Natural
Food
Antimicrobial
Systems

Natural
Food
Antimicrobial
Systems

Edited by

A.S. Naidu

Director, Center for Antimicrobial Research
Department of Food, Nutrition & Consumer Sciences
California State Polytechnic University
Pomona, California

CRC Press
Taylor & Francis Group
Boca Raton London New York

CRC Press is an imprint of the
Taylor & Francis Group, an **informa** business

CRC Press
Taylor & Francis Group
6000 Broken Sound Parkway NW, Suite 300
Boca Raton, FL 33487-2742

First issued in paperback 2019

ISBN-13: 978-0-8493-2047-7 (hbk)
ISBN-13: 978-0-367-39845-3 (pbk)
Library of Congress Card Number 00-036053

Library of Congress Cataloging-in-Publication Data

Naidu, A. S.
    Natural food antimicrobial systems / A. S. Naidu.
        p. cm.
    Includes bibliographical references and index.
    ISBN 0-8493-2047-X (alk. paper)
      1. Food—Microbiology. 2. Natural products. 3. Anti-infective agents.
4. Antibiosis. I. Title.
QR115 N.33 2000
664'.001'579—dc21
                                             00-036053
                                                  CIP

Visit the Taylor & Francis Web site at
http://www.taylorandfrancis.com

and the CRC Press Web site at
http://www.crcpress.com

Contents

SECTION-III   PHYTO-ANTIMICROBIALS

SECTION-IV   BACTO-ANTIMICROBIALS

About the Editor

A.S. 'Narain' Naidu is a medical microbiologist with more than 20 years of experience in studying the structure-function relationship of antimicrobial agents. He is considered a leading expert on protective and therapeutic applications of natural antimicrobials. As a public health expert, he has served various agencies, including the World Health Organization, the Hungarian Ministry of Health, the Directorate of Medical and Health Services of the Government of India, and the Royal Swedish Academy of Sciences. Dr. Naidu is currently the Director of the Center for Antimicrobial Research, California State Polytechnic University, Pomona.

Dr. Naidu received his Ph.D. in Microbiology from the Faculty of Medicine, Osmania University, India, in 1985. His research on staphylococcal enterotoxicosis, toxic shock syndrome and milk lactoferrin in Sweden during the mid-1980s has brought him international recognition. He has published over 50 peer-reviewed scientific papers and about 30 book chapters in this area. In 1997, he moved from the University of North Carolina at Chapel Hill to Cal Poly Pomona, to create a new university research center. Dr. Naidu is principal investigator for a number of scientific projects in the areas of food safety, public health, medicine and dentistry. He also supervises a multi-disciplinary team of scientists and research students.

Contributors

Bal'a, F.A.	Department of Food Science and Technology, Mississippi State University, P.O. Box 9805, Mississippi State, Mississippi 39762, USA.
Bidlack, W.R.	College of Agriculture, California State Polytechnic University, 3801 West Temple Avenue, Pomona, California 91768, USA.
Bogaert, J-C.	Research and Development, n.v. Glactic s.a., 7760 Escanaffles, Belgium.
Bostwick, E.F.	Research and Development, GalaGen Inc., P.O. Box 64314, Saint Paul, Minnesota 55164, USA.
Braun, S.	Research and Development, DMV International Nutritionals, 1712 Deltown Plaza, Fraser, New York 13753, USA.
Charter, E.A.	Research and Development, Canadian Inovatech Inc., 31212 Peardonville Rd., Abbotsford, British Columbia V2T 6K8, Canada.
Clarkson, M.R.	Research and Development, Aplin and Barrett Ltd., 15 North Street, Beaminster, Dorset DT8 3DZ, UK.
Clemens, R.A.	Department of Food Science and Nutrition, California State Polytechnic University, San Luis Obispo, California 93407, USA.
Cotton, L.N.	Department of Food Science and Technology, Mississippi State University, P.O. Box 9805, Mississippi State, Mississippi 39762, USA.
Crecelius, A.T.	Department of Food, Nutrition and Consumer Sciences, California State Polytechnic University, 3801 West Temple Avenue, Pomona, California 91768, USA.
Davidson, P.M.	Department of Food Science and Technology, University of Tennessee, P.O. Box 1071, Knoxville, Tennessee 37901, USA.
Debevere, J.M.	Department of Food Technology and Nutrition, Faculty of Agricultural and Applied Biological Sciences, Ghent University, Coupure 653, B-9000 Ghent, Belgium.
Delves-Broughton, J.	Research and Development, Aplin and Barrett Ltd., 15 North Street, Beaminster, Dorset DT8 3DZ, UK.
De Vuyst, L.	Research Group of Industrial Microbiology, IMDO, Department of Applied Biological Sciences, Vrije Universiteit Brussels (VUB), Belgium.
El-Ziney, M.G.	Department of Dairy Technology, Faculty of Agriculture, El-Satby, Alexandria University, 5-Aflatone St., Alexandria, Egypt.
Halaweish, F.	Department of Chemistry and Biochemistry, South Dakota State University, Brookings, South Dakota 57007, USA.
Hung, P.	Research and Development, Nutritional Foods Division, Taiyo Kagaku Co., Ltd., Yokkaichi, MIE 510-0825 Japan.

Ibrahim, H.R. Department of Biochemical Science and Technology, Kagoshima University, Kagoshima 890, Japan.

Isaacs, C.E. Department of Developmental Biochemistry, New York State Institute for Basic Research, 1050 Forest Hill Road, Staten Island, New York 10314, USA.

Jakobsen, M. Department of Dairy and Food Science, The Royal Veterinary and Agricultural University, Rolighedsvaj 30, DK-1958 Frederiksberg C, Copenhagen, Denmark.

Juneja, L.R. Research and Development, Nutritional Foods Division, Taiyo Kagaku Co., Ltd., Yokkaichi, MIE 510-0825 Japan.

Juneja, V.K. Microbial Food Safety Research Unit, USDA-ARS, Eastern Regional Research Center, 600 E. Mermaid Lane, Wyndmoor, Pennsylvania 19038, USA.

Khanna, N. Research and Development, Bio-cide International, Inc., P.O. Box 722170, Norman, Oklahoma 73070, USA.

Lampe, M.F. Department of Medicine, University of Washington, Box 356523, Seattle, Washington 98195, USA.

Lee, E.N. Department of Agricultural, Food and Nutritional Science, University of Alberta, Edmonton, Alberta T6G 2P5, Canada.

Leroy, F. Research Group of Industrial Microbiology, IMDO, Department of Applied Biological Sciences, Vrije Universiteit Brussels (VUB), Belgium.

Losso, J.N. Department of Food Science, Louisiana State University, Baton Rouge, Louisiana 70803, USA.

Marshall, D.L. Department of Food Science and Technology, Mississippi State University, P.O. Box 9805, Mississippi State, Mississippi 39762, USA.

Miller, K. W. Department of Molecular Biology, University of Wyoming, P.O. Box 3354, Laramie, Wyoming 82071, USA.

Mine, Y. Department of Food Science, Ontario Agricultural College, University of Guelph, Guelph, Ontario, Canada N1G 2W1.

Muthukumarappan, K. Department of Agricultural and Biosystems Engineering, South Dakota State University, Brookings, South Dakota 57007, USA.

Naidu, A.S. Center for Antimicrobial Research, California State Polytechnic University, 3801 West Temple Avenue, Pomona, California 91768, USA.

Nakai, S. Food, Nutrition and Health Program, University of British Columbia, 6650 NW Marine Drive, Vancouver, British Columbia CV6T 1Z4, Canada.

Okubo, T. Research and Development, Nutritional Foods Division, Taiyo Kagaku Co., Ltd., Yokkaichi, MIE 510-0825 Japan.

Oleszek, W. Department of Biochemistry and Crop Quality, Institute of Soil Science and Plant Cultivation, 24-100 Pulawy, Poland.

Prakash, A. Department of Food Sciences and Nutrition, Chapman University, 333 North Glassell Street, Orange, California 92866, USA.

Ravishankar, S.   National Center for Food Safety and Technology, Illinois Institute of Technology, Moffett Campus, 6502 S. Archer, Summit-Argo, Illinois 60501, USA.

Ray, B.   Department of Animal Science, University of Wyoming, P.O. Box 3354, Laramie, Wyoming 82071, USA.

Sharma, R.K.   Department of Food Science, Michigan State University, East Lansing, Michigan 48823, USA.

Sim, J. S.   Department of Agricultural, Food and Nutritional Science, University of Alberta, Edmonton, Alberta T6G 2P5, Canada.

Sofos, J.   Department of Animal Sciences, Colorado State University, Fort Collins, Colorado 80523, USA.

Steijns, J.   Research and Development, DMV International, NCB-laan 80, 5460 BA Veghel, The Netherlands

Sunwoo, H.H.   Department of Agricultural, Food and Nutritional Science, University of Alberta, Edmonton, Alberta T6G 2P5, Canada.

Thomas, L.V.   Research and Development, Aplin and Barrett Ltd., 15 North Street, Beaminster, Dorset DT8 3DZ, UK.

Whitmore, B.B.   Department of Foods, Nutrition and Consumer Sciences, California State Polytechnic University, 3801 West Temple Avenue, Pomona, California 91768, USA.

Preface

Food preservation is undoubtedly one of the most significant breakthroughs in human civilization. The technology has been explored for centuries, integrating culinary art with ingredient availability, social necessity, religion and ethnicity. The multifunctionality of natural preservatives, such as herbs, oils and spices, that led to designing a variety of health foods in ancient Asia, Europe and the Mediterranean was well recognized. The evolution of antimicrobial technology, from fermentation to pasteurization, started in wine and cheese factories and extended to revolutionize medicine. Despite our traditional knowledge of natural antimicrobial agents in milk, eggs, plants, probiotics, salts and acids, the structure-function relationship of these bioactive compounds has been unraveled only in the past few decades. We now understand the molecular mechanisms of the ability of a charged cleft in the C-lobe of a milk lactoferrin to bind iron and cause microbial stasis; or the muraminidase activity of a hen egg lysozyme triggering a cidal effect on the cell wall of a Gram-positive bacteria; or the capacity of a galactose-rich polysaccharide in common food-grade agar to detach microorganisms from meat surfaces; or the microbial interference phenomenon and competitive exclusion of pathogens by probiotic bacteria and their bacteriocins; or the pathways of potent oxidative events to break nucleic acid strands and trigger disinfection of viral pathogens by chloro-cides. There is a wealth of knowledge continuously being added to the ocean-depth of common sense we inherited from our ancestors on these fascinating natural food antimicrobials.

Emergence of the neo-breed of food-borne pathogens, concerns about the use of synthetic antimicrobials, possibly resulting in the development of microbial resistance, and the global consumer's awareness of the nutraceutical benefits of natural foods have catapulted the interests of food scientists and food technologists into the ancient art of food preservation. This volume, *Natural Food Antimicrobial Systems*, is an effort to consolidate the current developments in this area of food science.

Although many individuals have extended their help or suggestions in the preparation of this book, several deserve special mention. I am very grateful to my colleague Dr. Wayne Bidlack, Dean, College of Agriculture, California State Polytechnic University, Pomona, for his genuine interest and generous support of my academic endeavors at this institution. I am indebted to Dr. Fergus Clydesdale, Editor, *Critical Reviews in Food Science and Nutrition*, for recommending this project to CRC Press. Thanks to Lourdes Franco, Associate Editor, Life Science, CRC Press LLC, Boca Raton, Florida, for her professionalism and unstinting support in bringing this book to publication. My appreciation to Gail Renard and associates at CRC Press for expediting the final stages of printing.

This reference volume is the result of a scientific vision ascending from the click of a mouse to a floppy disk from an illustrious team of scientific experts around the world, both from academia and industry—my sincere appreciation for their relentless effort in contributing the chapters! This project required an in-depth literature search at every level. I would like to commend Erik Ennerberg, Reference Librarian, CalPoly University, Pomona, for providing documentation backup. I also appreciate Thomas Kennedy for drawing the chemical structures. I sincerely thank Brian Dowdle for the desktop design and publishing work. A final note of acknowledgment to my Naidu family: Smita, Tezus, Sreus and Aishwarya for proof reading, cross-referencing and designing the cover. I hope that this volume will contribute to the research and development of the natural food antimicrobial systems and lead to potential applications in nutrition and health.

A.S. Naidu
JANUARY 1, 2000

A.S. Naidu

Overview

1

The effort to process and store food for later use is undoubtedly a quantum leap in the evolution of mankind. It is impossible to establish the exact origins of the development of food processing technology. Anthropological investigations have continuously documented the existence of cooking with fire or some type of food processing dating back to primitive ages. Our interaction with nature and observations of the environment have led to identification of various types of edible foods; antidotes or remedies that cure and protect; as well as an array of flavorants and colorants to please the palate. The integration of such traditional knowledge has turned culinary art into science.

The discovery of microorganisms, the so-called 'animalcules', in the late 18th century placed food science under the microscope. The power of a microbe such as a milk-borne *Coxiella burnetti* to transmit human disease or the ability of a lactobacillus to ferment milk became evident. Pasteurization became the first-ever science-based antimicrobial food safety measure in dairy processing. The microbiology of food-borne infections and intoxications, as well as food spoilage, has redefined our traditional approaches to food protection. The nutraceutical benefits and the healing power of certain bioactive food ingredients that had been noticed for centuries came under close scientific scrutiny. A multifaceted approach from food science, microbiology, pharmacology and medicine has established 'Food Antimicrobial Science and Technology' as an emerging discipline.

FOOD ANTIMICROBIAL TECHNOLOGY

Demands and concerns of the global consumer play a critical role in dictating the momentum of research and development in food protection. The demand for minimally processed foods and sustained functionality of naturally occurring bioactive ingredients is steadily rising. This appeal is based on growing concerns about the use of synthetic preservatives with limited documentation on safety and tolerance; the suspected link between the overuse of antibiotics and the development of multi-drug resistance in microbes; and the increased media dissemination of knowledge on diet and health. Furthermore, recent emergence of food-borne pathogens such as the enterohemorrhagic

Escherichia coli O157:H7, multi-drug resistant *Salmonella* typhimurium DT104, *Listeria monocytogenes*, and *Enterococcus faecium* has made food safety a high priority with the federal, state and local health and regulatory authorities. This is reflected in the President's National Food Safety Initiative enforced in 1997.

Protection of food from spoilers and pathogens can be achieved by various methods including aseptic handling to prevent or minimize microbial entry, removal by physical methods through washing, centrifugation or filtration, destroying with heat, gas or irradiation, and inhibiting growth by refrigeration, freezing, dry packaging, or adding chemical preservatives. Each method is useful as an individual factor in microbial control or is part of a hurdle in a multi-factor food preservation system.

This reference volume shall focus on various aspects of '*Natural Food Antimicrobial Systems*' including molecular aspects of their mechanism of action, influence of factors on structure-function relationships, spectrum of antimicrobial activity, safety and tolerance issues, and more importantly the mass production and feasible application as food ingredients.

NATURAL FOOD ANTIMICROBIAL SYSTEMS

Numerous antimicrobial agents exist in animals and plants where they evolved as host defense mechanisms. Naturally occurring antimicrobials are abundant in environment. These compounds may exhibit antimicrobial activity in the foods as natural ingredients or may be used as additives to other foods. The discussion in this reference volume has been limited to six different classes of natural systems.

I. LACTO-ANTIMICROBIALS

Milk is the first complete functional food devised by nature for the protection and development of a newborn mammal. The prophylactic and therapeutic benefits of milk and its bioactive components have been recognized for centuries. The antimicrobial and nutraceutical potential of many components in milk has been unraveled in recent years. The dairy industry has started isolation of these substances from a variety of sources including cheese whey, hyperimmunized colostrum from cows, and a spectrum of multifunctional applications is being developed in the food and health industry.

Lactoferrin (LF) is a metal-binding glycoprotein present in milk and various exocrine secretions that bathe the mucosal surfaces. This natural antimicrobial agent is a multifunctional bioactive molecule with a critical role in many important physiological pathways. LF could elicit a variety of inhibitory effects against microorganisms comprising stasis, cidal, adhesion-blockade, cationic, synergistic and opsonic mechanisms. Broad-spectrum activity against different bacteria, viruses, fungi and parasites, in combination with anti-inflammatory and immunomodulatory properties makes LF a potent innate host defense mechanism. The current commercial production of bovine milk LF is approximately 100 metric tons worldwide and this figure is continuously increasing. This protein is finding its applications as an active ingredient in infant formulae, and health foods in South-East Asian countries, in particular. LF is also in use as a therapeutic and prophylactic agent to control intestinal illnesses and mucosal infections. A number of efficacy studies and clinical trials are ongoing in various laboratories with over 100 patents filed on this molecule in the past 10 years. Undoubtedly, LF is emerging as one of the leading natural microbial blocking agents in food safety and preservation.

Lactoperoxidase (LP) is an oxidoreductase secreted into milk which plays an important role in protecting the lactating mammary gland and the intestinal tracts of the newborn infants against pathogenic microorganisms. LP catalyzes the oxidation of thiocyanate and iodide ions to generate highly reactive oxidizing agents. These products have a broad spectrum of antimicrobial effects against bacteria, fungi and viruses. Microorganisms show great variability in their response to peroxidase systems. The enzymatic properties of LP have found usage in a variety of applications. Its antimicrobial activity has been used to preserve raw milk in warm climates, through the addition of hydrogen peroxide and thiocyanate.

Lactoglobulins are potential natural food antimicrobials with a wide spectrum of microbial pathogens and ability to neutralize microbial virulence factors such as toxins. Since colostrum/milk-derived antibodies are polyclonal, they present a multi-faceted approach to pathogen control that transcends the microbial ability to develop resistance mechanisms. Lactoglobulins such as bovine immunoglobulin concentrates have many advantages over synthetic antibiotics: low cost, polyclonal nature in microbial inactivation, activity most likely limited to the intestinal lumen, and less burden on the microbial gut ecology. From a commercial point of view, their entry into the market is rather fast due to less stringent regulatory issues. Lactoglobulin use in dietary supplements or special food products is being considered.

Lactolipids are not only nutrients but also antimicrobial agents that are part of an innate defense system functioning at mucosal surfaces. Lipid-dependent antimicrobial activity in milk is due to medium-chain saturated and long-chain unsaturated fatty acids and their respective monoglycerides. The antimicrobial activity of fatty acids and monoglycerides is additive so that dietary-induced alterations in the fatty acid distribution in milk do not reduce lipid-dependent antimicrobial activity. Fatty acids and monoglycerides can rapidly destabilize the membranes of pathogens. The antimicrobial activity of milk lipids can be duplicated using monoglyceride ethers, which very effectively inactivate enveloped viruses, bacteria and protozoa. Monoglyceride ethers are more stable than the ester linkages found in milk lipids since they are not degraded by bacterial or mammalian lipases. Lactolipids and related GRAS (Generally Recognized As Safe) antimicrobial lipids could also be added to infant formulas to reduce the risk of bacterial growth during transport.

II. OVO-ANTIMICROBIALS

In contrast to the immune system of animals, which produce antimicrobial polypeptides when needed, the egg efficiently resists microorganisms over a long period in the absence of such innate and systemic host defense. Apparently, the main function of ovo-antimicrobials is to keep microorganisms away from the yolk, the nutrient reservoir of egg. The wide diversity in the nature of microorganisms implies that ovo-antimicrobials possess multiple functions. Of many different types of proteins found in the egg albumen, most of them appear to possess antimicrobial properties or certain physiological function to interfere with the growth and spread of the invading microorganisms.

Lysozyme is a nutraceutical and a value-added product from hen egg white. Singly or in combination with other natural antimicrobial compounds, it has desirable antimicrobial properties for use in minimally processed foods. Lysozyme demonstrates cidal effects on Gram-positive bacteria, however, heat denatured lysozyme has antimicrobial activity against Gram-negative bacteria. Several patents claim the effectiveness of lysozyme, alone or in combination with other synergists, as a food preservative for fruits and vegetables, meat, beverages, and for non-food uses. Food products containing lysozyme, such as cheese, chewing gums, candies, and mouthwashes are on the market. Potential applications in the produce, meat, wine and animal feed industries are increasingly being demonstrated.

Ovotransferrin is the major iron-binding protein of egg, which is an equivalent to LF in milk. It acts as a bactericidal agent and its bacterial killing activity can be decoupled from its iron-sequestration properties, in addition to its known bacteriostatic effect (nutritional immunity). Of special interest is that the chemical cleavage at the Asp-X sequence of ovotransferrin produces a strong bactericidal hydrolysate to both Gram-positive and Gram-negative bacteria, heralding a fascinating opportunity for the potential use of the crude hydrolysate in food applications or infant formulas.

Ovoglobulin IgY. It has been long recognized that the hen, like its mammalian counterparts, provides young chicks with antibodies as protection against pathogens. This system facilitates the transfer of specific antibodies from serum to egg yolk, and provides a supply of antibodies (IgY class) to the developing embryo and the hatched chicken. The protection against pathogens that the relatively immuno-incompetent newly hatched chick has, is through transmission of antibodies from the mother via the egg. Egg yolk, therefore can be loaded with large amounts of IgY against pathogens which can immobilize the existing or invading pathogens during embryo development or in day-old chicks. Thus, the immunization of laying hens to various pathogens results in production of different antigen-specific IgY in eggs. Egg yolk contains 8-20 mg of IgY per ml or 136 - 340 mg per yolk, suggesting that more than 30 g of IgY can be obtained from one immunized hen in a year. The IgY technology opens new potential market applications in medicine, public health, veterinary medicine and food safety. A broader use of IgY technology could be as a biological or diagnostic tool, nutraceutical or functional food development, oral-supplementation for prophylaxis, and as pathogen-specific antimicrobial agents for infectious disease control.

Avidin is a basic tetrameric glycoprotein of egg white. The protein binds up to four molecules of biotin with exceptionally high affinity. This high affinity of avidin-biotin binding has been exploited thoroughly for the development of various molecular tools. Avidin has also been viewed as antibiotic protein in egg albumen inhibiting bacterial growth such as a remedial protein, or as a lytic enzyme. Avidin could inhibit the growth of biotin-requiring yeast and bacteria. The antimicrobial activity also may arise from the ability of avidin to bind various Gram-negative and Gram-positive bacteria. The specific avidin may act as a host defense with antimicrobial and enzyme inhibitor properties. Detailed molecular information through three dimensional structure and protein engineering of avidin would enable us to better understand the biological role and development of avidin as a natural food antimicrobial agent.

III. PHYTO-ANTIMICROBIALS

Phytochemicals exhibiting various degrees of antimicrobial activity occur in plant stems, leaves, barks, flowers, and fruits. Spices and herbs and their essential oils have varying degrees of biological activity. In many cases, concentrations of the antimicrobial compounds in herbs and spices are too low to be used effectively without adverse effects on the sensory characteristics of a food. However, they may contribute to the overall 'hurdle' system present naturally in a food product. The most consistent antimicrobial activity among herbs and spices has been found with components of cloves, cinnamon, mustard seed, oregano, rosemary, sage, thyme, and vanillin. Certain phyto-phenolic agents such as oleoresins from olive oil seem to deliver multifunctional physiological benefits to the consumer, and therefore are highly attractive to the health food industry. Since phyto-phenols have been in the food supply and consumed for centuries, these natural phyto-antimicrobials appear to be safe when compared to new synthetic preservatives.

Saponins are compounds that protect plants from biotic stress and also elicit antibacterial, antifungal and antiviral activities. This broad-spectrum of antimicrobial activity can be useful in food protection. The ability of saponins to complex with cholesterol in particular may find potential application to lower the diet-induced hypercholesterolemia. Since structurally divergent saponins present different activities, isolation of large quantities of homogenous bioactive form is still a challenge.

Flavonoids are extensively characterized by many laboratories for their broad-spectrum antimicrobial activity and the lack of resistance induction in target organisms. Isoflavonoids, as natural products and as potent fungal growth inhibitors, could be useful in controlling plant diseases. Due to this high specific antimicrobial activity a possible application for isoflavonoids as alternative to conventional fungicides in the control of storage fungi to prevent loss of grains in developing countries has been suggested. Flavonoids are also of great interest in food technology as they contribute to the sensory and nutritional qualities of fruits and fruit products. The antioxidant activity of flavonoids and phenylpropanoids in various beverages including fruit juices is well recognized. Recently, there has been a resurgence of interest in the role of flavonoids in health and disease following the French paradox suggesting an association between reduction in mortality from coronary heart disease and higher wine consumption. Flavonoids are now emerging as potent nutraceutical bioactive food ingredients with sensory appeal to the global food consumer.

Thiosulfinates such as allicin and ajoene are the antimicrobial components of garlic. Garlic and its derivatives are multifunctional as well as flavorful food additives. Garlic has been used in the past and is presently being used as a medicine, curative, preservative and a flavoring agent. Thiosulfinates could elicit both cidal and statis effects. Although garlic and its thiosulfinate derivatives appear to be safe for consumption, assessment and protein determination for the allergen responsible for hypersensitivity reactions needs in-depth evaluation. Certain reports have indicated allergic reactions associated with garlic consumption. If thiosulfinates are to be used as preservatives or as antimicrobial agents, the purification techniques and quality control methods need a great degree of standardization.

Catechins in green tea (GTC) could prevent plaque formation by inhibiting bacterial growth, glucan synthesis and the adherence of carcinogenic streptococci. In addition, GTC could reduce gum disease by suppressing the growth and adherence of periodontal pathogens to buccal epithelial cells as well as by inhibiting the activity of collagenases. GTC also strongly inhibit the growth of intestinal pathogens but do not affect other useful probiotic bacteria. The antimicrobial studies strongly suggest a potential prophylactic role for GTC in human medicine. GTC intake is safe and has health benefits, in addition, green tea is cheap and readily available in fractionated and extracted derivatives. The huge worldwide consumption and safety of GTC may well do significant good and could be supported as an adjunctive therapy and a health ingredient to a good diet, exercise, and avoidance of high-risk behavior.

Glucosinolates are abundant in various cruciferous vegetables including cabbage, Brussels sprouts, broccoli, cauliflower, horseradish, mustard, watercress, turnips, radish, rutabaga and kohlrabi. These bioactive compounds have been used for centuries in common herbal preparations and remedies ranging from rheumatism to treating snake poison. As stimulants, mustard and horseradish were applied to oily or high fatty proteins such as fish and meats. Volatile glucosinolates such as the isothiocyanates elicit a wide range of antibacterial and antifungal effects. Application of gaseous allyl isothiocyanates as a preservative in packaged foods has been suggested at low concentrations. A protective anti-tumor association between intake of cruciferous vegetables and cancer of the colon and rectum has been suggested. Broccoli, cabbage and Brussels sprouts were also shown to be protective against mammary carcinogenesis. However, certain studies with purified glucosinolates indicated toxicity in some animal models.

Agar is a basic ingredient in the solid-state cultivation of bacteria over most of this century. The possible adsorption of streptococci to the water-soluble galactose fraction from agar warrants attention in interpreting data on bacterial cell surface charge of hydrophobic properties, bacterial adhesion to eucaryotic cells, and agglutination assays for rapid identification of bacteria. The ability to interact with specific mucosal defense factors such as LF and to bind various bioactive compounds makes GRP an excellent immobilizing agent and an antimicrobial delivery system in various food products.

IV. BACTO-ANTIMICROBIALS
Food fermentation using microbial starter cultures is one of the oldest known uses of biotechnology. Fermented foods and beverages continue to provide an important part of human diet and constitute 20-40% of food supply worldwide. The most common use of probiotics is as food in the form of fermented milk products. Bacteriocins are naturally produced small peptides with bactericidal activity usually against closely related bacteria. Antimicrobial activity of bacteriocins in foods is affected by its levels, type and number of microorganisms, condition of application, interaction/inactivation by food components, and pH and temperature of product.

Probiotic strains of lactic acid bacteria (LAB) are generally enteric flora, believed to play a beneficial role in the ecosystem of the human gastrointestinal tract. Recent developments in the role of probiotics (live) and their probioactive (cellular) substances in the intestinal and extra-intestinal physiology of the host are overwhelming. LAB are capable of preventing the adherence, establishment, replication, and/or pathogenic action of specific enteric pathogens. Probiotic LAB have been used to treat disturbed intestinal microflora and increased gut permeability which are characteristic of many intestinal disorders such as acute rotaviral diarrhea, food allergy, colon disorders, metabolic changes during pelvic radiotherapy and changes associated with colon cancer development. In recent years, consumers have become aware of probiotic properties of cultured milk and dosage specifications. A concentration of 10^5 cfu/g or ml of the final product has been suggested as the therapeutic minimum. However, the list of probiotic effects and health claims with the use of LAB is expanding.

Nisin, a polypeptide bacteriocin produced by *Lactococcus lactis* ssp. *lactis*, exhibits antimicrobial activity against a wide range of Gram-positive bacteria associated with food, and is particularly effective against heat-resistant bacterial spores. It is not active against Gram-negative bacteria, yeasts and molds because of its inability to penetrate these cells and gain access to the cytoplasmic membrane. Toxicological studies have found nisin safe and it is now allowed as a food preservative in over 50 countries worldwide. The major use of nisin is in processed cheese, pasteurized dairy products and canned vegetables. In this type of product heat processing eliminates all spoilage microflora except for the nisin-sensitive Gram-positive spore forming bacteria such as bacilli and clostridia. Nisin is also used effectively in combination with heat in crumpets, liquid egg, and fresh soups. Spoilage by LAB can be controlled by nisin in products such as wine and beer, pasteurized vacuum packaged continental sausages, ricotta cheese and acidic foods such as salad dressings and sauces. As part of a hurdle system of preservation nisin can also be used to prevent the growth of pathogens. Current interest is focused on the combination of nisin with novel non-thermal preservation processes such as ultrahigh pressure and electroporation.

Pediocins from several strains of *Pediococcus acidilactici* and *Pediococcus pentosaceus*, due to their bactericidal property against many Gram-positive bacteria, are explored for their potential use as food preservatives and antibacterial agents. In general, the bactericidal property of pediocins is relatively resistant to heat, storage conditions, pH, organic solvents and hydrostatic pressure, but is destroyed by proteolytic enzymes and neutralized by some anions. The bactericidal action of pediocin PA-1/AcH against target cells is produced through the ability of the molecules to form pores in the membrane and disrupt the proton motive force. Limited studies have shown that pediocin PA-1/AcH, or the producer strain(s), can be effectively used in food systems to control *Listeria monocytogenes*. Under appropriate processing conditions it can also be used to control Gram-positive and Gram-negative spoilage bacteria in foods.

Reuterin has been proven to be an effective broad-spectrum antimicrobial able to play a powerful role to improve food safety and prolong shelf life. Further, *Lactobacillus reuteri* combined with glycerol represents an example of a protective culture which could be added in combination with starter cultures in the manufacturing of fermented fish, sausage, cheese and pickles. This system could control pathogen and spoilage microorganisms during the fermentation process and storage. The efficacy of reuterin as a biopreservative agent and *L. reuteri* as a protective culture is currently being explored over a wide range of applications. For example, it would be interesting to apply the *L. reuteri*-glycerol system in highly enriched glycerol foods such as shrimps and other seafoods to extend shelf life. The successful use of reuterin as a decontaminating agent for meat and cows' teat surfaces also suggests more surface disinfectant applications.

   Sakacins are bacteriocins produced by several strains of *Lactobacillus sakei*. They are small, cationic, hydrophobic peptides with an antibacterial mode of action against species of lactobacilli and certain food-borne pathogens such as *Listeria monocytogenes*. Environmental conditions, particularly the pH, will affect sakacin stability and efficacy. Application possibilities are predominantly in the field of sakacin-producing starter or co-cultures for meat fermentations since addition of purified sakacin is not (yet) accepted in food additives legislation. Continued study of the physical and chemical properties and structure-function relationships of sakacins and similar compounds, as well as their *in situ* functionality either as food additive or being produced by starter, co- or protective cultures is necessary if their potential for future use in meat preservation is to be exploited.

V. ACID-ANTIMICROBIALS

   Several naturally occurring organic acids are added to or are formed in foods through microbial fermentation. Organic acids are amply found in a variety of fruits including apples, apricots, bananas, cherries, citrus varieties, cranberries, grapes, peaches, etc. Acids inhibit microbial growth by lowering the pH, affecting the proton gradient across biological membrane, acidifying the cytoplasm and interfering with the chemical transport across cell membrane. Acid environment not only limits microbial growth but also enhances destruction of microorganisms synergistically with other antimicrobial systems. Most of the approved organic acid antimicrobials, although found naturally in various products, are synthesized chemically as additives of inclusion in food formulations but have a long history of safe use. Organic acids also have been tested or used as washes, sprays, or dips to decrease the microbial load of meat and poultry carcasses.

   Lactic acid and its derivatives elicit a broad-spectrum of antimicrobial activity against Gram-positive bacteria including spore-forming bacilli and clostridia, as well as Gram-negative pathogens such as enterohemorrhagic *E.coli* O157:H7 and salmonella. Lactate compounds also demonstrate antifungal activity against aflatoxin-producing *Aspergillus* sp. Considering these attributes, lactic acid and lactates are widely used in the food industry for decontamination of meat foods such as beef, poultry and pork during processing and packaging. They are also used in shelf life enhancement of fresh and semi-processed foods. Consumer acceptance of such processed foods, however, is largely based on visual and organoleptic qualities. Currently, the worldwide utility of lactic acid and its derivatives amounts to 100,000 metric tons per year.

   Sorbic acid use as a preservative relies on its ability to inhibit the growth of various yeasts, molds, and bacteria. Inhibition of microorganisms by sorbates, however, is variable with microbial types, species, strains, food, and environmental conditions. Some microbial strains are resistant to inhibition by sorbate or even metabolize the compound under certain conditions. In addition to being considered less toxic, sorbic acid is also considered more effective than benzoate or propionate in preserving foods such as cheese, fish and bakery products. Sorbates are considered effective food preservatives when used under sanitary conditions and in products processed following GMP.

Acetic acid and its GRAS salts have a long history of functional use in foods. The ability to lower pH with acetic acid or vinegar provides pickled foods with unsurpassed keeping quality, flavor, and safety. Aside from this traditional use, acetic acid use as a decontaminating agent of fresh and minimally processed foods remains under-exploited. Food processors should find acetic acid a desirable functional ingredient or processing aid because of its wide availability, low cost, consumer acceptance, and clean label requirements.

Citric acid is produced by most plants and animals during metabolism. The tartness of citrus fruits is due to the citric acid. Lemon juice contains 4 - 8% and orange has about 1% citric acid. Citric acid is widely used in foods, pharmaceuticals, and cosmetics. Citric acid has gained universal acceptance as GRAS ingredient and the Joint FAO/WHO committee has not set any ADI value that makes it a common chemical commodity. Although not a direct antimicrobial agent, citrate is recognized for enhancing the antimicrobial property of many other substances, either by reducing pH or by chelation of metal ions. Its use in controlling thermophiles, in particular, is well established. Citric acid is also used as an antioxidant to prevent oxidative discoloration in foods such as meats, fruits, and potatoes. Citrates of sodium, potassium and calcium also are recognized as GRAS ingredients with a wide range of food applications.

VI. MILIEU-ANTIMICROBIALS

Traditional antimicrobials include common salt (sodium chloride), sugar and wood smoke that are abundant in the milieu. Common salt has been used as a flavoring or a preservative in foods since ancient times. Curing of meat with impure salt led to the use of nitrates and nitrites as additives for the curing and the preservation of meat products. Exposure to wood smoke is a traditional practice still used in processing. Wood smoke contributes flavor and incorporates antimicrobial compounds such as phenols, formaldehyde, acetate and creosote, into a product. Oxidative damage has been nature's way of killing microorganisms in phagosomes and an integrated mechanism for antigen processing in various life forms. Various enzyme systems such as oxidases and halide-based catalysts contribute both additive and synergistic effects to numerous natural antimicrobials in the innate host defense. Thus, chloro-cides and ozone could be considered as natural hurdle mechanisms in the physiological milieu. As potent milieu-antimicrobials, chloro-cides and ozone could open a powerful hurdle-approach to food safety if integrated with other natural food antimicrobial systems.

Sodium chloride, commonly called table salt or salt, is a vital part of human life. Salt enhances the flavor of foods and plays a preserving as well as functional role in food processing. The antimicrobial activity of salt can be both direct or indirect depending on the amount added and the purpose it serves. Since the amount of sodium chloride that must be added to foods to prevent microbial growth is large and will cause an unacceptable taste, it is usually added in combination with other hurdles. The mechanism of inhibition of microorganisms by sodium chloride is mainly by lowering the water activity of the substrate. Studies have also indicated that sodium chloride could have a role in interfering with substrate utilization in microorganisms. The influence of sodium chloride alone as well as its interactive effects with other factors on various microorganisms has been well established. However, excessive sodium intake in humans has been linked to hypertension and the related cardiovascular problems and stroke. Hence there are many consumer concerns and the food industry is trying to minimize the salt content of food products.

Polyphosphates and their effects on microorganisms in food products depend on various factors which influence their activity such as the concentration, pH, nature of substrate, heat treatment, type and level of microbial contamination, formulation, and storage temperature. Long chain alkaline polyphosphates are more effective as antimicrobial agents compared to short chain polyphosphates, although sodium acid pyrophosphate, a short chain acidic polyphosphate, and trisodium phosphate are very effective in certain applications. The inhibition by sodium acid pyrophosphate is related to the decrease in pH, although pH reduction by itself does not account for suppression of microbial growth. The presence of polyphosphates also appears to increase the sensitivity of several microorganisms to heat. Since polyphosphates function by chelating metal ions essential for microbial growth, in foods with high levels of cations, polyphosphates are not very effective. Gram-positive bacteria and mold are sensitive to polyphosphates and Gram-negative bacteria are resistant to the action of polyphosphates. Level of contamination and age of the culture can also affect polyphosphate action. Synergistic effects have been reported with combinations of polyphosphates and other preservatives such as nitrites and sorbates. Sodium chloride, in some cases, enhances antimicrobial action, and in other cases, decreases effectiveness. Presence of spices may enhance activity. Certain combinations of polyphosphates and other additives such as salt have an additive effect, while other combinations elicit synergism. Polyphosphates lack the broad spectrum of activity exhibited by primary antimicrobials such as sorbate and benzoate. However, as indirect antimicrobials, polyphosphates can impart considerable protection against microbial growth and spoilage while at the same time providing physical and chemical functionality.

Chloro-cides are broad-spectrum antimicrobial compounds and oxidative damage is their fundamental mechanism for biocidal action. The biocidal component of most chloro-cides is hypochlorous acid. Currently, the compounds that produce hypochlorous acid are subject to scrutiny because of risks of forming undesired by-products such as trihalomethanes. With major advancements in chemical technology, chlorine dioxide (CD) has emerged as a safe and effective class of chloro-cide. The use of CD as a disinfectant and sanitizer is growing due to its high antimicrobial efficacy and absence of any hazardous by-products. It is extensively used for the treatment of drinking water without adding chlorine taste to the water. CD is also a popular choice of antimicrobial used for the sanitation of produce, poultry, and beef. It is highly effective for the prevention of late blight on potatoes. The recent FDA approval for the use of CD on seafood had provided a useful tool to the food-processing industry for controlling persisting microbial pathogens.

Ozone has been used in water purification for decades due to its disinfectant effect on a wide-variety of water-borne pathogens. Numerous studies have indicated that ozone could destroy with extreme efficiency the spores of molds, amoebae, viruses, and bacteria as well as various other pathogenic and saprophytic organisms. Such broad-spectrum and successful use of ozone has created a renewed interest in its utilization as a general germicidal agent. Ozone shows much promise in the food and agriculture industry, however, more research and development is required before its multipurpose usage. Recent GRAS status has encouraged food processors to use ozone for sanitation. Limited data are available on the effect of different concentrations of ozone in reducing bacterial contamination of various food materials and the changes in physical and biochemical characteristics of these foods.

EFFICACY AND APPLICATION

Of the many natural antimicrobial systems only a few have been tested or applied to foods. The effectiveness of these antimicrobials depends on a number of factors associated with the target microorganism, the food product, as well as the storage and handling environment.

Microbial spectrum. Antimicrobial systems have evolved in nature as specific mechanisms of host defense. Most of these systems are designed to be effective in native existence, and their broad-spectrum of activity is dependent on various factors that regulate the structure-function properties of the antimicrobial agent. Thus, conditions to elicit a potent antibacterial activity such as an iron-depriviation stasis of lactoferrin may not be optimal for a competitive blocking of viral adhesion to an enterocyte. The mechanism of activity also defines the spectrum of an antimicrobial agent. Thus, muraminidase activity would limit the cidal activity of hen egg lysozyme against Gram-positive bacteria. Similarly, nisin is a cationic molecule that interacts with membrane phospholipids of Gram-positive bacteria. However, a synergistic activity with EDTA could make these two antimicrobials effective against Gram-negative bacteria.

Various microbial factors also affect the antimicrobial outcome. Actively multiplying vegetative cells at exponential growth phase are more susceptible to an antimicrobial than organisms at stationary growth phase or their sporulating forms. Thus, organisms in an active metabolic state provide ready access to their cellular targets to an antimicrobial agent. However, such proliferating organisms also could develop mechanisms to resist an antimicrobial by shielding cellular targets, producing enzymes that hydrolyze the agent, or induce mutations. Processing conditions could induce heat-shock/acid-shock proteins and convert bacterial outer membrane resistant to antimicrobials.

Therefore, the efficacy of a natural antimicrobial system is based on the milieu optima to restore its structure-function, processing conditions and composition of the food product intended for protection. The type of microbial load and intended food safety determines the effective dosage.

Food composition. Temperature and pH of the food are the most important factors that influence the antimicrobial activity. Temperature affects the fluidity of the biological membranes and facilitates the perturbation mechanism of hydrophobic antimicrobials. Ionic conditions and pH of the milieu could contribute to the charge alternations and affect antimicrobial interactions with the target surface. Antimicrobials such as lactoferrin, ovotransferrin and lysozyme are cationic in nature and are most effective in acidic pH, a milieu condition that facilitates antimicrobial penetration across the microbial cytoplasmic membrane. Organic acid antimicrobials are weak acids and are most effective in their undissociated form. Thus, weak acids are able to penetrate the cell membrane more effectively in the protonated form, therefore, the pK_a value of these compounds is important in selecting a particular compound for an application. Thus, the lower the pH of a food product, the greater the proportion of acid in its undissociated form and the greater the antimicrobial activity. Polarity of the antimicrobial molecule is also an important characteristic; thus, side groups such as the alkyl, benzyl moieties could enhance the ionization of phyto-antimicrobials such as the flavonoids, saponins, and glucosinolates.

Most food antimicrobials are hydrophobic in nature, a physical property that facilitates their interaction with surfaces. Presence of certain lipids or proteins in certain foods may bind to an antimicrobial via non-specific hydrophobic interaction and inactivate them by immobilization. The antimicrobial activity of thiosulfinates from garlic and certain bacteriocins is compromised in high-lipid milieu. Oxidative antimicrobial activity of chloro-cides and ozone decreases in organic environment, protein-rich in particular.

Delivery system. An antimicrobial is added to a food processing line by various techniques such as a direct incorporation to the product formulation; spray or immersion of the product in an antimicrobial solution; dusting a dry powder of the antimicrobial over the product; application to the packing material that comes in contact with the food; or applying a multi-compound formulation with carriers such as vegetable oil, ethanol or propylene glycol. The mode of application for a specific product depends on type of food product; properties of the antimicrobial; type of processing; and expected efficacy. The delivery system selected should be compatible with the bio-functionality of the antimicrobial agent.

Cost factor. The extraction of bioactive antimicrobials from natural sources can be complex and expensive. Selection of natural source, isolation and commercial application is challenging due to the complexity of food, the variety of factors influencing preservation, and the complex chemical and sensory properties of natural antimicrobials. Non-denaturing extraction procedures, efficient large-scale isolation methods, and techniques to retain bio-functional properties are being optimized for most natural antimicrobials described in this volume.

REGULATORY ISSUES

Toxicological problems associated with the use of certain synthetic food antimicrobials has generated interest in the food industry for use of naturally occurring compounds. Commonly used antimicrobials, such as organic acids that are produced in large quantities through chemical synthesis, are also found naturally in many food products and are GRAS according to regulatory authorities. Many natural food ingredients are time tested with proven benefits as food additives; however, this does not warrant a safety claim for specific components isolated from these materials.

Isolation of bioactive components from a natural source poses new parameters of safety evaluation for the following reasons: 1) the isolation process might enrich an undesirable allergen, mutagen or toxin from a natural source; 2) an otherwise non-toxic component might be activated during isolation; 3) the extraction/purification methods possibly may denature a bioactive component and create a new artifact with toxic manifestations; 4) residual solvents or elution chemicals might react and/or contaminate the final product and compromise the structure-functional properties; 5) the level of acceptable daily intake of an isolated biological compound could be higher than its consumption via natural source; 6) the purified natural antimicrobial when incorporated in a formulation might react with other ingredients to form toxic species.

Therefore, all naturally occurring antimicrobials should demonstrate safety either by animal testing or by continuous consumption by consumers as a food over a long period. In addition to lacking toxicity, they must be non-allergenic and be metabolized and excreted and should not lead to residue build-up. Food antimicrobials should not react either to make important nutrients unavailable or destroy these nutrients. Their potential impact on the sensory characteristics of a food must be considered. Compounds that negatively affect flavor and odor or contribute inappropriate flavor and odor would be unacceptable.

PERSPECTIVE

In an attempt to satisfy the rapidly changing consumption patterns of the global market, the food processing industry is continuously developing various new and modified products. Such product development requires cutting-edge technology to ensure organoleptic distinction as well as other qualities such as wholesomeness, sanitary grade, shelf-life, and safety. Development of convenience foods, functional diets, nutraceutical health supplements also has necessitated the use of additives and antimicrobials. Many of the natural food antimicrobial systems also demonstrate a multifunctional physiological advantage and are emerging as value-added ingredients in various food products. Milk lactoferrin is now being explored as an antioxidant, an iron-delivery system and an immuno-modulating agent in various sports drinks and infant formulations. Globulin-rich ingredients such as whey-protein concentrates and hen egg proteins seem to reduce gastric ulcers in controlled clinical trials. Phyto-antimicrobials including thiosulfinate concentrates from garlic and oleuropeins from olives are suggested to lower cholesterol; catechins from green tea were shown to prevent gum disease; and flavonoids from cranberry were reported to lower the incidence of urinary tract infections. Lactic acid bacteria are believed to induce a probiotic effect and thereby improve gut physiology that might even contribute to the prevention of intestinal carcinogenesis. Many such natural food antimicrobial systems are currently used in a variety of over-the-counter remedies. Antimicrobial-flavorants including oleoresins of spices, vanillin, diacetyl and organic acids; antimicrobial-colorants such as turmeric, kola, anthocyanins, diterpenes, and stilbenes could serve as dual-function food additive systems. The application potential of natural food antimicrobial systems is enormous. The scientific basis for incorporation into various foods and the technical knowledge for the biofunctional maxima of natural food antimicrobial systems are currently being explored worldwide.

LITERATURE RECOMMENDED

1. Board, R.G., and Dillon, V. 1994. *Natural Antimicrobial Systems in Food Preservation*. Wallingford, UK: CAB International.
2. Davidson, P.M., and Branen, A.L. 1993. *Antimicrobials in Foods*. 2nd Edition. New York: Marcel Dekker.
3. De Vuyst, L., and Vandamme, E.J. 1994. *Bacteriocins of Lactic Acid Bacteria*. New York: Chapman and Hall.
4. Doyle, M.P., Beuchat, L.R., and Montville, T.J. 1997. *Food Microbiology – Fundamentals and Frontiers*. Washington, D.C: ASM Press.
5. Gould, G.W. 1989. *Mechanisms of Action of Food Preservation Procedures*. London: Elsevier Applied Science.
6. Gould, G.W. 1995. *New Methods of Food Preservation*. Glasgow, UK: Blackie Academic and Professional.
7. Hoover, D.G., and Steenson, L.R. 1993. *Bacteriocins of Lactic Acid Bacteria*. San Diego, California: Academic Press.
8. Jay, J.M. 1998. *Modern Food Microbiology*. 5th Edition. Gaithersburg, Maryland: Aspen Publishers.
9. Naidu, A.S., Bidlack, W.R., and Clemens, R.A. 1999. Probiotic spectra of lactic acid bacteria. *Crit. Rev. Food. Sci. Nutr.* 39:13-126.
10. Ray, B., and Daeschel, M.A. 1992. *Food Biopreservatives of Microbial Origin*. Boca Raton, Florida: CRC Press.
11. Russel, N.J., and Gould, G.W. 1991. *Food Preservatives*. London: Blackie and Son Ltd.
12. Sofos, J.N., Beuchat, L.R., Davidson, P.M., and Johnson, E.A. 1998. *Naturally Occurring Antimicrobials in Food*. Task Force Report No.132. Ames, Iowa: Council for Agricultural Science and Technology.

Section-I

LACTO-ANTIMICROBIALS

Lactoferrin
Lactoperoxidase
Lactoglobulins
Lactolipids

A.S. Naidu

Lactoferrin

2

I. INTRODUCTION

Lactoferrin (LF) is an iron-binding glycoprotein present in milk and many exocrine secretions that bathe the mucosal surface. Though the term 'LF' implies an iron-binding component in milk, this molecule co-ordinately binds to various metal ions and occurs in divergent biological milieu including saliva, tears, seminal fluids, mucins, and the secondary granules of neutrophils. LF has a multifunctional role in a variety of physiological pathways and is considered a major component of the preimmune innate defense in mammals. The ability of LF to bind two Fe^{+3} ions with high affinity in cooperation with two HCO_3^- ions is an essential characteristic that contributes to its major structure-functional properties including antimicrobial activities.

Iron is a transition metal belonging to group VII elements in the periodic table; it occurs at an approximate level of 2-g in the body of normal human adults and is essential for all living organisms. The ability of iron to alternate between its two valency states is its most important biological property, and is used in many metabolic pathways including the bioenergetic coupling of inorganic phosphate to adenosine phosphate (ADP) to form adenosine triphosphate (ATP). Nevertheless, this property can be hazardous for the cell, as it can lead to generation of reactive oxygen species. Under controlled conditions, however, iron is used beneficially, e.g., by the phagocytes in intra-lysosomal killing of microorganisms.

Various iron-binding proteins have evolved in the animal physiological system to sequester iron from the milieu. Ovotransferrin (conalbumin) from the egg white was the first iron-binding protein to be purified (Osborne & Campbell, 1900). In 1939, Sörensen and Sörensen identified a red iron-binding protein in bovine milk. Later, Schafer (1951) also reported a similar protein in human milk termed 'siderophilin'. Schade and Caroline (1946) isolated an iron-binding protein from human serum (serotransferrin) which was later named 'transferrin (TF)'. In 1960, Groves from the United Kingdom, Johansson from Sweden, Montreuil and co-workers from France, and Gruttner and co-workers from Germany independently isolated the red milk protein 'lactosiderophilin' from milk.

0-8493-2047-X/00/$0.00+$.50
© 2000 by CRC Press, LLC

During the 1960s several investigators isolated and characterized this protein from various exocrine secretions and tissues of humans and animals (Gordon et al., 1963; Blanc et al., 1963; Masson et al., 1965a; 1965b; 1965c; 1966; 1969; Masson & Heremans, 1968). Based on structural and chemical homology with serum TF, Blanc and Isliker (1961a; 1961b) proposed the name 'lactoferrin' for this protein. Though LF is not exclusively found in milk, this name is now widely recognized in the scientific community, although the term 'lactotransferrin' can be found in earlier publications.

LF was isolated as a major component in the specific granules of the polymorphonuclear leukocytes (PMNLs) with an important role in the amplification of inflammatory responses (Masson et al., 1969; Baggiolini et al., 1970; Oseas et al., 1982; Boxer et al., 1982a; 1982b; 1982c; Lash et al., 1983; Ambruso et al., 1984; Brittigan et al., 1989). LF has also been reported to be a component of the sperm-coating antigen (Ashorn et al., 1986) and was found to have cross-reactivity and sequence homology with the major histocompatibility antigen (Aguas et al., 1990). Extensive work by Masson and his Belgian group has established a clear role for LF in cellular immunity and has led to the identification of specific LF-receptors on macrophages, intermediation of endotoxic shock and hyposideremia (Van Snick et al., 1974; Van Snick & Masson, 1976). Pioneering efforts by Spik, Montreuil and their French group unraveled the biological chemistry of LF (Mazurier et al., 1981; Spik et al., 1982; Metz-Boutigue et al., 1984). Lönnerdal has opened the nutritional role for LF in the absorption of metal ions in the intestinal tract (Davidson & Lönnerdal, 1986; Davidson et al., 1990; 1994). In 1978, Broxmeyer and co-workers reported a regulatory function for LF in myelopoiesis (Broxmeyer et al., 1978; 1984).

In 1961, Oram and Reiter reported the ability of milk LF to inhibit the growth (stasis effect) of *Bacillus* sp. and found that nutritional deprivation of the bacteria from iron accounted for the antimicrobial activity. The antimicrobial spectrum of LF was further elucidated by Brock's group (Brock et al., 1978; 1983; Brines & Brock, 1983). In 1977, Arnold and co-workers reported cidal activity for LF against a variety of microorganisms at acidic pH and noted that it is not inhibited by salts in the media (Arnold et al., 1977; 1980; 1981). Despite wide citation in literature, the cidal effect of native LF appears to be artifactual in nature, caused by reactants other than LF in the milieu (Lassiter, 1990). The research group at the Morinaga Milk Company in Japan found that acid/pepsin hydrolysis of bLF could generate cationic antimicrobial peptides 'lactoferricins (LFcins)' (Tomita et al., 1989; 1991; 1994; Bellamy et al., 1992; 1993). This broad-spectrum cationic microbial killing seems to be non-specific and readily inhibited by salts in the milieu at physiological concentrations. Erroneously, LFcins are widely described in the literature as bactericidal domains, implying they are peptides responsible for the antimicrobial effect reported by Arnold's group.

Interaction of LF with specific targets on the microbial surface causes an array of outcomes either to the advantage of the host (microbial blocking effects) or the microorganism (iron-acquisition and pathogenesis).

Naidu and co-workers have identified, isolated and characterized LF-binding microbial targets in a variety of Gram-positive and Gram-negative bacteria (Naidu et al., 1990; 1991a; 1991b; 1992; 1993). Specific high-affinity interaction of LF with pore-forming outer membrane proteins (OMPS) of *Escherichia coli*, in particular, has unraveled a molecular mechanism for antimicrobial activity which seems to be well conserved in

Gram-negative enterics (Kishore et al., 1991; Tigyi et al., 1992; Erdei et al., 1993; Naidu & Arnold, 1994). The LF-mediated outer membrane damage in Gram-negative bacteria reported by Ellison et al. (1988) has explained certain antimicrobial effects such as antibiotic potentiation, release of lipopolysaccharides (LPS) and alterations in microbial OM permeation.

The critical role of iron in the pathogenesis of many microbial infections has been widely advocated (see reviews: Weinberg et al., 1975; Bullen, 1981). Thus, the mobilization of iron from physiological milieu from LF, TF and ferritin by pathogens appeared to be an important virulence trait. Rogers and Synge (1978) reported a siderophore-mediated mobilization of iron from LF by *E.coli*. Another type of iron-acquisition mechanism involving specific receptors was reported by Alderete's group as well as Sparling and co-workers, as a virulence factor in various intracellular pathogens, in particular, among the etiological agents of sexually transmitted diseases (Mickelsen et al., 1982; Peterson & Alderete, 1984; Biswas & Sparling, 1995). Later, Schryvers' group identified and characterized a number of specific receptors for LF on various mucosal pathogens (Schryvers & Morris, 1988; Schryvers, 1989).

This chapter is mainly confined to the antimicrobial spectrum of milk LF and its possible role in food safety and preservation. The multifunctionality of LF is also described to emphasize its nutraceutical benefits as a value-added food ingredient.

II. OCCURRENCE

A. Normal levels

In the human body, LF occurs in two major reservoirs, a circulatory pool stored in the polymorphonuclear lymphocytes (PMNL) and a stationary pool on the mucosal surfaces. In PMNL, LF is associated with the secondary (specific) granules at a concentration of about 15 μg/10^6 cells (Baggiolini et al., 1970; Bennett & Kokocinski, 1978) and is released isochromously with other lysosomal proteins into the plasma during phagocytosis (Bennett & Skosey, 1977; Leffell & Spitznagel, 1975). LF content in plasma is low or undetectable during agranulocytosis (Bennett & Kokocinski, 1978) and neutropenia (Hansen et al., 1975; Olofson et al., 1977), thus suggesting that PMNL are the sole source of intravascular LF production. The concentration of LF in human plasma is about 0.2 to 1.5 μg/ml (Rumke et al., 1971); the values are comparatively lower in women than in men (Bennett & Mohla, 1976). The plasma LF levels are also low in children due to reduced PMNL secretion (Gahr et al., 1987). LF is rapidly eliminated from the plasma with a mean fractional catabolic rate of 5.7/day, by liver and spleen; however, apo-LF is removed at a slower rate of 1.22/day (Bennett & Kokocinski, 1979).

Hirai and co-workers (1990) measured LF concentration in human milk and colostrum from 1 to 60 days after parturition (125 samples) by rocket immuno-electrophoretic assay using anti-hLF antiserum. The LF concentrations in colostrum (1-3 days of puerperium, n = 35), the transitional milk (4-7 days, n = 60), and mature milk (20-60 days, n = 30) were about 6.7, 3.7, and 2.6 g/L, respectively. Both the LF and total protein (TP) concentrations showed inverse correlation with the days after parturition. The LF/TP ratio in the mature milk (16.1%) was significantly less than that in the colostrum (20.4 %) and the transitional milk (21.4%). Furthermore, iron concentration in human milk was also measured by the internal standard technique of the spiked method on atomic absorption, and the LF iron-saturation was calculated. Neither Fe nor iron-saturation % showed a significant difference among these three stages of lactation. The means (n = 125) of Fe and iron-saturation % were about 0.61 µg/ml and 11.8%, respectively. However, significant correlation was observed between LF and Fe or between LF/TP and both Fe and TP in the mature milk. These results suggest that the mechanism stimulating the synthesis and secretion of LF is different from those of other proteins and LF can play variable roles in iron nutrition of infants during different stages of lactation.

An immunoperoxidase staining technique was used for detecting LF, TF and ferritin in routine histological paraffin sections of human tissue. LF was found in lactating breast tissue, bronchial glands, PMNLs, and gastric and duodenal epithelial cells (Mason & Taylor, 1978). The expression of LF was studied in human gastric tissues displaying normal, benign hyperplastic or malignant histology (Luqmani et al., 1991). A single 2.5-kb mRNA was detected in only 14% (2/14) of normal resections. This was similar to the finding that 85% of tumors were also negative, with 4/27 positive. In contrast, samples with superficial or atrophic gastritis had a high frequency of expression, with 5/7 and 9/14 positive, respectively. The higher incidence of LF mRNA in antral samples was a reflection of the greater proportion of these compared with body resections of patients with gastritis. No expression was seen in any of 5 gastric carcinoma cell lines. High levels were observed in the cardia, in contrast to complete absence in the oesophagus. Immunocytochemistry showed localization of LF in cells of both antral and body glands. Chief cells, but not adjacent parietal cells, were strongly stained. In tissues exhibiting superficial or atrophic gastritis a greater degree and intensity of staining was observed as compared with samples with normal histology

LF accounts for about 11.5% of the total secretary proteins of the bronchial glands (Harbitz et al., 1984). In these glands, LF is uniformly distributed in large amounts in the serous cells and is restricted only to the basal part of cytoplasm in the mucus cells (Masson et al., 1966). LF is released into the nasal secretions by serous cells of submucosal glands under the influence of the parasympathetic nervous system (Raphael et al., 1989). In human tears, LF is synthesized in the lacrimal glands (Janssen & van Bijsterveld, 1982; 1983) at a concentration of 1 to 3 mg/ml that accounts for ~25% of the total tear protein (Kijlstra et al., 1983). In the male reproductive tract, LF is found in the prostate, seminal vesicles, and seminal plasma (0.2 to 1.0 mg/ml), but not in testis (Rumke, 1974; Wichmann et al., 1989). As the sperm-coating antigen, LF may suppress lymphocyte response against sperm (Ashorn et al., 1986). It may also partially account for the antimicrobial and immune-suppressive activity of the seminal fluid in the female reproductive tract before and during fertilization (Broer et al., 1977). In the female reproductive tract, LF has been detected in the cervical mucus and endometrium of the secretory uterus (Tourville et al., 1970)

B. Clinical levels

Acute phase host responses such as an inflammation or toxic shock result in the depletion of iron from plasma (Klasing, 1984). The interleukin 1 (IL-1) seems to mediate multiple aspects of acute phase response and also induce exocytosis of PMNLs to release intracellular granule contents, including LF (Klempner et al., 1978; Dinarello, 1984). Eventually, the plasma LF levels increase in various pathological conditions and its estimation may serve as a prognostic marker. Thus, the elevated plasma LF level is an early indicator of septicemia and endotoxemia (Gutteberg et al., 1988). During meningococcal septicemia, LF level in the serum and cerebrospinal fluid is elevated within 18 h of onset as to the same magnitude of the C-reactive protein (Gutteberg et al., 1984). Similarly, an increase in the plasma and serum LF levels during cystic fibrosis seems to correlate with the intensity of the cystic fibrosis protein (Barthe et al., 1989). The estimation of elevated LF concentration with cholecytokinin secretin (CCK-S) test has been suggested to improve the differential diagnosis of chronic pancreatitis (Dite et al., 1989). LF to lysozyme ratio in the crevicular fluid may be helpful in the diagnosis of localized juvenile periodontitis (Friedman et al., 1983).

In the pathogenesis of rheumatoid arthritis, iron has an important role in the degradation of intact cartilage matrices, due to its capacity to generate free radicals that activate latent collagenases (Blake et al., 1981; 1984; Bukhhardt & Schwingel, 1986). Elevated levels of LF in the synovial fluids and synovial membranes due to inflammatory tissue damage is a characteristic of this disease (Ahmadzadeh et al., 1989). This high LF concentration has been suggested in the regulation of subsequent inflammatory processes critical for articular damage.

Neoplastic cells have an increased iron-requirement for the initiation and maintenance of DNA synthesis and for cell multiplication (Gatter et al., 1983; Barresi & Tuccari, 1987). In different types of malignancies, iron uptake may be mediated by specific receptors for LF or TF. Immunohistochemical studies have demonstrated LF in adenocarcinomas of the parotid gland (Caselitz et al., 1981), well differentiated prostatic carcinomas, breast carcinomas (Charpin et al., 1985), thyroid carcinomas of follicular origin (Tuccari & Barresi, 1985; Barresi & Tuccari, 1987), renal cell carcinomas (Loughlin et al, 1987), intestinal-type carcinomas, and incomplete intestinal metaplasia (Tuccari et al., 1989).

III. ISOLATION AND PURIFICATION

LF from bovine milk was first isolated by Groves (1960). Various procedures for isolation of LF from mammalian milk have been reported. The most commonly used methods include chromatographic separation on CM-Sephadex, Cibacron Blue F3G-A-Sepharose, Heparin agarose and single stranded DNA agarose (*FIGURE 1*).

A. Size-exclusion chromatography on CM-Sephadex columns

Gel filtration is one of the earlier techniques used for isolation of LF from various biological secretions (Butler, 1973). Tsuji and co-workers (1989) described a method to isolate LF from bovine colostrum. Colostrum obtained within 24 h after parturition was centrifuged at 3000 rpm for 10 min and 1 liter of the skimmed colostrum was dialyzed against distilled water in the presence of 0.01% sodium azide for 3 days at 4°C with several changes of distilled water. The dialyzed skimmed colostrum was loaded on a CM-Sephadex C-50 (Pharmacia) column (2 x 20 cm) equilibrated with 0.05 M potassium phosphate buffer, pH 8.0 (buffer-A). LF was eluted with a linear gradient of NaCl (0.1 to 0.7 M) in buffer-A after unbound proteins and weakly bound proteins were washed out sequentially with 100 ml of buffer-A and 100 ml of buffer-A containing 0.05 M NaCl. The combined fraction containing LF was dialyzed against buffer-A at 4°C overnight and re-loaded onto a column of CM-Sephadex C-50 (1 x 5 cm). LF was eluted with a linear gradient of NaCl (0.2 to 0.5 M) in buffer-A, and at this step, 192 mg of nearly homogenous LF was obtained.

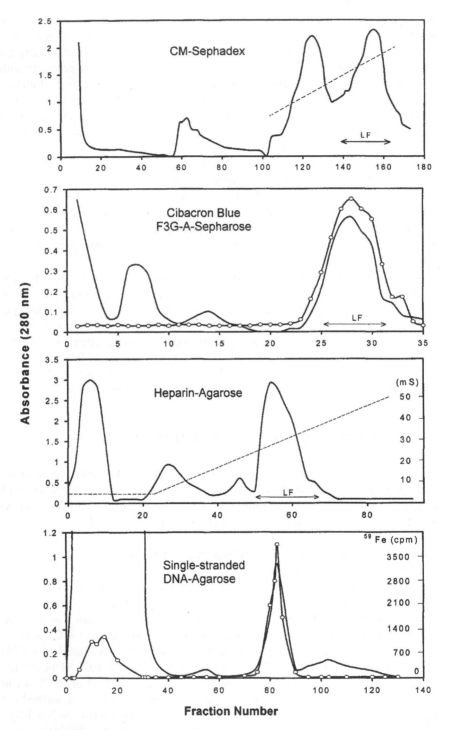

FIGURE 1. LF isolation methods - elution profiles on different chromatography columns [redrawn from Tsuji et al., 1989; Furmanski et al., 1989; Rejman et al., 1989; Hutchens et al., 1989].

B. Affinity-chromatography on Cibachron Blue-Sepharose columns

Various laboratories have adapted this method to isolate LF from human milk (Arnold et al., 1977; Bezwoda & Mansoor, 1986; Furmanski et al., 1989; Shimazaki & Nishio, 1991). Briefly, 5 ml of pooled human milk is skimmed and decaseinated by centrifugation (10,000 g, 40 min), acidification (to pH 4.7 with HCl), heating (40°C, 30 min) and recentrifugation (10,000 g, 40 min). The milk whey is dialyzed overnight against 500 ml veronal buffer. The whey was diluted to 25 ml with veronal buffer. The sample was applied to Cibachron Blue Sepharose CL-6B (Pharmacia) column (1 x 10 cm). The column is washed with 25-ml veronal buffer and eluted with a 30-ml linear gradient of 0.5-1 M NaCl in veronal buffer. A single protein peak corresponding to LF elutes at about 0.75 M NaCl. The radioisotope labeled ^{59}Fe-saturated LF demonstrated an elution pattern similar to apo-LF during this separation method (Furmanski et al., 1989).

C. Affinity-chromatography on heparin-cross linked columns

Heparin-agarose affinity chromatography has been used to isolate LF from human milk whey in a single chromatographic step (Blackberg & Hernell, 1980). Al-Mashikhi and Nakai (1987) have used heparin-sepharose affinity chromatography to isolate LF from cheddar cheese whey. In this procedure, whey is dialyzed against 0.05M NaCl in 0.005 M sodium barbital-HCl buffer, pH 7.4. Whey solution is applied to a heparin-agarose column equilibrated with the above dialysis buffer. Protein is eluted at a flow rate of 48 ml/h using a continuous gradient of 0.05M to 1.0 M NaCl constituted in the dialysis buffer. Fractions are collected and absorbance is read at 280 nm. Rejman et al. (1989) used this method to isolate bLF from mammary secretions collected during the nonlactating period. About 1600 absorbance units (280 nm) of whey protein were efficiently separated by the heparin-agarose column (packed with 2.0 x 16.5 cm of Affi-Gel heparin agarose from Bio-Rad) into four absorbance peaks. LF was identified in the fourth peak (eluting at a conductivity of 30 mS). Both iron-saturated and apo-forms of LF demonstrated similar elution profiles in this technique. IgG1 and secretory component are the trace contaminants with bovine milk whey fractionation. In contrast, serum albumin was identified as the contaminant with human milk whey separation on the heparin column.

D. Affinity-chromatography on DNA-agarose columns

Hutchens and co-workers (1989) reported that immobilized DNA is effective for a rapid and complete purification of apo-LF and holo-LF from colostrum in a single step. Urea was utilized as a mobile phase modifier to eliminate the interaction of other proteins such as serum albumin both with LF and with the immobilized DNA. Briefly, in this method, single-stranded DNA-agarose is packed into a 1.0 x 1.5 cm column to a bed volume of 5 to 10 ml. The column is washed with water, then equilibrated with 20 mM Hepes buffer, pH 8.0, with or without 6 M urea, at a flow rate of 30 ml/h. The separation procedure is performed at room temperature. Solid urea (up to 6 M) is added before the sample (5 to 10 ml) is applied to the DNA affinity column. The column is washed with equilibration buffer and prior to gradient elution, urea (if present) is removed with several column volumes of 20 mM Hepes buffer, pH 8.0 (HB). LF is eluted with a linear gradient of NaCl (0 to 1.0 M) in 20 mM HB. Fractions of 1 ml each are collected and the absorbance was measured at 280 nm. The radioisotope labeled ^{59}Fe-saturated LF demonstrated an elution pattern similar to apo-LF during this separation method. Finally, after each purification procedure the DNA-agarose column was washed extensively, first with 2 M NaCl in HB, followed by 8 M guanidine-HCl in HB and finally with water.

Iron saturation and desaturation: Various methods of iron saturation (holo-form) and desaturation (apo-form) of LF have been described. Briefly, apo-LF is prepared by dialysis against an acetic acid/sodium acetate buffer (pH 4.0), followed by exhaustive dialysis against deionized distilled water (Mazurier & Spik, 1980). Holo-LF is prepared by adding a large excess of citric acid (pH 2.5; 60 mol citric acid:1 mol iron). After incubation for 10 min, the pH is raised to 7.0 with 0.1 M NaOH. Excess sodium bicarbonate is added (2 mol bicarbonate:1 mol iron). Unbound iron is removed by gel filtration (Azari & Baugh, 1967).

Masson and Heremans (1968) described a method to prepare apo-LF. A concentration of 1% LF solution was deprived of iron by dialysis against 20 volumes of 0.1 M citric acid. After 36 h the citrate was eliminated by dialysis against 20 volumes of deionized water for 2-h at 4°C. The required amount of solid disodium phosphate was added to the dialysis flask. Stirring was discontinued and the flask was kept at 4°C. The pH of the suspension increased slowly as the crystals of disodium phosphate dissolved into the solution. This precaution is needed to prevent precipitation of the protein. Stirring is resumed as soon as the pH in the upper layer of buffer has reached a value of 5.0. The final pH is 7.6. After such treatment, the solution of LF turns completely colorless.

Fe(III) removal from LF by an Fe(III)-chelating resin with immobilized 3-hydroxy-2-methyl-4(1H)-pyridinone ligands at physiological pH in the presence of citrate has been described (Feng et al., 1995). The resin had a marked effect on the extent of iron removal. By using the Fe(III)-chelating resin, removal of iron from LF was nearly complete in < 24 h. Apo-LF with 4% iron saturation could be prepared conveniently from 100% or from 18% iron-saturated LF under mild conditions without affecting the iron-binding capacity of the protein.

IV. MOLECULAR PROPERTIES

A. Physico-chemistry

The physico-chemical characteristics of hLF and bLF are listed in *TABLE 1*. Like the transferrins of blood serum and egg white, LF is a single polypeptide chain with a molecular weight in the range 75 to 80-kDa. Dry weight determinations together with measurement of iron-binding capacities, showed combining weights per iron atom bound of 39,000 for bLF, and 40,000 for hLF (Aisen & Leibman, 1972). Accordingly, these correspond to molecular weights of 78,000 and 80,000 daltons, respectively, for protein molecules with two specific binding sites.

The isoelectric point (pI value) of bLF is reported at about 8.0 by free boundary electrophoretic methods (Groves, 1960; Szuchet-Drechin & Johanson, 1965); pI of 8.8 by Rotafors method (Shimazaki et al., 1993) and a pI of 8.2-8.9 by chromatofocusing (Shimazaki et al., 1993). On the otherhand, a wide range of pI values, from 5.5 to 10.0 have been reported for hLF by isoelectric focusing techniques [5.8 to 6.5 by Bezwoda & Mansoor, 1989; 6.9 by Malmquist & Johanson, 1971; 8.7 by Moguilevsky et al., 1985; 8.8

TABLE 1. Lactoferrin – Physico-chemical properties

Property	Human LF	Bovine LF	Reference
Molecular mass			
Sedimentation co-efficient	75,100	77,200 ± 1,300	Castellino et al., 1970
SDS-PAGE	76,800 ± 1,600	76,000 ± 2,400	Querinjean et al., 1971
Iron titration	80,000	78,500	Aisen & Leibman, 1972
Isoelectric point			Bezwoda & Mansoor, 1989
Chromato focusing	6.8 – 8.0	8.2 – 8.9	Yoshida & Xiuyun, 1991
Isoelectric focusing	5.8 – 6.5	9.5 – 10.0	Shimazaki et al., 1993
Absorption spectra			Aisen & Leibman, 1972
Apo-form at 280 nm	10.9	12.7	
Holo-form at 470 nm	0.510	0.400	
Glycosylation	Relatively high	Low	Metz-Boutigue et al., 1984
Protease sensitivity	Relatively low	High	Brines & Brock, 1983
IgA-complexes	Present	Absent	Watanabe et al., 1984
Iron-binding			
Equilibrium dialysis ($K_1 \times 10^{-4}$)	26.0	3.73	Aisen & Leibman, 1972
Thermal denaturation			Paulsson et al., 1993
Apo-LF denaturation (Tmax: °C)		71 ± .3 & 90 ± .3	
Apo-LF enthalpy (ΔH_{cal}: J/g)		12 ± .4 & 2 ± .5	
Holo-LF denaturation (Tmax: °C)		65 ± .3 & 93 ± .3	
Holo-LF enthalpy (ΔH_{cal}: J/g)		2 ± 1 & 37 ± 1	

to 8.9 by Birgens & Kristensen, 1990; and 8-10 by Kinkade et al., 1976]. By the Rotafors method hLF was focused at a pI of 8.7 and by chromatofocusing the hLF was eluted at pH 6.8-8.0 (Shimazaki et al., 1993).

Heat-induced enthalpy changes in different forms of bLF in water were examined by differential scanning calorimetry (Paulsson et al., 1993). Two thermal transitions with varying enthalpies were observed, depending on the iron-binding status of the protein. Iron-saturated holo-LF was more resistant to heat induced changes than the apo-form.

Investigations of metal-substituted hLF by fluorescence, resonance Raman, and electron paramagnetic resonance (EPR) spectroscopy confirm the close similarity between LF and serum TF (Ainscough et al., 1980). As in the case of Fe(III)- and Cu(II)-TF, a significant quenching of apo-LF's intrinsic fluorescence is caused by the interaction of Fe(III), Cu(II), Cr(III), Mn(III), and Co(III) with specific metal binding sites. Laser excitation of these metal-LFs produce resonance Raman spectral features at about 1605, 1505, 1275, and 1175 cm-1. These bands are characteristic of tyrosinate to the metal ions, as has been observed previously for serum TF, and permit the principal absorption band (l_{max} between 400 and 465 nm) in each of the metal-LFs to be assigned to charge transfer between the metal ion and tyrosinate ligands. Furthermore, as in serum TF the two metal binding sites in LF can be distinguished by EPR spectroscopy, particularly with the Cr(III)-substituted protein. Only one of the two sites in LF allows displacement of Cr(III) by Fe(III). LF is known to differ from serum TF in its enhanced affinity for iron. Accordingly, the kinetic studies show that the rate of uptake of Fe(III) from Fe(III)-citrate is 10 times faster for apo-LF than for apo-TF. Furthermore, the more pronounced conformational change which occurs upon metal binding to LF is corroborated by the production of additional EPR-detectable Cu(II) binding sites in Mn(III)-LF. The lower pH required for iron removal from LF causes some permanent change in the protein as judged by altered rates of Fe(III) uptake and altered EPR spectra in the presence of Cu(II). Thus, the common method of producing apo-LF by extensive dialysis against citric acid (pH 2) appears to have an adverse effect on the protein.

The anion binding properties of hLF, with Fe(III) or Cu(II) as the associated metal ion, highlight differences between the two sites, and in the anion binding behavior when different metals are bound (Brodie et al., 1994). Carbonate, oxalate and hybrid carbonate-oxalate complexes have been prepared and their characteristic electronic and EPR spectra recorded. Oxalate can displace carbonate from either one or both anion sites of $Cu_2(CO_3)_2$ LF, depending on the oxalate concentration, but no such displacement occurs for $Fe_2(CO_3)_2$ LF. Addition of oxalate and the appropriate metal ion to apo-LF under carbonate-free conditions gives dioxalate complexes with both Fe^{3+} and Cu^{2+}. Both the carbonate and oxalate ions bind in bi-dentate fashion to the metal, except that the carbonate ion in the N-lobe site of dicupric LF is mono-dentate. The hybrid copper LF complex shows that the oxalate ion binds preferentially in the C-lobe site in a bi-dentate mode. Overall these observations lead to a generalized model for synergistic anion binding by TF and allow comparisons to be made with non-synergistic anions such as citrate and succinate.

The amino acid composition of LF molecules isolated from the milk of different mammalian species is shown in *TABLE 2*. Tryptic peptide maps of hLF show some 40 spots, which is a much smaller number than would be predicted from the lysine and arginine content (Querinjean et al., 1971). LF is a glycoprotein containing two glycans attached through *N*-glycosidic linkages. The two *N*-acetyl-lactosaminic-type glycans are structurally heterogenous (Spik et al., 1982) and differ from those of other transferrins (Spik et al., 1975; Dorland et al., 1979; Van Halbeck et al., 1981). Analysis of bLF for carbohydrate content reveals 1 residue of terminal sialic acid, 10 to 11 residues of *N*-acetyl

TABLE 2. Lactoferrin - Amino acid composition of protein isolated from different species

Amino acid	Human	Bovine	Porcine	Monkey	Murine	Equine
Aspartate	71	71	57	72	71	72
Threonine	31	39	28	31	40	32
Serine	50	45	46	43	50	48
Glutamate	70	73	71	67	71	70
Proline	35	31	30	34	33	33
Glycine	56	43	50	61	51	50
Alanine	63	59	56	73	60	71
Cysteine	32	28	X	23	28	38
Methionine	6	4	4	2	5	3
Valine	49	43	42	50	45	44
Isoleucine	16	17	18	15	17	14
Leucine	61	61	53	62	57	60
Tyrosine	20	19	20	29	18	20
Phenylalanine	31	25	26	30	25	25
Tryptophan	11	9	X	X	10	12
Lysine	46	42	40	47	55	45
Histidine	9	10	8	9	10	10
Arginine	46	32	36	40	32	33

Adapted from Hutchens et al., 1989

glucosamine, 5 to 6 residues of galactose, and 15 to 16 residues of mannose per molecule (Castellino et al., 1970). In hLF, according to the sequence studies by Metz-Boutigue et al., 1984), asparagine residues 137 and 490 are glycosylated. Prediction of the secondary structure suggested that the two prosthetic sugar groups were linked to asparagine residues located in a β-turn conformation. The non-glycosylated asparagine residue 635 also occurs in a β-turn whereas asparagine residue 389 is located in a region of non-pre-dictable structure. Spik et al. (1982) elucidated the primary structure of glycans from hLF. The polypeptide chain of hLF consists of two glycoslyation sites to which glycans are linked through an *N*-(β-aspartyl)-*N*-acetylglucosaminylamine bond and which are struc-turally heterogenous. After chymotryptic or pronase digestions, glycopeptides with five different glycan structures were isolated. Three of these structures were determined by using methanolysis, methylation analysis, hydrazinolysis/nitrous deamination/ enzymatic cleavage and 1H-NMR spectroscopy at 360 MHz. *Glycopeptides-A/B:* NeuAc(α-2-6)Gal(β-1-4)GlcNAc(β-1-2)Man(α-1-3)[NeuAc(α-2-6)Gal(β-1-4)GlcNAc(β-1-2)Man(α-1-6)]Man(β-1-4)GlcNAc(β-1-4)[Fuc(α-1-6)]GlcNAc(β-1-)Asn;*Glyco-peptide-C:*NeuAc(α-2-6)(Gal(β-1-4)GlcNAc(β-1-2)Man(α-1-3)(Gal(β-1-4)[Fuc(α-1-3)] GlcNAc(β-1-2)Man(α-1-6))Man(β-1-4)GlcNAc(β-1-4)[Fuc(α-1-6)]GlcNAc(β-1-)Asn. *Glycopeptide-D:*NeuAc(α-2-6)Gal(β-1-4)GlcNAc(β-1-2)Man(α-1-3)[Gal(β-1-4)GlcNAc (β-1-2)Man(α-1-6)]Man(β-1-4)Glc-NAc(β-1-4)[Fuc(α-1-6)]GlcNAc(β-1)Asn. The other two glycopeptides were obtained in very low amounts with more com-plex structures.

It is generally believed that each iron-binding site contains two or three tyrosine residues (Windle et al., 1963) and one or two histidine residues (Krysteva et al., 1975; Mazurier et al., 1981); and concomitantly bound bicarbonate ion (Schlabach & Bates, 1975) may be held electrostatically to an arginyl side group (Rogers et al., 1978).

B. Structure

LF is a member of the iron-binding protein family collectively known as transfer-rins (TF). Human LF demonstrates amino acid sequence homology (more pronounced in the C-terminal region) with serum hTF (59%) and hen ovotransferrin (49%). Computer analysis has established an internal homology of the two lobes (residues 1-338, and 339-703), each containing a glycosylation site (aspargine residues 137 and 490) located in homologous position (Metz-Boutigue et al., 1984). Each lobe has a capacity to bind one Fe^{+3} ion with high affinity (K_d =10^{-20} M^{-1}) in the presence of a carbonate or bicarbonate anion (Harris, 1986). It has been suggested that the iron-binding site contain two or three tyrosine residues (Windle et al., 1963; Teuwissen et al., 1972) and one or two histidine residues (Mazurier et al., 1981); the concomitantly bound bicarbonate anion (Schlabach & Bates, 1975) may be held electrostatically to an arginyl side group (Rogers et al., 1978).

Baker and co-workers have extensively studied the three-dimensional structure of LF (*FIGURE 2*).The hLF molecule at 3.2-Å resolution has two-fold internal homology. The N- and C-terminal halves form two separate globular lobes, connected by a short alpha-helix, and carry one iron-binding site each (Anderson et al., 1987). The two lobes of the molecule have very similar folding; the only major differences being in surface loops. Each lobe is subdivided into two dissimilar α/β domains, one based on a six-strand-ed mixed β-sheet, the other on a five-stranded mixed β-sheet, with the iron site in the inter-domain cleft. The two iron sites appear identical at the present resolution. Each iron

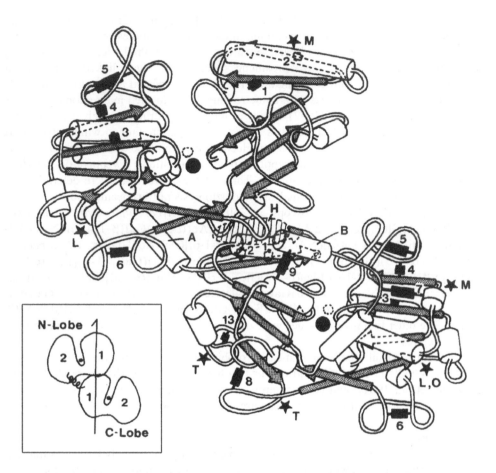

FIGURE 2. LF molecule: A schematic diagram. Helices are shown as cylinders, β-strands as arrows, iron atoms ●, probable anion sites ◯, disulfide bridges ■, and carbohydrate attachment sites ★ (labeled L for lactoferrin, T for serum transferrin, O for ovotransferrin, and M for melanotransferrin). The N-terminal half (N-lobe) is at top, the C-terminal half (C-lobe) at bottom; their relative orientations, related by a two-fold screw axis are shown in the inset. The two domains in each lobe are labeled 1 and 2. A region where hydrophobic interactions between the two lobes are made is indicated (H). Helices labeled are the connecting helix (A) and the C-terminal helix (B) [from Baker et al. (1987) with permission from the Elsevier Publications].

atom is coordinated to four protein ligands, 2 tyrosine, 1 aspartate, 1 histidine, and the specific CO_3^{2-}, which appears to bind to iron in a bi-dentate mode. The anion occupies a pocket between the iron and two positively charged groups on the protein, an arginine side-chain and the N terminus of helix 5, and may serve to neutralize this positive charge prior to iron binding. A large internal cavity, beyond the arginine side-chain, may account for the binding of larger anions as substitutes for CO_3^{2-}. Residues on the other side of the iron site, near the inter-domain crossover strands, could provide secondary anion binding sites, and may explain the greater acid-stability of iron binding by LF, compared with serum TF (Anderson et al., 1989).

X-ray structure analyses of four different forms of hLF (diferric, dicupric, an oxalate-substituted dicupric, and apo-LF), and of bovine diferric LF, have revealed various ways in which the protein structure adapts to different structural and functional states (Baker et al., 1991). Comparison of diferric and dicupric LFs revealed that different metals, through slight variations in the position, have different stereo-chemistry and anion coordination without any significant change in the protein structure. Substitution of oxalate for carbonate indicated that small side-chain movements in the binding site could accommodate larger anions. The multi-domain nature of LF also allows rigid body movements. The structure of apo-LF demonstrated the importance of large-scale domain movements for metal binding and release and suggested equilibrium between open and closed forms in solution, with the open form being the active binding species.

The crystal structure of a site-specific mutant of the N-terminal half-molecule of hLF(N), in which the iron ligand aspartate-60 has been mutated to serine, was studied to determine the effects of the mutation on iron binding and domain closure (Faber et al., 1996). At the mutation site the serine side-chain neither bound to the iron atom nor made any inter-domain contact similar to the substituted aspartate; instead a water molecule filled the iron coordination site and participated in inter-domain hydrogen bonding. The domain closure was also changed, with the mutant with a more closed conformation. Consideration of crystal packing suggested that the altered domain closure is a genuine molecular property but both the iron coordination and inter-domain contacts were consistent with weakened iron binding in the mutant.

The role of conserved histidine ligand in iron binding of LF was studied by site-directed mutagenesis and X-ray crystallographic analysis (Nicholson et al., 1997). *His-253* in the N-terminal half-molecule of hLF (residues 1–333) was changed to *Gly, Ala, Pro, Thr, Leu, Phe, Met, Tyr, Glu, Gln*, and *Cys* by oligonucleotide-directed mutagenesis. The mutant proteins were expressed in baby hamster kidney cells, at high levels, and purified. The study indicated that the *His* ligand is essential for the stability of the iron binding site. All of the substitutions destabilized iron binding irrespective of whether the replacements were potential iron ligands or not. Iron was lost below pH approximately 6.0 for the *Cys, Glu*, and *Tyr* mutants and below pH 7.0 or higher for the others, compared with pH 5.0 for the N-terminal half molecule. The destabilization was attributed to both steric and electronic effects. The decreased stability of the iron binding was attributed solely to the loss of the *His* ligand as the protein conformation and inter-domain interactions were unchanged.

	1		10		20		30
Human	G	RRRRS	VQWCA	VSQPE	ATKCF	QWQRN	MRKVR
Bovine		APRKN	VRWXT	ISQPE			
Porcine		APKKG	VRWCV	ISTAE	YSKCR	QWQSK	I RRTN
Murine		KATT	VRWCA	VSNSE	EEKCL	RWQNE	MRKVG
Equine		APRKS	VRWCT	ISPAE	AAKCA	K FQRN	MKK

FIGURE 3. Comparison of N-terminal sequences of LF molecule from different mammalian species.

The three-dimensional structure of diferric bLF and factors that influence its iron binding were reported (Moore et al., 1997). The final model comprised 5310 protein atoms (residues 5 to 689), 124 carbohydrate atoms (from ten monosaccharide units, in three glycan chains), 2 Fe^{3+}, 2 CO_3^{2-} and 50 water molecules. The folding of bLF molecule was similar to that of hLF, but bovine species differed in the extent of closure of the two domains of each lobe, and in the relative orientations of the two lobes. Differences in domain closure were attributed to amino acid changes in the interface, and differences in lobe orientations to slightly altered packing of two hydrophobic patches between the lobes. Changed inter-domain interactions were implied to the lesser iron affinity of bLF, compared with hLF, and two lysine residues behind the N-lobe iron site of bLF offer new insights into the 'dilysine trigger' mechanism proposed for iron release by TFs. The bLF structure was also notable for several well-defined oligosaccharide units that demonstrate the structural factors that stabilize carbohydrate structure. One glycan chain, attached to *Asn-545*, appears to contribute to inter-domain interactions and possibly modulate iron release from the C-lobe.

C. Heterogeneity

Among mammalian LFs, the human protein has been widely characterized. The amino acid and cDNA sequence data indicate that several animal LFs share extensive regions of primary sequence homology. Specifically, N-terminal sequences for porcine LF (Hutchens et al., 1989) indicate homology between LFs from human (Metz-Boutigue et al., 1984), bovine (Wang et al., 1984), equine (Jolles et al., 1984), monkey (Davidson & Lönnerdal, 1986) and murine (Pentecost & Teng, 1987) origins (*FIGURE 3*). X-ray diffraction studies also have demonstrated certain degrees of structural homology between hLF and bLF molecules (Norris et al., 1986). Peptide mapping also suggested structural homology between porcine and human LFs (Kokriakov et al., 1988).

LFs isolated from various sites of the human body demonstrate antigenic similarity. Considering the total amino acid sequence (Metz-Boutigue et al., 1984) and the polydispersity of the glycan structures (Spik et al., 1982), hLF has an estimated molecular mass of 82,400 $\pm$ 400. However, hLF in a number of human body fluids was found to possess different electrophoretic mobility due to its interaction with acidic macromolecules (Hekman, 1971). Several reports have also suggested polymerization of LF to a variable degree in biological fluids. Different forms of LF seem to appear at distinct stages of certain infections. Tabak et al. (1978) have detected LF polymers in the saliva of a patient with acute parotitis, however, apparent dimers and monomers were recovered when the inflammation gradually subsided. Similarly, bLF trimers appear in milk during acute stages of bovine mastitis, while dimers and eventually monomers emerge as predominant forms during the healing process (Harmon et al., 1976). The LF aggregation phenomenon in calcium containing fluids seems to inactivate certain biological activities of the molecule, such as the feedback control of granulopoiesis (Bennett et al., 1981).

Three isoforms of hLF with identical molecular mass, pI, partial proteolytic peptide patterns, and N-terminal amino acid sequence, but with distinct RNAse activity, were reported. The LF-alpha form binds iron; and, the other two, LF-beta and LF-gamma forms, express potent RNAse activity but lack the iron binding capacity (Furmanski et al., 1989). Two apparent forms of LF were also identified in bovine colostrum and the molecular heterogeneity seems due to a varying degree of protein glycosylation (Tsuji et al., 1989).

V. ANTIMICROBIAL EFFECTS

Structural characteristics and spatial orientation of the molecule are critical factors in the functionality of an antimicrobial compound. Occurrence in various milieu strongly emphasizes the significance of the structure-function relationship in the multifunctionality of the LF. As an exocrine secretory protein LF is present in different biological fluids of varying viscosity, pH, and ionic strength and co-exists with continuously changing ratios of other physiological substances. Thus, LF may be expected to perform a different antimicrobial function in the tear or saliva compared to its activity against an enteric bacteria at the intestinal mucosa. Moreover, as an acute-phase reactant LF also exists as a regulatory molecule in the cellular pool such as the neutrophils, and contributes to antigen processing in the phagosomes. Considering the diversity of LF's role in innate defense, a broad-spectrum of antimicrobial activity is expected. Accordingly, various modes of antimicrobial effects have been reported for LF.

A. Stasis effect

The iron-chelating capacity of LF in the metal-binding pockets in co-ordination with the bicarbonate anion has been suggested in the nutritional deprivation and a consequent inhibition of microbial growth in the stasis effect. This hypothesis was supported with various laboratory findings, such as exogenous addition of iron into the milieu could reverse the stasis effect or iron-saturated LF is non-inhibitory. During the early 1960s, Reiter's laboratory suggested the stasis mechanism of antimicrobial action for LF, which was further substantiated by Masson and co-workers. The stasis effect has been verified and validated by various laboratories during the past three decades. There are vast numbers of peer-reviewed publications in the scientific literature and this section will discuss the salient points.

Kirkpatrick et al. (1971) reported the fungistatic effect of apo-LF against *Candida albicans* and suggested a role for LF in the host-defense mechanism in chronic mucocutaneous candidiasis.

Reiter et al. (1975) found that two strains of *E. coli* were inhibited by colostral whey after dialysis or dilution in Kolmer saline and addition of precolostral calf serum or LF. Undiluted dialyzed milk was not inhibitory due to its low LF content but became inhibitory after addition of 1 mg/ml of LF. The lack of inhibition in undiluted whey is due to the high concentration of citrate in colostral whey (and milk) and it is suggested that citrate competes with the iron-binding proteins for iron and makes it available to the bacteria. Addition of bicarbonate, which is required for the binding of iron by TF and LF, could overcome the effect of citrate.

Bishop et al. (1976) tested the bacteriostatic effects of apo-bLF against strains of coliform bacteria associated with bovine mastitis. As low as 0.02 mg of apo-bLF per ml resulted in marked inhibition of growth of all coliforms. The stasis effect was lost if saturated LF or iron plus apo-LF was added to the synthetic medium. The inhibition of growth increased as the concentration of apo-LF increased from 0.02 to 0.2 mg/ml for *Klebsiella pneumoniae* and 2 mg/ml for *Aerobacter aerogenes*, and *E. coli*. As the concentration of apo-LF was increased above 0.2 or 2 mg/ml, there was less inhibition of growth except for *E. coli*. These results are compatible with the hypothesis that coliform bacteria respond to low-iron environments by production of iron-sequestering agents that compete effectively with apo-LF for free iron. Addition of apo-LF plus citrate resulted in loss of growth inhibition. The molar ratio (citrate to apo-LF) was found to be more important than the absolute concentration of either component. A ratio of 75 resulted in 50% growth inhibition, whereas ratios of 300 and greater resulted in less than 10% growth inhibition. These results suggest that the ratio of citrate to LF would be important in evaluating LF as a nonspecific protective factor of bovine mammary secretions.

An *in vitro* microassay was developed to evaluate bacteriostatic properties of apo bLF (Nonnecke & Smith, 1984a). The growth of coliform, staphylococcal, and streptococcal bacterial strains in a defined synthetic medium was inhibited by apo-bLF (0.5 to 30.0 mg/ml). Addition of holo-LF to the synthetic medium did not inhibit growth of test strains. Inhibition by apo-LF was greater for coliform than Gram-positive strains for all concentrations of apo-LF evaluated. No concentration of apo-LF proved bactericidal for either coliform or Gram-positive strains. Inhibition of two coliform strains by apo-LF (10 mg/ml) was abolished by addition of ferric iron to the assay system, indicating an iron-dependent nature of apo-LF induced inhibition of bacteria. Bicarbonate supplementation of the growth system containing apo-LF (1 mg/ml) increased inhibition of three coliform strains by apo-LF. Addition of increasing concentrations of citrate (2.0 mg/ml) to an assay system containing apo-LF (5 mg/ml) resulted in a concomitant reduction of growth inhibition of three coliform strains. These data confirmed a potential relationship between the molar ratio of citrate to LF of the lacteal secretion and its capacity to inhibit coliform strains associated with mastitis.

Mammary secretions were collected during physiologic transitions of the udder and were used in an *in vitro* microbiological assay to determine bacteriostasis of mastitis pathogens (Breau & Oliver, 1986). As mammary involution progressed, *in vitro* stasis of *Klebsiella pneumoniae, E. coli,* and *Streptococcus uberis* increased. Mammary secretions from concanavalin A (conA)- and phytohemagglutinin (PHA)-treated glands had significantly increased bacteriostasis. Secretions contained significantly increased concentrations of LF and a decreased citrate:LF molar ratio earlier in the dry period than did control mammary secretions. Greatest bacteriostasis was observed in mammary secretions obtained 7 days before parturition. However, differences in secretion composition or bacteriostasis were not found between conA- or PHA-treated and control udder halves during the prepartum period. Bacterial growth inhibition by mammary secretion decreased markedly during early lactation. A highly significant positive correlation was found between bacteriostasis and concentrations of LF, serum albumin, and IgG. A highly significant negative correlation was also reported in the citrate:LF molar ratio during early involution and the peripartum period.

The bacteriostasis effect bLF, TF and immunoglobulins against *E. coli* strain B117, acting alone or in combination, was investigated *in vitro* (Rainard, 1986a). Both LF and TF elicited a strong bacteriostasis without requirement for antibodies. After a short period of growth, the multiplication of bacteria was almost completely prevented by the iron-binding proteins. A significant but moderate additional stasis was achieved when IgG or IgM was added to TF, while addition of Ig to LF revealed no significant cooperative effect. All of 11 strains of *E. coli* isolated from bovine mastitis were sensitive to LF in the absence of Ig. It therefore appeared that antibodies were not required for LF to exert a potent bacteriostatic effect on mastitis isolates of *E. coli*. Rainard (1986b) also examined the bacteriostatic activity of bLF against mastitis pathogens using an *in vitro* microassay. The most susceptible species was *E. coli*; all of the 35 isolates tested were susceptible to bacteriostasis by apo-LF (0.1 mg/ml), although a few strains showed a lower degree of inhibition. Heterogeneity among strains was more pronounced among 10 isolates of *Staphylococcus aureus*, four of which were apparently unaffected by apo-LF (1 mg/ml). Under the same conditions, *Streptococcus agalactiae* (six isolates) and *Strep. uberis* (five isolates) resisted the bacteriostatic action of apo-LF.

The growth of *Streptococcus mutans* 6715-13 in a rich medium (Todd Hewitt broth) was drastically reduced by addition of apo-LF; this effect was bacteriostatic and reversible by saturation of LF with iron (Visca et al., 1989).

Dionysius et al. (1993) examined the *in vitro* antibacterial effects of various forms of LF on enterotoxigenic strains of *E. coli* using a microassay for bacterial growth. Native and apo-LF exhibited variable activity against 19 strains, whereas holo-LF had no effect. At a concentration of .2 mg/ml of apo-LF, strains could be distinguished as either sensitive or resistant to inhibition. Zinc-saturated LF was as bacteriostatic as apo-LF when sensitive and resistant strains were tested over the concentration range .04 to 1.0 mg/ml of LF. A bactericidal effect was observed for native, apo-, and Zn-saturated LF against some sensitive strains. The antibacterial activity of apo-LF depended on bacterial inoculum size and was not enhanced by the addition of lysozyme. Addition of holo-LF or cytochrome c diminished the antibacterial effect of apo-LF, whereas addition of BSA had no effect. Resistance to inhibition by LF was not related to the production of bacterial siderophores.

Paulsson et al. (1993) tested the effect of pasteurization- and UHT-treatments on the LF interaction with bacteria. The ability of native and iron-saturated LF to bind various bacterial species was unaffected by pasteurization. However, UHT treatment decreased this interaction capacity. Native LF, both unheated and pasteurized, showed similar bacteriostatic properties and moderately inhibited *E. coli*. However, this inhibitory capacity was lost after UHT treatment. Iron-saturated LF did not inhibit bacterial growth; neither pasteurization nor UHT could change this property. Thus, UHT seems to affect structural as well as certain biological properties (including bacteriostasis) of both native and iron-saturated bLF, and pasteurization seems to be a treatment of choice for products containing this protein.

The effect of LF on bacterial growth was tested by measuring conductance changes in the cultivation media by using a Malthus-AT system and was compared with the magnitude of ^{125}I-labeled LF binding in 15 clinical isolates of *E. coli* (Naidu et al., 1993). The binding property was inversely related to the change in bacterial metabolic rate and was directly related to the degree of bacteriostasis (*FIGURE 4*). The magnitude of LF-bacterium interaction showed no correlation with the MIC of LF. In certain strains, LF at supraoptimal levels reduced the bacteriostatic effect. Thus, the LF concentration in the growth media was critical for the antibacterial effect. The cell envelopes of *Salmonella typhimurium* 395MS with smooth lipopolysaccharide (LPS) and its five isogenic rough mutants revealed 38-kDa porin proteins as peroxidase-labeled-LF-reactive components in sodium dodecyl suLFate-polyacrylamide gel electrophoresis and Western blot (ligand blot) analysis. However, in the whole cell-binding assay, parent strain 395MS demonstrated a very low interaction with ^{125}I-LF. On the other hand, LF interaction gradually increased in correspondence with the decrease in LPS polysaccharide moiety in the isogenic rough mutants. Conductance measurement studies revealed that the low-level-LF-binding (low-LF-binding) strain 395MS with smooth LPS was relatively insusceptible to LF, while the high-LF-binding mutant Rd was more susceptible to LF (*FIGURE 5*). These

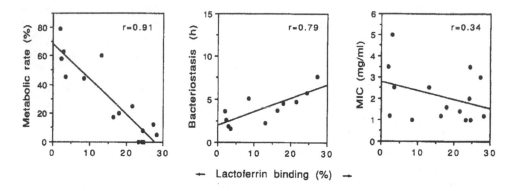

FIGURE 4. LF binding to *E.coli* and its relation to antimicrobial outcome. Correlation between parameters was made by linear regression analysis. The metabolic rate in the presence of LF (1 mg/ml) is expressed as a relative percentage, considering the change in the conductance rate of the control (without LF) as 100%. Bacteriostasis was estimated as the difference between bacteria growth in media with LF and the control. For MIC determinations, bacteria were grown in the presence of 15 different concentrations of LF within a range of 0.01 to 10 mg/ml. The lowest amount of LF that caused complete inhibition of bacterial growth at the time point when the metabolism of the control reached stationary phase was considered as MIC [reproduced from Naidu et al. (1993) with permission from the American Society for Microbiology].

data suggested a correlation between LF binding to porins and the LF-mediated antimicrobial effect. The polysaccharide moiety of LPS shielded porins from the LF interaction and concomitantly decreased the antibacterial effect.

Dial et al. (1998) have examined the *in vitro* and *in vivo* antimicrobial efficacy of bLF against *Helicobacter* species. LF was bacteriostatic to *H. pylori* when cultured at concentrations above or 0.5 mg/ml. Growth of *H. pylori* was not inhibited by another milk constituent, lysozyme, or by bLFcin, but growth was inhibited by the iron chelator deferoxamine mesylate. LF inhibition of growth could be reversed by addition of excess iron to the medium. LF in retail dairy milk was found to be more stable intra-gastrically than unbuffered, purified LF. Treatment of *H. felis*-infected mice with LF partially reversed mucosal disease manifestations. The authors concluded that bLF has a significant antimicrobial activity against *Helicobacter* species *in vitro* and *in vivo*.

The growth of *Bacillus cereus* was markedly inhibited by the addition of LF and was recovered by the addition of $FeCl_3$ (Sato et al., 1999). The bacteriostasis was also reversed by the addition of erythrocytes and hemoglobin. *B. cereus* could use heme or heme-protein complex (hemoglobin-haptoglobin and hematin-albumin complexes) as iron sources in iron deficient conditions. Thus, *B. cereus* seems to use such heme or heme-protein complexes to prevent the bacteriostasis of LF *in vivo*.

B. Cidal effect

In 1977, Arnold and co-workers reported a bactericidal effect for native LF molecule, which apparently was distinct from the stasis mechanism. These experiments were performed with microbial cells suspended in deionized water or buffer solutions at acid pH and the reported mechanism is controversial. Various laboratories have failed to demonstrate a similar bactericidal effect (Rainard, 1987; Gutteberg et al., 1990). Lassiter (1990), from Arnold's group, later published a doctoral thesis which indicated that conta-

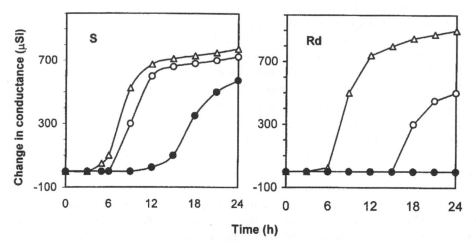

FIGURE 5. Growth of *S. typhimurium* 395MS(S) (S; parent with smooth LPS) and its isogenic mutant Rd (with rough LPS) in SPYE broth. Control (△), with 1 mg/ml bLF (○) and 5 mg/ml bLF (●). [Redrawn from Naidu, et al., 1993].

mination of EDTA during dialysis of LF could account for the cidal effect against *E.coli*. Furthermore, the cidal effects of LF against oral streptococci seem due to the acid pH of the test system. Degradation products in an LF preparation such as the cationic peptides could elicit membrane damage and kill microorganisms. In an antimicrobial milieu such as in the phagosome, LF could possibly elicit a cidal effect synergistic with oxidative events. However, clear evidence for a direct cidal effect with native (intact) LF molecule *in vitro* is still lacking. This section has reviewed the cidal effect in a chronological perspective with no endorsement for the mechanisms hypothesized in the literature.

LFs seem to elicit cidal effects against a variety of microorganisms including Gram-positive and Gram-negative bacteria, rods and cocci, facultative anaerobes, and aero-tolerant anaerobes. Similar morphological and physiological types are represented among the LF-resistant bacteria (Arnold et al., 1980). *S. mutans* was more resistant to LF when grown on a sucrose-containing medium than when it was grown on brain heart infusion broth without added sucrose. When an LF-sensitive, avirulent strain of *Streptococcus pneumoniae* was passed through mice, the resultant virulent culture demonstrated resistance to LF. Since organisms of the same species and even of the same strain such as *S. pneumoniae*, can differ in susceptibility to LF, it appears that accessibility to the LF target site may account for variations in susceptibility.

Influence of several physical conditions and the metabolic state of *Streptococcus mutans* on LF susceptibility were reported (Arnold et al., 1981). After exposure to LF, a 15-min lag period occurred before the initiation of killing, indicating that a two-step process is involved in LF killing. Cultures harvested during the early exponential phase were sensitive to LF, whereas cultures harvested in the early stationary phase were markedly resistant. The rate of killing was dependent on temperature; there was no loss of viability at 2°C. Killing occurred at pH 5.0 to 6.0 in water and 20 mM glycine, but not at any pH in 50 mM sodium phosphate or N-2 hydroxyethylpiperazine-N'-2-ethanesulfonic acid (HEPES) buffer. Addition of exogenous ferrous or ferric ions did not reverse or prevent LF killing, nor did 1 mM magnesium chloride.

Bactericidal effect of LF against *Legionella pneumophila* was reported (Bortner et al., 1986). Purified apo-hLF elicited cidal activity against *L. pneumophila* (serogroup 1), with a 4-log decrease in viability within 2 h at 37 °C. Guinea pig passage of this strain did not affect its sensitivity to LF. Addition of magnesium blocked the bactericidal activity. In addition, human milk was also cidal for *L. pneumophila*. Salts including $CaCl_2$, $Mg(NO_3)_2$, and $MgCl_2$, but not NaCl, blocked killing. Activity was pH dependent with the greatest activity at 5.0. Sensitivity of the organism was markedly affected by the growth conditions. Log-phase 12 h, broth-grown cells were most sensitive, with older cultures appearing more resistant. Plate-grown cells were completely resistant. LF binding, as detected by immunofluorescence microscopy, was temperature dependent (no binding was observed at 4°C), but was independent of killing (Bortner et al., 1989).

Actinobacillus actinomycetemcomitans is a fastidious, facultative Gram-negative rod associated with endocarditis, certain forms of periodontal disease, and other focal infections. Human LF is bactericidal for this pathogen (Kalmar & Arnold, 1988). This cidal activity required an unsaturated (iron- and anion-free) molecule that produced a 2-log reduction in viability within 120 min at 37°C at a concentration of 1.9 µM. Magnesium enhanced LF killing, while other cations, such as potassium and calcium, had no effect.

It was reported that selective anions were capable of inhibiting the expression of bactericidal activity by LF on *S. mutans* 10449 (Lassiter et al., 1987). The ability to block LF expression was directly related to the capacity of the anion to serve as a coordinate ion in iron-binding by the LF molecules. The authors hypothesized the presence of an anionic LF target site on the bacterial surface. Treatment of *S. mutans* with LF under anaerobic conditions abrogated the bactericidal effect. LF killing could be enhanced with thiocyanate and inhibited by catalase and lactoperoxidase; however, bovine serum albumin was equally effective as an inhibitor.

Antimicrobial effects of LF and human milk on *Yersinia pseudotuberculosis* was reported (Salamah & al-Obaidi, 1995a). Bacterial growth *in vitro* was inhibited by apo-but not holo-LF or human milk. Iron-free human milk and to a lesser extent normal human milk were bactericidal for *Y. pseudotuberculosis* cells that were suspended in deionized water. The *in vivo* studies also showed that iron-saturated LF enhanced growth, whereas the viable count was reduced by iron-free LF and EDTA. Nine envelope proteins were decreased or disappeared upon growth in iron-deficient medium, whereas one new high molecular weight protein appeared under the same conditions. The effect of pH, temperature, concentration of magnesium and calcium on the bactericidal activity of LF against *Yersinia pseudotuberculosis* was investigated (Salamah & al-Obaidi, 1995b). The bactericidal activity of LF was higher at acid pH, whereas the bactericidal activity of TF was higher at alkaline pH. Both were not efficient at 4°, 15°, and 25°C, but were efficient at 37°C. LF, but not TF, was very efficient at 42°C. The activity of both proteins were time and concentration dependent. Calcium did not effect their activity up to 60 mM, whereas magnesium reduced the activity of LF only.

C. Adhesion-blockade effect

E. coli is one of the major etiological agents of gastrointestinal illnesses in humans and animals. Bacterial adherence to intestinal epithelia is an important step in the pathogenesis of this disease. In the colonization process, bacterial adhesins such as fimbriae may recognize various mammalian subepithelial matrix components as receptors.

Substances that interfere in this host-pathogen interaction could be of therapeutic and pro-phylactic value, and nonimmunoglobulin fractions of milk, ileal mucus and mucin are among such potential inhibitors (Holmgren et al., 1981; Miedzobrodzki et al., 1989; Olusanya & Naidu, 1991; Cravito et al., 1991).

1. Adhesion-blockade of enteric pathogens. Several carbohydrates, such as 0.1% fucose or 0.5% glucose, as well as LPS (10 µg/ml) isolated from *Shigella flexneri* strong-ly inhibit the adherence of shigellae to guinea pig colonic cells. Fucose-containing pep-tides from hLF also inhibit the adhesion of *S. flexneri* to colonic epithelial cells (Izhar et al., 1987)

The non-immunoglobulin component of human milk responsible for the inhibi-tion of *E. coli* cell adhesion (hemagglutination) mediated by colonization factor antigen I (CFA-I) were identified by chromatographic fractionation of human whey proteins (Giugliano et al., 1995). Free secretory component (fSC) and LF were isolated and both compounds inhibited the hemagglutination by *E. coli* CFA1+. The lowest concentrations of fSC and LF able to inhibit the hemagglutination by *E. coli* strain TR50/3 CFA1+ were 0.06 mg/ml and 0.1 mg/ml, respectively. Commercial preparations of LF from human milk and TF from human serum also inhibited the hemagglutination, with MIC values of 0.03 mg/ml and 0.4 mg/ml, respectively.

Bovine LF mediated inhibition of hemagglutination activity of type 1 fimbriated *E. coli* has also been reported (Teraguchi et al., 1996). The agglutination reaction was specifically inhibited by glycopeptides derived from bLF or α-methyl-D-mannoside. These observations indicate that the glycans of bLF could serve as receptors for type 1 fimbrial lectin of *E. coli*.

The ability of LF to inhibit *in vivo* colonization of *E. coli* has been examined (Naidu et al., unpublished). Infection with *E. coli* strain F18 was established in strepto-mycin-treated mice by gastric intubation and bacterial excretion was estimated as colony forming units per gram (CFU/g) feces. The excretion of strain F18 in feces reached a steady-state (10^8 CFU/g) within 7 days, independent of challenge (dose: 8 x 10^8 or 10^3 CFU). Oral administration of bLF (20 mg/ml in 20% sucrose solution) caused a 1- to >3-log reduction in CFU/g feces with high and low dosages of strain F18. The bacterial mul-tiplication *in vivo* was markedly affected during the early 24 hours of infection, reflecting >3-log lower number of bacteria in the feces (2 x 10^3 CFU/g) than the control group. Oral administration of LF prior to infection reduced fecal excretion of *E. coli* from mouse intestine. *In vitro* effects of bLF on the molecular interactions of *E. coli* with subepithe-lial matrix proteins were examined. Bovine LF inhibited the binding of ^{125}I-labeled fibronectin, fibrinogen, collagen type-I, collagen type-IV and laminin to bacteria. This inhibitory effect was bLF dose-dependent, and was independent of coexistence (compet-itive) or preexistence (non-competitive) of bLF with the tissue matrix proteins. In dis-placement studies with bacteria-matrix protein complexes, bLF dissociated only collagen type-I and laminin interactions. Electron microscopy revealed the loss of type-1, CFA-I and CFA-II fimbria of *E. coli* grown in broth containing 10 µM LF. The inhibitory affect of LF on fimbrial expression was further confirmed by hemagglutination and yeast cell agglutination. The presence of 10 µM LF in the growth media, however, did not affect the P-fimbriation in *E.coli*. These data suggest a strong influence of LF on adhesion-colo-nization properties of *E.coli*.

2. Adhesion-blockade of oral pathogens. The influence of LF, salivary proteins (SP) and BSA on the attachment of *Streptococcus mutans* to hydroxyapatite (HA) was reported (Visca et al., 1989). Sorption of LF, SP, and BSA to HA was dependent on the protein concentration and reached the end-point at about 80 mg of proteins per gram of HA. Similarly, the number of streptococci adsorbed to HA was correlated to the amount of cells available up to at least 10^7 cells per mg of HA. The adsorption of LF, SP and BSA on HA reduced the number of attaching *S. mutans* cells. In particular, SP reduced the adsorption of *S. mutans* by 30%, whereas pre-coating of HA with apo- or iron-saturated LF resulted in a three orders of magnitude reduction of *S. mutans* adsorption to HA. The potent adherence-inhibiting effect of apo-LF together with its antibacterial activity against *S. mutans* suggests an important biological significance of these phenomena in the oral cavity.

Whole cells of *P. intermedia* demonstrate a high degree of binding to fibronectin, collagen type I and type IV and laminin, whereas a moderate interaction was detected with fibrinogen. The ability of bLF to affect the interactions of the above proteins with *P. intermedia* was examined (Alugupalli et al., 1994). In the presence of unlabeled bLF, a dose-dependent inhibition of binding was observed with all five proteins tested. Unlabeled bLF also dissociated the bacterial complexes with these proteins. The complexes with laminin or collagen type I were more effectively dissociated than fibronectin or fibrinogen, whereas the interaction with collagen type IV was affected to a lesser extent. A strain-dependent variation in the effect of bLF was observed.

The ability of hLF and bLF to inhibit adhesion of *A. actinomycetemcomitans* and *P. intermedia* to monolayers of fibroblasts, HEp-2, KB and HeLa cells was reported (Alugupalli & Kalfas, 1995). The inhibitory effect was dose-dependent in the concentration range 0.5-2500 µg/ml and not related to the bacterial growth phase. In the presence of LF, decreased association of bacteria with the cell monolayers was also found by microscopic examination of the preparations. These data suggested a possibility that LF could prevent the establishment of bacteria in periodontal tissues through adhesion-counteracting mechanisms in addition to its bacteriostatic and bactericidal properties.

LF also binds to fibroblast monolayers and matrigel, a reconstituted basement membrane, through ionic interactions. The adhesion of *A. actinomycetemcomitans* to these substrata was mainly dependent on the ionic strength of the environment. *P. intermedia* and *P. nigrescens* also adhere to fibroblasts mainly by ionic interactions, while their adhesion to matrigel seems to be mediated by specific mechanisms. Lectin-type interactions were not found involved in the binding of these bacteria to the substrata. Treatment of either *A. actinomycetemcomitans* or fibroblasts with LF decreased the adhesion in a dose-dependent manner, while LF treatment of matrigel alone had no adhesion-counteracting effect. Adhesion of *P. intermedia* and *P. nigrescens* to matrigel was not significantly affected by the ionic strength, but the presence of LF inhibited the adhesion. LF bound to matrigel, *P. intermedia* and *P. nigrescens* was rapidly released, while LF bound to *A. actinomycetemcomitans* and fibroblasts was retained. These findings indicate that LF-dependent adhesion-inhibition of *A. actinomycetemcomitans, P. intermedia* and *P. nigrescens* to fibroblasts and matrigel could involve binding of LF to both the bacteria and substrata. The decreased adhesion may be due to blocking of both specific adhesin-ligand as well as non-specific charge-dependent interactions (Alugupalli & Kalfas, 1997).

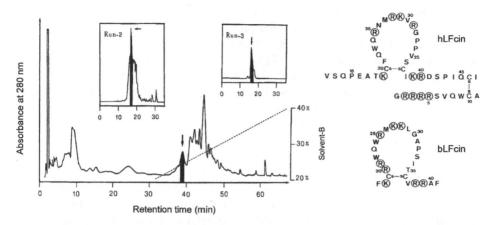

FIGURE 6. LFcin isolation by reverse-phase HPLC. Bovine LF was hydrolyzed with porcine gastric pepsin at pH 4.0 and the hydrolysate was fractionated on a Pep-S column. Shaded peaks are fractions with antimicrobial activity against *E.coli* H10407 in a microplate assay [Naidu & Erdei, unpublished data]. The amino acid sequence and the primary structures of bLFcin and hLFcin peptides are shown with basic residues encircled and sequence positions numbered [adapted from Bellamy et al., 1992].

D. Cationic effect

LF was found to contain an antimicrobial sequence near its N-terminus, which appears to function by a mechanism distinct from iron chelation. The identified domain contains a high proportion of basic residues, like various other antimicrobial peptides known to target microbial membranes and it appears to be located on the surface of the folded protein allowing its interaction with surface components of microbial cells (Tomita et al., 1994).

Hydrolysates prepared by cleavage of bLF with porcine pepsin, cod pepsin, or acid protease from *Penicillium duponti* showed strong activity against *E. coli* O111, whereas hydrolysates produced by trypsin, papain, or other neutral proteases were much less active (Tomita et al., 1991). Low molecular weight peptides generated by porcine pepsin cleavage of bLF showed broad-spectrum antibacterial activity, inhibiting the growth of a number of Gram-negative and Gram-positive species, including strains that were resistant to native LF. The antibacterial potency of the hydrolysate was at least eight-fold greater than that of undigested LF with all strains tested. The active peptides retained their activity in the presence of added iron, unlike native LF. The effect of the hydrolysate was bactericidal as indicated by a rapid loss of viability of *E. coli* O111.

A single active peptide representing antimicrobial domain was isolated following gastric pepsin cleavage of hLF, and bLF, and sequenced by automated Edman degradation. The antimicrobial sequence was found to consist mainly of a loop of 18 amino acid residues formed by a disulfide bond between cysteine residues 20 and 37 of hLF, or 19 and 36 of bLF (Bellamy et al., 1992). Synthetic analogs of this region similarly exhibited potent antibacterial properties. The active peptide of bLF was more potent than that of hLF having effectiveness against various Gram-negative and Gram-positive bacteria at concentrations between 0.3 µM and 3.0 µM, depending on the target strain. Effect of the isolated domain was lethal causing a rapid loss of colony-forming capability (*FIGURE 6*).

Human LF contains a 46 residue sequence named lactoferricin H (hLFcin) responsible for its cationic antimicrobial properties. Synthetic peptides HLT1, corresponding to the loop region of hLFcin (*FQWQR-NMRKVRGPPVS*) and HLT2, corresponding to its charged portion (*FQWQRNMRKVR*), exerted significant antibacterial effects against *E. coli* serotype O111 strains NCTC 8007 and ML35 (Odell et al., 1996). The corresponding sequences in native hLF were shown to adopt a charged helix and hydrophobic tail within the N-lobe remote from the iron binding site. Sequence similarities between LFcin and dermaseptin and magainins suggest that LFcin may act as an amphipathic α-helix.

The basic amino acid-rich region of bovine lactoferricin (bLFcin), *RRWQWRMKKLG* has many basic and hydrophobic amino acid residues. Using chemically synthesized bLFcin and its substituted peptides, the antimicrobial activities of the peptides were tested by determining the minimal inhibitory concentration (MIC) of *E. coli* and *Bacillus subtilis* and the disruption of the outer cell membrane of *E. coli*, and the peptide's toxicities were assayed by hemolysis (Kang et al., 1996). The short peptide (B3) composed of only 11 residues had similar antimicrobial activities while losing most of the hemolytic activities as compared with the 25 residue-long ones (B1 and B2). The short peptides (B3, B5 and B7) with double arginines at the N-termini had more potent antimicrobial activity than those (B4 and B6) with lysine. However, no antimicrobial and hemolytic activities were found in B8, in which all basic amino acids were substituted with glutamic acid, and in B9, in which all hydrophobic amino acids were substituted with alanine. The circular dichroism (CD) spectra of the short peptides in 30 mM SDS were correlated with their antimicrobial activities. These results suggested that the 11-residue peptide of bLFcin is involved in the interaction with bacterial phospholipid membranes and may play an important role in antimicrobial activity with little or no hemolytic activity.

To study the immunochemical and structural properties of bLFcin derived from N-lobe of bLF, monoclonal antibody (mAb) was prepared and the amino acid sequence concerned with binding to mAb identified (Shimazaki et al., 1996). Mice injected with bLFcin showed no production of antibody specific to this peptide, whereas those with bLFcin-KLH conjugate produced anti-bLFcin antibodies. None of the mAb reacted with bLF C-lobe, hLF or hLFcin. By the reactivity of the mAb against the peptides synthesized on cellulose membranes using spots and against chemically modified derivatives of bLFcin, the antigenic determinant was identified to be the sequence '*QWR*'.

Furthermore, three peptides with antibacterial activity toward enterotoxigenic *E. coli* have been purified from a pepsin digest of bLF (Dionysius & Milne, 1997). All peptides were cationic and originated from the N-terminus of the molecule in a region where a bactericidal peptide, bLFcin, had been previously identified. The most potent peptide, peptide I, was almost identical to bLFcin; the sequence corresponded to residues 17 to 42, and the molecular mass was 3195 as determined by mass spectrometry. A second, less active peptide, peptide II, consisted of two sequences, residues 1 to 16 and 43 to 48 (molecular mass of 2673), linked by a single disulfide bond. The third peptide, peptide III, also a disulfide-linked hetero-dimer, corresponded to residues 1 to 48 (molecular mass of 5851), cleaved between residues 42 and 43. Peptides I and II displayed antibacterial activity toward a number of pathogenic and food spoilage microorganisms, and peptide I inhibited the growth of *Listeria monocytogenes* at concentrations as low as 2 μM. Bacterial growth curves showed that bactericidal effects of peptides I and II were observable within 30 min of exposure. The results confirmed and extended those of earlier studies suggesting that the bactericidal domain of LF was localized in the N-terminus and did not involve iron-binding sites.

However, the antibacterial studies conducted by Hoek and co-workers (1997) indicated that the activity of LFcin is mainly, but not wholly, due to its N-terminal region. Several peptides sharing high sequence homology with bLFcin were generated from bLF with recombinant chymosin. Two peptides were co-purified, one identical to bLFcin and another differing from this cationic peptide by the inclusion of a C-terminal alanine. Two other peptides were copurified from chymosin-hydrolyzed LF, one differing from bLFcin by the inclusion of C-terminal alanyl-leucine and the other being a heterodimer linked by a disulfide bond. These peptides were isolated in a single step from chymosin-hydrolyzed LF by membrane ion-exchange chromatography and were purified by reverse-phase high-pressure liquid chromatography (HPLC). They were characterized by N-terminal Edman sequencing, mass spectrometry, and antibacterial activity determination. Pure LFcin, prepared from pepsin-hydrolyzed LF, was purified by standard chromatography techniques. This peptide was analyzed against a number of Gram-positive and Gram-negative bacteria before and after reduction of its disulfide bond or cleavage after its single methionine residue and was found to inhibit the growth of all the test bacteria at a concentration of 8 μM or less. Sub-fragments of LFcin were isolated from reduced and cleaved peptide by reverse-phase HPLC. Sub-fragment 1 (residues 1 to 10) was active against most of the test microorganisms at concentrations of 10 to 50 μM. Sub-fragment 2 (residues 11 to 26) was active against only a few microorganisms at concentrations up to 100 μM.

E. Synergistic effect

LF in combination with antibodies has powerful bacteriostatic effects *in vitro* and this phenomenon provides specific protection against many infections. LF appears to be essential for the antimicrobial function of polymorphonuclear leukocytes against *Pseudomonas aeruginosa* (Bullen, 1981).

Ellison et al. (1990) studied the ability of LF and TF to damage the Gram-negative outer membrane. Lipopolysaccharide (LPS) release by the proteins could be blocked by concurrent addition of Ca^{2+} and Mg^{2+}. Addition of Ca^{2+} also blocked the ability of LF to increase the susceptibility of *E. coli* to rifampicin. TF, but not LF, increased susceptibility of Gram-negative bacteria to deoxycholate, with reversal of sensitivity occurring with exposure to Ca^{2+} or Mg^{2+}. In transmission electron microscopy studies polymyxin B caused finger-like membrane projections, but no morphological alterations were seen in cells exposed to EDTA, LF or TF. These data provide further evidence that LF and TF act as membrane-active agents with the effects modulated by Ca^{2+} and Mg^{2+}.

Antimicrobial activities of LF were tested against 15 strains of 10 species of bacteria, and potent activities against *Staphylococcus aureus, E. coli, Klebsiella pneumoniae* and *Proteus spp.* were observed. Concomitant use of LF with antibiotic cefodoxime proxetil resulted in a synergistic activity against *S. aureus, E. coli, K. pneumoniae* and *Pseudomonas aeruginosa*; and an additive activity against *E. coli* strain NIHJ and *Providencia rettgeri*. The minimum inhibitory concentrations (MICs) of antibiotic in the presence of LF was reduced to < 1/64 with an efficacy rate of 53/57 (92.9%) in a patient group with infections (Chimura et al., 1993).

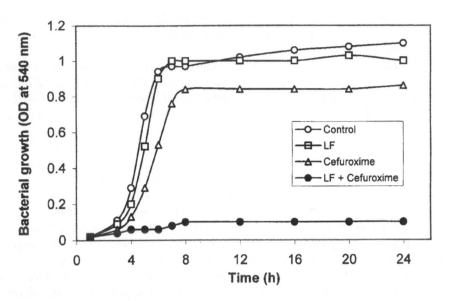

FIGURE 7. LF synergism with antibiotic cefuroxime. Growth of *S. typhimurium* strain ATCC13311 in special peptone yeast extract broth at 37°C was measured as change in optical density at 540 nm. Bacterial growth in media:
control (O); 500 µg/ml of bLF (□); 0.25 g/ml cefuroxime (△) and mixture of both agents at the above concentrations (●) [from Naidu & Arnold (1994) with permission from Elsevier Science Inc.].

The antibiotic susceptibility of *Salmonella* sp. in the presence of LF was examined (Naidu & Arnold, 1994). A mixture containing sub-minimum inhibitory concentration levels of LF (MIC/4) and cefuroxime (MIC/2) inhibited the bacterial growth. LF
strongly potentiated the action of erythromycin (eight-fold), whereas it increased the
activity only by two-fold for ampicillin, ciprofloxacin, chloramphenicol, and rifampicin;
similarly, these antibiotics also reduced the MIC of bLF by two-fold in *S. typhimurium.*
Such antimicrobial potentiation was not observed with bLF mixtures containing cefalex-
in, gentamycin, or polymyxin B against strain ATCC13311. BLF and cefuroxime also
demonstrated potentiation of varying degrees (2- to 16-fold) with nine other *Salmonella*
species. These data established the binding of LF to porins in salmonellae and a potentiation effect of LF with certain antibiotics (*FIGURE 7*).

The effects of LF and its peptides in combination with azole antifungal agents
against *Candida albicans* were investigated by a micro-broth-dilution method
(Wakabayashi et al., 1996). For pepsin hydrolysate of LF or the LF-derived antimicrobial
peptide bLFcin, the concentrations required to inhibit the growth of *Candida* decreased in
the presence of relatively low concentrations of clotrimazole (CTZ). The MIC of all azole
antifungal agents tested was reduced by 1/4-1/16 in the presence of a sub-MIC level of
each of these LF-related substances. Polyene and fluoropyrimidine antifungal agents did
not show such a combined effect with these LF-related substances. The anti-*Candida*
activity of LF or bLFcin in combination with CTZ was shown synergistic by checkerboard
analysis. These results indicate that LF-related substances function cooperatively with
azole antifungal agents against *C. albicans.*

Effects of apo-LF and lactoperoxidase system (lactoperoxidase, LP/SCN-/H$_2$O$_2$),
separately and together, on the viability of *Streptococcus mutans* (serotype c) *in vitro* was
tested (Soukka et al., 1991). Streptococci were incubated in buffered KCl (pH 5.5) with

and without the above components at concentrations normally present in human saliva. Both apo-LF and LP-system had a bactericidal effect against *S. mutans* at low pH. Together they showed an additive, but not a synergistic, antibacterial effect against *S. mutans*. Apo-LF enhanced the LP enzyme activity but decreased the yield of the antimicrobial component, hypothiocyanite (HOSCN/OSCN-), when incorporated into the reaction mixtures. This decrease, which was most pronounced at low pH, was due to an LP-independent reaction between apo-LF and HOSCN/OSCN-.

The effect of an antimicrobial protein, calprotectin, in combination with neutrophils on the growth of *C. albicans* was investigated (Okutomi et al., 1998). The growth inhibition of *C. albicans* by murine neutrophils was augmented by the addition of a low concentration of calprotectin prepared from rat peritoneal exudate cells. The concentrations of calprotectin causing 50% inhibition of growth of *C. albicans* in the absence or presence of neutrophils at an effector-to-target (E/T) ratio of 30 and 60 were estimated to be 0.45, 0.34 and 0.28 U/ml, respectively. The anti-*Candida* activity of calprotectin was completely inhibited by 2 µM of zinc ion, while it only partially lowered the activity of the combination of calprotectin and neutrophils. LF has strongly inhibited the growth of *C. albicans* in combination with calprotectin. These results suggest that calprotectin and LF released from neutrophils may cooperate to inhibit the growth of *C. albicans* at a local lesion of the infection where there is an accumulation of neutrophils.

F. Opsonic effect

The ability of hLF to stimulate the phagocytic and cytotoxic properties of macrophages was reported (Lima & Kierszenbaum, 1987). Fe-LF molecule was not required to increase the capacity of mouse peritoneal macrophages to take up *Trypanosoma cruzi* amastigotes, *Listeria monocytogenes*, or latex particles; it was necessary for LF to enhance intracellular killing of the two microorganisms. Thus, apo-LF, which did not increase macrophage cytotoxicity, after restoration of ferric ions prior to its use in treatments or when ferric citrate was added to the culture medium immediately after apo-LF treatment of the macrophages, does increase macrophage cytotoxicity. In that iron ions cannot be internalized as such, the latter observation suggested that apo-LF had taken up iron while membrane bound and then enhanced killing. Immunofluorescence studies revealed that comparable proportion of macrophage-bound apo-LF or LF at either 20 or 100% iron saturation were without appreciable differences in fluorescence intensity. Therefore, reduced binding of apo-LF compared with LF was not a likely explanation for the lack of effect of apo-LF on macrophage killing. LF did not enhance amastigote killing by macrophage in the presence of the iron chelator deferoxamine. Diethylaminetriaminepentaacetic acid, an iron chelator which is not incorporated into cells, had a similar effect. The iron-binding protein TF did not alter the capacity of macrophage to either take up or kill the amastigote, indicating that the noted LF effects were not shared by all iron-binding proteins. However, prior treatment of macrophages with TF enabled the cells to display a greater parasite killing capacity after apo-LF treatment, suggesting a role for iron in this activity.

Among the known life cycle stages of *Trypanosoma cruzi* only the amastigote form binds LF. This capacity was readily demonstrable by indirect immunofluorescence in amastigotes derived from mice, a mammalian cell culture, or grown in an axenic medium. No LF binding was detectable on trypomastigotes from blood or mammalian cells, insect-derived metacyclics or epimastigotes, or on epimastigotes grown in Warren's medium. Serum levels of LF were increased in mice acutely infected with *T. cruzi*, and amastigotes from the spleens of these animals were found to have the glycoprotein on their surface. The amastigote LF receptor may have biological significance in parasite-host interaction since mononuclear phagocytes also express an LF receptor, and treatment of these cells with LF has been shown to increase their capacities to take up and kill *T. cruzi* amastigotes *in vitro*. The LF receptor is the first marker for *T. cruzi* amastigotes for which a naturally occurring ligand has been described (Lima et al., 1988).

LF bound to *Streptococcus agalactiae* could interfere with the deposition of complement components on the bacterial surface (Rainard, 1993). Pretreatment of streptococci with LF shortened the lag phase preceding the deposition of C3 component on bacteria. The kinetics of C3 deposition was comparable to that obtained by adding antibodies against *S. agalactiae* to agammaglobulinaemic precolostral calf serum (PCS) heated at 56 °C for 3 min to inactivate the alternative pathway. Accelerated C3 deposition did not occur in the absence of calcium ions. Deposition of C4 on bacteria occurred only when either antibodies or LF were added to PCS. These results demonstrate that the interaction of LF with bacteria activated the classical pathway of complement in the absence of antibodies. The binding of purified C1q to bacteria was promoted in a dose-dependent manner by LF, suggesting that recruitment of classical pathway of complement resulted from the interaction of C1q with LF-adsorbed to the bacterial surface. Phagocytosis of bacteria opsonized with heated PCS (at 56°C for 3 min) and LF was comparable to that occurring in the presence of heated PCS and antibodies. These data suggested that LF was able to substitute for antibodies in order to activate the classical pathway of complement and to opsonize unencapsulated *S. agalactiae* efficiently.

VI. ANTIMICROBIAL SPECTRUM

Both LF-susceptible and -resistant organisms encompass a variety of types including Gram-positive and Gram-negative bacteria, rods and cocci, and aerobes and anaerobes; both DNA and RNA viruses; a variety of yeast as well as fungi; and parasites. Susceptibility depends on similarities in cell surface structure or the mode of LF action against individual organisms.

A. Antibacterial activity

The antibacterial properties of milk have been observed for a long time. Most of the relevant literature consists of observations that various pathogenic and saprophytic bacteria are killed or their growth temporarily inhibited by cow's milk. Bacteriostasis was the widely characterized inhibitory mechanism of LF with well-documented data. Over the past three decades, various laboratories have identified LF as a broad-spectrum antimicrobial and reported a variety of inhibitory mechanisms on both Gram-positive and Gram-negative bacteria (*TABLE 3*).

1. Gram-positive bacteria. In 1967, Reiter and Oram reported the antibacterial effects of LF against *Bacillus stearothermophilus* and *B. subtilis*. This study also observed that apo-LF was unable to inactivate bacterial spores but could inhibit their germination. A decade later, Arnold and co-workers (1977) reported cidal activity of LF against *Streptococcus mutans* and other oral streptococci.

The occurrence of LF in saliva has initiated many studies on antimicrobial activity against oral streptococci and control of caries. Apo-LF could cause a potent *in vitro* growth-inhibition of *Streptococcus mutans* and this effect could be reversed by iron (Visca et al., 1989). Furthermore, LF seems to reduce the adsorption of *S.mutans* cells to hydroxyapatite. This adherence-inhibiting effect of apo-LF together with bacteriostasis activity towards *S. mutans* suggests a possible patho-biological significance of caries control in the oral cavity *in vivo*. However, apo-bLF seems to elicit a low degree antimicrobial effect on mastitis-associated streptococci in bovine mammary secretions (Todhunter et al., 1985).

Naidu and co-workers (1990; 1991) have identified specific LF-binding proteins in *Staphylococcus aureus* isolated from human and animal infections as well as among various species of coagulase-negative staphylococci causing bovine mastitis. Apo-bLF at concentrations of 0.1%-0.4% could convert compact colonies of *Staphylococcus haemolyticus* transient to diffused in soft agar (Godo et al., 1997). This surface-active property of LF has prevented autoaggregation of cocci in compact ball-like colonies by hydrophobic interaction or trypsin-sensitive proteins. *In vivo* anti-staphylococcal activity of hLF, bLF and bLF hydrolysate was reported in an experimental mouse model (Bhimani et al., 1999). All the LF preparations demonstrated weak *in vitro* antibacterial activity while holo-LFs showed no activity. LF-treated mice (1 mg, i.v.) when injected with 10^6 staphylococci, showed 30-50% reduction in kidney infections, and viable bacterial counts in the kidney decreased 5- to 12-fold. The inhibitory effect was dose-dependent up to 1 mg LF. The LF preparations were effective when given 1 day prior to the bacterial challenge, after which there was no significant effect even at doses up to 5 mg. Apo- and Fe-saturated forms of hLF and bLF were all equally effective, while bLF hydrolysate was not protective. Different degrees of iron-saturation did not alter the *in vivo* antimicrobial property of either native LF preparation. Feeding mice with 2% bLF in drinking water also reduced the kidney infections by 40-60%, and viable bacterial counts, 5-12-fold.

Human LF was shown to be bactericidal *in vitro* for *Micrococcus luteus* but not for other *Micrococcus* species (*M. radiophilus, M. roseus* and *M. varians*) (de Lillo et al., 1997). A correlation between the binding of LF to the bacterial surface and the antimicrobial action was observed. Viability assays showed that ferric, but not ferrous, salts prevented binding and consequently *M. luteus* was not killed. The unsaturated form of LF showed a greater affinity than that of the iron-saturated molecule for lipomannan, a lipoglycan present on the cell wall of *M. luteus*, supporting the role for lipomannan as one of the possible binding sites of LF on *M. luteus*.

Custer and Hansen (1983) found that LF fragments could react with nitrite and cause inhibition of *Bacillus cereus* spore outgrowth. LF and lysozyme were shown to inhibit the growth of *Bacillus stearothermophilus* var. *calidolactis* spores (Carlsson et al., 1989). The growth of *Bacillus cereus* could be inhibited by LF and this effect could be reversed by the addition of iron (Sato et al., 1999). The growth inhibition was also reversed by the addition of erythrocytes and hemoglobin. *B. cereus* seems to use heme or heme-protein complex (hemoglobin-haptoglobin and hematin-albumin complexes) as iron sources in iron deficient conditions.

Oral administration of bLF with milk has been reported to inhibit various species of clostridia including *C. ramosum, C. paraputrificum* and *C. perfringens* in an experimental mouse model (Teraguchi et al., 1995).

TABLE 3. Inhibitory spectrum of hLF, bLF and LFcins against various bacteria.

Bacterial species	Form	Dose	Effect	Reference
Actino. actinomycetemcomitans	hLF	2 μM	Cidal (2-log reduction)	Kalmar & Arnold, 1988
Aeromonas hydrophila	bLF	0.1%	Adhesion-blockade (47%)	Paulsson et al., 1993
Bacillus cereus	bLFcin	6 μM	Cidal (4-log, 100%)	Hoek et al., 1997
Bacillus circulans	bLFcin	0.006%	Cidal (6-log, 100%)	Bellamy et al., 1992
Bacillus natto IFO3009	bLFcin	0.002%	Cidal (6-log, 100%)	Bellamy et al., 1992
Bacillus stearothermophilus	bLF	1:20	Stasis	Reiter & Oram, 1967
Bacillus subtilis	bLF	1:20	Stasis	Reiter & Oram, 1967
Bacillus subtilis ATCC6633	bLFcin	0.002%	Cidal (6-log, 100%)	Bellamy et al., 1992
Bifidobacterium longum	bLF	0.1%	Agglutination	Tomita et al., 1994
Corynebacterium diphtheriae	bLFcin	0.018%	Cidal (6-log, 100%)	Bellamy et al., 1992
Coryne. ammoniagenes	bLFcin	0.003%	Cidal (6-log, 100%)	Bellamy et al., 1992
Coryne. renale	bLFcin	0.001%	Cidal (6-log, 100%)	Bellamy et al., 1992
Clostridium innocuum	bLF	0.1%	Agglutination	Tomita et al., 1994
Clostridium perfringens	bLFcin	0.024%	Cidal (6-log, 100%)	Bellamy et al., 1992
Clostridium paraputrificum	bLFcin	0.003%	Cidal (6-log, 100%)	Bellamy et al., 1992
Enterococcus faecalis	bLFcin	0.06%	Cidal (6-log, 100%)	Bellamy et al., 1992
Escherichia coli E386	bLF	0.1%	Stasis (24-h, 100%)	Naidu et al., 1993
Escherichia coli	hLF	42 μM	Cidal (6-log reduction)	Arnold et al., 1980
Escherichia coli H10407	bLF	0.1%	Adhesion-blockade (50%)	Paulsson et al., 1993
Escherichia coli IID-861	bLFcin	10 μM	Cidal (3-log reduction)	Bellamy et al., 1992
Escherichia coli HB101	hLF	0.2%	Invasion-inhibition	Longhi et al., 1993
Escherichia coli CL99	bLF	20 μM	LPS release, OM damage	Yamauchi et al., 1993
Klebsiella pneumoniae	bLFcin	10 μM	Cidal (3-log reduction)	Bellamy et al., 1992
Lactobacillus casei	bLFcin	0.012%	Cidal (6-log, 100%)	Bellamy et al., 1992
Legionella pneumophila	hLF	0.03%	Cidal (4-log reduction)	Bortner et al., 1986
Listeria monocytogenes	bLF	10 μM	Cidal (4-log reduction)	Bellamy et al., 1992
L. monocytogenes NCTC7973	bLFcin	2 μM	Cidal (4-log, 100%)	Hoek et al., 1997
Micrococcus luteus	bLF	0.1%	Agglutination	Tomita et al., 1994
Proteus vulgaris JCM1668T	bLFcin	0.012%	Cidal (6-log, 100%)	Bellamy et al., 1992
Pseudomonas aeruginosa	hLF	42 μM	Cidal (7-log, 100%)	Arnold et al., 1980
Ps. aeruginosa IFO3446	bLFcin	10 μM	Cidal (3-log reduction)	Bellamy et al., 1992
Pseudomonas fluorescens	bLFcin	8 μM	Cidal (4-log, 100%)	Hoek et al., 1997
Salmonella abony	bLF	0.8%	Stasis (24-h, 100%)	Naidu & Arnold, 1994
Salmonella dublin	bLF	0.2%	Stasis (24-h, 100%)	Naidu & Arnold, 1994
Salmonella enteritidis	bLFcin	0.012%	Cidal (6-log, 100%)	Bellamy et al., 1992
Salmonella hartford	bLF	0.8%	Stasis (24-h, 100%)	Naidu & Arnold, 1994
Salmonella kentucky	bLF	0.2%	Stasis (24-h, 100%)	Naidu & Arnold, 1994
Salmonella panama	bLF	0.1%	Stasis (24-h, 100%)	Naidu & Arnold, 1994
Salmonella pullorum	bLF	0.2%	Stasis (24-h, 100%)	Naidu & Arnold, 1994
Salmonella rostock	bLF	0.2%	Stasis (24-h, 100%)	Naidu & Arnold, 1994
Salmonella salford	bLFcin	4 μM	Cidal (4-log, 100%)	Hoek et al., 1997
Salmonella montevideo	bLF	20 μM	LPS release, OM damage	Yamauchi et al., 1993
Salmonella thompson	bLF	0.1%	Stasis (24-h, 100%)	Naidu & Arnold, 1994
Salmonella typhimurium Rd	bLF	0.5%	Stasis (64%)	Naidu et al., 1993
Salm. typhimurium R10	bLF	0.1%	Adhesion-blockade (68%)	Paulsson et al., 1993
Salm. typhimurium SL696	bLF	20 μM	LPS release, OM damage	Yamauchi et al., 1993
Salmonella virchow	bLF	0.8%	Stasis (24-h, 100%)	Naidu & Arnold, 1994
Shigella flexneri	bLF	0.1%	Adhesion-blockade (30%)	Paulsson et al., 1993
Staphylococcus albus	bLF	0.5%	Stasis	Masson et al., 1966
Staphylococcus aureus	bLF	0.1%	Adhesion-blockade (54%)	Paulsson et al., 1993
Staph. aureus JCM2151	bLFcin	10 μM	Cidal (3-log reduction)	Bellamy et al., 1992
Staphylococcus epidermidis	bLFcin	0.006%	Cidal (6-log, 100%)	Bellamy et al., 1992
Staphylococcus haemolyticus	bLFcin	0.001%	Cidal (6-log, 100%)	Bellamy et al., 1992
Staphylococcus hominis	bLFcin	0.003%	Cidal (6-log, 100%)	Bellamy et al., 1992

Bacterial species	Form	Dose	Effect	Reference
Streptococcus bovis	bLFcin	0.006%	Cidal (6-log, 100%)	Bellamy et al., 1992
Streptococcus cremoris	bLFcin	0.003%	Cidal (6-log, 100%)	Bellamy et al., 1992
Streptococcus lactis	bLFcin	0.003%	Cidal (6-log, 100%)	Bellamy et al., 1992
Streptococcus mitior	hLF	42 µM	Cidal (6-log, 100%)	Arnold et al., 1980
Streptococcus mutans AHT	hLF	0.17%	Cidal (7-log, 100%)	Arnold et al., 1977
Strep. mutans LM-7	hLF	42 µM	Cidal (8-log, 100%)	Arnold et al., 1980
Strep. mutans ATCC25175	hLF	0.01%	Agglutination	Soukka et al., 1993
Streptococcus pneumoniae	hLF	42 µM	Cidal (7-log, 100%)	Arnold et al., 1980
Streptococcus salivarius	hLF	83 µM	Cidal (7-log, 100%)	Arnold et al., 1980
Streptococcus thermophilus	bLFcin	0.003%	Cidal (6-log, 100%)	Bellamy et al., 1992
Vibrio cholerae 569B	hLF	0.33%	Cidal (7-log, 100%)	Arnold et al., 1977

Groenink and co-workers (1999) reported a potent antimicrobial activity of synthetic cationic peptides derived from the N-terminal domain that comprises an amphipathic a-helix in hLF (hLF 18-31 and hLF 20-38) and bLF (bLF 17-30 and bLF 19-37). Peptide bLF 17-30, containing the largest number of positively charged amino acids, elicited the highest inhibitory spectrum against both Gram-positive and Gram-negative bacteria.

2. Gram-negative bacteria. Many studies have shown the antimicrobial activity of LF against Gram-negative bacteria, *E. coli*, in particular. Various antimicrobial effects of LF were demonstrated against *E.coli* and different mechanisms were postulated to elucidate these effects. LF elicits a bacteriostatic effect on *E. coli*. Based on the time required for *E.coli* O111 to reach one-half maximal cell density, Stuart and co-workers (1984) indicated that the *in vitro* effects of LF on the growth of *E. coli* was kinetic rather than bacteriostatic. Compared to a control, added apo-LF (0.25-1.0 mg/ml) produced only a delay effect indicating that these concentrations are probably within the sub-inhibitory concentration range. The kinetic delay effect of apo-LF also increased steadily in the presence of Zn^{2+} and Cu^{2+} cations. Cu^{2+}, Zn^{2+} and nitrilotriacetate did not affect the growth rate of this organism in the absence of LF compared to the control. These studies indicate that the mechanism by which LF alters the bacterial growth of *E. coli* O111 is more complex than simple iron deprivation.

Rainard (1987) reported the antibacterial activity of milk against a virulent strain of *E. coli* using milk fractions from normal or inflamed glands. Whey obtained from mastitis milk exhibited either bactericidal or bacteriostatic activities, depending on whether bacteria were enumerated by the pour plate technique or by surface plating onto sheep blood agar. The cidal activity, however, was not due to LF, even when assayed in distilled water. Milk whey ultra-filtrate was used to assay the ability of normal and mastitis milk to support the antibacterial activities of LF against *E. coli*. The addition of purified LF to ultra-filtrate from mastitis whey resulted in bacteriostasis, whereas LF was without effect in ultra-filtrate from normal whey. It was suggested that LF could inhibit the growth of LF-sensitive bacteria during mastitis depending on plasma exudation during mastitis. Dionysius et al. (1993) reported that the antibacterial activity of apo-LF depends on bacterial inoculum size and the addition of holo-LF or cytochrome-C could diminish the effect. Furthermore, the resistance to inhibition by LF was not related to the production of bacterial siderophores in *E.coli*.

Ellison et al. (1988) hypothesized that the iron-binding proteins could affect the Gram-negative outer membrane in a manner similar to that of the chelator EDTA. Further, both the whole protein and a cationic N-terminus peptide fragment directly damage the outer membrane of Gram-negative bacteria suggesting a mechanism for the supplemental effects. Several groups have also shown that LF could synergistically interact with immunoglobins, complement, and neutrophil cationic proteins against Gram-negative bacteria.

Klebanoff and Waltersdorf (1990) found that Fe^{2+} and apo-LF could generate hydroxy radicals via an H_2O_2 intermediate with toxicity to *E.coli*, and hypothesized that such a mechanism could possibly contribute to the microbicidal activity of phagocytes.

LF binds to surface structures expressed in *E. coli* K-12 strains grown under iron limitation (Visca et al., 1990). Both apo and holo forms of LF yielded a maximum of 1.6 X 10^5 bound molecules/*E. coli* K-12 cell. The amount of LF bound was independent of the expression of iron-regulated outer membrane proteins. However, LF did not bind to *E. coli* clinical isolates. Apo-LF (500 µg/ml) in a chemically defined medium inhibited the growth of *E. coli* K-12 strains but not of clinical isolates. These findings suggested that the antibacterial activity of the protein could be associated to its binding to the cell surface. Enterotoxigenic strains demonstrate higher LF interaction than enteropathogenic, enteroinvasive, enterohemorrhagic strains or normal intestinal *E. coli* isolates (Naidu et al., 1991). Also the enteropathogenic strains belonging to serotypes O44 and O127 demonstrate higher LF binding compared to O26, O55, O111, O119 and O126 serotypes. No significant differences in the degree of hLF or bLF binding were noticed between aerobactin-producing and non-producing strains. In later studies, Naidu and co-workers have identified and characterized porins in the outer membrane of Gram-negative bacteria as the specific receptors for LF interaction (Tigyi et al., 1992; Naidu et al., 1993; Erdei et al., 1994)

Using an *in vitro* model, Gutteberg and co-workers (1990) reported the early effect of *E. coli*, *Streptococcus agalactiae* (group B streptococci, GBS) and recombinant tumor necrosis factor alpha (TNF) on the release of LF and the generation of interleukin-1 (IL-1) due to *E. coli*, using heparinized whole blood from healthy full-term newborns. In a final concentration of 10^7 per ml both bacteria increased the release of LF markedly. The response to *E. coli* was immediate. GBS was a less potent stimulant than *E. coli* and the response was only apparent after 20 minutes. TNF in a concentration of 10 ng/ml as well as 1 ng/ml increased the release of LF significantly, whereas a concentration of 0.1 ng/ml had no effect. Whole blood incubated with different preparations of LF for 20 minutes did not increase the LF levels. No significant changes in IL-1 levels were observed. LF had bacteriostatic but no bactericidal effect on GBS and *Streptococcus mutans*.

Payne et al. (1990) demonstrated that apo-bLF had bacteriostatic activity against four strains of *L. monocytogenes* and an *E.coli* at concentrations of 15 to 30 mg/ml, in UHT milk. At 2.5 mg/ml the compound has no activity against *S. typhimurium, P. fluorescens* and limited activity against *E.coli* O157:H7 or *L. monocytogenes* VPHI (Payne et al., 1994).

Human LF and free secretory component (fSC) were shown to inhibit the hemagglutination induced by *E. coli* CFA1+ (Giugliano et al., 1995). The lowest concentrations of purified fSC and hLF to inhibit the hemagglutination by *E. coli* strain TR50/3 CFA1+ were 0.06 mg/ml and 0.1 mg/ml, respectively.

TABLE 4: Lactoferrin – Antiviral effects

Viral pathogen	Antiviral effect	Reference
Spleen focus forming virus (SFFV)	Decreases multiplication	Hangoc et al., 1987
Human influenza virus	Inhibits viral hemagglutination	Kawasaki et al., 1993
Human cytomegalovirus (HCMV)	Inhibits infection & replication	Hasegawa et al., 1994
	Inhibits MT4 cell cytopathy	Harmsen et al., 1995
Human herpes simplex virus (HSV-1)	Inhibits adsorption & penetration	Hasegawa et al., 1994
	Prevents plaque formation	Fujihara & Hayashi, 1995
Human immunodeficiency virus (HIV)	Inhibits MT4 cytopathic effect	Harmsen et al., 1995
	Inhibits vero cell cytopathy	Marchetti et al., 1996
Feline immunodeficiency virus (FIV)	Effects clinical outcome	Sato et al., 1996
Respiratory syncytial virus	Inhibits viral multiplication	Grover et al., 1997
Hepatitis C virus (HCV)	Binds E1 and E2 envelope proteins	Yi et al., 1997
Rotavirus	Inhibits HT-29 cell infection	Superti et al., 1997

The antimicrobial activities of bLF, and bLFcin against four clinical isolates of enterohemorrhagic *E. coli* O157:H7 were reported (Shin et al., 1998). The MICs against these isolates were 3 mg/ml for bLF, and 8-10 μg/ml for bLFcin in 1% Bacto-peptone broth. Transmission electron microscopy findings suggested that bLFcin acts on the bacterial surface and affects cytoplasmic contents. Furthermore, bLFcin affected the levels of verotoxins in the culture supernatant fluid of an *E. coli* O157:H7 strain.

The antimicrobial effect of LF against *Salmonella typhimurium* was tested by measuring conductance changes in the cultivation media by using a Malthus-AT system (Naidu et al., 1993). Conductance measurement studies revealed that the low-LF-binding strain 395MS with smooth LPS was relatively insusceptible to LF, while the high-LF-binding mutant Rd with rough LPS was more susceptible to LF suggesting an LPS shielding of antimicrobial effect. Later studies have led to the identification of porins as LF-binding outer membrane proteins in various species of *Salmonellae* (Naidu & Arnold, 1994).

Antimicrobial effects of LF against various Gram-negative bacterial pathogens, including *Aeromonas hydrophila, Yersinia enterocolitica, Campylobacter jejuni, Helicobacteri pylori, Pseudomonas aeruginosa, Vibrio sp.,* have also been reported (Arnold et al., 1977; Paulsson et al., 1993; Tomita et al., 1994).

B. Antiviral activity

LF demonstrates a broad-spectrum antiviral activity against both DNA and RNA viruses (*TABLE 4*). The ability of LF to interact with nucleic acids as well as its capacity to bind eucaryotic cells and prevent viral adhesion seem to be the possible antiviral mechanisms.

Abramson et al. (1984) suggested that the depressed chemotactic activity of PMNL infected with influenza virus could be due to changes occurring at the plasma membrane. Virus-treated PMNL stimulated with FMLP or *Staphylococcus aureus* exhibited a marked decrease for LF released into phagosomes, onto the cells' outer membrane, and into the extracellular medium as compared to control cells. Baynes et al. (1988) with the use of an immunoperoxidase stain for LF, showed that neutrophils in viral illness have reduced LF content compared to normal subjects. The authors suggested an acquired defect of neutrophil LF synthesis in viral infection. The LF levels in parotid saliva from individuals with different clinical stages of human immunodeficiency virus (HIV) infection were significantly decreased in parallel with their markedly reduced parotid secretory IgA output. This combined deficiency of parotid LF and secretory IgA may well contribute to the frequent oral infections seen in subjects with HIV infection (Muller et al, 1992).

Purified holo-hLF and recombinant murine IL-3 were assessed *in vivo* for their effects on replication of Spleen Focus Forming Viruses (SFFV) in spleens of DBA/2 mice injected with the polycythemia-inducing strain of the Friend Virus Complex (FVC) (Hangoc et al, 1987). LF and IL-3, inoculated 2 hr prior to the administration of the polycythemia-inducing strain of the FVC, respectively decreased and increased the replication of SFFV in mice as assessed by the spleen focus forming unit assay in primary and secondary DBA/2 mice. Since virus infectivity is associated with the DNA synthetic phase of the cell cycle and it has been shown that LF decreases and IL-3 increases the percent of hematopoietic progenitor cells in S-phase *in vivo*, the results suggest that the opposing actions of LF and IL-3 on replication of SFFV may reflect the actions of these molecules on cycling of the target cells for SFFV.

Human LF and bLF inhibit the infection of tissue culture cells with human cytomegalovirus (HCMV) and human herpes simplex virus-1 (HSV-1) (Hasegawa et al, 1994). The addition of LF inhibited both *in vitro* infection and replication of HCMV and HSV-1 in human embryo lung host cells. The maximum inhibition by more than six exponential of ID_{50} for HCMV and four exponential for HSV-1 was obtained at a concentration in a range from 0.5 to 1 mg of LF per ml of medium. The antiviral activity of LF was associated with its protein moiety, but not with its iron molecule or sialic acid. Furthermore, LF prevented virus adsorption and/or penetration into host cells, indicating an effect on the early events of virus infection. Preincubation of host cells with LF for 5 to 10 min was sufficient to prevent HCMV infection, even when LF was removed after addition of virus. These results suggest that LF possesses a potent antiviral activity and may be useful in preventing HCMV and HSV-1 infection in humans.

Native and chemically derivatized proteins purified from serum and milk were assayed *in vitro* to assess their inhibiting capacity on the cytopathic effect of human immunodeficiency virus (HIV)-1 and human cytomegalovirus (HCMV) on MT4 cells and fibroblasts, respectively (Harmsen et al, 1995). Only native and conformationally intact LF from bovine or human milk, colostrum, or serum could completely block HCMV infection (IC_{50} = 35-100 µg/mL). Moreover, native LF also inhibited the HIV-1-induced-cytopathic effect (IC_{50} = 40 µg/mL). When negatively charged groups were added to LF by succinylation, there was a four-fold stronger antiviral effect on HIV-1, but the antiviral potency for HCMV infection decreased. LF likely exerts its effect at the level of virus adsorption or penetration (or both), because after HCMV penetrated fibroblasts, the ongoing infection could not be further inhibited. A number of native and modified milk proteins from bovine or human sources were analyzed for their inhibitory effects on human immunodeficiency virus type 1 (HIV-1) and HIV-2 *in vitro* in an MT4 cell test system (Swart et al, 1996). The proteins investigated were LF, α-lactalbumin, β-lactoglobulin A, and β-lactoglobulin B. By acylation of the amino function of the lysine residues in the proteins, using anhydrides of succinic acid or cis-aconitic acid, protein derivatives were obtained that all showed a strong antiviral activity against HIV type 1 and/or 2. The *in vitro* IC_{50} values of the aconitylated proteins were in the concentration range of 0.3 to 3 nM. Succinylation or aconitylation of α-lactalbumin and β-lactoglobulin A/B also produced strong anti-HIV-2 activity with IC_{50} values on the order 500 to 3000 nM. All compounds showed virtually no cytotoxicity at the concentration used. Peptide-scanning studies indicated that the native LF as well as the charged modified proteins strongly binds to the V3 loop of the *gp120* envelope protein, with K_a values in the same concentration range as the above-mentioned IC_{50}. Therefore, shielding of this domain, resulting in inhibition of virus-cell fusion and entry of the virus into MT4 cells, may be the likely underlying mechanism of antiviral action.

LF prevents herpes simplex virus type-1 (HSV-1) plaque formation in Vero cell monolayer (Fujihara & Hayashi, 1995). Topical administration of 1% LF prior to the virus inoculation suppressed infection on ocular tissue, however it did not inhibit propagation of the virus in the mouse cornea. Marchetti et al. (1996) reported that both hLF and bLF are potent inhibitors of HSV-1 infection, the concentrations required to inhibit the cytopathic effect in Vero cells by 50% being 1.41 μM and 0.12 μM, respectively. Human LF and bLF exerted their activity through the inhibition of adsorption of virions to the cells independently of their iron withholding property showing similar activity in the apo- and iron-saturated form. The binding of [^{35}S]methionine-labelled HSV-1 particles to Vero cells was strongly inhibited when bLF was added during the attachment step. Bovine LF interacts with both Vero cell surfaces and HSV-1 particles, suggesting that the hindrance of cellular receptors and/or of viral attachment proteins may be involved in its antiviral mechanism.

The effect of LF on the growth of rotavirus and respiratory syncytial virus in cell culture was investigated (Grover et al., 1997). LF inhibited the growth of respiratory syncytial virus at a concentration tenfold lower than that normally present in human milk. The ability of LF to inhibit influenza virus hemagglutination was also reported (Kawasaki et al., 1993).

Hepatitis C virus (HCV) has two envelope proteins, E1 and E2, which form a hetero-oligomer. During dissection of interacting regions of HCV E1 and E2, Yi et al. (1997) found the presence of an interfering compound or compounds in skim milk identified as LF. The bindings of LF to HCV envelope proteins *in vitro* were confirmed by Western blotting and by the pull-down assay, with immuno-precipitated LF-bound protein A resin. Direct interaction between E2 and LF was proved *in vivo*, since anti-hLF antibody efficiently immuno-precipitated with secreted and intracellular forms of the E2 protein. The N-terminal loop of LF, the region important for the antibacterial activity, has only a little role in the binding ability to HCV E2 but affected the secretion or stability of LF. Taken together, these results indicate the specific interaction between LF and HCV envelope proteins *in vivo* and *in vitro*.

Different milk proteins were analyzed for their inhibitory effect on either rotavirus-mediated agglutination of human erythrocytes or rotavirus infection of the human enterocyte-like cell line HT-29 (Superti et al., 1997). Apo- and Fe-LF inhibited the replication of rotavirus in a dose-dependent manner, apo-LF being the most active. It was shown that apo-LF hinders virus attachment to cell receptors since it is able to bind the viral particles and to prevent both rotavirus hemagglutination and viral binding to susceptible cells. Moreover, LF markedly inhibited rotavirus antigen synthesis and yield in HT-29 cells when added during the viral adsorption step or when it was present in the first hours of infection, suggesting that this protein interferes with the early phases of rotavirus infection.

C. Antifungal activity

LF and lysozyme (muramidase), either singly or in combination, are fungicidal in nature and their combined activity is synergistic. Samaranayake and co-workers (1997) examined 20 oral isolates of *Candida krusei* and 5 isolates of *Candida albicans* for their susceptibility to human apo-LF and lysozyme, either singly or in combination, using an *in vitro* assay system. The two species exhibited significant interspecies differences in susceptibility to LF, but not for lysozyme; *C. krusei* was more sensitive to LF (1.4 times) than *C. albicans*. Both species revealed significant intraspecies differences in their susceptibility to lysozyme, but not for LF. No synergistic antifungal activity of the two proteins on either *Candida* species was noted.

LF could inhibit the growth of *C.albicans* in the absence of PMNL, and anti-LF antibodies reversed both this inhibition and the PMNL activation by MP-F2, GM-CSF, and LPS. Furthermore, PMNL may be activated by relevant candidal mannoproteins, and release of LF may add to other antimicrobial mechanisms of PMNL for the control of candidal infections (Palma et al., 1992).

Candida albicans was found highly susceptible to inhibition and inactivation by bLFcin, a peptide produced by enzymatic cleavage of bLF (Bellamy et al., 1993). Effective concentrations of the peptide varied within the range of 18 to 150 µg/ml depending on the strain and the culture medium used. Its effect was lethal, causing a rapid loss of colony-forming capability. ^{14}C-labeled bLFcin bound to *C. albicans* and the rate of binding appeared to be consistent with the rate of killing induced by the peptide. The extent of binding was diminished in the presence of Mg^{2+} or Ca^{2+} ions, which acted to reduce its anticandidal effectiveness. Binding occurred optimally at pH 6.0 and killing was maximal near the same pH. Such evidence suggests the lethal effect of bLFcin results from its direct interaction with the cell surface. Cells exposed to bLFcin exhibited profound ultrastructural damage that appeared to reflect its induction of an autolytic response.

D. Antiparasitic activity

LF could elicit defense against parasitic infections by phagocytic activity in the destruction of amastigotes, an intracellular parasitic form of *Trypanosoma cruzi* in macrophages. The effect of bLF on the intracellular growth *Toxoplasma gondii* parasites was examined in murine macrophage and embryonic cells (Tanaka et al., 1996). Co-cultures of host cells with the parasites were supplemented with LF, apo-LF, holo-LF or TF in the culture media for varying periods. The growth activity of intracellular parasites in the host cells was determined by the measurement of selective incorporation of ^{3}H-uracil. Supplement of LF had no effect on the penetration activity of the parasites, while development of intracellular parasites was inhibited linearly in concentration of LF. Supplement of apo-LF and holo-LF, but not TF showed similar effects. These suggest that LF induce the inhibitory effects on the development of intracellular parasites. Pretreatment of LF to the macrophages, however, did not show any inhibitory effects, whereas mouse embryonic cells preincubated with LF suppressed the intracellular growth. Thus, the action of LF to macrophages would be different from that of mouse embryonic cells.

The trophozoites of *Giardia lamblia* could be killed by nonimmune human milk in a time- and concentration-dependent manner. Removal of greater than 99% of the S-IgA from milk did not decrease its *Giardia*-cidal activity. Thus, the killing was not antibody dependent. Studies by Gillin and co-workers (1983) showed that in the presence of milk, trophozoites lost motility, swelled, and lysed. The *Giardia*-cidal activity may be specific to human milk, since unheated cow and goat milk were virtually devoid of activity. Human and bLFs and their derived N-terminal peptides were giardicidal *in vitro* (Turchany et al., 1995). Fe^{3+}, but not Fe^{2+}, protected trophozoites from both native LF and peptides, although the latter lack iron-binding sites. Other divalent metal ions protected only against native LF. Log-phase cells were more resistant to killing than stationary-phase cells. These studies suggest that LF, especially in the form of the N-terminal peptides, may be an important nonimmune component of host mucosal defenses against *Giardia lamblia*.

VII. INFLUENCING FACTORS

A. Citrate / bicarbonate ratio

The requirement of bicarbonate for formation of the red complex of LF was first pointed out by Blanc and Isliker (1963). In a later study, Masson and Heremans (1968) clearly established the involvement of bicarbonate in the metal-combining properties of LF. It was found that one molecule of bicarbonate (carbon dioxide) is taken up per atom of iron or copper during the formation of the colored LF-metal complex. It was observed that the color development proceeded very slowly when a gas-free sample of apo-LF was exposed to air in the presence of copper ions, whereas it took place instantly when bicarbonate was added. This suggested that the carbon dioxide from the air first had to become converted to bicarbonate ions before its participation in the reaction.

Fresh human and bovine milk are bacteriostatic *in vitro* for certain (milk-sensitive) strains of *E. coli*. Dolby et al. (1977) reported that the addition of bicarbonate to the test system could potentiate the bacteriostasis which also results in the inhibition of milk-resistant strains. The concentration of bicarbonate needed for such an effect is lower for human milk than for cow milk and is reduced even further by the addition of more LF. *In vivo* studies with infants and data deduced from the ratio of milk-sensitive to milk-resistant strains of *E. coli* isolated from fecal samples suggested that the neonatal intestinal secretions may contribute to the bacteriostatic activity of their feeds so that (i) in fully breastfed babies all strains of *E. coli* are inhibited to the same extent; there is no selection on the basis of milk sensitivity and equal numbers of strains resistant and sensitive to milk are found in the feces; (ii) in fully bottle-fed babies *E. coli* is not inhibited since the milk is non-bacteriostatic and again there is no selection; (iii) in babies fed at the breast but bottle-milk supplemented, only milk-sensitive strains are inhibited; milk-resistant strains are not, and preferentially colonized the large intestines.

Undiluted dialyzed milk was not inhibitory because of its low content of LF and the lack of inhibition in undiluted whey is due to the high concentration of citrate in colostral whey (and milk) (Reiter et al., 1975). It was suggested that citrate competes with the iron-binding proteins for iron and makes it available to the bacteria. Addition of bicarbonate, which is required for the binding of iron by LF, can overcome the effect of citrate. Coliform bacteria respond to low-iron environments by production of iron-sequestering agents that compete effectively with apo-LF for free iron. Addition of apo-LF plus citrate could reverse growth inhibition. The molar ratio (citrate to apo-LF) is more important than the absolute concentration of either component. Bishop et al. (1976) found that a ratio of

75 resulted in 50% growth inhibition, whereas ratios of 300 and greater resulted in less than 10% growth inhibition. These results suggest that the ratio of citrate to LF is important in the nonspecific protection of bovine udder. Inhibition of *E.coli* by apo-LF (10 mg/ml) was abolished by addition of ferric iron to the assay system, indicating an iron-dependent nature of apo-LF induced inhibition of bacteria (Nonnecke & Smith, 1984a). Bicarbonate supplementation of the growth system containing apo-LF (1 mg/ml) increased inhibition of three coliform strains by apo-LF. Addition of increasing concentrations of citrate (2.0 mg/ml) to an assay system containing apo-LF (5 mg/ml) resulted in a concomitant reduction of growth inhibition of three coliform strains. These data further indicate a potential relationship between the molar ratio of citrate to LF of the lacteal secretion and its capacity to inhibit coliform strains associated with mastitis. Changes in pH, concentration of serum albumin, immunoglobulin G, citrate, LF, and number of leukocytes in secretions were typical of milk from glands undergoing physiological transitions. Whey from different glands of the same cow differ markedly in their capacity to inhibit growth of coliforms. Bacteriostatic activity of whey increases markedly during the dry period and reaches maximal in whey collected on day-15 of the dry period and at parturition in the subsequent lactation (Nonnecke & Smith, 1984b).

Griffiths and Humphreys (1977) conducted a series of experiments to elucidate the importance of bicarbonate and milieu pH in the bacteriostatic activity of LF. At pH 7.4 and in the presence of bicarbonate, human milk and bovine colostrum inhibit the growth of *E. coli* O111. Adding sufficient iron to saturate the iron-binding capacity of the LF present in the milk or colostrum prevents bacteriostasis. At pH 6.8, neither milk nor colostrum could inhibit *E. coli* 0111. Adjusting the pH to 7.4 with bicarbonate resulted in the development of bacteriostatic activity. Adjusting the pH to 7.4 with NaOH was ineffective. Dialyzed colostrum and milk inhibited bacterial growth at pH 6.8 in the absence of added bicarbonate; addition of citrate or iron abolished bacteriostasis. The chromatographic elution profile of tyrosyl-tRNA from iron-replete *E. coli* differs significantly from that of tyrosyl-tRNA from iron-deficient organisms. Examination of the elution profile tyrosyl-tRNA from *E. coli* 0111 growing in colostrum without added bicarbonate showed that such bacteria were fully replete in iron. The nature of the elution profile of tyrosyl-tRNA also showed that iron was freely available to the bacteria when citrate was added to dialyzed colostrum but not available in its absence, even at pH 6.8. These data support the idea that the bacteriostatic action of milk and colostrum, due to the combined action of antibody and LF, depends on the addition of bicarbonate to counteract the iron-mobilizing effect of the citrate normally present in these secretions.

Thomas and Fell (1985) reported that in lactating cows hormones [oxytocin or ACTH (Synacthen)] affect the citrate and LF concentrations in the direction that would improve the antibacterial properties of milk, but that this was accompanied by adverse effects on milk secretion. However, the extent of the change was not sufficient to produce inhibition of coliform bacteria.

B. Milieu pH

In order to apply functionally active bLF to food products, the effect of pH on the heat stability of bLF was studied (Saito et al., 1994). Bovine LF was easily denatured to an insoluble state by heat treatment under neutral or alkaline conditions, above pH 6.0. In contrast, it remained soluble after heat treatment under acidic conditions at pH 2.0 to 5.0, and the HPLC pattern of LF heat-treated at pH 4.0 at 100°C for 5 min was the same as that of native bLF. Bovine LF was thermostable at pH 4.0, and could be pasteurized or sterilized without any significant loss of its physicochemical properties. Bovine LF was hydrolyzed by heat treatment at pH 2.0 to 3.0 at above 100°C, and its iron binding capacity and antigenicity were lost.

Acceleration of the autoxidation of Fe(II) by apoTF or apo-LF at acid pH is indicated by the disappearance of Fe(II), the uptake of oxygen, and the binding of iron to TF or LF. The product(s) formed oxidize iodide to an iodinating species and are bactericidal to *E. coli* (Klebanoff & Waltersdorph, 1990) Toxicity to *E. coli* by $FeSO_4$ (10 µM) and human apo-TF (100 µg/ml) or apo-hLF (25 µg/ml) was optimal at acid pH (4.5-5.0) and with log-phase organisms. Both the iodinating and bactericidal activities were inhibited by catalase and the hydroxyl radical (OH.) scavenger mannitol, whereas superoxide dismutase was ineffective. NaCl at 0.1 M inhibited bactericidal activity, but had little or no effect on iodination. Iodide increased the bactericidal activity of Fe(II) and apoTF or apo-LF. The formation of OH.was suggested by the formation of the OH.spin-trap adduct (5,5-dimethyl-1-pyroline N-oxide [DMPO]/OH)., with the spin trap DMPO and the formation of the methyl radical adduct on the further addition of dimethyl sulfoxide. (DMPO/OH).formation was inhibited by catalase, whereas superoxide dismutase had little or no effect. These data suggest that Fe(II) and apoTF or apo-LF can generate OH. via an H_2O_2 intermediate with toxicity to microorganisms, and raise the possibility that such a mechanism may contribute to the microbicidal activity of phagocytes.

Salamah and al-Obaidi (1995b) studied the effect of pH, temperature, magnesium and calcium on the bactericidal activity of LF and TF against *Yersinia pseudotuberculosis*. The bactericidal activity of LF was higher at acid pH, whereas the bactericidal activity of TF was higher at alkaline pH. Neither was efficient at 4°, 15°, and 25°C, but both were effective at 37°C. LF, but not TF, was very efficient at 42°C. The activity of both were time and concentration dependent. Calcium did not effect their activity up to 60 mM, whereas magnesium reduced the activity of LF only.

LF release from the secondary granules of activated PMNs is markedly lower at pH 7.2 than at pH 6.7 or 8.2 (Leblebicioglu et al., 1996). Moreover, phagocytosis of opsonized bacteria is lower at pH 7.2 than at pH 7.7. In addition to these effects on functional activation, extracellular pH influences the magnitude of intracellular Ca^{2+} mobilization. These findings suggest that the pH of an inflammatory milieu can selectively influence PMN activation, thereby altering the balance between bacteria and the host response.

C. Proteases

Holo-bLF is more resistant to proteolysis than the apo-form (Brock et al., 1976; 1978). In the trypsin digests of bLF, up to five different fragments with molecular weights ranging from 25-kDa to 53-kDa were detected, with no obvious qualitative difference between digests of holo and apo forms. The susceptibility of apo-bLF to tryptic digestion was only slightly reduced when the protein was complexed with β-lactoglobulin, suggesting that complex-formation is not a mechanism for protecting LF against intestinal degradation.

The susceptibility of hLF and bLF to digestion by trypsin and chymotrypsin has been compared (Brines & Brock, 1983). Neither enzyme had much effect on the hLF-mediated antimicrobial activity of human milk, and the iron binding capacity of hLF in the milk was only slightly reduced. Both enzymes had only a slight effect on the iron binding capacity of purified hLF. In contrast, trypsin destroyed the antimicrobial activity of bovine colostrum, and, in line with earlier studies, appreciably reduced the iron binding capacity of both colostrum and purified apo-bLF. Holo-bLF was more resistant to digestion. The unusual resistance of apo-hLF to proteolysis may reflect an evolutionary development designed to permit its survival in the gut of the infant.

Proteolytic hydrolysis of milk proteins in the mammary gland during involution could influence the bio-functionality of LF and other innate defense factors. Aslam and Hurley (1997) examined the activities of plasmin, plasminogen, and plasminogen activator on proteolysis of LF in mammary gland secretions collected during involution. Activities of plasmin, plasminogen, and plasminogen activator were significantly higher on day-7, -14, and -21 of involution than were those on day-7 postcalving. Protein fragments resulting from hydrolysis were detected by SDS-PAGE in samples collected on day-7, -14, and -21 of involution, but few protein fragments were observed in samples collected on day-7 postcalving when plasmin activity was low. Immunoblot analysis showed that a number of peptides observed during involution were generated from α-s-casein (CN), β-CN, κ-CN, or LF. The appearance of peptides from proteins of mammary secretions during early involution was generally correlated with increased plasmin activity. Elevated plasmin activity during mammary involution may be primarily responsible for the observed concurrent hydrolysis of milk proteins in mammary secretions.

The ability of periodontal pathogens *Porphyromonas gingivalis, Prevotella intermedia* and *Prevotella nigrescens* to degrade LF was reported (de Lillo et al., 1996). Strains of *P. gingivalis* completely degraded LF *in vitro*, whereas *P. intermedia* and *P. nigrescens* showed only partial degradation. It was suggested that LF binds to a high-affinity receptor on all these bacteria and, particularly in the case of *P. gingivalis*, is then degraded by cell-associated proteases. This property may provide protection to the cell against the effects of LF in periodontal sites and so is a possible virulence factor in disease. However, there was no association between the ability to degrade LF and whether the strains had originated from healthy or diseased oral sites.

Bacterium- and neutrophil-derived proteases have been suggested to contribute to tissue injury at sites of *Pseudomonas aeruginosa* infection. *Pseudomonas* elastase cleavage of LF and TF enhances *in vitro* iron removal from these proteins by the *P. aeruginosa* siderophore pyoverdin. Britigan et al. (1993) detected TF and LF cleavage products in bronchoalveolar lavage (BAL) samples from 21 of 22 and 20 of 21 cystic fibrosis (CF) patients, respectively. Three of eleven and two of nine BAL samples from individuals with other forms of chronic inflammatory lung disease had TF and LF cleavage products, respectively. Each patient in whom such products were detected was also infected with *P. aeruginosa*. No such products were detected in normal individuals. These data provide evidence that *P. aeruginosa*- and/or human-derived protease cleavage of TF and LF occurs *in vivo* in the airways of individuals with CF and other forms of chronic lung disease, suggesting that this process could contribute to *P. aeruginosa*-associated lung injury in these patients.

The effect of *Vibrio cholerae* non-O1 protease on LF was studied in relation to bacterial virulence mechanism (Toma et al., 1996). The proteins treated with the protease were analyzed by SDS-PAGE. The protease has cleaved LF into two fragments of 50-kDa and 34-kDa. The N-terminal amino acid sequencing of these fragments revealed that the cleavage site was near the hinge region, between *serine 420* and *serine 421*. This cleavage could affect the transition from open to closed configuration, which is involved in iron binding and release. However, the anti-bacterial activity of LF was not affected by protease treatment.

D. Microbial iron acquisition

Iron is essential to all microorganisms. The low concentration of free iron in body fluids creates bacteriostatic conditions for many microorganisms and is therefore an important defense factor of the body against invading bacteria. Iron-binding proteins, such as LF, TF, and ferritin, play a central role in human ferrokinetics. These iron-binding proteins also participate in the process of decreasing iron availability for the microorganisms. LF and TF restrict the amount of ionic iron available in body fluids to 10^{-18} M (Bullen, 1981). They do so by decreasing iron re-utilization. Anemia of inflammation (previously called anemia of chronic disease) is seen in the setting of infectious, inflammatory, and neoplastic diseases. It results, in part, from changes in the intracellular metabolism of iron. Alterations of iron physiology seen in many clinical circumstances make excess iron available to microorganisms, thus enhancing their pathogenicity. Understanding the molecular basis of iron withholding by the human host, both in the absence of and during infection, and that of iron acquisition by microorganisms may provide us with new and innovative antimicrobial agents and vaccines.

Pathogenic bacteria have developed several mechanisms for acquiring iron from the host (see reviews: Otto et al., 1992; Crosa, 1989). Siderophore-mediated iron uptake involves the synthesis of low molecular weight iron chelators called siderophores which compete with the host iron-binding glycoproteins LF and TF for iron. Other ways to induce iron uptake, without the mediation of siderophores, are the possession of outer membrane protein receptors that actually recognize the complex of TF or LF with iron, resulting in the internalization of this metal, and the use of heme-compounds released into the circulation after lysis of erythrocytes.

Rogers and Synge (1978) reported that enterochelin, an iron transporting compound of *E.coli* could abolish the bacteriostatic effect of human milk. The bacteriostatic phase in human milk could be abolished by adding sufficient iron to saturate the LF in human milk, and also by adding supernatant from a 24-h milk culture or by adding enterobactin, an enterobacterial iron chelator (Brock et al., 1983). Growth in the presence of enterobactin was even more rapid than in the presence of excess iron. Partial loss of bacteriostatic activity could be achieved by absorbing the milk with bacterial antigens, but no clear correlation with removal of antibodies to O, K, or H antigens was apparent.

Many strains of *E. coli* are able to synthesize two siderophores, aerobactin and enterochelin. Although aerobactin has a dramatically lower affinity for iron than enterochelin, it has been shown to provide a significant selective advantage for bacterial growth in conditions of iron limitation, such as in the body fluids and tissues. Differential regulation of the genetic determinants of the two siderophores resulted in preferential induction of the aerobactin system in the presence of unsaturated levels of TF and LF (Williams & Carbonetti, 1986).

Pathogenic *Neisseriae* have a repertoire of high-affinity iron uptake systems to facilitate acquisition of this essential element in the human host. They possess surface receptor proteins that directly bind the extracellular host iron-binding proteins TF and LF (see review: Schryvers & Stojiljkovic, 1999). Alternatively, they have siderophore receptors capable of scavenging iron when exogenous siderophores are present. Released intracellular heme iron present in the form of hemoglobin, hemoglobin-haptoglobin or free heme can be used directly as a source of iron for growth through direct binding by specific surface receptors. Although these receptors may vary in complexity and composition, the key protein involved in the transport of iron (as iron, heme or iron-siderophore) across the outer membrane is a TonB-dependent receptor with an overall structure presumably similar to that determined recently for *E. coli* FhuA or FepA. The receptors are potentially ideal vaccine targets in view of their critical role in survival in the host.

Helicobacter pylori is known to be an etiologic agent of gastritis and peptic ulcer disease in humans. Human LF supported full growth of the *H. pylori* in media lacking other iron sources, but neither human TF, bovine LF, nor hen ovoTF served as a source for iron (Husson et al., 1993). Since hLF is found in significant amounts in human stomach resections with superficial or atrophic gastritis, the iron acquisition system of *H. pylori* by the hLF receptor system may play a major role in the virulence of *H. pylori* infection. Most *H. pylori* strains also seem to produce extracellular siderophores (Illingworth et al., 1993).

The ability of malleobactin to mobilize iron from LF and TF was examined in an equilibrium dialysis assay in the absence of bacteria (Yang et al., 1993). Malleobactin was capable of removing iron from both LF and TF at pH values of 7.4, 6.0, and 5.0. However, the levels of iron mobilization were greater for TF than for LF at all the pH values used in the assay. *Bordetella bronchiseptica* uses a hydroxamate siderophore for removal of iron from LF and TF rather than relying upon a receptor for these host iron-binding proteins (Foster & Dyer, 1993).

Moraxella (Branhamella) catarrhalis, a mucosal pathogen closely related to *Neisseria* species, is a prominent cause of otitis media in young children and lower respiratory tract infections in adults. Campagnari et al. (1994) demonstrated that *M. catarrhalis* obtains iron from LF and TF and also maintains growth with ferric nitrate *in vitro*. Furthermore, when *M. catarrhalis* is grown under iron-limited conditions, the bacteria express new outer membrane proteins that are not detected in membranes of organisms cultured in an iron-rich environment. These iron-repressible proteins may be important for the acquisition and utilization of iron *in vivo*, which could allow *M. catarrhalis* to colonize and survive on human mucosal surfaces.

Vulnibactin, a siderophore produced by *Vibrio vulnificus*, has been shown to sequester TF- or LF-bound iron for growth (Okujo et al., 1996). Comparative studies with the strain producing vulnibactin and its exocellular protease-deficient mutant revealed the involvement of the protease in addition to vulnibactin could be effective in the utilization of Fe(III) bound to TF and LF. It appears that the protease causes cleavage of these proteins, thereby making bound iron more accessible to vulnibactin. In response to environmental iron stress, *Vibrio cholerae* produces the siderophore vibriobactin as well as a number of iron-induced outer membrane proteins (Tashima et al., 1996).

Leishmania chagasi, the cause of South American visceral leishmaniasis, requires iron for its growth. Wilson et al. (1994) reported the ability of promastigote forms of *L. chagasi* to take up ^{59}Fe chelated to either TF or LF, although uptake from ^{59}Fe-LF occurred more rapidly. ^{59}Fe uptake from either ^{59}Fe-TF or ^{59}Fe-LF was inhibited by a 10-fold excess of unlabeled holo-LF, holo-TF, apo-LF, apo-TF, or iron nitrilotriacetate but not ferritin or bovine serum albumin. There was no evidence for a role for parasite-derived siderophores or proteolytic cleavage of holo-LF or holo-TF in the acquisition of iron by promastigotes. This capacity to utilize several iron sources may contribute to the organism's ability to survive in the diverse environments it encounters in the insect and mammalian hosts.

Gardnerella vaginalis could acquire iron from hLF, but not from hTF (Jarosik et al., 1998). Siderophore production was detected in *G. vaginalis* strains and SDS-PAGE of the cytoplasmic membrane proteins isolated from *G. vaginalis* grown under iron-replete and iron-restricted conditions revealed several iron-regulated proteins ranging in molecular mass from 33- to 94-kDa.

VIII. MICROBIAL INTERACTIONS

LF could bind to microbial surface via an array of specific and non-specific interactions. Interactions of LF with specific microbial target sites could lead to events either for promoting host defense (microbial elimination) or microbial virulence (iron acquisition by pathogen) (Naidu & Arnold, 1997). This chapter shall focus the role of LF-binding microbial targets in antimicrobial susceptibility. Various LF-binding microbial targets have been listed in *TABLE 5*.

A. LF-binding targets

Specific LF-binding targets were identified on a variety of bacterial pathogens. The binding components in most microorganisms were proteins, and a few lectin-type interactions were also reported.

1. Staphylococcus spp. Bovine LF could bind to the following staphylococcal species associated with bovine intramammary infections: *S. epidermidis, S. warneri, S. hominis, S. xylosus, S. hyicus*, and *S. chromogenes* (Naidu et al., 1990). The bLF-binding mechanism was specific, with affinity constants (K_a values) ranging between 0.96 mM and 11.9 mM. The numbers of bLF-binding sites per cell, as determined by using Scatchard analysis, were as follows: *S. epidermidis*, 3,600; *S. warneri*, 1,900; *S. hominis*, 4,100; *S. xylosus*, 4,400; *S. hyicus*, 6,100; and *S. chromogenes*, 4,700. The bLF- binding receptors of the six coagulase-negative staphylococcal species demonstrated marked differences in patterns of susceptibility to proteolytic or glycolytic enzyme digestion and to heat or periodate treatment. These data suggest that the bLF-binding components in *S. epidermidis* and *S. warneri* are proteins containing glycosidyl residues. In the remaining four species, the proteinaceous nature of the bLF-binding component was evident, but the involvement of glycosidyl residues was not clear.

Naidu and co-workers (1991) investigated the hLF binding property of 489 strains of *S. aureus* isolated from various clinical sources. The hLF binding was common among *S. aureus* strains associated with furunculosis (94.3%), toxic shock syndrome (94.3%), endocarditis (83.3%) and septicaemia (82.8%) and other (nasal, vaginal or ocular) infections (96.1%). Naidu et al. (1992) also characterized the hLF-staphylococcal interaction in *S. aureus* strain MAS-89. The binding of ^{125}I-hLF to strain MAS-89 reached saturation in less than 90 min and was maximal between pH 4 and 9. Unlabelled hLF displaced ^{125}I-HLF binding. Various plasma and subepithelial matrix proteins, such as IgG, fibrinogen, fibronectin, collagen and laminin, which are known to interact specifically with *S. aureus*, did not interfere with hLF binding. The Scatchard plot was non-linear; that

TABLE 5: Specific binding of LF to various microbial cell surfaces - density, mass, binding-affinity (association constant) of cellular target sites

Microorganism	LF type	LF-binding cellular target site	Sites/cell	Mass	Affinity (K_A)	Reference
Actinobacillus actinomycetemcomitans	bLF/hLF	Heat-modifiable OMPs	ND	29-, 16.5-kDa	880 /1,800 nM	Alugupalli et al.,1995
Aeromonas hydrophila CCUG14551	bLF/hLF	Outer membrane proteins / porins	ND	30-, 40-, 60-kDa	ND	Kishore et al., 1991
Bordetella bronchseptica	hLF	Cell surface protein	ND	32-kD	ND	Menozzi et al., 1991
Bordetella pertussis	hLF	Cell surface protein	ND	27-kD	ND	Menozzi et al., 1991
Clostridium sp.	bLF	Surface layer protein	ND	33-kDa	ND	Tomita et al., 1998
Escherichia coli E34663	bLF/hLF	Outer membrane proteins (OMPs)	5,400	ND	140 nM	Naidu et al., 1991
Escherichia coli E34663	bLF/hLF	Porins Omp-C, Omp-F & Pho-E		37-kDa		Erdei et al., 1994
Escherichia coli O55B5	bLF/hLF	Lipopolysaccharide	ND		4 / 390 nM	Elas-Rochard et al., 1995
Haemophilus influenza KC548	hLF	Membrane protein	ND	106-kD, 105-kD	ND	Schryver, 1989
Helicobacter pylori	bLF/hLF	Heat-shock protein	ND	60-kDa	2,880 nM	Amini et al., 1996
	hLF	Outer membrane protein	ND	70-kDa	ND	Dhaenens et al., 1997
Mycoplasma pneumoniae	hLF	Cell membrane	10,000	ND	20 nM	Tryon & Baseman, 1987
Niesseria gonorrhoeae	hLF	Outer membrane receptor	ND	103-kD	ND	Biswas & Sparling, 1995
Neisseria meningitidis	hLF	Outer membrane receptor	ND	105-kD	ND	Schryver & Morris, 1988
	hLF	OMP - *IroA* protein	ND	ND	ND	Pettersson et al., 1994
Porphyromonas gingivalis	hLF	Fimbriae, lectin-type interaction	ND	ND	ND	Sojar et al., 1998
Prevotella intermedia 4H	bLF/hLF	Cell surface protein	45,000	62-kDa	550 nM	Kalfas et al., 1991; 1992
Pseudomonas aeruginosa	hLF	Outer membrane protein	ND	48-kD, 25-kD	ND	Carnoy et al., 1994
Salmonella abony NCTC6017	bLF/hLF	Outer membrane proteins / porins	ND	38-, 35-, 32-kDa	ND	Naidu & Arnold, 1994
Salmonella dublin NCTC9676	bLF/hLF	Outer membrane proteins / porins	ND	38-kDa	ND	Naidu & Arnold, 1994
Salmonella hartford HNCMB10063	bLF/hLF	Outer membrane proteins / porins	ND	37-, 35-, 25-kDa	ND	Naidu & Arnold, 1994
Salmonella panama NCTC5774	bLF/hLF	Outer membrane proteins / porins	ND	37-kDa	ND	Naidu & Arnold, 1994
Salmonella pullorum NCTC5776	bLF/hLF	Outer membrane proteins / porins	ND	38-, 35-kDa	ND	Naidu & Arnold, 1994
Salmonella rostock NCTC5767	bLF/hLF	Outer membrane proteins / porins	ND	39-kDa	ND	Naidu & Arnold, 1994
Salmonella kentucky NCTC5799	bLF/hLF	Outer membrane proteins / porins	ND	38-, 35-, 33-kDa	ND	Naidu & Arnold, 1994
Salmonella thompson NCTC5740	bLF/hLF	Outer membrane proteins / porins	ND	38-, 35-, 33-kDa	ND	Naidu & Arnold, 1994
Salmonella typhimurium ATCC13311	bLF/hLF	Outer membrane proteins / porins	ND	38-kDa	ND	Naidu & Arnold, 1994
Salmonella virchow NCTC 5742	bLF/hLF	Outer membrane proteins / porins	ND	38-, 35-, 33-kDa	ND	Naidu & Arnold, 1994
Shigella flexneri M90T	hLF	OMPs, porins	4,800	39-, 22-, 16-kDa	690 nM	Tigyi et al., 1992
Shigella flexneri M90T	bLF	OMPs, porins	5,700	39-, 22-, 16-kDa	104 nM	Tigyi et al., 1992
Staphylococcus aureus MAS89	hLF	Proteoglycan component	5,700	67-kDa / 62-kDa	27 nM	Naidu et al., 1992
Staphylococcus chromogenes AD1	bLF	Cell wall protein	4,700	ND	2,500 nM	Naidu et al., 1990
Staphylococcus epidermidis AF9	bLF	Peptidoglycan component	3,600	ND	11,900 nM	Naidu et al., 1990
Staphylococcus hominis AF93	bLF	Cell wall protein	4,100	ND	3,800 nM	Naidu et al., 1990

TABLE 5 (CONT.): Specific binding of LF to various microbial cell surfaces – density, mass, binding-affinity (association constant) of cellular target sites

Microorganism	LF type	LF-binding cellular target site	Sites/cell	Mass	Affinity (K_a)	Reference
Staphylococcus hyicus AC166	bLF	Cell wall protein	6,100	ND	960 nM	Naidu et al., 1990
Staphylococcus warneri AF101	bLF	Peptidoglycan component	1,900	ND	3,100 nM	Naidu et al., 1990
Staphylococcus xylosus AG12	bLF	Cell wall protein	4,400	ND	3,300 nM	Naidu et al., 1990
Treponema denticola	hLF	Cell surface proteins	ND	50-, 35-kDa	ND	Staggs et al., 1994
Treponema pallidum	hLF	Cell surface proteins	ND	49-, 34-, 29-kDa	ND	Staggs et al., 1994
Tritrichomonas foetus	hLF	Cell wall receptor	170,000	ND	3,600 nM	Tachezy et al., 1996
Trichomonas vaginalis	hLF	Cell surface receptor	90,000	ND	1,000 nM	Peterson & Alderete, 1984
Trypanosoma cruzi	hLF	Amastigote cell wall	1,000,000	ND	ND	Lima & Kierszenbaum, 1985

implied a low affinity (K_a:155 nM) and a high affinity (K_a:270 nM) binding mechanism. About 5,700 hLF-binding sites/cell were estimated. The staphylococcal hLF- binding protein (hLF-BP) was partially susceptible to proteolytic enzymes or periodate treatment and was resistant to glycosidases. An active hLF-BP with an apparent M_r of about 450-kDa was isolated from strain MAS-89 cell lysate by ion- exchange chromatography on Q-sepharose. In SDS-PAGE, the reduced hLF-BP was resolved into two components of 67- and 62-kDa. The two components demonstrated a positive reaction with hLF-HRPO in a Western blot. These data established a specific receptor for hLF in *S. aureus*.

2. Streptococcus spp. The exposure of *Streptococcus agalactiae* to bLF resulted in the binding of this protein to all the 12 strains of bovine origin tested, and also, although to a lesser degree, to the five tested strains of human origin (Rainard, 1992). The binding of LF was slightly affected by cultivation conditions, and appeared to be heat-stable. The binding of biotinylated LF was inhibited by unlabeled-LF but not by BSA.

Hammerschmidt et al. (1999) demonstrated specific binding of hLF to *Streptococcus pneumoniae*. Pretreatment of pneumococci with proteases reduced hLF binding markedly, indicating that the hLf receptor is proteinaceous. Binding assays performed with 63 clinical isolates belonging to different serotypes showed that 88% of the tested isolates interacted with hLF. Scatchard analysis showed the existence of two hLF-binding proteins with dissociation constants of 57 nM and 274 nM. The receptors were purified by affinity chromatography, and internal sequence analysis revealed that one of the *S. pneumoniae* proteins was homologous to pneumococcal surface protein A (PspA). The function of PspA as an hLF-binding protein was confirmed by the ability of purified PspA to bind hLF and to competitively inhibit hLF binding to pneumococci. *S. pneumoniae* may use the hLF-PspA interaction to overcome the iron limitation at mucosal surfaces, and this might represent a potential virulence mechanism.

3. Vibrio spp. Binding of LF to non-invasive *Vibrio cholerae* was reported (Ascencio et al., 1992). Iron-binding glycoproteins such as ferritin, TF, haemoglobin, and myoglobin moderately inhibited the LF interaction with the vibrios. Monosaccharides (N-acetyl glucosamine, mannose, galactose, and fucose), and other glycoproteins such as fetuin and orosomucoid also moderately inhibited the binding. *V. cholerae* showed a cell surface associated proteolytic activity which cleaved the cell-bound ^{125}I-labeled LF.

4. Helicobacter pylori: The interactions of *Helicobacter pylori* spiral and coccoid forms with LF was reported by Khin et al. (1996). The coccoid forms of 14 strains of *H. pylori* showed significant hLF binding (median 26%), found to be specific and was inhibited by unlabeled hLF and bLF.

Amini et al. (1996) reported the binding of bLF to a 60-kDa heat shock protein of *H. pylori*. Binding ability was related to human immunoglobulin G because bLF binding proteins were isolated by extraction of cell surface associated proteins with distilled water, applied on IgG-Sepharose and nickel sulfate chelate affinity chromatography. Binding was demonstrated by Western blot after purified protein was digested with alpha-chymotrypsin and incubated with peroxidase-labeled bLF. Binding was inhibited by unlabeled bLF, lactose, rhamnose, galactose, and two iron-containing proteins, ferritin and haptoglobin. Carbohydrate moieties of bLF seem to be involved in binding because glycoproteins with similar carbohydrate structures strongly inhibited binding. Scatchard plot analysis indicated a binding affinity (K_a) of 2.88 μM. In addition, binding of *H. pylori* cells to bLF was enhanced when bacteria treated with pepsin or α-chymotrypsin after isolation from iron-restricted and iron-containing media.

5. Oral bacteria. Interaction of LF with *Actinobacillus actinomycetemcomitans* was reported (Alugupalli et al., 1995). The binding of hLF and bLF reached maximum within 1 h. LF binding to the bacterium was pH-dependent and reversible. Scatchard analysis indicated the existence of two different types of binding sites on the bacterium, one with a high affinity constant (K_a:0.880 μM) and the other with a low affinity (K_a:1.8 μM). Bacteria in the exponential phase of growth showed higher binding than cells in the stationary phase. Bacteria grown in medium containing serum and/or lysed erythrocytes bound LF to a lesser extent. Heat-inactivated serum, lysed erythrocytes and other proteins such as mucin and laminin inhibited LF binding to *A. actinomycetemcomitans* in a competitive binding assay. SDS-PAGE and Western blot analysis revealed LF-reactive protein bands at 29-kDa and 16.5-kDa in the CE and OM of *A. actinomycetemcomitans*. The 29-kDa band displayed a heat-modifiable LF-reactive form with a molecular weight of 34-kDa. Neither proteinase K-treated cell envelope nor LPS of this bacterium showed reactivity with LF. These data suggest a specific LF interaction with OMPs of *A. actinomycetemcomitans*.

A LF-binding protein with an estimated molecular mass of 57-kDa was identified in the cell envelope of *Prevotella intermedia* by SDS-PAGE and Western- blot analysis (Alugupalli et al., 1994). Peroxidase-labeled bLF and hLF showed similar specific binding to this protein.

B. Porins

Porins are a well conserved heat-modifiable, pore-forming outer membrane proteins (OMPs) of the family *Enterobacteriaceae*. Porins are also reported in certain Gram-negative bacteria. Porins exist as trimers in the outer membrane usually surrounded by nine molecules of LPS. Porins are suggested to play an essential role in the transport of various solutes across the Gram-negative OM. Porins also serve as receptors for certain bacteriophages and colicins (see reviews: Lugtenberg & van Alpen, 1983; Nikaido & Vaara, 1985; Nikaido, 1989). Naidu and co-workers have reported the role of porins as LF anchorages in *E.coli* and other bacterial members of the family *Enterobacteriaceae* (Kishore et al., 1991; Tigyi et al., 1992; Erdei et al., 1993; Naidu & Arnold, 1994).

1. Escherichia coli. The degrees of hLF and bLF binding in 169 *E. coli* strains isolated from human intestinal infections, and in an additional 68 strains isolated from healthy individuals, were examined in a ^{125}I-labelled protein binding assay (Naidu et al., 1991). The binding was expressed as a percentage calculated from the total labelled ligand added to bacteria. The hLF and bLF binding to *E. coli* was in the range 3.7 to 73.4% and 4.8 to 61.6%, respectively. Enterotoxigenic (ETEC) strains demonstrated a significantly higher hLF binding than enteropathogenic (EPEC), enteroinvasive (EIEC), enterohaemorrhagic (EHEC) strains or normal intestinal *E. coli* isolates. EPEC strains belonging to serotypes O44 and O127 demonstrated significantly higher hLF binding compared to O26, O55, O111, O119 and O126 serotypes. No significant differences in the degree of hLF or bLF binding were found between aerobactin-producing and non-producing strains. The interaction was further characterized in a high LF-binding EPEC strain, E34663 (serotype O127). The binding was stable in the pH range 4.0 to 7.5, did not dissociate in the presence of 2M NaCl or 2M urea, and reached saturation within 2-h. Unlabelled hLF and bLF displaced the 125I-hLF binding to E34663 in a dose-dependent manner. Apo- and iron- saturated forms of LF demonstrated similar binding to E34663. Among various unlabelled subepithelial matrix proteins and carbohydrates tested (in 4-log excess) only fibronectin and fibrinogen caused a moderate inhibition of 125I-hLF binding. According to Scatchard plot analysis, 5,400 hLF-binding sites/cell, with an affinity constant (K_a) of 140 nM, were estimated in strain E34663. These data establish the presence of a specific LF-binding mechanism in *E. coli*.

Gado et al. (1991) reported that *E. coli* with low hLF binding are insusceptible to group A (A, E1, E2, E3, E6, and K) and group B (B, D, Ia, Ib, and V) colicins. Conversely, a spontaneous hLF high-binding variant demonstrated an increased suscepti-bility to both colicin groups. Colicin-insusceptible *E. coli* wild-type strains 75ColT, 84ColT, and 981ColT showed a low degree of hLF binding, i.e., 4, 8, and 10%, respec-tively. The hLF binding capacity was high in the corresponding colicin-susceptible mutants 75ColS (43%), 84ColS (32%), and 981ColS (43%). Furthermore, hLF low- (< 5%) and high- (> 35%) binding *E. coli* clinical isolates (10 in each category) were tested for susceptibility against 11 colicins. Colicin V susceptibility did not correlate with hLF binding in either categories. However, with the remaining colicins, three distinct hLF-binding, colicin susceptibility patterns were observed; (i) 10 of 10 hLF low-binding strains were colicin insusceptible, (ii) 6 of 10 hLF high-binding strains were also colicin insusceptible, and (iii) the remaining hLF high binders were highly colicin susceptible. Certain proteins in the cell envelope and outer membrane of wild-type H10407 (hLF low binder, colicin insusceptible) showed a lower mobility in SDS-PAGE compared to the cor-responding proteins of mutant H10407(LF) (hLF high binder, colicin susceptible). These mobility differences were also associated with hLF-binding proteins in Western blot (lig-and blot) analysis. The wild type showed a smooth form of LPS with a distinct ladder of O-chains, compared to the rough LPS of the mutant.

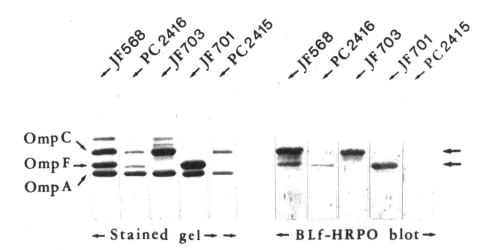

FIGURE 8. LF interaction with porins of *E.coli*. Outer membrane analyses of wild-type (JF568) and porin-defi-cient mutants (PC2416 express *PhoE* only; JF703 express *OmpC* only; JF701 express *OmpF* only; and PC2415 express none of the three porins) of *E.coli* by urea-SDS-PAGE and western blotting with HRPO-labeled bLF [from Erdei et al. (1994) with permission from the American Society for Microbiology].

Certain strains of *E. coli* (bacterial whole cells) demonstrate specific interaction with ^{125}I-LF. A band with a mass of approximately 37-kDa, which was reactive with horseradish peroxidase (HRPO)-labeled LF, was identified in the boiled cell envelope and outer membrane preparations of an LF-binding *E. coli* strain, E34663, and a non-LF-binding strain, HH45, by SDS-PAGE and Western blotting (Erdei et al., 1994). Such a band was not detected in the unboiled native cell envelope and outer membrane preparations. The molecular mass and the property of heat modifiability suggested that the LF-binding proteins were porins. The native trimeric form of porin OmpF isolated from strain B6 and its dissociated monomeric form both reacted with HRPO-labeled LF and with monoclonal antibodies specific for OmpF. Furthermore, by using *E. coli* constructs with defined porin phenotypes, OmpF and OmpC were identified as the LF-binding proteins by urea-SDS-PAGE and Western blotting and by ^{125}I-LF binding studies with intact bacteria (*FIGURE 8*). These data established that LF binds to porins, a class of well-conserved molecules common in *E. coli* and many other Gram-negative bacteria. However, in certain strains of *E. coli* these pore-forming proteins are shielded from LF interaction.

2. *Salmonella* spp. Interaction of LF with the CE and OM of *Salmonella typhimurium*-type strain ATCC13311 was tested by SDS-PAGE and Western-blot analyses (Naidu & Arnold, 1994). The HRPO-labeled bLF and hLF both recognized a heat-modifiable protein with an estimated molecular mass of 38-kDa in the OM. Simultaneous immunoblotting with an antiporin monoclonal antibody specific for a conserved porin domain in members of *Enterobacteriaceae* confirmed that the LF-binding protein is a porin. Such LF-binding porin proteins (37- to 39-kDa range) were readily detected in nine other common *Salmonella* species: *S. dublin, S. panama, S. rostock, S. abony, S. hartford, S, kentucky, S. pullorum, S. thompson,* and *S. virchow*. The latter six species also demonstrated one to three weak LF-reactive bands of low molecular weight in their CE.

3. *Shigella flexneri*. Tigyi et al. (1992) reported the interaction of LF with dysentry pathogen, *Shigella flexneri*. The interaction was specific, and approximately 4,800 hLF binding sites (K_a:690 nM) or approximately 5,700 bLF binding sites (K_a:104 nM) per cell were estimated in strain M90T by a Scatchard plot analysis. The native CE and OM did not reveal LF-binding components in SDS-PAGE. However, after being boiled, the CE and OM preparations showed three distinct horseradish peroxidase-LF reactive bands of about 39-, 22-, and 16-kDa. The 39-kDa component was also reactive to a monoclonal antibody specific for porin (PoI) proteins of members of the family *Enterobacteriaceae*. The LF binding protein pattern was similar with bLF or hLF, for Crb+ and Crb- strains. The protein-LF complex was dissociable by KSCN or urea and was stable after treatment with NaCl. Variation (loss) in the O chain of LPS markedly enhanced the LF-binding capacity in the isogenic rough strain SFL1070-15 compared with its smooth parent strain, SFL1070. These data establish that LF binds to specific components in the bacterial OM; the heat-modifiable, anti-PoI-reactive, and LPS-associated properties suggested that the LF-binding proteins are porins in *S. flexneri*.

4. *Aeromonas hydrophila*. The interaction of LF with *Aeromonas hydrophila* was tested in a ^{125}I-labeled protein- binding assay (Kishore et al., 1991). The LF binding was characterized in type strain *A. hydrophila* ssp. *hydrophila* CCUG 14551. The hLF and bLF binding reached a complete saturation within 2 h. Unlabeled hLF and bLF displaced [125I]- hLF binding in a dose-dependent manner, and more effectively by the heterologous (1 µg for 50% inhibition) than the homologous (10 µg for 50% inhibition) ligand. Apo- and holo-forms of hLF and bLF both inhibited more than 80%, while mucin caused approx. 50% inhibition of the hLF binding. Various other proteins (including TF) or car- bohydrates did not block the binding. Two hLF-binding proteins with an estimated mole- cular masses of 40-kDa and 30-kDa were identified in a boiled-CE preparation, while the unboiled CE demonstrated a short-ladder pattern at the top of the separating gel and a sec- ond band at approx. 60-kDa position. These data establish a specific interaction of LF and the LF-binding proteins seem to be porins in *A. hydrophila*.

C. Lipopolysaccharides (LPS)

1. Interactions. LF was reported to bind lipid A and intact LPS of *E. coli, Klebsiella pneumoniae, Pseudomonas aeruginosa, Neisseria meningitides* and *Haemophilus influenzae* (Appelmelk et al. 1994). LF binding to LPS was inhibitable by lipid A and polymyxin B but not by KDO (3-deoxy-D-manno-octulosonate), glycoside residues present in the inner core of LPS. Binding of LF to lipid A was saturable, and with an affinity constant of 2 nM.

Elass-Rochard et al (1995) reported the presence of two *E. coli O55B5* LPS-bind- ing sites on hLF: a high-affinity binding site (K_a:3.6 nM) and a low-affinity binding site (K_a:390 nM). Bovine LF, which shares about 70% amino acid sequence identity with hLF, showed similar interaction with LPS. Human serum TF, which is known to bind LPS, caused only 12% inhibition of hLF-LPS interaction, suggesting different binding domains on LPS. Binding and competitive binding experiments performed with the N-tryptic frag- ment (residues 4-283), the C-tryptic fragment (residues 284-692) and the N2-glycopeptide (residues 91-255) isolated from hLF have demonstrated that the high-affinity binding site is located in the N-terminal domain I of hLF, and the low-affinity binding site is present in the C-terminal lobe. The inhibition of hLF-LPS interaction by a synthetic octade- capeptide corresponding to residues 20-37 of hLF and bLFcin (residues 17-41), a prote- olytic fragment from bLF, revealed the importance of the 28-34 loop region of hLF and the homologous region of bLF for LPS binding. Direct evidence that this amino acid sequence is involved in the high-affinity binding to LPS was demonstrated by assays car- ried out with EGS-loop hLF, a recombinant hLF mutated at residues 28-34.

After differentiation, HL-60 cells showed a twofold increase of LF- binding sites with no difference in the specificity or affinity of LF between pre- and post-differentiated cells (Miyazawa et al, 1991). CD11a, CD11b, and CD11c Ag, which have been associat- ed with specific binding sites for LPS on monocytes/macrophages, were also increased 3- to 4-fold after differentiation. With the use of this system, the effect of LPS on LF bind- ing was tested. At 37°C, LPS enhanced LF binding on HL-60 cells, especially after dif- ferentiation. Conversely, at 4°C, LPS inhibited LF binding. There was little effect of tem- perature on LF binding in the absence of LPS. In the presence of polymyxin B sulfate, the enhanced LF binding by LPS was abrogated. In addition, pretreatment with mAbCD11 and/or mAb5D3, which are associated with or directed against candidate LPS receptors reduced LF binding. Cross-linking studies using an iodinated, photoactivatable LPS deriv- ative ([125I] ASD-LPS) demonstrated specific binding of LPS to LF. These data indicated a dichotomous nature of LF binding on monocyte/macrophage-differentiated HL-60 cells;

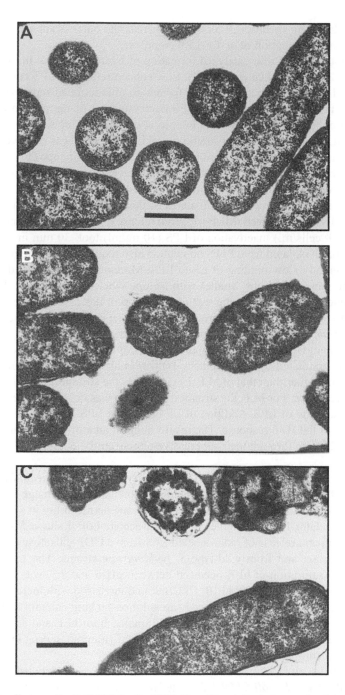

FIGURE 9. LFcin effects on *E.coli* CL99 I-2 cells (TEM: Bars, 50 nm). Bacteria were incubated for 2-h in 1% Bacto-peptone broth (A); or broth containing 0.1 mg/ml of bLFcin for 0-h (B) or 2-h (C). Bacteria exposed to bLFcin show altered cell membrane morphology with the appearance of membrane blisters. After 2-h incubation a large amount of cell debris was present and a number of remaining cells appear to have a clumping or coagulation of cytoplasmic elements in addition to membrane blistering [from Yamauchi et al., (1993) with permission from the American Society for Microbiology].

one being mediated by specific LF receptors whereas the other is apparently mainly via LPS receptors after formation of an LF-LPS complex.

LF could inhibit iron-catalyzed formation of hydroxyl radical in the presence of LPS at pH 7.4 and 4.5 (Cohen et al, 1992). Low concentrations of LPS could prime neutrophils toward enhanced function, such as formation of stimulated superoxide anion. LF inhibited LPS priming of neutrophils if LPS contamination of the protein (provided by commercial suppliers) was first reduced. Inhibition of LPS priming was observed whether apo-LF or Fe-LF was used. Similar inhibition of LPS priming was observed when neutrophils were incubated with other serum proteins (e.g., albumin, apoTF, or iron-saturated TF). These results indicated that LPS should not be expected to affect the free radical biology of LF, which is a crucial physiologic function of this protein.

PMN-induced LF could inactivate LPS, thereby blocking the ability of LPS to prime fresh PMNs for enhanced fMLP-triggered release of superoxide (Wang et al, 1995). Neutrophils (10^6 cells/ml) inactivation of LPS (10 ng/ml) took 30 min in a kinetic fashion. In addition, LF isolated from PMN population also required 30 min to inactivate LPS, indicating inherently slow binding of LF to LPS. Mononuclear cells failed to inactivate LPS under the same conditions. Studies with isotope-labeled LPS showed that inactivated LPS remained in the medium and was not taken up or destroyed by the PMNs during inactivation. Bovine LF was also reported to diminish the inflammatory reactions induced by *Mycobacterium bovis* (BCG) in a mouse model (Zimecki & Machnicki, 1994). Human LF, bLF and LFcin-B were found to suppress the IL-6 response in a monocytic cell line (THP-1) when stimulated by LPS (Mattsby-Baltzer et al, 1996). The suppression of bLF was similar to or higher than that of hLF. LFcin-B was the strongest inhibitor of the LPS-induced IL-6 response. For hLF, the strongest inhibition was observed when added 15-30 min after the addition of LPS. Addition of LF before the LPS induced an approximately 45% reduction of the IL-6 response. The results suggest an anti-inflammatory activity of hLF, bLF, and bLFcin through their suppressive effects on the cytokine release.

2. LF-induced LPS release - effects on microbial OM permeability. Ellison and co-workers (1988) reported that the iron-binding proteins could damage the gram-negative outer membrane and alter bacterial outer membrane permeability in a manner similar to that of the chelator EDTA. Studies in barbital-acetate buffer showed that EDTA and hLF cause significant release of radio-labeled LPS from a UDP-galactose epimerase-deficient *E. coli* mutant and from wild-type *S. typhimurium* strains. The LPS release was blocked by iron saturation of hLF, occurred between pH 6 and 7.5, was comparable for bacterial concentrations from 10^4 to 10^7 CFU/ml, and increased with increasing hLF concentrations. Studies using Hanks balanced salt solution lacking calcium and magnesium showed that TF could cause LPS release. Additionally, both hLF and TF increased the antibacterial effect of a sub-inhibitory concentration of rifampin, a drug excluded by the bacterial outer membrane.

Bovine LF and LFcin also reported to release intrinsically labeled [³H] LPS from three bacterial strains, *E. coli CL99 1-2*, *S. typhimurium SL696*, and *Salmonella montevideo SL5222* (Yamauchi et al, 1993). Under most conditions, more LPS are released by LFcin, the peptide fragment than by whole bLF. In the presence of either, LF or LFcin there is increased killing of *E. coli CL99 1-2* by lysozyme. Bovine LF and LFcin have the ability to bind to free intrinsically labeled [³H] LPS molecules, similar to hLF. In addition to these effects, whereas bLF was at most bacteriostatic, LFcin demonstrated consistent bactericidal activity against gram-negative bacteria. This bactericidal effect is modulated by the cations Ca^{2+}, Mg^{2+}, and Fe^{3+} but is independent of the osmolarity of the medium. Transmission electron microscopy of bacterial cells exposed to LFcin show the immediate development of electron-dense membrane blisters (*FIGURE 9*).

LF shows synergism with lysozyme; each protein alone is bacteriostatic, however, the mixture elicit a bactericidal effect for strains of *E. coli, S. typhimurium*, and *Vibrio cholerae* (Ellison & Giehl, 1991). The cidal activity is dose dependent, blocked by iron saturation of LF, and inhibited by high calcium levels, although LF does not chelate calcium. Growth media conditions inhibit completely or partially the effect of LF and lysozyme; the degree of inhibition correlate with media osmolarity. Dialysis chamber studies indicated that bacterial killing required direct contact with LF, and experiments with purified LPS suggested that this related to direct LPS- binding by the protein.

3. LPS induction of LF production in vivo. When incubated with *Salmonella typhimurium* LPS at 37°C, human PMN suspended in serum-free buffer releases the specific granule constituent LF into the surrounding medium (Koivuranta-Vaara et al., 1987). Release of LF from PMN vary with the concentration of LPS as well as with the duration of incubation and is not accompanied by significant release of the cytoplasmic enzyme lactate dehydrogenase. LPS-induced release of LF from PMN was augmented significantly when cell suspensions were supplemented with additional monocytes and lymphocytes. Only monocytes, however, secreted significant amounts of LF-releasing activity (in a time- and concentration- dependent manner) when incubated separately with LPS. LF-releasing activity was heat (80°C for 15 min) labile, eluted after chromatography on Sephadex G-100 with an apparent molecular weight of approximately 60-kDa, and was inhibited by antibodies to TNF-α. Thus, LPS-induced noncytotoxic release of LF from human PMN suspended in serum-free buffer is mediated, at least in part, by TNF-α derived from contaminating monocytes.

Injection of *Salmonella typhimurium* or LPS into mice resulted in a dose-dependent increase in plasma LF. Endotoxin challenge of normal and neutropenic mice showed a direct correlation of plasma LF level with the granulocyte count in peripheral blood. Physiological neutropenia did not inhibit the LF release (Sawatzki & Rich, 1989). *In vivo* studies by Gutteberg et al. (1989) measured the total serum iron, plasma LF and circulating leukocytes in piglets during the early phase of severe gram-negative septicemia and endotoxemia following infusion of LPS or *E. coli*. Iron dropped significantly during the first 30 min of LPS infusion from a median of 32 mM to 13.4 mM. A similar decrease in serum iron was observed at 120 min after the *E. coli* infusion. Plasma levels of LF increased significantly 120 min after the LPS infusion (6 mg/L compared to preinfusion value of 0.25 mg/L). After intravenous infusion of *E. coli* a significant rise of plasma LF was demonstrated in 30 min after bacterial infusion (2.1 mg/L compared to preseptic value of 0.8 mg/L). This increase was accompanied with a significant drop of circulating leukocytes (7.3 x 10⁹/L compared to preinfusion, 17 x 10⁹/L) in the piglets receiving *E. coli* by intravenous route. However, intraperitoneal inoculaltion of *E. coli* did not show any significant change of plasma LF. The rapid onset of hyposideremia during endotoxemia and *E. coli* septicemia appeared to correlate with the release of LF from granulocytes and the clearance of iron-bound LF from blood or peritoneal cavity. LPS-stimulation of heparinized human blood resulted in a significant rise of LF, TNF-α, and thromboplastin levels (Gutteberg et al, 1990). LPS induced the secretion of LF from granulocytes; the levels were 2.1 mg/L (in 5 min) and 5.3 mg/L (in 60 min) compared to the control values of 0.2 mg/L and 1.3 mg/L, respectively. Systemic administration of bLF into mice 24 hours prior to intravenous challenge of LPS (50 mg), significantly lowered the serum concentration of TNF-α and IL-6 (Machnicki et al, 1993). Doses of bLF (<0.1 mg) and pretreatment of mice with bLF (days 6-2 or 12-2 hours) before LPS challenges were not effective.

Lipopolysaccharides (1 to 100 ng/ml) from *E. coli*, *S. typhimurium*, and *Serratia marcescens*, strongly enhanced growth inhibition of *Candida albicans* by human PMNs *in vitro*. Flow cytometry analysis demonstrated that LPS markedly augmented phagocytosis of *Candida* cells by increasing the number of yeasts ingested per neutrophil as well as the number of neutrophils capable of ingesting fungal cells. LPS activation caused augmented release of LF, and antibodies against LF effectively and specifically reduced the anti-*C. albicans* activity of both LPS-stimulated and unstimulated PMN. Northern (RNA blot) analysis showed enhanced production of mRNAs for IL-1b, TNF-α, and IL-6 and in neutrophils within 1 h of stimulation with LPS. The cytokines were also detected in the supernatant of the activated PMN, and their synthesis was prevented by pretreatment of LPS-stimulated PMN with protein synthesis inhibitors, such as emetine and cycloheximide. These inhibitors, however, did not block either LF release or the anti-*Candida* activity of LPS-stimulated PMN. These results demonstrate the ability of various bacterial LPS to augment neutrophil function against *C. albicans* and may contribute to the antifungal effect of PMN. Moreover, the ability to produce cytokines upon stimulation by ubiquitous microbial products such as the endotoxins points to an extraphagocytic, immunomodulatory role of PMN during infection.

D. DNA

LF was reported to interact with deoxyribonucleic acid (DNA) in both its double-stranded and single-stranded configurations (Bennett & Davis, 1982). Additional evidence for a preferential reactivity with dsDNA was provided by the enzymatic treatment of preformed dsDNA-LF and ssDNA-LF complexes with S1 endonuclease, and DNAse 1. DNase digestion alone liberated free LF. The interaction of LF with DNA partially inhibited the binding of anti-DNA antibodies from patients with SLE in a standard Farr assay. Furthermore, DNA-anti-DNA (labeled with ^{125}I-IgG) complexes could be dispersed *in vitro* by the addition of LF. It is hypothesized that the release of LF by neutrophils chemotactically attracted to DNA-anti-DNA complexes may act as a feedback loop to modulate the inflammatory response in SLE.

Studies indicated that the cell membrane DNA might have a novel role as a receptor for LF. Bennett et al (1983) demonstrated the binding of human ^{125}I-labeled LF to a population of adherent mononuclear cells (ADMC) and nonrosetting (E-) lymphocytes, which was abolished by prior treatment of the cells with deoxyribonuclease (DNase), but not ribonuclease (RNase). Saturable binding of LF at 0 °C was demonstrated for both E- and ADMC, with equilibrium dissociated constants of 0.76 x 10^{-6} M and 1.8 x 10^{-6} M, respectively. E- cells bound 2.5 x 10^7 and ADMC bound 3.3 x 10^7 molecules of LF at saturation. Cell membranes isolated from ADMC and E- reacted with ^{125}I-labeled LF and prior treatment of the membranes with DNase abolished the binding. This study provided evidence that cell membrane DNA acts as a binding site for exogenous LF.

Further investigations showed that binding of LF to the surface of human neutrophils was also dependent upon the presence of cell surface DNA (Bennett et al, 1986). This evidence of a DNA-LF interaction was obtained by the co-isolation of LF with DNA by both gel chromatography and affinity chromatography using Heparin Sepharose CL 6B. The interaction of LF with neutrophils was a saturable phenomenon with a K_a of 6.2 x 10^{-6} M and a maximum binding of 9.2 x 10^6 molecules per cell.

Various groups have consistently observed that LF interacts avidly with nucleic acids (Bennett & Davis, 1982; Bennett et al., 1986; Hutchens et al., 1989). To determine which region of the molecule is important in these interactions, solid-phase ligand binding assays were performed with LF from human milk (natural hLF) and N-terminally deleted hLF variants (van Berkel et al., 1997). Iron-saturated and natural hLF bound equally well to heparin, lipid A, human lysozyme and DNA. Natural hLF lacking the first two N-terminal amino acids (*Gly1-Arg2*) showed reactivities of one-half, two-thirds, one-third and one-third towards heparin, lipid A, human lysozyme and DNA respectively compared with N-terminally intact hLF. A lack of the first three residues (*Gly1-Arg2-Arg3*) decreased binding to the same ligands to one-eighth, one-quarter, one-twentieth and one-seventeenth respectively. No binding occurred with a mutant lacking the first five residues (*Gly1-Arg2-Arg3-Arg4-Arg5*). An anti-hLF monoclonal antibody (E11) that reacts to an N-lobe epitope including *Arg5* completely blocked hLF-ligand interaction. These results showed that the N-terminal stretch of four consecutive arginine residues, *Arg2-Arg3-Arg4-Arg5*, has a decisive role in the interaction of hLF with DNA, heparin, lipid A, and lysozyme.

Incubation of LF-Cu(II) complexes with supercoiled plasmid Bluescript II SK DNA rapidly form open circular or linear forms of DNA. LF with bound Cu(II) caused extensive degradation of yeast tRNA molecules in the presence of hydrogen peroxide. Covalent modification of surface-exposed histidyl residues by carboxyethylation with diethylpyrocarbonate abolished the LF-associated hydrolytic activity. These results indicated that LF-bound Cu(II) could facilitate the hydrolysis of DNA and RNA molecules. Copper-binding sites on LF possibly served as centers for repeated production of hydroxyl radicals via a Fenton-type Haber-Weiss reaction. Thus, it was speculated that an enhanced nuclease activity associated with elevated local concentrations of LF could potentiate microbial degradation (Zhao & Hutchens, 1994).

Recent studies showed that LF enters the cell from the serum and is transported into the nucleus where it binds DNA. Specific binding sites for LF were reported on K562 cells (estimate the number of binding sites and the dissociation constant). Western blotting analysis of K562 lysates identified a 120-kDa molecule as specific LF-binding receptor. This binding at the cell surface caused a temperature-dependant internalization of LF, which was immunologically detectable as a DNA-linked protein in nuclear extracts (Garre et al, 1992). Specific DNA sequences that could confer LF-induced gene transcription of a reporter gene have now been identified. RNA and DNA could inhibit LF enhancement of the activity of natural killer and lymphokine-activated killer cells *in vitro*. LF taken up by K562 human myelogenous leukemia cells appears in the nucleus where it is bound to DNA. This binding of LF to DNA occurs under stringent conditions with distinct sequence specificity, and that interaction between LF and these sequences intracellularly leads to transcriptional activation (He & Furmanski, 1995). This potential direct transcriptional function of LF is unique and remains to be confirmed in whole cells or tissues.

TABLE 6: Lactoferrin – Multifunctional activities

Cellular function	Reference
Inflammatory amplification and neutrophil aggregation	Oseas et al., 1981
Inhibition of antibody-mediated cytotoxicity	Nishiya and Horwitz, 1982
Specific growth stimulation of lymphocytes	Hashizume et al., 1983
Down regulation of myelopoiesis	Pelus et al., 1981; Broxmeyer et al., 1987
Complement activation by C3 convertase inhibition	Kulics & Kijlstra, 1987
Intestinal iron absorption	Cox et al., 1979; Davidson & Lönnerdal, 1988
Enterocyte proliferation and gut maturation	Nichols et al., 1990
Up-regulation of thymocyte maturation	Zimecki et al., 1991
Up-regulation of monocyte cytotoxicity	McCormick et al., 1991
Regulation of antibody production	Zimecki et al., 1991
Regulation of cytokine production	Crouch et al., 1992
Down-regulation of tumor necrosis factor	Machnicki et al., 1993
Prevention of hydroxy radical-mediated tissue injury	Britigan et al., 1994

IX. MULTIFUNCTIONAL ACTIVITIES *IN VIVO*

LF plays a critical regulatory role in various physiological pathways (*TABLE 6*). Though iron-binding is considered an important molecular property of LF, a number of cellular functions are independent of this metal binding property. Specific and non-specific interactions of LF with cells, co-existence with a variety of bio-molecules at different milieu, molecular heterogeneity and structural flexibility confers a spectrum of multifunctionality to the LF molecule *in vivo*.

A. Pathogen control *in vivo*

1. Bacteria. Zagulski et al. (1989) reported a protective effect of bLF when administered intravenously to mice 24 h before a challenge with a lethal dose of *E. coli*. About 70% of mice pretreated with bLF survived challenge. The survival rates in control mice treated with *E. coli* alone and pretreated with bovine serum albumin were 4 and 8%, respectively. Similar protection was also observed with hLF administration. Sufficient amounts of ferric ions were given to mice, in single and multiple doses, for full serum TF saturation 30 min before or after *E. coli* administration. The multiple dose of ferric ions did not change considerably the survival rate of mice pretreated with bLF. In contrast, a single dose of ferric ions gradually decreased the survival rate of the mice after the first week of experiment. From day-14 this decrease was statistically significant in all groups of mice treated with a single dose of ferric ions when compared with mice pretreated only with bLF, and the difference ranged from 25 to 35% on day-30.

Denisova et al. (1996) reported a synergistic effect of bLF, lactoperoxidase, and lactoglobulin obtained from cow milk, as well as hLF obtained from human milk, in protection against mice infected intranasally with *Shigella sonnei*. The combined administration of these preparations in doses, each one having no protective action, contributed to the elimination of the bacteria from the lungs and prevented the death of the animals. This phenomenon was found to be nonspecific and was suggested for use in the development of preparations for passive immunization against a wide spectrum of microorganisms.

Oral administration of bLF with milk in mice could suppress the proliferation of *Clostridium ramosum* C1 *in vivo* and decreased the numbers of *C. ramosum* and other bacteria in the feces (Teraguchi et al., 1995a). This bacteriostatic effect of bLF was dependent on the concentration of bLF, the duration of feeding, and the administered dose of *C. ramosum* C1. Compared with bovine serum albumin, ovalbumin, bovine whey protein isolate, or bovine casein, only bLF showed this specific activity. A similar effect of bLF was observed after oral inoculation with *C. ramosum* JCM 1298, *C. paraputrificum* VPI 6372, or *C. perfringens* ATCC 13124. A hydrolysate prepared by digestion of bLF with porcine pepsin showed the same inhibitory effect on proliferation of *C. ramosum in vivo* as occurred with undigested bLF. These results indicate that ingested bLF could exert a bacteriostatic effect against clostridia in the gut even after enzymatic digestion in gut.

Furthermore, supplementation of the milk diet with bLF or a pepsin-generated hydrolysate of bLF also resulted in significant suppression of bacterial translocation from the intestines to the mesenteric lymph nodes, and the bacteria involved were mainly members of the family *Enterobacteriaceae* (Teraguchi et al., 1995b). This ability of LF to inhibit bacterial translocation may be due to its suppression of bacterial overgrowth in the guts of milk-fed mice.

Wada et al (1999) suggested that bLF exerts an *in vivo* inhibitory effect on colonizing *Helicobacter pylori* by detaching the bacterium from the gastric epithelium and by exerting a direct anti-bacterial effect. Germfree BALB/c mice were orally inoculated with *H. pylori* to induce infection. Three weeks after infection the mice were given bLF orally once daily for 2 or 4 weeks and were then killed to examine the bacterial number in the stomach and the serum antibody titer to *H. pylori*. To count the number of epithelium-bound *H. pylori*, the resected stomach was agitated in phosphate-buffered saline to remove non-bound *H. pylori* before bacterial enumeration. The administration of 1% bLF for 3 to 4 weeks decreased the number of *H. pylori* in the stomach to one-tenth and also exerted a significant inhibitory effect on the attachment of *H. pylori* to the stomach. Furthermore, the serum antibody titer to *H. pylori*, decreased to an undetectable level.

2. Viruses. Lu et al. (1987) examined the *in vivo* effects of holo-hLF on the survival rates of mice and titers of spleen focus-forming viruses (SFFV) in mice inoculated with the polycythemia-inducing strain of the Friend virus complex (FVC-P). LF prolonged the survival rates and decreased the titers of SFFV in mice infected with FVC-P. Titers of SFFV, assayed 14 days after administration of FVC-P, were measured by the spleen focus-forming unit assay in secondary mouse recipients. Decreases in titers of SFFV were apparent when LF was given *in vivo* as a single bolus dose of 200 μg within 2 h of the FVC administration, or as a total dosage of 200 μg given on days 1, 2, 4, 7, 9, and 11 after FVC-P, and to a lesser degree when LF was given as a total dosage of 200 μg on days 3, 4, 7, 9, and 11 after FVC-P. No decreases in titers of SFFV were detected when LF was given up to 3 days before or more than 3 days after FVC-P. LF did not appear to be directly inactivating the viruses as it did not inactivate the SFFV or the Friend murine leukemia helper virus *in vitro*. The results suggested that the protective effect of LF *in vivo* was probably due to an action on cells responding to the FVC or to an action on cells which influence the cells responding to the FVC or which influence the virus.

LF could inhibit bacterial growth in the murine conjunctival sac (Fujihara & Hayashi, 1995). The antiviral activity of LF against herpes simplex virus type-1 (HSV-1) *in vitro* (infectivity of Vero cell monolayers) and inhibit infection *in vivo* in the mouse cornea was also reported. LF effectively prevented HSV-1 plaque formation.

Administration of topical 1% LF prior to the virus inoculation suppressed infection on ocular tissue; however, it did not inhibit propagation of the virus.

The administration of bLF (1 mg/g body weight) before the murine cytomegalovirus (MCMV) infection completely protected the BALB/c mice from death due to the infection (Shimizu et al., 1996). The LF-treated mice showed a significant increase in the NK cell activity but not of the cytolytic T lymphocytes that recognize an MCMV-derived peptide. Moreover, the elimination of the NK cell activity by an injection with anti-asialo GM1 antibody abrogated such augmented resistance, thus supporting the hypothesis that the LF-mediated antiviral effect *in vivo* is performed through the augmentation of NK cell activity. No such LF-mediated antiviral effect *in vivo* with the increased NK cell activity was found in athymic nude mice, whereas it was restored completely by the transfer of splenic T cells from LF-treated donors. These findings suggest that T lymphocytes induce both the augmentation of NK cell activity and the resultant antiviral effect in the LF-treated hosts.

Effects of oral administration of bLF on intractable stomatitis in feline immunodeficiency virus (FIV)-positive and FIV-negative cats, and phagocytosis of neutrophils in healthy and ill cats was tested (Sato et al., 1996). Bovine LF (40 mg/kg ot body weight) was applied topically to the oral mucosa of cats with intractable stomatitis daily for 14 days and improvement of clinical signs of disease (pain-related response, salivation, appetite, and oral inflammation) was evaluated. Assay of neutrophil phagocytosis was examined before and 2 weeks after starting LF treatment, using nonopsonized hydrophilic polymer particles. Bovine LF could improve intractable stomatitis and concurrently enhanced the host defense system. Topical application of bLF to oral mucus membrane was suggested useful as a treatment for intractable stomatitis also for FIV-positive cats.

3. Mycoplasma. The protective effect of LFcin against *Toxoplasma gondii* infection was examined in experimental murine toxoplasmosis (Isamida et al., 1998). All the mice that were dosed orally with LFcin (5.0 mg), and challenged with cysts of *T. gondii* at a dose of LD_{90} survived until the end of the experiment (35 days post challenge). Intraperitoneal administration of LFcin (0.1 mg) also prevented death in 100% of treated mice challenged with *T. gondii* cysts. In contrast, 80% of untreated mice died of acute toxoplasmosis within 14 days post challenge. In the mice treated per-orally with LFcin, the number of cysts in the brain was significantly lower than that in untreated mice. Levels of IFN-g in the serum of infected mice treated per-orally with LFcin showed a tendency to lower than those in the infected mice without treatment.

B. Immuno-modulatory functions

Natural killer (NK) and lymphokine-activated killer (LAK) cell cytotoxic functions are strongly augmented by LF (Shau et al., 1992). LF significantly enhances NK and LAK activities when added at the beginning of NK or LAK cytotoxicity assays. LF is effective in augmenting cytotoxic activities at concentrations as low as 0.75 µg/ml, and higher concentrations of LF induce greater augmentation of NK and LAK. Iron does not appear to be essential for LF to increase NK and LAK, as depleting iron from LF with the chelator deferoxamine does not affect the capacity of LF to increase cytotoxicity. LF is known to have RNase enzymatic activity and LF enhancement of NK and LAK can be blocked by RNA. However, LFs from two different sources with over 100-fold difference in RNase activity are equally effective in enhancing NK and LAK. Furthermore, purified non-LF RNase does not modulate NK or LAK activity and DNA is as effective as RNA in blocking LF augmentation of NK or LAK cytotoxicity. Therefore, the RNase activity is unlikely to be responsible for LF enhancement of the cytotoxicity. Newborn infants are known to have low NK activity and NK and LAK cells have been implicated in host defense against microbial infections. Thus, maternal milk- derived LF may have a role in boosting antimicrobial immunity in the early stages of life. In adults, LF released from neutrophils may enhance NK and LAK functions in the inflammatory process induced by microbial infections.

Debbabi et al. (1998) investigated the nature of immune responses induced by repeated oral administration of bLF to mice (1998). Mice were fed daily for 4 weeks with two doses of protein antigen: a low (0.05 mg/g body weight per d) or high (1 mg/g body weight per d) dose of LF, or water as a control. A fourth group was immunized intramuscularly with 0.01 mg LF in complete Freund's adjuvant. Anti-LF IgA and IgG were detected in the intestinal fluid and serum of mice given LF. Total immunoglobulins were higher in the intestinal fluid in LF groups than in the control group. No difference could be detected in the serum. IgA and IgG secretion was enhanced in Peyer's patches and spleen from LF-fed mice, in comparison with controls. [3H]thymidine uptake into Peyer's patch and spleen cells from both control and LF-fed mice was enhanced by 75 μg of LF/ml *in vitro*, but LF groups had a greater proliferation rate than the control group. These findings suggested that LF could act as an immunostimulating factor on the mucosal immune system and that activation of the mucosal immune system is dependent on the ability of LF to bind to the intestinal mucosa.

In vivo administration of bLF seems to inhibit delayed type hypersensitivity (DTH) reactions in mice (Zimecki & Machnicki, 1994). Administration of bLF at 48 or 24 h before eliciting the DTH reaction was not effective; however, bLF suppressed the reaction when given at the peak of the inflammatory process. The effects of bLF were strongest when the protein was injected intravenously. Intraperitoneal or intramuscular administrations of bLF were less inhibitory. In addition, bLF diminished, although to a much lesser degree, the inflammatory reactions induced by BCG antigen. The inhibitory action of bLF does not involve liver since treatment of mice with galactosamine does not reverse the inhibition. Studies on cytokine production revealed that peritoneal macrophages, derived from mice pretreated with LF, have an increased ability to produce IL-6 *in vitro* after induction with LPS. In addition, the inhibition of macrophage migration, mediated by migration inhibition factor, was abolished by bLF. However, the inhibitory effect of bLF could not be transferred with serum from donors treated with bLF.

Miyauchi et al. (1997) evaluated the immunomodulatory effects of bLF hydrolysate on the proliferation of murine splenocytes. The hydrolysate enhanced [3H]thymidine uptake by splenocytes, but undigested bLF exerted an inhibitory effect. The hydrolysate had the ability to inhibit the blastogenesis induced by mitogens such as concanavalin A, phytohemagglutinin, and LPS; inhibition was similar to that with undigested bLF. These results suggested that the hydrolysate contained both immunostimulatory and immunoinhibitory peptides. The stimulatory effect of the hydrolysate in the absence of mitogens was then explored in more detail using nonadherent splenocytes. The proliferative response of splenocytes to the hydrolysate was much greater in the fraction that was enriched with B cells than in the fraction that was enriched with T cells. The hydrolysate did not affect thymocyte proliferation. These data indicated that the adherent cells resembling macrophages found among the splenocytes were not the target cells of the hydrolysate. The stimulatory effect of the hydrolysate was due to the activation of B cells by the hydrolysate and enhanced immunoglobulin production by splenocytes. Because the hydrolysate also enhanced the proliferation and immunoglobulin A production of Peyer's Patch cells, the immunostimulatory effect of the hydrolysate *in vivo* was examined using mice that had been orally immunized with cholera toxin. The concentrations of immunoglobulin A conjugated against cholera toxin in bile and in the intestinal contents of mice fed liquid diets containing 1% (wt/vol) LF hydrolysate were greater than those of mice fed control diets. These results suggest a beneficial effect of LF on the mucosal immunity.

Zimecki et al. (1998a) also investigated the effect of bLF oral treatment on carrageenan-induced inflammation in rats. Animals were given 5 oral doses of bLF (10 mg each) on alternate days and 24 h after the last dose a carrageenan inflammation was induced in the hind foot. Control rats were given 0.9% saline or BSA. The magnitude of the reaction was measured after 2 h (optimal response) and expressed as an increase of the foot pad thickness in millimeters. The evaluation of bLF effects on carrageenan reaction was supplemented by determination of the ability of spleen cell cultures to produce IL-6 and TNF-α upon LPS induction. Bovine LF inhibited the carrageenan-induced inflammation in by 50 and 40% as compared to saline and BSA control groups, respectively. The inhibition was also associated with a substantial decrease in the ability of splenocytes to produce IL-6 in bLF-treated rats (94 and 83% as compared to saline- and BSA-treated groups). The LPS-induced TNF-α production was also decreased, although to a lesser degree (48 and 35%, respectively). The decreased ability of spleen cells to produce inflammatory cytokines in bLF-treated rats indicates that hypo-reactivity of the immune system cells may be the basis for the inhibition of carrageenan-induced inflammation.

Intravenous (i.v.) administration of bLF (10 mg/mouse), 24 h before thymectomy, reduced, on average, the level of serum IL-6 by 70% as measured 4 h after surgical procedure (Zimecki et al., 1998b). The inhibiting effect of bLF on TNF-α production was low with a mean 30% reduction. The effects of bLF (i.v.) administration on the cytokine levels following splenectomy were less inhibitory. Bovine LF caused an approximate 35% fall in IL-6 levels and even weaker effects (20% inhibition) on TNF-α release. Application of much lower (1-0.2 mg) doses of bLF was even more effective in lowering IL-6 levels after thymectomy (up to 90%) after 5 bLF doses, and by 55% of TNF-α. The authors suggest that LF may find therapeutic application for diminishing manifestations of shock caused by clinical insults.

Zimecki et al. (1998c) conducted a clinical trial in Poland to monitor several immune parameters in 17 healthy volunteers taking commercially available capsules containing bLF (40 mg/ daily) orally for 10 days. Leukocyte number and content of main blood cell types, spontaneous and phytohemagglutinin A (PHA)-induced proliferation of lymphocytes, plasma levels of IL-6 and TNF-α, as well as spontaneous and LPS-induced production of these cytokines in peripheral blood cell cultures, were evaluated. All measurements were performed before, one day and 14 days following cessation of bLF treatment. A transient drop was observed in the percentage of neutrophils accompanied by an opposite phenomenon with regard to lymphocyte levels. More profound changes were registered in the percentage of other cell types, such as a 100% increase in the level of immature cell forms. At the same time the percentages of eosinophils and monocytes markedly declined. All these changes were, however, more individual and regulatory, the direction of these changes depended on initial profile of blood cells. Although the proliferative response of lymphocytes showed, on average, a transient decrease, differentiated effects of bLF treatment were observed depending on initial ability of lymphocytes to proliferate. TNF-α serum levels showed a tendency to decrease during the monitoring time, the changes of IL-6 levels were, however, not significant. As in the case of the proliferative response, the treatment with bLF was regulatory with respect to serum TNF-α levels. The influence of other ingredients such as selenium or vitamins, contained in the capsules, cannot be excluded, although the data indicated that orally taken bLF alone can induce identical changes as the capsules containing bLF. The study suggests that oral administration of bLF-containing capsules may regulate certain immune responses in healthy individuals.

In a clinical trial conducted in Japan, Yamauchi et al. (1998) examined the effects of orally administered bLF on the immune system of healthy volunteers. Ten healthy male volunteers (age range: 31-55) were given bLF (2 g/body/day) for 4 weeks. Blood samples were drawn before, during and after LF administration. Phagocytic activity and superoxide production activity of PMN were evaluated from the number of polymer particles phagocytosed by PMN and by the dichlorofluorescein (DCFH) oxidation assay, respectively. The expression levels of CD11b, CD16 and CD56 molecules on leukocytes were quantified using flow cytometry. The phagocytic activity of PMN increased during the period of LF administration in 3 of the 10 volunteers. In 2 of the 3 volunteers in which the phagocytic activity increased, PMN expressed CD16 at higher levels corresponding to the increase in 3 of the 10 volunteers, whereas the CD11b+ lymphocytes and CD56+ lymphocytes increased in 4 volunteers including the same 3 volunteers who showed an increase in CD16+. These results suggest that the proportion of NK cells among the lymphocytes might have increased in these subjects. It was demonstrated that the phagocytic activity or superoxide production activity of PMN or the proportions of CD11b+, CD16+ and CD56+ in lymphocytes was influenced by LF administration in 7 of the 10 volunteers, while the effects of LF on the immune system differed in individual cases. These results suggest that LF administration may influence primary activation of the host defense system.

C. Nutritional functions

LF has been suggested to play an important role in the intestinal absorption of iron (de Vet & van Vugt, 1971; de Vet & van Gool, 1974). During early life, infants usually consume a diet that is heavily dominated by milk. The bio-availability of the trace elements iron, zinc, copper and manganese from human milk is high compared to cow's milk and infant formulas. This high bio-availability may be explained by the presence of LF in human milk, which may facilitate iron and manganese uptake via an intestinal receptor for this protein (Lönnerdal, 1989).

1. Specific LF-binding receptors on intestinal brush border. Several lines of evidence have been suggested the occurrence of a specific LF receptor in the small intestinal brush-border membrane in several animal species, which is thought to be involved in LF-mediated intestinal iron absorption. Hu et al. (1990) isolated and partially characterized an LF receptor from mouse intestinal brush border. The receptor has been purified to homogeneity by affinity chromatography on an immobilized hLF column. The purified receptor was found to be active in that it binds apo- and holo-LF with a K_a of 0.1 μM. Anti-receptor antibodies were prepared, and the receptor was further isolated by immunoaffinity chromatography in higher yield but in a denatured form. The purified receptor has a molecular mass of about 130-kDa, and consisted of a single polypeptide chain with an isoelectric point at 5.8. The receptor was also shown to bear concanavalin A and phytohemagglutinin binding glycans.

Kawakami and Lönnerdal (1991) have isolated a hLF receptor from solubilized human fetal intestinal brush-border membranes. The molecular weight of the receptor was 110-kDa by SDS-PAGE under non-reducing conditions and 37-kDa under reducing conditions. Competitive binding studies demonstrated specific binding of hLF. The binding was pH dependent, with an optimum between pH 6.5 and 7.5. Scatchard plot analysis indicated 4.3×10^{14} binding sites/mg membrane protein with an affinity constant of 0.3 μM for hLF. Both half-LF and deglycosylated LF bound to the receptor with an affinity similar to intact LF. In contrast, little binding of bLF or human TF to human brush border membrane vesicles occurred.

Rosa and Trugo (1994) investigated iron uptake from hLF by brush-border membrane vesicles (BBMV) obtained from the small intestine of human neonates. Uptake experiments were performed by incubation of ^{55}Fe-citrate or ^{55}Fe-LF with BBMV, followed by rapid filtration through microporous filters. ^{55}Fe uptake from LF by BBMV was dependent on pH, with a maximum at 7.5, and increased with incubation time, reaching a maximum at 1 min. When ^{55}Fe was bound to citrate, uptake was slower (maximum at 5 min) and not dependent on pH. In both experiments, the maximum uptake of iron bound to LF was about twice that of iron bound to citrate (230 pmol and 105 pmol/mg microvillus protein, respectively). Partial degradation of LF in two fragments resulted in the loss of its capacity to increase iron uptake by BBMV. The study suggested that LF could increase iron absorption during the neonatal period, contributing to the high bio-availability of this mineral in human milk.

Differentiated Caco-2 colon carcinoma cell monolayers grown in bicameral chambers have been used as an *in vitro* model to study the effect of different carrier molecules on mucosal iron transport (Sanchez et al., 1996). Transfer of iron across the monolayers in the apical-to-basolateral direction was greater from holo-LF than from iron citrate, while very little transport occurred from holo-TF. However, a greater proportion of iron was retained by the cells when Fe-citrate was the donor. Caco-2 cells expressed TF receptors (1.3×10^5/cell; K_a:20 nM), but binding of LF, though substantial in quantity, had an affinity too low to measure. When monolayers were incubated with ^{125}I-labelled LF or TF some ^{125}I-activity was transported, but almost all was TCA-soluble, suggesting that degradation products rather than intact protein were being transported. Addition of 10 μM S-nitroso-N-acetyl-D,L-penicillamine, which produces nitric oxide (NO) in solution, caused a significant increase in iron transport from ferric citrate, but not from Fe-LF or Fe-TF. It was concluded that in this *in vitro* system LF but not TF enhances mucosal iron transport, and that NO may play a regulatory role in iron absorption.

2. LF-mediated intestinal iron uptake. Various clinical trials and *in vivo* animal experiments on the role of LF in intestinal iron absorption are inconclusive. The efficacy of supplementing iron bound to LF to iron-deficient and iron-sufficient young mice was evaluated in comparison with supplementation of iron as iron chloride (Fransson et al., 1983). Mice fed a non-supplemented milk diet (approximately 1 mg Fe/L) for 4 weeks had a microcytic, hypochromic anemia and low tissue iron concentrations. Iron supplementation of the diet with LF-iron, or iron chloride at a level of 5 mg Fe/L prevented the anemia and resulted in tissue iron levels similar to levels found for mice fed a stock commercial diet. There was no significant difference in any of the parameters analyzed between the groups of mice receiving the two iron supplements following a diet deficient in iron. Apo-LF when supplemented to the diet had no negative effect on the iron status of the mice. These results suggested that LF could be a useful vehicle for supplementation of iron.

Davidson et al. (1990) examined the iron retention from human milk, milk-based infant formula (IF) with and without supplemental ferrous sulfate, and IF supplemented with either hLF or bLF in infant rhesus monkeys. There was no significant difference in iron retention from the experimental diets: human milk, IF, IF+Fe, IF+hLF, or IF+bLF. The authors concluded that the infant monkeys absorb and retain iron similarly from human milk and infant formula. Supplementation of infant formula with hLF or bLF resulted in similar iron retention to that of ferrous sulfate-supplemented infant formula.

Iron balance studies were performed in 16 term infants from week-3 through week-17 of life (Schulz-Lell et al., 1991). The balance studies were performed at home and comprised five periods with an interval of 3 to 4 weeks, each consisting of three 24-h collections of milk and stool samples. Seven infants were fed an adapted infant formula supplemented with bLF (1 mg/ml) and nine received the same formula without bLF. The LF supplemented group received 169 µg iron/kg b.w/day and retained 63 µg/kg b.w./day. The mean iron intake of infants fed with the adapted formula without supplementation of bLF was 118 µg/kg b.w./day. The retention of iron was 43 µg/kg b.w./day. Mean percentage retention of iron in the supplemented group was 36%, in the non-supplemented group 28%.

In a study by Chierici et al. (1992) infant formulae supplemented with various amounts of bLF were given to two groups of infants. These infants were compared with infants receiving unsupplemented formula and breast-fed infants. The effects of these diets on levels of hemoglobin, hematocrit, serum iron, ferritin and zinc were examined for a study period of 150 days. At birth, concentrations of iron, hemoglobin, hematocrit and zinc were comparable in all four feeding groups. The serum zinc level was not altered by LF supplementation. Ferritin levels of breast-fed infants were significantly higher than in non-supplemented formula-fed infants at day 30 and day 90. This difference was seen only at day 30, when comparing breast-fed infants to LF-supplemented formula-fed infants. Comparing the infants receiving formulae, the formula supplemented with the higher amount of bLF induced significantly higher serum ferritin levels compared to the unsupplemented formula at day 90 and day 150. These observations favor the hypothesis that LF is possibly involved in iron absorption.

The following studies, however, do not support the efficacy of LF in facilitating iron uptake in the gastrointestinal tract. McMillan and co-workers (1977) compared iron availability from human milk with that from other formulas supplemented with LF. The study concluded that manipulation of the protein, fat, lactose, calcium, phosphorus, or LF content of proprietary milk did not reproduce the iron absorption demonstrated with human milk. Studies by Fairweather-Tait et al. (1987) with ^{59}Fe-labels suggested that bLF has no effect on iron absorption in rats. The results were compared with those obtained from a group of infants fed a similar level of iron as ferric chloride, labeled with ^{59}Fe, together with 30 mg ascorbic acid. There was a very wide variation in percent iron retention amongst the infants but no overall difference between the LF and ferric chloride groups.

The potential effect of LF on iron absorption was investigated by measuring iron absorption in infants fed breast milk (with its native content of LF) and the same milk from which LF had been removed (> 97%) by treatment with heparin-Sepharose. Eight breast-fed infants (2-10 mo; mean age 5 mo) were fed 700 to 1000 g of each milk in a randomized, cross-over design with each child acting as his/her own control. The milk was labeled with 8.6 μM (0.5 mg) of ^{58}Fe and iron absorption was measured by quantifying the incorporation of the isotope into red blood cells 14 d after intake using thermal ionization mass spectrometry. Fractional iron absorption was significantly lower from breast milk than from LF-free breast milk. The geometric mean (range) was 11.8% (3.4-37.4%) for breast milk and 19.8% (8.4-72.8%) for LF-free breast milk. These results do not support a direct role for LF in the enhancement of iron absorption from human milk at this age. In addition, iron absorption (11.8%) from human milk fed over several feeds was lower than that previously reported for single feed studies.

D. Physiological functions

The germfree, colostrum-deprived, immunologically 'virgin' piglet model was used to evaluate the ability of bLF to protect against lethal shock induced by intravenously administered endotoxin (Lee et al., 1998). Piglets were fed bLF or BSA prior to challenge with intravenous *E. coli* LPS, and temperature, clinical symptoms, and mortality were observed for 48 h following LPS administration. Prefeeding with LF resulted in a significant decrease in piglet mortality compared to feeding with BSA (16.7 versus 73.7% mortality). Protection against the LPS challenge by LF was also correlated with both resistance to induction of hypothermia by endotoxin and an overall increase in wellness, as quantified by a toxicity score developed for these studies. *In vitro* studies using a flow cytometric assay system demonstrated that LPS binding to porcine monocytes was inhibited by LF in a dose-dependent fashion, suggesting that the mechanism of LF action *in vivo* may be inhibition of LPS binding to monocytes/ macrophages and, in turn, prevention of induction of monocyte/macrophage-derived inflammatory-toxic cytokines.

Antibody to an estrogen inducible mouse uterine protein (Teng et a., 1986) has been used to isolate cDNA to the messenger RNA. Analysis of the deduced primary structure and additional biochemical characterization indicated that the protein is LF. An increase in the level of LF mRNA of at least 300-fold can be induced in the mouse uterus by estrogen (Pentecost & Teng, 1987).

Neutrophils can inactivate LPS and block its ability to prime fresh neutrophils for enhanced fMLP-triggered release of superoxide. Wang et al. (1995) showed that the inactivation of LPS by neutrophils was primarily due to LF. A time course for inactivating LPS showed that neutrophils (5 million/ml) took 30 min to inactivate 10 ng/ml LPS. Mononuclear cells could not inactivate LPS under the same conditions. Experiments with radioactive LPS showed that inactivated LPS remained in the medium and was not taken up or destroyed by the neutrophils during inactivation. Inactivated LPS still gelled Limulus lysate and primed monocytes. Cell-free medium from neutrophil suspensions also inactivated LPS. A single LPS-inactivating factor was purified from medium by heparin-agarose chromatography. SDS-PAGE showed a single band at 80-kDa, which was identified as LF by immunoblotting. Anti-LF immunoglobulin G removed the LPS-inactivating activity from purified LF and cell-free medium. Purified neutrophil LF required 30 min to inactivate LPS, indicating inherently slow binding of LF to LPS.

Burrin et al. (1996) examined the anabolic effect of orally administered bLF on visceral organ growth and protein synthesis in newborn pigs. We studied a total of 18 unsuckled newborn pigs. Pigs were randomly assigned to one of three dietary treatment groups: bottle-fed (10 ml/h) formula, formula containing physiologic levels (1 mg/ml) of added bLF, or colostrum. After 24 h of feeding, the visceral organ protein synthesis *in vivo* was measured using a flooding dose of [^{3}H]phenylalanine. The visceral organ protein and DNA mass, as well as intestinal hydrolase activities and villus morphology was also measured. Hepatic protein synthesis in pigs fed either formula containing bLF or colostrum was similar and in both groups was significantly higher than in pigs fed formula. Splenic protein synthesis was not significantly different in pigs fed either formula or formula containing bLF, but was significantly higher in colostrum-fed animals. There were no significant differences in small intestinal growth, protein synthesis, or hydrolase activities between newborn pigs fed formula, formula containing bLF, or colostrum. These results indicate that feeding formula containing physiologic concentrations of added bLF could increase hepatic protein synthesis in newborn pigs, suggesting that colostrum-borne LF serves an anabolic function in neonates.

Antitumor activity. Yoo et al. (1997) studied the effect of bLF and bLFcin on inhibition of metastasis in murine tumor cells, B16-BL6 melanoma and L5178Y-ML25 lymphoma cells, using experimental and spontaneous metastasis models in syngeneic mice. Subcutaneous (s.c.) administration of apo-bLF (1 mg/mouse) and bLFcin (0.5 mg/mouse) 1 day after tumor inoculation significantly inhibited liver and lung metastasis of L5178Y-ML25 cells. However, apo-hLF and holo-bLF at the dose of 1 mg/mouse failed to inhibit tumor metastasis of L5178Y-ML25 cells. Similarly, the s.c. administration of apo-bLF as well as bLFcin, but not apo-hLF and holo-bLF, 1 day after tumor inoculation resulted in significant inhibition of lung metastasis of B16-BL6 cells in an experimental metastasis model. Furthermore, *in vivo* analysis for tumor-induced angiogenesis, both apo-bLF and bLFcin inhibited the number of tumor-induced blood vessels and suppressed tumor growth on day 8 after tumor inoculation. However, in a long-term analysis of tumor growth for up to 21 days after tumor inoculation, single administration of apo-bLF significantly suppressed the growth of B16-BL6 cells throughout the examination period, whereas bLFcin showed inhibitory activity only during the early period (8 days). In spontaneous metastasis of B16-BL6 melanoma cells, multiple administration of both apo-bLF and bLFcin into tumor-bearing mice significantly inhibited lung metastasis produced by B16-BL6 cells, though only apo-bLF exhibited an inhibitory effect on tumor growth at the time of primary tumor amputation (day-21) after tumor inoculation. These data suggest that apo-bLF and bLFcin inhibit tumor metastasis through different mechanisms, and that the inhibitory activity of bLF on tumor metastasis may be related to iron-saturation.

The influence of bLF on colon carcinogenesis was investigated in male F344 rats treated with azoxymethane (Sekine et al., 1997a). After three weekly injections of azoxymethane, the animals received 2 or 0.2% bLF for 36 weeks. No effects indicative of toxicity were noted, but significant reduction in both the incidence and number of adenocarcinoma of the large intestine was observed with both doses. Thus, the incidences of adenocarcinoma in the groups receiving 2% and 0.2% bLF were 15% and 25%, respectively, in contrast to the 57.5% control value. These results suggest a possible application for bLF in the chemoprevention of colon cancer. Sekine et al. (1997b) also reported that the administration of bLF (2%) and *Bifidobacterium longum* (3%) significantly decreased the numbers of aberrant crypt foci. Most importantly large size foci composed of four or more crypts were always significantly decreased by 2% bLF. Studies on the natural killer activity of spleen cells demonstrated enhancement by bLF and *B. longum* in line with the levels of influence on foci induction, indicating a possible role for elevated immune cytotoxicity in the observed inhibition.

Iigo et al. (1999) reported that oral administration of bLF and the bLF hydrolysate also demonstrated significant inhibition of lung metastatic colony formation from s.c. implanted tumors without appreciable effects on tumor growth. bLFcin displayed a tendency for inhibition of lung metastasis. On the other hand, bLF did not exert marked anti-metastatic activity in athymic nude mice bearing Co 26Lu, though bLF had a tendency to inhibit the lung metastatic colony formation associated with anti-asialoGM1 antibody treatment. AsialoGM1+ and CD8+ cells in white blood cells were increased after treatment with bLF. *In vitro*, the viability of Co26Lu-F55 cells was markedly decreased when co-cultured with white blood cells from mice administrated bLF, but recovered on treatment with anti-asialoGM1 antibody or anti-CD8 mAb and complement. The results suggest bLF and related compounds might find application as tools in the control of metastasis and that asialoGM1+ and CD8+ cells in the blood are important for their inhibitory effects.

X. APPLICATIONS

The emerging knowledge about diseases and the role of natural foods in lowering the risk of such disease processes, and research efforts to identify and develop nutritionally important supplemental foods have been steadily increasing. Milk contains the widest range of biologically active ingredients. LF is one such bioactive protein present in milk (milk-derived whey or derivatives). It is a complex molecule with a number of potential functional properties (Naidu & Bidlack, 1998).

A. Potential additive for food safety
Consumption of cow's milk has been an integral part of human civilization since antiquity. LF is now recognized as a significant constituent of milk, responsible for nutraceutical benefits and innate protection against intestinal illnesses. A vast amount of literature has been published on the prophylactic, therapeutic, and regulatory role of LF in various physiological functions during the past four decades. The following points validates milk LF as a potent natural antimicrobial for food safety.

Ultimately, an antimicrobial must be non-toxic to test animals and humans. It is also important that the antimicrobial be metabolized and excreted by the body. *In vivo* studies have demonstrated that the liver plays a central role in the elimination of LF from plasma. When injected by intravenous route, LF is taken up rapidly by the receptors in the liver reticuloendothelial system. The integrity of the protein moiety of LF is required for its effective uptake by the liver. Histological and cytochemical investigations indicate that sinusoidal and Kupfer cells are responsible for hepatic uptake of LF (Courtoy et al., 1984). Studies with iodinated LF identified parenchymal cells as the prime source of LF catabolizing activity (Ziere et al., 1992). The liver degrades LF with a half-life of 2.7 h (Regoeczi et al., 1985). LF seems to remain associated with the plasmalemma for prolonged periods of time (Ziere et al., 1992) which could be interpreted as a sign of slow internalization. However, LF is readily detected in both endosomes and lysosomes as early as 15 min after administration (Regoeczi et al., 1985). LF appears in bile 35 min after intravenous injection (Regoeczi et al., 1994).

Water solubility of an antimicrobial agent is probably the most important physical property. Hydrophilic properties are necessary to assure that the antimicrobial agent is soluble in aqueous phase where microbial growth occurs. LF is water-soluble, yet dissolves in a wide range of inorganic as well as organic solvents.

Thermal behavior can also directly influence the activity of an antimicrobial agent, especially with its carry-through properties. If the food is heated during processing, a highly volatile agent such as a phenolic compound can be vaporized and lost. Paulsson et al. (1993) examined the heat-induced enthalpy changes in apo- and iron-saturated bLFs by differential scanning calorimetry. Two thermal transitions with varying enthalpies were observed, depending on the iron-binding status of the protein. Iron-saturated LF was more resistant to heat-induced changes than was the apo-LF. Native LF had two transitional peaks and pasteurization affected only the low temperature transition. Iron-saturated LF revealed a single transitional peak that was resistant to denaturation by pasteurization temperatures. Native LFs, both unheated and pasteurized, showed similar antibacterial properties suggesting temperature-resistance of the molecule.

The ability of an antimicrobial compound to ionize can alter its biological activity depending on the pH of the food system in which it is used. LF exerts antimicrobial activity at a wide pH range in different ionic milieus. Natural compounds, occurring as the major food components, can also influence activity of an antimicrobial depending on the properties of the antimicrobial and those of the naturally occurring compound. The cationic nature enables LF to interact and complex with other proteins such as casein, α-lactalbumin (Hekman, 1971; Lampreave et al., 1990), lysozyme (Perraudin et al., 1974), and immunoglobulin-A (Watanabe et al., 1984). Various regulatory functions were attributed to such LF complexes. Interactions with acidic polysaccharides (Mann et al., 1994) which are ubiquitous components of cell membranes. The arginine-rich N-terminus of the LF molecule (Mann et al., 1994) allows LF to bind with varying degrees of affinity to almost any cell, as well as to low-density lipoproteins.

Knowledge of the mode of action of an antimicrobial and of the ability of the organisms to overcome this mode of action can be helpful in determining the efficiency and usefulness of an antimicrobial. The protective role of LF in host defense against various microbial illnesses has been established (Naidu & Arnold, 1997). This chapter has identified several of the molecular mechanisms related to LF antimicrobial activity.

Valid assay methods for antimicrobial agents are essential so that the levels can be easily determined. Various laboratories have developed rapid, sensitive, enzyme-linked immunosorbant assays (ELISA) for quantitation of LF in different biological systems. Shinmoto et al. (1997) reported a highly sensitive competitive ELISA for bLF. Culliere et al. (1997) recently developed a microparticle-enhanced nephlometric immunoassay for LF quantitation. This assay is sensitive (detection limit in reaction mixture, 0.2 mg/L) and can be performed in diluted milk (1/300 in reaction mixtures), excluding any interference or sample pretreatment. It allows the quantification of LF on a large range of concentrations (0.675-21.6 g/L) with accuracy (linear-recovery in dilution-overloading assay) and precision (3-6% variation). A rapid immunoluminometric assay and a non-competitive avidin-biotin immunoassay were also developed for LF quantitation. The LactoCard, a new commercial assay, allows rapid determination of LF concentrations in 10-15 min.

TABLE 7. In vivo effects of LF - human clinical trials and consumption studies in different experimental animal models.

Test parameter	Model	Test design (Treated/Total)	LF	Dosage	Carrier	Route	Reference
Effect on intestinal iron absorption	Mouse	Suckling mice	bLF	4 g/L or 50 µM for 4 wks	Milk	Oral	Fransson et al., 1983
Effect on tissue mineral content	Mouse	Suckling mice	bLF	4 g/L or 50 µM for 4 wks	Milk	Oral	Keen et al., 1984
Prophylaxis during neutropenia	Human	Cancer patients (5/14)	hLF	800 mg/day for 10 d	Capsule	Oral	Trumpler et al., 1989
Effect on fecal microflora	Human	Newborn term infants (38/69)	bLF	2.8 g/L for 14 d	Formula	Oral	Balmer et al., 1989
Iron absorption in the newborn	Monkey	Newborn monkeys (6/6)	bLF	1 g/L one time per diet	Formula	Oral	Davidson et al., 1990
Effect on muscle growth	Rats	Adult animals (48/80)	bLF	0.25, 1 or 4 mg/d for 14-d	Saline	S.C.	Byatt et al., 1990
		Neonatal animals (40/60)	bLF	0.01-0.1 mg/kg b.wt.	Saline	S.C.	Byatt et al., 1990
Effect on iron balance	Human	3-week old infants (7/16)	bLF	1 g/L for 3.5 mo.	Formula	Oral	Schulz-Lell et al., 1991
Effect on fecal microflora	Human	New born term infants (29/55)	bLF	0.1 or 1.0 g/L for 3 mo.	Formula	Oral	Roberts et al., 1992
Serum Zn, Fe & ferritin levels	Human	New born term infants (28/51)	bLF	0.1 or 1.0 g/L for 5 mo.	Formula	Oral	Chierici et al., 1992
Control of M. paratuberculosis	Mouse	Gnotobiotic mice (8/32)	bLF	2 mg/L for 10 mo.	Water	Oral	Hamilton & Czuprynski, 1992
Effects on fecal flora	Mouse	Gnotobiotic mice (5/75)	bLF	2 g/L for 14 days	Formula	Oral	Hentges et al., 1992
Effects on induced pancreatitis	Human	Adult subjects (6/35)	bLF	100 mg/kg b. wt./ dose	Saline	I.P	Koike & Makino, 1993
Effect on iron balance	Bovine	New born calves (7/53)	bLF	11.4 g /day for 5 d	Colostrum	Oral	Kume et al., 1994
Effect on iron balance	Bovine	1-week old calves (12/36)	bLF	5 g/day for 5-10 d	Formula	Oral	Kume et al., 1995
Antitumor activity	Human	Adult cancer patients (7/7)	bLF	0.7% per day for 6 mo.	Formula	Oral	Kennedy et al., 1995
Bacteriostasis of Clostridium spp.	Mouse	(9/9); (15/25) & (30/40)	bLF	0.5 – 2.0% for 7-14 d	Milk	Oral	Teraguchi et al., 1995a
Effect on bacterial translocation	Mouse	5-week old mice (20/30)	bLF	0.5 – 2.0% for 7 d	Milk	Oral	Teraguchi et al., 1995b
Antiviral activity against stomatitis	Cats	Cats 5-10 yr. (1/7) (5/12)	bLF	40 mg/kg b.wt for 14 d	None	Oral	Sato et al., 1996
Protein synthesis de novo	Porcine	New born pigs (6/18)	bLF	1 g/L for 24 h	Formula	Oral	Burrin et al., 1996
Inhibition of induced colon tumors	Rats	6-wk rats (70/115)	bLF	0.2 to 2.0% for 30 wks	Diet	Oral	Sekine et al., 1997a
Inhibition of induced colon tumors	Rats	6-wk rats (16/30) (8/15)	bLF	0.2 to 2.0% for 4-13 wks	Diet	Oral	Sekine et al., 1997b
Control of Toxoplasma gondii	Mouse	8-wk mice (25/30)	LFcin	0.2 g/kg b.wt for 8-d	Saline	I.P	Isamida et al., 1998

XI. SAFETY AND TOLERANCE

LF is already used in a wide range of products including infant formulas, sport and functional foods, personal care products, as well as veterinary and feed specialties. In United States, LF is yet to be considered for a GRAS status. On the other hand, LF products have found a niche in various parts of Europe and southeast Asia controlled by the following regulations.

A. Legal status

In Europe, bLF is manufactured according to the Dairy Hygiene Directive 92/46 of the EC and meets the corresponding requirements. As per the EC regulations, bLF can be added to food for nutritional reasons. However, for allowed use, the relevant food standards should be considered.

In Japan, LF concentrates are allowed under number 438 as 'a substance composed mainly of LF obtained from mammal's milk' in food according to the Ministry of Health and Welfare Announcement No.160 of August 10, 1995.

According to the Public Code of Food Additives in South Korea, LF concentrates are allowed in food products. No special provisions are laid down.

In Taiwan, LF may be used in special nutritional food stuffs under the condition 'only for supplementing foods with an insufficient nutritional content and may be used in appropriate amounts according to actual requirements'.

B. *In vivo* metabolism and turnover

Bennett and Kokocinski (1979) measured the turnover of ^{125}I-LF in ten adult human subjects by simultaneous organ radioactivity counting procedure. Ferrokinetic studies were performed in three adults after the intravenous injection of [^{59}Fe]LF. 3. LF was rapidly eliminated from the plasma with a mean fractional catabolic rate of 5.7/day. Apo-LF (one subject) was eliminated at a slower rate (fractional catabolic rate 1.22/day). Of the administered ^{125}I label 99% was recovered in the urine, as free iodine, within the first 24 h. In the ^{59}Fe studies no appreciable activity was found in the urine. Organ radioactivity counting showed that LF was rapidly taken up by the liver and spleen. In the ^{125}I studies the rapid excretion of free ^{125}I suggested catabolism at these sites. In the ^{59}Fe studies, the radioactivity persisted in the liver and spleen for several weeks and was slowly transferred to the bone marrow before appearing in circulating erythrocytes. From the values of fractional catabolic rate, plasma LF, neutrophil LF and plasma volume, a 'derived neutrophil turnover' was calculated for each subject. The mean value was 8×10^8 neutrophils/day. This is about 1% the value obtained from the actual measurement of labeled cells. It is postulated that this 'derived value' represents only that portion of neutrophil turnover accounted for by intravascular senescence.

Ziere et al. (1992) characterized the hepatic recognition of LF. Intravenously injected ^{125}I-LF was cleared rapidly from the circulation by the liver (93% of the dose at 5 min after injection). Parenchymal cells contained 97% of the hepatic radioactivity. Internalization, monitored by measuring the release of liver-associated radioactivity by the polysaccharide fucoidin, occurred slowly. Only about 40% of the liver-associated LF was internalized at 10 min after injection, and it took 180 min to internalize 90%. Subcellular fractionation indicated that internalized LF is transported to the lysosomes.

Binding of LF to isolated parenchymal liver cells was saturable with a dissociation constant of 10 μM (20 x 10^6 binding sites/cell).

Regoeczi et al. (1994) studied the ability of liver to transfer LF from the plasma to the bile by injecting a dose (10 to 20 μg/100 gm) of labeled bLF intravenously and following its appearance in bile over 3 h. Both diferric- and apo-LF peaked in the bile 35 min after administration (i.e., the same time as bovine lactoperoxidase and diferric rat TF). However, only a small portion of the LF dose (approximately 1%) was recovered with the bile in 3 h. On the basis of autoradiographic evidence, the excreted LF appeared intact. The biliary excretion profile of albumin, a protein thought to reach the canaliculus by para-cellular diffusion, was notably devoid of a peak. This, together with competition observed between LF and lactoperoxidase on one hand and diferrin-TF and LF on the other for transfer to bile, suggests that LF is routed through the hepatocyte in vesicles. The process is initiated by binding to a plasma membrane component to which lactoperoxidase and diferric-TF can also bind. Most ^{59}Fe bound to LF accompanied the protein carrier to the bile. The study concluded that under normal circumstances (i.e., when concentration of LF in the plasma is very low), LF transferred from plasma by the liver is probably not the major source of this protein in bile.

C. Consumption studies and safety data

The *in vivo* efficacy as well as safety of LF has been documented in human clinical trials and in several consumption studies in experimental animal models (*TABLE 7*). Infant food formulae have been supplemented with bLF in four different clinical trials at levels of 0.1 to 1.0 g/L for three to five months and at 2.8 g/L for fourteen days to term human infants with no adverse effects. Also, bLF has been safety administered orally to cancer patients at 1.6 g/day for six months.

Much higher levels of bLF have been administered orally to mice and rats, as high as 20 g/L of milk for fourteen days and 20 g/kg diet for thirty weeks, respectively, with no known side effects. Subcutaneous or intraperitoneal administration of bLF also support the safety of bLF when it is given by more sensitive routes of delivery. Bovine LF has been given as a single intraperitoneal dose to rats at 100 mg/kg body weight with no known adverse effect. Other animal species, including cats, pigs, calves, and monkeys have been given bLF orally with no detrimental effects.

Bovine LF demonstrates structural homology and functional similarity to hLF. However, cow milk contains only about one-tenth the amount of LF compared to its occurrence in human milk. The levels of hLF in colostrum and mature milk is about 30 g/L and 2 g/L. Thus, a safe upper level of LF consumption by children, adolescents, and adults would be expected at these levels.

The literature describes no reports on toxicity of LF. No standard toxicology tests (acute, subchronic, chronic, carcinogenicity, reproductive, developmental, etc.) of LF were reported, Furthermore, LF has not been assigned a Chemical Abstract Services (CAS) registry number.

XII. SUMMARY

LF is an antimicrobial glycoprotein present in milk and various exocrine secretions that bathe the mucosal surfaces. This metal-binding protein is a multifunctional bioactive molecule with a critical role in many important physiological pathways. LF could elicit a variety of inhibitory effects against microorganisms comprising stasis, cidal, adhesion-blockade, cationic, synergistic and opsonic mechanisms. Broad-spectrum activity against different bacteria, viruses, fungi and parasites, in combination with anti-inflammatory and immunomodulatory properties makes LF a potent innate host defense mechanism. The current global production of bovine milk LF is approximately 100 metric tons and this figure is continuously increasing. This protein is finding it applications as an active ingredient in infant formulae, and health foods in South-East Asian countries, in particular. LF is also in use as a therapeutic and prophylactic agent to control intestinal illnesses and mucosal infections. Certain oral hygiene products, skin care products, and animal feed supplements contain LF. Recent advances in LF research to elucidate the structural function relationships, antimicrobial mechanisms, cost-effective technologies for large-scale protein isolation and biotechnology is opening unlimited opportunities for this natural antimicrobial in the development of new products and formulation. A number of efficacy studies and clinical trials are ongoing in various laboratories with over 100 patents filed on this molecule in the past 10 years. Undoubtedly, LF is emerging as one of the leading natural microbial blocking agent in food safety and preservation.

XIII. REFERENCES

1. Abramson, J.S., Parce, J.W., Lewis, J.C., Lyles, D.S., Mills, E.L., Nelson, R.D., and Bass, D.A. 1984. Characterization of the effect of influenza virus on polymorphonuclear leukocyte membrane responses. *Blood* 64:131-138.
2. Aguas, A., Esaguy, N., Sunkel, C.E., and Silva, M.T. 1990. Cross reactivity and sequence homology between the 65-kilodalton mycobacterial heat shock protein and human lactoferrin, transferrin, and DR beta subsets of major histocompatibility complex class II molecules. *Infect. Immun.* 58:1461-1470.
3. Ahmadzadeh, N., Shingu, M., and Nobunaga, M. 1989. Iron-binding proteins and free iron in synovial fluids of rheumatoid arthritis patients. *Clin. Immunol.* 8:345-351.
4. Ainscough, E.W., Brodie, A.M., Plowman, J.E., Bloor, S.J., Loehr, J.S., and Loehr, T.M. 1980. Studies on human lactoferrin by electron paramagnetic resonance, fluorescence, and resonance Raman spectroscopy. *Biochemistry* 19:4072-4079.
5. Aisen, P., and Leibman, A. 1972. Lactoferrin and transferrin: A comparative study. *Biochim. Biophys. Acta* 257:314-323.
6. Al-Mashikhi, S.A., and Nakai, S. 1987. Isolation of bovine immunoglobulins and lactoferrin from whey proteins by gel filtration techniques. *J. Dairy Sci.* 70:2486-2492.
7. Alugupalli, K.R., Kalfas, S., Edwardsson, S., Forsgren, A., Arnold, R.R., and Naidu, A.S. 1994. Effect of lactoferrin on interaction of *Prevotella intermedia* with plasma and subepithelial matrix proteins. *Oral Microbiol. Immunol.* 9:174-179.
8. Alugupalli, K.R., and Kalfas, S. 1995. Inhibitory effect of lactoferrin on the adhesion of *Actinobacillus actinomycetemcomitans* and *Prevotella intermedia* to fibroblasts and epithelial cells. *APMIS* 103:154-160.
9. Alugupalli, K.R., Kalfas, S., Edwardsson, S., and Naidu, A.S. 1995. Lactoferrin interaction with *Actinobacillus actinomycetemcomitans*. *Oral Microbiol. Immunol.* 10:35-41.
10. Alugupalli, K.R., and Kalfas, S. 1997. Characterization of the lactoferrin-dependent inhibition of the adhesion of *Actinobacillus actinomycetemcomitans*, *Prevotella intermedia* and *Prevotella nigrescens* to fibroblasts and to a reconstituted basement membrane. *APMIS* 105:680-688.
11. Ambruso, D.R., Sasada, M., Nishiyama, H., Kubo, A., Komiyama, A., and Allen, R.H. 1984. Defective bactericidal activity and absence of specific granules in neutrophils from a patient with recurrent bacterial infections. *J. Clin. Immunol* . 4:23-30.
12. Amini, H.R., Ascencio, F., Ruiz-Bustos, E., Romero, M.J., and Wadstrom, T. 1996. Cryptic domains of a 60 kDa heat shock protein of *Helicobacter pylori* bound to bovine lactoferrin. *FEMS Immunol. Med. Microbiol.* 16: 247-55.
13. Andersson, B.F., Baker, H.M, Dodson, E.J., Norris, G.E., Rumball, S.V., Waters, J.M, and Baker E.N. 1987. Structure of human lactoferrin at 3.2-Å resolution. *Proc. Natl. Acad. Sci. USA.* 84:1769-1773.

14. Anderson, B.F., Baker, H.M., Norris, G.E., Rice, D.W., and Baker, E.N. 1989. Structure of human lactoferrin: crystallographic structure analysis and refinement at 2.8 A resolution. *J. Mol. Biol.* 209:711-734.
15. Appelmelk, B.J., An, Y.Q., Geerts, M., Thijs, B.G., de Boer, H.A., MacLaren, D.M., de Graaff, J., and Nuijens, J.H. 1994. Lactoferrin is a lipid A-binding protein. *Infect. Immun.* 62: 2628-2632.
16. Arnold, R.R., Cole, M.F., and McGhee, J.R. 1977. A bactericidal effect for human lactoferrin. *Science* 197:263-265.
17. Arnold, R.R., Brewer, M., and Gauthier, J.J. 1980. Bactericidal activity of human lactoferrin: sensitivity of a variety of microorganisms. *Infect. Immun.* 28:893-898.
18. Arnold, R.R., Russell, J.E., Champion, W.J., and Gauthier, J.J. 1981. Bactericidal activity of human lactoferrin: influence of physical conditions and metabolic state of the target microorganism. *Infect. Immun.* 32:655-660.
19. Ascencio, F., Ljungh, A., and Wadstrom, T. 1992. Lactoferrin binding properties of *Vibrio cholerae*. *Microbios.* 70:103-117.
20. Ashorn, R.G., Eskola, J., Tuohimaa, P.J., and Krohn, K.J. 1986. Effect of progesterone-inducible proteins human lactoferrin and avidin on lymphocyte proliferation. *Human Reproduct.* 1:149-151.
21. Aslam, M., and Hurley, W.L. 1997. Proteolysis of milk proteins during involution of the bovine mammary gland. *J. Dairy Sci.* 80:2004-2010.
22. Azari, P., and Baugh, R.F. 1967. A simple and rapid procedure for preparation of large quantities of pure ovotransferrin. *Arch. Biochem. Biophys.* 118:138-145.
23. Baggiolini, M., De Duve, C., Masson, P.L., and Heremans, J.F. 1970. Association of lactoferrin with specific granules in rabbit heterophil leukocytes. *J. Exp. Med.* 131:559-570.
24. Baker, E.N., Rumball, S.V., and Andersson, B.F. 1987. Transferrins: insight into structure and function from studies on lactoferrin. *Trends Biochem. Sci.* 12:350-353.
25. Baker, E.N., Anderson, B.F., Baker, H.M., Haridas, M., Jameson, G.B., Norris, G.E., Rumball, S.V., and Smith, C.A. 1991. Structure, function and flexibility of human lactoferrin. *Int. J. Biol. Macromol.* 13:122-129.
26. Balmer, S.E., Scott, P.H., and Wharton, B.A. 1989. Diet and feacal flora in the newborn: lactoferrin. *Arch Dis Child.* 64:1685-1690.
27. Barresi, G., and Tuccari, G. 1987. Iron-binding proteins in thyroid tumours: an immunocytochemical study. *Pathol. Res. Pract.* 182:344-351.
28. Barthe, C., Galabert, C., Guy-Crotte, O., and Figarella, C. 1989. Plasma and serum lactoferrin levels in cystic fibrosis. Relationship with the presence of cystic fibrosis protein. *Clin. Chim. Acta* 181:183-188.
29. Batish, V.K., Chander, H., Zumdegni, K.C., Bhatia, K.L., and Singh, R.S. 1988. Antibacterial effect of lactoferrin against some common food-borne pathogenic organisms. *Austral. J. Dairy Technol.* 5:16-18.
30. Baynes, R.D., Bezwoda, W.R., and Mansoor. N. 1988. Neutrophil lactoferrin content in viral infections. *Am. J. Clin. Pathol.* 89:225-228.
31. Bellamy, W., Takase, M., Yamauchi, K., Wakabayashi, H., Kawase, K., and Tomita, M. 1992. Identification of the bactericidal domain of lactoferrin. *Biochim. Biophys. Acta* 1121:130-136.
32. Bellamy, W., Wakabayashi, H., Takase, M., Kawase, K., Shimamura, S., and Tomita, M. 1993. Killing of *Candida albicans* by lactoferricin B, a potent antimicrobial peptide derived from the N-terminal region of bovine lactoferrin. *Med. Microbiol. Immunol. (Berl)* 182:97-105.
33. Bennett, R.M., and Kokocinski, T. 1978. Lactoferrin content of peripheral blood cells. *Br. J. Haematol.* 39:509-521.
34. Bennett, R.M., and Mohla, C. 1976. A solid phase radioimmunoassay for the measurement of lactoferrin in human plasma: variation with age, sex and disease. *J. Lab. Clin. Med.* 88:156-166.
35. Bennett, R.M., and Skosey, J.L. 1977. Lactoferrin and lysozyme levels in synovial fluid: differential indices of articular inflammation and degradation. *Arthrit. Rheumat.* 20:84-90.
36. Bennett, R.M., and Kokocinski, T. 1979. Lactoferrin turnover in man. *Clin. Sci.* 57:453-460.
37. Bennett, R.M., Bagby, G.C., and Davis, J. 1981. Calcium-dependent polymerization of lactoferrin. *Biochem. Biophys. Res. Commun.* 101:88-95.
38. Bennett, R.M., and Davis, J. 1982. Lactoferrin interacts with deoxyribonucleic acid: a preferential reactivity with double-stranded DNA and dissociation of DNA-anti-DNA complexes. *J. Lab. Clin. Med.* 99:127-38.
39. Bennett, R.M., Davis, J., Campbell, S., and Portnoff, S. 1983. Lactoferrin binds to cell membrane DNA. Association of surface DNA with an enriched population of B cells and monocytes. *J. Clin. Invest.* 71:611-618.
40. Bennett, R.M., Merritt, M.M., and Gabor, G. 1986. Lactoferrin binds to neutrophilic membrane DNA. *Br. J. Haematol.* 63:105-117.

41. Bezwoda, W.R., and Mansoor, H. 1986. Isolation and characterization of lactoferrin separated from human whey by adsorption chromatography using Cibacron Blue F3G-A linked affinity adsorbent. *Clin. Chim. Acta* 157:89-95.

42. Bezwoda, W.R., and Mansoor, N. 1989. Lactoferrin from human breast milk and from neutrophil granulocytes. Comparative studies of isolation, quantitation, characterization and iron binding properties. *Biomed. Chromat.* 3:121-126.

43. Bhimani, R.S., Vendrov, Y., and Furmanski, P. 1999. Influence of lactoferrin feeding and injection against systemic staphylococcal infections in mice. *J. Appl. Microbiol.* 86:135-144.

44. Birgens, H.S., and Kristensen, L. O. 1990. Impaired receptor binding and decrease in isoelectric point of lactoferrin after interaction with human monocytes. *Eur. J. Haemat* 45:31-35.

45. Bishop, J.G., Schanbacher, F.L., Ferguson, L.C., and Smith, K.L. 1976. *In vitro* growth inhibition of mastitis-causing coliform bacteria by bovine apo-lactoferrin and reversal of inhibition by citrate and high concentrations of apo lactoferrin. *Infect. Immun.* 14:911-918.

46. Biswas, G.B., and Sparling, P.F. 1995. Characterization of *lbpA*, the structural gene for a lactoferrin receptor in *Neisseria gonorrhoeae. Infect. Immun.* 63:2958-2967.

47. Blackberg, L., and Hernell, O. 1980. Isolation of lactoferrin from human whey by a single chromatography step. *FEBS Letts.* 109:180-184.

48. Blake, D.R., Hall, N.D., Bacon, P.A. 1981. The importance of iron in rheumatoid disease. *Lancet* 21:1142-1143.

49. Blake, D.R., Gallagher, P.J., and Potter, A.R. 1984. The effect of synovial iron on the progression of rheumatoid disease. *Arthritis Rheum.* 27:495-501.

50. Blanc, B., and Isliker, H.C. 1961a. Isolement et caracterisation de la proteine rouge siderophile dulait maternel: la lactoferrine. *Helv. Physiol. Pharmacol. Acta* 19:C-13-14.

51. Blanc, B., and Isliker, H.C. 1961b. Isolement et caracterisation de la proteine rouge siderophile dulait maternel: la lactoferrine. *Bull. Soc. Chim. Biol (Paris)* 43:929-943.

52. Bortner, C.A., Miller, R.D., and Arnold, R.R. 1986. Bactericidal effect of lactoferrin on *Legionella pneumophila. Infect. Immun.* 51:373-377.

53. Bortner, C.A., Arnold, R.R., and Miller, R.D. 1989. Bactericidal effect of lactoferrin on *Legionella pneumophila*: effect of the physiological state of the organism. *Can. J. Microbiol.* 35:1048-1051.

54. Boxer, L.A., Bjorksten, B., Bjork, J., Yang, H.H., Alen, J.M., and Baehner, R.L. 1982a. Neutropenia induced by systemic infusion of lactoferrin. *J. Lab. Clin. Med.* 99:866-872.

55. Boxer, L.A., Haak, R.A., Yang, H.H., Wolach, J.B., Whitcomb, J.A., Butterick, C.J., and Baehner, R.L. 1982b. Membrane-bound lactoferrin alters the surface properties of polymorphonuclear leukocytes. *J. Clin. Invest.* 70:1049-1057.

56. Breau, W.C., and Oliver, S.P. 1986. Growth inhibition of environmental mastitis pathogens during physiologic transitions of the bovine mammary gland. *Am. J. Vet. Res.* 47:218-222.

57. Brines, R.D., and Brock, J.H. 1983. The effect of trypsin and chymotrypsin on the *in vitro* antimicrobial and iron-binding properties of lactoferrin in human milk and bovine colostrum. Unusual resistance of human apolactoferrin to proteolytic digestion. *Biochim. Biophys. Acta* 759:229-235.

58. Britigan, B.E., Hayek, M.B., Doebbeling, B.N., and Fick, R.B. 1993. Transferrin and lactoferrin undergo proteolytic cleavage in the *Pseudomonas aeruginosa*-infected lungs of patients with cystic fibrosis. *Infect. Immun.* 61:5049-5055.

59. Britigan, B.E., Serody, J.S., and Cohen, M.S. 1994. The role of lactoferrin as an anti-inflammatory molecule. *Adv. Exp. Med. Biol.* 357:143-156.

60. Brock, J.H., Arzabe, F., Lampreave, F, and Pineiro, A. 1976. The effect of trypsin on bovine transferrin and lactoferrin. *Biochim. Biophys. Acta* 446:214-225.

61. Brock, J.H., Pineiro, A., and Lampreave, F. 1978. The effect of trypsin and chymotrypsin on the antibacterial activity of complement, antibodies, and lactoferrin and transferrin in bovine colostrum. *Ann. Rech. Vet.* 9:287-294.

62. Brock, J.H., Pickering, M.G., McDowall, M.C., and Deacon, A.G. 1983. Role of antibody and enterobactin in controlling growth of *Escherichia coli* in human milk and acquisition of lactoferrin- and transferrin-bound iron by *Escherichia coli. Infect. Immun.* 40:453-459.

63. Brodie, A.M., Ainscough, E.W., Baker, E.N., Baker, H.M., Shongwe, M.S., and Smith, C.A. 1994. Synergism and substitution in the lactoferrins. *Adv. Exp. Med. Biol.* 357:33-44.

64. Broer, K.H., Dauber, U., Herrmann, W.P., and Hirshhauser, C. 1977. The presence of lactoferrin (SCA) on the human sperm head during *in vitro* penetration through cervical mucus. *IRCS J. Med. Sci.* 5:362.

65. Broxmeyer, H.E., Smithyman, A., Eger, R.R., Meyers, P.A., and de Sousa, M. 1978. Identification of lactoferrin as the granulocyte-derived inhibitor of colony-stimulating activity production. *J. Exp. Med.* 148:1052-1067,

66. Broxmeyer, H.E., and Platzer, E. 1984. Lactoferrin acts on I-A and I-E/C antigen subpopulations of mouse peritoneal macrophages in the absence of T lymphocytes and other cell types to inhibit production of granulocyte-monocyte colony stimulatory factors *in vitro. J. Immunol.* 133:306-314.

67. Broxmeyer, H.E., Williams, D.E., Hangoc, G., Cooper, S., Gentile, P., Shen, R.N., Ralph, P., Gillis, S., and Bicknell, D.C. 1987. The opposing action *in vivo* on murine myelopoiesis of purified preparations of lactoferrin and the colony stimulating factors. *Blood Cells* 13:31-48.

68. Bukhhardt, H., and. Schwingel, M. 1986. Oxygen radicals as effectors of cartilage destruction. *Arthritis Rheum.* 30:57-63.

69. Bullen, J.J. 1975. Iron-binding proteins in milk and resistance to *Escherichia coli* infection in infants. *Postgrad. Med. J.* 51:67-70.

70. Bullen, J.J. 1976. Iron-binding proteins and other factors in milk responsible for resistance to *Escherichia coli. Ciba Found. Symp.* 42:149-169.

71. Bullen, J.J. 1981. The significance of iron in infection. *Rev. Infect. Dis.* 3:1127-1138.

72. Burrin, D.G., Wang, H., Heath, J., and Dudley, M.A. 1996. Orally administered lactoferrin increases hepatic protein synthesis in formula-fed newborn pigs. *Pediatr. Res.* 40(1):72-76.

73. Butler, J.E. 1973. The occurrence of immunoglobulin fragments, two types of lactoferrin and a lactoferrin-IgG2 complex in bovine colostral and milk whey. *Biochim. Biophys. Acta* 295:341-351.

74. Byatt, J.C., Schmuke, J.J., Comens, P.G., Johnson, D.A., and Collier, R.J. 1990. The effect of bovine lactoferrin on muscle growth *in-vivo* and *in-vitro. Biochem. Biophys. Res. Comm.* 173:548-553.

75. Byrd, T.F., and Horwitz, M.A. 1991. Lactoferrin inhibits or promotes *Legionella pneumophila* intracellular multiplication in nonactivated and interferon gamma-activated human monocytes depending upon its degree of iron saturation. Iron-lactoferrin and nonphysiologic iron chelates reverse monocyte activation against *Legionella pneumophila. J. Clin. Invest.* 88:1103-1112.

76. Campagnari, A.A., Shanks, K.L., and Dyer, D.W. 1994. Growth of *Moraxella catarrhalis* with human transferrin and lactoferrin: expression of iron-repressible proteins without siderophore production. *Infect. Immun.* 62:4909-4914.

77. Carlsson, A., Bjorck, L., and Persson, K. 1989. Lactoferrin and lysozyme in milk during acute mastitis and their inhibitory effect in Delvotest P. *J. Dairy. Sci.* 72:3166-3175.

78. Carnoy, C., Scharfman, A., van Brussel, E., Lamblin, G., Ramphal, R., and Roussel, P. 1994. *Pseudomonas aeruginosa* outer membrane adhesins for human respiratory mucus glycoproteins. *Infect. Immun.* 62:1896-1900.

79. Caselitz, J., Jaup, T., and Seifer, G. 1981. Lactoferrin and lysozyme in carcinomas of the parotid gland. *Virchows Arch. A.* 394:61-73.

80. Castellino, F.J., Fish, W.W., and Mann, K.G. 1970. Structural studies on bovine lactoferrin. *J. Biol. Chem.* 245:4269-4275.

81. Charpin, C., Lachard, A., and Pourreau-Schneider, N. 1985. Localization of lactoferrin and nonspecific cross-reacting antigen in human breast carcinomas. *Cancer* 55:2612-2617.

82. Chierici, R., Sawatzki, G., Tamisari, L., Volpato, S., and Vigi, V. 1992. Supplementation of an adapted formula with bovine lactoferrin. 2. Effects on serum iron, ferritin and zinc levels. *Acta Paediatr.* 81:475-479.

83. Chimura, T., Hirayama, T., Nakahara, M. 1993. *In vitro* antimicrobial activities of lactoferrin, its concomitant use with cefpodoxime proxetil and clinical effect of cefpodoxime proxetil. *Jpn. J. Antibiot.* 46:482-485.

84. Cohen, M.S., Mao, J., Rasmussen, G.T., Serody, J.S., and Britigan, B.E. 1992. Interaction of lactoferrin and lipopolysaccharide (LPS): effects on the antioxidant property of lactoferrin and the ability of LPS to prime human neutrophils for enhanced superoxide formation. *J. Infect. Dis.* 166:1375-1378.

85. Courtoy, P.J., Moguilevsky, N., Retegui, L.A., Castracane, C.E., and Masson, P.L. 1984. Uptake of lactoferrin by the liver. II. Endocytosis by sinusoidal cells. *Lab. Invest.* 50:329-334.

86. Cox, T.M., Mazurier, J., Spik, G., Montreuil, J., and Peters, T.J. 1979. Iron-binding proteins and influx of iron across the duodenal brush border. Evidence for specific lactotransferrin receptors in the human small intestine. *Biochim. Biophys. Acta* 588:120-128.

87. Cravito, A., Tello, A., Villafan, H., Ruiz, J., del Vedovo, S., and Neeser, J.R. 1991. Inhibition of localized adhesion of enteropathogenic *Escherichia coli* to Hep-2 cells by immunoglobulin and oligosaccharide fractions of human colostrum and breast milk. *J. Infect. Dis.* 163:1247-1255.

88. Crosa, J.H. 1989. Genetics and molecular biology of siderophore-mediated iron transport in bacteria. *Microbiol. Rev.* 53:517-530.

89. Crouch, S.P.M., Slater, K.J., and Fletcher, J. 1992. Regulation of cytokine release from mononuclear cells by the iron-binding protein lactoferrin. *Blood* 80:235-240.

90. Culliere, M.L., Montagne, P., Mole, C., Bene, M.C., and Faure, G. 1997. Microparticle-enhanced nephlometric immunoassay of lactoferrin in human milk. *J. Clin. Lab. Anal.* 11:239-243.
91. Custer, M.C., and Hansen, J.N. 1983. Lactoferrin and transferrin fragments react with nitrite to form an inhibitor of *Bacillus cereus* spore outgrowth. *Appl. Environ. Microbiol.* 45:942-949.
92. Dalmastri, C., Valenti, P., Visca, P., Vittorioso, P., and Orsi, N. 1988. Enhanced antimicrobial activity of lactoferrin by binding to the bacterial surface. *Microbiologica* 11: 225-230.
93. Davidson, L.A., and Lönnerdal, B. 1986. Isolation and characterization of rhesus monkey lactoferrin. *Pediatr. Res.* 20:197-201.
94. Davidson, L., and Lönnerdal, B. 1988. Specific binding of lactoferrin to brush border membrane: Ontogeny and effect of glycan chain. *Am. J. Physiol.* 257:930-934.
95. Davidson, L.A., Litov, R.E., and Lönnerdal, B. 1990. Iron retention from lactoferrin-supplemented formulas in infant rhesus monkeys. *Pediatr. Res.* 27:176-180.
96. Davidson, L., Kastenmayer, P., Yuen, M., Lönnerdal, B., and Hurrell, R.F. 1994. Influence of lactoferrin on iron absorption from human milk in infants. *Pediatr. Res.* 35:117-124.
97. de Lillo, A., Teanpaisan, R., Fierro, J.F., and Douglas, C.W. 1996. Binding and degradation of lactoferrin by *Porphyromonas gingivalis, Prevotella intermedia* and *Prevotella nigrescens. FEMS Immunol. Med. Microbiol.* 14:135-143.
98. de Lillo, A., Quiros, L.M., and Fierro, J.F. 1997. Relationship between antibacterial activity and cell surface binding of lactoferrin in species of genus *Micrococcus. FEMS Microbiol. Lett.*150:89-94.
99. de Vet, B.J., and van Gool, J. 1974. Lactoferrin and iron absorption in the small intestine. *Acta Med. Scand.* 196:393-402.
100. de Vet, B.J., and van Vugt, H. 1971. Lactoferrin, in particular the relation with iron absorption. *Ned. Tijdschr. Geneeskd.* 115:961-967.
101. Debbabi, H., Dubarry, M., Rautureau, M., and Tome, D. 1998. Bovine lactoferrin induces both mucosal and systemic immune response in mice. *J. Dairy Res.* 65:283-293.
102. Denisova, I.I., Gennadeva, T.I., Zhdanova, T.V., Khristova, M., and Khazenson, L.B. 1996. The action of lactoperoxidase, lactoferrin and lactoglobulin on *Shigella sonnei* in an *in vivo* experiment. *Zh. Mikrobiol. Epidemiol. Immunobiol.* 1:66-69.
103. Dhaenens, L., Szczebara, F., and Husson, M.O. 1997. Identification, characterization, and immunogenicity of the lactoferrin-binding protein from *Helicobacter pylori. Infect. Immun.* 65:514-518.
104. Dial, E.J., Hall, L.R., Serna, H., Romero, J.J., Fox, J.G., and Lichtenberger, L.M. 1998. Antibiotic properties of bovine lactoferrin on *Helicobacter pylori. Dig. Dis. Sci.* 43:2750-2756.
105. Dinarello, C.A. 1984. Interleukin-1. *Rev. Infect. Dis.* 6:51-95.
106. Dionysius, D.A., and Milne, J.M. 1997. Antibacterial peptides of bovine lactoferrin: purification and characterization. *J. Dairy Sci.* 80:667-674.
107. Dionysius, D.A., Grieve, P.A., and Milne, J.M. 1993. Forms of lactoferrin: their antibacterial effect on enterotoxigenic *Escherichia coli. J. Dairy Sci.* 76:2597-2600.
108. Dite, P., Bartova, M., and Skaunic, V. 1989. Assessment of lactoferrin in the diagnosis of pancreatic disease. *Vnitrni. Lek.* 35:132-136.
109. Dolby, J.M., Stephens, S., and Honour, P. 1977. Bacteriostasis of *Escherichia coli* by milk. II. Effect of bicarbonate and transferrin on the activity of infant feeds. *J. Hyg. (Lond).* 78:235-242.
110. Dorland, L., Haverkamp, J., Vliegenthart, J.F.G., Spik, G., Fournet, B., and Montreuil, J. 1979. Investigation by 300-MHz 1H-nuclear magnetic resonance spectroscopy and methylation analysis of the single glycan chain of chicken ovotransferrin. *Eur. J. Biochem.* 100:569-574.
111. Elass-Rochard, E., Roseanu, A., Legrand, D., Trif, M., Salmon, V, Motas, C., Montreuil, J., and Spik G. 1995. Lactoferrin-lipopolysaccharide interaction: involvement of the 28-34 loop region of human lactoferrin in the high-affinity binding to *Escherichia coli* 055B5 lipopolysaccharide. *Biochem. J.* 312:839-45
112. Ellison, R.T., Giehl, T.J., and LaForce, F.M. 1988. Damage of the outer membrane of enteric gram-negative bacteria by lactoferrin and transferrin. *Infect. Immun.* 56:2774-2781.
113. Ellison, R.T., LaForce, F.M., Giehl, T.J., Boose, D.S., and Dunn, B.E. 1990. Lactoferrin and transferrin damage of the gram-negative outer membrane is modulated by Ca^{2+} and Mg^{2+}. *J. Gen. Microbiol.* 136:1437-1446.
114. Ellison, R.T., and Giehl, T.J. 1991. Killing of gram-negative bacteria by lactoferrin and lysozyme. *J. Clin. Invest.* 88:1080-1091.
115. Erdei, J., Forsgren, A., and Naidu, A.S. 1994. Lactoferrin binds to porins OmpF and OmpC in *Escherichia coli. Infect. Immun.* 62:1236-1240.
116. Faber, H.R., Bland, T., Day, C.L., Norris, G.E., Tweedie, J.W., and Baker, E.N. 1996. Altered domain closure and iron binding in transferrins: the crystal structure of the *Asp60Ser* mutant of the amino- terminal half-molecule of human lactoferrin. *J. Mol. Biol.* 256:352-363.

117. Fairweather-Tait, S.J., Balmer, S.E., Scott, P.H., and Minski, M.J. 1987. Lactoferrin and iron absorption in newborn infants. *Pediatr. Res.* 22:651-654.
118. Feng, M., van der Does, L., and Bantjes, A. 1995. Preparation of apolactoferrin with a very low iron saturation. *J. Dairy Sci.* 78:2352-2357.
119. Foster, L.A., and Dyer, D.W. 1993. A siderophore production mutant of *Bordetella bronchiseptica* cannot use lactoferrin as an iron source. *Infect. Immun.* 61:2698-2702.
120. Fransson, G.B., Keen, C.L., and Lonnerdal, B. 1983. Supplementation of milk with iron bound to lactoferrin using weanling mice: Effects on hematology and tissue iron. *J. Pediatr. Gastroenterol. Nutr.* 2:693-700.
121. Friedman, S.A., Mandel, I.D., and Herrera, M.S. 1983. Lysozyme and lactoferrin quantitation in the crevicular fluid. *J. Periodontol.* 54:347-350.
122. Fujihara, T., and Hayashi, K. 1995. Lactoferrin inhibits herpes simplex virus type-1 (HSV-1) infection to mouse cornea. *Arch. Virol.* 140:1469-72
123. Furmanski, P., Li, Z-P., Fortuna, M.B., Swamy, C.V.B., and Das, M.R. 1989. Multiple molecular forms of human lactoferrin. Identification of a class of lactoferrins that possess ribonuclease activity and lack iron-binding capacity. *J. Exp. Med.* 170:415-429.
124. Gado, I., Erdei, J., Laszlo, V.G., Paszti, J., Czirok, E., Kontrohr, T., Toth, I., Forsgren, A., and Naidu, A.S. 1991. Correlation between human lactoferrin binding and colicin susceptibility in *Escherichia coli*. *Antimicrob. Agents Chemother.* 35:2538-2543.
125. Gahr, M., Schulze, M., Schefeczyk, D., Speer, C.P., and Peters, J.H. 1987. Diminished release of lactoferrin from polymorphonuclear leukocytes in human neonates. *Acta Hematol.* 77:90-95.
126. Garre, C., Bianchi-Scarra, G., Sirito, M., Musso, M., and Ravazzolo, R. 1992. Lactoferrin binding sites and nuclear localization in K562(S) cells. *J. Cell. Physiol.* 153:477-482.
127. Gatter, K.C., Brown, G., Trowbridge, I.S., Woolston, R.E., and Mason, D.Y. 1983. Transferrin receptors in human tissues: their distribution and possible clinical relevance. *J. Clin. Pathol.*36:539-545.
128. Gillin, F.D., Reiner, D.S., and Wang, C.S. 1983. Killing of *Giardia lamblia* trophozoites by normal human milk. *J. Cell Biochem.* 23:47-56.
129. Giugliano, L.G., Ribeiro, S.T., Vainstein, M.H., and Ulhoa, C.J. 1995. Free secretory component and lactoferrin of human milk inhibit the adhesion of enterotoxigenic *Escherichia coli*. *J. Med. Microbiol.* 42:3-9.
130. Godo, Z.I., Magyar, E., Andirko, I., and Rozgonyi, F. 1997. Compact growth of *Staphylococcus haemolyticus* in soft agar is not due to hydrophobic interaction between the cocci. *Acta Microbiol. Immunol. Hung.* 44:343-349.
131. Gorden, W.G., Groves, M.L., and Basch, J.J. 1963. Bovine milk 'red protein': amino acid composition and comparison with blood transferrin. *Biochemistry* 2:817-820.
132. Griffiths, E., and Humphreys, J. 1977. Bacteriostatic effect of human milk and bovine colostrum on *Escherichia coli*: importance of bicarbonate. *Infect. Immun.* 15:396-401.
133. Groenink, J., Walgreen-Weterings, E., van't Hof, W., Veerman, E.C., and Nieuw Amerongen, A.V. 1999. Cationic amphipathic peptides, derived from bovine and human lactoferrins, with antimicrobial activity against oral pathogens. *FEMS Microbiol. Lett.* 179:217-222.
134. Grover, M., Giouzeppos, O., Schnagl, R.D., and May, J.T. 1997. Effect of human milk prostaglandins and lactoferrin on respiratory syncytial virus and rotavirus. *Acta. Paediatr.* 86:315-316.
135. Groves, M. 1960. The isolation of a red protein from milk. *J. Am. Chem. Soc.* 82:3345-3350.
136. Gruttner, R., Schafer, K.H., and Schroter, W. 1960. Zur reindarstellung eines eisenbindenden proteins in der frauenmilch. *Klinische Wochenschrift* 38:1162-1165.
137. Gutteberg, T.J., Haneberg, B., and Jørgensen, T. 1984. The latency of serum acute phase proteins in meningococcal septicemia, with special emphasis on lactoferrin. *Clin. Chim. Acta* 136:173-178.
138. Gutteberg, T.J., Røkke, O., Andersen, O., and Jørgensen, T. 1988. Lactoferrin as an indicator of septicemia and endotoxaemia in pigs. *Scand. J. Infect. Dis.* 20:659-666.
139. Gutteberg, T.J., Røkke, O., Andersen, O., and Jørgensen, T. 1989. Early fall of circulating iron and rapid rise of lactoferrin in septicemia and endotoxemia: an early defence mechanism. *Scand. J. Infect. Dis.* 21: 709-715.
140. Gutteberg, T.J., Osterud, B., Volden, G., and Jørgensen, T. 1990a. The production of tumour necrosis factor, tissue thromboplastin, lactoferrin and cathepsin C during lipopolysaccharide stimulation in whole blood. *Scand. J. Clin. Lab. Invest.* 50: 421-427.
141. Gutteberg, T.J., Dalaker, K., and Vorland, L.H. 1990b. Early response in neonatal septicemia. The effect of *Escherichia coli*, *Streptococcus agalactiae* and tumor necrosis factor on the generation of lactoferrin. *APMIS* 98:1027-1032.

142. Hamilton, H.L., and Czuprynski, C.J. 1992. Effects of mycobactin J and lactoferrin supplementation of drinking water on the *in vivo* multiplication of *Mycobacterium paratuberculosis* in gnotobiotic mice. *Can J Vet Res.* 56:70-73.

143. Hammerschmidt, S., Bethe, G., Remane, P., and Chhatwal, G.S. 1999. Identification of pneumococcal surface protein A as a lactoferrin-binding protein of *Streptococcus pneumoniae*. *Infect. Immun.* 67:1683-1687.

144. Hangoc, G., Lu, L., Oliff, A., Gillis, S., Hu, W., Bicknell, D.C., Williams, D., and Broxmeyer, H.E. 1987. Modulation of Friend virus infectivity *in vivo* by administration of purified preparations of human lactoferrin and recombinant murine interleukin-3 to mice. *Leukemia* 1: 762-764.

145. Hansen, N.E., Malmquist, J., and Thorell, J. 1975. Plasma myeloperoxidase and lactoferrin measured by radioimmunoassay: relations to neutrophil kinetics. *Acta Med. Scand.* 198:437-443.

146. Harbitz, O., Jensen, A.O., and Smidsrod, O. 1984. Lysozyme and lactoferrin in sputum from patients with chronic obstructive lung disease. *Eur. J. Res. Dis.* 65:512-520.

147. Harmon, R.J., Schanbacher, F.L., Fergusson, L.C., and Smith, K.L. 1976. Changes in lactoferrin, immunoglobulin G, bovine serum albumin, and alpha-lactalbumin during acute experimental and natural coliform mastitis in cows. *Infect. Immun.* 13:533-542.

148. Harmsen, M.C., Swart, P.J., de Bethune, M.P., Pauwels, R., De Clercq, E., The, T.H., and Meijer D.K. 1995. Antiviral effects of plasma and milk proteins: lactoferrin shows potent activity against both human immunodeficiency virus and human cytomegalovirus replication *in vitro*. *J. Infect. Dis.* 172:380-388.

149. Harris, W.R. 1986. Thermodynamics of gallium complexation by human lactoferrin. *Biochemistry* 25:803-808.

150. Hasegawa, K., Motsuchi, W., Tanaka, S., and Dosako, S. 1994. Inhibition with lactoferrin of *in vitro* infection with human herpes virus. *Jpn. J. Med. Sci. Biol.* 47:73-85

151. Hashizume, S., Kuroda, K., and Murakami, H. 1983. Identification of lactoferrin as an essential growth factor for human lymphocytic cell lines in serum-free medium. *Biochim. Biophys. Acta* 763:377-382.

152. He, J., and Furmanski, P. 1995. Sequence specificity and transcriptional activation in the binding of lactoferrin to DNA. *Nature* 373: 721-724

153. Hekman, A. 1971. Association of lactoferrin with other proteins, as demonstrated by changes in electrophoretic mobility. *Biochim. Biophys. Acta* 251:380-387.

154. Hentges, D.J., Marsh, W.W., Petschow, B.W., Thal, W.R., and Carter, M.K. 1992. Influence of infant diets on the ecology of the intestinal tract of human flora-associated mice. *J Ped Gastroenterol. Nutr.* 14:146-152.

155. Hirai, Y., Kawakata, N., Satoh, K., Ikeda, Y., Hisayasu, S., Orimo, H., and Yoshino, Y. 1990. Concentrations of lactoferrin and iron in human milk at different stages of lactation. *J. Nutr. Sci. Vitaminol. (Tokyo)* 36:531-544.

156. Hoek, K.S., Milne, J.M., Grieve, P.A., Dionysius, D.A., and Smith, R. 1997. Antibacterial activity in bovine lactoferrin-derived peptides. *Antimicrob. Agents Chemother.* 41:54-59.

157. Holmgren, J., Svennerholm, A.M., and Ahren, C. 1981. Nonimmunoglobulin fraction of human milk inhibits bacterial adhesion (hemagglution) and enterotoxin binding of *Escherichia coli* and *Vibrio cholerae*. *Infect. Immun.* 33:136-141.

158. Honour, P., and Dolby, J.M. 1979. Bacteriostasis of *Escherichia coli* by milk. III. The activity and stability of early, transitional and mature human milk collected locally. *J. Hyg. (Lond.).* 83: 243-254.

159. Hu, W.L., Mazurier, J., Montreuil, J., and Spik, G. 1990. Isolation and partial characterization of a lactotransferrin receptor from mouse intestinal brush border. *Biochemistry* 29:535-541.

160. Husson, M.O., Legrand, D., Spik, G., and Leclerc, H. 1993. Iron acquisition by *Helicobacter pylori*: importance of human lactoferrin. *Infect. Immun.* 61:2694-2697.

161. Hutchens, T.W., Magnuson, J.S., and Yip, T.T. 1989. Rapid purification of porcine colostral whey lactoferrin by affinity chromatography on single-stranded DNA-agarose. Characterization, amino acid composition, and N-terminal aminoacid sequence. *Biochim. Biophys. Acta* 999:323-329.

162. Iigo, M., Kuhara, T., Ushida, Y., Sekine, K., Moore, M.A., and Tsuda, H. 1999. Inhibitory effects of bovine lactoferrin on colon carcinoma 26 lung metastasis in mice. *Clin. Exp. Metastasis* 17:35-40.

163. Illingworth, D.S., Walter, K.S., Griffiths, P.L., and Barclay, R. 1993. Siderophore production and iron-regulated envelope proteins of *Helicobacter pylori*. *Zentralbl. Bakteriol.* 280:113-119.

164. Isamida, T., Tanaka, T., Omata, Y., Yamauchi, K., Shimazaki, K., and Saito, A. 1998. Protective effect of lactoferricin against *Toxoplasma gondii* infection in mice. *J. Vet. Med. Sci.* 60:241-244.

165. Izhar, M., Nuchamowitz, Y., and Mirelman, D. 1982. Adherence of *Shigella flexneri* to guinea pig intestinal cells is mediated by a mucosal adhesin. *Infect. Immun.* 35:1110-1118.

166. Janssen, P.T., and van Bijsterveld, O.P. 1982. Immunochemical determination of human tear lysozyme (muramidase) in keratoconjunctivitis sicca. *Clin. Chim. Acta.* 121:251-260.

167. Janssen, P.T., and van Bijsterveld, O.P. 1983. A simple test for lacrimal gland function: a tear lactoferrin assay by radial immunodiffusion. *Grafes. Arch. Clin. Exp. Opthalmol.* 220:171-174.
168. Jarosik, G.P., Land, C.B., Duhon, P., Chandler, R., and Mercer, T. 1998. Acquisition of iron by *Gardnerella vaginalis. Infect. Immun.* 66:5041-5047.
169. Johanson, B. 1960. Isolation of an iron-containing red protein from human milk. *Acta Chem. Scand.* 14:510-512.
170. Jolles, J., Donda, A., Amiguet, P., and Jolles, P. 1984. Mare lactotransferrin: purification, analysis and N-terminal sequence determination. *FEBS Lett.* 176:185-188.
171. Kaltas, S., Andersson, M., Edwardsson, S., Forsgren, A., and Naidu, A.S. 1991. Human lactoferrin binding to *Porphyromonas gingivalis, Prevotella intermedia,* and *Prevotella melaninogenica. Oral Microbiol. Immunol.* 6:350-355.
172. Kalfas, S., Tigyi, Z., Wikstrom, M., and Naidu, A.S. 1992. Laminin binding to *Prevotella intermedia. Oral Microbiol. Immunol.* 7:235-239.
173. Kalmar, J.R., and Arnold, R.R. 1988. Killing of *Actinobacillus actinomycetemcomitans* by human lactoferrin. *Infect. Immun.* 56:2552-2557.
174. Kang, J.H., Lee, M.K., Kim, K.L., and Hahm, K.S. 1996. Structure-biological activity relationships of 11-residue highly basic peptide segment of bovine lactoferrin. *Int. J. Pept. Protein Res.* 48:357-363.
175. Kawakami, H., and Lönnerdal, B. 1991. Isolation and function of a receptor for human lactoferrin in human fetal intestinal brush-border membranes. *Am. J. Physiol.* 261:G841-846.
176. Kawasaki, Y., Isoda, H., Shinmoto, H., Tanimoto, M., Dosako, S., Idota, T., and Nakajima, I. 1993. Inhibition by kappa-casein glycomacropeptide and lactoferrin of influenza virus hemagglutination. *Biosci. Biotechnol. Biochem.* 57:1214-1215.
177. Keen, C.L., Fransson, G.B., and Lonnerdal, B. 1984. Supplementation of milk with iron bound to lactoferrin using weanling mice: II. Effects on tissue manganese, zinc and copper. *J. Ped. Gastroenterol. Nutr.* 3:256-261.
178. Kennedy, R.S., Konok, G.P., Bounous, G., Baruchel, S., and Lee, T.D.G. 1995. The use of a whey protein concentrate in the treatment of patients with metastatic carcinoma:A phase I-II clinical study. *Anticancer Res.* 15:2643-2650.
179. Khin, M.M., Ringner, M., Aleljung, P., Wadstrom, T., and Ho, B. 1996. Binding of human plasminogen and lactoferrin by *Helicobacter pylori* coccoid forms. *J. Med. Microbiol.* 45:433-439.
180. Kijlstra, A.S., Jeurissen, H.M., and Koning, K.M. 1983. Lactoferrin levels in normal human tears. *Br. J. Opthalmol.* 67:199-202.
181. Kinkade, J.M., Miller III, W.W.K., and Segars, F.M. 1976. Isolation and characterization of murine lactoferrin. *Biochim. Biophys. Acta* 446:407-418.
182. Kioke, D., and Makino I. 1993. Protective effect of lactoferrin on caerulein-induced acute pancreatitis in rats. *Digestion.* 54:84-90.
183. Kirkpatrick, C.H., Green, I., Rich, R.R., and Schade, A.L. 1971. Inhibition of growth of *Candida albicans* by iron-unsaturated lactoferrin: relation to host-defense mechanisms in chronic mucocutaneous candidiasis. *J. Infect. Dis.* 124:539-544.
184. Kishore, AR., Erdei, J., Naidu, S.S., Falsen, E., Forsgren, A., Naidu, A.S. 1991. Specific binding of lactoferrin to *Aeromonas hydrophila. FEMS Microbiol. Lett.* 67:115-119.
185. Klasing, K.C. 1984. Effect of inflammatory agents and interleukin 1 on iron and zinc metabolism. *Am. J. Physiol.* 247:R901-R904.
186. Klebanoff, S.J., and Waltersdorph, A.M. 1990. Prooxidant activity of transferrin and lactoferrin. *J. Exp. Med.* 172:1293-1303.
187. Klempner, M.S, Dinarello, C.A., and Gallin, J.I. 1978. Human leukocytic progen induces release of specific granule contents from human neutrophils. *J. Clin. Invest.* 61:1330-1336.
188. Koivuranta-Vaara, P., Banda, D., and Goldstein, M. 1987. Bacterial-lipopolysaccharide-induced release of lactoferrin from human polymorphonuclear leukocytes: role of monocyte-derived tumor necrosis factor alpha. *Infect. Immun.* 55: 2956-61
189. Kokriakov, V.N., Alishina, G.M., Slepenkov, S.V., Lakovleva, M.F., and Pigarevskii, V.E. 1988. O Stepeni strukturnoi gomologii laktokerrinov moloka i neitrofil'nykh granulositov. *Biokhimiia* 53:1837-1843.
190. Krysteva, M., Mazurier, J., Spik, G., and Montreuil, J. 1975. Comparative study on histidine modification by diethylpyrocarbonate in human serotransferrin and lactotransferrin. *FEBS Lett.* 56:337-340.
191. Kulics, J. and Kijlstra, A. 1987. The effect of lactoferrin on complement mediated modulation of immune complex size. *Immunol. Lett.* 14:349-353.
192. Kume, S., and Tanabe, S. 1994. Effect of twinning and supplemental iron-saturated lactoferrin on iron status of newborn calves. *J. Dairy Sci.* 77:3118-3123.

193. Kume, S., and Tanabe, S. 1995. Effect of supplemental lactoferrin with ferrous iron on iron status of newborn calves. *J. Dairy Sci.* 79:459-464.

194. Lampreave, F., Pineiro, A., Brock, J.H., Castillo, H., Sanchez, L., and Calvo, M. 1990. Interaction of bovine lactoferrin with other proteins of milk whey. *Int. J. Biol. Macromol.* 12:2-5.

195. Lash, J.A., Coates, T.D., Lauze, J., Baehner, R.L., and Boxer, L.A. 1983. Plasma lactoferrin reflects granulocyte activation *in vivo. Blood* 61:885-888.

196. Lassiter, M.O., Newsome, A.L., Sams, L.D., and Arnold, R.R. 1987. Characterization of lactoferrin interaction with *Streptococcus mutans. J. Dent. Res.* 66:480-485.

197. Lassiter, M.O. 1990. Ethylenediaminetetraacetate (EDTA): Influence on the antimicrobial activity of human lactoferrin against *Escherichia coli* and *Streptococcus mutans*. Doctoral Thesis. Department of Pathology, School of Medicine. Atlanta, Georgia: Emory University.

198. Leblebicioglu, B., Lim, J.S., Cario, A.C., Beck, F.M., and Walters, J.D. 1996. pH changes observed in the inflamed gingival crevice modulate human polymorphonuclear leukocyte activation *in vitro. J. Periodontol.* 67:472-477.

199. Lee, W.J., Farmer, J.L., Hilty, M., and Kim, Y.B. 1998. The protective effects of lactoferrin feeding against endotoxin lethal shock in germfree piglets. *Infect. Immun.* 66:1421-1426.

200. Leffell, M.S., and. Spitznagel, J.K. 1975. Fate of human lactoferrin and myeloperoxidase in phagocytosing human neutrophils: effects of immunoglobulin G subclasses and immune complexes coated on latex beads. *Infect. Immun.* 12:813-820.

201. Li, Y.M., Tan, A.X., and Vlassara, H. 1995. Antibacterial activity of lysozyme and lactoferrin is inhibited by binding of advanced glycation-modified proteins to a conserved motif. *Nat. Med.* 1:1057-1061.

202. Lima, M.F., and Kierszenbaum, F. 1985. Lactoferrin effects on phagocytic cell function. I. Increased uptake and killing of an intracellular parasite by murine macrophages and human monocytes. *J. Immunol.* 134:4176-4183.

203. Lima, M.F., and Kierszenbaum, F. 1987. Lactoferrin effects of phagocytic cell function. II. The presence of iron is required for the lactoferrin molecule to stimulate intracellular killing by macrophages but not to enhance the uptake of particles and microorganisms. *J. Immunol.* 139:1647-1651.

204. Lima, M.F., Beltz, L.A., and Kierszenbaum, F. 1988. *Trypanosoma cruzi*: a specific surface marker for the amastigote form. *J. Protozool.* 35: 108-110.

205. Longhi,C., Conte, M.P., Seganti, L., Polidoro, M., Alfsen, A., and Valenti, P. 1993. Influence of lactoferrin on the entry process of *Escherichia coli* HB101 (pRI203) in HeLa cells. *Med. Microbiol. Immunol.* 182:25-35.

206. Lönnerdal, B. 1989. Trace element absorption in infants as a foundation to setting upper limits for trace elements in infant formulas. *J. Nutr (Suppl.)* 119:1839-1844.

207. Loughlin, K.R., and Gittes, R.F., Partridge, D., and Stelos, P. 1987. The relationship of lactoferrin to the anemia of renal cell carcinoma. *Cancer* 59:566-571.

208. Lu, L., Hangoc, G., Oliff, A., Chen, L.T., Shen, R.N., and Broxmeyer, H.E. 1987. Protective influence of lactoferrin on mice infected with the polycythemia-inducing strain of Friend virus complex. *Cancer Res.* 47:4184-4188.

209. Lugtenberg, B.,and van Alphen, L. 1983. Molecular architecture and functioning of the outer membrane of *Escherichia coli* and other gram-negative bacteria. *Biochim. Biophys. Acta* 737:51-115.

210. Luqmani, Y.A., Campbell, T.A., Bennett, C., Coombes, R.C., and Paterson, I.M. 1991. Expression of lactoferrin in human stomach. *Int. J. Cancer* 49:684-687.

211. Machnicki, M., Zimecki, M., and Zagulski, T. 1993. Lactoferrin regulates the release of tumour necrosis factor alpha and interleukin 6 *in vivo. Int. J. Exp. Pathol.* 74: 433-439.

212. Malmquist, J., and Johansson, B.G. 1971. Interaction of lactoferrin with agar gels and with trypan blue. *Biochim. Biophys. Acta* 236:38-46.

213. Mann, D.M., Romm, E., and Migliorini, M. 1994. Delineation of the glycosaminoglycan binding site in the human inflammatory response protein lactoferrin. *J. Biol. Chem.* 269:23661-23667.

214. Marchetti, M., Longhi, C., Conte, M.P., Pisani, S., Valenti, P., and Seganti, L. 1996. Lactoferrin inhibits herpes simplex virus type 1 adsorption to Vero cells. *Antiviral. Res.* 29:221-231.

215. Mason, D.Y., and Taylor, C.R. 1978. Distribution of transferrin, ferritin, and lactoferrin in human tissues. *J. Clin. Pathol.* 31:316-327.

216. Masson, P.L., Carbonara, A.O., and Heremans, J.F. 1965a. Studies on the proteins of human saliva. *Biochim. Biophys. Acta* 107:485-500.

217. Masson, P.L., Heremans, J.F., and Prignot, J., 1965b. Immunohistochemical localization of the iron-binding protein lactoferrin in human bronchial glands. *Experimentia (Basel)* 21:604-610.

218. Masson, P.L., Carbonara, A.O., and Heremans, J.F. 1965c. Studies on the proteins of human bronchial secretions. *Biochim. Biophys. Acta* 111:466.

219. Masson, P.L., Heremans, J.F., Prignot, J.J., and Wauters, G. 1966. Immunohistochemical localization and bacteriostatic properties of an iron-binding protein from bronchial mucus. *Thorax* 21:538-544.

220. Masson, P.L., Heremans, J.F., and Schonne, E. 1969. Lactoferrin, an iron-binding protein in neutrophilic leukocytes. *J. Exp. Med.* 130:643-658.

221. Masson, P.L., and Heremans, J.F. 1968. Metal-combining properties of human lactoferrin (red milk protein). 1. The involvement of bicarbonate in the reaction. *Eur J Biochem.* 6: 579-584.

222. Masson, P.L., and Heremans, J.F. 1971. Lactoferrin in milk from different species. *Comp. Biochem. Physiol.* 39B:119-129.

223. Mattsby-Baltzer, I., Roseanu, A., Motas, C., Elverfors, J., Engberg, I., and Hanson, L.A. 1996. Lactoferrin or a fragment thereof inhibits the endotoxin-induced interleukin-6 response in human monocytic cells. *Pediatr. Res.* 40:257-262.

224. Mazurier, J., and Spik, G. 1980. Comparative study of the iron-binding properties of human transferrins. I. Complete and sequential iron saturation and desaturation of the lactotransferrin. *Biochim. Biophys. Acta* 629:399-407.

225. Mazurier, J., Léger, D., Tordera, V., Montreuil, J., and Spik, G. 1981. Comparative study of the iron-binding properties of transferrins: Differences in the involvement of histidine residues as revealed by carbethoxylation. *Eur. J. Biochem.* 119:537-543.

226. McCormick, J.A., Markey, G.M., and Morris, T.C.M. 1991. Lactoferrin-inducible monocyte cytotoxicity for K562 cells and decay of natural killer lymphocyte cytotoxicity. *Clin. Exp.Immunol.* 83:154-156.

227. McMillan, J.A., Oski, F.A., Lourie, G., Tomarelli, R.M., and Landaw, S.A. 1977. Iron absorption from human milk, simulated human milk, and proprietary formulas. *Pediatrics* 60:896-900.

228. Menozzi, F.D., Ganteiz, C., and Locht, C. 1991. Identification and purification of transferrin and lactoferrin-binding proteins of *Bordetella pertussis* and *Bordetella bronchoseptica*. *Infect. Immun.* 59:3982-3988.

229. Metz-Boutigue, M.H., Jolles, J., Mazurier, J., Schoentgen, F., Legrand, D., Spik, G., Montreuil, J., and Jolles, P. 1984. Human lactotransferrin: amino acid sequence and structural comparisons with other transferrins. *Eur. J. Biochem.* 145:659-676.

230. Mickelsen, P.A., Blackman, E., and Sparling, P.F. 1982. Ability of *Neisseria gonorrhoeae, Neisseria meningitidis* and commensal *Neisseria* species to obtain iron from lactoferrin. *Infect. Immun.* 35:915-920.

231. Miedzobrodzki, J., Naidu, A.S., Watts, J.L., Ciborowski, P., Palm, K., and Wadstrom, T. 1989. Effect of milk on fibronectin and collagen type-I binding to *Staphylcoccus aureus* and coagulase-negative staphylococci isolated from bovine mastitis. *J. Clin. Microbiol.* 27:540-544.

232. Miyauchi, H., Kaino, A., Shinoda, I., Fukuwatari, Y., and Hayasawa, H. 1997. Immunomodulatory effect of bovine lactoferrin pepsin hydrolysate on murine splenocytes and Peyer's patch cells. *J. Dairy Sci.* 80:2330-2339.

233. Miyazawa, K., Mantel, C., Lu, L., Morrison, D.C., and Broxmeyer, H.E. 1991. Lactoferrin-lipopolysaccharide interactions. Effect on lactoferrin binding to monocyte/macrophage-differentiated HL-60 cells. *J. Immunol.* 146:723-729.

234. Moguilevsky, N., Retegui, L.A., and Masson, P.L. 1985. Comparison of human lactoferrins from milk and neutrophilic leukocytes: relative molecular mass, isoelectric point, iron-binding properties and uptake by the liver. *Biochem. J.* 229:353-359.

235. Montreuil, J., Tonnelat, J., and Mullet, S. 1960. Preparation and properties of lactoferrin of human milk. *Biochim. Biophys. Acta* 45:413-421.

236. Moore, S.A., Anderson, B.F., Groom, C.R., Haridas, M., and Baker, E.N. 1997. Three-dimensional structure of diferric bovine lactoferrin at 2.8 A resolution. *J. Mol. Biol.* 274:222-236.

237. Muller, F., Holberg-Petersen, M., Rollag, H., Degre, M., Brandtzaeg, P., and Froland, S.S. 1992. Nonspecific oral immunity in individuals with HIV infection. *J. Acquir. Immune Defic. Syndr.* 5:46-51.

238. Naidu, A.S., Miedzobrodzki, J., Andersson, M., Nilsson, L.E., Forsgren, A., and Watts, J.L. 1990. Bovine lactoferrin binding to six species of coagulase-negative staphylococci isolated from bovine intramammary infections. *J. Clin. Microbiol.* 28:2312-2319.

239. Naidu, A.S., Miedzobrodzki, J., Musser, J.M., Rosdahl, V.T., Hedstrom, S.A., and Forsgren, A. 1991a. Human lactoferrin binding in clinical isolates of *Staphylococcus aureus*. *J. Med. Microbiol.* 34:323-328.

240. Naidu, S.S., Erdei, J., Czirok, E., Kalfas, S., Gado, I., Thoren, A., Forsgren, A., and Naidu, A.S. 1991b. Specific binding of lactoferrin to *Escherichia coli* isolated from human intestinal infections. *APMIS.* 99:1142-1150.

241. Naidu, A.S., Andersson, M., and Forsgren, A. 1992. Identification of a human lactoferrin-binding protein in *Staphylococcus aureus*. *J. Med. Microbiol.* 36:177-183.

242. Naidu, S.S., Svensson, U., Kishore, A.R., and Naidu, A.S. 1993. Relationship between antibacterial activity and porin binding of lactoferrin in *Escherichia coli* and *Salmonella typhimurium*. *Antimicrob. Agents Chemother.* 37:240-245.

243. Naidu, A.S., and Arnold, R.R. 1994. Lactoferrin interaction with salmonellae potentiates antibiotic susceptibility in vitro. *Diagn. Microbiol. Infect. Dis.* 20:69-75.

244. Naidu, A.S. and Arnold, R.R. 1997. Influence of lactoferrin on host-microbe interactions. In *Lactoferrin - Interactions and Biological Functions*, ed. T.W. Hutchens, and B. Lonnerdal, 259-275. Totowa, NJ: Humana Press.

245. Naidu, A.S., and Bidlack, W.R. 1998. Milk lactoferrin - Natural microbial blocking agent (MBA) for food safety. *Environ. Nutr. Interact.* 2:35-50.

246. Nichols, B.L., McKee, K.S., and Huebers, H.A. 1990. Iron is not required in the lactoferrin stimulation of thymidine incorporation into the DNA of rat crypt enterocytes. *Pediatr. Res.* 27:525-528.

247. Nicholson, H., Anderson, B.F., Bland, T., Shewry, S.C., Tweedie, J.W., and Baker, E.N. 1997. Mutagenesis of the histidine ligand in human lactoferrin: iron binding properties and crystal structure of the histidine-253 methionine mutant. *Biochemistry* 36:341-346.

248. Nikaido, H., and Vaara, M. 1985. Molecular basis of bacterial outer membrane permeability. *Microbiol. Rev.* 49:1-31.

249. Nikaido, H. 1989. Outer membrane barrier as a mechanism of antimicrobial resistance. *Antimicrob. Agents Chemother.* 33:1831-1836.

250. Nishiya, K. and Horwitz, P.A. 1982. Contrasting effects of lactoferrin on human lymphocyte and monocyte natural killer cell activity and antibody-dependent cell mediated cytotoxicity. *J. Immunol.* 129:2519-2523.

251. Nonnecke, B.J., and Smith, K.L. 1984a. Inhibition of mastitic bacteria by bovine milk apo-lactoferrin evaluated by *in vitro* microassay of bacterial growth. *J. Dairy Sci.* 67:606-613.

252. Nonnecke, B.J., and Smith, K.L. 1984b. Biochemical and antibacterial properties of bovine mammary secretion during mammary involution and at parturition. *J. Dairy Sci.* 67:2863-2872.

253. Norris, G.E., Andersson, B.F., Baker, E.N., Baker, H.M., Gartner, A.L., Ward, J., and Rumball, S.V. 1986. Preliminary crystallographic studies on bovine lactoferrin. *J. Mol. Biol.* 191:143-145.

254. Odell, E.W., Sarra, R., Foxworthy, M., Chapple, D.S., and Evans, R.W. 1996. Antibacterial activity of peptides homologous to a loop region in human lactoferrin. *FEBS Lett.* 382:175-178.

255. Okujo, N., Akiyama, T., Miyoshi, S., Shinoda, S., and Yamamoto, S. 1996. Involvement of vulnibactin and exocellular protease in utilization of transferrin- and lactoferrin-bound iron by *Vibrio vulnificus*. *Microbiol. Immunol.* 40:595-598.

256. Okutomi, T., Tanaka, T., Yui, S., Mikami, M., Yamazaki, M., Abe, S., and Yamaguchi, H. 1998. Anti-*Candida* activity of calprotectin in combination with neutrophils or lactoferrin. *Microbiol. Immunol.* 42:789-793.

257. Olofson, T., Olsson, I., Venge, P., and Elgefors, B. 1977. Serum myeloperoxidase and lactoferrin in neutropenia. *Scand. J. Haematol.* 18:73-80.

258. Olusanya, O., and Naidu, A.S. 1991. Influence of mucin on serum and connective tissue protein binding to *Staphylococcus aureus* isolated from nasal carriage and clinical sources. *Afr. J. Med. Sci.* 20:89-95.

259. Oram, J.D. and Reiter, B. 1968. Inhibition of bacteria by lactoferrin and other iron-chelating agents. *Biochim. Biophys. Acta* 170:351-365.

260. Osborne, T.B., and Campbell, G.F. 1900. The protein constituents of egg white. *J. Am. Chem. Soc.* 22:422-426.

261. Oseas, R., Yang, H.H., Baehner, R.L., and Boxer, L.A. 1981. Lactoferrin: a promoter of polymorphonuclear leukocyte adhesiveness. *Blood* 57:939-945.

262. Otto, B.R., Verweij-van Vught, A.M., and MacLaren, D.M. 1992. Transferrins and heme-compounds as iron sources for pathogenic bacteria. *Crit. Rev. Microbiol.* 18:217-233.

263. Palma, C., Serbousek, D., Torosantucci, A., Cassone, A., and Djeu, J.Y. 1992. Identification of a mannoprotein fraction from *Candida albicans* that enhances human polymorphonuclear leukocyte (PMNL) functions and stimulates lactoferrin in PMNL inhibition of candidal growth. *J. Infect. Dis.* 166:1103-1112.

264. Paulsson, M.A., Svensson, U., Kishore, A.R., and Naidu, A.S. 1993. Thermal behavior of bovine lactoferrin in water and its relation to bacterial interaction and antibacterial activity. *J. Dairy Sci.* 76:3711-3720.

265. Payne, K.D., Davidson, P.M., Oliver, S.P., and Christian, G.L. 1990. Influence of bovine lactoferrin on the growth of Listeria monocytogenes. *J. Food Prot.* 53:468-472.

266. Payne, K.D., Davidson, P.M., and Oliver, S.P. 1994. Comparison of EDTA and apo-lactoferrin with lysozyme on the growth of food-borne pathogenic and spoilage bacteria. *J. Food Prot.* 57:62-65.

267. Pelus, L.M., Broxmeyer, H.E., and Moore, M.A. 1981. Regulation of human myelopoiesis by prostaglandin E and lactoferrin. *Cell Tissue Kinet.* 14:515:526.

268. Pentecost, B.T., and Teng, C.T. 1987. Lactotransferrin is the major estrogen inducible protein of mouse uterine secretions. *J. Biol. Chem.* 262:10134-10139.
269. Perraudin, J.P., Prieels, J.P., and Leonis, J. 1974. Interaction between lysozyme and some lactoferrin complex in human milk. *Arch. Int. Physiol. Biochim.* 82: 1001.
270. Peterson, K.M., and Alderete, J.F. 1984. Iron uptake and increased intracellular enzyme activity follow host lactoferrin binding by *Trichomonas vaginalis* receptors. *J. Exp Med.* 160:398-410.
271. Pettersson, A., Maas, A., and Tommassen, J. 1994. Identification of the iroA gene product of *Neisseria meningitidis* as a lactoferrin receptor. *J. Bacteriol.* 176:1764-1766.
272. Querinjean, P., Masson, P.L., and Heremans, J.F. 1971. Molecular weight, single-chain structure and amino acid composition of human lactoferrin. *Eur. J. Biochem.* 20:420-425.
273. Rainard, P. 1986a. Bacteriostasis of *Escherichia coli* by bovine lactoferrin, transferrin and immunoglobulins (IgG1, IgG2, IgM) acting alone or in combination. *Vet. Microbiol.* 11:103-115.
274. Rainard, P. 1986b. Bacteriostatic activity of bovine milk lactoferrin against mastitic bacteria. *Vet. Microbiol.* 11:387-392.
275. Rainard, P. 1987. Bacteriostatic activity of bovine lactoferrin in mastitic milk. *Vet. Microbiol.* 13: 159-166.
276. Rainard, P. 1992. Binding of bovine lactoferrin to *Streptococcus agalactiae*. *FEMS Microbiol. Lett.* 77:235-239.
277. Rainard, P. 1993. Activation of the classical pathway of complement by binding of bovine lactoferrin to unencapsulated *Streptococcus agalactiae*. *Immunology* 79:648-652.
278. Raphael, G.D., Jeney, E.V., Baranuik, J.N., Kim, I., Meridith, S.D., and Kaliner, M.A. 1989. Pathophysiology of rhinitis: lactoferrin and lysozyme in nasal secretion. *J. Clin. Invest.* 84:1528-1535.
279. Regoeczi, E., Chindemi, P.A., Debanne, M.T., and Prieels, J.P. 1985. Lactoferrin catabolism in the rat liver. *Am.J. Physiol.* 248:G8-G14.
280. Regoeczi, E., Chindemi, P.A., and Hu, W.L. 1994. Transport of lactoferrin from blood to bile in the rat. *Hepatology* 19:1476-1482.
281. Reiter, B., and Oram, J.D. 1967. Bacterial inhibitors in milk and other biological fluids. *Nature* 216:328-330.
282. Reiter, B., Brock, J.H., and Steel, E.D. 1975. Inhibition of *Escherichia coli* by bovine colostrum and post-colostral milk. II. The bacteriostatic effect of lactoferrin on a serum susceptible and serum resistant strain of *E. coli*. *Immunology* 28:83-95.
283. Rejman, J.J., Hegarty, H.M., and Hurley, W.L. 1989. Purification and characterization of bovine lactoferrin from secretions of the involuting mammary gland: Identification of multiple molecular forms. *Comp. Biochem. Physiol.* 93B:929-934.
284. Roberts, A.K., Chierici, R., Hill, M.J., Volpata, S., and Vigi, V. 1992. Supplementation of an adapted formula with bovine lactoferrin: I. Effect on the infant fecal flora. *Acta Paediatr.* 81:119-124.
285. Rogers, H.J., and Synge, C. 1978. Bacteriostatic effect of human milk on *Escherichia coli*: the role of IgA. *Immunology* 34:19-28.
286. Rogers, T.B., Borresen, T., and Feeney, R.E. 1978. Chemical modification of the arginines in transferrins. *Biochemistry* 17:1105-1109.
287. Rosa, G., and Trugo, N.M. 1994. Iron uptake from lactoferrin by intestinal brush-border membrane vesicles of human neonates. *Braz. J. Med. Biol. Res.* 27:1527-1531.
288. Rumke, P., Visser, D., Kwa, H.G., and Hart, A.A.M.. 1971. Radioimmunoassay for lactoferrin in blood plasma of breast cancer patients, lactating, and normal women: prevention of false high levels caused by leakage from neutrophil leukocytes in vitro. *Folia Med. Netherland.* 14:156-168.
289. Rumke, P. 1974. The origin of immunoglobulins in semen. *Clin. Exp. Med.* 14:1068-1084.
290. Saito, H., Takase, M., Tamura, Y., Shimamura, S., and Tomita, M. 1994. Physicochemical and antibacterial properties of lactoferrin and its hydrolysate produced by heat treatment at acidic pH. *Adv. Exp. Med. Biol.* 357:219-226.
291. Salamah, A.A., and al-Obaidi, A.S. 1995a. *In vivo* and *in vitro* effects of lactoferrin on *Yersinia pseudotuberculosis*. *New Microbiol* 18:267-274.
292. Salamah, A.A., and al-Obaidi, A.S. 1995b. Effect of some physical and chemical factors on the bactericidal activity of human lactoferrin and transferrin against *Yersinia pseudotuberculosis*. *New Microbiol* 18:275-281.
293. Samaranayake, Y.H., Samaranayake, L.P., Wu, P.C., and So, M. 1997. The antifungal effect of lactoferrin and lysozyme on *Candida krusei* and *Candida albicans*. *APMIS* 105:875-883.
294. Sanchez, L., Ismail, M., Liew, F.Y., and Brock, J.H. 1996. Iron transport across Caco-2 cell monolayers. Effect of transferrin, lactoferrin and nitric oxide. *Biochim. Biophys. Acta* 1289:291-297.

295. Sato, N., Kurotaki, H., Ikeda, S., Daio, R., Nishinome, N., Mikami, T., and Matsumoto, T. 1999. Lactoferrin inhibits *Bacillus cereus* growth and heme analogs recover its growth. *Biol. Pharm. Bull.* 22:197-199.

296. Sato, R., Inanami, O., Tanaka, Y., Takase, M., and Naito, Y. 1996. Oral administration of bovine lactoferrin for treatment of intractable stomatitis in feline immunodeficiency virus (FIV)-positive and FIV-negative cats. *Am. J. Vet. Res.* 57:1443-1446.

297. Sawatzki, G., and Rich, I.N. 1989. Lactoferrin stimulates colony stimulating factor production *in vitro* and *in vivo*. *Blood Cells* 15: 371-385.

298. Schade, A.L., and Caroline, L. 1946. An iron-binding component in human blood plasma. *Science* 104:340.

299. Schafer, K.H. 1951. Elekrophoretische untersuchungen zum milcheiweissproblem. *Montsssschrift fur Kindrheilkunde* 99:69-71.

300. Schlabach, M.R. and Bates, G.W. 1975. The synergistic binding of anions and Fe(III) by transferrin. Implications for the interlocking sites hypothesis. *J. Biol. Chem.* 250:2182-2188.

301. Schryvers, A.B., and Morris, L.J. 1988. Identification and characterization of the transferrin receptor from *Neisseria meningitidis*. *Mol. Microbiol.* 2:281-288.

302. Schryvers, A.B. 1989. Identification of the transferrin- and lactoferrin-binding proteins in *Haemophilus influenzae*. *J. Med. Microbiol.* 29:121-130.

303. Schryvers, A.B., and Stojiljkovic, I. 1999. Iron acquisition systems in the pathogenic *Neisseria*. *Mol. Microbiol.* 32:1117-1123.

304. Schulz-Lell, G., Dorner, K., Oldigs, H.D., Sievers, E., and Schaub, J. 1991. Iron availability from an infant formula supplemented with bovine lactoferrin. *Acta Paediatr. Scand.* 80:155-158.

305. Sekine, K., Watanabe, E., Nakamura, J., Takasuka, N., Kim, D.J., Asamoto, M., Krutovskikh, V., Baba-Toriyama, H., Ota, T., Moore, M.A., Masuda, M., Sugimoto, H., Nishino, H., Kakizoe, T., and Tsuda, H. 1997a. Inhibition of azoxymethane-initiated colon tumor by bovine lactoferrin administration in F344 rats. *Jpn. J. Cancer Res.* 88:523-526.

306. Sekine, K., Ushida, Y., Kuhara, T., Iigo, M., Baba-Toriyama, H., Moore, M.A., Murakoshi, M., Satomi, Y., Nishino, H., Kakizoe, T., and Tsuda, H. 1997b. Inhibition of initiation and early stage development of aberrant crypt foci and enhanced natural killer activity in male rats administered bovine lactoferrin concomitantly with azoxymethane. *Cancer Lett.* 121:211-216.

307. Shau, H., Kim, A., Golub, S.H. 1992. Modulation of natural killer and lymphokine-activated killer cell cytotoxicity by lactoferrin. *J. Leukoc. Biol.* 51:343-349.

308. Shimazaki, K., and Nishio, N. 1991. Interacting properties of bovine lactoferrin with immobilized Cibacron blue F3GA in column chromatography. *J. Dairy Sci.* 74:404-408.

309. Shimazaki, K., Kawaguchi, A., Sato, T., Ueda, Y., Tommimura, T., and Shimamura, S. 1993. Analysis of human and bovine milk lactoferrins by rotafor and chromatofocusing. *Int. J. Biochem.* 25:1653-1658.

310. Shimazaki, K., Nam, M.S., Harakawa, S., Tanaka, T., Omata, Y., Saito, A., Kumura, H., Mikawa, K., Igarashi, I., and Suzuki, N. 1996. Monoclonal antibody against bovine lactoferricin and its epitopic site. *J. Vet. Med. Sci.* 58:1227-1229.

311. Shimizu, K., Matsuzawa, H., Okada, K., Tazume, S., Dosako, S., Kawasaki, Y., Hashimoto, K., and Koga, Y. 1996. Lactoferrin-mediated protection of the host from murine cytomegalovirus infection by a T-cell-dependent augmentation of natural killer cell activity. *Arch. Virol.* 141:1875-1889.

312. Shin, K., Yamauchi, K., Teraguchi, S., Hayasawa, H., Tomita, M., Otsuka, Y., Yamazaki, S. 1998. Antibacterial activity of bovine lactoferrin and its peptides against enterohaemorrhagic *Escherichia coli* O157:H7. *Lett. Appl. Microbiol.* 26:407-411.

313. Shinmoto, H., Kobori, M., Tsushida, T., and Shinohara, K. 1997. Competitive ELISA of bovine lactoferrin with bispecific monoclonal antibodies. *Biosci. Biotechnol. Biochem.* 61:1044-1046.

314. Smith, K.L., Conrad, H.R., and Porter, R.M. 1971. Lactoferrin and IgG immunoglobulins from involuted bovine mammary glands. *J Dairy Sci.* 54:1427-1435.

315. Sojar, H.T., Hamada, N. and Genco, R.J. 1998. Structures involved in the interaction of *Porphyromonas gingivalis* fimbriae and human lactoferrin. *FEBS Lett.* 422:205-208.

316. Sørensen, M., and Sørensen, S.P.L. 1939. The proteins in whey. *CR. Trav. Lab. Carlsberg.* 23:55-99.

317. Soukka, T., Lumikari, M., and Tenovuo, J. 1991. Combined inhibitory effect of lactoferrin and lactoperoxidase system on the viability of *Streptococcus mutans*, serotype c. *Scand. J. Dent. Res.* 99: 390-396.

318. Spik, G., Bayard, B., Fournet, B., Strecker, G., Bouquelet, S., and Montreuil, J. 1975. Studies on glycoconjugates. LXIV. Complete structure of two carbohydrate units of human serotransferrin. *FEBS Lett.* 50:296-299.

319. Spik, G., Strecker, G., Fournet, B., Bouquelet, S., Montreuil, J., Dorland, L., Van Halbeek, H., and Vliegenthart, J.F.G. 1982. Primary structure of glycans from human lactotransferrin. *Eur. J. Biochem.* 121:413-419.

320. Staggs, T.M., Greer, M.K., Baseman, J.B., Holt, S.C., and Tyron, V.V. 1994. Identification of lactoferrin-binding proteins from *Treponema pallidum* subspecies *pallidum* and *Treponema denticola*. *Mol. Microbiol.* 12:613-619.

321. Stuart, J., Norrell, S., and Harrington, J.P. 1984. Kinetic effect of human lactoferrin on the growth of *Escherichia coli* O111. *Int. J. Biochem.* 16:1043-1047.

322. Superti, F., Ammendolia, M.G., Valenti, P., and Seganti, L. 1997. Antirotaviral activity of milk proteins: lactoferrin prevents rotavirus infection in the enterocyte-like cell line HT-29. *Med. Microbiol. Immunol. (Berl.)* 186: 83-

323. Swart, P.J., Kuipers, M.E., Smit, C., Pauwels, R., deBethune, M.P., de Clercq, E., Meijer, D.K., and Huisman, J.G. 1996. Antiviral effects of milk proteins: acylation results in polyanionic compounds with potent activity against human immunodeficiency virus types 1 and 2 *in vitro*. *AIDS Res. Hum. Retroviruses.* 12:769-775.

324. Szuchet-Derechin, S., and Johnson, P. 1965. The 'Albumin' fraction of bovine milk-II. The isolation and physicochemical properties of the red proteins. *Eur. Polymer J.* 1:283-291.

325. Tabak, L., Mandel, I.D., Herrera, M., and Baurmash, H. 1978. Changes in lactoferrin and other proteins in a case of chronic recurrent parotitis. *J. Oral Path.* 1:97-99.

326. Tachezy, J., Kulda, J., Bahnikova, I., Suchan, P., Razga, J., and Schrevel, J. 1996. *Tritrichomonas foetus:* iron acquisition from lactoferrin and transferrin. *Exp. Parasitol.* 83:216-228.

327. Tanaka, T., Omata, Y., Saito, A., Shimazaki, K., Igarashi, I., and Suzuki, N. 1996. Growth inhibitory effects of bovine lactoferrin to *Toxoplasma gondii* parasites in murine somatic cells. *J. Vet. Med. Sci.* 58: 61-65.

328. Tashima, K.T., Carroll, P.A., Rogers, M.B., and Calderwood, S.B. 1996. Relative importance of three iron-regulated outer membrane proteins for *in vivo* growth of *Vibrio cholerae*. *Infect. Immun.* 64:1756-1761.

329. Teng, C. T., Walker, M. P., Bhattacharyya, S. N., Klapper, D. G., DiAugustine, R. P., and McLachlan, J. A. 1986. Purification and properties of an oestrogen-stimulated mouse uterine glycoprotein (approx. 70 kDa). *Biochem. J.* 240:413-422.

330. Teraguchi, S., Shin, K., Ozawa, K., Nakamura, S., Fukuwatari, Y., Tsuyuki, S., Namihira, H., and Shimamura, S. 1995a. Bacteriostatic effect of orally administered bovine lactoferrin on proliferation of *Clostridium* species in the gut of mice fed bovine milk. *Appl. Environ. Microbiol.* 61:501-506.

331. Teraguchi, S., Shin, K., Ogata, T., Kingaku, M., Kaino, A., Miyauchi, H., Fukuwatari, Y., and Shimamura, S. 1995b. Orally administered bovine lactoferrin inhibits bacterial translocation in mice fed bovine milk. *Appl. Environ. Microbiol.* 61:4131-4134.

332. Teraguchi, S., Shin, K., Fukuwatari, Y., and Shimamura, S. 1996. Glycans of bovine lactoferrin function as receptors for the type 1 fimbrial lectin of *Escherichia coli*. *Infect. Immun.* 64:1075-1077.

333. Teuwissen, B., Masson, P.L., Osinski, P., and Heremans, J.F. 1972. Metal-combining properties of human lactoferrin. The possible involvement of tyrosyl residues in the binding sites. Spectrophotometric titration. *Eur. J. Biochem.* 31:239-245.

334. Thomas, A.S., and Fell, L.R. 1985. Effect of ACTH and oxytocin treatment on lactoferrin and citrate in cows' milk. *J. Dairy Res.* 52:379-389.

335. Tigyi, Z., Kishore, A.R., Maeland, J.A., Forsgren, A., and Naidu, A.S. 1992. Lactoferrin-binding proteins in *Shigella flexneri*. *Infect. Immun.* 60:2619-2626.

336. Todhunter, D.A., Smith, K.L., and Schoenberger, P.S. 1985. *In vitro* growth of mastitis-associated strepto-cocci in bovine mammary secretions. *J. Dairy Sci.* 68:2337-2346.

337. Toma, C., Honma, Y., and Iwanaga, M. 1996. Effect of *Vibrio cholerae* non-O1 protease on lysozyme, lactoferrin and secretory immunoglobulin A. *FEMS Microbiol. Lett.* 135:143-147.

338. Tomita, M., Bellamy, W., Takase, M., Yamauchi, K., Wakabayashi, H., and Kawase, K. 1991. Potent antibacterial peptides generated by pepsin digestion of bovine lactoferrin. *J. Dairy Sci.* 74:4137-4142.

339. Tomita, M., Takase, M., Wakabayashi, H., and Bellamy, W. 1994. Antimicrobial peptides of lactoferrin. *Adv. Exp. Med. Biol.* 357: 209-218.

340. Tomita, S., Shirasaki, N., Hayashizaki, H., Matsuyama, J., Benno, Y., and Kiyosawa, I. 1998. Binding char-acteristics of bovine lactoferrin to the cell surface of *Clostridium* species and identification of the lacto-ferrin-binding protein. *Biosci. Biotechnol. Biochem.* 62:1476-1482.

341. Tourville, D.R., S.S. Ogra, J. Lippes, and T.B. Tomassi Jr. 1970. The human female reproductive tract:immunohistological localization of gammaA, gammaG, gammaM "secretory piece" and lactoferrin. *Am. J. Obstet. Gynecol.* 108:1102-

342. Trumpler, U., Straub, P.W., and Rosenmund, A. 1989. Antibacterial prophylaxis with lactoferrin in neu-tropenic patients. *Eur. J. Clin. Microbiol. Infect. Dis.* 8:310-313.

343. Tryon, V.V., and Baseman, J.B. 1987. The acquisition of human lactoferrin by *Mycoplasma pneumoniae*. *Microb. Pathog.* 3:437-443.

344. Tsuji, S., Y. Hirata, and K. Matsuoka. 1989. Two apparent molecular forms of bovine lactoferrin. *J. Dairy Sci.* 72:1130-1136.

345. Tuccari, G., and G. Barresi. 1985. Immunohistochemical demonstration of lactoferrin in follicular adenomas and thyroid carcinomas. *Virchows Arch. A.* 406:67-74.

346. Tuccari, G., Barresi, G., Arena, F., and Inferrera, C. 1989. Immunocytochemical detection of lactoferrin in human gastric carcinomas and adenomas. *Arch. Pathol. Lab. Med.*113:912-915.

347. Turchany, J.M., Aley, S.B., and Gillin, F.D. 1995. Giardicidal activity of lactoferrin and N-terminal peptides. *Infect. Immun.* 63:4550-4552.

348. van Berkel, P.H., Geerts, M.E., van Veen, H.A., Mericskay, M., de Boer, H.A., and Nuijens, J.H..1997.N-terminal stretch *Arg2, Arg3, Arg4* and *Arg5* of human lactoferrin is essential for binding to heparin, bacterial lipopolysaccharide, human lysozyme and DNA. *Biochem. J.* 15:145-451.

349. Van Halbeck, H., Dorland, L., Vliegenthar, J.F.G., Spil, G., Cheron, A., and Montreuil, J. 1981. Structure determination of two oligomannoside-type glycopeptides obtained from bovine lactotransferrin, by 500 MHz 1H-NMR spectroscopy. *Biochim. Biophys. Acta* 675:293-296.

350. Van Snick, J., Masson, P.L., and Heremans, J.F. 1974. The involvement of lactoferrin in the hyposideremia of acute inflammation. *J. Exp. Med.* 140:1068-1084.

351. Van Snick, J., and Masson, P.L. 1976. The binding of human lactoferrin to mouse peritoneal cells. *J. Exp. Med.* 144:1568-1580.

352. Visca, P., Berlutti, F., Vittorioso, P., Dalmastri, C., Thaller, M.C., and Valenti, P. 1989. Growth and adsorption of *Streptococcus mutans* 6715-13 to hydroxyapatite in the presence of lactoferrin. *Med. Microbiol. Immunol. (Berl)* 178:69-79.

353. Visca, P., Dalmastri, C., Verzili, D., Antonini, G., Chiancone, E., Valenti, P. 1990. Interaction of lactoferrin with *Escherichia coli* cells and correlation with antibacterial activity. *Med. Microbiol. Immunol. (Berl)* 179:323-333.

354. Wada, T., Aiba, Y., Shimizu, K., Takagi, A., Miwa, T., and Koga, Y. 1999. The therapeutic effect of bovine lactoferrin in the host infected with *Helicobacter pylori*. *Scand. J. Gastroenterol.* 34:238-243.

355. Wakabayashi, H., Abe, S., Okutomi, T., Tansho, S., Kawase, K., and Yamaguchi, H. 1996. Cooperative anti-Candida effects of lactoferrin or its peptides in combination with azole antifungal agents. *Microbiol. Immunol.* 40:821-825.

356. Wang, C.S., Chan, W.Y., and Kloer, U.II. 1984. Comparative studies on the chemical and immunochemical properties of human milk, human pancreatic juice and bovine milk lactoferrin. *Comp. Biochem. Physiol.* (B) 78:575-580.

357. Wang, D., Pabst, K.M., Aida, Y., and Pabst, M.J. 1995. Lipopolysaccharide-inactivating activity of neutrophils is due to lactoferrin. *J. Leukoc. Biol.* 57:865-874.

358. Watanabe, T., Nagura, H., Watanabe, K., and Brown, W.R. 1984. The binding of human milk lactoferrin to immunoglobulin A. *FEBS Lett.* 168:203-207.

359. Weinberg, E.D. 1975. Nutritional immunity. Host's attempt to withold iron from microbial invaders. *JAMA.* 231 (1): 39-41.

360. Wichmann, L., Vaalasti, A., Vaalasti, T., and Tuohimaa, P. 1989. Localization of lactoferrin in the male reproductive tract. *Int. J. Androl.* 12:179-186.

361. Williams, P.H., and Carbonetti, N.H. 1986. Iron, siderophores, and the pursuit of virulence: independence of the aerobactin and enterochelin iron uptake systems in *Escherichia coli. Infect. Immun.* 51:942-947.

362. Wilson, M.E., Vorhies, R.W., Andersen, K.A., and Britigan, B.E. 1994. Acquisition of iron from transferrin and lactoferrin by the protozoan *Leishmania chagasi. Infect. Immun.* 62:3262-3269.

363. Windle, J.J., Wiersema, A.K., Clark, J.R., and Feeney, R.E. 1963. *Biochemistry* 2:1341-1346.

364. Wu, H.F., Monroe, D.M., and Church, F.C. 1995. Characterization of the glycosaminoglycan binding region of lactoferrin. *Arch. Biochem. Biophys.* 317:85-92.

365. Yamauchi, K., Tomita, M., Giehl, T.J., and Ellison, R.T. 1993. Antibacterial activity of lactoferrin and a pepsin-derived lactoferrin peptide fragment. *Infect. Immun.* 61:719-728.

366. Yamauchi, K., Wakabayashi, H., Hashimoto, S., Teraguchi, S., Hayasawa, H., and Tomita, M. 1998. Effects of orally administered bovine lactoferrin on the immune system of healthy volunteers. *Adv. Exp. Med. Biol.* 443:261-265.

367. Yang, H., Kooi, C.D., and Sokol, P.A. 1993. Ability of *Pseudomonas pseudomallei* malleobactin to acquire transferrin-bound, lactoferrin-bound, and cell-derived iron. *Infect. Immun.* 61:656-662.

368. Yi, M., Kaneko, S., Yu, D.Y., and Murakami, S. 1997. Hepatitis C virus envelope proteins bind lactoferrin. *J. Virol.* 71:5997-6002.

369. Yoo, Y.C., Watanabe, S., Watanabe, R., Hata, K., Shimazaki, K., and Azuma, I. 1997. Bovine lactoferrin and lactoferricin, a peptide derived from bovine lactoferrin, inhibit tumor metastasis in mice. *Jpn. J. Cancer Res.* 88:184-190.

370. Yoshida, S. 1989. Preparation of lactoferrin by hydrophobic interaction chromatography from milk acid whey. *J. Dairy Sci.* 72:1446-1450.
371. Yoshida, S., and Ye-Xiuyun. 1991. Isolation of lactoperoxidase and lactoferrins from bovine milk acid whey by carboxymethyl cation exchange chromatography. *J. Dairy Sci.* 74:1439-1444.
372. Zagulski, T., Lipinski, P., Zagulska, A., Broniek, S., and Jarzabek, Z. 1989. Lactoferrin can protect mice against a lethal dose of *Escherichia coli* in experimental infection *in vivo. Br. J. Exp. Pathol.* 70:697-704.
373. Zhao, X.Y., and Hutchens, T.W. 1994. Proposed mechanisms for the involvement of lactoferrin in the hydrolysis of nucleic acids. *Adv. Exp. Med. Biol.* 357: 271-278.
374. Ziere, G.J., van Dijk, M.C., Bijsterbosch, M.K., and van Berkel, T.J. 1992. Lactoferrin uptake by the rat liver. Characterization of the recognition site and effect of selective modification of arginine residues. *J. Biol. Chem.* 267:11229-11235.
375. Zimecki, M., Mazurier, J., Machnicki, M., Wieczorek, Z., Montreuil, J., and Spik.G. 1991. Immunostimulatory activity of lactotransferrin and maturation of CD4- and CD8- murine thymocytes. *Immunol. Lett.* 30:119-124.
376. Zimecki, M., and Machnicki, M. 1994. Lactoferrin inhibits the effector phase of the delayed type hypersensitivity to sheep erythrocytes and inflammatory reactions to *M. bovis* (BCG). *Arch. Immunol. Ther. Exp. (Warsz).* 42:171-177.
377. Zimecki, M., Miedzybrodzki, R., and Szymaniec, S. 1998a. Oral treatment of rats with bovine lactoferrin inhibits carrageenan-induced inflammation; correlation with decreased cytokine production. *Arch. Immunol. Ther. Exp. (Warsz)* 46:361-365.
378. Zimecki, M., Wlaszczyk, A., Cheneau, P., Brunel, A.S., Mazurier, J., Spik, G., and Kubler, A. 1998b. Immunoregulatory effects of a nutritional preparation containing bovine lactoferrin taken orally by healthy individuals. *Arch. Immunol. Ther. Exp. (Warsz)* 46:231-240.
379. Zimecki, M., Wlaszczyk, A., Zagulski, T., and Kubler, A. 1998c. Lactoferrin lowers serum interleukin 6 and tumor necrosis factor alpha levels in mice subjected to surgery. *Arch. Immunol. Ther. Exp. (Warsz)* 46:97-104.

A.S. Naidu

Lactoperoxidase

3

I. INTRODUCTION

Lactoperoxidase (LP) is a hemoprotein present in milk, tears, and saliva (Tenovuo & Pruitt, 1984)). LP constitutes one of the nonimmunoglobulin defense factors in the mucosal secretions (Mandel & Ellison, 1985). In 1881, Arnold demonstrated the peroxidase activity in bovine milk. Later in 1943, Theorell and Åkeson, isolated the protein responsible for this enzyme activity, by precipitation methods, and termed 'lactoperoxidase'. The purification of LP has been considerably refined in the later years by several groups. Subsequent studies have analyzed the amino terminus, along with amino acid and carbohydrate composition (Rombauts et al., 1967; Carlstrom, 1969; Sievers, 1981; Paul & Ohlsson, 1985). These studies further suggested that LP could exist in multiple forms, probably reflecting heterogeneity in glycosylation. LP contains at least two domains differing in thermostability (Pfeil & Ohlsson, 1986). In addition, a peroxidase with similar biochemical and immunological properties, which could originate from the same gene as LP, has been found in human saliva (Mansson-Rahematulla, 1988). The gene sequences encoding several peroxidases including LP has been published (Dull et al., 1990).

The involvement of LP in the inhibition of microbial growth (stasis) was first suggested by Hanssen (1924) and was ultimately demonstrated by Wright and Trammer (1958). Involvement of hydrogen peroxide (Jago & Morrison, 1968) and thiocyanate (Reiter et al., 1963) in the inhibitory phenomenon was observed. Earlier, this stasis effect was considered as a limitation in the manufacture of cultured dairy products from normal cow milk, which contains large amounts of LP and thiocyanate. Later, it was suggested that the presence of LP in the mammary, salivary, lacrimal and harderian glands may constitute a naturally occurring defense mechanism in the body to control microbial proliferation (Klebanoff & Luebke, 1965; Slowey et al., 1968).

Hydrogen peroxide is cidal to a variety of life forms inspite of their highly developed defense mechanisms against this oxidizing agent (Chance et al., 1979). Thus, free hydrogen peroxide may pose a threat to both human and microbial cells. Hydrogen peroxide is a substrate for LP in oxidizing thiocyanate (SCN-) to hypothiocyanate (OSCN) (Thomas, 1985). Several studies have suggested that the peroxidation reaction serves at least two important functions in the mucosal secretions: 1) the products of the reaction lead to antimicrobial activity (Pruitt & Reiter, 1985); and 2) the reaction prevents the accumulation of hydrogen peroxide excreted by many microorganisms (Carlsson et al., 1983) and host cells (Pruitt et al., 1983). Hydrogen peroxide is highly toxic to mammalian cells and is consumed rapidly by the peroxidase reaction into nontoxic byproducts (Hanström, 1983).

The LP system also inhibits hexokinase (Adamson & Pruitt, 1981) and glyceraldehye-3-phosphate dehydrogenase (Carlsson et al., 1983). The enzyme inhibitory activity could contribute to the antimicrobial function of the LP system. The LP system has been suggested to protect the gut of the calf from enteric pathogens (Reiter et al., 1980) and a possible role for LP in the defense of the mammary gland has been postulated (Reiter & Bramley, 1975). LP has been used to preserve milk without refrigeration (Bjorck et al., 1979). The LP system could inhibit bacteria, fungi, parasites and viruses; therefore, it is considered as a broad-spectrum natural antimicrobial for food safety and preservation.

II. OCCURRENCE

LP is the most abundant enzyme in bovine milk, constituting about 1% of the whey proteins or 10-30 µg/ml of milk (Reiter, 1985). Feeding practices could influence the LP levels (Kiermeier & Kuhlmann, 1972), and elevated somatic cell counts correlate with increased LP levels in bovine and human milk (Gothefors & Marklund, 1975). The concentration of LP is low in the bovine colostrum and increases rapidly to reach a peak at 4-5 days of postpartum. It declines rapidly afterwards to reach a constant, rather high, plateau during the lactation. Human LP concentration is high in colostrum and declines rapidly within 1 week (Kiermeier & Kuhlmann, 1972; Gothefors & Marklund, 1975). The concentrations of LP and SCN- in the bovine mammary secretions during the early dry period are similar to the milk before drying off. However, the free cysteine levels progressively increase in the secretions beginning 3 to 5 days after the last milking. The secretions, when diluted in steamed milk, show greater stimulation of *Streptococcus agalactiae* growth as the drying-off period progresses. This effect has been attributed to the increased concentrations of cysteine that counteracts the LP/SCN-/ H_2O_2 inhibitory system for *S. agalactiae*. This effect on the LP system has been suggested to increase the susceptibility of bovine udder to *S. agalactiae* infection during the dry period (Brown & Mickelson, 1979).

Milk from infected udders shows higher levels of LP and SCN- than milk from normal udders. Udder irritation could increase bacteriostatic activity of milk (Janota-Bassalik et al., 1977). Estrogen could also influence the LP concentration in bovine milk. Thus, the peroxidative activity of milk varies periodically in non-pregnant but not in pregnant cows. The milk of cows in estrus shows increased LP activity and turns normal after ovulation (Kern et al., 1963). LP concentration in cervical mucus significantly drops after ovulation (Linford, 1974). Immunofluorescence and hybridization techniques revealed the localization of LP in the goat lactating mammary gland (Cals et al., 1994). *In situ* hybridization experiments have demonstrated the presence of LP mRNA within the cytoplasm of alveolar epithelial cells (acini).

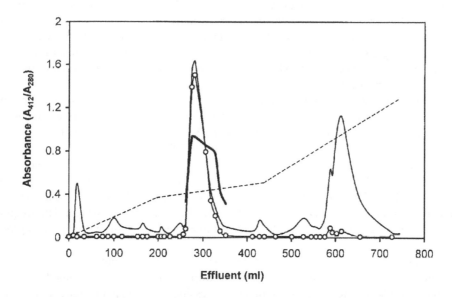

FIGURE 1: Isolation of lactoperoxidase on Bio-Rex 70 resin column at 4 °C by ion-exchange chromatography. The column was equilibrated with starting buffer (0.02 M sodium acetate, pH 7.35) at a flow rate of 30 ml/h. The absorbance of fractions was measured at 280 mμ (thin line) and 412 mμ (line with circles). At the end of the first gradient (dotted line) the green LPO was located on the top of the column and a more gradual gradient with 0.35 M sodium acetate, pH 7.9 (dotted line) completely eluted the enzyme. The fractions with an A412:A280 ratio (thick line) above 0.9 were pooled and concentrated (redrawn from Rombauts et al., 1967).

Pruitt et al. (1990) described an assay to measure peroxidase concentration in biological fluids. Regression equations were used to calculate concentrations of bovine LP, human salivary peroxidase, and human myeloperoxidase. Electron donors: pyrogallol, guaiacol, 2,2'-azinobis (3-ethylbenzylthiazoline-6-sulfonic acid, ABTS), and thiocyanate (SCN-) were used to measure the peroxidase activity. The peroxidation rates of these donors depend upon the concentrations of H_2O_2, thus, should be carefully controlled for accurate, and reproducible results.

III. MOLECULAR PROPERTIES

A. Isolation and Purification

Casein precipitation with rennet and adsorption of whey proteins to ion exchangers are basic steps in the isolation and purification of LP (*FIGURE 1*). Paul and co-workers (1980) described a method that minimizes extreme ionic fluctuations, and avoids over exposure to potassium phosphate. Unpasteurized skim milk is coagulated with rennin (2 mg/L). After cooling with ice, the whey is stirred for 3-h with the ion exchanger CG-50-NH_4 (20 g/L). The LP-resin complex after two washes with water, three washes with 50 mM sodium acetate, is transferred to a column and allowed to settle in 50 mM sodium acetate buffer under cautious stirring to prevent channeling. LP is eluted with 2 M sodium acetate and transferred to a phenyl-Sepharose column. LP is further eluted with decreasing, linear gradient of sodium acetate (2.0-0.5M, pH 7.0). The eluate is concentrated in a Diaflow XM-50 cell with slow stirring, dialyzed against 10 mM sodium phosphate (pH 6.0). The dialysate is fractionated by chromatography on CM-52 with a sodium phosphate gradient (10-130 mM, pH 6.0). The final purification and separation of LP subfractions is performed on DEAE-Sephadex. Borate (10 mM, pH 9.0 and 8.0) resolves LP into a number of fractions.

Human LP with stability and immunoreactive properties similar to that of bovine milk LP has been partially purified from human colostrum by hydrophobic interaction chromatography (Langbakk & Flatmark, 1984). A 10-fold purification with an apparent recovery of about 45% was obtained by using Phenyl-Sepharose C1-4B.

A procedure for isolation and purification of LP from cow milk was described by Denisova et al. (1986). This procedure involves isolation of casein from milk, sorption of LP on CM-Sephadex, concentration of the eluate using ultrafiltration, salting out with ammonium sulfate and a final isoelectric focusing in borate-polyol system. Highly purified, active preparation of LP was obtained within a relatively short period by this procedure.

Ferrari and co-workers (1995) developed an improved purification method for LP using preparative chromatography and electrophoresis methods combined with analytical electrophoresis techniques and image processing. Electron paramagnetic resonance (EPR) analysis has been used to evaluate LP homogeneity against lactoferrin and minor LP components to define the final steps of purification. Two samples of LP (from farm and commercial milk) were purified and characterized by EPR spectroscopy to clarify the Fe(III)-heme high-spin nature of the native LP. These data indicated the presence of an iron(III)-heme high-spin catalytic site in the native enzyme. In contrast, LP from commercial milk, a low-spin iron(III)-heme species, was observed due to nitrite impurities. The LP/SCN- activity in phosphate and acetate buffer was optimal at pH 5.5.

A fractionation scheme for isolation LP, lactoferrin, and IgG, based on cation exchangers was developed (Hahn et al., 1998). Four different cation-exchange media (S-HyperD-F, S-Sepharose FF, Fractogel EMD SO3- 650 (S) and Macro-Prep High S Support) were compared for dynamic binding capacity to IgG and different elution behaviors with sequential step gradients of NaCl buffers. Peak fractions were analyzed by size-exclusion chromatography and sodium dodecyl sulfate polyacrylamide gel electrophoresis (SDS-PAGE). LP activity was monitored by the oxidation of O-phenylenediamine. Fractogel EMD had the highest binding capacity for IgG (3.7-mg/ml gel at a linear flow-rate of 100 cm/h), however, the resolution was low compared to the other three media. S-Hyper D and S-Sepharose FF showed lower capacities (3.3 and 3.2-mg/ml gel, respectively) but exhibited better protein resolution. These effects could be partially explained by the k' versus salt concentration plots. The binding capacity of Macro-Prep S was considerably lower compared to that of the other resins. S-Sepharose FF and S-Hyper D combine relatively high dynamic capacity for IgG and provided good resolution. Compared to studies with standard proteins, such as 100-mg/ml bovine serum albumin for S-Hyper D, their binding capacities were very low. Even after removal of low-molecular-mass compounds, the capacity could not be improved significantly. The running conditions (low pH) were responsible for the low protein binding capacity, since low-molecular-mass compounds in the feed do not compete with the adsorption of whey protein. The dynamic capacity did not decrease to a large extent within the range of flow-rates (100-600 cm/h) investigated. The dynamic capacity of HyperD and Fractogel was at least five times higher when pure bovine IgG was used for determination. In conclusion, S-Sepharose FF, S-Hyper D-F and Fractogel EMD SO3- 650 (S) are considered as successful candidates for the large-scale purification of bovine whey proteins.

B. Chemistry and structure

Early ORD studies with LP in the far ultraviolet region indicated a 17% alpha-helical structure (Maguire & Dunford, 1971). Later evaluation of optical and CD spectra revealed 65% beta-structure, 23% alpha-helix, and 12% unordered structure for LP (Sievers, 1980). Reducing SDS-PAGE with 1% SDS/mercaptoethanol showed a single component, indicating only one peptide chain. Dansylation revealed only one N-terminal amino acid, leucin. LP contains 8 disulfide bonds that contribute to the rigidity of the molecule.

The single peptide chain of bovine LP contains 612 amino acid residues, including 15 half-cystines and 4 or 5 potential *N*-glycosylation sites. The corresponding peptide segments of human myeloperoxidase, eosinophil peroxidase and thyroperoxidase display 55%, 54% and 45% identity with bovine LP, respectively, with 14 out of the 15 half-cysteines present in each of the four enzymes being located in identical positions (Cals et al., 1991). The occurrence of an odd number of half-cysteines in bovine LP supports the finding of a heme thiol released from this enzyme by a reducing agent, suggesting that the heme is bound to the peptide chain via a disulfide linkage. The amino acid composition of LP is compared with salivary peroxidase and myeloperoxidase in *Table 1*.

Peptide sequences obtained from cyanogen bromide fragments of bovine LP were used to design oligonucleotide probes for cDNA library screening constructed from bovine mammary tissue (Dull et al., 1990). Three overlapping clones were obtained, the longest of which (T3) contained a reading frame of 712 amino acid residues. The encoded amino acid sequence was homologous to those reported for myelo-, thyro-, and eosinophil peroxidases. Two possible amino termini of the mature enzyme were identified, and the predicted mature protein matched previous molecular weight estimates of 78,500. Of eight bovine tissues tested, transcription of T3 sequences were detected in mammary tissue only. Using the bovine LP cDNA as a probe, a single hybridizing clone was found in a human mammary gland cDNA library. This clone encoded the carboxy-terminal 324 residues of a peroxidase distinct from the other three known human peroxidases, and was closely related to bovine LP. Ueda and co-workers (1997) reported the molecular cloning of human LP chromosomal gene and determined its gene organization. The human LP gene is arranged with the myeloperoxidase gene in a tail-to-tail manner. Similar to the human myeloperoxidase and eosinophil peroxidase genes, the human LP gene was split by 11 introns and spans 28 kb. Unlike most introns in mammalian gene, the 5' splice donor sequence of intron 11 starts with GC instead of GT. When the minigene comprised of exon 11, intron 11 and exon 12 of the human LP gene was introduced into COS cells, the correct splicing of the intron was found, suggesting the intron 11 of the human LP gene was functional. The coding sequence of human LP consists of 2136 bp, and codes for a protein of 712 amino acids. The amino acid sequence of human LP showed 51% homology with human myeloperoxidase and eosinophil peroxidase, suggesting that these peroxidase genes have evolved from a common ancestral gene. On the other hand, the nucleotide sequences of the 5' promoter regions of these peroxidase genes exhibit no similarity, which agrees with their tissue-specific expression.

TABLE 1: Comparison of aminoacid composition in lactoperoxidase, salivary peroxidase and myeloperoxidases

Amino acid	Bovine LP	Human SP	Human MP
Aspartic acid	127	118	132
Threonine	50	49	56
Serine	60	77	54
Glutamic acid	107	105	136
Proline	80	97	84
Glycine	77	77	76
Alanine	69	74	72
Cysteine	15	16	24
Valine	40	38	48
Methionine	16	5	24
Isoleucine	32	31	32
Leucine	113	117	60
Tyrosine	27	22	24
Phenylalanine	52	50	48
Lysine	55	50	24
Histidine	23	18	12
Arginine	61	57	96
Tryptophan	ND	ND	12

(Adapted from Cals et al., 1991; Matheson et al., 1981; Mansson-Rahemtulla et al., 1988)

C. Stability

During pasteurization, whole milk loses about three-quarters of LP activity, whereas partially purified LP is stable for 15 min at this temperature (Wutrich et al., 1964). Kinetic studies of the rate of formation of Compound-I indicated that LP is stable when stored at neutral pH 7.0 but deactivates by storage at acidic pH 3.0 (Maguire et al., 1971). The deamidization of LP by glycine at pH 10.3 for 48 hr at room temperature seemed not to inactivate LP.

No heme was released by acid butanone after digestion of LP with pepsin followed by trypsin (Morell, 1953) or chymotrypsin (Sievers, 1979). Also trypsin and thermolysin did not inactivate native LP, while chymotrypsin caused a moderate effect. Pronase rapidly digests native LP to fragments from which heme could be extracted (Sievers, 1979). LP is not inactivated by the gastric juice (pH 5.0), whereas pepsin at pH 2.5 inactivated LP (Gothefors & Marklund, 1975). The prosthetic group of milk LP has been isolated from a pronase hydrolysate of the enzyme and identified as protoheme IX (Sievers, 1979). Partial degradation of the heme occurs during the proteolysis, possibly because of coupled oxidation in the presence of glycine and oxygen. The heme seems to be buried in the protein molecule in a crevice that allowed ligands to bind to the iron on one side only.

Pfeil and Ohlsson (1986) studied the thermal unfolding of LP by differential scanning calorimetry and optical methods. The protein demonstrated at least two domains differing in thermostability. The prosthetic group belonged to the domain of lower thermostability. Thermodynamic parameters of protein unfolding were found similar to the globular proteins.

Marcozzi and co-workers (1998) investigated the activity and the stability of bovine LP in the presence of different surfactants. The cationic benzalkonium chloride was effective in preserving the enzymatic activity for over 10 days, while the native LP completely lost its activity within 3-4 days. In the presence of benzalkonium chloride, LP could preserve its secondary structure for a long time. This condition creates an enhanced hydrophobic environment for LP, as indicated in fluorescence studies. Moreover, this surfactant at a concentration of 0.01% (0.3 mM) increased the LP activity in the first 2-h of incubation at 37°C. Both hydrophobic and electrostatic interactions of the cationic surfactant seem to be responsible for the enzyme activation and stabilization.

D. Heterogeneity

Two different molecular forms of LP were reported, probably representing a monomer and an aggregate (Tenovuo, 1981). The isoelectric point of LP (pI, 8.1) did not change during aggregation. The presence of substrate (SCN-) caused some disaggregation. Purified milk LP showed a similar chromatographic pattern to salivary LP.

Two distinct LP forms, a non-heme and an enzymatically active LP were isolated from bovine milk by Dumontet and Rousset (1983). Both forms of LP demonstrated similar molecular weight of 85-kDa in the denatured (SDS-PAGE) and native form (velocity sedimentation on sucrose gradient). However, the non-heme form was devoid of light absorption properties in the Soret region and of enzyme activity. Both forms showed similar carbohydrate content and peptide maps after limited proteolysis by subtilisin or trypsin. The two forms showed immunological relatedness in immunodiffusion and radioimmuno assays. The non-heme LP was readily detected in milks from cow, goat, sheep, and human species. In bovine milk, LP and non-heme LP were found in comparable amounts.

Antibodies raised against bovine LP moderately cross-react with human salivary peroxidase. In double-immunodiffusion experiments, the two enzymes show partial identity, and in competitive radioimmunoassay and enzyme-linked immunosorbent assay, Western blot analysis differentiate salivary peroxidase from human and rat saliva samples and the purified enzyme in its non-reduced, reduced, and de-glycosylated forms. The major activity of these antibodies was directed against the protein core of the antigen. Immunodetection of the peptide fragments of bovine LP and human salivary peroxidase revealed structural differences in the two enzymes (Masson-Rahematullah et al., 1990).

E. Molecular interactions

LP interacts with other components of milk and saliva, and this interaction could be significant *in vivo*. LP binds to secretory IgA, IgM, myeloma IgA_1, and myeloma IgA_2 (Tenovuo et al., 1982). The interactions between LP and the immunoglobulins are non-specific. The LP-Ig complex was more stable than the free enzyme. However, LP does not bind to IgG or lactoferrin. The interaction of LP with lysozyme and ribonuclease from cow milk has also been reported (Hulea et al., 1989). LP is slightly activated when complexed to lysozyme, while IgA and IgM were inhibitory. On the other hand, IgG and ribonuclease had no effect on the enzyme activity although the latter did form a complex with the LP. The interaction between LP and lysozyme appears to be lectin-type since the alteration of LP sugar moiety by periodate oxidation, prevented the formation of the LP-lysozyme complex.

The interaction of bovine LP with several inorganic species including SCN-, I-, Br-, Cl-, F-, NO_2-, N_3-, CN- was reported (Ferrari et al., 1997). Based on the ability of LP to form 1:1 complexes with the above ligands and the dissociation constant values (K_d) of the adducts, three groups were observed: 1) SCN-, I-, Br-, and Cl- (K_d increases along the series); 2) F- (which shows a singular behavior); 3) NO_2-, N_3-, and CN- (that bind at the iron site). K_d values for the LP/SCN- adduct were altered in the presence of other inorganic species; a strong competition between the substrate and all other anions (with the exception of F-) was observed. Binding studies on the natural substrates SCN- and I- indicated protonation of a common site in proximity of the iron (possibly distal histidine). Computer-assisted docking simulations suggested the ability of all ligands to penetrate the heme pocket.

IV. ENZYME ACTIVITY

Peroxidases belong to a group of enzymes that catalyze the oxidation of numerous organic and inorganic substrates by H_2O_2. Most peroxidases, including LP contain ferriprotoporphyrin IX as a prosthetic group. A characteristic feature of hemoprotein peroxidases is their ability to exist in various oxidation states. There are five known enzyme intermediates. In increasing order of their oxidative equivalents, these are ferrous enzyme, ferric or native enzyme, Compound-II, Compound-I, and Compound-III. They are readily distinguished from each other by their absorbance in the Soret region (380-450 nm) and visible range (450-650 nm). In the course of Compound III and Compound-II conversion back to the native peroxidase, oxygen-derived free radicals such as O^{2-}, $HO^{2\cdot}$, and $\cdot OH$ are generated. Simultaneously the enzyme is irreversibly damaged. In the presence of an exogenous electron donor, such as I-, the interconversion between the various oxidation states of the peroxidase is markedly affected. Compound-II and/or Compound-III formation is inhibited, depending on the H_2O_2. In addition, the enzyme is largely protected from irreversible inactivation. These effects of I- are readily explained by a) the two-electron oxidation of I- to I^{ox} by Compound-I, which bypasses Compound-II as an intermediate and b) the rapid oxidation of H_2O_2 to O_2 by the oxidized species of I- which prevents the generation of oxygen derived free radicals (Kohler & Jenzer, 1989).

Piatt and O'Brien (1979) reported the evidence for singlet oxygen formation for the $LP/H_2O_2/Br$- system by monitoring 2,3-diphenylfuran and diphenylisobenzofuran oxidation, O_2 evolution, and chemiluminescence. This mechanism provided an explanation for the cytotoxic and microbicidal activity of peroxidases and polymorphonuclear leukocytes. Evidence for singlet oxygen formation included:

- Chemiluminescence accompanying the enzymatic reaction was doubled in a deuterated buffer and inhibited by singlet oxygen traps.
- Singlet oxygen traps, diphenylfuran and diphenylisobenzofuran, were oxidized to their known singlet oxygen oxidation products in the presence of $LP/H_2O_2/Br$-.
- Rate of oxidation of diphenylfuran and diphenylisobenzofuran was inhibited when monitored in the presence of known singlet oxygen traps or quenchers.
- Singlet oxygen traps inhibited oxygen evolution from the enzymatic reaction but not by singlet oxygen quenchers.
- Traps or quenchers which were effective inhibitors in the experiments above did not inhibit peroxidase activity, were not competitive peroxidase substrates and did not react with the hypobromite intermediate since they did not inhibit hydrogen peroxide consumption by the enzyme.

Pruitt and co-workers (1982) examined the LP-catalyzed oxidation of SCN- by two different polarographic techniques: direct current polarography and linear sweep voltametry. The main oxidation product at pH 6.5, with a half-wave potential (Efi) of -0.39 to -0.44 V, was identified as OSCN- ion. All three components of the LP system (i.e. LP/SCN-/ H_2O_2) were required to produce OSCN-. Addition of excess H_2O_2 or H_2O_2/LP to an OSCN-/SCN- mixture generated a new peak associated with a simultaneous decrease of OSCN- concentration. This indicates a possible reaction between H_2O_2 or H_2O_2/LP and OSCN-. This new peak possibly represents higher oxy acids of SCN- (O_2SCN-, O_3SCN-), formed in the oxidation of OSCN- by H_2O_2 or by H_2O_2/LP. This reaction might explain why, in solutions that already contain OSCN- (e.g., in saliva), the addition of H_2O_2 results in the formation of highly reactive, short-lived antimicrobial products in addition to OSCN-.

The heme environmental structures of LP have been studied by the use of hyperfine-shifted proton NMR and optical absorption spectra (Shiro & Morishima, 1986). The NMR spectra of the enzyme in native and cyanide forms in water indicated that the fifth ligand of the heme iron is the histidyl imidazole with an anionic character and that the sixth coordination site is possibly vacant. These structural characteristics are quite similar to those of horseradish peroxidase (HRPO), suggesting that these may be prerequisite to peroxidase activity.

LP-catalyzed H_2O_2 metabolism proceeds through one of three different pathways, depending on the nature and the concentration of the second substrate as an electron donor and/or on pH conditions (Jenzer et al., 1986). In the LP/H_2O_2, at low H_2O_2 levels and/or alkaline conditions the peroxidative cycle involves ferric LP/Compound-I/Compound-II/ferric LP conversion, whereas high H_2O_2 levels and/or acidic conditions favor the ferric LP/Compound-I/Compound-II/Compound-III/ferrous LP/ferric LP pathway. The Compound-III/ferroperoxidase states are associated with irreversible enzyme inactivation by cleavage of the heme moiety and liberation of iron. It is likely that either singlet oxygen or superoxide and hydroxyl radicals are involved in the attack on heme iron, because inactivation correlates with oxygen production and can be decreased to a certain degree by scavengers such as ethanol, 1-propanol, 2-propanol, or mannitol. In the LP/ H_2O_2/I- system, the enzyme may also be inactivated by iodide generated in the course of enzymatic I- oxidation (i.e. during ferric LP/Compound-I/ferric LP cycles).

The LP-catalyzed oxidation of glutathione (GSH) and SCN- was studied (Lovaas, 1992). One or two moles of GSH were oxidized per mole of H_2O_2, depending on the reaction conditions. Omission of SCN- prevented the oxidation of GSH. The oxidation of GSH required only catalytic amounts of SCN-, which was therefore recycled. Iodide (I-) could replace SCN-, while chloride or bromide was ineffective. The apparent Michaelis constant for SCN- was 17 μM. Oxidation of SCN- generated two reactive intermediates, one stable and one unstable. The stable intermediate (-OSC.) decayed by a second-order reaction. The decay of the unstable radical was very fast. These data elucidate the short- and long-term antibacterial effects of LP/halide/ H_2O_2; point to possible deleterious effects due to glutathione depletion; and support observations on lipid peroxidation/halogenation in biological membranes, liposomes, and unsaturated fatty acids.

Products formed from the LP catalyzed oxidation of SCN- with H_2O_2 was studied by ^{13}C-NMR at pH 6.0 and 7.0 (Pollock & Goff, 1992). Ultimate formation of OSCN-as the major product correlates well with the known optical studies. The oxidation rate of SCN- appears to be greater at pH < or = 6.0. At H_2O_2/SCN- ratios of < or = 0.5, OSCN- was not formed immediately, but an unidentified intermediate was produced. At H_2O_2/SCN- ratios of > 0.5, SCN- appears to be directly oxidized to OSCN-. Once formed, OSCN- slowly degrades over a period of days to CO_2, HCO_3-, and HCN. An additional product also appears after formation of OSCN-. On the basis of carbon-13 chemical shift information this new species was suggested to result from rearrangement of OSCN- to yield the thiooxime isomer, SCNO- or SCNOH.

It should be noted that the final product of LP reaction depends on the method of H_2O_2 generation and relative proportions of the substrates (Dionysius et al., 1992).

V. ANTIMICROBIAL ACTIVITY

Peroxidase enzymes catalyze the oxidation of electron donors by peroxide to generate highly reactive products with a wide range of antimicrobial properties. The actual structures of these products and the chemistry of their reactions depend upon the specific peroxidase-electron donor pair. The SCN- ion is the most significant electron donor *in vivo*, which is a normal constituent of bovine milk and is derived from the diet.

The major end product of the LP/SCN- at neutral pH is the OSCN- anion, which is in equilibrium with hypothiocyannous acid (HOSCN, pK 5.3) (Thomas, 1981). There are other products of the LPS/SCN- that are generated in small amounts and are relatively unstable at neutral pH compared to SCN- (Aune & Thomas, 1977; Bjorck & Clasesson, 1980; Pruitt & Tenovuo, 1982).

A. Mechanism(s) of action

The antibacterial activity of the LP/SCN-/ H_2O_2 system has been suggested due to the OSCN- anion. Relatively pure solutions of OSCN- could be prepared using an immobilized enzyme. To support the above hypothesis pure OSCN- preparations were directly tested for antimicrobial effects on *Escherichia coli* and *Streptococcus lactis* (Marshal & Reiter, 1980). *E. coli* was killed in the presence of OSCN- anion whereas the effect on *Streptococcus lactis* was only bacteriostatic. These studies supported the similarities of LP/SCN-/ H_2O_2 and OSCN- systems. The cellular targets for OSCN- and HOSCN interaction are sulfhydryl groups (Aune & Thomas, 1978) and reduced nicotinamide nucleotides (Oram & Reiter, 1966). The sulfhydryl groups could be the cysteine residues of specific proteins or reduced GSH. Sulfhydryl groups are oxidized to disulfides (-S-S-), sulfenyl thiocyanates (-S-SCN-), and sulfinic acids (-S-OH). Disulfides, sulfenyl thiocyanates, and sulfinic acids are readily converted back to sulfhydryls by an excess of reducing agents such as cysteine, mercaptoethanol, dithiothreitol, sodium hydrosulfite and GSH. The nucleotides NADH and NADPH are oxidized to NAD^+ and $NADP^+$, and this reaction is also reversible. LP is effective at concentrations as low as 0.5 µg/L (about 6.5 x 10^{-12} M, assuming a M_r of 77 kDa).

LP catalyzes the oxidation of SCN- by H_2O_2 into OSCN-, a reaction that could protect bacterial and mammalian cells from killing by H_2O_2. Carlsson and co-workers (1984) demonstrated, however, that LP in the presence of SCN- could potentiate the bactericidal and cytotoxic effects of H_2O_2hydrogen under specific conditions, such as when

H_2O_2 is present in the reaction mixtures in excess of SCN-. The toxic agent was also formed in the absence of LP in a reaction between OSCN- and H_2O_2. Sulfate, sulfite, cyanate, carbonate, and ammonia, generated during the OSCN- oxidation by H_2O_2 were not bactericidal and did not potentiate the bactericidal effect of H_2O_2. The authors suggested that cyanosulfurous acid, the only other postulated product of the chemical oxidation of OSCN- by H_2O_2, as the possible cidal agent.

Adsorption of LP to microbial cell surface seems necessary to elicit antimicrobial effects (Tenovuo & Knuuttila, 1977). There have been conflicting reports regarding the binding of LP to bacterial cell surfaces. Effects of bacterial cell-bound LP on acid production by suspensions of *Streptococcus mutans* in the presence of H_2O_2 and SCN- was reported (Pruitt et al., 1979). Addition of H_2O_2 and SCN- to bacterial cells treated with LP (0.1 mg/ml and thoroughly washed) caused marked reduction in acid production by *S.mutans*. After a 3-h incubation in saline, however, the LP-treated bacteria produced acid in the presence of H_2O_2 and SCN-. The authors concluded that LP is initially bound to the bacterial cell surface in an enzymatically active form at a concentration sufficient to inhibit acid production. The LP seems to slowly degrade or desorb as the bacteria stand in saline suspension.

Mickelson (1979) reported that low amounts of LP adsorb to the cell surface of *S. agalactiae* and are not removed by washing. A diffusible antibacterial product of LP/H_2O_2/SCN-reaction was not detected. Incubation of *S. agalactiae* cells with LP/H_2O_2 and ^{14}C-labeled sodium SCN resulted in the incorporation of SCN into the bacterial protein. Most of the LP-catalyzed, incorporated SCN was released from the bacterial protein with dithiothreitol. Cells that had their membrane permeability changed by treatment with Cetab or 80% ethanol incorporated more SCN than did untreated cells (about 1 mol of SCN for each mol of sulfhydryl group present in the reaction mixture). Alteration of membrane permeability caused protein sulfhydryls, normally protected by the cytoplasmic membrane, to become exposed to oxidation. It was suggested that LP/H_2O_2-catalyzed the incorporation of SCN into the proteins of *S. agalactiae* by a mechanism similar to that reported for bovine serum albumin. It was speculated that the removal of reactive protein sulfhydryls from a functional role in membrane transport and in glycolysis has a possible role in eliciting antibacterial effect for *S. agalactiae*. LP is enzymatically effective even when fewer than 100 molecules are adsorbed per cell of *Streptococcus mutans* (Pruitt et al., 1979).

B. Antimicrobial target sites

Cytoplasmic membrane and/or cytoplasm are suggested to be major targets for antimicrobial products of the LP system. The cell wall and outer membrane may partially limit accessibility but do not totally exclude the products of the LP/SCN- from the cell interior. The cell wall of *E. coli* may also be altered by exposure of cells to the LP/SCN- (Reiter, 1978).

Marshall and Reiter (1980) reported the ability of OSCN- to alter bacterial cytoplasmic membranes. Treatment of *E. coli* suspensions with OSCN- caused extensive leakage of [14C] amino acids and [42]K in 10 min. Damage to the inner membrane of *E. coli* was also suggested by the increase in hydrogen ion permeability of cells treated with the LP/SCN- (Law & John, 1981). The cell walls of gram-positive bacteria are less susceptible to damage by the products LP system than of gram-negative bacteria.

The cytoplasmic membrane of gram-positive bacteria is also altered by the LP system. Treatment with the LP system or OSCN- could inhibit amino acid transport in *Lactobacillus acidophilous* (Clem & Klebanoff, 1966; Slowey et al., 1968) and *S. aureus* (Hamon & Klebanoff, 1973), and glucose and oxygen uptake in numerous species. Mickelson (1977) found that the LP/SCN- inhibits glucose transport system of *S. agalactiae* by modification of essential membrane sulfhydryl groups. Later studies indicated that the LP system incorporated $S^{14}CN^-$ into bacterial protein which could be released by treatment with reducing agents (Mickelson, 1979).

Antimicrobial products of LP system could inhibit several glycolytic enzymes including hexokinase, aldolase, 6-phosphogluconate dehydrogenase and glucose-6-phosphate dehydrogenase activities in *Streptococcus cremoris* (Oram & Reiter, 1966). Similar enzyme inhibition was also observed with *S. mutans*. Thus, LP-treated cells show marked reduction in pyruvate kinase, hexokinase, and aldolase activities. However, the levels of enolase phophoglycerate kinase and phosphoglycerate mutase were slightly affected (Hoogendoorn, 1974).

Glyceraldehyde-3-phosphate dehydrogenase activity of *S. pyogenes* is inhibited by LP/H_2O_2 mixtures in the absence of SCN- (Mickelson, 1966). Later studies also showed that OSCN- could directly oxidize glyceraldehyde-3-phosphate-dehydrogenase (Carlsson et al., 1983). Treatment of cell extracts of this enzyme (NAD linked) from *S. mitis, S. sanguis, S, mutans* and *S. salivarius* with OSCN- solutions resulted in complete inhibition of enzyme activity.

C. Regulating factors

Concentration of LP is not a limiting factor for antibacterial activity in bovine milk or in human saliva. However, the SCN- concentration is a limiting factor in milk and sometimes in saliva, but usually it is the availability of H_2O_2 that determines the magnitude of antibacterial activity of peroxidase systems (Pruitt et al., 1982).

Susceptibility of microorganisms to the antimicrobial activity of LP is dependent on the state of their metabolic growth. In general, resting cells or cells in the stationary growth phase are more susceptible to killing or inhibition than are metabolically active or growing cells. Bacteria grown anaerobically are more susceptible to LP than are those grown aerobically (Carlsson et al., 1983). Also the increased permeability of the cell envelope is associated with increased susceptibility to LP-mediated antibacterial effects (Purdy et al., 1983).

Under aerobic conditions, LP and SCN- could protect *E. coli* and oral streptococci from the bactericidal effect of H_2O_2 (Adamson & Carlsson, 1982). LP in the absence of SCN- was protective but potentiated H_2O_2 toxicity for certain other bacterial species under similar conditions. The products of reaction between H_2O_2 and SCN- in the presence of LP were not bactericidal except for *E. coli*.

Inhibition of bacterial metabolism by the $LP/SCN-/H_2O_2$ system was studied with serotypes A through G of *S. mutans* (Thomas et al., 1983). When the washed, stationary-phase cells were incubated aerobically with LP/SCN- and glucose, > 90% inhibition of sugar utilization and lactate production was obtained with strains that released large amounts of H_2O_2; 20 to 50% inhibition was obtained with strains that released about half as much H_2O_2; and no inhibition was obtained with strains that released only small amounts of H_2O_2. Inhibition was most effective at pH 5.0, whereas release of H_2O_2 and accumulation of the inhibitor (OSCN- ion) were highest at pH 8.0. With H_2O_2-releasing cells from early stationary phase, preincubation with glucose abolished inhibition, though it did not influence H_2O_2 release. Cells harvested 24 h later were depleted of sulfhydryl compounds. Inhibition of these cells was abolished by preincubation with glucose and certain sulfhydryl or disulfide compounds (reduced or oxidized GSH, cysteine or cystine). This preincubation increased cell sulfhydryl content but had no effect on H_2O_2 release. All strains were inhibited when incubated with LP/SCN- and added (exogenous) H_2O_2. Smaller amounts of H_2O_2 were required to inhibit at pH 5.0, and larger amounts were required to inhibit cells preincubated with glucose alone or in combination with sulfhydryl or disulfide compounds. The results indicate that pH, amount of H_2O_2, cell sulfhydryl content, and stored-carbohydrate content determine susceptibility to inhibition.

Cystine reduction in *S. agalactiae*, resulting in sulfhydryl formation, may account for antagonism of the antibacterial effect of LP/SCN-/H_2O_2 system when cystine is present in excess (Mickelson & Anderson, 1984). The reduction of cystine seems to occur by a coupling reaction between GSH reductase and GSH-disulfide transhydrogenase activity, both of which were found in the supernatant fraction from cell homogenates. NADPH-specific GSH reductase activity was found in the pellet and supernate fractions from cell homogenates. Two sulfhydryls were formed for each mole of NADPH used during cystine reduction.

Excess amounts of I- (10 mM) and H_2O_2 (0.1-10 mM) could cause rapid, irreversible inactivation of LP without forming Compound III (Huwiler et al., 1986). In contrast, in the absence of I- Compound III was formed and inactivation proceeded in a slow fashion. Increasing the LP concentration accelerated the rate of inactivation. Irreversible inactivation of LP involved cleavage of the prosthetic group and liberation of heme iron. The rate of enzyme destruction was correlated with the production of molecular oxygen (O_2), which originated from the oxidation of excess H_2O_2. Since H_2O_2 and O_2 *per se* do not affect the heme moiety of the peroxidase, it was suggested that the damaging species may be a primary intermediate of the H_2O_2 oxidation, such as oxygen in its excited singlet state, superoxide radicals, or consequently formed hydroxyl radicals.

Jenzer and Kohler (1986) have investigated irreversible inactivation of LP in the presence of excess H_2O_2. Serial overlay absorption spectra of the Soret region indicated that the rate and total amount of enzyme inactivation depend on the proton concentration. Perhydroxyl or superoxide radicals could not be established as the inactivating species in this mechanism; however, these radicals could influence the rate of reconversion of the intermediate LP-compound III back to the resting ferric form.

The irreversible inactivation of bovine LP by thiocarbamide goitrogens was reported, and the kinetics were consistent with a mechanism-based (suicide) mode (Doerge, 1986). Sulfide ion-inactivated, 2-mercaptobenzimidazole-inactivated, and 1-methyl-2-mercaptoimidazole-inactivated LP demonstrated different visible spectra, suggesting the formation of various products. These data support a mechanism in which reactive intermediates are formed by S-oxygenation reactions catalyzed by LP Compound II. It was proposed that the reaction of electron-deficient intermediates with the heme prosthetic group was responsible for the observed spectral changes and inactivation by thiocarbamides.

The oxidation of SCN- by H_2O_2 in the presence of LP does not take place at pH greater than 8.0. Since SCN- does not bind to LP above this pH, the binding of SCN- to LP is considered to be prerequisite for the oxidation of SCN-. Maximum inhibition of oxygen uptake by *Streptococcus cremoris* was observed when H_2O_2 and SCN- were present in equimolar amounts and the pH was below 6.0 (Modi et al., 1991).

Fluoride (F-) ions at concentrations present *in vivo* at the plaque/enamel interface (0.05-10 mM) inhibit the activities of LP, myeloperoxidase and total salivary peroxidase in a pH- and dose-dependent manner (Hannuksela et al., 1994). Furthermore, the generation of antimicrobial products *in vivo*, hypothiocyanite (HOSCN/OSCN-), of the oral peroxidase systems was inhibited by F-, again at low pH (5.0-5.5) both in buffer (by 45%) and in saliva (by 15%). Small molecular weight media components in brain-heart infusion broth could also interfere with the antimicrobial activity of LP (Hoogendorn & Moorer, 1973).

D. Resistance

Several strains of *Streptococcu cremoris* demonstrate a low degree of susceptibility to the antimicrobial effects of LP system. Extracts from a resistant strain were able to reverse the inhibition of a susceptible strain. It was suggested that the resistant strains utilize an $NADH_2$-oxidizing enzyme to catalyze the oxidation of $NADH_2$ by OSCN-, and thereby lower the inhibitory levels (Oram & Reiter, 1966a). Under aerobic conditions, certain strains of *S. agalactiae* develop a cyanide-sensitive respiratory system resistant to LP/SCN- (Mickelson, 1979).

Carlsson and co-workers (1983) reported that *Streptococcus mitis* and *Streptococcus sanguis*, but not *Streptococcus salivarius* or *S. mutans*, had a high capacity for recovering from inhibition by OSCN-. This resistance is due to the presence of a streptococcal oxidoreductase that catalyzes the reduction of OSCN- to SCN-. The activity of this enzyme was much higher in *S. mitis* and *S. sanguis* than in *S. salivarius* and *S. mutans*. The relative resistance of these organisms to OSCN- was suggested due to relative difference in activity of this oxidoreductase enzyme.

Hoogendoorn (1976) suggested the potential role of NADPH in *S. mutans* resistance to the LP/SCN-. The inhibited streptococcal cells possibly recover by utilizing NADPH to reduce the compounds oxidized by OSCN-. When NADPH levels are exceeded by H_2O_2 (from which OSCN- is generated), streptococci fail to reverse the antimicrobial effects of LP/SCN-.

Resistance of oral bacteria to OSCN- inhibition seems to correlate with their peroxidogenicity (Mansson-Rahemtulla et al., 1982). Thomas and co-workers (1983) identified three major types of peroxide-generating *S. mutans*, Class-I strains produced large amounts of H_2O_2; Class-II produced moderate amounts of H_2O_2; and Class-III produced little H_2O_2 or none. Class I and class II strains were auto-inhibited due to the utilization of streptococcal H_2O_2 by LP/SCN- to generate antimicrobial OSCN-. Among Class-I and -II strains the H_2O_2 and OSCN- accumulation increased with culture age. Detectable H_2O_2 was excreted by class III cells but did not accumulate due to the high levels of peroxide-reducing enzymes present in these strains. However, cells of all classes were inhibited by the LP/SCN- when H_2O_2 was added exogenously.

Preincubation of cells with glucose and with certain sulfhydryl compounds could lead to elevated cell sulfhydryl content and to increased resistance (Thomas et al., 1983). A two-step mechanism to elucidate LP/SCN- resistance was suggested. The first step postulates a reductase-dependent NADPH generation of sulfhydryl compounds from intracellular disulfides. In the second step, these sulfhydryl compounds rapidly reduce OSCN- to SCN- in an enzyme-independent reaction. The net result is an NADPH-dependent reduction of OSCN- to SCN- with conservation of intracellular sulfhydryl concentration.

VI. ANTIMICROBIAL SPECTRUM

The LP system could elicit stasis and/or cidal activity on a variety of susceptible microorganisms including bacteria, fungi, viruses and parasites (*TABLE 2*). The molecular mechanism(s) of such inhibitory effects depend on the type of electron donor, test media, temperature, pH, etc. and could range from oxidative killing to blockade of glycolytic pathways or interference in cytopathic effects (*FIGURE 2*).

A. Antibacterial effects

In 1894, Hesse found that gram-negative, catalase-positive species of *Vibrio* and *Salmonella* were killed in raw milk. Gram-negative bacteria seem to be more easily killed than are gram-positive cells by extended incubation with OSCN-, especially at low temperatures. The pH-dependence of killing also seems to be greater for gram-negative species. *E. coli* and *S. typhimurium* are difficult to kill above pH 7.0 and at temperatures over 20°C, when SCN- is the electron donor. However, under similar conditions, a cidal activity could be achieved with I- as electron donor. Both SCN- and I- promote killing by the LP system when incubated with bacteria at pH 5.5 or less and at 0-5°C (Pruitt & Reiter, 1985). The inner membrane of gram-negative bacteria appears to be more extensively damaged by LP treatment than is that of gram-positive species (Marshall & Reiter, 1980).

LP present in various secretions oxidizes SCN- in the presence of H_2O_2 to an unstable oxidation product (OSCN-), which is bactericidal for enteric pathogens including multiple antibiotic resistant strains of *E. coli*. The system damages the inner membrane causing leakage and cessation of uptake of nutrient, leading eventually to death of the organisms and lysis. The antimicrobial activity of $LP/H_2O_2/SCN$- system against *E. coli* seems related to the oxidation of bacterial sulfhydryls (Thomas & Aune, 1978). LP catalyzed oxidation of SCN- results in accumulation of OSCN-. A portion of the bacterial sulfhydryls was oxidized by OSCN- to yield sulfenic acid and sulfenyl thiocyanate derivatives. The remaining sulfhydryls were not oxidized, although OSCN- was present in large excess. The oxidation of sulfhydryls to sulfenyl derivatives inhibits bacterial respiration. Adding sulfhydryl compounds to reduce the sulfenyl derivatives and the excess OSCN- could reverse this inhibition. Also, the removal of excess unreacted OSCN- by washing cells could reverse this inhibition. After washing, the bacteria demonstrated a time-dependent recovery of their sulfhydryl content and resulted in cellular restoration of the ability to respire. The inhibited cells were viable if diluted and plated shortly after the incubation with the $LP/H_2O_2/SCN$- system. On the other hand, long-term incubation in the presence of the excess OSCN- results in loss of viability. Also, the inhibition of respiration is irreversible. During this long-term incubation, the excess OSCN- was consumed and the sulfenyl derivatives disappear.

The LP/SCN-$/H_2O_2$ system could inhibit the growth of enterotoxigenic *E. coli* strains that cause scouring in neonatal and post-weaning piglets (Grieve et al., 1992). An enzymatic system for H_2O_2 generation (glucose oxidase; 0.1 U/ml) and a chemical source

TABLE 2: Antimicrobial activity of lactoperoxidase system

Microorganism	Donor	H$_2$O$_2$ source	Inhibitory effect	Reference
Gram +ve bacteria				
Bacillus cereus	SCN-	Reagent	Collagenase production	Tenovuo et al., 1983
B. megatherium	SCN-	Reagent	Stasis	Klebanoff et al., 1966
Lactobacillus casei	SCN-	Bacteria	Stasis	Iwamoto et al., 1972
L. acidophilus	I-	Bacteria	Lysine accumulation	Zeldow, 1963
L. plantarum	SCN-	Bacteria	Stasis	Hogendron & Moorer, 1973
L. bulgaricus	SCN-	Bacteria	Acid production	Portman et al., 1962
L. helvaticus				
L. jugurti				
Sarcina lutea	SCN-	Reagent	Stasis	Reiter, 1978
Staph. albus	SCN-	Reagent	Stasis	Klebanoff et al., 1966
Staph. aureus	I- or SCN-	*S. mitis*	Amino acid uptake	Harmon & Klebanoff, 1973
Strep. cremoris	Milk/ SCN-	Bacteria	Glycolysis	Reiter et al., 1963
S. lactis	SCN-	Reagent	Stasis	Marshall & Reiter, 1980
S. pyogenes	Milk	Bacteria	Cidal / production	Mickelson, 1979
S. agalactiae	Milk/ SCN-		Lactic acid production	Brown & Mickelson, 1979
			Sugar transport	Mickelson, 1977
S. fecalis	SCN-	Bacteria	Stasis	Klebanoff et al., 1966
S. mitis	SCN-	Reagent	Peroxide excretion	Carlsson et al., 1983
S. mutans	SCN-	Bacteria	Sugar uptake	Thomas et al., 1983
S. salivarius	SCN-	Reagent	Acid production	Carlsson et al., 1983
S. sanguis	SCN-	Bacteria	Stasis / acid production	Carlsson et al., 1983
Gram -ve bacteria				
Escherichia coli	SCN-	Reagent	Stasis	Klebanoff et al., 1966
	I-	Reagent	Stasis	Klebanoff, 1967
	Whey	Oxidase	Amino acid uptake	Harmon & Klebanoff, 1973
	Milk	Reagent	Stasis	Stephens et al., 1979
L. pneumophila	Cl-	Reagent	Cidal	Lochner et al., 1983
S. typhimurium	SCN-	Oxidase	Cidal	Reiter et al., 1976
		Reagent	Cidal / stasis	Purdy et al., 1983
P. aeruginosa	SCN-	Oxidase	Cidal	Reiter et al., 1976
		Reagent	Cidal	Pruitt & Reiter, 1985
P. fluorescens	Milk/ SCN-	Oxidase	Cidal	Björck et al., 1975
		Reagent	Cidal	Björck, 1978
Fungi				
Candida tropicalis	I-	*S.mitis*	Cidal	Harmon & Klebanoff, 1973
Candida albicans	I-	Reagent	Cidal	Lehrer, 1969
Rhodotorula rubra	SCN-	Oxidase	Stasis	Popper & Knorr, 1997
Sacch. cerevisiae				
Aspergillus niger				
Byssochlamys fulva				
Viruses				
Polio virus	I- / Br-	Reagent	Cidal	Belding & Klebanoff, 1970
Vaccinia virus				
HSV-1	SCN-	Reagent	Delay/loss of cytopathy	Courtois et al., 1990
HIV-1	SCN-/I-	Oxidase	Cell infectivity	Yamaguchi et al., 1993

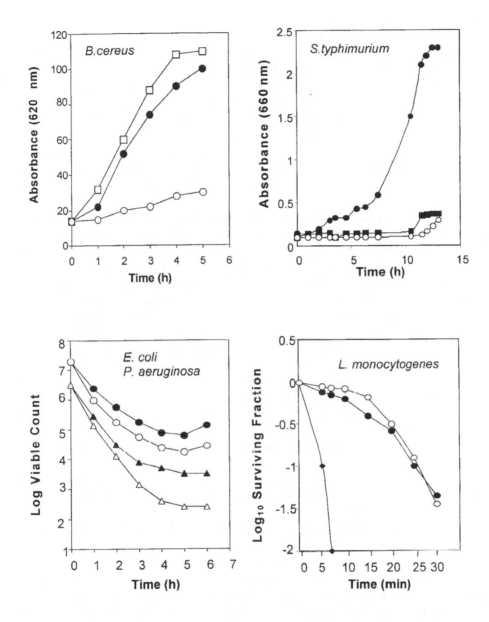

FIGURE 2: Antimicrobial spectrum of lactoperoxidase system. (i) *Bacillus cereus* growth was inhibited by 250 mU/ml LP and 1 mM I- in the presence of H_2O_2 either at 0.1 mM (●) or 0.2 mM (O) concentration compared to the control (□) (adapted from Tenovuo et al., 1985). (ii) *Salmonella typhimurium* rough mutant TA1535 growth in LP system containing 300 μM OSCN- either with 50 μg/ml of catalase (■) or without catalase (O) and compared to control (●) (adapted from Purdy et al., 1983). (iii) Cidal effects of LP system against *Escherichia coli* in the presence of SCN- at concentrations of 0.15 mM (▼) or 0.30 mM (▽); also against *Pseudomonas aeruginosa* with SCN- at concentrations of 0.15 mM (●) or 0.30 mM (O) (adapted from Reiter et al., 1976). (iv) *Listeria monocytogenes* survival curves at 57.8 °C in milk (O), in milk containing 0.6 mM H_2O_2 (●) and milk containing lactoperoxidase system, 2.4 mM SCN- / 0.6 mM H_2O_2 / 9.2 μg of inherent LP/ml (♦) (adapted from Kamau et al., 1990).

(sodium carbonate peroxyhydrate; 90 mg/L) were used in the LP system to test 19 strains in a 6-h growth assay at 37°C. Only three strains were highly sensitive to the LP/glucose oxidase system, while all showed marked growth inhibition with the LP/peroxyhydrate system. H_2O_2 alone had less effect than the complete system.

The interaction between milk xanthine oxidase and LP in a model system and antimicrobial action of these enzymes on *E. coli O111* was reported (Al'perovich et al., 1992). Bacterial superoxide dismutase (SOD), which transforms O^{2-}. (reaction product of xanthine oxidase) into H_2O_2 was necessary for the reaction. It was suggested that the combination of these enzymes possibly protect neonates against intestinal infections. Bacterial SOD could act as the key factor, creating the antimicrobial LP system. Antimicrobial activity of LP system against verotoxigenic *E.coli* O157:H7 has also been reported (Heuvelink et al., 1998).

The LP/SCN-/H_2O_2 system exerts both bacteriostatic and bactericidal activities against strains of *S. typhimurium* (Purdy et al., 1983). The bactericidal activity is dependent on the permeability of the bacterial cell envelope. The deep rough mutant TA1535, with the most permeable cell envelope, is killed both at neutral and acid pH, whereas very little or no killing is observed with the intact cells of the wilds-type parent strain. The delta gal mutant, TA1530, representing an intermediate in cell envelope permeability, is inhibited to a much lesser extent than TA1535. Bacteria in log phase of growth are more sensitive to the bactericidal effects than are those in stationary phase. Growth phase has little influence on the bacteriostatic effects. The wild-type strain produced significant quantities of acid in the presence of glucose. This acid production is inhibited by the LP/SCN-/H_2O_2 system. In contrast, this inhibition is not reversed by addition of a reducing agent (2-mercaptoethanol) in several strains of streptococci.

The inhibition of *Bacillus cereus* by LP and myeloperoxidase antimicrobial systems was reported (Tenovuo et al., 1985). With the LP/SCN-/H_2O_2 system, the growth inhibition is directly proportional to the amount of OSCN- ions present. This inhibition is associated with reduced extracellular release of collagenase activity from the cells. With LP, the antimicrobial efficiency of the oxidizable substrates SCN- is greater than I-, and with myeloperoxidase, the efficiency was I- greater than Cl- greater than SCN-, respectively. LP failed to oxidize Cl-.

The LP/SCN-/ H_2O_2 system consisting of LP (0.37 U/ml), KSCN (0.3 mM), and H_2O_2 (0.3 mM), delayed but did not prevent growth of *Listeria monocytogenes* at 30°C in broth and at 20°C in milk (Siragusa & Johnson, 1989).

LP (10-200 μg/ml) could elicit significant reduction of *S. mutans* adherence to hydroxyapatite in a dose-dependent manner (Roger et al., 1994). The strongest inhibition of adhesion was found when both saliva-coated apatite and bacteria were pretreated with LP. The inhibition of adherence of a serotype c strain of *S. mutans* to saliva-coated hydroxyapatite is a novel antibacterial mechanism for LP.

Stephens and co-workers (1979) examined the LP activity in milk of sows suckling piglets infected with *E. coli*. A 5-fold increase of LP activity in milk was observed during the 3-4 weeks of lactation. The antibacterial activity of milk from sows suckling normal young increased with the LP, and this bactericidal activity could be reversed by LP inhibitors such as penicillamine and cysteine but not by iron supplementation to saturate the lactoferrin. In milk from sows suckling infected young, bacteriostatic activity was observed in 14 days after infection and needed iron or both iron and penicillamine (or cysteine) for reversal, indicating the role of both the antibody-lactoferrin and the LP systems.

In vivo antimicrobial activation of the LP system (LP/SCN-/ H_2O_2) in the calf abomasum was reported (Reiter et al., 1980). Milk provided the two essential factors, namely the LP and the SCN-, while the third factor was rendered either by an H_2O_2 generating system (glucose oxidase and glucose) or by H_2O_2-producing lactobacilli (natural gut flora in the abomasum of the calf).

Marshall and co-workers (1986) suggested an important role for the LP/SCN-/H_2O_2 system in protecting lactating mammary glands from *Streptococcus uberis* infections. However, this protective barrier seemed to be ineffective as the involution progressed.

Products of SCN- oxidation by LP could inhibit peroxide producing gram-positive bacteria. Such products demonstrate cidal activity against gram-negative bacteria including *Pseudomonas spp.* and *E. coli* (Bjorck et al., 1975). The LP system could inhibit most strains of lactobacilli and streptococci; however, the sensitivity varies from strain to strain. Exposure of susceptible cells to the LP system could cause a rapid (<1 min) inhibition of bacterial metabolism resulting in leakage of amino acids and potassium. Various microbial cellular events including carbohydrate transport and utilization; oxygen uptake; amino acid and purine transport (protein and nucleic acid syntheses); production and excretion of extracellular products (i.e. lactic acid, H_2O_2, collagenase); and growth may be inhibited.

LP/SCN-/ H_2O_2 complex inhibits transport of 2-deoxyglucose or glucose in *Streptococcus agalactiae*. The inhibition is reversible with dithiothreitol. *N*-ethylmaleimide, *p*-chloromercuribenzoate and dithiothreitol. The inhibition of glucose transport by the LP-complex and its reversibility with dithiothreitol suggest the modification of functional sulfhydryl groups in the cell membrane as a cause of transport inhibition (Mickelson, 1977). The effects of the LP/SCN-/H_2O_2 system on *S. mutans, S. sanguis, S. mitis,* and *S. salivarius,* in particular, the rate of acid production and oxygen uptake by intact cells of, the activity of glycolytic enzymes in cell-free extracts, and the levels of intracellular glycolytic intermediates were reported (Carlsson et al., 1983). All strains consumed oxygen in the presence of glucose. *S. sanguis, S. mitis,* and anaerobically grown *S. mutans* excreted H_2O_2. There was higher NADH oxidase and NADH peroxidase activity in aerobically grown cells than in anaerobically grown cells. NADPH oxidase activity was low in all species. Acid production, oxygen uptake, and, consequently, H_2O_2 excretion were inhibited in all the strains by LP/SCN-/H_2O_2 system. *S. sanguis* and *S. mitis* had a higher capacity than *S. mutans* and *S. salivarius* to recover from this inhibition. Higher activity in the former strains of an NADH-OSCN oxidoreductase, which converted OSCN- into SCN-, could be accounted for this difference. The change in levels of intracellular glycolytic intermediates after inhibition of glycolysis by OSCN- and the actual activity of glycolytic enzymes in cell-free extracts in the presence of OSCN- indicated that the primary target of OSCN- in the glycolytic pathway was glyceraldehyde 3-phosphate dehydrogenase.

The susceptibility of *Capnocytophaga ochracea, Eikenella corrodens, Eubacterium yurii, Fusobacterium nucleatum, Peptostreptococcus micros, Prevotella intermedia, Selenomonas sputigena,* and *Wolinella recta* to OSCN- generated by the LP system was reported (Courtois et al., 1992). Bacterial growth was inhibited after OSCN-exposure, with an intra- and inter-species variability from 0 to 95% for *C. ochracea,* 34-

100% for *E. corrodens*, 0-83% for *E. yurii*, 1-15% for *F. nucleatum*, 8-61% for *P. micros*, 0-100% for *P. intermedia*, 0-44% for *S. sputigena* and 0-8% for *W. recta*.

B. Antifungal effects

Candida tropicalis could be killed by the LP system when I- was used as electron donor (Hamon & Klebanoff, 1973). In this study, H_2O_2 was provided by a strain of peroxidogenic oral streptococci (*S. mitis*) inoculated into the test medium along with the *C. tropicalis*. Lehrer (1969) reported the fungicidal activity of peroxidase/ H_2O_2/I- system against several strains of *C. albicans* as well as species of *Saccharmoyces*, *Rhodoturula*, *Geotrichum*, and *Aspergillus*.

The ability of LP (5, 50, and 500 U/ml) to degrade aflatoxin in the presence of NaCl (225 µM) and H_2O_2 (50 µM) was reported (Doyle and Marth, 1978). Increasing the amount of LP from 50 to 500 units/ml of reaction mixture resulted in augmenting the rate of degradation of aflatoxin B1 from 3.6 to 5.1%/24 h. When comparable amounts of LP were present, aflatoxin G1 was degraded approximately 1.5 times faster than was aflatoxin B1.

Popper and Knorr (1997) reported the antifungal activity of LP system with glucose oxidase as H_2O_2 source in salt solution and in apple juice. *Rhodutorula rubra* and *Saccharomyces cerevisiae* cultivated aerobically in apple juice; and agar-grown *Mucor rouxii*, *Aspergillus niger* and *Byssochlamys fulva* were tested. The antifungal activity of LP was tested with initial counts of about 10^5 cfu/ml (yeast cells or spores) in salt solution supplemented with SCN- (25 mg/L) and glucose (20 g/L) or in apple juice supplemented with the same amount of SCN-. Antifungal activity was observed against all test organisms in both media. The yeast strains were found to be least stable while *B. fulva* was most resistant. However, a combination of LP (5 U/ml) with glucose oxidase (0.5 to 1 U/ml) caused total inactivation of this mold in salt solution within 2 h. The LP system also showed antifungal activity in apple juice at acid pH (3.2), although its effectiveness was reduced. In this medium, *B. fulva* was inactivated by LP (20 U/ml) and glucose oxidase (1 U/ml) within 4 h. Strains of *R. rubra* and *S. cerevisfiae* were also inhibited in apple juice by LP (5 U/ml) and glucose oxidase (1 U/ml).

C. Antiviral effects

LP could kill both RNA (poliovirus) and DNA (vaccina) virus with halides (I-, Br-) as electron donors (Belding & Klebanoff, 1970). These particular viruses are more resistant than are most others to the effects of drying, heat, and disinfectants. Optimum virucidal activity could be obtained with LP at pH 4.5.

Preincubation with LP/SCN- oxidizing system before inoculation in MRC5 fibroblast cell line could result either in delay or loss of cytopathic potential of HSV1 (Courtois et al., 1990). This antiviral effect was found to be time-dependent.

Pourtois et al. (1990) reported the inhibition of human immunodeficiency virus type-1 (HIV-1) infectivity of ARV-4 cell line by LP generated OSCN-. LP and glucose oxidase are also virucidal to HIV-1 in the presence of sodium iodide, as assessed by the loss of viral replication in a syncytium-forming assay or by the inhibition of cytopathic effects on infected cells (Yamaguchi et al., 1993). In the presence of low concentrations of sodium iodide, HIV-1 was susceptible to virucidal effects of LP at low enzyme concentrations. The loss of viral replication was linearly related to the time of incubation in the enzyme solutions, with an inactivation rate of 1 log unit every 30-min. These *in vitro* findings demonstrate that the $LP/H_2O_2/halide$ system provides potent virucidal activity against HIV-1.

VII. MULTIFUNCTIONALITY

The $LP/SCN-/H_2O_2$ system was found to inhibit lipoprotein lipase (LPL) in milk (Ahrne & Bjorck, 1985). This inhibition was related to the amount of SCN-/ H_2O_2 added to milk. Lipolysis was unaltered until LPL activity was decreased to less than 40% of the original activity. When LPL was inhibited further, decrease in lipolysis was observed. LP also inhibits purified LPL at similar concentrations of SCN-/ H_2O_2 as in milk. Non-enzymatically generated OSCN- inhibited LPL at concentrations comparable to those produced by the LP. Cysteine could eliminate the effect of LP on LPL. The sulphydryl reagents dithiobisnitrobenzoic acid and sodium tetrathionate did not affect LPL activity.

The ability of LP to activate peritoneal macrophages from C57BL/6 mice and the resulting cytotoxic activity toward 3T12 cells *in vitro* was reported (Wei et al., 1986). A 50% cytotoxic effect on 3T12 cells by macrophages stimulated with LP (1.6 μM) was observed. On a molar basis the peroxidases are much less potent macrophage activators than interferon (alpha + beta) and endotoxin. Nevertheless, the data clearly indicated that peroxidases are a group of enzymes capable of inducing macrophage activation, resulting in cytostatic and/or cytocidal activity.

A. Tumoricidal activity

The toxicity of various combinations of $LP/SCN-/H_2O_2$ was reported against cell lines of human epidermoid carcinoma of the cervix (HeLa), a Chinese hamster ovary cell line (CHO), and human gingival fibroblasts (Hanstrom et al., 1983). The capacity of cells to proliferate, exclude trypan blue and produce acids from glucose was evaluated. One-hour exposure of the cells to 10 μM H_2O_2 reduced cell proliferation. After 1-h exposure to 100 μM H_2O_2 the cells were irreversibly damaged. In the presence of LP and SCN- the cells were protected from the toxic effects of H_2O_2 as evidenced by their capacity to proliferate and exclude trypan blue. The rate of acid production of cells in the presence of glucose was inhibited, however, by the combination $LP/SCN-/H_2O_2$. The kinetics of the cytolytic activity of LP on erythrocytes in the presence of H_2O_2 and I- was investigated at physiological pH (McFaul et al., 1986). Optimal concentrations of H_2O_2 and I- were found to be 40 and 25 μM, respectively, while higher concentrations of H_2O_2 inhibited the cytolytic activity. The cytolytic effect was maximal at pH 6.3, and was inhibited by tyrosine, tryptophan, cysteine, and to a lesser extent by histidine.

Stanislawski and co-workers (1989) reported the tumoricidal potency of enzyme immunotoxins constructed of antibodies conjugated to glucose oxidase and to LP. Murine plasmacytoma cells were targeted *in vitro* with the use of affinity-purified rabbit anti-plasmacytoma membrane antibodies (conjugated to glucose oxidase or LP) or rabbit serum raised against plasmacytoma microsome membranes followed by goat anti-rabbit immunoglobulin conjugates (to glucose oxidase or LP). Cytotoxicity was generated subsequently by incubation of the washed cells in a medium supplemented with glucose and sodium iodide, which were the substrates of these enzymes. This resulted in the presumed metabolic release of highly toxic reduced oxygen species and iodinated derivatives. Targeting of tumor cells with both conjugates produced a synergistic killing effect.

Halide-dependent cytolysis of B-16 melanoma (black tumor) cells mediated by LP and myeloperoxidase systems was reported (Odajima et al., 1996). LP could replace myeloperoxidase in systems containing bromide, but not chloride. A significant suppression of black tumors was detected in the groups of mice inoculated with melanoma cells exposed to the systems containing LP or myeloperoxidase.

VIII. APPLICATIONS

A. Preservation of milk

In dairy practice, milk is often cooled and stored for long periods prior to distribution. Such milk is susceptible to spoilage by psychrotrophs (mainly pseudomonads) that survive pasteurization. In many developing countries, raw, uncooled milk is taken to collecting centers, also without cooling facilities, and transported for many hours to the creameries. Depending on the original hygienic quality, ambient temperature, and distance from the collecting center, a high proportion of the milk becomes unfit for further handling, distribution as liquid milk, or for manufacture. The addition of SCN- and sodium percarbonate, as a source of H_2O_2 could increase the shelf life of milk sufficiently to be acceptable even after transportation for several hours at high ambient temperatures. Milk does not normally contain sufficient SCN- for the LP system, and therefore SCN- or sodium percarbonate must be added as an H_2O_2 source.

LP catalyzes the oxidation of SCN- by H_2O_2 and generates an intermediary product with antibacterial properties. The components of this system, with the exception of H_2O_2, are present in milk. Thus, generation of H_2O_2 by enzymatic process in milk could make the milk LP system complete. Accordingly, a two-enzyme coupled system consisting of ß-galactosidase and glucose oxidase has been developed to activate the LP system to inactivate milk-borne bacteria (Bjorck, 1976). The use of LP system for improvement of raw milk storage at low temperatures was suggested (Bjorck, 1978). Activation of the antibacterial LP system in milk by increasing SCN- concentration to 0.25 mM and adding an equimolar amount of H_2O_2, results in a substantial reduction of the bacterial flora and prevents the growth of psychrotrophic bacteria for up to 5 days. This treatment neither alters physico-chemical properties of milk nor develops LP-resistant bacteria.

The antimicrobial ability of the LP system is increased by the addition of larger amounts of SCN- and H_2O_2. Laboratory and field trials revealed that such activated systems could preserve poor-quality raw milk for longer periods of time, at tropical temperatures (Aparicio et al., 1986). It is possible to preserve certain milks at 20°C for more than one day, without diminishing the overall keeping quality. At 36°C, milks did not show acidity development for about 10 hours. Tests conducted under real collection and transportation conditions validate these findings. It was therefore proved that the system can be used practically and that its bactericidal/bacteriostatic effect on the spoilage flora of milk can be increased to overcome adverse conditions of milk handling in the tropics.

Conventional methods of ensuring the safety of cow milk for human consumption, such as pasteurization, are not always practical in poor socioeconomic conditions or in rural communities that lack modern amenities. Activation of the LP system and souring of milk were investigated as potential alternative methods to sustain the safety of milk by controlling potential pathogens (Kangumba et al., 1997). The activation of the LP system could inhibit the growth of *S. aureus* and *E. coli* by 2-log values. The inhibition of *Brucella abortus* is negligible. The multiplication of *Coxiella burnetti* in milk is not disturbed even after 17 h of LP-system activation at 20°C. Souring inhibited the growth of *S. aureus* and *E. coli* also by 2-log values.

In acidified raw milk, supplementation of LP with an exogenous supply of H_2O_2, rapidly decreases the salmonella counts. Different salmonella serotypes are inhibited; rough strains, however, are more susceptible than smooth strains. When calves are fed on fresh milk, containing LP, and challenged with *Salmonella typhimurium* in doses of either 10^9 or 10^{10} cfu, the clinical findings and salmonella excretion patterns were similar to those of control calves fed on heated milk. The role of LP as a possible non-antibiotic system to control salmonellosis in dairy farms was suggested (Wray & McLaren, 1987).

B. Control of *Listeria* sp.

The LP system seems to enhance the thermal destruction of *Listeria monocytogenes* and *S. aureus* (Kamau et al., 1990). After LP activation, biphasic survival curves were observed for *L. monocytogenes* at 57.8 °C and for *S. aureus* at 55.2°C. The more heat-sensitive fractions (93% of the *L. monocytogenes*, 92% of the *S. aureus*) were killed almost instantly. The most rapid killing of *L. monocytogenes* occurred when samples were heated soon after activation of the LP. It was suggested that LP activation followed by heating could increase the margin of safety with respect to milk-borne pathogens.

Activity of the raw milk LP/SCN-/H_2O_2 system on four *Listeria monocytogenes* strains at refrigeration temperatures after addition of 0.25 mM sodium thiocyanate and 0.25 mM H_2O_2 was reported (Gaya et al., 1991). The LP system elicited cidal activity against *L. monocytogenes* at 4 and 8°C; the activity was dependent on temperature, length of incubation, and strain of *L. monocytogenes* tested. D values in activated LP-system milk for the four strains tested ranged from 4.1 to 11.2 days at 4°C and from 4.4 to 9.7 days at 8°C. The LP level in raw milk declined during a 7-day incubation, the decrease being more pronounced at 8°C than at 4°C and in control milk than in activated-LP system milk. The SCN- concentration decreased considerably in activated-LP system milk at both temperatures during the first 8-h of incubation. LP system activation was shown to be a feasible procedure for controlling growth of *L. monocytogenes* in raw milk at refrigeration temperatures.

The listeriocidal activity of bacteriocin-producing lactic acid bacteria alone and in combination with the LP system in refrigerated raw milk was reported (Rodriguez et al., 1997). After 4 days at 4°C, *L. monocytogenes* counts in milk were reduced by 0.21-0.24 log units when inoculated with bacteriocin-producing *Lactococcus lactis ssp. lactis* ATCC 11454, *L. lactis ssp. lactis* ESI 515 or *Enterococcus faecalis* INIA 4. Activation of the LP system did not enhance inhibition at this temperature. After 4 days at 8°C, *L. monocytogenes* counts in the non-activated LP system milk inoculated with lactic acid bacteria were reduced by 1.87, 1.54 and 1.11 log units compared to control milk, whereas in the activated LP system milk, this reduction was 1.99, 2.10 and 1.06, respectively. The higher nisin production by *L. lactis ssp. lactis* ESI 515 in milk with activated LP system than in non-activated LP system milk was credited with the enhanced listeriocidal effect.

Addition of nisin (10 or 100 IU/ml) to ultra-high temperature (UHT) processed skim milk did not inhibit *L. monocytogenes* after 24 h at 30°C. However, under similar conditions, addition of the LP system resulted in decrease in listerial counts by three log units. When the two preservatives were added to actively growing cells of *L. monocytogenes* in two steps with a 2-h interval, their synergistic effect was enhanced (Zapico et al., 1998).

C. Control of oral pathogens

LP/SCN-/ H_2O_2 system-generated OSCN- and hypothiocyanous acid (HOSCN) inhibits a number of oral bacteria, including mutans streptococci. The effects of LP on model biofilms of *Staphylococcus aureus, Staphylococcus epidermidis, Pseudomonas fluorescens,* and *Pseudomonas aeruginosa* on steel and polypropylene substrata; and plaque-resembling biofilms of *Streptococcus mutans, Actinomyces viscosus,* and *Fusobacterium nucleatum* on saliva-coated hydroxyapatite was examined (Johansen et al., 1997). Fluorescence microscopy and an indirect conductance assay (measuring carbon dioxide evolution) were used to evaluate the antimicrobial activity of enzymes against bacterial cells in biofilm. LP combined with glucose oxidase was bactericidal against biofilm bacteria but did not remove the biofilm from the substrata. A complex mixture of polysaccharide-hydrolyzing enzymes was able to remove bacterial biofilm from steel and polypropylene substrata but did not have a significant bactericidal activity. However, combining oxidoreductases with polysaccharide-hydrolyzing enzymes resulted in bactericidal activity as well as removal of the biofilm. Toothpaste (Biotene™) that comprises the complete LP system is commercially available.

D. Animal trials

In vivo protective effect of LP in calves was suggested based on four animal trials (>200 calves). Calves are susceptible to scour (diarrhea) early in life, particularly during transfers to large calf units. Scouring is either dietary when milk replacers are used or caused by infection with *E. coli* or viruses. Depending on the severity and period of infection, afflicted animals lose weight, are retarded in growth development and may die. Reiter and co-workers (1981) demonstrated that feeding of LP supplemented milk could prevent incidence of scouring in calves and also improve weight-gain in LP-treated animals.

The clinical efficacy of a preparation based on the LP system and lactoferrin was tested in calves experimentally infected with enterotoxigenic *E. coli* (Still et al., 1990). Mortality, occurrence and duration of diarrhea were significantly lower (P <0.05) and general clinical status significantly improved (P <0.05) in infected calves treated with LP and lactoferrin preparation than in infected but non-treated calves. The study suggested that LP and lactoferrin are effective in the treatment of enteric colibacillosis in calves.

Oral administration of LP and passive enteral immunization with lactobacilli sera has been suggested in protection against intestinal infections with salmonellae and shigellae (Denisova et al., 1987). Combined therapy with LP, lactoferrin and lactoglobulin could elicit synergistic antimicrobial effect against *Shigella sonnei* during intranasal challenge in an experimental mouse model (Denisova et al., 1996). This phenomenon was nonspecific and the therapy was suggested for passive immunization against shigellosis.

IX. SAFETY AND TOLERANCE

In vivo clearance

In vivo uptake of bovine LP in a calcium-dependent, saturable process was reported using a rat liver model (Hildenbrandt & Aronson, 1985). Both hepatocytes and Kupffer or other nonparenchymal demonstrated the endocytosis of LP. The mediating receptors were the Gal/GalNAc lectin of hepatocytes and the Man/GlcNAc lectin of nonparenchymal cells. Blocking either one of these receptors caused a large shift in distribution of accumulated LP into the cell type whose receptor was left unblocked, but the extent of uptake was unaffected and the rate was only moderately reduced. Effective inhibition of overall uptake into the perfused organ required the presence of competitors for both receptors. Conversely, LP was an effective competitor of other ligands (asialo orosomucoid or mannan) for either of the two receptors. The major LP clearance was associated with hepatocytes by a process completely inhibited by asialofetuin. A faster cycling time for Gal/GalNAc receptors when bound to LP was suggested. The glycoprotein selectively lost its affinity for Man/GlcNAc receptors when digested by endoglycosidase H, suggesting that LP contain mannose-rich oligosaccharides.

X. SUMMARY

LP is an oxidoreductase secreted into milk, and plays an important role in protecting the lactating mammary gland and the intestinal tract of the newborn infants against pathogenic microorganisms. LP catalyzes the oxidation of thiocyanate and iodide ions to generate highly reactive oxidizing agents. These products have a broad spectrum of antimicrobial effects against bacteria, fungi and viruses. The molecular components of cells that are oxidized are sulfhydryl groups, NADH, HADPH, and under some conditions aromatic amino acid residues. Oxidation of these molecular components alters the functions of cellular systems. The cytoplasmic membrane, sugar, and amino acid transport systems, and glycolytic enzymes may be damaged. The result of such damage may be cell death or inhibition of growth, respiration, active transport, or other vital metabolic functions. The major products responsible for these effects are the hypothiocyanite ion (OSCN-), hypothiocyanous acid (HOSCN), and iodine (I_2). However, minor products are also generated. These minor products are highly reactive and short-lived but have even more potent antimicrobial properties. Microorganisms show great variability in their response to peroxidase systems. Killing of cells is usually greater at low pH, low temperature, and with I- as the electron donor. Gram-negative bacteria are more susceptible to killing and to cytoplasmic membrane damage than are gram-positive species. However, the killing of both species requires long incubation periods (on the order of an hour) and high concentrations ([OSCN-]) > 10 µM at pH 7.0) of the active products. Some species of bacteria are able to resist the antibacterial effects by generating substances that reduce OSCN- to SCN-. These same reducing agents can also enable cells to recover from inhibition by reversing the oxidation of sulfhydryl groups of cellular components caused by peroxidase system products.

The enzymatic properties of LP have found usage in a variety of applications. Its antimicrobial activity has been used to preserve raw milk in warm climates, through the addition of hydrogen peroxide and thiocyanate. Other studies have targeted the cytotoxic effects of LP to specific pathogenic cell types by chemically linking to monoclonal anti-bodies. Also, LP itself has been used as an iodinating agent and is considered to be less disruptive to protein structure.

XI. REFERENCES

1. Adamson, M., and Carlsson, J. 1982. Lactoperoxidase and thiocyanate protect bacteria from hydrogen peroxide. *Infect. Immun.* 35:20-24.
2. Ahrne, L., and Bjorck, L. 1985. Effect of the lactoperoxidase system on lipoprotein lipase activity and lipolysis in milk. *J. Dairy Res.* 52:513-520.
3. Al'perovich, D.V., Kesel'man, E.V., Shepelev, A.P., and Chernavskaia, L.N. 1992. Free-radical mechanism of antimicrobial action of xanthine oxidase and lactoperoxidase. *Biull. Eksp. Biol. Med.* 114:272-274.
4. Aparicio, M.A., Peralta, L.M., and Garcia, H.S. 1986. Potentialization of the lactoperoxidase system for preservation of raw milk in the tropics. *Arch. Latinoam. Nutr.* 36:725-733.
5. Aune, T.M., and Thomas, E.L. 1977. Accumulation of hypothiocyanite ion during peroxidase-catalyzed oxidation of thiocyanate ion. *Eur. J. Biochem.* 80:209-214.
6. Aune, T.M., and Thomas, E.L. 1978. Oxidation of protein sulfhydryls by products of peroxidase-catalyzed oxidation of thiocyanate ion. *Biochemistry* 17:1005-1010.
7. Belding, M.E., Klebanoff, S.J., and Ray, G.C. 1970. Peroxidase-mediated virucidal systems. *Science* 167:195-196.
8. Bjorck, L. 1976. An immobilized two-enzyme system for the activation of the lactoperoxidase antibacterial system in milk. *Biotechnol. Bioeng.* 18:1463-1472.
9. Bjorck, L. 1978. Antibacterial effect of the lactoperoxidase system on psychrotrophic bacteria in milk. *J. Dairy Res.* 45:109-118.
10. Bjorck, L., Claesson, O., and Schulthess, W. 1979. The lactoperoxidase/thiocyanate/hydrogen peroxide system as a temporary preservative for raw milk in developing countries. *Milchwissenschaft* 34:726-729.
11. Bjorck, L., Rosen, C., Marshall, V., and Reiter, B. 1975. Antibacterial activity of the lactoperoxidase system in milk against pseudomonads and other gram-negative bacteria. *Appl. Microbiol.* 30:199-204.
12. Brown, R.W., and Mickelson, M.N. 1979. Lactoperoxidase, thiocyanate, and free cystine in bovine mammary secretions in early dry period and at the start of lactation and their effect on *Streptococcus agalactiae* growth. *Am. J. Vet. Res.* 40:250-255.
13. Cals, M.M., Guillomot, M., and Martin, P. 1994. The gene encoding lactoperoxidase is expressed in epithelial cells of the goat lactating mammary gland. *Cell. Mol. Biol.* 40:1143-1150.
14. Cals, M.M., Mailliart, P., Brignon, G., Anglade, P., and Dumas, B.R. 1991. Primary structure of bovine lactoperoxidase, a fourth member of a mammalian heme peroxidase family. *Eur. J. Biochem.* 198:733-739.
15. Carlsson, J., Edlund, M.B., and Hanstrom, L. 1984. Bactericidal and cytotoxic effects of hypothiocyanite hydrogen peroxide mixtures. *Infect. Immun.* 44:581-586.
16. Carlsson, J., Iwami, Y., and Yamada, T. 1983. Hydrogen peroxide excretion by oral streptococci and effect of lactoperoxidase-thiocyanate-hydrogen peroxide. *Infect. Immun.* 1983 40:70-80.
17. Carlstrom, A. 1969. Physical and compositional investigations of the subfractions of lactoperoxidase. *Acta Chem. Scand.* 23:185-202.
18. Chang, C.S., Sinclair, R., Khalid, S., Yamazaki, I., Nakamura, S., and Powers, L. 1993. An extended X-ray absorption fine structure investigation of the structure of the active site of lactoperoxidase. *Biochemistry* 32:2780-2786.
19. Clem, W.H., and Klebanoff, S.J. 1966. Inhibitory effect of saliva on glutamic acid accumulation by *Lactobacillus acidophilus* and the role of the lactoperoxidase-thiocyanate system. *J. Bacteriol.* 91:1848-1853.
20. Courtois, P., Majerus, P., Labbe, M., Vanden Abbeele, A., Yourassowsky, E., and Pourtois, M. 1992. Susceptibility of anaerobic microorganisms to hypothiocyanite produced by lactoperoxidase. *Acta. Stomatol. Belg.* 89:155-162.
21. Courtois, P., van Beers, D., de Foor, M., Mandelbaum, I.M., and Pourtois, M. 1990. Abolition of herpes simplex cytopathic effect after treatment with peroxidase generated hypothiocyanite. *J. Biol. Buccale* 18:71-74.
22. Denisova, I.I., Gennad'eva, Tia, Zhdanova, T.V., Khristova, M., and Khazenson, L.B. 1996. The action of lactoperoxidase, lactoferrin and lactoglobulin on *Shigella sonnei* in an in-vivo experiment. *Zh. Mikrobiol. Epidemiol. Immunobiol.* 1:66-69.

23. Denisova, I.I., Krasheniuk, A.I., Azhitskii, G.I., Sharaeva, T.K., and Umovskaia, E.A. 1986. Isolation and purification of lactoperoxidase from cow's milk. *Vopr. Med. Khim.* 32:116-119.

24. Denisova, I.I., Vasser, N.R., Bel'kova, E.I., Velichko, L.N., and Gennad'eva, Tia. 1987. Action of the lactoperoxidase system on *Salmonellae* and *Shigellae*. *Zh. Mikrobiol. Epidemiol. Immunobiol.* 11:99-103.

25. Dionysius, D.A., Grieve, P.A., and Vos, A.C. 1992. Studies on the lactoperoxidase system: reaction kinetics and antibacterial activity using two methods for hydrogen peroxide generation. *J. Appl. Bacteriol.* 72:146-153.

26. Doerge, D.R. 1986. Mechanism-based inhibition of lactoperoxidase by thiocarbamide goitrogens. *Biochemistry* 25:4724-4728.

27. Doyle, M.P., and Marth, E.H. 1978. Degradation of aflatoxin by lactoperoxidase. *Z. Lebensm. Unters. Forsch.* 166:271-273.

28. Dull, T.J., Uyeda, C., Strosberg, A.D., Nedwin, G., and Seilhamer, J.J. 1990. Molecular cloning of cDNAs encoding bovine and human lactoperoxidase. *DNA Cell. Biol.* 9:499-509.

29. Dumontet, C., and Rousset, B. 1983. Identification, purification, and characterization of a non-heme lactoperoxidase in bovine milk. *J. Biol. Chem.* 258:14166-14172.

30. Ferrari, R.P., Ghibaudi, E.M., Traversa, S., Laurenti, E., De Gioia, L., and Salmona, M. 1997. Spectroscopic and binding studies on the interaction of inorganic anions with lactoperoxidase. *J. Inorg. Biochem.* 68:17-26.

31. Ferrari, R.P., Laurenti, E., Cecchini, P.I., Gambino, O., and Sondergaard, I. 1995. Spectroscopic investigations on the highly purified lactoperoxidase Fe(III)-heme catalytic site. *J. Inorg. Biochem.* 58:109-127.

32. Gaya, P., Medina, M., and Nunez, M. 1991. Effect of the lactoperoxidase system on *Listeria monocytogenes* behavior in raw milk at refrigeration temperatures. *Appl. Environ. Microbiol.* 57:3355-3360.

33. Goff, H.M., Gonzalez-Vergara, E., and Ales, D.C. 1985. High-resolution proton nuclear magnetic resonance spectroscopy of lactoperoxidase. *Biochem. Biophys. Res. Commun.* 133:794-799.

34. Gothefors, L., and Marklund, S. 1975. Lactoperoxidase activity in human milk and in saliva of newborn infants. *Infect. Immun.* 11.1210-1215.

35. Grieve, P.A., Dionysius, D.A., and Vos, A.C. 1992. In vitro antibacterial activity of the lactoperoxidase system towards enterotoxigenic strains of *Escherichia coli*. *Zentralbl. Veterinarmed.* [B] 39:537-545.

36. Hahn, R., Schulz, P.M., Schaupp, C., and Jungbauer, A. 1998. Bovine whey fractionation based on cation-exchange chromatography. *J. Chromatogr.* 795:277-287.

37. Hamon, C.B., and Klebanoff, S.J. 1973. A peroxidase-mediated, *Streptococcus mitis*-dependent antimicrobial system in saliva. *J. Exp. Med.* 137:438-450.

38. Hannuksela, S., Tenovuo, J., Roger, V., Lenander-Lumikari, M., and Ekstrand, J. 1994. Fluoride inhibits the antimicrobial peroxidase systems in human whole saliva. *Caries Res.* 28:429-234.

39. Hanssen, F.S. 1924. *Br. J. Exp. Pathol.* 5:271.

40. Hanstrom, L., Johansson, A., and Carlsson, J. 1983. Lactoperoxidase and thiocyanate protect cultured mammalian cells against hydrogen peroxide toxicity. *Med. Biol.* 61:268-274.

41. Heuvelink, A.E., Bleumink, B., van den Biggelaar, F.L., Te Giffel, M.C., Beumer, R.R., and de Boer, E. 1998. Occurrence and survival of verocytotoxin-producing *Escherichia coli* O157 in raw cow's milk in The Netherlands. *J. Food Prot.* 61:1597-1601.

42. Hildenbrandt, G.R., and Aronson, N.N. Jr. 1985. Endocytosis of bovine lactoperoxidase by two carbohydrate-specific receptors in rat liver. *Arch. Biochem. Biophys.* 237:1-10.

43. Hoogendoorn, H. 1974. "The Effect of lactoperoxidase-thiocyanate-hydrogen peroxide on the metabolism of cariogenic microorganisms in vitro and in the oral cavity" (Thesis). Delft, Mouton, The Hague, Netherlands.

44. Hoogendoorn, H., and Moorer, W.R. 1973. Lactoperoxidase in the prevention of plaque accumulation, gingivitis and dental caries. I. Effect on oral streptococci and lactobacilli. *Odontol. Revy* 24:355-366.

45. Hulea, S.A., Mogos, S., and Matei, L. 1989. Interaction of lactoperoxidase with enzymes and immunoglobulins in bovine milk. *Biochem. Int.* 19:1173-1181.

46. Huwiler, M., Jenzer, H., and Kohler, H. 1986. The role of compound III in reversible and irreversible inactivation of lactoperoxidase. *Eur. J. Biochem.* 158:609-614.

47. Iwamoto, Y., Nakamura, R., Watanabe, T., and Tsunemitsu, A. 1972. Heterogeneity of peroxidase related to antibacterial activity in human parotid saliva. *J. Dent. Res.* 51:503-8.

48. Jago, G.R., and Morrison, M. 1962. *Proc. Soc. Exp. Biol. Med.* 111:585.

49. Janota-Bassalik, L., Zajac, M., Pietraszek, A., and Piotrowska, E. 1977. Bacteriostatic activity in cow's milk from udders infected with *Streptococcus agalactiae* and *Staphylococcus aureus*. *Acta Microbiol. Pol.* 26:413-419.

50. Jenzer, H., and Kohler, H. 1986. The role of superoxide radicals in lactoperoxidase-catalysed H_2O_2-metabolism and in irreversible enzyme inactivation. *Biochem. Biophys. Res. Commun.* 139:327-332.

51. Jenzer, H., Jones, W., and Kohler, H. 1986. On the molecular mechanism of lactoperoxidase-catalyzed H_2O_2 metabolism and irreversible enzyme inactivation. *J. Biol. Chem.* 261:15550-15556.

52. Johansen, C., Falholt, P., and Gram, L. 1997. Enzymatic removal and disinfection of bacterial biofilms. *Appl. Environ. Microbiol.* 63:3724-3728.

53. Kamau, D.N., Doores, S., and Pruitt, K.M. 1990. Enhanced thermal destruction of *Listeria monocytogenes* and *Staphylococcus aureus* by the lactoperoxidase system. *Appl. Environ. Microbiol.* 56:2711-2716.

54. Kangumba, J.G., Venter, E.H., and Coetzer, J.A. 1997. The effect of activation of the lactoperoxidase system and souring on certain potential human pathogens in cows' milk. *J. S. Afr. Vet. Assoc.* 68:130-136.

55. Kern, V.R., Wildbrett, G., and Kiermeier, F. 1963. Dependence of peroxidase activity in milk on the sexual cycle in cows. *Zeitschrift Naturforschung* 18:1082-1084.

56. Kiermeier, F., and Kuhlmann, H. 1972. Lactoperoxidase activity in human and in cows' milk. Compartive studies. *Munch. Med. Wochenshr.* 114:2144-2146.

57. Klebanoff, S. J. 1967. Iodination of bacteria: a bactericidal mechanism. *J. Exp. Med.* 126:1063-78.

58. Klebanoff, S.J., Clem, W. H., and Luebke, R. G. 1966. The peroxidase-thiocyanate-hydrogen peroxide antimicrobial system. *Biochim. Biophys. Acta.* 117:63-72.

59. Kohler, H., and Jenzer, H. 1989. Interaction of lactoperoxidase with hydrogen peroxide. Formation of enzyme intermediates and generation of free radicals. *Free Radic. Biol. Med.* 6:323-339.

60. Langbakk, B., and Flatmark, T. 1984. Demonstration and partial purification of lactoperoxidase from human colostrum. *FEBS Lett.* 174:300-303.

61. Law, B.A., and John, P. 1981. Effect of lactoperoxidase bactericidal system on the formation of the electrochemical proton gradient in *Escherichia coli*. *FEMS Microbiol. Lett.* 10:67-70.

62. Lehrer, R.I. 1969. Antifungal effects of peroxidase systems. *J. Bacteriol.* 99:361-365.

63. Linford, E. 1974. Cervical mucus: an agent or a barrier to conception? *J. Reprod. Fertil.* 87:239-250.

64. Lochner, J. E., Friedman, R. L., Bigley, R. H., and Iglewski, B. H., 1983. Effect of oxygen-dependent antimicrobial systems on *Legionella pneumophila*. *Infect. Immun.* 39:487-9.

65. Lovaas, E. 1992. Free radical generation and coupled thiol oxidation by lactoperoxidase/SCN-/H_2O_2. *Free Radic. Biol. Med.* 13:187-195.

66. Maguire, R.J., and Dunford, H.B. 1971. The effect of ligand binding on the optical rotatory dispersion of lactoperoxidase. *Can. J. Biochem.* 49:666-670.

67. Maguire, R.J., Dunford, H.B., and Morrison, M. 1971. The kinetics of the formation of the primary lactoperoxidase-hydrogen peroxide compound. *Can. J. Biochem.* 49:1165-1171.

68. Mansson-Rahemtulla, B., Rahemtulla, F., and Humphreys-Beher, M.G. 1990. Human salivary peroxidase and bovine lactoperoxidase are cross-reactive. *J. Dent. Res.* 69:1839-1846.

69. Marcozzi, G., Di Domenico, C., and Spreti, N. 1998. Effects of surfactants on the stabilization of the bovine lactoperoxidase activity. *Biotechnol. Prog.* 14:653-656.

70. Marshall, V.M., and Reiter, B. 1980. Comparison of the antibacterial activity of the hypothiocyanite anion towards *Streptococcus lactis* and *Escherichia coli*. *J. Gen. Microbiol.* 120:513-516.

71. Marshall, V.M., Cole, W.M., and Bramley, A.J. 1986. Influence of the lactoperoxidase system on susceptibility of the udder to *Streptococcus uberis* infection. *J. Dairy. Res.* 53:507-514.

72. McFaul, S.J., Lin, H., and Everse, J. 1986. The mechanism of peroxidase-mediated cytotoxicity. I. Comparison of horseradish peroxidase and lactoperoxidase. *Proc. Soc. Exp. Biol. Med.* 183:244-249.

73. Mickelson, M.N. 1966. Effect of lactoperoxidase and thiocyanate on the growth of *Streptococcus pyogenes* and *Streptococcus agalactiae* in a chemically defined culture medium. *J. Gen. Microbiol.* 43:31-43.

74. Mickelson, M.N. 1977. Glucose transport in *Streptococcus agalactiae* and its inhibition by lactoperoxidase-thiocyanate-hydrogen peroxide. *J. Bacteriol.* 132:541-548.

75. Mickelson, M.N. 1979. Antibacterial action of lactoperoxidase-thiocyanate-hydrogen peroxide on *Streptococcus agalactiae*. *Appl. Environ. Microbiol.* 38:821-826.

76. Mickelson, M.N., and Anderson, A.J. 1984. Cystine antagonism of the antibacterial action of lactoperoxidase-thiocyanate-hydrogen peroxide on *Streptococcus agalactiae*. *Appl. Environ. Microbiol.* 47:338-342.

77. Modi, S., Deodhar, S.S., Behere, D.V., and Mitra, S. 1991. Lactoperoxidase-catalyzed oxidation of thiocyanate by hydrogen peroxide: 15N nuclear magnetic resonance and optical spectral studies. *Biochemistry* 30:118-124.

78. Odajima, T., Onishi, M., Hayama, E., Motoji, N., Momose, Y., and Shigematsu, A. 1996. Cytolysis of B-16 melanoma tumor cells mediated by the myeloperoxidase and lactoperoxidase systems. *Biol. Chem.* 377:689-693.

79. Oram, J.D., and Reiter, B. 1966a. The inhibition of streptococci by lactoperoxidase, thiocyanate and hydrogen peroxide. The effect of the inhibitory system on susceptible and resistant strains of group N streptococci. *Biochem. J.* 100:373-381.

80. Oram, J.D., and Reiter, B. 1966. The inhibition of streptococci by lactoperoxidase, thiocyanate and hydrogen peroxide. The oxidation of thiocyanate and the nature of the inhibitory compound. *Biochem. J.* 100:382-388.

81. Paul, K.G., Ohlsson. P.I., and Henriksson, A. 1980. The isolation and some liganding properties of lactoperoxidase. *FEBS Lett.* 110:200-204.

82. Pfeil, W., and Ohlsson, P.I. 1986. Lactoperoxidase consists of domains: a scanning calorimetric study. *Biochim. Biophys. Acta* 872:72-75.

83. Piatt, J., and O'Brien, P.J. 1979. Singlet oxygen formation by a peroxidase, H_2O_2 and halide system. *Eur. J. Biochem.* 93:323-332.

84. Popper, L., and Knorr, D. 1997. Inactivation of yeast and filamentous fungi by the lactoperoxidase-hydrogen peroxide-thiocyanate-system. *Nahrung* 41:29-33.

85. Portman, A., Gate, Y., and Auclair, J. 1962. *Sixteenth International Dairy Congress.* Vol. B. Copenhagen, Denmark.

86. Pourtois, M., Binet, C., Van Tieghem, N., Courtois, P., Vandenabbeele, A., and Thiry, L. 1990. Inhibition of HIV infectivity by lactoperoxidase-produced hypothiocyanite. *J. Biol. Buccale* 18:251-253.

87. Pruitt, K.M. Tenovuo, J., Fleming, W., and Adamson, M. 1982. Limiting factors for the generation of hypothiocyanate ion, an antimicrobial agent, in human saliva. *Caries Res.* 16:315-323.

88. Pruitt, K.M., Adamson, M., and Arnold, R. 1979. Lactoperoxidase binding to streptococci. *Infect. Immun.* 25:304-309.

89. Pruitt, K.M., and Reiter, B. 1985. Biochemistry of peroxidase system - antimicrobial effects. In *The Peroxidase System – Chemistry and Biological Significance,* ed. K.M. Pruitt, and J.O. Tenovuo, pp.143-177. New York: Marcel Dekker.

90. Pruitt, K.M., and Tenovuo, J. 1982. Kinetics of hypothiocyanite production during peroxidase-catalyzed oxidation of thiocyanate. *Biochem. Biophys. Acta* 704:204-214.

91. Pruitt, K.M., Kamau, D.N., Miller, K., Mansson-Rahemtulla, B., and Rahemtulla, F. 1990. Quantitative standardized assays for determining the concentrations of bovine lactoperoxidase, human salivary peroxidase, and human myeloperoxidase. *Anal. Biochem.* 191:278-86.

92. Pruitt, K.M., Tenovuo, J., Andrews, R.W., and McKane, T. 1982. Lactoperoxidase-catalyzed oxidation of thiocyanate: polarographic study of the oxidation products. *Biochemistry* 21:562-567.

93. Purdy, M.A., Tenovuo, J., Pruitt, K.M., and White, W.E. Jr. 1983. Effect of growth phase and cell envelope structure on susceptibility of *Salmonella typhimurium* to the lactoperoxidase-thiocyanate-hydrogen peroxide system. *Infect. Immun.* 39:1187-1195.

94. Reiter, B. 1978. The lactoperoxidase-thiocyanate-hydrogen peroxide antibacterium system. *Ciba Found. Symp.* 65:285-294.

95. Reiter, B. 1978. Review of nonspecific antimicrobial factors in colostrum. *Ann. Rech. Vet.* 9:205-224.

96. Reiter, B. 1985. The lactoperoxidase system of bovine milk. In *The Peroxidase System - Chemistry and Biological Significance,* ed. K.M. Pruitt, and J.O. Tenovuo, pp.123-141. New York: Marcel Dekker.

97. Reiter, B., Marshall, V.M., and Philips, S.M. 1980. The antibiotic activity of the lactoperoxidase-thiocyanate-hydrogen peroxide system in the calf abomasum. *Res. Vet. Sci.* 28:116-122.

98. Reiter, B., Marshall, V.M.E., Bjorck, L., and Rosen, C.G. 1976. Nonspecific bactericidal activity of the lactoperoxidases-thiocyanate-hydrogen peroxide system of milk against *Escherichia coli* and some gram-negative pathogens. *Infect. Immun.* 13:800-807.

99. Reiter, B., Pickering, A., Oram, J.D., and Pope, G.S. 1963. *J. Gen. Microbiol.* 33:xii.

100. Rodriguez, E., Tomillo, J., Nunez, M., and Medina, M. 1997. Combined effect of bacteriocin-producing lactic acid bacteria and lactoperoxidase system activation on Listeria monocytogenes in refrigerated raw milk. *J. Appl. Microbiol.* 83:389-395.

101. Roger, V., Tenovuo, J., Lenander-Lumikari, M., Soderling, E., and Vilja, P. 1994. Lysozyme and lactoperoxidase inhibit the adherence of *Streptococcus mutans* NCTC 10449 (serotype c) to saliva-treated hydroxyapatite in vitro. *Caries Res.* 28:421-428.

102. Rombauts, W.A., Schroeder, W.A., Morrison, M. 1967. Bovine lactoperoxidase. Partial characterization of the further purified protein. *Biochemistry* 6:2965-2977.

103. Sandholm, M., Ali-Vehmas, T., Kaartinen, L., and Junnila, M. 1988. Glucose oxidase (GOD) as a source of hydrogen peroxide for the lactoperoxidase (LPO) system in milk: antibacterial effect of the GOD-LPO system against mastitis pathogens. *Zentralbl. Veterinarmed.* [B] 35:346-352.

104. Shiro, Y., and Morishima, I. 1986. Structural characterization of lactoperoxidase in the heme environment by proton NMR spectroscopy. *Biochemistry* 25:5844-5849.
105. Sievers, G. 1979. The prosthetic group of milk lactoperoxidase is protoheme IX. *Biochim. Biophys. Acta* 579:181-190.
106. Sievers, G. 1980. Structure of milk lactoperoxidase. A study using circular dichroism and difference absorption spectroscopy. *Biochim. Biophys. Acta* 624:249-259.
107. Sievers, G. 1981. Structure of milk lactoperoxidase. Evidence for a single polypeptide chain. *FEBS Lett.* 127:253-256.
108. Siragusa, G.R., and Johnson, M.G. 1989. Inhibition of *Listeria monocytogenes* growth by the lactoperoxidase-thiocyanate- H_2O_2 antimicrobial system. *Appl. Environ. Microbiol.* 55:2802-2805.
109. Slowey, R.R., Eidelman, S., and Klebanoff, S.J. 1968. Antibacterial activity of the purified peroxidase from human parotid saliva. *J. Bacteriol.* 96:575-579.
110. Stanislawski, M., Rousseau, V., Goavec, M., and Ito, H. 1989. Immunotoxins containing glucose oxidase and lactoperoxidase with tumoricidal properties: in vitro killing effectiveness in a mouse plasmacytoma cell model. *Cancer Res.* 49:5497-504
111. Steele, W.F., and Morrison, M. 1969. Antistreptococcal activity of lactoperoxidase. *J. Bacteriol.* 97:635-639.
112. Stephens, S., Harkness, R.A., and Cockle, S.M. 1979. Lactoperoxidase activity in guinea-pig milk and saliva: correlation in milk of lactoperoxidase with bactericidal activity against *Escherichia coli*. *Br. J. Exp. Pathol.* 60:252-258.
113. Still, J., Delahaut, P., Coppe, P., Kaeckenbeeck, A., and Perraudin, J.P. 1990. Treatment of induced enterotoxigenic colibacillosis (scours) in calves by the lactoperoxidase system and lactoferrin. *Ann. Rech. Vet.* 21:143-152.
114. Tenovuo, J. 1981. Different molecular forms of human salivary lactoperoxidase. *Arch. Oral Biol.* 26:1051-1055.
115. Tenovuo, J., and Knuuttila, M.L. 1977. Antibacterial effect of salivary peroxidases on a cariogenic strain of *Streptococcus mutans*. *J. Dent. Res.* 56:1608-1613.
116. Tenovuo, J., Makinen, K. K., Sievers, G. 1985. Antibacterial effect of lactoperoxidase and myeloperoxidase against *Bacillus cereus*. *Antimicrob. Agents Chemother.* 27:96-101.
117. Tenovuo, J., Makinen, K.K., and Sievers, G. 1985. Antibacterial effect of lactoperoxidase and myeloperoxidase against *Bacillus cereus*. *Antimicrob. Agents Chemother.* 27:96-101.
118. Tenovuo, J., Modoveanu, Z., Mestecky, J., Pruitt, K.N., and Mansson-Rahemtulla, B. 1982. Interaction of specific and innate factors of immunity: IgA enhances the antimicrobial effect of the lactoperoxidase system against *Streptococcus mutans*. *J. Immunol.* 128:726-731.
119. Thomas, E.L. 1981. Lactoperoxidase-catalyzed oxidation of thiocyanate: equilibria between oxidized forms of thiocyanate. *Biochemistry* 20:3273-80.
120. Thomas, E.L., and Aune, T.M. 1978. Lactoperoxidase, peroxide, thiocyanate antimicrobial system: correlation of sulfhydryl oxidation with antimicrobial action. *Infect. Immun.* 20:456-463.
121. Thomas, E.L., and Aune, T.M. 1978. Susceptibility of Escherichia coli to bactericidal action of lactoperoxidase, peroxide, and iodide or thiocyanate. *Antimicrob. Agents Chemother.* 13:261-265.
122. Thomas, E.L., Pera, K.A., Smith, K.W., and Chwang, A.K. 1983. Inhibition of *Streptococcus mutans* by the lactoperoxidase antimicrobial system. *Infect. Immun.* 39:767-778.
123. Ueda, T., Sakamaki, K., Kuroki, T., Yano, I., and Nagata, S. 1997. Molecular cloning and characterization of the chromosomal gene for human lactoperoxidase. *Eur. J. Biochem.* 243:32-41.
124. Wei, R.Q., Lefkowitz, S.S., Lefkowitz, D.L., and Everse, J. 1986. Activation of macrophages by peroxidases. *Proc. Soc. Exp. Biol. Med.* 182:515-521.
125. Wray, C., and McLaren, I. 1987. A note on the effect of the lactoperoxidase systems on salmonellas in vitro and in vivo. *J. Appl. Bacteriol.* 62:115-118.
126. Wright, R.C., and Tramer, J. 1958. Factors influencing the activity of cheese starters. The role of milk peroxidase. *J. Dairy Res.* 25: 104-118.
127. Wutrich, S., Richterich, R., and Hostettler, H. 1964. *Zeitschrift Lebensmittel-Untersuch. Forsch* 124: 345.
128. Yamaguchi, Y., Semmel, M., Stanislawski, L., Strosberg, A.D., and Stanislawski, M. 1993. Virucidal effects of glucose oxidase and peroxidase or their protein conjugates on human immunodeficiency virus type 1. *Antimicrob. Agents Chemother.* 37:26-31.
129. Zapico, P., Medina, M., Gaya, P., and Nunez, M. 1998. Synergistic effect of nisin and the lactoperoxidase system on *Listeria monocytogenes* in skim milk. *Int. J. Food Microbiol.* 40:35-42.
130. Zeldow, B. J. 1963. Studies on the antibacterial action of human saliva. III. Cofactor requirements of a lactobacillus bactericidin. *J. Immunol.* 90:12-16.

E.F. Bostwick
J. Steijns
S. Braun

Lactoglobulins

4

I. INTRODUCTION

One of the best known natural antimicrobial systems is the production of immunoglobulins (Ig). Ig or antibodies are produced by all vertebrates in lymphocytes or immunocytes in response to an antigen. Ig have the ability to recognize, precipitate or otherwise neutralize bacteria, viruses, polysaccharides, nucleotides, peptides and proteins. Ig can provide extremely specific protections such that single molecules are targeted or a single type of virus or bacterium. This specific protection has long been recognized and was the principle on which vaccines were developed, long before there was an understanding of the immune system or Ig.

It was clear as early as 430 BC that surviving the plague provided immunity against future attack, but only against that specific disease. Specific reasons for this immunity and elucidation of the immune system would not occur until the end of the 19th century, but by then the principle had been used for a considerable period of time in vaccines such as smallpox (Stryer, 1975).

The benefit of colostrum, which contains high levels of Ig (lactoglobulins), was well known more than one hundred years ago (Ehrlich, 1892). Colostrum was known as a means of transporting immune factors from mother to newborn. In pigs, horses, sheep and calves this is the primary source of immunity, since there is limited transfer of maternal Ig *in utero* via the placenta. In humans, rabbits, rats and mice, passive immunity is conveyed by transfer via the uterus, although colostrum still plays an important role in these species. There have been extensive reviews on colostrum and its ability to provide immunity to offspring (Roy, 1956; Butler, 1994; Reddy et al., 1988; Yolken, 1985; Hilpert, 1987; Stephan, 1990).

The use of Ig as antimicrobials and specifically generating, purifying and commercializing the fractions for prevention and treatment of various microbial diseases is relatively recent. A number of human trials have been performed and reviewed (Weiner, 1999; Davidson, 1996). Use in food systems as antimicrobials for preservative reasons has been suggested (Korhonen, 1998; Korhonen et al., 1998) but has been limited due to costs,

0-8493-2047-X/00/$0.00+$.50
© 2000 by CRC Press, LLC

stability and the need for either broader specificity or higher activity against a single organism, neither of which is obtainable without purification or hyper-immunization of the animal source.

Although it is possible to derive Ig from a number of commercial sources, including cows, sheep, goats, cell culture and eggs among others, the most practical and therefore more probable commercial sources with current technology are bovine milk and colostrum.

II. OCCURRENCE

Ig can be characterized a number of ways, but one method is to characterize them as secretory and serum fractions. Secretory Ig appears in external secretions, particularly in intestinal mucosa, lacteal secretions, tears, and bronchial mucosa. Since the focus of this chapter is on Ig (lactoglobulin) as natural antimicrobial systems in foods, the primary emphasis will be on non-serum fractions. The most commonly consumed foods containing Ig are milk and eggs. In milk, colostrum and other foods such as eggs, Ig have a role in preserving the food, although the primary role is in providing passive immunity to the newborn.

Edelman (1959) first described that Ig structure was made up of two types of polypeptide chains. Porter (1973) showed that the fragments after proteolysis still retained the ability to bind antigens. This work led to the characterization of Ig which revealed existence of a similar structural basis for all immunoglobulin (Edelman, 1970; 1973).

Generally, there are five classes of immunoglobulin, known as IgG, IgA, IgM, IgE and IgD, with an additional class, IgY contained in eggs. Milk contains IgG, IgA and IgM. Bovine IgG occurs in two subclasses, IgG_1, and IgG_2 of which IgG_1 comprises 75% of the IgG in milk (Butler, 1994). In human milk all four subclasses of IgG have been detected, with IgG_4 being in relatively higher concentrations.

IgA is the principal antibody in external secretions such as mucus, saliva and mammary secretions, except for ruminant animals that have IgG. Since ruminant milk and sometimes colostrum are regularly consumed, especially bovine milk, this is an important difference. In human milk, IgA is the primary Ig and in human colostrum IgA is almost 90% of the Ig content (Hanson et al., 1993). IgA has been suggested to provide passive protection against intestinal pathogens for the newborn until the development of an active immune system. IgA is present in a dimer form called SIgA, which is resistant to digestion by proteolytic enzymes. SIgA serves as a first line of defense against bacterial and viral antigens, although it does not bind complement or act as an opsonin in phagocytosis. The primary role appears in the agglutination and antigen binding and thereby prevents interaction of viruses and bacterial toxins to cell walls. Without this binding to cell wall, most bacteria and viruses lose their virulence. Therefore, it is not unexpected that SIgA is the primary source of Ig defense in the gut of infants and adults (Hanson et al., 1988).

The activity of SIgA against intestinal pathogens, in specific and microflora, in general, is probably one of the reasons SIgA is not in high concentrations in bovine colostrum or milk. High concentration of SIgA could interfere with the development of ruminant digestive flora. IgG, which is found in high levels in bovine colostrum, is found in lower levels in human milk and colostrum probably because Ig can be transferred *in utero* via the placenta (Porter, 1979).

TABLE 1. Protein and immunoglobulin content in bovine and human milk and colostrum

	Colostrum (mg/ml)	Milk (mg/ml)
Human		
IgA	17.4 - 32	3.7 - 4.6
IgG	0.4 - 0.53	0.28 - 0.65
IgM	1.6 - 1.13	0.24 - 0.46
Total Ig Fraction		
Protein	20 - 32	12 -16
Bovine		
IgA	4.5 - 5.4	0.1 - 0.08
IgG	50 - 90	0.8
IgM	6.0- 6.8	0.05 - 0.086
Total Ig Fraction	60 - 120	0.7 - 1.0
Protein	140	29 - 38

Based on Hanson et al., 1993; Mickleson & Moriarty, 1982; Kulkarni & Pimpale, 1989

In human colostrum IgA concentrations can be as high as 32 mg/ml, effectively constituting the entire protein fraction (*TABLE 1*). The IgA concentration drops substantially in breast milk as does the overall immunoglobulin fraction, which may only constitute 35% of the total protein. The protein content of bovine colostrum and milk is higher than in human milk and colostrum and the Ig fraction is proportionally higher in colostrum. Although, IgG rather than IgA predominates in bovine colostrum, IgA levels can approach 25% of those in human colostrum. Ig levels in human milk are substantially higher than those in bovine milk (Butler, 1994; Kulkarni & Pimpale, 1989; Mickleson & Moriarty, 1982).

III. CHEMISTRY AND STRUCTURE

A. Structure

The basic structure of all the major Ig classes is similar. Ig are multi-chain proteins primarily made of heavy and light polypeptide chains. The IgG class can be considered as a prototype. IgG is a symmetrical glycoprotein consisting of four polypeptide chains, two identical glycosylated heavy (H) chains and two non-glycosylated light chains (*FIGURE 1*). The glycosylation varies between 2 and 12% and consists primarily of mannoses and N-acetylglucosamine (Vasilov, 1982).

Differences in the heavy chains account for the differences in Ig classes and subclasses. There are four subclasses of heavy chains in human Ig and two classes in light chains. The Ig molecule is always symmetrical; all the light chains and heavy chains are of the same class. Within the heavy and light chains there are regions that are variable (V) and invariant or constant (C). The variable section consists of about 100 amino acids near the N-terminal region, while the invariant region is at the C-terminal (Edelman, 1973). The variable regions determine the antigen specificity of the Ig.

IgG has a molecular mass of about 180 kD. The heavy chains in IgG are about 50 kD and consist of about 440 amino acids. Differences between the heavy chains give each class of antibodies its distinctive characteristics. The heavy chains can range up to 80 kD in other Ig classes. The light chain in IgG consists of about 220 amino acids and is about 25 kD (Edelman, 1970; 1973). The heavy chains are disulfide bonded to each other and in turn the light chains are disulfide bonded to this heavy chain construction (see

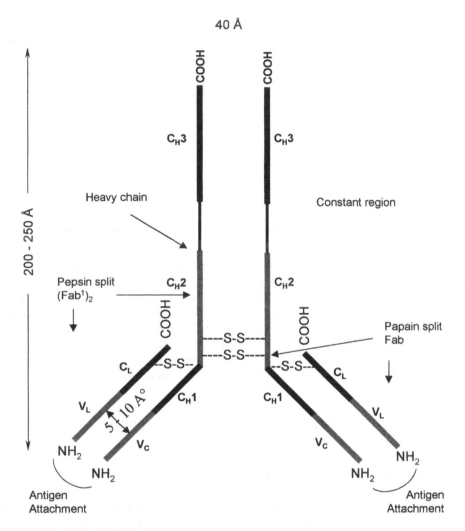

FIGURE 1. Basic structure of immunoglobulin. IgG is a symmetrical glycoprotein with four polypeptide chains, two identical heavy (H) and two light (L) chains.

FIGURE 1). The basic four-chain structure can be assembled into higher order structures such as pentamers for IgM with a molecular mass of 900 kD. IgA can be found as a monomer, dimer or trimer, or tetramer, with a molecular mass of 170 - 720 kD (Hanson et al., 1988).

Monomeric IgA is found primarily in serum, but in milk IgA is present as a dimer bond through a polypeptide J chain. An additional 75 kD polypeptide glycopeptide secretory component is also bound to IgA found in milk and the resulting dimer is known as secretory IgA, which is more resistant to proteolysis and therefore is more stable in the intestinal tract. SIgA from milk primarily ends up on the mucosal membranes of the gastrointestinal tract. (Hanson et al., 1988). IgA and IgG are similar except for differences in the heavy chain and an octapeptide added to the C-terminal heavy chain of IgA.

IgM has the highest molecular mass (950 kD) of any class of Ig. IgM consists of five subunits similar to IgA, also linked by disulfide bonds and a J chain. IgM is produced in low levels but it is the first Ig fraction to be produced in response to an antigen followed by production of much higher levels of IgG. IgM antibodies are more efficient at complement fixation, opsonization, and agglutination of bacteria and viruses. The complete role of IgM is not understood.

The roles of IgD and IgE are also not well understood. IgD and IgE are similar in structure to IgG and have similar molecular mass of 175 kD and 200 kD, respectively. The primary structural differences are in the heavy chains. IgE is known to play a role in mediating allergic reactions, which is well established. Since these classes of Ig are not found in milk or colostrum, they will not be further discussed.

Much of the structure of Ig has been determined by enzymatic digestion resulting in fragments that can more easily be analyzed. Papain digestion produces Fc (see *FIGURE 1*)) which is inactive (Fab for the univalent fraction and (Fab^1)$_2$ for the bivalent fraction from pepsin treatment). The loss of the Fc fraction does not affect the antibody activity. The Fc fragment contains most of the carbohydrate fraction. The diameter of an IgG molecule is about 40 Å and about 200-250 Å in length. The arms of the Y are flexible (Parham, 1984; Reilly et al., 1997). The Fab fragments retain some of their neutralizing activity locally in the gastrointestinal tract (Reilly et al., 1997).

B. Biosynthesis

Since there are several protein chains that make up Ig as well as two types of structural forms (membrane-bound [receptor] and secretory forms) the synthesis of Ig is very complex. A number of genes on different chromosomes are involved in the production of an Ig molecule. The genes for the L and H chains are located on different chromosomes. The variable and constant regions each have separate genes. There are four different genes that code for the constant regions in IgG known as $\gamma 1$, $\gamma 2$, $\gamma 3$ and $\gamma 4$. These differences in the constant regions account for the differences in IgG fractions as well as molecular mass differences, although the homology between the encoded amino acid sequences is very high. It may be that membrane and secretory Ig have a single source (Kemp et al., 1983) and later cleavage differentiates them; however it has also been suggested that the secretory form may result from termination of transcription before the membrane exons. Since the membrane form contains an additional 25 amino acid sequence that is predominantly hydrophobic it would seem likely that termination of transcription is the route of synthesis. This additional hydrophobic sequence aids in retention in the membrane.

The genes on the same chromosome have discrete coding sections (exons) separated by non-coding sections (introns). Tonegawa et al. (1976) showed that the variable regions of the gene segment are moved close to the constant region, which results in a functional H or L chain gene. A series of splices to the mRNA are required to remove the introns, which results in mRNA with only coding regions.

Shortly after synthesis the H and L chains are glycosylated primarily at asparagine. Glycoslyation of serine and threonine, which is common in other proteins, is found occasionally in IgA and IgD. The glycosylation is primarily mannose oligosaccharides with some N-acetylgalactosamine linked to the proximal monosaccharide (Kemp et al., 1983).

The assembly process for Ig is no less complex than the synthesis. There is a repeating pattern of amino acid sequences approximately 110 amino acids long in both the H and L chains. This sequence is repeated four times in the H chain (440 amino acids). These 110 amino acid length segments form a β-pleated sheet structure (Baumal & Scharff, 1973). Cystein groups form disulfide bonds. In the cell, there normally is an excess pool of L chains. It has been suggested that this drives the assembly, but the initial assembly step can either be the formation and crosslinking of the two heavy chains followed by the addition of an L chain or the combination of an H and L chain. This process is specific for the antibody class (Parkhouse, 1971; Buxbaum et al., 1971).

Immunoglobulins are synthesized in a class of lymphocytes known as B cells. When activated, beta cells differentiate to plasma cells that secrete antibodies. This is a rather over simplification of the process but further description here will be confined mainly to production of antibodies in exocrine secretion. In human breast milk that is primarily SIgA and in bovine milk IgG.

The process of generation of either IgA as mucosal antibodies or SIgA excreted in milk starts in the Peyer's patches in the small intestine or the lamina propria of the bronchi (Cebra et al., 1979). In the gut, an antigen or pathogen is recognized by the M cells in the Peyer's patches and then transported across the mucosal epithelia to lymphoid tissue containing B cells. This same process also occurs in the lamina propria of the bronchi (BALT), and the appendix and tonsils. The presentation of the antigen to the B cell activates or primes it. The B cell contains both IgM and as much as 10-times more IgD. IgD probably is the major receptor for the antigen and has a role in the activation of B cells. After activation, IgD levels on the surface decrease substantially and the IgM is secreted. After presentation of an antigen, IgM first increases and then falls off, followed by an increase in the IgG fraction (Lamm et al., 1979). The expression may switch to IgG at this point. Non-secreting cells with membrane IgG are memory cells. These memory cells do not need antigen stimulation to remain active (Swain et al., 1999).

Primed B cells that produce IgA move from the Peyer's patches or BALT into the mesenteric lymph nodes where they proliferate and differentiate into plasma blasts. These are transferred to the vena cava and vascular circulation. The plasma blasts move to mucosal or glandular sites and the mature plasma cells secrete polymeric IgA. Lactation hormones direct these plasma blasts to the mammary gland where they differentiate into plasma cells and produce large amounts of mucosal antibodies in the colostrum and milk. The result is that antibodies to a mucosal antigen are transferred from the mother to infant resulting in passive immunity (Lamm et al., 1979).

C. Assays

By far the simplest test for antibody, which can be made quantitative, is a precipitation test. Since most antibody-antigen complexes are insoluble under appropriate conditions this can be used to determine concentration. If there is excess antigen the antigen-antibody complex can be resolubilized. This precipitation process is the basis for analysis (and for antimicrobial activity) and was first proposed in agar gels (Oudin, 1952) and later using double diffusion from two wells in the agar plate, one holding the antigen and another filled with the antibody (Ouchterlony, 1962). This method is commonly used, but is more qualitative than quantitative and requires dilutions to determine concentrations. Concentrations of antibody may sometimes be expressed in precipitating or agglutinating titers obtained through serial dilutions. Through use of anti-IgG or other anti-Ig classes it is possible to measure the specific Ig class. Antibody based methods of detection remain among the most sensitive methods of detection.

There is a wide range of chromatography methods to isolate and quantitate Ig classes. These methods do not quantitate the specific activity for a single antigen, but rather measure all Ig molecules in a certain preparation. Analysis by chromatography often involves analysis of fragments of Ig generated by reducing agents such as β-mercaptoethanol, which breaks disulfide bonds, and enzymatic hydrolysis that can yield classic cleavage fragments such as Fab and Fc. These can be used to identify the antibody class. The use of SDS polyacrylamide gel electrophoresis allows quick analysis of the fragments when mercaptoethanol is used or the complete polypeptide (Laemmli, 1970). SDS-PAGE can also be used as a qualitative measure to evaluate purification steps (Kaneko et al., 1985).

Other non-immunological methods developed for identification of IgG fractions in milk include centrifugation and turbidity measurements using a colostrometer (Kummer et al., 1992).

Many different types of chromatographic techniques have been used to separate Ig from other proteins as well as to quantitate the fractions. Typically these methods do not offer as much resolution as SDS-PAGE, but can be quicker and more quantitative. Gel filtration using Sepharcyl S-300 to separate and quantitate Ig in colostrum and crude Ig fractions was reported (Al Mashikhi et al., 1988). Fractogel TSK HW-55 was also evaluated in this study with similar results. Heparin-Sepharose chromatography also provided good results (Boseman-Finkelstein & Finkelstein, 1982). Other types of gel chromatography include use of Sepharose 6B (Al Mashikhi & Nakai, 1987). Limited success has been obtained using HPLC and reverse-phase HPLC.

Immunological methods developed for Ig detection in milk include double and single radial immunodiffusion, immuno-nephelometry, immuno-electrophoresis and direct and indirect ELISA (Kummer et al., 1992)

Quantitative and highly specific immunoassays are available to measure specific activity against a single antigen or measure the whole class of antibody. However, a majority of the developed assays have been specifically developed for the IgG fraction in bovine milk. Since bovine milk primarily contains IgG, it is a logical starting point for measuring commercial samples used for antimicrobial purposes. Radial immunodiffusion (RID) has been commonly used, but it has low sensitivity (Leung et al., 1991). Combining immunoassays with fluorescence can provide assays with detection limits down to 5 ng of IgG/ml (Losso et al., 1993; Losso & Nakai, 1994). Rabbit anti-bovine IgG is covalently bound to a carboxyl-polystyrene particle that binds to IgG in milk. This in turn is detected by a fluorescent-labeled antibody and after filtration and washing is measured by epifluorometry. This method has some advantages over ELISA methods that are sensitive to the matrix.

The above methods primarily describe ways to analyze IgG and IgA fractions in bovine and human milk. However, other methods have also been developed for IgE and IgM antibodies. There is certain cross reactivity between anti-IgM and IgE with IgG as might be expected, but cross reactivity of IgG with IgM is limited. Highly specific non cross-reacting methods for detection of IgE have been developed (Glenn et al., 1999).

IV. ISOLATION AND PURIFICATION

There are a number of sources for the isolation of Ig. However, large-scale isolation for use as antimicrobials requires relatively large quantities of feed streams containing Ig. In particular, there are bovine-derived sources (both colostrum and milk), isolation and purification from blood sources and isolation and purification from large-scale production of (monoclonal) antibodies.

For certain applications, Ig may require extensive purification even from sources including colostrum. However, many applications such as oral antimicrobial use of antibodies does not require high purification, since sufficient amounts can be consumed safely even by infants (Reilly et al., 1997; Korhonen et al., 1998). Furthermore, Ig from bovine colostrum and milk do not restrict use or create safety concerns unlike Ig from cell culture, blood or other sources. Thus, the type of Ig application would dictate the purity. There are various commercial and non-commercial methods for isolation of Ig to different degrees of purity and associated cost for the final product.

Bovine milk and colostrum provide ready commercial sources of Ig fractions. Many of the efficacy trials reviewed later in this chapter are based on unpurified dried colostrum or partially purified fractions. An advantage of this source is that it is a recognized food source and there are lower safety and regulatory concerns about introducing a purified or semipurified fraction into a food source.

Relatively high concentration of Ig makes colostrum perhaps the ready source to purify or partially purify lactoglobulins. However, lower availability of colostrum compared to milk supply is a limitation. Furthermore, the Ig fraction in colostrum falls quickly from as much as 60 mg/ml to 24 mg/ml after 3 days (Kulkarni & Pimpale, 1989). In addition, the total volume during the first few days of lactation may only account for 30-60 kg of colostrum resulting in maximum production of Ig fraction of 3.6 kg. In contrast, the Ig fraction in milk can vary from 0.8-1.2 mg/ml, a factor of 20-75 times lower (Kulkarni & Pimpale, 1989). However, the volume of milk available can easily be more than two orders of magnitude that of colostrum resulting in production of 7-20 kgs of Ig. Accordingly, the relatively low concentration in milk makes it more difficult to isolate and purify Ig.

Drying of colostrum results in a final product containing 18-25% of Ig. Although the antibodies in colostrum are heat-sensitive, a substantial fraction of the activity will survive pasteurization and drying (Chen & Chang, 1998). Special methods have been developed to heat treat colostrum and preserve most of its Ig activity. Removal of fat fraction from colostrum by centrifugation could increase the Ig fraction over 30%. A simple process to produce semi-purified colostrum is shown in *FIGURE 2*. Fractions resulting from this process contain more than 50% Ig. These methods are fairly simple and cost-effective. Ultra-filtration or a combination with ion-exchange chromatography (*FIGURE 2*) can be used to further purify the fraction ((Kanamaru et al., 1993, Fukumoto et al., 1994, Cordle et al., 1994). The ultra-filtration step can be used to remove lactose, the non-protein nitrogen fraction and ash resulting in a product containing more than 80% Ig. This semipurified colostrum/Ig fraction is similar to the process of making whey protein concentrates.

Most of the above steps are also applicable to colostrum-based process collected from hyperimmunized cows. In this case, the Ig faction could constitute a higher overall percentage of the protein fraction. It also means that if further fractionation is done a

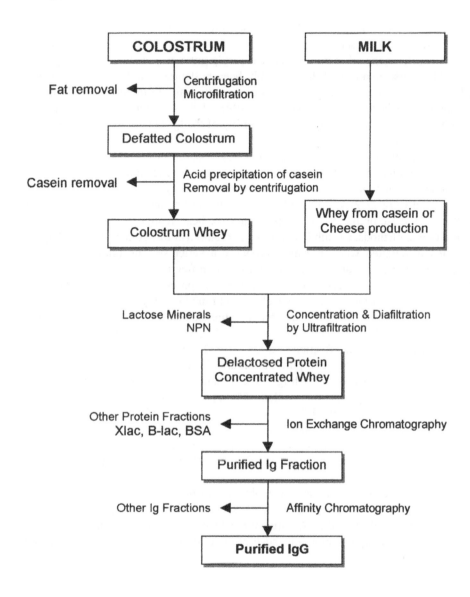

FIGURE 2. Schematic flow diagram of immunoglobulin isolation from bovine colostrum and cheese whey.

preparation specific for a particular antigen—bacteria or virus—can be obtained. This approach has been adapted by Galagen Inc. to produce a fraction with specific antimicrobial activity (Weiner et al., 1999).

Down-stream purification to IgG can be done by precipitation chromatography and other methods including immobilized metal chelate chromatography (Al-Mashikhi & Nakai, 1988; Boesman-Finklestain & Finkelstein, 1982; Li-Chan et al., 1990), gel filtration chromatography (Al-Maskikhi et al., 1988), or ion exchange (Ahmend et al., 1998). Further purification, although possible, is not practical on a commercial scale. Thus, affinity chromatography can be used to further purify the IgG fractions or isolate IgA or IgM fractions (Lihme et al., 1997).

Although the potential supply of Ig from bovine milk is large, concentration and purification from milk present two difficulties: 1) initial concentrations are substantially lower than in colostrum and 2) from a practical standpoint, whey is the only starting material available, consequently milk has probably been pasteurized and the whey also heat treated by some method. This results in loss of activity of the Ig fraction and the antimicrobial quality may be lowered. Li-Chan et al. (1995) showed a decrease of 10-30% in the Ig fraction after HTST pasteurization. Heat stability of the Ig fraction has been extensively evaluated (Chen & Chang, 1998).

The protein fraction of the starting whey contains from 10-20% Ig, with the majority of the protein being β-lactoglobulin, albumin and α-lactalbumin. The amount of Ig in commercial whey protein concentrates (WPC) and whey protein isolates (WPI) can vary substantially (*TABLE 2*). It is possible to isolate the Ig fraction from commercial WPC and WPI. Since these preparations undergo extensive heat treatment and are purified by chromatographic methods, a solubilization step is required at low concentrations. Any liquid delactosed, defatted whey stream is probably a better starting point. Commercial processes for complete separation of whey into protein fractions including the IgG fractions are performed at Carbarry and New Zealand. A process developed by Sepragen Corporation employs a radial-flow ion exchange chromatography (Ahmed et al., 1998). Similar chromatographic processes are not commercialized yet. Many such methods result in isolation of β-lactoglobulin as the classical IgG fraction. A further purification step is needed to separate this fraction (Pearce, 1988; Delespesse, 1986).

Although a high degree of purification is difficult, precipitation and removal of selected protein fractions could enrich Ig in the remaining whey fraction (Pearce, 1988). The resulting fraction may contain more than 50% Ig, suitable for many applications. This fraction could be further purified (Lihme et al., 1997). A more traditional method is to precipitate the protein fraction containing the Ig and to leave the rest in solution. Precipitation with ammonium sulfate, ethanol, potassium phosphate, and polyethylene glycol could be the methods of choice. However, these precipitation methods present some problems for further use in food products.

High degree of purification can be obtained using protein G and protein A affinity chromatography. These are expensive methods suitable for lab-scale purification or for isolation of monoclonal antibodies. Additional purification methods include hydrophobic-interaction chromatography, thiophilic adsorption chromatography and other ligands (Delespesse, 1986).

TABLE 2. Composition of commercial whey protein samples

WHEY PROTEIN SAMPLES: COMPOSITION									
	% on Product				Protein				
Production method	Manufactor	Protein	Lactose	Fat	aLac	Ig	blg	BSA	GMP
Ultrafiltration/MF	1	90	<5	1%	22	23	37	11	0
UF/MF	2	91	<1	0.5	20	16	43	5	23
UF/MF	3	90	<3	1	20	12	42	6	24
Ion Exchnage	4	92	<1	1	15	10	59	21	0
Ultrafiltration	5	70	11	7	18	9	23	13	1.05

V. ANTIMICROBIAL ACTIVITY

A. Mechanism of action

The initial binding step of an antibody to its complementary antigen target is a simple, reversible noncovalent interaction. Various types of noncovalent interactions, including hydrogen bonds, hydrophobic bonds, coulombic interactions and van der Waals forces, may work alone or in combination to stabilize antibody binding. Target antigens may consist of amino acids, nucleic acids, carbohydrates or lipids, but it is the degree of complementarity between the combining site on the antibody molecule and the shape of the antigen that determines the overall bonding strength. The exquisite specificity of this interaction has often been described as a 'lock and key' mechanism.

Once bound to a target, different antibody subclasses may trigger events that lead to inflammatory processes or increase the effectiveness of other components of the immune system. Antibody versatility is the result of both this initial interaction and its sequelae. Milk/colostrum-derived and egg yolk antibodies developed for food use are typically polyclonal, meaning that there are many antibody specificities present that are directed against multiple binding sites on a single microorganism or its extracellular products. Further, many of the naturally occurring (background, broad-spectrum) antibodies from animal sources are cross-reactive and therefore, capable of binding to related antigen targets on more than one strain within a class of pathogens, and could also affect human pathogens. To maximize the milk or egg's antibody content specific for human pathogens, animals could be immunized against specific microbial targets.

The following are all mechanisms by which an antibody could affect a pathogen target. They may function individually or in combination in response to the appropriate target(s).

1. Agglutination. Antibody molecules are flexible enough to permit cross-linking of cells with common surface antigens. With bacteria, for example, this effect causes a network of cell formation that could be removed mechanically from the body. Although direct agglutination of cell membrane antigens does occur, more common targets of agglutinating antibodies are the flagella and pili that extend from the microbial surface and are typically involved in motility or epithelial adherence. Agglutinated microorganisms are usually unable to colonize or bind to receptors and release toxins.

2. Cidal activity. Antibodies are rarely cidal. Antibody-mediated microbial killing due to membrane lysis typically requires the presence of complement and the synergistic activity of phagocytic cells. Non-specific factors such as lysozyme could enhance bacterial killing.

3. Complement activation. Under appropriate conditions, antibody binding could activate the complement enzyme cascade. When multiple bovine IgG antibodies, for example, bind to an appropriately spaced repeating antigen sequence, C1q, the first component in the complement sequence, fixes or binds to a portion of the IgG molecule. This fixation step initiates the cascade and activation of other complement components. Ultimately, this process generates a 'membrane attack complex' that has the ability to lyse, or punch holes, in the cell membranes of microorganisms. Certain complement proteins could also function as opsonins to attract phagocytic cells and facilitate ingestion of antibody-coated microorganisms. Antibody/complement mediated lysis is effective against certain bacteria and some virus-infected cells.

4. Stasis activity. Most commonly, antibody binding to microorganisms is the initial step in a slow cytotoxic process. Bound antibodies may, however, cause structural alterations at the cell surface or disturb cellular processes resulting in immobilization, metabolic disturbances or increases in membrane permeability. Antibody-mediated enzyme inhibition impacts certain key viral (e.g. neuraminidase) and bacterial activities. Impairment of cell growth and multiplication follow until other anti-infective immune processes can assist in destruction or removal of microorganisms from the body.

5. Adherence-blockade. Specific antibodies may also bind to certain cell surface components (e.g., adhesins) that are responsible for initiating contact with the host cell surface receptors. Once coated with antibodies, microbial adherence to the receptor cannot be established. For many microorganisms, such as bacteria and fungi, adherence is an obligatory first step to establishing colonization and infection. Gut pathogens, in particular, may have receptors for epithelial surfaces that are readily blocked by antibodies.

6. Opsonization. Antibodies could signal other immune cells to help target and destroy invading pathogens, and enhance phagocytic activity. Aggregation of multiple IgG antibody molecules bound to the surface of a pathogen could trigger binding of Fc portion of the antibody molecule to receptors on the surface of macrophages. When present in sufficient concentration, Fc binding stimulates a conformational change in the macrophage that leads to pathogen engulfment and destruction. This method of attack is used against viruses, bacteria and fungi.

7. Toxin neutralization. Bacterial toxins often must be actively transported inside the host cells, an event triggered by receptor binding, to cause cell death. Antibodies could prevent this internalization by either binding to various portions of the toxin molecules, thereby preventing the toxin from attaching to its specific cellular receptor or by binding to the receptor itself to block toxin attachment. Antibody-bound toxins are also more easily recognized by macrophages, enhancing clearance. Alternatively, antibody binding to soluble toxin molecules may cause formation of an insoluble complex or precipitate. Complexed toxin antigens are generally unable to bind to their receptors or are absorbed across cell membranes.

8. Virus neutralization. Viruses fail to replicate until receptor-mediated endocytosis permits fusion of the virus with host cell DNA. Antibodies neutralize virus activity (render the virus non-infective) by blocking initial binding of the virus to its cell surface receptor or by inhibiting the fusion event itself.

VI. INFLUENCING FACTORS

Newborn animals receive antibodies in a matrix of other bioactive proteins by various delivery mechanisms via the placenta or through colostrum ingestion or in egg yolk components. As an immensely capable and versatile molecule, Ig is designed to function best in concert with other immune enhancing factors. Some of the more common bio-logically active components that are also found in eggs, milk and colostrum are discussed below:

1. Lactoferrin (LF). LF is the major iron-transport protein that exerts antimicro-bial effects by sequestering iron and limiting bacterial growth and metabolism (for an extensive review refer to *Chapter 2*). Both LF and Ig are naturally present in bovine milk and colostrum to provide local GI protection to the weaning mammal. Both proteins appear to survive the human GI transit. In combination with antibody binding, LF can synergistically inhibit the growth of certain bacteria or provide a 'one-two punch' that results in a more effective reduction in disease symptoms and infectivity. For example, LF alone has been shown to elicit a bactericidal effect against *Helicobacter pylori in vitro* and in animal model studies, while bovine antibodies can block the cytotoxic effects of Cag A and Vac A toxins of this gastric pathogen. Lactoferricin, a pepsin-digest of LF has even greater antibacterial activity than the native molecule (Dial et al., 1998). It has also been reported that LF has antiviral activity. *In vitro* LF prevents binding of rotavirus to sus-ceptible cultured enterocyte-like cells. Therefore, LF present in immune colostrum prepa-rations evaluated in prophylactic clinical studies in infants may have contributed to their successful outcomes. In experimentally infected mouse model studies, for example, skimmed immune colostrum was effective, while the purified Ig from the same colostrum was not.

2. Lactoperoxidase (LP). LP is one of the enzymes found in milk and colostrum known to elicit cidal components to assist in the destruction of microorganisms (for an extensive review refer to *Chapter 3*). *In vitro*, preincubation with LP, thiocyanate and hydrogen peroxide enhances the ability of specific milk antibodies to inhibit glucose uptake by *Streptococcus mutans* or *Streptococcus sobrinus*. The antibacterial activities of LF and lysozyme may also be enhanced in the presence of LP.

3. Lysozyme (LZ). LZ demonstrates both bactericidal and fungistatic activities by cleaving peptidoglycans in the cell wall of Gram-positive bacteria and by antibacterial mechanisms other than lytic effects (for an extensive review refer to *Chapter 6*). LZ is a common constituent of body secretions including human breast milk. It is highly resistant to proteolytic digestion and may work together with specific antibodies to destroy bacte-ria at the mucosal surface. In eggs, LZ is found in the albumin or egg white fraction; sep-aration of egg yolks from whites during fractionation of IgY may, therefore, eliminate potential synergistic effects.

4. Complement. Presence of complement plus opsonized antibody-coated bacte-ria results in more efficient ingestion by phagocytic cells. Enveloped viruses are also more effectively neutralized by the combination of antibody plus complement. Unfortunately, complement activity in milk is more sensitive to heat denaturation than antibody mole-cules and may be reduced to undetectable levels in many processed foods. Host comple-ment components are available, however, within the GI tract, especially in the presence of infectious inflammation. *In vitro* evidence suggests that bovine colostral IgG can interact with human complement to stimulate opsonization and phagocytosis of specific antibody-coated bacteria (unpublished observations).

5. Secretory component. Secretory component is mainly responsible for the transport of IgA and IgM across epithelial cells and appears as a critical component of the secretory IgA. Free secretory component has been identified in bovine colostrum and milk. This protein molecule has been shown to independently inhibit adhesion of *E. coli* to cell receptors *in vitro* and may provide non-specific defense support for antibody-mediated inhibition of pathogen binding.

6. Probiotics. The gastrointestinal tract, where the antibodies and other bioactive proteins have their major effect, is populated by various commensal microorganisms. Probiotics are supplemental bacteria, typically *Lactobacillus* and/or *Bifidobacteria* strains, whose presence enhances desirable health conditions in the gut (for an extensive review refer to *Chapter 17*). *In vitro*, the combination of these bacteria with antibodies could cause synergistic reductions in colonization by GI pathogens. Ingestion of fermented foods and/or live probiotic cultures is also reported to have direct stimulatory effects on the host immune system, in particular increasing the magnitude of Ig response.

VII. SPECTRUM

Antibodies may be unique as food antimicrobials due to their potential effects on disease-causing organisms across the pathogenic spectrum and neutralize microbial virulence factors such as toxins. Since colostrum/milk and egg-derived antibodies are polyclonal, they present a multi-faceted approach to pathogen control that transcends the microbial ability to develop resistance mechanisms.

A. Antibacterial activity

Pathogenic bacteria rely on two basic properties, invasiveness and toxicity, to colonize their hosts. Antibodies are particularly useful in blocking both these mechanisms. Toxin-producing bacteria can be further subdivided into groups that have toxin as a major virulence factor, and those where the toxin alone is responsible for manifestation of disease. Some of the earliest published clinical uses of bovine milk antibodies were prophylactic and therapeutic against infantile diarrhea caused by *Escherichia coli*. Polyclonal anti-pili antibodies could block motility and adherence of bacteria to GI epithelium to limit infectivity, and anti-toxin antibodies seem to prevent the cellular damage that results in secretory diarrhea. Bacterial antigen targets are usually composed of proteins or polysaccharides (e.g. LPS or endotoxin).

B. Antiviral activity

Antibodies are most effective in the early stages of viral infection when the infectious virions can be found in the extracellular matrix as they spread from cell to cell. Viral antigens are typically proteins or glycoproteins. Those antibodies that bind to surface features involved in attachment (e.g. hemagglutinin) are the most protective. Addition of milk-derived anti-rotavirus antibodies to infant formula was more effective in preventing initial infection and in limiting the spread of disease (reduced viral shedding) than in treating symptoms of rotaviral diarrhea.

C. Antifungal activity

Antifungal antibodies at mucosal surfaces such as the GI tract epithelial layer function much like antibacterial antibodies to prevent attachment and invasion. Orally administered bovine antibodies may prevent colonization and overgrowth of *Candida albicans*, a normal inhabitant of the gastric mucosa.

D. Antiprotozoal activity

Protozoal-specific Ig in the gastrointestinal tract prevents infection largely by blocking receptor attachment and blocking parasite entry into cells. Antibody binding could cause direct alteration of the outer cell membrane, damaging or killing the protozoa. Bound antibodies also enhance phagocytosis and activate complement to cause lysis. Anti-protozoal antibodies are stage-specific; in the gut, antibodies against oocyst and early larval stages are important anti-infective tools. Protozoal antigens are typically carbohydrates, proteins or lipids.

E. Antitoxin activity

Inhibition of hemolysis due to *Staphylococcus aureus* toxin and reduction of cytotoxic effects on Vero cell monolayers has been demonstrated *in vitro* using colostrum from non-immunized cows. Colostrum from cows immunized against enterotoxigenic *E. coli* cells and heat-labile enterotoxin was 100% protective in a human volunteer challenge study. Anti-toxin A colostral antibodies were also effective in preventing fatal diarrhea in Golden Syrian hamsters challenged with *Clostridium difficile*.

VIII. APPLICATIONS

Whereas in man the major part of the passively acquired Ig is transmitted prenatally, via the uterus to the newborn, in ruminants this occurs postnatally, predominantly via the colostrum in the first 36 hours after birth. Since it became clear in the 1970s that the immunoglobulins were largely responsible for the protective activity of colostrum against enteric infections, the feed industry has taken up the concept of 'immune milk' for newborn farm animals like calves, piglets and lambs. Field trials with calves have shown that feeding colostrum with high levels of Ig (>60 g/L) had a better impact on weight gain and duration and severity of scours than the low Ig colostrum (< 45 g/L) (Nocek et al., 1984). Soon vaccination methods were developed to stimulate the protective activity of the colostrum passed on to the newborn with respect to specific infective agents or virulence factors. Rotaviruses and enterotoxic *E. coli*, the causative agents of diarrhea, could be effectively controlled (Saif & Smith, 1985; Acres, 1985).

In the 1980s ideas were developed to use the Ig in cow's colostrum, or even in cow's milk, with 50-100 times lower levels for prevention and/or treatment of diarrheal outbreaks in humans (Reddy et al., 1988; Ebina, 1992). This was partly due to an increased awareness of the types of pathogens and their modes of action, based on the experience from animal farm trials. Also the increasing knowledge on the composition of specific and non-specific protective factors in human breast milk, with its well-documented health benefits, will provide scientific basis for the nutraceutical potential of cow based 'immune milks'.

However, several issues have to be addressed in order to judge the commercial attractiveness of bovine Ig enriched products for human application. These will be dealt with in the following sections.

A. Ig from hyperimmunized cows

Bovine milk and colostrum may contain antibodies directed against pathogens and virulence factors occurring in the environment of the cows. Yolken and co-workers (1985) detected anti-human rotavirus antibodies in raw and pasteurized cow milk. The IgG1 class had geometric mean levels of 0.54 µg/L (n=31) and 0.23 µg/L (n=37), respectively. Hilpert et al. (1987) also described milk Ig concentrates with anti-human rotavirus activity. The neutralizing titers, as measured with an ELISA, for non-immunized colostral milk, hyperimmune milk collected up to 40 or 10 days post-calving were 1:330, 1:1100 and 1:6000, respectively. Janson et al. (1994) compared the titers of specific human rotaviruses in colostra from immunized and non-immunized cows and found a 100-fold increase in favor of the hyperimmune colostrum.

Stephan et al. (1990) prepared an Ig preparation from the first colostrum of more than 100 cows after calving. About 84% of the total protein comprised Ig with IgG as the predominant form in addition to some IgA and IgM. The preparation displayed antibody activity against strains of *E.coli, Klebsiella spp., Pseudomonas spp., Proteus spp., Serratia spp., Salmonella spp., Staphylococcus spp., Streptococcus spp.* and *Candida spp.* Moreover, the product contained Ig against hemolytic toxins of *S. aureus*. The product was tested in a 4-year-old boy with HIV (Shield et al., 1993) and in a prospective, open and uncontrolled study with HIV positive patients (Plettenberg et al., 1993). Evidence for the presence of antibodies directed against *Cryptosporidium* was obtained.

Facon et al. (1995) reported the presence of 0.55-5.2 mg/L of the human ETEC colonization factor antigen I (CFA-I) from colostrum of non-vaccinated cows. This represented 0.002-0.007% of the total Ig amount. Cow's milk was analyzed as well. The level found was 0.1 mg/L, which represented 0.012% of the total Ig. Vaccination increased the ratio of specific Ig to total Ig about 10-25 times in the milk.

Cordle et al. (1991) used porcine ETEC and human rotavirus immunogens to obtain hyperimmune colostra; compared to non-immunized controls hyperimmune colostra contained 10- and 10-16 fold higher levels of the specific antibodies, respectively. Similarly, Tacket et al. (1992) found 64-1024 fold enhancement in the levels of Ig against LPS of *Shigella flexneri* in the hyperimmunized colostrum compared to the preparations from non-immunized cows.

Syväoga et al. (1994) reported antibodies against *Campylobacter jejuni* in hyperimmune colostrum but not in colostrum from non-immunized cows. There was a correlation (R=0.8) between anti-*Campylobacter* antibodies and Ig content, suggesting that the first colostral samples have the highest specific titers.

Clostridium difficile toxins A and B antibodies were also markedly increased, when assayed with ELISA, in Ig concentrates from colostra of hyperimmunized cows (Kelly et al., 1996). Challenge studies with *C.difficile* culture filtrates in rat ileal loop model has indicated that administration of anti-*C-difficile* bovine Ig concentrate could protect the architecture of ileal mucosa with minimal disruption. In contrast, the controls showed complete destruction of normal villous architecture and submucosal vascular congestion, neutrophil margination and infiltration (*FIGURE 3*).

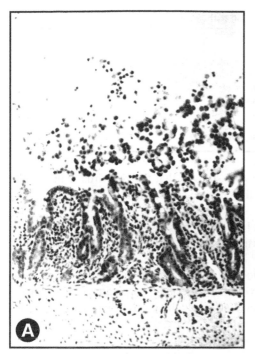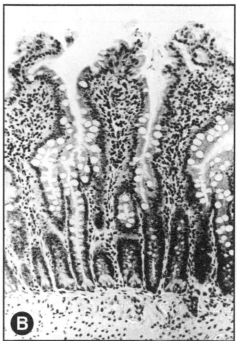

FIGURE 3. Histological sections of mucosa from rat ileal loop exposed to *Clostridium difficile* culture filtrate. (A) Demonstrates complete destruction of normal villous architecture and submucosal vascular congestion, neutrophil margination and infiltration. (B) In the presence of anti *C-difficile*

Greenberg and Cello (1996) reported a 16-20 greater activity against *Cryptosporidium parvum* in hyperimmune bovine colostrum.

Loimaranta and co-workers (1998a; 1998b) used *Streptococcus mutans* and *Streptococcus sobrinus* for vaccination. Ig from the control colostrum (only adjuvant injection) had no activity against these strains in an ELISA, whereas the hyperimmune preparation showed high specific activity. For *Helicobacter pylori,* a 100-fold higher antibody titer in the hyperimmune colostrum was obtained (Caswell et al., 1998).

From most of these studies it appears that the benefit of immunization is a higher specific activity of the desired Ig. Indeed, most clinical studies have used Ig preparations from hyperimmune colostrum or milk. So, currently most companies active in the field have developed a proprietary technology for vaccination in order to increase the levels of the desired antibodies in the milk or colostrum, from which further concentration or purification is possible.

The spectrum of organisms towards specific antibodies raised for commercial human application is summarized in *TABLE 3*.

B. Survival of Ig in the GI tract

A prerequisite for biological activity of Ig enriched products is that they are not digested by the enzymes present in the stomach or the duodenum/ileum. Several *in vitro* and *in vivo* studies have addressed this issue.

TABLE 3. Bovine Ig - Passive immunity against human pathogens

Bacteria	*E.coli* (ETEC, EPEC)
	Shigella flexneri
	Campylobacter jejuni
	Clostridium difficile
	Helicobacter pylori
	Streptococcus mutans / S. sobrinus
Yeast and molds	*Candida albicans*
Viruses	Rotavirus
Zooparasites	*Cryptosporidium parvum*

1. In vitro studies. Hilpert et al. (1987) showed that the rotavirus neutralizing ability of milk Ig concentrate is affected by the enzyme-substrate ratio and the pH when digested with pepsin and trypsin.

Petchow and Talbot (1994) purified Ig from the first six milkings after calving in hyperimmunized cows against simian rotavirus strain SA-11. The final product contained 80-85% whey proteins comprising about 35% IgG. Gastric juice aspirates were collected from (fasted) children or adults, whereas duodenal fluids were derived from fasted children only. Furthermore, the Ig concentrates were also digested with commercial gastric or pancreatic enzymes. In the absence of duodenal enzymes the gastric digestion by pepsin was extensive below pH 3.0, as evidenced from the reduction in rotavirus neutralizing titer. Above pH 4.0, the neutralizing titer remained essentially unchanged versus the control. However, in a two-phase digestion model, the neutralizing titer was further reduced above pH 4.0 due to the action of duodenal enzymes like trypsin and chymotrypsin. Combined digestion under pH conditions favoring pepsin or trypsin activity resulted in at least 50-fold reduction of virus neutralizing titer. Surprisingly, addition of skim milk to the digestion samples, in order to affect pH or provide substrate competition, did not reduce the degradative action of the gastro-intestinal enzymes on the biological activity of the immunoglobulins.

2. In vivo studies. Hilpert et al. (1987) conducted a study with infants receiving Ig concentrate 2 grams per kg body weight for 5 days. The stool samples of these infants contained intact bovine IgG1 and various Ig fragments of bovine antibodies. Furthermore, a correlation was established between the presence of bovine IgG1 and virus neutralizing activity. Based on the amount of ingested concentrate by infants and recovery after GI passage, the virus neutralizing activity in the fecal samples was measured. It was estimated that about 10% of the specific antibodies survived GI transit in these infants.

Roos et al. (1995) used ^{15}N-labeled Ig from a non-immunized bovine colostrum to investigate its resistance to digestion in the human adult intestine. Healthy individuals drank 400 ml with about 6 grams of Ig, containing about 84% of IgG. During 8 h, ileal samples were collected through naso-intestinal intubation. Ig in the aspirate was measured with radial immunodiffusion for active Ig. The isotope enrichment was determined by isotope ratio mass spectrometry. The mean digestibility of the exogenous nitrogen was 79%, which was significantly lower than the ileal digestibility of most other milk proteins by at least 90%. This demonstrates that the Ig preparation, wherein 90% of the protein comprises Ig, has increased resistance to gut enzymes. A 19% recovery of immunologically active IgG and IgM was observed.

Kelly et al. (1997) studied the survival of *Clostridium difficile* toxin specific antibodies and non-specific Ig (IgG) from a single oral dose of a hyperimmune bovine Ig concentrate (BIC) to human volunteers. The BIC was administered at different conditions: in the fasted or fed state, in combination with antacid or an acid secretion blocker (omeprazole), or encapsulated allowing release only above pH 6.0. With 45 g of BIC, containing about 14 g of bovine IgG, omeprazole increased the level of IgG in fecal samples versus the fasted, fed or antacid subjects, which did not differ significantly from each other. With 8 g of BIC, containing about 3 g of bovine IgG, the subjects that received encapsulated BIC had significantly higher fecal levels of bovine IgG compared to fed subjects. In the omeprazole treated subjects, an inverse relation was established between the first appearance of carmine red in the feces and the stool level of bovine IgG, indicating that higher GI transit time increased the survival of IgG. Control subjects receiving no BIC did not have toxin neutralizing activity in the stools, whereas the stools from subjects who ingested BIC contained neutralizing activity towards *C.difficile* toxins A and B.

In a follow-up study, ileostomy patients received 5 g of BIC with 2.1 g of bovine IgG on four different occasions: standardized diet with 4 h fasting prior to oral BIC administration, inclusion of antacid, during omeprazole treatment and BIC within enteric coated capsules (Warny et al., 1999). Effluent was sampled at 30 min intervals during 6 h. Total bovine IgG was measured with radial immunodiffusion, specific anti- *C. difficile* toxin A antibodies with ELISA. Rounding of Chinese hamster ovary cells in monolayer culture was also used to determine the ability of ileal samples to neutralize cytotoxicity of *C.difficile* toxin A. Contrary to the previous study, the enteric coated BIC showed a low IgG recovery (4%) in the ileal effluent. However, many capsules remained intact or partially unopened. In the other treatments the mean recoveries of free bovine IgG were 30% (antacid), 49% (BIC alone) and 50% (omeprazole), yet variance analysis did not reveal statistical difference. There were considerable differences in transit time between individuals; higher IgG in intestinal samples reflected shorter transit times. The time course of total bovine IgG and toxin specific IgG paralleled. A positive correlation (r=0.95; p<0.001) between total bovine IgG amount and ability to neutralize cytotoxicity was obtained for the intestinal fluids.

Taken together the *in vitro* and *in vivo* studies indicated that immunoglobulins are degraded by digestive enzymes, however, a part (20% or higher) of the ingested antibodies seem to survive and retain their biological activity. To increase survival rate, it is important to protect against gastric digestion. Encapsulation of bovine Ig with acid resistant coatings may be a simple way to achieve this GI protection.

3. Clinical trials. Recently several reviews have summarized various clinical studies performed with bovine Ig concentrates from non-immunized and vaccinated cows (Davidson, 1996; Bogstedt et al., 1996; Weiner et al., 1999). Weiner et al. (1999) provided an extensive overview of the human clinical trials conducted on oral pathogens (*Streptococcus mutans, Candida albicans*), the human-specific gastric pathogen *Helicobacter pylori*, rotavirus infections in children, *C. parvum* induced diarrhea in immunocompromised hosts and enteropathogenic *E. coli* infections in children.

The suppression of *H. pylori* by bovine Ig concentrates (BIC) has so far yielded only modest results or no effect (Opekun et al., 1999; Casswall et al., 1998). Weiner et al. (1999) refer to a study where 2 out of 8 patients cleared the infection after taking 4 g of hyperimmune BIC twice daily for 4 weeks.

Positive outcome has been reported in trials with infants suffering from rotavirus infection. Prophylactic studies showed that the clinical effect might vary from total protection to mitigated disease state. The daily dose of IgG ranged from 600 to 2000 mg in these studies; the BIC preparations were taken for 1-10 days. Reduced viral shedding and mitigated disease were also reported from therapeutic treatments. The daily required dose was between 2000-9000 mg of IgG for 3-5 days.

Cryptosporodiosis is a common infection in humans, which is self-limiting in immunologically competent individuals. In immunocompromised situations, like AIDS infection, chemotherapy treatment and malnutrition, the *Cryptosporidium*-induced diarrhea may become chronic and even life-threatening. Currently, there is no effective therapy and in general a combination of treatments is in practice, including anti-diarrheal agents, fluid replacement, nutritional support and careful drug treatment affecting the immune performance.

Ig preparations from bovine colostrum produced by non-immunized cows were succesfully used in oral immunotherapy of HIV-positive patients with chronic diarrhea (Plettenberg et al., 1993). Seven out of 25 patients in this prospective, open and uncontrolled study had confirmed cryptosporidiosis. Patients were treated with a drink containing about 4.5 g IgG, 0.5 g IgM and 1.4 g IgA, given daily for 10 days. The cryptosporidiosis group had 2 non-responders, 2 with partial and 3 with complete remission. The non-cryptosporidiosis patients (n=18) had 7 non-responders, 4 with partial and 7 with complete remission. In 8 of the 9 non-responders the dose was subsequently doubled for another 10 days. Five patients showed complete or partial remission. Treatment resulted in at least partial remission of the diarrhea in 2/3 of the patients, which was similar to results as reported in an earlier study (Stephan et al., 1990).

Greenberg and Cello (1996) treated 16 patients with confirmed *C. parvum* infection using hyperimmune BIC. Treatment lasted for 21 days with 24 g of BIC per day, either provided as drinks from powder (n=8) or in capsules (n=8). A significant decrease in stool weight was only observed in the group using the powder supplement. No satisfactory explanation for the difference with the capsules was given.

Tacket and co-workers (1988) conducted a double blind controlled trial comprising human volunteers challenged with enterotoxigenic *E.coli* producing colonization factor antigen I (CFA I) and heat-labile and heat-stabile enterotoxins. The volunteers ingested 3 servings per day of 3 g of hyperimmune colostrum derived Ig concentrate, containing about 1.15 g of Ig, after each meal for seven days. The cows that produced the colostrum had been vaccinated with several ETEC sero- and fimbria types, *E.coli* heat-labile enterotoxin and cholera toxin. The control preparation did not contain anti-*E.coli* activity. The challenge was given after the seventh dose, on day-3. Nine of the ten volunteers receiving the control developed diarrhea, whereas the 10 volunteers receiving the BIC had no problems. All volunteers excreted the challenge organism in their feces. It was concluded that the BIC could be a good prophylactic against traveller's diarrhea. In a recent study, a hyperimmune bovine milk antibody concentrate was tested derived from cows vaccinated with purified colonization factor antigens of ETEC (Freedman et al., 1998). The product contained 78% of protein of which 76% was IgG; the specific activity of the CFA antibodies in the hyperimmune product was 32-fold higher than that of either labile toxin, LPS or type I pillus. The placebo comprised a lactose free infant formula. The human adult volunteers ingested a drink, supplemented with bicarbonate to increase stomach pH, with either the placebo (n=10), a low dose of BIC (n=4) or a high dose of BIC (n=11); however, the study did not describe the precise amounts of Ig given to the volunteers. The drinks were taken 3 times daily for 7 days; the challenge with ETEC occurred on day-3. Seven out of the 10 individuals taking the placebo developed diarrhea; only 1 (high dose) out of 15 persons in the treatment groups got ill, although some clinical symptoms like headaches, gas development and cramps were reported in a number of cases. Prevention of shigellosis by BIC was reported by Tacket et al. (1992). The hyperimmune product, with a 1000-fold higher titer of anti-*Shigella flexneri* LPS antibodies versus the non-immunized BIC, was given to human volunteers for 7 days, three times a day after meals. The challenge with *Shigella flexneri* took place on day-3. Per serving 10 grams of a chocolate flavored specific BIC were administered; no clear data were provided on the IgG content. The hyperimmune BIC was compared to a control without Ig; bicarbonate was included in both suspensions. The result of this double-blind, randomized study was that 45% of the control group developed illness whereas the BIC-treated individuals had no disease symptoms.

Both the above studies provide evidence from serum immune responses that the luminal inactivation of pathogens by passively acquired Ig is reflected in a less pronounced active immune response.

C. Specific applications

For feed applications the market has not really grown due to the expensive nature of the bovine Ig concentrates (BIC) in a low cost feed product. However, recently the use of antibiotics as growth promotors is under critical evaluation, and in some countries certain antibiotics are already banned due to possible risk of developing multi-resistant bacterial strains. Therefore, natural antimicrobials such as BIC, LF, LP or antibiotic peptides could play a prominent role.

Control of infantile diarrhea in developing countries is one of the earliest human applications suggested for BIC (Reddy et al., 1988). BIC may have a positive effect in minimizing high health risks in the developed world including hospital-acquired infections, neonatal nurseries, day care centers, immunocompromised situations, and traveller's diarrhea (Davidson, 1996; Bogsted et al., 1996).

Specific dosage of BIC for an effective prophylactic and therapeutic usage has been suggested. Weiner et al. (1999) have elucidated the purpose of administration of antibodies for passive immunity against oral pathogens such as streptococci and candida; enteric infectious agents like rotaviruses and *E.coli*; and *Clostridium difficile* or *Cryptosporidium* in immunocompromised condition. Antibiotic use is indeed an effective way of treating infections. However, their use is often expensive, sometimes causes side-effects and possibly contributes to the incidence of multi-resistant pathogens. Therefore use of BIC is an effective alternative.

IX. PRODUCTION TECHNOLOGY

Many patents are available describing the isolation and/or purification of Ig from bovine milk, whey or colostrum. Products described therein range from whey protein concentrates with about 10% of Ig to preparations with more than 80% Ig. Colostrum is often used as the starting raw material due to its rich source of Ig, however, cheese whey is also widely used. Typical processing steps involve skimming to remove fat, coagulation of casein (if applicable) and filtration steps to concentrate the (colostral) whey. To further increase the Ig concentration one may exploit the use of sequential ultrafiltration (UF) steps or combinations of ultrafiltration with ion exchange chromatography. As UF is used in the dairy industry for the preconcentration of milk before cheesemaking or for the production of whey protein concentrates, UF membranes with cut off at e.g., 100 kD were used in the 1980s to produce products with about 12% IgG. Anion exchange resins may bind non-Ig whey proteins and thus, in combination with UF, result in Ig enrichment. Fukumoto et al. (1994) used UF to concentrate cheddar cheese whey that was subsequently loaded to a column packed with chelating Sepharose equilibrated with copper ions. This immobilized metal affinity chromatography (IMAC) column was compared to an anion exchange resin (AIEX). IMAC proved to give higher recovery and purity of IgG than AIEX. Depending on the conditions chosen, the purity of Ig on protein with IMAC was in the range from 50-60%, whereas this was 20-35% with AEIX.

Concentration of IgG at $\geq 80\%$ could be achieved by passing the immune (colostral) whey through immunoaffinity columns (Konecny et al., 1996). However, this process is relatively expensive and could as well be disadvantageous by removing bioactive components, like e.g., lactoferrin, that may have synergy with IgG. For cost-effective production membrane technologies are currently in use; the commercial preparations contain Ig roughly between 60 and 75%.

Li-Chan et al. (1995) investigated the thermal stability of bovine IgG in phosphate buffered saline and in commercially processed milk products. Slightly increased heat stability was found for the IgG in a milk matrix between 60 and 80 ºC. The IgG content in HTST pasteurized milks was between 59 and 76% of the level found in raw milks. Specific antibodies did not differ in response compared to total IgG. However, UHT sterilized or evaporated canned milk had hardly any IgG. Chen and Chang (1998) studied bovine IgG model systems and found stabilizing or protective effects upon heating when adding components like fructose, maltose, sorbitol, glutamic acid or glycine. It also appeared that IgG in colostrum is more heat stable in the temperature range between 75 and 100˚C than in whey or phosphate buffered saline.

X. SUMMARY

Lactoglobulins are potential natural food antimicrobials with a wide-spectrum of microbial pathogens and ability to neutralize microbial virulence factors such as toxins. Since colostrum/milk-derived antibodies are polyclonal, they present a multi-faceted approach to pathogen control that transcends the microbial ability to develop resistance mechanisms. Lactoglobulins such as BIC have many advantages over synthetic antibiotics: low cost, polyclonal nature in microbial inactivation, activity most likely limited to the intestinal lumen, and less burden on the microbial gut ecology. From a commercial point of view, their entry into the market is faster due to less stringent regulatory issues. Lactoglobulin use in dietary supplements or special food products is under consideration. However, BIC for human application is only at its infancy due to uncertainties with respect to dosage and timing of administration. Furthermore, Ig are heat-sensitive molecules, subject to human digestive action and may present difficulties with respect to palatability or shelf life of food products due to contaminating enzymes of a proteolytic nature. Encapsulation of immunoglobulins with acid resistant coating is one approach under investigation. However, these issues need further studies including clinical trials and sophisticated product development.

XI. REFERENCES

1. Acres, S.D. 1985. Enterotoxigenic *Escherichia coli* infections in newborn calves: A review. *J. Dairy Sci.* 68: 229-256.
2. Ahmed, S. H., Saxena, V., Mozaffar, Z., and Miranda Q. R. 1998. Sequential separation of whey proteins and formulations thereof. *US Patent No.* 5,756,680.
3. Al-Mashikhi, S. A., Li-Chan, E., and Nakai, S. 1988. Separation of Ig and lactoferrin from cheese whey by chelating chromoatography. *J. Dairy Sci.* 74:1747-1755.
4. Al-Mashikhi, S. A., and Nakai, S. 1987. Isolation of bovine Ig and lactoferrin from whey protein by gel filtration technique. *J. Dairy Sci.* 70:2486-2492.
5. Altekruse, S.F., Stern, N.J., Fields, P.I and Swerlow, D.L. 1999. *Campylobacter jejuni:* An emerging food-borne pathogen. *Emerging Infect. Dis.* 5:28-35.
6. Baumal R., and Scharff, M. D. 1973. Synthesis, assembly, and secretion of mouse Ig. *Transplant Rev.* 14:163.
7. Boesman-Finkelstein, M., and Finkelstein, R.A. 1982. Sequential purification of lactoferrin, lysozyme and secretory Ig A from human milk. *FEMS Microbiol. Lett.* 144:1-5.
8. Bogstedt, A.K., Johansen, K., Hatta, H., Kim, M., Casswall, T., Svensson, L. and Hammerström, L. 1996. Passive immunity against diarrhea. *Acta Paediatr.* 85:125-128.
9. Butler, J. E. 1983. Bovine Ig: and augmented review. *Vet. Imm Immunopathol.* 4:43-152.
10. Butler, J. E. 1994. Passive immunity and Ig diversity. In: *Indigenous Antimicrobial Agents of Milk: Recent Developments,* pp.14-47. International Dairy Federation Press.
11. Buxbaum, J., Zolla, S., Scharff, M., and Franklin, E. 1971. Syntheses and assembly of Ig by malignant human plasmacytes and lymphocytes. *J. Exp. Med.* 133:1118.
12. Casswall, T.H. , Sarker, S.A., Albert, M.J., Fuchs, G.J., Bergström, M., Björck, L. and Hammerström, L. 1998. Treatment of *Helicobacter pylori* infection in infants in rural Bangladesh with oral Ig from hyper-immune bovine colostrum. *Aliment Pharmacol. Ther.* 12:563-568.
13. Cebra, J. J., Crandall, C. A., Gearhart, P. J., Robertson, S. M. Tseng, J., and Watson, P. M 1979. Cellular events concerned with the initiation, expression, and control of the mucosal immune response. *Immunology of Breast Milk*:1-7.
14. Chen, C-C., and Chang, H-M. 1998. Effect of thermal protectants on the stability of bovine milk Ig G. *J. Agric. Food Chem.* 46:3570-3576.
15. Cordle, C.T., Schaller, J.P., Winship, T.R., Candler, E.L., Hilty, M.D., Smith, K.L., Saif, L.J., Kohler, E.M. and Krakowka, S. 1991. Passive immune protection from diarrhea caused by rotavirus or *E.coli*: animal model to demonstrate and quantitate efficacy. *Adv. Exp. Med. Biol.* 310:317-327.
16. Cordle, C.T., Thomas, R. L., Griswell, L. G., Westfall, P. H. 1994. Purification of active Ig from whey using iso-sieving filtration, In: *Indigenous Antimicrobial Agents of Milk: Recent Developments,* pp.88. International Dairy Federation Press.
17. Davidson G.P. 1996. Passive protection against diarrheal disease. *J Pediatr. Gastroenterol. Nutr.* 23:207-212.
18. Delespesse, G. 1986. Purified Ig-related factor, novel monoclonal antibodies, hybridoma cell lines, processes and applications, *US Pat. Application* 86810244.3.
19. Dial, E.J., Hall, L.R., Serna, H., Romero, J.J., Fox, J.G., and Lichtenberger, L.M. 1998. Antibiotic properties of bovine lactoferrin on *Helicobacter pylori*. *Dig. Dis. Sci.* 43:2750-2756.
20. Ebina, T. 1992. Passive immunizations of suckling mice and infants with bovine colostrum containing antibodies to human rotavirus. *J. Med. Virol.* 38:117-123.
21. Edelman, G. M. 1973. Antibody structure and molecular immunology. *Science* 180:830-840.
22. Edelman, G. M. 1970. The structure and function of antibodies. *Sci. Amer.* 223:34-42.
23. Ehrlich, P. 1982. ‹ber Immunit%ot durch Verebung und Zeugung, *Z. Hyg. Infekt Krankh* 12:183-203.
24. Facon, M., Skura, B.J., and Nakai, S. 1995. Antibodies to a colonization factor of human enterotoxic *Escherichia coli* in cowís milk and colostrum. *Food Res. Intl.* 28:387-391.
25. Freedman, D.J., Tacket, C.O., Delehanty, A., Maneval, D.R., Nataro, J. and Crabb, J.H. 1998. Milk Ig with specific activity against purified colonization factor antigens can protect against oral challenge with enterotoxigenic *Escherichia coli*. *J. Infect. Dis.* 177:662-667.
26. Fukumoto, L.R., Li-Chan, E., Kwan, L., and Nakai, S. 1994. Isolation of Ig from cheese whey using ultra-filtration and immobilized metal affinity chromatography. *Food Res. Intl.* 27:335-348.
27. Fukushima, Y., Kawata, Y., Hara, H., Tereda, A., and Mitsuoka, T. 1998. Effect of a probiotic formula on intestinal IgA production in healthy children. *Int. J. Food Microbiol.* 42:39-44.

28. Glen, F. R., Porter, J. P., Rushlow, K. E., and Wassom, D. L. 1999. Method to detect IgE. *Intl. Patent* 05945294.

29. Greenberg, P.D., and Cello, J.P. 1996. Treatment of severe diarrhea caused by *Cryptosporidium parvum* with oral bovine Ig concentrate in patients with AIDS. *J. AIDS Human Retrovirol.* 13:348-354.

30. Hanson, L. A., Ashraf, R. N., Hahn-Zoric, M., Carlsson, B., Herias, V., Wiedermann, U., Dahlgren, U., Motas, C., Mattsby-Baltzer, I., Gonzales-Cossio, T., Cruz, J. R., Karlberg, J., Lindblad, B. S., and Jalil, F. 1993. Breast milk: roles in neonatal host defense, In: *Immunophysiology of the Gut*, pp. 248-249.

31. Hanson, L. A., Carlsson, B., Jalil, F., Hahn-Zoric, M., Hermodson, S., Karlberg, J., Mellander, L., Khan, S. R., Lindblad, B., Thiringer, K., and Zaman, S. 1988. Antiviral and antibacterial factors in human milk, *Biology of Human Milk* 15:141-143.

32. Hilpert, H., Brüssow, H., Mietens, C., Sidoti, J., Lerner, L. and Werschau, H. 1987. Use of bovine milk concentrate containing antibody to rotavirus to treat rotavirus gastroenteritis in infants. *J. Infect. Dis.* 156:158-166.

33. Janson, A., Nava, S., Brüssow, H., Mahanalabis, D. and Hammerström, L. 1994. Titers of specific antibodies in immunized and non-immunized colostrum. Implications for their use in patients with gastrointestinal infections. In: *Indigenous Antimicrobial Agents of Milk: Recent Developments*, pp.221-228. International Dairy Federation Press.

34. Kanamaru, Y., Ozeki, M., Nagaoka, S., Kuzuya, Y., and Niki, R. 1993. Ultrafiltration and gel filtration methods for separation of Ig with secretory component from bovine milk - Separation of Ig. *Milchwissenschaft* 48:247-251.

35. Kaneko, T., Wu, B.T., and Nakai, S. 1985. Selective concentration of bovine Ig and β-Lactalbumin from acid whey using FeCl₃. *J. Food Sci.* 50:1531-1536.

36. Kelly, C.P., Chetham, S., Keates, S., Bostwick, E.F., Roush, A.M., Castagliuolo, I., Lamont, J.T. and Pothoulakis, C. 1997. Survival of anti-*Clostridium difficile* bovine Ig concentrate in the human gastrointestinal tract. *Antimicrob. Agents Chemother.* 41:236-241.

37. Kelly, C.P., Pothoulakis, C., Vavva, F., Castagliuolo, I., Bostwick, E.F., O'Keane, J.C., Keates, S., and, J.T. 1996. Anti-*Clostridium difficile* bovine Ig concentrate inhibits cytotoxicity and enterotoxicity of *C.difficile* toxins. *Antimicrob. Agents Chemother.* 40:373-379.

38. Kemp, D. J., Morahan, G., Cowman, A. F., and Harris, A. W. 1983. Production of RNA for secreted Ig μ chains does not require transcriptional termination 5' to the μm exons. *Nature* 301:84.

39. Konecny, P., Brown, R.J., and Scouten W.H. 1996. Purification of monospecific polyclonal antibodies from hyperimmune bovine whey using immunoaffinity chromatography *Prep. Biochem. Biotechnol.* 26:229-243.

40. Konecny, P., Brown, R. J., and Scouten, W. H. 1994. Chromatographic purification of Ig G from bovine milk whey. *J. Chromatogr.* 673: 45-53.

41. Korhonen, H. 1998. Colostrum Ig and the complement system - potential ingredients of functional foods. *Bull. IDF* 336:36-40.

42. Korhonen, H., Syvaoja, E.L., Ahola-Luttila, H., Sivela, S., Kopola, S., Husu, J., and Kosunen, T.U. 1995. Bactericidal effect of bovine normal and immune serum, colostrum and milk against *Helicobacter pylori*. *J. Appl. Bacteriol.* 78:655-662.

43. Korhonen, H., Pihlanto-Leppala, A., Rantamaki, P., and Tupasela, T. 1998. The functional and biological properties of whey proteins: prospects for the development of functional foods. *Agri. Food Sci. (Finland)* 7:283-296.

44. Kulkarni, P. R., and Pimpale, N. V. 1989. Colostrum - a review, *Indian J. Dairy Sci.* 42: 216-224.

45. Kummer, A., Kitts, D. D., Li-Chan, E., Losso, J. N., Skkura, B. J., and Nakai, S. 1992. Quantification of Bovine IgG in milk using enzyme-linked immunosorbent assay. *Food Agri. Immunol.* 4:93-102.

46. Laemmli, U. K. 1970. Cleavage of structural proteins during the assembly of the head of bacteriophage T4. *Nature* 227:680.

47. Lamm, M. E., Weisz-Carrington, P., Roux E. M., McWilliams, M., and Phillips-Quagliata, D. 1979. Mode of induction of an IgA response in the breast and other secretory sites by oral immunization In: *Immunology of Breast Milk*, pp.105-109.

48. Leung, C., Kuzmanoff, K. M., and Beattie, C. W. 1991. Isolation and characterization of monoclonal antibody directed against bovine a$_{s2}$-casein, *J. Dairy Sci.* 74:2872-2878.

49. Li-Chan, E., Kummer, A., Losso, J.N., Kitts, D.D. and Nakai S. 1995, Stability of bovine Ig to thermal treatment and processing. *Food Res. Intl.* 28:9-16.

50. Li-Chan, E., Kwan, L., and Nakai, S. 1990. Isolation of whey proteins during metal chelate interaction chromatography. *J. Dairy Sci.* 73:2075-2086.

51. Lihme, A. O. F. 1998. Isolation of Ig. *Pat. Application* PCT/DK97/00359.

52. Loimaranta, V., Carlèn, A., Olsson, J., Tenovuo, J., Syväoja, E-L. and Korhonen, H. 1998. Concentrated bovine colostral whey proteins from *Streptococcus mutans/Strep. Sobrinus* immunized cows inhibit the adherence of *Streptococcus mutans* and promote the aggregation of mutans streptococci, *J. Dairy Res.* 65:599-607.

53. Loimaranta, V., Tenovuo, J., and Korhonen, H. 1998. Combined inhibitory effect of bovine immune whey and peroxidase-generated hypothiocyanite against glucose uptake by *Streptococcus mutans*. *Oral Microbiol. Immunol.* 13:378-381.

54. Losso, J. N., Kummer, A., Li-Chan, E., and Nakai, S. 1993. Development of a particle concentration fluorescence immunoassay for the quantitative determination of IgG in bovine milk, *J. Agric. Food Chem.* 41:682-686.

55. Losso, J. N., and Nakai, S. 1994. Fluorescence quenching of heat-treated Ig and determination of their biological activity by PCFIA. *Food Agric. Immunol.* 6:287-295.

56. Mickleson, K. N. P., and Moriarty, K. M. 1982. Ig levels in human colostrum and milk. *J. Ped. Gastroenterol. Nutrit.* 1:381-384.

57. Nocek, J.E., Braund, D.G., and Warner R.G. 1984. Influence of neonatal colostrum administration, Ig, and continued feeding of colostrum on calf gain, health and serum protein. *J. Dairy Sci.* 67: 319-333.

58. Opekun, A.R., El-Zaimaity, H.M.T., Osato, M.S., Gilger, M.A., Malaty, H.M., Terry, M., Headon, D.R. and Graham, D.Y. 1999. Novel therapies for *Helicobacter pylori* infection. *Aliment Pharmacol. Ther.* 13:35-41.

59. Ouchterlony, Ö. 1962. Quantitative immunoelectrophoresis. *Acta Path. Microbiol. Scand.* 154:252.

60. Oudin, J. 1952. Specific precipitation in gels and its application to immunochemical analysis. *Methods Med. Res.* 5:335-378.

61. Parham P. 1983. Monoclonal antibodies against HLA products and their use in immunoaffinity purification. *Methods Enzymol.* 92:110-138.

62. Parkhouse, R. M. E. 1971. Ig M biosyntheses: production of intermediates and excess of light chain in mouse myeloma MOPC-104E. *Biochem. J.* 123:635.

63. Pearce, R. J. 1988. Whey protein fractions. *Patent Application* PCT/AU88/00141.

64. Petschow, B.W., and Talbott, R.D. 1994. Reduction in virus-neutralizing activity of a bovine clostrum Ig concentrate by gastric acid and digestive enzymes. *J. Pediatr. Gastroenterol. Nutr.* 19:228-235.

65. Pierre, A., and Fauquant, J. 1986. Principes pour un procede industriel de fractionement de proteines du lactoserum. *Lait* 66: 405-419.

66. Pirro, F., Wieler, L.H., Failing, K., Bauerfeind, R., and Baljer, G. 1995. Neutralizing antibodies against Shiga-like toxins from *Escherichia coli* in colostra and sera of cattle. *Vet. Microbiol.* 43:131-141.

67. Plettenberg, A., Stochr, A., Stellbrink, H-J., Albrecht, H. and Meigel, W. 1993. A preparation from bovine colostrum in the treatment of HIV-positive patients with chronic diarrhea. *Clin. Invest. 71: 42-45*

68. Porter, P. 1979. Adoptive immunization of the neonate by breast factors. *Immunology of Breast Milk.* 197-206.

69. Porter, R. R. 1973. Structural studies of Ig. *Science* 180:713-716.

70. Reddy, N.R., Roth, S.M., Eigel, W.N. and Pierson, M.D. 1988, Foods and food ingredients for prevention of diarrheal disease in children in developing countries, *J. Food Prot.* 51:66-75.

71. Reilly, R. M., Domingo, R., and Sandhu, J. 1997. Oral delivery of antibodies future pharmacokinetic trends, *Clin. Pharmacokinet.* 313-323.

72. Roos, N., Mahé, S., Benamouzig, R., Sick, H., Rautureau, J. and Tomé, D. 1995. 15N-labeled Ig from bovine colostrum are partially resistant to digestion in human intestine. *J. Nutr.* 125:1238-1244.

73. Roy, J. H. B. 1956. Studies in calf intestine with special reference to the protective action of colostrum, Ph.D. Thesis, University of Reading, UK.

74. Saif, L.J. and Smith, K.L. 1985. Enteric viral infections of calves and passive immunity. *J Dairy Sci.* 68:206-228.

75. Shield, J., Melville, C., Novelli, V., Anderson, G., Scheimberg, I., Gibb, D. and Milla, P. 1993. Bovine colostrum Ig concentrate for cryptosporidiosis in AIDS. *Arch. Dis. Child* 69: 451-453.

76. Stephan, W., Dichtelmüller, H. and Lissner, R. 1990. Antibodies from colostrum in oral immunotherapy. *J. Clin. Chem. Clin. Biochem.* 28:19-23.

77. Stryer, L. 1975. Ig, In: *Biochemistry,* 730-731

78. Superti, F., Ammendolia, M.G., Valenti, P., and Seganti, L. 1997. Antirotaviral activity of milk proteins: lactoferrin prevents rotavirus infection in the enterocyte-like cell line HT-29. *Med. Microbiol. Immunol.* 186:83-91.

79. Swain, L. S., Hu, H., and Huston, G. 1999. Class II-independent generation of CD4 memory of T cells form effectors. *Science* 286:1381-1383.

80. Syväoja, E-L., Ahola-Luttila, H.K., Kalsta, H., Matilainen, M.H., Laakso, S., Husu, J.R. and Kosunen, T.U. 1994. Concentration of *Campylobacter*-specific antibodies in the colostrum of immunized cows. *Milchwissenschaft* 49:27-31.

81. Tacket, C. O., Binion, S.B., Bostwick, E., Losonsky, G., Roy, M.J. and Edelman, R. 1992. Efficacy of bovine milk Ig concentrate in preventing illness after *Shigella flexneri* challenge. *Am. J. Trop. Med. Hyg.* 47:276-283.

82. Tacket, C. O., Losonsky, G., Link, H., Hoang, Y., Guesry, P., Hilpert, H. and Levine, M.M. 1988. Protection by milk Ig concentrate against oral challenge with enterotoxigenic *Escherichia coli*. *N. Eng. J. Med.* 318:1240-1243.

83. Tejada-Simon, M.V., Lee, J.H., Ustunol, Z., and Pestka, J.J. 1999. Ingestion of yogurt containing *Lactobacillus acidophilus* and *Bifidobacterium* to potentiate Ig A responses to cholera toxin in mice. *J. Dairy Sci.* 82:649-660.

84. Tonegawa, S., Hozumi, N., Matthyssens, G., and Schuller, R. 1976. Somatic changes in the content and context of Ig genes. *Cold Spring Harb. Symp. Quant. Biol.* XLI:877.

85. Vasilov, R. G., and Ploegh, H. L. 1982. Biosynthesis of murine IgD: heterogenecity of glycosylation. *Eur. J. Immunol.* 12:804.

86. Warny, M., Fatimi, A., Bostwick, E.F., Laine, D.C., Lebel, F., Lamont, J.T., Pothoulakis, C., and Kelly C.P. 1999. Bovine Ig concentrate-*Clostridium difficile* retains *C. difficile* toxin neutralizing activity after passage through the human stomach and small intestine. *Gut* 44:212-217.

87. Weiner, C., Pan, Q., Hurtif, M., Borén, T., Bostwick, E., and Hammerström L. 1999. Passive immunity against human pathogens using bovine antibodies. *Clin. Exp. Immunol.* 116:193-205.

88. Yolken, R.H., Losonsky, G.A., Vonderfect, S., Leister, F., and Wee, S-B. 1985. Antibody to human rotavirus in cow's milk. *N. Engl. J. Med.* 312:605-610.

89. Yoshida, S., and Xiuyun, Y.X. 1991. Isolation of lactoperoxidase and lactoferrins from bovine milk acid whey by carboxymethyl cation exchange chromatography. *J. Dairy Sci.* 74:1439-1444.

C.E. Isaacs

M.F. Lampe

Lactolipids

5

I. INTRODUCTION

Mammalian milk is the natural functional food. It is not only a vehicle which delivers nutrients to the newborn but also transports molecules and cells which are part of the secretory immune system and play an important role in innate defense by protecting mucosal surfaces from infection (Goldman & Goldblum, 1990). Milk lipids are a class of molecules that have both a nutritional and protective function. According to the reviews by Kabara (Kabara, 1978; Kabara, 1980), the antimicrobial activity of lipids, in particular fatty acids, has been studied since the late 19th century. The lipid-dependent antimicrobial activity delivered in milk to the neonate results from the release of antimicrobial fatty acids and monoglycerides from milk triglycerides. These triglycerides are present in milk fat globules and represent 98% of the milk fat (Hamosh, 1991; Jensen, 1996; Hamosh, 1995). The triglyceride core of the milk fat globule is surrounded by a membrane of polar lipids, which prevents hydrolysis of the triglycerides by lipases in milk and therefore the lipid-dependent antimicrobial activity is released by lipases in the infant's gastrointestinal tract.

Milk lipids are not unique in possessing antimicrobial activity. Human epidermis-derived skin lipids, especially free fatty acids, have been found to inactivate *Staphylococcus aureus* (Bibel et al., 1989; Kearney et al., 1984; Miller et al., 1988; Aly et al., 1972). The removal of skin lipids by acetone extraction permitted added bacteria to persist on the skin of volunteers whereas application of the skin lipids, in particular fatty acids, inactivated the bacteria. Studies at other mucosal surfaces have shown that lung surfactant from humans, dogs, rats, and guinea pigs contained free fatty acids that were bactericidal for pneumococci (Coonrod & Yoneda, 1983; Coonrod, 1987; Coonrod, 1986). Inhaled pneumococci were killed extracellularly in rats (Coonrod, 1987) and the antibacterial activity was found to reside in the long-chain polyunsaturated free fatty acids purified from lung surfactant, providing further support for the potential antimicrobial role of surfactant free fatty acids. Fractionation of porcine intestinal lipids showed that the

0-8493-2047-X/00/$0.00+$.50

© 2000 by CRC Press, LLC

antibacterial activity against *Clostridium welchii* was primarily due to the activity of long-chain unsaturated fatty acids (Fuller & Moore, 1967).

Free fatty acids have also been found to have antimicrobial activity in nine species of brown algae (Rosell & Srivastava, 1987). Both gram-positive and gram-negative bacteria were inactivated by algal lipids. Additionally, in another study of Caribbean marine algae (Ballantine et al., 1987), over 70% of the lipid extracts from approximately 100 algae tested had antibacterial activity. Fatty acids and their antimicrobial derivatives not only have a protective role at the mucosal surfaces of vertebrates, but also appear to be utilized by simpler eucaryotes for antibacterial activity.

Lipids are one of a number of nonspecific protective factors present in milk (Welsh & May, 1979). These nonspecific protective factors are part of an innate defense system that phylogenetically precedes the adaptive immune system, e.g. antibodies, which is found solely in vertebrates (Fearon, 1997). In addition to lipids, these nonspecific protective factors include lactoferrin, lactoperoxidase, lysozyme, receptor oligosaccharides, and antimicrobial peptides (Goldman et al., 1985; Ogra & Losonsky, 1984; Welsh & May 1979; Newburg, 1997; Newburg et al., 1992; Kabara, 1980; Boman, 1998). Nonspecific protective factors such as lipids can prevent or inhibit the establishment, multiplication, and spread of pathogenic microorganisms in the host (Mandel & Ellison, 1985). In fact, microbial inactivation measured using individual purified antimicrobial factors may underestimate their *in vivo* effectiveness because antimicrobial factors in milk interact, e.g. sIgA, peroxidase, and lactoferrin produce synergistic antimicrobial activity (Maldoveanu et al., 1983; Watanabe et al, 1984; Reiter, 1981).

A. Related antimicrobial systems

1. Antimicrobial peptides. Antimicrobial peptides, the first novel class of antimicrobial agents since the introduction of nalidixic acid in 1970 (Hancock, 1997), are, in general, short polypeptides and are produced by prokaryotes, plants, and a wide variety of animals (Hancock, 1997; Nissen-Meyer & Nes, 1997; Boman, 1998). In mammals, they are expressed in a variety of cells. In insects, bacterial infection induces the release of these antibacterial peptides into the haemolymph. They are key effectors of innate immunity due to their more rapid synthesis than proteins and faster diffusion rate. There are generally three main groups of antimicrobial peptides: those with a high content of 1 or 2 amino acids, often proline and arginine; those with intramolecular disulfide bonds often stabilizing a predominantly β-sheet structure; and those with a linear structure, no cysteines, and often amphiphilic regions (Boman, 1998). Their exact mechanism of bacterial membrane permeabilization is not known. In general, they directly cause lysis of microorganisms. Several peptides may form a complex of membrane spanning helices, which results in transient pore formation. The first step of binding may involve initial competitive displacement of divalent cations from lipopolysaccharides (LPS). Several authors have shown that a variety of antimicrobial peptides have good activity against sexually transmitted pathogens (Yasin et al, 1996; Qu et al, 1996; Lampe et al, 1996). Both antimicrobial lipids and peptides function by disrupting the integrity of the membranes of pathogenic microorganisms (Noseda, et al., 1989; Hancock, 1997).

2. Biocides. There are a wide variety of chemical agents that are able to inactivate microorganisms. Disinfectants are active on inanimate objects or surfaces while antiseptics are active in or on tissues. They have a wide spectrum of activity against bacteria, viruses, fungi, and protozoa. In general, biocides interact with the cell surface, penetrate into the cell, and are active at an intracellular site (McDonnell & Russell, 1999). Several reports have been published that describe the activity of a variety of biocides on sexually transmitted pathogens (Howett et al, 1999; Lampe et al, 1998b; Lyons & Ito, 1995; Yong et al, 1982).

II. OCCURRENCE

A. Biosynthesis

The lipid-dependent antimicrobial activity in milk is primarily present in two groups of fatty acids and their derivatives: the long-chain unsaturated fatty acids and the medium-chain saturated fatty acids. The origin of fatty acids in milk is three-fold: the diet, the mobilization of endogenous stored fatty acids, and synthesis in the mammary gland (Jensen, 1996). The relative proportion of milk fatty acids from each of these sources is dependent upon the mother's diet (Silber et al., 1988; Francois et al., 1998). Maternal diets which are high energy and low fat stimulate the production of milks with relatively high amounts of medium-chain fatty acids (Spear et al., 1992; Thompson & Smith, 1985; Francois et al., 1998) as the result of increased medium-chain fatty acid synthesis in the mammary gland. Therefore, even when the fatty acid composition of human milk triglyceride varies as the result of diet (Silber et al, 1988), there will always be a sufficient concentration of antimicrobial fatty acids to inactivate susceptible pathogens.

B. Screening/Identification

To find biological sources of antimicrobial lipids, the lipid fraction can be extracted using chloroform-methanol (2:1), the extract dried and taken up in ethanol (Thormar et al., 1987), and the extracted lipids can be tested against susceptible microorganisms.

C. Quantitative assays

Lipids present in human milk can be quantified using methods developed and modified by Jensen et al. (1997). Lipids from other sources can be extracted as described above and quantified using thin layer chromatography and flame ionization detection (Peyrou et al., 1996). The method used to determine antimicrobial activity varies with the organism of interest. Antimicrobial lipids in milk inactivate enveloped viruses (Isaacs et al., 1986; Isaacs et al., 1990; Isaacs & Thormar, 1991; Isaacs et al., 1994). Virucidal activity can be determined by incubating infectious virus particles in the presence of lipid and then determining the reduction in virus titer by plating the virus on susceptible cell cultures (Isaacs et al, 1986). The decrease in virus titer is determined either by determining the cytopathic endpoint in 96 well tissue culture plates (Thormar et al., 1987) or by using a plaque assay (Prichard et al., 1990).

1. Measurement of activity against Chlamydia trachomatis. In early studies, *in vitro* minimal inhibitory concentration (MIC) and minimal bactericidal concentration (MBC) assays using tissue culture cells in antibiotic-free conditions were developed to examine the anti-chlamydial activity of various antibiotics (Stamm et al, 1986; Stamm et al, 1991). These studies required extensive development of techniques to grow *Chlamydia trachomatis* in antibiotic free media and to determine antibiotic MIC/MBCs versus *C. trachomatis*. In addition, staining with a fluorescein-labeled anti-chlamydial monoclonal antibody, rather than other staining methods, was found to be superior because it allows visualization of abnormal *C. trachomatis* inclusions which are frequently produced in tissue culture cells in the presence of antimicrobials and thus refined the MIC value.

In subsequent studies, these methods were adapted to develop the minimal cidal concentration (MCC) assay. This assay examines the direct effect of microbicides on extracellular chlamydial EBs, mimicking the conditions found in the vaginal tract at the time of sexual contact. *C. trachomatis* is transmitted from an infected to an uninfected individual in genital secretions via sexual contact. Although *C. trachomatis* is an obligate intracellular bacterium, the elementary body (EB) or infectious form of the organism is found extracellularly in secretions. EBs are adapted for survival in cell-free conditions, but are susceptible *in vitro* to various antimicrobials such as detergents (Lampe et al, 1998b), peptides (Yasin et al, 1996), whole human milk, and fractions of milk (Elbagir et al, 1990). Extracellular EBs would be introduced into the genital tract of an uninfected individual in their partner's genital secretions. These newly introduced organisms either would then infect genital epithelial cells very rapidly or be killed by a previously applied topical microbicide. After various times of direct exposure to the microbicide in the in vitro assay, the organisms are tested for viability by inoculating them onto tissue culture cells. The cultures are incubated for the appropriate length of time and any surviving *C. trachomatis* stained and counted. Killing by the microbicide is calculated by comparing the numbers of surviving organisms after treatment with organisms that were not exposed to any microbicide. This assay was also adapted to examine other conditions that may be found in the female vaginal tract both during normal and disease conditions. Human blood (10%) was added to examine its effect on microbicide activity during the menstrual cycle. The pH was altered to examine the activity of microbicides at a normal pH of 4, pH 5 or 6 in women with bacterial vaginosis, pH 7 in women who are post-menopausal or bleeding, and pH 8 after semen is deposited.

2. Minimal Cidal Concentration (MCC) assay against C. trachomatis.

C. trachomatis strains: Two standard strains of *C. trachomatis* are used in these studies, including serovars D (UW-3/Cx) and F (UW-6/Cx). These strains are propagated in McCoy mouse fibroblast cells (ATCC CRL 1696), purified on Renografin density gradients (Howard et al, 1974), and stored at -70 °C in SPG (219 mM sucrose, 3.8 mM KH_2PO_4, 8.6 mM Na_2HPO_4, 4.9 mM glutamic acid, pH 7.0) (Jackson & Smadel, 1951) until needed. Their serotype is confirmed with monoclonal antibodies in an inclusion typing method prior to use (Suchland & Stamm, 1991). Immediately prior to use, the purified organisms are thawed and diluted in SPG.

MCC Assay: This assay examines the direct effect of microbicides on extracellular chlamydial EBs, which would occur in the vaginal tract at the time of sexual contact. One day prior to the assay, 9×10^4 low passage (<15 passages) McCoy cells are added to individual wells in a 96 well microtiter plate and incubated for 24-h to permit the cells to form a confluent monolayer. These cells are maintained in antibiotic-free Eagle's minimal essential medium with 10% fetal calf serum (EMEM) and checked routinely once per month for mycoplasma contamination prior to the MCC assays. 1×10^6 *C. trachomatis* inclusion forming units (IFU) in SPG are added to dilutions of each microbicide or positive and negative controls and incubated for 0, 30, 60, 90, or 120 minutes. After each time period, a 5 ml aliquot of the organism and drug mixture is diluted 1:40 in 195 ml of SPG. 100 μl of this dilution is then added to a McCoy mouse fibroblast cell monolayer in a 96 well microtiter plate and centrifuged for 1 hour to inoculate the tissue culture cells. The organism/drug mixture is then removed and replaced with EMEM with 1 μg/ml cyclohexamide. The cultures are incubated for 48 hours, stained with the *Chlamydia* genus-specific fluorescein isothiocyanate (FITC)-labeled monoclonal antibody CF2 (Stamm et al., 1983), and chlamydial inclusions counted in three fields. All assays are performed in triplicate, the inclusion counts averaged, and positive and negative controls run in parallel. Additional wells are inoculated with either the *C. trachomatis* organisms only, which serves as an organism control, or SPG only, to monitor McCoy cell morphology. Percent killing in tests with the microbicide and the positive and negative controls are calculated by the following formula: (mean IFU of organism control - mean IFU of test/mean IFU of organism control) X 100. The lowest concentration of microbicide, which shows 100% killing is defined as the Minimal Cidal Concentration (MCC). 100% killing represents a decrease of at least 1×10^4 organisms, from the original 1×10^6 IFU in the inoculum to 160 IFU, the minimum number of IFU which can be counted in our assay.

    Alamar-Blue™ Cytotoxicity Assay: The MCC assay is duplicated with the exception that no *C. trachomatis* are added. The various lipid dilutions are further diluted 1:40 in EMEM as in the MCC assay, then added to 24 hour monolayers of McCoy cells in 96 well microtiter plates and incubated for 1 hour. The lipid dilutions are removed from the wells and replaced with EMEM with 1 μg/ml cyclohexamide and the cell cultures incubated for 48 hours. The growth indicator alamarBlue (Alamar Biosciences, Inc., Sacramento, CA) is next added to each well and the cells further incubated for 4 more hours. The metabolic activity of growing cells chemically reduces alamarBlue, which causes the fluorometric/colorimetric REDOX indicator to change from a non-fluorescent blue oxidized form to a fluorescent red reduced form. The optical density is read at 570 nm and 600 nm and the percent inhibition of McCoy cells calculated compared to EMEM negative controls using the following formula: % inhibition = ((Mean OD_{570} - Mean OD_{600} of negative cell control) - (Mean OD_{570} - Mean OD_{600} of test)/ (Mean OD_{570} - Mean OD_{600} of negative cell control) to 100.

    MCC Assay in the presence of 10% human blood: Because blood found during the menstrual period may alter antimicrobial activity, 10% whole human blood was added to the MCC assay described above. Blood is collected from an individual who has no anti-chlamydial antibodies as measured by microimmunofluorescence (Wang et al, 1977) and who is not receiving antibiotics. 10% human blood is also added to the controls and only 0 and 120 minute time points are taken.

    MCC Assay with pH Alterations: Because the pH can vary in the human vagina and alter the activity of topical antimicrobials, the MCC assay described above is followed with the exception that lipid dilutions are adjusted to pH 4, 5, 6, 7, or 8 with 1M Na_2HPO_4 or 1M KH_2PO_4 before *C. trachomatis* IFU are added. All controls are tested at each different pH value and only 0 and 120 minute time points are taken.

III. MOLECULAR PROPERTIES

A. Isolation and purification

The primary commercial source of antimicrobial fatty acids that are also found in milk is plant oils. These are primarily coconut, soy, palm, and high oleic safflower oils. Coconut oil is a major source of lauric acid (12 carbons) which is a medium-chain fatty acid with strong antimicrobial activity (Isaacs et al., 1992; Kabara, 1978). The other plant oils are excellent sources of long-chain unsaturated fatty acids with antimicrobial activity. Antimicrobial fatty acids can be purchased commercially from a number of different suppliers. Antimicrobial monoglycerides in milk have ester linkages between the fatty acid and the glycerol backbone. This linkage can be replaced with an ether linkage using well-established methods of synthesis and it provides a more stable compound with equal or greater antimicrobial activity (Baumann, 1972; Isaacs et al., 1994; Lampe et al., 1998a).

B. Physico-chemistry

The antimicrobial activity of short-chain fatty acids is due to the undissociated molecule and not the anionic form of the molecule (Kabara, 1978; Lundblad & Seng, 1991) and therefore their antimicrobial activity is pH dependent. Work by Kabara (1978) shows that as the degree of dissociation of the acid increases, the activity decreases. The minimum inhibitory concentration of short-chain acids (6, 8 and 10 carbons) increases with increasing pH while the minimum inhibitory concentration of medium-chain acids (12 and 14 carbons) decreased with increasing pH. The activity of long-chain unsaturated fatty acids was unaffected by changing pH values. Monoglyceride esters and ethers of short-chain fatty acids (see Section IV-A2) as opposed to the free acids are unaffected by pH alterations due to the attachment of the fatty acid carboxyl group to the glycerol backbone.

C. Structure

The structures of fatty acids, monoglyceride esters and monoglyceride ethers can be seen in *FIGURE 1*.

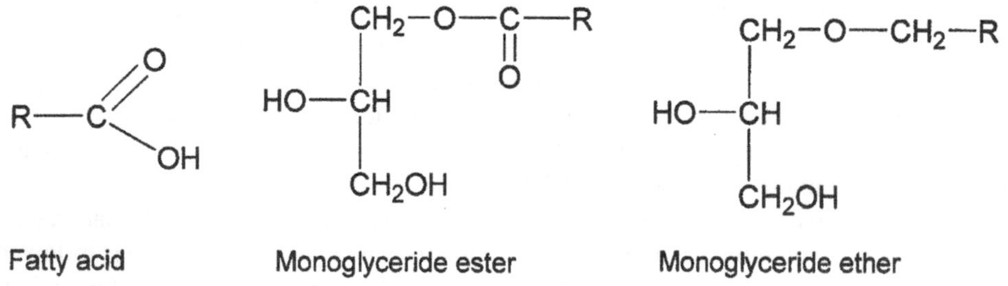

FIGURE 1. The structures of fatty acids, monoglyceride esters, and monoglyceride ethers.

TABLE 1. Fatty acids with ability to inactivate enveloped viruses[1]

Fatty acid	C-atom: double bonds
Butyric	(4:0)
Caproic	(6:0)
Palmitic	(16:0)
Stearic	(18:0)
Caprylic	(8:0)
Capric	(10:0)
Myristic	(14:0)
Palmitoleic	(16:1)
Lauric	(12:0)
Oleic	(18:1)
Linolenic	(18:3)
Linoleic	(18:1)
Arachidonic	(20:4)

[1]Fatty acids are listed from the least effective to the most effective.

IV. ANTIMICROBIAL ACTIVITY

A. Mechanism of action

The antimicrobial activity produced by the hydrolysis of milk triglycerides can be duplicated using purified fatty acids and monoglycerides (Thormar et al., 1987). Short-chain (butyric, caproic, and caprylic) and long-chain saturated (palmitic and stearic) fatty acids have no or minimal activity against three enveloped viruses even at exceedingly high concentrations (Thormar et al., 1987). On the other hand medium-chain saturated and long-chain unsaturated fatty acids are strongly antiviral against enveloped viruses regardless of whether the virus is a DNA, RNA, or a retrovirus (*TABLE 1*). In contrast, incubation of the nonenveloped poliovirus with capric, lauric, myristic, palmitoleic, oleic, linoleic, and arachidonic acids did not significantly reduce viral infectivity.

This data suggests that the envelope of viruses such as HSV and human immunodeficiency virus (HIV) (Isaacs & Thormar, 1990) is the target for lipid-dependent viral inactivation. Electron microscopic studies done in our laboratory confirmed this and showed that the envelope of VSV is completely destroyed by linoleic acid (Thormar et al., 1987). Other purified lipids, which are also the product of milk triglyceride hydrolysis by lipases, e.g. 1-monoglycerides of milk fatty acids, also effectively inactivate enveloped viruses (Thormar et al., 1987). In most instances the monoglyceride derivative of the fatty acid is antiviral at a concentration that is 5 to 10 times lower (mM) than the corresponding fatty acid (Thormar et al., 1987). Results with purified monoglycerides indicate that monoglycerides produced by hydrolysis of milk triglycerides may have an antimicrobial role in the infant's gastrointestinal tract.

Further evidence that antimicrobial lipids destabilize membranes is provided by electron microscopic studies of leukemic cells treated with ester and ether lipids (Noseda et al., 1989; Verdonck & van Heugten, 1997). These experiments showed that the plasma membrane of the leukemic cells was the primary target of the lipids and that morphological damage induced by the lipids consisted of the formation of blebs, formation of holes, and increased porosity of the membrane. *In vivo*, cells in the infant's gastrointestinal tract are protected from lipid damage by the mucosal barrier.

Depending upon the concentration, lipid dependent inactivation of enveloped viruses can occur within a few seconds or over the period of a few hours. The antiviral effect of active fatty acids and monoglycerides is additive (Isaacs & Thormar, 1990). When medium-chain saturated and long-chain unsaturated fatty acids are tested together, each at a concentration previously shown to be below that needed for antiviral activity, the mixture is antiviral because the total concentration of antiviral lipids is at or above a level sufficient to destabilize the lipid envelope of the virus. Regardless of whether fatty acid mixtures consisted of two medium-chain saturated fatty acids, a medium-chain saturated and a long-chain unsaturated fatty acid, or two long-chain unsaturated fatty acids, they were antiviral. Mixtures were made with as many as seven different fatty acids each at a suboptimal concentration, but the lipid mixture itself was antiviral.

Milk lipids inactivate all enveloped viruses that have been examined including herpes simplex virus (Isaacs et al., 1986; Sands et al., 1979), influenza virus (Kohn et al., 1980a; Kohn et al., 1980b), respiratory syncytial virus (Laegreid et al., 1986), measles virus (Isaacs et al., 1986), vesicular stomatitis virus (Isaacs et al., 1986), visna virus (Isaacs et al., 1986), mouse mammary tumor virus (Sarkar, 1973), dengue virus types 1-4 (Falker et al., 1975), cytomegalovirus (Welsh et al., 1979), Semliki forest virus (Welsh et al., 1979; Welsh & May, 1979), Japanese B encephalitis virus (Fieldsteel, 1974), and human immunodeficiency virus (Isaacs & Thormar, 1990). Milk lipids also inactivate both gram positive, e.g. *Staphylococcus epidermidis*, and gram negative, e.g. *Escherichia coli*, bacteria (Isaacs et al., 1990) as well as the protozoal pathogen *Giardia lamblia* (Hernell et al., 1986; Rohrer et al., 1986; Reiner et al., 1986).

As mentioned above, the antimicrobial activity of free fatty acids may be altered by pH depending upon the chain length of the molecule, but monoglycerides maintain their antimicrobial activity across a wide pH range. The temperature at which lipids are incubated does, however, affect the rate at which they inactivate pathogens. The 8 carbon monoglyceride ether 1-*O*-octyl-*sn*-glycerol can decrease the titer of VSV by more than 10,000 fold at 37 °C in 1 hour but requires 18 hours for the same reduction at 4°C (Isaacs et al., 1994).

The antibacterial and antifungal activity of lipids is by the same membrane destabilizing mechanism as shown for enveloped viruses (Kabara, 1978; Shibasaki & Kato, 1978; Gershon & Hanks, 1978). Mild heating which makes bacterial membranes more fluid increases the lipid dependent killing of the gram negative bacterial pathogens *E. coli* and *Pseudomonas aeruginosa* by 1,000-100,000 fold (Shibasaki & Kato, 1978) providing further evidence that the bacterial membrane is the site of antimicrobial lipid activity.

1. Effect against STD pathogens. Up to 12 million new cases of sexually trans-mitted diseases (STDs) occur each year in the US, and many of these infections result in serious reproductive sequellae (Eng & Butler, 1997). Given the lack of a vaccine or other effective means to prevent most sexually transmitted diseases, new approaches to pre-vention are urgently needed. One new approach to reducing the transmission of *C. tra-chomatis* and other STD pathogens would be the topical application of antimicrobials intravaginally before sexual contact. Topical microbicides are particularly attractive in that they are broad spectrum in their activity and could be self administered by women before sexual contact. An ideal microbicide would prevent STDs caused by bacterial pathogens such as *C. trachomatis* and *Neisseria gonorrhoeae*, as well as those caused by human immunodeficiency virus (HIV), human papillomavirus (HPV), herpes simplex viruses (HSV), and the parasite *Trichomonas vaginalis*. In addition, such preparations should be spermicidal but not disrupt the normal flora or be toxic to the vaginal epithelium.

Development of a topical microbicide that is safe, inexpensive, easy to use, and effective against the most common STD pathogens would help reduce the transmission of STDs. The lipids we have examined in these studies may be good candidates for such a topical microbicide.

A number of topical microbicides, including nonoxynol-9 (Benes & McCormack, 1985, Kelly et al, 1985), chlorhexidine (Lyons & Ito, 1995), defensins, and protegrins (Yasin et al. 1996) have been examined for their in vitro effect on *C. trachomatis* and other STD pathogens (Qu et al, 1996).

To this end, we examined the activity of five synthetic lipids adapted from naturally occurring compounds found in human breast milk. *C. trachomatis* serovar D or F elementary bodies were added to serial dilutions of the lipids and incubated for various times. Aliquots were then cultured in monolayers of McCoy cells and inclusions were counted.

2. Effect against *C. trachomatis*. Chlamydiae are atypical gram-negative bacteria that have evolved a biphasic life cycle that facilitates their efficient transmission. The bacterium infects columnar epithelial cells of the human genital and respiratory tracts and replicates within these cells in membrane bound inclusions in a form designated the reticulate body. When the infected host cell has been depleted of its nutrients by the newly replicated bacteria, the bacterium changes its morphology to that of the infectious, non-metabolically active form designated the elementary body (EB). EBs are released from the depleted infected cell and can either infect adjacent cells or be transmitted to uninfected individuals through sexual contact. The EB surface contains several cysteine-rich proteins that provide rigidity to the cell wall through intra- and inter-molecular disulfide bonds (Hatch et al, 1984). Although the bacterium does not produce peptidoglycan, its cell wall is sufficiently strengthened by these sulfhydryl bonds such that the organism is environmentally resistant. If the organism in its EB form can be killed with a topical microbicide before it infects the target columnar epithelial cell, the chlamydial infectious process would be blocked.

Human breast milk contains numerous antimicrobial compounds (Isaacs & Thormar, 1991), some of which are lipid based. Some of these antimicrobial lipids were modified to increase their stability and aqueous solubility while maintaining their antimicrobial activity, and then synthesized for use in antimicrobial assays. These novel lipids were shown to have inhibitory activity against *E. coli*, *Salmonella enteriditis*, and *Staph. epidermidis* (Isaacs et al, 1990). We examined whether genital strains of *C. trachomatis* were killed by these novel antimicrobial lipids adapted from human breast milk.

Lipids. Five lipids, including 1 and 2-*O*-hexyl, 1-*O*-heptyl, and 1 and 2-*O*-octyl-*sn*-glycerol which were designed by Dr. C. E. Isaacs and synthesized by Deva Biotech (Hatboro, PA), were tested. 1 and 2-*O*-hexyl-*sn*-glycerol were diluted to 15 mM (2.64 mg/ml) and 1-*O*-heptyl and 1 and 2-*O*-octyl-*sn*-glycerol were diluted to 50 mM (10.2 mg/ml) in SPG or EMEM. The lipids were mixed by vortexing for 15-30 seconds to ensure their even distribution after each dilution.

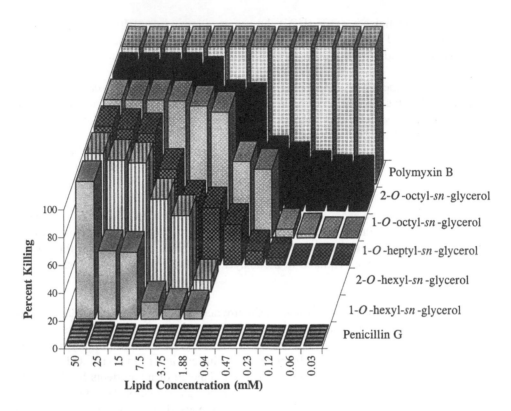

Figure 2. Comparison of the anti-chlamydial activity of five novel lipids. The MCC assay was used and percent killing of *C. trachomatis* serovar D after 120 minutes of exposure to each lipid or the controls were plotted. Polymyxin B and penicillin G initial concentrations were 2 mg/ml.

Controls. Polymyxin B and penicillin-G were used as positive and negative controls at initial concentrations of 2 mg/ml in the MCC assays. In the alamar-Blue™ cell toxicity assay, EMEM and EtOH were used as controls. All controls were prepared the day of the assay.

The five lipids described above were each tested in the MCC assay. A summary of the anti-chlamydial activity of the five lipids against *C. trachomatis* serovar D at 120 minutes is shown in *Figure 2*.

The 2 and 1-*O*-octyl-*sn*-glycerol derivatives showed 100% killing at 7.5 mM and 15 mM respectively. However, the 1-*O*-heptyl and 1 and 2-*O*-hexyl-*sn*-glycerol derivatives showed 100% killing only at higher concentrations of 25 and 50 mM. A summary of the MCC values of each lipid against *C. trachomatis* serovars D and F is shown in *Table 2*.

To ensure that the diluted lipid contained no residual toxicity for the McCoy cells used to culture *C. trachomatis*, the MCC assay described above includes a 1:40 dilution of the lipid-organism mixture prior to adding the inoculum. To directly test whether this dilution eliminated all toxicity, all dilutions of the lipids were then tested in the alamar-Blue™ cytotoxicity assay, which can detect low amounts of toxicity. We found that none of the dilutions of the lipids used in the MCC assay showed significant cytotoxicity. The

TABLE 2. MCC values for all four lipids against *C. trachomatis* serovars D and F.

Lipid	Time (min)[a]	Serovar D[b] MCC (mM)	Time (min)[a]	Serovar F[b] MCC (mM)
2-0-octyl-*sn*-glycerol	90	7.5	120	7.5
1-0-octyl-*sn*-glycerol	30	15	120	7.5
1-0-heptyl-*sn*-glycerol	90	25	ND[c]	ND
2-0-hexyl-*sn*-glycerol	90	50	120	12.5
1-0-hexyl-*sn*-glycerol	120	>50	120	50

a) Time (minutes) organisms were exposed to lipid prior to inoculation.
b) Concentration (mM) of lipid, which caused 100% killing of *C. trachomatis* at this time point.
c) ND, not done.

results with 2-*O*-octyl-*sn*-glycerol are shown in TABLE 3. All other lipids showed similar negative results.

Since urogenital infections can be caused by as many as at least eight different *C. trachomatis* serovars, we examined whether these antimicrobial lipids were active against different serovars. All dilutions of the lipids were tested against serovars D and F for up to 120 minutes. The results with 2-*O*-octyl-*sn*-glycerol at 0 and 120 minutes are shown in FIGURE 3A and show that its MCC was 7.5 mM at 120 minutes for both organisms. The other lipids tested showed more activity against serovar F than serovar D but both serovars were killed after 120 minutes. 1 *O*-hexyl-*sn*-glycerol was the least active because it never completely killed serovar D even after 120 minutes.

Since human blood found during the menstrual cycle may reduce antimicrobial activity, we tested the lipids in the MCC assay in the presence and absence of 10% whole human blood. The results comparing 2-*O*-octyl-*sn*-glycerol activity against *C. trachomatis* serovar D with and without 10% human blood after 0 and 120 minutes of exposure are shown in FIGURE 3B. Since we found that the MCC with blood (15 mM) was one dilution higher than without blood (7.5 mM) after 120 minutes, we conclude that blood has a small but most likely insignificant effect on lipid anti-chlamydial activity.

Because the pH can range from 4 to 8 in the female vagina and because different pHs can affect antimicrobial activity, the activity of the lipids at pH 4, 5, 6, 7, and 8 was tested against *C. trachomatis* serovar D for 0 and 120 minutes. FIGURE 3C shows that some 2-*O*-octyl-*sn*-glycerol pH/concentration combinations showed increased (pH 8, 3.75

TABLE 3. 2-0-octyl-*sn*-glycerol cytotoxicity under pre-inoculation MCC assay conditions.

Concentration (mM)[a]	mean OD_{570} - mean OD_{600}	SD	% cytotoxicity
0.3750	0.405	0.064	2.0
0.1875	0.366	0.045	11.5
0.0938	0.410	0.052	1.0
0.0469	0.406	0.039	1.9
0.0234	0.414	0.041	0.0
0.0118	0.414	0.007	0.0
0.0058	0.414	0.045	0.0
0.0029	0.414	0.038	0.0
0.0014	0.406	0.058	1.9
0.0008	0.405	0.032	2.2
No lipid	0.414	0.052	0.0

a) Values represent lipid concentration (mM), diluted 1:40 and plated directly on McCoy cell monolayers.

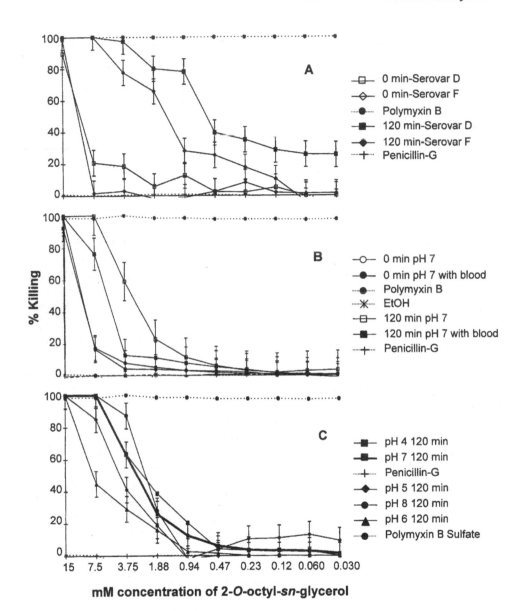

FIGURE 3. Antimicrobial activity of 2-*O*-octyl-*sn*-glycerol activity against *C. trachomatis*

A) Comparison of activity against *C. trachomatis* serovars D and F. Percent killing of each serovar after lipid exposure at 0 and 120 minutes in the MCC assay. Results for the polymyxin B positive and penicillin-G negative controls are also shown.

B) Percent cidal values for *C. trachomatis* serovar D exposed to the indicated lipid concentrations at 0 and 120 minutes in the presence and absence of 10% whole human blood. Values for the organism, polymyxin B, and penicillin G controls were also determined in the MCC assay with and without the addition of 10% whole human blood.

C) Activity of lipid at pH 4, 5, 6, 7, 8 at 120 minutes. Percent killing at each pH value is depicted, including standard deviations for each concentration. Values were calculated compared to the organism control at the same pH value.

mM) or decreased (pH 6, 7.5 mM) killing of *C. trachomatis* serovar D after 120 minutes. Thus, in summary, different pHs did not have a significant effect on lipid activity.

In summary, the 2-*O*-octyl-*sn*-glycerol lipid was the most active of the five closely related lipids tested. 7.5 mM of this lipid killed *C. trachomatis* after 90 minutes of exposure to the organism. 15 mM of 1-*O*-octyl-*sn*-glycerol also killed *C. trachomatis* after 30 minutes of contact. The 1-*O*-heptyl and 1 and 2-*O*-hexyl-*sn*-glycerol derivatives were less active, either requiring higher concentrations or longer exposure times to kill *C. trachomatis* or showing incomplete killing at the concentration and times examined. 7.5 mM 2-*O*-octyl-*sn*-glycerol can easily be synthesized and incorporated into a vehicle suitable for self administration and is currently being formulated. Since this concentration is well above the minimum necessary to kill *C. trachomatis* and can readily be synthesized, mM lipid concentrations are physiologically relevant for a topical microbicide. A further advantage of such a lipid preparation is its markedly lesser cost in comparison to other antimicrobials such as peptides.

Lipids disrupt the membranes of pathogens and thus might also disrupt eukaryotic cell membranes they contact. This might also include the epithelial cells lining the human vagina. However, cellular toxicity caused by lipids applied to mucosal surfaces would probably be partly prevented by the mucous which covers the cells. In our assay with one hour of contact between a 1:40 dilution of each lipid concentration, the lipids were not found to cause toxicity to McCoy cells measured in the alamar-Blue™ test. Toxicity of these lipids is important and should, however, be examined further in more appropriate systems.

Because there are at least eight different *C. trachomatis* serovars which can cause urogenital disease, we determined whether these antimicrobial lipids were active against multiple serovars. The same concentration of 2-*O*-octyl-*sn*-glycerol necessary to kill serovar D (7.5 mM) was also completely active against serovar F. We thus found that none of the lipids tested showed significant differences in activity against the two *C. trachomatis* serovars D and F. Additional strains and serovars should be tested but these results suggest that a topical lipid preparation would be broadly active against the different *C. trachomatis* serovars which cause STDs.

Vaginally applied microbicides would come in contact with serum proteins contained in menstrual blood and some microbicides show reduced activity in the presence of serum (Lehrer et al., 1993). The lipids examined in these studies could potentially bind to serum proteins with fatty acid binding sites found in blood (Isaacs et al, 1990). However, these lipids did not show decreased activity in the presence of 10% whole human blood, indicating that a lipid preparation would be active during the menstrual period when other antimicrobials might have reduced activity.

Because the optimal topical antimicrobial would be effective at different pH values found in the vagina, the activity of the lipids was examined at pH 4, 5, 6, 7, and 8. Some differences in activity of the lipids at pH values other than 7 were found but these differences are not significant, indicating that a vaginal lipid preparation would be active under different pH conditions.

In summary, these lipids kill *C. trachomatis* directly with 2-*O*-octyl-*sn*-glycerol showing the most activity of those tested (Lampe et al., 1998a). Lipid activity was not affected by 10% human blood or pH differences from 4.0 to 8.0, conditions that can be found in the vagina. After further testing of toxicity and efficacy in humans, these lipids, especially 2-*O*-octyl-*sn*-glycerol, may function as an effective topical microbicide to kill STD pathogens including *C. trachomatis* at the time of sexual transmission, before infection occurs. These lipids which are naturally found at mucosal membranes in humans may be relatively nontoxic and ideal for their proposed use.

3. Effect against other sexually transmitted pathogens. The lipid ether 1-*O*-octyl-*sn*-glycerol was tested against the sexually transmitted pathogens *N. gonorrhoeae* and *T. vaginalis* as well as organisms that are part of the vaginal microflora (Rabe et al., 1997) at pH's 4-8 in the presence and absence of blood. *N. gonorrhoeae* and *T. vaginalis* were killed by 5-7.5 mM 1-*O*-octyl-*sn*-glycerol over a pH range of 4-8 while *Lactobacillus crispatus* and other members of the vaginal microflora were not killed by lipid concentrations cidal to sexually transmitted pathogens. 1-*O*-octyl-*sn*-glycerol effectively inactivates HIV-1 and HIV-2 (Isaacs et al., 1994). HIV-1 and HIV-2 titers were decreased by more than one million fold (6 logs) after 10 minutes of incubation with the lipid. Studies done in our laboratories in conjunction with the Division of AIDS (NIH) found complete inactivation of HIV-1 in the presence of mucin indicating that antimicrobial lipid ethers should be able to effectively inactivate HIV intravaginally.

4. Outcome. Microbicides can have at least four modes of action; they can disrupt the bacterial cell wall or cell membrane, they can block receptor-ligand interaction, they can inhibit intracellular replication of an obligate intracellular organism such as chlamydiae, and they can inhibit an intracellular target.

Antimicrobial lipids may be active because they disrupt membranes. In electron micrographs of *C. trachomatis* treated with these lipids, not shown here, there was disruption of the chlamydial inner membrane, allowing the cytoplasmic contents to leak out of the cell. This confirms that the lipids act directly on the organisms. Our finding that there was no eukaryotic cellular toxicity as measured by the alamar-Blue™ assay also confirms our conclusion that the antichlamydial activity of these lipids is due to direct disruption of the chlamydial membrane and not due to toxicity to the McCoy cells.

5. Antimicrobial kinetics. These lipids are immediately active on *C. trachomatis*. Aliquots were removed from the lipid-organism MCC assay immediately after mixture and assayed for chlamydial viability. As shown in Figure 3, serovar D showed 89.6% and serovar F 100% killing by the most active lipid, 2-*O*-octyl-*sn*-glycerol, at 0 minutes. Both serovars showed 100% killing by 15 mM of this lipid after only 30 minutes. These results demonstrate the rapid action of these lipids on *C. trachomatis*.

B. Influencing factors

The antimicrobial activity of fatty acids can be reduced in the presence of proteins especially albumin through specific and nonspecific binding (Shibasaki & Kato, 1978; Kato & Shibasaki, 1975). Microbicidal inhibition by unsaturated fatty acids can also be reduced by other surface active agents such as cholesterol (Kabara, 1978; Ammon, 1985). Staphylococci, which are inactivated *in vivo* by long-chain unsaturated fatty acids, can develop resistance to fatty acid inactivation by producing an esterifying enzyme, which binds the fatty acids to cholesterol (Kapral & Mortensen, 1985). The use of monoglycerides would prevent this particular mechanism of inactivation.

1. The effect of strain variation. The *C. trachomatis* trachoma, urogenital, and lymphogranuloma venereum serovars differ not only in their serological reactivities but biologically as evidenced by the different types of infections they cause. The different urogenital serovars have not been shown, however, to differ in their antibiotic sensitivities. To ensure, nevertheless, that no strain to strain variations occurs in lipid MCC assay results, two urogenital serovars, D and F, were examined. As demonstrated in *FIGURE 3A*, there was little difference in lipid activity on these two serovars. Thus, these lipids will most likely be active against the eight *C. trachomatis* serovars, which cause urogenital infections.

2. Exposure in the presence of 10% human blood. Because the activity of topical microbicides would be reduced in the presence of human blood found in the vagina during the menstrual cycle, the lipids were tested in the pre-inoculation MCC assay in the presence and absence of 10% whole human blood. Blood was collected from an individual who was not receiving antibiotics and who has no antibodies against the test organism. 10% human blood was also added to the organism and antibiotic controls. *FIGURE 3B* shows that 10% human blood had an insignificant effect on the 2-*O*-octyl-*sn*-glycerol activity. Similar results were obtained for the four other lipids.

3. Activity at different pH values. Since the pH can vary from 4 to 8 in the human vagina, the pre-inoculation MCC assay was followed except that microbicide dilutions were adjusted to pH 4, 5, 6, 7, or 8 with $1M$ Na_2HPO_4 or $1M$ KH_2PO_4 before the pathogen was added. All controls were tested at each different pH value. *FIGURE 3C* shows the effect of different pH values on the activity of 2-*O*-octyl-*sn*-glycerol. Generally pH differences did not have a significant effect on lipid activity.

C. Spectrum

Antimicrobial lipids can inactivate both gram positive and gram negative bacteria. Lauric acid (12:0) as well as the monounsaturated palmitoleic acid (16:1) and the polyunsaturated linoleic acid (18:2) have been shown to be the most active against gram positive bacteria and those gram negative bacteria which can be inactivated by antimicrobial lipids (Kabara, 1978). A number of gram negative bacteria, including *E. coli*, are extremely resistant to fatty acid activation, especially by long-chain unsaturated fatty acids (Shibasaki & Kato, 1978), while others including *N. gonorrhea* and *Haemophilus* species are as susceptible to lipid inactivation as gram positive bacteria (Knapp & Melly, 1986; Rabe et al., 1997). *C. trachomatis*, a common sexually transmitted pathogen, is also gram negative and, as can be seen by the results presented in this section, is inactivated by 8 carbon fatty acid ethers. Antimicrobial fatty acids present in milk also inactivate protozoa, fungi, yeast, and all enveloped viruses tested thus far. The intestinal protozoan *Giardia lamblia* (Hernell et al., 1986; Reiner et al., 1986) is susceptible to inactivation by both medium-chain saturated and long-chain unsaturated fatty acids as well as fatty acid derivatives including monoglycerides and lysophosphatidylcholine (Reiner et al., 1986). The fungi *Aspergillus niger* and *Trichoderma viride* are inactivated primarily by short-chain fatty acids (3 - 5 carbon) and their derivatives (Gershon & Shanks, 1978) whereas the yeast *Saccharomyces cerevisiae* and *Kluyveromyces marxianus* are inhibited by medium chain (8 and 10 carbon) fatty acids (Viegas et al., 1989). Enveloped viruses are more readily inactivated by lipids than other microbes. They are susceptible to medium-chain saturated and long-chain unsaturated fatty acids and their monoglyceride derivatives.

1. Chlamydia trachomatis. *C. trachomatis* is the most common bacterial sexually transmitted pathogen worldwide, causing millions of infections each year (CDC, 1997). These infections frequently result in serious sequelae, especially in women, including pelvic inflammatory disease, ectopic pregnancy, infertility, and chronic pelvic pain. In addition, chlamydial infections may increase the risk of human immunodeficiency virus transmission (Laga et al, 1993). Mechanical barriers such as condoms were recently found to be protective in men but not in women (Holmes, 1998). Currently available spermicides have not been shown in clinical trails to reduce the transmission of chlamydial or other STD infections (NIAID, 1998). A consensus is emerging that the use of topical microbicides may be an effective approach for the prevention of chlamydial infections and STDs in general.

Chlamydiae are obligate intracellular bacteria which can only grow within eukaryotic cells (Schachter, 1999). Infectious elementary bodies are found in genital secretions and are adapted for survival in this environment. The organisms are only extracellular for a brief period of their life cycle but this extracellular phase is essential for their transmission from person to person. It is during this brief time that chlamydiae are susceptible to killing by topical microbicides. The development of new methods for prevention of sexually transmitted *C. trachomatis* infection is a top public health priority. Topical self administered vaginal microbicides represent one such approach in which the organism is eradicated at the time of initial exposure. A further goal is to develop antimicrobials that are not only active against *C. trachomatis* but also *N. gonorrhoeae*, HIV, HSV, HPV, and *T. vaginalis,* are non-toxic to vaginal tissues and the normal flora, can be spermicidal, are inexpensive, and can be self-administered topically by women.

V. BIOLOGICAL ADVANTAGE

A. Additive advantage to foods

Antimicrobial fatty acids and monoglycerides have been used as food preservatives (Shibasaki & Kato, 1978) and in cosmetics (Kabara, 1991). Many of these lipids are considered as GRAS (generally regarded as safe) chemicals by the U.S. Food and Drug Administration (Anonymous, 1973) and can therefore be used in food products to reduce food borne infections. One potential new area where these lipids could be used is in infant formulas (Isaacs et al., 1995). At present the formulas contain triglycerides which are hydrolysed in the infant's gastrointestinal tract to produce antimicrobial fatty acids and monoglycerides in a similar fashion to milk triglycerides (Canas-Rodrigues & Smith, 1966; Smith, 1966). However, the formula is not protected from contamination between the time it is put into a bottle and ingested by the infant allowing time for potentially significant bacterial growth. The addition of these GRAS compounds to infant formulas could extend the time during which they can be safely transported.

B. Physiological advantage

The use of antimicrobial lipids at mucosal surfaces would provide increased protection from infection using molecules that are naturally occurring *in vivo* and could be used in human and animals. The transfer of lipid based protection is already known to be successful because of the transfer of lipid based protection in milk (Canas-Rodriguez

& Smith, 1966) and food (Fuller & Moore, 1967). The increased protection that would be provided by antimicrobial lipids at the host's mucosal surfaces will be especially important for stopping the spread of viruses such as HIV and HSV for which it has not been possible to develop a vaccine. A recent clinical trial shows that a recombinant subunit vaccine containing 2 major HSV surface proteins induced neutralizing antibody production but did not prevent HSV infection (Corey et al., 1999; Mascola, 1999). HSV and HIV establish latent infections in the immuno-competent host in contrast to a virus such as poliovirus which does not persist in the host and is eventually removed from host tissue by an immune system primed by vaccination. Vaccination does not prevent initial virus replication in host tissue (Clements-Mann, 1998). Lipids provided to mucosal surfaces could inactivate HIV and HSV prior to their infection of cells at mucosal surfaces and prevent establishment of latent infection.

STD pathogens and topical activity. Most STD pathogens are transmitted in sexual secretions at the time of contact. Many of them, including *C. trachomatis*, *N. gonorrhoeae*, HIV, and HPV, are intracellular pathogens. Most of them can only replicate within eukaryotic cells and once they enter their target host cells, they may be resistant to killing by topical microbicides. However, for the short period of time before they are internalized, they may be susceptible to topical microbicide killing. The MCC assay, described above, is designed to test immediate, topical killing of *C. trachomatis* by these lactolipids. Our results showed that the lipids were directly active on the exposed extracellular organisms. Thus, the lactolipids may be effective against STD pathogens in general during the brief time after they are introduced into the genital tract and before they become intracellular. Once they have been internalized and initiate infection, they are then more effectively treated by standard antibiotics. Topical activity of lactolipids can potentially reduce the vertical transmission of sexually transmitted pathogens. It is estimated that 40-80% of the children who acquire HIV-1 from their mothers become infected during delivery (Mostad et al., 1997; Goldenberg et al., 1997) by exposure of the skin or mucous membranes to virus (Baba, 1994). HSV is also transmitted from mother to infant and infection in the perinatal period is responsible for approximately 80% of all instances of neonatal HSV infection (Whitely, 1994). Lipid protection during birth would not only be against vertically transmitted viruses but also against bacteria present in the birth canal. *C. trachomatis* has one of the highest incidences of infection of any sexually transmitted disease and infants have a 33-50% risk of contracting the infection from an infected mother during vaginal delivery (Eng & Butler, 1997). The antibacterial activity of lactolipids and related lipids could also be used to prevent bacterial vaginosis which leads to preterm delivery of low-birth-weight infants (Hillier et al., 1995; Goldenberg et al., 1997). As a result of the negative impact of vaginal infections on the mother and infant a simple, inexpensive, non-toxic intervention that can be used routinely is required (Hofmeyr & McIntyre, 1997). Lactolipids meet all of these criteria.

VI. APPLICATIONS

A. Suitability/Adaptability

Antimicrobial lipids found in milk can be developed for use in a microbicide delivered to mucosal surfaces. Medium-chain (6 - 12 carbon) fatty acids and their monoglyceride derivatives are potentially more adaptable for use in a topical microbicide than the long-chain unsaturated fatty acids, e.g. oleic acid, which provide most of the lipid dependent antimicrobial activity in human milk. Medium-chain fatty acids have greater solubility in an aqueous environment than longer-chain length fatty acids. In milk, lipid is transported to the intestinal mucosa as emulsified globules, which are surrounded by a membrane (Jensen, 1996; Jensen et al., 1990; Lammi-Keefe & Jensen, 1984). It would be much simpler to use a lipid with an increased aqueous solubility than to develop an artificial system to mimic the human milk delivery system. Another advantage to using antimicrobial medium-chain fatty acids is that they are saturated (no double bonds) whereas long-chain unsaturated lipid are susceptible to oxidation with possible loss of activity and production of undesirable oxidation products.

As mentioned above, monoglyceride esters are more strongly antimicrobial on a concentration basis than the corresponding fatty acid (Thormar et al., 1987). We have found that changing the linkage in a monoglyceride between the fatty acid and the glycerol backbone from an ester bond to an ether linkage can increase the antimicrobial activity of some medium-chain monoglycerides. The inactivation studies described above with *C. trachomatis* show that the use of ether linkages provides a stable effective microbicide.

B. Possible applications

Currently lactolipids are being used in various cosmetic products (Kabara, 1991). These lactolipids could also be effective against STD pathogens because they are directly active against *C. trachomatis*, *N. gonorrhoeae*, HIV, HSV, HPV and *T. vaginalis*.

C. Limitations

Because many STD pathogens are quickly internalized into their target host cells, topical preparations containing these lactolipids need to be present at or before sexual contact to ensure their contact with the infectious agents. Topical preparations thus need to be actively inserted by women at the appropriate time to be effective. Failure to insert the topical agents before sexual contact will not prevent STDs. This limitation results from inappropriate application by the users, not failure of the topical preparations themselves.

VII. SAFETY/TOLERANCE AND EFFICACY

Fatty acids and monoglycerides are already present in food and are considered GRAS compounds (Anonymous, 1973). Monoglyceride esters, however, can be hydrolysed by lipases of either mammalian or microbial origin producing the free fatty acid which is less active than the monoglyceride and can be inactivated by bacterial esterifying enzymes (Kapral & Mortensen, 1985). A major advantage of using an ether linkage is that this bond is stable toward enzymatic and chemical hydrolysis (Paltauf, 1983). Therefore, intravaginally, for example, the monoglyceride ether structure will be maintained in the presence of lipases, whether of human or microbial origin, and antimicrobial activity will not be diminished. Ether lipids are fairly innocuous compounds which have been shown to be nontoxic even at high doses and can be used by mammals to synthesize membrane alkyl glycerolipids and plasmalogens (Mangold, 1983; Mangold, 1972). Ether lipids are also currently being developed as anticancer drugs for use against leukemic cells

(Verdonck & vanHeugten, 1997). Catabolism of ether lipids takes place primarily in the intestine and liver by oxidative cleavage of the ether bond (Weber, 1985). Glycerol ether lipids are present in human and bovine milk and colostrum (Oh & Jadav, 1994; Hallgren et al., 1974; Hallgren & Larsson, 1962). Previous studies have shown that ether lipids are not transported across the placenta from mother to infant (Das et al., 1992). Ether mono-glycerides will potentially remain active longer *in vivo* than the comparable ester linked compound but not produce any increase in toxicity. Studies performed in our laboratory in conjunction with the National Institute of Allergy and Infectious Diseases have found that the lipid ether 1-*O*-octyl-*sn*-glycerol does not cause irritation in a rabbit model and is acceptable for vaginal administration.

VIII. BIOTECHNOLOGY

The synthesis of alkyl glycerol ethers is well-established (Baumann, 1972; Paltauf, 1983). These compounds can be synthesized and then matched with water soluble polymers for delivery to mucosal surfaces. It is also potentially important that biotransformations of ether lipids, which occur in animal cells also proceed in cultured plant cells (Mangold & Weber, 1987). Complex ether lipids could be synthesized from simple precursors using plant cells that do not contain endogenous ether lipids.

IX. SUMMARY

Milk lipids are not only nutrients but also antimicrobial agents that are part of an innate defense system functioning at mucosal surfaces. Lipid-dependent antimicrobial activity in milk is due to medium-chain saturated and long-chain unsaturated fatty acids and their respective monoglycerides. The antimicrobial activity of fatty acids and mono-glycerides is additive so that dietary induced alterations in the fatty acid distribution in milk do not reduce lipid dependent antimicrobial activity. Fatty acids and monoglycerides can rapidly destabilize the membranes of pathogens. The antimicrobial activity of milk lipids can be duplicated using monoglyceride ethers, which very effectively inactivate enveloped viruses, bacteria and protozoa. Monoglyceride ethers are more stable than the ester linkages found in milk lipids since they are not degraded by bacterial or mammalian lipases. The sexually transmitted pathogens *C. trachomatis*, HIV, HSV, *N. gonorrhoeae*, and *T. vaginalis* are effectively inactivated by the lipid ether 1-*O*-octyl-*sn*-glycerol. This compound could be used as part of a vaginal microbicide to reduce both the spread of sexually transmitted pathogens between adults as well as the vertical transmission of infectious agents from mother to infant during birth. Lactolipids and related GRAS antimicrobial lipids could also be added to infant formulas to reduce the risk of bacterial growth during transport.

X. REFERENCES

1. Aly, R., Maibach, H.I., Shinefield, H.R., and Strauss, W.G. 1972. Survival of pathogenic microorganisms on human skin. *J. Invest. Dermatol.* 58:205-210.
2. Ammon, H.V. 1985. Effects of fatty acids on intestinal transport. p. 173-181. In J.J. Kabara (ed.) *The Pharmacological Effect of Lipids II,* The American Oil Chemists Society, Champaign, IL.
3. Anonymous, GRAS (Generally recognized as safe) Food Ingredients - Glycerine & Glycerides ordering No. PB-221227, National Technical Information Service, U.S. Department of Commerce, Springfield, VA, 1973.

4.	Baba, T.W., Koch, K.J., Mittler, E.S., Greene, M., Wyand, M., Penninck, D., and Ruprecht, R.M. 1994. Mucosal infection of neonatal rhesus monkeys with cell-free SIV. *AIDS Res. Human Retroviruses* 10:351-357.
5.	Ballantine, D.L., Gerwick, H.W., Velez, S.M., Alexander, E., and Guevara, P. 1987. Antibiotic activity of lipid-soluble extracts from Caribbean marine algae. In *Hydrobiologia*, ed. M.A. Ragan, and C.J. Bird, *pp*. 463-469. Twelfth International Seaweed Symposium. Dordecht, Netherlands: Dr. W. Junk Publishers.
6.	Baumann, W.J. 1972. The chemical syntheses of alkoxylipids. In *Ether Lipids Chemistry and Biology*, ed, F. Snyder, *pp*. 51-79. New York: Academic Press.
7.	Benes, S., and McCormack, W.M. 1985. Inhibition of growth of *Chlamydia trachomatis* by nonoxynol-9 in vitro. *Antimicrob. Agents Chemother.* 27:724-726.
8.	Bibel, D.J., Miller, S.J., Brown, B.E., Pandey, B.B., Elias, P.M., Shinefield, H.R., and Aly, R. 1989. Antimicrobial activity of stratum corneum lipids from normal and essential fatty acid-deficient mice. *J. Invest. Dermatol.* 92:632-638.
9.	Boman, H.G. 1998. Gene-encoded peptide antibiotics and the concept of innate immunity: An update review. *Scand. J. Immunol.* 48:15-25.
10.	Canas-Rodriguez, B., and Smith, H.W. 1966. The identification of the antimicrobial factors of the stomach contents of sucking rabbits. *Biochem. J.* 100:79-82.
11.	CDC. 1997. *Chlamydia trachomatis* Genital Infections-United States, 1995. *MMWR.* 46:193-198.
12.	Clements-Mann, M.L. 1998. Lessons for AIDS vaccine development from non-AIDS vaccines. *AIDS Res. Hum. Retroviruses* 14 (Suppl. 3):S197-S203.
13.	Coonrod, J.D. 1986. The role of extracellular bactericidal factors in pulmonary host defense. *Sem. Respir. Infect.* 1:118-129.
14.	Coonrod, J.D. 1987. Role of surfactant free fatty acids in antimicrobial defenses. *Eur. J. Respir. Dis.* 71:209-214.
15.	Coonrod, J.D., and Yoneda, K. 1983. Detection and partial characterization of antibacterial factor(s) in alveolar lining material of rats. *J. Clin. Invest.* 71:129-141.
16.	Corey, L., Langenberg, A.G.M., Ashley, R., Sekulovich, R.E., Izu, A.E., Douglas, J.M., Handsfield, H.H., Warren, T., Marr, L., Tyring, S., DiCarlo, R., Adimora, A.A., Leone, P., Dekker, C.L., Burke, R.L., Leong, W.P., and Straus, S.E. 1999. Recombinant glycoprotein vaccine for the prevention of genital HSV-2 infection. *JAMA* 4:331-340.
17.	Das, A.K., Holmes, R.D., Wilson, G.N., and Hajra, A.K. 1992. Dietary ether lipid incorporation into tissue plasmalogens of humans and rodents. *Lipids* 27:401-405.
18.	Elbagir, A., Petterson, M., Lindahl, M., Genc, M., Froman, G., and Mardh, P.A. 1990. Influence of whole human milk, and fractions thereof, on inclusion-formation of *Chlamydia trachomatis* in McCoy cells. *APMIS* 98:609-614.
19.	Eng, T.R., and Butler, W.T. (eds.) 1997. Medicine. In *The Hidden Epidemic: Confronting Sexually Transmitted Diseases*, *pp*. 1-432. Washington DC: National Academy Press.
20.	Falkler, W.A., Jr., Diwan, A.R., and Halstead, S.B. 1975. A lipid inhibitor of Dengue virus in human colostrum and milk; with a note on the absence of anti-Dengue secretory antibody. *Arch. Virol.* 47:3-10.
21.	Fearon, D.T. 1997. Seeking wisdom in innate immunity. *Nature* 388:323-324.
22.	Fieldsteel, A.H. 1974. Nonspecific antiviral substances in human milk active against arbovirus and murine leukemia virus. *Cancer Res.* 34:712-715.
23.	Francois, C.A., Connor, S.L., Wander, R.C., and Connor, W.E. 1998. Acute effects of dietary fatty acids on the fatty acids of human milk. *Am. J. Clin. Nutr.* 67:301-308.
24.	Fuller, R., and Moore, J.H. 1967. The inhibition of the growth of *Clostridium welchii* by lipids isolated from the contents of the small intestine of the pig. *J. Gen. Microbiol.* 46:23-41.
25.	Gershon, H., and Shanks, L. 1978. Antifungal activity of fatty acids and derivatives: Structure activity relationships. In *The Pharmacological Effect of Lipids,* ed. J.J. Kabara, *pp*. 51-62. St.Louis, Mo: The American Oil Chemists Society.
26.	Goldenberg, R.L., Andrews, W.W., Yuan, A.C., MacKay, H.T., and Michael, E. 1997. Sexually transmitted diseases and adverse outcomes of pregnancy. *Infect. Perinatol.* 24:23-41.
27.	Goldman, A.S., and Goldblum, R.M. 1990. Human milk: Immunologic-nutritional relationships. *Annals NY Acad. Sci.* 587:236-245.
28.	Goldman, A.S., Ham Pong, A.J., and Goldblum, R.M. 1985. Host defenses: Development and maternal contributions. In *Advances in Pediatrics*, ed. L.A. Barnes, *pp*. 71-100. Chicago: Year Book Publication.
29.	Hallgren, B., and Larsson, S. 1962. The glyceryl ethers in man and cow. *J. Lipid Res.* 3:39-38.
30.	Hallgren, B., Niklasson, A., Stallberg, G., and Thorin, H. 1974. On the occurrence of 1-0-alkylglycerols and 1-0(2-Methoxyalkyl) glycerols in human colostrum, human milk, cow's milk, sheep's milk, human red bone marrow, red cells, blood plasma and a uterine carcinoma. *Acta Chem. Scand.* B 28:1029-1034.

31. Hamosh, M. 1991. Lipid metabolism. In *Neonatal Nutrition and Metabolism*, ed. W.W. Hay, Jr., *pp.* 122-142. St. Louis, Mo: Mosby Year Book.

32. Hamosh, M. 1995. Lipid metabolism in pediatric nutrition. *Ped. Clinics N.A.* 42:839-859.

33. Hancock, R. E. 1997. Antibacterial peptides and the outer membranes of gram-negative bacilli. *J Med. Microbiol.* 46(1):1-3.

34. Hatch, T. P., Allan, I., and Pearce, J.H. 1984. Structural and polypeptide differences between envelopes of infective and reproductive life cycle forms of *Chlamydia* spp. *J. Bacteriol.* 157:13-20.

35. Hernell, O., Ward, H., Blackberg, L., and Pereira, M.E.A. 1986. Killing of *Giardia lamblia* by human milk lipases: An effect mediated by lipolysis of milk lipids. *J. Infect. Dis.* 153:715-720.

36. Hillier, S.L., Nugent, R.P., Eschenback, D.A., Krohn, M.A., Gibbs, R.S., Martin, D.H., Cotch, M.F., Edelman, R., Pastorek, J.G., Rao, A.V., McNellis, D., Regan, J.A., Carey, C., Klebanoff, M.A. 1995. Association between bacterial vaginosis and preterm delivery of a low-birth-weight infant. *N. Engl. J. Med.* 333:1373-1742.

37. Hofmeyr, G.J., and McIntyre, J. 1997. Preventing perinatal infections. *BMJ* 315:199-200.

38. Holmes, K. K. 1998. Personal Communication.

39. Howard, L., Orenstein, N.S., and King, N.W. 1974. Purification on renografin density gradients of *Chlamydia trachomatis* grown in the yolk sac of eggs. *Appl. Microbiol.* 27:102-106.

40. Howett, M.K., Neely, E.B., Christensen, N.D., Wigdahl, B., Krebs, F.C., Malamud, D., Patrick, S.D., Pickel, M.A., Welsh, P.A., Reed, C.A., Ward, M.G., Budgeon, L.R., and Kreider, J.W. 1999. A broad-spectrum microbicide with virucidal activity against sexually transmitted viruses. *Antimicrob. Agents Chemother.* 43:314-321.

41. Isaacs, C. E., Kashyap, S., Heird, W.C., and Thormar, H. 1990. Antiviral and antibacterial lipids in human milk and infant formula feeds. *Arch. Dis. Child.* 65:861-864.

42. Isaacs, C.E., and Thormar, H. 1990. Human milk lipids inactivate enveloped viruses. In *Breastfeeding, Nutrition, Infection and Infant Growth in Developed and Emerging Countries*, ed. S.A. Atkinson, L.A. Hanson, and R.K. Chandra, *pp.* 161-174. Newfoundland, Canada: ARTS Biomedical Publishers and Distributors.

43. Isaacs, C.E., and Thormar, H. 1991. The role of milk-derived antimicrobial lipids as antiviral and antibacterial agents. In *Immunology of Milk and the Neonate*, ed. J. Mestecky et al., *pp.* 159-165. New York: Plenum Press.

44. Isaacs, C.E., Kim, K.S., and Thormar, H. 1994. Inactivation of enveloped viruses in human bodily fluids by purified lipids. *Slow Inf. Central Nerv. System* 724:457-464.

45. Isaacs, C.E., Litov, R.E., and Thormar, H. 1995. Antimicrobial activity of lipids added to human milk, infant formula, and bovine milk. *J. Nutr. Biochem.* 6:362-366.

46. Isaacs, C.E., Litov, R.E., Marie, P., and Thormar, H. 1992. Addition of lipases to infant formulas produces antiviral and antibacterial activity. *J. Nutr. Biochem.* 3:304-308.

47. Isaacs, C.E., Thormar, H., and Pessolano, T. 1986. Membrane-disruptive effect of human milk: Inactivation of enveloped viruses. *J. Infect. Dis.* 154:966-971.

48. Jackson, E. B., and Smadel, J.E. 1951. Immunization against scrub typhus. II. Preparation of lyophilized living vaccine. *Amer. J. Hyg.* 53:326-331.

49. Jensen, R.G. 1996. The lipids in human milk. *Prog. Lipid Res.* 35:53-92.

50. Jensen, R.G., Ferris, A.M., Lammi-Keefe, C.J., and Henderson, R.A. 1990. Lipids of bovine and human milks: A comparison. *J. Dairy Sci.* 73:223-240.

51. Jensen, R.G., Lammi-Keefe, C.J., and Koletzko, B. 1997. Representative sampling of human milk and extraction of fat for analysis of environmental lipophilic contaminants. *Toxicol. Environ. Chem.* 62:229-247.

52. Kabara, J.J. 1978. Fatty acids and derivatives as antimicrobial agents - A review. In *The Pharmacological Effect of Lipids*, ed. J.J. Kabara, *pp.* 1-14. St. Louis, Mo: The American Oil Chemists Society.

53. Kabara, J.J. 1980. Lipids as host-resistance factors of human milk. *Nutr. Rev.* 38:65-73.

54. Kabara, J.J. 1991. Chemistry and biology and monoglycerides in cosmetic formulations. *Cosmetic Science Technol. Ser.* 311-344.

55. Kapral, F.A., and Mortensen, J.E. 1985. The inactivation of bactericidal fatty acids by an enzyme of *Staphylococcus aureus*. In *The Pharmacological Effect of Lipids II*, ed. J.J. Kabara, *pp.* 103-109. Champaign, IL: The American Oil Chemists Society.

56. Kato, N., and Shibasaki, I. 1975. Comparison of antimicrobial activities of fatty acids and their esters. *J. Ferment. Technol.* 53:793-801.

57. Kearney, J.N., Ingham, E., Cunliffe, W.J., and Holland, K.T. 1984. Correlations between human skin bacteria and skin lipids. *Br. J. Dermatol.* 110:593-599.

58. Kelly, J. P., Reynolds, R.B., Stagno, S., Louv, W.C., and Alexander, W.J. 1985. *In vitro* activity of the spermicide nonoxynol-9 against *Chlamydia trachomatis*. *Antimicrob. Agents Chemother.* 27:760-762.

59. Knapp, H.R., and Melly, M.A. 1986. Bactericidal effects of polyunsaturated fatty acids. *J. Infect. Dis.* 154:84-94.

60. Kohn, A., Gitelman, J., and Inbar, M. 1980a. Interaction of polyunsaturated fatty acids with animal cells and enveloped viruses. *Antimicrob. Agents Chemother.* 18:962-968.

61. Kohn, A., Gitelman, J., and Inbar, M. 1980b. Unsaturated free fatty acids inactivate animal enveloped viruses. *Arch. Virol.* 66:301-307.

62. Laegreid, A., Kolstootnaess, A.-B., Orstavik, I., and Carlsen, K.H. 1986. Neutralizing activity in human milk fractions against respiratory syncytial virus. *Acta Paediatr. Scand.* 75:696-701.

63. Laga, M., A. Manoka, M. Kivuvu, B. Malele, M. Tuliza, N. Nzila, J. Goeman, F. Behets, V. Batter, M. Alary, and et al. 1993. Non-ulcerative sexually transmitted diseases as risk factors for HIV-1 transmission in women: results from a cohort study. *AIDS* 7(1):95-102.

64. Lammi-Keefe, C.J., and Jensen, R.G. 1984. Lipids in human milk: A review. 2: Composition and fat-soluble vitamins. *J. Ped. Gastroenter. Nutr.* 3:172-198.

65. Lampe, M.F., Ballweber, L.M., Borders, M.A., Isaacs, C.E., Stamm, W.E. 1996. A new approach to control sexually transmitted diseases: Effect of novel topical microbicides on *Chlamydia trachomatis*, Proceedings, Third Meeting of the European Society for *Chlamydia* Research, Stary, A (ed.), Societa Editrice Esculapio, Bologna, Italy.

66. Lampe, M.F., Ballweber, L.M., Isaacs, C.E., Patton, D.L., and Stamm, W.E. 1998a. Inhibition of *Chlamydia trachomatis* by novel antimicrobial lipids adapted from human breast milk. *Antimicrob. Agents Chemother.* 42:1239-1244.

67. Lampe, M.F., Ballweber, L.M., and Stamm, W.E. 1998b. Susceptibility of *Chlamydia trachomatis* to chlorhexidine gluconate gel. *Antimicrob. Agents Chemother.* 42:1726-1730.

68. Lehrer, R. I., Lichtenstein, A.K. and Ganz, T. 1993. Defensins: antimicrobial and cytotoxic peptides of mammalian cells. *Ann. Rev. Immunol.* 11:105-128.

69. Lundblad, J.L., and Seng, R.L. 1991. Inactivation of lipid-enveloped viruses in proteins by caprylate. *Vox Sang* 60:75-81.

70. Lyons, J. M., and Ito, J.I.J. 1995. Reducing the risk of *Chlamydia trachomatis* genital tract infection by evaluating the prophylactic potential of vaginally applied chemicals. *Clin. Infect. Dis.* 21:S174-177.

71. Mandel, I.D., and Ellison, S.A. 1985. The biological significance of the nonimmunoglobulin defense factors. In *The Lactoperoxidase System Chemistry and Biological Significance*, ed. K. M. Pruitt, and J.O. Tenovuo, *pp.* 1-14. New York: Marcel Dekker.

72. Mangold, H.K. 1972. Biological effects and biomedical applications of alkoxylipids. In *Ether Lipids Chemistry and Biology*, ed. F. Snyder, *pp.* 157-176. New York: Academic Press.

73. Mangold, H.K. 1983. Ether lipids in the diet of humans and animals. In *Ether Lipids Biochemical and Biomedical Aspects*, ed. H.K. Mangold and F. Paltauf, pp. 231-238. New York: Academic Press.

74. Mangold, H.K., and Weber, N. 1987. Biosynthesis and biotransformation of ether lipids. *Lipids* 22:789-799.

75. Mascola, J.R. 1999. Herpes simplex virus vaccines - Why don't antibodies protect? *JAMA* 4:379-380.

76. McDonnell, G., and Russell, A.D. 1999. Antiseptics and disinfectants: activity, action, and resistance. *Clin. Microbiol. Rev.* 12:147-179.

77. Miller, S.J., Aly, R., Shinefeld, H.R., and Elias, P.M. 1988. *In vitro* and *in vivo* antistaphylococcal activity of human stratum corneum lipids. *Arch. Dermatol.* 124:209-215.

78. Moldoveanu, Z., Tenovuo, J., Pruitt, K.M., Mansson-Rahemtulla, B., and Mestecki, J. 1983. Antibacterial properties of milk: IgA-peroxidase-lactoferrin interactions. *Ann. NY Acad. Sci.* 409:848-850.

79. Mostad, S.B., Overbaugh, J., DeVange, D.M., Welch, M.J., Chohan, B., Mandaliya, K., Byange, P., Martin, H.L., Jr., Ndinya-Achola, J., Bwayo, J.J., and Kreiss, J.K. 1997. Hormonal contraception, vitamin A deficiency, and other risk factors for shedding of HIV-1 infected cells from the cervix and vagina. *Lancet* 350:922-927.

80. Newburg, D.S. 1997. Do the binding properties of oligosaccharides in milk protect human infants from gastrointestinal bacteria? *J. Nutr.* 127(5 Suppl):980S-984S.

81. Newburg, D.S., Viscidi, R.P., Ruff, A., and Yolken, R.H. 1992. A human milk factor inhibits binding of human immunodeficiency virus to the CD4 receptor. *Pediatr. Res.* 31(1):22-28.

82. NIAID. 1998. Research on Topical Microbicides for Prevention of STDs/AIDS. RFA AI-98-011.

83. Nissen-Meyer, J., and Nes, I.F. 1997. Ribosomally synthesized antimicrobial peptides: their function, structure, biogenesis, and mechanism of action. *Arch. Microbiol.* 167(2-3):67-77.

84. Noseda, A., White, J.G., Godwin, P.L., Jerome, W.G., and Modest, E.J. 1989. Membrane damage in leukemic cells induced by ether and ester lipids: An electron microscopic study. *Exper. Mol. Pathol.* 50:69-83.

85. Ogra, P.L., and Losonsky, G.A. 1984. Defense factors in products of lactation. In *Nutritional and Immunological Interactions*, ed. P.L. Ogra, *pp.* 67-87. Orlando, FL: Grune and Stratton.

86. Oh, S.Y., and Jadhav, L.S. 1994. Effects of dietary alkylglycerols in lactating rats on immune responses in pups. *Ped. Res.* 36:300-305.
87. Paltauf, F. 1983. Ether lipids as substrates for lipolytic enzymes. In *Ether Lipids Biochemical and Biomedical Aspects*, ed. H.K. Mangold and F. Paltauf, pp. 211-229. New York: Academic Press.
88. Peyrou, G., Rakotondrazafy, V., Mouloungui, Z., and Gaset, A. 1996. Separation and quantitation of mono-, di-, and triglycerides and free oleic acid using thin-layer chromatography with flame-ionization detection. *Lipids* 31:27-32.
89. Prichard, M.N., Turk, S.R., Coleman, L.A., Engelhardt, S.L., Shipman, C., and Drach, J.C. 1990. A microtiter virus yield reduction assay for the evaluation of antiviral compounds against human cytomegalovirus and herpes simplex virus. *J. Virol. Methods* 28:106-106.
90. Qu, X. D., Harwig, S.S., Oren, A.M., Shafer, W.M. and Lehrer, R.I. 1996. Susceptibility of *Neisseria gonorrhoeae* to protegrins. *Infect. Immun.* 64:1240-1245.
91. Rabe, L.K., Coleman, M.S., Hillier, S.L., and Isaacs, C. 1997. The *in vitro* activity of a microbicide, octyl-glycerol, against *Neisseria gonorrhoeae, Trichomonas vaginalis,* and vaginal microflora. International Congress of Sexually Transmitted Diseases, Seville, Spain.
92. Reiner, D.S., Wang, C.-S., and Gillin, F.D. 1986. Human milk kills *Giardia lamblia* by generating toxic lipolytic products. *J. Infect. Dis.* 154:825-832.
93. Reiter, B. 1981. The contribution of milk to resistance to intestinal infection in the newborn. In *Immunological Aspects of Infection in the Fetus and Newborn*, ed. H.P. Lambert and C.B.S. Wood. London: Academic Press.
94. Rohrer, L., Winterhalter, K.H., Eckert, J., and Kiohler, P. 1986. Killing of *Giardia lamblia* by human milk is mediated by unsaturated fatty acids. *Antimicrob. Agents Chemother.* 30:254-257.
95. Rosell, K.-G., and Srivastava, L.M. 1987. Fatty acids as antimicrobial substances in brown algae. In *Hydrobiologia*, ed. M.A. Ragan and C.J. Bird, Twelfth International Seaweed Symposium, *pp.*471-475. Dordrecht, Netherlands: Dr. W. Junk Publishers.
96. Sands, J., Auperin, D., and Snipes, W. 1979. Extreme sensitivity of enveloped viruses, including herpes simplex, to long-chain unsaturated monoglycerides and alcohols. *Antimicrob. Agents Chemother.* 15:67-73.
97. Sarkar, N.H., Charney, J., Dion, A.S., and Moore, D.H. 1973. Effect of human milk on the mouse mammary tumor virus. *Cancer Res.* 33:626-629.
98. Schachter, J. 1999. Biology of *Chlamydia trachomatis*. In *Sexually Transmitted Diseases*, 3rd edition, ed. K. K. Holmes et. al., *pp.*391-405. NewYork: McGraw-Hill.
99. Shibasaki, I., and Kato, N. 1978. Combined effects on antibacterial activity of fatty acids and their esters against gram-negative bacteria. In *The Pharmacological Effect of Lipids,* ed. J.J. Kabara, *pp.* 15-24. St.Louis, Mo: The American Oil Chemists Society.
100. Silber, G.H., Hachey, D.L., Schanler, R.J., and Garza, C. 1988. Manipulation of maternal diet to alter fatty acid composition of human milk intended for premature infants. *Am. J. Clin. Nutr.* 47:810-814.
101. Smith, H.W. 1966. The Journal of Pathology and Bacteriology. *J. Path. Bact.*91:1-9.
102. Spear, M.L., Bitman, J., Hamosh, M., Wood, D.L., Gavula, D., and Hamosh, P. 1992. Human mammary gland function at the onset of lactation: Medium-chain fatty acid synthesis. *Lipids* 27:908-911.
103. Stamm, W.E., and Wong, K.G. 1991. Antimicrobial Activity of Sparfloxacin (AT-4140, PD 131501, CI-978) against *Chlamydia trachomatis* in Cell Culture. *Eur. J. Clin. Microbiol. Infect. Dis.* Supp. 566-567.
104. Stamm, W.E., Tam, M., Koester, M. and Cles, L. 1983. Detection of *Chlamydia trachomatis* inclusions in McCoy cell cultures with fluorescein-conjugated monoclonal antibodies. *J. Clin. Microbiol.* 17:666-668.
105. Stamm, W.E., and Suchland, R. 1986. Antimicrobial activity of U-70138F (Paldimycin), roxithromycin (RU 965), and ofloxacin (ORF 18489) against *Chlamydia trachomatis* in cell culture. *Antimicrob. Agents Chemother.* 30: 806-807,
106. Suchland, R. J., and Stamm, W.E. 1991. Simplified microtiter cell culture method for rapid immunotyping of *Chlamydia trachomatis*. *J. Clin. Microbiol.* 29:1333-1338.
107. Thompson, B.J., and Smith. S. 1985. Biosynthesis of fatty acids by lactating human breast epithelial cells: An evaluation of the contribution to the overall composition of human milk fat. *Ped. Res.* 19:139-143.
108. Thormar, H., Isaacs, C.E., Brown, H.R., Barshatzky, M.R., and Pessolano, T. 1987. Inactivation of enveloped viruses and killing of cells by fatty acids and monoglycerides. *Antimicrob. Agents Chemother.* 31:27-31.
109. Verdonck, L.F., and van Heugten, H.G. 1997. Ether lipids are effective cytotoxic drugs against multidrug-resistant acute leukemia cells and can act by the induction of apoptosis. *Leukemia Res.* 21:37-43.
110. Viegas, C.A., Rosa, M.F., Sa-Correia, I., and Novais, J.M. 1989. Inhibition of yeast growth by octanoic and decanoic acids produced during ethanolic fermentation. *Appl. Environ. Microbiol.* 55:21-28.

111. Wang, S.-P., Grayston, J.T., Kuo,C-C., Alexander, E.R., and Holmes, K.K. 1977. Serodiagnosis of *Chlamydia trachomatis* infection with the micro-immunofluorescence test. In *Nongonococcal Urethritis and Related Infections*, eds. D. Hobson and K.K. Holmes, *pp*. 237-248. Washington DC: American Society for Microbiology.
112. Watanabe, T., Nagura, H., Watanabe, K., and Brown, W.R. 1984. The binding of human milk lactoferrin to immunoglobulin A. *FEBS Letts*. 168:203-207.
113. Weber, N. 1985. Metabolism of orally administered rac-1-O[^{14}C]dodecylglycerol and nutritional effects of dietary rac-1-O-dodecylglycerol in mice. *J. Lipid Res*. 26:1412-1420.
114. Welsh, J.K., and May, J.T. 1979. Anti-infective properties of breast milk. *J. Pediatr*. 94:1-9.
115. Welsh, J.K., Arsenakis, M., Coelen, R.J., and May, J.T. 1979. Effect of antiviral lipids, heat, and freezing on the activity of viruses in human milk. *J. Infect. Dis*. 140:322-328.
116. Whitley, R.J. 1994. Herpes simplex virus infections of women and their offspring: Implications for a developed society. *Proc. Natl. Acad. Sci*. 91:2441-2447.
117. Yasin, B., Harwig, S.S., Lehrer, R.I., and Wagar, E.A. 1996. Susceptibility of *Chlamydia trachomatis* to protegrins and defensins. *Infect Immun*. 64:709-713.
118. Yong, E.C., Klebanoff, S.J., and Kuo, C.C. 1982. Toxic effect of human polymorphonuclear leukocytes on *Chlamydia trachomatis*. *Infect. Immun*. 37:422-426.

Section-II

OVO-ANTIMICROBIALS

Lysozyme
Ovotransferrin
Ovoglobulin IgY
Avidin

J.N. Losso
S. Nakai
E.A. Charter

Lysozyme

6

I. INTRODUCTION

Lysozymes were discovered and named by Alexander Fleming (1922). Later Alderton et al. (1945) identified lysozyme in hen's egg albumen and found it to be the same as a previously known protein, globulin G1, discovered by Longworth et al. (1940) using moving-boundary electrophoresis. The lysozymes of avian egg albumen, which is the most plentiful source, shell and viteline membrane, belong to a class of enzymes that lyse the cell walls of gram-positive bacteria by hydrolyzing the β-1,4 linkage between N-acetylmuramic acid (NAM) and N-acetyl-glucosamine (NAG) of gigantic polymers $(NAM-NAG)_n$ in the peptidoglycan (murein). They are sometimes termed muraminidases or more precisely N-acetylmuramideglycanohydrolases (EC 3.2.1.17). 'Lysozyme' is the recommended common name. Lysozymes are ubiquitous in both the animal and plant kingdoms, and play an important role in the natural defense mechanism. Lysozyme is attractive as a natural food preservative because it is endogenous to many foods, specific to bacterial cell walls, and harmless to humans. There are microbial, viral, phage, insect, plant, and animal tissue lysozymes. Body fluids such as tears, saliva, urine and human milk contain 2.6, 0.13, trace, and 0.2-0.4 mg/ml of lysozyme, respectively (Grossowicz & Ariel, 1983). The lysozyme content in cow's milk was reported to be much lower (<1/100) than the above values in human milk (Packard, 1982).

II. OCCURRENCE

A. Sources

Egg-white lysozyme is the classic representative of the lysozyme family and the related enzymes are called c type (chicken- or conventional-type) lysozymes. One egg contains about 0.3-0.4 g of lysozyme. Until now, eggs have been the easiest and most economical source for recovering lysozyme. The presence of a radically different lysozyme, called g type after the Embden goose, was discovered. Three of only four lysozyme g sequences known to date have been reported. There is no evidence for the occurrence of lysozyme g or its gene in living systems other than birds (Prager & Jolles, 1996).

Furthermore, a *v* type (viral) family has also been defined for phage lysozymes. Phage Lambda is by definition the prototype of lambdoid phage. This λ family is different from other lysozymes, as it is not a hydrolase but a transglycosidase. This enzyme is active on *Escherichia coli* cell walls but not on *Micrococcus luteus* (another name *M. lysodeikticus* is also used) or on chitin (β-1,4-polyNAG). Several other phage lysozymes are totally unrelated to the above phage lysozymes, and named after the first member of the family investigated. It is called CH type phage lysozyme as it is produced by the fungus *Chaloropsis* (Fastrez, 1996). In plants, *h* type (hevamine type) and *b* type (barley type) families have been identified (Bientema & van Scheltinga, 1996).

To restrict our discussion to food applications of lysozymes, *c* and *g* family lysozymes only will be herein discussed, with an exception of phage T4 lysozyme (*v* type) for comparison because of its antimicrobial activity against a Gram-negative bacterium, *E. coli*.

B. Screening / Identification

Major techniques utilize the characteristic properties of lysozyme, i.e. relatively low molecular weight ($\cong$14,500 daltons) among other proteins and high isoelectric point (pI $\cong$ 10.7). Accordingly, cation (or anion) exchange chromatography, gel filtration chromatography, electrophoresis, and MALDI-TOF-MS have been used as effective separation as well as identification methods. Gel electrophoresis has been the most popular identification method. A potential exists, however, to use affinity chromatography for more specific separation or identification by constructing affinity columns of immobilized anti-lysozyme antibody or lysozyme substrates on chromatographic columns (Ahvenainen et al., 1980).

C. Quantitative assays

Methods of identification stated above in *Section IIB* could be used for molecular quantification. The methods for determining lysozyme activity have been reviewed by Grossowicz and Ariel (1983), which include turbidimetric and agar plate methods using *Micrococcus luteus* (Smoelis & Hartsell, 1949), immunoassays, fluorometric and colorimetric methods, as well as histochemical and cytochemical localization. Pitotti et al. (1992) proposed the use of *Leuconostoc oenos* as a suitable substrate for lysozyme activity at acidic pH since the enzyme is unable to lyse *Micrococcus luteus*, the substrate used in the classical lysozyme assay, at acidic pH. The method is very rapid, reproducible, with good precision, but the suggestion has been made to use a more sensitive strain, such as *Leuconostoc oenos* L294. This has the potential to make it as sensitive as the *Micrococcus luteus* method carried out at pH 6.2. Recently, as a general and direct method for detection and quantification of cell lysis, leakage of cell-specific enzymes has been proposed, in particular a sensitive enzymatic luminometric inorganic pyrophosphatase detection assay (Nyrén & Edwin, 1994). This assay detects lysozyme at concentrations <5 ng/ml.

III. MOLECULAR PROPERTIES

A. Isolation / Purification

The classical preparation of lysozyme involves direct crystallization from egg white by addition of 5% NaCl at pH 9.5, with a lysozyme recovery of 60-80% (Alderton & Fevold, 1946). However, the functionality of the remaining egg white is lost. To solve

```
        1        10          20        30        40        50
Avian   KVFGRCELAAAMKRHGLDNYRGYSLGNVVCAAKFESNFNTQATNRNT-DGS
Human      E       RTL   L M G    I   A   M L   W GY   R    Y AG R
Bovine   K Q       RTL   KL    G K     A    L LTKW  SY   K    Y PSSE

        51       60          70        80        90        100
Avian   TDYG I LQ I NSRWWCNDGRTPGSRNL CN I PCSALLSSD I TASVNCAKK I VS
Human        F      Y       K    AV A  H LS       QDN ADA A    R V R
Bovine       F      K       K    NAVDG HVS     E MEND AKA A    H

        101      110         120       129
Avian   DGNGMNAWVAWRNRCKGT DV QAW I RGCRL
Human   PQ  I R              QNR  RQY V Q   GV
Bovine  E-Q  I T      KSH RDH   SSK V E   T
```

FIGURE 1. Amino acid sequences of lysozymes of chicken, human and cow. Sites 1-40 of bovine are from cow's milk.

this problem, separation of lysozyme from egg white by ultrafiltration, ion exchange or affinity chromatography has been suggested (Ahvenainen et al., 1980). A variety of cation exchangers were compared and a macroporous weak acid type resin Duolite C-464 was selected on the basis of a high lysozyme recovery of 90-95%, retention of whipping and gelling properties of the lysozyme-free egg white, and ease of column manipulation (Li-Chan et al., 1986). The purity of lysozymes separated by salting-out crystallization, ultra-filtration and cation exchange were compared by Kijowski et al. (1999) using elec-trophoresis, calorimetry and amino acid analysis. Polymerization, including the formation of active or inactive reversible dimers, upon denaturation especially heating was also reported by the same authors.

B. Physico-chemistry

The molecular weight of chicken lysozyme computed from the amino acid sequence of 129 residues is 14,307 and the isoelectric point is 10.7. Molecular weight obtained by matrix assisted laser desorption/ionization mass spectrometry (MALDI-TOF-MS) is 14,308 (Yang et al., 1998). The precision of MALDI-TOF-MS is 0.05%.

The isoelectric points of lysozymes are not always high; those of lysozymes secreted in the stomach are much lower with pH values of 6-8. This variation is due to the difference of arginine residues in lysozyme molecules in the range of 3-14. Chicken lysozyme has a higher extinction coefficient $E^{1\%}$ at 280 nm (26.4 vs. 4-15) than most pro-teins because of higher contents of aromatic amino acids in the lysozyme molecule. There is no sequence homology of chicken lysozyme (129 residues) with g type lysozyme (185 residues). The g type is three times more active than c type (Canfield & McMurry, 1967). The amino acid sequences of chicken, human and cow lysozymes are shown in *FIGURE 1*. The latter two are mammalian lysozymes, and therefore, more homologous to each other than to chicken lysozyme.

Human lysozyme is about four times more active in bacteriolysis than chicken lysozyme although the hydrolytic activity against glycol chitin is about the same. Human lysozyme has more arginine residues (14 vs. 11) than chicken lysozyme, especially near *Asp52* (*FIGURE 1*), but has a lower K_a of 10000 vs. 71400 (M^{-1}) (Imoto, 1996). This fact might imply that there is no need of an excessively strong binding ability with substrates

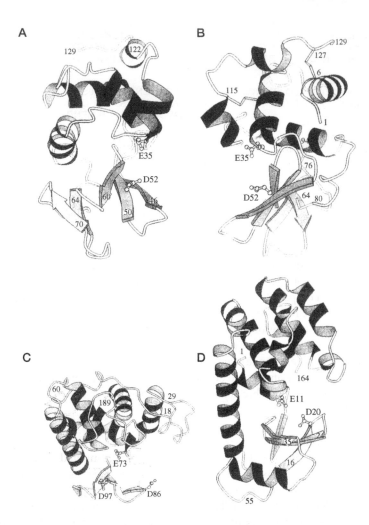

FIGURE 2. Ribbon diagrams of *c, g,* and *v* type lysozymes. (a) Chicken lysozyme facing into the active site. (b) Chicken lysozyme horizontally vertical to (a) highlighting the four-disulfide bridges. (c) Goose lysozyme. (d) T4 lysozyme [reprinted with permission from Strynadka and James (1996); copyright Birkhäuser].

for catalysis. Whether this unusual presence of *Arg49* near *Asp52* in human lysozyme is a cause of higher lysis activity than that of chicken lysozyme is unknown. The conversion of *Trp62* to *Tyr* enhanced the bacteriolytic activity of hen egg-white lysozyme by two-fold (Kumagi & Miura, 1989). In contrast, *Tyr62Trp* of human lysozyme decreased lytic activity (Muraki et al., 1987). Meanwhile, the double mutant *Val73Arg/Gln125Arg* of human lysozyme increased lytic activity at higher ionic strength and higher pH, whereas *Arg41Gln/Arg100Ser* showed the same behavior at lower ionic strength and lower pH (Muraki et al., 1988). It was postulated therefore that lysozyme might be designed to have an optimal balance of charges so as to express a proper lytic activity under conditions where it is set to work.

Since lysozyme has four disulfide bonds, thereby making the molecule unusually compact, it is considerably heat stable. In dried powder forms, lysozyme can be stored

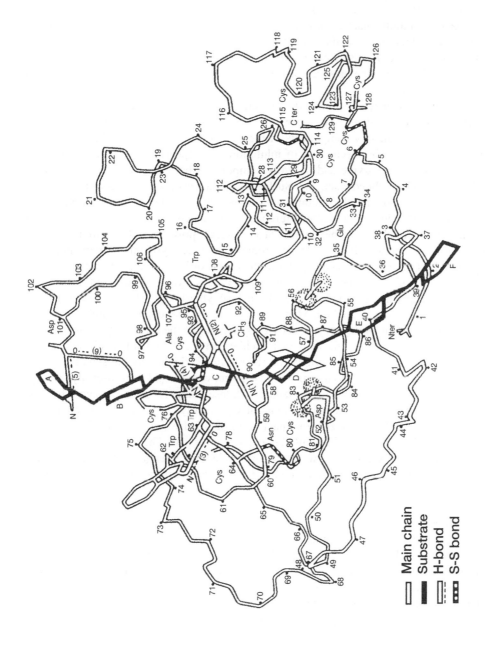

FIGURE 3. Estimated structure of lysozyme-(NAG)₆ complex by X-ray analysis [reprinted with permission from Imoto (1996); copyright Birkhäuser].

for a long time (>6 months) at temperatures up to 30 °C without losing lytic activity. Chicken lysozyme is stable at 100 °C for 2 min at pH 4.5 and at 100 °C for 30 min at pH 5.29. But the enzyme is inactivated quickly in alkaline solutions. Proctor and Cunningham (1988) reviewed the effect of heat, pH and ionic strength on lysozyme activity. Lysozyme is rapidly inactivated in the presence of thiol compounds that split the disulfide linkages, which are essential for maintaining the tight structure especially near the substrate-binding site (*FIGURE 2 & FIGURE 3*). This fact explains the instability of chicken lysozyme attributable to the presence of free sulfhydryl groups in other albumen proteins such as ovalbumin.

Proteolytic enzymes such as trypsin, chymotrypsin, and papain do not hydrolyze lysozyme, but pepsin does; but these enzymes do hydrolyze denatured lysozyme. Lysozyme is inactivated by components of the egg yolk such as lipovitellin (Proctor & Cunningham, 1988). The mechanism of inhibition was reported to involve electrostatic interaction between lysozyme and the yolk components because salt and urea partially reactivate the yolk-inhibited lysozyme. Surface active agents such as SDS form complexes with lysozyme and strongly inhibit lysozyme activity. Iodine, concentrated urea, and anhydrous formic acid at room temperature reversibly inactivate lysozyme. Lysozyme is stable in 20% alcohol, and stable against heat in sugar, polyols, and salt solutions. The reaction of lysozyme with lipid oxidation products produced lysozyme dimers and polymers and a loss of enzymatic activity. More than 60% of lysozyme activity is destroyed when 0.35 Mrad of irradiation was applied to a 0.3% lysozyme solution, but in the presence of other egg white proteins the activity of lysozyme is only slightly reduced when 0.85 Mrad irradiation was applied. Manganese, mercury, and copper at high concentrations (10^{-3} to 10^{-2} M) inhibit lysozyme activity by binding to the carboxyl groups of *Glu35* and *Asp52* at the active site of the enzyme (Teichberg et al., 1974; Gallo et al., 1971). Iron (III) binds to and inactivates lysozyme; iron (II) requires oxidation by phosphate buffer to inactivate lysozyme at pH 7.4 and the activity is recovered at pH 4; iron-catalyzed oxygen radicals irreversibly inactivate lysozyme; catalase competes with lysozyme for iron binding; benzoate, formate and mannitol partially protect lysozyme from iron complexation (Sellak et al., 1992). Lysozyme is inactive in distilled water, activated by low concentration of salt and inhibited by high concentration of salt. The activation at low salt concentration is mostly correlated with a non-specific ionic strength and the inhibition at high salt concentration is mostly correlated with cationic concentration and charge (Chang & Carr, 1971).

C. Structure

Chicken lysozyme contains four disulfide bonds at 6-12, 30-115, 64-80 and 76-94, with no free sulfhydryl (SH-) group (*FIGURE 2 & FIGURE 3*). The lysozyme molecule consists of two domains or lobes, i.e. α- and β-domains, linked by a long α-helix between which lies the active site. Helix A (4-15), helix B (24-36), helix C (88-99), helix D (108-115) and 3_{10} helix (120-125) form the α-domain. Triple-stranded β-sheet (41-60), central 3_{10} helix (79-84), and the large loop (61-78) form the β-domain (*FIGURE 2*). Striking resemblance in 3D structure determined by X-ray crystallography led to the proposal that lysozymes of the *c* and *g* types as well as bacteriophage T4 lysozyme might be derived from a common ancestral protein (Prager & Jollès, 1996). The active site of hen egg white lysozyme consists of six subsites A, B, C, D, E, and F (*FIGURE 3*), which are sufficient to

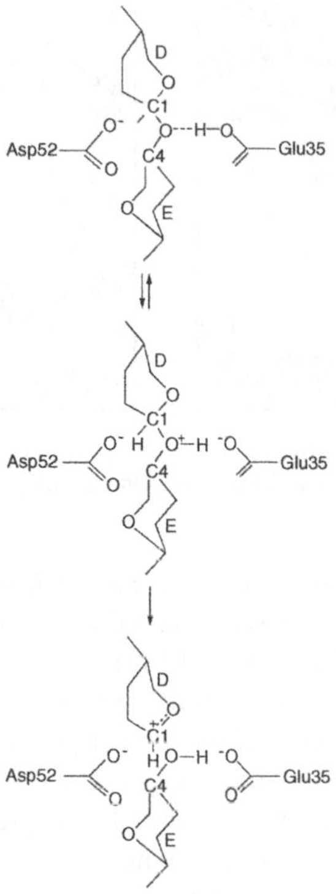

FIGURE 4. Reaction mechanism of *c* type lysozyme [reprinted with permission from Imoto (1996); copyright Birkhäuser].

bind six sugar residues (Imoto, 1996). The six subsites along the active site cleft position the catalytic groups Glu-35 and Asp-52 between subsites D and E.

IV. ANTIMICROBIAL ACTIVITY

A. Mechanism of action

Lysozyme promotes catalysis by inducing steric stress in the substrate. The reaction mechanism of lysozyme is illustrated in *FIGURE 4* (Imoto, 1996). *Asp52* and *Glu35* are the precise amino acids that participate in the catalysis. *Glu35*, which lies in a hydrophobic environment, participates in catalysis in the protonated form, and *Asp52*, which lies in a hydrophilic environment, does so in the dissociated form. The bond between this oxygen and C1 on the D sugar is cleaved. A carbonium ion is formed on the D sugar and this is stabilized by the formation of an oxocarbonium ion. The distortion of the D sugar from the chair form to the sofa form favors this process. The negative charge on *Asp52* stabilizes the formation of a positive charge on the D sugar. *Glu35*, which participates in catalysis in the protonated form, has an abnormally high pKa of 6.1 (normally 4.3), thus enhancing the catalytic efficiency. This abnormality is produced by the negative charge of *Asp52* and by the surrounding hydrophobic environment, especially of *Trp108* (Inoue et al., 1992a; 1992b). Site-directed mutagenesis studies in which *Asp52* and

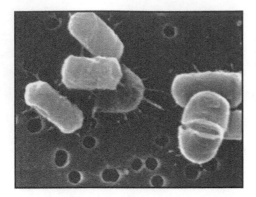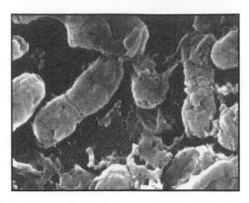

FIGURE 5. Scanning electron micrograph showing the effect of lysozyme *C. tyrobutyricum.* (left) Control bacterial cells and (right) lysozyme-treated bacterial cells [reprinted with permission from Canadian Inovatech, Inc. (1999); copyright Inovatech, Inc.].

Glu35 were converted to the corresponding *Asn* and *Gln* demonstrated the critical requirement for *Glu35* (0% activity) and lesser essentiality of *Asp52* (0.5% substrate specificity) against cell wall substrate (Malcom et al., 1989).

In the case of *g* type lysozyme with 185 amino acid residues, it was suggested that *Glu73* and *Asp86* correspond to *Glu35* and *Asp52* of *c* type, respectively. However, Weaver et al. (1995) found no counterpart in *g* to *Asp52* in *c* and suggested that only one acidic residue is essential for the catalytic activity of lysozyme *g* and, further, that *Asp52* might not be required for catalysis by lysozyme *c*.

T4 lysozyme shares 74 common amino acids with chicken lysozyme and 91 common residues with goose lysozyme. In 164 residues of T4 lysozyme, *Asp20* was proposed to play the same role as *Asp52* in chicken lysozyme, namely stabilization of the positive charge of the oxocarbonium intermediate. As well, *Glu11* was suggested to act as the proton donor facilitating departure of the leaving sugar E. *Glu11* and *Asp20* in T4 lysozyme lie on opposite side of the activity site cleft and superimposed closely with the analogous *Glu35* and *Asp52* in chicken lysozyme (*FIGURE 2*) (Strynadka & James, 1996). Unlike chicken lysozyme and goose lysozyme, the environment of *Glu11* is relatively polar and solvent exposed. *Glu11* does not interact with a helix dipole as in the case of chicken lysozyme, but rather forms a strong ion pair with the adjacent guanidium group of *Arg145*. *Asp20* carboxylate does not interact directly with other protein side chains as in the case of the D52-S50-N46-N59 platform observed in chicken lysozyme (*FIGURE 3*) (Strynadka & James, 1996).

The rate of lysozyme catalysis depends upon the pH of the medium. The pH profile is bell-shaped with a maximum at pH 5.0 and inflections at pH 3.8 and 6.7. The osmotic uptake of water leads to the expansion, eventual rupture of the cytoplasmic membrane, and the rupture of the cytoplasmic membrane causes cell death. Water then reacts with the acylium ion, and the enzyme is ready for another round of catalysis. Gram positive bacteria are very susceptible to lysozyme because the cell wall of gram-positive bacteria is about 90% peptidoglycan. In some gram-positive bacteria such as *Bacillus cereus* the absence of NAG residues confers resistance to lysozyme (Board, 1995). A scanning electron micrograph (SEM), depicting the *in vitro* effect of lysozyme on cells of *Clostridium tyrobutyricum* is shown in *FIGURE 5*.

In Gram-negative bacteria, the peptidoglycan that makes only 5-10% of the cell wall lies beneath the outer membrane of the cell envelope. The lipopolysaccharide layer of the outer membrane acts as a barrier against macromolecules and hydrophobic compounds. The lipid component of the inner core of the lipopolysaccharide molecules through their content of phosphate and carboxyl groups and their electrostatic interactions with divalent cations reinforce the stability of gram negative bacteria cell walls towards microbicidal agents such as lysozyme.

The antiviral activity of lysozyme is not associated with its lytic activity and was reported to be associated with the positive charge of lysozyme (Cisani et al., 1984). Addition of negative charges on lysozyme molecule, by succinylation, did not induce antiviral activity against influenza virus (Schoen et al., 1997). It is possible that a purely electrostatic effect of a protein does not suffice for antiviral activity. Possibly, a much more specific interaction is involved.

Recently, Düring et al. (1999) unexpectedly discovered that heat denatured T4 lysozyme as well as hen egg white lysozyme, in which the enzymatic activity was abolished, preserved its antimicrobial activity. The membrane perturbing activity of denatured lysozyme was demonstrated on bacterial, fungal, and plant cells. The amphiphatic C-terminal domains of the lysozymes seem to have mediated the bactericidal and fungicidal activities of the denatured lysozymes. A synthetic peptide, with sequence homology to the amphiphatic C-terminal of T4 or hen egg white lysozyme, showed bactericidal and fungicidal activities similar to the microbicidal activity of the C-terminal domain of heat denatured T4 and hen egg white lysozyme. This finding suggests that lysozyme has both enzymatic and non-enzymatic microbicidal activity in native and denatured state, respectively. Therefore, the antimicrobial activity of lysozyme may not be limited by heat treatment encountered during food processing operations.

B. Spectrum of activity

Lysozyme is most effective against some specific gram-positive bacteria including *B. stearothermophilus*, *Clostridium thermosaccharolyticum*, and *Clostridium tyrobutyricum*. The spectrum of activity of lysozyme is easily broadened to other spoilage and pathogenic organisms and even to some gram-negative bacteria when lysozyme is used in combination with other compounds. Johnson (1994) demonstrated that lysolecithin enhanced lysozyme inhibitory activity against yeasts. Gram-negative bacteria became susceptible to lysozyme after the outer membrane of the bacteria has been disrupted by compounds such as EDTA, aprotinin, organic acids or when lysozyme was conjugated to carbohydrates (Johnson, 1994; Pellegrini et al., 1992). It has also been shown that within some species, lysozyme may have different inhibitory activity against bacteria. Lysozyme was shown to be effective against some *S. aureus* and ineffective against some others (Johnson, 1994). *Salmonella senftenberg* 775 W is sensitive to lysozyme while *S. typhimirium* is not. The spectrum of antimicrobial activity of lysozyme against gram-negative and gram-positive bacteria is shown in *TABLE 1 and TABLE 2*, respectively. Results are mostly from *in vitro* studies and have been compiled from different sources (Cisani et al., 1984; Carini et al., 1985; Johnson, 1994).

TABLE 1. Antimicrobial spectrum of lysozyme against Gram-positive bacteria

Family	Species	Sensitivity
MICROCOCCACEAE	*Micrococcus lysodeikticus*	+
	Staphylococcus aureus	-/+
	Sarcina lutea	+
LACTOBACILLACEAE	*Lactobacillus bulgaricus*	nds/-
	Lactobacillus helveticus	+
	Lactobacillus acidophilus	-
	Lactobacillus lactis	-/+
	Lactobacillus casei	-
	Lactobacillus plantarum	-
	Lactobacillus fermentum	-
STREPTOCOCCACEAE	*Streptococcus thermophilus*	+
	Streptococcus lactis	-/+
	Streptococcus cremoris	-/+
	Streptococcus diacetilactis	-
	Streptococcus faecalis	-
	Streptococcus faecium	+
	Streptococcus faecalis var. liquefaciens	-/nds
	Streptococcus faecalis var. zymogenes	-
	Streptococcus bovis	-
LEUCONOSTOC	*Leuconostoc citrovorum*	-
PEDIOCOCCUS	*Pediococcus cerevisiae*	ndr
BIFIDOBACTERIUM	*Bifidobacterium spp*	+
BACILLACEAE	*Bacillus spp.*	+
	Bacillus subtilis	+/nds
	Bacillus cereus	-/+
	Bacillus megaterium	+
	Bacillus stearothermophilus var. calidolactis	+
CLOSTRIDIUM	*Clostridium welchii*	-
	Clostridium tyrobutyricum	+
	Clostridium butyricum	nds
	Clostridium sporogenes	+
	Clostridium perfringens	-
	Clostridium botulinum A, B, E	m+
PROPIONOBACTERIACEAE	*Propionobacterium spp.*	-/ndr
CORYNEBACTERIUM	*Corynebacterium. betae*	+
	Corynebacterium flaccumfaciens	+
	Corynebacterium poinsettae	+
	Corynebacterium tritici	-

+ = sensitive, ▬ = resistance, m+ = moderately sensitive, nds = not demonstrated sensitivity, ndr = not demonstrated resistance (modified from Cisani et al., 1984; Carini et al., 1985; Johnson, 1994).

V. BIOLOGICAL ADVANTAGE

A. Additive advantage to foods

Currently, new safe processes that enhance food safety are receiving top regulatory approval around the world. Lysozyme is a protein endogenous to most foods, is safe, and is stable under pH conditions mostly encountered in foods. It is available in ready to use form such as a liquid or spray dried powder. The spray-dried form is often granulated to reduce the formation of aerosols during its use in food processing. Lysozyme is not

TABLE 2. Antimicrobial spectrum of lysozyme against Gram-negative bacteria

Family	Species	Sensitivity
ENTEROBACTERIACEAE	*Escherichia coli*	-/ndr/+
	Salmonella typhimurium	-/+
	Shigella	+
	Proteus vulgaris	-
	Serratia marcescens	-
	Erwinia spp.	+
	Yersinia enterocolitica	m+
PSEUDOMONADIACEAE	*P. aeruginosa*	+
	P. fluorescens	ndr
	Achromobacter spp.	-
PASTEURELLA	*Pasteurella spp.*	+
NEISSERIACEAE	*Neisseria spp.*	+
VIBRIONACEAE	*Vibrio cholerae*	-
CAMPYLOBACTERIACEAE	*Campylobacter jejuni*	m+
KLEBSIELLEAE	*Klebsiella pneumoniae*	-

+ = sensitive, ‑ = resistance, m+= moderately sensitive, nds= not demonstrated sensitivity, ndr = not demonstrated resistance (modified from Cisani et al., 1984; Carini et al., 1985; Johnson, 1994).

known to undergo any adverse chemical or organoleptic changes that would be of concern under normal food processing conditions.

Lysozyme has proven to be an economically viable alternative to chemical preservatives in a number of food applications, most notably in cheese making.

Lysozyme is selective, at the level used industrially, for certain gram-positive spoilage bacteria and may not interfere with a number of organisms that are considered to be beneficial for human health.

B. Physiological advantage to host

Since its discovery, lysozyme has never lost its role in the control of bacterial infection and the modification of host immunity (Sava, 1996). Lysozyme added to foods is destroyed in the human stomach, making it harmless to humans. Lysozymes of both human and avian origin kill gram-positive bacteria, binds to most of the lipopolysaccharides (LPS) produced by various gram-negative bacteria, independent of variations in the structure of lipid A and /or polysaccharide portions of the LPS molecule, with a high binding affinity to produce a complex, LPS-lysozyme, which results in lipopolysaccharide detoxification and lysozyme inactivation. As a consequence of the detoxification, the biological activities of LPS such as the mitogenic and tumor necrosis factor (TNF) production are significantly reduced by the LPS-lysozyme complex. The biological activity of LPS, such as the mitogenic activity of murine splenic B lymphocytes and the TNF production of RAW264.7 cells *in vitro* and of mice *in vivo,* were significantly reduced by lysozyme (Takada et al., 1994).

Lysozyme, orally administered, has proven to be an effective immunostimulant, antiviral and anti-inflammatory in cancer patients and individuals with herpetic lesions (Inoue, 1987; De Douder & Marias, 1974; Satoh et al., 1980).

VI. APPLICATIONS

A. Suitability/ Adaptability

Consumer demand for fresh, healthy, and natural foods that contain less or no chemical preservatives is growing exponentially. Accordingly, the food industry has opted to investigate and potentially offer some naturally occurring preservatives as alternatives to traditional preservation agents. At the same time, there is a widespread concern, within the scientific community, that minimally processed foods are prone to the growth or survival of pathogenic bacteria such as *Listeria monocytogenes, Yersinia enterocolitica, Aeromonas hydrophila, Campylobacter jejuni*, and the toxigenic *Clostridium botulinum* which can grow at temperature as low as 3.3°C (Johnson, 1994). Hen egg white lysozyme has several desirable properties as a food preservative. The FAO/WHO and several countries including Austria, Australia, Belgium, Denmark, Finland, France, Germany, Italy, Japan, Spain, and The United Kingdom have acknowledged the non-toxicity of lysozyme and have therefore approved its use in some foods. The GRAS tentative final ruling has been granted to lysozyme in the US, and Canada is presently considering a regulatory approval.

B. Existing applications

In vitro studies of the effectiveness of lysozyme on both gram-positive and gram-negative bacteria have been the subject of excellent academic publications and will not be reviewed here (Satoh et al., 1980; Ibrahim et al., 1994; 1997; Razavi-Rohani & Griffiths, 1996). For pharmacological and therapeutic applications, the publication by Sava (1996) is recommended. Japan has filed the greatest number of patents and *in vivo* applications of lysozyme as a food preservative.

1. Dairy products: Cheese, particularly Italian cheeses, and Edam or Gouda cheeses, are classical examples of dairy products that have shown the effectiveness and selectivity of lysozyme as a food antimicrobial. Butyric acid bacteria are commonly found in milk and if not properly eliminated, they metabolize lactic acid into butyric acid, acetic acid, carbon dioxide, and hydrogen. Accumulation of butyric acid makes cheese unpalatable. Accumulation of carbon dioxide and hydrogen can cause cheese to blow, making it undesirable for human consumption. In the cheese making process, lysozyme is added to cheese vats in either a liquid or spray dried and granulated form. At the pH of the milk, lysozyme is positively charged, and as the curd forms, lysozyme is electrostatically attracted and adsorbed to the negatively charged casein molecules that form the curd. After separation of the curd from the whey, greater than 90% of the lysozyme activity resides with the curd. After molding and brining, the cheese rounds are aged for up to two years. As the center of the round develops anaerobic conditions, the contaminating spores of *Clostridium tyrobutyricum* slowly begin to germinate. In the vegetative form, the organism is susceptible to the lytic action of the enzyme and the lysozyme present in the cheese continues, for the duration of the aging process, to inhibit growth. Comparison between good Grana cheese and *C. tyrobutyricum* contaminated Grana cheese is shown in *FIGURE 6*. Lysozyme, at concentration of 21 mg/L, successfully killed 99% of 5×10^5 resting vegetative cells of *C. tyrobutyricum* within 24 h incubation at 25 °C, and severely inhibited actively growing vegetative cells (Wasserfall & Teuber, 1979; Hughey et al.,

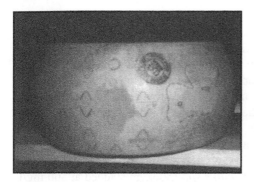 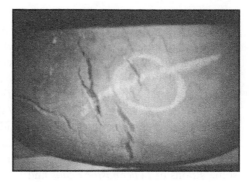

FIGURE 6. Comparison between good Grana cheese and *C. tyrobutyricum* contaminated Grana cheese [reprinted with permission from Canadian Inovatech, Inc. (1999); copyright Inovatech, Inc.].

1989). Lysozyme at 20 to 35 mg/L successfully replaced commonly used preservatives such as formaldehyde, nitrate, nisin, and hydrogen peroxide in protecting cheese against *C. tyrobutyricum*. Formaldehyde, nitrate, and nisin may be inhibitory to the starter and secondary cultures required for the ripening of cheese. There was no significant difference, organoleptically, between lysozyme stabilized cheese and nitrate stabilized cheese. The effectiveness of lysozyme in providing a natural enzymatic protection against *C. tyrobutyricum* is directly associated to its specificity. In general, lysozyme inhibits the spoilage organism, while not interfering with the starter culture. The activity of four starter cultures, in Gouda cheese, was not inhibited by lysozyme at concentration up to 2500 units/ml while growth of coliform isolates was inhibited by 1000 units/ml of lysozyme (Bester & Lombard, 1990). In specific cases the tolerance level of the starter culture may be somewhat lower and special care must be taken to inhibit spoilage without inhibiting the culture. *Lactobacillus helveticus* in milk was inhibited by a lysozyme concentration higher than 50 ppm (Makki & Durance, 1996).

Several patents claim the effectiveness of lysozyme at concentration as low as 50 ppm to prevent the development of undesirable microorganisms in butter and cheese for more than 24 months. A patent obtained by Dell'Acqua et al. (1989) reported enhancement of lysozyme activity around pH 5.2 in milk. Large volumes of lysozyme, about 100 tons, are used annually by the cheese industry to prevent the growth of *C. tyrobutyricum* from germinated spores.

Listeria monocytogenes is of greatest concern in soft cheese and milk (Farber & Peterkin, 1991; Schuchat et al., 1992). Lysozyme, at 20 to 200 mg/L delayed the growth of all four strains of *L. monocytogenes* isolated during a food poisoning outbreak (Johnson, 1994). The sensitivity of the pathogens to lysozyme depended mostly on the physiological state of the microbe and on the growth medium. Milk and dairy products have contributed the most to the outbreak of listeriosis. The rate of survival of *L. monocytogenes* in milk is very high and the resistance to lysozyme has been associated with the presence of minerals or mineral-associated components. In Camembert cheese, lysozyme was bacteriostatic but did not rid the cheese of viable *L. monocytogenes* (Hughey et al., 1989).

2. Alcoholic beverages: Microbes commonly found in wines are non-pathogenic because of the alcohol content and pH of wines. The most common bacteria and yeast responsible for wine spoilage, in the cellar or in the bottle, are *Lactobacillus* spp., *Leuconostoc* spp., *Pediococcus* spp., and the *Saccharomyces* spp. The control of malolactic fermentation, a metabolic process by which lactic acid bacteria convert L-malic acid into L(+)-lactic acid and carbon dioxide, is an important process in the production of low acid (high pH) wines. Until recently, sulfur dioxide was used, worldwide, to control malolactic fermentation, inhibit spoilage bacteria and yeast and provide antioxidative properties to wines during storage. Unfortunately the use of sulfur dioxide in wines is associated with numerous side effects. Sulfur dioxide, at concentration above 220 ppm in wines, causes allergic reactions in asthmatic individuals, generates hydrogen sulfide, is not effective against the growth of lactic acid bacteria in low acid wines, and is not active against some strains of *Leuconostoc oenos* and *Lactobacillus sp.* Lysozyme, added at 500 mg/L of grape must, inhibited malolactic fermentation, and addition of lysozyme at 125-250 mg/L of red wines after malolactic fermentation promoted microbial stabilization, and prevented an increase in the content of acetic acid and biogenic amines (Gebraux et al., 1997). Villa (1996) confirmed the antibacterial efficacy of lysozyme with regards to total or partial inhibition of malolactic fermentation in red and white wines. It was suggested that lysozyme may not completely replace SO_2 since it has no antioxidant properties but a combination of lysozyme and less SO_2 may allow good control of bacterial flora and the production of high quality wines with low sulfite levels. Makki and Durance (1996) studied the potential of lysozyme as an antimicrobial in unpasteurized beer containing 5% alcohol at pH 4.7. The growth of *Lactobacillus brevis* strain B-12 and *Pediococcus damnosus* strain B-130 was delayed but not prevented using 50 ppm of lysozyme. Japanese sake, a well-known beverage, has been reported to be preserved with a mixture of lysozyme or lysozyme salt plus β-glucopyranose aerodehydrogenase (Eisai, 1980).

3. Fish and seafood products: Edwardsiellosis is a serious bacterial infection, caused by *Edwardsiella tarda*, that threatens the farming industry as the disease often occurs in fish grown in farming ponds (Hatta et al., 1994). In the past, antibiotics have been used to control the infection, which has resulted in the emergence of drug-resistant strains and the potential of antibiotics in the food chain. Lysozyme is widely distributed in fish tissues as a protective antibacterial. Lysozyme alone, lysozyme conjugated to galactomannan or lysozyme conjugated to palmitic acid was administered orally to carp, *Cyprinus carpio* L., infected with a virulent strain of gram-negative *Edwardsiella tarda* (Nakamura et al., 1996). The survival rate was 20% for fish treated with lysozyme used singly or lysozyme-palmitic acid conjugate and 30% for fish treated with lysozyme-galactomannan conjugate while all the control fish died within 3 days.

The use of lysozyme in seafoods has been patented by Eisai (1971; 1972a; 1972b). Lysozyme, at a concentration of 0.005 - 0.03% in combination with 0.25 to 0.35% NaCl has been used to preserve low quality salmon or trout roe. A combination of lysozyme and glycine has been reported by Eisai (1973) to be an effective preservative of fish paste.

4. Fruits and vegetables: In the past two decades, there has been a noticeable increase in the consumption of fresh fruits and vegetables mostly because research has shown an inverse relationship between fruit and vegetable consumption and the incidence of cancer and coronary heart disease. However, public health officials have documented significant increase in the number of minimally processed produce-associated foodborne disease outbreaks mostly in the US. Outbreak data have linked *Salmonella, Shigella, E. coli O157:H7, Enterotoxigenic E.coli, Cryptosporidium parvum, Cyclospora, Bacillus cereus,* hepatitis A virus, and *L. monocytogenes.* Fruits and vegetables are low-acid foods and are very susceptible to spoilage by *C. botulinum* strains and *L. monocytogenes.* Three US patents obtained by Dell'Acqua et al. (1989), Johnson et al. (1991), and Ueno et al. (1996) report on bacterial decontamination of vegetables using hen egg white lysozyme. The synergistic effect of chelators, in combination with lysozyme, delayed toxin production in potato suspensions (Cunningham et al., 1991). Fruits and vegetables may also be contaminated by non-pathogenic bacteria and fungi that may alter the quality of the product. Fresh vegetables, tofu, kimchi, Japanese potato salad, sushi, Chinese noodles, and creamed custards have been preserved using lysozyme or a combination of lysozyme and amino acids (Cunningham et al., 1991). Patents filed, in Japan, by Eisai (1971; 1972a; 1972b) report the use of lysozyme in combination with glycine to preserve fruit juice, uncooked noodles, bean jam, and custard cream. Fruit and vegetables were preserved by a combination of lysozyme chloride, lower fatty acid monoglyceride, phytic acid and /or sodium acetate (Q.P. CORP., 1985).

   5. Chewing gum and toothpaste: Lysozyme, in chewing gum formulations, was effective against periodontitis-causing bacteria, and gingivitis associated inflammation and bleeding (Sava, 1996). Lysozyme is used in toothpaste, in combination with chloride, fluoride, thiocyanate, and bicarbonate to lyse *Streptococcus mutans* (Goodman et al., 1981). A US patent reports the use of lysozyme in combination with EDTA to control gum infections (Rabussay, 1982).

C. Synergism - the hurdle effect

   Lysozyme is not totally effective against all gram-positive bacteria and is ineffective against most gram-negative bacteria. Hurdle technology has been defined as a combination of existing and novel preservation techniques in order to establish a series of preservative factors (hurdles) that any organism present in the food system should not be able to overcome (Leistner et al., 1995). Each preservative factor individually affects the growth and survival of microorganisms and concertedly controls microbial spoilage and food poisoning and makes it unlikely for any microorganism to survive. The hurdles may be a combination of temperature, water activity, pH, redox potential, enzyme antimicrobial, salt, sugar, pasteurization, bacteriocin, ultra high pressure treatment, edible coatings, gas packaging, bioconservation, etc. The preservative factors may disturb several or just one of the homeostatic mechanisms of microorganisms, such as pH habituation of resistant cells, and as a result the microorganism will not multiply but instead remain inactive or even die. Hurdle technology as applied to lysozyme has shown significant improvement in lysozyme activity against a wide range of bacteria. The combination of natural preservatives such as nisin, lactoferrin, glycine, organic acids, egg white, trypsin, aprotinin, gelatin, ultrahigh pressure, and electroporation has in most cases performed better against a wide range of bacteria than lysozyme used singly (Hauben et al, 1996; De Douder & Marias, 1974).

   Lysozyme and nisin, used singly or in combination with EDTA, was incorporated into the structure of biodegradable packaging films made from corn zein or soy protein isolate and was evaluated for inhibition against *Lactobacillus plantarum* NCDO 1752 and *E. coli* ATCC 25922 (Padgett et al., 1998). Packaging films were made using the heat-press and casting methods, respectively. The cast films yielded higher antimicrobial activity in films than heat pressing, lysozyme and nisin retained their inhibitory effect against *Lactobacillus plantarum* and the addition of EDTA to either lysozyme or nisin to the mixture increased the inhibitory effects of films against gram negative *E.coli*. Payne et al. (1994) used a combination of lysozyme, EDTA, and apo-lactoferrin to evaluate their antimicrobial effectiveness in ultra-high temperature pasteurized milk against *Pseudomonas fluorescens, S. typhimurium, E. coli* 0157:H7, and *L. monocytogenes*. Lysozyme, apo-lactoferrin, alone slowed the growth of *L. monocytogenes*, but had no effect on *E.coli*. The combination of EDTA (1.5- 2.5 mg/ml)-lysozyme (150 µg/ml) was bactericidal to *L. monocytogenes* and slowed the growth of *E.coli*. Hughey et al. (1989) reported that the combination of EDTA and lysozyme was more bactericidal against *L. monocytogenes* Scott A when added to fresh corn, fresh green beans, shredded cabbage, shredded lettuce, and carrots during storage at 5 °C than lysozyme alone.

The major disadvantages associated with the use of trisodium phosphate in food decontamination procedures are the high concentration of trisodium phosphate (approximately 10% w/v) which results in a high pH values (> 12.5), the changes in the organoleptic properties of the food, the effectiveness against gram negative bacteria only, and the disposal of the phosphate in the environment. Carneiro et al. (1998) investigated the effect of lysozyme and nisin on the viability of *Campylobacter jejuni, E. coli, Pseudomonas fluorescens*, and *Salmonella enteritidis* cells on chicken skin followed by chicken skin treatment with trisodium phosphate. The data suggested that the high trisodium phosphate concentration could be reduced from 250 mM to 2-100 mM and by following triphosphate treatment with exposure of chicken skin to low concentration of lysozyme (10 µg/ml) or nisin (1 µM).

Wang and Shelef (1991) reported that sensitivity of *L. monocytogenes* strain Scott A and Brie-1 to lysozyme was enhanced by other egg white proteins such as ovomucoid, conalbumin, and by alkaline pH (pH 9) conditions. The synergistic effect of egg white proteins enhanced the antimicrobial activity of lysozyme against wine spoilage microbes *Lactobacillus kunkeei, Pediococcus*, and *Kloeckera* (Gao et al., 1999). Lysozyme activity was shown to be additive to the antimicrobial effect of apolactoferrin on *Staphylococcus epidermis* in particular and on gram-negative bacteria, in general (Leitch & Wilcox, 1998). Lactoferrin interferes with the LPS layer of gram-negative bacteria by increasing its permeability, in a similar way to that of EDTA, which allows greater accessibility of lysozyme into the bacteria.

High hydrostatic pressure is receiving increased attention as a preservation method that could induce lethal or sublethal injury to bacteria without affecting the organoleptic attributes of the minimally processed foods. Cell suspensions of *E.coli* MG1655 were treated by hydrostatic high pressure alone (180-320 MPa or in combination with EDTA (1mM-5mM), lysozyme (10-100 µg/ml), and nisin (100 µM) in phosphate buffer at pH 7.0 (Hauben et al., 1996). Cell inactivation was not dose-dependent for lysozyme (10 to 100 µg/ml) or for nisin (100 to 200 IU/ml). Inactivation of cell suspensions occurred only when lysozyme was present during pressure treatment. This demonstrates that cell permeabilization, which facilitated lysozyme incorporation into the cell, was not maintained after pressure release. Kim (1984) also reported similar enhanced lysozyme activity against gram-negative bacteria at increased hydrostatic pressure.

D. Potential applications

1. Meat products: Decontamination of meat and poultry carcasses during or at the end of the production line is a recommended procedure as part of the HACCP approach to help food processing facilities minimize the risk of contamination. In Europe, chemical or physical treatment of carcasses are not allowed, while these treatments are approved in the USA (Bolder, 1997). There is no report of meat carcass decontamination using hen egg white lysozyme. The use of lysozyme as natural sanitizer to prevent bacterial load on meat surfaces would fit the trend towards utilizing natural preservatives for safety in foods. The shelf life of raw meat, following removal from carbon dioxide storage, is always threatened by food pathogens and spoilage bacteria such as *L. monocytogenes, Pseudomonas, Lactobacillus, Brochothrix thermosphacta,* and *E.coli.* Lysozyme delayed the growth of *L. monocytogenes* in fresh pork sausage but did not eliminate the bacteria (Hughey et al., 1989). *L. monocytogenes* has been implicated in the contamination of pâté, sausages and frankfurters (De Douder & Marias, 1974). Extensive work carried out in Johnson's laboratory at the University of Wisconsin, USA, has demonstrated that lysozyme, alone, was less effective in controlling *L. monocytogenes* in pork sausage (bratwurst), pork, beef, or turkey frankfurters than lysozyme plus EDTA. It was suggested that lysozyme may have inhibitory activity against *L. monocytogenes* in meat products, but more conclusive work needs to be done. The synergistic effect of chemical preservatives such as EDTA, polyphosphates, phytic acid, parabens, benzoic acid, sorbic acid, hydrogen peroxide, and *p*-hydroxybenzoate was essential to delay growth and kill a significant number of *L. monocytogenes* cells in minimally processed meat products mentioned above. Nisin, in combination with lysozyme, was more effective than lysozyme alone on meat products. *In vitro* studies have also shown delayed growth of *L. monocytogenes* cells but not a total growth inhibition. Lysozyme alone or in combination with other biological preservatives offers an additional and acceptable processing parameter for improving the safety and assuring the quality of meat products. Cunningham et al. (1991) reported that vienna sausages were best preserved by a combination of dipping the casings in 0.05 % lysozyme plus phosphate buffer at pH 6.5, adding 0.055% lysozyme to the cured meat, and dipping the sausage in 0.05% lysozyme in phosphate buffer at pH 6.5 after cooking.

Lysozyme contains a pentadecapeptide sequence I-V-S-D-G-N-G-M-N-A-W-V-A-W-R (amino acids 98-112), obtained by enzymatic hydrolysis using clostripain at pH 7.5 and 37 °C. The pentadecapeptide induced non-catalytic bacterial death, against *E.coli, Klebsiella pneumoniae, Serratia marcescens* and *Streptococcus zooepidemicus,* while the cell membranes of these bacteria were not enzymatically attacked by native lysozyme (Pelligrini et al., 1997).

2. Wine spoilage: The potential application of lysozyme as a partial replacement of SO_2 to control malolactic fermentation in wines has reached industrial trials as the enzyme has shown its inhibitory effect on the growth of lactic acid bacteria (Green et al., 1995a; 1995b; 1995c; Amati et al., 1995). Two other enzymes considered as effective alternatives for reducing SO_2 use in wines are glucose oxidase and catalase. Glucose oxidase and catalase are safe and effective anti-oxidizing systems. Glucose oxidase has been used successfully to stabilize flavor and color in bottled white and rosé wines (Gomez et al., 1995; Pickering, 1998). The addition of lysozyme alone or in combination with an oxygen scavenger such as glucose oxidase may have significant merit. This could be attractive to the wine industry since the enzymes are natural and offer continuing protection to the wine.

3. Plants: The majority of bacterial plant pathogens, about 95%, including *Agrobacterium tumefaciens*, *Pseudomonas syringae*, and *Erwinia carotovora*, are gram-negative. Hen egg white lysozyme in combination with compounds, such as EDTA, that help permeate the cell walls of gram-negative bacteria may provide protection against bacterial spoilage. Ruminant lysozyme has been genetically engineered and was shown to provide protection to plants against fungal pathogens (Mirov et al., 1998).

4. Animal feed: Probiotics are reported to improve the good enteric microflora in human as well as animal gastrointestinal tract and promote health benefits. Lysozyme (5-20 mg/kg feed), administered simultaneously with 100,000 to 5,000,000 viable cells of *Bacillus cereus* CIP 5832 in feed has improved the growth of animals and chicks (Nguyen et al., 1999). The authors acknowledged that *Bacillus cereus* is sensitive to lysozyme, *in vitro*, but once combined with feed an unexplainable growth was observed.

5. AIDS: Atkinson et al. (1989) reported that the concentration of four salivary antimicrobial proteins, lactoferrin, lysozyme, secretory IgA, and histatins, were found to be increased in the stimulated submandibular/sublingual saliva of all HIV-1 positive patients as well as the subset of unmediated HIV-positive patients. The highest concentration of lysozyme and histatins, potent antifungal proteins in saliva, were found in patients with evidence of oral candidiasis. The review by Sava (1996) is a good reference for the immunomodulating and antiviral properties ascribed to lysozyme. Humanization of hen egg-white lysozyme using genetic engineering may provide opportunity for egg white lysozyme, which is plentiful to be explored by the AIDS drug research.

E. Limitations

The activity of lysozyme against bacterial spores may be limited by the ability of the enzyme to penetrate the cell wall of gram-negative bacteria and the intact bacterial spore coat. The hurdle is sometimes overcome when lysozyme is used in combination with metal chelators which disrupt the integrity of the cell wall thus allowing lysozyme to lyse the remaining of the cell wall. Alternatively, a smaller lysozyme, in which the conserved disulfide bridged loop region of 65-79 was replaced with a short loop of *AsnGlySerAsn*, was engineered, for the food industry, in order to improve access to large substrate and understand the minimum requirement for protein folding, stability and activity (Pickersgill et al., 1994). The engineered small hen lysozyme folded correctly, had a reduced antigenicity, and a pH-activity profile similar to that of native lysozyme, but showed only 25% of the activity of the wild-type enzyme against *Micrococcus lysodeikticus* cell walls.

The use of lysozymes, including a genetically modified lysozyme with increased heat resistance, and other lytic enzyme preservatives increased the risk of growth of non-proteolytic *C. botulinum* (Ueno et al., 1996). Non-proteolytic strains of *C. botulinum* (type B, E, F) are capable of growth and toxin production at chilled temperature and are potential hazard in minimally processed foods. Lysozyme, at 5-50 µg/ml added prior to heating has been shown to increase the apparent heat resistance of spores of non-proteolytic *C. botulinum*. A 6-D inactivation could not be achieved with a heat treatment equivalent to 19.8 min at 90 °C and growth of *Clostridium* spores was detected in less than 93 days as compared to samples containing no lysozyme, which did not show presence of *Clostridium* for at least 93 days (Peck et al., 1993; 1992). Peck and Fernandez (1995) suggested that for processed foods containing lysozyme at concentration up to 50 μg/ml prior to heating, with an intended shelf-life of no more than 4 weeks and temperature of exposure approaching 12 °C, incubation at temperature <12 °C would be required to ensure safety with respect to non-proteolytic *C. botulinum*.

Lysozyme is probably not used extensively in processed cheese making because the processing temperature was thought to be typically high enough to cause a significant reduction in enzyme activity. However, the finding by Düring et al. (1999) on lysozyme will probably find applications where food is processed at temperature above lysozyme denaturation temperature.

Cost has been another factor that has limited the use of lysozyme in applications where the use of standard chemical preservatives, such as potassium sorbate, and nitrates, is less expensive. However, as consumer demands for more natural and healthy food products increase over time this hurdle may be overcome as the 'nutraceutical attributes' of lysozyme alone or in combination with other antimicrobials and preservatives become more evident.

VII. SAFETY/ TOLERANCE AND EFFICACY

A. Metabolism: Physiological turnover/clearance

Lysozyme, like many other low molecular proteins, is easily filtered through the kidney, reabsorbed by the proximal tubular cells and catabolized in the lysosomes. This property has opened the door to using lysozyme as anti-infective and carrier for renal delivery of various drugs including anti-bacterial agents (Haas et al., 1997; Mcijer et al., 1996).

B. Cell culture and animal studies

Egg-white lysozyme bound to LPS from *E.coli* 0111:B4 suppressed TNF-α production from macrophage-like cell lines resulting in a reduction of the lethal toxicity associated with LPS (Kurasawa et al., 1996). The binding of lysozyme to LPS reduces the mitogenic activity and TNF production by LPS and may be an important therapeutic modulator of inflammatory response during sepsis and septic shock (Takada et al., 1994).

Immunization of Balb/c mice with hen egg white lysozyme stimulated T-cell activation while a conjugate lysozyme-PEG lowered the T-cell activating capacity of hen egg white lysozyme (So et al., 1996).

C. Clinical trials

Clinical trials of lysozyme have been reported with mixed results. Yamada et al. (1993) reported clinical studies that showed the development of specific IgE antibody titers to hens's egg white lysozyme in children allergic to egg. A careful examination of several publications, such as by Urisu et al. (1997), Bernhisel-Broadbent et al. (1994), Aabin et al. (1996), Holen et al. (1990), Picher and Campi (1992), Anet et al. (1985), and

Hoffman (1982) showed that researchers who used commercially available egg white proteins reported lysozyme as a major allergen. When individual egg white proteins were prepared to the highest purity available, lysozyme ranked last among egg white protein allergens. Ovomucoid, ovotransferrin, and ovalbumin were the major immunodominant proteins in egg white. Proteinase treatment of allergenic proteins is the most popular strategy for reducing allergenicity associated with food proteins and has shown reduced allergenicity.

VIII. BIOTECHNOLOGY

A. Large-scale production

Despite the substantial amount of genetic engineering work carried using hen egg white lysozyme, the enzyme is still commercially isolated almost exclusively from hen egg white. Large-scale isolation procedures such as direct crystallization, membrane separation, affinity chromatography, and ion exchange have been used. Column ion exchange chromatography with carboxymethyl cellulose or Duolite C-464 resins seem to be superior to other isolation methods in terms of cost, ease of reusability, purity and yield of lysozyme (Li-Chan et al., 1986). Purity and yield of lysozyme are better with a column ion exchange than a stirred tank since there are a larger number of theoretical plates in a column as compared to a single equilibration process in a stirred tank.

Systems for expression and secretion of hen egg white lysozyme have been constructed using yeast and *E.coli*. Unfortunately, until now the amount of lysozyme secreted into the medium has been low, ranging from 50 µg/L to 500 mg/L (Arima et al., 1997; Inoue et al., 1992b).

The world supply/demand of lysozyme is stable. As more opportunities are found for lysozyme there will be a need for genetic manipulation to enhance production. Enhancement of lysozyme bactericidal, thermal and chemical stability, and yield by genetic engineering have been attempted by several investigators with mixed results. Approaches have involved (1) introduction of disulfide bridges; (2) increasing internal hydrophobic packing; (3) reducing solvent-exposed nonpolar surface area (4) replacing the conserved long disulfide loop containing residue 65-79 by a short loop, and (5) replacement in the *hs* variant (A31V and 155L/S91 T) (Matsumura et al., 1989; Malcolm et al., 1990; Pakula & Sauer, 1990; Shih & Kirsch, 1995).

1. Modification of lytic activity: By modifying lysozyme chemically or by genetic engineering, we can achieve better understanding of the protein behavior, as shown in *Section IIIA* as derived from mutants, and enhanced lysozyme properties.

2. Glycosylation by mild Maillard reaction: Glycosylation of lysozyme by mild heating in the presence of carbohydrates was reviewed by Kato et al. (1994). The chicken lysozyme-dextran conjugate thus prepared had antimicrobial activity against gram-negative bacteria. In addition, the conjugation improved the surface activity of lysozyme, with regards to its emulsifying, foaming and gelling properties, as well as heat stability. The controlled heating conditions used were mild: typically involving dry heating of a powder mixture of protein and polysaccharide at 60 °C, which was held at a relative humidity of 65-79%, for one to two weeks.

A drawback of the above controlled Maillard reaction is the very slow reaction rate. As an alternative to the atmospheric treatment, it was found that shortening of the reaction time was successful by using a high-pressure treatment of a solution (150 Mpa, 60 °C for 1 h). A remarkable 200% increase in the lysozyme activity compared to the untreated lysozyme may have been due to a decrease in heat denaturation of the enzyme protein rather than the promotion of reaction rate. As a result, high solubility was maintained for lysozyme without melanoidin formation (Nakamura et al., 1997). The same group later found that from a lysozyme/dextran mixed solution treated at pH 4.5 with NaCl (0.1 M) under 192 MPa and 19.3 °C, 10.4% of the conjugate was obtained after 88-min treatment. This yield should still need improvement compared to the dry-powder process with higher than 70% yield. The high-pressure treated conjugate exhibited an increased antimicrobial activity against *E. coli* and *S. typhimurium.* Much higher activity against gram-negative bacteria was revealed by a lysozyme-lactoferrin complex conjugated with dextran (unpublished result).

Through chemical modification of lysozyme by linking a saturated fatty acid or hydrophobic peptide (*PhePheValAlaPro*) to the C-terminus, the activity against gram-negative bacteria was detected (Ibrahim et al., 1991; 1997). The gram-negative cells have an additional membrane outside the cell wall. The cells manipulate the biochemistry of the wall in a way that causes a local change in the biophysical parameter analogous to surface tension. A possible explanation of the inactivity of lysozyme is that the enzyme is entrapped in the outer membrane through specific binding to the LPS and hence becomes arrested at this site. The specific hydrophobic binding and the resultant inactivation of lysozyme by LPS isolated from a variety of gram-negative organisms were reported (Ibrahim et al., 1994).

3. Genetic engineering: Kumagai and Miura (1989) were successful in enhancing the cell-lysing activity of chicken lysozyme three-fold by simultaneously mutating N37G, W62Y and D101G, despite the fact that individual mutants did not show increased lysing activity.

High-level expression of lysozyme (550 mg/L) in *Pichia pastoris* was reported (Digan et al., 1989). It is desirable to produce the enzyme with different pH optimum, altered specificity, improved activity and stability (Imoto, 1996). Frequently, mutations that boost enzyme activity occur in unpredictable sites for simultaneous mutation of more than one site (Rawls, 1999). The application of a regulated random optimization for site-directed mutagenesis is recommended (Nakai et al., 1998). Selection of amino acid residues to replace at site, which are randomly chosen, is made by using amino acid scales for hydrophobicity, propensities for helix and strand or bulkiness. The Windows computer program written for genetic optimization could accommodate any new amino acid scale, if it is effective in searching for important functions of proteins.

It is worth noting that only one amino acid substitution (*Ile55Thr* or *Asp66His*) turned human lysozyme to an amyloid fibril protein (Pepys et al., 1993). Most proteins may have an inherent ability to form fibrils like those found in the brains of Alzheimer's disease victims (Borman, 1999).

IX. SUMMARY

Lysozyme, a nutraceutical and value-added product from hen egg white, singly or in combination with other natural antimicrobial compounds, has desirable antimicrobial properties for use in minimally processed foods. Heat denatured lysozyme has antimicrobial activity against gram-negative bacteria. Japan leads the world for the variety of industrial uses of lysozyme. Outside Japan, lysozyme is mostly used by the dairy industry to control the growth of *C. tyrobutyricum*. Several patents claim the effectiveness of lysozyme, alone or in combination with other synergists, as a food preservative for fruits and vegetables, meat, beverages, and for nonfood uses. Food products containing lysozyme, such as cheese, chewing gums, candies, and mouthwashes are on the market. Potential applications in the produce, meat, wine and animal feed industries are increasingly being demonstrated (Canadian Inovatech, 1999). Until genetic engineering makes it possible to produce lysozyme with both gram-positive and gram-negative antimicrobial activity, hurdle technology will provide lysozyme more opportunities for use as a natural food preservative in the future.

Acknowledgment: The technical contribution of Angela Stessen and John Masuhara, QC Manager and Production Manager at the Canadian Inovatech, respectively, is greatly appreciated.

X. REFERENCES

1. Aabin, B., Poulsen, L.K., Ebbehoj, K., Norgaard, A., Frokiær, H., Bindslev-Jensen, C., and Barkholt, V. 1996. Identification of IgE-binding egg white proteins: Comparison of results obtained by different methods. *Int. Arch. Allergy Immunol.* 109: 50-57.
2. Ahvenainen, R., Heikonen, M., Kreula, M., Linko, M., and Linko, P. 1980. Separation of lysozyme from egg white. *Food Process. Eng.* 2: 301-310.
3. Alderton, G. and Fevold, H.L. 1946. Direct crystallization of lysozyme from egg white and some crystalline salts of lysozyme. *J. Biol. Chem.* 164: 1-5.
4. Alderton, G., Ward, W.H., and Fevold, H.L. 1945. Isolation of lysozyme from egg white. *J. Biol. Chem.* 157: 43-58.
5. Amati, A., Arfelli, G., Dell'Acqua, E., Ferrarini, R., Gerbi, V., Riponi, C., and Zironi, R. 1995. Lysozyme applications to control malolactic fermentation: Industrial trials. 46th Annual meeting of the American Society for Enology and Viticulture. Portland, Oregon, June 22-24.
6. Anet, J., Back, J.F., Baker, R.S., Barnett, D., Burley, R.W., and Howden, M.E.H. 1985. Allergens in the white and yolk of hen's egg. A study of IgE binding by egg proteins. *Int. Archs. Allergy Appl. Immun.* 77: 364-371.
7. Arima, H., Ibrahim, H.R., Kinoshita, T., and Kato, A. 1997. Bactericidal action of lysozymes attached with various sizes of hydrophobic peptides to the C-terminal using genetic modification. *FEBS Lett.* 415:114-118.
8. Atkinson, J.C., Yeh, C.K., Oppenheim, F.G., Bermudez, D., Baum, B.J., and Fox, P.C. 1989. Elevation of salivary antimicrobial proteins following HIV-1 infection. *J. Acquired Immune Deficiency Syndromes* 3:41-48.
9. Beintema, J.J. and van Scheltinga, A.C.T. 1996. Plant lysozyme. In: *Lysozymes: Model Enzymes in Biochemistry and Biology.* ed. P. Jollès, pp. 75-86. Basel: Birkhäuser Verlag.
10. Bernhisel-Broadbent, J. Dintzis, H.M., Dintzis, R., and Sampson, H.A. 1994. Allergenicity and genicity of chicken egg ovomucoid (Gal d III) compared with ovalbumin (Gal d I) in children with egg allergy and in mice. *J. Allergy Clin Immunol.* 93:1047-1059.
11. Bester, B.H., and Lombard, S.H. 1990. Influence of lysozyme on selected bacteria associated with Gouda cheese. *J. Food Prot.* 53: 306-311.
12. Board, R.G. 1995. New methods of food protection preservation, ed. G.W. Gould, pp. 41-57. London: Blackie Academic and Professional.

13. Bolder, N.M. 1997. Decontamination of meat and poultry carcasses. *Trends Food Sci.Technol.* 8: 221-227.

14. Borman, S. 1999. Shedding light on amyloid diseases. *Chem. Eng. News* (April 5) p. 7.

15. Canadian Inovatech. 1999. Lysozyme-bacteriocin combination provides improved food protection. *Food Technol.* 43(6):151.

16. Canfield, R.E. and McMurry, S. 1967. Purification and characterization of a lysozyme from goose egg white. *Biochem. Biophys. Res. Commun.* 26; 38-42.

16a. Carini, S., Muchetti, G., and Neviani, E. 1985. Lysozyme: Activity against clostridia and use in cheese production - A Review *Microbiologie Aliments Nutrition* 3:299-320.

17. Carneiro de Melo, A.M.S.,Cassar, C.A., and Miles, R.J. 1998. Trisodium phosphate increases sensitivity of gram-negative bacteria tp lysozyme and nisin. *J. Food Prot.* 61: 839-844.

18. Chang, K.Y., and Carr, C.W. 1971. Studies on the structure and function of lysozyme. I. The effect of pH and cation concentration on lysozyme activity. *Biochim. Biophys. Acta* 229:496-503.

19. Cisani, G., Varaldo, P.E., Ingianni, A., Pompei, R., and Satta, G. 1984. Inhibition of herpes simplex virus-induced cytopathic effect by modified hen egg-white lysozymes. *Curr. Microbiol.* 10: 35-40.

20. Cunningham, F.E., Proctor, V.A., and Gretsch, S.J. 1991. Egg-white lysozyme as a food preservative: a review. *World's Poultry Science Journal* 47: 141-163.

21. De Douder, C., and Marias, J. 1974. On lysozyme therapy. I. Lysozyme tablets in treatment of some localized and generalized viral skin diseases. *Medikon* 3: 19-20.

22. Dell' Aqua, E., Bruzzese, T., and Van Den Heuvel, H.H. 1989. *US Patent* 4,810,508.

23. Digan, M.E., Lair, S.V., Brierley, R.A., Siegel, R.S., Williams, M.E., Ellis, S.B., Kellaris, P.A., Provow, S.A., Craig, W.S., Velicelebi, G., Harpold, M.M., and Thill, G.P. 1989. Continuous production of a novel lysozyme via secretion from the yeast, *Pichia pastoris. Bio/Technol.* 7: 160-164.

24. Düring, K., Porsch, P., Mahn, A., Brinkmann, O., and Gieffers, W. 1999. The non-enzymatic microbicidal activity of lysozymes. *FEBS Lett.* 449: 93-100.

25. Eisai Co. Ltd 1973. Preservation of foods-hy adding a bacteriolytic enzyme and glycine. *Japanese Patent* JP73016613.

26. Eisai Co. Ltd. 1971. Ikura and suzuko by lysozyme treatment of inferior roe. *Japanese Patent* JP 7103732.

27. Eisai Co. Ltd. 1972a. Preserving raw fishery product by saline lysozyme solution. *Japanese Patent* JP 7203732

28. Eisai Co. Ltd. 1972b Fish product preservation. *Japanese Patent* JP 72005710.

29. Eisai KK. 1980. Storing Japanese sake-with addition of lysozyme or its salt and β-glucopyranose aeodehydrogenase. *Japanese Patent* JP 80035105.

30. Farber, J.M., and Peterkin, P.I. 1991. *Listeria monocytogenes*, a food-borne pathogen. *Microbiol. Rev.* 55: 476-511.

31. Fastrez, J. 1996. Phage lysozymes. In: *Lysozymes: Model Enzymes in Biochemistry and Biology,* ed. P. Jolles, pp. 35-64. Basel: Birkhäuser Verlag.

32. Fleming, A. 1922. On a remarkable bacteriolytic element found in tissue and secretions. *Proc.Royal Soc. London* B39: 306-317.

33. Gallo, A.A., Swift, T.J., and Sable, J.Z. 1971. Magnetic resonance study of the Mn^{+2}-lysozyme complex. *Biochem. Biophys. Res. Comm.* 43: 1232-1238.

34. Gao, Y.C., Krentz, S., Power, J., Lagarde, G., and Charter, E. 1999. Use of lysozyme and other egg white proteins to inhibit wine spoilage and barley infestation microorganisms. 41st Annual Conference of the Canadian Institute of Food Science and Technology. Kelowna, BC. June 6-9.

35. Gebraux, V.V., Monamy, C., and Bertrand, A. 1997. Use of lysozyme to inhibit malolactic fermentation and to stabilize wine after malolactic fermentation. *Am. J. Enol. Vit.* 48: 49-54.

36. Gomez, E., Martinez, A., and Laencina, J. 1995. Prevention of oxidative browning during wine storage. *Food Res. Intl.* 28: 213-217.

37. Goodman, H., Pollock, J.J., Katona, L.I., Iacono, V.J., Cho, M., and Thomas, E. 1981. Lysis of *Streptococcus mutans* by hen egg white lysozyme and inorganic sodium salts. *J. Bacteriol.* 146: 764-774.

38. Green, J.L., and Daeschel, M.A. 1995. The effects of wine components on the activity of lysozyme in a model wine system. 46th Annual meeting of the American Society for Enology and Viticulture. Portland, Oregon, June 22-24.

39. Green, J.L., Watson, B.T., and Daeschel, M.A. 1995. Efficacy of lysozyme in preventing malolactic fermentation in Oregon Chardonnay and Pinot Noir wines (1993 and 1994 vintages). 46th Annual meeting of the American Society for Enology and Viticulture. Portland, Oregon, June 22-24.

40. Green, J.L., Watson, B.T., and Daeschel, M.A. 1995. Efficacy of lysozyme in preventing malolactic fermentation in Pinot Noir wines. 46th Annual meeting of the American Society for Enology and Viticulture. Portland, Oregon, June 22-24.

41. Grossowicz, N. and Ariel, M. 1983. Methods for determination of lysozyme activity. *Methods Biochem. Anal.* 29: 435-446.

42. Haas, M., Kluppel, A.C.A., Wartna, E.S., Moolenaar, F., Meijer, D.K.F., De Jong, P.E., and Zeuw, D. 1997. Drug-targeting to the kidney: Renal delivery and degradation of a naproxen-lysozyme conjugate *in vivo*. *Kidney International* 52: 1693-1699.

43. Hatta, H., Mabe, K., Kim, M., Yamamoto, T., Gutierrez, M.A., and Miyazaki, T. 1994. Prevention of fish disease using egg yolk antibody. In *Egg Uses and Processing Technologies. New Developments*, eds. J.S. Sim and S. Nakai. Wallingford (UK): CAB International.

44. Hauben, K.J.A., Wuytack, E.Y., Soontjens, C.C.F., and Michiels, C.W. H. 1996. High-pressure transient sensitization of *Escherichia coli* to lysozyme and nisin by disruption of outer membrane permeability. *J. Food Prot.* 59: 350-355.

45. Hoffman, D.R. 1982. Immunochemical identification of the allergens in egg white. *J. Allergy Clin. Immunol.* 71: 481-486.

46. Holen, E., and Elsayed, S. 1990. Characterization of four major allergens of hen egg-white by IEF/SDS-PAGE combined with electrophoretic transfer and IgE-immunoautoradiography. *Int. Arch. Allergy Appl. Immunol.* 91: 136-141.

47. Hughey, V.L., Wilger, P.A., and Johnson, E.A. 1989. Antibacterial activity of hen egg white lysozyme against *Listeria monocytogenes* Scott A in foods. *Appl. Envir. Microbiol.* 55: 631-638.

48. Ibrahim, H.R., Higashiguchi, S., Sugimoto, Y., and Aoki, T. 1997. Role of divalent cations in the novel bactericidal activity of the partially unfolded lysozyme. *J. Agric. Food Chem.* 45: 89-94.

49. Ibrahim, H.R., Kato, a., and Kobayashi, K. 1991. Antimicrobial effects of lysozyme against gram-negative bacteria due to covalent binding of palmitic acid. *J. Agrtic. Food Chem.* 39:2077-2082.

50. Ibrahim, H.R., Yamada, M., Matsushita, K., Kobayashi, K., and Kato, A. 1994. Enhanced bacterial action of lysozyme to *Escherichia coli* by inserting a hydrophobic pentapeptide into C terminus. *J. Biol. Chem.* 269: 5059-5063.

51. Ibrahim, H.R., K., Kobayashi, K., and Kato, A. 1993. Length of hydrocarbon chain and antimicrobial action to Gram-negative bacteria of fatty acylated lysozyme. *J. Agri. Food Chem.* 41:1164-1168.

52. Imoto, T. 1996. Engineering of lysozyme. In *Lysozymes: Model Enzymes in Biochemistry and Biology*. P. Jollès (ed.) Birkhäuser Verlag, Bassel, Switzerland, pp. 163-181.

53. Inoue, K. 1987. The effect of lysozyme chloride on the immune response in patients with head and neck cancer. *Gan No Rinsho* 33:627-632.

54. Inoue, M., Yamad, H., Hashimoto, Y., Yasukochi, T., Hamaguchi, K., Miki, T., Horiuchi, T., and Imoto, T. 1992a. Stabilization of a protein by removal of unflavorable abnormal pHa: Substitution of undissociable residue for glutamic acid-35 in chicken lysozyme. *Biochemistry* 31: 8816-8821.

55. Inoue, M., Yamada, H., Yasukochi, T., Kuroki, R., Miki, T., Horiuchi, T., and Imoto, T. 1992b. Multiple role of hydrophobicity of typtophan-108 in chicken lysozyme: Structural stability, saccharide binding ability, and abnormal pHa of glutamic acid-35. *Biochemistry* 31: 5545-5553.

56. Johnson, E. 1994. Egg white lysozyme as a preservative for use in foods. In *Egg Uses and Processing Technologies. New Developments,* eds. J.S. Sim and S. Nakai, pp. 177-191. Wallingford (UK): CAB International.

57. Johnson, E.A., Dell'Acqua, E., and Ferrari, L. 1991. Process for bacterial decontamination of vegetable foods. *US Patent* 5,019,411.

58. Kato, A., Ibrahim, H.R., Nakamura, S., and Kobayashi, K. 1994. In *Egg Uses and Processing Technologies.* eds. Sim, J.S. and Nakai, S., pp. 250-268, Wallingford (UK): CAB International.

59. Kijowski, J., Lesnierowski, G., and Fabisz-Kijowska, A. 1999. Lysozyme polymer formation and functionality of residuals after lysozyme extraction. In *Egg Nutrition and Biotechnology*, eds. Sim, J., Nakai, S. and Guenter, W., Chapter 21. Wallingford (UK): CAB International.

60. Kim, J. 1984. Increased lytic activity of lysozyme at increased hydrostatic pressure. *Dev. Ind. Microbiol.* 25: 689-694.

61. Kumagai, I., and Miura, K. 1989. Enhanced bacteriolytic activity of hen egg-white lysozyme due to conversion of Trp62 to other aromatic amino acid residues. *J. Biochem.* 105: 946-948.

62. Kurasawa, T., Takada, K., Ohno, N., and Yadome, T. 1996. Effect of murine lysozyme on lipopolysaccharide-induced biological activities. *FEMS Immunol. Med. Microbiol.* 13: 293-301.

63. Leistner, L., and Gorris, L.G.M. 1995. Food preservation by hurdle technology. *Trends Food Sci. Technol.* 6:41-47.

64. Leitch, E.C., and Willcox, D.P. 1998. Synergic antistaphylococcal properties of lactoferrin and lysozyme. *J. Med. Microbiol.* 47: 837-842.

65. Li-Chan, E., Nakai, S., Sim, J., Bragg, D.B., and Lo, K.V. 1986. Lysozyme separation from egg white by cation exchange column chromatography. *J. Food Sci.* 51: 1032-1036.

66. Longthworth, L.G., Cannan, R.K., and MacInnes, D.A. 1940. An electrophoretic study of the proteins of egg white. *J. Am. Chem. Soc.* 62: 2580-2590.
67. Makki, F., and Durance, T.D. 1996. Thermal inactivation of lysozyme as influenced by pH, sucrose, and sodium chloride and inactivation and preservative effect in beer. *Food Res. Int.* 29: 635-645.
68. Malcolm, B.A., Wilson, K.P., Matthews, B.W., Kirsch, J.F., and Wilson, A.C. 1990. Ancestral lysozyme reconstructed, neutrality tested, and thermostability linked to hydrocarbon packing. *Nature* 345:86-89.
69. Malcom, B.A., Rosenberg, S., Corey, M.J., Allen, J.S., Baetselier, A.D., and Kirsch, J.F. 1989. Site-directed-mutagenesis of the catalytic residue Asp-52 and Glu-35 of chicken egg white Lysozyme. *Proc. Nat. Acad. Sci. USA.* 86: 133-136.
70. Matsumura, M., Signor, G., and Matthews, B.W. 1989. Substantial increase of protein stability by multiple disulfide bonds. *Nature* 342:291-293.
71. Meijer, D.K.F., Molema, G., Moolenaar, F., Zeeuw, D., and Swart, P.J. 1996. Glyco-protein drug carriers with an intrinsic therapeutic activity: The concept of dual targeting. *J. Control.Release* 39: 163-172.
72. Mirkov, T.E., Tex, H., and Fitzmaurice, L.C. 1998. Protection of plants against plant pathogens. *US Patent* 5,850,025.
73. Muraki, M., Jigami, Y., Morikawa, M., and Tanaka, H. 1987. Engineering of the active site of human lysozyme: Conversion of aspartic acid 53 to glutamic acid and tyrosine 63 to tryptophan or phenylalanine. *Biochim. Biophy. Acta* 911: 376-380.
74. Muraki, M., Morikawa, M., Jigami, Y., and Tanaka, H. 1988. Engineering of human lysozyme as a polyelectrolyte by the alteration of molecular surface charge. *Protein Eng.* 2: 49-54.
75. Nakai, S., Nakamura, S., and Scaman, C.H. 1998. Application of random-centroid optimization to one-site mutation of *Bacillus stearothermophilus* neutral protease to improve thermostability. *J. Agric. Food Chem.* 46: 1655-1661.
76. Nakamura, K., Furukawa, N., Matsuoka, A., Takahashi, T., and Yamanaka, Y. 1997. Enzyme activity of lysozyme-dextran complex prepared by high pressure treatment. *Food Sci. Technol. Tokyo* 3: 235-238.
77. Nakamura, S., Gohya, Y., Losso, J.N., Nakai, S., and Kato, A. 1996. Protective effect of lysozyme-galactomannan or lysozyme-palmitic acid conjugates against *Edwardsiella tarda* infection in carp, *Cyprinus carpio* L. *FEBS Lett.* 383: 251-254.
78. Nguyen, T.H., Guyon-Varch, A., and Brongniart, I. 1999. International Patent Application, except US. WO 99/00022.
79. Nyrén, P. and Edwin, V. 1994. Inorganic pyrophosphatase-based detection system. II. Detection and quantification of cell lysis and cell-lysing activity. *Anal. Biochem.* 220: 46-52.
80. Packard, V.S. 1982. *Human Milk and Infant Formula.* New York: Academic Press.
81. Padgett, T., Han, I.Y., and Dawson, P.I. 1998. Incorporation of food-grade antimicrobial compounds into biodegradable packaging films. *J. Food Prot.* 61: 1330-1335.
82. Pakula, A.A., and Sauer, R.T. 1990. Reverse hydrophobic effects relieved by amino acid substitutions at protein surface. *Nature* 344: 363-364.
83. Payne, K.D., Oliver, S.P., and Davidson, P.M. 1994. Comparison of EDTA and apo-lactoferrin with lysozyme on the growth of foodborne pathogenic and spoilage bacteria. *J. Food Prot.* 57: 62-65.
84. Peck, M.W., Fairbairn, D.A., and Lund, B.M. 1993. Heat-resistance of spores of non-proteolytic *Clostridium botulinum* estimated on a medium containing lysozyme. *Lett. Appl. Microbiol.* 16: 126-131.
85. Peck, M.W., Fairbairn, D.A., and Lund, B.M. 1992. Factors affecting growth from heat-treated spores of non-proteolytic *Clostridium botulinum.* *Lett. Appl. Microbiol.*15: 152-155.
86. Peck, M.W., and Fernandez, P.S. 1995. Effect of lysozyme concentration, heating at 90 °C, and then incubating at chilled temperature on growth from spores of non-proteolytic *Clostridium botulinum. Lett. Appl. Microbiol.* 21: 50-54.
87. Pellegrini, A., Thomas, U., von Fellenberg, R., and Wild, P. 1992. Bactericidal activities of lysozyme and aprotinin against Gram-negative and Gram-positive bacteria related to their basic character. *J. Appl. Bacteriol.* 72: 180-187.
88. Pelligrini, A., Thomas, U., Bramaz, N., Klauser, S., Hunziker, P., and von Fellenberg, R. 1997. Identification and isolation of a bactericidal domain in chicken egg white lysozyme. *J. Appl. Microbiol.* 82: 372-378.
89. Pepys, M.B., Hawkins, P.N., Booth, D.R., Vigushin, D.M., Tennent, G.A., Soutar, A.K., Totty, N., Nguyen, O., Blake, C.C.F., Terry, G.J., Feest, T.G., Zalin, A.M., and Hsuan, J.J. 1993. Human lysozyme gene mutation cause hereditary systemic amyloidosis. *Nature* 362:553-557.
89a. Phillips, D.C. 1966. The three-dimensional structure of an enzyme molecule. Sci. Amer. 215:78-90.
90. Pichler, W.J., and Campi, P. 1992. Allergy to lysozyme/egg white-containing vaginal suppositories. *Annal. Allergy* 69:521-525.
91. Pickering, G. 1998. The use of enzymes to stabilise colour and flavour in wine- an alternative to SO$_2$? *The Australian Grapegrower & Winemaker.* September 1998:101-102.

92. Pickersgill, R., Varvill, K., Jones, S., Perry, B., Fischer, B., Henderson, I., Garrard, S., Summer, I., and Goodenough, B. 1994. Making a small enzyme smaller; removing the conserved loop structure of hen lysozyme. *FEBS Lett.* 347: 199-202.

93. Pitotti, A., Dal Bo, A., and Boschelle, O. 1992. Assay of lysozyme by its lytic action on *Leuconostoc oenos*: a suitable substrate at acidic pH. *J. Food Biochem.* 15: 3393-403.

94. Prager, E.M. and Jollés, 1996. Animal lysozymes *c* and *g*: An overview. In *Lysozymes: Model Enzymes in Biochemistry and Biology*, ed. P. Jollés, pp. 9-31. Basel: Birkhäuser Verlag.

95. Proctor, V.A., and Cunningham, F.E. 1988. The chemistry of lysozyme and its use as a food preservative and a pharmaceutical. *Crit. Rev. Food. Sci. Nutr.* 26: 359-395.

96. Q.P. CORP. 1985. Preservative for food-comprise lysozyme lower fatty acid monoglyceride and phytic acid and/or sodium acetate. *Japanese Patent* JP 60037965.

97. Rabussay, D.P. 1982. Method of dental treatment. *US Patent* 4355022

98. Rawls, R.L. 1999. Taking a clue from biology. Chem. Eng. News (April 12): 38-43.

99. Razavi-Rohani, S.M. and Griffiths, M.W. 1996. Inhibition of spoilage and pathogenic bacteria associated with foods by combinations of antimicrobial agents. *J. Food Safety* 16: 87-104.

100. Satoh, J., Okano, T., Kikuchi, T., Suzuki, N., and Goto, Y. 1980. In vitro and in vivo effect of lysozyme chloride on mtigenic responses of lymphocytes of mice. *Tokohu Igaku Zasshi* 93:32-37.

101. Sava, G. 1996. Pharmacological asspects and therapeutic applications of lysozymes. In *Lysozymes: Model Enzymes in Biochemistry and Biology.* ed. P. Jollés, Basel: Birkhäuser Verlag.

102. Schoen, P., Corver, J., Meijer, K.F., Wilschut, J., and Swart, P. 1997. Inhibition of influenza virus fusion by polyanionic proteins. *Biochem. Pharmacol.* 53: 995-1003.

103. Schuchat, A., Deaver, K.A., Wenger, J.D., Plikaytis, B.D., Mascola, L., Pinner, R.W., Reingold, A.L., and Broome, C.V., and the Listeria Study Group. 1992. Role of foods in sporadic listeriosis I. Case-control study of dietary risk factors. *JAMA.* 267: 2041-2045.

104. Sellak, H., Franzini, E., Hakim, J., and Pasquier, C. 1992. Mechanism of lysozyme inactivation and degradation by iron. *Arch. Biochem. Biophys.* 299: 172-178.

105. Shih, P., and Kirsch, J.F. 1995. Design and structural analysis of an engineered thermostable chicken lysozyme. *Protein Sci.* 4: 2063-2072.

106. Smoelis, A.N. and Hartsell, S.E. 1949. The determination of lysozyme. *J. Bacteriol.* 58: 731-736.

107. So, T., Ito, H-O., Koga, T., Ueda, T., and Imoto, T. 1996. Reduced immunogenicity of monomethoxypoly ethylene glycol-modified lysozyme for activation of T cells. *Immunol. Lett.* 49: 91-97.

108. Strynadka, K.C.J. and James, M.N.G. 1996. Lysozyme: A model enzyme in protein crystallography. In *Lysozymes: Model Enzymes in Biochemistry and Biology,* ed. P. Jollés, pp. 185-222. Basel: Birkhäuser Verlag.

109. Takada, K., Ohno, N., and Yadome, T. 1994. Detoxification of lipopolysaccharide (LPS) by egg white lysozyme. *FEMS Immunol. Med. Microbiol.* 9: 255-264.

110. Teichberg, V.I., Sharon, N., Moult, J., Smilansky, A., and Yonath, A. 1974. Binding of divalent copper ions to aspartic acid residue 52 in hen egg-white lysozyme. *J. Mol. Biol.* 87: 357-368.

111. Ueno, R., Fujita, Y., Nnagamura, U., Kamino, Y., and Tabata, A. 1996. Preservation of foods by the combined action of a natural antimicrobial agent and separately packaged deoxidizing agent. *US Patent* 5,549,919.

112. Urisu, A., Ando, H., Morita, Y., Wada, E., Yasaki, T., Yamada, K., Komada, K., Tori, S., Goto, M., and Wakamatsu, T. 1997. Allergenic activity of heated and ovomucoid-depleted egg white . *J. Allergy Clin. Immunol.* 100: 171-176.

113. Villa, A. 1996. Controlling malolactic fermentation (MLF) with lysozyme: Applications and Results. Wine spoilage Microbiology Conference. California State University, Fresno, CA. March 8, 1996

114. Wang, C., and Shelef, L.A. 1991. Factors contributing to antilisterial effects of raw egg albumen. *J. Food Sci.* 56: 1251-1254.

115. Wasserfall, F.E., and Teuber, M. 1979. Action of egg-white lysozyme on *Clostridium tyrobutyricum. Appl. Environ. Microbiol.* 38: 197-199.

116. Weaver, L.H., Grütter, M.G., and Mattews, B.W. 1995. The refined structures of goose lysozymes and its complex with a bound trisaccharide show that the 'goose-type' lysozyme 1 lacks a catalytic aspartic residue. *J. Mol. Biol.* 245: 54-68.

117. Yamada, T., Yamada, M., Sasamoto, K., Nakamura, H., Mishima, T., Yasueda, H., Shida, T., and Iikura, Y. 1993. Specific IgE antibody titers to hens' egg white lysozyme in allergic children to egg. *Japanese J. Allerg.* 42:136-141.

118. Yang, H.H., Li, X. C., Amft, M., and Grotemeyer, J. 1998. Protein conformational changes determined by matrix assisted laser desorption mass spectrometry. *Anal. Biochem.* 258:118-126.

H. R. Ibrahim

Ovotransferrin

7

I. INTRODUCTION

With the increased public interest in natural foods, it is pertinent to consider natural antimicrobial systems with the hope of assessing their potential utility in food preservation and formulated drug systems. Interpretation of the molecular basis of such natural antimicrobials and their interaction with cell envelopes of bacteria will provide a crucially needed theoretical basis upon which food technologists and microbiologists can establish systematic investigations to assess the potential use of these natural antimicrobial molecules in formulated food and drug systems.

Of the many natural antimicrobial systems, that of the avian egg (which produces life within twenty-one days) is the least understood. It is one in which enzymes play little role and cells other than the immunologically incompetent ones of the embryo are absent (Banks et al., 1986). The yolk, the principal food reserve of the avian egg, is protected from microbial infection by the interplay between the physical defense provided by the egg's integument (cuticle, calcite shell, and shell membranes) and the chemical defense of the albumin (the egg white). Egg-white constitutes the second defense line against invading bacteria next to the shell and shell-membranes. Egg-white proteins have been studied by many scientists from different disciplines using them as a source of typical globular proteins. The biological functions of egg white proteins have been thought to prevent the penetration of microorganisms into the yolk and the embryo during development. Of different proteins found in the egg albumen, most appear to possess antimicrobial properties or certain physiological functions which interfere with the growth and spread of invading microorganisms; essentially the proteins act as the chemical defense system (Ibrahim, 1997a). These antimicrobial activities include: the bacteriolytic lysozyme, which hydrolyses the peptidoglycan in the cell walls of bacteria; ovotransferrin which chelates metal ions, specially iron; the vitamin binding proteins, i.e., avidin and ovoflavoprotein which chelate biotin and riboflavin, respectively; the proteinase inhibitors, i.e., ovoinhibitor, ovomucoid, ovomacroglobulin, and cystatin.

0-8493-2047-X/00/$0.00+ $.50

From review of the literature, it was concluded that ovotransferrin is the major component of the avian egg's antimicrobial defense system (Tranter & Board, 1982) as it renders the iron unavailable to microbial growth within the albumin, the so called nutritional immunity of albumin. Indeed there is no compelling evidence of the starvation strategy in albumin by ovotransferrin since saturation of ovotransferrin with iron does not convert albumin into an optimal medium for the growth of *Escherichia coli* (Banks et al., 1986) nor affect its *in vitro* bactericidal activity against *Staphylococcus aureus* (Ibrahim et al., 1998).

This chapter summarizes the current information on ovotransferrin of the hen's egg albumin, to clarify its structure-antimicrobial relations and the possible role of its existence in the avian eggs, which may allow it to be placed in a clearer structural functional perspective among the natural food antimicrobial agents.

II. STRUCTURE

Ovotransferrin (also called conalbumin) is a monomeric glycoprotein consisting of a 686-residue single polypeptide chain with a molecular mass of 78-80 kDa and an isoelectric point of 6.0 (*TABLE 1*). It has a capacity to bind reversibly two Fe^{3+} ions per molecule con-

TABLE 1: Physico-chemical properties of ovotransferrin from hen's egg white

Property	Value	Reference
% of total egg white proteins:	12%	Vadehra & Nath, 1973
Amino Acid Residues:	686	Williams et al., 1982
Molecular Weight:	77,770 Da	Williams et al., 1982
Metal Binding:	2 mol/mol of OTf	Williams et al., 1982
(Fe^{3+}/Zn^{2+}/Al^{3+}/Cu^{2+}/Mn^{2+})		
Iron Affinity (pH 7.5):		
OTf	10^{30} M^{-1}	Tranter & Board, 1982
N-terminal lobe	1.5×10^{14} M^{-1}	Kurokawa et al., 1995
C-terminal lobe	1.5×10^{18} M^{-1}	Kurokawa et al., 1995
Optical Extinction Coefficient:		
$^{1\%}E_{280nm}$	11.6	Vadehra & Nath, 1973
(2Fe) $^{1\%}E_{465nm}$	0.58	Komatsu & Feeney, 1967
Isoelectric point (*p*I):		
apo-protein	6.73	Wenn & Williams, 1968
2Fe-protein	5.78	Wenn & Williams, 1968
Carbohydrate Chain (Asn 473):		
	4 mannose	Williams, 1968
	8 N-acetylglucosamine	
Disulfide Bonds:		
OTf	15 bridges (30 Cysteine)	Williams et al., 1982
N-terminal lobe	6 bridges	Williams et al., 1982
C-terminal lobe	9 bridges	Williams et al., 1982
Heat Stability (Tm):		
Apo-OTf pH 7.0	61.0°C	Burley & Vadehra, 1989
pH 9.0	62.0°C	
2Fe-OTf pH 7.5	82.5°C	Mason et al., 1996
Al^{3+}-OTf pH 7.0	73.5°C	Burley & Vadehra, 1989
pH 9.0	72.5°C	

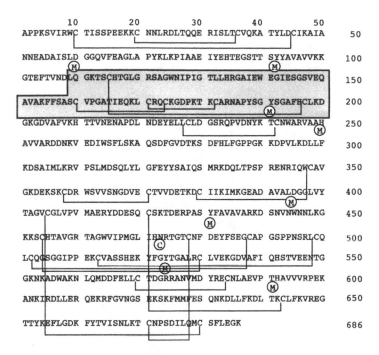

FIGURE 1. The amino acid sequence of ovotransfcrrin. Disulfide bridges (solid lines), N-linked glycosylation site (C), and the residues involved in the iron-binding sites (M) are indicated under the sequence. The sequence of OTAP-92 antibacterial peptide is indicated by light shading. The mature amino acid sequence of ovotransferrin and disulfide bridges was generated by using Brookhaven PDB file (P02789, TRFE-CHICK) retrieved from Swiss-Prot protein data base.

comitantly with two bicarbonate anions. It is a member of three homologous and closely related proteins called the transferrin family. The transferrins are a group of iron-binding proteins which are widely distributed in various biological fluids including serum transferrin (iron transport proteins; transferrin, siderophilin, β1-metal binding protein), lactoferrin (in milk and other secretions), and ovotransferrin (in avian egg white) (Baker et al., 1987). The amino acid sequence of ovotransferrin (*FIGURE 1*) depicts the position of disulfide bonds, amino acid residues involved in the iron binding sites (M), and sites of carbohydrate moieties (C) of the protein. The transferrins show a remarkable similarity in the amino acid sequence as well as the global folding of the molecules (*FIGURE 2*).

The avian transferrin gene is expressed both in liver and oviduct, where its protein products are known as serum transferrin and ovotransferrin, respectively. The modulation of the expression of the avian transferrin gene by iron level and steroid hormones is very different in the two organs. The protein part of avian serum transferrin and ovotransferrin is identical but differs only in the nature of their attached carbohydrate groups (Thibodeau et al., 1978). There are two putative glycosylation sites on the ovotransferrin polypeptide chain. Both sites are located in the C-terminal lobe (473-476, NRTG and 618-621, NGSE), but a single carbohydrate chain, attached to Asn 473 (*TABLE 1*) lacking sialic acid, was detected by glycopeptide mapping (Williams, 1968; Thibodeau et al., 1978). The oligosaccharide attached to this basic glycosylation site (Asn 473) of ovotransferrin

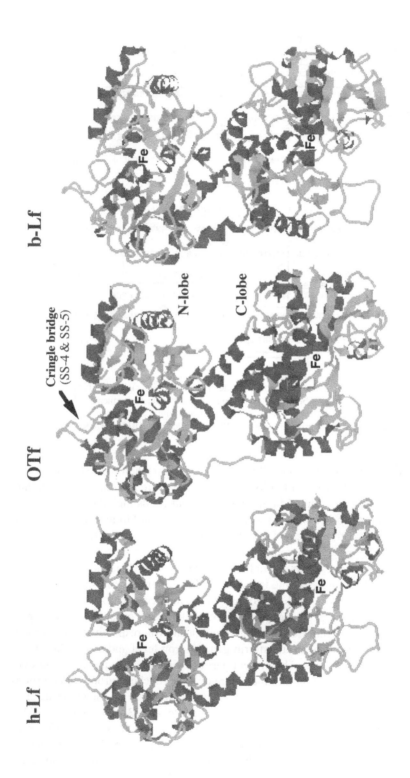

FIGURE 2. Stereo view of the folding of human lactoferrin (h-Lf), hen egg-white ovotransferrin (OTf), and bovine lactoferrin (b-Lf). The half-molecules N-lobe (upper) and C-lobe (lower), location of iron ion (Fe) the binding cleft of each lobe, and the spatial location of the cringle-bridges region are indicated. The ribbon structures were drawn by RasMol ver. 2.6 software (Biomolecular Structures Group, Glaxo R. & D., Greenford, UK) using the Brookhaven PDB files (TRFL-HUMAN (1LCF), TRFEñCHICK (1OVT), and TRFL-BOVIN (1BLF) for h-Lf, OTf and b-Lf, respectively) from Swiss-Prot data base.

is composed of 4 residues of mannose and 8 residues of N-acetylglucosamine. Avian serum transferrin has most of its carbohydrate moiety also in a single unit composed of 2 residues of mannose, 2 residues of galactose, 3 residues of N-acetylglucosamine, and 1 or 2 residues of sialic acid. The glycosylation sites are the same in ovotransferrin and serum transferrin (Williams, 1968). However, avian transferrins are not homogeneous with respect to its main glycan group, since the proteins appear in two electrophoretic variants (Jacquinot et al., 1994; Iwase & Hotta, 1977). These forms are probably derived from different content of sialic acid on serum transferrin and the variable glycan constituents of ovotransferrin. Another possible explanation of this may be the synthesis of the protein in more than one tissue, each of which may attach a characteristic oligosaccharides group to the molecule. The functional significance of the difference in the glycan's structure of ovotransferrin and serum transferrin is unknown although the two proteins are immunologically similar.

A. Iron-binding domains

Like the mammalian transferrins, the polypeptide chain of ovotransferrin is folded into two homologous halves (N-terminal lobe, residues 1-332 and C-terminal lobe, residues 342-686) each with a single iron site. Each lobe is further divided into two domains (N-terminal lobe, N1 and N2 domains; C-terminal lobe, C1 and C2 domains) of about 160 residues each, with an iron site in between (Baker et al., 1987; Kurokawa et al., 1995). The two domains are linked by two antiparalleled β-strands that allow them to open and close by a hinge (*FIGURE 2*). The other contact area between the two lobes includes residues 89 to 91 and 308 to 326 in the N-terminal lobe and residues 378 to 388, 597, and 674 to 686 in the C-terminal lobe (Kurokawa et al., 1995). The iron binding sites comprise the two phenolate oxygens of two tyrosines (Tyr 92 and Tyr 191 in N-terminal and Tyr 431 and Tyr 524 in C-terminal lobe) and a carboxylate oxygen from aspartic acid (Asp 60 in N-terminal and Asp 395 in C-terminal lobe), and one neutral nitrogen, the imidazole nitrogen of a single histidine (His 250 in N-terminal and His 592 in C-terminal lobe). The overall structure of hen ovotransferrin (OTf) is similar to human (hLf), bovine (bLf) lactoferrins (*FIGURE 2*) and human and rabbit serum transferrins (Kurokawa et al., 1995).

B. Variations in two half-molecules

The two lobes show marked similarities in their sequences with 37% identical residues. The positions of the 15 disulfide bridges are shown in *FIGURE 1*; there are 6 homologous bridges in each half of the molecule and 3 extra bridges which occur only in the C-terminal half. When the sequences of the two lobes of hen ovotransferrin are compared with those of human transferrin, all four half-molecules are homologous. The pairwise comparison indicated that the N-terminal domains of the two species and the two C-terminal domains of the two species are more similar than an N-terminal domain and a C-terminal domain (Williams et al., 1982). Furthermore, the three-dimensional superposition of the C-terminal lobe on the N-terminal lobe of ovotransferrin gives high identity and most of the secondary structural elements are comparable between the two lobes (Kurokawa et al., 1995). The main differences between the two lobes were found in loop regions.

The N-terminal lobe of ovotransferrin shows a much lower transition temperature for thermal denaturation than the C-terminal lobe (Nakazato et al., 1988; Lin et al., 1994), and this may be due to the greater number of disulfide bridges in the latter (*TABLE 1*).

These structural differences are associated with different iron-binding properties that the affinity of the C-terminal lobe for iron is also greater than the N-terminal lobe (Kurokawa et al., 1995). In addition, C-terminal lobe of ovotransferrin retains iron to lower pH values than the N-terminal lobe. Oe and co-workers (1988) have shown that the two half molecules of ovotransferrin can be obtained by limited proteolysis procedure. The two fragments corresponded to the N-terminal and C-terminal lobes and each contains an intact ferric ion-binding site. The isolated half molecules had the ability to re-associate non-covalently in solution. It would be of great interest to know whether this structural differentiation of the two lobes is accompanied by different biological functions for ovotransferrin. This remains an open question.

III. IRON-BINDING AND RECEPTOR-INTERACTION

Hen transferrin is unique in that a single protein of the same gene serves two functions (Thibodeau et al., 1978). Like all serum transferrins, hen serum transferrin is synthesized in the liver and implicated in the transport of iron, in a soluble form, to the target cells by membrane bound receptor-mediated endocytosis (Mason et al., 1996). Hen ovotransferrin is synthesized in the oviduct and delivered to newly formed egg albumin which is believed to primarily function as a bactericidal protein in the second defense line, the albumin, of hen's egg. The binding of ovotransferrin to the chick-embryo red blood cells (CERBC) has been reported and thus is believed to secondarily function as an iron transporter to the chick embryo (Brown-Mason and Woodworth, 1984; Mason et al., 1987). Studies from different laboratories indicated that both the N- and C-lobes, isolated either by trypsin cleavage of the connecting peptide of OTf (Mason et al., 1987; Brown-Mason & Woodworth, 1984) or the recombinant lobes (Mason et al., 1996), must be both present and pre-associated in order to bind to the chick reticulocyte receptor (a co-operative interaction) and to allow the delivery of iron (Alcantara & Schryvers, 1996; Jacquinot et al., 1994; Mason et al., 1996). Moreover, it has been documented that when one lobe is loaded with iron, the presence or absence of iron in the contralateral lobe makes little difference to receptor binding (Oratore et al., 1989). The basic conclusion that arises from these studies is that the ovotransferrin receptor must possess specific binding sites for both the N- and C-terminal lobes of ovotransferrin (Alcantara & Schryvers, 1996). It is apparent, therefore, that the mode of the binding of ovotransferrin to chicken receptor is different from the mode of mammalian transferrins. The isolated C-terminal lobe of human serum transferrin can donate iron to cells in a similar manner to intact transferrin (Zak et al., 1994). Furthermore, ovotransferrin does not bind to mammalian transferrin receptors, while most of the mammalian transferrin receptors can interact with other transferrins (Shimooka et al., 1986).

IV. ANTIMICROBIAL ACTIVITY

The high affinity of ovotransferrin for iron ($\sim 10^{30}$ M^{-1}) implies that in the presence of unsaturated ovotransferrin (apo-OTf), iron will be sequestered and rendered unavailable for the growth of microorganisms. Since ovotransferrin is largely free of iron in egg albumin, it has been suggested that its bacteriostatic action in egg white would be caused by depriving the bacteria of iron essential for their growth (Valenti et al., 1983a;

Tranter & Board, 1982). Although the studies attributed the antimicrobial effects of ovotransferrin to be bacteriostatic (growth inhibition), the bactericidal (bacterial killing regardless of iron deprivation) action of ovotransferrin against a wide spectrum of microorganisms (e.g. *Pseudomonas, E. coli, Streptococcus mutans, S. aureus, Proteus,* and *Klebsiella* sp.) has also been suggested (Valenti et al., 1982; 1983b; 1984; 1985; 1987). The antibacterial activity of metal complexes of ovotransferrin was studied *in vitro* against different bacterial species and the Zn^{2+} -loaded ovotransferrin appeared to be more active by comparison with the apo-protein and other metal complexes. The effect was neither due to Zn^{2+} ions, nor to iron deprivation, but to a specific activity of the Zn^{2+} -ovotransferrin complex. This antibacterial activity required a direct contact of Zn^{2+} -ovotransferrin with the bacterial surface (Valenti et al., 1987). It was also reported that ovotransferrin has antifungal activity, and the authors suggested an interaction between the protein and *Candida* cells, and again, its effect is decoupled from iron sequestration (Valenti et al., 1985).

In a compelling review, Tranter and Board (1982) concluded that ovotransferrin was the essential component of the egg's antimicrobial defense system. The antimicrobial potency of hen egg albumin was boosted at alkaline pH condition (pH 9.5) and by temperature near 40°C, the physiological temperature of birds (Tranter & Board, 1984). It was evident that saturation of ovotransferrin, by iron supplement of egg albumin at 39.5°C, does not convert hen egg albumin into an optimal medium for the growth of *E. coli*. It is now clearly established that ovotransferrin possesses both the bacterial growth inhibition (static) and the iron withholding-independent (cidal) antimicrobial actions. Exertion of a particular antimicrobial action seems to be environment-dependent and microorganism-specific.

However, many microorganisms are able to acquire iron bound to transferrins, including ovotransferrin (Law et al., 1992). Microorganisms accomplish this through two kinds of inducible mechanisms. One operates through production of high affinity iron-chelators (called siderophores) capable of removing iron from transferrin and delivering it to the cell via siderophore-receptors on the cell surface (Crosa, 1989; Lindsay & Riley, 1994; Lindsay et al., 1995; Trivier et al., 1995; Lim et al., 1996). The other mechanism is by expressing a receptor for the transferrin itself by which the bacteria can then remove the iron through a mechanism usually involving outer membrane and periplasmic iron-transport proteins (Chart et al., 1986; Chart & Rowe, 1993; Husson et al., 1993; Keevil et al., 1989; Modun et al., 1994; Ogunnariwo & Schryvers, 1992; Redhead et al., 1987; Tranter & Board, 1982). As a consequence, the bacterio-static / -cidal activity of ovotransferrin remained obscure for many years. From the literature, the antimicrobial activity of ovotransferrin has repeatedly been suggested to be decoupled from the iron-sequestration effect and shown to involve a direct contact of the molecule with the bacterial surface. However, the molecular basis that accounts for such bacterio-static or -cidal action remains an enigma. Attempts to elucidate the structural basis and antimicrobial mechanism of ovotransferrin would be an important contribution for modern biotechnology and medicine. It is particularly remarkable that OTf plays a key role in the antimicrobial defense of hen's egg and constitutes 12 % of the total proteins of albumin (15 g per liter of egg white) thus having great promise, as a natural and safe protein, for potential use in formulated food and drug systems.

TABLE 2: Bactericidal activity ovotransferrin and its half-molecules against *Staphylococcus aureus* and *Escherichia coli* as a function of Iron-saturation

	S. aureus		E. coli K-12	
	CFU/mL	**Δ Log killing/h**	**CFU/mL**	**Δ Log killing/h**
Control	7.2×10^5		1.0×10^6	
OTf -apo	3.5×10^4	1.32	9.9×10^5	0.01
-2Fe	4.1×10^4	1.24	6.2×10^5	0.21
Nt-apo	8.3×10^3	1.94	ND	
-Fe	9.9×10^3	1.86	ND	
Ct-apo	3.6×10^5	0.31	ND	
-Fe	6.4×10^5	0.05	ND	

Bactericidal activity was assessed by incubating bacteria with 20 µg/mL protein for 1 h at 37°C with gentle shaking. The colony forming units (CFU) were determined on nutrient agar plates after incubation for 24 h at 37°C. Cidal activity (Δ Log killing) was calculated by using the formula: Δ Log killing=Log nc/np, where nc and np are the CFU/mL of control and protein-treated cells, respectively. Iron-free (-apo) proteins were prepared by dialyzing against 0.1 M citrate buffer (pH 4.5) for 48h. Iron-saturated (Fe) proteins were prepared by incubating (over night) proteins with excess molar concentration of ferric chloride in tris buffer (pH 8.0) containing 5 mM sodium bicarbonate. Preparations were dialyzed against water and freeze dried. ND, not determined.

A. Antimicrobial mechanism

Many protein functions can be inferred from the known functions of homologous proteins; however, the existence of multifunctional proteins complicates such an approach (Jeffery, 1999). The bactericidal (killing) action of lactoferrin (Lf) has been reported against a wide spectrum of microorganisms which was found to operate regardless of iron-withholding properties (Bortner et al., 1989; Bellamy et al., 1992; Yamauchi et al., 1993). Briefly, the bactericidal action of Lf was attributed to the presence of a basic sequence, 25 amino acid residues long, in the N-terminal region of lactoferrin molecule, isolated by pepsin proteolysis and called lactoferricin. The biological activities of lactoferrin and the mode of action of lactoferricin are described in detail elsewhere in this volume.

A recent work from our laboratory (Ibrahim, 1997b) demonstrated a direct bacterial killing of ovotransferrin against Gram-positive *Staphylococcus aureus* and *Bacillus cereus* and the Gram-negative *E. coli* whether saturated or not with iron (*FIGURE* 3). Testing the antimicrobial activity of the iron-free (OTf-apo) and diferric- (OTf-2Fe) ovotransferrin and its half-molecules against *Staphylococcus aureus* and *E. coli* K12 indicated an interesting finding. Regardless of the degree of iron saturation, ovotransferrin exhibited a strong antimicrobial action against the Gram-positive *S. aureus*. On the other hand, though weak effect, the Gram-negative *E. coli* was more susceptible to the iron-saturated ovotransferrin, OTf-2Fe, than to the iron-free ovotransferrin, apo-OTf (*TABLE* 2). Most probably, the large-scale conformational changes occurring when the iron is bound to the ovotransferrin molecule (Grossmann et al., 1992) is responsible for the antimicrobial activity against *E. coli*. Both of the isolated half molecules of OTf showed remarkable reduction in the CFU of *S. aureus* in iron-free form (Nt-apo and Ct-apo), but the N-terminal lobe exhibited stronger bactericidal activity than the C-terminal lobe (*TABLE* 2). Strikingly, iron-saturation of the half molecules abolished the bactericidal activity of C-terminal lobe (Ct-Fe), while the N-lobe (Nt-Fe) restored its bactericidal effect almost equal in magnitude to the iron-free N-lobe (Nt-apo). Based on weight concentration of the

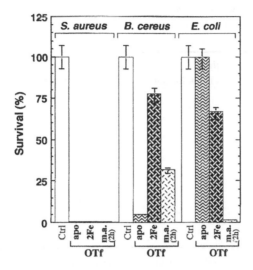

FIGURE 3. Antibacterial activity of hen egg-white ovotransferrin as a function of iron saturation and mild-acid hydrolysis against *Staphylococcus aureus*, *Bacillus cereus* and *Escherichia coli*. The bactericidal test was performed by incubating bacteria (initial levels of 3.2×10^8, 4.5×10^7, and 1.2×10^8 CFU / mL, respectively) with 20 μg / mL protein for 2 hr at 37°C. The results are presented as percentage to control cells incubated in the absence of protein. Ctrl, control cells; apo, iron-free OTf; 2Fe, iron-saturated OTf; and m.a, mild-acid total hydrolysate (2 hours) of OTf.

protein, the bacterial killing of the iron-saturated N-lobe was approximately two-fold greater than that of the iron-saturated intact OTf. But on the molar basis, bacterial killing of N-lobe (38.5 kDa) and OTf (78 kDa) were almost equal (*TABLE 2*). Indeed these results and previously reported observations (Valenti et al., 1987; 1985; 1983b; 1982) confirm that the bactericidal activity of ovotransferrin is operationally independent of its iron binding properties (nutritional deprivation).

These promising results have suggested that ovotransferrin possesses antibacterial domain located in the N-terminal lobe. As the comparative analysis of the amino acid sequence covering the bactericidal peptides found in lactoferrins (bLf 18-41, and hLf 12-47) indicated a low percentage of homology, we have hypothesized that the antibacterial domain may lies in a different location of the N-terminal region of ovotransferrin. Therefore, selection of the cleavage specificity for this investigation would be of great importance in understanding the structure-related antibacterial action of OTf.

B. Antibacterial peptide, OTAP-92

The isolation and structure determination of the bactericidal domain of OTf was achieved by utilizing the partial-acid proteolysis technique (Zhou & Smith, 1990), which produces relatively large peptides via specific cleavage at Asp-X sequence. The bactericidal domain of OTf was purified from the active hydrolysate by gel filtration and subsequent reversed-phase HPLC (Ibrahim, 1997b; Ibrahim et al., 1998). *TABLE 3* summarizes the antibacterial activity of OTf, its partial-hydrolysates, the two antibacterial fractions (A and B) on Sephadex G-50 column of the OTf hydrolysate for 2.5 h, and the purified bactericidal peptide (OTAP-92) by RP-HPLC, peak 4. Hydrolysis for 2.5 h produced the most

TABLE 3: Bactericidal activity against *Staphylococcus aureus* and *Escherichia coli* of the Asp-X cleaved ovotransferrin and its separated peptide on Sephadex G-50 column and reverse phase-HPLC.

Treatment	S. aureus		E. coli K-12	
	CFU/mL	Δ Log killing/5h	CFU/mL	Δ Log killing/5h
Control	3.2×10^9		1.3×10^9	
OTf Hydrolysate				
0 min	1.2×10^6	2.42	7.5×10^8	0.23
120 min	0.9×10^6	2.55	1.0×10^7	2.11
150 min	1.7×10^2	**6.26**	1.4×10^6	**2.96**
Sephadex G-50				
peak A	3.4×10^7	0.96	5.1×10^7	1.40
peak **B**	8.1×10^5	**2.59**	1.7×10^6	**2.88**
RP-HPLC				
peak 4(**OTAP-92**)	0.6×10^2	**6.72**	7.4×10^7	**1.24**

Ovotransferrin was cleaved at aspartyl residues by mild-acid hydrolysis for different lengths of time. The active OTf hydrolysate (150 min) was separated into four peptide peaks (A, B, C, and D) on Sephadex G-50 column. The maximal bactericidal fraction (peak B) was further separated into six peptide peaks (I-6) on a Wakosil-II 5C18-200 column (RP-HPLC) whereas peak 4 was the most bactericidal fraction. N-terminal microsequencing of peak 4 corresponded to the peptide Leu 109-Asp 200 of OTf (referred to as OTAP-92, Ovotransferrin Antibacterial Peptide-92 residues). The bactericidal activity was assessed by incubating bacteria with 20 µg/mL peptide fraction for 5 h at 37°C. The CFU value and Δ Log killing were obtained as described in TABLE 2.

potent bactericidal peptides against both test strains. Strikingly, the specific cleavage at the aspartyl residues produced more potent bactericidal component than the intact OTf (*FIGURE 3* and *TABLE 3*), providing a great opportunity for food technologists to consider the mild-acid hydrolysate of OTf as a potent antimicrobial agent. The active hydrolysate (2.5 h) were separated into four fractions (A-D) on Sephadex G-50 column, where both fraction A and B were bactericidal when screened against *S. aureus*. However, fraction B exhibited greater bacterial killing than A (*TABLE 3*). The decreased antibacterial activity of peak B than the total hydrolysate (150 min) may be due to antagonism of other peptides present in peak B to the bactericidal peptide of OTf. This was supported by the recovery of the antimicrobial action after complete isolation of the bactericidal peptide, peak 4 (OTAP-92), by reversed-phase HPLC on a Wakosil-II 5C18 column, using linear gradient of acetonitrile in acidified water. The purified peptide was more bactericidal than the intact OTf (*TABLE 3*) or its half-molecules (*TABLE 2*). Furthermore, this peptide showed strong bactericidal activity against both Gram-positive *S. aureus* and Gram-negative *E. coli* strains (*TABLE 3*), where more than six $\log_{10}$ orders of killing against *S. aureus* and more than one $\log_{10}$ order of killing against *E. coli* was observed. It was the first report on the identification and isolation of a bactericidal domain from ovotransferrin.

Structural features of OTAP-92: Combination of electrophoretic analysis and protein sequencing revealed a bactericidal peptide with average M_r of 9.9 kDa, corresponding to a surface exposed cationic peptide (ca. pI 8.84) consisting of 92 residues (Leu 109-Asp 200) located at the lip of the iron-binding cleft of the N-lobe of OTf (shaded sequence in Figure 1), designated OTAP-92 (Ibrahim et al., 1998). The OTAP-92 com-

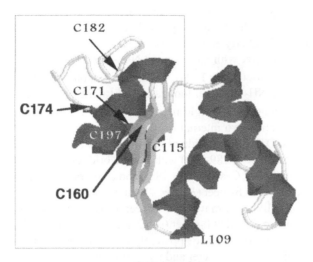

OTAP-92

FIGURE 4. Stereo view of the folding of the ovotransferrin antibacterial peptide (OTAP-92). The six cysteine residues involved in the three disulfide linkages (C115-C197, C160-C174, and C171-C182) are shown. The structural motif (helix-sheet motif) in OTAP-92 that resembles the 3D structure of insect defensins is indicated by boxed region.

prised six cysteines engaged in three intrachain disulfide bridges, Cys 115-Cys 197, Cys 160-Cys 174, and Cys 171-Cys 182. Of the three disulfide bridges in OTAP-92, two (Cys 160-Cys 174, and Cys 171-Cys 182) form cringle bridges, the two disulfides SS-IV and SS-V of OTf. The OTAP-92 domain packed into three helices with a two-stranded ß-sheet together with the three disulfide linkages (*FIGURE 4*), thus projecting the cringle bridges and a set of basic residues to the exterior of the N-lobe (*FIGURE 2*).

Several important structural features were observed in OTAP-92. It is scattered with an array of positively charged residues (Lys 169, Lys 175, Lys 179, Lys 181, Arg 135, Arg 172 and Arg 184), which are prerequisites for most antimicrobial peptides. The OTAP-92 peptide does not bind iron, but it contributes by the Tyr 191 to the tetrapartite protein ligands (Asp 60, Tyr 92, Tyr 191, and His 250) of the iron-binding site of N-lobe. The cringle lies on the opposite side of the iron-binding site from the interdomain cleft (*FIGURE 2*). This structural motif of OTAP-92 was found in other transferrins, including human, bovine, and pig lactoferrins, and all showed the same pattern of cysteines of the two disulfide bridges in the cringle region (Ibrahim et al., 1998). This unique structural motif of OTAP-92, strongly suggests an important role in the biological function of OTf, particularly when it is surface exposed and located in the structurally unstable domain (N-lobe) (Nakazato et al., 1988).

Multiple alignment showed remarkable sequence similarities between the optimum region of OTAP-92 (Lys 160-Arg 184) and several insect defensins, such as agitoxin (AGTX-1, Swiss-Prot P46110), iberiotoxin (IBTX, Swiss-Prot P24663) and kaliotoxin (KTX-1, Swiss-Prot P24662), which are active in killing both Gram-positive and Gram-negative bacteria (Ibrahim et al., 1998). Although the insect defensins exhibit great structural diversities, the cysteine-rich defensins contain a characteristic cysteine motif analogous to the cringle region of OTAP-92. The cysteine-rich defensins contain six cysteine residues involved in three disulfide bridges with molecular mass ranging from 2 to 6 kDa (Charlet et al., 1996; Lamberty et al., 1999). A characteristic fold of insect defensins is a central amphipathic α-helix with an extended NH$_2$-terminal loop and a COOH-terminal antiparallel ß-sheet (Charlet et al., 1996). The helix is stabilized via two disulfide bridges to the ß-sheet, and the NH$_2$-terminal loop is linked to one of the strands of the sheet via the third disulfide bridge. The cringle region of OTAP-92 showed sequence similarity, including the conservative two internal disulfide bridges, of these defensins (Ibrahim et al., 1998). The entire OTAP-92 peptide of OTf folds into three helices (helix-1, 121-136; helix-2, 147-156; and helix-3, 189-200) with cringle bridges maintained by the two disulfides (Cys 160-Cys 174 and Cys 171-Cys 182), and a central two-stranded (strand-1, 112-116; and strand-2, 157-161) ß-sheet (*FIGURE 4*). Interestingly, a similar 3D structural motif of insect defensins (helix-sheet motif) can be seen (boxed region of *FIGURE 4*) in OTAP-92 consisting of the two-stranded ß-sheet (sequence 112-116 and 157-161) and the helix-3 (residues 189-200). Therefore, one apparently universal antibacterial action of OTAP-92 is its relatively high basic character and the strikingly high degree of conservative cysteine array, together with the helix-sheet motif, a characteristic of many native antibacterial peptides. These structural similarities between OTAP-92 (cringle region) and insect defensins may allow anticipating a similar mechanism of antimicrobial action. A broader analysis of the structural requirement and the antimicrobial mechanism of the sequences of the other transferrins overlapping the OTAP-92 region will certainly be rewarding.

V. SUMMARY

In contrast to the immune system of animals, which produce antimicrobial polypeptides when needed, the egg albumin efficiently resists microorganisms over a long period in the absence of such innate and systemic host defense. Apparently, the main function of albumin is to keep microorganisms away from the yolk, the nutrient reservoir of the egg. The wide diversity in the nature of microorganisms implies that albumin would possess multiple antimicrobial functions. Of many different types of proteins found in the egg albumin, most appear to possess antimicrobial properties or certain physiological functions which interfere with the growth and spread of the invading microorganisms. Indeed there is a strong belief that only one of the factors plays an important role in the chemical defense of the egg albumin, and the literature supports the view that ovotransferrin is the cardinal factor (Board & Hornsey, 1978). In egg albumin and in mammalian serum as well, there is no free iron available for bacterial growth. In mammalian serum, transferrin is believed to act as an antimicrobial agent, where the iron-deprived bacteria are killed as a result of lysis caused by complement and lysozyme (Tranter & Board, 1982). Tranter and Board (1982) were unable to detect lysozyme-induced lysis of bacteria (*E. coli* 0111) denied Fe^{3+}. Hence, there must be an unknown mechanism responsible for the bactericidal activity of egg albumin.

The present review has attempted a broad survey of the chemistry and antimicrobial activity of ovotransferrin and provides key evidence that ovotransferrin acts as a bactericidal agent and that its bacterial killing activity can be decoupled from its iron-

TABLE 4: Putative motif sites in ovotransferrin sequence spatially located around antimicrobial peptide (OTAP-92)

OTf sequence		Motif site	Description
99-102	KKGT	CAMP-PHOSPHO-SITE	cAMP-and cGMP-dependent protein kinase phosphorylation site
13-16	SSPE		
14-17	SPEE		
27-30	TQQE		
41-44	TYLD	CK2-PHOSPHO-SITE	Casein kinase II phosphorylation site
105-108	TVND		
212-215	TVNE		
87-92	GSTTSY		
101-106	GTEFTV		
118-123	GLGRSA	MYRISTYL-MODIFICATION	N-myristoylation site
142-147	GIESGS		
193-198	GAFHCL		

The scanning of ovotransferrin sequence was performed by the on-line PROSITE tool against the pattern entries of Swiss-prot (the Swiss Institute of Bioinformatics).

sequestration properties beside its known bacteriostatic effect (nutritional immunity). The bactericidal activity has been ascribed to the presence of a peptide (OTAP-92), which can be released by specific cleavage at Asp-X sequence, recognized as the bactericidal domain of the OTf molecule.

Structural similarity between ovotransferrin and lactoferrin with various biological functions provides interesting information about the actual role of ovotransferrin in the egg. The multiple roles of mammalian lactoferrins discovered so far (including stimulation of a variety of immune cell lines, regulation of normal cell growth, inhibition of tumor growth and spread of cancer, antiviral activity) imply that ovotransferrin still has unknown function(s) in the egg particularly during embryogenesis. Clearly, this hypothesis has the attraction of embracing the key features of lactoferrin to explain the yet unknown physiological functions of ovotransferrin. The phenomenon is of great interest which may explain the function of ovotransferrin in the avian egg. The comparison between the structure and the mode of various biological actions of lactoferrin and ovotransferrin needs to be re-evaluated given that one multiple functional protein can perform two different functions in two different locations within the cell, homologous proteins may have one activity as a monomer and another as a multimer, and complex interactions among the various components make up a certain biological system. The occurrence of multiple conformational states of proteins in the egg albumin is expected to have important consequences for interpretation of their biological roles and thus provide a clue to elucidate the so far unknown physiological functions of ovotransferrin. This assumption can be verified if some allosteric and/or proteolytic activation were detected in the egg albumin. It is worth noting that ovotransferrin was found to possess various putative posttranslational modification sites, such as phosphorylation and fatty acylation motifs, particularly in the vicinity of the newly discovered antibacterial peptide (*TABLE 4*). The role of these sites on the physiological function, if any, of ovotransferrin is an enigma.

In conclusion, this review explored the presence of a cationic antibacterial peptide in the N-terminal half-molecule of OTf, thus introducing a good candidate to control pathogenic bacteria known to associate with foods, and to prolong shelf life. Perhaps the greatest roadblock to the use of naturally occurring antimicrobials could be economics. Of special interest is that the chemical cleavage at Asp-X sequence of OTf produces a strong bactericidal hydrolysate to both Gram-positive and Gram-negative bacteria, heralding a fascinating opportunity for the potential use of the crude hydrolysate in food applications or infant formulas. Knowledge of the antimicrobial mechanism of a compound may allow selection of combinations of antimicrobials with different mechanisms that could be utilized against the microorganisms in a food product. Therefore, the structural motif of the antibacterial peptide of OTf (OTAP-92), which shows structural and functional similarities with known bactericidal defensins, opens the door for its future candidacy as a novel antibiotic peptide in food and in the pharmaceutical industries as well. For this, the sequence coding for the OTAP-92 is now being isolated from the cDNA of ovotransferrin in our laboratory with the objective of understanding its structure-antimicrobial relations and also to better design a novel antimicrobial molecule using our previously adopted approaches for lysozyme (Ibrahim et al., 1992; 1994b; 1993; 1994a). Eventually, the unique location of this bactericidal domain on the exterior lip of the N-lobe thus paves the way to future studies on clarifying whether this domain has other biological function(s) in ovotransferrin.

Acknowledgment: The work on ovotransferrin and its antibacterial peptide was supported in part by a grant-in-aid for scientific research (No. 8064) from the Ministry of Education, Science and Culture of Japan.

VI. REFERENCES

1.   Alcantara, J. and Schryvers, A. B. 1996. Transferrin binding protein two interacts with both the N-lobe and C-lobe of ovotransferrin. *Microb. Pathog.* 20:73-85.

2.   Baker, E. N., Rumball, S. V. and Anderson, B. F. 1987. Transferrins: insights into structure and function from studies on lactoferrin. *Trends Biochem. Sci.* 12:350-353.

3.   Banks, J. G., Board, R. G. and Sparks. 1986. N. H. C. Natural antimicrobial systems and their potential in food preservation of the future. *Biotechnol. Appl. Biochem.*8:103-147.

4.   Bellamy, W., Takase, M., Yamauchi, K., Wakabayashi, H., Kawase, K. and Tomita, M. 1992. Identification of the bactericidal domain of lactoferrin. *Biochim. Biophys. Acta.* 1121:130-136.

5.   Board, R. G. and Hornsey, D. J. 1978. In *Chemical Zoology,* ed. M. Florkin and B. J. Scheer, pp.37-74. London: Academic Press.

6.   Bortner, C. A., Arnold, R. R. and Miller, R. D. 1989. Bactericidal effect of lactoferrin on *Legionella pneumophila*: effect of the physiological state of the organism. *Can. J. Microbiol.* 35:1048-1051.

7.   Brown-Mason, A. and Woodworth, R. C. 1984. Physiological levels of binding and iron donation by com plementary half-molecules of ovotransferrin to transferrin receptors of chick reticulocytes. *J. Biol. Chem.* 259:1866-1873.

8.   Burley, R. W. and Vadehra, D. V. 1989. *The Avian Egg: Chemistry and Biology.* New York: John Wiley & Sons, Inc.

9.   Charlet, M., Chernysh, S., Philippe, H., Hetru, C., Hoffmann, J. A. and Bulet, P. 1996 Innate immunity: isolation of several cysteine-rich antimicrobial peptides from the blood of a mollusc, *Mytilus edulis.* *J. Biol. Chem.* 271:21808-21813.

10. Chart, H., Buck, M., Stevenson, P. and Griffiths, E. 1986. Iron regulated outer membrane proteins of *Escherichia coli*: variations in expression due to the chelator used to restrict the availability of iron. *J. Gen. Microbiol.* 132:1373-1378.

11. Chart, H. and Rowe, B. 1993. Iron restriction and the growth of *Salmonella enteritidis. Epidemiol. Infect.* 110:41-47.

12. Crosa, J. H. 1989. Genetics and molecular biology of siderophore-mediated iron transport in bacteria. *Microbiol. Lett.* 53:517-530.

13. Grossmann, J. G., Neu, M., Pantos, E., Schwab, F. J., Evans, R. W., Townes-Andrews, E.,Lindley, P. F., Appel, H., Thies, W. G. and Hasnain, S. S. 1992. X-ray solution scattering reveals conformational changes upon iron uptake in lactoferrin, serum and ovotransferrin. *J. Mol. Biol.* 255:811-819.

14. Husson, M. O., Legrand, D.,Spik, G. and Leclerc, H. 1993. Iron acquisition by *Helicobacter pylori*: importance of human lactoferrin. *Infect. Immun.* 61:2694-7.
15. Ibrahim, H. R. 1997a. Insights into the structure-function relationships of ovalbumin, ovotransferrin, and lysozyme. In *Hen Eggs: Their Basic and Applied Science*, ed. T. Yamamoto, pp. 37-56. New York: CRC Press, Inc.
16. Ibrahim, H. R. 1997b. Isolation and characterization of the bactericidal domain of ovotransferrin. *Nippon Nogeikagaku Kaishi* 71:39-41.
17. Ibrahim, H. R., Hatta, H., Fujiki, M., Kim, M. and Yamamoto, T. 1994a. Enhanced antimicrobial action of lysozyme against Gram-negative and Gram-positive bacteria due to modification with perillaldehyde. *J. Agric. Food Chem.* 42:1813-1817.
18. Ibrahim, H. R., Iwamori, E., Sugimoto, Y. and Aoki, T. 1998. Identification of a distinct antibacterial domain within the N-lobe of ovotransferrin. *Biochim. Biophys. Acta*, 1401: 289-303.
19. Ibrahim, H. R. and Kobayashi, K. Kato. 1993. A. Length of hydrocarbon chain and antimicrobial action of gram-negative bacteria of fatty acylated lysozyme. *J. Agric. Food Chem.* 41:1164-1168.
20. Ibrahim, H. R., Yamada, M., Kobayashi, K. and Kato, A. 1992. Bactericidal action of lysozyme against Gram-negative bacteria due to insertion of a hydrophobic pentapeptide into its C-terminus. *Biosci. Biotech. Biochem.* 56:1361-1363.
21. Ibrahim, H. R., Yamada, M., Matsushita, K., Kobayashi, K. and Kato, A. 1994b. Enhanced bactericidal action of lysozyme to *Escherichia coli* by inserting a hydrophobic pentapeptide into its C-terminus. *J. Biol. Chem.* 269:5059-5063.
22. Iwase, H. and Hotta, K. 1977. Ovotransferrin subfractionation dependent upon carbohydrate chain differences. *J. Biol. Chem.* 252:5437-5443.
23. Jacquinot, P. M., Leger, D., Wieruszeski, J. M., Coddeville, B., Montreuil, J. and Spik, G. 1994. Change in glycosylation of chicken transferrin glycans biosynthesized during embryogenesis and primary culture of embryo hepatocytes. *Glycobiology* 4:617-624.
24. Jeffery, C. J. 1999. Moonlighting proteins. *Trends Biochem. Sci.*24:8-11.
25. Keevil, C. W., Davies, D. B., Spillane, B. J. and Mahenthiralingam, E. 1989. Influence of iron-limited and replete continuous culture on the physiology and virulence of *Neisseria gonorrhoeae. J. Gen. Microbiol.* 135:851-863.
26. Komatsu, S. K. and Feeney, R. E. 1967. Role of tyrosyl groups in metal binding properties of transferrins. *Biochemistry* 6:1136-1141.
27. Kurokawa, H., Mikami, B. and Hirose, M. 1995. Crystal structure of diferric hen ovotransferrin at 2.4 ≈ resolution. *J. Mol. Biol.* 254:196-207.
28. Lamberty, M., Ades, S., Uttenweiler-Joseph, S., Brookhart, G., Bushey, D., Hoffmann, J. A. and Bulet, P. 1999. Insect immunity: Isolation from the *Lepidopteran heliothis virescens* of a novel insect defensin with potent antifungal activity. *J. Biol. Chem.* 274: 320-9326.
29. Law, D., Wilkie, K. M., Freeman, R. and Gould, F. K. 1992. The iron uptake mechanisms of enteropathogenic *Escherichia coli*: the use of haem and haemoglobin during growth in an iron-limited environment. *J. Med. Microbiol.* 37:15-21.
30. Lim, Y., Lee, M. Y., Shin, S. H., Yang, N. U., Lee, S. E., Rhee, J. H., Park, Y. and Kim, I. S. 1996. Effect of iron limitation on the production of siderophore and hemolysin in *Staphylococcus aureus. J. Korean Soc. Microbiol.* 31:331-337.
31. Lin, L. N., Mason, A. B., Woodworth, R. C. and Brandts, J. F. 1994. Calorimetric studies of serum trans ferrin and ovotransferrin. Estimates of domain interactions and study of the kinetic complexities of ferric ion binding. *Biochemistry* 33:1881-1888.
32. Lindsay, J. A. and Riley, T. V. 1994. Staphylococcal iron requirements, siderophore production, and iron-regulated protein expression. *Infect. Immun.* 62:2309-2314.
33. Lindsay, J. A., Riley, T. V. and Mee, B. J. 1995. *Staphylococcus aureus* but not *Staphylococcus epidermidis* can acquire iron from transferrin. *Microbiology* 141:197-203.
34. Mason, A. B., Brown, S. A. and Church, W. R. 1987. Monoclonal antibodies to either domain of ovo transferrin block binding to transferrin receptors on chick reticulocytes. *J. Biol. Chem.* 262:9011-9015.
35. Mason, A. B., Woodworth, R. C., Oliver, R. W., Green, B. N., Lin, L. N., Brandts, J. F., Savage, K. J., Tam, B. M. and MacGillivray, R. T. 1996. Association of the two lobes of ovotransferrin is a prerequisite for receptor recognition. Studies with recombinant ovotransferrins. *Biochem. J.* 319:361-368.
36. Modun, B., Kendall, D. and Williams, P. 1994. *Staphylococci* express a receptor for human transferrin: identification of a 42-kilodalton cell wall transferrin-binding protein. *Infect. Immun.* 62:3850-3858.
37. Nakazato, K., Yamamura, T. and Satake, K. 1988. Different stability of N- and C-domain of diferric ovo transferrin in urea and application to the determination of iron distribution between the two domains. *J. Biochem. (Tokyo)* 103:823-828.

38. Oe, H., Doi, E. and Hirose, M. 1988. Amino-terminal and carboxyl-terminal half-molecules of ovotransferrin: preparation by a novel procedure and their interactions. *J. Biochem. (Tokyo)* 103:1066-1072.

39. Ogunnariwo, J. A. and Schryvers, A. B. 1992. Correlation between the ability of *Haemophilus paragallinarum* to acquire ovotransferrin-bound iron and the expression of ovotransferrin-specific receptors. *Avian Dis.* 36:655-663.

40. Oratore, A., D'Andrea, G., Moreton, K. and Williams, J. 1989. Binding of various ovotransferrin fragments to chick-embryo red cells. *Biochem. J.* 257:301-304.

41. Redhead, K., Hill, T. and Chart, H. 1987. Interaction of lactoferrin and transferrins with the outer membrane of *Bordetella pertussis*. *J. Gen. Microbiol.* 133:891-898.

42. Shimooka, T., Hagiwara, Y. and Ozawa, E. 1986. Class specificity of transferrin as a muscle trophic factor. *J. Cell. Physiol.* 126:341-351.

43. Thibodeau, S. N., Lee, D. C. and Palmiter, R. D. 1978. Identical precursors for serum transferrin and egg white conalbumin. *J. Biol. Chem.* 253:3771-3774.

44. Tranter, H. S. and Board, R. G. 1982. The antimicrobial defense of avian eggs: Biological perspective and chemical basis. *J. Appl. Biochem.* 4:295-338.

45. Tranter, H. S. and Board, R. G. 1984. The influence of incubation temperature and pH on the antimicrobial properties of hen egg albumen. *J. Appl. Bacteriol.* 56:53-61.

46. Trivier, D., Davril, M., Houdret, N. and Courcol, R. J. 1995. Influence of iron depletion on growth kinetics, siderophore production, and protein expression of *Staphylococcus aureus*. *Fems. Microbiol. Lett.* 127:195-200.

47. Vadehra, D. V. and Nath, K. R. 1973. Eggs as a source of protein. *CRC Crit. Rev. Food Technol.* 4: 193-309.

48. Valenti, P., Antonini, G., Fanelli, M. R., Orsi, N. and Antonini, E. 1982. Antibacterial activity of matrix-bound ovotransferrin. *Antimicrob. Agents Chemother.* 21:840-1.

49. Valenti, P., Antonini, G., Von-Hunolstein, C., Visca, P., Orsi, N. and Antonini, E. 1983a. Studies of the antimicrobial activity of ovotransferrin. *Int. J. Tissue React.* 5:97-105.

50. Valenti, P., Antonini, G., Von-Hunolstein, C., Visca, P., Orsi, N. and Antonini, E. 1984. Studies of the antimicrobial activity of ovotransferrin. *J. Appl. Bacteriol.* 56:53-61.

51. Valenti, P., Antonini, G., Von-Hunolstein, C., Visca, P., Orsi, N. and Antonini, E. 1987. Studies of the antimicrobial activity of ovotransferrin. *Int. J. Tissue React.* 5:97-105.

52. Valenti, P., Visca, P., Antonini, G. and Orsi, N. 1985. Antifungal activity of ovotransferrin towards genus Candida. *Mycopathologia* 89:169-175.

53. Valenti, P., Visca, P., von-Hunolstein, C., Antonini, G., Creo, C. and Orsi, N. 1983b. Importance of the presence of metals in the antibacterial activity of ovotransferrin. *Ann. Ist Super Sanita* 18:471-472.

54. Wenn, R. V. and Williams, J. 1968. The isoelectric fractionation of hen's-egg ovotransferrin. *Biochem. J.* 108:69-74.

55. Williams, J. 1968. A comparison of glycopeptides from ovotransferrin and serum transferrin of the hen. *Biochem. J.* 108:57-67.

56. Williams, J., Elleman, T. C., Kingston, I. B., Wilkins, A. G. and Kuhn, K. A. 1982. The primary structure of hen ovotransferrin. *Eur. J. Biochem.* 122:297-303.

57. Yamauchi, K., Tomita, M.,Giehl, T. J. and Ellison, R. T. 1993. Antibacterial activity of lactoferrin and a pepsin-derived lactoferrin peptide fragment. *Infect. Immun.* 61:719-728.

58. Zak, O., Trinder, D. and Aisen, P. 1994. Primary receptor-recognition site of human transferrin is in the C-terminal lobe. *J. Biol. Chem.* 269:7110-7114.

59. Zhou, Z. and Smith, D. L. 1990. Assignment of disulfide bonds in proteins by partial acid hydrolysis and mass spectrometry. *J. Protein Chem.* 9:523-532.

J.S. Sim

H.H. Sunwoo

E.N. Lee

Ovoglobulin IgY

8

I. INTRODUCTION

Domestic avian species such as chickens, turkeys, and ducks produce antibodies in blood and eggs against factors which cause avian diseases, as well as against other antigens. Immunoglobulin (Ig) in avian blood is transferred to the yolk of eggs to give acquired immunity to the offspring (Rose & Orlans, 1981). The protection against pathogens is provided for the relatively immuno-incompetent newly hatched chick. In the egg, the white contains IgA and IgM at relatively low concentrations of 0.7 and 0.15 mg/ml, respectively, while yolk contains a considerably higher concentrations of 25 mg IgG/ml (Rose et al., 1974). The antibody in egg yolk has been referred to as IgY (Leslie & Clem, 1969) due to its difference from mammalian IgG in structure and immunological properties. Thus, IgY is larger in size (Kobayashi & Hirai, 1980), slightly more acidic, and lower in molecular rigidity (Higgins, 1975) than mammalian IgG. Unlike mammalian IgG, IgY does not fix mammalian complement. It neither binds to protein A, G nor F_C receptors (Jensenius et al., 1981; Akerstrom et al., 1985). Since IgY is mainly composed of γ-livetin, which is a larger molecule than any other α-, β-livetin in egg yolk, it is relatively easy to separate from other proteins in the water-soluble fraction of egg yolk. Several methods have been used for the isolation and large-scale purification of IgY from egg yolk (Polson et al., 1980, 1985; Jenenius et al., 1981; Bade & Stegemann, 1984; Hassl & Aspock, 1988; McCannel & Nakai, 1989, 1990; Hatta et al., 1990; Akita & Nakai, 1992).

A hen lays an average of 240 eggs a year (Canadian Egg Marketing Agency, 1986). The annual production of IgY by a hen is about 24 g, which is high productivity compared to antibody production by mammals such as rabbits, mice and goats. Chicken egg yolk has, therefore, received much attention as a good source of antibodies (Leslie & Clem, 1969). As a food ingredient and/or a reagent, IgY is a versatile molecule for prophylactic, therapeutic and diagnostic purposes (Losch et al., 1986; Shimizu et al., 1988; Bar-Joseph et al., 1980; Vieira et al., 1984; Yolken et al., 1988). The purified IgY is practically applicable for passive immunization by oral administration. The effectiveness of passive immunization by oral administration of IgY to prevent infection has been reported for rotavirus diarrhea in human and animals (Bishop et al., 1973; DeMol et al., 1986; Yolken et al., 1988; Kuroki et al., 1993, 1994, 1997; Hatta, et al., 1993; Ebina 1996), dental caries (Otake et al., 1991; Hatta et al., 1997b), enteric colibacillosis (Yokoyama et al., 1992; Imberechts et al., 1997; Zuniga et al., 1997), and salmonellosis (Peralta et al., 1994; Sunwoo et al., 1996; Yokoyama et al., 1998ab).

Passive immunization of IgY could be achieved by antimicrobial activity against essential virulence determinants of pathogens. Antibodies interfere with the adhesion of pathogens to the intestinal wall and neutralize partially, or completely, their colonization potential. In the case of porcine enterotoxigenic *Escherichia coli* (ETEC), antibodies against the adhesive fimbriae were shown to protect against intestinal colonization and disease (Moon, 1976; Hampson, 1994). The activity of IgY may change with pH, temperature, and digestive enzymes in the intestinal tract and the dose of antibodies. These variables, therefore, enhance the effectiveness of antibody application. Shimizu et al. (1988) reported that the activity of IgY specific to *E. coli* is quite stable on incubation with trypsin or chymotrypsin but sensitive to pepsin, especially at pH lower than 4.5. Dose-dependent effects of specific IgY in controlling diarrhea of newborn calves have been demonstrated (Ozpinar et al., 1996).

In this chapter, the knowledge of egg yolk antibody, namely IgY, is extensively reviewed with an emphasis on its antimicrobial activity and potential prophylaxis or treatment for microbial infections. Finally, the role of IgY as a natural food antimicrobial in designing safe and healthy foods is discussed.

II. OCCURRENCE

A. Biosynthesis

The avian circulatory system consists of at least three kinds of immunoglobulins, IgG, IgA, and IgM, which are distinguishable in concentration, structure, and immunological function. The major immunoglobulin is IgG, which comprises about 75% of the total immunoglobulin pool. The concentrations have been reported to be 5.0, 1.25, and 0.61 mg/ml serum for IgG, IgA, and IgM, respectively (Leslie & Martin, 1973). Chickens produce immunoglobulins against many antigens including bacteria, viruses, and foreign substances in host defense.

In mammals, the transfer of maternal antibody occurs after birth via the mammary secretions and the neonatal gut, whereas in birds, maternal antibody must be present in the egg to protect the newly hatched chick. Specific antibody, especially IgG, is transferred from hen serum to yolk and then to the circulation of the chick via the endoderm of the yolk sac. Antibody is secreted into the ripening egg follicle (Patterson et al., 1962) and is incorporated into the egg white in the oviduct along with the egg albumen secretion. The concentrations of IgM and IgA in egg white and IgG in egg yolk are about 0.15 mg, 0.7 mg, and 25 mg per ml, respectively (Rose et al., 1974). Compared to other plasma proteins, IgG is selectively secreted into egg follicle and the IgG secretion from the hen's circulatory system into the ovarian follicle increase at a specific stage in its development. This transfer to the ovarian follicle is a receptor dependent and the ovarian IgG receptor allows the selective transport of all IgG subpopulations presented by the maternal blood (Locken & Roth, 1983). Maternal IgA and IgM, present in oviduct secretions, are acquired by the egg as it passes down the oviduct where the egg white is laid down.

At this time the yolk, which is fully formed on leaving the ovary, is surrounded by the vitelline membrane. The subsequent transfer of these immunoglobulins to the embryonic gut via swallowed amniotic fluid, at a time when neither is present in the serum, resembles the transfer of certain of the colostral or milk immunoglobulin to the newborn mammal. Immunoglobulins in eggs provide passive immunity so that the antibodies in eggs originated from the mother hen protect the newly hatched chick from a variety of infectious diseases (Losch et al., 1986).

B. IgY production

Several laboratories have compared the antibody production in hen egg to rabbit serum. The total antibody activity of the eggs laid by a hen in a month is equivalent to that produced in a half liter of serum from an immunized rabbit (Jensenius et al., 1981). Gottstein and Hemmeler (1985) compared the productivity of IgY from the eggs laid by a hen over a year with that of IgG from the whole serum of a rabbit in which both animals were immunized with the same antigens. Hens usually lay approximately 240 eggs in a year. The amount of serum collected from a rabbit is about 40 ml. One gram of egg yolk laid by the immunized hen contains about 10 mg of IgY whereas 1 ml of rabbit serum yields about 35 mg of IgG. Therefore, an immunized hen yields 40 g of IgY compared to 1.4 g of IgG produced by an immunized rabbit. The antibody production in hens is nearly thirty times greater than that of rabbits based on the weight of antibody produced per head per year (Hatta et al., 1997a).

The effects of egg and yolk weights on IgY production in laying chickens were reported by Li et al. (1998b). The percentage of hen-day production was approximately twice higher in the Single Comb White Leghorn (SCWL) hens than in the Rhode Island Red (RIR) hens, and both egg and yolk weight were 1.3 times greater in the SCWL hens than in the RIR hens. The ratio of yolk weight to egg white is similar in both strains of chicken. The total content of IgY in the egg yolk is relatively constant (average 0.6%) among the chickens regardless of the strain, egg weight or egg production per day during the experimental period. However, the total content of IgY in the yolk is approximately 1.3 times greater, and the total IgY produced during the 18 week experimental period is three times greater in the SCWL hens than in the RIR hens. Therefore, the egg yolk weight and the percentage hen-day production are considered to be important factors for the efficient production of IgY.

C. Quantitative assays

1. IgY-ELISA: Egg yolk antibodies raised against antigens can be assayed by an enzyme-linked immunosorbent assay (ELISA). The assay is based on the principle that specific antibody could bind to antigen, especially the antigenic binding site. Antibodies bound to antigen could be quantified by the reaction of substrate and enzyme conjugated with secondary antibody. The procedures are as follows. Microtiter plates are used as a solid support and are coated with antigens. A 10 µg/ml concentration of antigen in carbonate-bicarbonate buffer (0.05 M, pH 9.6) is added to each well and incubated for 24 h at room temperature. The plates are washed three times with phosphate buffered saline (PBS). After washing, 150 µl of 1 % (wt/vol) bovine serum albumin (BSA) in carbonate-bicarbonate buffer (0.05 M, pH 9.6) is added to each well, and incubated at 37 °C for 30 min. The BSA solution is discarded and each well is washed three times with phosphate buffered saline-Tween 20 (0.05%) (PBS-Tween). Diluted samples containing egg yolk antibodies (IgY) are added to the plate and incubated at 37°C for 1 h. The plates are washed three times with PBS-Tween and 150 µl of rabbit anti-chicken IgG conjugated with horseradish peroxidase (1:1,000 in PBS-Tween) is added to each well. After incubation at 37°C for 1 h, plates are washed three times with PBS-Tween, followed by addition of 100 µl of freshly prepared substrate solution, 2-2'-azino-bis (3-ethylbenzthiazoline-6-sulfonic acid) in 0.05 M phosphate citrate buffer (pH 5.0) containing 0.03 % sodium perborate. The reaction is continued for 30 min. Absorbance of chromophore produced in the reaction mixture is read at 405 nm using a microplate reader (Hornbeck, 1991).

   2. Antimicrobial assay: The inhibitory effect of IgY on the growth of bacteria could be measured by turbidity (OD at 550 nm) and standard plate counts. Sugita-konishi et al. (1996) showed the effects of specific IgY on the growth of three bacterial strains by turbidity assay. Each type of bacteria (10^4 cfu/ml) in tryptic soy broth (TSB) was incubated with different concentrations of specific and control IgY at 37°C. The turbidity of the culture was measured at OD 550 nm at 2 h intervals. Plate counting was used as another method for the measurement of antimicrobial activity (Shimizu et al., 1988). IgY (1-10 mg/ml) was added to a test culture (10^6 cfu *E. coli*/ml in TSB) and samples of the culture were taken after 1, 3, and 5 h. A viable cell count was made on tryptic soy agar (TSA) plates after appropriate dilution of the bacteria-IgY mixtures. The inoculated plates were incubated at 37 °C overnight.

III. MOLECULAR PROPERTIES

A. Isolation and purification

   Egg yolk is a complex food which can be separated by centrifugation into particles, 'the granules' and a clear fluid supernatant, 'the plasma' (Stadelman & Cotterill, 1977). Granules are composed of 70% α- and β- lipovitellins, 16% phosvitin, and 12% low-density lipoproteins (LDL) (Burley & Cook, 1961). Plasma is about 78% of the total yolk and composed of a lipid-free globular protein, livetin (α-, β-, and γ-) which represents about 10.6% of the total yolk solids and LDL (MaCully et al., 1962). IgY is known as γ-livetin and exists in egg yolk together with two other water-soluble proteins, α-, and β- livetin, and lipoprotein; therefore, separation of IgY or γ- livetin requires extraction of the water-soluble fraction (WSF) from yolk lipoprotein followed by purification from other livetins (Polson et al., 1980).

   Based on the aggregation of yolk lipoproteins at low ionic strengths as reported by Jensenius et al. (1981), several researchers used water dilution followed by centrifugation or filtration to fractionate WSF from water-insoluble components of egg yolk (Kwan et al., 1991; Akita & Nakai, 1992). For extraction of WSF from egg yolk with water, two factors are critical: the pH and the extent of egg yolk dilution (Akita & Nakai, 1992). The pH is extremely important to obtain the highest recovery of IgY. Fichtali et al. (1993) obtained a maximum recovery (54%) of IgY in WSF by 10-fold dilution and pH 5.5. IgY recovery was increased to 93-96% under the condition of 6 times dilution and pH 5.0 after incubation at 4 °C for 6 h followed by centrifugation or filtration (Akita & Nakai, 1992).

   After separation of WSF containing livetins from the egg granule, the next step

TABLE 1. Comparison of avian IgY and mammalian IgG

Character	Avian IgY	Mammalian IgG
Molecular weight	180 kDa	150 kDa
Isoelectric point	> acidic	< acidic
Heat stability	> sensitive	< sensitive
pH stability	> sensitive	< sensitive
F_c receptor binding activity	Low	High
Protein A/ protein G binding	no	yes
Interference with mammalian IgG	no	yes
Interference with rheumatoid factor	no	yes
Complement activation	no	yes

is to isolate IgY from the other water-soluble proteins, α-, and β- livetin, and LDL. A variety of methods has been used for the purification of IgY: ultracentrifugation (McBee & Coteterill, 1979), organic solvents (Bade & Stegemann, 1984; Polson et al., 1980), precipitation of lipoproteins by polyethyleneglycol (Polson et al., 1985), precipitation using sodium dextran sulphate (Jensenius et al., 1981), or natural gums (Hatta et al., 1988, 1990), ultrafiltration (Akita & Nakai, 1992), ion exchange chromatography (McCannel & Nakai, 1990) and metal chelate interaction chromatography (Greene & Holt, 1997).

B. Physico-chemistry

IgY is different from the IgG of mammals in molecular weight, isoelectric point, and binding behavior with complements, etc. as summarized in TABLE 1.

1. Stability to pH and heat: The heat and pH stability of IgY and rabbit IgG specific to human rotavirus were compared by measuring the antibody activity by ELISA (Hatta et al., 1993). IgY is more sensitive than IgG to temperature higher than 70°C. The study also showed that the maximum temperature of denaturation endotherm (T_{max}) of IgY was 73.9°C while that of IgG was 77.0°C according to differential scanning calorimetry. The IgY activity under various acidic conditions (pH 2 and 3) was more sensitive than that of rabbit IgG. These observations were supported by Otani et al. (1991) in studies comparing anti-αsl-casein IgY and rabbit IgG specific to mouse IgG. These differences in heat and acid sensitivity between IgY and IgG may be attributed to the variations in their protein structures.

Shimizu et al. (1988, 1992, 1993) reported on the molecular stability of IgY antibodies in comparison with that of mammalian IgG antibodies and found that heat (>75°C) or acid (<pH 3.0) treatment reduced the antibody activity of IgY. The activity of IgY was decreased by incubating at pH 3.5 or lower and almost completely lost at pH 3.0. The activity of IgY was decreased by heating for 15 min at 65°C or higher. Under alkaline conditions, changes in the activity of IgY did not occur until the pH is increased to 11 but was markedly diminished by incubation at pH 12 or higher.

2. Stability to proteolytic enzymes: IgY is relatively resistant to trypsin or chymotrypsin digestion, but is sensitive to pepsin digestion. However, the IgY is more susceptible to pepsin, trypsin or chymotrypsin digestion than the rabbit IgG antibody. Otani et al. (1991) showed the digestion profiles of egg yolk IgY and rabbit serum IgG antibodies specific to αsl-casein by determination of a percentage decrease in undigested heavy chain over a time period by pepsin, trypsin or chymotrypsin digestion.

Susceptibility of IgY against *E. coli* (Simizu et al., 1993), and human rotavirus (Hatta et al., 1993) was also examined. SDS-PAGE profiles of IgY after incubation with pepsin revealed that IgY at pH 2.0 is hydrolyzed into small peptides, and no bands corresponding to IgY were detected (Hatta et al., 1993). Conversely, IgY when incubated with pepsin at pH 4.0 resulted in the clear heavy (H)-and light (L)-chains after 4 h, although certain new bands appeared between H- and L-chains.

The behavior of IgY with trypsin and chymotrypsin was also examined (Hatta et al., 1993). Changes in the neutralization titer of IgY were almost similar for trypsin and chymotrypsin. After 8 h incubation, the activity of IgY in neutralization titer remained 39% and 41% for the mixtures with trypsin and chymotrypsin, respectively. On incubation with trypsin, the IgY H-chain disappeared, and several bands between H- and L-chains appeared on SDS-PAGE. Both H- and L-chains of IgY remained unchanged with chymotrypsin, although a small band below H-chain was observed.

C. Structure

The structure of IgY is identical to the major immunoglobulin found in serum, but is distinguishable from that of the mammalian IgG. IgY consists of two heavy chains (H) and two light chains (L) and has a molecular mass of ~180 kDa. The H chains of IgY possess one variable (V) region domain, four constant (C) region domains and no genetic hinge, unlike mammalian IgG which has three constant region domains and a hinge region (Parvari et al., 1988). The molecular structure of IgY, therefore, is similar to mammalian IgM or IgE, which consists of four domains. Comparisons of C-region sequences in IgG and IgY revealed that the Cγ2 and Cγ3 domains of IgG are most closely related to the Cv3 and Cv4 domains of IgY, respectively, and that the equivalent of the Cv2 domain is absent in γ chains. The Cv2 domain is probably 'condensed' to form the IgG hinge region (Burton, 1987; Parvari et al., 1988; Fellah et al., 1993; Magor et al., 1994).

Shimizu et al. (1992) reported that the content of β-sheet structure in the constant domains of IgY is lower than that of rabbit IgG, and the flexibility of the boundary region between Cv1 and Cv2 domains corresponding to the hinge region of IgG was less than that of rabbit IgG. Ohta et al. (1991) examined the whole structure of the various sugar chains of IgY, and indicated that 27.1% of asparagine-linked carbohydrate chains of IgY have glucose as the nonreducing end residue of the glycolic chains.

The structural properties of IgY (e.g. molecular size, conformation of domains, intramolecular bonding, lack of disulfide linkage in the IgY L-chain, and lower flexibility of the hinge region) could influence the overall properties of IgY molecule and are structural factors that might have bearing on the lower molecular stability of IgY, compared with mammalian IgG (Pilz et al., 1977; Shimizu et al., 1992).

IV. ANTIMICROBIAL ACTIVITY

A. Mechanism of action

1. Antigen-antibody interactions: The antigen-antibody interaction is a biomolecular association similar to an enzyme-substrate interaction but with the important distinction that it does not lead to an irreversible chemical alteration in either the antibody or antigen and therefore is reversible. The interaction between an antibody and an antigen involves various noncovalent interactions between the antigenic determinant, or epitope, of the antigen and the variable-region (V_H/V_L) domain of the antibody molecule, particularly the hypervariable regions, or complementarity-determining regions (CDR_s) (Kuby, 1997).

2. Anti-viral effects: The fundamental process of viral infection is the expression of the viral replicative cycle (partial or complete) in a host cell. Viral pathogenesis refers to the interaction of viral and host factors that lead to disease production. Rotaviruses are major cause of gastroenteritis and infect cells in the villi of the small intestine. They multiply in the cytoplasm of enterocytes, damage the transport mechanisms and lead to diarrhea. Coronaviruses also infect the gastrointestinal tract, but this infection is usually asymptomatic and has narrow host ranges. Human coronaviruses cause common colds and have been implicated in gastroenteritis in infants. Animal coronaviruses cause diseases of economic importance in domestic animals. Disease occurs in young animals and is marked by epithelial cell destruction and loss of absorptive capacity. Newcastle virus, the prototype avian parainfluenza virus, is the agent of respiratory infections, which initiate infection via the replication in the epithelia of the respiratory tract (Brooks et al., 1998).

Mechanism: Antibodies specific for viral surface antigens are often crucial in controlling the spread of a virus during acute infection and in protecting against re-infection. Most viruses express surface receptor molecules that enable them to initiate infection by binding specifically to host cell membrane molecules. If antibody is produced to the viral receptor, it could block infection by preventing binding of viral particles to host cells. Viral neutralization by antibody sometimes involves mechanisms that operate following viral attachment to host cells. In some cases antibodies may block viral penetration by binding to epitopes with the plasma membrane and can also agglutinate viral particles (Kuby, 1997).

3. Antibacterial effects: The pathogenesis of bacterial infection includes initiation of the infectious process and mechanisms that lead to the development of disease symptoms. The infection can be transient or persistent (Brooks et al., 1998).

Many factors determine bacterial virulence, or ability to cause infection and disease. Virulence factors are characteristics that enable bacteria to cause disease. Pathogenic bacteria may have one or several virulence factors. Many bacteria with capsules adhere to mucosal cells as the first step in causing disease (Bene & Schmidt, 1997). Adherence, which is only one step in the infectious process, is followed by development of microcolonies and subsequent steps in the pathogenesis of infection. The interaction between bacteria and host cell surfaces are complicated on the adhesion process. Bacteria have specific surface molecules that interact with host cells (Brooks et al., 1998). Adhesion factors are surface structures, including pili, lipopolysaccharide (LPS) side chains (O antigens), and M protein. In most cases, adhesion factors are pili [also designated as fimbriae and, in specific cases, as colonization factor antigens (CFAs)], hair like appendages that extend from the bacterial cell surface and mediate adherence of the bacteria to host cell surfaces (Bene & Schmidt, 1997). For example, certain *E. coli* strains have type I pili, which adhere to epithelial cell receptors. The *E. coli* that cause diarrheal diseases have pilus-mediated adherence to intestinal epithelial cells, although the pili and specific molecular mechanisms of adherence appear to be different depending on the form of the *E. coli* that induce the diarrhea (Brooks et al., 1998). LPS have also been implicated to play a role in adhesion. Adhesion factors interact with cells and tissues depending on the expression of receptors. Differences in tissue and organ tropism exhibited by bacteria that reach the systemic circulation reflect the different affinities of adhesion factors expressed by various bacteria (Bene & Schmidt, 1997).

For many pathogens, invasion of the host epithelium is critical to the infectious process. The invasiveness of bacteria seems to be a complex process. Some bacteria (e.g. *Salmonella* species) invade tissues through the junctions between epithelial cells. Once inside the host cell, bacteria may remain enclosed in a vacuole composed of the host cell membrane, or the vacuole membrane may be dissolved and bacteria may be dispersed into the cytoplasm (Brooks et al., 1998). In addition, many bacteria produce and secrete enzymes, which may play an important pathogenic role by a variety of mechanisms. These include a wide array of enzymes that degrade collagen, fibrin, and cellular material or modify and inactivate antibiotics.

Bacterial toxins can be classified into two broad categories. Exotoxins are proteins produced and released from the cell to cause toxicity, whereas endotoxins are part of the bacterial cell wall. Exotoxins can be common to all bacteria of a given genus or unique to pathogenic strains. These toxins are categorized into enterotoxins, neurotoxins, and cytotoxins. Enterotoxins affect the gastrointestinal tract and include heat-labile toxins (LT-I, LT-II), heat-stable toxin (ST), and cholera toxin. Endotoxins are lipopolysaccharides (LPS) derived from Gram-negative bacterial cell walls. All Gram-negative bacteria have endotoxin in their outer membranes, but endotoxins of different bacteria vary in their potency and ability to cause characteristic symptoms (Bene & Schmidt, 1997).

E. coli that cause diarrhea are extremely common worldwide. These *E. coli* are classified by the characteristics of their virulence properties, and each group causes disease by a different mechanism. There are five groups, enteropathogenic *E. coli* (EPEC), enterotoxigenic *E. coli* (ETEC), enterohemorrhagic *E. coli* (EHEC), enteroinvasive *E. coli* (EIEC), and enteroaggressive *E. coli* (EAEC). Enterotoxigenic *E. coli* (ETEC) is a common cause of 'traveler's diarrhea' in many developed countries and a very important cause of diarrhea in infants in developing countries (Brooks et al., 1998). The pathogenesis of diarrhea caused by ETEC strains has two main steps: intestinal colonization mediated by CFA and hypersecretion of water and electrolytes caused by ST or LT enterotoxins, or both. ETEC are limited to the proximal small intestine. Attachment to the intestinal wall is mediated by fimbriae. At least eight different colonization factors have now been identified in human strains, with all ETEC possessing one or more of these antigens. In addition to the colonization factors, virulence in ETEC strains is dependent on the formation of plasmid-mediated enterotoxins. Two major types of toxins have been described, which are readily distinguishable on the basis of heat stability, molecular weight and mode of action; that include LT and ST (Moon et al., 1978; Parry & Rooke, 1985; Holmgren, 1985; Saeed et al., 1983). The basis of oral administration of antibodies could cause interference with bacterial adherence and neutralization of toxins produced by the pathogens.

There are several determinants related to the pathogenicity of *Salmonella*; namely, endotoxin, an envelope containing LPS with an antigenic polysaccharide (O antigen) and invasins, outer membrane proteins (OMP) that mediate adherence to, and penetration of, intestinal epithelial cells. The portal of entry (small intestine epithelium) is common to all species. All virulent species apparently can survive gastric acidity and penetrate the intestinal epithelium and subepithelium (Virella & Schmidt, 1997). In the distal ileum, *Salmonella* adhere to and pass through intestinal epithelial cells, primarily the M cells of the follicle associated epithelium (Clark et al., 1994; Jones et al., 1994; Kohbata et al., 1986). *Salmonella* has various surface components, which are virulence-related. Lipopolysaccharide (LPS), flagella, or outer membrane proteins (OMP) play a role as pathogenicity determinants (Galdiero et al., 1990).

Mechanism: Anti-F18ab fimbriae antibodies interfere with the adhesion of bacteria to the intestinal wall. By neutralizing the colonization factor, the onset of disease was consequently reduced (Imberechts et al., 1997). Antibodies against the fimbriae of *E. coli* K88+ prevented the attachment of *E. coli* to the mucosal receptor (Jin et al., 1998). The attachment of bacterial cells was strongly inhibited when homologous anti-fimbrial antibody solutions were used in the *in vitro* adhesion assay (Yokoyama et al., 1992; Jungling et al., 1991). The *in vivo* prophylactic effect of IgY on piglet diarrhea could be due to their ability to inhibit the adhesion of *E. coli* to the intestinal mucus. The *in vitro* binding studies suggest that IgY interferes with the binding of ETEC to small intestine mucins, a step that may be necessary for this organism to colonize the small intestine (Wanke et al., 1990).

The passive transfer of anti-OMP antibodies may influence colonization by *Salmonella enteritidis* and *S. typhimurium* by binding to these surface proteins. The mechanism by which anti-OMP antibodies protect against *Salmonella* invasion of the host is unclear. On the basis of the present knowledge on OMP of Gram-negative bacteria and their biological functions as studies on adhesion to and invasion of the mucosa by *Salmonella,* Yokoyama et al. (1998a) hypothesized that OMP, since exposed on the bacterial surface are easily recognized by the antibodies. The subsequent binding may lead to impairment of the biological functions of OMP. The end result is a reduced invasiveness of *Salmonella* with loss of ability to colonize the intestinal tract.

Streptococcus mutans synthesize large polysaccharides such as dextrans or levans from sucrose and contribute importantly to the genesis of dental caries. IgY has specificity to insoluble glucans surrounding the cell surface of *S. mutans* and inhibits bacteria cell adherence properties. Although the mechanism by which immune antibodies protected against *S. mutans* colonization in humans was not elucidated, Ma et al. (1987) and Lehner et al. (1985) suggested that anti-*S. mutans* antibody bound to the antigens on the surface of the bacterium prevented adherence to the tooth, and that these opsonized bacteria may be eliminated by local host defenses. Filler et al. (1991) suggested that the binding of antibodies to the surface of *S. mutans* led to an inhibition of metabolic events important for the growth of *S. mutans*. Hatta et al. (1997b) also suggested that the IgY inhibits *S. mutans* adherence and its colonization in humans.

Antibodies may bind to the bacterial pili and block microbial colonization of the small intestine. IgY was detected in the intestine of neonatal pigs and, to a lesser degree, in weaning pigs. Since IgY could be destroyed by pepsin in the stomach (pH 2.3) of older pigs, fewer antibodies could pass through the intestine without hydrolysis or with intact structure and function. In young pigs, antibodies could readily pass the stomach without any damage by the immature gastric environment, thus allowing absorption of structurally functional antibodies in the small intestine and transfer to the circulation system (Yokoyama et al., 1993).

B. Influencing factors

The effectiveness of IgY for inhibiting adhesion of ETEC is influenced by two factors, i.e. dose of antibodies and the concentration of ETEC. The supply of antibodies in the intestine should be sufficiently high to prevent binding of *E. coli* to the mucosal receptors. Jin et al. (1998) showed that IgY when diluted 50- and 100-fold had a very strong inhibition activity against *E. coli* K88 at a concentration of 10^9 cfu/ml. However, dilution of 100 times for IgY was insufficient to inhibit the adhesion of *E. coli* to intestinal mucus when the concentration of *E. coli* K88 was 10^{10} cfu/ml.

V. ANTIMICROBIAL SPECTRUM

Since the use of IgY has been developed to prevent infectious diseases, many investigators have reported that IgY obtained from egg yolk of hens possesses the antimicrobial activity. This specific IgY activity against infectious pathogens such as rotavirus, *E. coli*, and *Salmonella spp.* raised from hens immunized with the corresponding pathogens appears to inhibit the growth of microorganisms *in vitro*. In addition, the passive immunization of the specific IgY against infectious pathogens reduced the rate of disease in animals. The antimicrobial spectrum of IgY is shown in *TABLE 2*.

A. Anti-rotavirus IgY

Rotaviruses are major causes of diarrhea illness in human infants and young animals, including calves and piglets. Infections in adult humans and animals are also common. Rotaviruses possess common antigens located on most, if not all, the structural proteins. These could be detected by immunofluorescence, ELISA, and immune electron microscopy. At least, nine serotypes have been identified among human rotaviruses, and about five more serotypes exist among animal isolates. Some animal and human rotaviruses share serotype specificity (Brooks et al., 1998).

Rotaviruses have a wide host range. Most isolates have been recovered from newborn animals with diarrhea. In experimental studies, human rotavirus can induce diarrhea illness in newborn colostrum-deprived animals (e.g. piglets, calves). Homologous infections may have a wider age range. Swine rotavirus infects both newborn and weaning piglets. Newborns often exhibit subclinical infection perhaps due to the presence of maternal antibody, whereas overt disease is more common in weaning animals (Brooks et al., 1998).

1. Anti-human rotavirus IgY: Human rotavirus (HRV) was found in 1973 (Bishop et al., 1973) and was identified as a major causative agent of infectious gastroenteritis in infants and young children. HRV infection is characteristically localized in the epithelial cells of the intestinal tract and leads to severe diarrhea with vomiting. Vaccination trials for HRV infection were unsuccessful due to the difficulty of introducing active antibody into the intestinal tract of infants whose immunity is generally under developed (DeMol et al., 1986). Thus, oral passive immunization has been investigated to prevent HRV infectious disease. There have been several reports on effects of IgY on the inhibition of the HRV growth *in vitro* and *in vivo*.

Yolken et al. (1988) investigated that commercially available eggs prevented the infection of HRV. Oral administration of anti-HRV IgY from chickens immunized with

TABLE 2. Specific anti-IgY against various bacterial and viral pathogens

Pathogens	Antimicrobial effect	References
Salmonella	Preventing gut colonization and organ invasion in chicks infected with S. *enteritidis*	Opitz et al.,1993
	Protecting mice challenged with *S.enteritidis* or *S.typhimurium* from experimental salmonellosis	Yokoyama et al., 1998a
	Preventing fatal salmonellosis in neonatal calves exposed with *S.typhimurium* or *S.dublin.*	Yokoyama et al., 1998b
	Inhibiting adhesion of *S.enteritidis* to human intestinal cells	Yoshiko et al., 1996
Escherichia coli	Preventing K88+, K99+, 987P+ enterotoxigenic *E.coli* (ETEC) infection in neonatal piglets	Yokoyama et al., 1992
	Protecting neonatal calves from fatal enteric colibacillosis by K99-pilated ETEC	Ikemori et al., 1992
	Preventing diarrhea in rabbits challenged with ETEC	O'Farrelly et al., 1992
	Reducing intestinal colonization with F18+ ETEC in weaned pigs	Zuniga et al., 1997
	Inhibiting shedding of F18+ *E.coli* by infected pigs	Imberechts et al., 1997
	Inhibiting adhesion of ETEC K88 to piglet intestinal mucus	Jin et al., 1998
	Curing diarrhea affected piglets in a field study	Wiedemann ct al., 1991
	Reducing diarrhea incidences of neonatal calves in field trial	Ozpinar et al., 1996
	Protecting pigs challenged with K88+ ETEC from *E.coli*-induced enterotoxemia	Yokoyama et al., 1993
	Neutralizing ETEC heat-labile toxin	Akita et al., 1998
Streptococcus	Reducing caries formation in rats infected with S. *mutans*	Hamada et al., 1991
	Preventing the establishment of S. *mutans* in dental plaque of humans	Hatta et al., 1997b
Edwardsiella	Preventing Edwardsiellosis of Japanese eels infected with *Edwardsiella tarda*	Hatta et al., 1994
Staphylococcus	Inhibiting the production of *Staphylococcus aureus* enterotoxin-A	Yoshiko et al., 1996
Pseudomonas	Inhibiting the growth of *Pseudomonas aeruginosa*	Yoshiko et al., 1996
Rotavirus	Preventing bovine rotavirus (BRV) induced diarrhea in murine model	Kuroki et al., 1993
	Protecting calves from BRV disease	Kuroki et al., 1994
	Protecting neonatal calves from BRV diarrhea under field conditions.	Kuroki et al., 1997
	Preventing human rotavirus induced gastroenteritis in mice	Ebina, 1996
Coronavirus	Protecting neonatal calves from bovine coronavirus-induced diarrhea	Ikemori et al., 1997
IBD-virus	Protecting chicks from infectious bursal disease	Eterradossi et al., 1997

HRV completely prevented the diarrhea in suckling mice (Hatta et al., 1993). These results indicated that the activity of IgY administration before challenging mice with HRV was highly effective, emphasizing its prophylactic role. Similarly, Ebina (1996) demonstrated that IgY against HRV (MO strain) prevented diarrhea in mice.

2. Anti-bovine rotavirus IgY: Rotavirus is also a potent pathogen to neonatal calves. The passive immunization of anti-bovine rotavirus (BRV) IgY has been reported to prevent diarrhea in animal models including mice and calves.

Kuroki et al. (1993; 1994) reported a dose-dependent protection of IgY (oral administration) in animal models infected with BRV serotypes G6 (strain Shimane) and G10 (strain KK-3). Furthermore, the oral efficacy of IgY specific for BRV serotypes G6 and G10 in protecting neonatal calves was examined in a herd of cattle under field conditions. Parameters such as body weight gain, mortality, decrease of BRV titer in stool and fecal score were evaluated. In one of the three trials, IgY-treated calves tested under high relative humidity (RH) showed a significantly increased mean body weight and a decrease in number of calves shedding high titers of BRV (G6) in stool compared to control calves. The observation that IgY-treated calves had a marked advantage in overall body weight gain and virus excretion profile concluded that IgY product was effective in a field condition with an epidemic outbreak of BRV diarrhea (Kuroki et al., 1997).

B. Anti-*Salmonella* IgY

Salmonella spp. are often pathogenic to humans or animals. The pathobiological symptoms include nausea, vomiting, abdominal cramps, diarrhea, and enteric fever. The immune response of chickens against LPS antigens resulted in the production of antibodies specific to LPS, a major constituent of the outer membrane of Gram-negative bacteria (Sunwoo et al., 1996; Yokoyama et al., 1998). Sunwoo et al (1996) found that the activity of antibodies against LPS fraction containing lipid-A was higher than that of those against a LPS fraction lacking lipid-A. These results indicated that IgY specific for the LPS fraction might be useful in the prevention of *Salmonella* adhesion and progression of the disease. Other fractions of 14 kDa fimbriae (Peralta et al., 1994) and outer membrane proteins (Yokoyama et al., 1998) of *Salmonella* were investigated for the possible control of salmonellosis.

IgY specific against 14 kDa fimbriae of *S. enteritidis* was orally administered to mice infected with the corresponding bacteria. The result showed a decrease of bacterial virulence (Peralta et al., 1994). Yokoyama et al. (1998) demonstrated that passive immunization with IgY specific for *S. typhimurium* and *S. dublin* could prevent fatal salmonellosis in calves.

Animal feed formulated with IgY against *S. enteritidis* phage type 13A showed significant differences from the control feed with other treatment such as multiple probiotics, organic acid and *Lactobacillus spp.* (Opitz et al., 1993). Anti-*S. enteritidis* IgY was effective in preventing gut colonization and organ invasion in chicks infected with *S. enteritidis* phage type 13A. This indicates that egg powder containing IgY against the microbial pathogen may be used for the feed additives, which provide prophylactic and therapeutic functions.

C. Anti-ETEC IgY

ETEC is the major cause of diarrhea and death in neonatal calves and piglets (Moon et al., 1976; Myers & Guinee, 1976; Hampson, 1994). Diarrhea in neonatal and post-weaning pigs has become a serious problem due to the increasing trend towards large intensive herds and early weaning (2 weeks of age). Such conditions lead to fatal enteric colibacillosis and considerable economic loss (Hampson, 1994). Passive protection of specific IgY against K88, K99, and 987P fimbrial adhesins of ETEC prevented infection in piglets deprived of colostrum (Yokoyama et al. 1992). The reaction of IgY against ETEC resulted in the strong resistance to bacterial adhesion to intestine. A similar study was conducted by Jin et al. (1998) showing the binding ability of specific IgY against K88+ MB *E. coli* in the mucus isolated from the intestine of piglets. Imberechts et al. (1997) also demonstrated that anti-F18ab fimbriae IgY inhibited the attachment of F18ab+ *E. coli* to the intestinal mucosa *in vitro* and diminished the onset of diarrhea and death in piglet infected with F18ac+ *E. coli*. Another field study showed the suppression of F18 fimbriated *E. coli* in newly weaned pigs treated with dietary specific IgY powder (Zuniga et al., 1997).

The K99 pilus antigen is one of the major adherence factors among ETEC isolated from neonatal calves (Guinee et al., 1976; Isaacson et al., 1978; Gaastra & de Graaf, 1982). The protective effects of egg yolk powder prepared from hens immunized with heat-extracted antigens from K99-piliated ETEC strain 431 were evaluated in a colostrum-fed calf model of ETEC-induced diarrhea caused by a heterologous strain (B44). Results indicate that the orally administered egg yolk powder protected against ETEC-induced diarrhea in neonatal calves. The protective components seem to be the antibodies raised by vaccination of chickens against ETEC (Ikemori et al., 1992).

O'Farrelly et al. (1992) examined the protective effects of IgY against ETEC B16-4 with LT and CFA-I in the rabbit reversible ileal tie model of diarrhea. Oral ingestion of egg yolks containing IgY for 4 days prior to inoculation protected rabbits from diarrhea after challenge with the same strain of *E. coli*. Rabbits showed no adverse effects from the ingestion of the egg yolks. *In vitro* experiments showed that IgY interfered with the binding of *E. coli* to purified small bowel mucins. These findings indicate that eggs from hens immunized with appropriate antigens have potential as a useful source of passive immunity.

The neutralization effect of IgY and its antigen-binding Fab fragment against LT of ETEC was evaluated by Akita et al. (1998). Both IgY and Fab were found to be effective in neutralizing LT produced by ETEC strain H10407.

D. Anti- *S. mutans* IgY

S. mutans is a major etiologic agent of human dental caries (Bratthall & Kohler, 1976; Loesche et al., 1975). Oral passive immunization with IgY specific to *S. mutans* seems to be effective in protecting against dental caries. Otake et al. (1991) showed that oral administration of immune IgY specific to *S. mutans* resulted in a statistically significant reduction in caries development in an experimental animal model. Another study also indicated that rats infected with *S. mutans* had a significant reduction in dental plaque accumulation and caries formation (Hamada et al., 1991). Furthermore, a mouth rinse containing IgY specific to *S. mutans* prevented the establishment of this bacterium in dental plaque of humans *in vitro* and *in vivo*. These data support the effectiveness of IgY with specificity to *S. mutans* grown in the presence of sucrose as an efficient method to control the colonization of *S. mutans* in the oral cavity of humans (Hatta et al., 1997b). These studies provide evidence for potential advantages of using IgY with specificity to *S. mutans* for controlling plaque and subsequent oral health problems associated with plaque accumulation.

E. Other IgY activities

Bovine coronavirus (BCV) is an important agent of neonatal calf diarrhea and acute diarrhea of adult cattle, referred to as 'winter dysentery'. BCV is known to cause a more severe disease and higher mortality than by BRV since BCV multiplies in both the small intestine and the large intestine whereas BRV infects only the small intestine (Kapil et al., 1990). Passive immunization of IgY specific against BCV prevented the diarrhea in neonatal calves infected with BCV (Ikemori et al., 1997).

Yoshiko et al. (1996) studied the immune functions of IgY obtained from hens immunized with a mixture of formalin-treated *Pseudomonas aeruginosa*, *Staphylococcus aureus*, enterotoxin-A, and *Salmonella enteritidis*. The results indicated that specific IgY effectively prevented the pathogenesis.

Newcastle disease has now been effectively controlled by immunizing hens with an attenuated Newcastle disease virus (vaccination). In the vaccination (active immunization), it takes about a week before rise in sufficient antibody activity in sera for protection. Since IgY specific to Newcastle disease virus is produced from egg yolk, and the use of the IgY for hens might not have immunogenicity, the risk period could be covered by intramuscular administration of the IgY to hens at the same time hens are vaccinated. Stedman et al. (1969) demonstrated the effectiveness of IgY isolated from egg yolk in conferring passive protection against challenge by Newcastle disease.

The infection of eels with *Edwardsiella tarda* is known to cause the most severe mortality. Hatta et al. (1994) demonstrated that Edwardsiellosis in Japanese eels could be prevented by oral administration of anti-*E. tarda* IgY. Many pathogens of fish have been reported to spread by infection through intestinal mucosa. The oral administration of specific IgY against fish pathogens in feed could be an alternative to use of antibiotics or chemotherapeutics for controlling bacterial infections in fish. Furthermore, effective drugs have not so far been developed against such bacterial fish diseases.

VI. ADVANTAGES

Antigen-specific IgG are conventionally isolated from sera of animals, such as rabbits, mice, and goats that have been superimmunized with an aimed antigen. IgY could be isolated from the egg yolk laid by a previously superimmunized hen. Chickens are potent antibody producers, and their immunological responsiveness is similar to that of mammals. The IgY production from hen's egg yolk is relatively simple and economical way to raise polyclonal antibodies. Therefore, IgY can be used as a successful substitution of mammalian antibodies. Several advantages in the production of IgY from hen's egg yolk are summarized as follows:

- Large-scale feeding of hens for egg production is currently in practice for collecting the source of a specific antibody. Maintenance of a large flock of layers is economical, which makes IgY feasible for large scale-production (Hamada et al., 1991).
- The number of animals used for antibody production is reduced because chickens produce larger amounts of antibodies than mammals (Schade et al., 1996).
- Immunization of hens (vaccination) has long been used to control various avian infectious diseases, indicating that immunization of hens is much more likely to be effective than with animals.
- Conventional methods for raising polyclonal antibodies inevitably sacrifices animals after harvesting specific IgG from their circulation system. On the other hand, the method of using hens requires only collecting the eggs laid by immunized hens. For separation of IgY, large-scale methods are applicable by automatic separation of the egg yolk using a machine.
- Egg yolk as the source of IgY is much more hygienic than mammalian sera from which IgG is separated.
- As egg yolk contains only IgY, the isolation of IgY from the yolk is much easier than that of IgG from animal sera.
- Because it is taxonomically different, the hen can produce antibodies whose formation is difficult or impossible in mammals (Hatta et al., 1997a).
- Antibody titers of concentrated IgY and WSF, and high immune response in the hen are stable for a long period (Tsunemitsu et al., 1989; Kuroki et al., 1993).

VII. APPLICATIONS

A. Suitability/ Adaptability

The protective effect of IgY is not destroyed by removal of the Fc fragment, and the Fab fragment is stable against further peptic digestion at pH above 4.2. Akita and Nakai (1993) suggest that IgY could be effective in young infants under 6 months of age. The gastric pH of infants under 6 months is normally in the range of 4-5 even 2-3 h after intake of milk (Nakai, 1962). Consequently, peptic digestion of IgY in the stomach of such infants leads to a stable functional Fab fragment.

IgG is often digested and inactivated by gastric juice. Hydroxypropyl methylcellulose phthalate (HPMCP) for the separation of yolk lipids has been used as an enteric coating substance for some drugs (Dressman & Amidon, 1984; Takada et al., 1989). Apparently, HPMCP-coated drugs are resistant to gastric juice, and dissolution in intestinal fluid is pH-dependent. It is likely that the antibody powder coated with HPMCP, although not perfect, may confer enteric resistance properties against low pH, thereby allowing safe passage through the stomach and ensuring the ultimate release of functioning antibodies in the small intestine. In studies by O'Farrelly et al. (1992), carbonate-bicarbonate buffer was used to suspend the egg yolks during oral inoculation to protect IgY in its passage through the stomach so that it could reach the small intestine.

B. Existing applications

1. Microstatic and prophylatic use: IgY can be applied to prevent the growth of various pathogenic bacteria such as *E. coli* and *Salmonella* using specific polyclonal antibodies, which effectively neutralize or reduce bacterial proliferation in meat products and prevent the growing threat of food-borne illnesses. The bacterial pathogens in contact with IgY may lose their mobility and reduce their colony-forming abilities. The inhibitory effects of IgY on the growth of porcine ETEC 987P were examined *in vitro* (Sunwoo et al. 1999). Bacteria (1.2×10^7 cfu./ml) were cultured with specific IgY or control IgY at concentrations of 0.625 –40 mg/ml at 37°C for 1-6 h in TSB. Turbidity measurements at 550 nm indicated the inhibition of bacterial growth at IgY concentrations of 5 and 10 mg/ml. Thus, anti-ETEC IgY could be useful in the control of bacterial growth as well as blocking the attachment of bacteria to the intestinal epithelium.

2. Passive immunization: Bacteria invade the body either through a number of natural entry routes (i.e., the respiratory tract, the gastrointestinal tract, and the genitourinary tract) or through unnatural routes opened up by breaks in mucus membranes or skin. The gastrointestinal (GI) tract continuously processes large quantities of foreign material (Tomasi & McNabb, 1987). It is important that the gut immunologically tolerates ingested dietary antigens. There is a general consensus that IgA, which is produced by GI tract-associated lymphoid tissue and transported into the lumen of the gut, assists in protection against bacterial, viral, and parasitic infections by inhibiting binding, preventing colonization, and neutralizing toxins (Williams & Gibbons, 1972). In certain circumstances, the secretory IgA system is absent or inadequate; thus enhancing the possible colonization of the GI tract by bacterial or viral pathogens. A particularly vulnerable period is the early stages of infancy, when the immune system is immature (Stiehm & Fudenberg, 1966). Numerous studies have documented the role of colostrum and breast milk in protecting the newborn against GI tract infections (Glass et al., 1983; Jason et al., 1984; Welsh & May, 1979). Recent studies have examined the possible use of nonmaternal antibodies for passive immunization of the GI tract.

Since *E. coli* is a major cause of infectious diseases in domestic animals, strategies to reduce the incidence and severity of disease have been considered as important priorities to reduce large economic losses resulting from these diseases (Moon & Bunn, 1993). The principle of vaccination, discovered in 1796 by Jenner, has provided the means to effectively reduce diseases caused by pathogens. In particular, there are numerous commercial vaccines for the prevention of neonatal diarrhea in pigs, calves, and lambs. While vaccines had a major impact on disease control, other management strategies that employ improved sanitation, use of antibiotics, and assurance that neonates receive colostrum early in life are also important in the reduction of disease and should remain as an integral part of disease prevention. Colonization of the intestine with *E. coli* leads to strong anti-colonization immunity associated with the appearance of anti-*E. coli* antibody of the IgA class in the serum. However, so far no effective vaccination protocol to combat this problem has been developed (Bianchi et al., 1996). Passive immunization may therefore be an attractive alternative. It may be achieved by the ingestion of antibody against essential virulence determinants. In the case of porcine ETEC, colostral antibodies against the adhesive fimbriae had effects on the prevention of the intestinal colonization and disease (Rutter & Jones, 1973; Moon, 1981).

Passive immunization is an important application of IgY, in which the specific binding ability of antibodies to the antigens (pathogens, venoms, etc) neutralize the pathobiological activities. Passive immunization seems to be one of the most valuable applications of antibodies in which pathogen-specific IgG is administered to individuals to result in prevention of infectious diseases. Passive immunization differs from active immunization (vaccination) in that the former employs an antibody obtained from other animals. The administration of this antibody specific to certain antigens (bacteria, virus, toxin, etc.) to individuals orally or systemically works to neutralize infectious activity or toxicity of the antigens. For practical application of passive immunization, an effective method of preparation of the antibody will be necessary, because large amounts of antibody may be required to administer the antibody for the passive immunization. The antigen-specific IgY could be prepared on an industrial scale from eggs laid by hens immunized with selected antigens. Passive immunization using IgY could find a wider practice in the near future (Hatta et al., 1997a).

IgY can be used for both therapeutic as well as prophylactic purposes. Wiedemann et al. (1991) found that IgY was as successful as a common antibiotic therapy in curing piglets suffering from diarrhea. Yokoyama et al. (1992) showed that antibodies prepared from the yolk of eggs from hens immunized with fimbrial antigens of *E. coli* were protective in newborn piglets against a challenge with homologous ETEC strains.

3. Immunological tool: Antigen-specific IgG isolated from sera of superimmunized animals, such as rabbits, mice, cows, and goats has been widely applied as an immunological tool in the field of diagnosis as well as basic research. The antigen-specific IgY is useful in its binding specificity comparable to the mammalian IgG. They both aid in detecting antigens with high specificity, which will never be achieved by any other method (Hatta et al., 1997a).

Diagnosis: Altschuh et al. (1984) reported that IgY against human antibody (IgG and IgM) was applicable in determining their concentration in biological fluid by rocket-immunoeletrophoresis. While using rabbit IgG chemical modification, such as carbamylation, of the IgG is generally needed to change its isoelectric point from that of the human antibody. However, in application of IgY, the carbamylation is not necessary since its isoelectric point is different from that of the human antibody. Fertel et al. (1981) demonstrated the application of IgY in determining prostaglandin in serum using radioimmunoassay using prostaglandin conjugated with hemocyanin (hapten) as an antigen for immunization of hens. Gardner and Kaye (1982) prepared IgY specific to rotavirus, adenovirus, and influenza virus, and performed immunological detection of viruses using IgY as the first antibody, and fluorescein isothiocyanate (FITC)-conjugated rabbit IgG specific to IgY as the second antibody. The preparation of IgY to these viruses was achieved with less effort compared to the conventional rabbit IgG, since the purification of the virus as antigen was unnecessary. Since these viruses can be cultivated using fertilized eggs, the contaminants in the virus culture that are components of egg must not show any immunogenicity to hens. It was suggested that IgY is a suitable antibody for detecting pathogens in stool samples, because it does not bind to protein A derived from *Staphylococcus aureus* as a common fecal contaminant. Thus, this characteristic of IgY could diminish identifying false positives in stool. Many researchers have also demonstrated the application of IgY for determination of various biological compounds that occur at relatively low concentrations such as plasma kallikrein (Burger et al., 1985), 1.25-dihydroxyvitamin D (Bauwens et al., 1988), hematoside (NeuGc) (Hirabayashi et al., 1983), human transferrin (Ntakarutimana et al., 1992), ochratoxin A (Clarke et al., 1994), human dimeric IgY (Polson et al., 1989), and high-molecular weight mucin-like glycoprotein-A (Shimizu, 1995).

Another advantage of IgY as an immunological tool over using rabbit IgG is the sensitivity of hens to antigens originating from mammals. There are a number of mammalian proteins with well-conserved amino acid sequences and many of these molecules have low or no antigenicity in mammals. Antibody production against such protein antigens in the hen is highly promising as an alternative animal, due to its immunological diversity from mammals. Carroll and Stollar (1983) succeeded in preparing IgY against RNA polymerase II that generally fails to elicit specific antibody response in mammals. Many laboratories have succeeded in producing IgY specific to low immunogenic mammalian antigens (*TABLE 3*).

Immunoadsorbent ligand: Immuno-affinity chromatography has been a useful tool for purification of proteins (antigens). Rabbit IgG has been conventionally used as a ligand to immobilize to adsorbents such as cellulose or agarose. However, the rabbit IgG poses several disadvantages because acidic pH (< 2.0) conditions are necessary for eluting the protein bound to the rabbit IgG column. Therefore, the eluted protein is often denatured depending on its physico-chemical nature. Moreover, production of rabbit IgG in large amounts is generally expensive. The immuno-affinity chromatography using rabbit IgG as a ligand has thus been applied for isolation of only certain proteins.

It has been demonstrated that IgY could be an effective alternative antibody ligand for such immunoadsorbent techniques. Specific IgY and rabbit IgG against mouse IgG were immobilized on Sepharose 4B and compared for their eluting efficacy in affinity purification of antibody from mouse sera. The adsorbent was eluted, stepwise, with a

TABLE 3. Production of IgY specific to less immunogenic antigens against mammals.

Antigen	Reference
Proliferating cell nuclear antigen of calf thymus	Gassmann et al., 1990
Heat-shock protein (Hsp 70)	Gutierrez & Guerriero, 1991
Human insulin	Lee et al., 1991
Rat glutathion peroxide	Yoshimura, 1991
Peptidylglycine α-amidating enzyme	Sturmer et al., 1992
von Willebrand factor	Toti et al., 1992
Platelet glycoprotein Iib-IIIa	Toti et al., 1992
Parathyroid hormone related protein	Rosol et al., 1993
Mouse erythroprotein receptor	Morishita et al., 1996
Proteoglycan	Li et al., 1998a

buffer solution of pH 4.0 and 2.0. The mouse IgG was dissociated at pH 4.0 only half of that applied, and the remaining IgG was eluted with pH 2.0 buffer solution with rabbit IgG as a ligand. On the other hand, 97% of the mouse IgG was dissociated even at pH 4.0 using IgY as a ligand (Hatta et al., 1997a).

C. Possible applications

The use of antibodies in combination with other preventive measures could improve survival rates of calves during *Salmonella* epizootic infections. The value of oral antibody administration in a combination treatment acts as the first line of defense in mucosal protection causing rapid elimination of invading bacteria. Oral passive immunization, using IgY could be considered as an adjunct to vaccination or antimicrobial treatment, or both, when there is a risk of exposure to S*almonella* (Yokoyama et al., 1998b). Hatta et al. (1997b) indicated that the immune IgY could be a novel ingredient for foods and mouth rinses to prevent the colonization of *S. mutans*, especially in the presence of dextrans.

The protective effect of IgY could be enhanced by producing a cocktail of antibodies against several important antigens including the various CFAs, enterotoxins and the more important O antigens endemic to a particular region. Although anti-adherence antibodies may offer better protection because of serotype specificity, anti-enterotoxin antibodies may also confer cross-protection against heterologous strains producing the enterotoxins (Akita et al., 1998).

D. Limitations

There are limitations to use IgY as an oral prophylactic treatment due to its susceptibility to proteolysis. *In vitro* studies on pepsin digestion of IgY showed both rapid and complete digestion at low pH (Shimizu et al., 1988; Schmidt et al., 1989). IgY has a serum half-life of 1.85 days in newborn pigs. This is considerably shorter than the reported serum half-life of 12 to 14 days for homologous IgG (colostrum antibodies) (Curtis & Bourne, 1971, 1973). The fraction of IgY absorbed into the circulation when administered in pigs decreased with increasing age of pigs (Yokoyama et al., 1993).

The effect of IgY is not always prevention of a disease but rather a reduction of infectious agents (Kuroki et al., 1993). Although IgY inhibits the adhesion of *E. coli* K88[+] MB to piglet mucus, prolonged incubation of IgY and *E. coli* cause no further reduction in the degree of *E. coli* adhesion to mucus. It was also shown that antibodies were not able to displace *E. coli* K88[+] MB once bound to the mucus of the small intestine. This indicates that the antibodies have a lower affinity for the receptor than that of K88[+] fimbriae. Since IgY could not remove previously bound *E. coli* K88[+] MB from the mucus, a prophylactic use for IgY is suggested (Jin et al., 1998). However, therapeutic effects of IgY in diarrhea of piglets caused by *E. coli* K88 have been reported (Yokoyama et al., 1992; Wiedemann et al., 1991). Actively acquired local anti-fimbrial immunity has been shown to be cross-protective against the other antigenic variants of fimbriae F18 (Sarrazin & Bertschinger, 1997), and Imberechts et al. (1995) has observed such cross-protection with a high dose of anti-fimbrial IgY.

While both IgY and whole egg yolk appear to have the same protective effect, various theoretical and practical concerns should be considered. Though IgY precipitation may be more stable for long periods of storage, it is more time-consuming and expensive to prepare. However, egg yolk precipitation carries the risk of transmission of *Salmonella* or other bacterial contamination and, in the developed world, the concern of cholesterol content (O'Farrelly et al., 1992).

VIII. SAFETY AND EFFICACY

Ikemori et al. (1997) showed that significant protection against bovine coronavirus (BCV) using specific IgY was achieved in calves treated with high titers of IgY and colostrum antibodies. Survival against BCV-induced mortality was 100% when 1:2,560 antibody titer was used for treatment of calves with the egg yolk powder. The calves treated with the IgY did not have severe diarrhea and had higher weight gains. About 8 g of egg yolk powder were obtained from one chicken egg and only one egg yolk was needed to protect one calf from diarrhea over a 7-day course of treatment. This resulted in an earlier protection trial using IgY specific for ETEC in ETEC-infected calves (Ikemori et al., 1992). It appears that greater amount of colostrum powder than egg yolk powder is necessary to prevent diarrhea. The difference in the minimal protective titers of the antibody powder between the egg yolk and the colostrum may have two possible explanations. Firstly, the avidity of antibodies derived from colostrum is lower than that of antibodies obtained from egg yolk (Ikemori et al., 1993). Compared to colostral antibodies, the BCV specific antibody from egg yolk may have reacted more strongly and stably with coronavirus epitopes *in vivo* in the neutralization reaction. Secondly, less chicken antibody may have been digested and inactivated by the gastric juice. The antibodies that escaped digestion in the stomach are still functional in the small intestine. The antibody of the egg yolk is considered to be of almost the same stability or slightly more susceptible to gastric juice than mammalian antibody (Shimizu et al., 1992). However, Ikemori et al. (1997) suggested that the antibody from egg yolk was more effective compared to colostral antibody in that yolk components in the egg yolk powder such as protein and fat may have protected the immunoglobulin fraction from digestive enzymes and allowed safe passage of IgY through the stomach enough to confer protection in the target areas of the small intestine of calves (Ikemori et al., 1997).

In the study of oral administration of IgY, pigs did not survive despite the high serum chicken IgG concentration detected. It can be assumed that most of the antibodies might have been transferred to the blood, with only a few antibodies remaining in the intestinal tract to fight the infection. To achieve protective antibody activity in the intestine, the supply of antibodies in the area must be sufficient to bind with the infective antigens. IgY is innocuous, and adverse effect was not observed in association with its use for treatment of infectious intestinal diseases in pigs (Yokoyama et al., 1993).

IX. SUMMARY

It has been long recognized that the hen, like its mammalian counterparts, provides young chicks with antibodies as protection against hostile invaders. This system facilitates the transfer of specific antibodies from serum to egg yolk, and provides a supply of antibodies called immunoglobulin Y (IgY) to the developing embryo and the hatched chick. The protection against pathogens that the relatively immuno-incompetent newly hatched chick has, is through transmission of antibodies from the mother via the egg. Egg yolk, therefore, can be loaded with a large amount of IgY against pathogens which can immobilize the existing or invading pathogens during the embryo development or in day-old chicks. Thus, the immunization of laying hens to various pathogens results in production of different antigen-specific IgY in eggs. Egg yolk contains 8-20 mg of immunoglobulins (IgY) per ml or 136 - 340 mg per yolk, suggesting that more than 30 g of IgY can be obtained from one immunized hen in a year. By immunizing laying hens with antigens and collecting IgY from egg yolk, low cost antibodies at less than $10 per g compared to more than $20,000 per g of mammalian IgG can be obtained.

This IgY technology opens new potential market applications in medicine, public health, veterinary medicine and food safety. A broader use of IgY technology could be applied as biological or diagnostic tool, nutraceutical or functional food development, oral-supplementation for prophylaxis, and as pathogen-specific antimicrobial agents for infectious disease control. This chapter has emphasized that when IgY-loaded chicken eggs are produced and consumed, the specific antibody binds, immobilizes and consequently reduces or inhibits the growth or colony forming abilities of microbial pathogens. This concept could serve as an alternative agent to replace the use of antibiotics, since today, more and more antibiotics are less effective in the treatment of infections, due to the emergence of drug-resistant bacteria.

X. REFERENCES

1. Akerstrom, G., Brodin, T., Reis, K., and Borg, L. 1985. Protein G: a powerful tool for binding and detection of monoclonal and polyclonal antibodies. *J. Immunol.* 135:2589-2592.
2. Akita, E.M., and Nakai, S. 1992. Immunoglobulins from egg yolk: isolation and purification. *J. Food Sci.* 57:629-634.
3. Akita, E.M., and Nakai, S. 1993. Comparison of four purification methods for the production of immunoglobulins from eggs laid by hens immunized with an enterotoxigenic *E.coli* strain. *J. Immunol. Methods.* 160:207-214.
4. Akita, E.M., Li-Chan, E.C.Y., and Nakai, S. 1998. Neutralization of enterotoxigenic *Escherichia coli* heat-labile toxin by chicken egg yolk immunoglobulin Y and its antigen-binding fragments. *Food Agri. Immunol.* 10:161-172.
5. Altschuh, D., Hennache, G., and Regenmortel, M.H.V. 1984. Determination of IgG and IgM levels in serum by rocket immunoelectrophoresis using yolk antibodies from immunized chickens. *J. Immunol. Methods.* 69:1-7.
6. Bade, H., and Stegemann, H. 1984. Rapid method of extraction of antibodies from hen egg yolk. *J. Immunol. Methods.* 72:421-426.
7. Bar-Joseph, M., and Malkinson, M. 1980. Hen egg yolk as a source of antiviral antibodies in the enzyme linked immunosorbent assay (ELISA): a comparison of two plant viruses. *J. Virol. Methods.* 1:179-183.

8.   Bauwens, R.M., Devos, M.P., Kint, J.A., and Leenheer, A.P. 1988. Chicken egg yolk and rabbit serum compared as sources of antibody for radioimmunoassay of 1,25-dihydroxyvitamin D in serum or plasma. *Clin. Chem*. 34:2153-2154.

9.   Bene, V.D., and Schmidt, M.G. 1997. Bacterial virulence factors. In *Microbiology and Infectious Diseases*. ed. Virella, G., pp.66-70. Williams & Wilkins.

10. Bianchi, A.T.J., Scholten, J.W., van Zijderveld, A.M., van Zijderveld F.G., and Bokhout, B.A. 1996. Parenteral vaccination of mice and piglets with F4⁺ *Escherichia coli* suppresses the enteric anti-F4 response upon oral infection. *Vaccine*. 14:199-206.

11. Bishop, R.F., Davidson, G.P., Holmes, I.H., and Ruck, B.J. 1973. Virus particles in epithelial cells of duodenal mucosa from children with acute nonbacterial gastroenteritis. *Lancet*. 2:1281-1283.

12. Bratthall, D., and Kohler, B. 1976. *Streptococcus mutans* serotypes: Some aspects of their identification, distribution, antigenic shifts and relationship to caries. *J. Dent. Res*. 55:C15-C21.

13. Brooks, G.F., Butel, J.S., and Morse, S.A. 1998. *Medical Microbiology*. Appleton & Lange.

14. Burger, D., Ramus, M.A., and Schapira, M. 1985. Antibodies to human plasma kallikrein from egg yolks of an immunized hen: Preparation and characterization. *Thromb. Res*. 40:283-288.

15. Burley, R.W., and Cook, W.H. 1961. Isolation and composition of avian yolk granules and their constituents alpha- and beta-lipovitellins. *Can. J. Biochem. Phyisiol*. 39:1295-1307.

16. Burton, D.R. 1987. *Molecular Genetics of Immunoglobulins*. ed. Calabi, F., and Neuberger, M.S., pp. 1-50. Elsevier Science Publishers.

17. Canadian Egg Marketing Agency. 1986. The Amazing Egg. Canadian Egg Marketing Agency, Ottawa, Ontario.

18. Carroll, S.B., and Strollar, B.D. 1983. Antibodies to calf thymus RNA polymerase II from egg yolks of immunized hens. *J. Biol. Chem*. 258:24-26.

19. Clark, M.A., Jepson, M.A., Simmons, N.L., and Hirst, B.H. 1994. Preferential interaction of *Salmonella typhimurium* with mouse Peyer's patch M cells. *Res. Microbiol*. 145:543-552.

20. Clarke, J.R., Marquardt, R.R., Frohlich, A.A., Oosterveld, A., and Madrid, F.J. 1994. Isolation, characterization, and application of hen egg yolk polyclonal antibodies. In *Egg Uses and Processing Technologies-New Developments*, ed. Sim, J.S., and Nakai, S., pp.207. CAB International, Oxon.

21. Curtis, J., and Bourne, F.J. 1971. Immunoglobulin quantitation in sow serum, colostrum, and milk and in the serum of young pigs. *Biochem. Biophys. Acta*. 236:319-332.

22. Curtis, J., and Bourne, F.J. 1973. Half-lives of immunoglobulins IgG, IgA, and IgM in the serum of newborn pigs. *Immunol*. 24:147-155.

23. DeMol, P., Zissis, G., Butzler, J.P., Mutwewingabo, A., and Andre, F.E. 1986. Failure of live, attenuated oral rotavirus vaccine. *Lancet*. 2:108.

24. Dressman, J.B., and Amidon, G.L. 1984. Radiotelemetric method for evaluating enteric coatings *in vivo*. *J. Pharm. Sci*. 73:935-938.

25. Ebina, T. 1996. Prophylaxis of rotavirus gastroenteritis using immunoglobulin. *Arch.Virol*.12: 217-223.

26. Eterradossi, N., Toquin, D., Abbassi, H., Rivallan, G., Cotte, J.P., and Guttet, M. 1997. Passive protection of specific pathogen free chicks against infectious bursal disease by *In-ovo* injection of semi-purified egg-yolk antiviral immunoglobulins. *J. Vet. Med*. B44:371-383.

27. Fellah, J.S., Kerfourn, F., Wiles, M.V., Schwager, J., and Charlemagne, J. 1993. Phylogeny of immunoglobulin heavy chain isotypes: structure of the constant region of *Ambystoma mexicanum* upsilon chain deduced from cDNA sequence. *Immunogenetics*. 38:311-317.

28. Fertel, R., Yetiv, J.Z., Coleman, M.A., Schwarz, R.D., Greenwald, J.E., and Bianchine, J.R. 1981. Formation of antibodies to prostaglandins in the yolk of chicken eggs. *Biochem. Biophys. Res. Commun*. 102:1028-1033.

29. Fichtali, J., Charter, E.A., Lo, K.V., and Nakai, S. 1993. Purification of antibodies from industrially separated egg yolk . *J. Food Sci*. 58:1282-1285, 1290.

30. Filler, S.J., Gregory, R.L., Michalck, S.M., Katz, J., and McGhee, J.R. 1991. Effect of immune bovine milk on *Streptococcus mutans* in human dental plaque. *Arch. Oral. Biol*. 36:41-47.

31. Gaastra, W., and de Graaf, F.D. 1982. Host-specific fimbrial adhesions of non-invasive enterotoxigenic *Escherichia coli* strains. *Microbiol. Rev*. 46:129-161.

32. Galdiero, F., Tufano, M.A., Galdiero, M., Masiello, S., and Rosa, M.D. 1990. Inflammatory effects of *Salmonella typhimurium* porins. *Infect. Immun*. 58:3183-3186.

33. Gardner, P.S., and Kaye, S. 1982. Egg globulins in rapid virus diagnosis. *J. Virol. Methods*. 4:257-262.

34. Gassmann, M., Thommes, P., Weiser, T., and Hubscher, U. 1990. Efficient production of chicken egg yolk antibodies against a conserved mammalian protein. *FASEB J*. 4:2528-2532.

35. Glass, R.I., Svennerholm, A.M., and Stoll, B.J. 1983. Protection against cholera in breast-fed children by antibodies in breast-milk. *N. Engl. J. Med*. 308:1389-1392.

36. Gottstein, B., and Hemmeler, E. 1985. Egg yolk immunoglobulin Y as an alternative antibody in the serology of echinococcosis. *Z. Parasitenkid.* 71:273-276.

37. Greene, C.R., and Holt, P.S. 1997. An improved chromatographic method for the separation of egg yolk IgG into subpopulations utilizing immobilized metal ion (Fe^{3+}) affinity chromatography. *J. Immunol. Methods.* 209:155-164.

38. Guinee, P.A.M., Jansen, W.H., and Agterberg, C.M. 1976. Detection of the K99 antigen by means of agglutination and immunoelectrophoresis in *Escherichia coli* isolates from calves and its correlation with enterotoxigenicity. *Infect. Immun.* 13:1369-1377.

39. Gutierrez, J.A., and Guerriero, V. 1991. Quantitation of Hsp 70 in tissues using a competitive enzyme-linked immunosorbent assay. *J. Immunol. Methods.* 143:81-88.

40. Hamada, S., Horikoshi, T., Minami, T., Kawabata, S., Hiraoka, J., Fujiwara, T., and Ooshima, T. 1991. Oral passive immunization against dental caries in rats by use of hen egg yolk antibodies specific for cell-associated glucosyltransferase of *Streptococcus mutans*. *Infect. Immun.* 59:4161-4167.

41. Hampson, D.J. 1994. Postweaning *Escherichia coli* diarrhea in pigs. In *Escherichia coli in Domestic Animals and Humans,* ed. Gyles, C.L., pp.171-191. CAB International.

42. Hassl, A., and Aspock, H. 1988. Purification of egg yolk immunoglobulins: a two-step procedure using hydrophobic interaction chromatography and gel filtration. *J. Immunol. Methods.* 110:225-228.

43. Hatta, H., Kim, M., and Yamamoto, T. 1990. A novel isolation method for hen egg yolk antibody "IgY". *Agri. Biol. Chem.* 54:2531-2535.

44. Hatta, H., Mabe, K., Kim, M., Yamamoto, T., Gutierrez, M.A., and Miyazaki, T. 1994. Prevention of fish disease using egg yolk antibody. In *Egg Uses and Processing Technologies-New Developments*, ed. Sim, J.S., and Nakai, S., pp.241. CAB International.

45. Hatta, H., Ozeki, M., and Tsuda, K. 1997a. Egg yolk antibody IgY and its application. In *Hen Eggs: Their Basic and Applied Science*, ed. Yamamoto, T., Juneja, L.R., Hatta, H., and Kim, M., pp.151-178. Boca Raton: CRC Press.

46. Hatta, H., Sim, J. S., and Nakai, S. 1988. Separation of phospholipids from egg yolk and recovery of water-soluble proteins. *J. Food Sci.* 53:425-427, 431.

47. Hatta, H., Tsuda, K., Akachi, S., Kim, M., Yamamoto, T., and Ebina, T. 1993. Oral passive immunization effect of anti-human rotavirus IgY and its behavior against proteolytic enzymes. *Biosci. Biotech. Biochem.* 57:1077-1081.

48. Hatta, H., Tsuda, K., Ozeki, M., Kim, M., Yamamoto, T., Otake, S., Hirasawa, M., Katz, J., Katz, J., Childers, N.K., and Michalek, S.M. 1997b. Passive immunization against dental plaque formation in humans: Effect of a mouth rinse containing egg yolk antibodies (IgY) specific to *Streptococcus mutans*. *Caries Res.* 31:268-274.

49. Higgins, D.A. 1975. Physical and chemical properties of fowl immunoglobulins. *Vet. Bull.* 45:139-154.

50. Hirabayashi, Y., Suzuki, T., Suzuki, Y., Taki, T., Matumoto, M., Higashi, H., and Kato, S. 1983. A new method for purification of antiglycospingolipid antibody. Avian antihematoside (NeuGc) antibody. *J. Biochem.* 94:327-330.

51. Holmgren, J. 1985. Toxins affecting intestinal transport processes. In *The Virulence of Escherichia coli,* ed. Sussman, M., pp. 177-192. London: Academic Press Inc.

52. Hornbeck, P. 1991. Assays for antibody production. In *Current Protocols in Immunology.* ed. Coligan, J.E., Krisbeek, A.M., Shevach, E.M., and Strober, W. John Wiley & Songs, Inc.

53. Ikemori, Y., Ohta, M., Umeda, K., Icatlo, F.C., Kuroki, M., Yokoyama, H., and Kodama, Y. 1997. Passive protection of neonatal calves against bovine coronavirus-induced diarrhea by administration of egg yolk or colostrum antibody powder. *Vet. Microbiol.* 58:105-111.

54. Ikemori, Y., Peralta, R.C., Kuroki, M., Yokoyama, H., and Kodama, Y. 1993. Research note: avidity of chicken yolk antibodies to enterotoxigenic *Escherichia coli* fimbriae. *Poultry Sci.* 72:2361-2365.

55. Ikemori,Y., Kuroki, M., Peralta,R.C., Yokoyama, H., and Kodama, Y. 1992. Protection of neonatal calves against fatal enteric colibacillosis by administration of egg yolk powder from hens immunized with K99-piliated enterotoxigenic *Escherichia coli*. *Am. J. Vet. Res.* 53:2005-2008.

56. Imberechts, H., Deprez, P., Van Driessche, E., and Pohl, P. 1997. Chicken egg yolk antibodies against F18ab fimbriae of *Escherichia coli* inhibit shedding of F18 positive *E.coli* by experimentally infected pigs. *Vet. Microbiol.* 54:329-341.

57. Imberechts, H., Pohl, P., and Deprez, P. 1995. Passive protection of weaned pigs against experimental infection with F18 positive *E. coli*. First International Rushmore Conference on Mechanisms in the Pathogenesis of Enteric Diseases. Abstract 39. Rapid City, SD.

58. Isaacson, R.E., Moon, H.W., and Schneider, R.A. 1978. Distribution and virulence of *Escherichia coli* in the small intestines of calves with and without diarrhea. *Am. J. Vet. Res.* 39:1750-1755.

59. Jason, J.M., Nieburg, P., and Marks, J.S. 1984. Mortality and infectious disease associated with infant-feeding practices in developing countries. *Pediatrics.* 74:702-727.

60. Jensenius, J.C., Andersen, I., Hau, J., Crone, M., and Koch, C. 1981. Eggs: conveniently packaged antibodies. Methods for purification of yolk IgG. *J. Immunol. Methods.* 46:63-68.

61. Jin, L.Z., Baidoo, S.K., Marquardt, R.R., and Frohlich, A.A. 1998. *In vitro* inhibition of adhesion of enterotoxigenic *Escherichia coli* K88 to piglet intestinal mucus by egg-yolk antibodies. *FEMS Immunol. Med. Microbiol.* 21:313-321.

62. Jones, B.D., Ghori, N., and Falkow, S. 1994. *Salmonella typhimurium* initiates murine infection by penetrating and destroying the specialized epithelial M cells of the Peyer's patches. *J. Exp. Med.* 180:15-23.

63. Jungling, A., Wiedemann, V., Kuhlmann, R., Erhard, M., Schmidt, P., and Losch, U. 1991. Chicken egg antibodies for prophylaxis and therapy of infectious intestinal diseases IV. *In vitro* studies on protective effects against adhesion of enterotoxigenic *E. coli* to isolated enterocytes. *J. Vet. Med.* B38:373-381.

64. Kapil, S., Trent, A.M., and Goyal, S.M. 1990. Excretion and persistence of bovine coronavirus in neonatal calves. *Arch. Virol.* 115:127-132.

65. Kobayashi, K., and Hirai, H. 1980. Studies on subunit components of chicken polymeric immunoglobulins. *J. Immunol.* 124:1695-1704.

66. Kohbata, S., Yokoyama, H., and Yabuuchi, E. 1986. Cytopathogenic effect of *Salmonella typhi* GIFU 10007 on M cells of murine Peyer's patches in ligated ileal loops: and ultrastructural study. *Microbiol. Immunol.* 30:1225-1237.

67. Kuby, J. 1997. *Immunology*, 3rd ed. W.H. Freeman.

68. Kuroki, M., Ikemori, Y., Yokoyama, H., Peralta, R.C., Icatlo, F.C., and Kodama, Y. 1993. Passive protection against bovine rotavirus-induced diarrhea in murine model by specific immunoglobulins from chicken egg yolk. *Vet. Microbiol.* 37:135-146.

69. Kuroki, M., Ohta, M., Ikemori, Y., Icatlo, F.C., Kobayashi, C., Yokoyama, H., and Kodama, Y. 1997. Field evaluation of chicken egg yolk immunoglobulins specific for bovine rotavirus in neonatal calves. *Arch. Virol.* 142:843-851.

70. Kuroki, M., Ohta, M., Ikemori, Y., Peralta, R.C., Yokoyama, and H., Kodama, Y. 1994. Passive protection against bovine rotavirus in calves by specific immunoglobulins from chicken egg yolk. *Arch. Virol.* 138:143-148.

71. Kwan, L., Li-Chan, E., Helbig, N., and Nakai, S. 1991. Fractionation of water-soluble and –insoluble components from egg yolk with minimum use of organic solvents. *J. Food Sci.* 56:1537-1541.

72. Lehner, T., Caldwell, J., and Smith, R. 1985. Local passive immunization by monoclonal antibodies against streptococcal antigen I/II in the prevention of dental caries. *Infect. Immun.* 50:796-799.

73. Lee, K., Ametani, A., Shimizu, M., Hatta, H., Yamamoto, T., and Kaminogawa, S. 1991. Production and characterization of anti-human insulin antibodies in the hen's egg. *Agric. Biol. Chem.* 55:2141-2143.

74. Leslie, G.A., and Clem, L.W. 1969. Phylogeny of immunoglobulin structure and function. III. Immunoglobulins of the chicken. *J. Exp. Med.* 130:1337-1352.

75. Leslie, G.A., and Martin, L.N. 1973. Studies on the secretory immunologic system of fowl. III. Serum and secretory IgA of the chicken. *J. Immunol.* 110:1-9.

76. Li, X., Nakano, T., Chae, H.S., Sunwoo, H.H., and Sim, J.S. 1998a. Production of chicken egg yolk antibody (IgY) against bovine proteoglycan. *Can. J. Anim. Sci.* 78:287-291.

77. Li, X., Nakano, T., Sunwoo, H.H., Paek, B.H., Chae, H.S., and Sim, J.S. 1998b. Effects of egg and yolk weights on yolk antibody (IgY) production in laying chickens. *Poultry Sci.* 77:266-270.

78. Locken, M.R., and Roth, R.E. 1983. Analysis of maternal IgG subpopulations which are transported into the chicken oocyte. *Immunol.* 49:21-28.

79. Loesche, W.J., Rowan, J., Straffon, L.H., and Loos, P.J. 1975. Association of *Streptococcus mutans* with human dental decay. *Infect. Immun.* 11:1252-1260.

80. Losch, Y., Schranner, I., Wanke, R., and Jurgens, L. 1986. The chicken egg, an antibody source. *J. Vet. Med.* B33:609-619.

81. Ma, J.K.C., Smith, R., and Lehner, T. 1987. Use of monoclonal antibodies in local passive immunization to prevent colonization of human teeth by *Streptococcus mutans*. *Infect. Immun.* 55:1274-1278.

82. Magor, K.E., Higgins, D.A., Middleton, D., and Warr, G.W. 1994. One gene encodes the heavy chains for three different forms of IgY in the duck. *J. Immunol.* 153:5549-5555.

83. McBee, L.E., and Cotterill, O.J. 1979. Ion-exchange chromatography and electrophoresis of egg yolk proteins. *J. Food Sci.* 44:656-660.

84. McCannel, A.A., and Nakai, S. 1989. Isolation of egg yolk immunoglobulin-rich fractions using copper-loaded metal chelate interaction chromatography. *Can. Inst. Food Sci. Techonol. J.* 22:487-490.

85. McCannel, A.A., and Nakai, S. 1990. Separation of egg yolk immunoglobulins into subpopulations using DEAE-ion exchange chromatography. *Can. Inst. Food Sci. Techonol. J.* 23:42-46.

86. McCully, K.A., Mok, C.C., and Common, R.H. 1962. Paper electrophoretic characterization of proteins and lipoproteins of hen's egg yolk. *Can. J. Biochem. Physiol.* 40:937-952.
87. Moon, H.W. 1978. Mechanism in the pathogenesis of diarrhea: a review. *J. Am. Vet. Med. Assoc.* 172:443-448.
88. Moon, H.W. 1981. Protection against enteric colibacillosis in pigs suckling orally vaccinated dams: Evidence for pili as protective antigens. *Am. J. Vet. Res.* 42:173-177.
89. Moon, H.W., and Bunn, T.O. 1993. Vaccines for preventing enterotoxigenic *Escherichia coli* infections in farm animals. *Vaccine.* 11:213-220.
90. Moon, H.W., Whipp, S.C., and Skartvcdt, S.M. 1976. Etiologic diagnosis of diarrheal diseases of calves: frequency and methods for detecting enterotoxin and K99 antigen production by *Escherichia coli*. *Am. J. Vet. Res.* 37:1025-1029.
91. Morishita, E., Narita, H., Nishida, M., Kawashima, N., Yamagishi, K., Masuda, S., Nagao, M., Hatta, H., and Sasaki, R. 1996. Anti-Erythroprotein receptor monoclonal antibody: Epitope mapping, quantification of the soluble receptor, and detection of the soublized transmembrane receptor and the receptor-expressing cells. *Blood* 88:465-471.
92. Myers, L.L., and Guinee, P.A.M. 1976. Occurrence and characteristics of enterotoxigenic *Escherichia coli* isolated from calves with diarrhea. *Infect. Immun.* 13:1117-1119.
93. Nakai, S. 1962. Study on digestion of milk protein and its improvement, PhD thesis, University of Tokyo.
94. Ntakarutimana, V., Demedts, P., Sande, M.V., and Scharpe, S. 1992. A simple and economical strategy for downstream processing of specific antibodies to human transferrin from egg yolk. *J. Immunol. Methods.* 153:133-140.
95. O'Farrelly, C., Branton, D., and Wanke, C.A. 1992. Oral ingestion of egg yolk immunoglobulin from hens immunized with an enterotoxigenic *Escherichia coli* strain prevents diarrhea in rabbits challenged with the same strain. *Infect. Immun.* 60:2593-2597.
96. Ohta, M., Hamako, J., Yamamoto, S., Hatta, H., Kim, M., Yamamoto, T., Oka, S., Mizuochi, T., and Matuura, F. 1991. Structures of asparagine-linked oligosaccharides from hen egg-yolk antibody (IgY). Occurrence of unusual glucosylated oligomannose type oligosaccharides in a mature glycoprotein. *Glycoconjugate J.* 8:400-413.
97. Opitz, H.M., El-begearmi, M., Flegg, P., and Beane, D. 1993. Effectiveness of five feed additives in chicks infected with *Salmonella enteritidis* phage type 13A. *J. Appl. Poultry Res.* 2:147-153.
98. Otake, S., Nishihara, Y., Makimura, M., Hatta, H., Kim, M., Yamamoto, T., and Hirasawa, M. 1991. Protection of rats against caries by passive immunization with hen-egg-yolk antibody (IgY). *J. Dent. Res.* 70:162-166.
99. Otani, H., Matsumoto, K., Saeki, A., and Hosono, A. 1991. Comparative studies on properties of hen egg yolk IgY and rabbit serum IgG antibodies. *Lebensm. Wiss. Techonol.* 24:152-158.
100. Ozpinar, H., Erhard, M.H., Aytug, N., Ozpinar, A., Baklaci, C., Karamuptuoglu, S., Hofmann, A., and Losch, U. 1996. Dose-dependent effects of specific egg-yolk antibodies on diarrhea of newborn calves. *Preventive Vet. Med.* 27:67-73.
101. Parry, S.H., and Rooke, D.M. 1985. Adhesins and colonization factors of *Escherichia coli*. In *The Virulence of Escherichia coli,* ed. Sussman, M., pp. 79-156. London: Academic Press Inc.
102. Parvari, R., Avivi, A., Lentner, F., Ziv, E., Tel-Or, S., Burstein, Y., and Schechter, I. 1988. Chicken immunoglobulin γ-heavy chains: Limited V_H gene repertoire, combinatorial diversification by D gene segments and evolution of the heavy chain locus. *EMBO J.* 7:739-744.
103. Patterson, R., Younger, J.S., Weigle, W.O., and Dixon, F.J. 1962. The metabolism of serum proteins in the hen and chick and secretions of serum proteins by the ovary of the hen. *J. Physiol.* 45:501-513.
104. Peralta, R.C., Yokoyama, H., Ikemori, Y., Kuroki, M., and Kodama, Y. 1994. Passive immunization against experimental salmonellosis in mice by orally administered hen egg yolk antibodies specific for 14-kDa fimbriae of *Salmonella enteritidis*. *J. Med. Microbiol.* 41:29-35.
105. Pilz, I., Schwarz, E., and Palm, W. 1977. Small-angle X-ray studies of the human immunoglobulin molecule. *Eur. J. Biochem.* 75:195-199.
106. Polson, A., and Von Wechmar, M.B. 1980. Isolation of viral IgY antibodies from yolks of immunized hens. *Immunol. Comm.* 9:476-493.
107. Polson, A., Coetzer, T., Kruger, J., Von Maltzahn, E., and Vac der Merwe, K. J. 1985. Improvements in the isolation of IgY from the yolks of eggs laid by immunized hens. *Immunol. Invest.* 14:323-327.
108. Polson, A., Maass, R., and Van Der Merwe, K.J. 1989. Eliciting antibodies in chickens to human dimeric IgA removal of factors from human colostrum depressing anti-IgA antidbody production. *Immunol. Invest.* 18:853-877.
109. Rose, M.E., and Orlans, E. 1981. Immunoglobulins in the egg, embryo and young chick. *Devel. Comp. Immunol.* 5:15-20, 371-375.

110. Rose, M.E., Orlans, E., and Buttress, N. 1974. Immunoglobulin classes in the hen's egg: their segregation in yolk and white. *Eur. J. Immunol.* 4:521-523.

111. Rosol, T.J., Steinmeyer, C.L., McCauley, L.K., Merryman, J.I. Werkmeister, J.R., Grone, A., Weekmann, M.T., Swayne, D.E., and Capen, C.C. 1993. Studeies on chicken polyclonal anti-peptide antibodies specific for parathyroid hormone related protein (1-36). *Vet. Immunol. Immunopathol.* 35:321-337.

112. Rutter, I.W., and Jones, G.W. 1973. Protection against enteric disease caused by *Escherichia coli*- a model for vaccination with a virulence determinant. *Nature.* 242:531-532.

113. Saeed, A.M.K., Sriranganathan, N., and Cosand, W. 1983. Purification and characterization of heat stable enterotoxin from bovine enterotoxigenic *Escherichia coli. Infect. Immun.* 40:781-787.

114. Sarrazin, E., and Bertschinger, H.U. 1997. Role of fimbriae F18 for actively acquired immunity against porcine enterotoxigenic *Escherichia coli. Vet. Microbiol.* 54:133-144.

115. Schade, R., Staak, C., Hendriksen, C., Erhard, M., Hugl, H., Koch, G., Larsson, A., Pollmann, W., Regenmortel, M., Rike, E., Spielmann, H., Steinbusch, H., and Straughan, D. 1996. The production of avian (egg yolk) antibodies : IgY. *ALTA.* 24:925-934.

116. Shimizu, M., Fitzsimmons, R.C., and Nakai, S. 1988. Anti-*E.coli* immunoglobulin Y isolated from egg yolk of immunized chickens as a potential food ingredient. *J. Food Sci.* 53:1360-1366.

117. Shimizu, M., Nagashima, H., and Hashimoto, K. 1993. Comparative studies on molecular stability of immunoglobulin G from different species. *Comp. Biochem. Physiol.* 106B:255-261.

118. Shimizu, M., Nagashima, H., Sano, K., Hashimoto, K., Ozeki, M., Tsuda, K., and Hatta, H. 1992. Molecular stability of chicken and rabbit immunoglobulin G. *Biosci. Biotech. Biochem.* 56:270-274.

119. Shimizu, M., Watanabe, A., and Tanaka, A. 1995. Detection of high-molecular weight mucin-like glycoprotein-A (HMGP-A) of human milk by chicken egg yolk antibody. *Biosci. Biotech. Biochem.* 59:138-139.

120. Shmidt, P., Wanke, R., Linckh, E., Wiedemann, V., Kuhlmann, R., and Losch, U. 1989. Chicken egg antibodies for prophylaxis and therapy of infectious intestinal diseases. II. *In vitro* studies on gastric and enteric digestion of egg yolk antibodies specific against pathogenic *Escherichia coli* strains. *J. Vet. Med.* B36:619-629.

121. Stadeklman, W. J., and Cotterill, O.J. 1977. *Egg Science and Technology*, 2ⁿᵈ ed. pp. 65-91. AVI Pub. Co. Westport, CT.

122. Stedman, R.A., Singleton, L., and Box, P.G. 1969. Purification of Newcastle disease virus antibody from the egg yolk of the hen. *J. Comp. Path.* 79:507-516.

123. Stiehm, E.R., and Fudenberg, H.H. 1966. Serum levels of immune globulins in health and disease: a survey. *Pediatrics.* 37:715-720.

124. Sturmer, A.M., Driscoll, D.P., and Jackson-Matthews, D.E. 1992. A quantitative immunoassays using chicken antibodies for detection of native and recombinant α-amidating enzyme. *J. Immunol. Methods.* 146:105-110.

125. Sugita-konishi, Y., Shibata, K., Yun, S.S., Hara-kudo, Y., Yamaguchi, K., and Kumagai, S. 1996. Immune functions of immunoglobulin Y isolated from egg yolk of hens immunized with various infectious bacteria. *Biosci. Biotech. Biochem.* 60:886-888.

126. Sunwoo, H.H., Li, X., Lee, E.N., and Sim, J.S. 1999. Preparation of antigen-specific IgY for food application. In *Egg Nutrition and Biotechnology*, ed. Sim, J.S., and Guenter, W., pp.311-322. CAB International 2000.

127. Sunwoo, H.H., Nakano, T., Dixon, W.T., and Sim, J.S. 1996. Immune responses in chickens against lipopolysaccharide of *Escherichia coli* and *Salmonella typhimurium. Poultry Sci.* 75:342-345.

128. Takada, K., Ohhashi, M., Furuya, Y., Yoshikawa, H., and Muranishi, S. 1989. Enteric solid dispersion of ciclosporin A (CiA) having potential to deliver CiA into lymphatics. *Chem. Pharm. Bull.* 37:471-474.

129. Tomasi, T.B., and McNabb, P.C. 1987. The secretory immune system. In *Basic and Clinical Immunology*, ed. Stobo, J.D., and Wells, J.V. Norwalk, Conn.: Appleton & Lange.

130. Toti, F., Gachet, C., Ohlamann, P., Stierle, A., Grunebaum, L., Wiesel, M.L., and Cazenave, J.P. 1992. Electrophoretic studies on molecular defects of von Willebrand factor and platelet glycoprotein Iib-IIIa with antibodies produced in egg yolk from laying hens. *Haemostasis.* 22:32-40.

131. Tsunemitsu, H., Shimizu, M., Hirai, T., Yonemichi, H., Kudo, T., Mori, K., and Onoe, S. 1989. Protection against bovine rotaviruses in newborn calves by continuous feeding of immune colostrum. *Jap. J. Vet. Sci.* 51:300-308.

132. Vieira, J.G.H., Oliveira, M.A.D., Russo, E.M.K., Maciel, R.M.B., and Pereira, A.B. 1984. Egg yolk as a source of antibodies for human parathyroid hormone (hPTH) radioimmunoassay. *J. Immunoassay.* 5:121-129.

133. Virella, G., and Schmidt, M.G. 1997. Gram-negative rods II: *Enterobacteriaceae* and other enteropathogenic Gram-negative rods. In *Microbiology and Infectious Diseases,* ed. Virella, G., pp.141-157. Williams & Wilkins.

134. Wanke, C.A., Cronan, S., Goss, C., Chadee, K., and Guerrant, R.L. 1990. Characterization of binding of *Escherichia coli* strains which are enteropathogens to small-bowel mucin. *Infect. Immun.* 58:794-800.
135. Welsh, J.K., and May, J.T. 1979. Anti-infective properties of breast-milk. *J. Pediatr.* 94:1-9.
136. Wiedemann, V., Linckh, E., Kuhlmann, R., Schmidt, P., and Losch, U. 1991. Chicken egg antibodies for prophylaxis and therapy of infectious intestinal diseases. *J. Vet. Med.* B38:283-291.
137. Williams, R.C., and Gibbons, R.J. 1972. Inhibition of bacterial adherence by secretory immunoglobulin A: a mechanism of antigen disposal. *Science.* 177:697-699.
138. Yokoyama, H., Peralta, R.C., Diaz, R., Sendo, S., Ikemori, Y., and Kodama, Y. 1992. Passive protective effect of chicken egg yolk immunoglobulins against experimental enterotoxigenic *Escherichia coli. Infect. Immun.* 60: 998-1007.
139. Yokoyama, H., Peralta, R.C., Umeda, K., Hashi, T., Icatlo, F.C., Kuroki, M., Ikemori, Y., Kodama, Y. 1998b. Prevention of fatal salmonellosis in neonatal calves, using orally administered chicken egg yolk *Salmonella*-specific antibodies. *Am. J. Vet. Res.* 59:4, 416-420.
140. Yokoyama, H., Peralta., R.C., Sendo, S., Ikemori, Y., and Kodama, Y. 1993. Detection of passage and absorption of chicken egg yolk immunoglobulins in the gastrointestinal tract of pigs by use of enzyme-linked immunosorbent assay and fluorescent antibody testing. *Am. J. Vet. Res.* 54:867-872.
141. Yokoyama, H., Umeda, K., Peralta, R.C., Hashi, T., Icatlo, F.C., Kuroki, M., Ikemori, Y., and Kodama, Y. 1998a. Oral passive immunization against experimental salmonellosis in mice using chicken egg yolk antibodies specific for *Salmonella enteritidis* and *S. typhimurium. Vaccine.*16:388-393.
142. Yolken, R.H., Leister, F., Wee, S.B., Miskuff, R., and Vonderfecht, S. 1988. Antibodies to rotaviruses in chickens'eggs: A potential source of antiviral immunoglobulins suitable for human consumption. *Pediatrics.* 81:291-295.
143. Yoshiko, S.K., Shibata, K., Yun., S.S., Yukiko, H.K., Yamaguchi, K., and Kumagai, S. 1996. Immune functions of immunoglobulin Y isolated from egg yolk of hens immunized with various infectious bacteria. *Biosci. Biotech. Biochem.* 60:886-888.
144. Yoshimura, S., Watanabe, K., Suemizu, H., Onozawa, T., Mizoguchi, J., Tsuda, K., Hatta, H., and Moriuchi, T. 1991. Tissue specific expression of the plasma glutathione peroxidase gene in rat kidney. *J. Biochem.* 109:918-923.
145. Zuniga, A., Yokiayama, H., Albicker-Rippinger, P., Eggenberger, E., and Bertschinger, H.U. 1997. Reduced intestinal colonisation with F18-positive enterotoxigenic *Escherichia coli* in weaned pigs fed chicken egg antibody against the fimbriae. *FEMS Immunol. Med. Microbiol.* 18:153-161.

Y. Mine

Avidin

9

I. INTRODUCTION

The early history of avidin studies is closely associated with the discovery of biotin. Hen egg avidin was recognized as a biological factor in egg white in the late 1920s during the discovery and isolation of the vitamin, biotin (Boas, 1924; Eakin et al., 1941). Avidin is a glycosylated and positively charged protein which can bind up to four biotin and form stable complexes (Green, 1964). Avidin activity was found in various avian eggs and egg jelly of invertebrates but could not be found in many other tissues or organisms. This protein is a trace component (0.05%) of egg white, but it is well studied due to its ability to bind to biotin (Green, 1975). Its function in the egg was presumed to be that of an antibiotic along with several other proteins of similar functions, but very different mechanism with no general biological function, and no apparent clinical use as an antibiotic. Avidin was essentially ignored for about two decades until the discovery of the coenzyme function of covalently bound biotin and detailed study of avidin was conducted by several goups (Green et al., 1975). Chaiet et al. (1964) reported a streptavidin, avidin like activity from various species of streptomyces in the series of antibiotic materials produced by these bacteria, while similar to egg avidin in many ways. In 1970s, avidin has been viewed as an antimicrobial factor in egg albumen. The discovery of streptavidin confirmed this; however, biological function of avidin has not been clearly established yet in spite of numerous applications of avidin and development of avidin-biotin technology.

II. MOLECULAR PROPERTIES

A. Biosynthesis

Avidin is a minor protein in the avian albumen. Avidin is synthesized in the goblet cells of the epithelium of the oviduct in hens and this synthesis is specifically stimulated by progesterone. Avidin production is found in the oviducts of laying hens, but not in those of non-laying adult hens or immature chickens, which indicates a regulation by ovarian function. Avidin production for albumen formation seems to be regulated by several steroids and the progesterone probably acts by control of transcription of the avidin mRNA (Elo & Korpela, 1984).

B. Isolation and purification

Avidin is found in the egg white at a maximum concentration of about 0.05% of the total proteins. Due to the low content and positive charge of avidin, the isolation was complicated by the formation of slightly soluble complexes between the basic avidin and components of egg white. The earliest purification methods of Eakin et al (1941) and Dhyse (1954) made use of selective solubilization of avidin by dilute salt from alcohol-precipitated egg proteins and this yielded about 50% of active avidin (7 units/mg). Fraenkel-Conrat (1952) developed an alternative approach in which egg proteins were adsorbed on bentonite. Avidin was eluted with 1M dipotassium phosphate and purified by ammonium sulfate fractionation to a specific activity of 8-10 units/mg. Subsequently, the availability of cellulose ion exchange led to the development of an improved procedure based on adsorption of the basic proteins on CM-cellulose at high pH, followed by elution of avidin with ammonium carbonate. Melamed and Green (1963) acheived a yield of 93% active using a preliminary batch adsorption on carboxymethylcellulose (CMC) and Amberlite (G-50) ion exchange resin which was later crystallized at pH 5.0 from ammonium sulfate (3M) or potassium phosphate (3M) (Green & Toms, 1970). Affinity chromatographic methods have also been suggested. Cautrecasa and Wilchek (1968) bound avidin to biocytin sepharose column and eluted with 6M guanidine-HCl at pH 1.5. Iminobiotin, which binds avidin at pH 11 but not at pH 4, was also employed as an affinity ligand for avidin. With this approach, yield of 95% and 99% avidin purity was reported (Heney & Orr, 1981). Durance and Nakai (1988a: 1988b) reported a single column cation exchange method that allowed simultaneous recovery of lysozyme and avidin from undiluted egg white. Avidin recovery was better than that of ion exchange methods (74-80%) and the purity was up to 40.9%.

C. Physico-chemistry

Avidin is a basic tetrameric glycoprotein with isoelectric point at pH 10.0 (Woolley & Longsworth, 1942). The protein is composed of four identical subunits with a molecular weight reported between 15 kDa to 15.8 kDa, giving a total molecular weight of 66 - 69 kDa for avidin (Kopela, 1984a). The amino acid composition of avidin is shown in *TABLE 1*. There are 0-2 residues each of histidine, proline, cysteine and methionine, and the contents of asparatate and threonine are very high. The complete primary sequence established by DeLange and Huang (1971) is shown in *TABLE 1*. The amino acid sequence of the 128 amino acids residues in each subunit indicates a single intramolecular disulfide bond between *Cys-4* and *Cys-83*. Heterogenecity of the polypeptides chain has been indicated by presence of either isoleucine or threonine at residue 34. The carbohydrate moiety, which comprises about 10% of the molecular weight of avidin, consists of a single oligosaccharide chain with four or five mannose and three N-acetylglucosamine residues per subunit linked to *Asp-17* of the polypeptide chain (Bruce & White, 1982). Avidin is not inactivated by treatment with iodine at neutral pH, nor by acetylation of 60% of the amino groups, nor by esterification of 20% of the carboxyl groups. Significant (>70%) inactivation resulted from oxidation with H_2O_2 in the presence of Fe^{2+} and from treatment

TABLE 1. Amino acid composition and sequence of avidin from hen egg white.

Amino acid	Residues/subunit	Sequence
Lysine	9	
Histidine	1	
Arginine	8	5 10 15 20 25
Aspartate	15	**ARKCSLTGKWTNDLGSNMTIGAVNS**
Threonine	19	
Serine	9	
Glutamine	10	30 35 40 45 50
Proline	2	**RGEFTGTYITAVTATSNEIKEESPLH**
Glycine	11	
Alanine	5	
Cysteine	2	55 60 65 70 75
Valine	7	**GTENTINKRTQPTFGFTVNWKFSESTT**
Methionine	2	
Isoleucine	8	
Leucine	7	80 85 90 95 100 105
Tyrosine	1	**VFTGQCFIDRNGKEVLKTMWLLRSSVNDI**
Phenylalanine	7	
Tryptophan	4	
		110 115 120 125
TOTAL	128	**GDDWKATRVGINIFTRLRTQKE**
Amide	16	
Mannose	4	
Glucosamine	3	
Subunit weight	15,600	
Molecular weight	67,000	

The amino acid composition and the molecular weight is based on the sequence (DeLange & Huang, 1971).

with formaldehyde in the presence of alanine or with hydroxylamine at 50°C (Fraenkel-Conrat, 1952). All four tryptophans are rapidly oxidized at pH 4.0 and biotin-binding activity is lost when an average of the two had been destroyed (Green, 1963). The involvement of a tryptophan residue and an amino group at the biotin-binding site of avidin has been investigated in relation to structure of avidin, biotin and biotin analogs which contribute to the highly specific and tight binding of biotin to avidin (Green, 1975). One tryptophan could be destroyed with loss of only 25% of activity, but all activity disappeared when a second residue was oxidized. Furthermore, when half the activity was lost about half of tryptophans 10 and 70 were destroyed, *Trp 97* was intact, and the evidence on *Trp* 110 was inconclusive. Either or both *Trp 10* and *Trp 70* are essential to the binding (Huang & Delange, 1971).

Raman spectroscopic analysis has indicated that avidin has 10% of α-helix and 55% of β-strands (Honzatko & Williams, 1982). This was consistent with the low CD of avidin below 215 nm (Green, 1975). Crystallization of avidin has been carried out by several groups (Green & Toms, 1970; Pinn et al., 1982; Gatti et al., 1984). These data indicated that each molecule of avidin (M_r 66-kDa) contained four identical subunits. The three dimensional structure of avidin, when refined to a crystallographic R-factor of 0.164, indicated that each protomer is organized in an eight-stranded antiparallel orthogonal β-barrel with extended loop regions that define the biotin binding pocket in the promoter core (Pugliese et al., 1994) (*FIGURE 1*).

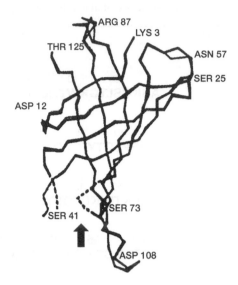

FIGURE 1. Stereo view of avidin subunit with defined biotin cavity. The biotin-binding site is the central cavity enclosed by the 8 β-strands marked by an arrow in the figure [reproduced from Pugliese et al., (1994) with permission from the Academic Press].

D. Biotin-avidin binding properties

Avidin binds four biotin molecules one per subunit (Green, 1964) and the biotin-avidin interact is one of the strongest noncovalent protein-ligand interactions found in nature, with a dissociation constant K_a of about $10^{15}M^{-1}$ (Green, 1963). Avidin is remarkably stable over a wide range of pH and temperature with regard to its biotin-binding ability, however, its stability is decreased at low ionic strength. The subunit of avidin is dissociated in 0.1 M HCl, 0.1 M sodium dodecyl sulfate or 6 M guanidine HCl with complete loss of biotin-binding ability (Green, 1963). The avidin-biotin complex is resistant to denaturation and proteolytic enzymes. It was shown that avidin was irreversibly denatured at temperature higher than 70°C, but that the complex was stable at 100°C. In the absence of biotin, the binding site is only partly occupied by water molecule. The structure of the binding site residues is highly complementary to that of the incoming biotin molecule, accounting for prompt and specific recognition (Gitin et al., 1988; Pugliese et al., 1994). The crystal structure of the holo-avidin complex has shown that biotin sits in a deep protein-core pocket displaying polar residues at its dead-end recognizing the ureido group of the biotin, while the remaining part of the pocket is essentially hydrophobic (Pugliese et al., 1993). The high specific and affinity of biotin binding of avidin has many uses in applications such as affinity chromatography, histochemistry, pathological probe, diagnostics, signal amplification, immunoassay, hybridoma technology and blotting technology (Wilckel & Bayer, 1990).

III. ANTIMICROBIAL ACTIVITY

Egg white contains several antibacterial proteins such as lysozyme, a flavin-binding protein and ovotransferrin. Avidin has also been suggested to play a role as an antimicrobial agent (Kopela, 1984b). The view is not surprising. The discovery of strep-

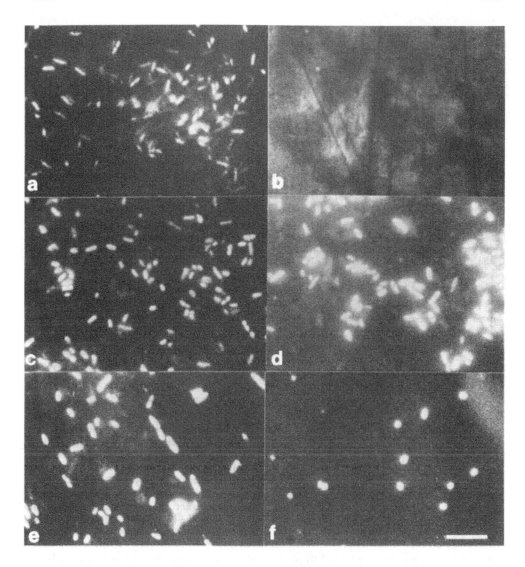

FIGURE 2. Fluorescence microscopy of bacteria treated with TRITC-labeled avidin (5 µg/ml). The following bacteria are visualized: (a) *E. coli* (ATCC25922), (b) *E. coli* fluorescence blocked with unlabeled avidin (50 µg/ml), (c) *K. pneumoniae*, (d) *S. marcescens*, (e) *P. aeruginosa*, (f) *S. epidermidis*. [reproduced from Korpela et al., (1984) with permission from the Federation of European Microbiological Societies].

tavidin is perhaps the most convincing support of this view in the light of the general versatility of streptomyces in deriving antibiotic system. This is further supported by the induction of avidin synthesis at the site of tissue injury in chicken (Elo et al., 1984). However, the biological function of avidin in terms of antimicrobial activity has not been clearly established, but it has been generally thought that avidin inhibits the growth of biotin-requiring microbes (Eakin et al., 1941; Herts, 1946), since it has a high affinity for biotin and the complex is very stable. Avidin inhibits *in vitro* the growth of biotin-requiring yeasts and bacteria (Board & Fuller, 1974; Green, 1975).

The induction of avidin in inflammation and cellular damage and its secretion by macrophage has suggested a role for avidin as a possible host-defence factor (Korpela, 1984b). Elo et al., (1978) proposed that avidin may act as an inducible antimicrobial substance in the inflammatory reaction of chicken tissues caused by bacterial and viral infections. This is further supported by the finding that marked amounts of avidin are present in chicken tissues after intra-peritoneal and intra-venous administration of *E. coli*. On the other hands, avidin may also have effects in tissues other than the possible antibacterial activity. By binding extra cellular/intracellular biotin, avidin might render it unavailable for the animal's own cells (Miller & Tauig, 1964). This should result in a decreased activity of biotin enzymes such as carboxylase and transcarboxylases (Wood & Barden, 1977) and thereby in an altered cell metabolism and growth (Messmer & Young, 1977). The unusually high isoelectric point of avidin at pH 10.0 suggests that its biological effects are not based on biotin-binding only. Basic proteins are essentially involved in the regulation of inflammation reaction (Elo & Korpela, 1984) and exhibit antibacterial activity.

Korpela et al. (1984) found that avidin binds *in vitro* to different Gram-negative and Gram-positive bacteria. Using tetramethyl-rhodamine isothiocyanate (TRITC)-conjugated avidin in fluorescence microscopy, all Gram negative bacteria tested *E. coli, K. pneumoniae, S. marcescens, P. aeruginosa* were found the binding of avidin. Among Gram-positive bacteria, both *S. aureus* and *S. epidermidis* also bound avidin, but no binding with *S. pyogenes* or *B. subtilis* (*FIGURE 2*). The binding of avidin was dose-dependent below 10 μg avidin/ml. The binding was also dose-dependent on the number of bacteria (*FIGURE 3A*). The binding kinetics reach maximum in 10 min and are optimal at physiological pH (7.0-8.0) and at 4°C, 22°C and 37°C. Showing no further increase during prolonged incubation for up to 120 min. (*FIGURE 3B*), bacterial binding of avidin is independent of the saturation of its biotin-binding site. The avidin receptor of *E. coli* was shown to be the porin protein of the outer membrane with the major 36 kDa (Korpela et al., 1984). Porins extend to the outer surface of the outer membrane; they are among the most abundant proteins of the cell and a major outer membrane of a similar structure and function is found in most Gram-negative bacteria (Lugtenberg & Van Alphen, 1983). No proteins analogous to porins are present in the cell envelope of Gram-positive bacteria. This indicates that the porin is the only avidin-binding component of the cell envelope and no avidin is bound by the other major abundant OM protein or by lipopolysaccharide. It remains to be seen whether this binding involves specific receptors or if the acidic character of the porin protein (pI 4.8-5.0) plays a role. Avidin could be antimicrobial by binding biotin tightly enough to make it unavailable to microbes. Avidin at the cell surface might effectively trap biotin and prevent its entry into the cell. However, there is limited data on the possible inhibition of growth of pathogenic bacteria by avidin. Furthermore, many bacteria such as *E. coli* and Salmonellae are not normally dependent on exogeneous biotin, and *E. coli* can grow in the presence of avidin (Campbell et al., 1972). It suggests avidin also has a role in the mechanism of infection or in host defence not related to its inter-defence with biotin utilization.

Recently, the mature hen avidin encoded by a synthetic cDNA was expressed in *E. coli* in an insoluble form (Nardone et al., 1998). The protein showed unaltered biotin-binding activity. The procedure yields about 20 mg pure protein/L cell culture. Site-directed mutagenesis to create mutated proteins, which have desirable amino acid substitution, is a powerful technique for obtaining further information on the structure function rela-

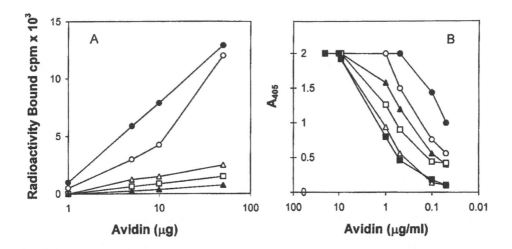

FIGURE 3. (A) Binding of avidin to bacteria attached to solid phase. (●) *P. aeruginosa,* (○) *S. marcescens,* (▲) *S. epidermidis,* (❑) *E. cloacae,* (△) *E. coli,* and (■) *K. pneumoniae* (1.5 x 10⁸ bacteria cells/ml) were attached to microtiter plates and incubated with different concentration of avidin. Bound avidin was detected with enzyme immunoassay.

(B) Binding of [¹⁴C]biotin labeled avidin to bacteria. (●) *P. aeruginosa,* (○) *S. aureus,* (△) *S. epidermidis,* (▲) *E. coli,* and (❑) *K. pneumoniae* (5 x 10⁹ bacteria cells/ml) were attached to microtiter plates and incubated with different amounts of labeled avidin for 2-h. The attached avidin was solubilized by 1% SDS and radioactivity was counted [reproduced from Korpela et al., (1984) with permission from the Federation of European Microbiological Societies].

tionship of proteins. The protein engineering technique could be a powerful tool for the studies on further avidin bactericidal defensins to clarify a defence mechanism of avidin and its potential application as antimicrobial agent.

IV. SUMMARY

Avidin is a basic tetrameric glycoprotein isolated from egg white. The protein binds up to four molecules of biotin with exceptionally high affinity. This high affinity of avidin-biotin binding has been exploited thoroughly for the development of various molecular tools. Avidin has also been viewed as antibiotic protein in egg albumen inhibiting bacterial growth such as a remedial protein, in reproduction or as a lytic enzyme. It has been found avidin inhibits *in vitro* the growth of biotin-requiring yeast and bacteria. The antimicrobial activity also may arise from the ability of avidin to bind the various Gram-negative and Gram-positive bacteria. The specific avidin induction by tissue injury and inflammation in chickens has also suggested that avidin may act as a host defense with antimicrobial and enzyme inhibitor properties. The detailed molecular information through three dimensional structure and protein engineering of avidin would enable us to better understand the biological role and development of avidin as a natural food antimicrobial agent.

V. REFERENCES

1.   Boas, M.A. 1924. An observation on the value of egg white as the sole source of nitrogen for young growing rats. *Biochem. J.* 18:422-424.
2.   Board, R.G., and Fuller, R. 1974. Non-specific antimicrobial defences of the avian egg, embryo and neonate. *Bio. Rev.* 49:15-49.

3. Bruch, R.C., and White, H.B. 1982. Compositional and structural heterogeneity of avidin glycopeptides. *Biochemistry*, 21:5334-5341.
4. Cautrecasas, P., and Wilchek, M. 1968. Single-step purification of avidin from egg white by affinity chromatography on biocytin-Sepharose columns. *Biochem. Biophys. Res. Commun.* 33:235-237.
5. Campbell, A., del Campillo-Campbell, A., and Chang, R. 1972. A mutant of *Escherichia coli* that requires high concentration of biotin. *Proc. Natl. Acad. Sci. USA* 69:676-680.
6. Chaiet, L., and Wolf, F.J. 1964. Properties of streptavidin, a biotin-binding protein produced by *Streptmycetes*. *Arch. Biochem. Biophys.* 106:1-5.
7. Delange, R., and Huang, T.S. 1971. Egg white avidin 3. Seuqence of the 78-residue middle cyanogen bromide peptides. Complete amino acid sequence of the protein subunit. *J. Biol. Chem.* 246:698-709.
8. Dhyse, F.G. 1954. A practical laboratory preparation of avidin concentrates for biological investigation. *Proc. Soc. Exp. Biol. Mod.* 85:515-518.
9. Durance, T.D. and Nakai, S. 1988a. Purification of avidin by cation exchange, gel filtration, metal chelate interaction and hydrophobic interaction chromatography. *Can. Inst. Food Sci. Technol.* 21:279-286.
10. Durance, T.D., and Nakai, S. 1988b. Simultaneous isolation of avidin and lysozyme from egg albumen. *J. Food Sci.* 53:1096-1102.
11. Eakin, R.E., Snell, E.E., and Williams, R.J. 1941. The concentration and assay of avidin, the injury-producing protein in raw egg white. *J. Biol. Chem.* 140:535-543.
12. Elo, H.A., Kulomaa, M. and Tuohimaa, P. 1978. Avidin induction by tissue injury and inflammation in male and female chickens. *Comp. Biochem. Physiol.* 62:237-240.
13. Elo, H.A., and Korpela, J. 1984. The occurrence and production of avidin: a new conception of the high-affinity biotin-binding protein. *Comp. Biochem. Physiol. B.* 78:15-20.
14. Fraenkel-Contrat, J. and Fraenkel-Contrat, H. 1952. Metabolic fate of biotin and of avidin-biotin complex upon parenteral administration. *Biochim. Biophys. Acta.* 8:66-70.
15. Gatti, G., Bolognes, M., Coda, A., Chiolerio, F., Fillippini, E., and Malcovati, M. 1984. Crystalization of hen egg-white avidin in a tetragonal form. *J. Mol. Biol.* 178:787-789.
16. Gitlin, G., Bayer, E.D. and Wilchek, M. 1988. Studies on the biotin-binding site of avidin. *Biochem. J.* 250:291-294.
17. Green, N.M. 1963. Avidin 3. The nature of the biotin-binding site. *Biochem. J.* 89:599-609.
18. Green, N.M. 1964. The moleculer weight of avidin. *Biochem J.* 92:16-17.
19. Green, N.M., and Toms, E.J. 1970. Purification and crystallization of avidin. *Biochem. J.* 118:67-70.
20. Green, N.M., and Joynson, M.A. 1970. Preliminary crystallographic investigation of avidin. *Biochem. J.* 118:71-72.
21. Green, N.M. 1975. Avidin. *Adv. Protein Chem.* 29:85-133.
22. Heney, G., and Orr, G.A. 1981. The purification of avidin and its derivatives on 1-iminobiotin-6-amino-hexyl-Sepharose 4B. *Anal. Biochem.* 114:92-96.
23. Hertz, R. 1946. Biotin and the avidin-biotin complex. *Physiol. Rev.* 26:479-494
24. Honzatko, R.B., and Williams, R.W. 1982. Raman spectroscopy of avidin: Secondary structure, disulfide conformation, and the envioronment of tyrosine. *Biochemistry* 21:6201-6205.
25. Huang, T.S., and DeLange, R.J. 1971. Egg white avidin 2. Isolation, composition and amino acid sequence of the tryptic peptides. *J. Biol. Chem.* 246:686-697.
26. Korpela, J. 1984a. Avidin, a high affinity biotin-binding protein as a tool and subject of biological research. *Med. Biol.* 62:5-26.
27. Korpela, J. 1984b. Chicken macrophages synthesize and secrete avidin in culture. *Eur. J. Cell Biol.* 33:105-111.
28. Korpela, J., Salonen, E.M., Kuusela, P., Sarvas, M., and Vaheri, A. 1984. Binding of avidin to bacteria and to the outer membrane porin of *Escherichia coli*. *FEMS Microbiology Letters.* 22:3-10.
29. Lugtenberg, B., and Van Alphen, L. 1983. Molecular architecture and functionins of the outer membrane of *Escherichia coli* and other Gram-negative bacteria. *Biochim. Biophys. Acta.* 737:51-115.
30. Miller, A.K., and Tausig, F. 1964. Biotin-binding by parenterally-administered streptavidin or avidin. *Biochem. Biophys. Res. Commun.* 14: 210-214.
31. Messmer, T.O., and Young, D.V. 1977. The effect of biotin and fatty acids on SV3T3 cell growth in the presence of normal calf serum. *J. Cell Physiol.* 90:265-270.
32. Melamed, M.D., and Green, N.M. 1963. Avidin II. Purification and composition. *Biochem. J.* 89:591-599.
33. Nardone, E., Rosano, C., Santambrogio, P., Curnis, F., Corti, A., Magni, F Siccardi, A., Paganelli, G., Losso, R., Apreda, B., Bolognesi, M., Sidoli, A., and Arosio, P. 1998. Biochemical characterization and crystal structure of a recombinant hen avidin and its acidic mutant expressed in *Escherichia coli*. *Eur. J. Biochem.* 256:453-460.

34. Pinn, E., Pahler, A., Saenger, W., Petsko, G.A., and Green, M. 1982. Crystallization and preliminary X-ray investigation of avidin. *Eur. J. Biochem.* 123:545-546.

35. Pugliese, L., Coda, A., Malcovait, M. and Bolognesi, M. 1993. Three demensional structure of the tetragonal crystal form of egg white avidin in its functional complex with biotin at 2.7 Å resolution. *J. Mol. Biol.* 231:698-710.

36. Pugliese, L., Malcovati, M., Coda, A., and Bolognesi, M. 1994. Crystal structure of apo-avidin from hen egg-white. *J. Mol. Biol.* 235:42-46.

37. Wilchek, E.A., and Bayer, M. 1990. Introduction to avidin-biotin technology. *Methods Enzymol.* 184:5-13.

38. Woolley, D.W., and Longsworth, L.G. 1942. Isolation of an antibiotic factor from egg white. *J. Biol. Chem.* 142:285-290.

39. Wood, H.G. and Barden, R.E. 1977. Biotin enzyme. *Rev. Biochem.* 46:385-413.

Section-III

PHYTO-ANTIMICROBIALS

Phyto-phenols
Saponins
Flavonoids
Thiosulfinates
Catechins
Glucosinolates
Agar

P.M. Davidson

A.S. Naidu

Phyto-phenols

10

I. INTRODUCTION

Food preservation dates back to prehistoric times and has refined into a culinary art in various parts of the world. The potential benefits of edible plants, as well as their phytochemicals, in food preservation and improvement of organoleptic qualities of certain traditional foods has been practiced for centuries. As long as 5000 years ago, the Chinese used herbs as medicines (Farrell, 1985). Reports on the early use of spices or herbs as antimicrobial preservatives can be traced to 1550 BC, when the ancient Egyptians used cinnamon, cumin and thyme both for food preservation and mummification (Webb & Tanner, 1944; Hirasa & Takemasa, 1998). The storage of yogurt under olive oil has been practiced since Biblical times, and it was assumed that the oil has a preservative role.

Scientific evidence on the preservation potential of spices emerged in the early 19th century. Chamberland (1887) first reported the antimicrobial activity of cinnamon oil against spores of anthrax bacilli (Webb & Tanner, 1944). Grove (1918) observed the ability of aqueous and alcoholic extracts of ground cinnamon to preserve tomato sauce. Prasad and Joshi (1929) developed a method in India, for preserving native fruits with ground cloves and salt. Fabian et al. (1939) found that cinnamon inhibits microbial growth at a 1:50 dilution (extract of 10 g in 100-ml water); and cloves inhibit *Bacillus subtilis* at 1:100 and *S. aureus* at 1:800 dilutions, respectively.

Spices and herbs are used for adding desirable sensory properties to food products. According to Lindsay (1996), herbs may be defined as portions of aromatic, soft-stemmed plants and aromatic shrubs and trees, while spices are rhizomes, roots, bark, flower buds, fruits and seeds (*TABLE 1*). Herbs are generally from sub-tropical and temperate plants while spices are from tropical plants. Spices may be marketed fresh or dried. Dried ground spices may be extracted with a solvent, such as ethanol, to obtain the spice oleoresin. They may also be subjected to distillation, cold pressed, or extracted with supercritical carbon dioxide to produce the essential oil (Hirasa & Takemasa, 1998).

Although ancient civilizations acknowledged the antiseptic and antimicrobial potential of many plant extracts, it was not until recently that the implied phytochemicals

TABLE 1. Spices and herbs with antimicrobial activity and their flavor components

	Plant part	Major flavor component
Herbs:		
Basil, sweet (*Ocimum basilicum*)	Leaves	Linalool / Methyl chavicol
Oregano (*Origanum vulgare*)	Leaves/ Flowers	Carvacrol / Thymol
Rosemary (*Rosmarinus offinicalis*)	Leaves	Camphor / 1,8-cineole / α-pinene / Borneol / Bornyl acetate / Verbenone
Sage (*Salvia officinalis*)	Leaves	Thujone / 1,8-cineole / Borneol / Camphor
Thyme (*Thymus vulgares*)	Leaves	Thymol / Carvacrol
Spices:		
Allspice, pimento (*Pimenta dioica*)	Berry/ leaves	Eugenol / β-caryophyllene
Cinnamon (*Cinnamomum zeylanicum*)	Bark	Cinnamic aldehyde / Eugenol
Clove (*Syzygium aromaticum*)	Flower bud	Eugenol
Mustard (*Brassica*)	Seed	Allyl isothiocyanate
Nutmeg (*Myristica fragrans*)	Seed	Myristicin / α-pinene / Sabinene
Vanilla (*Vanilla planifola, V. pompona, V. tahitensis*)	Fruit/ seed	Vanillin (4-hydroxymethoxybenzaldehyde)/ *p*-OH-benzyl methyl ether

Based on Farrell, 1985; Charalambous, 1994; Lindsay, 1996; Lachowicz et al., 1998; Mangena & Muyima, 1999

were characterized. Advances in molecular separation techniques led to the isolation of various phyto-phenolic compounds. The increased demand for minimally processed, extended shelf-life foods has further revived interest in exploitation of these natural antimicrobial agents.

This chapter reviews the antimicrobial properties of phyto-phenolic compounds from essential oils of spices, herbs, edible grains and seeds. The multifunctional benefits of selected phyto-phenols as food additives are also described.

II. CHEMISTRY

The compounds responsible for antimicrobial activity of spices are primarily phenolic components of the essential oil fraction (Beuchat, 1994). The commonly occurring phenolic compounds associated with various plant sources are listed in *TABLE 2*. Antimicrobial activity of cinnamon, allspice, and cloves is attributed to eugenol (2-methoxy-4-allyl phenol) and cinnamic aldehyde, which are major constituents of the volatile oils of these spices. Cinnamon contains 0.5-1.0% volatile oil, which contains 65-75% cinnamic aldehyde and 8% eugenol. Allspice contains up to 4.5% volatile oil, of which 80% is eugenol. Clove buds have an average essential oil content of 17% that is 93-95% eugenol (Farrel, 1985).

Oregano, savory, and thyme demonstrate antifungal activity. Terpenes carvacrol, *p*-cymene, and thymol are the major volatile components of oregano, thyme, and savory and likely account for the antimicrobial activity. The essential oil of oregano contains up to 50% thymol; thyme has 43% thymol and 36% *p*-cymene, and savory, 30-45% carvacrol and 30% *p*-cymene (Farag, et al., 1989).

The active antimicrobial fraction of sage and rosemary has been suggested to be the terpene fractions of the essential oils. Rosemary contains borneol, camphor, 1,8-cineole, α-pinene, camphene, verbenone and bornyl acetate while sage contains thujone [4-methyl-1-(1-methylethyl)bicyclo-3-1-O-hexan-3-one].

TABLE 2: Phenolic compounds associated with various plant sources

Phenolic compound	Plant source
Apigenin-7-glucoside	Oleaceae sp.
Benzoic acid	Spices
Berbamine	Beriberis
Caffeine	Tea
Caffeic acid	Oleaceae sp. / avocado / artichoke / apple
O-Caffeylquinic acid	Artichoke / plum / carrots
Caryophelene	Hops
Catechin	Skin and seeds of wine grapes / tea
Cinnamic acid	Brassica oil seeds
Chlorogenic acid	Brassica oil seeds / artichoke / apple
Chelldonic acid	Berberis
Chicorin	Chicory
Columbamine	Berberis
Coumarin	Spices
p-coumaric acid	Oleaceae sp. / avocado / brassica oil seeds / apple
Cyanarine	Artichoke
Dihydrocafeic acid	Oleaceae sp.
Dimethyloleuropein	Oleaceae sp.
Esculin	Chicory
Ferulic acid	Oleaceae sp. / avocado / brassica oil seeds
Gallic acid	Oleaceae sp. / grapes / avocado / brassica oil seeds
Gingerols	Spices
Humulon	Hops
Hydroxytyrosol	Oleaceae sp.
4-hydroxybenzoic acid	Avocado / brassica oil seeds / vanilla / carrots
4-hydroxycinnamic acid	Oleaceae sp. / carrots
Isovanillin	Avocado
Linalool	Boldo
Lupulon	Hops
Luteoline-5-glucoside	Oleaceae sp.
Ligustroside	Oleaceae sp.
Myricetin	Tea
3-methoxybenzoic acid	Oleaceae sp.
Oleoside	Oleaceae sp.
Oleuropein	Oleaceae sp.
Paradols	Spices
Procatechuic acid	Oleaceae sp. / avocado
O-pyrocatechuic	Avocado
Quercetin-3-rutinoside	Tea
Quercetin	Oleaceae sp. / tea
Resocrylic acid	Avocado
Sesamol	Sesame oil
Shogoals	Spices
Syringic acid	Brassica oil seeds
Sinapic acid	Avocado / brassica oil seeds
Tannins	Wine grapes / spices / brassica oil seeds / boldo / herbs
Tannic acid	Skin and seeds of wine grapes
Thymol	Thyme
Trimethoxybenzoate	Oleaceae sp.
Trihydroxyphyenylacetate	Oleaceae sp.
Tryrosol	Oleaceae sp.
Verbascoside	Oleaceae sp.
Vanillin	Spices / vanilla
Vanillic acid	Oleaceae sp. / avocado / brassica oil seeds / vanilla

Adapted from Nychas, 1994.

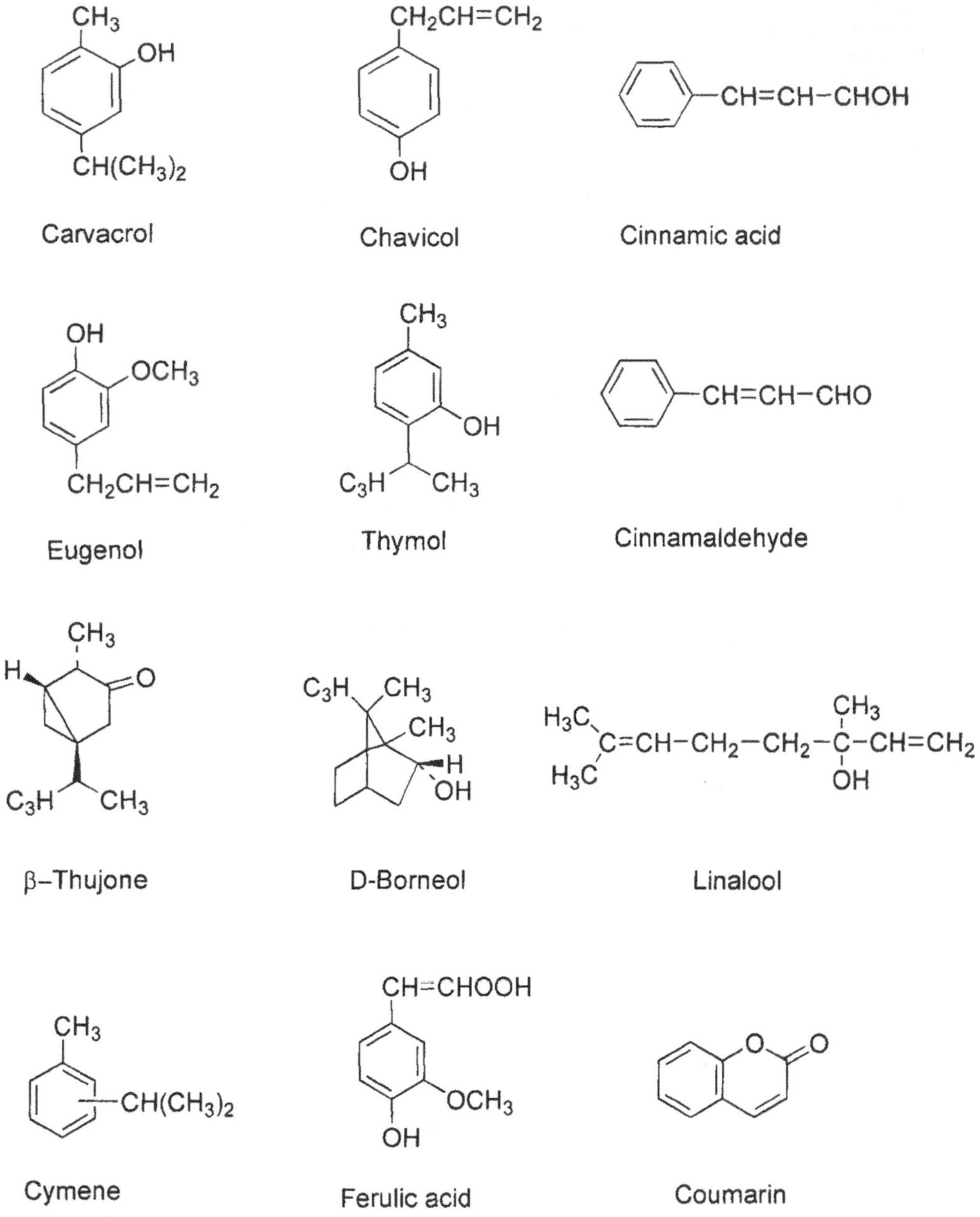

FIGURE 1. Chemical structures of the commonly occurring phytophenols of essential oils.

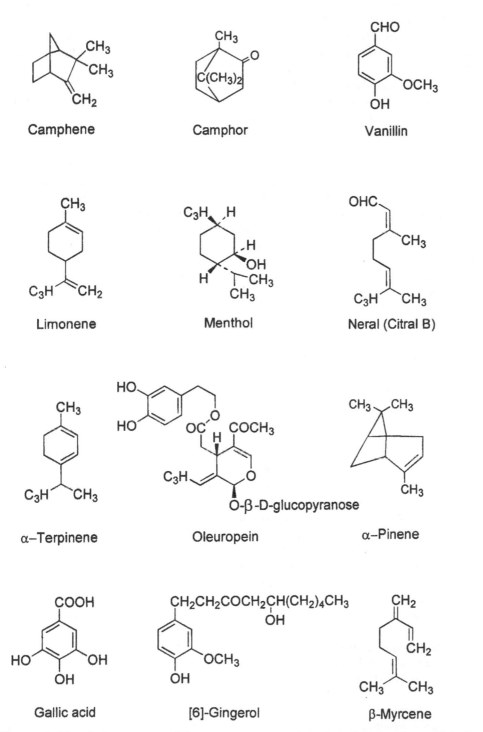

FIGURE 1. Chemical structures of the commonly occurring phytophenols of essential oils.

Recently, a number of potential phyto-phenolic compounds have been isolated from olives and virgin olive oil, and amongst these are polyphenols and glycosides. Some of these phyto-phenols are effective against lactic acid bacteria. The compounds identified are tyrosol, *p*-hydroxyphenylacetic acid, *p*-coumaric acid and ferulic acid (Keceli, et al., 1998).

The essential oil from herb of *Ducrosia anethifolia (DC.) Boiss.* consists mainly of aliphatic compounds (Janssen, et al., 1984). α-pinene, myrcene and limonene are main components of the hydrocarbons present in the oil, while *N*-decanal, *N*-dodecanal, *N*-decanol, *trans*-2-dodecenal, and *cis*-chrysanthenyl acetate are the major oxygen-containing constituents. The oil and the main oxygen-containing aliphatic components show a potent antimicrobial activity against Gram-positive bacteria, yeast and dermatophytes. The volatile oil of *Ducrosia ismaelis Asch.* is a light yellow volatile liquid with a strong aromatic odor and a specific gravity of 0.9573 (Al-Meshal, 1986). Spectrometry studies revealed the presence of free alcohols, alkenes and highly conjugated alkenes, aromatic functions, alicyclic structures and cyclic ketones. The pharmacological studies of this oil showed a highly significant and dose dependent central nervous system depressant and marked neuromuscular blocking actions. Experiments on smooth muscles and heart show a parasympatholytic activity. It also exhibits significant antimicrobial activity against *S. aureus, B. subtilis* and *C. albicans*.

Chemical structures of the commonly occurring phyto-phenols of essential oils are shown in *FIGURE 1*.

Isolation methods. Screening and isolation of antimicrobial phyto-phenols require a multidisciplinary approach. The primary screen, any *in vitro* or *in vivo* detection system (bioassay), is determining factor for successful isolation of a bioactive compound. In chosing a solvent for extraction, its ability to extract components of a solute has to be considered. The more efficient the extraction step, the greater is the range of compounds present in the extract. The need to use pure solvents for extractions is obvious. Less obvious, is the need to carry out the extraction under mild conditions, utilizing, whenever possible, solvents of low reactivity. The possibility of generating artifacts should never be discounted. The following techniques to isolate bioactive phytochemicals have been reviewed by Ghisalberti (1993).

Dry plant materials could be extracted with a variety of solvents, and sometimes sequentially from low to high polarity, if a crude fractionation of metabolites is sought. Generally, however, a polar solvent such as ethyl acetate or methanol is used. With dried material, ethyl acetate or low-polarity solvents will only rinse or leach the sample, whereas alcoholic solvents presumably rupture cell membranes and extract a greater amount of endocellular materials.

The water content of the fresh plant materials makes dichloro-methane-methanol solvent mixture ideal for extraction purposes. Once the extract has been partitioned and the methanol removed from the aqueous phase, the latter can be back extracted with ethyl acetate and then with butanol, to separate the lipophilic material from the water solubles.

Borneol (I) and isoborneol (II) are separated by dry-column chromatography (DCC) (Wu et al., 1989). The sample mixed with a small amount of adsorbent is transferred to the top of the column packed with silica gel (200 mesh) and developed with petroleum ether-ethyl acetate-chloroform (20:2:6) until the solvent has spread from the top to the bottom of column. The zones of I and II can be located by the corresponding Rf

TABLE 3: Phyto-phenolic compounds from various plant sources and their antimicrobial spectrum

Plant source	Phyto-phenolic	Susceptible microorganism	Reference
Nigella sativa L.	Thymohydroquinone	Gram positive bacteria	El-Fatatry, 1975
Anethum graveolens L	Volatile oil	*Saccharomyces vini /Lacobacillus buchneri*	Shcherbanovsky & Kapelev, 1975
Ducrosia anethifolia	α-pinene / limonene	Gram-positive bacteria / yeast / dermatophytes	Janssen et al., 1984
Ducrosia ismaelis Asch	Cyclic ketones	*S. aureus / B. subtilis / C. albicans.*	Al Meshal, 1986
Thymus vulgaris	Volatile oils	Enteropathogenic bacteria	Ramanoelina et al., 1987
Ocimum sp.	Volatile oils	*E. coli / B. subtilis, S. aureus / Trichophyton mentagrophytes*	Janssen et al., 1989
Lauraceae sp.	Volatile oils	Fungi	Raharivelomanana et al., 1989
Bystropogon. sp.	Pulegone	Bacteria / Fungi	Economou & Nahrstedt, 1991
Achillea fragrantissima	Terpinen-4-ol	Gram-positive & -negative bacteria/ *C. albicans*	Barel et al., 1991
Hoslundia opposita	Sesquiterpenes	*Aspergillus niger/ Acinetobacter calcoacetica / Brochothrix thermosphacta/*	Gundidza et al., 1992
Cedronella canariensis	Volatile oils	*Bordetella bronchiseptica / Cryptococcus albidus*	Lopez-Garcia et al., 1992
Humulus lupulus L	Chloroform extract	*B. subtilis/ S. aureus/ E.coli / T. mentagrophytes/ C. albicans*	Langezaal et al., 1992
Jasonia candicans	Intermediol/ borneol	*B. subtilis / Cryptococcus neoformans / Trichophyton mentagrophytes /*	Hammerschmidt et al., 1993
Jasonia montana	Camphor	*C. albicans*	
Thymus vulgaris L	Volatile oils	Fungi	Panizzi et al., 1993
Schinus molle L	Volatile oils	*Klebsiella pneumoniae/ Alcaligenes faecalis/ P. aeruginosa/ E.coli/*	Gundidza, 1993
		Enterobacter aerogenes/ Proteus vulgaris/ Clostridium sporogenes/	
		Acinetobacter calcoacetica/ Citrobacter freundii/ Serratia marcescens/	
		B. subtilis/ Brochothrix thermosphacata/ Aspergillus sp.	
Melaleuca alternifolia	Linalool/ terpinenes	*C. albicans/ E. coli/ S. aureus*	Carson & Riley, 1995
Helianthus annuus	Precocenes	*Pyricularia oryzae*	Satoh et al., 1996
Piper angustifolium	Camphor/ camphene	*P. aeruginosa/ C. albicans/ Cryptococcus neoformans/ Aspergillus sp./ E.coli.*	Tirilini et al., 1996
Bellis perennis L	Polyacetylenes	Gram-positve & -negative bacteria	Avato et al., 1997
Cistus creticus	Cadinenes	Gram-positve & -negative bacteria	Demetzos et al., 1997
Pelargonium sp.	Methanol extract	*S. aureus / S. epidermidis / Proteus vulgaris / B. cereus*	Lis-Balchin et al., 1998
Ocimum basilicum L.	Linalool / chavicol	*Lb. curvatus / S. cerevisiae*	Lachowicz et al., 1998
Cistus creticus	Sclareol	*S. aureus/ S. epidermidis / S. hominis*	Demetzos et al., 1999
Calamintha nepeta	Pulegone	*Listeria monocytogenes/ B. cereus/ Salmonella sp./ Aspergillus niger*	Flamini et al., 1999
Peumus boldus	Monoterpenes	*Streptococcus pyogenes/ Micrococcus sp./ Candida sp*	Vila et al., 1999
Leptospermum scoparium	Monoterpenes/	Bacteria / Fungi	Porter & Wilkins, 1999
Kunzea ericoides	sesquiterpene		

of TLC. The compounds are then eluted separately with absolute alcohol. After the concentration of the eluent by vacuum distillation, the crystals of I and II are isolated from the residual solution on standing. The purities of I and II are tested with known specimens by TLC. Though the absolute stereochemistry of D-borneol is described as 1S, 2R-form, the opposite optical rotation for the same structure is described in other literatures. The application of improved Mosher's method to D-borneol resulted in 1R, 2S-form for its absolute stereochemistry (Ezaki et al., 1997).

A simple, rapid high-performance liquid chromatography method has been devised to separate and quantify the xanthophylls capsorubin and capasanthin present in red pepper (*Capsicum annuum* L.) fruits and preparations made from them (paprika and oleoresin) (Weissenberg et al., 1997). A reversed-phase isocratic non-aqueous system allows the separation of xanthophylls within a few minutes, with detection at 450 nm, using methyl red as internal standard to locate the various carotenoids and xanthophylls found in plant extracts. The method has been proposed for rapid screening of large plant populations, plant selection, as well as for paprika products and oleoresin, and also for nutrition and quality control studies.

A method to extract oleoresins from dried guajillo peppers with four different solvents: ethanol, acetone, ethyl acetate and hexane with further fractionation into red and paprika extracts was described (Amaya Guerra et al., 1997). As the polarity of the solvent increased, the amount of pigments extracted also increased. Acetone had good affinity for pungent (capsaicin) compounds. Utilization of these solvents alone did not produce red and paprika oleoresins that meet commercial specifications. Fractionation of acetone extracted oleoresins with ethanol: water (90:10) yielded a precipitate and a solution. The precipitate and solution produced red and paprika extracts that meet pungency and color specifications.

III. ANTIMICROBIAL SPECTRUM

A myriad of factors influences the antimicrobial activity of phyto-phenols. For example, chemical composition, as influenced by geographic origin and crop to crop variations, could significantly affect the activity of whole spices or essential oils (Moyler, 1994). Furthermore, the type of assay method could influence the antimicrobial outcome of these compounds (Hammer et al., 1999a). Certain antimicrobial components of spices are hydrophobic, therefore, disk assays or agar diffusion methods may not be appropriate for evaluating the activity. Co-existence of other food components (protein, lipid), surfactants, minerals, pH, time and temperature also control the biological activity of phyto-phenols. Finally, the microbial targets including the organism (e.g., bacteria, fungi), form (e.g., vegetative cell, spore), genus, species and strain, stage of growth, and previous exposure to stress and injury are critical factors that contribute to the antimicrobial spectrum. Unless these factors are controlled, studies on the antimicrobial activity of phyto-phenols from oils, spices or herbs may vary considerably. Phyto-phenolic compounds from various plant sources and their antimicrobial spectrum is shown in *TABLE 3*.

A. Oleuropein

The hydrolysis products of olive oil are reported to elicit a wide spectrum of antimicrobial activity (Juven et al., 1972; Fleming et al., 1973). Oleuropein, the bitter principle of olives, is one of the major antimicrobial components of the polyphenol fraction of olive oil.

1. Antimicrobial effects. The phenolic compounds extracted from olives with ethyl acetate inhibit the germination and outgrowth of *Bacillus cereus* T spores (Tassou et al., 1991). Purified oleuropein from the olive extract also cause strong inhibition. The addition of oleuropein and olive extracts 3 or 5 min after germination, immediately decreased the rate of change of phase bright to phase dark spores and significantly delayed overgrowth.

The presence of low concentrations (0.1% w/v) of oleuropein from olives, delayed the growth of *Staphylococcus aureus* as indicated by changes in conductance *in vitro*, whilst higher concentrations (0.4-0.6% w/v) inhibited growth completely (Tranter et al., 1993). Oleuropein, at concentration higher than 0.2% inhibited growth and production of enterotoxin B *in vitro*. Lower levels (0.1%) did not affect the final viable count and production of toxin but decreased the number of viable organisms. An increase in the concentration of oleuropein resulted in a decrease in the amount of glucose assimilated and consequently the amount of lactate produced. In addition, oleuropein prevented the secretion of a number of exo-proteins. Finally, the addition of oleuropein during the exponential phase appeared to have no effect on the growth of *S. aureus*.

The inhibitory effect of commercial oleuropein against *Salmonella* Enteritidis in a coliform broth and in reconstituted milk (model food system) was also reported (Tassou & Nychas, 1995). The inhibition of *Salmonella* Enteritidis in the broth was influenced by the initial inoculum size, pH of the medium and concentration of the additive. The inhibition was more pronounced in samples with low pH and low inoculum size.

Aziz et al. (1998) have tested the antimicrobial potential of eight phenolic compounds isolated from olive cake against the growth of *Escherichia coli, Klebsiella pneumoniae, Bacillus cereus, Aspergillus flavus* and *Aspergillus parasiticus*. The phenolic compounds included *p*-hydroxy benzoic, vanillic, caffeic, protocatechuic, syringic, and *p*-coumaric acids, oleuropein and quercetin. Caffeic and protocatechuic acids (0.3 mg/ml) inhibited the growth of *E. coli* and *K. pneumoniae*. The same compounds apart from syringic acid (0.5 mg/ml) completely inhibited the growth of *B. cereus*. Oleuropein, and *p*-hydroxy benzoic, vanillic and *p*-coumaric acids (0.4 mg/ml) completely inhibited the growth of *E. coli, K. pneumoniae* and *B. cereus*. Vanillic and caffeic acids (0.2 mg/ml) completely inhibited the growth and aflatoxin production by both *A. flavus* and *A. parasiticus*, whereas the complete inhibition of the molds was attained with 0.3 mg/ml *p*-hydroxy benzoic, protocatechuic, syringic, and *p*-coumaric acids and quercetin.

2. Antioxidant effects. Natural antioxidants contained in the Mediterranean diet have been suggested to play a role in the prevention of cardiovascular diseases, through inhibition of low density lipoprotein (LDL) oxidation. Oleuropein has been shown to be a potent antioxidant endowed with anti-inflammatory properties. Visioli and Galli (1994) tested this hypothesis *in vitro* by inducing LDL oxidation with copper sulfate and preincubating the samples with oleuropein. Oleuropein 10 μM effectively inhibited $CuSO_4$-induced LDL oxidation, as assessed by various parameters. These findings suggested a new link between the Mediterranean diet enriched with oleuropeins and prevention of coronary heart disease.

Visioli et al. (1998a) have also investigated the effects of oleuropein on nitric oxide (NO) release in cell culture and its activity toward nitric oxide synthase (NOS) expression. Oleuropein enhanced nitrite production, in a dose-dependent manner, in LPS-challenged mouse macrophages. Also, Western blot analysis of cell homogenates showed that oleuropein increased NOS expression in such cells. These data suggested that, during endotoxin challenge, oleuropein potentiates the macrophage-mediated response, resulting in higher NO production, currently believed to be beneficial for cellular protection.

The free radical-scavenging activity of some olive oil phenolic compounds, namely hydroxytyrosol and oleuropein, with respect to superoxide anion generation, neutrophils respiratory burst, and hypochlorous acid has been investigated (Visioli et al., 1998b). The low EC_{50} indicate that both compounds are potent scavengers of superoxide radicals and inhibitors of neutrophils respiratory burst; whenever demonstrated *in vivo*, these properties may partially explain the observed lower incidence of coronary heart disease and cancer associated with the Mediterranean diet.

3. Multifunctionality. Petroni et al. (1995) have investigated the *in vitro* effects of phenolic compounds extracted from olive oil and from olive-derived fractions. More specifically, the studies were performed on platelets of 2-(3,4-di-hydroxyphenyl)-ethanol (DHPE), a phenol component of extra-virgin olive oil with potent antioxidant properties. The variables studied include i) aggregation of platelet rich plasma (PRP) induced by ADP or collagen; ii) thromboxane B2 (TxB2) production by collagen or thrombin-stimulated PRP. In addition, TxB2 and 12-hydroxy-eicosatetraenoic acid (12-HETE) produced during blood clotting were measured in serum. Preincubation of PRP with DHPE for at least 10 min resulted in maximal inhibition of the various measured variables. The IC_{50} (concentration causing 50% inhibition) of DHPE for ADP or collagen-induced PRP aggregations were 23 and 67 μM, respectively. At 400 μM DHPE, a concentration which completely inhibited collagen-induced PRP aggregation, TxB2 production by collagen- or thrombin-stimulated PRP was inhibited by over 80 percent. At the same DHPE concentration, the accumulation of TxB2 and 12-HETE in serum was reduced by over 90 and 50 percent, respectively. The authors also tested the effects of PRP aggregation of oleuropein, and of selected flavonoids (luteolin, apigenin, quercetin) and found them to be less active. On the other hand, a partially characterized phenol-enriched extract obtained from aqueous waste from olive oil showed rather potent activities.

Kohyama et al. (1997) have investigated the effects of olive fruit extract on arachidonic acid lipoxygenase activities using rat platelets and rat polymorphonuclear leukocytes (PMNL). Olive extract strongly inhibited both 12-lipoxygenase (12-LO) and 5-lipoxygenase (5-LO) activities. One of the compounds responsible for this inhibition was purified and identified as DHPE. DHPE inhibited platelet 12-LO activity (IC_{50}, 4.2 μM) and PMNL 5-LO activity (IC_{50}, 13 μM) but not cyclo-oxygenase activity in cell-free conditions. It also inhibited 12-LO activity in intact platelets (IC_{50}, 50 μM) and reduced leukotriene B4 production in intact PMNL stimulated by A23187 (IC_{50}, 26 μM). The inhibition by DHPE of both lipoxygenase activities was stronger than that by oleuropein, caffeic acid, or 7 other related phenolic compounds, especially in intact cells. These results suggest that DHPE is a potent specific inhibitor of lipoxygenase activities.

Recently, de la Puerta and co-workers (1999) determined the anti-eicosanoid and antioxidant effects of the principal phenolic compounds from the 'polar fraction': oleuropein, tyrosol, hydroxy-tyrosol, and caffeic acid. In intact rat peritoneal leukocytes stimulated with calcium ionophore, all four phenolics inhibited leukotriene B4 generation at the 5-lipoxygenase level with effectiveness hydroxytyrosol > oleuropein > caffeic acid > tyrosol (approximate EC_{50} values: 15, 80, 200, and 500 µM, respectively). In contrast, none of these compounds caused substantial inhibition of thromboxane generation via the cyclo-oxygenase pathway. Hydroxytyrosol, caffeic acid, oleuropein, and tyrosol (decreasing order of effectiveness) also quenched the chemiluminescence signal due to reactive oxygen species generated by phorbol myristate acetate-stimulated rat leukocytes. None of these compounds were toxic to leukocytes at the concentrations tested. The authors have concluded that the phenolics found in virgin olive oil possess an array of potentially beneficial lipoxygenase-inhibitory, prostaglandin-sparing, and antioxidant properties.

Saenz et al. (1998) have tested oleuropein, tyrosol, squalene and the fraction of sterols and triterpenoid dialcohols from the unsaponifiable fraction obtained from virgin olive oil for possible cytostatic activity against McCoy cells, using 6-mercaptopurine as a positive control. Samples of sterols and triterpenic dialcohols showed a strong activity.

B. Oleoresins

1. Antimicrobial activity. Conner and Beuchat (1984) reported that an oleoresin of cinnamon was extremely inhibitory against eight yeasts, i.e. *Candida lipolytica, Debaryomyces hansenii, Hansenula anomala, Kloeckera apiculata, Lodderomyces elongisporus, Rhodotorula rubra, S. cerevisiae,* and *Torulopsis glabrata.* Essential oil of clove dispersed (0.4% v/v) in a concentrated sugar solution, had a marked germicidal effect against various bacteria and *C. albicans* (Briozzo, *et al.*, 1989). *S. aureus* (five strains), *Klebsiella pneumoniae, P. aeruginosa, Clostridium perfringens,* and *E. coli* inoculated at a level of 10^7 cfu/ml, and *C. albicans* (inoculum 4.0 x 10^5 cfu/ml) were killed (>99.9%) after 2-7 min in broth supplemented with 63% (v/w) of sugar, and containing 0.4% (v/w) of essential oil of clove. Presence of organic matter (i.e. human or bovine serum) did not impair its antimicrobial activity. Sugar was not necessary for the antimicrobial activity of clove oil, but the concentrated sugar solution provided a good vehicle for obtaining uniform oil dispersion that is relatively stable for certain practical applications.

2. Multifunctionality. The oleoresin from Brazilian *Copaifera* species yielded copalic acid and sesquiterpenes and showed marked anti-inflammatory activity (Basile et al., 1988). The oleoresin significantly inhibited carageenan-induced pedal edema following oral doses from 0.70 to 2.69 ml/kg, but was somewhat less effective than 50 mg/kg calcium phenylbutazone. Repeated administration of the oleoresin at a dose of 1.26 ml/kg for a 6-day period reduced granuloma formation with a response comparable to that of 20 mg/kg of calcium phenylbutazone. This same dose of oleoresin also reduced the vascular permeability to intra-cutaneous histamine. The LD_{50} value of the oleoresin in rats was estimated to be 3.79 (3.21-4.47) ml/kg.

Consumption of carotenoids has frequently been inversely correlated with cancer incidence. Sharoni and co-workers (1997) have used the 7,12-dimethyl-benz[a]anthracene (DMBA)-induced rat mammary tumor model to compare the effect of lycopene-enriched tomato oleoresin on the initiation and progression of these tumors with that of β-carotene. Rats were injected i.p. with lycopene-enriched tomato oleoresin or β-carotene (10 mg/kg, twice per week) for 2 weeks prior to tumor induction by DMBA and for an additional 16 weeks after carcinogen administration. HPLC analysis of carotenoids extracted from several tissues showed that both carotenoids were absorbed into blood, liver, mammary gland, and mammary tumors. The tomato oleoresin-treated rats developed significantly fewer tumors, and the tumor area was smaller than that of the unsupplemented rats. Rats receiving β-carotene showed no protection against the development of mammary cancer.

In vitro studies on the effect of alcoholic extracts of turmeric, turmeric oil and turmeric oleoresin (TOR), on the incidence of micronuclei (MN) in lymphocytes from normal healthy subjects showed that the test compounds did not cause any increase in the number of MN as compared with those found in untreated controls (Hastak et al., 1997). Further it was observed that all three compounds offered protection against benzo[a]pyrene induced increase in MN in circulating lymphocytes. In subsequent studies, patients suffering from submucus fibrosis were given a total oral dose of TO (600 mg turmeric oil mixed with 3 g turmeric/day). TOR (600 mg + 3 g turmeric/day) and 3 g turmeric/day as a control for 3 months. It was observed that all three treatment modalities decreased the number of micronucleated cells both in exfoliated oral mucosal cells and in circulating lymphocytes. TOR was found to be more effective in reducing the number of MN in oral mucosal cells, but in circulating lymphocytes the decrease in MN was comparable in all three groups.

C. Thymol and carvacrol

The antimicrobial activity of oregano and thyme has been attributed to their essential oils which contain the terpenes: carvacrol [2-methyl-5-(1-methylethyl)phenol] and thymol [5-methyl-2-(1-methylethyl)phenol], respectively.

1. Antimicrobial effects. Katayama and Nagai (1960) tested thymol and carvacrol against *Bacillus subtilis*, *Salmonella* Enteritidis, *S. aureus*, *Pseudomonas aeruginosa*, *Proteus* and *E. coli* and found inhibition of all microorganisms at dilutions as low as 1:2000. Essential oil from oregano had the highest activity of a number of essential oils tested against both fungi and bacteria (Maruzella & Henry, 1958; Maruzella & Liguori, 1958). Ting and Deibel (1992) determined the MIC of oregano against *Listeria monocytogenes* at 24°C in a microbiological medium to be 0.5-0.7% w/v. Oregano at 0.5% or 1.0% was bacteriostatic to *L. monocytogenes* in trypticase soy agar at 4 or 24°C but had little effect growth of the microorganism in beef at 1.0% (Ting & Deibel, 1992). Growth and lactic acid production of *Lactobacillus plantarum* and *Pediococcus cerevisiae* were inhibited by 4 mg/ml oregano for 7 days (Zaika & Kissinger, 1981). Lower concentrations of the spice stimulated acid production. Rosemary, sage and thyme inhibited the microorganisms to a lesser extent than oregano but also caused stimulation of acid production (Zaika et al., 1983). Kim et al. (1995) evaluated the antibacterial activity of essential oil constituents against *E. coli*, *E. coli* O157:H7, *Salmonella* Typhimurium, *L. monocytogenes*, and *Vibrio vulnificus* at 5, 10, 15, and 20% using disk and dilution assays. Carvacrol showed strong bactericidal activity against all strains, while limonene, nerolidol, and β-ionone were mostly inactive. Carvacrol was highly bactericidal at 250 μg/ml

TABLE 4. Minimum lethal concentrations (μg/ml; ≥ 99.9% decrease in viability) of thymus essential oil and its components against various bacteria and yeasts.

Microorganism	Thyme oil M.L.C. Range	Active Component	Component M.L.C (μg/ml)
Gram-negative bacteria			
Escherichia coli	450-900	Thymol / Carvacrol	225
Escherichia coli O157:H7	450-900	Carvacrol	225
Pseudomonas aeruginosa	>900	-	-
Salmonella Typhimurium	450-900	Thymol	56.25
Yersinia enterocolitica	450-900	Thymol / Carvacrol	225
Gram-positive bacteria			
Bacillus cereus	225-900	Thymol	450
Listeria monocytogenes	225-900	Thymol / Carvacrol	450
Staphylococcus aureus	225-900	Thymol	225
Yeasts			
Candida albicans	225-400	Thymol / Carvacrol	112.5
Saccharomyces cerevisiae	225-400	Thymol / Carvacrol	112.5

Adapted from Cosentino et al. (1999)

against *Salmonella* Typhimurium and *Vibrio vulnificus*. Pol and Smid (1999) reported that carvacrol enhanced the antimicrobial activity of nisin against growth of both *Bacillus cereus* and *L. monocytogenes*.

Karapinar and Aktug (1986) determined the inhibitory activity of thyme, mint, and bay leaves on growth of *Salmonella* Typhimurium, *S. aureus* and *Vibrio parahaemolyticus*. Thyme was the most effective antimicrobial of the spices and bay leaves the least active. *Vibrio parahaemolyticus* was the most sensitive bacteria and *Salmonella* Typhimurium the most resistant. Beuchat (1976) also showed that oregano and thyme were bactericidal to *Vibrio parahaemolyticus* at 0.5% as whole spices and bacteriostatic at 100 μg/ml as essential oils. Firouzi et al. (1998) found that thyme essential oil was the most effective antimicrobial against *L. monocytogenes* 4b growth compared to a number of other spice and herb extracts. Cosentino et al. (1999) reported on the composition and antimicrobial activity of thyme from different sources. They found that the composition of the essential oil of thymus varied greatly. Among the four samples analyzed, all of which were from Sardinia, the major component in all was thymol (29-50%) followed by variable amounts of carvacrol, *p*-cymene and α-pinene. The researchers then determined the minimum lethal concentration of the essential oil and components of the essential oil against various bacteria and yeasts (TABLE 4). While thymol and carvacrol were the most active components of the essential oils, the activity varied depending upon the source and the type of microorganism. Cosentino et al. (1999) suggested that components of the essential oil may have acted synergistically (e.g., thymol and carvacrol) or antagonistically (e.g., *p*-cymene) with other components to alter overall antimicrobial effectiveness. Juven et al. (1994) evaluated thyme essential oil and its components against *Salmonella* Typhimurium to determine their activity and assess the interactions of environmental factors on antimicrobials activity. They first determined that 175 μg/ml or 225 μg/ml thymol or carvacrol, respectively, completely inhibited *Salmonella* Typhimurium recovery under aerobic or anaerobic conditions. *p*-Cymene up to 500 μg/ml had no effect on the microorganism. Factors reducing the activity of thymol included presence of Tween 80, aerobic incubation, reduced pH and presence of bovine serum albumin.

TABLE 5. Minimum inhibitory concentrations (% v/v) of selected essential oils of herbs and spices against various bacteria and yeasts as determined by agar dilution.

Spice	Enterococcus faecalis	E. coli	Pseudomonas aeruginosa	Salmonella Typhimurium	S. aureus	Candida albicans
Basil	>2.0	0.5	>2.0	2.0	2.0	0.5
Bay	0.5	0.12	1.0	0.25	0.25	0.12
Black pepper	1.0	>2.0	>2.0	>2.0	>2.0	>2.0
Clove	0.5	0.25	>2.0	>2.0	0.25	0.12
Coriander	0.25	0.25	>2.0	1.0	0.25	0.25
Fennel	>2.0	0.5	>2.0	1.0	0.25	0.5
Ginger	>2.0	>2.0	>2.0	>2.0	>2.0	>2.0
Lemongrass	0.12	0.06	1.0	0.25	0.06	0.06
Marjoram	2.0	0.25	>2.0	0.5	0.5	0.25
Oregano	0.25	0.12	2.0	0.12	0.12	0.12
Peppermint	2.0	0.5	>2.0	1.0	1.0	0.5
Rosemary	>2.0	1.0	>2.0	>2.0	1.0	1.0
Sage	2.0	0.5	>2.0	2.0	1.0	0.5
Spearmint	2.0	0.25	>2.0	0.5	0.25	0.12
Tea tree	2.0	0.25	>2.0	0.5	0.5	0.5
Thyme	0.5	0.12	>2.0	>2.0	0.25	0.12
Wintergreen	>2.0	0.5	>2.0	0.5	2.0	0.25

Adapted from Hammer et al. (1999a)

Hammer et al. (1999a) found that the method for determining antimicrobial effectiveness of essential oils may have a significant effect on identifying active components. For example, they showed that, using an agar dilution assay, the most effective of 52 spice and herb essential oils and plant extracts against a variety of bacteria and a yeast were lemongrass, oregano and bay (*TABLE 5*). Little activity was demonstrated for thyme. However, when a different assay was used, such as the broth micro-dilution assay, thyme was the most effective essential oil among 20 herbs, spices and plant extracts tested against *E. coli*, *S. aureus* and *Candida albicans* (*TABLE 6*). This shows that the type of assay may influence the results of antimicrobial activity analyses. Similar results were demonstrated by Kubo et al. (1991), who found no inhibition of selected microorganisms with a series of volatile components from cardamom using a disk diffusion assay but significant inhibition using a broth dilution assay. In contrast, Hao et al. (1998) found no antimicrobial activity by thyme against *Aeromonas hydrophila* or *L. monocytogenes* in cooked beef. This could have been due to interference from food components, especially lipids.

Both growth and aflatoxin production of *A. parasiticus* were inhibited by 500 μg/ml thymol for 10 days at 28°C (Buchanan & Sheperd, 1981). At 100 μg/ml, thymol nearly eliminated aflatoxin production by the mold. Conner and Beuchat (1984) tested 32 essential oils from spices and plant extracts at 1 and 10% for growth inhibition activity against 13 different yeast species. The most active growth inhibitors were oregano and thyme followed by clove and cinnamon. Karapinar (1985) confirmed that thyme was an effective inhibitor of *Aspergillus parasiticus* growth.

Aqueous and ethanolic extracts of *Thymus capitatus* (10-200 mg/ml), saponin, resin and essential oil of the plant (10-5000 μg/ml) inhibit the growth of several bacteria and fungi (Kandil et al., 1994). The essential oil of thyme or its constituent thymol, especially under anaerobic conditions reduce the viable counts of *Salmonella* Typhimurium (Juven et al., 1994). Antagonistic effects of thymol against *S. aureus* were also greater

TABLE 6. Comparison of minimum inhibitory concentrations (% v/v) of essential oils of herbs and spices against *Escherichia coli, Staphylococcus aureus* and *Candida albicans* in agar dilution (AD) and broth micro-dilution (BD) assays.

Spice	Escherichia coli		Staphylococcus aureus		Candida albicans	
	AD	BD	AD	BD	AD	BD
Bay	0.12	0.12	0.25	0.12	0.12	0.06
Clove	0.25	0.12	0.25	0.12	0.12	0.12
Ginger	>2.0	>4.0	2.0	>4.0	>2.0	>4.0
Lemongrass	0.06	0.12	0.06	0.06	0.06	0.06
Peppermint	0.5	0.12	1.0	.012	0.5	0.12
Thyme	0.12	0.03	0.25	0.03	0.12	0.03

Adapted from Hammer et al. (1999a)

under anaerobic conditions. In contrast to the phenolic constituents, the chemically related terpenes *p*-cymene and γ-terpinene had no antagonistic effects against *Salmonella* Typhimurium. The addition of Desferal counteracted the antibacterial effects of both thyme oil and thymol, however, any possible role of iron in the oxygen-related antibacterial action was not established. In the presence of thymol, the viable counts of *Salmonella* Typhimurium obtained on a minimal medium were lower than those obtained on nutrient agar. Addition of bovine serum albumin (BSA) neutralized the antibacterial action of thymol, suggesting that the effects of BSA or Desferal are due to their ability to bind phenolic compounds through their amino and hydroxylamine groups, respectively, thus preventing complexation reactions between the oil phenolic constituents and bacterial membrane proteins. This hypothesis was supported by the marked decrease in the viable counts of *Salmonella* Typhimurium caused by either thyme oil or thymol when the pH of the medium was changed from 6.5 to 5.5 or the concentration of Tween 80 in the medium was reduced.

The antifungal activity of several components of essential oils structurally related to eugenol were evaluated using a paper-disk method (Pauli & Knobloch, 1987). Equimolar amounts were tested on more than ten fungal strains known to contaminate food. Iso-eugenol, cinnamaldehyde, carvacrol, eugenol and thymol revealed the strongest antifungal activity. The most resistant strain appeared to be *Penicillium verrucosum* var. *cyclopium*, and the most sensitive was *P. viridicatum*. Some of the structural effects were considered, including a free hydroxyl group in connection with an alkyl substituent which seemed to represent an especially active configuration of phenolic compounds and which rendered antimicrobial activity.

The inhibitory effects of 10 selected Turkish spices, oregano essential oil, thymol and carvacrol towards growth of 9 foodborne fungi were investigated in culture media at pH 3.5 and 5.5 (Akgul & Kivanc, 1988). The antifungal effects of sodium chloride, sorbic acid and sodium benzoate and the combined use of oregano with sodium chloride were also tested under the same conditions for comparison. Of the spices tested, only oregano at 1.0, 1.5, 2.0% (w/v) levels showed effect on all fungi. 8% (w/v) sodium chloride was less effective than oregano. Oregano essential oil, thymol or carvacrol at concentrations of 0.025% and 0.05% completely inhibited the growth of all fungi, showing greater inhibition than sorbic acid at the same concentrations. The combined use of oregano and sodium chloride exhibited a synergistic antifungal effect. Organic and aqueous solvent extracts and fractions of the plant *Micromeria nervosa* (Labiatae) consists of carvacrol and thymol as main ingredients. These extracts strongly inhibit *Candida albicans* and various bacteria including *Proteus vulgaris, Pseudomonas aeruginosa* (Ali-Shtayeh et al., 1997).

2. Antioxidant effects. Antioxidants minimize oxidation of the lipid components in foods. There is an increasing interest in the use of natural and/or synthetic antioxidants in food preservation, but it is important to evaluate such compounds fully for both antioxidant and pro-oxidant properties. The antioxidant properties of thymol, carvacrol, 6-ginerol, hydroxytyrosol and zingerone are well characterized (Aeschbach et al., 1994). These monoterpenoids decrease peroxidation of phospholipid liposomes in the presence of iron(III) and ascorbate, however, zingerone has only a weak inhibitory effect on the system. The compounds were good scavengers of peroxyl radicals (CCl_3O_2; calculated rate constants $> 10^6$ M/sec) generated by pulse radiolysis. Thymol, carvacrol, 6-gingerol and zingerone were not able to accelerate DNA damage in the bleomycin-Fe(III) system. Hydroxytyrosol promoted deoxyribose damage in the deoxyribose assay and also promoted DNA damage in the bleomycin-Fe(III) system. This promotion was inhibited strongly in the deoxyribose assay by the addition of BSA to the reaction mixtures. These data suggest that thymol, carvacrol and 6-gingerol possess useful antioxidant properties and may become important in the search for 'natural' replacements for 'synthetic' antioxidant food additives.

3. Therapeutic effects. The antimicrobial action of thymol-camphor against *S. aureus* was compared with 21 pharmaceutical preparations *in vitro* and was found superior (Trebitsch, 1978). The essential oil of *Bupleurum fruticosum* was investigated qualitatively and quantitatively together with the anti-inflammatory activity of the whole essential oil and its major components (Lorente et al., 1989). In addition, antispasmodic activity was determined in rat uterus preparations using acetylcholine and oxytocin as agonists. The anti-inflammatory activity shown by the essential oil was attributed to α-pinene and β-pinene, and the presence of thymol and carvacrol as minor components, have potentiated the action of these hydrocarbons. Didry et al. (1993) have tested the antimicrobial activity of thymol, carvacrol and cinnamaldehyde against bacteria involved in upper respiratory tract infections. The broad spectrum of activity and the synergistic effect observed with some combinations (specially thymol and carvacrol) could allow the use of the three compounds alone or, like thymol an carvacrol, combined during the treatment of respiratory infections.

Thymol and carvacrol, as the essential oils of two labiatae plants, *Mosla chinensis Maxim.* and *Pogostemon cablin Benth* were shown to elicit antibacterial activity against periodontopathic bacteria, including *Actinobacillus, Capnocytophaga, Fusobacterium, Eikenella* and *Bacteroides* species. (Osawa et al., 1990). Didry et al. (1994) have also tested the antimicrobial activity of thymol, carvacrol, cinnamaldehyde and eugenol alone or combined on eight oral bacteria. The compounds inhibited seven microorganisms and a synergistic effect was observed with certain combinations. The four compounds were suggested for therapeutic applications alone or in combination, as eugenol and thymol, eugenol and carvacrol, thymol and carvacrol, during the treatment of oral infectious diseases. Antimicrobial effects of thymol on oral pathogens *Porphyromonas gingivalis, Selenomonas artemidis* and *Streptococcus sobrinus* were reported (Shapiro & Guggenheim, 1995). The extremely rapid efflux of intracellular constituents evoked by thymol is consistent with its postulated membrane-tropic effects. Correlations between leakage-inducing concentrations of thymol and MIC and minimal bactericidal concentrations suggest that membrane perforation is a principal mode of action of this substance. The thymol-induced decline in intracellular ATP in *S. sobrinus* appears to be entirely attributable to leakage, whereas in *P. gingivalis* thymol may also inhibit ATP-generating pathways. Relative changes in the transmembrane potential of resting cells of *S. sobrinus* pulsed with glucose are as sensitive to thymol as is leakage from this organism. The effects of thymol on transmembrane potential are probably secondary to those arising from leakage of intracellular substances.

In vitro leishmanicidal and trypanocidal activities of a petroleum ether extract of *Oxandra espintana* have been investigated (Hocquemiller et al., 1991). Four aromatic monoterpenes were isolated, of which two are novel: espintanol, responsible for the antiparasite activity, and *O*-methylespintanol. Espintanol was tested *in vitro* on 20 strains of *Trypanosoma cruzi* and 12 strains of *Leishmania spp*. Its structure was determined by spectroscopic methods and confirmed by its preparation starting from carvacrol.

Thymol and carvacrol are also shown to inhibit *Cryptococcus neoformans*, an opportunistic fungal pathogen in AIDS patients. The MIC was 100-200 µl/L *in vitro* (Viollon & Chaumont, 1994). Thymol and carvacrol are also the main essential oil components of *Lippia sidoides Cham*. These essential oils show antibacterial and antifungal properties against microorganisms on the skin of feet and armpits (Lacoste et al., 1996). These essential oils elicit strong antagonistic activities against *Corynebacterium xerosis* that contributes to axillary odor.

D. Borneol

The active antimicrobial fraction of sage and rosemary has been suggested to be borneol and other phyto-phenols of the terpene fraction. At 2% in bacterial growth medium sage and rosemary were more active against Gram-positive than Gram-negative strains (Shelef et al., 1980). The inhibitory effect of these two spices at 0.3% was bacteriostatic while at 0.5% they were bactericidal to the Gram-positive strains. Pandit and Shelef (1994) studied the antimicrobial effectiveness of 18 spices against the growth of *L. monocytogenes* in culture medium. The most effective compound was 0.5% rosemary which was bactericidal. The fraction of rosemary essential oil that was most inhibitory to *L. monocytogenes* was α-pinene. Smith-Palmer et al. (1998) demonstrated that rosemary (0.02-0.05%) and sage (0.02-0.075%) were inhibitory to Gram-positive *L. monocytogenes* and *S. aureus* but not to Gram-negative bacteria. Further, Pandit and Shelef (1994) showed that *L. monocytogenes* Scott A growth in refrigerated fresh pork sausage was delayed by 0.5% ground rosemary or 1% rosemary essential oil. In contrast, Mangina and Muyima (1999) indicated little antimicrobial activity with essential oil of rosemary against bacteria and yeast in an agar diffusion assay. They found that, with a 1:2 dilution of the oil, 25 of 29 bacterial strains and 10 of 12 yeast strains tested showed zones of inhibition about 3.5 mm. The lack of activity could be due to the assay utilized. Shelef et al. (1984) found that sensitivity of *Bacillus cereus*, *S. aureus* and *Pseudomonas* to sage was greatest in microbiological medium and significantly reduced in rice and chicken and noodles. It was theorized that loss of activity was due to solubilization of the antimicrobial fraction in the lipid of the foods. Hefnawy et al. (1993) evaluated 10 herbs and spices against two strains of *L. monocytogenes* in tryptose broth. The most effective spice was sage which at 1% decreased viable *L. monocytogenes* by 5-7 logs after 1 day at 4°C. Allspice was next most effective inactivating the microorganism in 4 days.

Hammerschmidt et al. (1993) analyzed the essential oils of the aerial parts of *Jasonia candicans* and *J. montana* by gas chromatography-mass spectrometry (GC/MS). Of twenty-one components identified in the volatile oil of *J. candicans*, intermediol was the main constituent. Fifty-eight components were characterized in the essential oil of *J. montana*. Borneol, bornyl acetate, camphor, chrysanthemol, intermediol, and 1,8-cineole were the major constituents in this oil. The two oils showed antibacterial activity against *Bacillus subtilis*. They also showed a marked antifungal activity against *Trichophyton mentagrophytes, Cryptococcus neoformans*, and *Candida albicans*.

Liu (1990) reported the therapeutic effects of borneol. In a clinical trial, 170 patients with purulent otitis media were treated with borneol-walnut oil of various concentrations, and the controls (108 patients) were treated with neomycin. The total effective rates were 98.1% and 84.1% respectively (p <0.001) indicating that the therapeutic effects of borneol-walnut oil for the treatment of purulent otitis media is superior to that of neomycin compound. The most optimal concentration of borneol-walnut oil was 20% through clinical and laboratory observations. Due to its simple composition, significant therapeutic effects and nontoxic reactions, the borneol-walnut oil has been proved a promising external remedy for the treatment of purulent otitis media.

E. Eugenol and cinnamic aldehyde

The major antimicrobial components of cinnamon and clove are cinnamic aldehyde (3-phenyl-2-propenal) and eugenol [2-methoxy-4-(2-propenyl)-phenol], respectively. Cinnamon contains 0.5-1.0% volatile oil of which 75% is cinnamic aldehyde and 8% eugenol while cloves contain 14-21% volatile oil, 95% of which is eugenol (Bullerman et al., 1977). Zaika and Kissinger (1979) showed that clove at 0.4% in a Lebanon bologna formulation inhibited growth and acid production by a lactic acid bacterial starter culture. Cinnamon at 0.8% in the same system has moderately inhibited growth. Both spices stimulated acid production at low concentrations. Ting and Deibel (1992) reported that up to 3.0% cinnamon had no effect on growth of *L. monocytogenes* at 24°C. In contrast, Bahk et al. (1990) found that 0.5% cinnamon was the most effective spice against growth of *L. monocytogenes* strain V7 in tryptose broth at 37°C. Smith-Palmer et al. (1998) determined that the 24-h MIC of cinnamon essential oil against *Campylobacter jejuni, E. coli, Salmonella* Enteritidis, *L. monocytogenes*, and *S. aureus* were 0.05, 0.05, 0.05, 0.03, and 0.04%, respectively, in an agar dilution assay.

Bullerman (1974) determined that 1.0% cinnamon in raisin bread inhibited growth and aflatoxin production by *Aspergillus parasiticus*. A 0.2% alcoholic extract of cinnamon was 95% effective in inhibiting toxin production by the same organism in microbiological media. In a related study, Bullerman et al. (1977) found that the essential oils of clove and cinnamon, eugenol and cinnamic aldehyde, were more effective inhibitors of *A. parasiticus* growth and toxin production than the parent spices. Azzouz and Bullerman (1982) evaluated 16 ground herbs and spices at 2% w/v against nine mycotoxin producing *Aspergillus* and *Penicillium* species. The most effective antifungal spice evaluated was clove, which inhibited growth initiation at 25°C by all species for over 21 days. Cinnamon was the next most effective spice, inhibiting three *Penicillium* species for over 21 days.

Aqueous clove infusions (0.1-1.0% w/v) inhibited outgrowth of germinated spores of *Bacillus subtilis* in nutrient agar (Al-Khayat & Blank, 1985). The MIC of cloves against *L. monocytogenes* growth at 24°C in a microbiological medium was 0.5-0.7% w/v (Ting & Deibel, 1992). Similar results were reported by Bahk et al. (1990). Cloves at 0.5% were shown to be bactericidal to *L. monocytogenes* in trypticase soy agar at 4 or 24°C but had little effect growth of the microorganism in beef at 1.0% w/v (Ting & Deibel, 1992). Smith-Palmer et al. (1998) determined the MICs of clove oil against *Campylobacter jejuni, E. coli, Salmonella* Enteritidis, *L. monocytogenes*, and *S. aureus* and they were 0.05, 0.04, 0.04, 0.03, and 0.04%, respectively. Essential oil of clove at 500 µg/ml inhibited *Aeromonas hydrophila* growth in microbiological media and in vacuum-packaged and air-packed cooked pork (Stecchini et al., 1993). Similarly, Wendakoon and Sakaguchi (1993) found that 0.5% clove oil in combination with 2% NaCl completely inhibited growth and biogenic amine formation by *Enterobacter aerogenes* in Mackerel broth. Eugenol at 0.06% v/v was found to prevent outgrowth of 10^5 CFU/ml *Bacillus subtilis* after 72-h at 37°C. Eugenol was also the most effective antibacterial spice component tested against *Salmonella* Typhimurium, *S. aureus* and *Vibrio parahaemolyticus* compared to thymol, anethole and menthol (Karapinar & Aktug, 1987). Hao et al. (1998) added nine spice extracts (0.1 ml of a 20% ethanol extract) to 25 g cooked beef inoculated with either *Aeromonas hydrophila* or *L. monocytogenes* and incubated at 5 or 15°C for up to 14 days. Of the extracts, eugenol was most inhibitory against both microorganisms although it did not completely prevent the growth of either.

Deans et al. (1995) tested the antimicrobial activity of clove essential oils against a collection bacteria and molds. The essential oil was most effective as an antifungal agent against *Aspergillus* and *Fusarium* species. Eugenol at 300 µg/ml was completely inhibitory to growth of mycotoxin- and non-mycotoxin-producing strains of *Aspergillus* and *Penicillium* in Sabouraud dextrose agar (Mansour et al., 1996). In addition, aflatoxin B_1 production by *A. parasiticus* was reduced 58% by 100 µg/ml at while, as indicated by mycelial dry weight, was inhibited only 5.3%. Jayashree and Subramanyam (1999) determined that the antioxidant properties of eugenol allow its inhibition of aflatoxin production by *Aspergillus parasiticus* through inhibition of the final steps of aflatoxin biosynthesis involving lipid peroxidation.

Smith-Palmer, et al., (1998) tested the antimicrobial properties of 21 plant essential oils and two essences against five important food-borne pathogens including *Campylobacter jejuni, Salmonella* Enteritidis, *E. coli, S. aureus* and *L. monocytogenes*. The oils of bay, cinnamon, clove and thyme were the most inhibitory, each having a bacteriostatic concentration of 0.075% or less against all five pathogens. In general, Gram-positive bacteria were more sensitive to inhibition by plant essential oils than the Gram-negative bacteria. *Campylobacter jejuni* was the most resistant of the bacteria investigated to plant essential oils, with only the oils of bay and thyme having a bactericidal concentration of less than 1%. *L. monocytogenes* was extremely sensitive to the oil of nutmeg. A concentration of less than 0.01% was bacteriostatic and 0.05% was bactericidal, but when the temperature was reduced to 4 degrees, the bacteriostatic concentration was increased to 0.5% and the bactericidal concentration to >1%.

Shcherbanovsky and Kapelev (1975) reported the antimicrobial activity of 25 volatile oils from aerial parts and seeds of dill (*Anethum graveolens L.*) against yeast *Saccharomyces vini* and *Lactobacillus buchneri*. Yousef and Tawil (1980) evaluated the bacteriostatic and fungistatic activities of 22 volatile oils, wherein, the cinnamon oil showed the highest activity against the tested bacteria and fungi.

F. Linalool and chavicol

Sweet basil (*Ocimum basilicum* L.) essential oil has some antimicrobial activity due primarily to the presence of linalool and methyl chavicol (Wan et al., 1998). Lachowicz et al. (1998) evaluated essential oils of sweet basil extracted by distillation against 33 bacteria, yeasts and molds in an agar well assay. The essential oils were active against certain fungi including *Mucor* and *Penicillium* species but demonstrated little antimicrobial activity against bacteria in the assay system. A strain of *Bacillus cereus* and *Salmonella* Typhimurium were most sensitive to the basil. Addition of 0.1% or 1.0% basil oil to tomato juice medium inactivated 10^4 CFU/ml of *Lactobacillus curvatus* (7 days) or *Saccharomyces cerevisiae* (immediate) at 30°C, respectively. Wan et al. (1998) did a similar study by screening the essential oil components of sweet basil, linalool and methyl chavicol, against 35 strains of bacteria, yeasts and molds using an agar well diffusion assay. Again, the compounds demonstrated limited activity against most microorganisms except the fungi, *Mucor* and *Penicillium* with the well assay system employed. In an system using impedance to monitor growth, 0.125% (v/v) methyl chavicol inhibited *Aeromonas hydrophila* growth for 48-h at 30°C while 1% (v/v) of linalool was required. Against *Pseudomonas fluorescens*, a minimum of 2% (v/v) of the essential oil components were required for inhibition. Methyl chavicol (0.1%) in filter-sterilized fresh lettuce supernatant reduced viable *Aeromonas hydrophila* by 5-logs and, as a wash for lettuce leaves, the compound was as effective as 125 µg/ml chlorine (Wan et al., 1998). Smith-Palmer et al. (1998) reported MICs of basil essential oil of 0.25, 0.25, 0.1, 0.05, and 0.1% for *Campylobacter jejuni*, *E. coli*, *Salmonella* Enteritidis, *L. monocytogenes* and *S. aureus*, respectively. Yin and Cheng (1998) detected no antifungal activity against *Aspergillus flavus* or *Aspergillus niger* with water extracts of basil or ginger in a paper disk assay.

Pattnaik, et al., (1997) tested antimicrobial activity of five aromatic constituents of essential oils (cineole, citral, geraniol, linalool and menthol) against eighteen bacteria (including Gram-positive cocci and rods, and Gram-negative rods) and twelve fungi (three yeast-like and nine filamentous). Linalool was the most effective and inhibited seventeen bacteria, followed by cineole, geraniol (each of which inhibited sixteen bacteria), menthol and citral aromatic compounds, which inhibited fifteen and fourteen bacteria, respectively. Against fungi the citral and geraniol oils were the most effective (inhibiting all twelve fungi), followed by linalool (inhibiting ten fungi), cineole and menthol (each of which inhibited seven fungi) compounds.

Lachowicz, et al., (1998) examined essential oils from five varieties of *Ocimum basilicum* L. plants (anise, bush, cinnamon, dark opal and dried basil) for antimicrobial activity against a wide range of foodborne Gram-positive and Gram-negative bacteria, yeasts and moulds. All five essential oils showed antimicrobial activity against most of the bacteria tested, except *Flavimonas oryzihabitans* and *Pseudomonas* species. Synergistic effects were observed between anise oil, low pH (4.2) and salt (5% NaCl). Anise oil demonstrated antimicrobial effect in tomato juice medium and inhibited the growth of *Lactobacillus curvatus* and *Saccharomyces. cerevisiae*.

G. Vanillin

Vanillin (4-hydroxy-3-methoxybenzaldehyde) is a major constituent of vanilla beans, the fruit of an orchid (*Vanilla planifola*, *Vanilla pompona*, or *Vanilla tahitensis*). It has been shown to possess antimicrobial activity (Beuchat & Golden, 1989). Jay and

Rivers (1984) reported that vanillin was most active as against molds and non-lactic Gram positive bacteria. López-Malo et al. (1995) prepared fruit based agars containing mango, papaya, pineapple, apple and banana with up to 2000 μg/ml vanillin and inoculated each with *Aspergillus flavus*, *A. niger*, *A. ochraceus*, or *A. parasiticus*. Vanillin at 1500 μg/ml significantly inhibited all strains of *Aspergillus* in all media. The compound showed least effectiveness in banana and mango agars. Cerrutti and Alzamora (1996) demonstrated complete inhibition of growth for 40 days at 27°C of *Debaryomyces hansenii*, *Saccharomyces cerevisiae*, *Zygosaccharomyces bailii* and *Zygosaccharomyces rouxii* in laboratory media and apple puree at a_w of 0.99 and 0.95 by 2000 μg/ml vanillin. In contrast, 2000 μg/ml vanillin was not effective against the yeasts in banana puree. These researchers attributed this loss of activity to binding of the phenolic vanillin by protein or lipid in these banana and mango, a phenomenon demonstrated for other antimicrobial phenolic compounds (Rico-Muñoz & Davidson, 1983). López-Malo et al. (1997,1998) studied the effect of vanillin concentration, pH and incubation temperature on *Aspergillus flavus*, *A. niger*, *A. ochraceus*, and *A. parasiticus* growth on potato dextrose agar. Depending upon species, vanillin in combination with reduced pH had an additive or synergistic effect on growth of the molds. The most sensitive was *A. ochraceus* (MIC = 500 μg/ml at pH 3.0, ≤ 25°C or pH 4.0, ≤ 15°C) and the most resistant was *A. niger* (MIC = 1000 μg/ml, pH 3.0, 15°C). Cerrutti et al. (1997) utilized vanillin to produce a shelf-stable strawberry puree. A combination of 3000 μg/ml vanillin, 1000 μg/ml calcium lactate, and 500 μg/ml ascorbic acid preserved strawberry puree acidified with citric acid to pH 3.0 and adjusted to a_w 0.95 with sucrose for over 60 days against growth of natural microflora and inoculated yeasts.

H. Terpenes

Tea-tree oil (an essential oil of the Australian native tree *Melaleuca alternifolia*) has long been regarded as a useful topical antiseptic agent in Australia and has been shown to have a variety of antimicrobial activities; however, only anecdotal evidence exists for its efficacy in the treatment of various skin conditions. Gustafson et al. (1998) reported on the antimicrobial activity of Australian tea tree (*Melaleuca alternifolia*) essential oil. The major active components of tea tree oil include terpinen-4-ol, γ-terpinene, α-terpinene, α-terpineol, α-pinene, 1,8-cineole and linalool (Gustafson et al., 1998; Hammer et al., 1999b). Against *E. coli* AG100, the MIC and MBC were 0.25% in microbiological media (Gustafson et al., 1998). Addition of tea tree oil at 0.25% to log and stationary phase cells caused a reduction of > 8 logs of viable *E. coli* in 30 min and 3-4 logs after 2-h, respectively. Hammer et al. (1999b) determined the antimicrobial effectiveness of tea tree oil and characterized factors affecting its activity. The MICs and MLCs, respectively, determined for the oil against selected bacteria and a yeast were (% v/v): *E. coli*, 0.12-0.25 and 0.12-0.25, *Salmonella* Typhimurium, 0.25-0.5 and 0.25-0.5, *S. aureus*, 0.5 and 0.5-1.0, and *Candida albicans*, 0.25-0.5 and 0.5. The presence of surfactants, such as Tween 20 or 80, generally increased the MICs against microorganisms, while the presence of minerals or organic matter, such as blood, bovine serum albumin, baker's yeast, or skim milk, slightly increased or had no effect on the MIC depending upon microorganism. Cox et al. (1998) showed that tea tree essential oil caused leakage and inhibited glucose dependent respiration of *E. coli* AG100. This indicated a mechanism of action involving disruption of the cytoplasmic membrane.

Bassett, et al., (1990) conducted a single-blind, randomized clinical trial on 124 patients to evaluate the efficacy and skin tolerance of 5% tea-tree oil gel in the treatment of mild to moderate acne when compared with 5% benzoyl peroxide lotion. The results of this study showed that both 5% tea-tree oil and 5% benzoyl peroxide had a significant effect in ameliorating the patients' acne by reducing the number of inflamed and non-inflamed lesions (open and closed comedones), although the onset of action in the case of tea-tree oil was slower.

Aromatic plants from the Labiatae family (*Thymus vulgaris, Ocimum gratissimum*), the Myrtaceae family (*Eugenia caryophyllata, Melaleuca viridiflora*) and the Compositae (*Helichrysum lavanduloides, H. bracteiferum, H. gymnocephalum, Psiadia altissima*) show antimicrobial activity against enteropathogenic and food spoilage organisms (Ramanoelina, et al., 1987). Three oils from *Thymus vulgaris, Ocimum gratissimum* and *Eugenia caryophyllata* demonstrated broad-spectrum activity. The essential oil of *Melaleuca viridiflora* had a high inhibitory effect especially on Gram-positive bacteria.

Essential oil from *Achillea fragrantissima* exerts cidal effect on several Gram-positive and Gram-negative bacteria as well as on *C. albicans*. The active phyto-phenolic compound was identified as terpinen-4-ol (Barel, et al., 1991). Essential oils from *Cedronella canariensis (L.) W. et B.* inhibit respiratory tract pathogens *Bordetella bronchiseptica* and *Cryptococcus albidus* (Lopez-Garcia, et al., 1992). The essential oil from the leaves of *Hoslundia opposita* contains largely the sesqui-terpenes and sesquiterpene alcohols. These phyto-phenolic compounds show significant activity against *Aspergillus niger, Acinetobacter calcoacetica, Brochothrix thermosphacta* and *Flavobacterium suaveolens*. (Gundidza, et al., 1992). Essential oils from *Satureja montana L., Rosmarinus officinalis L., Thymus vulgaris L.,* and *Calamintha nepeta (L.) Savi* demonstrate potent antimicrobial and fungicide activities (Panizzi, et al., 1993). Camphor and camphene are the major essential oil constituents of *Piper angustifolium Lam*. These phyto-phenolic compounds are bacteriostatic and fungistatic against *Trichophyton mentagrophytes, P. aeruginosa, C. albicans, Cryptococcus neo-formans, Aspergillus flavus, Aspergillus fumigatus,* and *E. coli* (Tirillini, et al., 1996).

I. Compounds with limited activity

Many herbs and spices have demonstrated limited or no antimicrobial activity. They include anise, bay (laurel), black pepper, cardamom, cayenne (red pepper), celery seed, chili powder, coriander, cumin, curry powder, dill, fenugreek, ginger, juniper oil, mace, marjoram, mint, nutmeg, orris root, paprika, sesame, spearmint, tarragon, and white pepper (Fabian et al., 1939; Webb & Tanner, 1944; Marth, 1966; Zaika & Kissinger, 1979; Tassou et al., 1995). For example, chili, paprika, parsley, red pepper and black pepper at up to 3.0% w/v did not inhibit the growth of *L. monocytogenes* Scott A at 24°C (Ting & Deibel, 1992). Hao et al. (1998) showed it (pimento) had little effect on the microorganism in cooked beef. This was suggested to be due to the reduction of activity of the essential oil in the presence of lipids in the food (Hao et al., 1998). Hao et al. (1998) showed that angelica, bay leaf, caraway and marjoram had no activity against *L. monocytogenes* in cooked beef. Tassou et al. (1995) evaluated the effect of mint (*Mentha piperita*) essential oil on *Salmonella* Entertidis and *L. monocytogenes* in culture medium and three foods. The effect of the essential oil was variable and depended on pH, temperature and the food. Eucalyptus, marjoram and peppermint were inhibitory primarily to the Gram-positive bacteria. Bay (laurel) was found by these researchers to have inhibitory and bactericidal activity in the ranges of 0.002-0.075% and 0.04-0.5%, respectively, against the five bacteria. This is in agreement with some reports (Hammer et al., 1999a) and in contrast to other reports that bay has little or no activity (Karapinar & Aktug, 1986; Hao et al., 1998). Perhaps the most unique finding by Smith-Palmer et al. (1998) was that nutmeg essential oil was inhibitory to *L. monocytogenes* at < 0.01% and bactericidal at 0.05% at 35°C. The activity of nutmeg was highly dependent upon temperature as the MIC and MBC increased to 0.5% and > 1% at 4°C. The antimicrobial effectiveness of bay, cinnamon and thyme was not influenced by a temperature shift from 35 to 4°C. Huhtanen (1980) showed that 10% ethanol extracts of selected spices were inhibitory to *Clostridium botulinum* including (MIC in μg/ml in parenthesis): mace (31), bay (125), black pepper (125), nutmeg (125), and white pepper (125). Ejechi et al. (1998) found that high concentrations (15-20% v/v) of water extracts of ginger and nutmeg could inhibit the growth of natural spoilage microflora of mango juice for 1 month at 30°C. This high concentration of extract was not acceptable on a sensory basis. However, 4% of each extract combined with heating the mango juice to 55°C prevented growth of both yeast and mold in the juice and produced a product with satisfactory sensory attributes. Kubo et al. (1991) determined the antimicrobial activity of cardamom essential oil and 10 components of the *n*-hexane extract of cardamom using a broth dilution assay. Against food-related bacteria, eugenol (also present in cloves) was most effective against the Gram-positive bacteria *Bacillus subtilis* and *S. aureus* (MIC 400-800 μg/ml) while, with the exception of *E. coli* which was inhibited by 400 μg/ml eugenol, none of the components was inhibitory to the other Gram-negative bacteria, *Pseudomonas aeruginosa* or *Enterobacter aerogenes*. The most effective antimicrobial components against fungi were geraniol, limonene and α-terpinene. Cardamom essential oil was very limited in its antimicrobial activity generally requiring 800 to > 1,600 μg/ml to inhibit food-related bacteria and fungi (Kubo et al., 1991).

Turmeric was shown to inhibit a variety of bacteria, including *B. cereus, S. aureus, E. coli,* and *Lactobacillus plantarum* (Bhavani Shankar & Sreenivasa Murthy, 1979). Whereas, nutmeg, curry powder, mustard, black pepper, and sassafras could moderately inhibit *V. parahaemolyticus* (Beuchat, 1976). The spice, *Aframomum danielli*, on a wet weight basis with a moisture content of 10.5%, protein content of 8.2% (dry matter basis) could inhibit the growth of *Salmonella enteriditis, Pseudomonas fragi, P. fluorescens, Proteus vulgaris, Streptococcus pyogenes, S. aureus, Aspergillus flavus, A. parasiticus, A. ochraceus* and *A. niger*. The MIC for *Klebsiella pneumoniae* and *P. aeruginosa* was 1 in 32 whilst the MIC for *S. aureus* was 1 in 8,000 (Adegoke & Skura, 1994).

J. Mechanisms of action

Some studies have focused on the mechanism by which spices or their essential oils inhibit microorganisms. Since it has been concluded that the terpenes in essential oils of spices are the primary antimicrobials, the mechanism most likely involves these compounds. Many of the most active terpenes, e.g., eugenol, thymol and carvacrol, are phenolic in nature. Therefore, it would seem reasonable that their modes of action might be related to other phenolic compounds. The mode of action of phenolic compounds is generally thought to involve interference with functions of the cytoplasmic membrane including proton motive force and active transport (Eklund, 1985; Davidson, 1993). In fact, Juven et al. (1994) hypothesized that inhibition by thymol of *Salmonella* Typhimurium was due to a reaction of the compound with proteins in the cytoplasmic membrane of the microorganism. This caused changes in permeability of the membrane resulting in possible leakage and affected the proton motive force. In addition, terpenes may have other antimicrobial mechanisms. Conner and Beuchat (1984) suggested that spice essential oils may inhibit enzyme systems in yeasts, including those involved in energy production and synthesis of structural components. Conner et al. (1984) demonstrated energy depletion in yeasts caused by essential oils of several spices including clove, thyme, oregano and cinnamon. In addition, ethanol production, respiratory activity and sporulation of yeast cells was also influenced by these essential oils.

IV. SUMMARY

Spices and herbs and their essential oils have varying degrees of biological activity. Of the 70 herbs and spices officially recognized as useful for food ingredients (Lindsay, 1996), only a handful have demonstrated significant antimicrobial activity. In many cases, concentrations of the antimicrobial compounds in herbs and spices are too low to be used effectively without adverse effects on the sensory characteristics of a food. However, they may contribute to the overall 'hurdle' system present naturally in a food product. The most consistent antimicrobial activity among herbs and spices has been found with components of cloves, cinnamon, mustard seed, oregano, rosemary, sage, thyme, and vanillin. Effective phyto-phenolic agents preserve food by various mechanisms including stasis (growth inhibition of microorganisms) and cidal (direct destruction of microorganisms) effects. Certain phyto-phenolic agents such as from olive oil seem to deliver multifunctional physiological benefits to the consumer, therefore are highly attractive to the health food industry. Since phyto-phenolic compounds have been in the food supply and consumed for many years, these natural phytochemicals appear to be safe when compared to new synthetic preservatives.

V. REFERENCES

1. Adegoke, G.O., and Skura, B.J. 1994. Nutritional profile and antimicrobial spectrum of the spice *Aframomum danielli K. Schum. Plant Foods Hum. Nutr.* 45:175-182.
2. Aeschbach, R., Loliger, J., Scott, B.C., Murcia, A., Butler, J., Halliwell, B., and Aruoma, O.I. 1994. Antioxidant actions of thymol, carvacrol, 6-gingerol, zingerone and hydroxytyrosol. *Food Chem. Toxicol.* 32:31-36
3. Akgul, A., and Kivanc, M. 1988. Inhibitory effects of selected Turkish spices and oregano components on some foodborne fungi. *Int. J. Food Microbiol.* 6:263-268.
4. Ali-Shtayeh, M.S., Al-Nuri, M.A., Yaghmour, R.M., and Faidi, Y.R. 1997. Antimicrobial activity of *Micromeria nervosa* from the Palestinian area. *J. Ethnopharmacol.* 58:143-147.
5. Al-Khayat, M.A., and Blank, G. 1985. Phenolic spice components sporostatic to *Bacillus subtilis. J. Food Sci.* 50:971-974.
6. Al-Meshal, I.A. 1986. Isolation, and characterization of a bioactive volatile oil from *Ducrosia ismaelis Asch. Res. Commun. Chem. Pathol. Pharmacol.* 54:129-132.
7. Amaya Guerra, C.A., Othon Serna Saldivar, S.R., Cardenas, E., and Nevero Muunoz, J.A. 1997. Evaluation of different solvent systems for the extraction and fractionation of oleoresins from guajillo peppers. *Arch. Latinoam. Nutr.* 47:127-130.
8. Avato, P., Vitali, C., Mongelii, P., and Tava, A. 1997. Antimicrobial activity of polyacetylenes from *Bellispernnis* and their synthetic derivatives. *Planta Med.* 63:503-507.

9.   Aziz, N.H., Farag, S.E., Mousa, L.A., and Abo-Zaid, M.A. 1998. Comparative antibacterial and antifungal effects of some phenolic compounds. *Microbios* 93:43-54.

10. Azzouz, M.A., and Bullerman, L.B. 1982. Comparative antimycotic effects of selected herbs, spices, plant components and commercial fungal agents. *J. Food Sci.* 45:1298-1301.

11. Bahk, J., Yousef, A.E., and Marth, E.H. 1990. Behavior of *Listeria monocytogenes* in the presence of selected spices. *Lebensmittel-Wissenschaft-und-Technologie* 23:66-69.

12. Barel, S., Segal, R., and Yashphe, J. 1991. The antimicrobial activity of the essential oil from *Achillea fragrantissima*. *J. Ethnopharmacol.* 33:187-191.

13. Basile, A.C., Sertie, J.A., Freitas, P.C., and Zanini, A.C. 1988. Anti-inflammatory activity of oleoresin from Brazilian Copaifera. *J. Ethnopharmacol.* 22:101-109.

14. Bassett, I.B., Pannowitz, D.L., and Barnetson, R.S. 1990. A comparative study of tea-tree oil versus benzoylperoxide in the treatment of acne. *Med. J. Aust.* 153:455-458.

15. Beuchat, L.R. 1976. Sensitivity of *Vibrio parahaemolyticus* to spices and organic acids. *J. Food Sci.* 41:899-902.

16. Beuchat, L.R. 1994. Antimicrobial properties of spices and their essential oils. In *Natural Antimicrobial Systems and Food Preservation*, eds. V.M. Dillon., and R.G. Board, pp.167-179. Wallingford, UK: CAB Intl..

17. Beuchat, L.R., and Golden, D.A. 1989. Antimicrobials occurring naturally in foods. *Food Technol.* 43(1):134-142.

18. Bhavani Shankar, T.N., and Sreenivasa Murthy, V. 1979. Effect of tumeric (*Curcuma longa*) fractions on the growth of some intestinal and pathogenic bacteria in vitro. *J. Exp. Biol.* 17:1363.

19. Briozzo, J., Nunez, L., Chirife, J., Herszage, L., and D'Aquino, M. 1989. Antimicrobial activity of clove oil dispersed in a concentrated sugar solution. *J. Appl. Bacteriol.* 66:69-75.

20. Buchanan, R.L., and Sheperd, A.J. 1981. Inhibition of *Aspergillus parasiticus* by thymol. *J. Food Sci.* 46:976-977.

21. Bullerman, L.B. 1974. Inhibition of aflatoxin production by cinnamon. *J. Food Sci.* 39:1163-1165.

22. Bullerman, L.B., Lieu, F.Y., and Seier, S.A. 1977. Inhibition of growth and aflatoxin production by cinnamon and clove oils, cinnamic aldehyde and eugenol. *J. Food Sci.* 42:1107-1109.

23. Carson, C.F., and Riley, T.V. 1995. Antimicrobial activity of the major components of the essential oil of *Melaleuca alternifolia*. *J. Appl. Bacteriol.* 78:264-269.

24. Cerrutti, P., Alzamora, S.M., and Vidales, S.L. 1997. Vanillin as an antimicrobial for producing shelf-stable strawberry puree. *J. Food Sci.* 62:608-610.

25. Cerrutti, P., and Alzamora, S.M. 1996. Inhibitory effects of vanillin on some food spoilage yeasts in laboratory media and fruit purées. *Intl. J. Food Microbiol.* 29:379-386.

26. Charalambous, G. (ed.) 1994. *Spices, Herbs and Edible Fungi*. Amsterdam: Elsevier.

27. Conner, D.E. and Beuchat, L.R. 1984. Effects of essential oils from plants on growth of food spoilage yeasts. *J. Food Sci.* 49:429-434.

28. Conner, D.E., Beuchat, L.R., Worthington, R.E., and Hitchcock, H.L. 1984. Effects of essential oils and oleoresins of plants on ethanol production, respiration and sporulation of yeasts. *Int. J. Food Microbiol.* 1:63-74.

29. Cosentino, S., Tuberoso, C.I.G., Pisano, B., Satta, M., Mascia, V., Arzedi, E., and Palmas, F. 1999. *In-vitro* antimicrobial activity and chemical composition of Sardinian *Thymus* essential oils. *Lett. Appl. Microbiol.* 29:130-135.

30. Cox, S.D., Gustafson, J.E., Mann, C.M., Markham, J.L., Liew, Y.C., Hartland, R.P., Bell, H.C., Warmington, J.R., and Wyllie, S.G. 1998. Tea tree oil causes K$^+$ leakage and inhibits respiration in *Escherichia coli*. *Lett. Appl. Microbiol.* 26:355-358.

31. Davidson, P.M. 1993. Parabens and phenolic compounds, p. 263-306, In P.M. Davidson and A.L. Branen (ed.), *Antimicrobials in Foods*, 2nd ed. New York: Marcel Dekker, Inc.

32. de la Puerta, R., Ruiz Gutierrez, V., and Hoult, J.R. 1999. Inhibition of leukocyte 5-lipoxygenase by phenolics from virgin olive oil. *Biochem. Pharmacol.* 57:445-449.

33. Deans, S.G., Noble, R.C., Hiltunen, R.,Wuryani, W., and Penzes, L.G. 1995. Antimicrobial and antioxidant properties of *Syzygium aromaticum* (L.) Merr. & Perry: impact upon bacteria, fungi and fatty acid levels in ageing mice. *Flavour and Fragrance J.* 10:323-328.

34. Demetzos, C., Katerinopoulous, H., Kouvarakis, A., Stratigakis, N., Loukis, A., Ekonomakis, C., Spiliotis, V., and Tsaknis, J. 1997. Composition and antimicrobial activity of the essential oil of *Cistus creticus* ssp. ericephalus. *Planta Med.* 63:477-479.

35. Demetzos, C., Stahl, B., Anastassaki, T., Gazouli, M., Tzouvelekis, L.S., and Rallis, M. 1999. Chemical analysis and antimicrobial activity of the resin *Ladano*, of its essential oil and of the isolated compounds. *Planta Med.* 65:76-78.

36. Didry, N., Dubreuil, L., and Pinkas, M. 1993. Antibacterial activity of thymol, carvacrol and cinnamaldehyde alone or in combination. *Pharmazie* 48:301-304.

37. Didry, N., Dubreuil, L., and Pinkas, M. 1994. Activity of thymol, carvacrol, cinnamaldehyde and eugenol on oral bacteria. *Pharm. Acta Helv.* 69:25-28.

38. Economou, D., and Nahrstedt, A. 1991. Chemical, physiological and toxicological aspects of the essential oil of some species of the genus Bystropogon. *Planta Med.* 57:347-351.

39. Ejechi, B.O., Souzey, J.A., and Akpomedaye, D.E. 1998. Microbial stability of mango (*Mangifera indica* L.) juice preserved by combined application of mild heat and extracts of two tropical spices. *J. Food Prot.* 61:725-727.

40. Eklund, T. 1985. Inhibition of microbial growth at different pH levels by benzoic and propionic acids and esters of *p*-hydroxybenzoic acid. *Inter. J. Food Microbiol.* 2:159.

41. El-Fatatry, H.M. 1975. Isolation and structure assignment of an antimicrobial principle from the volatile oil of *Nigella sativa* L. seeds. *Pharmazie* 30:109-111.

42. Ezaki, K., Sekita, S., Kawahara, N., Shirota, O., Kamakura, H., and Satake, M. 1997. Stereochemical structure of D-borneol in 'The Japanese Standards of Food Additives'. *Kokuritsu Iyakuhin Shokuhin Eisei Kenkyusho Hokoku* 115:125-129.

43. Fabian, F.W., Krehl, C.F., and Little, N.W. 1939. The role of spices in pickled food spoilage. *Food Res.* 4:269.

44. Farag, R.S., Daw, Z.Y., Hewedi, F.M., and El-Baroty, G.S.A. 1989. Antimicrobial activity of some Egyptian spice essential oils. *J. Food Prot.* 52:665.

45. Farbood, M.I., MacNeil, J.H., and Ostovar, K. 1976. Effect of rosemary spice extractive on growth of microorganisms in meat. *J. Milk Food Technol.* 39:675.

46. Farrell, K.T. 1985. *Spices, Condiments, and Seasonings.* AVI Publ. Co., Westport, CT.

47. Firouzi, R., Azadbakht, M., and Nabinedjad, A. 1998. Anti-listerial activity of essential oils of some plants. *J. Appl. Ani. Res.* 14:75-80.

48. Flamini, G., Cioni, P.L., Puleio, R., Morelli, I., and Panizzi, L. 1999. Antimicrobial activity of the essential oil of *Calamintha nepeta* and its constituent pulegone against bacteria and fungi. *Phytother. Res.* 13:349-351.

49. Fleming, H.P., Walter, W.M. Jr, and Etchells, J.L. 1973. Antimicrobial properties of oleuropein and products of its hydrolysis from green olives. *Appl. Microbiol.* 26:777-782.

50. Ghisalberti, E.L. 1993. Detection and isolation of bioactive natural products. In *Bioactive Natural Products: Detection, Isolation and Structural Determination,* ed. S.M. Colgate and R.J. Molyneux, pp. 9-57. Boca Raton, FL: CRC Press.

51. Grove, O. 1918. The preservative action of various spices and essential oils. *Annual Report Agric. Hort. Res. Station,* pp. 29. Bristol, England: Long Ashton,

52. Gundidza, G.M., Deans, S.G., Svoboda, K.P., and Mavi, S. 1992. Antimicrobial activity of essential oil from *Hoslundia opposita. Cent. Afr. J. Med.* 38:290-293.

53. Gundidza, M., Deans, S.G., Kennedy, A.I., Mavi, S., Waterman, P.G., and Gray, A.I. 1993. The essential oil from *Heteropyxis natalensis Harv:* its antimicrobial activities and phytoconstituents. *J. Sci. Food Agri.* 63:361-364.

54. Gundidza, M., Sibanda, M. 1991. Antimicrobial activities of *Ziziphus abyssinica* and *Berchemia discolor. Cent. Afr. J. Med.* 37:80-83.

55. Gustafson, J.E., Liew,Y.C., Chew, S., Markham, J., Bell, H.C., Wyllie, S.G., and Warmington, J.R. 1998. Effects of tea tree oil on *Escherichia coli. Lett. Appl. Microbiol.* 26:194-198.

56. Hammer, K.A., Carson, C.A., and Riley, T.V. 1999a. Antimicrobial activity of essential oils and other plant extracts. *J. Appl. Microbiol.* 86:985-990.

57. Hammer, K.A., Carson, C.F., and Riley, T.V. 1999b. Influence of organic matter, cations and surfactants on the antimicrobial activity of *Melaleuca alternifolia* (tea tree) oil *in vitro. J. Appl. Microbiol.* 86:446-452.

58. Hammerschmidt, F.J., Clark, A.M., Soliman, F.M., el-Kashoury, E.S., Abd el-Kawy, M.M., and el-Fishawy, A.M. 1993. Chemical composition and antimicrobial activity of essential oils of *Jasonia candicans* and *J. montana. Planta Med.* 59:68-70.

59. Hao, Y.Y., Brackett, R.E., and Doyle, M.P. 1998. Inhibition of *Listeria monocytogenes* and *Aeromonas hydrophila* by plant extracts in refrigerated cooked beef. *J. Food Prot.* 61:307-312.

60. Hastak, K., Lubri, N., Jakhi, S.D., More, C., John, A., Ghaisas, S.D., and Bhide, S.V. 1997. Effect of turmeric oil and turmeric oleoresin on cytogenetic damage in patients suffering from oral submucous fibrosis. *Cancer Lett.* 116:265-269.

61. Hefnawy, Y.A., Moustafa, S.I., and Marth, E.H. 1993. Sensitivity of *Listeria monocytogenes* to selected spices. *J. Food Prot.* 56:876-878.

62. Hirasa, K. and Takemasa, M. 1998. *Spice Science and Technology.* New York: Marcel Dekker.

63. Hocquemiller, R., Cortes, D., Arango, G.J., Myint, S.H., Cave, A., Angelo, A., Munoz, V., and Fournet, A. 1991. Isolation and synthesis of espintanol, a new antiparasitic monoterpene. *J. Nat. Prod.* 54:445-452.

64. Huhtanen, C.N. 1980. Inhibition of *Clostridium botulinum* by spice extracts and aliphatic alcohols. *J. Food Prot.* 43:195-196.

65. Janssen, A.M., Scheffer, J.J., Baerheim Svendsen, A., and Aynehchi, Y. 1984. The essential oil of *Ducrosia anethifolia (DC.) Boiss.* Chemical composition and antimicrobial activity. *Pharm. Weekbl.* 6:157-160.

66. Janssen, A.M., Scheffer, J.J., Ntezurubanza, L., and Baerheim Svendsen, A. 1989. Antimicrobial activities of some *Ocimum* species grown in Rwanda. *J. Ethanopharmacol.* 26:57-63.

67. Jay, J.M. and Rivers, G.M. 1984. Antimicrobial activity of some food flavoring compounds. *J. Food Safety* 6:129-139.

68. Jayashree, T. and Subramanyam, C. 1999. Antiaflatoxigenic activity of eugenol is due to inhibition of lipid peroxidation. *Lett. Appl. Microbiol.* 28:179-183.

69. Juven, B., Henis, Y., and Jacoby, B. 1972. Studies on the mechanism of the antimicrobial action of oleuropein. *J. Appl. Bacteriol.* 35:559-567.

70. Juven, B.J., Kanner, J., Schved, F., and Weisslowicz, H. 1994. Factors that interact with the antibacterial action of thyme essential oil and its active constituents. *J. Appl. Bacteriol.* 76:626-631.

71. Kandil, O., Radwan, N.M., Hassan, A.B., Amer, A.M., el-Banna, H.A., and Amer, W.M. 1994. Extracts and fractions of *Thymus capitatus* exhibit antimicrobial activities. *J. Ethnopharmacol.* 44:19-24.

72. Karapinar, M. 1985. The effects of citrus oils and some spices on growth and aflatoxin production by *Aspergillus parasiticus* NRRL 2999. *Intl. J. Food Microbiol.* 2:239-246.

73. Karapinar, M. and Aktug, S.E. 1986. Sensitivity of some common food-poisoning bacteria to thyme, mint and bay leaves. *Intl. J. Food Microbiol.* 3:349-354.

74. Karapinar, M. and Aktug, S.E. 1987. Inhibition of foodborne pathogens by thymol, eugenol, menthol and anethole. *Int J. Food Microbiol.* 4:161-166.

75. Katayama, T. and Nagai, I. 1960. Chemical significance of the volatile components of spices from the food preservation standpoint. IV. Structure and antibacterial activity of some terpenes. *Nippon Suisan Gakkaishi* 26:29.

76. Keceli, T., Robinson, R.K., and Gordon, M.H. 1998. The antimicrobial activity of some selected polyphenols from virgin olive oil. IFT Annual Meeting, Atlanta, Abstract 46-B-20.

77. Kim, J., Marshall, M.R., and Wei, C.I. 1995. Antibacterial activity of some essential oil components against five foodborne pathogens. *J. Ag. Food Chem.* 43:2839-2845.

78. Kohyama, N., Nagata, T., Fujimoto, S., and Sekiya, K. 1997. Inhibition of arachidonate lipoxygenase activities by 2-(3,4-dihydroxyphenyl)ethanol, a phenolic compound from olives. *Biosci. Biotechnol. Biochem.* 61:347-350.

79. Kubo, I., Himejima, M., and Muroi, H. 1991. Antimicrobial activity of flavor components of cardamom *Elattaria cardamomum* (Zingiberaceae) seed. *J. Agric. Food Chem.* 39:1984-1986.

80. Lachowicz, K.J., Jones, G.P., Briggs, D.R., Bienvenu, F.E., Wan, J., Wilcock ,A., and Coventry, M.J. The synergistic preservative effects of the essential oils of sweet basil (*Ocimum basilicum* L.) against acid-tolerant food microflora. *Lett. Appl. Microbiol.* 26:209-214.

81. Lacoste, E., Chaumont, J.P., Mandin, D., Plumel, M.M., and Matos, F.J. 1996. Antiseptic properties of essential oil of *Lippia sidoides* Cham. Application to the cutaneous microflora. *Ann. Pharm. Fr.* 54:228-230.

82. Langezaal, C.R., Chandra, A., and Scheffer, J.J. 1992. Antimicrobial screening of essential oils and extracts of some *Humulus lupulus* L. cultivars. *Pharm. Weekbl. Sci.* 14:353-356.

83. Lindsay, R.C. 1996. Food additives, p. 767-823. In O.R. Fennema. *Food Chemistry, 3rd ed.* New York: Marcel Dekker.

84. Lis-Balchin, M., Buchbauer, G., Hirtenlehner, T., and Resch, M. 1998. Antimcrobial activity of *Pelargonium* essential oils added to a quiche filling as a model food system. *Lett. Appl. Microbiol.* 27:207-210.

85. Liu, S.L. 1990. Therapeutic effects of borneol-walnut oil in the treatment of purulent otitis media. *Chung Hsi I Chieh Ho Tsa Chih* 10:93-95.

86. Lopez-Garcia, R.E., Hernandez-Perez, M., Rabanal, R.M., Darias, V., Martin-Herrera, D., Arias, A., and Sanz, J. 1992. Essential oils and antimicrobial activity of two varieties of *Cedronella canariensis (L.) W. et B. J. Ethnopharmacol.* 36:207-211.

87. López-Malo, A., Alzamora, S.M., and Argaiz, A. 1995. Effect of natural vanillin on germination time and radial growth of moulds in fruit-based agar systems. *Food Microbiol.* 12:213-219.

88. López-Malo, A., Alzamora, S.M., and Argaiz, A. 1997. Effect of vanillin concentration, pH and incubation temperature on *Aspergillus flavus*, *Aspergillus niger*, *Aspergillus ochraceus*, and *Aspergillus parasiticus* growth. *Food Microbiol.* 14:117-124.

89. López-Malo, A., Alzamora, S.M., and Argaiz, A. 1998. Vanillin and pH synergistic effects on mold growth. *J. Food Sci.* 63:143-146.

90. Lorente, I., Ocete, M.A., Zarzuelo, A., Cabo, M.M., and Jimenez, J. 1989. Bioactivity of the essential oil of *Bupleurum fruticosum. J. Nat. Prod.* 52:267-272.

91. Mangena, T. and Muyima, N.Y.O. 1999. Comparative evaluation of the antimicrobial activities of essential oils of *Artemisia afra*, *Pteronia incana* and *Rosmarinus officinalis* on selected bacteria and yeast strains. *Lett. Appl. Microbiol.* 28:291-296.

92. Mansour, N., Yousef, A.E., and Kim, J.G. 1996. Inhibition of surface growth of toxigenic and nontoxigenic aspergilli and penicillia by eugenol, isoeugenol and monolaurin. *J. Food Safety* 16:219-230.

93. Marth, E.H. 1966. Antibiotics in foods—naturally occurring, developed and added. *Residue Rev.* 12:65-161.

94. Maruzella, J.C., and Henry, P.A. 1958. The *in vitro* antibacterial activity of essential oils and oil combinations. *J. Amer. Pharm. Assn.* 47:294.

95. Maruzella, J.C., and Liguori, L. 1958. The *in vitro* antifungal activity of essential oils. *J. Amer. Pharm. Assn.* 47:331-336.

96. Moyler, D. 1994. Spices - recent advances. p. 1-70. In G. Charalambous (ed.) 1994. *Spices, Herbs and Edible Fungi*. Amsterdam: Elsevier.

97. Nychas, G.J. 1994. Natural antimicrobials from plants. In *Natural Antimicrobial Systems and Food Preservation,* ed. V.M. Dillon., and R.G. Board, pp. 58-83. Wallingford, UK:CAB Intl..

98. Osawa, K., Matsumoto, T., Maruyama, T., Takiguchi, T., Okuda, K., and Takazoe, I. 1990. Studies of the antibacterial activity of plant extracts and their constituents against periodontopathic bacteria. *Bull. Tokyo Dent. Coll.* 31:17-21.

99. Pandit, V.A., and Shelef, L.A. 1994. Sensitivity of *Listeria monocytogenes* to rosemary (*Rosmarinus officianalis* L.). *Food Microbiol.* 11:57-63.

100. Panizzi, L., Flamini, G., Cioni, P.L., and Morelli, I. 1993. Composition and antimicrobial properties of essential oils of four Mediterranean Lamiaceae. *J. Ethnopharmacol.* 39:167-170.

101. Pattnaik, S., Subramanyam, V.R., Bapaji, M., and Kole, C.R. 1997. Antibacterial and antifungal activity of aromatic constituents of essential oils. *Microbios.* 89:39-46.

102. Pauli, A., and Knobloch, K. 1987. Inhibitory effects of essential oil components on growth of food-contaminating fungi. *Z. Lebensm. Unters. Forsch.* 185:10-13.

103. Petroni, A., Blasevich, M., Salami, M., Papini, N., Montedoro, G.F., and Galli, C. Inhibition of platelet aggregation and eicosanoid production by phenolic components of olive oil. *Thromb. Res.* 78:151-160.

104. Pol, I.E. and Smid, E.J. 1999. Combined action of nisin and carvacrol on *Bacillus cereus* and *Listeria monocytogenes. Lett. Appl. Microbiol.* 29:166-170.

105. Porter, N.G., and Wilkins, A.L. 1999. Chemical, physical and antimicrobial properties of essential oils of *Leptospermum scoparium* and *Kunzea ericoides. Phytochemistry* 50:407-415.

106. Prasad, M., and Joshi, N. 1929. The preservative value of spices used in pickling raw fruits in India. *Agric. J. India* 24:202.

107. Raharivelomanana, P.J., Terrom, G.P., Bianchini, J.P., and Coulanges, P. 1989. Study of the antimicrobial action of various essential oils extracted from Malagasy plants. II. Lauraceae. *Arch. Inst. Pasteur Madagascar* 56:261-271.

108. Ramanoelina, A.R., Terrom, G.P., Bianchini, J.P., and Coulanges, P. 1987. Antibacterial action of essential oils extracted from Madagascar plants. *Arch. Inst. Pasteur Madagascar* 53:217-226.

109. Rico-Muñoz, E., and Davidson, P.M. 1983. The effect of corn oil and casein on the antimicrobial activity of phenolic antioxidants. *J. Food Sci.* 48:1284-1288.

110. Saenz, M.T., Garcia, M.D., Ahumada, M.C., and Ruiz, V. 1998. Cytostatic activity of some compounds from the unsaponifiable fraction obtained from virgin olive oil. *Farmaco* 53:448-449.

111. Satoh, A., Utamura, H., Ishizuka, M., Endo, N., Tsuji, M., and Nishimura, H. 1996. Antimicrobial benzopyrans from the receptacle of sunflower. *Biosci. Biotechnol. Biochem.* 60:664-665.

112. Shapiro, S., and Guggenheim, B. 1995. The action of thymol on oral bacteria. *Oral Microbiol. Immunol.* 10:241-246.
113. Sharoni, Y., Giron, E., Rise, M., and Levy, J. 1997. Effects of lycopene-enriched tomato oleoresin on 7,12-dimethyl-benz[a]anthracene-induced rat mammary tumors. *Cancer. Detect. Prev.* 21:118-23
114. Shcherbanovsky, L.R., and Kapelev, I.G. 1975. Volatile oil of *Anethum Graveolens L.* as an inhibitor of yeast and lactic acid bacteria. *Prikl. Biokhim. Mikrobiol.* 11:476-477.
115. Shelef, L.A., Jyothi, E.K., and Bulgarelli, M.A. 1984. Growth of enteropathogenic and spoilage bacteria in sage-containing broth and foods. *J. Food Sci.* 49:737-740.
116. Shelef, L.A., Naglik, O.A., and Bogen, D.W. 1980. Sensitivity of some common foodborne bacteria to the spices sage, rosemary, and allspice. *J. Food Sci.* 45:1042-1044.
117. Smith-Palmer, A., Stewart, J., and Fyfe, L. 1998. Antimicrobial properties of plant essential oils and essences against five important food-borne pathogens. *Lett. Appl. Microbiol.* 26:118-122.
118. Stecchini, M.L., Sarais, I., and Giavedoni, P. 1993. Effect of essential oils on *Aeromonas hydrophila* in a culture medium and in cooked pork. *J. Food Prot.* 56:406-409.
119. Tassou, C.C., and Nychas, G.J. 1995. Inhibition of *Salmonella enteritidis* by oleuropein in broth and in a model food system. *Lett. Appl. Microbiol.* 20:120-124.
120. Tassou, C.C., Drosinos, E.H., and Nychas, G.J.E. 1995. Effects of essential oil from mint (*Mentha piperita*) on *Salmonella enteritidis* and *Listeria monocytogenes* in model food systems at 4 and 10°C. *J. Appl. Bacteriol.* 78:593-600.
121. Tassou, C.C., Nychas, G.J., and Board, R.G. 1991. Effect of phenolic compounds and oleuropein on the germination of *Bacillus cereus* T spores. *Biotechnol. Appl. Biochem.* 13:231-237.
122. Ting, W.T.E. and Deibel, K.E. 1992. Sensitivity of *Listeria monocytogenes* to spices at two temperatures. *J. Food Safety* 12:129-137.
123. Tirillini, B., Velasquez, E.R., and Pellegrino, R. 1996. Chemical composition and antimicrobial activity of essential oil of *Piper angustifolium. Planta. Med.* 62:372-373.
124. Tranter, H.S., Tassou, S.C., and Nychas, G.J. 1993. The effect of the olive phenolic compound, oleuropein, on growth and enterotoxin B production by *Staphylococcus aureus. J. Appl. Bacteriol.* 74:253-259.
125. Trebitsch, F. 1978. Antimicrobial action of thymol-camphor compared with 20 pharmaceutical preparations against *Staphylococcus aureus. Aust. Dent. J.* 23:152-155.
126. Vila, R., Valenzuela, L., Bello, H., Canigueral, S., Montes, M., and Adzet, T. 1999. Composition and antimicrobial activity of the essential oil of *Peumus boldus* leaves. *Planta Med.* 65:178-179.
127. Viollon, C., and Chaumont, J.P. 1994. Antifungal properties of essential oils and their main components upon *Cryptococcus neoformans. Mycopathologia* 128:151-153
128. Visioli, F., and Galli, C. 1994. Oleuropein protects low density lipoprotein from oxidation. *Life Sci.* 55:1965-1971.
129. Visioli, F., Bellosta, S., and Galli, C. 1998a. Oleuropein, the bitter principle of olives, enhances nitric oxide production by mouse macrophages. *Life Sci.* 62:541-546.
130. Visioli, F., Bellomo, G., and Galli, C. 1998b. Free radical-scavenging properties of olive oil polyphenols. *Biochem. Biophys. Res. Commun.* 247:60-64.
131. Walter, W.M. Jr, Fleming, H.P., and Etchells, J.L. 1973. Preparation of antimicrobial compounds by hydrolysis of oleuropein from green olives. *Appl. Microbiol.* 26:773-776.
132. Wan, J., Wilcock, A., and Coventry, M.J. 1998. The effect of essential oils of basil on the growth of *Aeromonas hydrophila* and *Pseudomonas fluorescens. J. Appl. Microbiol.* 84:152-158.
133. Webb, A.H. and Tanner, F.W. 1944. Effect of spices and flavoring materials on growth of yeasts. *Food Res.* 10:273-282.
134. Weissenberg, M., Schaeffler, I., Menagem, E., Barzilai, M., and Levy, A. 1997. Isocratic non-aqueous reversed-phase high-performance liquid chromatographic separation of capsanthin and capsorubin in red peppers (*Capsicum annuum* L.), paprika and oleoresin. *J. Chromatogr.* (A) 757:89-95.
135. Wendakoon, C.N. and Sakaguchi, M. 1993. Combined effect of sodium chloride and clove on growth and biogenic amine formation of *Enterobacter aerogenes* in mackerel muscle extract. *J. Food Prot.* 56:410-413.
136. Wu, C.Y., Yang, S.H., and Wu, S.C. 1989. Preparative separation of borneol and isoborneol in bingpian by dry-column chromatography. *Hua Hsi I Ko Ta Hsueh Hsueh Pao* 20:327-330.
137. Yin, M.C., and Cheng, W.S. 1998. Inhibition of *Aspergillus niger* and *Aspergillus flavus* by some herbs and spices. *J. Food Prot.* 61:123-125.
138. Yousef, R.T., and Tawil, G.G. 1980. Antimicrobial activity of volatile oils. *Pharmazie* 35:698-701.
139. Zaika, L.L., and Kissinger, J.C. 1979. Effect of some spices on acid production by starter cultures. *J. Food Prot.* 42:572.

140. Zaika, L.L., and Kissinger, J.C. 1981. Inhibitory and stimulatory effects of oregano on *Lactobacillus plantarum* and *Pediococcus cerevisiae. J. Food Sci.* 46:1205-1210.
141. Zaika, L.L., Kissinger, J.C., and Wasserman, A.E. 1983. Inhibition of lactic acid bacteria by herbs. *J. Food Sci.* 48:1455-1459.

W.A. Oleszek

11

Saponins

I. INTRODUCTION

Saponins are a group of naturally occurring glycosides which are predominantly found in the plant kingdom, however, certain lower marine animals also contain these compounds. They can be found in different plant parts including roots, shoots, flowers and seeds. The common feature of most of the saponins is the formation of soapy lather when shaken in water solution. The height of the froth and the time of its disappearance have been often used as a semiquantitative test for characterization of plant species with respect to saponin content. Most literature concerning saponin distribution is based on this test. In the Orient, saponins were used in traditional folk medicine or as a soap and thus, in many cases trivial names of saponin-rich plant species are derived from this feature e.g. soapwort (*Saponaria officinalis*), soaproot (*Chlorogalum pomeridianum*), soapbark (*Quillaja saponaria*), soapberry (*Sapindus saponaria*), soapnut (*Sapindus mukurossi*) (Hostettmann & Marston, 1995).

Depending on the structure saponins may show hemolytic, antimicrobial, membrane depolarizing, cholesterol binding and allelopathic activities. Hemolytic activity of saponins *in vitro* was reported by Kobert as early as 1887 and was based on the destruction of erythrocyte membranes (lysis) and release of hemoglobin. From the nutritional point of view hemolytic activity does not seem to create any problems as saponins neither cross membrane barrier nor enter the blood stream. This characteristic was applied in a number of hemolytic semiquantitative tests for saponin determination and has been reviewed by Birk (1969).

Cholesterol binding and permeation of the epithelial barrier may have nutritional consequences and much work has been recently devoted to these effects (Gee et al., 1998; Malinow et al., 1978, 1981). These aspects generated dramatic progress in the development of isolation, structure determination and testing techniques of individual saponins. Antimicrobial activities are important to the plant for protection against pathogens. The localization of saponins with higher antimicrobial activity in plant parts previously exposed to pathogens suggests such a function. Aerial parts of some plants may

0-8493-2047-X/00/$0.00+$.50
© 2000 by CRC Press, LLC

cumulate bitter taste saponins, a feature important in plant protection against herbivores (Oleszek et al., 1999).

II. OCCURRENCE

Saponins are widely distributed in plant species, about 100 families. As reported by Gubanov et al. (1970) 76% of Asian plant families contain saponins. They are found in edible plants such as soya, beans, peas, oat, *Solanum* and *Allium* species, tomato, asparagus, tea, peanut, spinach, sugar beet, yam and blackberry; in plants used as animal feed such as alfalfa, clover, legumes, forage and cover crops, sunflower, horse chestnut, guar and lupine; and in plants used as flavorings, health foods, tonics etc. including fenugreek, liquorice, nutmeg, *Quillaya*, *Yucca*, *Gypsophila*, herbs, edible seeds, health foods, and ginseng (Price et al., 1987). Their concentration in plants ranges from traces up to 10% of dry matter as in the *Yucca schidigera* trunk. The concentration depends on the cultivar, age, physiological stage and geographical location of the plant. Considerable qualitative and quantitative variation also exists between plant parts. For example alfalfa seeds contain only low toxicity soyasapogenol glycosides (0.2% of DM), while during germination glycosides of medicagenic acid, zanhic acid and hederagenin are synthesized (Oleszek, 1998). Roots of alfalfa contain medicagenic acid, zanhic acid, hederagenin and soyasapogenol glycosides with a total concentration of saponins 2.5% DM while aerial parts possess similar glycosides but at lower concentration 0.1- 1.5% DM (Nowacka & Oleszek, 1994). Depending on the origin of the variety in a saponin mixture, soyasapogenol glycosides, medicagenic acid glycosides or zanhic acid tridesmoside may dominate.

As mentioned above the presence of saponins in plant extract can be simply demonstrated by the froth formation after shaking in water solution. Nevertheless, some saponins, especially those with two or three branched sugar chains do not form stable froth. On the other hand, certain plant extracts not containing saponins may also produce froth, and may mislead. Some saponins show hemolytic activity and this feature in some cases could be used as a semiquantitative test, with several modifications (Birk, 1969). In general, saponin-containing material or its water extract is mixed with ox blood or with washed erythrocytes in isotonic buffered solution (0.9% NaCl). After 20-24 h incubation, samples are centrifuged and hemolysis is indicated by the presence of hemoglobin in the supernatant. According to the European Pharmacopoeia, a unit hemolytic index (HI) is defined as the amount (in milliliters) of ox blood (2%, v/v) hemolyzed by 1-g of crude saponins or plant material. The saponin mixture of *Gypsophila paniculata* L. (HI = 30 000) or Saponin white (Merck, HI = 15 000) are usually used as reference. Hemolytic indices of saponins are calculated according to the following equation: $HI = HI_{std} \times a/b$; where HI_{std} is the hemolytic index of standard saponin, and a and b are the means of lowest concentration from each replicate of standard and saponins, respectively, at which full hemolysis occurred. Mackie et al. (1977) measured the absorbance at 545 nm of the supernatant after hemolysis and defined one unit of activity as the quantity of hemolytic material that caused 50% hemolysis.

The other modification of the above method is a hemolytic micromethod. In this assay, bovine blood is stabilized with sodium citrate (3.65% w/v) and mixed with gelatin solution. Gelatin (4.5 g) is dissolved in 100 ml of isotonic buffered solution and 75 ml of

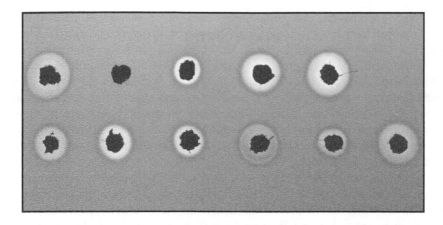

FIGURE *1*. Hemolytic test performed on blood-gelatin covered plates. Black spots – mashed alfalfa plant material; transparent rings – hemolytic zones; background in original – red.

this solution is mixed with 20 ml of the blood. Gelatin/blood mixture is spread on a glass plate (10 x 20 cm) to a thickness of 0.5 mm. After coagulation, plates are used for hemolytic tests. Saponin samples (10 ml) or mashed fresh plant material is placed in localized areas on the gelatin/blood-covered plates and after 20 h the width of the resulting hemolytic rings are measured (*FIGURE 1*). Standard saponin is measured as a reference control on each plate. Detailed descriptions of HI and microhemolytic methods are described by Oleszek (1988).

Saponins can also be quantified in bio-assays using organisms such as *Trichoderma viride* (Zimmer et al., 1967), *Tribolium castaneum* (Shany et al., 1970), lettuce seed germination (Pedersen et al., 1967) or with gravimetry (van Atta et al., 1961), spectrophotometry (Hiai, 1976; Baccou, 1977; Gestetner, 1966) and chromatography (TLC, GC, HPLC) (Besso et al., 1979; Domon et al., 1984; Kitagawa et al., 1984; Slacanin et al., 1988; Oleszek et al., 1990; Tava et al., 1993; Xin et al., 1994).

Qualitative assay for saponins can be performed with TLC on silica gel or reversed phase C18 plates. Most often used solvents for silica gel TLC include chloroform-methanol-water (7:3:1; 65:35:10 and similar combinations), ethyl acetate-acetic acid-water (7:2:2), n-butanol-ethanol-ammonia (7:2:5), n-butanol-acetic acid-water (BAW 4:1:5, upper layer), chloroform-methanol-acetic acid-water (60:32:12:8), n-butanol saturated with water, n-butanol-ethyl acetate-water (100:25:50), ethyl acetate-pyridine-water (30:10:30, upper phase, for glycoalkaloids). For reversed phase (C18, C8) plates almost exclusively different proportion of water-methanol and water-acetonitrile are used depending on the polarity of analyzed saponins. Several sprayers are used for visualization of saponins and depending on the structure, they may produce different colored spots. The most common sprayers are vanillin-sulfuric acid (a 1% solution of vanillin in ethanol, mixed with 3% solution of perchloric acid in water in a 1:1 ratio and sprayed onto the TLC plate followed by the 10% solution of sulfuric acid in ethanol); Liebermann-Burchard (methanol:acetic anhydride:sulfuric acid, 50:5:5 v/v) and Ehrlich reagent (1 g *p*-dimethylaminobenzaldehyde in 50 ml 36% HCl mixed with 50 ml ethanol). All listed sprayers require heating of the plates for several minutes at 105-120°C. The vanillin-sulfuric acid reagent generates color product with most of the saponin aglycones but the reaction is not very specific and certain other classes of substances may also react. Liebermann-Burchard reagent produces color spots with aglycones (blue, green, pink, brown, yellow) and with glycosides e.g. green for medicagenic acid, pink for hederagenin, bricky for soyasapogenol, yellowish for zanhic acid, brown for steroidal furostanol and spirostanol saponins. The Ehrlich reagent is highly specific for furostanol steroidal saponins and produces pink-red spots, while spirostanol glycosides are not visualized. This reagent could effectively distinguish these two groups of compounds. Many of listed sprayers are successfully used for spectrophotometric determination of saponins. This has, however, to be performed with caution, since certain compounds present in the sample matrices cause interference.

Gas chromatography has been used for determination of aglycones liberated after hydrolysis of saponins. This technique has been used for determination of trimethylsilyl ethers of soyasapogenols and medicagenic acid (Jurzysta & Jurzysta, 1978) trimethylsilyl ethers of soyasapogenol A and B in soybean (Kitagawa et al., 1984), dimethyl ester, di(trimethylsilyl)ether of medicagenic acid (Brawn et al., 1981), methylated-acetylated medicagenic acid, hederagenin and soyasapogenols B, C, D, E and F (Tava et al., 1993), trimethylsilyl ethers of panaxadiol and panaxatriol from ginseng (Sakamoto et al., 1975) and methyl ester TMS ether of glycyrrhetinic acid from *Glycyrrhiza radix* and human serum (Bombardelli et al., 1976, 1979; Itoh et al., 1985). Formation of artifacts during saponin hydrolysis and artifact formation during derivatization and analysis is a limitation to this technique. Derivatized compounds are often poorly separated by this method.

High performance liquid chromatography (HPLC) seems to be the most reliable technique for quantification. The main problem in chromatographic determination is the lack of chromophores in saponin molecules, which creates a problem for UV detection. The non-specific detection at 200-210 nm is possible, however, this requires a careful pre-column purification of extracts, and removal of other components from the matrix (Domon et al., 1984). Pre-column derivatization of saponins with reagents introducing chromophores is an alternative (Besso et al., 1979; Slacanin et al., 1988).

III. MOLECULAR PROPERTIES

Saponins consist of non-sugar aglycone coupled to sugar chain units. These sugars could be attached to the aglycone either as one, two or three side chains, and the terms monodesmoside, bidesmoside and tridesmoside has been given to these saponins, respectively (Greek *desmos* = chain) (Wulf 1968). According to the nature of aglycone they could be divided into the groups of saponins containing steroidal or triterpene (*FIGURE 2*) aglycones - sapogenins. Some authors also include steroidal glycoalkaloids of solanidans and spirosolan class in the group of saponins (*FIGURE 3*). Steroidal skeletons have in most cases furostanol or spirostanol form; furostanol glycosides usually have a bidesmosidic and spirostanol monodesmosidic nature. Both steroidal and triterpene saponins may have a number of functional groups (-OH, -COOH, -CH3) causing great natural diversity only because of aglycone structure. Over 100 steroidal and probably an even higher number of triterpene saponins have been known (Hostettmann & Marston, 1995). This diversity could be further multiplied by the composition of sugar chains, their number, branching patterns and substitution. It is well known that even within one plant species, distinct parts may harbor structurally different saponins. This diversity makes isolation and identification of single saponins quite difficult.

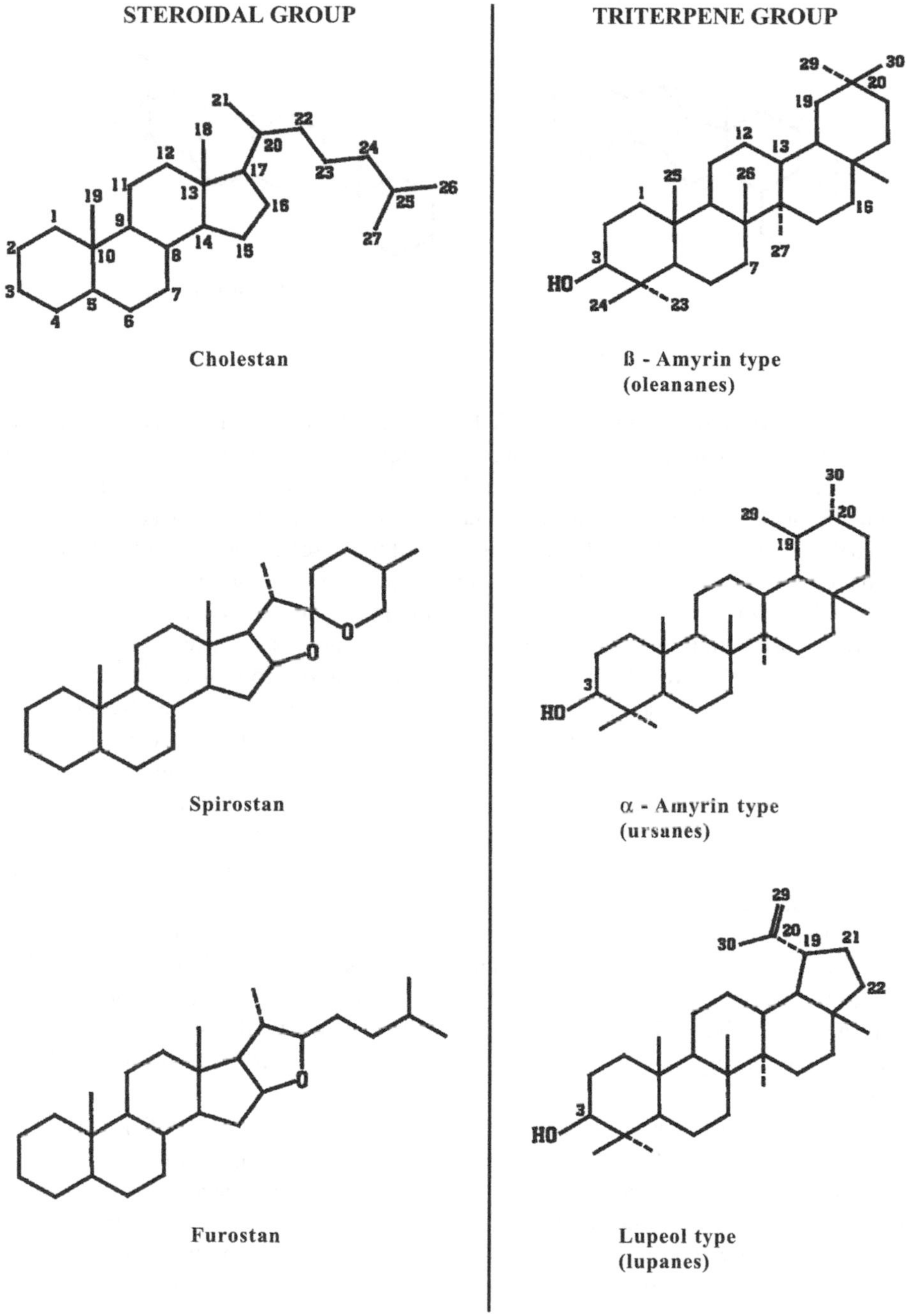

STEROIDAL GROUP

Cholestan

Spirostan

Furostan

TRITERPENE GROUP

ß - Amyrin type
(oleananes)

α - Amyrin type
(ursanes)

Lupeol type
(lupanes)

FIGURE 2. Sapogenin skeletons – Steroidal and Triterpene

GLYCOALKALOIDS GROUP

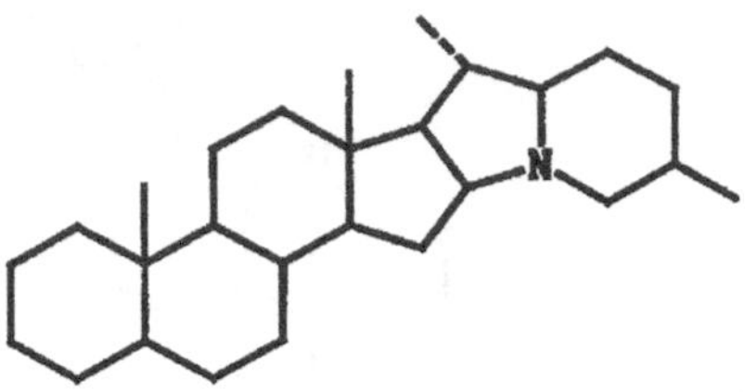

Solanidan 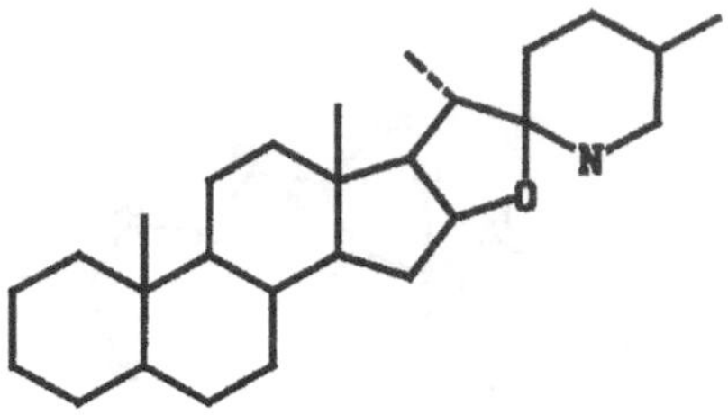   Spirosolan

FIGURE 3. Sapogenin skeletons – Glycoalkaloids

Early work on saponins included hot extraction with aqueous methanol or ethanol of defatted plant material, followed by evaporation of alcohol and extraction of saponins into n-butanol (Wall et al., 1952). In this liquid-liquid extraction, highly polar components present in extract matrix e.g. saccharides stay in the water solution giving some purification of the extract. However, in most of the cases evaporation of butanol produces dark-brown solid rich solution with many interfering compounds. Some authors propose dialysis for removal of small water-soluble molecules (Massiot et al., 1988). The next step of purification may include precipitation of saponins from water solution with lead acetate, barium hydroxide or cholesterol or precipitation from alcohol solution with diethyl ether or acetone. Such treatments produce crude saponin mixtures. Nevertheless, these procedures also have many disadvantages and some, such as cholesterol precipitation, are inconvenient due to difficulties in ensuring the complete washing of non-precipitated saponin from the viscous precipitate. The extraction with hot aqueous methanol may lead to the formation of $-OCH_3$ derivatives, lactones, both in triterpene and steroidal saponins or may result in the degradation of genuine saponins into their artifacts as in the case of soybean saponins (Massiot et al., 1996). Such transformations occur on the aglycone skeleton or on the sugar part, or consist in the loss of labile substituents. In liquid-liquid extraction, some highly polar saponins (bidesmosides and tridesmosides) could be lost or extraction may not be quantitative. This has been observed in zanhic acid tridesmoside present in alfalfa aerial parts (Oleszek et al., 1992). In the Wall's procedure, tridesmoside, in spite of its very high concentration in some cultivars, remained in water with carbohydrates.

The alternative for Wall's procedure is solid-phase extraction (SPE) on reversed phase (C8, C18) sorbent (Oleszek, 1988). In the SPE method, saponin extract (aqueous 10-20% methanol) can be loaded on a short column and then washed with methanol-water. Depending on the concentration of methanol different classes of matrix components can be removed. The water can easily remove all carbohydrates while water-methanol mixtures (up to 40% of methanol) may remove phenolics, including most of the flavonoids. Saponins can be washed out with solvents containing more than 40% methanol; depending on the concentration selective extraction of different classes of saponins can be achieved. This, however, requires preliminary tests for different classes of saponin extract that can be performed on ready to use 1-2 cm³ cartridges. Selectivity of the C18 sorbent to retain saponins strongly depends on the pH of the loaded sample, which is especially convenient in separation of saponins having COOH groups. The SPE

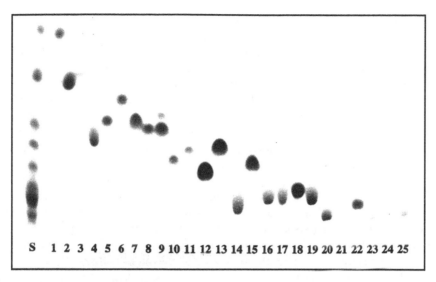

FIGURE 4. TLC chromatogram of alfalfa root saponins; s – crude saponin mixture; 1-25 individual saponins.

method is especially convenient with steroidal saponins. In many cases, this is a good procedure for separation of highly purified crude saponin mixtures for evaluation of biological activity.

The separation of crude mixtures into individual saponins is quite a challenge. The wide range of polarities of individual glycosides makes it almost impossible to separate individual saponins in one run. Open column or low-pressure liquid chromatography on silica gel and reversed phase sorbents has to be applied in combination. The first separation usually provides saponin fractions that consist of several glycosides with similar retention properties. For further separation, it is necessary to switch to another sorbent or solvent system. In the case of triterpene alfalfa saponins, separation on C18 sorbent in water-methanol gradient is very convenient and further purification of individual saponins can be accomplished on silica gel columns (Oleszek et al., 1990). However, initial separation on silica gel column with n-butanol saturated with water followed by C18 column also yields good separation (*FIGURE 4*) (Bialy 1998). For steroidal saponins, best separations can be achieved by silica gel column chromatography washed with chloroform-methanol mixtures followed by C18 separation with methanol-water mixtures.

IV. ANTIMICROBIAL ACTIVITY

A. Antibacterial activity

The general principle of the action of saponins on microorganism is their interaction with membrane sterols. Thus, the bacteria should not be sensitive to saponins, as their membranes are low in cholesterol. However, recent report on the saponin-sensitive and insensitive *Trichoderma viride* strains indicate that fatty acid composition of membranes may equally be important in this respect (Gruiz, 1966). Thus, a number of saponins, both triterpene and steroidal, show antibacterial activities (*TABLE 1*). Antibacterial activity of saponins has been extensively reviewed by Takeya (1997) and Tamura (1996).

The degree of activity depends on the saponin structure, target microorganism and may range from 5 ppm as reported for gitogenin glycoside (I) (Masamitsu et al., 1995) up to 5-10 mg/ml found for saponins from common ivy (hederagenin glycosides) (Cioaca et al., 1978). In general, the common ivy saponins were more active against Gram-positive than Gram-negative bacteria with minimal inhibitory concentrations (MIC) of 0.3-1.25 and 1.25-5 mg/ml, respectively. Positive reactions between antibiotics (ampicillin, kanamycin, cephalexin, oxytetracycline and chloramphenicol) and ginseng saponins were studied by checkerboard method (Kim et al., 1987). It was documented that ampicillin, kanamycin, oxytetracycline and chloramphenicol against *Bacillus subtilis* and ampicillin and cephalexin against *Staphylococcus aureus* were synergistic with ginseng saponin. No antagonism between saponins and antibiotics could be observed. Quillaya saponins show emulsifying properties enhancing the antibacterial effectivity of fumaric acid in food preservation. Antibacterial agent was prepared by mixing 100 g of fumaric acid and 0.5 g Quillaya saponin powder (Nishina et al., 1992). Mixing saponins with acetic acid was shown to be an effective antimicrobial composition for the control of bacteria grown on the skin after perspiration (Tatebe, 1986).

The major effect of saponins on bacteria is the leakage of protein and certain enzymes from their cells (Zablotowicz et al., 1996, Hoagland et al., 1996). The ß-aescin, glycyrrhetic acid and medicagenic glycosides increased extracellular protein leakage from three bacterial strains including *Pseudomonas fluorescens* RA-2, *Curtobacterium flaccumfaciens* JM-1011 and *Bacillus thuringiensis* UZ404, as measured 24 h after treatment with 500 µM saponin solutions. Betulin and hecogenin had no influence or slightly reduced protein leakage, while soyasaponin I increased leakage in case of *C. flaccumfaciens* and reduced it by 20 and 50% in *P. fluorescens* and *B. thuringiensis*, respectively. Different effects of structurally divergent alfalfa saponins were also observed on enzymatic activity, FDA-esterase and TTC-dehydrogenase in the *B. thuringiensis* but not in *P. fluorescens* (Oleszek et al., 1999). These data clearly indicate that bacterial membranes can be affected by certain saponins, resulting in a significant loss of vital activity, especially in some Gram-positive genera such as the *Bacillus sp.*

B. Antifungal activity

Among the saponins, those with triterpene and spirostanol skeleton generally demonstrate antifungal activities (*TABLE 2*). Furostanol saponin with bisdesmosidic nature lack bacteriostatic and fungicidal effects, and does not complex with cholesterol. Complexing with membrane sterols, proteins and phospholipids seems to be a major mechanism that predispose sensitivity of a microbe to saponins, but the effect is relatively nonspecific. In the case of fungi, as shown by Gruiz (1996), ergosterol content of saponin sensitive and insensitive *Trichoderma viride* strains did not differ significantly. In both cases, however, endogenous saponin treatment significantly increased ergosterol level. Considerable changes after saponin treatment were noticed in the lipid composition of sensitive and insensitive strains. It was concluded that metabolic changes in fatty acid composition of membrane phospholipids discriminate fungus sensitivity to exogenous saponins.

Regarding the structural features of saponin, their fungicidal activities are determined by the aglycone structure and also by the number, quality and branching patterns of sugar substitution. The aglycone by itself may show very high antifungal activity. Medicagenic acid (*FIGURE 5*) in a free form may totally inhibit *T. viride* at the concentra-

TABLE 1. Antimicrobial acitivity of saponins from different plant sources

Plant source	Saponin	Microorganism	MIC	Reference
Aesculus hippocastanum	triterpene	*Agrobacterium tumefaciens*	100 μM	Oleszek et al., 1999
		Rhizobium meliloti	100 μM	
		Bradyrhizobium japonicum	100 μM	
		Pseudomonas fluorescens	25 μM	
Astragalus melanophrurius	triterpene	Several species	modest	Calis et al., 1997
Beta vulgaris	triterpene	*Streptococcus mutans*	10 μg/ml	Sasazuka et al., 1995
Capsicum annum	steroidal	*Saccharomyces cerevisiae*		Gal, 1968
		Bacillus cereus, B. subtilis		
Hedera helix	triterpene	22 different bacteria		Cioaca et al., 1978
Hedyotis molicaulis	triterpene	*Bacillus subtilis M45*		Masataka et al.,1998
Henricia laeviuscola	steroidal	*Staphylococcus aureus*		Andersson et al., 1989
Hosta sieboldiana	steroidal	*Debaryomyces polymorphus*	10 ppm	Masamitsu et al., 1997
		Zygosaccharomyces rouxii	5 ppm	
Medicago sativa	triterpene	*Agrobacterium tumefaciens*		Timbekova et al., 1996
		Corynebacterium insidiosum		
		Pseudomonas lachrymans		
		Coryno. michiganense		
		Xanthomonas campestris		
Medicago sativa	triterpene	*Staphylococcus aureus*		Boguslavsky et al, 1992
		Pseudomonas aeruginosa		
Sanicula europaea	triterpene	*C. diphtheriae*		Przyborowski et al., 1967
Sapindus saponaria	triterpene	*Pseudomonas aeruginosa*		Lemos et al., 1992
		Bacillus subtilis		
		Cryptococcus neoformans		
Smilax aspera	steroidal	*Cryptococcus neoformans*		Petritic et al., 1969
	steroidal	8 bacterial strains		Bruno 1968
Solenostemma argel	unspec.	*Streptococcus* spp.		El Hady et al., 1994
		Escherichia coli		
		Bacillus anthracis		
		Staphylococcus aureus		
		Klebsiella pneumoniae		
Symphytum officinale	triterpene	*Salmonella typhi*	20 μg/ml	Ahmad et al., 1993
		Staphylococcus aureus		
		Streptococcus faecalis		
		Escherichia coli		

tion as low as 2.2 μg/ml, but the activity of hederagenin differing slightly in the structure from medicagenic acid, has been reduced, while soyasapogenol B structurally related to these two sapogenins is lacking in any activity. In the case of medicagenic acid, the maximum activity has been found when all functional groups (-COOH, -OH) were free. Blockage of the two carboxylic groups led to the impaired growth, so that 50-60% of the growth was maintained even at a concentration of 100 μg/ml. Also blocking of the -OH groups in the aglycone produced a very significant lowering of antifungal activity (Oleszek et al., 1988). Glycoalkaloid saponin, α-tomatine owes also its antimicrobial effect to the tomatidine moiety (Arneson & Durbin, 1968, Jadhav et al., 1981) and sugar beet saponins to oleanolic acid (Sasazuka et al., 1995). It is suggested that glycosides are ineffective against microorganisms. However, aglycones released from them by the enzymes present on the cell surface are antimicrobial. In some *in vitro* tests, the aglycones could elicit marginal activity, and this may happen due to their poor water solubility, as sapogenins are usually hydrophobic compounds. This problem could be circumvented by

TABLE 2. Antifungal acitivity of saponins from different plant sources

Plant source	Saponin	Fungal species	MIC	Reference
Agave sisalana	steroidal	*Aspergillus* spp., *C. albicans, Rodotorula glutinis, S.cerevisiae*		Uikawa & Purchio, 1989
Albizzia lebbek	unspec.	*Macrophomina phaseolina*	32.8 µg/ml	Ahmad et al., 1990
Allium ampeloprasum	steroidal	*Mortierella ramanniana*	10 µg/disc	Sata et al., 1998
Allium sativum	steroidal	*Candida albicans*	100 µg/ml	Matsura et al., 1996
Asparagus officinalis	steroidal	*Trichophyton rubrum, C. albicans*	2-3 µg/ml	Magota et al., 1994
Asparagus officinalis	steroidal	*Cryptococcus* spp., *Trichophyton* spp. *Microsporum* spp., *Epidermophyton* spp.	0.5-8 µg/ml	Shimoyamada et al., 1990
Asparagus officinalis	steroidal	Several fungi		Shimoyamada et al., 1996
Avena sativa	triterpene	*Fusarium avenaceum*		Luening et al., 1978
Avena sativa	triterpene	*Botritis cinerea, Fusarium lini*	2.5 mg/ml	Singh et al., 1992
		Altenaria spp.		
Bellis perennis	triterpene	*Candida* spp., *Microsporum* spp. *Trichophyton* spp., *Aspergillus niger*		Willigmann et al., 1992
Camellia japonica	triterpene	*Botritis cinerea, Cochliobolus miyabeanus*		Nagata et al., 1985
		Pyricularia oryzae, Pestalotia longiseta		
Camellia japonica	triterpene	*T. mentagrophytes, Epidermophyton floccosum*	25 µg/ml	Uemura et al., 1997
Camellia oleifera	triterpene	*Trichopyton* spp., *Epidermophyton flocosum*	0.1-3 µg/ml	Jin et al., 1993
		Microsporum audouinii	10 µg/ml	
Camelia spp.	triterpene	*Pilicularia oryzae, Cochliobolus miyabeanus*		Hamaya et al., 1984
Celmisia petriei	triterpene	*Cladosporium cladosporoides*		Rowan & Newman, 1984
Chisocheton paniculatus	steroidal	*Curvularia verruciformis, Altenaria solani*		Bordoloi et al., 1993
		Dreschlea oryzae		
Clerodendrum wildii	triterpene	*Cladosporium cucumerinum*	3.3 µg/plate	Toyota et al., 1990
Dolichos kilimandscharicus	triterpene	*Cladosporium cucumerinum*	2-5 µg/plate	Hostettmann et al., 1996
Dracaena mannii	steroidal	*Trichophyton* spp., *Cladosporium carrionii*	6-12.5 µg/ml	Okunji et al., 1990, 1996
		Trichosporon cutaneum, Geotrichum candidum	6.25 µg/ml	
		Microsporum spp., *Phialophora verrucosa*	12-50 µg/ml	
		Ramichloridium subulatum, Exophiala jeanselmei	25 µg/ml	

Plant species	Type	Target organism(s)	Concentration	Reference
Echallium elaterium	triterpene	*C.albicans, Rhodotorula* sp., *Fonsecaes pedrosoi, Candida tropicalis*	25 μg/ml	Bar-Nun & Mayer, 1990
Hedera helix	triterpene	*Botritis cinerea*	100 μg/ml	Timon-David et al., 1980
		Candida albicans CS60	100 mg/kg	
		Microsporum canis	0.5 mg/ml	
Hedera helix	triterpene	*Candida glabrata*		Favel et al., 1994
Heteropappus altaicus	triterpene	*Candida* spp.		Bader & Hiller, 1995
Heteropappus biennis	triterpene	*Candida* spp.		Bader & Hiller, 1995
Lycopersicon esculentum	steroidal	*Aspergillus* spp., *C. albicans, Trichophyton* spp.	10mM	Jadhav et al., 1981
Lycopersicon esculentum	steroidal	*Gaeumannomyces graminis*		Osbourn et al., 1995
Lycopersicon esculentum	steroidal	*Septoria lycopersici, Cladosporium fulvum*		Melton et al., 1998
Lycopersicon esculentum	steroidal	*Corticium rolfsii*		Schlosser, 1976
Maesa ramentacea	triterpene	10 fungal species	150 μg/ml	Phongpaichit et al., 1995
Malus meliana	unsp.	Food preservation	10-50 μg/ml	Xie et al., 1998
Medicago lupulina	triterpene	*Trichoderma viride*	17-50 μg/ml	Oleszek et al., 1988
Medicago sativa	triterpene	*Gaeumannomyces graminis*	3.5-7 μg/ml	Martyniuk et al., 1996
Medicago sativa	triterpene	*Sclerotium rolfsii, Rhizoctonia solani, A. niger*	50 μg/ml	Zehavi et al., 1993
		Fusarium oxysporum, Cephalosporium gramineum		
Medicago sativa	triterpene	*T. mentagrophytes, Epidermophyton floccosum*	7-25 μg/ml	Zehavi et al., 1996
		Microsporum canis, Cryptococcus sp., *C.albicans*	5-10 μg/ml	
Medicago sativa	triterpene	*Trichoderma viride*	17-50 μg/ml	Oleszek, 1996
Medicago sativa	triterpene	*Trichoderma viride, Candida albicans*		Boguslavsky et al., 1992
Medicago sativa	triterpene	*Phytophthora cinnamomi*		Zentmeyer & Thompson, 1967
Medicago spp.	triterpene	*Trichoderma viride*		Jurzysta & Waller, 1996
Mollugo pentaphylla	triterpene	*Cladosporium cucumerinum*	1.5 μg/plate	Hamburger et al., 1989
Nicotiana tabacum	steroidal	*Cladosporium cucumerinum, Puccinia recondita*		Grunweller et al., 1990
Pisum sativum	triterpene	*Fusarium solani*	150-300 μg/ml	Christian et al., 1989
Polemonium caeruleum	triterpene	yeasts, dermatophytes, molds		Hiller et al., 1981
Polygala vulgaris	triterpene	*Ceratocystis ulmi*		Chesne et al., 1983
Primula acaulis	triterpene	*Candida* spp.		Margineau et al., 1976
Rapanea melanophloeos	triterpene	*Cladosporium cucumerinum*	1 μg/plate	Ohtani et al., 1993
Rudgea viburnioides	triterpene	*Cladosporium cladosporioides*		Young et al., 1998
Ruscus aculeatus	steroidal	*Aspergillus fumigatus, T. mentagrophytes*		Guerin & Reveillere, 1984
Sapindus mukurossi	steroidal	*Cladosporium cucumerinum*	1.5 μg/plate	Tamura et al., 1990
Serjania salzmanniana	triterpene	*Cryptococcus neoformans, Candida albicans*	8-16 μg/ml	Ekabo et al., 1996

(Contd.)

TABLE 2. (Contd.)

Plant source	Saponin	Fungal species	MIC	Reference
Silphium perfoliatum	triterpene	*Drechslera graminea, R. nodosus, R. nigricans*	10-1000 µg/ml	Davidyants et al., 1997
Solanum tuberosum	steroidal	*Trichoderma viride, Cladosporium fulvum*		Jadhav et al., 1981
		Helmintosporium carbonum, Fusarium caeruleum		
Solenostemma argel	unspec.	*Asperillus, Penicillim, Chrisoporium, Candida spp*		El Hady et al., 1994
		C. neoformans, Mucor, Rhodotorula		
Solidago virgaurea	triterpene	*Candida spp.*	200 µg/ml	Bader and Hiller, 1995
Trigonella foenum-graecum	steroidal	*Rosellinia necatrix, T. harzianum, T. viride*	50-100 µg/ml	Sauvaire et al., 1996
		Candida albicans		
Trillium glandiflorum	steroidal	*C. albicans, S. cerevisiae, T. mentagrophytes*	1.5-12.5µg/ml	Hufford et al., 1988
		Cryptococcus neoforman, Aspergillus spp		
Yucca schidigera	steroidal	*Saccharomyces spp., Candida albicans*	31-62 µg/ml	Tanaka et al., 1996
		C. farmata, Hansenula spp., Cryptococcus spp.	31.3 µg/ml	
		Pichia spp., Debaromyces spp., Trichophyton spp.	31.3 µg/ml	
		Zygosaccharomyces spp., Sabouraudites canis	31.3 µg/ml	
		Epidermophyton floccosum	31.3 µg/ml	
Zygophyllum spp.	triterpene	*Verticillium albo-atrum, Fusarium oxysporum*		Oui et al., 1994

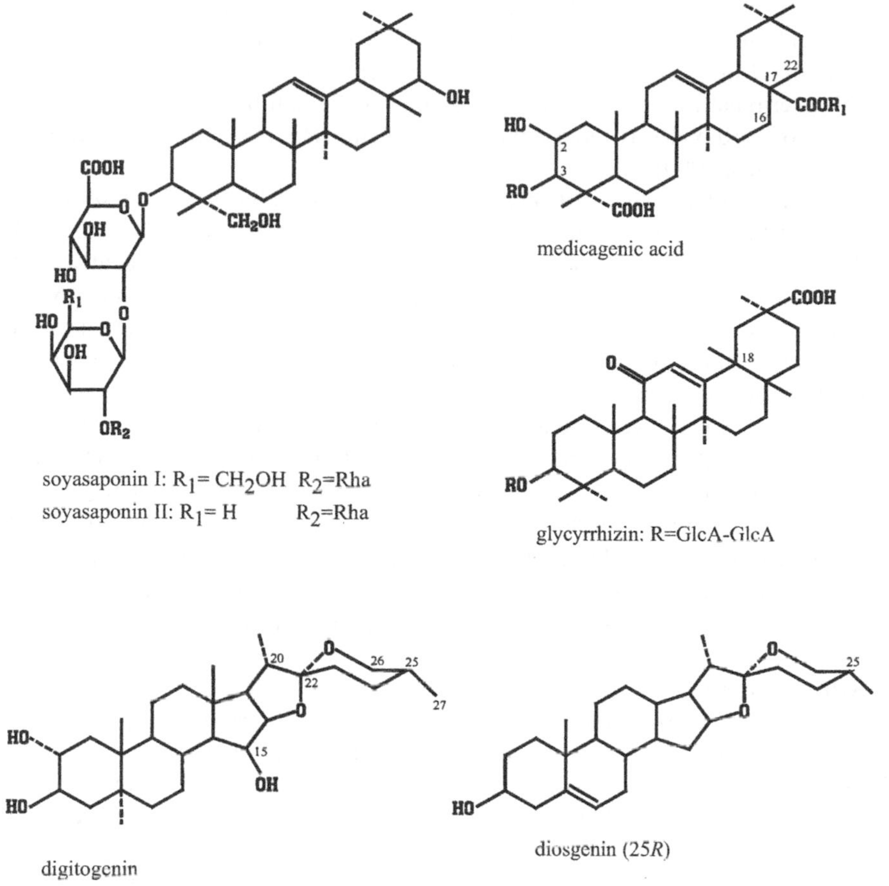

soyasaponin I: R_1= CH$_2$OH R_2=Rha
soyasaponin II: R_1= H R_2=Rha

medicagenic acid

glycyrrhizin: R=GlcA-GlcA

digitogenin

diosgenin (25*R*)

FIGURE 5. Structures of selected saponins and sapogenins with extensively elucidated biological activities.

the incorporation of the aglycone to cyclodextrin (CD) (Sasazuka et al., 1995). The antimicrobial activity of oleanolic acid (OA) incorporated into CD was similar to that of OA. The solubility of ß-CD was 1.85 g/100g of water at 25°C, while that of OA-ß-CD was 3.85 g/100 g of water and increased with the temperature.

Structurally divergent glycosides of the same aglycone molecule may show a wide range of activities. In this respect, monoglycosides with one sugar attached demonstrate similar activity to the aglycone. This was proven with medicagenic acid glucoside (Oleszek et al., 1990b) and diosgenin glucoside and galactoside (Takechi & Tanaka 1991).

Medicagenic acid glucoside with glucose at C-3 showed however, much higher activity than similar glycoside substituted with glucuronic acid. Prolongation of the sugar chain in monodesmosides could lead to a gradual decrease in activity. Comparison of hemolytic and antifungal activities of partial acid hydrolysates of dioscin and dioscinin showed that the activities were in general proportional to number of sugar residues, but compounds with branched sugar chain showed higher activities than those with straight chains. The substitution of aglycone with glucose at position C-3 was indicated to be important for antifungal activity. It is of interest to notice that even pattern of linkage of two sugars had influence on activity with Rha(1-2)Glc contributed more that Rha(1-4)Glc (Takechi et al., 1991). In the studies of α-hederin from *Hedera rhombea* it was documented that terminal rhamnose is very important for the antifungal activity of saponin. Methylation of carboxyl was more important for hemolytic than for antifungal activity (Takechi & Tanaka, 1990).

A number of structurally divergent saponins from alfalfa (*Medicago sativa*) have been tested for their activity against *Trichoderma viride* (*TABLE 3*) (Oleszek et al., 1990b, Bialy, 1998). It was shown that bidesmosidic analogues of medicagenic acid were less active than monodesmosides with the activity gradually declining as the number of sugar residue in saponin molecule increased. The same trend was found for the 3-O-glucuronide medicagenic acid and its bidesmoside. However, all glucuronic acid substituted at C3 glycosides were much less active than their analogues with glucose linked at C3. However, no straight correlation between sugar number and activity could be found. The 16-OH derivative of medicagenic acid, the zanhic acid tridesmoside was totally inactive in the range up to 100 mg/ml in spite of the fact that bidesmosidic medicagenic acid glycosides with the same number of sugars had IA_{50} 104 µg/ml (*TABLE 3*). Monodesmosidic saponin, 3-Glc-Glc-Glc zanhic acid, released after basic hydrolysis of tridesmoside showed IA_{50} around 65 mg/ml. Similar effect of reduction of antifungal and hemolytic activity has been observed for 17-OH derivatives of dioscinin (Takechi et al., 1991).

C. Antiviral activity

Most of the work that has been performed on antiviral activity of plant saponins did not exceed the experimental level. Some of the studies are listed in *TABLE 4*. The most frequently studied compound was glycyrrhizin (*FIGURE 5*), which was shown to inhibit some DNA and RNA viruses *in vitro* and to have therapeutic and prophylactic effects on some chronic viral hepatitis (Pompei et al., 1979). This compound was shown to inhibit the growth of several enveloped viruses including varicella-zoster virus (VZV), giving 100% inhibition of cytopathic effect (CPE) at the concentration 5.33 µg/ml, human immunodeficiency virus (HIV) (IC_{50} = 404 mg/ml) and VZV (IC_{50} = 584 µg/ml). This did not affect naked virus (poliovirus type 1). Glycyrrhizin had no direct inactivation effect on virus particles, and no interferon inducing activity *in vitro* or *in vivo*. Its biological activity is believed to be related to the action of one or more steps of the viral replication cycle. Similar results were obtained by Aquino et al. (1996) for the number of triterpene saponins including quinovic acid and oleanolic acid glycosides. The infection by enveloped virus (VSV) was generally more sensitive to these glycosides than infection by naked virus (HRV). However, much lower concentrations were required for plaque reduction (MIC_{50} ranged from 20 to 60 µg/ml). The most effective of quinovic acid glycosides was the compound with unsubstituted -COOH groups at C-27 and C-28 (MIC_{50} = 20 µg/ml, CD_{50} = 100 mg/ml). No correlation was found between the activity against VSV and the number of sugars in glycoside molecule. Several oleanolic acid glycosides were tested against HRV and VSV infection on HeLa and CER cells, respectively. The maximum non-toxic concentration ranged from 4 to 12 µg/ml was found for compounds having free carboxyl groups, while their esters showed reduced activity. The HRV was reduced affected only by one glycoside (De Tomassi et al., 1991).

TABLE 3: Inhibitory activity (IA$_{50}$) of alfalfa root medicagenic acid glycosides against *T. viride*

Compound	IA$_{50}$ (µg/ml)
3-Glc	1.6
3-Glc, 28-Glc	33.0
3-Glc-Glc, 28-Glc	68.0
3-Glc, 28-Ara-Rha	42.0
3-Glc-Glc-Rha	40.0
3-Glc-Glc-Glc, 28-Glc	94.0
3-Glc, 28-Ara-Rha-Xyl	13.5
3-Glc-Glc, 28-Ara-Rha-Xyl	35.0
3-Glc-Glc-Glc, 28-Ara-Rha-Xyl	64.0
3-Glc-Glc-Glc, 28-Ara-Rha-(Api)-Xyl	104.0
3-GlcA	9.1
3-GlcA, 28-Ara-Rha-Xyl	47.5

Based on Oleszek et al., 1990b, Bialy, 1998

The steroidal spirostane saponins, dioscin and gracillin, both having diosgenin (*FIGURE 5*) as an aglycone are cyctotoxic above 4 and 20 mg/ml, respectively and do not give interesting results below these concentration in the antiviral screening using VSV and HRV-1B. Furostanol glycosides are less cytotoxic and the compounds with 25R configuration gave 100% plaque reduction at 100 mg/ml but were less active against HRV. The opposite was found for compounds with 25S configuration, which were active against HRV, which cytopathic effect is reduced to 25% at 4 mg/ml. It was strongly stressed that 25R and 25S epimers give an inverted intensity of action against enveloped and naked viruses (Aquino et al., 1996).

In the studies of Akhov et al., (1999) on the effect of steroidal furostanol saponin known under its common name as deltoside, isolated from underground parts of *Allium nutans* was tested against tobacco mosaic virus (TMV). The virus was treated with 0.6, 1.25 and 2.5 µg/ml solutions of saponin and after 30 min it was put on the grid and observed with electronic microscope (*FIGURE 6*). It was shown that saponin caused the agglutination of virons and creation of knots and fibrils on the surface of the virus.

Interesting data were published by Okubo et al., (1994) on the activity of saponins having soyasapogenol as an aglycone. These saponins are quite common in legume seeds and most dominant is soyasaponin I (*FIGURE 5*), the triterpene saponin related in the structure to glycyrrhizin. They have partial inhibitory effect on HIV- induced cytopathology in infected MT-2 lymphocyte cultures (Nakashima et al., 1989). They also completely inhibited HIV- induced cytopathic effects and virus specific antigen expression 6 days after infection at the concentration higher than 0.25 mg/ml (Okubo et al., 1994). This saponins inhibit virus replication during an early stage of replicative cycle, i.e. adsorption or penetration of HIV, and in this respect might be a very promising specific for preventing a large number of healthy carriers from developing AIDS. It seems that soyasaponin I causes no side effects and is naturally present in legume seeds, a very important source of protein in the diet of oriental people.

TABLE 4. Antiviral activity of saponins

Saponin	Plant Source	Virus type	Reference
Chikusetsusaponin III	*Panax japonicum*	HSV-1	Fukushima et al., 1995
Deltoside	*Allium nutans*	TMV	Akhov et al., 1999
Extract MeOH	*Mimosa hamata*		Jain et al., 1997
Extract	*Verbascum thapsiforme*	HSV-1	Slagowska et al., 1987
Glycyrrhizin	*Glycyrrhiza glabra*	HAV	Crance et al., 1990
		VZV	Baba & Shigeta, 1987
		HIV-1	Ito et al., 1987
		HIV-1, HSV-1	Hirabayashi et al., 1991
		HIV	Hattori et al., 1989
		HIV-1	Hasegawa et al., 1994
Holoturinosides	*Holothuria forskalii*	VSV	Rodrigues et al., 1991
Oleanolic acid glycosides	*Calendula arvensis*	VSV, HRV	De Tomassi et al., 1991
		HRV-1B, VSV	Aquino et al., 1996
Protoprimulagenin glycosides	*Anagalis arvensis*	HSV-1, AV-6, VSV	Amoros et al., 1987 Amoros et al., 1988
Quinovic acid glycosides	*Uncaria tometosa* *Guettarda platypoda*	HRV-1B, VSV	Aquino et al., 1996
Rosamultin, Kajiichigoside F1			Rucker et al., 1991
Saikosaponin a	*Bupleurum falcatum*	HSV	Ushio & Abe, 1992
Soyasaponin (Bb)	*Glycine max*	HIV-1	Okubo et al., 1994
Soyasaponin I and II	*Glycine max*	HSV-1	Hayashi et al., 1997
Soyasaponin B1 and B2	*Glycine max*	HIV	Nakashima et al., 1989
Spirostane and furostane glyc.	*Tamus communis* *Asparagus cochinchinensis*	HRV-1B, VSV	Aquino et al., 1996
Taurosid I		HIV-1	Krivorutchenko et al., 1997
Tormentic acid, euscafic acid		HSV-1, poliovirus	Simoes et al., 1990
Oleanolic acid glycosides			
Triterpene saponins		Influenza A2	Rao et al., 1974
Triterpene saponins	*Chenopodium anthelminticum*	Influenza A2	Vichkanova and Goryunova, 1971
	Callistephus chinensis *Glycyrrhiza glabra*		
Zingibroside R1	*Panax zingiberensis*	HIV-1	Hasegawa et al., 1994

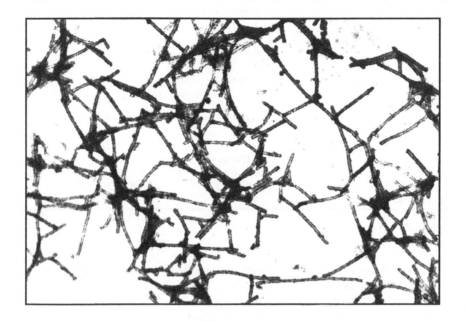

FIGURE 6. A micrograph of tobacco mosaic virus treated with 1.25 µg/ml solution of deltoside, furostanol saponin from *Allium mutans*.

V. BIOLOGICAL ADVANTAGE

A. Additive advantage to foods

Saponins can be found as ingredients of many plants used as a human food. Some of the most important have been listed in *TABLE 5*. The knowledge about these compounds as food components has increased substantially during last few years. It has been documented that they are generally not absorbed in the digestive tract and thus, do not create serious toxicological problems. Hemolytic and toxic effects in fish and high toxicity to invertebrate pests are well known. Oral toxicity of saponins has been estimated to be low (Price et al., 1987). Ingestion of saponin containing food by man and animals has been associated with both deleterious and beneficial effects; the former resulting in reduced weight gain in animals fed for example alfalfa, and the latter their purported hypocholesterolemic effect. It is not clear what mechanism reduces weight gain in animals, but it is generally agreed that the responsible factor is not toxicity but rather bitterness of saponins, resulting in reduced feed consumption. However, some of these effects may be attributed to the ability of saponins to affect membrane integrity. Studies of a number of saponins showed that the features which appear to be necessary to produce a permeabilizing effect are a combination of type and the number of sugar chains per molecule in addition to the nature of the aglycone to which these sugars are attached (Gee et al.,1998). Preliminary data obtained by computer modeling showed that saponin molecules can have cage-like shapes, which appear to be related to their ability to reduce

TABLE 5. Saponin content (%) of various food plants

Plant source	Saponin content (%)
Peanuts	1.3 - 1.6
Millet	0.02
Alfalfa sprouts	0.6
Spinach root	4.7
Horse chesnut	3 - 6
Guar	10
Oats	0.1
Sesame seeds	0.3
Asparagus	1.5
Garlic	0.3
Lentils	0.11-0.46
Pea products	0.01-0.51
Bean products	0.02-3.3
Soybean products	0.15-0.6

Adapted from Price et al., 1987

potential difference across the intestinal epithelium, which is thought to reflect a loss of membrane integrity affecting some or all of the mucosal cells, possibly as a result of formation of stable micelle-like structures in cell membrane. Such intestine "holes" may permit food toxins and xenobiotics to be easily absorbed.

The beneficial effect of saponins reflects their abilities to bind diet cholesterol. Saponin-cholesterol sandwich-like insoluble, nondigestible macromolecules are formed in digestive tract, which in turn significantly prevent the absorption of bile salts from the perfused small intestine (Oakenfull, 1981). It has been documented that saponins reduce diet-induced hypercholesterolemia in rats and lead to the significant increases in fecal exertion of neutral sterols and bile acids (Oakenfull et al., 1983). The authors suggest that particular saponins reduce hypocholesterolemia by different mechanisms. It has been suggested that alfalfa and Quillaya saponins reduce the primary absorption of dietary cholesterol, while saponins derived from *Saponaria* and soybean increase fecal excretion of bile salts. The discovery that some saponins are hypocholesterolemic has provoked considerable research and clinical interest. Malinow and co-workers (1981) observed a regression of aortic and coronary atherosclerosis in macaques (*Cynomolgas macaques*) fed 1% isolated alfalfa root or 0.6% alfalfa top saponins without any side effects.

The main problem in application of saponins as a cholesterol reducing formula is the fact that many inconsistent opinions can be found in literature on their effects as well as on the toxicity. The main reason is the fact that much research has been performed on extracts or very poorly defined mixtures due to difficulties with separation of single pure compounds.

B. Physiological advantage to host

The glycosylation process that takes place in plants is generally regarded as a means for increasing solubility of the compound, allowing saponin transport. From the other side, it can be regarded as deactivation of active aglycone to protect cell organelles. The activation can be performed when the producing organism needs protection. Thus, damage of the plant tissue may release appropriate enzymes that are able to hydrolyze bidesmosidic saponins to more active monodesmosidic forms. In this respect saponins appear to be an exception among plant antifungal compounds since even monodesmosidic compounds are polar, while the majority of other antifungal natural compounds tend to be strongly lipophilic and inactive in glycosidic forms. The *Hedera helix* which is rich with inactive bidesmosidic hederasaponin C contains also enzyme hydrolyzing ester linked sugar moiety to produce highly active monodesmosidic α-hederin (Schlosser, 1973). *Avena sativa* leaves contain inactive bidesmosidic furostanol saponins avenocoside A and B, which on tissue damage are hydrolyzed by ß-glucosidase removing glucose at C-26 and converting them to active monodesmosidic forms (Luening & Schloesser, 1976). However, the plant cell can be disrupted by the endogenous saponin as well, the case we face in *in vitro* activity tests. In the studies on membranolytic action of different saponins on mycelium of *Botrytis cinerea* and *Rhisoctonia solani* it was documented that digitonin, α-hederin and tomatine caused considerable leakage of free amino acids while aescin and thea saponin were less effective (Segal & Schloesser, 1975). Another saponin, cyclamin, also significantly damaged cell membranes but the effect was selective to some microorganisms. Disruption of cell membrane was accompanied by enzymatic conversion of saponins into their corresponding aglycones. The hydrolysis of the aglycone can, however, in some cases provide a protection for the fungi. Intramural ß-glucosidase in *Drechslera avenacea* can hydrolyse monodesmosidic 26-deglucoavenacosides A and B into aglycone nuatigenin, which due to the low water solubility precipitates and is not able to reach fungal plasma membrane (Luening & Schloesser, 1976). The same was true for tomatidine released from tomatine by *Altenaria* species pathogenic to the tomato fruit; non-pathogenic *Altenaria* species do not have such an enzymatic inactivation system (Schloesser, 1977).

Some plant fungi do not produce extracellular enzymes that degrade saponins, but still can grow on the medium containing such a saponin. The best examples are *Phyllostica concentrica* and *Pestalotia microspora,* microspora of which can grow in ivy leaves in spite of not having such an enzyme. It was suggested that these fungi produce other substance(s) inhibitory to enzymes converting bi- to monodesmosidic forms (Schloesser, 1973). Some fungi can avoid saponin toxicity by invading plant parts low with these compounds. Artificial inoculation of *Botrytis cinerea* revealed that the fungus can only develop on the stems but not on the leaves of *Cyclamen persicum* (Schloesser, 1971). This difference can be explained by the cyclamin content of the respective organs. The ability of a phytopathogenic fungi to detoxify plant saponins may suggest that saponin detoxification determines the host range of these fungi (Bowyer et al., 1995).

The cucurbitacin I, triterpene saponin from some *Cucurbitaceae*, was shown to play a protective function against *Botrytis cinerea*. Cucurbitacine I, when applied to cucumber fruits or cabbage leaves prior to inoculation, prevented fungus infection of these tissues. The protective effect was not due to the induction of lignification, but appeared to be caused by inhibition by cucurbitacin I of laccase formation by *Botrytis* (Bar-Nun & Mayer, 1990).

VI. APPLICATIONS

Several plant sources have been commercially used (*TABLE* 6) but only two of them are actually permitted as food additives. These are *Quillaya saponaria* and *Yucca schidigera* extracts, containing triterpene and steroidal saponins, respectively. They are regarded as GRAS (generally recognized as safe) products. In the UK the use of Quillaya extracts in food was permitted as early as 1962; in beverages, the use of such extracts as emulsifiers and foaming agents was restricted to not more than 20 ppm by weight. These two extracts are also permitted in the USA by the FDA for use in food and beverages. They are used as a flavor enhancer in some categories of beverages including root beer, flavored teas and juices and additions to non-fat dairy products such as chocolate milk and vanilla pudding. As natural emulsifiers they have been used to emulsify oil-based flavors for candy, to prevent precipitation in a protein containing liquid composition, to prevent oil separation in mayonnaise, for use as a leavening agent in the bakery industry, to disperse oil in water type emulsion composition, to increase stability of cream when added to coffee and as a natural dispersing agent for waxes used in food coatings. They can also play an important function as natural preservatives against certain strains of bacteria (*E. coli, Staphylococcus aureus, Salmonella, Pseudomonas aeruginosa*) and fungi (*Aspergillus niger, Penicilium chrysogenum*). A recent patent claims that a simple process using *Quillaya* and *Yucca* saponins removes up to 83% of the cholesterol from the milk, 77% from cream, 80% from butter oil and reduces the cholesterol from liquefied eggs, or any other foodstuffs in aqueous systems.

In Japan the use of *Quillaya* and *Yucca* extracts, soya bean saponins and enju saponins as emulsifiers has also been permitted, as a food additive.

The important application of many steroidal saponins is their usage as starting materials for steroid hormone semi-synthesis. A majority of the steroids produced by the pharmaceutical industry and used as contraceptives are obtained by semi-synthesis from natural substances including saponins. The main saponin that has been used for a long time for this purpose was diosgenin extracted from different plant sources (see TABLE 6).

A number of plant sources containing saponins are used as drugs for different ailments including:
- in phlebology and proctology - Buchers broom (*Ruscus aculeatus* L.), Figwort (*Ranunculus ficaria* L. = *Ficaria ranunculoides* Roth.)
- cough treatment - Snakeroot (*Polygala senega* L.), Common ivy (*Hedera helix* L.), Primrose (*Primula veris* L. = *Primula officinalis* (L.) Hill.)
- anti-inflammatory - Licorice (*Glycyrrhiza glabra* L.), Common horse chestnut (*Aesculus hippocastanum* L.)
- adaptogens - Ginseng (*Panax ginseng* C.A. Meyer), Siberian ginseng (*Eleutherococcus enticosus* Maxim)
- in dermatology - Hydrocotyle (*Centella asiatica* L.), Urban ginseng [*Mimosa tenuiflora* (Willd.)], Marigold (*Calendula officinalis* L.)
- kidney and gall stone formation - Mohave Yucca (*Yucca schidigera* Roezl ex Ortgies)
- hypertension - green tea (*Camelia sinensis* L.)
- premenstrual tension syndrome - soybean (*Glycine max* L.)
- nasal and ocular delivery of insulin (*Quillaya saponaria* Molina)
- detergents - Quillaya (*Quillaya saponaria* Molina), Soapworth (*Saponaria officinalis* L.), Soaproot (*Gypsophila* spp.), Mohave Yucca (*Yucca schidigera* Roezl ex Ortgies)

Saponin extracts e.g. *Yucca* and *Quillaya* can be successfully used in animal feeding. Experiments with chicken show that addition of Yucca extract to the drinking water for broilers can reduce ammonia release from feces (Podgorski et al., 1996). In some other animals these saponins can modify nitrogen metabolism, alter rumen fermentation, show

TABLE 6: Plant sources for industrially utilized saponins

Plant source	Common name	Use
Agave sisalana	Sisal, Henequen	Hecogenin source
Balanites aegyptica		Diosgenin source
Chlorogalum pomeridianum	Soaproot, California soap plant	Amolonin source
Costus speciosus		Diosgegnin source
Digitalis lanata	Purple Foxglove	Digitonin source
Digitalis purpurea	Digitalis	Digitonin source
Dioscorea composita	Yams, Barbasco	Diosgenin source
Dioscorea terpinapensis		
Dioscorea spp		
Glycine max	Soybean	Soyasaponin source
Quillaya saponaria	Quillaya	Soaps, foaming agents
Smillax spp.		Smilagenin/sarsapogenin source
Solanum spp.		Solasodine source
Trigonella faenum-graecum	Fenugreek	Diosgenin source
Yucca schidigera	Mohave Yucca, Joshua Tree	Soaps, foaming agents

antiprotozoal activity, modify fat digestion and absorption and generally improve animal performance (Makkar & Becker, 1996). Industrial uses of saponins may include:

- bleaching or de-inking agent in recycled paper industry
- preventing pitch stain during the paper making process
- reducing odors in municipal sewage pond systems
- acceleration of microbial growth in both aerobic and anaerobic systems
- coating and dispersing fat particles in waste water
- suspending the silver halides in photographic films

VII. SAFETY

Cytotoxicity of plant saponins precludes their practical application and some of them may show quite high cytotoxic activity (*in vitro*), anti-tumor (*in vivo*) and chemo-preventive (*in vitro* and *in vivo*). These activities show concentration dependent effects and change on the structure of the aglycone and sugar substitutions. Some examples of these activities are listed in TABLE 7.

Along with cytotoxic activities a number of antimutagenic activities have also been performed. Generally, saponins show no or little mutagenic activity and some antimutagenic activities. Thirteen saponins, oleanolic acid and hederagenin glycosides, isolated from *Calendula officinalis*, *Calendula arvensis* and *Hedera helix* were tested for toxicity and mutagenicity using a modified liquid technique of the Salmonella/microsomal assay (Elias et al., 1990). All tested saponins showed no toxicity and mutagenicity for doses of 400 μg. Screening for the antimutagenic activity was performed with two known promutagenic substances benzo[a]pyrene (MaP) and mutagenic urine concentrate from smokers (SU). Saponins showed antimutagenic activity against BaP (1μg) and SU (5μl) and the activity showed dose-response relationship.

The medicagenic acid (Na⁺), its 3-O-glucopiranoside and soyasaponin I have been studied for their mutagenicity with the Ames test in the concentration range of 0-500 mg per plate. None of the tested compounds increased the number of his revertants in *Salmonella typhimurium* strains TTA97, TA98, TA100, TA102, neither in the presence or absence of S9 fraction from rat liver. Thus, according to Ames criterion none of the compounds was mutagenic (Czeczot et al., 1994).

Some saponins e.g. glycyrrhizin and glycyrrhetinic acid, when applied at high doses for a prolonged time, show a side effect causing edema and hypertension in some patients. This side effect, called pseudoaldosteronism, is induced by the inhibitory effect of these two saponins against the reductase for mineral cortocoid in the liver resulting in retention of Na^+ and water and excretion of K^+.

VIII. BIOTECHNOLOGY

Industrial demand for the *Quillaya*, *Yucca* and ginseng saponins requires development of the new technologies for their production. For decades, *Quillaya* and *Yucca* extracts have been produced from trees. Every year about 60,000 trees were cut causing serious ecological damage and a shortage of resources. One of the solutions is pruning the trees on the plantations. For ginseng, the predominant source is field-cultivated material. However, the field cultivation is a time-consuming and labor-intensive process subject to several additional conditions like soil, climate, pathogens and pests.

The alternative to field production is plant cell culture technology. Trials with *Panax notoginseng* performed in China showed that bubble column reactors, air-lift bioreactors and shake flask techniques can be used for high density (23-25 g/l) cell cultivation. The biomass productivity was 1083, 146 and 916 and production of saponins was 32, 43 and 50 mg per liter per day (Zhong & Yao, 1999). Callus and cell suspension cultures of American ginseng (*Panax quinquefolium*) were compared for biomass growth and ginsenoside production over a 35-day culture cycle on modified MS medium. The biomass yields in suspension and callus cultures were maximal on the 25 and 30 day of growth. Both types of culture were able to produce ginsenosides in amounts and quality comparable to the cultivated plants. Appreciable amounts of ginsenosides, particularly Rg 1, were found to leach out in the culture medium of 30- to 35 day-old suspension culture (Archana et al., 1994). The production of saponins can be increased by appropriate sugar type and concentration. The culture media of *Panax pseudoginseng* containing 30, 60, or 100 g sucrose per liter produced on average 4.55, 37.25 and 91.83 embryoids (derived from ginseng shoots), respectively. The HPLC profile of saponin composition of the ginseng embryoids was similar to that found for the roots of naturally grown plants (Asaka et al., 1994).

Cell suspension culture technique has been applied to the production of antifungal spirostanol saponins from *Solanum chrysotrichum* (Schldl.). Hypocotyl-derived calluses grown for 25 days in shake flasks can yield 14 mg saponins per gram, which is 50 times that of field grown plants. When grown in 10 l airlift reactor with 3g/l dry inoculum in batch culture, higher levels of biomass were achieved with cotyledon derived (14.6 g/l) than with hypocotyl derived (7.7 g/l) cells. The maximum productivity of saponins in bioreactors was 0.025 g/l/day after 9 days in culture (Villarreal et al., 1997).

Oleanolic acid on an industrial scale can be obtained from sugar beet pulp after extraction of sucrose. Oleanolic acid is present in this solution in glycosidic form (glucuronide, glucoside or methylated glucoside). The ß-glucuronidase from *Pseudomonas paucimobilis* H-3 hydrolyzed almost all beet saponin containing glucuronic acid to produce oleanolic acid and 73% of beet saponin was hydrolyzed to oleanolic acid in 24 hours (Ishikawa, 1996).

TABLE 7. Cytotoxic activity of saponins

Saponin	Plant Source	Cytotoxicity	Concentration
Diosgenin, dihydrodiosgenin	*Dracena afromontana*	KB cells	10 µg/ml
Diosgenin glycosides	*Paris polyphylla*	KB cells	0.16-0.29 µg/ml
		P-388	0.22-0.44 µg/ml
		L-1210	0.14-0.43 µg/ml
Diosgenin glycosides	*Balanites aegyptica*	P-388	0.21-2.4 µg/ml
Echinocystic acid octaglyc.	*Entada phaseoloides*	L-5178 Y	0.83 µg/ml
Ginsenoside R_{b2}	*Panax ginseng*	RLE, B16	not effective
		HRA	10-100 µmol
α-Hederin	*Hedera helix*	B 16	5 mg/ml
Holestane glycosides	*Holothuria forskalii*	P-388	0.38-0.46 µg/ml
Oleanolic acid diglycoside	*Panax zingiberensis*	MT-4	46.2 µmol
Pectiniosides A-F	*Asterina pectinifera*	L 1210	8.8-11 µg/ml
		KB	10-11.5 µg/ml
Saikosaponins		PLC/PRF/5	20-50 µg/ml
Sho-saiko-to medicine		Hep-G2	
Saikosaponi-acetylglycosyl	*Buphlerum wenchuanese*	P-388	<5 µg/ml
Sarasinoside A1	*Asteropus sarasinosum*	P-388	2.8 µg/ml
Sarasinosides D-G	*Asteropus sarasinosum*	A-549	7.4 µg/ml
		HT-29	4.55 µg/ml
		P-388	3.62 µg/ml
		B 16-F 10	9.62 µg/ml
Solamargine		PLC/PRF/5	1.53 µg/ml
Khasianine	*Solanum* spp.	PLC/PRF/5	8.60 µg/ml
Spirostanol glycosides	(*Chamadeorca linearis*)	L 1210	32 µmol
Trillin	*Dracaena afromontana*	KB cells	100 µg/ml
Triterpene glycosides	*Dysoxylum cumingianum*	MOLT-4	0.0045-0.0062 µg/ml
Triterpene glycosides	*Myrsine australis*	P-388	0.85 µg/ml
Tubeimoside 1	*Bobolstemma paniculatum*	GOTO	0.24 µmol/L
		A-172	0.15 µmol/L
		PANC-1	0.27 µmol/L
		COLO 320 DM	0.43 µmol/L
Yamogenin glycosides	*Balanites aegyptiaca*	P-388	0.21-2.40 µg/ml
Zingiberoside R1	*Panax zingiberensis*	MT-4	84.4 µmol

Based from Lacaille-Dubois & Wagner, 1996; Hostettmann & Marston, 1995

IX. SUMMARY

Saponins as natural phytoantimicrobial agents elicit antibacterial, antifungal and antiviral effects. They constitute an important defense mechanism for the host plant and protect against biotic stress factors. These activities can be utilized in natural food preservation against susceptible strains of bacteria and fungi, as well as therapeutics for certain chronic diseases. Saponins, in general, are not mutagenic or cytotoxic compounds. Their abilities to complex diet cholesterol may find wide application in reduction of diet-induced hypercholesterolemia. Prophylactic application of saponins can be achieved by the consumption of food rich in these compounds or by commercial preparation of beverages or food additives enriching daily saponin intake. However, since structurally divergent saponins show different activities, and isolation of large quantities of purified, single saponins is still a challenge, more research is needed to identify active principles and to find out the conditions for enhanced rates of synthesis by plant.

X. REFERENCES

1. Ahmad, R., Ata, A., Islam, B., Ashfaq, M. 1990. Studies on saponins from some indigenous plants. *Pak. Vet. J.,* 10:146-148

2. Ahmad, V.U., Noorwala, M., Mohammad, F.V., Aftab, K., Sener, B., Giliani, A.U. 1993. Triterpene saponins from the roots of *Symphytum officinale. Fitoterapia.* 64:478-479.

3. Akhov, L.S., Musienko, M.M., Shishova, Y., Polishuk, V.P., Oleszek W. 1999. Biological activity of deltoside from *Allium nutans* L. *Int. Conf. on "Saponins in Food, Feedstuffs and Medicinal Plants"*, Book of Abstr. p.

4. Amoros, M., Fauconnier, B., Girre, R.L. 1987. In vitro antiviral activity of a saponin from *Agalinis arvensis*, Primulaceae, against herpex simplex virus and polivirus. *Antiviral Res.* 8:13-25.

5. Amoros, M., Fauconnier, B., Girre, R.L. 1988. Effect of saponins from *Agalinis arvensis* on experimental herpes simplex keratitis in rabbits. *Planta Med.* 54:128-31.

6. Andersson, L., Bohlin, L., Iorizzi, M., Riccio, R., Luigi, M., Washington, M.L. 1989. Biological activity of saponins and saponin-like compounds from starfish and brittle-stars. *Toxicon.* 27:179-188.

7. Aquino, R., De Simone, F., Pizza, C., Conti, C., Stain, M.L. 1989. Plant methabolites. Structure and *in vitro* antiviral activity of quinovic acid and glycosides from *Uncaria tomentosa* and *Guettarda platypoda. J. Nat. Prod.* 52:679-685.

8. Aquino, R., De Simone, F. De Tomassi, N., Piacente, S., Pizza, C. 1996. New biologically active steroidal and triterpenoid glycosides from medicinal plants. In *Saponins Used in Traditional and Modern Medicine.* eds. Waller, G.R. Yamasaki, K., Plenum Press, New York, pp. 401-414.

9. Arhana, M., Shukla, Y.N., Mahesh, P., Ahuja, P.S., Uniyal, G.C., Mathur, A., Pal, M. 1994. Saponin production in callus and cell suspension cultures of *Panax quinquefolium. Phytochemistry*, 35:1221-1225.

10. Arneson, P.A., Durbin, R.D. 1968. The mode of action of tomatine as a fungitoxic agent. *Plant Physiol.* 43:683-686.

11. Asaka, I., Ii, I., Hirotani, M., Asada, Y., Yoshikawa, T., Furuya, T. 1994. Mass production of ginseng (*Panax ginseng*) embryoids on media containing high concentrations of sugar. *Planta Medica*, 60:146-148.

12. Atsuro, N., Yoshitaka, N., Eiji, O., Hiroshi, K. 1992. Antibacterial agents from fumaric acid solubilized by emulsifiers for foods. JP 04187066 A2 920703 Hesei, Patent written in Japanese.

13. Baba, M., Shigeta, S. 1987. Antiviral activity of glycyrrhizin against varicella-zoster virus *in vitro. Antiviral Res.* 7:99-107.

14. Baccou, J.C., Lambert, F., Sauvaire, Y. 1977. Spectrophotometric method for the determination of total steroidal sapogenin. *Analyst.* 102:458-460.

15. Bader, G., Kulhanek, Y., Ziegler-Boehme, H. 1990. Antifungal action of glycosides of polygalactic acid. *Pharmazie*, 45:618-620.

16. Bader, G., Hiller, K. 1995. Antitumoral and antifungal actions of different saponins of polygalacic acid. *Book of Abstracts, 210th ACS National Meeting*, Chicago, IL, AGFD-203.

17. Bar-Nun, N., Mayer, A.M. 1990. Cucurbitacins protect cucumber tissue against infection by *Botrytis cinerea. Phytochemistry*, 29:787-791.

18. Besso, H., Saruwatari, Y., Futamura, K., Kunihiro, K., Fuwa, T., Tanaka, O. 1979. High performance liquid chromatographic determination of ginseng saponin by ultraviolet derivatisation. *Planta Medica.* 37:226-233.

19. Bialy, Z. 1998. Chemical composition and fungistatic activity of alfalfa (*Medicago sativa* L. var Radius) root saponins. *PhD thesis, IUNG, Pulawy* pp. 64 (in Polish)

20. Bombardelli, E., Bonati, A., Gabetta, B., Martinelli, E.M., Mustich, G. 1976. Gas-liquid chromatographic and mass spectrometric investigation on saponins in *Panax ginseng* extracts. *Fitoterapia.* 47:99-106.

21. Bombardelli, E., Gabetta, B., Martinelli, E.M., Mustich, G. 1979. Gas chromatographic - mass spectrometric (GC-MS) analysis of medicinal plants. Part III. Quantitative evaluation of glycyrrhetic acid and GC-MS investigation of licorice triterpenoids. *Fitoterapia.* 50:11-24.

22. Bowyer, P., Clarke, B.R., Lunnes, P., Daniels, M.J., Osbourn, A.E. 1995. Host range of plant pathogenic fungus determined by a saponin detoxifying enzyme. *Science.* 267:371-374.

23. Birk, Y. 1969. Saponins. In *Toxic Constituents of Foodstuffs,* ed. I.E. Liener, pp. 169-210. New York: Academic Press.

24. Boguslavsky, V.M., Shirshova, T.I., Burtseva, S.A., Krepis, Ye.S.Obrezha, V.M., Tsigulaya, T.Y., Borisova, T.A., Panyushkina, K.A. 1992. Antimicrobial and antifungal activities of alfalfa saponins. *Stiinte Biol. Chim.* 3:42-44.

25. Bordoloi, M., Saikia, B., Mathur, R.K., Goswami, B.N. 1993. A meliacin from *Chisocheton paniculatus. Phytochemistry*, 34:583-584.

26. Brawn, P.R., Lindner, N.M., Miller, J.M., Telling, G.M. 1981. A gas chromatographic method for the determination of medicagenic acid in lucerne (alfalfa) leaf protein concentrate. *J. Sci. Food Agric.* 32:1157-1162

27. Calis, I., Aysen, Y., Deniz, T., Wright, A.D., Sticher, O., Luo, Y.D., Pezzuto, J.M. 1997. Cycloartane triterpene glycosides from the roots of *Astragalus melanophrurius*. *Planta Med*. 63:183-186.

28. Czeczot, H., Radhen-Staron, I., Oleszek, W., Jurzysta, M. 1994. Isolation and studies of the mutagenic activity of saponins in the Ames tes. *Acta Poloniae Pharmac*. 51:133-136.

29. Chesne, C., Amoros, M., Girre, L. 1983. Study of the antifungal activity of higher plants. III. Comparative study of techniques for isolation of an antifungal saponoside from aerial parts of *Polygala vulgaris* L. *Fr. Plant Med. Phytother*. 17:191-201.

30. Christian, D.A., Hadwiger, L.A. 1989. Pea saponins in the pea-*Fusarium solani* interaction. *Exp. Mycol*., 13:419-427.

31. Cioaca, C., Margineanu, C., Cucu, V. 1978. The saponins of *Hedera helix* with antibacterial activity. *Pharmazie*. 33:609-610.

32. Crance, J.M., Biziagos, E., Passagot, Van Cuyck-Gandre, H., Deloince, R. 1990. Inhibition of hepatitis A virus replication *in vitro* by antiviral compounds. *J. Med. Virol*. 31:155-160.

33. Davidyants, E.S., Kartasheva, I.A., Neshin, I.V. 1997. The effect of triterpene glycosides of *Silphium perfoliatum* on phytopathogenic fungi. *Rastit. Resur*. 33:93-98.

34. De Tomassi, N., Conti, C., Stein, M.L., Pizza, C. 1991. Structure and *in vitro* antiviral activity of triterpenoid saponins from *Calendula arvensis*. *Planta Med*. 57:250-253.

35. Domon, B.; Dorsaz, A.C., Hostettmann, K. 1984. High-performance liquid chromatography of oleane saponins. *J. Chromatogr*. 315:441-446.

36. Ekabo, O.A., Farnsworth, N.R., Henderson, T.O., Mao, G., Mukherjee, R. 1996. Antifungal and molluscicidal saponins from *Serjania salzmanniana*. *J. Nat. Prod*. 59:431-435.

37. El Hady, F.K., Gegazi, A.G., Ata, N., Enbaawy, M.L. 1994. Studies on determining antimicrobial activity of *Solenostemma argel* (Del) Hayne. 2. Extraction with chloroform/methanol in different proportions. *Quatar Univ. Sci. J*. 14:143-146.

38. Elias, R., De Meo, M., Vidal-Olivier, E., Laget, M., Balansard, G., Dumenil, G. 1990. Antimutagenic activity of some saponins isolated from *Calendula officinalis* L., *C. arvensis* L. and *Hedera helix* L. *Mutagenesis*. 5:327-331.

39. Favel, A., Steinmetz, M.D., Regli, P., Vidal-Olivier, E., Elias, R., Balansard, G. 1994. *In vitro* activity of triterpenoid saponins. *Planta Med*., 60:50-53.

40. Fukushima, Y., Bastov, K.F., Ohba, T., Lee, K.H. 1995. Antiviral activity of dammarane saponins against herpes simplex virus I. *Int. J. Pharmacogn*. 33:2-6.

41. Gal, I.E. 1968. Antibacterial activity of the spice, paprika. Testing of capsicidin and capsaicin activity. *Z. Lebensm.-Unters. Forsch*. 138:86-92.

42. Gee, J.M., Price, K.R., Ng, K., Johnson, I.T., Rhodes, M.J. 1998. The relationship between saponins structure and bioactivity - a preliminary study. COST 98, 4: 8-14.

43. Gestetner, B., Birk, Y., Bondi, A., Tencer, Y. 1966. Soybean saponin. VII. A method for the determination of sapogenin and saponin contents in soybean. *Phytochemistry*. 5:803-805.

44. Gruiz, K. 1996. Fungitoxic activity of saponins: practical use and fundamental principles. In *Saponins Used in Traditional and Modern Medicine*, ed. Waller, G.R., Yamasaki, K., Plenum Press, New York, pp. 527-534.

45. Grunweller, S., Schroder, E., Kesselmeier, J. 1990. Biological activities of furostanol saponins from *Nicotiana tabacum*. *Phytochemistry*, 29:2485-2490.

46. Gubanov, I.A., Libizov, N.I., Gladkikh, A.S. 1970. Search for saponin containing plants among flora of Central Asia and Southern Kazakhstan. *Farmatsiya (Moscow)*, 19:23-31.

47. Guerin, J.C., Reveillere, H.P. 1984. Activite antifongique diextraits vegetaux a usage therapeutique. I. Etude de 41 extraits sur 9 souches fongiques. *Ann. Pharm. Fr*. 42:553-559.

48. Hamaya, E., Tsushida, T., Nagata, T., Nishino, C., Enoki, N., Manabe, S. 1984. Antifungal components of camellia plants. *Nippon Shokubutsu Byori Gakkaiho*, 50:628-636.

49. Hamburger, M., Dudan, G., Ramachandran, N., Jayaprakasam R., Hostettmann, K. 1989. An antifungal triterpenoid from *Mollugo pentaphylla*. *Phytochemistry*, 28:1767-1768.

50. Hasegawa, H., Matsumiya, S., Uchiyama, M., Kurokawa, T., Inouye, Y., Kasai, R., Ishibashi, S., Yamasaki, K. 1994. Inhibitory effect of some triterpenoid saponins on glucose transport in tumor cells and its application to *in vitro* cytotoxic and antiviral activities. *Planta Med*. 60:240-243.

51. Hattori, T., Ikematsu, S., Koito, A., Matsushita, S., Maeda, Y., Hada, M., Fujimaki, M., Takatsuki, K. 1989. Preliminary evidence for inhibitory effect of glycyrrizin on HIV replication in patients with AIDS. *Antiviral Res*. 11:255-262.

52. Hayashi, K., Hayashi, H., Hiroaka, N., Ikeshiro, Y. Inhibitory activity of soyasaponin II on virus replication in vitro. *Planta Med.*, 63:102-105.

53. Hiai, S., Oura, H., Nakajima, T. 1976. Color reaction of some sapogeninsand saponins with vaniline and sulfuric acid. *Planta Med.* 29:116-119.

54. Hiller, K., Paulick, A., Friedrich, E. 1981. Inhibitory activity of *Polemonium* saponin against fungi. *Pharmazie*, 36:133-134.

55. Hirabayashi, K., Iwata, S., Matsumoto, H., Mori, T., Shibata, S., Baba, M., Ito, M., Shigeta, S., Nakashima, H., Yamamoto, N. 1991. Antiviral activities of glycyrrizin and its modified compounds against human immunodeficiency virus type 1 (HIV-1) and herpers simplex virus type 1 (HSV-1) *in vitro*. *Chem Pharm. Bull.* 39:112-115.

56. Hoagland, R.E., Zablotowicz, R.M., Oleszek, W., Jurzysta, M. 1996. Effect of alfalfa saponins on rhizosphere bacteria. *Phytopathology*. 86, S97.

57. Hostettman, K., Marston, A. 1995. *Saponins*. Cambridge University Press.

58. Hostettmann, K., Marston, A., Maillard, M., Wolfender, J.L. 1996. Search for molluscicidal and antifungal saponins from tropical plants. In *Saponins Used in Traditional and Modern Medicine*. ed. Waller, G.R., Yamasaki K. Plenum Press, New York, pp. 117-128.

59. Hufford, C.D., Liu, S., Clark, A.M. 1988. Antifungal activity of *Trillium grandiflorum* constituents. *J. Nat. Prod.* 51:94-98.

60. Ishikawa, H. 1996. Enzymatic production of oleanolic acid from sugar beet saponin. *Proc. Res. Soc. Japan Sugar Raf. Techn.* 44:29-34

61. Ito, M., Nakashima, H., Baba. M., 1987. Inhibitory effect of glycyrrizin on the *in vitro* infectivity and cytopathic activity of the human immunodeficiency vrus [HIV(HTLV-III/LAV)]. *Antiviral Res.* 7:127-137.

62. Itoh, M., Asakawa, N., Hashimoto, Y., Ischibashi, M., Miyazaki H. 1985. Quantitative analysis of glycyrrhizin and glycyrrhetinic acid in plasma after administration of FM-100 using gas chromatography - selected ion monitoring. *Yakugaku Zasshi*, 105:1150-1155.

63. Jadhav, S.J., Sharma, R.P., Salunkhe, D.K. 1981. Naturally occurring toxic alkaloids in foods. *Crit. Rev. Toxicol.* 11:21-35.

64. Jain, R., Arora, R., Alam, S., Jain, S. 1997. Pharmacological evaluation of *Mimosa hamata* roots. *Fitoterapia*. 68:377-378.

65. Jin, J., Du, S., Zhong, M. 1993. Studies on antifungal active constituents in the oil cakes of *Camellia oleifera* Abel seeds. *Tianran Chanwu Yanjiu Yu Kaifa*, 5:48-52.

66. Jurzysta, M., Jurzysta A. 1978. Gas-liquid chromatography of trimethylsilyl ethers of soyasapogenols and edicagenic acid. *J. Chromatogr.* 148:517-520.

67. Jurzysta, M., Waller, G.R. 1996. Antifungal and haemolytic activity of aerial parts of alfalfa (*Medicago*) species in relation to saponin composition. In *Saponins Used in Traditional and Modern Medicine*, ed. Waller, G.R. and Yamasaki, K., Plenum Press, New York. pp. 565-574.

68. Khan, M.M.A.A., Singh, N., Dhavan, K.N. 1996. Occurrence and identification of a new antiviral saponin from *Lawsonia alba* fruits. Nat. Acad. Sci. Lett. 19:7-7.

69. Kim, H.S., Seong, S., Oh, K.W., Jeong, T.S., Nam, K.Y. 1987. Effect of ginseng saponin on the antimicrobial activities of some antibiotics. *Haniguk Kyunhakhoechi*. 15:87-91.

70. Kitagawa, I., Yoshikawa, M., Hayashi, T., Taniyama, T. 1984. Characterization of saponin constituents in soybeans of various origins and quantitative analysis of soyasaponins by gas-liquid chromatography. *Yakugaku Zasshi*, 104:162-168.

71. Kobert, R. 1887. Über Quillajasäure. Ein Beitrag zur Kenntnis der saponingruppe. *Arch. Exper. Pathol. Pharmakol.* 23, 233-272.

72. Krivorutchenko, Y.L., Andronovskaya, I.B., Hinkula, J., Krivoshein, Y.S., Ljungdahl-Stahle, E., Pertel, S. S., Grishkovets, V.I., Zemlyakov, A.E., Wahren, B. 1997. Study of adjuvant activity of new MDP derivatives and purified saponins and their influence on HIV-1 replication *in vitro*. *Vaccine*, 15:1479-1486.

73. Lacaille-Dubois, M.A., Wagner, H. 1996. A review of the biological and pharmaceutical activities of saponins. *Phytomedicine*, 2:363-386.

74. Lemos, T.L.G., Mendes, A.L., Sousa, M.P., Brazfilho, R. 1992. New saponin from *Sapindus saponaria*. *Fitoterapia*. 63:515-517.

75. Luening, H.U., Schloesser, E. 1975. Role of saponins in antifungal resistance. V. Enzymatic activation of avenacosides. *Z. Pflanzenkr. Pflanzenschutz.*, 82:699-703.

76. Luening, H.U., Schloesser, E. 1976. Role of saponin in antifungal resistance. VI. Interaction between *Avena sativa* and *Drechslera avenacea*. *Z. Pflanzenkr. Pflanzenschutz.* 83:317-327.

77. Luening, H.U., Waiyaki, B.G., Schlosser, E. 1978. Role of saponins in antifungal resistance. VIII. Interactions *Avena sativa-Fusarium avenaceum*. *Phytopathol. Z.*, 92:338-345.

78. Mackie, A.M., Singh, H.T., Owen, J.M. 1977. Studies on the distribution, biosynthesis and function of steroidal saponins in echinoderms. *Comp. Biochem. Physiol.* 56:9-12.

79. Magota, H., Okubo, K., Shimoyamada, M., Suzuki, M., Maruyama, M. 1994. Isolation of steroidal saponins as antifungal agent. *Jpn. Kokai Tokkyo Koho*, 5 pp. JP 03048694 A2 910301 (Patent written in Japanese)

80. Makkar, H.P.S., Becker, K. 1996. Effect of Quillaya saponins on in vitro rumen fermentation. In *Saponins Used in Food and Agriculture*, eds. Waller, G.R., Yamasaki, K., Plenum Press, New York, pp. 387-394.

81. Malinow, M.R., McLaughlin, P., Stafford, C. 1978. Prevention of hypocholesterolemia in monkeys (*Mucaca fascicularis*) by digitonin. *Am. J. Clin. Nutr.* 32:814-818.

82. Malinow, M.R., Connor, W.E., McLaughlin, P., Stafford, C., Lin, D.S., Livingston, A.L., Kohler, G.O., McNulty, W.P. 1981. Cholestero;l and bile acid balance in *Mucaca fascicularis*. Effects of alfalfa saponins. *J. Clin. Invest.* 67:156-164.

83. Martyniuk, S., Wroblewska, B., Jurzysta, M., Bialy, Z. 1996. Saponin as inhibitors of cereal pathogens: *Gaeumannomyces graminis* v. *tritici* and *Cephalosporium gramineum*. *Proc. Int. Symp. Mod. Fungic.Antif. Comp.* Intercept, Andover, UK. pp. 193-197.

84. Masamitsu, O., Etsuzo, T. 1998. Novel steroidal saponin and antimicrobial agents and antitumor agents containing it. JP 10158295 A2 980616 Heisei, Patent written in Japanese

85. Masataka, K., Yoshio, H., Misuto, T., Taro, N., Sazali, H.A., Bte, A.R., Halila, J. 1998. Triterpeneid saponins from *Dedyotis nudicaulis*. *Phytochemistry*, 48:525-528.

86. Massiot, G., Lavaud, C., Guillaume, D., Le Men-Olivier, L. 1988. Reinvestigation of sapogenins and prosapogenins from alfalfa (*Medicago sativa*). *J. Agric. Food Chem.* 36:902-909.

87. Massiot, G., Dijoux, M.G., Lavaud, C. 1996. Saponins and artifacts. In *Saponins Used in Food and Agriculture*, ed. G.R. Waller and K. Yamasaki, pp. 183-192. New York: Plenum Press.

88. Margineau, C., Cucu, V., Grecu, L., Parvu, C. 1976. Anticandida action of a saponin from *Primula Planta Med.* 30:35-38.

89. Matsura, H., Morita, T., Gokuchi, T., Itakura, Y.,Hayashi, N. 1996. Antifungal steroid saponins from *Allium*. *Jpn Kokai Tokkyo Koho* 8 pp., JP 01224396 A2 890907. (Patent written in Japanese)

90. Melton, R.E., Flegg, L.M., Brown, J.K.M., Olivier, R.P., Daniels, M.J., Osbourn, A. 1998. Heterologous expression of *Septoria lycopersici* tomatinase in *Cladosporium fulvum*: effects of compatabile and incompatabile interactions with tomato seedlings. *Mol. Plant-Microbe Interact.* 11:228-236.

91. Nagata, T., Tsushida, T., Hamaya, E., Enoki, N., Manabe, S., Nishino, C. 1985. Camellidins, antifungal saponins isolated from *Camellia japonica*. *Agric. Biol. Chem.* 49:1181-1186.

92. Nakashima, H., Okubo, K., Honda, Y., Tamura, T., Matsuda, S., Yamamoto, N. 1989. Inhibitory effect of glycosides like saponin from soybean on the infectivity of HIV *in vitro*. *AIDS*. 3:655-658.

93. Nishina, A., Nozaki, Y., Oohara, E., Kihara, H. 1992. Antibacterial agents from fumaric acid solubilized by emulsifiers for foods. Jpn. Kokai Tokkyo Koho, 6 pp., JP 04187066 A2 920703 (Patent written in Japanese).

94. Nowacka, J., Oleszek W. 1994. Determination of alfalfa (*Medicago sativa*) saponins by high-performance liquid chromatography. *J. Agric. Food Chem.* 42:727-730.

95. Oakenfull, D.G. 1981. Saponins in food. *Food Chem.*, 6:19-40.

96. Oakenfull, D.G., Topping, D.L., Illman, R.J., Fenwick, D.E. 1984. Prevention of dietary hyperholesterolaemia in the rat by the soya bean and quillaya saponins. *Nutr. Rep. Int.* 29:1039-144.

97. Ohtani, K., Mavi, S., Hostettmann, K. 1993. Molluscicidal and antifungal triterpenoid saponinsfrom *Rapanea melanophloeos*. *Phytochemistry*, 33:83-86.

98. Okubo, K., Kudou, S., Uchida, T., Yoshiki, Y., Yoshikoshi, M., Tonomura, M. 1994. Soybean saponin and isoflavonoids: structure and antiviral activity against human immunodeficiency virus *in vitro*. *ACS Symp. Ser.* 546:330-339.

99. Okunji, C.O., Okeke, C.N., Gugnani, H.C., Iwu, M.M. 1990. An antifungal spirostanol saponin from fruit pulp of *Dracaena mannii*. *Int. J. Crude Drug Res.* 28:193-199.

100. Oleszek, W. 1988. Solid-phase extraction-fractionation of alfalfa saponins. *J. Sci. Food Agric.* 44:43-49.

101. Oleszek, W. 1990. Structural specificity of alfalfa (*Medicago sativa*) saponin haemolysis and its impact on two haemolysis-based quantification methods. *J. Sci. Food Agric.* 53:477-485.

102. Oleszek, W. 1996. Alfalfa saponins: Structure, biological activity and chemotaxonomy. In *Saponins Used in Food and Agriculture*, ed. Waller, G.R. Yamasaki, K., Plenum Press, New York, pp. 155-170.

103. Oleszek, W. 1998. Composition and quantitation of saponin in alfalfa (*Medicago sativa*) seedlings. *J. Agric. Food Chem.* 46:960-962.

104. Oleszek W., Price, K.R., Fenwick, G.R. 1988. Sensitivity of Trichoderma viride to medicagenic acid, its natural glucosides (saponins) and derivatives. *Acta Soc. Bot. Pol.* 57:361-370.

105. Oleszek, W., Price, K.R., Colquhoun, I.J., Jurzysta, M., Ploszynski, M., Fenwick, G.R. 1990b. Isolation and identification of alfalfa (*Medicago sativa* L.) root saponins: Their activity in relation to a fungal bioassay. *J. Agric. Food Chem.* 38:1810-1817.

106. Oleszek, W., Jurzysta, M., Price, K.R., Fenwick, G.R. 1990. High-performance liquid chromatography of alfalfa root saponins. *J. Chromatogr.* 519:109-116.

107. Oleszek, W., Jurzysta, M., Ploszynski, M., Colquhoun, I.J., Price, K.R., Fenwick, G.R. 1992. Zanhic acid tridesmoside and other dominant saponins from alfalfa (*Medicago sativa* L.) aerial parts. *J. Agric. Food Chem.* 40:191-196.

108. Oleszek, W., Hoagland, R.E., Zablotowicz, R.M. 1999. Ecological significance of plant saponins. In *Principles and Practicies of Plant Ecology*, ed. Inderjit,Dakshini, K.M.M., Foy, C.L. CRC Press LLC, pp.451-465.

109. Osbourn, A., Bowyer, P., Lunness, P. Clark, B., Daniels, M. 1995. Fungal pathogens of oat roots and tomato leaves employ closely related enzymes to detoxify different host plant saponins. *Mol. Plant-Microbe Interact.* 8:971-978.

110. Oui, S.A., Abdel Hady, F.K., ElGamal, M.H., Shaker, K.H. 1994. Isolation of antifungal compounds from some *Zygophyllum* species and their bioassay against two soil-borne plant pathogens. *Folia Microbiol.* 39:215-221.

111. Pedersen, M.W., Zimmer, D.E., McAllister, D.R., Anderson, J.O., Wilding, M.D., Taylor, G.A., McGuire, C.F. 1967. Comparative studies of saponins on several alfalfa varieties using chemical and biochemical assays. *Crop Sci.*, 7:349-351.

112. Petricic, J., Radosevic, A. 1969. Asperosid, a new bidesmosine 22-hydroxyfurostanol saponin from *Smilax aspera*. *Farm. Glas.* 25:91-95.

113. Phongpaichit, S., Schneider, E.F., Picman, A.K.Tantiwachwuttikul, P., Wiriyachitra, P., Arnason, J.T. 1995. Inhibition of fungal growth by an aqueous extract and saponins from leaves of *Maesa ramentacea* Wall. *Biochem. Syst. Ecol.* 23:17-25.

114. Podgorski, W., Trawinska, B., Mardarowicz, L., Polonis, A. 1996. The influence of saponins on broilers and level of ammonia and urea in their faeces. *Ann. Univ. M.Curie-Sklod.,* 14:167-171 (in Polish).

115. Pompei, R., Flore, O., Marciallis, M.A., Pani, A., Loddo. B. 1979. Glycyrizzic acid inhibits virus growth and inactivates virus particles. *Nature.* 281:689-690.

116. Price, K.R., Johnson, I.T., Fenwick, G.R. 1987. The chemistry and biological significance of saponins in foods and feedingstuffs. *CRC Crit. Rev. Food Sci. Nutr.* 26:27-135.

117. Rao, G.S., Sinsheimer, J.E., Cochran, K.W. 1974. Antiviral activity of triterpenoid saponins containing ß-amyrin aglycones. *J Pharm. Sci.*, 63:471-473.

118. Rodriguez, J., Castro, R., Riguera, R. 1991. Holothurinosides: new antitumor non sulphated triterpenoid glycosides from the sea cucumber *Holothuria forskalii. Tetrahedron*, 47:4753-4762.

119. Rowan, D.D., Newman, R.H. 1984. Noroleanane saponins from *Celmisia petriei. Phytochemistry*, 23:639-644.

120. Rucker, G., Mayer, R.,Shin-Kim, J.S. 1991. Triterpene saponins from the Chinese drug 'Daxueteng' (Calius sargentodoxae). *Planta Med.* 57:468-470.

121. Sakamoto, I., Morimoto, K., Tanaka, O. 1975. Quantitative analysis of dammarane type saponins of ginseng and its application to the evaluation of the commercial ginseng tea and ginseng extrcats. *Yakugaku Zassi*, 95:1456-1460.

122. Sasazuka, T., Kameda, Y., Endo, M., Suzuki, H., Hiwatachi, K. 1995. Water soluble oleanolic acid. Production, inhibition of insoluble glucan synthesis and antibacterial action. *Seito Gijutsu Kenkyu Kaishi.* 43:63-67.

123. Sata, N., Matsunaga, S., Fusetani, N., Nishikawa, H., Takamura, S., Saito, T. 1998. New antifungal and cytotoxic steroidal saponins from the bulbs of an elephant garlic mutant. *Biosci., Biotechnol., Biochem.* 62: 1904-1911.

124. Sauvaire, Y., Baissac, Y., Leconte, O., Petit, P., Ribes, G. 1996. Steroid saponins from fenugreek and some their biological properties. In *Saponins Used in Food and Agriculture*, ed. Waller, G.R., Yamasaki, K., Plenum Press, New York, pp 37-46.

125. Schloesser, E. 1973. Role of saponins in antifungal resistance. II. The hederasaponins in leaves of English ivy. *Z. Pflanzenkr. Pflanzenschutz*, 80:704-710.

126. Schloesser, E. 1976. Role of saponins in antifungal resistance. VII. Significance of tomatin in species-specific resistance of tomato fruits against fruit rotting fungi. *Meded. Fac. Landbouwwet., Rijksuniv.Gent.* 41:499-503.

127. Schloesser, E. 1977. Role of saponins in antifungal resistance. IV. Tomatin-dependent development of species of *Altenaria* on tomato fruits. *Proc. Ger. Curr. Top. Plant Pathol.*, ed. Kiraly, Z., Akad. Kiado, Budapest, Hung. pp. 77-87.

128. Segal, R., Schloesser, E. 1975. Role of glycosidases in the membranolytic, antifungal action of saponins. *Arch. Microbiol.* 104:147-150.
129. Shany, S., Gestetner, B., Birk, Y., Bondi, A. 1970. Lucerne saponins. III. Effect of lucerne saponins on larval growth and their detoxification by various sterols. *J. Sci. Food Agric.* 21:508-512.
130. Shimoyamada, M., Suzuki, M., Sonta, H., Maruyama, M., Okubo, K. 1990. Antifungal activity of the saponin fraction obtained from *Asparagus officinalis* L. and its active principle. *Agric. Biol. Chem.* 54:2553-2557.
131. Shimoyamada, M., Suzuki, M., Maruyama, M., Watanbe, K. 1996. An antifungal saponin from white asparagus (*Asparagus officinalis* L) bottoms. *J. Sci. Food Agric.* 72:430-434.
132. Simoes, C.M.O., Amoros, M., Schenkel, E.P., Shin-Kim, J.S., Ruker, G., Girre, L. 1990. Preliminary studies of antiviral activity of triterpenoid saponins: relationship between their chemical structure and antiviral activity. *Planta Med.* 56:652-653.
133. Singh, U.O., Srivastava, B.P., Singh, K.P., Pandey, V.B. 1992. Antifungal activity of steroid saponins and sapogenins from *Avena sativa* and *Costus speciosus*. *Naturalia Sao-Paulo*, 17:71-77.
134. Slacanin, I., Marston, A., Hostettmann, K. 1988. High-performance liquid chromatographic determination of molluscicidal saponins from *Phytolacca dodecandra* (Phytolaccaceae) *J. Chromatogr.* 448:265-274.
135. Slagowska, A., Zgorniak-Nowosielska, I., Grzybek, J. 1987. Inhibition of herpes simplex virus replication by Flos verbasci infusion. *Polish J. Pharmacol. Pharm.* 39:55-61.
136. Takechi, M., Tanaka, Y. 1990. Structure-activity relationships of saponin a-hederin. *Phytochemistry*, 29:451-452.
137. Takechi, M., Tanaka, Y. 1991. Structure-activity relationships of synthetic diosgenyl monoglycosides. *Phytochemistry*, 30:2557-2558.
138. Takechi, M., Shimada, S., Tanaka, Y. 1991. Structure-activity relationships of the saponin dioscin and dioscinin. *Phytochemistry*, 30:3943-3944.
139. Takeya, K. 1997. Antimicrobial activities of natural medicines. *Kagaku Kogyo*. 48:888-897.
140. Tamura, K., Mizutani, K., Yamamoto, S. 1990. Isolation of monodesmoside saponins from rind of *Sapindus trifoliatus* or - *mukurosii* fruits as skin fungicides. 02 160798 (Patent written in Japanese)
141. Tamura, Y. 1996. Antimicrobial activities of extracts from yucca. *Kurin Tekunoroji*, 6:67-71.
142. Tanaka, O., Tamura, Y., Masuda, H., Mizutani, K. 1996. Application of saponins in foods and cosmetics: saponins of Mohave yucca and *Sapindus mukurossi*. In *Saponins Used in Food and Agriculture*. ed. Waller G.R. Yamasaki, K. Plenum Press, New York, pp. 1-11.
143. Tatebe, M. 1986. Antimicrobial compositions containing acid-saponin mixtures for the skin. *Jpn. Kokai Tokkyo Koho*, 3 pp. (Patent written in Japanese).
144. Tava, A., Oleszek, W., Jurzysta, M., Berardo, N., Odoardi, M. 1993. Alfalfa saponins and sapogenins: Isolation and quantification in two different cultivars. *Phytochem. Anal.* 4:269-274.
145. Timbekova, A.E., Isaev, M.I., Abubakirov, N.K. 1996. Chemistry and biological activity of triterpenoid glycosides from *Medicago sativa*. In *Saponins Used in Food and Agriculture*, eds. Waller, G.R., Yamasaki, K., Plenum Press, New York, pp. 171-182.
146. Timon-David, P., Julien, J., Gasquet, M., Balansard, G., Bernard, P. 1980. Research of antifungal activity from several active principle extracts from climbing-ivy: *Hedera helix* L. Ann. Pharm. Fr., 38:545-552.
147. Toyota, M., Msonthi, J.D., Hostettmann, K. 1990. A molluscicidal and antifungal triterpenoid saponin from the roots of *Clerodendrum wildii*. *Phytochemistry*, 29:2849-2851.
148. Trzesznik, U., Przyborowski, R., Holler, K., Linzer, B. 1967. Components of some Saniculoidae. VII. Antimicrobial properties of *Sanicula* saponins. Pharmazie. 22:715-717.
149. Tschese, R., Wulf, G. 1972. Chemie und Biologie der Saponine. *Prog. Chem. Org. Nat. Prod.* 30:462-606.
150. Uemura, T., Sagesaka, H., Kodama, K. 1997. Antifungal agents containing tea saponins. *Jpn. Kokai Tokkyo Koho*, 4 pp, JP 09110712 A2 970428 (Patent written in Japanese)
151. Ujikawa, K., Purchio, A. 1989. Substancias antifungicas, inhibidoras de *Aspergillus flavus* e de outras fungicas, isoladas de *Agave sisalana* (sisal). *Cienc. Cult.* 41:1218-1224.
152. Ushio, Y., Abe, H. 1992. Inactivation of measles virus and herpes simplex virus by saikosaponin a. *Planta Med.* 58:171-173.
153. van Atta, G.R., Guggolz, J., Thompson, C.R. 1961. Determination of saponins in alfalfa. *Agric. Food Chem.* 9:77-81.
154. Vichkanova, S.A., Goryunova, L.V. 1971. Antiviral activity of some saponins. *Tr. Vses. Nauch. Issled. Inst. Lek. Rast.* 14:204-212.
155. Villarreal, M.L., Arias, C., Feria-Velasco, A., Ramirez, O.T., Quintero, R. 1997. Cell suspension culture of *Solanum chrysotrichum* (Schldl.) - a plant producing an antifungal spirostanol saponin. *Plant Cell, Tissue Organ Cult.* 50:39-44.

156. Villarreal, M.L., Arias, C., Vega, J., Feria-Velasco, A., Ramirez, O.T., Nicasio, P., Rojas, G., Quintero, R. 1997. Large-scale cultivation of *Solanum chrysotrichum* cells. Production of the antifungal saponin SC-1 in 10-L airlift bioreactors. *Plant Cell Rep.*, 16:653-656.

157. Wall, M.E., Eddy, C.R., McClennan, M.L., Klumpp, M.E. 1952. Detection and estimation of steroidal sapogenins in plant tissue. *Anal. Chem.* 24:1337-1341.

158. Walters, B. 1968. Antibiotic effect of saponins. IV. Antibiotic effect of neutral steroid glycosides with and without saponin characteristics. *Planta Med.* 16:114-119.

159. Willigmann, I., Schnelle, G., Bodinet, C., Beuscher, N. 1992. Antimycotic compounds from different *Bellis perennis* varieties. *Planta Med.* 58:7-12.

160. Wulf, G. 1968. Neuere Entwicklungen auf dem Saponingebiet. *Dt Apoth. Ztg* 108:797-808.

161. Xie, D., Zhao, S., Zeng, X., Lei, S., 1998. Identification of effective mycostatic constituents in *Malus meliana* leaf. *Linchan Huaxue Yu Gongye*. 18:33-38 (in Chinese).

162. Xin, D., Wen, W., Yuqing, S. 1994. Separation and quantitation of ginseng saponins by two-dimensional TLC-densitometry. *Chinese J. Chromat.* 12:173-174.

163. Young, M.C.M., Araujo, A.R., da Silva, C.A., Lopez, M.N., Trvisan, L.M.V., Bolzani, V. 1998. Triterpenes and saponins from *Rudgea viburnioides*. *J. Nat. Prod.* 61:936-938.

164. Zablotowicz, R.M., Hoagland, R.E., Wagner, S.C. 1996. Effects of saponins on the growth and activity of rhizosphere bacteria, In *Saponins Used in Food and Agriculture*, eds. Waller, G.R., Yamasaki, K., Plenum Press, New York, pp. 83-95.

165. Zehavi, U., Ziv-Fecht, O., Levy, M., Naim, M., Evron, R., Polacheck, I. 1993. Synthesis and antifungal activity of medicagenic acid saponins on plant pathogens: modification of saccharide moiety and the 23a substitution. *Carbohydr. Res.* 244:161-169.

166. Zehavi, U., Polacheck, I. 1996. Saponins as antimycotic agents: glycosides of medicagenic acid. In *Saponins Used in Traditional and Modern Medicine*, ed. Waller, G.R., Yamasaki, K., Plenum Press, New York, pp. 535-546.

167. Zentmyer, G.A., Thompson, C.R. 1967. The effect of saponins from alfalfa on *Phytophthora cinnamomi* in relation to control of root rot of avocado. *Phytopathology.* 57:1278-1280.

168. Zhong, J.J., Yao, H. 1999. Production of ginseng saponins by cell suspension cultures of *Panax notoginseng* in bioreactors. *Phytochem. Soc. Europe Meeting on "Saponins in Food Feedstuffs and Medicinal Plants"* Pulawy, Book of Abstracts pp.

169. Zhou, J., 1989. Some bioactive substances from plants of West China. *Pure Appl. Chem.*, 61:457-460.

170. Zimmer, D.E., Pedersen, M.W., McGuire, C.F. 1967. A bioassay for alfalfa saponins using fungus *Trichoderma viride*. *Crop Sci.* 7:223-224.

A.S. Naidu

W.R. Bidlack

A.T. Crecelius

Flavonoids

12

I. INTRODUCTION

The term flavonoids (Lat. *Flavus* = yellow) was first used for the family of yellow-colored compounds with a flavone moiety (2-phenyl-chromone) (Swain, 1976). It was later extended to various plant polyphenols to include less intensely colored flavanones (with the saturated 2,3-double bond), colorless flavon-3-ols (catechins, without a C4 carbonyl group), and even more intensely colored red and blue anthocyanidins. In plants, the flavonoids occur as white and yellow pigments in flowers, fruits, bark and roots, astringent parasite deterrents, and, because of their favorable UV-absorbing properties, they also protect the plant from harmful UV radiation from the sun (Swain, 1976; Swain et al., 1979). Thus, flavonoids are a group of C_{15} aromatic plant pigments, which are biosynthesized via a confluence of the acetate/malonate and shikimate pathways. They are found virtually in all land-based green plants and, although not produced by animals may occasionally be accumulated from food sources.

A single plant may contain different flavonoids, and their distribution within a plant family could be useful in the taxonomy. Flavonoids play different roles in the ecology of plants. Due to their attractive colors, flavonols, flavones, and anthocyanidins are likely to be a visual signal for pollinating insects. Catechins and other flavanols possess astringent characteristics and act as feed repellents, while isoflavones are important plant-protective phytoalexins. Phenolic acids are of great interest to humans in different aspects as they contribute to the sensory and nutritional qualities of fruits and fruit products. Acylation of anthocyanins with *p*-coumaric and caffeic acids is common in fruits, and it is responsible for a better color stability in fruit products (Mazza & Miniati, 1993).

Among numerous substances identified in medicinal plants, flavonoids represent one of the most interesting class of bioactive compounds. Approximately 40 plant species are currently in use as phytomedicines because of their flavonoid content. In these plants, flavonoids commonly occur as glycosides, while free aglycones are less frequent. The most common classes are flavonols, flavones and their dihydroderivatives followed by

0-8493-2047-X/00/$0.00+$.50
© 2000 by CRC Press, LLC

anthocyanins, flavans and isoflavones. This chapter is designed to describe structure-function relationship of flavonoids as phytoantimicrobial agents with an emphasis on their nutraceutical advantages.

II. OCCURRENCE

Flavonoids are comprised of a range of C_{15} aromatic compounds, including chalcones, dihydroflavones (flavanones), flavones, biflavonoids, dihydroflavonols, flavonols, anthocyanidins, and (often) proanthocyanidin tannins, together with numerous derivatives of the basic forms (*FIGURE 1*). Predominant among naturally occurring derivatives are the glycosidic forms located in cell vacuoles of the plant. More lipophilic forms such as methylated, acylated, and prenylated aglycones are found in or on the cuticular waxes, and biflavonoids are located in the cuticle (Gadek et al., 1984).

A. Distribution in fruits

Grapes are rich sources of phenolic compounds including flavonoids and non-flavonoids. The most abundant classes of flavonoids include the flavan-3-ols, anthocyanins (in red grapes), and flavonols, while the most abundant class of nonflavonoids are the hydroxycinnamates; however, other phenolics also exist. Flavonols comprise another major flavonoid component of grapes, although these are found at lower levels than either flavan-3-ols or anthocyanins. The specific flavonols found in grapes are all glycosides (Cheynier & Rigaud, 1986) and include glycosides of myricetin, quercetin, kaempferol, and isorhamnetin. The glycosidic substituents are glucose, glucuronic acid, glucosylgalactose, glucosylarabinose, and glucosylxylose. Rutin has also been found in wine (Alonso et al., 1986), presumed to originate from the grape. The levels of total flavonols in grapes vary widely, from 20 to 95 mg/kg fresh weight.

Macheix et al. (1990) characterized flavonoids and phenylpropanoids from apple, orange, and black currant fruit. A variation in phenolic composition exists based on the ripeness as well as the variety of the fruit. In apple, chlorogenic acid, the 5'-quinic acid ester of caffeic acid, is the major phenylpropanoid, together with the quinic acid esters and glucosides of *p*-coumaric acid. Quercetin is the predominant flavonol, and epicatechin is the most abundant flavanol. The dihydrochalcones phloretin glucoside (phloridzin) and phloretin xyloglucoside are found in apple flesh, while the skin of red apples is rich in cyanidin-3-galactoside and other anthocyanins. The ratio of *p*-coumaric acid esters to chlorogenic acid vary considerably, and the ratio of phenylpropanoids to flavonoids decline during maturation of the fruit.

In orange (sweet orange or citrus sinensis used in juice manufacture), the phenylpropanoids include *p*-coumaric, ferulic, sinapic, and chlorogenic acids, together with their glycosides. Cyanidin-3-glucoside is the predominant anthocyanin, while flavones such as sinensetin, nobiletin, and tangeretin are present. Citrus fruits are characterized by their high concentrations of flavanone glycosides, especially hesperidin (hesperetin 7-rutinoside), narirutin, and eriocitrin.

In black currant fruit, chlorogenic acid is present, but not as the major phenylpropanoid. 3'-caffeoylquinic acid is the main hydroxycinnamate, together with *p*-hydroxybenzoic acid, salicylic (2-hydroxybenzoic) acid, gallic acid, and their esters. The anthocyanins cyanidin and delphinidin are present as the glucosides and rutinosides, giving

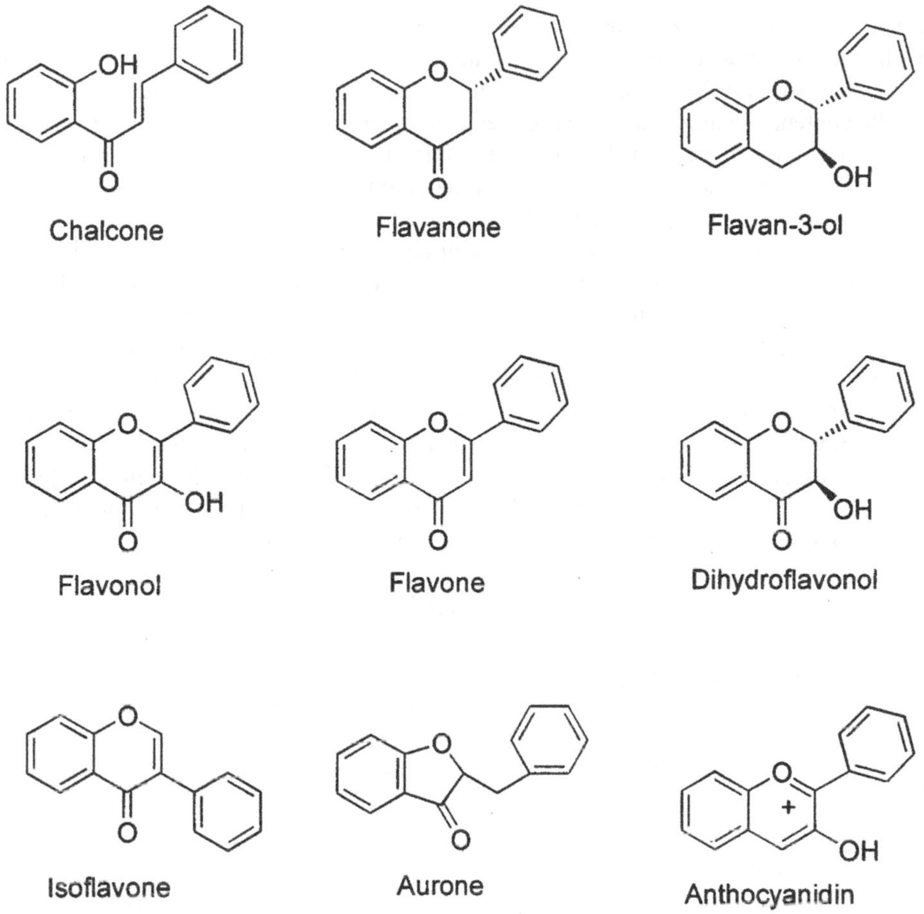

FIGURE 1. Basic structural classes of flavonoids.

black currant its characteristic color. Kaempferol, myricetin, and quercetin are the major flavanols present, while catechin, epicatechin, and the gallocatechins are the flavanols located in black currant.

B. Influencing factors

Phenolic acids are directly implied in the response of fruit to different kinds of stress (Harborne, 1995): mechanical (wounding), chemical (various types of treatment), or microbiological (pathogen infection). Phenolic acids are involved in resistance in two ways (Macheix et al., 1990): i) they contribute to the healing of wounds by lignification of cell walls around the wounded zone (Herrmann, 1989); ii) they elicit antimicrobial properties (Harborne, 1995; Lattanzio et al., 1994). The compounds involved can be classified in three groups (Macheix et al., 1990). Some are already present in the plant, and their level generally increases following stress (Herrmann, 1989). Others are formed only after injury but are derived from existing substances by hydrolysis or oxidation; (Tobias et al., 1993) still others are biosynthesized *de novo* and can be classified as phytoalexins.

The effect of wounding has been well studied, tomato in particular (Fleuriet & Macheix, 1984a). The most immediate response to wounding is the oxidation of pre-existing phenolic compounds and their degradation. Thus, the chlorogenic acid content of tomato fruit pericarp falls in 6 h after wounding. Afterward there is an increase in total phenolic content: accumulation of chlorogenic acid, feruloyl-, p-coumaroyl, and sinapoyl glucose in wounded pericarp. Finally, as the third response to wounding, formation of healing tissues ('wound lignin') occurs to protect the fruit from water loss and also form a physiological barrier to prevent possible penetration of pathogens.

The increase in phenolic content with pathogen infection has been well documented at a molecular level (Dixon & Paiva, 1995). As a defense mechanism, ferulic and p-coumaric acids may be esterified to wall polysaccharides, possibly rendering the wall resistant to fungal enzymes either by masking the substrate or altering the solubility properties of these wall polysaccharides (Iiyama et al., 1994; Dixon & Paiva, 1995).

Certain phenolic compounds can be biosynthesized *de novo* after infection. These compounds that do not exist before infection and that elicit antimicrobial activity are called phytoalexins. They are produced by plants as a defense mechanism in response to microbial infection (Dixon & Paiva, 1995; Harborne, 1995; Kuc, 1995), the accumulated compounds are often flavonoids (Swinburne & Brown, 1975).

III. CHEMISTRY

Phenolic acids belong to two different classes, HBA and HCA, which derive from two nonphenolic molecules: benzoic and cinnamic acids, respectively. Nevertheless, in most cases, phenolic acids are not found in a free state but as combined forms either soluble, and then accumulated in the vacuoles, or insoluble when linked to cell-wall components.

A. Hydroxybenzoic acid (HBA)

HBA have a common structure of C6-C3 type derived from benzoic acid. These are present as glucosidic forms in cherry and plum or different spices (Herrmann, 1989). Different new glycosides of HBA with potent radical-scavenging activity have been identified in the fruits of *Boreava orientalis* (Sakushima et al., 1995a; 1995b). Gallic acid, hexahydroxydiphenic acid, and pentagalloylglucose are constituents of hydrolyzable tannins. In addition, low concentrations of gallic acid are found in fruits in the form of esters with quinic acid (theogallin) or glucose (glucogallin) and in the form of glucosides. Glucogallin has also been identified in persimmon and isolated only from astringent immature fruit, whereas free gallic acid was found in immature fruit of both astringent and nonastringent varieties (Nakabayashi, 1971). Glucogallin has been proposed as a good index for distinguishing between astringent and nonastringent varieties. Gallic acid is also found combined with naringenin in fruits of *Acacia farnesiana* or with (-)epicatechin to form epicatechin-3-*O*-gallate, a constituent of unripe grapes (Macheix et al., 1990).

p-hydroxybenzoic and vanillic acids are also present in numerous fruits (Herrmann, 1989; Sakushima et al., 1995a), and the native forms are frequently simple combinations with glucose. Other derivatives have been detected in certain fruits (Macheix et al., 1990; Herrmann, 1989): the methyl ester of p-hydroxybenzoic acid in passion fruit, 3-4-dihydroxybenzoic aldehyde in banana, a phenylpropene benzoic acid derivative in fruits of Jamaican piper species, and benzoyl esters and other derivatives in the fruits of *Aniba riparia*. Syringic acid has only been reported in grape, and has a limited distribution in fruits.

Protocatechuic acid is found in a number of soft fruits in the form of glucosides, less abundant than those of *p*-hydroxybenzoic acid (Macheix et al., 1990; Herrmann, 1989). Salicylic and gentisic acids have been reported in low quantities in the fruits of certain Solanaceae (tomato, eggplant, pepper), Cucurbitaceae (melon, cucumber), and other species (e.g., kiwi fruit, grapefruit, grape).

B. Hydroxycinnamic acid (HCA)

Among fruit phenolic compounds, HCA derivatives play an important role due to both their abundance and diversity. They all derive from cinnamic acid and are essentially present as combined forms of four basic molecules: coumaric, caffeic, ferulic, and sinapic acids (*FIGURE 2*). Two main types of soluble derivatives have been identified: i) those involving an ester bond between the carboxylic function of phenolic acid and one of the alcoholic groups of an organic compound (i.e., quinic acid, flucose, etc.), for example, chlorogenic acid, which has been identified in numerous fruits; and ii) those that involve a bond with one of the phenolic groups of the molecule, i.e., *p*-coumaric acid *O*-glucoside in tomato fruit. Although native compounds are mainly of the *trans* form, isomerization occurs during extraction/purification and under the effect of light.

HCA are generally present in fruits in combined forms. However, free forms of HCA might accumulate in tomato and grape during extreme extraction conditions, physiological disturbances, contamination by microorganisms, and anaerobiosis. Free *p*-coumaric and caffeic acids may also appear during fruit processing (eg. preparation of fruit juices and wine making).

Quinic esters of HCA have long been reported in fruits. The first were chlorogenic acid (5-*O*-caffeoylquinic acid) and *p*-coumaroylquinic acid in apples (Bradfield et al., 1952). Chlorogenic acid was subsequently found in many other fruits, often accompanied by other caffeoylquinic isomers such as neochlorogenic acid (3-*O*-caffeoylquinic acid) and cryptochlorogenic acid (4-*O*-caffeoylquinic acid), isochlorogenic acid (a mixture of several di-*O*-caffeoylquinic acids) in coffee bean, apple, pineapple, cherry, and peach (Macheix et al., 1990). The presence of 1,3,5-tri-*O*-caffeoylquinic acids in the fruit of *Xanthium stumarium* seems to be uncommon (Agata, et al., 1993). Quinic derivatives of other HCA have also been identified in numerous fruits, e.g., several isomers of *p*-coumaroylquinic acid in apple and 5-*O*-feruloylquinic acid in tomato (Macheix et al., 1990). Quinic derivatives are generally abundant in fruits, however, certain fruits lack these compounds (e.g., grape, cranberry, and strawberry). Tartaric esters are limited to certain fruits of *Vitis* species. Caffeoylshikimic esters are not widespread in plants, but abundant in dates.

Many HCA derivatives with other hydroxy acids have also been reported in plants (Herrmann, 1989), but not been identified in fruits, with the exception of *p*-coumaroylmalic acid in pear skin (Oleszek et al., 1994), 2'-*O*-*p*-coumaroyl-, 2'*O*-feruloylgalactaric acids and 2'-*O*-*p*-coumaroyl-, 2'-*O*-feruloyl-, and 2',4'-*O*-diferuloylglucaric acids in the peel of citrus fruits (Naim et al., 1992).

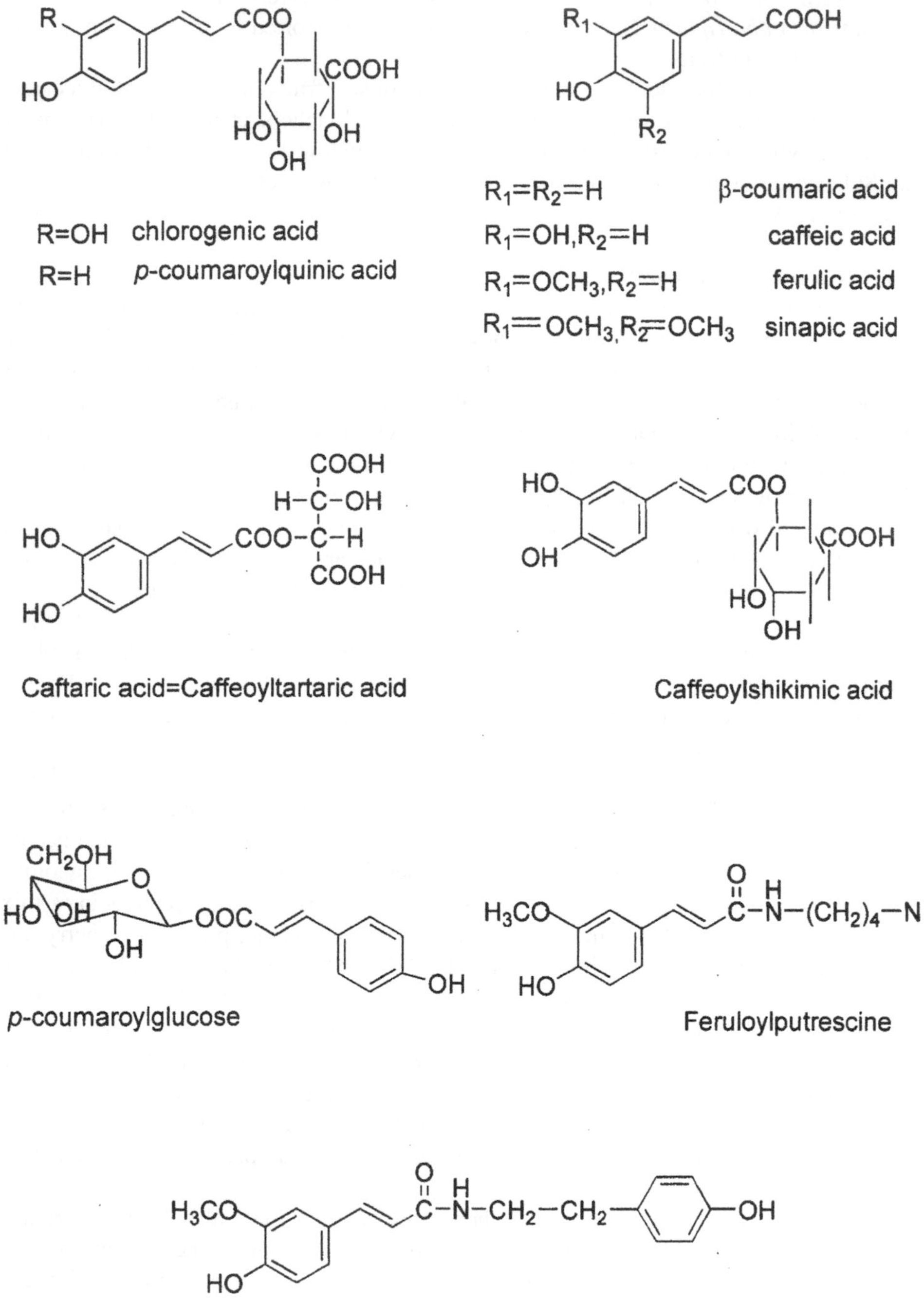

R=OH   chlorogenic acid
R=H   *p*-coumaroylquinic acid

$R_1=R_2=H$ β-coumaric acid
$R_1=OH, R_2=H$ caffeic acid
$R_1=OCH_3, R_2=H$ ferulic acid
$R_1=OCH_3, R_2=OCH_3$ sinapic acid

Caftaric acid=Caffeoyltartaric acid

Caffeoylshikimic acid

p-coumaroylglucose

Feruloylputrescine

Feruloyltyramine

FIGURE 2. Chemical structure of hydroxycinnamic acids and some common derivatives identified in fruits.

Since the identification of 1-*O-p*-coumaroylglucose and caffeic acid 3-*O*-gluco-side in potato berry (Harborne & Corner, 1961), numerous derivatives of HCA with sim-ple sugars have been identified in various fruits (Macheix et al., 1990; Herrmann, 1989; Shimazaki et al., 1991). Glucose esters and glycosides may be present simultaneously in the same fruit, even though HCA glycosides are often encountered in plants, and are rarely reported in fruits (Herrmann, 1989; Sukrasno & Yeoman, 1993). In tomato fruit, *p*-coumaric and ferulic acids are present both as glucosides and as glucose esters, whereas caffeic acid is only represented by caffeoylglucose (Fleuriet & Macheix, 1981b). Glucose esters of sinapic acid have been occasionally reported in fruits: they are present in toma-to and in *Boreava orientalis*, in addition to a glucosinolate salt (Fleuriet & Macheix, 1981c; Sakushima et al., 1994; 1995). Different new phenylpropanoid derivatives with simple sugars have been shown in the fresh fruit of *Piscrama quassiodes* (Yoshikawa et al., 1995). Verbascoside is an example of a rather more complex chemical combination that was identified in olives, and several other caffeoyl glycosides of dihydrox-yphenylethanol have been identified in the fruits of various species of *Forsythia*, which is also a member of the Oleaceae family (Fleuriet & Macheix, 1984c).

Although HCA derivatives with sugars and hydroxy acids are present simultane-ously in numerous fruits (e.g., apple, tomato, and cherry), several exceptions have been reported. Glucose derivatives of HCA are not present or only present as traces in pear and grape, whereas HCA are only present in the form of derivatives with sugars in strawber-ry and cranberry (Macheix et al., 1990; Herrmann, 1989).

The presence of hydroxycinnamoyl amides in fruits has rarely been reported. Ferulyputrescine occurs in grapefruit and orange juice (Naim ct al., 1992) but has not been found in tangerine or lemon. Likewise, two new phenolic amides were isolated from the fruit of white pepper (*Piper nigrum* L.): N-*trans*-feruloyltyramine, and N-*trans*-feru-loylpiperidine together with some other derivatives of piperidine and phenolics (Inatani et al., 1981).

Acylatioin of anthocyanins with certain phenolic acids has been known for a long time. Grape has been studied extensively (Macheix et al., 1990; Macheix et al., 1991; Mazza & Miniati, 1993), and it has been shown that *p*-coumaric acid plays a major role in the acylation of malvidin and other anthocyanins. Whereas caffeic acid only combines with malvidin 3-glucoside, a situation common in fruits and vegetables (Mazza & Miniati, 1993). In eggplant, delphinidin is acylated with coumaric and caffeic acids, delphinidin 3-(*p*-coumaroylrutinoside)-5-glucoside being a major pigment in purple-skinned varieties. In the fruit of *Solanum guineese* (garden huckleberry), petunidin 3-(*p*-coumaryl-ruti-noside)-5-glucoside forms at least 70% of anthocyanins and is accompanied by very small quantities of several other acylated derivatives (Price & Wrolstad, 1995). Flavonoid gly-cosides other than anthocyanins can also be acylated with HCA, but are rarely reported in fruits: e.g., kaempferol-*p*-coumaroylglycosides in *Tribulus terrestris*, 7-*O-p*-coumaroyl-glycoside-naringenin in *Mabea caudata*, or rhamnetin-3-*p*-coumaroyl-rhamninoside from *Rhamnus petiolaris* (Ozipek et al., 1994). *p*-Coumaric and ferulic acids are present in combination with betanidin monoglucoside in fruits from *Basella rubra* (Glassgen et al., 1993).

IV. ISOLATION

Since some flavonoids are unstable and could be degraded by enzyme action in undried plant material, caution is required during processing. It is safe to freeze-dry the plant material. The resulting powder material can be stored for future use in sealed containers in a freezer. For flavonoid quantification, snap freezing in liquid nitrogen immediately after harvest is recommended. In practice, if anthocyanins or tannins are not involved, drying of well-spread-out plant material in an oven at 100°C is an acceptable method. Subsequent storage of the dried material sealed in a plastic bag under refrigeration will prevent significant flavonoid losses for many months. Air-drying plant material at room temperature is not recommended, since this method could cause enzymatic degradation, and convert glycosides to aglycones. A satisfactory alternative is to extract the plant material in its undried form, for example, by chopping up the sample in a blender with an appropriate solvent. Enzyme activity can be avoided if alcohol is included in the extraction medium.

A. Extraction methods

The appropriate extraction procedure is determined by the types of flavonoids to be extracted and the purpose. Anthocyanins are readily extracted at room temperature from ground or mashed plant material such as petals or berries using acidified solvents. Use of formic, acetic, or trifluoroacetic acid (TFA) as the acid component reduces the likelihood of hydrolysis or deacylation. Typical solvents include methanol:water:acetic acid (70:23:7) or methanol:water:TFA (70:29.9:0.1). Flavonoid glycosides can be extracted with methanol:water (70:30), although glycuronides and C-glycosides are better extracted with a higher proportion of water. Excessive chlorophyll content in the resulting extracts may be diminished by extraction with chloroform or diethyl ether after the removal of methanol. Pre-extraction of plant material with one or other of these solvents is another option. Quantitative yields of the constituent flavonoid glycosides are only obtained when two to three sequential extractions of the original plant material are pooled. Fewer polar aglycones are obtained from the leaf surface by rinsing the intact plant material in an organic solvent such as ether or ethyl acetate. In contrast, proanthocyanidin tannins and their precursors are normally extracted from ground plant material using solvents such as acetone or acetone-water (Porter, 1994), while biflavonoids are extracted with solvents ranging from chlorinated hydrocarbons to ether, acetone, or methanol, depending upon their structural type (William & Harborne, 1989).

B. Large-scale fractionation

Further separation of the flavonoid-containing fraction(s) can be accomplished by column chromatography. Polyamide (i.e. MN SC-6) and Sephadex LH-20 are useful column beds for this purpose. Cellulose beds can also be used but the elution is slow.

In a typical separation using polyamide, the concentrated flavonoid fraction is applied to a column previously equilibrated in acid (HCl, formic, or acetic acids):water (pH 2.0) (Griesbach & Asen, 1990; Markham, 1975). At least one column volume of acidic water is passed through the column before a stepwise increase in alcohol content is started. Anthocyanin glycosides without acyl groups are eluted with acidic water alone, flavonol and flavone glycosides require 10-60% methanol, while nonpolar flavonol aglycones are eluted with 60-100% methanol. Polyamide chromatography is not the method of choice for proanthocyanidin tannins due to their strong adsorption. Fractions may be collected in set volumes corresponding to column size, or smaller fractions can be collected using an automatic fraction collector. Similar fractions can later be combined on the basis of TLC or HPLC analysis. At this stage some of the fractions may contain essentially pure compounds, which require only minor subsequent clean-up.

Separation by gel permeation may be used for further purification of fractions from larger column separations. Sephadex LH-20 is the most commonly used method to separate substances based not only on molecular size but also on the H-bonding interaction. Most flavonoids can be isolated to homogeneity on LH-20 column, although in some cases the chromatography needs repetition to obtain fractions of sufficient purity for specific applications. These columns are usually run with a single solvent system. For anthocyanins, methanol:acetic acid:water (10:1:9) is useful (Lu et al., 1991), while for other flavonoids water-methanol (or ethanol) or alcohol alone provide good separations (Markham, 1975). Acetone can be added to assist the elution of some tannins (Porter, 1989).

V. ANTIMICROBIAL ACTIVITY

Lipophilic flavonoids display antimicrobial activity, a property associated with their ability to penetrate biological membranes (Harborne, 1983). Methylation of 5-OH has been considered a structural feature essential for such antimicrobial activity (Tomas-Barberan et al., 1988).

A. Antibacterial activity

Wyman and Van Etten (1977) tested six isoflavonoids in both semisolid agar and liquid media for antibacterial activity against 20 phytopathogenic and saprophytic isolates representing the genera *Pseudomonas, Xanthomonas*, and *Achromobacter*. Phaseollin and pisatin were slightly to moderately inhibitory to a few isolates. Phaseollin isoflavan exhibited slight to moderate activity against several pseudomonads but strongly inhibited the xanthomonads and *Achromobacter* spp. Kievitone was the most inhibitory compound tested, significantly affecting the growth of a large proportion of the isolates. The xanthomonads and *Achromobacter* isolates tended to be more strongly inhibited by the active compounds than the pseudomonads. Phaseollin isoflavan and kievitone elicited bactericidal activity against some isolates. The sensitivity of the bacterial isolates to the compounds tested varied widely. The pseudomonads generally were more tolerant of the isoflavonoids than the xanthomonads or *Achromobacter* isolates. No clear correlation between pathogenicity on leguminous species and tolerance to the isoflavonoids was evident. The reaction of the *P. phaseolicola* isolates could not be correlated with race.

Species of the genus *Psiadia* (Compositae) are used in African traditional medicine as an expectorant for the treatment of bronchitis and asthma, a poultice for rheumatoid arthritis and as an analgesic for brain and nerves. From a dichloromethane extract and a hydrolysed methanolic extract from the leaves of *Psiadia trinervia*, 13,3-methylated flavonols (Chrysosplenol-D, isokaempferide, 5,7,4'-trihydroxy-3,3'-dimethoxyflavone and 5,7,4'-trihydroxy-3,8-dimethoxyflavone) displayed antibacterial activity (Wang et al., 1989). Twenty-nine derivatives were prepared by permethylaiton and selective methylation of the free hydroxyl group at C5. The antimicrobial activities of the isolates and derivatives were determined by bioautographic assays using *Cladosporium cucumerinum* and *Bacillus cereus* as test organisms.

Among the Polynesian medicinal plants *Aleurites moluccana* extracts showed antibacterial activity against *Staphylococcus aureus* and *Pseudomonas aeruginosa*, while *Pipturus albidus* and *Eugenia malaccensis* extracts showed growth inhibition of *Staphylococcus aureus* and *Streptococcus pyogenes* (Locher et al., 1995).

A series of 100 Rwandese medicinal plants (267 plant extracts), used by traditional healers to treat infections, were screened for antibacterial, antifungal and antiviral properties (Vlietinck et al., 1995). About 45% of the extracts were active against *Staphylococcus aureus*, 2% against *Escherichia coli*, 16% against *Pseudomonas aeruginosa*, 7% against *Candida albicans*, 80% against *Microsporum canis* and 60% against *Trichophyton mentagrophytes*.

B. Antiviral activity

Flavonoids are widespread in nature and some of them are used in the treatment of different human diseases. Cutting and coworkers (1949; 1953) were the first to report that plant flavonoids have antiviral activities. Since then, several flavones (e.g., quercetin and morin) found to have virucidal activity against enveloped viruses, including herpes simplex virus, but not against non-enveloped viruses (Beladi et al., 1977). The same flavonoids also proved to be significantly protective against lethal cardiovirus infection in mice (Veckenstedt & Pustzai, 1981). Certain flavonoids are effective in inhibiting the reverse transcription of retroviruses and in decreasing the incidence of RSV-induced sarcoma in chickens. Flavonol (+)-cyanidanol-3 (Catergen) and the flavanolignan silymarin have been shown to be effective against active viral hepatitis in man (Blum et al., 1977; Pelloni et al., 1977).

Flavonoids display potent antiviral effects both *in vitro* and *in vivo* (Beladi et al., 1981); flavonoids such as rutin, hesperidin, citrus bioflavonoids complex are used in the therapy of viral diseases (Verbenko et al., 1979; Berenge & Exposite, 1975; Kaul et al., 1985). Flavonoids such as phelodendrozide, pentamethylquercetin, herbacitin, rhamnetin and rutin show no antimicrobial activity. All of these compounds contain a glucose moiety or methyl group. Substitution of hydroxyl group with a sugar moiety decrease or even completely abolish the antiviral effect in quercetin and luteolin. Presence of OH group at C3 was not necessary since luteolin (flavon) showed the highest activity. Similarly to quercetin derivatives, introduction of sugar moiety or methyl group to lutcolin resulted in a decrease of the antiviral activity.

Lycorine, one of the main alkaloids of the Amaryllidaceae family, could elicit a pronounced antiviral activity associated with the crude extracts from the roots and leaves of *Clivia miniata* Regel (Ieven et al., 1982). The inhibitory activity of lycorine against DNA and several RNA viruses on vero cells was not virucidal. Poliomyelitis virus inhibition occurred at lycorine concentrations as low as 1 µg/ml, whereas concentrations exceeding 25 µg/ml caused cytotoxicity.

Ro 09-0179 (4',5-dihydroxy-3,3',7-trimethoxyflavone), isolated from a Chinese medicinal herb showed potent antiviral activity. It selectively inhibited the replication of human picornaviruses, such as rhino and coxsackie viruses in tissue culture, but not other DNA and RNA viruses. Ro 09-0298 (4',5-diacetyloxy-3,3',7-trimethoxyflavone), an orally active derivative of Ro 09-0179, prevented coxsackievirus (B1) infection in mice

(Ishitsuka et al., 1982a). The critical time for the inhibition of rhinovirus replication by Ro 09-0179 was 2 to 4 h after virus adsorption, i.e., in the early stages of virus replication. It markedly inhibited coxsackievirus and rhinovirus RNA synthesis in infected HeLa cells, but not in a cell-free system using the RNA polymerase complex isolated from the infected cells. In the infected cells, the RNA polymerase complex was not formed in the presence of Ro 09-0179. It was suggested that Ro 09-0179 interferes with certain processes of viral replication which occurs between viral uncoating and the initiation of viral RNA synthesis.

Further studies on various analogs related to Ro 09-0179 led to the identification of 4'-ethoxy-2'-hydroxy-4,6'-dimethoxychalcone (Ro 09-0410), a new and a different type of antiviral agent (Ishitsuka et al., 1982b). Ro 09-0410 had a high activity against rhinoviruses but no activity against other picornaviruses. Of 53 rhinovirus serotypes tested, 46 were susceptible to Ro 09-0410 in HeLa cell cultures. The concentration of Ro 09-0410 inhibiting 50% of the types of rhinovirus was about 0.03 µg/ml, whereas the 50% cytotoxic concentration was 30 µg/ml. Ro 09-0410 inactivated rhinoviruses in direct dose, time-, and temperature-dependent fashion. Ro 09-0410 seems to bind to some specific site(s) on the virion of susceptible virus strains and prevents viral replication in the cell. Ro 09-0410, 4',6-dichloroflavan, and RMI-15,731 exert their activities by binding to or interaction with some specific site on the viral capsid protein. Furthermore, the binding or interaction sites for these three agents either are the same or close to each other (Ninomiya et al., 1984; 1985).

3-Methylquercetin (3MQ), a natural compound isolated from *Euphorbia grantii* selectively inhibits poliovirus replication, but has no effect on encephalomyocarditis virus (Castrillo et al., 1986). When the compound is present from the beginning of infection, the bulk of viral protein synthesis is prevented, but the shut-off of host protein synthesis continued. Addition of 3MQ 3-h after infection has a slight effect on viral protein synthesis, suggesting that this compound block a step of viral replication different from translation. Poliovirus RNA synthesis is potently blocked by 3MQ, i.e., 50% inhibition at 2 µg/ml. In poliovirus-infected cells, the viral protein and RNA synthesis were severely reduced, provided 3-methylquercetin was present between 1 and 2 h post-infection. Under these conditions, the virally induced host shut-off remained in effect. On the other hand, in uninfected HeLa cells, protein and RNA synthesis was inhibited moderately by 3-methylquercetin (Vrijsen et al., 1987).

Salicin and salireposide, the phenolic glucosides of *Populus* cultivar *Beaupre*, were shown to inhibit poliomyelitis and Semliki forest viruses (Van Hoof et al., 1989). 4'-Hydroxy-3-methoxyflavones are natural compounds with known antiviral activities against picornaviruses such as poliomyelitis and rhinoviruses. In order to establish a structure-activity relationship a series of analogs were synthesized, and their antiviral activities and cytotoxicities were compared with those of flavones from natural origin (De Meyer et al., 1992). The 4'-hydroxyl and 3-methoxyl groups, a substitution in the 5 position and a polysubstituted A ring appeared to be essential for a potent antiviral activity.

Several flavonoids have been shown to inhibit the replication of picornaviruses. Two classes of compounds can be distinguished according to their antiviral spectrum and mechanism of action. Some chalcones and flavans inhibit selectively different serotypes of rhinoviruses (Fujui et al., 1980; Bauer et al., 1981; Ishitsuka, et al., 1982a; 1982b). Compounds such as 4'-ethoxy-2'-hydroxy-4, 6'-dimethoxychalcone and 4', 6-dichloroflavan interact directly with specific sites on the viral capsid proteins, thereby uncoating and consequently lineration of viral RNA (Ninomiya, et al., 1984; 1985; Tisdale & Selway, 1984). Sensitivity of a virus depends on its serotype and the compound. Aza analogs such as 2H-pyrano [2,3-b]pyridines and N-benzylbenzamides show similar activity (Bargar et al., 1986). A second class of compounds consists of flavones. Various substituted derivatives active against a wide range of picornaviruses, except Mengo virus, and vesicular stomatitis virus were isolated from several plants (Ishitsuka, et al., 1982a; 1982b; Van Hoof et al., 1984; Tsuchiya et al., 1985; Vleitinck et al., 1988). These compounds interfere with an early step in the viral RNA synthesis. Although the molecular mechanism is not completely understood, they possibly inhibit the formation of minus-strand RNA of poliovirus by interacting with one of the proteins involved in the binding of the virus replication complex to vesicular membranes where the poliovirus replication takes place (Castrillo, et al., 1986; Castrillo & Carasco, 1987; Vrijsen et al., 1987; Lopez Pila et al., 1989).

Selected plants having a history of use in Polynesian traditional medicine for the treatment of infectious disease were investigated for antiviral, antifungal and antibacterial activity *in vitro* (Locher et al., 1995). Extracts from *Scaevola sericea, Psychotria hawaiiensis, Pipturus albidus* and *Eugenia malaccensis* showed selective antiviral activity against herpes simplex virus-1 and 2 and vesicular stomatitis virus.

Of the 100 Rwandese medicinal plants tested, about 27% of the plant species exhibited prominent antiviral properties against one or more test viruses, more specifically 12% against poliomyelitis, 16% against coxsackie, 3% against Semliki forest, 2% against measles and 8% against herpes simplex virus (Vlietinck et al., 1995).

A reactive mechanism of action, pronounced and broad-spectrum antiviral activity, and the lack of resistance-induction (Ninomiya et al., 1985) by flavones prompted many laboratories to explore this class of flavonoids and to establish a link between structure and activity. Studies with natural flavones have shown that the 3-methoxyl and 4'-hydroxyl groups are important for potent antiviral activity (Van Hoof et al., 1984; Tsuchiya et al., 1985; Vleitinck et al., 1988).

The small number of potent antiviral drugs with specific antiviral effects applicable for human therapy led to the idea that in combinations the already known substances would exert effects with lower non-toxic concentration; this would also decrease the possibility of the emergence of drug-resistant viruses occurring with the use of nucleoside analogs.

Quercetin: Quercetin is a bioflavonoid which is widely distributed in many fruits, vegetables and tea (Kuhnau, 1976). The biological effects of quercetin and some of its derivatives have been reported. For example: inhibitory effects on growth of malignant cells and glycolysis (Suolinna et al., 1975); macromolecule synthesis (Graziani et al., 1979); activity of protein kinase (Graziani et al., 1981); activity of ATPases (Kuriki et al., 1976); replication of viruses (Ishitsuka et al., 1982a; 1982b; Castrillo et al., 1986; Vrijsen et al., 1987); and induction of heat shock proteins (Hosokawa et al., 1990). Quercetin also seem to enhance the antiviral activity of recombinant human TNF-α (Ohnishi & Bannaj, 1993).

Quercetin showed a significant protective effect in systemic picornavirus infections in animal models (Cutting et al., 1953; Guttner et al., 1982., Ishitsuka et al., 1982a; 1982b; Veckenstedt & Pusztai, 1981; Veckenstedt et al., 1987). Likewise, interferon has proven to be efficient against such diseases (Gresser et al., 1968; Olsen et al., 1976; Weck et al., 1982; 1983). Combinations of antiviral compounds and interferon may produce enhanced efficacy against encephalomyocarditis virus infections in mice (Chany & Cerutti, 1977; Werner et al., 1976). Quercetin provided a macrophage-dependent protection against murine cardiovirus infections (Guttner et al., 1982; Veckenstedt & Pusztai, 1981; Veckenstedt et al., 1981; 1987) but failed to protect Sindbis virus-infected mice. This confirms earlier observations suggesting that the antiviral activity of flavonoids in animal models is limited to picornaviruses (Ishitsuka et al., 1982a; 1982b; Tisdale & Selway, 1984).

Beladi et al. (1981) described that quercetin and quercitrin inhibit the replication of herpesviruses in cell culture. 5-ethyl-2'-deoxyuridine (EDU) is one of the potent nucleoside analogs could diminish the HSV yield in cell culture and its inhibitory effect is dependent on the virus-induced thymidine kinase (De Clercq et al., 1980). It has been reported that quercetin raises the cAMO level in Ehrlich ascites tumour cells (Graziani & Chayoth, 1977), and also that flavonoids inhibit cAMO phosphodiesterase, which is responsible for the breakdown of cAMP (Beretz et al., 1978; Ferrel et al., 1979). In addition, it has been shown that other cAMP-enhancers, such as dibutyryl cAMP, inhibit the multiplication of HSV-1 and HSV-2 (Stanwick et al., 1977). Mucsi (1984) studied the combined antiviral effects of quercetin, quercitrin and 5-ethyl-2'-deoxyuridine (EDU) on the multiplication of herpes simplex virus type 1 (HSV-1) and pseudorabies virus were studied *in vitro* by the yield reduction test. Quercetin in combination with EDU had an additive antiviral effect on either herpes-virus. The combined application of quercitrin and 5-ethyl-2'-deoxyuridine showed a synergic effect on pseudorabies virus.

The effect of different substituents of quercetin and luteolin on the ability to inhibit the herpes simplex virus (HSV-1) replication in RK-13 cells were studied (Wleklik et al., 1988). It seems that flavonoid compounds with free hydroxyl groups at C5, C7, C3', C4' and additionally at C3 have the highest activity. Substitution of those groups caused decrease of or completely abolished the antiviral activity of the tested compounds.

C. Antifungal activity

In view of phytoalexin properties, the fungicidal activity of isoflavonoids has attracted the attention of many research groups (Van Etten, 1976). Natural products, as potent fungal growth inhibitors, could be very useful substances in controlling plant disease. Fruit phenolic acids possess antimicrobial properties, and their content may increase after infection. Thus, chlorogenic and *p*-coumaroylquinic acids in apple are inhibitors of both *Botrytis* spore germination and mycelial growth (Swinburne & Brown, 1975). When their effects on growth of certain fungi associated with infections in apple are compared, *p*-coumaroylquinic is more inhibitory than chlorogenic at the same concentrations for *B. cinerea* and *Alternaria* spp., whereas *P. expansum* is less sensitive. Both quinic derivatives are stimulatory at low concentrations for *Botrytis* and *Penicillium*.

The acquisition of antimicrobial properties by phenolic compounds may derive from oxidation or hydrolysis. Firstly, *O*-quinones are generally more active than *o*-diphenols, and browning intensity is often greatest in highly resistant plants, suggesting that black and brown pigments contribute to resistance (Bell, 1981). Hydrolysis may be carried out by fungal pectic enzymes, as suggested by the appearance of free *p*-coumaric, caffeic, and ferulic acids in apple infected with *Penicillum expansum*. In this case, the damage does not cause much browning around the infection site due to the inhibition of the phenolase system by acids released after hydrolysis of chlorogenic and *p*-

coumaroylquinic acids in the fruit. Again in apple, antifungal compounds, such as 4-hydroxybenzoic acid produced after infection with *Sclerotinia fructigena*, is thought to be derived from the transformation of chlorogenic acid by the fungus. Free phenolic acids are the best inhibitors of growth of the fungi appearing during the postharvest storage and their structure/activity relationships have been studied *in vitro* (Lattanzio et al., 1994). An additional methoxy group caused increased activity of HBA and HCA derivatives. Thus, ferulic and 2,5-methoxybenzoic acids showed a strong inhibition against all fungi tested.

Kramer and co-workers (1984) tested isoflavones from food sources for fungicidal activity against *Aspergillus ochraceus, Penicillium digitatum* and *Fusarium culmorum*. These fungal pathogens demonstrated differential susceptibilities to isoflavonoids. The growth of *A. ochraceus* was markedly inhibited by some isoflavonoids, *P. digitatum* and *F. culmorum* are inhibited as well as stimulated in their growth depending on the substances and their applied concentrations. Among the isoflavonoids, the isoflavone-1 stimulated the growth of *P. digitatum* and *F. culmorum* at higher concentrations and showed a low inhibitory effect on *A. ochraceus*. Isoflavanone-5 was antifungal only to *F. culmorum*. The isoflavan-9 actively inhibited the growth of *A. ochraceus* and *F. culmorum* and *P. digitatum*. Diadzein inhibited *F. culmorum* at all concentrations tested. The isoflavonone-6 stimulated *F. culmorum*, but *P. digitatum* was strongly inhibited. The isoflavan-10 inhibited *P. digitatum* and *F. culmorum* only at higher concentrations.

Flavonoids, besides other biological activities, have been shown to be active against microorganisms (Thomas-Barberan et al., 1988; Harborne, 1983). Since these natural compounds possess highly specific antimicrobial activity, they could used as alternative to conventional fungicides in the control of storage fungi which cause great losses in the storage of grains in the developing countries (Neergaard, 1977). Weidenborner and co-workers (1989; 1990) studied the ability of isoflavonoids to inhibit growth of fungi. Four flavones, three flavanones and one catechin were tested for fungicidal activity in malt extract broth against five storage fungi of the genus *Aspergillus* (Weidenborner et al., 1990). Unsubstituted flavone and flavanone were highly active while the hydroxylated flavonoids demonstrated a weak activity. Accordingly, unsubstituted flavone and flavanone caused mycelial inhibition up to 90% in the case *A. glaucus*, while *A. flavus* and *A. petrakii* were inhibited only up to 70% at 5 mM concentration. In contrast, the activity of the oxygenated flavonoids was low. Flavonol inhibited mycelial growth of *A. repens* to 33.5% at 20 mM concentration, while this substance stimulated growth of *A. chevalieri* to 29% at 20 and 80 mM concentrations. An inhibition of 8.7% was caused by 7-hydroxyflavone in the case of *A. flavus* at 80 mM concentration. The flavanones naringenin and hesperetin were only mildly active against fungi of the *A. glaucus* group. The highest inhibition was 20.7% with naringenin at 80 mM concentration using *A. chevalieri* as the test organism.

The isoflavone biochanin A with hydroxyl groups at 5,7 positions, showed potent antifungal activity against *Rhizoctonia solani* and *Sclerotium rolfsii* (Weidenborner et al., 1989). In flavones, at least two methoxy groups in ring A at the 5-position seems essential for potent antifungal activity (Thomas-Barberan et al., 1988). Nobiletin, 5,6,7,8,3',4'-hexamethoxyflavone is fungistatic against *Deuterophoma tracheiphila* and 5,4'-dihydroxy-6,7,8,3'-tetramethoxyflavone is also fungistatic. Flavones methoxylated in ring A and in rings A and C are fungicidal to *C. cucmerinum* while 5-hydroxy-6,7-dimethoxy-and the 5-hydroxy-6,7,8-trimethoxyflavone are ineffective (Tomas-Barberan et al., 1988).

In general, the high antifungal activity of the flavone skeleton can be ascribed to the absence of polar groups in the molecule.

5,8-dihydroxy-6,7-dimethoxy- and the 7-hydroxy-5,6,8-trimethoxy-flavonone active against sclerotia germination of *S. rolfsii, Thanatephorus cucumerinum* and *Rhizoctonia solani*. The former flavanone also inhibits spore germination of *Helmithosporium oryzae, Rhizopus artocarpi* and *Fusarium oxysporum ciceri*. In both substances ring A is fully substituted and contains at least two methoxy groups. The low activity of hesperetin and naringenin is possibly due to partially substituted ring A and absence of methoxy groups. However, 7-hydroxy- and 6,7-dihydroxyflavanone, are fungicidal to *Alternarai solani* and *Curvularia lunata*.

Phaseolus species produce an array of fungitoxic isoflavonoids in response to either fungal attack or $CuCl_2$ treatment. *Phaseolus vulgaris* L. has been investigated previously, with the isolation of some 20 isoflavonoids (Wandward, 1980). Phytochemical investigation of the species *Phaseolus aureus, Phaseolus mungo* L., *Phaseolus coccineus* L. and *Phaseolus lunatus* L. has led to the isolation of significant amounts of a large range of isoflavones, isoflavanones, pterocarpans and isoflavans and has made possible the further assessment of fungitoxic structure-activity relationships within these groups of isoflavonoids (Adesanya et al., 1984; 1985; O'Neil et al, 1982; 1983; 1986; O'Neill, 1983).

Polynesian medicinal plant extracts from *Psychotria hawaiiensis* and *Solanum niger* inhibited growth of the fungi *Microsporum canis, Trichophyton rubrum* and *Epidermophyton floccosum*, while *Ipomoea sp., Pipturus albidus, Scaevola sericea, Eugenia malaccensis, Piper methysticum, Barringtonia asiatica* and *Adansonia digitata* extracts showed antifungal activity to a lesser extent. *Eugenia malaccensis* was also found to inhibit the classical pathway of complement, suggesting an immunological basis for its *in vivo* activity (Locher et al., 1995).

The fungicidal activity of Daidzein for *F. culmorum* or of the isoflavanones 5 and 6 for *P. digitatum* and *F. culmorum*, respectively shows a deviation indicating that other factors may be important for the antifungal property of the isoflavonoids. It seems that the fungicidal property of the isoflavonoids is specific to individual fungi, substances, and their concentrations and no absolute generalization is possible in this context.

Harborne and Inoham (1978) studied the isoflavone lutcone and suggested that a lipophilic side chain was important for fungitoxicity. Investigations by Smith (1976; 1978) indicated that the antifungal nature of kievitone (Van Etten, 1967) is dependent on two molecular features, the phenolic hydroxyl groups and the lipophilic dimethylallyl substituent. Adesanya et al. (1986) tested a series of isoflavones, isoflavanones, pterocarpans, isoflavans and the coumestan aureol, obtained from five *Phaseolus* species for fungitoxicity against *Aspergillus niger* and *Cladosporium cucumerinum*. A high level of lipophilicity and the presence of at least one phenolic function appear important to isoflavonoid fungitoxicity. Compared to the antibacterial flavonols, the antifungal compounds tend to be more lipophilic. These lipophilic compounds are deacylated in the fungal cell to yield an active phenol that inhibits fungi (Wang et al., 1989).

In the past there has been much speculation as to the possible isoflavonoid structural requirements for fungitoxicity. Perrin and Cruickshank (1969) postulated that the isoflavans, due to a planar conformation, fit into the probable receptor sites in the cells of sensitive fungi (Van Etten, 1976).

VI. MULTIFUNCTIONALITY

The ability of flavonoids to act as antioxidants in biological systems was recognized in the 1930s (Benthsath et al, 1936; 1937; Rusnyak & Szent-Gyorgy, 1936); however, the antioxidant mechanism was largely neglected for many years (Bors et al., 1990; Jovanovic et al., 1994). Several investigators have attributed the predominant mechanism of the chemopreventive protective action of naturally occurring plant compounds to their antioxidant activity and the capacity to scavenge free radicals (Ames, 1983; Caragay, 1992; Stavric, 1994). Among the most investigated chemopreventers are plant polyphenols, flavonoids, catechins, and some spice components present in fruits, grains, tea, red wine, and vegetables.

Phenylpropanoids (such as the hydroxycinnamates) are the metabolic precursors of flavonoids in higher plants through their role in th eshikimic acid pathway. Since the phenylpropanoids are a major part of the plant constituents (Macheix et al., 1990) that can be identified as antioxidants (Rice-Evans et al., 1995), they are also significant antioxidant components of fruit juices, as is the case with chlorogenic acid in apple juice (Miller et al., 1995).

The antioxidant activity of flavonoids and phenylpropanoids has been comprehensively ranked (Rice-Evans et al., 1995; 1996) and can be demonstrated in various beverages in addition to fruit juices, for example, the catechin gallate esters in green and black teas (Salah et al., 1995) and the anthocyanins in wines (Frankel et al., 1995; Miller & Rice-Evans, 1995). Their antioxidant potential is dependent on the number and arrangement of the hydroxyl groups across the structure, as well as the presence of electron-donating and electron-withdrawing substituents in the ring structure (Bors et al., 1990). While relatively little is known about the bioavailability of the flavonoids in humans, their excretion has been demonstrated through the bile (Ueno et al., 1983). In addition, flavonoids derived from citrus fruits have been shown in animal studies to undergo degradation in the intestine to low molecular weight compounds (such as hydroxybenzoates and hydroxycinnamates), which are readily absorbed and subsequently excreted in the urine (Booth et al., 1957).

Flavonoids have been shown to inhibit platelet aggregation (Tzeng et al., 1991). The major antiplatelet effect of fisetin, kaempferol, morin, myricetin, and quercetin were reportedly due to both inhibition of thromboxane formation and thromboxane receptor antagonism. In a study using rat models of thrombosis, the most effective antithrombotic tested was a combination of flavonoids and acetylsalicylic acid, a combination that eliminates some of the untoward effects of the latter drug (Hladovec, 1972).

Investigations by Fitzpatrick et al. (1993; 1995) showed that wines, grape juices, and grape skin extracts relaxed pre-contracted smooth muscle of intact rat aortic rings but had no effect on aortas in which the endothelium had been removed.

Among the pharmacological activities of flavonoids is their role in arachidonic acid metabolism. The mechanisms of 'cytoprotection' of the 3-palmitoyl-(+)-catechin (Palm-cat), a new flavonoid compound ($C31\ H44\ O7$), in experimental hepatitis induced in the rat by galactosamine (GalN) and *E.coli* O55:B5 endotoxin (LPS), hepatic cAMP and cGMP, transaminases, bilirubin, and endotoxemia were investigated (Scevola et al., 1984). Palm-cat significantly increased cyclic-GMP levels in the liver and reduced or slightly modified the cAMP. Flavonoid significantly decreased the frequency of endotoxemia. It was suggested that reticuloendothelial system and hepatocyte functions and immune and inflammatory responses can be affected in liver disease by flavonoids via cyclic nucleotide regulation.

A multi-center hospital-based case-control study was simultaneously performed in a high-risk and a low-risk area for stomach cancer in Germany (Boeing et al., 1991). Increased consumption of processed meat and of beer was positively associated with risk, whereas increased wine and liquor consumption was negatively associated with risk. The association of alcoholic beverages with stomach cancer risk may reflect a particular life style rather than being causally related to risk. In a prospective study of stomach cancer in relation to diet, cigarette, and alcohol consumption, the consumption of alcohol, either from beer, spirits, or wine, did not affect the incidence (Nomura et al., 1990).

Recently, there has been a resurgence of interest in the role of flavonoids in health and disease. Much of this interest centers around a phenomenon now well known as the French paradox. A population-based epidemiological study in 1979 (St. Leger et al., 1979) showed an association between reduction in mortality from coronary heart disease (CHD) and higher wine consumption. In 1987, it was recognized that the French paradox lies in the contrast between a food rich in saturated fatty acids and a moderate coronary mortality rate, fairly similar to that observed in Mediterranean countries where the dietary fat intake is much smaller than in France. The high mean level of alcohol consumption in France might be one of the factors responsible for this French peculiarity (Richard, 1987). Renaud and de Lorgeril (1992), referring to the French paradox, suggested that this paradox might, in part, be attributable to high wine consumption. Epidemiological studies indicate that consumption of alcohol at the level of intake in France (20-30g per day) can reduce risk of CHD by at least 40%. Alcohol is believed to protect from CHD by preventing atherosclerosis through the action of high-density-lipoprotein cholesterol, but serum concentrations of this factor are no higher in France than in other countries. Re-examination of previous results suggests that, in the main, moderate alcohol intake does not prevent CHD through an effect on atherosclerosis, but rather through a hemostatic mechanism. The authors suggested that the inhibition of platelet reactivity by alcohol in wine might explain in part the protection against CHD in France.

Many of the alleged effects of pharmacological doses of flavonoids are linked to their known functions as strong antioxidants, free-radical scavengers, and metal chelators and their interaction with enzymes, adenosine receptors, and biomembranes (Middleton & Kandaswami, 1994; Formica & Regelson, 1995; Ji et al., 1996; Saija et al., 1995; Bors et al., 1996). However, it is likely that the therapeutic value of most flavonoid medicinal plants rests not on the flavonoid fraction alone, but on a complex mixture of chemically different compounds. This aspect is common to many phytomedicines, whose activity cannot be assigned to specific constituents, since other components may either directly contribute or play an adjuvant role that strengthens the action of the active principles.

VII. SAFETY AND TOLERANCE

Due to their presence, both in edible plants as well as in foods and beverages derived from plants, flavonoids are important constituents of the non-energetic part of human diet with an average daily intake (ADI) of about 600 mg/day. Also, a dozen flavonoid containing species are known and have long been used in traditional medicine.

During the past two decades an increased effort in pharmacognosy has led to validating a number of these phytomedicines for the long-term treatment of mild and chronic diseases or to attain and maintain a condition of well-being.

Despite the potential significant effects of flavonoids, limited information is available regarding the absorption, metabolism, and excretion of most flavonoid classes in humans (Scheline, 1991; Sawai et al., 1987). Indeed, few studies have been performed on individual aglycones taken at pharmacological doses (Cova, 1992), and the results are somewhat controversial. Morevover, most studies concerning flavonoid activities have been performed *in vitro* and the results can be only partly extrapolated to *in vivo* conditions, as many flavonoids are rapidly metabolized to phenolic acids.

VIII. SUMMARY

A reactive mechanism of action, pronounced and broad-spectrum antimicrobial activity, and the lack of resistance induction prompted many laboratories to explore flavonoids as food antimicrobials. Isoflavonoids, as natural products and as potent fungal growth inhibitors could be extremely useful in controlling plant diseases. Due to this high specific antimicrobial activity a possibile application for isoflavonoids as alternative to conventional fungicides in the control of storage fungi to prevent loss of grains in developing countries has been suggested. Flavonoids are also of great interest in food technology as they contribute to the sensory and nutritional qualities of fruits and fruit products. The antioxidant activity of flavonoids and phenylpropanoids in various beverages in addition to fruit juices is well recognized.

Recently, there has been also a resurgence of interest in the role of flavonoids in health and disease following the French paradox suggesting an association between reduction in mortality from coronary heart disease and higher wine consumption. Incidentally, flavonoid preparations have long been used in medical practice to treat disorders of peripheral circulation, to lower blood pressure, and to improve aquaresis. Numerous phytomedicines containing flavonoids are marketed in different countries as anti-inflammatory, antispasmodic, antiallergic, and antiviral remedies. In recent years, many flavonoids with potent antimicrobial activity in cell cultures and in experimental animals have been detected. Some of these compounds are currently undergoing either preclinical or clinical evaluation, and perspectives for finding new antimicrobial drugs are promising. Some of these flavonoids exhibit a unique antiviral mechanism of action and are good candidates for further clinical research. The importance of the plant kingdom as a source of new antiviral flavonoids against rhinovirus and HIV is well recognized. Flavonoids are now emerging as potent nutraceutical bioactive food ingredients with sensory appeal to the global food consumer.

IX. REFERENCES

1.   Adesanya, S.A., O'Neil, M.J. and Roberts, M.F. 1984. Induced and constitutive isoflavonoids in *Phaseous mungo* Leguminosae. *Zeitschrift fuer Naturfarschung.* 39c:881-893.
2.   Adesanya, S.A., O'Neill, M.J. and Roberts, M.F. 1986. Structure-related fungitoxicity of isoflavonoids. *Physiol. Molecular Plant Pathol.* 29:95-103.
3.   Adesanya, S.A., O'Neill, M.J., and Roberts, M.F. 1985. Isoflavonoids from *Phaseolus coccineus* I. *Phytochemistry* 24:2699-2706.

4. Agata, I., Goto, S., Hatano, T., Nishibe, S., and Okuda, T. 1993. 1,3,5-Tri-*O*-caffeoylquinic acid from *Xanthium strumarium. Phytochemistry* 33:508-509.

5. Alonso, E., Estrella, M.I., and Revilla, E. 1986. HPLC Separation of flavonol glucosides in wine. *Chromatography* 22:268-270.

6. Ames, B.N. 1983. Dietary carcinogens and aniti-carcinogens. *Science* 221-1256-1260.

7. Bargar, T., Dulworth, J., Kenny, M., Massad, R., Daniel, J., Wilson, T., and Sargent, R.J. 1986. 3,4dihydro-2 phenyl-2H-pyrano (2,3-b) pyridines with potent antirhinovirus activity. *J. Med. Chem.* 29:1590-1595.

8. Bauer, D., Selway, J., Batchelor, J., Tisdaele, M., Caldwell, I., and Young, D. 1981. 4',6-Dichloroflavon (BW683C), a new anti-rhinovirus compound. *Nature.* 292:369-370.

9. Beladi, I., Mucsi, I., Pusztai, R., Bakay, M., Rosztoezy, I., Gabor, M., and Veckenstedt, A. 1981. *In vitro* and *in vivo* antiviral effects of flavonoids. In *Proceedings of the International Bioflavonoid Symposium* (6th Hungarian Bioflavonoid Symposium), Munich, FRG, ed. L. Farkas, M. Gabor, F. Kallay, and H. Wagner, pp.443-450. Budapest: Akademiai Kiado.

10. Beladi, I., Pusztai, R., Mucsi, I., Bakay, M., and Gabor, M. 1977. Activity of some flavonoids against viruses. *An. N.Y. Acad. Sci.* 234:358-364.

11. Bell, E.A. 1981. Biochemical mechanisms of disease resistance. *Annu. Rev. Plant Physiol.* 32:21-81.

12 Benthsath, A., Rusznyak, S., and Szent-Gyorgy, A. 1936. Vitamin nature of flavones. *Nature* 798.

13. Benthsath, A., Rusznyak, S., and Szent-Gyorgy, A. 1937. Vitamin P. *Nature* 326.

14. Berenge, A. and Exposite, R. 1975. A double-blind trial of (+) cyanidol-3 in viral hepatitis. In *New Trends in the Therapy of Liver Diseases*, pp.201-203. Intern. Symp. Tirrena, Basel:Karger.

15. Beretz, A., Anton, R., and Stoclet, J.C. 1978. Flavonoid compounds are potent inhibitors of cyclic AMP phosphodiesterase. *Experientia.* 34:1054-1055.

16. Blum, A.L., Doelle, W., Kortum, K., Peter, P., Strohmeyer, G., Berthet, P., Goebell, H., Pelloni, S., Poulsen, H., and Tygstrup, N. 1977. Treatment of acute viral hepatitis with (+) cyanidanol-3. *Lancet.* 2:1153-1155.

17. Boeing, H., Frentzel-Beyme, R., Berger, M., Berndt, V., Gores, W., Korner, M., Lohmeier, R., Menarcher, A., Mannl, H.F., Meinhardt, M., Muller, R., Ostermeier, H., Paul, F., Schwemmle, K., Wagner, K.H., and Wahrendorf, J. 1991. Case-control study on stomach cancer in Germany. *Int. J. Cancer* 47:858-864.

18. Booth, A.N., Jones, F.T., and DeEds, F. 1957. Metabolic fate of hesperidin, eriodictyol, homoeriodictyol, and diosmin. *J. Biol. Chem.* 230:661-668.

19. Bors, W., Heller, W., Michel, C., and Stettmaier, K. 1996. Flavonoids and polyphenols:chemistry and biology. In *Handbook of Antioxidants*, eds. Cadenas, E., and Packer, L., pp. 409-466. New York: Marcel Dekker.

20. Bors, W., Heller, W., Michel, S., and Saran, M. 1990. Flavonoids as antioxidants: Determination of radical scavenging efficiencies. *Methods Enzymol.* 186:343-355.

21. Calomme, M., Pieters, L., Vlietinck, A., and Vanden Berghe, D. 1996. Inhibition of bacterial mutagenesis by Citrus flavonoids. *Planta Med.* 62:222-226.

22. Caragay, A.B. 1992. Cancer preventive foods and ingredients. *Food Technol.* 46:65-68.

23. Castrillo, J.,and Carasco, L.1987. Action of 3-methylquercetin on poliovirus RNA replication. *J. Virol.* 61:3319-3321.

24. Castrillo, J.L., Vanden Berghe, D., and Carrasco, L. 1986. 3-Methylquercetin is a potent and selective inhibitor of poliovirus RNA synthesis. *Virology* 152:219-227.

25. Chany, C. and Cerutti, I. 1977. Enhancement of antiviral protection against encephalomyocarditis virus by a combination of isoprinosine and interferon. *Arch. Virol.* 55:225-231.

26. Cheynier, V., and Rigaud, J. 1986. HPLC separation and characterization of flavonols in the skins of *Vitis vinfera* var. Cinsault. *Am. J. Enol. Vitic.* 37:248-252.

27. Cova, D., De Angelis, L., Giavarin, F., Palladini, G., and Perego, A. 1992. Pharmacokinetics and metabolism of oral diosmin in healthy volunteers. *Int. J. Clin. Pharmacol. Ther. Toxicol.* 30:29-33.

28. Cutting, W.C., Dreisbach, R.H., and Matsushima, F. 1953. Antiviral chemotherapy VI. Parenteral and other effects of flavonoids. *Stanford Med. Bull.* 11:227-229.

29. Cutting, W.C., Dreisbach, R.H., and Neff, B.J. 1949. Antiviral chemotherapy III. Flavonos and related compounds. *Stanford Med. Bull.* 7:137-138.

30. De Clercq, E., Descamps, J., Verhelst, G., Walker, R.T., Jones, A.S., Torreneo, P.F., and Shugar, D. 1980. Comparative efficacy of antiherpes drugs against different strains of herpes simplex cirus. *J. Infect. Dis.* 141:563-574.

31. De Meyer, N., Haemers, A., Mishra, L., Pandey, H.K., Pieters, L.A., Vanden Berghe, D.A., and Vlietinck, A.J. 1992. 4'-Hydroxy-3-methoxyflavones with potent antipicornavirus activity. *J. Med. Chem.* 34:736-746.

32. Dixon, R.A., and Paiva, N.L. 1995. Stress-induced phenylpropanoid metabolism. *Plant Cell* 7:1085-1097.
33. Ferrell, J.E. Jr., Chang Sing, P.D.G., Loew, G., King, R., Mansour, J.M. and Mansour, T.E. 1979. Structure activity studies of flavonoids as inhibitors of cyclic AMP phosphodiesterase and relationship to quantum chemical indices. *Mol. Pharmacol.* 16:556-568.
34. Fitzpatrick, D.F., Hirschfield, S.L., and Coffey, R.G. 1993. Endothelium-dependent vasorelaxing activity of wine and other grape products. *Am. J. Physiol.* 265:H774-H778.
35. Fitzpatrick, D.F., Hirschfield, S.L., Ricci, T., Jantzen, P., and Coffey, R.G. 1995. Endothelium-dependent vasorelaxation caused by various plant extrancts. *J. Cardiovasc. Pharmacol.* 26:90-95.
36. Fleuriet, A., and Macheix, J.J. 1984c. Mise en evidence et dosage par chromatographie liquide a haute performance du verbascoside dans le fruit de six cultivars d'*Olea europea* L. *CR Acad Sci Paris* 299:253-256.
37. Fleuriet, A., and Macheix, J.J. 1984b. Orientation nouvelle du metabolism des acides hydroxycinnamiques dans les fruits de tomates blesses (*Lycopersicon esculentum*). *Physiol. Plant* 61:64-68.
38. Fleuriet, A., and Macheix, J.J. 1984a. Quinyl esters and glucose derivatives of hydroxycinnamic acids during growth and ripening of tomato fruit. *Phytochemistry* 20:667-671.
39. Formica, J.V., and Regelson, W. 1995. Review of the biology of quercetin and related bioflavonoids. *Food Chem. Tox.* 33:1061-1080.
40. Frankel, E.N., Waterhouse, A.L., and Teissedre, P.L. 1995. Principal phenolic phytochemicals in selected California wines and their antioxidant activity in inhibiting oxidation of human low-density lipoproteins. *J. Agric. Food Chem.* 43:890-894.
41. Fujui, M., Suhara, Y., and Ishitsuka, H. 1980. *Eur. Patent* 0013960 Al.
42. Gadek, P.A., Quinn, C.J., and Ashford, A.E. 1984. Localization fo the biflavonoid fraction in plant leaves, with special reference to *Agathis robusta* (C. Moore ex F Muell.) F.M. Bail. *Aust. J. Bot.* 32:15-31.
43. Glassgen, W.E., Metzger, J.W., Heuer, S., and Strack, D. 1993. Betacyanins from fruits of *Basella rubra*. *Phytochemistry* 33:1525-1527.
44. Graziani, Y., and Chayoth, R. 1977. Elevation of cyclic AMP levels in Ehrlich ascites tumor cells by quercetin. *Biochem. Pharmac.* 26:1259-1261.
45. Graziani, Y., and Chayoth, R. 1979. Regulation of cyclic AMP level and synthesis of DNA, RNA and protein by quercetin in Ehrlich ascites tumor cells. *Biochem. Pharmacol.* 28:397-403.
46. Graziani, Y., Chayoth, R., Karny, N., Feldman, B. and Levy, J. 1981. Regulation of protein kinases activity by quercetin in Ehrlich ascites tumor cells. *Biochim. Biophys. Acta* 714:415-421.
47. Greissman, T.A. 1962. *The Chemistry of Flavonoid Compounds*. Oxford: Pergamon Press.
48. Gresser, J., Bourali, C., Thomas, M.T. and Falcoff, E. 1968. Effect of repeated inoculation of interferon preparations on infection of mice with encephalomyocarditis virus. *Proc. Soc. Exp. Biol. Med.* 127:491-496.
49. Griesbach, R.J., and Asen, S. 1990. Characterization of the flavonol glycosides in Petunia. *Plant Sci.* 70:49-56,
50. Guttner, J., Veckenstedt, A., Heinecke, H. and Pusztai, R. 1982. Effect of quercetin on the course of Mengo virus infection in immunodeficient and normal mice. A histologic study. *Acta Virol.* 26:148-155.
51. Harborne, J.B. 1983. Plant and Fungal toxins. In *Handbook of Natural Toxins*, eds. Keeler, R.F., and Tu, A.T., pp. 743-750. New York: Marcel Dekker.
52. Harborne, J.B. and Inoham, J. 1978. Biochemical aspects of coevolution of higher plants with their fungal parasites. In *Biochemical Aspects of Coevolution in Higher Plants*, ed. J.B. Harborne, pp. 343-405. London:Academic Press.
53. Harborne, J.B., and Corner, J.J. 1961. Plant polyphenols. 4-Hydroxycinnamic acid-sugar derivatives. *Biochem. J.* 81:242-250.
54. Harborne, J.B.1995. Plant polyphenols and their role in plant defense mechanisms. In *Polyphenols*, eds. Brouillard, R., Jay, M., and Scalbert, A., pp. 19-26. Paris: INRA Editions.
55. Havsteen, B. 1983. Flavonoids, a class of natural products of high pharmacological potency. *Biochem. Pharmacol.* 32(7):1141-1149.
56. Herrmann, K. 1989. Occurrence and content of hydroxycinnamic and hydroxybenzoic acid compounds in foods. *Crit. Rev. Food Sci. Nutr.* 28:315-347.
57. Hladovec, J. 1972. Antithrombotic drugs and experimental thrombosis. *Cor. Vasa.* 20:135-141.
58. Hosokawa, N., Hirayoshi, K., Nakai, A., Hosokawa, Y., Marui, N., Yoshida, M., Sakai, T., Nishino, H., Aoike, A., Kawai, K. and Nagata, K. 1990. Flavonoids inhibit the expression of heat shock proteins. *Cell Struct. Funct.* 15:393-401.
59. Hu, J.P., Calomme, M., Lasure, A., De Bruyne, T., Pieters, L., Vlietinck, A., and Vanden Berghe,D.A. 1995. Structure-activity relationship of flavonoids with superoxide scavenging activity. *Biol. Trace Elem. Res.* 47:327-331.

60. Ieven, M., Vlietinck, A.J., Vanden Berghe, D.A., Totte, J., Dommisse, R., Esmans, E., and Alderweireldt, F. 1982. Plant antiviral agents. III. Isolation of alkaloids from *Clivia miniata* Regel (Amaryllidaceae). *J. Nat. Prod.* 45:564-573.

61. Iiyama, K., Lam, T.B.T., and Stone, B.A. 1994. Covalent cross-links in the cell wall. *Plant Physiol.* 104:315-320.

62. Inatani, R., Nakatani, N., and Fuw, H. 1981. Structure and synthesis of new phenolic amides from *Piper nigum* L. *Agr. Biol. Chem. Tokyo* 45:667-673.

63. Ishitsuka, H., Ninomiya, Y.T., Ohsawa, C., Fujiu, M., and Suhara, Y. 1982b. Direct and specific inactivation of rhinovirus by chalcone Ro 09-0410. *Antimicrob. Agents Chemother.* 22:617-621.

64. Ishitsuka, H., Ohsawa, C., Ohiwa, T., Umeda, I., and Suhara, Y. 1982a. Antipicornavirus flavone Ro 09-0179. *Antimicrob. Agents Chemother.* 22:611-616.

65. Ji, X.D., Melman, N., and Jacobson, K.A. 1996. Interactions of flavonoids and other phytochemicals with adenosine receptors. *J. Med. Chem.* 39:781-788.

66. Jovanovic, S.V., Steenken, S., Tosic, M., Marjanovic, B., and Simic, M.G. 1994. Flavonoids as antioxidants. *J. Am. Chem. Soc.* 116:4846.

67. Kaul, T.N., Middleton, J. and Ogra, R.J. 1985. Antiviral activity of flavonoids on human viruses. *J. Med. Virol.* 15:71-79.

68. Kramer, P.R., Hindorf, H., Jha, C.H., Kallage, J., and Zilliken, F. 1984. Antifungal Activity of soybean and chickpea isoflavones and their reduced derivatives. *Phytochemistry.* 23:2203-2205.

69. Kuc, J. 1995. Phytoalexins, stress metabolism, and disease resistance in plants. *Annu. Rev. Phytopathol.* 33:275-297.

70. Kuhnau, J. 1976. The flavonoids. A class of semi-essential food components: Their role in human nutrition. *World Rev. Nutrit. Dietetics* 24:117-191.

71. Kuriki, Y and Racker, E. 1976. Inhibition of (Na+, K+) adenosine triphosphatase and its partial reactions by quercetin. *Biochemistry.* 15:4951-4956,

72. Lattanzio, V., De Cicco, V., Di Venere, D., Lima, G., and Salerno, M. 1994. Antifungal activity of phenolics against fungi commonly encountered during storage. *Ital. J. Food Sci.* 6:23-30.

73. Locher, C.P., Burch, M.T., Mower, H.F., Berestecky, J., Davis, H., Van Poel, B., Lasure, A., Vanden Berghe, D.A., and Vlietinck, A.J. 1995. Antimicrobial activity and anti-complement activity of extracts obtained from selected Hawaiian medicinal plants. *J. Ethnopharmacol.* 49:23-32.

74. Lopez Pila, J., Kopecka, H.,and Vanden Berghe, D. 1989. Lack of evidence for strand-specific inhibition of poliovirus RNA synthesis by 3-methylquercetin. *Antiviral Res.* 1:47-53.

75. Lu, T.S., Saito, N., Yokoi, M., Shigihara, A., and Honda, T. 1991. An acylated peonidin glycoside in the violet-blue flowers of *Pharbitis nil. Phytochemistry* 30:2387-2390.

76. Macheix, J.J., Fleuriet, A., and Billot, J. 1990. *Fruit Phenolics.* Boca Raton, FL: CRC Press.

77. Macheix, J.J., Sapis, J.C., and Fleuriet, A. 1991. Phenolic compounds and polyphenoloxidase in relation to browning in grapes and wines. *Crit. Rev. Food Sci. Nutr.* 30:441-486.

78. Markham, K.R. 1975. Isolation techniques for flavonoids. In: *Flavonoids*, eds. Harborne, J.B., Mabry, T.J., and Mabry, H., pp.1-44. London: Chapman and Hall.

79. Mazza, G., and Miniati, E. 1993. *Anthocyanins in Fruits, Vegetables and Grains.* Boca Raton, FL:CRC Press.

80. Middleton, E. Jr., and Kandaswami, C. 1994. The impact of plant flavonoids in mammalian biology: influence for immunity, inflammation and cancer. In *The Flavonoids-Advances in Research Since 1986*, ed. Harborne, J.B., pp.619-652. London: Chapman and Hall.

81. Miller, N.J., and Rice-Evans, C.A. 1995. Antioxidant activity of resveratrol in red wine. *Clin. Chem.* 41:1789.

82. Miller, N.J., Diplock, A.T., and Rice-Evans, C.A. 1995. Evaluation of the total antioxidant activity as a marker of the deterioration of apple juice on storage. *J. Agric. Food Chem.* 43:1794-1801.

83. Mucsi, I. 1984. Combined antiviral effects of flaovnoids and 5-ethyl-2'-deoxyuridine on the multiplication of herpes virus. *Acta Virol.* 28:395-400.

84. Naim, M., Zehavi, U., Nagy, S., and Rouseff, R.L. 1992. Hydroxycinnamic acids as off-flavor precursors in Citrus fruits and their products. In *Phenolic Compounds in Food and Their Effects on Health I. Analysis, Occurrence, and Chemistry,* eds. Ho, C.T., Lee, C.Y., and Huang, M.T., Series 506, pp.180-191. Washington, DC: ACS.

85. Nakabayashi, T. 1971. Studies on tannin of fruits and vegetables. VII. Difference of the compound of tannin between the astringent and non astringent persimmon fruits. *J. Jpn. Soc. Technol.* 18:33-37.

86. Neergaard, P. 1977. *Seed Pathology.* London: MacMillan Press.

87. Ninomiya, Y., Aoyama, M., Umeda, I., Suhara, Y., and Ishitsuka, H. 1985. Comparative studies on the modes of action of the antirhinovirus agents Ro 09-0410, Ro 09-0179, RMI-15,731, 4',6-dichloroflavan, and enviroxime. *Antimicrob. Agents Chemother.* 27:595-599.
88. Ninomiya, Y., Ohsawa, C., Aoyama, M., Umeda, I., Suhara, Y., and Ishitsuka, H. 1984. Antivirus agent, Ro 09-0410, binds to rhinovirus specifically and stabilizes the virus conformation. *Virology* 134:269-276.
89. Nomura, A., Grove, J.S., Stemmermann, G.N., and Severson, R.K. 1990. A prospective study of stomach cancer and its relation to diet, cigarettes, and alcohol consumption. *Cancer Res.* 50:627-631.
90. O'Neil, M.J, Adesanya, S.A., Roberts, M.F., and Pantry, I.R. 1986. Novel inducible isoflavonoids from the lima bean *Phaseolus lunatus L. Phytochemistry.* 25:1315-1322.
91. O'Neill, M.J. 1983. Aureol and Phaseol, two new coumestans from *Phaseolus aureus* Roxb. *Zeitschrift fuer Naturforschung.*38:698-700.
92. O'Neill, M.J., Adesanya, S.A. and Roberts, M.F. 1983. Antifungal phytoalexins in Phaseolus aureus Roxb. *Zeitschrift fuer Naturforschung.* 38:693-697.
93. O'Neill, T.M. and Mansfield, J.W. 1982. Antifungal activity of hydroxyflavans and other flavonoids. *Trans. Br. Mycol. Soc.* 79:229-237.
94. Ohnishi, E. and Bannaj, H. 1993. Short communication quercetin potentiates TNF-α induced antiviral activity. *Antiviral Res.* 22:327-331.
95. Oleszek, W., Amiot, M.J., and Aubert, S.Y. 1994. Identification of some phenolics in pear fruit. *J. Agric. Food Chem.* 42:1261-1265.
96. Olsen, G.A., Kern, E.R., Glasgow, L.A. and Overall, J.C., Jr. 1976. Effect of treatment with exogenous interferon, polyinosinic acid-polycytidylic acid, or polyinosinic acid-polycytidylic acid-poly-L-lysine complex on encephalomyocarditis virus infections in mice. *Antimicrob. Agents Chemother.* 10:668-676.
97. Ozipek, M., Calis, I., Ertan, M., and Ruedi, P. 1994. Rhammetin 3-*p*-coumaroyl-rhamninoside from *Rhamnus petiolaris. Phytochemistry* 37:249-253.
98. Pelloni, S., Berthet, P., Blum, A.L., Doelle, Goebell, H., Kortum, K., Peter, P., Poulsen, H., Strohmeyer,G., and Tygstrup, N. 1977. Treatment of acute viral hepatitis with (+) cyanidanol-3. *Schweiz. Med. Wschr.* 107:1859-1861.
99. Perrin, D.R. and Cruickshank, I.A.M. 1969. The antifungal activity of pterocarpans towards *Monilinia fracticola. Phytochemistry.* 8:971-978.
100. Porter, L.J. 1989. Tannins. In *Methods in Plant Biochemistry. Vol. 1. Plant Phenolics*, ed. Harborne, J.B., pp.389-419. London: Academic Press.
101. Porter, L.J. 1994. Flavans and proanthocyanidins. In *The Flavonoids-Advances in Research Since 1986*, ed. Harborne, J.B., pp.23-53. London: Chapman and Hall.
102. Price, C.L., and Wrolstad, R.E. 1995. Anthocyanin pigments of royal Okanogan huckleberry juice. *J. Food Sci.* 60:369-374.
103. Renaud, S., and de Lorgeril, M. 1992. Wine, alcohol, platelets and the French paradox for coronary heart disease. *Lancet* 339:1523-1526.
104. Rice-Evans, C.A., Miller, N.H., and Paganga, G. 1996. Structure-antioxidant activity relationships of flavonoids and phenolic acids. *Free Rad. Biol. Med.* 20:933-956.
105. Rice-Evans, C.A., Miller, N.J., Bolwell, P.G., Bramley, P.M., and Pridham, J.B. 1995. The relative antioxidant activities fo plant-derived polyphenolic flavonoids. *Free Rad. Res.* 22:375-383.
106. Richard, J.L. 1987. [Coronary risk factors. The French paradox]. *Arch. Maladies Coeur Vaisseaux* 80 Spec No:17-21.
107. Rusznyak, S., and Szent-Gyorgy, A. 1936. Vitamin P: flavonols as vitamins. *Nature* 27.
108. Saija, A., Salese, M., Lanza, M., Mazzullo, D., Bonina, F., and Castelli F. 1995. Flavonoids as antioxidant agents: importance of their interaction with biomembranes. *Free Rad. Biol. Med.* 19:481-486.
109. Sakushima, A., Coskun, M., and Maoka, T. 1995a. Hydroxybenzoic acids from *Boreava orientalis. Phytochemistry* 40:257-261.
110. Sakushima, A., Coskun, M., and Maoka, T. 1995b. Sinapinyl but-3-enylglucosinolate from *Boreava orientalis. Phytochemistry* 40:483-485.
111. Sakushima, A., Coskun, M., Tanker, M., and Tanker, N. 1994. A sinapic acid ester from *Boreava orientalis. Phytochemistry* 35:1481-1484.
112. Salah, N., Miller, N.H., Paganga, G., Tijburg, L., and Rice-Evans, C.A. 1995. Polyphenolic flavanols as scavengers of aqueous phase radicals and as chain-breaking antioxidants. *Arch. Biochem. Biophys.* 322:339-346.
113. Sawai, Y., Kahsaka, K., Nishiyama, Y., and Ando, K. 1987. Serum concentration of rutoside metabolites after oral administration of a rutoside formulation to humans. *Arzneim-Forsch/Drug Res.* 1987:729-732.

114. Scevola, D., Barbarini, G., Grosso, A., Bona, S., and Perissoud, D. 1984. Flavonoids and hepatic cyclic monophosphates in liver injury. *Boll. Ist Sieroter. Milan* 63:77-82.
115. Scheline, R.R. 1991. *Handbook of Mammalian Metabolism of Plant Compounds.* Boca Raton, FL: CRC Press.
116. Shimazaki, N., Mimaki, Y., and Sashida, Y. 1991. Prunasin and acetylated phenylpropanoic acid sucrose esters, bitter principles from the fruits of *Prunus jamasakura* and *P. maximowiczii. Phytochemistry* 30:1475-1480.
117. Smith, D.A. 1976.Some effects of the phytoalexin, kievitone, on the vegetative growth of *Aphanomyces euteiches, Rhizoetenin solani and Fusarium solani* f. sp. *Phaseoli. Physiol. Plant Pathol.* 9:45-55.
118. Smith, D.A. 1978. Observations on the fungitoxicity of the phytoalexin, kievitone. *Phytopathology.* 48:81-87.
119. St. Leger, A.S., Cochrane, A.L., and Moore, F. 1979. Factors associated with cardiac mortality in developed countries with particular reference to the consumption of wine. *Lancet* I:1017-1020.
120. Stanwick, T.L., Anderson, R.W., and Nahmias, A.J. 1977. Interaction between cyclic nucleotides and Herpes simplex viruses: productive infection. *Infect. Immun.* 18:342-347.
121. Stavric, B. 1994. Role of chemopreventers in human diet. *Clin. Biochem.* 27:319-332.
122. Sukrasno, N., and Yeoman, M.M. 1993. Phenylpropanoid metabolism during growth and development of *Capsicum frutescens* fruits. *Phytochemistry* 32:839-844.
123. Suolinna, E.M., Buchsbaum, R.N. and Racker, E. 1975. The effect of flavonoids on aerobic glycolysis and growth of tumor cells. *Cancer Res.* 35:1865-1872.
124. Swain, T. 1976. Nature and Properties of Flavonoids. In *Chemistry and Biochemistry of Plant Pigments. Vol. 1.*, ed. Goodwin, T.W., pp.425. London: Academic Press.
125. Swain, T., Harborne, J.B., Van Sumere, C.F. 1979. *Biochemistry of Plant Phenolics.* Vol. 112. Plenum Press: New York.
126. Swinburne, T.R., and Brown, A.E. 1975. Biosynthesis of benzoic acids in Bramley's seedling apples infected by *Nectria gallignea. Physiol. Plant Pathol.* 6:259-264.
127. Tisdale, M., and Selway, J. 1984. Effect of dichloroflavan (BW683C) on the stability and uncoating of rhinovirus type 1B. *J. Antimicrob. Chemother.* 14: Suppl A 97-105.
128. Tobias, R.B., Conway, W.S., Sams, C.E., Gross, K.C., and Whitaker, B.D. 1993. Cell wall composition of calcium-treated apples inoculated with *Botrytis cinerea. Phytochemistry* 32:35-39.
129. Tomas-Barberan, F.A., Maillard, M., and Hostettmann, K. 1988. In *Plant Flavonoids in Biology and Medicine.* eds. Cody, V., Middleton, Jr. E., Harborne, J.B., and Beretz, A., pp.61-79. New York: Alan, R. Liss.
130. Tsuchiya, Y., Shimizu, M., Hiyama, Y., Iton, K., Hashimota, Y., Nakayama, M., Horie, and T., Morita, N. 1985. Antiviral activity of natural occurring flavonoids *in vitro. Chem. Pharm. Bull.* 33:3881.
131. Tzeng, S.H., Ko, W.C., Ko, F.N., and Teng, C.M. 1991. Inhibition of platelet aggregation by some flavonoids. *Thromb. Res.* 64:91-100.
132. Ueno, I., Nakano, N., and Hirono, I. 1983. Metabolic fate of [^{14}C]quercetin in the ACI rat. *Jpn. J. Exp. Med.* 53:41-50.
133. Van Etten, H.D. 1967. Antifungal activity of pterocarpans and other selected isoflavonoids. *Phytochemistry.* 15:655-659.
134. Van Hoof, L., Totte, J., Corthout, J., Pieters, L.A., Mertens, F., Vanden Berghe, D.A., Vlietinck, A.J., Dommisse, R., and Esmans, E. 1989. Plant antiviral agents, VI. Isolation of antiviral phenolic glucosides from *Populus* cultivar *Beaupre* by droplet counter-current chromatography. *J. Nat. Prod.* 52:875-878.
135. Van Hoof, L., Vanden Berghe, D.A., Hatfield, G.M., and Vlietinck, A.J. 1984. Plant antiviral agents; V. 3-Methoxyflavones as potent inhibitors of viral-induced block of cell synthesis. *Planta Med.* 50:513-517.
136. Veckenstedt, A. and Pusztai, R. 1981. Mechanism of antiviral action of quercetin aginast cardiovirus infection in mice. *Antiviral Res.* 1:249-261.
137. Veckenstedt, A., Guttner, J., and Beladi, I. 1987. Synergistic action of quercetin and murine alph/beta interferon in the treatment of Mengo virus infection in mice. *Antiviral Res.* 7:169-178.
138. Verbenko, E., Petrova, I., Goncharova, L., and Dudarev, A. 1979. Chelepine treatment of some viral diseases of the skin. *Vest. Dermatol. Venerol.* 6:51-53.
139. Vleitinck, A.J., Vanden Berghe, D.A., Haemers, A. 1988. In *Plant Flavonoids in Biology and Medicine II: Biochemical, Cellular and Medicinal Properties,* eds. Cody, V., Middleton., Jr. Harborne, J.B., Beretz, A., pp. 293. New York: Allan Liss Inc.
140. Vlietinck, A.J., Van Hoof, L., Totte, J., Lasure, A., Vanden Berghe, D., Rwangabo, P.C., and Mvukiyumwami, J. 1995. Screening of hundred Rwandese medicinal plants for antimicrobial and antiviral properties. *J. Ethnopharmacol.* 46:31-47.

141. Vrijsen, R., Everaert, L., Van Hoof, L.M., Vlietinck, A.J., Vanden Berghe, D.A., and Boeye, A. 1987. The poliovirus-induced shut-off of cellular protein synthesis persists in the presence of 3-methylquercetin, a flavonoid which blocks viral protein and RNA synthesis. *Antiviral Res.* 7:35-42.

142. Wandward, M.D. 1980. Phaseollin formation and metabolism in *Phaseolus vulgaris*. *Phytochemistry.* 19:921-927.

143. Wang, Y., Hamburger, M., Gueho, J., Hostettmann, K. 1989. Antimicrobial flavonoids from *Psiadia trinervia* and their methylated and acetylated derivatives. *Phytochemistry.* 28:2323-2327.

144. Weck, P.K., Harkins, R.N. and Stebbing, N. 1983. Antiviral effects of human fibroblast interferon from *Escherichia coli* against encephalomyocarditis virus infection of squirrel monkeys. *J. Gen. Virol.* 64:415-419.

145. Weck, P.K., Rinderknecht, E., Estell, D.A. and Stebbing, N. 1982. Antiviral activity of bacteria-derived human alpha interferons against encephalomyocarditis virus infection of mice. Infect. *Immun.* 35:660-665.

146. Weidenborner, M., Hindorf, H., Jha, C.H., and Tsotsonos, P. 1990. Antifungal activity of flavonoids against storage fungi of the genus *Aspergillus*. *Phytochemistry.* 29:1103-1105.

147. Werner, G.H., Jasmin, C. and Chermann, J.C. 1976. Effect of ammonium 5-tungsto-2-antimoniate on encephalomyocarditis and vesicular stomatitis virus infections in mice. *J. Gen. Virol.* 31:59-64.

148. William, C.A., and Harborne, J.B. 1989. Biflavonoids. In *Methods in Plant Biochemistry. Vol. 1. Plant Phenolics*, ed. Harborne, J.B., pp.357-388. London: Academic Press.

149. Wleklik, M., Luczak, M., Panasiak, W., Kobus, M., Lammer-Zarawska, E. 1988. Structural basis for antiviral activity of flavonoids-naturally occcuring compounds. Acta virol. 32:522-525.

150. Wyman, J.G., and Van Etten, H.D. 1977. Antibacterial activity of selected isoflavonoids. *Phytopathology* 68:583-589.

151. Yoshikawa, K., Sugawara, S., and Arihara, S. 1995. Phenylpropanoids and other secondary metabolites from fresh fruits of *Picrasma quassioides*. *Phytochemistry* 40:253-256.

B.B. Whitmore

A.S. Naidu

13

Thiosulfinates

I. INTRODUCTION

Garlic (*Allium sativum* L), a native of central Asia, is a perennial herb with flat, spear-like leaves (up to 35 cm long and 2.5 cm wide) and a white membranous covered bulb composed of about 3 to 15 bulblets or cloves. The English name is of Anglo-Saxon origin, being derived from the words '*Gar*' (spear) and '*Lac*' (a plant). The botanical name of garlic, *Allium,* may be derived from the Celtic word '*All*,' meaning pungent, and *sativum* from the Latin word *'sative'*, meaning sown from seeds.

Garlic has been used throughout history as a flavoring agent in food, and as a medicinal herb for thousands of years. There are references to the medicinal potential of garlic in ancient Chinese texts, Egyptian papyrus, as well as Greek and Roman scriptures. Inscriptions on the pyramids of Giza show that the workers in ancient Egypt used garlic for medicinal powers to complete their tasks by increasing physical stamina and spiritual integrity. Pharaohs were entombed with clay and woodcarvings of garlic and onions to ensure the meals in their afterlives were well seasoned. The Codex Ebers (Ebers Papyrus), an Egyptian holy book dating from about 1550 BC, lists more than 800 therapeutic formulas in use at that time, with 22 containing garlic (Block, 1984). Hippocrates, in ancient Greece, used garlic for treating infections, wounds and intestinal disorders. Aristotle and Aristophanes also recommended garlic for medicinal effects. Pliny the Elder in ancient Rome tabulated many therapeutic uses for *Allium*. It was postulated that the first Olympic athletes used garlic as a stimulant.

During the 1720s, garlic was used in Europe in a potion called 'Four Thieves Vinegar' (*vinaigre des quatre voleurs*) for prophylaxis against the plague. It is still used in France today. In 1858, Louis Pasteur noted the antimicrobial effects of garlic. (Pasteur, 1858). In England, garlic was used during the First World War as an antiseptic to prevent gangrene. The crushed cloves were diluted with water, placed on sterilized Sphagnum moss and applied to the wound. Animals were also fed garlic to prevent sickness. Cattle were fed with garlic to prevent anthrax and chickens were fed garlic to prevent gapes (*Syngamus trachealis*).

0-8493-2047-X/00/$0.00+ $.50
© 2000 by CRC Press, LLC

In the early 1900s, researchers began to identify the antimicrobial substances in garlic. German chemists Theodor Wertheim and F.W. Semmler were the first to identify allyl compounds in garlic oil by steam distillation. Steam distillation causes some of the volatile oil to yield diallyl disulfide and related compounds (Tyler, 1993). Cavallito et al. (1944) used ethyl alcohol (instead of steam distillation) to produce an oil containing allyl 2-propenethiosulfinate, which he named 'allicin'. (Block, 1984).

II. OCCURRENCE

Garlic is related to onions (*Allium cepa*), leeks (*Allium porum*), shallots (*Allium ascalonicum*), rocambole (sand leeks) (*Allium scorodoprasum*), chives (*Allium schoenoprasum*), and garlic chives (Chinese chives or Oriental garlic) (*Allium tuberosum*). Many of these species also contain active antimicrobial compounds found in garlic. Yoo and Pike (1998) analyzed the contents and relative proportion of thiosulfinates in different *Allium* species. Their data revealed different species of *Allium* contained different proportions of thiosulfinates, with *A. sativum* containing the most S-2-propenyl(allyl)-L-cysteine sulfoxide, a 83.7% relative proportion. Giant garlic (*A. ampeloprasum*) contained 62.7% relative proportion of S-2-propenyl(allyl)-L-cysteine sulfoxide.

Allium sativum L plants grow easily in both cool and warm climates. The cloves are planted and grown for 5-6 months. The cloves are then harvested and the bulbs are dried for a week. After drying, it is trimmed and sometimes braided and hung in the shade to dry.

Continuous controversies exist regarding the taxonomic status of garlic. Certain botanists classify it in the *Amaryllidaceae* (Amaryllis family) based on the umbel flower infloration. Some place it in the *Lilliacae* family, while still others place it under the family *Alliaceae* (Martin, 1999).

A. Biosynthesis

The site of natural biosynthesis of the extract is located in the bulb of the garlic plant. An intact bulb contains alliin (+S-allyl-L-cysteine-S-oxide), the precursor to allicin. Alliin is hydrolyzed by alliinase when the bulb is damaged and becomes allicin, ammonia and pyruvate. In an intact bulb, alliin and alliinase are stored in separate cellular compartments (Feldberg, 1988). Damage to the bulb destroys the barriers between compartments, causing the subsequent reaction.

The alliinase enzyme catalyzes the conversions in several of the sulfur compounds in garlic (as well as conversion into the lacrimatory factors in *Allium cepa*-the brown onion), one of which is the conversion of alliin to allicin. Pyridoxal phosphate, a cofactor, acts on the substrate, forming a complex with the enzyme. This binding complex also includes an electrostatic interaction of the substrate with a metal ion. An alkaline group on the enzyme removes a proton in the substrate, causing the dissolution of the substrate and subsequent release of 2-propenesulfenic acid, ammonia and pyruvate. The 2-propenesulfenic acid dimerizes with a second molecule of 2-propenesulfenic acid to make allicin (Stoll & Seebeck, 1949).

The types of sulfur compounds extracted from garlic depend on the extraction conditions. Steam distillation is a severe technique, yielding diallyl disulfide from the condensed steam. Ethyl alcohol and water at 25°C yields allicin, whereas ethyl alcohol at

0°C yields the compound alliin. Because alliin has optical isomerism, four forms are possible; however only one form is found in garlic (Block, 1984).

Feldberg et al. (1988) demonstrated the formation of allicin by the reaction of alliin with alliin lyase (alliinase). Alliin lyase (6.25 U) was added to 13mM alliin in a 40-ml volume of 0.05M potassium phosphate buffer (pH 6.5) containing 10^{-5} M pyridoxal 5'-phosphate, and 5% glycerol. After a 4-hour incubation the reaction was measured by the formation of pyruvate according to a dinitrophenylhydrazine assay. The reaction was 90-95% complete at this concentration of alliin. The allicin was concentrated by extraction of the reaction mixture, with an equal volume of ethyl ether. The ether was removed with a nitrogen stream, and the oil residue dissolved in 4 ml of deionized water. The allicin concentration was 35mM according to the N-ethylmaleimide assay.

Allicin decomposition can also form ajoene. Three molecules of allicin combine to form two molecules of ajoene. Based on studies with methyl methanethiosulfinate, *trans* and *cis* forms of ajoene were identified, with *cis* being the more potent of the two. It was also shown that ajoene could be formed by heating allicin with a mixture of water and organic solvent such as acetone (Block, 1984). This was recently substantiated in a study by Yoshida et al. (1999) which also determined that the double bound position or the *trans* structure reduces the antimicrobial activity.

B. Screening/Identification:

1. Analytical assays: Barone and Tansey (1977) used thin layer chromatography (TLC) for separation and bioassay of the anticandidal activity of the natural and synthetic preparations. TLC was done on precoated silica gel of 0.25mm thickness. Several contaminants were identified in the commercial preparations of allyl disulfide.

Column chromatographic methods are used to separate the active ingredients in raw garlic extract. The anticandidal component of garlic extract binds to Sephadex G-10 gel and the active component could be separated from other small molecules by gel filtration (Barone & Tansey, 1977).

Naganawa (1996) separated ajoene E and Z by high-performance liquid chromatography (HPLC) using a Supelcosil LC-Si column. The homogeneity of the preparation was further determined by nuclear magnetic resonance (NMR) spectroscopy.

Gas chromatographic techniques are also frequently used to determine sulfur groups in garlic extract (Davis et al., 1990).

Indirect spectrophotometric assays based on the use of excess cysteine and allicin's ability to react rapidly can also be used to determine allicin concentration. This resulting reaction forms S-allylmercaptocysteine. Since the reaction proceeds with one molecule of allicin reacting with 2 molecules of cysteine, the decrease of cysteine can be utilized as a measure of allicin. This decrease of cysteine is measured with thiol-reactive reagents. A single-step spectrophotometric assay for both allicin and alliinase activity utilizes 2-nitro-5-thiobenzoate (NTB) (Miron, et al., 1998). Alliin reacts with alliinase to yield allicin and pyruvate. The allicin further reacts with the NTB forming allylmercapto-NTB. The reaction between allicin and NTB was monitored by decreasing absorbance at 412 nm. This rate was found to be <30 minutes.

Keusgen (1998) developed a flow injector analyzer (FIA) to determine the alliin contents. This method is based on immobilized allinase, and utilizes ammonium instead of pyruvic acid in the alliin/alliinase reaction. Simultaneous determination of alliin and ammonium occurs in channel 1 and 2 respectively. In channel 1, ammonium is formed enzymatically and converted to ammonia by adding a NaOH solution before the buffer stream reaches the diffusion module. There, the ammonia passes through the internal membrane, affecting a color change in the bromocresol purple indicator solution. Colorimetric measurements were made at 605 nm. Channel 2 monitors the 'native' ammonium concentration. The concentration of alliin can be calculated from the peak area or evaluation of peak height. This method is not suitable for alliin determination in other *Allium* species containing cysteine sulfoxides in large amounts unless an enzyme with a higher specificity for alliin can be located.

2. Biological sources: Allicin is found in the damaged garlic bulb as well as in the leaves (Mütsch-Eckner et al., 1993) Alliin is the precursor of allicin and is converted to allicin with allinase. Fresh garlic extracts have been shown to contain the most antimicrobial activity. Most commercial preparations show no antimicrobial activity (Barone & Tansey, 1977; Block, 1985; Caporaso, 1983; Tyler, 1993).

III. ISOLATION AND PURIFICATION

A. Preparation of the extract

In vitro synthesis of allicin involves oxidation of diallyl disulfide by an equimolar amount of organic per-acid. Diallyl disulfide (0.1 M) is dissolved in 500 ml of chloroform in an ice bath. To this solution, meta-chloro perbenzoic acid (0.1 M) dissolved in 1150 ml of chloroform is slowly added at a rate of 5 ml per minute. After 1 hour incubation at room temperature, 0.1M aqueous sodium bicarbonate was added to this mixture as a 5% solution. After thorough mixing, 25 ml of 2% aqueous sodium bicarbonate is added to neutralize acids. Chloroform is removed by distillation at 20°C in a rotary evaporator under reduced pressure. N-hexane (30 ml) is added to the residue and the hexane mixture is extracted with 50 ml of water until small aliquots of the water show no turbidity when added to a few drops of 10% aqueous NaOH. Development of turbidity would indicate that alkyl thiosulfinates are present, thus further extraction is warranted. Water extracts are combined and extracted 3 times with 200-ml of N-hexane. The remaining water solution should be extracted five times with 200 ml of anhydrous sodium sulfate and evaporated at 20°C in a rotary evaporator under low pressure.

Garlic extract: Moore and Atkins, (1976) used a modification of Tynecka and Gos (1973), to extract garlic juice from garlic cloves. One kilogram of garlic is hulled, pressed through a garlic press and blended at high speed for 2 minutes, until thick. The mash is squeezed through eight layers of gauze, and the resulting liquid is centrifuged for 10 minutes at 8,000x*g*. The supernatant is decanted and sterilized using Millipore filters with successive smaller pore sizes (0.8 μ, 0.45 μ, and 0.22 μ). About 200 ml of garlic juice is obtained using this procedure.

Caporaso et al. (1983) using a modification of Fromtling and Bulmer (1978) method prepared garlic extract from fresh cloves. The garlic cloves are hulled and 100 grams of garlic cloves are homogenized in 10 ml of distilled water in a high-speed blender. Two-minute bursts are performed in a Waring™ blender for a total of 10 minutes. The vessel is placed in an ice bath between bursts. The mixture is centrifuged at 2,000x*g* for 20 min. The mixture is filtered and sterilized by passing through a 0.2 μ Nalgene filter. The filtrate was then tested for antifungal activity and stored at -25°C until used.

A Sephadex G-10 gel column (Pharmacia) was used by Barone and Tansey (1977), for fractionation of garlic extract by molecular sieve chromatograpy. The column bed size was 10 X 190 mm, and calibrated with NaCl-blue dextran solution. Two ml of aqueous garlic extract was pumped through the column. Fractions were eluted with phosphate buffer and the chromatography was performed at 5°C.

Barone and Tansey (1977) also used thin layer chromatography (TLC) for separation of anticandidal activity of natural and synthetic preparations. TLC was performed on 20 x 20 cm precoated silica gel 60 glass plates having a layer thickness of 0.25mm. Two different solvent systems: 1) carbon tetrachloride: methanol: water, 20:10:2 and 20:10:1; and 2) benzene: ethyl acetate, 90:10, were tested. Substances were diluted with cold chloroform if needed. Spots of 30-50 λ were applied to plates with a micropipette. TLC plates were developed using several different methods as follows: 1) plates were incubated with iodine crystals at 40°C, 2) plates were sprayed with a mixture of 0.1 M copper acetate and 0.1 M silver nitrate in 3 M alcoholic ammonia. (Prepared by dissolving 1.7 g silver nitrate and 1.8 copper acetate in 20-ml conc. ammonia and then adding absolute ethanol to achieve a 100-ml total volume), and 3) plates were sprayed with a 1:5 mixture of 0.1 N silver nitrate and 5 N ammonium hydroxide, and heated at 105°C for 20 minutes. Finally, copies of the chromatograms were made by tracing the plates immediately after the staining and heat treatments. All precipitates were removed by filtration.

B. Isolation of allicin

Raw garlic was fractionated into allicin and ajoene by a process described by Yoshida et al. (1987). Raw garlic was homogenized with water. Half of the homogenate was steam distilled and acidified to pH 3.0-4.0. Ethanol was added and the mixture was heated and extracted with EtOAc silica gel chromatography. This process produced ajoene. The other half of the homogenate was filtered and extracted with EtOAc that formed an organic and an aqueous layer. The organic layer was concentrated and passed through silica gel chromatography to separate allicin (*FIGURE 1*).

Feldberg et al. (1988) demonstrated the allicin formation by reacting alliin with alliin lyase (alliinase). Alliin lyase (6.25 U) was added to 13 mM alliin, a 0.05 M potassium phosphate buffer (pH 6.5), 10-5 M pyridoxal 5'-phosphate, and 5% glycerol for a total volume of 40 ml, and incubated for 4 hours. The extent of the reaction was measured by the formation of pyruvate according to the dinitrophenylhydrazine assay, and was found to be 90-95% complete at this concentration of alliin. The allicin was concentrated by extraction of the reaction mixture, with equal volume of ethyl ether. The ether was removed with a nitrogen stream, and the oil residue was dissolved in 4 ml of deionized water. The allicin concentration determined by N-ethylmaleimide assay, was 35mM. The product was further purified through a Bio-Rad AG50W column to obtain pure allicin (determined by paper chromatography). The final concentration was 1.1 to 1.5 M and was adjusted to pH 6.5 before use.

Mutsch-Eckner et al. (1993), separated allicin from garlic leaves by column chromatography with Amberlite IR-120. The leaves were freeze-dried (3.2 kg) and macerated with 80% methanol, four times for 2 days, each step with 12 liters at 2°C. When mixture was concentrated to 4 liters, the aqueous phase was extracted with CH_2Cl_2, and

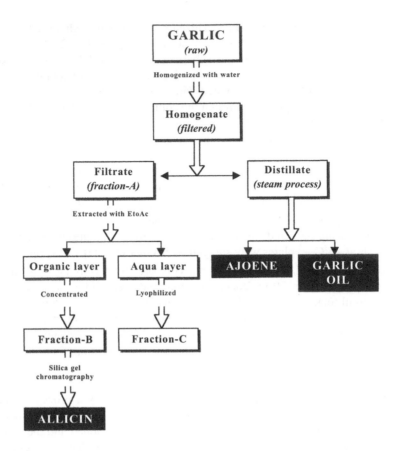

FIGURE 1. Fractionation of phytoantimicrobial thiosulfinates from raw garlic (adapted from Yoshida et al., 1987)

EtOAc three times using 4 liters of each. The remaining aqueous layer was freeze-dried and measured at 670.3 g. The amino acids of 120 g of the water extract was eluted with a 0.75 M NH$_4$OH solution. With gel-permeation chromatography on Sephadex LH-20 the amino acid fraction (2.1 g) was separated into two fractions. This was accomplished by increasing the amount of methanol in H$_2$O from 0% to 20% for 20 h. The process was repeated to purify the fractions. Fraction-1 was found to be (-)-N-(1'-Deoxy-1'-beta-D-fructopyranosyl)-S-allyl-L-cysteine sulfoxide and fraction-2 (allicin), and fraction-3 (+)-S-methyl-L-cysteine sulfoxide. An amount of 300-mg was used in a semi-preparative HPLC (H$_2$O) yielding pure amounts of (27 mg) of fraction-3, and (140 mg) of allicin. HPLC on RP-18 column, the Sephadex fraction-1 further resolved a main fraction after freeze-drying and crystallization from Me$_2$CO/H$_2$O, and resulted in 245 mg of allicin.

C. Isolation of ajoene

Three molecules of allicin combine to form two molecules of ajoene. Block (1984) found *trans* and *cis* forms of ajoene, with *cis* being the more potent of the two. Ajoene was synthesized by heating allicin with a mixture of water and an organic solvent such as acetone. Yoshida et al. (1986) reported the separation of both allicin and ajoene

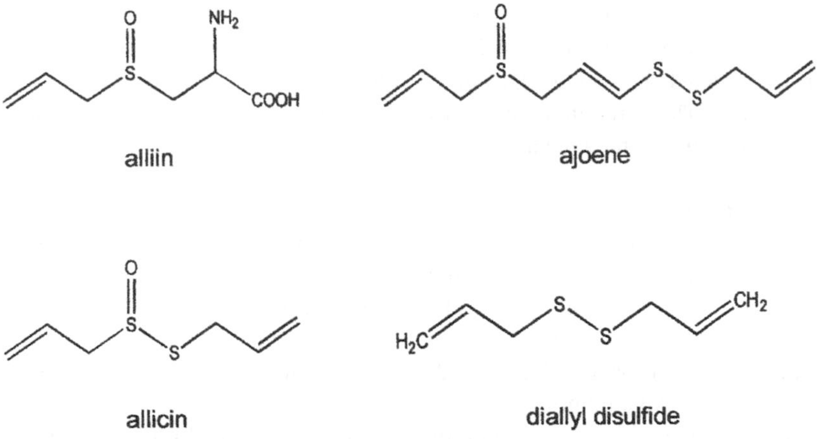

FIGURE 2. Chemical structure of major thiosulfinate compounds from garlic

from raw garlic. The purified compound was identified by comparison to a sample by ^{1}H- and ^{13}C-nuclear magnetic and mass spectroscopy. The purity of the isolated material was estimated to 99.0%. A process for the manufacture of synthetic ajoene (M_r 234) has also been developed (Apitz-Castro & Jain, U.S. patent 4665099, May 1987).

D. Isolation of alliinase

A method for purification of alliinase has been described by Feldberg and co-workers (1988). Two-kilograms of fresh garlic cloves were processed through a juicer and mixed with 0.05 M potassium phosphate (pH 6.5), 10-5 M pyridoxal 5'-phosphate, and 5% glycerol. The mixture was centrifuged at 20,000 x g for 1 hour and the pellet was sus-pended in 0.05 M potassium phosphate buffer and dialyzed against the same buffer. Alliinase was isolated by negative chromatography on a DEAE-trisacryl M column. After precipitation by 50% ammonium sulfate, the pellet was re-suspended and applied to a Sephacryl S-300 column, equilibrated with a 0.01 M potassium phosphate buffer. The eluted active material was applied to a hydroxyapatite-Ultrogel column, washed and equi-librated with 0.01 phosphate buffer and a 0.07 M phosphate buffer. Enzyme activity was eluted with a 0.20 M phosphate buffer. This method yielded 25 U/ml of the enzyme. A molecular mass of 41 kDa was determined for alliinase by SDS-PAGE. Krest and Keusgen (1999) reported alliinase from garlic powder with two different molecular weights of 53 and 54 kDa, whereas fresh garlic consists of two identical subunits.

IV. MOLECULAR PROPERTIES

A variety of thiosulfinate compounds have been isolated and identified from gar-lic (*Allium sativium L.*). These include allicin, alliin, deoxyalliin, alliin, thio-2-propene-1-sulfinic acid S-allyl ester, +S-allyl-L-cysteine sulfoxide, ajoene, organic trisulfur, diallyl thiosulfate, diallyl sulfate, diallyl sulfide (DAS), diiallyl disulfide (DADS), diallylthio-sulfinate, diallyl trisulfide, ethanolic garlic extract, Z-4,5,9-trithiadeca-1,6-diene-9-oxide, Z-10-devinylajoene; Z-10-DA, E-4,5,9-trithiadeca-1,7-diene-9-oxide (iso-E-10-devinyla-joene, iso-E-10-DA). The structure of major thiosulfinates is shown in *FIGURE 2*.

The chemical formula and molecular mass of alliin ($C_6H_{10}O_3SN$), allicin ($C_6H_{10}OS_2$ =162), ajoene ($C_9H_{14}OS_3$=234), and diallyl disulfide (DADS; $C_6H_{10}S_2$ = 146) are well documented. The molecular mass of alliinase enzyme was estimated to 41 kDa (Feldberg, 1988). Thiosulfinate compounds are extensively characterized for their physiochemical properties in the recent years.

A. Sensitivity to enzymes

Alliin was one of the first natural chemicals to display optical isomerism. Such configurations in alliin are possible at both the sulfur and carbon atoms. Alliinase enzyme [alliin alkyl-sulfinate-lyase (Feldberg et al., 1988)] hydrolyzes alliin to 2-propenesulfenic acid and acts preferentially on the isomer of alliin that produces the clockwise rotation of polarized light. The 2-propenesulfenic acid dimerizes with a second molecule of 2-propenesulfenic acid to make allicin.

The enzyme alliinase also requires a cofactor, pyridoxal phosphate, which acts on the substrate, forming a complex with the enzyme. This binding complex also includes an electrostatic interaction of the substrate with a metal ion. An alkaline group on the enzyme removes a proton, in the substrate causing the dissolution of the substrate and subsequent release of 2-propenesulfenic acid, ammonia and pyruvate. The 2-propene-sulfenic acid dimerizes with a second molecule of 2-propenesulfenic acid to make allicin (Block, 1985; Feldberg et al., 1988).

Allinase acts on a number of molecules besides the alliin. All such molecules contain cysteine, one of the sulfur containing amino acids. Allinase along with the cofactor pyridoxal phosphate produces several sulfenic acids with the generic formula RSOH, where R is a radical. These radicals may be allyl, 1-propenyl, methyl or propyl. The byproducts of the reactions are pyruvate and ammonia and an unstable sulfenic acid (Block, 1985).

Allicin also affects other kinds of enzymes from hosts and parasites. Ankri et al. (1997) determined that allicin inhibits the ability of *Entamoeba histolytica* trophozoites to destroy monolayers of baby hamster kidney cells. It was also found that cysteine proteinases (an amoebic virulence factor) as well as alcohol dehydrogenase are strongly inhibited by allicin.

Chung (1998) reported a dose-dependant effect of DAS or DADS treatment with viability studies and decreased K_m and V_{max} of arylamine N-acetyltransferase (NAT) enzymes from the gastric pathogen *Helicobacter pylori*.

Chen et al.(1998) demonstrated a decreased NAT activity in *Klebsiella pneumoniae* with DAS and DADS. It was also shown that DADS and DAS could inhibit NAT activity in a human colon tumor cell line (adenocarcinoma).

B. Thermal behavior

Allicin is heat labile, but stable in low concentrations of acid or alkali. Allicin is somewhat soluble in water (2.5% at 10°C) and insoluble in aliphatic hydrocarbon skellysolee-B (n-hexanes) (Cavallito & Bailey, 1944). Barone and Tansey, (1977) demonstrated the heat labile property of garlic extract by autoclaving (121°C for 10 minutes) which abolished anticandidal activity. In addition, a gradual loss of anti-candidal activity was observed when the extract was incubated for longer times at 37°C.

C. Effects of milieu

The anticandidal activity of garlic extract (1 mg/ml) is stable at acidic conditions (pH 2.0). However, this antimicrobial activity was lost after exposure to alkaline conditions (pH 12.0) (Barone & Tansey, 1977).The sulfur-oxygen interaction of allyl disulfinate confers a more positive charge on the sulfur, making it more susceptible to a nucleophilic attack in an alkali environment by a hydroxyl group. This may cause the unstable sulfinic acid to decompose, liberating sulfur dioxide. The mercaptan salt easily oxidizes to allyl disulfide in the presence of air or oxygen.

Anticandidal activity may involve inactivation of essential thiol groups by combining with the allyl sulfide with the sulfhydryl group to form a stable disulfide or that simple oxidation of the adjacent thiol to the corresponding disulfide by the labile oxygen of allicin. Cavallito and Bailey (1944) showed that sulfhydryl compounds such as cysteine stoichiometrically react with the polar thiosulfinate producing S-thioallyl cysteine. The addition of dithioerythritol also liberate the mercaptan, probably by reducing the thiosulfinate. Because there is such low amounts of reducing agent, disulfide should also be formed. (Barone & Tansey, 1977)

Sulfhydryl compounds such as L-cysteine or a reducing compound (dithioerythritol) interferes with the anticandidal activity of synthetic allicin and garlic extract at low concentrations (5 x 10^{-5} moles of each). Thus, addition of L-cysteine or dithioerythritol could abolish the anticandidal activity of garlic extract (Barone & Tansey, 1977). Naganawa, (1996) reported that addition of twice the amount of cysteine to the media caused a 20% reduction in the ajoene concentration and abolished the antimicrobial properties. San-Blas et al. (1997), also described this inhibitory phenomenon with thiosulfinates and the fungi *P. brasiliensis*.

Inhibition of alliinase activity directly effects the production of allicin. Compounds such as hydroxylamine, penicillamine and allylcysteine inhibit alliinase activity. (Miron et al., 1998).

D. Effects of storage conditions

A gradual loss of anticandidal activity was observed in agar diffusion assays after incubation at room temperature for several days, suggesting that the active compound (allyl thiosulfinate) gradually degrade over time (Barone & Tansey, 1977).

Studies with oral doses of garlic show that the antifungal activity decreases in serum after 60 minutes after the first presence. Urine did not appear to inhibit the antifungal activity of garlic extract. These findings suggest that in biological systems certain factors in serum and urine may cause degradation of garlic's active ingredients. It is possible that high serum protein may bind allicin and may inactivate the molecule. The reduction in antifungal activity in serum could also be attributed to its high pH in the absence of CO_2. Thus, when the assay was performed in 5% CO_2 the antifungal activity of the extract in the serum was found to be slightly elevated (Caporaso et al., 1983).

Yamada and Azuma (1977) showed that addition of allicin to SG broth medium retained its activity for at least 10 days. Possibly, allicin bound to the fungal spore or cell is more stable than the free allicin in the medium. It was observed that more allicin was required to inhibit growth in liquid than in the solid media, suggesting that allicin could deteriorate more rapidly in the liquid media at pH 5.6.

Moore and Atkins, (1976) also described the loss in antimicrobial activity of garlic to *C. albicans* at 4 °C. The MIC values decreased from 1:128 to 1:16 in a two week period. The antimicrobial activity of garlic extract stored at –10°C and –60°C decreased from 1:128 to 1:64 within the first week and remained at this level for the rest of the test period.

Allicin decomposition occurs in several ways. Three molecules of allicin combine to form two molecules of ajoene (Block, 1985). Allicin also decomposes into 2-propensulfenic acid and thioacrolein, however, the former compound self-condenses back into allicin. The two thioacrolein molecules proceed through a Diels-Alder reaction to form 2-vinyl-[4H]-1,3-dithiin and 3-vinyl-[4H]-1,2-dithiin.

V. ANTIMICROBIAL ACTIVITY

The juice and vapors of onions, garlic, and horseradish have been evaluated for their antimicrobial activity since the early 1900s. Walker and co-workers (1925) reported the fungistatic properties of garlic/onion juice and vapors. Walton et al. (1936) developed a simple method for evaluating the antimicrobial activity of garlic vapors. An agar plate is inverted and minced garlic is placed inside the top lid to expose the media to vapors. After exposure for varying lengths of time, the media is streaked with the test strains. Strains, including *Bacillus subtilis, Serratia marcescens,* and two *Mycobacterium* species, were inhibited to varying extents according to this method.

Helicobacter pylori is the causative agent of gastric ulcers and is implicated in the etiology of stomach cancer. The incidence of gastric cancer is lower in individuals and populations with high allium vegetable intakes. Standard antibiotic regimens against *H. pylori* are frequently ineffective in high-risk populations. Wong, et al. (1996) reported the inhibitory activity of garlic extract on *H. pylori*. You et al. (1998) suggested that infection with *H. pylori* was a risk factor for pre-cancerous lesions and garlic may be protective. Sivam et al. (1997) investigated the role of allium vegetable intake on cancer prevention, and tested its antimicrobial activity against *H. pylori*. An aqueous extract of garlic cloves was standardized for its thiosulfinate concentration and tested for its antimicrobial activity on *H. pylori* grown on chocolate agar plates. MIC was determined at 40 μg/ml of thiosulfinate. *S. aureus* tested under the same conditions was not susceptible to garlic extract up to the maximum thiosulfinate concentration tested (160 μg/ml). The authors suggested that the sensitivity of *H. pylori* to garlic extract at such low concentration may be related to the reported lower risk of stomach cancer in those with a high allium vegetable intake. Graham et al. (1999) performed a prospective crossover study on healthy individuals infected with *H. pylori*. Ten subjects were fed 10 sliced gloves of garlic, 6 sliced fresh jalapeños (capsaicin), and eleven people with 2 tablets of bismuth subsalicylate (Pepto-Bismol). The study found neither garlic or capsaicin had any *in vivo* effect on *H. pylori*, whereas the bismuth had a marked inhibitory effect. However, alliinase enzyme degradation in the stomach (inhibition of the alliin to allicin or ajoene conversion) or possible complex formation with the sulfur amino acids (from proteins) in the food should be considered.

Researchers in Iran, while assessing oils from various herbs and spices, found that the oil from garlic produced a zone of inhibition against *Listeria monocytogenes* serotype 4a (foodborne) in agar diffusion tests (Firouzi et al., 1998).

Chang et al. (1997) found that garlic was responsible for the removal of coliforms in kimchi (pickled cabbage) and Paludan-Muller et al. (1999) suggested that garlic has a dual role in fermented Thai fish (som-fak). Paludan-Muller et al. (1999) found that garlic could inhibit gram-negative bacteria and yeast, and it provided a better fermentation source than rice starch. The more acid tolerant strains of *Lactobacillus* species *(L. casei. L. pentosus and L. plantarum)* are found toward the end of fermentation, while *Leuconostoc* spp., *Lactobacillus curvatus* and *Lactobacillus lactis* are found toward the beginning of the process. Initial presence and growth of LAB such as *L. plantarum* that could ferment garlic rapidly (thus lowering the pH of the fish), may also be a factor.

Thiosulfinates from garlic demonstate a wide spectrum of antimicrobial activity against bacteria, fungi, viruses and parasites.

TABLE 1: Antibacterial activity of thiosulfinate compounds from garlic

Bacterial species	MIC (compound)	Effect	Reference
Bacillus cereus	300 µg/ml (80% Z + 20% E-ajoene)	A	Naganawa et al., 1996
	<100 µg/ml (E-trithiadecadieneoxide)	B	Yoshida et al., 1999
Bacillus subtilis	≥500 µg/ml (80% Z + 20% E-ajoene)	A	Naganawa et al., 1996
	<100 µg/ml (E-trithiadecadieneoxide)	B	Yoshida et al., 1999
Coliforms	2% garlic extract	C	Chang et al. 1997
Escherichia coli	400 µg/ml (80% Z + 20% E-ajoene)	A	Naganawa et al., 1996
Gram +ve and Gram -ve	1:125,000 titer (allicin)	C	Cavallito & Bailey, 1944
Gram -ve	4% minced garlic	E	Paludan-Muller et al.,1999
Helicobacter pylori	40 µg/ml (thiosulfinate)	C	Sivam et al., 1997
		F	You et al., 1998
		G	Chung et al., 1998
Klebsiella pneumoniae	DAS and DADS	G	Chen et al., 1999
	>500 µg/ml (80% Z + 20% E-ajoene)	A	Naganawa et al., 1996
Listeria monocytogenes	Garlic oil	C	Firouzi et al. 1998
Micrococcus luteus	>500 µg/ml (80% Z + 20% E-ajoene)	A	Naganawa et al., 1996
Mycobacterium phlei	>500 µg/ml (80% Z + 20% E-ajoene)	A	Naganawa et al., 1996
M. smegmatis	>500 µg/ml (80% Z + 20% E-ajoene)	A	Naganawa et al., 1996
Pseudomonas aeruginosa	>500 µg/ml (80% Z + 20% E-ajoene)	A	Naganawa et al., 1996
Salmonella typhimurium	0.4 mM (49 µg/ml; allicin)	D	Feldberg, et al., 1988
Staphylococcus aureus	<100 µg/ml (E-trithiadecadieneoxide)	B	Yoshida et al., 1999
	400 µg/ml (80% Z + 20% E-ajoene)	A	Naganawa et al., 1996
	160 µg/ml (thiosulfinate)	C	Sivam et al., 1997
Streptococcus spp.	>500 µg/ml (80% Z + 20% E-ajoene)	A	Naganawa et al., 1996
Streptomyces griseus	>500 µg/ml (80% Z + 20% E-ajoene)	A	Naganawa et al., 1996
Xanthomanas maltophilia	>500 µg/ml (80% Z + 20% E-ajoene)	A	Naganawa et al., 1996

A) Ajoene effects are elicited by the presence of both the disulfide bond and the sulfinyl group.
B) Trans configuration or position of double bond reduces antimicrobial activity.
C) Mechanism not elaborated/explored in this study.
D) Allicin inhibits DNA synthesis, cellular replication as well as mRNA degradation and RNA synthesis.
E) Garlic is a C-source for LAB fermentation. It increases pH and inhibits gram-ve bacteria and yeast.
F) Garlic consumption protects against *H. pylori* and precancerous lesions.
G) Decreased arylamine N-transferase (NAT) activity.

TABLE 2: Antifungal activity of thiosulfinate compounds from garlic

Fungal species	MIC (compound)	Effect	Reference
Aspergillis niger –M phase	1:512 in broth, 1:8 serum	A	Caporaso et al.1983
A. niger (ATCC 16404)	20 μg/ml (ajoene, 99% pure)	C	Yoshida et al. 1987
A. flavis-M phase	1:256 in broth, 1:8 in serum	A	Caporaso et al. 1983
A. fumigatus-M phase	1:128 in broth, 1:16 in serum	A	Caporaso et al. 1983
Candida albicans	1:5,000 titer (garlic extract)	A	Appleton & Tansey, 1975
C. albicans	1:128 titer (garlic extract)	A	Moore & Atkins, 1976
C. albicans	50-300 μg/ml-static	B	Barone & Tansey, 1977
	400 μg/ml – cidal (ajoene).		
C. albicans Y phase	1:512 in broth, 1:64 in serum		Caporaso et al. 1983
C. albicans (ATCC 10231)	20 μg/ml (ajoene, 99% pure)	A	Yoshida et al. 1987
C. albicans	≥500 μg/ml (ajoene)	C	Naganawa et al. 1996
Candida glabrata Y-phase	1:64 in broth, 1:4 in serum	D	Caporaso et al. 1983
Candida guilliermondii	1:5,000 titer (garlic extract)	A	Appleton & Tansey, 1975
C. guilliermondii -Y-phase	1:1,024 in broth, 1:128 in serum	A	Caporaso et al. 1983
Candida krusei Y-phase	1:128 in broth, 1:16 in serum	A	Caporaso et al. 1983
Candida krusei	1:128 titer (garlic extract)	A	Moore & Atkins, 1976
Candida parapsilosis	1:5,000 titer (garlic extract)	A	Appleton & Tansey, 1975
C. parapsilosis	1:128 titer (garlic extract)	A	Moore & Atkins, 1976
C. parapsilosis Y-phase	1:512 in broth, 1:64 in serum	A	Caporaso et al. 1983
Candida stellatoidea	1:512 in broth, 1:1,228 in serum	A	Caporaso et al. 1983
Candida tropicalis	1:5,000 titer (garlic extract)	A	Appleton & Tansey, 1975
C. tropicalis Y-phase	1:128 titer (garlic extract)	A	Caporaso et al. 1983
Cryptococcus albidus	1:512 in broth, 1:64 in serum	A	Caporaso et al. 1983
Cryptococcus neoformans	1:512 titer	A	Fromtling & Bulmer, 1978
C. neoformans Y- phase	1:128 in broth, 1:16 in serum	A	Caporaso et al. 1983
C. neoformans	1 mg/per day (allitridium)	A	Davis et al. 1990
C. neoformans	100 μg/ml (allitridium)	A	Shen et al. 1996
Dactulomyces crustaceus	1:5,000 titer (garlic extract)	A	Appleton & Tansey, 1975
Dactulomyces floccosum	1:5,000 titer (garlic extract)	A	Appleton & Tansey, 1975
Hanseniaspora valbyensis	400 μg/ml (ajoene)	D	Naganawa et al. 1996
Humicola lanuginosa	1:5,000 titer (garlic extract)	A	Appleton & Tansey, 1975
Microsporum spp	1:5,000 titer (garlic extract)	A	Appleton & Tansey, 1975
Mucor pusillus	1:1,024 in broth, 1:16 in serum	A	Caporaso et al. 1983
Paracoccidioides brasiliensis	50 μM (synthetic ajoene)	E	San-Blas et al. 1989
Paracoccidioides brasiliensis	25 & 50 mM for Y & M phase	F	San Blas et al. 1997
Penicillium marneffei	1:5,000 (garlic extract)	A	Appleton & Tansey, 1975
Phialophora pedrosoi	1:5,000 (garlic extract)	A	Appleton & Tansey, 1975
Pichia anomala	≥500 μg/ml (ajoene)	D	Naganawa et al. 1996
Rhizopus arrhizus	1:64 in broth, 1:16 in serum	A	Caporaso et al. 1983
Saccharomyces cerevisiae	300 μg/ml (ajoene)	D	Naganawa et al. 1996
Schizosaccharomyces pombe	300 μg/ml (ajoene)	D	Naganawa et al. 1996
Scopulariopsis brevicaulis,	1:5,000 (garlic extract)	A	Appleton & Tansey, 1975
Tinea pedis	0.4% (w/w ajoene & cream)	A	Ledezma et al. 1996
Torulopsis glabrata	1:1,024 (garlic extract)	A	Moore & Atkins, 1976
Trichophyton spp.	1:5,000 (garlic extract)	A	Appleton & Tansey, 1975

A) No data from the study.

B) Interferes with cell metabolism by: 1) inactivation of proteins by oxidation of essential thiols to a disulfide. 2) Competitive inhibition of -SH compounds (cysteine or glutathione). 3) Noncompetitive inhibition of enzyme function by oxidation of the binding to the –SH groups at the active site.

C) Severe damage to hyphae, thickening/morphological changes of cell wall and destruction of organelles.

D) Antimicrobial effects may result from the presence of both the disulfide bond and the sulfinyl group.

E) Ajoene perturbs the cell membrane leading to cell lysis, which probably affects lipids in the bilayer.

F) Ajoene induces alterations in phospholipid and fatty acid moieties. Saturated fatty acids are reduced in Y phase. Blocks thermal induced transition from M to Y phase and leads to decay of fungal structures.

A. Allicin

The antimicrobial activity of garlic (*Allium sativum L*) was reported by Cavallito (1944) and the active component diallylthiosulfinate was named 'allicin'. Stoll et al. (1951) confirmed that the allicin is derived from the alliin-alliinase system. Dankert et al. (1979) examined the crude juices of garlic in an agar diffusion test for their growth inhibitory effect on five gram negative and three gram positive bacterial species and two yeast species. All test organisms were inhibited by garlic juice. Addition of complex-forming agents and organic matter to the crude juice reduced its activity on all test organisms. Volatile substances showed a strong inhibitory activity after exposure for 8 hours or longer at 23°C or 37°C. Minimal inhibition concentrations (MIC) determined in a dilution test were found to be high for gram negative bacteria and low for both yeast species. The D-values of different test organisms in undiluted garlic juice were calculated. *Pseudomonas aeruginosa* had a very low D-value, whilst the bacteriostatic concentration was high. This indicates a large concentration exponent of crude garlic juice for this organism. The opposite was found for *Staphylococcus aureus*. The antimicrobial activity of garlic extract on the oral flora of volunteers was investigated by Elnima et al. (1983). A mouthwash containing 10% garlic in quarter Ringer solution elicited a significant reduction in the number of oral bacteria.

The effect of bacteriostatic concentrations of allicin (0.2 to 0.5 mM) on the growth of *Salmonella typhimurium* revealed a pattern of inhibition characterized by: (i) a lag of approximately 15 min between addition of allicin and onset of inhibition, (ii) a transitory inhibition phase whose duration was proportional to allicin concentration and inversely proportional to culture density, (iii) a resumed growth phase which showed a lower rate of growth than in uninhibited controls, and (iv) an entry into stationary phase at a lower culture density (*FIGURE 3*). Whereas DNA and protein syntheses showed a delayed and partial inhibition by allicin, inhibition of RNA synthesis was immediate and total, suggesting that this is the primary target of allicin action (Feldberg et al., 1988).

Barone and Tansey (1977) demonstrated that serial dilutions of synthesized allicin showed complete fungistasis in shake cultures of *C. albicans* at allicin dilutions of 1×10^{-4}.

The aqueous extract of garlic and allicin both show a potent *in vitro* antibacterial activity against isolates of multiple drug-resistant *Shigella dysenteriae 1, Shigella flexneri Y, Shigella sonnei* and enterotoxigenic *E. coli* (Chowdhury et al., 1991). The minimum inhibitory concentrations of the aqueous extract and allicin against *Shigella flexneri* were 5 and 0.4 µl/ml, respectively. The two agents also showed potent *in vivo* antibacter-

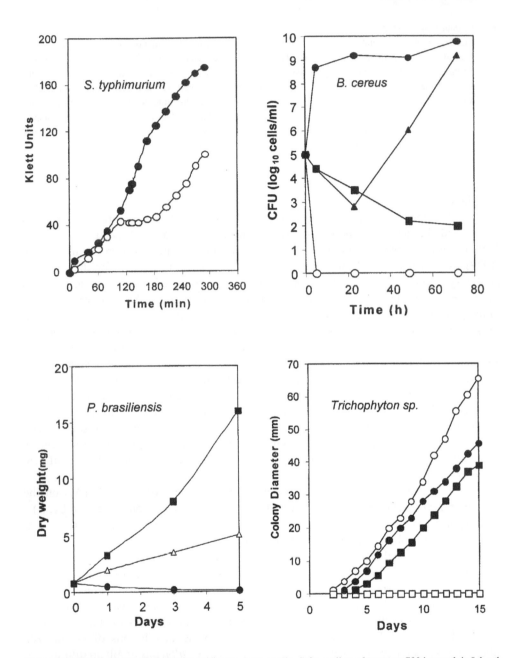

FIGURE 3. **Antimicrobial effects of thiosulfinates from garlic.** *Salmonella typhimurium* 7004 growth in L broth at 37 °C with no treatment [●] or with allicin [○] added at 91 min to a final concentration of 0.3 mM (redrawn from Feldberg et al., 1988). *Bacillus cereus* growth in the presence of ajoene, ● = control; ▲ = 10 µg/ml ajoene; ■ = 20 µg/ml ajoene; ○ = 30 µg/ml ajoene (redrawn from Naganawa et al., 1996). *Trichophyton mentagrophytes* radial growth on SG medium in the presence of allicin, ○ = control; ● = 0.78 µg/ml allicin; ■ = 1.57 µg/ml allicin; □ = 3.13 µg/ml allicin (redrawn from Yamada & Azuma, 1977). *Paracoccidioides brasiliensis* growth in the presence of increasing concentrations of ajoene, ■ = Control; ▲ = 50 µM ajoene; ● = 100 µM ajoene (redrawn from San-Blas et al., 1989).

ial activity against experimental shigellosis in a rabbit model. Oral administration of the two agents completely cured the infected rabbits within 3 days. On the contrary, 4 of the 5 rabbits in the control group died within 48 h after challenge. The experimental groups were pathogen-free on the second day of treatment. The antibacterial activity against the challenge strain was observed in the sera of the treated rabbits with 30-60 min of administration of the agents. The LD_{50} values of the aqueous extract and allicin in mice were 174 ml/kg and 204 µl/kg of body weight, respectively. At the therapeutic dose, the two agents did not show any adverse effects on the standard biochemical profile of blood.

Using direct pre-infection incubation assays, Weber et al. (1992) reported the *in vitro* virucidal effects of fresh garlic extract, its polar fraction, and the following garlic associated compounds: diallyl thiosulfinate (allicin), allyl methyl thiosulfinate, methyl allyl thiosulfinate, ajoene, alliin, deoxyalliin, diallyl disulfide, and diallyl trisulfide. Activity was determined against selected viruses including, herpes simplex virus type 1, herpes simplex virus type 2, parainfluenza virus type 3, vaccinia virus, vesicular stomatitis virus, and human rhinovirus type 2. The order for virucidal activity generally was: ajoene > allicin > allyl methyl thiosulfinate > methyl allyl thiosulfinate. Ajoene was found in oil-macerates of garlic but not in fresh garlic extracts. No activity was found for the garlic polar fraction, alliin, deoxyalliin, diallyl disulfide, or diallyl trisulfide. Fresh garlic extract, in which thiosulfinates appeared to be the active components, was virucidal to each virus tested. The predominant thiosulfinate in fresh garlic extract was allicin. Lack of reduction in yields of infectious virus indicated undetectable levels of intracellular antiviral activity for either allicin or fresh garlic extract. Furthermore concentrations that were virucidal were also toxic to HeLa and Vero cells. Virucidal assay results were not influenced by cytotoxicity since the compounds were diluted below toxic levels prior to assaying for infectious virus. These results indicate that virucidal activity and cytotoxicity may have depended upon the viral envelope and cell membrane, respectively. However, the authors concluded that the activity against non-enveloped virus may have been due to inhibition of viral adsorption or penetration.

Diallyl trisulfide is a chemically stable final transformation product of allicin was synthesized in 1981 in China and used for treatment of bacterial, fungal and parasitic infections in humans. Lun et al. (1994) investigated the activity of diallyl trisulfide in several important protozoan parasites *in vitro*. The IC_{50} (concentration which inhibits metabolism or growth of parasites by 50%) for *Trypanosoma brucei brucei*, *T.b. rhodesiense*, *T.b. gambiense*, *T. evansi*, *T. congolense* and *T. equiperdum* was in the range of 0.8-5.5 µg/ml. IC_{50} values were 59 µg/ml for *Entamoeba histolytica* and 14 µg/ml for *Giardia lamblia*. The cytotoxicity of the compound was evaluated on two fibroblast cell lines (MASEF, *Mastomys natalensis* embryo fibroblast and HEFL-12, human embryo fibroblast) *in vitro*. The maximum tolerated concentration for both cell lines was 25 µg/ml. These results indicated that diallyl trisulfide has potential to be used for treatment of several human and animal parasitic diseases. In a recent study, allicin was shown to inhibit the ability of *Entamoeba histolytica* trophozoites to destroy monolayers of baby hamster kidney cells (Ankri et al., 1997). Allicin has strongly inhibited cysteine proteinases, an important contributor to amoebic virulence, as well as alcohol dehydrogenase system of the parasite.

Mechanism of action: Barone and Tansey (1977) proposed that garlic and allicin interfered with *C. albicans* cell metabolism by inactivation of proteins, competitive inhibition of sulfhydryl compounds, or by noncompetitive inhibition of enzyme function by oxidation. It is hypothesized that at the cidal or static levels allicin disrupts cell metabolism in *Candida* by inactivating proteins by the oxidation of essential thiols to a disulfide. This competitively inhibits the activity of sulfhydryl compounds by interaction with glutathione or cysteine. Noncompetitive inhibition of enzyme function is from the oxidation binding to the –SH groups at the enzyme's allosteric sites. *Candida* cells in the yeast phase are more pathogenic than the mycelial form. Allicin may interfere with the electron flow through the disulfhydryl reductase system and antagonize the reductase function by oxidation of the sulfhydryl groups within the *Candida* cell wall. In addition, direct interference with the reductase molecule may occur. The actions may uncouple cell division from cell metabolism, resulting in an increase of mycelial forms of *Candida*. This effect of allicin on yeast division may decrease virulence of *Candida* by enhancing the conversion of the yeast form to the mycelial form (Barone & Tansey, 1977). Yamada and Azuma (1977) found allicin inhibits swelling and germination of spores at concentrations of 3.13 µg/ml or more. Normal hyphae were not observed after allicin administration in *Candida albicans, Cryptococcus neoformans, Aspergillis fumigatus, Trichophyton mentagrophytes, T. ferrugineum, T. rubrum, Microsporum gypseum,* and *Epidermophyton floccosum.* The MIC's for *Candida albicans, Cryptococcus neoformans, Trichophyton mentagrophytes, T. ferrugineum, T. rubrum, Microsporum gypseum, Epidermophyton floccosum* were as low as 3.13 to6.25 µg/ml on SG agar media and 1.57-6.25 µg/ml in SG broth medium, even after 5 to 15 days after incubation (*FIGURE 3*). The MIC's for *Aspergillis fumigatus* was 12.5 to 25 µg/ml (both broth and agar) after 5 days of incubation. It appears that allicin interferes with normal hyphae production.

Allicin may affect cellular replication involving DNA and or RNA synthesis. Cellular lesions or depletion of nutrients may reset the initiator protein mass/DNA ratio required for cellular replication, in addition to being a reversible inhibitor of RNA synthesis in the cells. It is possible that allicin may affect RNA polymerase, and or possibly inhibit mRNA degradation as well as RNA synthesis (Feldberg et al., 1988).

Ankri et al. (1997) determined that allicin inhibits the ability of *Entamoeba histolytica* trophozoites to destroy monolayers of baby hamster kidney cells. Ankri also found that cysteine proteinases (a major contributor to the amoebic virulence) as well as alcohol dehydrogenase are strongly inhibited by allicin.

B. Ajoene

Yoshida et al. (1987) reported the antifungal activity of six fractions derived from garlic in an *in vitro* system. Ajoene had the strongest activity in these fractions. The growth of both *Aspergillus niger* and *C. albicans* was inhibited by ajoene at less than 20 µg/ml. San-Blas et al. Demonstrated ajoene mediated inhibition of *Paracoccidioides brasiliensis* growth by 90% in the Y phase and 60% in the M form with just 50 mM of ajoene, and at 100 mM cell membrane thickened to an average of 60nm (control cells were 12 nm thick). In addition, cell membrane lysis was observed with 200 µM ajoene. During the M phase, 100 µM ajoene showed a moderate thickening of the membrane (20 nm) as well as some discontinuity and detachment of the membrane from the cell wall (*FIGURE 3; FIGURE 4A & 4B*).

Naganawa et al. (1996) studied the effect of ajoene and DAD effects on fungi and bacteria. Results indicated that the anti-microbial effects may result due to the presence of disulfide bond and sulfinyl group in each compound. Fungicidal effects were found against *Candida albicans, Hanseniaspora valbyensis, Pichia anomala, Schizosaccharomyces pombe, and Saccharomyces cerevisiae*, and bactericidal effects were observed against *Bacillus cereus, Bacillus subtilis, Staphylococcus aureus, Mycobacterium smegmatis, Mycobacterium phlei, Micrococcus luteus, Lactobacillus plantarum, Streptococcus sp., Streptomyces griseus, Escherichia coli, Klebsiella pneumoniae, Xanthomanas maltophilia, and Pseudomonas aeruginosa*. The minimal microbicidal concentrations (MMC) for the bacteria were found to be: *Bacillus cereus* (300 µg/ml) (*Figure 3*), *Bacillus subtilis* (>500 µg/ml) *Staphylococcus aureus* (400 µg/ml*)*, *Mycobacterium smegmatis* (>500 µg/ml*)*, *Mycobacterium phlei* (>500 µg/ml), *Micrococcus luteus* (>500 µg/ml) , *Lactobacillus plantarum (not determined), Streptococcus sp.* (> 500 µg/ml), *Streptomyces griseus* (>500 µg/ml), *Escherichia coli* (400 µg/ml*)*, *Klebsiella pneumoniae* (> 500 µg/ml), *Xanthomanas maltophilia* (> 500 µg/ml) *Pseudomonas aeruginosa* (no inhibition found up to 500 µg/ml concentration).

The MMC for yeast was found to be >500 µg/ml for *Candida albicans, Hanseniaspora valbyensis* (400 µg/ml*)*, *Pichia anomala* (>500 µg/ml), *Schizosaccharomyces pombe* (300 µg/ml), *Saccharomyces cerevisiae* (300 µg/ml). The effective mixture of ajoene consisted of a solution of 80% Z-ajoene and 20% E-ajoene.

Ajoene also exhibits a broad-spectrum antimicrobial activity (Naganawa et al., 1996). Growth of gram-positive bacteria, such as *Bacillus cereus, Bacillus subtilis, Mycobacterium smegmatis*, and *Streptomyces griseus* were inhibited at 5 µg/ml of ajoene. *S. aureus* and *Lactobacillus plantarum* also were inhibited below 20 µg/ml of ajoene. For gram-negative bacteria, such as *E. coli, Klebsiella pneumoniae,* and *Xanthomonas maltophilia*, MICs were between 100 and 160 µg/ml. Ajoene also inhibited yeast growth at concentrations below 20 µg/ml. The microbicidal effect of ajoene on growing cells was observed at slightly higher concentrations than the corresponding MICs. *B. cereus* and *Saccharomyces cerevisiae* (10^5 cfu/ml) were killed at 30 µg/ml of ajoene in 24 h. However, the MIC for resting cells were at 10 to 100 times higher. The disulfide bond in ajoene appears to be necessary for the antimicrobial activity of ajoene, since reduction by cysteine, which reacts with disulfide bonds, abolished its antimicrobial activity.

Recently, Yoshida et al. (1998) isolated a compound showing antimicrobial activity from an oil-macerated garlic extract by silica gel column chromatography and preparative TLC. On basis of the results of NMR and MS analyses, the compound was identified as Z-4,5,9-trithiadeca-1,6-diene-9-oxide (Z-10-devinylajoene; Z-10-DA). Z-10-DA exhibited a broad spectrum of antimicrobial activity against gram-positive and gram-negative bacteria as well as yeasts. The antimicrobial activity of Z-10-DA was comparable to that of Z-ajoene, but was superior to that of E-ajoene. Z-10-DA and Z-ajoene are different in respect of substitution of the allyl group by the methyl group flanking a sulfinyl group. This result suggests that substitution by the methyl group would also be effective for the inhibition of microbial growth. Further studies have shown newly identified organosulfur compound from oil-macerated garlic extract and its antimicrobial effect against *B. cereus, B. subtilis, S. aureus,* and yeast at concentrations less than 100 µg/ml, but not against gram negative bacteria. From the results of NMR, IR and MS analyses, the structure was determined to be E-4,5,9-trithiadeca-1,7-diene-9-oxide (iso-E-10-devinyl

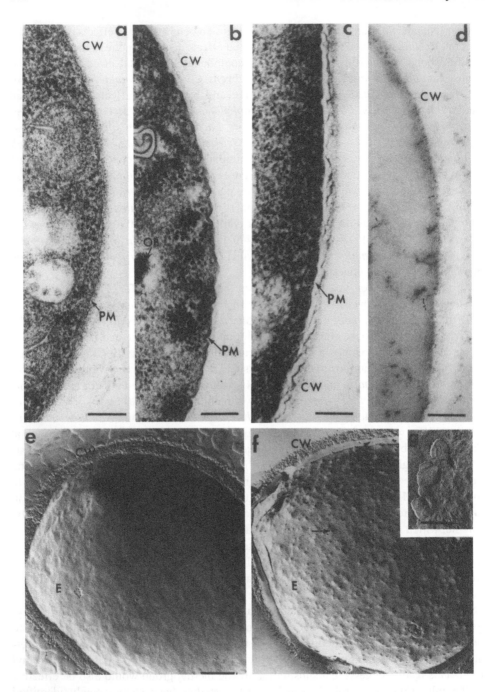

FIGURE 4A. Electron microscopy of *P. brasiliensis* Y Phase. Thin sections: (a) control, (b) 50 μM ajoene, (c) 100 μM ajoene (bar, 200 nm). Observe the thickening and disturbance of the cell membrane with increasing amounts of ajoene. Freeze-itching: (e) control, (f) 100 μM ajoene (bar, 600 nm). An irregular fracture plane is evident in the presence of ajoene. Abbreviations: CW, cell wall; PM, cytoplasmic membrane; PS, periplasmic space; OB, osmiophilic body; E, E face of the membrane (Reproduced with permission from the American Society for Microbiology; San-Blas et al., 1989).

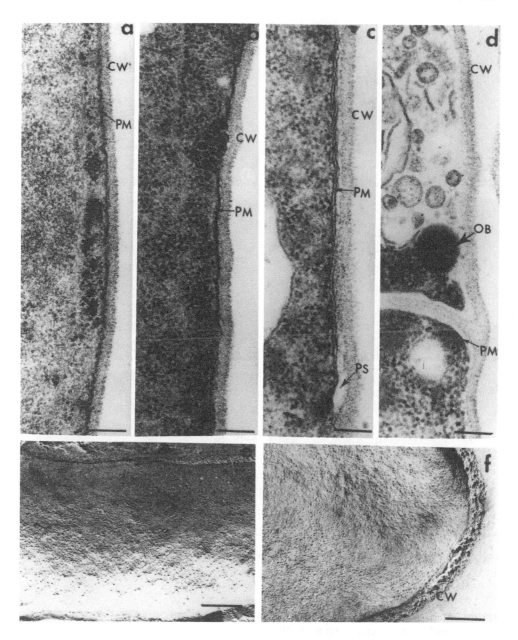

FIGURE 4B. Electron microscopy of *P. brasiliensis* M Phase. Thin sections: (a) control, (b) 50 μM ajoene, (c) 100 μM ajoene, (d) 200 μM ajoene. Thickening of the membrane is present at an ajoene concentration of 200 μM. Freeze-itching: (e) control, (f) 100 μM ajoene (bar, 200 nm). No important changes are observed in the membrane structure. Abbrebiations are as in Figure 4A. (Reproduced with permission from the American Society for Microbiology; San-Blas et al., 1989).

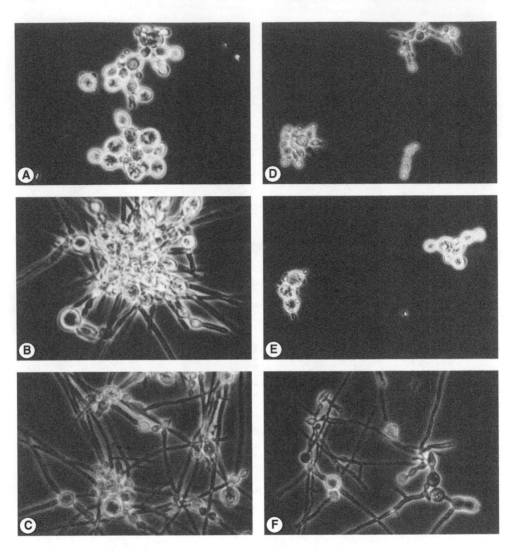

FIGURE 5A. Effect of ajoene on Y-to-M dimorphism of *P. brasiliensis.* (a) - (c) Control cultures transforming at 0, 24, and 48 h, respectively. (d) - (f) Culture left to transform in the presence of ajoene, at 0, 24 and 48 h, respectively. Bar = 25 μm. (Reproduced with permission from San-Blas et al., 1993).

ajoene, iso-E-10-DA). This organosulfur compound was found to be different than E-4,5,9-trithiadeca-1,6-diene-9-oxide (E-10-devinylajoene, E-10-DA). (The position of the double bond in the seventh position instead of the sixth position is the difference.) Yoshida et al. (1999), determined that compound is inferior to E-ajoene, Z-ajoene and Z-10-DA, and suggested that the position of the double bond reduced the antimicrobial activity.

  Mechanism of action: San-Blas and co-workers (1997) reported that ajoene could induce alterations in phospholipid and fatty acid moieties, reducing phosphatidyl-choline (PC) in both Y and phases of *Paracoccidioides brasiliensis*. PE increased to 38% in the yeast (Y) phase and 44% in the mycelial (M) phase, suggesting and inhibition of

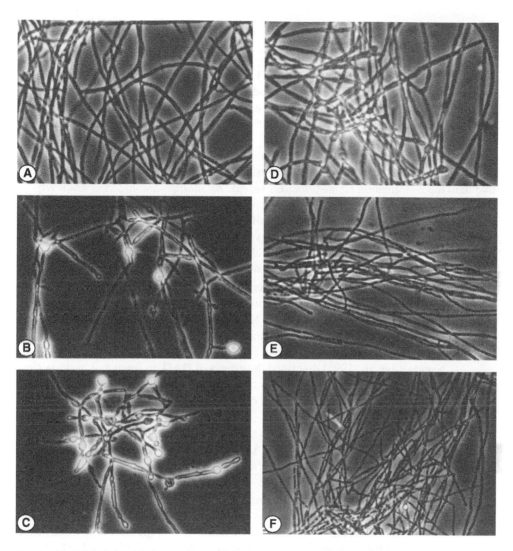

FIGURE 5B. Effect of ajoene on M-to-Y dimorphism of *P. brasiliensis.* Descriptions as in *FIGURE 5A.* (Reproduced with permission from San-Blas et al., 1993).

PC synthesis. The saturated fatty acids 16:0 and 18:0 were reduced in the Y phase from 67 to 35% with increase in unsaturated components. This phenomenon was not observed in the M phase. Ajoene also seems to block the thermally induced transition from M to Y phase, and disturb the membrane, leading to deterioration in the fungal structures (*FIGURE 5A & 5B*). In earlier studies, San-Blas et al. (1989) proposed that ajoene could cause perturbations in the cell membrane leading to cellular lysis and subsequent deterioration of all fungal structures. In addition, ajoene was taken up by the fungus, and seemed to affect the lipids in the bilayer. In eucaryotic studies with *Trypanosoma cruzi*, Urbina et al. (1997)

also indicated that ajoene altered the phospholipid composition of *T. cruzi* epimastigotes and amastigotes.

Yoshida et al. (1987) observed hyphae damage in *Aspergillis niger* (ATCC 16404) and *Candida albicans* (ATCC 10231) as well as damage to the cellular organelles. These morphological changes to the cell and destruction to the organelles suggests that ajoene has a multifactorial mechanism rather than a single mechanism in the cell. Ajoene was found to be more effective in inhibiting fungi growth than allicin at the concentration of 20 µg/ml.

Trans configuration or position of double bond reduces antimicrobial activity of garlic. Yoshida et al. (1999) found E-4.5.9-trithiadeca-1,7-diene-9-oxide to be less effective than the Z- formation against *Bacillus cereus, B. subtilis, Staphylococcus aureus*.

C. Diallyl disulfide (DADS) and diallyl sulfide (DAS)

DADS and DAS could decrease arylamine N-transferase (NAT) activity in bacteria. Chen (1999) reported a decreased NAT activity in *Klebsiella pneumoniae* with increased levels of DADS and DAS, decreasing K_m and V_{max} of NAT. Chung (1998) also reported dose-dependant effects with DAS or DADS treatment with viability studies and decreased K_m and V_{max} of arylamine NAT enzymes of *H. pylori*.

Klebsiella pneumoniae growth and NAT activities were decreased with increased levels of DAS or DADS. This was demonstrated in growth studies and cytosol examinations (Chen et al., 1999).

Naganawa.et al. (1996) described the effects of DAD on fungi and bacteria similar to that of ajoene (see above in ajoene section). Nok et al. (1996) found DADS inhibits procylic forms of *Trypanosoma brucei brucei* as well as inhibiting phospholipases from *T. conglense, T. vivax, and T. brucei brucei.*

VI. BIOLOGICAL ADVANTAGE

A. Additive advantage in foods

The U.S. Food and Drug Administration (FDA) has already approved DADS for food use and it has GRAS status (by the Flavor and Extracts Manufacturers Association) for use as a synthetic flavoring substance. The Council of Europe lists it as a substance that can be added to food without being a hazard to the public's health (Ford, 1988).

Paludan-Muller (1999) suggested that garlic has a dual role in Thai low-salt fermented fish products: 1) as a substrate for fermentation, and 2) as an inhibitor of gram-negative bacteria and yeasts. Presence and proliferation of LAB proved essential for rapid pH reduction. Garlic provided a more important role than rice starch during fermentation in 'som-fak', when a garlic-fermenting strain of *L. plantarum* was identified.

B. Physiological advantage to host

Garlic has attained a firm place in folk medicine for centuries. In addition to antimicrobial properties, garlic could elicit multifunctional effects to benefit human health. Garlic is capable of lowering blood cholesterol and reducing secondary vascular changes. It also raises fibrinolytic activity and inhibits thrombocyte aggregation. Therefore, garlic contains highly active therapeutic principles which appear to be particularly suitable for prophylaxis of arteriosclerosis (Ernest, 1981).

Allicin inhibits human platelet aggregation *in vitro* without affecting cyclooxygenase or thromboxane synthase activity or cyclic adenosine monophosphate levels (Mayeux et al., 1988). Allicin does not alter the activity of vascular prostacyclin synthase. However, it inhibits ionophore A23187-stimulated human neutrophil lysosomal enzyme release. *In vivo*, allicin dilates the mesenteric circulation independent of prostaglandin release or a beta-adrenergic mechanism.

Garlic has been claimed to be effective against diseases, in the pathophysiology of which oxygen free radicals (OFRs) have been implicated. Effectiveness of garlic could be due to its ability to scavenge OFRs. Prasad et al. (1995) investigated the ability of allicin contained in the commercial preparation 'garlicin' to scavenge hydroxyl radicals (•OH) using high pressure liquid chromatographic (HPLC) method. The •OH radical was generated by photolysis of H_2O_2 (1.25-10 µmoles/ml) with ultraviolet light and was trapped with salicylic acid which is hydroxylated to produce •OH adduct products 2,3- and 2,5-dihydroxybenzoic acid (DHBA). H_2O_2 produced a concentration-dependent •OH as estimated by •OH adduct products 2,3-DHBA and 2,5-DHBA. Allicin equivalent in 'Garlicin' (1.8, 3.6, 7.2, 14.4, 21.6, 28.8 and 36 µg) produced concentration-dependent decreases in the formation of 2,3-DHBA and 2,5-DHBA. The inhibition of formation of 2,3-DHBA and 2,5-DHBA with 1.8 µg/ml was 32.36% and 43.2% respectively while with 36.0 µg/ml the inhibition was approximately 94.0% and 90.0% respectively. The decrease in •OH adduct products was due to scavenging of •OH and not by scavenging of formed •OH adduct products. Allicin prevented the lipid peroxidation of liver homogenate in a concentration-dependent manner. These results suggest that allicin scavenges •OH and 'garlicin' has antioxidant activity.

Chen et al. (1999) demonstrated the decrease in arylamine N-acetyltransferase (NAT) activity in *Klebsiella pneumonae* as well as showing that DADS and DAS inhibited NAT activity in human colon tumor cells (adenocarcinoma).

Zheng et al. (1997) recently reported the inhibitory effects of allicin on proliferation of tumor cells. The effect was associated with the cell cycle blockage of S/G2M boundary phase and induction of apoptosis.

You et al. (1998), in their epidemiological studies found garlic consumption to be protective against *Helicobacter pylori* and precancerous lesions.

Gallwitz et al. (1999) proposed that the anti-parasitic and cytostatic actions of ajoene may be partially due to the many different effect of enzymes such as human glutathione reductase(GR) in the antioxidant thiol metabolism. It was found that ajoene is a covalent inhibitor and a substrate of GR. A crystal resolution structure of GR inhibited by ajoene demonstrated a mixed disulfide in the active site. This modified enzyme increased oxidase activity as compared to the free GR. This leads to the formation of single-electron reduced products resulting in superoxide anion radicals, which could damage the cells by oxidative damage.

Intraperitoneal administration of radioactive labeled DADS in mice indicated a rapid uptake of the compound by liver during the first 30 minutes and highest after 90 minutes. Liver cytosol contained more than 70% of the radioactivity after 2 hours, and 80% was found as sulfates and 8% as unchanged DADS. (Pushpendran et al., 1980).

VII. APPLICATIONS

Garlic is firmly entrenched in folk medicine. Garlic has been used, at one time or another to treat problems such as asthma, coughs, heart problems, regulation of blood pressure, as a cholesterol lowering agent, for increased circulation, impotence, diabetes, cancer, lung disorders (breathing in a combination of water, vinegar and garlic steam), epilepsy, dropsy (in a poultice), hysteria (held under the nose and sniffed), leprosy, smallpox (in a poultice), vermiferge (ingested orally), rheumatism, healing cuts (placing crushed garlic in the wound), baldness (in wine for a topical lotion), and as a digestion aid. Garlic is also widely used as a flavoring agent in cooking, and is used in the cuisine of most cultures.

A. Food preservative

Naganawa et al. (1996) suggested a potential role for ajoene as an antimicrobial food preservative based on studies with *B. cereus*. A concentration of 100 μg/ml of ajoene was required to inhibit growth compared to 1 mg/ml of sorbic acid.

An alternative method to preserve raw buffalo milk was reported by Jandal (1998) using lactoperoxidase, garlic extract and ethanol. This system uses the oxidation of sulfur compounds in garlic by oxygen released from the ethanol in the presence of lactoperoxidase. The oxidation products inhibit the spoilage organisms by affecting the respiratory system. Garlic juice in a 2-10% solution was effective in eliminating coliform bacteria in the first 7 days of 'kimchi' (fermented cabbage) fermentation (Chung et al., 1997).

Mielnik (1997) conducted a study to assess the use of fresh or dried garlic as an antioxidant in frozen, de-boned chicken meat. TBA increase was more in fresh garlic than in dried garlic, and increased in all samples during frozen storage, but increases were smaller in the garlic group. In addition to the antioxidant properties, the added garlic reduced the sensory rancidity of the meat during storage. Use of garlic in mechanically de-boned meat was suggested in products where flavor and aroma are acceptable.

B. Antibacterial agent

Thiosulfinates from garlic are effective antibacterial agents against a variety of pathogens including: *Salmonella, Listeria, E. coli, B. cereus, S. aureus, Streptococcus, Pseudomonas aeruginosa* and *H. pylori* cause food-borne illnesses. Garlic has also been shown to inhibit various strains of food spoilage organisms (Chung, 1998; Feldberg, 1988; Firouzi, 1998; Naganawa, 1996; Sivam, 1997; Yoshida, 1999). Perhaps garlic should be considered as both a food and a non-restricted preservative.

C. Antifungal agent

Moore and Atkins (1976) found that *C. albicans* was inhibited and killed in garlic extract diluted at 1:1024, at 37°C. It appears that garlic extract is effective at body temperature and may have potential as a medical alternative to Nystatin. Barone and Tansey (1977) also observed fungistatic conditions at 50, 150, and 300 μg/ml and fungicidal at 400 μg/ml against *C. albicans*. Shen et al. (1996) showed that a Chinese preparation, allitridium (diallyl trisulfide), in conjunction with amphotericin B has a synergistic effect *in vitro* for fungal infections.

Ledzema et al. (1996) demonstrated the efficacy of ajoene to treat short-term therapy of tinea pedia (Athlete's foot). After 7 days of treatment, 27 of 34 patients were shown to have complete clinical and mycological cure in seven days. The remainder of the patients (seven) was cured after 7 additional days of treatment. All patients evaluated after 90 days after treatment yielded negative cultures for mycological growth. In the People's Republic of China, commercial *Allium sativum* products are used as antifungal drugs to treat systemic fungal infections (Shen, et al. 1996).

D. Antiviral agent

Weber et al. (1992) reported *in vitro* virucidal effects of fresh garlic extract, its polar fraction, and the following garlic associated compounds: diallyl thiosulfinate (allicin), allyl methyl thiosulfinate, methyl allyl thiosulfinate, ajoene, alliin, deoxyalliin, diallyl disulfide, and diallyl trisulfide. Activity was determined against selected viruses including, herpes simplex virus type 1, herpes simplex virus type 2, parainfluenza virus type 3, vaccinia virus, vesicular stomatitis virus, and human rhinovirus type 2. The order for virucidal activity generally was: ajoene > allicin > allyl methyl thiosulfinate > methyl allyl thiosulfinate. Ajoene was found in oil-macerates of garlic but not in fresh garlic extracts. No activity was found for the garlic polar fraction, alliin, deoxyalliin, diallyl disulfide, or diallyl trisulfide. Fresh garlic extract, in which thiosulfinates appeared to be the active components, was virucidal to each virus tested. The predominant thiosulfinate in fresh garlic extract was allicin. Results of this study indicate ajoene may be evaluated as a possible virucide against herpes simplex virus type 1, herpes simplex virus type 2, parainfluenza virus type 3, vaccinia virus, vesicular stomatitis virus, and human rhinovirus type 2.

E. Antiprotozoal agent

Garlic and garlic derivatives have been shown to elicit anti-protozoal activity (Ankri 1997, Gallwitz, 1999; Nok, 1996; Urbina, 1997) in particular, against *Entamoeba histolytica* and *Trypanosoma cruzi, T. brucei, T. congolense, T. vivax,* and *Trypanosoma cruzi* epimastigotes and amastigotes.

F. Limitations

Storage of the garlic extract is a definite limitation. Many studies have shown that garlic extract loses its antimicrobial effects during storage. These losses vary as to the temperature as well as the length of time of storage.

Many dried garlic preparations contain only alliin, and no allicin or ajoene. Alliin must be converted to allicin, which requires alliinase. Stomach acid denatures this enzyme, rendering it ineffective. Thus, slight to no conversion to allicin or ajoene takes place in the stomach. Fresh whole garlic allows the conversion to take place while chewing, and possibly holding it in the mouth for a longer period of time would allow the conversion to be more efficient (as well as noticeably more flavorful). Even enteric-coated garlic may not convert to the active forms in the small intestine. Allicin created in the small intestine will invariably react with cysteine, if protein-containing foods are consumed with garlic. S-allylmer captocysteine binds to allicin, preventing uptake into the bloodstream (Tyler, 1993). Block (1985) failed to detect ajoene in garlic powder, pills, oils, or extracts. It was concluded that this was most likely due to steam distillation disintegration of the compound.

Commercially peeled cloves were found to have very high numbers of microbes as compared with freshly peeled cloves. These microbes include *Leuconostoc mesenteroides*, *L. dextranicum*, *Enterobacter intermedium*, *E. sakazakii*, *Serratia liquefaciens*, *Kluyvera cryocrescens*, and *Cryptococcus neoformans* per gram (Shim, 1999). Park et al. (1998) showed dry peeled garlic and root removal reduced microbial load more effectively than a wet peeling process. Washing also reduced microbial counts, and repeated washings with water at 5°C further reduced the contaminant load.

Garlic flavor may be a problem to some persons sensitive to strong flavors. Studies done with garlic show that a flavored solution of garlic in both mouth and overall were stronger with water, intermediate with 0.1% xanthan, and the lowest with 0.3% guar gum. It was found that it lowered the flavor release by 50% (Yven, 1998). It has not yet been shown if the antimicrobial or antioxidant properties are still active in the guar gum and garlic solution.

VIII. SAFETY AND TOLERANCE

Studies done with oral doses of garlic show that the antifungal activity decreases in serum after 60 minutes after first presence of antifungal activity. Urine did not appear to inhibit the antifungal activity of garlic extract. These findings suggest that some factors in serum and urine may cause degradation of garlic's active ingredients in those systems. It is possible that high serum protein (or possibly protein containing high amounts of sulfur or sulfur-containing amino acids) may bind allicin or it may inactivate it. The drop in antifungal activity in serum may be due to the high pH of the serum in the absence of CO_2. When those plates were incubated in 5% CO_2 the antifungal activity of the extract in the serum was noted to be slightly higher. Caporaso and co-workers (1983) also reported that the maximum tolerable dose in humans is 25 ml of the filtered extract (100 g garlic) homogenized in 10 ml of distilled water. Data from the National Toxicological Program indicates no reported toxicity in humans (Dausch & Nixon, 1990).

After ingestion of fresh garlic in a maximally tolerable dose to five volunteer patients, it was apparent that it caused significant discomfort to the oral and gastric mucosa (Caporaso et al., 1984). This could limit the effectiveness of raw garlic extract as a prophylactic substance. Possibilities may exist for placing fresh garlic in a capsule form, however gastric mucosa may still be affected.

A case of partial thickness burns from a garlic-petroleum jelly plaster was reported on a child, that was applied at the direction of a naturopath physician (Parish et al., 1987).

As with any protein product, allergy and hypersensitivity reactions may occur. Contact dermatitis caused by occupational contact with raw garlic has been documented. Jappe (1999), in a review of literature and in a case report of a cook who contracted garlic induced contact dermatitis, determined the garlic related adverse effects to be irritant contact dermatitis, with a rare variant of zosteriform dermatitis, allergic contact dermatitis (which includes the hematogenic variant) as well as combination reactions. Eight patients in Hong Kong developed contact dermatitis after rubbing the cut end of fresh garlic into the skin to treat infections of the groin, neck, lower limb, hands or face. The patients were treated with a fluorinated topical steroid. All treatments were successful (Lee & Lam, 1991).

Contact sensitivity of garlic and DADS of 155 patients in a patch-test showed 5.2% of all patients reacted to garlic. It was confirmed that diallyl disulfide was the sensitizing agent in the garlic (Lembo et al., 1991). It has been suggested by Brenner and Wolf (1994) that DADS and other compounds may possibly cause a type of relapsing chronic skin diseases known as pemphigus.

A. Animal studies

San-Blas et al. (1989) described a lack of toxicity of ajoene in mice and dogs—at levels up to 25 mg/kg. DADS toxicity in laboratory animals was determined by Ford et al. (1988) to be 0.26 g/kg (the acute oral LD_{50} in rats), and 3.6 g/kg (the acute dermal LD_{50} in rabbits). The undiluted solution produced skin abnormalities and irritant effects in the rabbit patch test at doses of 2.5, 3.75, or 5.0 g/kg for a span of 24 hours.

The effect of an oily extract, possibly DADS from garlic was shown to suppress the ability of the *Trypanosoma brucei* parasites to infect the host in mice studies. When administered to experimentally infected mice at a dosage of 120 mg/kg per day, the trypanosomiasis subsided in 4 days. The extract also showed inhibition of the procyclic forms of *Trypanosoma brucei* and phosholipases from *T. congolense* and *T. vivax*. It was postulated that DADS interferes with the membrane synthesis of the parasite.

B. Clinical trials

After ingestion of fresh garlic in a maximally tolerable dose, it was apparent that it caused significant discomfort to the oral and gastric mucosa of the volunteer patients in the study (Caporaso et al., 1984).

Adverse effects shown in Davis et al. (1990) was similar to those from eating fresh garlic. Fresh garlic effects include vomiting, diarrhea, anorexia, flatulence, weight loss, or garlicky body odor. There were no cases of anaphylaxis. One patient experienced abdominal discomfort and nausea, which depended on rate of drug delivery. The patient had symptoms at 30 mg/hour, but not at 15 mg/hour. Twenty five percent of the patients had abdominal discomfort, nausea and thrombophlebitis at the site of intravenous administration.

IX. BIOTECHNOLOGY

A. Large-scale production

Currently, many companies are marketing garlic extract in various forms. During large-scale production many of the thiosulfinates seem to lose their antimicrobial activity depending on the processing conditions (Block, 1985; Barone & Tansey, 1977; Caporaso et al., 1984; Tyler, 1993).

Synthetic allicin has been made by a method involving the oxidation of diallyl disulfide by an equimolar amount of organic per-acid. The steps for synthetic preparation of the extract may be found under the *Section: Molecular properties*, of this chapter. Stoll and Seebeck (1948) found that attaching an allyl group and an oxygen atom to the sulfur atom in the amino acid cysteine forms alliin.

B. Quality control

Processing and storage conditions of commercial preparations of garlic extract may compromise the antimicrobial activity of garlic products. A commercial preparation of garlic extract was assayed and no antifungal activity was detected for either yeast or mycelial fungi (Caporaso et al.,1983). Ajoene or allicin could not be detected in dehydrated garlic powder, pills, oils, extracts or other proprietary garlic preparations (Block, 1985; Barone & Tansey, 1977; Caporaso et al., 1983).

Krest and Keusgen (1999) showed two molecular forms of alliinase (53 and 54 kDa) from garlic powder, in contrast to two identical subunits in fresh garlic. This indicates that the drying process affects the alliinase in garlic powder, however, the preparation was capable of converting alliin to allicin.

Commercially peeled cloves in Korea were found to have much higher numbers of microbes than freshly peeled cloves (Shim, 1999). Raw garlic can also be contaminated by *Clostridium botulinum* (Roever, 1998) as can garlic flavored oils (LaGrange, 1998). Large-scale producers of fresh garlic should consider risks of microbial contamination.

Presently the US Food and Drug Administration does not control the quality of herbal preparations or supplements. Any quality control done in the U.S. is self-regulated.

X. SUMMARY

It has been demonstrated that garlic and its derivatives may be multifunctional as well as flavorful food additives. Garlic has been used in the past and is presently being used as a medicine, curative, preservative and a flavoring agent. The recent reports of antibiotic-resistant organisms and the threat of new emerging pathogens is a global public health problem. Microbes adapt to new environment by mutation, allowing them to survive adverse conditions by natural selection. Therefore, it is suggested that alternative or synergistic therapies that offer possible solutions or prevention should be investigated.

The antimicrobial properties of garlic and its derivatives have been documented for many years. It has been shown that garlic and its derivatives exhibit both cidal and statis effects. Many studies have indicated that garlic derivatives exhibit antimicrobial properties and fresh garlic appears to provide the most optimum and efficient cidal and inhibitory effects.

Although garlic and its derivatives appear to be safe for consumption, assessment and protein determination for the allergen responsible for hypersensitivity reactions needs in-depth evaluation. Certain reports have indicated allergic reactions associated with garlic consumption.

These preliminary indications are worthy of further research into the specific mechanisms and the effects of garlic derivatives on susceptible microorganisms. It is apparent that the conflicts reported in the studies (MIC values etc.) are due to different strains of organisms used. Determination of the most effective sulfinate concentration range to use for a specific organism or infection is another consideration. If garlic sulfinates are to be used as preservatives or as antimicrobial agents, the purification techniques and quality control methods should be standardized.

Acknowledgments. We thank Dr. Gioconda San-Blas, Centro de Microbiologia y Biologia Cellular, Venezuela for providing electron micrographs for this chapter.

XI. REFERENCES

1. Appleton, J.A., and Tansey, M.R. 1975. Inhibition of growth of zoopathogenic fungi by garlic extract. *Mycologia*. 67:882-885.
2. Barone, F.E., and Tansey, M.R. 1977. Isolation, purification, identification, synthesis, and kinetics of activity of the anticandidal component of *Allium sativum*, and a hypothesis for its mode of action. *Mycologia*. 69:793-824.
3. Block, E. 1985. The chemistry of garlic and onions. *Sci Am*. 252(3):114-119.
4. Brenner, S., and Wolf, R. 1994. Possible nutritional factors in induced pemphigus. *Dermatol*. 189:337-339.
5. Caporaso, N., Smith, S.M., and Eng, R.H.K. 1983. Antifungal activity in human urine and serum after ingestion of garlic (*Allium sativum*). *Antimicrob. Agents Chemother*. 23:700-702.
6. Cavallito, C.J., and Bailey, J.H. 1944. Allicin, the antibacterial principle of *Allium sativum*.I. Isolation, physical properties and antibacterial action. *J. Am. Chem. Soc*. 66:1950-1951.
7. Cavallito, C.J., Bailey, J.H., Buck, J.S., and Suter, C.M., 1944. Allicin, the antibacterial principle of *Allium sativum*. II. Determination of the chemical structure. *J. Am. Chem. Soc*. 66:1952-1954.
8. Chen, G.W., Chung, J.G., Ho, H.C., and Lin, J.G. 1999. Effects of the garlic compounds diallyl sulphide and diallyl disulphide on arylamine N-acetyltransferase activity in *Klebsiella pneumonae*. *J. Appl. Toxicol*. 19:75-81.
9. Chen, G.W., Chung, J.G., Hsieh, C.L., and Lin, J.G. 1998. Effects of the garlic components diallyl sulfide and diallyl disulfide on arylamine N-acetyltransferase activity in human colon tumour cells. *Food Chem. Toxicol*. 36:761-770.
10. Chung, C.H., Kim, Y.S., Yoo, Y.J., and Kyung, K.H. 1997. Presence and control of coliform bacteria in kimchi. *Korean J. Food Sci*.29:999-1005.
11. Chowdhury, A.K., Ahsan, M., Islam, S N., and Ahmed, Z.U. 1991. Efficacy of aqueous extract of garlic and allicin in experimental shigellosis in rabbits. *Indian J. Med. Res*. 93:33-36.
12. Chung, J.G., Chen, G.W., Wu, L.T., Chang, H.L., Lin, J.G., Yeh, C.C., and Wang, T.F. 1998. Effects of garlic compounds diallyl sulfide and diallyl disulfide on arylamine N-acetyltransferase activity in strains of *Helicobacter pylori* from peptic ulcer patients. *Am. J. Chin Med* 26:353-364.
13. Dankert, J., Tromp, T.F., de Vries, H., and Klasen, H.J. 1979. Antimicrobial activity of crude juices of *Allium ascalonicum, Allium cepa* and *Allium sativum*. *Zentralbl. Bakteriol*. 245:229-239.
14. Dausch, J.G., and Nixon, D.W. 1990. Garlic: A review of its relationship to malignant disease. *Prev Med* 19:346-361.
15. Elnima, E.I., Ahmed, S.A., Mekkawi, A.G., and Mossa, J.S. 1983. The antimicrobial activity of garlic and onion extracts. *Pharmazie* 38:747-748.
16. Ernest, E. 1981. Garlic therapy? Theories of a folk remedy. *MMW Munch Med. Wochenschr*. 123:1537-1538.
17. Feldberg, R.S., Chang, S.C., Kotik, A.N., Nadler, M., Neuwirth, Z., Sundstrom, D.C., and Thompson, N.H. 1988. In vitro mechanism of inhibition of bacterial cell growth by allicin. *Antimicrob. Agents Chemother*. 32:1763-1768.
18. Ford, R.A., Letizia, C., and Api, A.M. 1988. Diallyl disulfide. *Food Chem Tox*. 26:297-298.
19. Fromtling, R.A., and Bulmer, G.S. 1978. In vitro effect of aqueous extract of garlic (*Allium sativum*) on the growth and viability of *Cryptococcus neoformans*. *Mycologia*. 70:397-405.
20. Firouzi, R., Azadbakht, M., and Nabinediad, A. 1998.Anti-listerial activity of essential oils of some plants. *J. Appl. Animal Res*.14:75-80.
21. Gallwitz, H., Bonse, S., Martinez-Cruz, A., Schlichting, I., Schumacher, K., and Krauth-Siegel, R.L. 1999. Ajoene is an inhibitor and subversive substrate of human glutathione reductase and *Trypanosoma cruzi* trypanothione reductase: crystallographic, kinetic, and spectroscopic studies. *J Med Chem*. 42:364-372.
22. Graham, D.Y., Anderson, S.Y., and Lang, T. 1999. Garlic or jalapeno peppers for treatment of *Helicobacter pylori* infection. *Am J. Gastroenterol*. 94:1200-1202.
23. Jandal, J.M. 1998. Lactoperoxidase/garlic extract/ethanol system in the preservation of buffalo milk. *Buffalo Journal*. (1):95-101.
24. Jappe, U., Bonnekoh, B., Hausen, B.M., and Gollnick, H. 1999. Garlic-related dermatoses: case report and review of the literature. *Am. J. Contact Dermat*. 10:37-39.
25. Keusgen, M.1998. A high-throughput method for the quantitative determination of Alliin. *Planta Medica*. 64:736-740.

26. Krest, I., and Keusgen, M. 1999. Quality of herbal remedies from *Allium sativum*: differences between alliinase from garlic powder and fresh garlic. *Planta Med.* 65:139-43.

27. LaGrange, L.A. 1998. Botulism in flavored oils-a review. *Dairy Food Environ. Sanitation.* 18:438-441.

28. Ledezma, E., DeSousa, L., Jorquera, A., Sanchez, J., Lander, A., Rodriguez, E., Jain, M.K., and Apitz-Castro R. 1996. Efficacy of ajoene, an organosulphur derived from garlic, in the short-term therapy of tinea pedis. *Mycoses.* 39:393-395.

29. Lee, T.Y., and Lam, T.H. 1991. Contact dermatitis due to topical treatment with garlic in Hong Kong. *Contact Dermatitis.* 24:193-196.

30. Lembo, G., Balato, N., Patruno, C., Auricchio, L., and Ayala, F. 1991. Allergic contact dermatitis due to garlic (*Allium sativum*). *Contact Derm.* 25:330-331.

31. Lun, Z.R., Burri, C., Menzinger, M., and Kaminsky, R. 1994. Antiparasitic activity of diallyl trisulfide (Dasuansu) on human and animal pathogenic protozoa (*Trypanosoma sp., Entamoeba histolytica* and *Giardia lamblia*) in vitro. *Ann. Soc. Belg. Med. Trop.* 74:51-59.

32. Martin, B. 1999. Botany toxonomy inquiry. La Sierra University, Riverside, CA. April 3.

33. Mayeux, P.R., Agarwal, K.C., Tou, J.S., King, B.T., Lippton, H.L., Hyman, A.L., Kadowitz, P.J., and McNamara, D.B. 1988. The pharmacological effects of allicin, a constituent of garlic oil. *Agents Actions* 25:182-190.

34. Mielnik, M. 1997. Garlic as an antioxidant. Stabilization of frozen, mechanically deboned chicken meat. *InforMAT* 10:16-17.

35. Miron, T., Rabinkov, A., Mirekman, D., Weiner, L., Wilchek, M.1998. A spectrophotometric assay for allicin and alliinase activity: Reaction of 2-nitro-5-thiobenzoate with thiosulfinates. *Analytical Biochemistry.* 265:317-325.

36. Moore, G.S., and Atkins, R.D. 1976. The fungicidal and fungistatic effects of an aqueous garlic extract on medically important yeast-like fungi. *Mycologia.* 69:341-348.

37. Mutsch-Eckner, M., Erdelmeier, C.A.J., and Sticher, O. 1993. A novel amino acid glycoside and three amino acids from *Allium sativum.* *J. Nat Prod.* 56:864-869.

38. Naganawa, R., Iwata, N., Ishikawa, K., Fukuda, H., Fujino, T., and Suzuki, A. 1996. Inhibition of microbial growth by ajoene, a sulfur-containing compound derived from garlic. *Appl. Environ. Microbiol.* 62:4238-4242.

39. Naidu, A.S. 1999. Phytoantimicrobial (PAM) agents as multifunctional food additives. In *Phytochemicals as Bioactive Agents,* eds. W.R. Bidlack, S.T. Omaye, M.S. Meskin, and D. Jahner. Lancaster: Technomic Publishing Co. (in press)

40. Paludan-Muller, C., Huss, H.H., and Gram, L. 1999. Characterization of lactic acid bacteria isolated from a Thai low-salt fermented fish product and the role fo garlic as substrate for fermentation. *Int. J. Food Microbiol.* 46:219-229.

41. Parish, R.A., McIntire, S., and Heimbach, D.M. 1987. Garlic burns: a naturopathic remedy gone awry. *Pediatr. Emerg. Care.* 3:258-260.

42. Park, W.P., Cho, S.H., and Lee, P.S.1998. Effect of minimal processing operations on the quality of garlic, green onion, soybean sprouts and watercress. *J. Sci. Food Agric.* 77:282-286.

43. Pasteur, L. 1858. Memoire sur la fermentation appelle lactique. *Mem. Soc. Roy. Sci. Lille* 5:13-26.

44. Prasad, K., Laxdal, V.A., Yu, M., and Raney, B.I. 1995. Antioxidant effect of allicin, an active principle in garlic. *Mol. Cell. Biochem.* 148:183-189.

45. Pushpendran, C.K., Devasagayam, T.P.A., Chintalwar, G.J., Banerji, A., and Eapen, J. 1980. The metabolic fate of S-diallyl disulfide in mice. *Experientia.* 36:1000-1001.

46. Roever, C. de. 1998. Microbiological safety evaluations and recommendations on fresh produce. *Food Control* 9:321-347.

47. San-Blas, G., San-Blas F., Gil, F., Marino, L., and Apitz-Castro, R. 1989. Inhibition of growth of the dimorphic fungus *Paracoccidioides brasiliensis* by ajoene. *Antimicrob. Agents Chemother.* 33:1641-1644.

48. San-Blas, G., Urbina, J.A., Marchan, E., Contreras, L.M., Sorais, F., and San-Blas, F. 1997. Inhibition of *Paracoccidioides brasiliensis* is associated with blockade of phosphatidyl choline biosynthesis. *Microbiology.* 143:1583-1586.

49. Shen, J., Davis, L.E., Wallace, J.M., Cai, Y., and Lawson, L.D. 1996. Enhanced diallyl trisulfide has in vitro syngergy with amphotericin B against *Cryptococcus neoformans. Planta Med.* 62:415-418.

50. Shim, S.T., and Kyung, K.H. 1999.Natural microflora of pre-peeled garlic and their resistance to garlic antimicrobial activity. *Food Microbiol.* 16:165-172.

51. Sivam, G.P., Lampe, J.W., Ulness, B., Swansy, S.R., and Potter JD. 1997. *Helicobacter pylori* in vitro suceptibility to garlic (*Allium sativum*) extract. *Nutr. Cancer.* 27:118-121.

52. Stoll, A., and Seebeck, E. 1951. Chemical investigations on alliin, the specific principles of garlic. *Advan. Enzymol.* 11:377.
53. Tynecka, A., and Gos, Z. 1973. The inhibitory action of garlic (*Allium sativum L.*) on growth and respiration of some microrganisms. *Acta Microbiol. Polon Scr. B.* 5:51-62.
54. Tyler, V.E. 1993. Garlic and Other Alliums. *The Honest Herbal.* Pharmaceutical Products Press. N.Y. 139-143.
55. Tyler, VE.1994. Cardiovascular System Problems-Garlic. *Herbs of Choice.* Pharmaceutical Products Press.New York.104-108.
56. Walker, J.C., Lindegren, C.C., and Bachman, F.M. 1925. Further studies on the toxicity of juice extracted from succulent onion scales. *J. Agri. Res.* 30:175.
57. Walton, L., Herbold, M., and Lindegren, C.C. 1936. Bactericidal effects of vapors from crushed garlic. *Food Res.* 1:163.
58. Weber, N.D., Anderson, D.O., North, J.A., Murray, B.K., Lawson, L.D., and Hughes, B.G. 1992. In vitro virucidal effects of *Allium sativum* (garlic) extract and compounds. *Planta Med.* 58:417-423.
59. Wong, R.M., Kondo, Y., Bamba, H., Matsuzaki, S., and Sekine, S. 1996. Anti-*Helicobacter pylori* activity in the garlic extract. *Nippon Shokakibyo Gakkai Zasshi* 93:688.
60. Yamada, Y., and Azuma, K. 1977. Evaluation of the in vitro antifungal activity of allicin. *Antimicrob. Agents Chemother.* 11:743-749.
61. Yoo, K.S., and Pike, L.1998. Determination of flavor precursor compound S-alk(en)yl-L-cysteine sulfoxides by an HPLC method and their distribution in *Allium* species. *Scientia Horticulturae.*75:1-10.
62. Yoshida, S., Kasuga, S., Hayashi, N., Ushiroguchi, T., Matsura, H., and Nakagawa, S. 1987. Antifungal activity of ajoene derived from garlic. *Applied and Environmental Microbiology.* 53:615-617.
63. Yoshida, H., Katsuzaki, H., Ohta, R., Ishikawa, K., Fukuda, H., Fujino, T., and Suzuki, A. 1999. An organosulfur compound isolated from oil-macerated garlic extract, and its antimicrobial effect. *Biosci. Biotechnol Biochem.* 63:588-590.
64. You, W.C., Zhang, L., Gail, M.H., Ma, J.L., Chang, Y.S., Blot, W.J., Li, J.Y., Zhao, C.L., Liu, W.D., Li, H.Q., Hu, Y.R., Bravo, J.C., Correa, P., Xu, G.W., Fraumeni, J.F. Jr. 1998. *Helicobacter pylori* infection, garlic intake and precancerous lesions in a Chinese population at low risk of gastric cancer. *Int J Epidemiol.* 27:941-944.
65. Yven, C., Guichard, E., Giboreau, A., and Roberts, D.D. 1998. Assessment of interactions between hydrocolloids and flavor compounds by sensory, headspace, and binding methodologies. *J. Agri. Food Chem.* 46:1510-1514.

L.R. Juneja

T. Okubo

P. Hung

Catechins

14

I. INTRODUCTION

Tea has served as a popular beverage worldwide for centuries. Historians trace the first use of tea to China, in the twenty-eighth century BC. However, written documentation does not appear until the third century BC. Green tea has been considered a crude medicine in China for 4000 years. Today, tea is the second most consumed beverage in the world following water. Green tea is one of the most popular beverages in Japan and China, and its consumption accounts for about a half of all tea consumed worldwide. The tea plant is cultivated in more than 20 countries of Asia, Africa, and South America; India and China are the biggest growers, producing 672 and 600 million pounds a year, respectively (FAO, 1993).

The degree of fermentation greatly affects the quality and type of tea. According to the degree of fermentation, tea is classified into three kinds: green tea (unfermented), oolong tea (semi-fermented), and black tea (fermented). Research shows that the processing of black tea destroys certain beneficial health constituents such as catechins (polyphenolic compounds) that exist in green tea due to oxidation.

Green tea has been attracting many researchers from various fields exploring treatment for ailments ranging from acne to heart disease. Green tea is a powerful physiological antioxidant and a strong candidate for therapeutic and prophylactic applications. Green tea catechins (GTC) inhibit bacterial growth (Okubo et al., 1997; Fukai et al., 1991) and suppress tumors (Ji et al., 1997). Certain studies also indicated the ability of GTC to prevent oxidation of low-density lipoprotein (LDL), precursors to heart disease (Luo et al., 1997). GTC inhibits a wide range of microorganisms including cariogenic bacteria (Sakanaka, 1997), and enteric pathogens (Okubo & Juneja, 1997; Ishihara & Akachi, 1997). The volatile constituents of green tea also demonstrate antimicrobial activity against several bacteria, fungi, viruses and parasites. Several studies have shown that GTC strongly inhibits the growth of *Vibrio cholerae* 01, *Staphylococcus aureus*, *Vibrio parahaemoliticus*, *V. minicus*, *Staphylococcus epidermidis*, *Campylobacter jejuni*, *Plesiomonas shigellloides* (Toda et al., 1989); *Streptococcus mutans* (Sakanaka et al.,

TABLE 1. Green tea catechin composition of Sunphenon®

Catechin compound	Composition (%, w/w)
(+)-Catechin (C)	2.7
(+)-Gallocatechin (GC)	10.7
(-)-Epicatechin (EC)	5.9
(-)-Epicatechin gallate (ECg)	5.2
(-)-Epigallocatechin (EGC)	16.5
(-)-Epigallocatechin gallate (EGCg)	31.9
Total catechin content	74.8

1989) or viruses such as *Influenza* A, *Influenza* B (Nakayama et al., 1993) and more than 99% growth of viruses such as *Polio* type 3, *Echo* type 9, *Coxasackie* B6, *Reovirus* type 1, *Vaccinia, Herpes simplex,* (John et al., 1979).

The antimicrobial effect of the Sunphenon®, a commercial product of GTC containing 74.8% polyphenol prepared from green tea (*Camellia sinensis*) by *Taiyo Kagaku Co. Ltd.* (Yokkaichi, Japan), was used in the most of studies reported in this Chapter. The catechin composition of Sunphenon® is shown in *TABLE 1.*

II. OCCURRENCE

Tea leaves essentially contain catechins, theanine (γ-ethylamino-L-glutamic acid), and caffeine. Most of the interest has been focused on catechins based on the isoflavan structure, and these compounds comprise about 30% of the dry weight of tea flush (the growing point of the plant, consisting of the buds and immature leaves that are picked for processing). These compounds are infused with hot water or extracted with ethyl acetate from the aqueous solution. The slight astringent and bitter taste of green tea infusion is attributed to catechins.

About 5% of the dry weight of black tea (10% in case of green tea) and its aqueous extracts is made up of catechins, which are simple, well-characterized isoflavonoids. GTC mainly consists of six compounds: (+)-Catechin (C); (+)-Gallocatechin (GC); (-)-Epicatechin (EC); (-)-Epicatechin gallate (ECg); (-)-Epigallocatechin (EGC); and (-)-Epigallocatechin gallate (EGCg). These catechins are synthesized in tea leaves through malonic acid- and shikimic acid-metabolic pathways. The content of catechin varies depending on the species of tea plant and the season for harvesting, in which, EGCg is not found in other plants and is the major catechin in green tea (Nakabayashi, 1991; Bradfield & Penney, 1948).

III. MOLECULAR PROPERTIES

A. Structure

Tea catechins are a class of flavonols which are C-15 compounds, and their derivatives are composed of two phenolic nuclei (A ring and B ring) connected by three carbon units (C-2, C-3 and C-4). The flavonol structure of catechin (3, 3', 4', 5, 7-pentahydroxylflavan) contains two asymmetric carbon atoms at C-2 and C-3. Catechins and their derivatives also have nucleophilic centers at C-6 and C-8, which are reactive with electrophilic specimens. Chemically they are highly reactive, with properties of metal chelation, oxidative radical scavenging, nitrosation inhibition, etc.

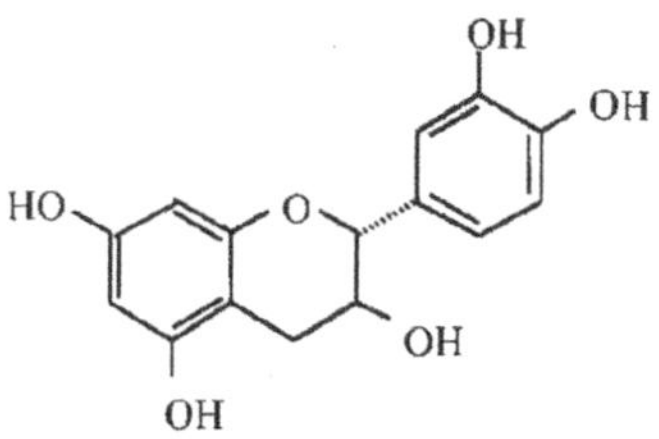

(+)-Catechin
$C_{15}H_{14}O_6$, mol.wt. 290
m.p. 176°C, $[\alpha]_D+18°$,

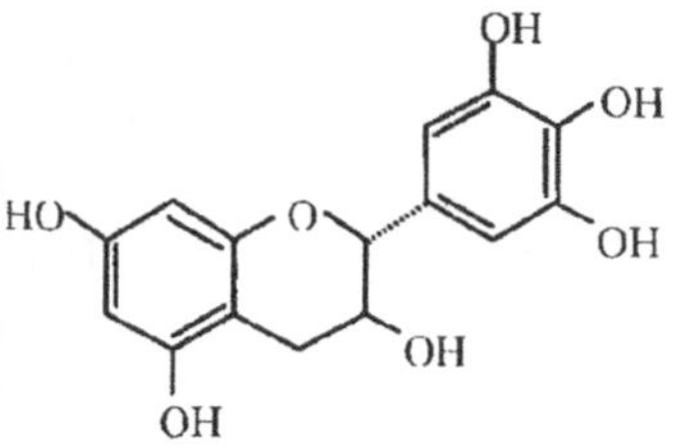

(+)-Gallocatechin
$C_{15}H_{14}O_7$, mol.wt. 306
m.p. 188°C, $[\alpha]_D+15°$,

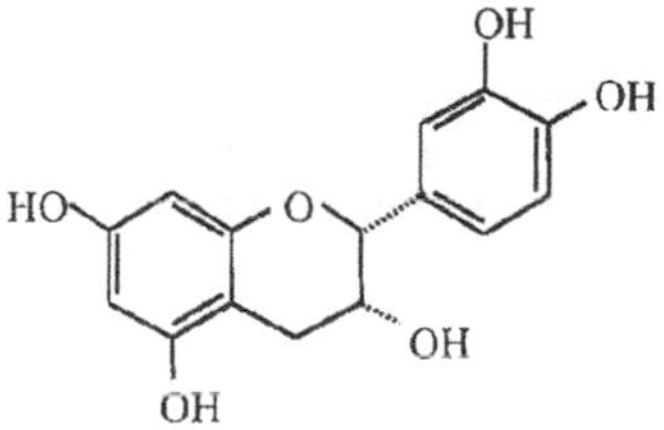

(−)-Epicatechin
$C_{15}H_{14}O_6$, mol.wt. 290
m.p. 242°C, $[\alpha]_D-69°$,

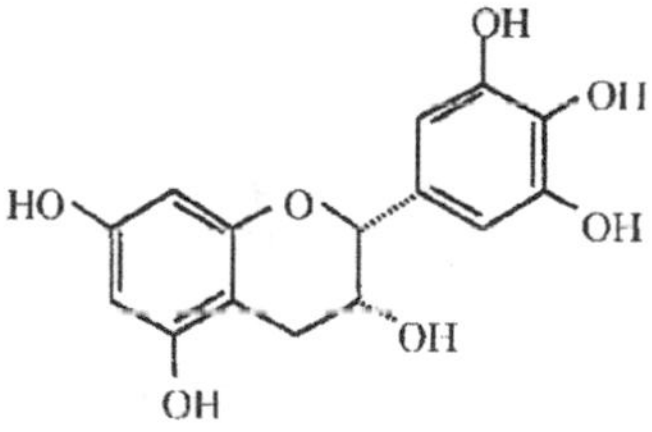

(−)-Epigallocatechin
$C_{15}H_{14}O_7$, mol.wt. 306
m.p. 218°C, $[\alpha]_D-50°$,

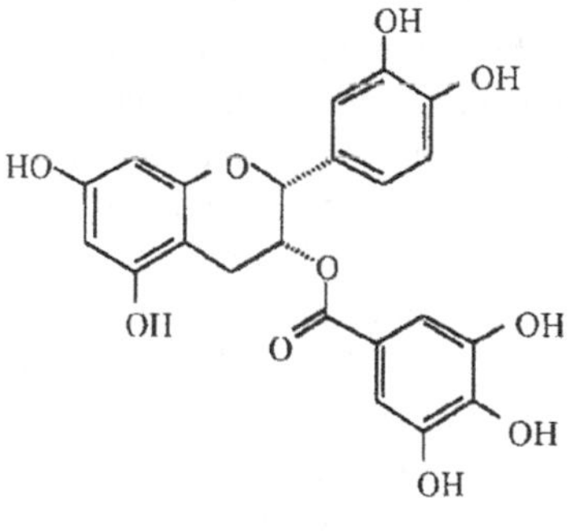

(−)-Epicatechin gallate
$C_{22}H_{18}O_{10}$, mol.wt. 442
m.p. 253°C, $[\alpha]_D-177°$,

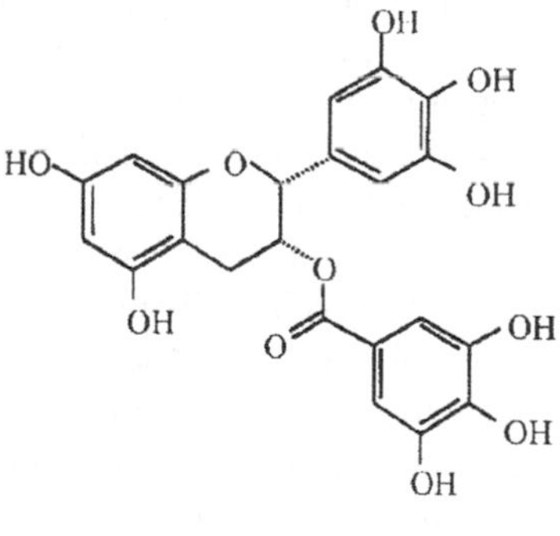

(−)-Epigallocatechin gallate
$C_{22}H_{18}O_{11}$, mol.wt. 458
m.p. 254°C, $[\alpha]_D-190°$,

FIGURE 1. Chemical structure of green tea catechins (GTC).

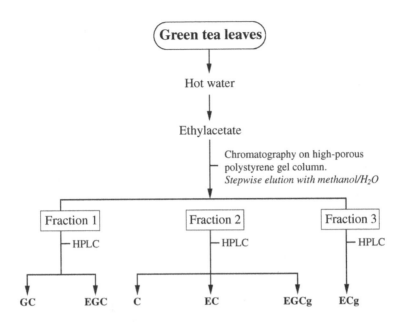

FIGURE 2. Isolation and purification of green tea catechins (GTC).

Structure, molecular formula, molecular weight (mol.wt), melting point (m.p), optical rotation $[(\alpha)D]$ of each component of GTC are shown in *FIGURE 1* (Hergert & Kurth, 1953; Bradfield & Penney, 1948; Birch et al., 1957).

B. Isolation and purification

The leaves of green tea were steeped into hot water (95 °C) for 30 min with gentle stirring. The ratio of tea and water is set at 1:10 (w/w). The mixture was then filtered and the tea infusion is concentrated to about quarter by volume with vacuum dryer. The catechins are extracted by adding ethylacetate three times by volume to the water extract. The ethylacetate layer is then dried with cyclone-vacuum dryer to obtain the catechins.

A schematic pathway of isolation and purification of GTC is shown in *FIGURE 2*. GTC is separated into three fractions using chromatography with a high porosity polystyrene gel column (Diaion HP-20, particle size 75–150 μm, *Mitsubishi Kasei Co.*, Japan). A sample of 100 mg catechins dissolved in 20 ml water is applied and eluted stepwise with 100 ml of methanol/H_2O = 1/5 (fractions 1 and 2) and with methanol/H_2O = 2/5 (fraction 3). Elution is performed at a flow rate of 5 ml/min. Each fraction obtained above is purified using recycling liquid chromatography equipped with GS-310 (particle size 9 μm, LC-908). The mobile phase consists of acetonitrile/H2O (3/7, v/v). Elution is performed at a flow rate of 10 ml/min with monitoring at 280 nm. These components are then analyzed using HPLC Model 600E equipped with auto-sampler model 717 and programmable multi-wavelength detector model 490E (*Millipore Co.*, USA). A sample volume of 5 μl of is injected to a Develosil ODS-P-5 column (*Nomura Chemical Co.*, Japan); eluted with a mixture of acetic acid:acetonitrile:N, N-dimethylformamide:H2O=3:1:15:81 (v/v/v/v) at a flow rate of 0.5 ml/min. Catechins are detected at the absorbance of 280 nm.

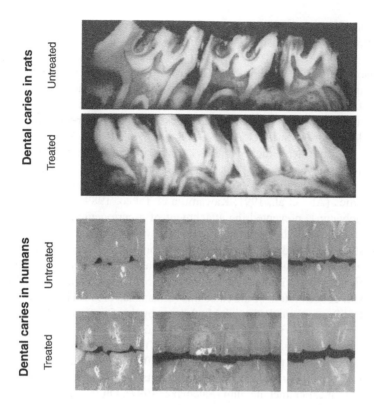

FIGURE 3. Anticaries activity of Sunphenon® - a commercial GTC formulation. *Dental caries in rats*: Animals were fed with chocolate and caramel diets containing 56% sucrose with or without 0.05% Sunphenon® for 40 days. The caries scores were then counted. (Source: Sakanaka et al., 1992). *Dental caries (plaque formation) in humans*: Twenty-six volunteers rinsed their mouths 3 times a day with 0.05% Sunphenon® solution for 3 days. Control group rinsed with water only. The subjects did not brush the teeth during the test period.

IV. ANTIMICROBIAL ACTIVITY

A. Effects on dental caries

Several epidemiological studies have suggested efficacy of green tea infusion for prevention of dental caries (Onishi et al., 1978; 1981a; & 1981b). Dental caries and periodontal diseases are the main infectious diseases in the oral cavity. *Streptococcus mutans* and *S. sobrinus* play an important role in the cause of dental caries. These bacteria synthesize water-soluble and insoluble glucans by glucosyltransferase (GTase) from sucrose in foods and drinks. Water-insoluble and highly branched glucan is responsible for the bacterial cell adherence to tooth surface (Hamada & Llade, 1980). Thus, the three major factors that play a critical role in the development of dental caries are host (tooth), bacteria, and substrate (sucrose) for the bacterial metabolism.

S. mutans and other microorganisms grow in the synthesized polysaccharides (glucan) and form dental plaque. These bacteria metabolize various saccharides and produce organic acids, especially lactic acid which is retained in dental plaque and the tooth is decayed gradually. Usually one cannot recognize a dental caries before it gets worse to a considerable extent. The tooth surface once plaqued is never back to the original state. Therefore, prevention of dental caries is much more important than cure of the plagued teeth.

The most traditional method of decreasing cariogenic bacteria and dental plaque is use of dental brush or chemicals such as antibiotics, antibacterial substances, as well as enzyme inhibitors and their combination (Ikeda et al., 1978; Ohshima et al., 1983; Backer et al., 1983). Various plant extracts have been investigated aiming prevention of dental caries by many laboratories (Saeki et al., 1983; Okami et al., 1981). Among such phytochemicals, GTC was considered highly effective for prevention of dental caries.

Several studies have confirmed that purified catechin fraction from green tea, and ECG and EGCg in particular, inhibit the growth of many bacterial species and elicit anticariogenic properties (Ahn et al., 1991; Kawamura & Takeo, 1989; Otake et al., 1991). Specifically, Sunphenon® prevented the attachment of a cariogenic *S. mutans* strain to hydroxyapatite and inhibited the bacterial GTase activity (Sakanaka et al., 1990; Otake et al., 1991). Our studies showed that Sunphenon® was not only effective in the inhibition the growth of *S. mutans*, but also against gingivitis, glucan synthesis and of cellular adherence of cariogenic streptococci. The inhibition of the growth of *S. mutans* and *S. sobrinus* was paralleled with concentration of the catechins (Sakanaka, 1997). The GC and EGC at concentration of 250 µg/ml markedly inhibited the growth of strains such as MT-8148, IFO-13955, MT-4502, MT-4532, MT-4245, MT-4251, 6715-DP, and Mfe-28 (Sakanaka, 1997). The minimum inhibitory concentration (MIC) against cariogenic bacteria of EGCg, a major component of tea catechins, was 250-1,000 µg/ml in the brain heart infusion (BHI) medium, and their growth inhibitory effects were enhanced by two-fold or more when tested in meat extract medium (Sakanaka, 1997). On the other hand, ECg and EGCg almost completely inhibited glucan synthesis at concentrations of 25-30 µg/ml, and the adherence of bacteria cells at concentration of 50 µg/ml (Sakanaka et al., 1990; 1992).

In vivo experiments in a rat model were performed to evaluate the inhibitory effect of GTC on dental caries. Rats were fed with chocolate and caramel diets containing 56% sucrose with or without 0.05% Sunphenon® for 40 days. The caries scores were then measured. The results indicated that Sunphenon® administration has significantly reduced the caries score by approximately 25% (*FIGURE 3 -TOP*).

Human clinical studies were also conducted on the preventive effect of GTC against dental plaque. The results showed that GTC significantly decreased the dental plaque formation in the volunteers who rinsed their mouth three times a day for 3 days with a solutions containing 0.05% GTC (*FIGURE 3 - BOTTOM*). The total number of bacteria and streptococci on the dental plaque also decreased in all the test groups. In addition, the concentration of lactic acid in saliva decreased after rinsing with the GTC solutions (Sakanaka et al., 1997; Oiwa et al., 1993). These results suggest that GTC is a potent inhibitor of dental plaque formation in the human oral cavity.

Furthermore, epidemiological studies carried out in the primary schools of two typical Japanese farm villages for 5 years which required drinking a cup of tea (*Bancha*) after lunch, resulted in a significant decrease in the average number of caries lesions in both village groups (Onishi, 1985).

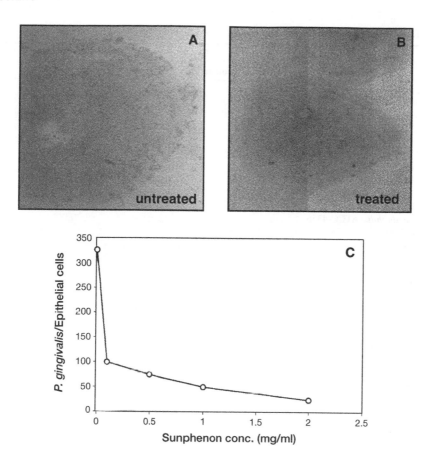

FIGURE 4. Inhibition adherence of *Porphyromonas gingivalis* adherence to oral epithelial cells Epithelial cells from oral cavity were pre-treated with Sunphenon® solution. Bacterial suspension of *P. gingivalis* was added to epithelial cells and incubated for 60 min. The epithelial cells and adherent bacteria were observed under an optical microscope (Source: Sakanaka et al., 1996).

Dose-response curve of adhesion-inhibition: Epithelial cells and *P. gingivalis* were incubated with Sunphenon® at various concentrations from 0.1 - 2.5 mg/ml at 37 °C for 60 min. Epithelial cells and bacteria were washed 5 times by centrifugation (1000 rpm, 3 min) with 0.1M PBS (pH 6.0). Cells were smeared on a glass slide, dried, and stained with a safranin solution. Adherent bacteria to 20 epithelial cells were counted by light microscope and the mean values were depicted.

B. Effects on periodontal disease

Periodontal diseases (periodontitis) occur during middle or old age and are considered one of the major geriatric disorders. The disease is characterized by inflammatory lesions in periodontal tissues and gingivitis. *Porphyromonas gingivalis*, the subgingival plaque bacteria, is one of the etiological agents of periodontitis. The adherence of *P. gingivalis* to oral epithelial cells is the initial step in the pathogenesis of periodontitis. *In vitro* experiments showed that Sunphenon® at 0.1 mg/ml and above strongly inhibit the adherence of *P. gingivalis* to epithelial cells (*FIGURE 4*). Furthermore, it was also observed that GTC, especially EGCg, GCg, and ECg, strongly inhibits the adherence of *P. gingivalis* (Katoh, 1995; and Sakanaka et al., 1996).

TABLE 2. Effects of GTC on growth and inhibition of enteric bacteria

Strain	Growth		Inhibition
	PYF medium	Gyorgy medium	
Bifidobacterium adolescentis E-194a	-	+	-
B. adolescentis E-319a	-	+	-
B. bifidum E-319a	-	+	-
B. bifidum S-28a	-	-	-
B. breve S-1	-	+	-
B. breve S-46	-	+	-
B. infantis S-12	-	+	-
B. infantis 1-10-5	-	+	-
B. longum E-194b	-	+	-
B. longum Kd-5-6	-	+	-
Lactobacillus acidophilus ATCC-4356	+	-	-
L. casei ATCC-7469	+	+	-
L. salivarius ATCC-11741	+	+	-
Bacteroides distasonis B-26	-	-	-
B. distasonis S-601	-	-	-
B. fragilis M-601	-	-	-
B. fragilis VI-23	-	-	-
B. thetaiotaomicron AS-126	-	-	-
B. vulgatus B-24	-	-	-
B. vulgatus F-62	-	-	-
Clostridium bifermentans B-1	-	-	+
C. bifermentans B-4	-	-	+
C. butyricum ATCC-14823	-	-	-
C. butyricum S-601	-	-	+
C. coccoides B-2	-	-	-
C. difficile ATCC-9689	-	-	+
C. innocuum M-601	-	-	+
C. paraputrificum B-3-4	-	-	+
C. paraputrificum B-78	-	-	+
C. paraputrificum VPI-6372	-	-	+
C. perfringens ATCC-13124	-	-	+
C. perfringens B-3-7	-	-	+
C. perfringens B-3-8	-	-	+
C. perfringens B-165-16	-	-	+
C. perfringens C-01	-	-	+
C. ramosum ATCC-25582	-	-	+
C. ramosum C-00	-	-	-
Eubacterium aerofaciens S-601	-	-	-
E. aerofaciens S-605	-	-	-
E. lentum M-601	-	-	-
E. limosum E-1	-	-	-
Escherichia coli E-605	-	-	-
E. coli M-602	-	-	-
E. coli O-601	-	-	-
E. coli V-603	-	-	-

Test strains were obtained from the RIKEN, cultured in peptone yeast fields (PYF) medium without C-source supplementation. Gyorgy medium was supplemented with 0.5% glucose and 1% methanol extract of green tea.. The inhibitory effect of GTC was assayed by paper-disc method (8mm diameter, Toyo Roshi, Japan) containing 10 mg of GTC per disc on Brucella agar. Tests were done three times and a mean of inhibition zone of 12 mm or larger was estimated as '+'.

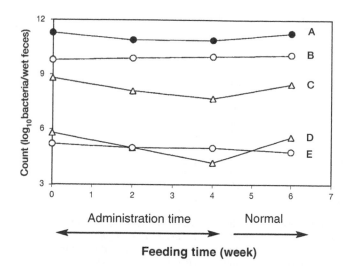

FIGURE 5. Effect of Sunphenon® on human enteric microflora. Human volunteers were administrated with 1.2g Sunphenon®/day/person for 4 weeks. Bacterial enumeration was performed from fecal samples (A - Total count; B - *Bifidobacterium* spp.; C - *Clostridium* spp., D - *Clostridium perfringens*; E - *Lactobacillus* spp.

Collagenase, which is produced by *P. gingivalis*, breaks down collagen in the gums, weakens the periodontal pocket and eventually recedes the gums leading to gingival and periodontal diseases. The test results showed that the activity of collagenase was significantly inhibited by 50 µg/ml Sunphenon®, and the enzyme activity was almost inhibited to zero at 100 µg/ml Sunphenon®.

These results suggested that Sunphenon® significantly decreased the periodontal disease by inhibiting the growth, adherence of *P. gingivalis* as well as suppressing the activity of collagenase.

C. Effects on human enteric microflora

In the human large intestine, the undigested and unabsorbed substances are subjected to react with various enteric bacteria. These bacteria and their metabolites are closely associated with nutrition, health, aging and pathogenesis of gastrointestinal diseases. For instance, lactic acid bacteria (LAB) such as *Bifidobacterium* and *Lactobacillus* spp. are known to be useful to human and animal health as probiotics. These organisms stimulate digestion and absorption of nutrients and inhibit the growth of unfavorable microbes (Hentges, 1983; Mitsuoka, 1984). On the other hand, certain clostridia such as *Clostridium perfringens*, *C. difficile*, etc. are closely associated with intestinal malfunction, that could lead to accelerated aging or causing tumors (Bokkenheuser, 1983; Godman, 1983; Gary & Sherwood, 1984). The microflora in the large intestine are always affected by external environmental factors, in which diets are the most important.

For measurement of the MIC, 0.1ml of the GTC at various concentrations was added to 9.9 ml of medium containing 1.5% agar at around 60°C in a petri dish. The mixture was agitated thoroughly and solidified at 37°C. One platinum loop of the microorganism suspension, obtained from a pre-culture at 37°C overnight, was inoculated on the agar medium, and incubated at 37°C, 48 h for bacteria or at 25°C, 96 h for fungi and yeast. Brain heart infusion (BHI) was used for bacteria, and malt extract broth was used for fungi and yeast.

TABLE 3. Antimicrobial spectrum of GTC against various bacteria (MIC in µg/ml)

Bacterial strain	MIC	Bacterial strain	MIC
Food-borne bacilli		*Escherichia coli strains*	
Bacillus subtilis IFO-3007	2,000	*E. coli* O157:H7 SAKAI212	250
B. cereus ATCC-14579	600	*E. coli* O157:H7 ATCC43895	250
B. brevis IFO-3331	700	*E. coli* O139-CH15	750
B. circulans IFO-13626	600	*E. coli* O139-CH20	500
B. macerans JCM-2500	800	*E. coli* K-88	750
B. polymyxa JCM-2507	900	*E. coli* IFO-3545	750
Thermophilic spore formers		*Gram +ve / -ve strains*	
B. stearothermophilus IAM-1035	100	*Proteus vulgaris* IFO-3581	2,000
B. stearothermophilus IFO-12550	150	*Pseudomonas aeruginosa* IFO-3080	250
B. stearothermophilus ATCC-12980	150	*P. fluorescens* IFO-13922	300
B. coagulans JCM-2257	200	*Serratia marcescens* IFO-3046	4,000
Clostridium thermoaceticum No-5801	400	*Staphylococcus aureus* IFO 12732	500
C. thermoaceticum No-5802	400	*Lactobacillus acidophilus* IAM-1043	1,000
C. thermoaceticum No-5809	400	*Streptococcus lactis* IFO-12546	2,000
C. thermosaccharolyticum No-5604	100	*Helicobacter pylori* ATCC-43504	500
		Propionibacterium acnes ATCC-6919	500
		P. acnes ATCC-11828	600

For MIC determination, 0.1 ml of the GTC at various concentrations was added to 9.9 ml of medium containing 1.5% agar. One platinum loop of overnight-grown bacterial suspension was inoculated on the BHI agar, and incubated at 37 ˚C for 48 h. For thermophilic sporeformers, BHI and the modified thioglycollate medium were used. Culture conditions: were aerobic for *Bacillus stearothermophilus* (50 ˚C, 48 h), *Bacillus coagulans* (37 ˚C, 48 h), and anaerobic for *Clostridium* bacteria. (55 ˚C, 72 h).

Growth response and inhibition of bacteria isolated from human intestines were examined. The results showed that GTC promoted the growth of useful probiotic flora while it inhibited the growth of harmful bacteria such as *C. bifermentan, C. difficile, C. innocuum, C. paraputrificum, C. perfringens,* and *C. ramosum.* Ahn et al. (1991) showed that ECg and EGCg (5 mg/disk) significantly inhibited the growth of *C. difficile* and *C. perfringens (TABLE 2).*

In vivo studies on human volunteers also revealed that intake of GTC 1.2 g/day/person for 4 weeks significantly inhibited the growth of harmful bacteria but did not affect the growth of useful probiotic bacteria (*FIGURE 5*).

Other tests were also carried out to determine the MIC of GTC against various bacteria strain isolated from foods, human sources, animals, and fish. Thermophilic spore forming bacteria were also tested for GTC susceptibility (*TABLE 3 & 4*).

The effect of GTC on the heating time required for decreasing 10% CFU of spores (D-value) of *Bacillus stearothermophilus* was tested. In the presence of 500 ppm of GTC, the D-value significantly decreased at all tested temperatures as compared to control sample. The inhibition rate became greater following the increase of treated temperature.

The antifungal activity of GTC was also tested on the fungi and yeast such as *Saccharomyces cerevisiae* IFO-0203, *Candida albicans* IFO-1061, *Rhodotorula rubra* IFO-0001, *Hansenula anomala* IFO-0136, *Kluyveromyces fragilis* IFO-1963, *K. lactis* IFO-1090, *Metschnikowia pulcherrima* IFO-1678, *Aspergillus niger* ATCC-3275, *Mucor mucedo* IFO-7684, *Rhizopus chinensis* IFO-4745, *Penicillium chrysogenum* IFO-5809, *P. citrinum* IFO-7784, *Trycophyton rubrum* IFO-5807, *T. mentagrophytes* IFO-5466, *T. mentagrophytes* IFO-5809, and *T. tonsurans* IFO-5946. However, no marked inhibitory effect was observed, even at 4 mg/ml concentration of GTC in the media.

TABLE 4: MIC (µg/ml) values for GTC against animal- and fish-borne pathogens

Animal-borne pathogens	MIC	Fish-borne pathogens	MIC
Cattle		**Ayu**	
Salmonella dublin L-729	4,000	*Vibrio anguillarum* IFO-13266	200
Pseudomonas aeruginosa KK-1001	500	*V. anguillarum* UP-1	200
S. aureus KK-103	500	*V. anguillarum* PT-213	150
S. aureus spp.	1,000	*V. anguillarum* PT-493	150
S. aureus KK-101	1,000	*Vibrio* spp.	200
S. epidermidis KK-108	500		
S. epidermidis spp.	500	**Crawfish**	
Escherichia coli K-99	>4,000	*Vibrio* sp. PJ	50
		V. alguillarum	400
Pig		*V. fluvialis*	200
Salmonella enteritidis ZK-2a	4,000	*V. damsela* ATCC-33539	200
E. coli K-88(Abbotstown)	4,000		
E. coli K-88(G-1253)	4,000		
E. coli K-88(MN-1)	4,000	**Yellowtail**	
Staphylococcus pyogenes spp.	1,000	*Streptococcus* spp.	1,000
		S. KS-8903	700
		S. KS-8930	800
Chicken		*S.* KS-8982	900
Salmonella enteritidis L-58	4,000	*S. b-hemlosys*	900
S. typhimurium L-413	4,000	*Pasteurella piscicida* spp.	100
S. infantis L-164 4,000	4,000	*P. piscicida* OT-8447	200
S. thompson L-131	4,000	*P. piscicida* 5866	150
S. softa L-59	2,000		
S. mbandaka L-743	4,000	**Eel**	
S. mbandaka spp.	4,000	*Edwardjella tarda* SH-89133	400
S. huderberg spp.	4,000	*E. tarda* E-2812	300
		E. tarda SY-84006	400
		Vibrio vulnificus ATCC-33147	200
		V. parahaemolyticus ATCC-17802	200
		Salmon	
		Vibrio vulnificus	200
		Vibrio spp.	100
		Other fish	
		Aeromonas salmonisida	800
		A. hydrophila	200
		Vibrio alguillarum ATCC-19264	400
		V. harveyi	100

For MIC determination, 0.1 ml of the GTC at various concentrations was added to 9.9 ml of medium containing 1.5% agar. One platinum loop of overnight-grown bactcrial suspension was inoculated on the BHI agar, and incubated at 37 °C for 48 h.

D. Effects on *Escherichia coli* O157:H7

Escherichia coli O157:H7 is one of the enterohemorrhagic pathogens. In 1993, the outbreak of hemorrhagic colitis caused by *E. coli* O157:H7 in uncooked hamburger meats was reported in Mermelstein (United States). The seriousness of this strain generated interest during the sensational outbreak in Japan in 1996. More than 3,000 people infected with *E. coli* O157:H7 seriously suffered from diarrhea with a high rate of mortality. *E.coli* O157:H7 produces a cytotoxin known as verotoxin or shiga-like toxin, which shows strong toxicity on vero cells. This cytotoxin causes hemorrhagic colitis and hemolytic uremic syndrome in humans (Karmali et al. 1985). Various laboratories have tested the efficacy of various antimicrobial systems to control *E.coli* O157:H7. Such systems included a wide variety of antibiotics (Hathcox et al., 1996; Podolak et al., 1996), antioxidants (Ogunrinola et al., 1996), and treatment with acids and temperature (Tsai et al., 1996). Recently, it was reported that green tea extract prevented gnotobiotic mice from the *E. coli* infection (Isogai et al., 1998). We have conducted further investigation regarding the growth inhibitory effect of GTC on various *E. coli* including the serotype O157:H7.

1. Bacterial strains and culture conditions. The bacterial strains used in this study were *E. coli* O157:H7 Sakai strain which was isolated from a patient involved in the Sakai-city outbreak during the summer of 1996 (Izumiya et al. 1997), *E. coli* O157:H7 43895, *E. coli* O139-CH15, *E. coli* O55-CH20, K99-B41 and IFO-3545. Bacterial strains were transferred from stock cultures by inoculating a loopful of each in trypticase soy broth (TSB) and incubated at 37 ˚C for 24 h. Bacteria were serially transferred to culture for an additional 24 h. Bacteria were collected by centrifugation at 12,000 rpm x 10 min at 4°C. The cell pellet was then washed with 0.01M phosphate-buffer saline (PBS) (pH 7.2) and resuspended in the same buffer. The concentration of cells was then adjusted to 10^4 cells/ml. The MIC of GTC against *E. coli* was measured by the method described in previous sections.

2. Effects on the growth-multiplication. A volume of 0.5 ml cell suspension (10^4 CFU/ml) was added to 4.5 ml PBS (0.01M) solution or Muller Hilton medium, incubated at 37 ˚C for 0, 2, 4, 6 and 24 h, with or without GTC at various concentrations from 0 to 1000 µg/ml. After incubation, 0.5 ml of bacterial suspension was plated on TS agar medium, incubated at 37 ˚C for 24 h, and the colony counts were performed for viable cells.

The MIC of GTC against various strains of *E. coli* were determined (*TABLE 3*). GTC has completely inhibited the growth of two strains of *E. coli* O157:H7 at the concentration of 250 µg/ml, and at 500 or 750 µg/ml against other strains of *E. coli*. These data suggest that, *E. coli* O157:H7 was more susceptible to GTC than the other strains of *E. coli*.

The inhibitory effects of GTC against *E. coli* O157:H7 Sakai strain (*FIGURE 6A*) and *E. coli* IFO-3545 (*FIGURE 6B*) incubated in PBS at 37 ˚C for various time periods was studied. GTC inhibited the growth of both *E. coli* strains, and the inhibition was increased in a dose-dependent manner with the GTC concentration. The growth of bacteria was completely inhibited by in the buffer solution containing 1,000 µg/ml of GTC for 6 h or at > 200 µg/ml concentration after 24 h incubation.

The effect of GTC in Muller Hilton medium on the growth of *E.coli* was tested. The survival of *E. coli* IFO-3545 in culture medium after 24 hr incubation was not affected even at the concentration of 1,000 µg/ml of GTC and the viable cell counts of this strain increased from initial count of 10^5 CFU/ml to 10^{11} CFU/ml (*Figure 6D*). On the other hand, GTC at 1000 µg/ml inhibited the growth of *E. coli* O157:H7 Sakai strain after 24 h of cultivation, and the viable of bacteria decreased from 10^4 CFU/ml to <10 CFU/ml (*FIGURE 6C*).

These observations indicated that the inhibitory effect of GTC on the growth of *E. coli* O157:H7 was stronger than on the other *E. coli* strains. Therefore, the habitual intake of green tea or green tea catechins might be useful for control of *E. coli* O157:H7 in the digestive tract.

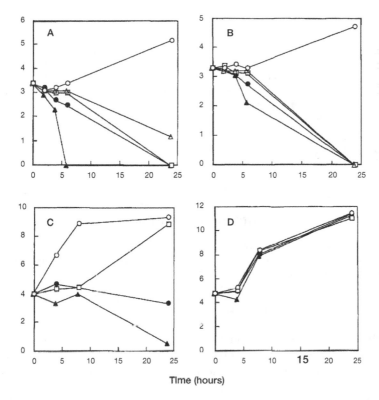

FIGURE 6. Effect of GTC on growth of *E. coli* in PBS (A - Sakai strain O157:H7; B - strain IFO-3545) and Mueller-Hinton broth (C - Sakai strain O157:H7; D - strain IFO-3545). GTC was added to the culture medium at various concentrations: 0 μg/ml (O), 100 μg/ml (△), 200 μg/ml (□), 500 μg/ml (●), and 1000 μg/ml (▲).

E. Effects on *Helicobacter pylori*

Helicobacter pylori causes severe inflammation and plays a critical role in the disorders of gastric mucosa. Cancer of the distal stomach, both of the intestinal and diffuse type, is strongly associated with *H. pylori* colonization. This bacterium causes chronic active inflammation of the gastric mucosa in the majority of colonized subjects. In a considerable number of subjects, this will eventually lead to a loss of gastric glands, and thus the establishment of atrophic gastritis, which is associated with the development of intestinal metaplasia and dysplasia. Development of atrophy and metaplasia of the gastric mucosa are thus strongly associated with *H. pylori* infection, instead of a direct and inevitable consequence of aging. The presence of these consecutive disorders leads to a 5-90-fold increased risk for cancer of the distal stomach, in particular of the intestinal type. It was reported that, ranitidine bismuth citrate is a novel salt that seems to be useful in eradication of *H. pylori* from gastric mucosa when it is combined with antibiotics (Alarcon et al., 1999). Shinchi et al. (1997) suggested that consumption of fresh vegetables could protect against *H. pylori* infection. National Cancer Center Hospital (Japan) has investigated the effect of GTC on the growth and adhesion-colonization of *H. pylori* to gastric mucin.

For the determination of *H. pylori* adhesion, *H. pylori* ATCC-43504 cells were cultured in mucin-coated plates for 1 h at 37°C, with or without GTC at various concentrations ranging from 10 to 100 ppm to fix the *H. pylori*. The cells were then rinsed four

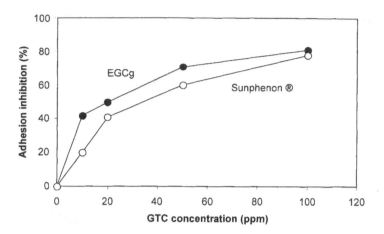

FIGURE 7. Effect of GTC on the adhesion of *Helicobacter pylori* to gastric mucin. Bacterial cells were cultured in mucin-coated plates for 1h at 37°C, with or without GTC at various concentrations (10 to 100 ppm) to fix the *H. pylori*. Cells were rinsed four times with PBS-tween 20 solution, anti-*H. pylori* IgG was added and incubated for 2 h. Bound IgG was detected with anti-mouse IgG-IgY-HRPO reagent after incubation for 2h at 37 °C. Wells were rinsed four times and substrate solution (*o*-phenylenediamine) was added. The absorbance at 450 nm was measured using a microplate reader. EGCg (●); Sunphenon® (O).

times with PBS-Tween 20 solution, anti-*H. pylori* IgG was added and incubated for 2 h. Bound IgG was treated with anti-mouse IgG IgY-HRP for 2 h at 37°C. Wells were rinsed four times and substrate solution (*o*-phenylenediamine) was added. The absorbance was measured at 450 nm using a microplate reader.

The MIC of various GTC compounds including C (>600 ppm), EC (600 ppm), GC (600 ppm), EGC (550 ppm) was low. Much lower and effective MIC values were obtained for Ecg (350 ppm), EGCg (200 ppm), and Sunphenon® (400 ppm) against *H. pylori*.

In the presence of GTC, the adhesion of *H. pylori* to gastric mucin was strongly inhibited, and the inhibition increased following the increase of GTC concentration (*FIGURE 7*). The results suggested that GTC is one of the effective inhibitors against *H. pylori* and may be beneficial for prevention of gastric cancer.

In a clinical study, seven patients positively infected with *H. pylori*, volunteered for therapeutic experiments with Sunphenon®. The patients were dosed with GTC every day using intake jellies or capsules consisting of 600 mg Sunphenon® for 4 weeks. The clearance activity of Sunphenon® against *H. pylori* colonization in their stomachs was measured by comparing the relative *H. pylori* number before and after administration. GTC was proven effective for 4 out of 7 patients. These results suggested that GTC might play an important role in the clearance of *H. pylori* in human gastric milieu.

F. Antiviral activity

Based on nucleic acid content, viruses can be classified as DNA or RNA types. *Hepadna*, *Popova*, *Adeno*, *Herpes*, *Pox*, and *Parvo* are DNA-viruses pathogenic to humans. *Picorna*, *Entero*, *Calici*, *Toga*, *Flavi*, *Orthomyxo*, *Paramyxo*, *Corona*, *Arena*, *Bunya*, *Retro*, *Rhabdo*, and *Reo* are the RNA-viruses. These viruses infect humans through skin, sexual organs, respiratory organs and digestive tracts during any tissue damage. The general symptoms caused by viral infections include toxemia, fever, fatigue and loss of appetite etc. Some of *retro* and *herpes* viruses are associated with certain cancers. The prophylaxis measures used to control viral diseases include hygiene, sanitation, vector control, vaccination, physical and chemical treatments.

Black tea extract has been reported to inhibit the multiplication of influenza A virus in embryonated eggs (Green, 1949). The extract of green tea leaves has been reported to show antiviral activity against influenza virus A and B (Nakayama et al., 1993), *vaccinia*, *herpes simplex*, *coxsackie* B6 and *polio* 1 viruses (John & Mukundan, 1979). Mukoyama et al. (1991) also reported that tea extracts prevented rotavirus and entero-virus from infecting rhesus monkey kidney cells. They also observed that EGCg inhibited the hemaglutination activity of the influenza viruses.

Rotaviruses belong to *Reoviridae* family, the causal viruses of gastroenteritis in infants and children (Barnett, 1983; Cukor & Blacklow, 1984). Every year, millions of infants and children have been reported to suffer from diarrhea globally, and about a half of such diseases are of viral etiology, mainly the *rotaviruses*. Ebina (1991) reported that ECg and EGCg are strong inhibitors as compared to other compounds of tea catechins. At concentrations of 1 µg/ml, GTC strongly inhibited the viral propagation by 84.4% and 96.2%, respectively. Gallic acid and its derivatives were not effective. These results suggested that the galloyl moiety linked by ester linkage in catechin molecule is important for the antiviral activity.

It was reported that GTC and flavonoids are strong inhibitors of reverse transcriptase of HIV (human immuno-deficiency virus) and several DNA- and RNA-polymerases (Ono & Nakane, 1991; Nakane & Ono, 1990). HIV is a retrovirus and its reverse transcriptase is important for infection. ECg and EGCg are strong inhibitors of HIV-reverse transcriptase, however, C, EGC, and gallic acid were ineffective. ECg and EGCg could inhibit by 50% of HIV-reverse transcriptase at 0.017 mg/ml and 0.012 mg/ml, respectively (Nakane & Ono, 1990). It is likely that the effective catechins mentioned above will be developed to use for prevention or treatment of various viral diseases.

V. APPLICATIONS

Tea catechins have been used widely in confectioneries such as chewing gums, candies, caramels, jelly beans, beverages, and several other food products in Japan and other countries. It was reported that approximately 50-100 mg catechins are contained in 100 ml of green tea infusion (Maeda & Nakagawa, 1977), and with the MIC range from 250 µg to 1000 µg/ml, green tea is quite effective on dental caries prevention as well as prevention of other antimicrobial activities. Onishi (1985) reported that a cup of green tea after lunch resulted in a reduction of dental caries risk in schoolchildren. Green tea has been applied in the treatment of diarrhea in China. This suggests that GTC is effective for prevention or treatment of certain diseases caused by rotavirus or bacteria. The GTC products could offer the food industry a tremendous opportunity for processed foods targeted in prevention of diseases as well as in terms of nutrition. The GTC products, Sunphenon® and Sunkatol®, are now used not only in processing of general foods but also for physiological functional applications all over the world.

VI. SAFETY AND TOLERANCE

The application of antibacterial agents of green tea catechins is especially attractive, because it focuses on the eradication of the causal microorganisms, in contrast to fluoridation or dietary control. In contrast, the use of antibiotics such as penicillin, erythromycin, and tetracycline are accompanied by a potential risk of undesirable and unacceptable side effects (Fitzerald, 1972). Green tea and its extracts have been recognized as a safe food and as food ingredients for a long time all over the world. Since green tea has been widely consumed by people as a daily beverage, extracts or purified flavor compounds of green tea may be safe, or risk-free, for use in oral care products. The safety of Sunphenon® has been confirmed with acute toxicity, subacute toxicity and mutagenecity tests. In 1993, Sunphenon® was approved as a "Food for Specified Health Use" (FOSHU) in Japan. Many studies have been conducted to reconfirm the safety of GTC. One of the studies estimated that the dose of LD_{50} of Sunphenon® would be 3.09 g per kg for female and more than 5.0 g per kg for male of *ddy* mice. In an oral administration study, the daily dose of 15 and 75 mg per kg for 88 days showed no toxic symptoms in rats (Takahashi & Nimomiya, 1997).

VII. SUMMARY

Studies showed that GTC could prevent plaque formation by inhibiting bacteria growth, glucan synthesis and the adherence of carciogenic streptococci. In addition, GTC reduced periodontal disease by suppressing the growth and adherence of *P. gingivalis* on buccal epithelial cells as well as by inhibiting the activity of collagenase. GTC strongly inhibited the growth of harmful intestinal bacteria but did not affect other useful probiotic bacteria both *in vitro* and *in vivo*.

These results indicate that the antimicrobial activities of GTC support its practical use for caries control, periodontal therapy, and management of disturbed intestinal conditions caused by microorganisms. The results strongly suggest a potential prophylactic role in human medicine for green tea catechins. GTC intake is safe and has health benefits; in addition, green tea is cheap and readily available in fractionated and extracted derivatives. However, the use of green tea is still a long way from pharmaceutical application. The huge worldwide consumption and safety of green tea catechins may well do significant good and could be supported as an adjunctive therapy and a healthy ingredient to a good diet, exercise, and avoidance of high-risk behavior.

VIII. REFERENCES

1. Ahn, A.J., Kawamura T., Kim, M., Yamamoto, T., and Mitsuoka, T. 1991. The catechins: Selective growth inhibitors of *clostridium* spp., *Agric. Biol. Chem.* 55:1425.
2. Alarcon, T., Domingo, D., Prieto, N., and Lopez-Brea, M. 1999. Bacteriostatic and bactericidal activity of ranitidine bismuth citrate in *Helicobacter pylori* clinical isolates. *Rev. Esp. Quimioter.* 12(1):64.
3. Albert, Z.K., and Robert, M.C. 1985 In *Rotavirus Virology*, ed. B.N. Fields, pp.863. New York: Raven Press.
4. Barnett, B. 1983. Viral gastroenteritis, *Med. Clin. North Am.* 67:1031.
5. Baker, P.J., Slots, J., Genco, R.J., and Evans, R.T. 1983. Minimal inhibitory concentrations of various antimicrobial agents for human oral anaerobic bacteria. *Antimicrob. Agents Chemother.* 24:420.
6. Birch, A.J., Clark-Lewis, J.W., and Robertson, A.V. 1957. The relative and absolute configurations of catechins and epicatechins. *J. Chem. Soc.* 3586.

7.   Bokkenheuser, V.D. 1983. Biotransformation of steroids. In *Human Intestinal Microflora in Health and Disease*, ed., D.J. Hentges, pp.215. New York: Academic Press.

8.   Bradfield, A.E., and Penney, M. 1948. The catechins of green tea, Part II. *J. Chem. Soc.* 2249.

9.   Cukor, G., and Blacklow, N.R. 1984. Human viral gastroenteritis. *Microbiol. Rev.* 48:157.

10. Ebina, T. 1991. Infantile gastroenteritis: prevention and treatment of rotaviral diarrhea. *J. Antibact., Antifung. Agents.* 19:349.

11. Fitzgerald, R.J. 1972. Inhibition of experimental dental caries by antibiotics. *Antimicrob. Agents Chemother.* 1:296.

12. FAO, 1993. *Production Year Book* 47:174.

13. Fukai, K., Ishigami, T., and Hara, Y. 1991. Antibacterial activity of tea catechins against phytopathogenic bacteria. *Agric. Biol. Chem.* 55(7):1895.

14. Gary, L.S., and Sherwood, L.G. 1984. Intestinal flora in Health and Disease. *Gastroenterology* 86:174.

15. Goldman, P. 1983. Biochemical pharmacology and toxicology involving the intestinal flora. In *Human Intestinal Microflora in Health and Disease*, ed. D.J. Hentges, pp.241. New York: Academic Press.

16. Green, R.H. 1949. Inhibition of multiplication of influenza virus by extracts of tea. *Proc. Soc. Exp. Biol. Med.* 71:84.

17. Hamada, S., and Llade, H.D. (1980) Biology, immunology and cariogenecity of *Streptococcus mutans*. *Microbiol. Rev.* 44:443.

18. Hathcox, A.K., and Beuchat, L.R. 1996. Inhibitory effects of sucrose fatty acid esters, alone and in combination with etylenediaminetetraacetic acid and other organic acids, on viability of *Escherichia coli* O157:H7. *Food Microbiol.*13:213.

19. Hatta, H., Sakanaka, S., Tsuda, N., Kanatake, N., Yamamoto, T. and Ebina, T. 1989. In *37ᵗʰ Annual Meeting of Japanese Society of Virologists in Osaka*, pp. 327. Osaka.

20. Hentges, D.J. 1983. Role of the intestinal microflora in host defense against infection. In *Human Intestinal Microflora in Health and Disease*, ed. D.J. Hentges, pp.311. New York: Academic Press.

21. Hergert, H.L., and Kurth, E.T. 1953. The isolation and properties of catechol from white fir bark. *J. Org. Chem.*, pp.521.

22. Ikeda, T., Shiota, T., Mcghee, J.R., Otake, S., Michalek, S.M., Ochiai, K., Hirasawa, M., and Sugimoto, K. 1978. Virulence of *Streptococcus mutans*, Comparison of the effects of a coupling sugar and sucrose on certain metabolic activities and cariogenicity. *Infect. Immun.* 19:477.

23. Ishikawa, N., and Akachi, S. 1997. Green tea extract as a remedy for diarrhea in farm-raised calves. In *Chemistry and Applications of Green Tea*, eds. T. Yamamoto, L.R. Juneja, D.C. Chu, and M. Kim, pp.137. New York: CRC Press.

24. Isogai, E., Isogai, H., Takeshi, K. and Nishikawa, T. 1998. Protective effect of Japanese green tea extract on gnotobiotic mice infected with an *Escherichia coli* O157:H7 strain. *Microbiol. Immunol.* 42(2):125.

25. Izumiya, H., Terajima, J., Wada, A., Isogai, Y., Ito, K., Tamura, K., and Watanabe, H. 1997. Molecular typing of Enterohemorrhagic *Escherichia coli* O157:H7 isolates in Japan by using pulsed-field gel electrophoresis. *J. Clin. Microbiol.* 35:1675.

26. Ji, B.T., Chow, W.H., Hsing, A.W., McLaughlin, J.K., Dai, Q., Gao, Y.T., Blot, W.J., and Fraumeni, J.F.Jr. 1997. Green tea consumption and the risk of pancreatic and colorectal cancers. *Int'l J. Cancer* 70:255.

27. John, T.J, and Mukundan, P. 1979. Virus inhibition by tea, caffeine and tannic acid. *Ind. J. Med. Res.* 69:542.

28. Karmali, M.A., Petric, M., Lim, C., Fleming, P.C., Arbus, G.S., and Lior, H. 1985. The association between idiopathic hemolytic uremic syndrome and infected by Verotoxin-producing *Escherichia coli*. *J. Infect. Dis.* 151:775.

29. Katoh, H. 1995. Prevention of mouse experimental periodontal disease by tea catechins. *Nihon Univ. J. Oral Sci.* (in Japanese), 21:1.

30. Kawamura, J., and Takeo, T. 1989. Antibacterial activity of tea catechin to *Streptococcus mutans*. *J. Jpn. Soc. Food Sci. Technol.* 36:463.

31. Luo, M., Kannar, K., Wahlqvist, M.L., O'Brien, R.C. 1997. Inhibition of LDL oxidation by green tea extract. *Lancet* 349:360.

32. Maeda, S., and Nakagawa, M. 1977. General chemical and physical analysis of various kinds of green tea. *Chagyo Kenkyu Hokoku* (in Japanese) 45:85.

33. Mitsuoka T. 1984. *A Color Atlas of Anaerobic Bacteria*, Shobunsha, Tokyo.

34. Mukoyama, A., Ushijima, H., Nishimura, S., Koike, H., Toda, M., Hara, Y., and Shimamura, T. 1991. Inhibition of rotavirus and entero-virus infections by tea extracts. *Jpn. J. Med. Sci. Biol.* 44:181.

35. Nakabayashi, T., Chemical components in tea leaves. 1991. In *Chemistry and Function of Green Tea, Black Tea, and Oolong Tea* (in Japanese), eds. T. Nakabayashi, K. Ina, and K. Sakata, pp.20. Kawasaki, Japan: Kogaku Shuppan.

36. Nakane, H., and Ono, K. 1990. Differential inhibitory effects of some catechin derivatives on the activities of human immunodeficiency virus reverse transcriptase and cellular deoxyribonucleotic and ribonucleotic acid polymerases. *Biochemistry* 29:2841.

37. Nakayama, M., Toda, M., Okubo, S., and Shimamura, T. 1990. Inhibition of influenza virus infection by tea. *Lett. Appl. Microbiol.* 11:38.

38. Nakayama, M., Suzuki, K., Toda, M., Okubo, S., Hara, Y., and Shimamura, T. 1993. Inhibition of the infectivity of influenza virus by tea catechins. *Antiviral Res.* 21:289.

39. Ogunrinola, O.A., Fung, D.Y.C., and Jeon, I.J. 1996. *Escherichia coli* O157:H7 growth in laboratory media as affected by phenolic antioxidants. *J. Food Sci.* 61:1017.

40. Ohshima, T., Izumitani, A., Sobue, S., Okahashi, N., and Hamada, S. 1983. Non-cariogenicity of the disaccharide palatinose in experimental dental caries of rats. *Infect. Immunol.* 39:43.

41. Oiwa, T., Sakanaka, S., Kashiwagi, M., Hasegawa, Y., Yoshihara, Y., and Yoshida, S. 1993. Inhibitory effect of human plaque formation by green tea catechins (Sunphenon). *Jpn. J. Prd. Dent.* (in Japanese) 31:247.

42. Okubo, T., and Juneja, L.R. 1997. Effect of green tea catechins on human intestinal microflora. In *Chemistry and Applications of Green Tea*, eds. T. Yamamoto, L.R. Juneja, D.C. Chu, and M. Kim, pp.109. New York: CRC Press.

43. Okami, Y., Takashio, M., and Umezawa, H. 1981. Ribocitorin, A new inhibitor of dextransucrase. *J. Antibiot.* 24:344.

44. Onishi, M., Okaumura, F., and Murakami, Y. (1978) *In vitro* screening of tea leaves effective against dental caries. *J. Dent. Health.* 27:279.

45. Onishi, M., Shimura, N., Nakamura, C., and Sato, M. (1981a) A field test on the caries preventive effect of tea drinking. *J. Dent. Health.* 31:13.

46. Onishi, M., Ozaki, F., Yoshino, F., and Murakami, Y. (1981b) An experimental evidence of caries preventive activity of non-fluoride component in tea. *J. Dent. Health.* 31:158.

47. Onishi, M. 1985. The feasibility of tea drinking program for dental public health in primary schools. *J. Dent. Health*, 35:402.

48. Ono, K., and Nakane, H. 1991. In *Proceedings of the International Symposium on Tea Science in Japan*, pp.277.

49. Otake, S., Makimura, M., Kuroki, T., Nishihara Y., and Hirasawa M. 1991. Anticaries effects of polyphenolic compounds from Japanese green tea. *Caries Res.* 25:438.

50. Podolak, R.K., Zayas, J.F., Kastner, C.L., and Fung, D.Y. 1996. Inhibition of *Listeria monocytogenes* and *Escherichia coli* O157:H7 on beef application of organic acids. *J. Food Prot.* 59:370.

51. Saeki, Y., Ito, Y., Shibata, M., Sato, Y., Okuda, K., and Takazoe, I. 1989. Antibacterial action of natural substances on oral bacteria. *Bull. Tokyo Dent. Coll.* 30:129.

52. Sakanaka, S., Sato, T., Kim, M., and Yamato, T. 1990. Inhibitory effects of green tea catechins on glucan synthesis and cellular adherence of cariogenic streptococci. *Agric. Biol. Chem.* 54:2925.

53. Sakanaka, S., Shimura, N., Aizawa, M., Kim, M., and Yamato, T. 1992. Effect of green tea catechins against dental caries in conventional rats. *Biosci. Biotech. Biochem.* 56:592.

54. Sakanaka, S., Aizawa, M., Kim, M., and Yamamoto, T. 1996. Inhibitory effects of green tea catechins on growth and cellular adherence of an oral bacterium, *Porphyromonas gingivalis. Biosci. Biotech. Biochem.* 60:745.

55. Sakanaka, S. 1997. Green tea catechins for prevention of dental caries. In *Chemistry and Applications of Green Tea*, eds. T. Yamamoto, L.R. Juneja, D.C. Chu, and M. Kim, pp.87. New York: CRC Press.

56. Shinchi, K., Ishii, H., Imanishi, K., and Kono, S. 1999. Relationship of cigarette smoking, alcohol use, and dietary habits with Helicobacter pylori infection in Japanese men. *Scand. J. Gastroenterol.* 32(7):651.

57. Takahashi, H., and Ninomiya, M. 1997. Metabolism of tea catechins. In *Chemistry and Applications of Green Tea*, eds. T. Yamamoto, L.R. Juneja, D.C. Chu, and M. Kim, pp.87. New York: CRC Press.

58. Toda, M., Okubo, S., Ohnishi, R., and Shimamura, T. 1989. Antibacterial and bactericidal activities of Japanese green tea. *Nippon saikingaku zassh.* 44:669.

59. Toda, M., Okubo, S., Ikigai, H., and Shimamura, T. 1990. Antibacterial and anti-hemolysin activities of tea catechins and their structural relatives. *Nippon saikingaku zassh.* 45(2):561.

60. Tsai, S.H., and Chou, C.C. 1996. Injury, inhibition and inactivation of *Escherichia coli* O157:H7 by potassium sorbate and sodium nitrite as affected by pH and temperature. *J. Sci. Agric.* 71:10.

B.B. Whitmore

A.S. Naidu

Glucosinolates

15

I. INTRODUCTION

Glucosinolates occur in many species of plants of the family Brassicaceae (formerly named Cruciferae). Members of the Brassica family contain four sepals, four petals, and a superior ovary with 1 to many seeds contained in a capsule (silique), with two deciduous valves. There are 300+ genera with 3,000 or more species worldwide. Many of the common species are used as food (Hickman, 1993).

Plants of Brassica include mustards, cabbage (derived from *Brassica oleracea*, Capitata, Pekinensis, Chinensis and Acephala groups), cauliflower (derived from *B. oleracea* var. *botrytis*), brussel sprouts (*B. oleracea* var. *gemmifera* Zenker), broccoli (derived from *B. oleracea*), kohlrabi (*B. oleracea* L. *var. gongylodes* L.), kale (Acephala group-ornamental), horseradish (*Armoracia rusticana* G. M. Sch.), wasabi or Japanese horseradish (*Wasabia japonica* Matsum or *Wasabia tenuis* Matsum), and radishes (*Raphanus sativus* L. var. *radicula* Pers.). Some common mustards include Black mustard (*B. nigra* [L] W. Koch), *B. juncea* (formerly *B. rugosa*, brown, Indian mustard, leaf or mustard greens, Oriental or Asian mustard), *B. hirta* Moench (also referred to as *B. alba*, *Sinapis alba* or white mustard), *B. carinata*, *B. napus* (Rape, or rapeseed, Colza, Argentine rape, from which canola oil is made), and *B. rapa* L. (turnips, Rape, or field mustard).

Many other plant families besides Brassicaceae also contain glucosinolates. Capparaceae (Capparidaceae) and the subfamily Cleomoideae, form glucosinolates when injured. Capers are considered as a member of this family. The Order Capparales contain both Brassicaceae and Capparaceae (Cronquist, 1981; Thorne 1992; Dahlgren 1980; Takhtajan 1980; Martin 1999). Capers differ from Brassicaceae by exhibiting slightly zygomorphic flowers, exerted stamens, and an ovary on a gynophore or stipe (Martin, 1999). Unopened caper blossoms are used in many ethnic foods as a condiment and in sauces (commonly prepared fish tartar sauces). Tropaeolaceae contains 'mustard oil'. One example of this family is the Garden Nasturtium (*Tropaeolum majus*). Bataceae (Batidaceae), Akaniaceae (Bretschneidera-ceae), Limnanthaceae (*Limnanthes douglasii*),

Moringaceae (*Moringa oleifera* Lam.), Euphorbiaceae (*Drypetes roxbarghii* Hurusawa), Resedaceae, Salvadoraceae, Tovariaceae (sometimes placed in Capparaceae), Gyrostemonaceae (*Codonocarpus cotinifolius*), Plantaginaceae (Plantain-*Plantago major*), and Caricaceae seeds (*Carica* L.), also contain isothiocyanates (mustard oils) when damaged (Mitchel & Rook, 1979).

II. OCCURRENCE

Glucosinolates are abundant in cruciferous vegetables, including cabbage, Brussels sprouts, broccoli, cauliflower, horseradish, mustard, watercress, turnips, radish, rutabaga, and kohlrabi (Fenwick et al., 1983; Mitchel & Rook, 1979). Cabbage varieties contain 4 to 146 ppm allyl isothiocyanate (AITC); 0 to 2.8 ppm benzyl isothiocyanate (BITC); and 1 to 6 ppm phenylethyl isothiocyanate (PEITC). Other isothiocyanates are present as components of total glucosinolate content in cabbages (Van Etten et al., 1976). Ripe papaya fruit contains 4 ppm BITC in the pulp and 2910 ppm in the seed. During growth, BITC concentration increases in the seeds and decreases in the pulp; when ripe, seeds contain 500 times more BITC than pulp (Tang, 1971). Three-day-old sprouts of cultivars of broccoli and cauliflower contain 10 to 100 times higher levels of glucoraphanin (the glucosinolate of suforaphane) than the corresponding mature plants. Extracts of 3-day-old broccoli cultivars are highly effective in reducing the incidence, multiplicity, and rate of development of mammary tumors in DMBA-treated rats. Sprouts of many broccoli cultivars contain negligible quantities of indole glucosinolates, which predominate in the mature vegetable and may give rise to degradation products (eg. Indole-3-carbinol) that could possibly enhance tumorigenesis (Fahey et al., 1997).

A. Biosynthesis

1. Site of production. Injury to plants in Brassica causes hydrolysis of existing glucosinolates by the enzyme myrosinase (thioglucosidase). Myrosinase produces AITC as the major product, as well as nitriles, epinitriles, and glucose. Myrosinase has been suggested to be a cell wall or membrane-bound enzyme. Recent studies by Kelly and co-workers (1998) have established that sinigrin is localized to protein bodies in aleurone-like cells, and is absent from myrosin cells. Glucosinolates are composed of a sulfonated oxime, a thioglucose group, and a variable carbon R-group, which determines the name of the glucosinolate compound (*TABLE 1*).

The endproduct of a glucosinolate-myrosinase reaction is determined by the physical, chemical and environmental conditions in which the hydrolysis reaction takes place. More than 80% of all glucosinolates identified are in the family of Brassicaceae, considered as a chemotaxonomic criterion for Brassica classification (Kjaer, 1976).

Sinigrin, a glucosinolate is hydrolyzed by thioglucosidase (myrosinase) producing 1-cyano-2, 3-epithiopropane (CETP), allyl cyanide (AC), AITC, and allyl thiocyanate (ATC) (Shofran et al., 1998) (*FIGURE 1*). Mustard seed glucosinolate contains about 93% sinigrin (Shahidi & Gabon, 1990). Glucobrassicin (an indole glucosinolate) hydrolyzes enzymatically to yield 3,3'-diindolylmethane, 3-indoleacetonitrile, and 3-indolemethanol (Labague et al., 1991). Isothiocyanates have a hydroxyl group at position-2, cyclise, producing oxazolidine-2-thiones. Goitrin is an example of this process, which is produced

TABLE 1. Common terms associated with R-Glucosinolates

Name	R-Glucosinolate
Glucobrassicanapin	4-pentenyl
Glucobrassicin	3-indolymethyl
Glucobrassicin-1-sulfonate	1-sulfonate-3-indolylmethyl
Glucoberteroin	5-methylthiopentyl
Glucocheirolin	3-methylsulfonylpropyl
Glucoerucin	4-methylthiobutyl
Glucoiberin	3-methylsulfinylpropyl
Gluconapin	3-butenyl
Gluconapoleiferin	2-hydroxy-4-pentenyl
4-Hydroxyglucobrassicin	4-hydroxy-3-indolylmethyl
4-Methoxyglucobrassicin	4-methoxy-3-indolylmethyl
Neoglucobrassicin	1-methoxy-3-indolylmethyl
Progoitrin	2-hydroxy-3-butenyl
Sinigrin	2-propenyl

(Adapted from Ettlinger & Kjaer, 1968; Kjaer, 1976)

from 2-hydroxy-but-2-enylglucosinate (Hanley et al., 1990). Allyl-, benzyl-, and 4-methylthio-butylglucosinolates are enzymatically degraded to thiocyanates. Many mechanisms have been proposed in thiocyanate formation with no conclusive data.

Hydrolysis of alkenyl glucosinolates occurs with myrosinase and epithiospecifier protein (Tookey, 1973). Epithiospecifier protein (ESP) interacts with myrosinase to promote sulfur transfer from S-glucose moiety to the terminal alkenyl moiety (Macleod & Rossiter, 1985).

The pH conditions affect the production of glucosinolate compounds. AC production proceeds at pH 4.0, whereas AITC is produced at pH 7.0, and nitrile and elemental sulfur production occurs at lower pH (Bones & Rossiter, 1996).

Different grades of mustard exhibit different amounts of AITC. The Grain Research Laboratory and the Canada Grain Commission obtains quality data on Oriental and Brown (*B. juncea*) mustards to determine the amounts of isothiocyanates in different grades (*TABLE 2*). Mayton (1996) determined that different cultivars of the same species show different levels of AITC.

It has also been shown that plants, in response to damage by insects or injury, alter their glucosinolate metabolism (Olsson & Jonasson, 1994). Indoles and aromatic glucosinolate production increase while the aliphatic glucosinolates (like sinigrin) decrease. The indole glucosinolates stimulate feeding and oviposition in the Diamondback Moth (*Plutella xylostella* L. or *Plutella maculipennis*) and the White Cabbage Butterfly (*Pieris brassicae*). Sinigrin repels adult Black Swallowtail Butterflies (*Papilio polyxenes*) (Hughes & Renwick, 1997; David & Gardner, 1962).

Methyl methanethiosulfinate (MMTSO) is the primary breakdown product of S-methyl-L-cysteine sulfoxide by SMCSO lyase (Kyung & Fleming, 1994). It degrades to other sulfur compounds including methyl methanethiosulfonate (MMTSO$_2$), dimethyl disulfide (DMDS) and dimethyl trisulfide (DMTS). They also found that the principle antibacterial activity of fresh cabbage juice was MMTSO. Kyung et al. (1997) determined that heating cabbage juice resulted in the formation of MMTSO$_2$, a potent antibacterial agent.

F_{IGURE} *1.* Myrosinase-glucosinolate pathway of sinigrin hydrolysis (Based on Bones & Rossiter, 1996).

2. Distribution. Whole mustard seed is ground into flour, which is used to produce condiment mustard or sold as a spice. Ground, powdered, seed and prepared mustard can be purchased at grocery markets. Commonly white, black and brown mustard seeds are carefully blended to achieve precise flavor or aroma, depending on the desired product.

SMCSO can be synthesized by oxidizing S-methyl-L-cysteine. Fractional crystallization is used to separate the D(+) from the L(-) form using either an acetone-water or water-ethanol mixture. MMTSO can be synthesized by oxidizing DMDS with peracetic acid and vacuum distillation at 2 mm Hg. The fraction, which boils at 65°C, is collected and can be stored at -83°C (Kyung & Fleming, 1997).

T_{ABLE} 2. Variations in AITC levels in *Brassica juncea*

Type	Grade	No. of samples	AITC mg/g
Oriental-*B. juncea*	No.1	102	12.8
	No.2	2	11.4
	No.3	2	11.6
	Sample	4	13.0
Brown- *B. juncea*	No.1	93	10.4
	No.2	8	10.2
	No.3	9	10.8
	Sample	5	9.1

(Adapted from the Quality data, Grain Research Laboratory, 1998)

III. MOLECULAR PROPERTIES

Karcher and El Rasso (1998) reported various techniques to separate glucosides from cabbage using HPLC, high-performance capillary electrophoresis (HPCE), and capillary electrochromatography (CEC). For the detection of glucosinolates and their acid hydrolysis products, labeled fluorescent tags (linked to the carboxyl groups) were used. The hydrolysis products were reacted with 7-aminonaphthalen-1,3-disulfonic acid (ANDSA) fluorescent tag in the presence of carbo-diamide. This produced stable fluorescent derivatives detected by HPCE-laser induced fluorescence at 25 nM, with little to no side products and high yields. Capillary zone electrophoresis with n-octylk-β-D-glucoside (OG) micelle separated the ANDSA products using 50 mM sodium phosphate, pH 6.5, containing 50 nM OG as the running electrolyte. In a second separation approach, the derivatives were separated by micellar electro-kinetic capillary chromatography (MECC) with in situ charged OG-borate complexed micelles. Since the micellular system is highly flexible, the selectivity of the separation can be manipulated by varying the surfactant concentration. Due to high ionic nature, glucosinolates are not easily amenable to gas chromatography and require a precolumn derivitization.

A. Preparation of cabbage extract

Kyung et al. (1997) used Green Boy cabbage that was quartered, and passed through a Fitzmill. The pulp was pressed to extract the juice, and filtered through an ultra-filtration unit. The juice was frozen and stored at -20°C in glass containers. Unheated juice was extracted using a centrifuge-like juicer. Quartering and steaming cabbage at about 100°C in an autoclave at atmospheric pressure resulted in boiled cabbage juice with MMTSO$_2$.

B. Isolation of 1-cyano-2, 3-epithiopropane (CEPT)

Shofran ct al. (1998) adapted the method of Luthy and Benn (1979) for isolation of CEPT. About 2.9 g of 1-cyano-2, 3-epoxypropane is dissolved in 9 ml of acetone. The mixture was added drop-wise to a stirred combination mixture of 2.6 g thiourea, 4.3 g benzoic acid, in 22 ml of acetone. A white precipitate (formed in 2 h) was collected by filtration, and washed with diethyl ether. The precipitate was re-dissolved in a 3:7 acetone/water solution (190-ml). A solution of sodium carbonate (2 g in 10 ml water) was added drop-wise to the solution and extracted with 30 ml of benzene. Sodium carbonate solution (2 ml) was added to the aqueous phase and re-extracted with 30 ml benzene. The benzene extracts were combined, washed with water and saturated sodium chloride, and dried over magnesium sulfate. After filtration, 10 mg of hydroquinone was added to prevent polymerization. Reduced pressure was used to concentrate the solvent into a yellow liquid. This liquid was then distilled under reduced pressure (68°C, 1 mm Hg) to produce 1.4 g of CETP. The resulting CETP was estimated to be >99% pure by HPLC analysis.

C. Isolation of allyl thiocyanate (ATC)

Shofran et al. (1998) synthesized ATC by mixing 30 mM of allyl bromide in 10 ml of 95% ethanol, 30 mM sodium pentahydrate in 10 ml of water, and reflux stirred for 1 hour. The mixture was cooled to room temperature followed by extraction with diethyl ether, three times. The aqueous residue was cooled on ice, and stirred with 30 mM of potassium cyanide for 15 min. After 20 min, the cold mixture was extracted with pentane (4-times), dried over anhydrous sodium sulfate, and concentrated at 0°C. This method yielded 1.56 grams of ATC, determined to be >99% pure by HPLC.

D. Effects of milieu

Ascorbic acid may modulate myrosinase activity by inducing a conformational change, possibly in the active site (Shikita et al, 1999). It could activate the process at low concentrations and inhibit it at high concentrations (Bones & Slupphaug, 1989). It is also inactivated by 2-deoxy-2-fluoroglucotropaeolin (Cottax, et al. 1996).

Epithiospecifier protein (ESP) interacts with myrosinase to promote transfer of sulfur from the S-glucose moiety to the terminal alkenyl moiety. ESP is also inactive in the absence of ferrous ions. Addition of ferrous ions cause the hydrolysis reaction products to change from oxazolidine-2-thione to epithionitrile (Macleod & Rossiter, 1985).

Different pH environments effect production of the compounds. AC production proceeds at pH 4.0, whereas AITC is produced at pH 7.0, and nitrile and elemental sulfur production proceeds at lower pH (Bones & Rossiter, 1996).

AITC interacts with proteins containing high amounts of cysteine. Kawakishi and Kaneko (1987) reported that AITC has a 'high reactivity' with bovine serum albumin (BSA). A 50% decrease of cysteine was found to occur in insulin. Polymerization occurred in BSA, as well as ovalbulmin, and insulin (from bovine pancreas) with AITC exposure. This is possibly due to protein sulfenic acid formed by scission of the disulfide bonds. Further studies indicated that a minor part of insulin was depolymerized to a lower molecular peptide.

Due to its non-ionizable structure, the antimicrobial activity of MMTSO is not markedly influenced by pH changes. Thus, AITC and MMTSO may have application as food preservatives in acidic as well as non-acidic, and minimally processed foods (Kyung & Fleming, 1997).

IV. ANTIMICROBIAL ACTIVITY

Thiocyanates demonstrate a wide range of antimicrobial and allelopathic effects. Methanethiol, and its oxidation product (CH_3SSCH_3), as well as methyl isothiocyanate suppress the pea root-rot disease (*Aphanomyces euteiches*) when used as a soil fumigant (Lewis & Papavizas, 1971). These compounds adversely affect the morphological properties, mycelial growth, and oospore development. Addition to the soil with leaf tissue of cabbage and other *Brassicas* have a similar effect on pea, bean, sesame wheat root-rot fungi (Kirkegaard et al., 1996; Angus et al., 1994). Thiocyanates are also effective against wire-worms, nematodes and a number of insects (Beekhuis, 1975).

Volatile isothiocyanates are considered as the major allelochemicals responsible for the inhibitory effects of *Brassica* on growth of other plants. Accordingly, methyl isothiocyanate inhibits germination of a number of weeds such as pigweed, dandelion, crabgrass and common ragweed (Beekhuis, 1975; Teasdale & Taylorson, 1986). Not all weeds are inhibited by *Brassica* alleochemicals. Choesin and Boerner (1991) found that *Brassica napus* is not alleopathic (inhibitory) to *Medicago sativa* (alfalfa).

TABLE 3. Glucosinolates and their antibacterial spectrum

Bacterial species	Glucosinolate	MIC (ppm)	Reference
Aeromonas hydrophilia 7966	AITC	200	Shofran et al, 1998
Bacillus subtilis IS75	AITC	200	Shofran et al, 1998
Bacillus subtilis B96	MMTSO	50	Kyung & Fleming, 1997
	AITC	50	
	MMTSO$_2$	20	
Enterobacter aerogenes B146	MMTSO	200	Kyung & Fleming, 1997
	AITC	300	
	MMTSO$_2$	50	
E coli ATCC 43895 (O157:H7)	AITC (gas)	1,500	Delaquis & Sholberg, 1997
E. coli B34 (ATCC 33625)	MMTSO	50	Kyung & Fleming, 1997
	AITC	50	
E coli NC101	AITC	100	Shofran et al, 1998
E. coli 33625	AITC	200	Shofran et al, 1998
Lactobacillus brevis LA25	MMTSO	100	Kyung & Fleming, 1997
	AITC	300	
	MMTSO$_2$	50	
Lactobacillus brevis LA200	MMTSO	100	Kyung & Fleming, 1997
	AITC	300	
	MMTSO$_2$	50	
Lactobacillus brevis MD42	AITC	1,000	Shofran et al, 1998
Lactobacillus plantarum LA97	MMTSO	50	Kyung & Fleming, 1997
	AITC	300	
	MMTSO$_2$	50	
Lactobacillus plantarum LA70	MMTSO	50	Kyung & Fleming, 1997
Leuconostoc mesenteroides FFL44	AITC	500	Shofran et al, 1998
Leuconostoc mesenteroides LA10	MMTSO	100	Kyung & Fleming, 1997
	AITC	500	
	MMTSO$_2$	100	
Leuconostoc mesenteroides LA113	MMTSO	200	Kyung & Fleming, 1997
	AITC	500	
	MMTSO2	20	
Listeria monocytogenes 81-861	AITC (gas)	500	Delaquis & Sholberg, 1997
Listeria monocytogenes B67	MMTSO	50	Kyung & Fleming, 1997
	AITC	200	
	MMTSO$_2$	50	
Listeria monocytogenes B70	MMTSO	50	Kyung & Fleming, 1997
	AITC	200	
	MMTSO$_2$	50	
Pediococcus pentosaceous FFL48	AITC	1,000	Shofran et al, 1998
Pedicoccus pentasaceus LA3	MMTSO	200	Kyung & Fleming, 1997
	AITC	300	
	MMTSO$_2$	100	
Pediococcus pentosaceus LA76	MMTSO	200	Kyung & Fleming, 1997
	AITC	300	
	MMTSO$_2$	100	
Pseudomonas fluorescens MD13	AITC	200	Shofran et al, 1998
Salmonella typhimurium ATCC 14028	AITC (gas)	1,500	Delaquis & Sholberg, 1997
Salmonella typhimurium B38	MMTSO	50	Kyung & Fleming, 1997
	AITC	100	
	MMTSO$_2$	50	
	MMTSO	50	
Staphylococcus aureus B31	MMTSO	50	Kyung & Fleming, 1997
	AITC	100	
	MMTSO$_2$	50	
S. aureus 4220	AITC	200	Shofran et al, 1998
S. aureus B31	SMCSO	0.075%	Kyung et al., 1997

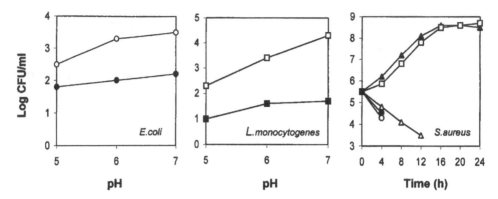

FIGURE 2. Inhibitory activity of glucosinolates against different food-borne bacteria. MIC (ppm) of AITC (●) and sodium benzoate (○) against *Escherichia coli* [redrawn from Shofran et al., 1998]. MIC (ppm) of MMTSO (■) and sodium benzoate (❑) against *Listeria monocytogenes* F5069 at different pH [Kyung & Fleming, 1997]. Effect of MMTSO2 on the growth of *Staphylococcus aureus* B31 in nutrient broth. Concentrations of MMTSO2 were 5 ppm (❑), 10 ppm (△), 20 ppm (●), 50 ppm (○) and control (▲) [redrawn from Kyung et al., 1997].

A. Antibacterial effects

Glucosinolates elicit a wide spectrum of antimicrobial activity against a variety of bacteria (*FIGURE 2*; *TABLE 3*). Shofran et al. (1998) reported that sinigrin, AC, and CETP were not inhibitory to any of the bacteria or yeasts tested. Since, the ATC converts to AITC, the MIC values reported may not be accurate. However, AITC had a MIC of 50 to 1,00 ppm for the bacteria tested (*E. coli* 33625, *E coli* NC101, *Pseudomonas fluorescens* MD13, *Aeromonas hydrophilia* 7966, *Staphylococcus aureus* 4220, *Bacillus subtilis* IS75, *Pediococcus pentosaceous* FFL48, *Leuconostoc mesenteroides* FFL44, *Lactobacillus brevis* MD42, but not for *Lactobacillus plantarum* MOP3), and 1 to 4 ppm for selected yeasts (*Saccharomyces cerevisiae* MD7, *Pichia membranefaciens* Y1617, *Torulopsis etchellsii* Y6651, *Hansenula mrakii* Y1617, *Torulaspora delbruekii* Y2547) but not against certain zero-tolerant yeasts (*Saccharomyces rouxii* Y2547, *Zygosacchraomyces globiformis* SPY9, *Z. bisporus* T12626).

Delaquis and Sholberg (1997) tested gaseous form of AITC for stasis and cidal activity against *Salmonella* Typhimurium (ATCC 14028), *Listeria monocytogenes* (strain 81-861), *E. coli* O157:H7 (ATCC 43895), *Pseudomonas corrugata* (isolated from lettuce), *Penicillium expansum* (isolated from apple), *Aspergillis flavus* (wheat isolate), and *Bortytis cinerea* (isolated from apple). Colony formation in *Salmonella* Typhimurium (ATCC 14028), *Listeria monocytogenes* (strain 81-861), *E. coli* O157:H7 (ATCC 43895), were completely inhibited at concentrations of 1,500 µg/L or more. Variations at different incubation temperatures were also observed. It was determined than *E. coli* was the most resistant of the bacteria and *Penicillum expansum* was more resistant to the sporicidal effects of AITC.

Methyl methanethiosulfinate (MMTSO) is the primary breakdown product of S-methyl-L-cysteine sulfoxide by SMCSO lyase (Kyung & Fleming, 1994). It degrades to other sulfur compounds including methyl methanethiosulfonate (MMTSO$_2$), dimethyl disulfide (DMDS) and dimethyl trisulfide (DMTS). It was reported that MMTSO is the principle antibacterial activity of fresh cabbage juice. Kyung et al. (1997) found that heating cabbage juice results in the formation of MMTSO$_2$, a potent antibacterial agent. In a later study, Kyung and Fleming (1997) examined susceptibility of 15 bacterial strains and 4 yeast species to different sulfur compounds. Results indicated that sinigrin and SMCSO were not inhibitory to all strains of bacteria and yeast up to 1,000 ppm, while MMTSO, AITC, and $MMTSO_2$ elicited inhibitory effects.

Kyung et al. (1997) tested the growth-multiplication of *S. aureus* (B31) in auto-claved and unheated cabbage juice, autoclaved SMCSO and $MMTSO_2$ added to nutrient broth. Autoclaved juice either undiluted or diluted with distilled water to 60% juice, com-pletely inhibited staphylococcal growth. When SMCSO was autoclaved in the broth, the concentration for MIC was 0.075 and 0.1%. When SMCSO was autoclaved separately and then added to nutrient broth, concentrations of 0.05, 0.075 and 0.1% caused growth inhi-bition. The MIC values for $MMTSO_2$ added to the broth were in the range of 10-50 ppm.

Hashem and Saleh (1999) studied the effects of *Diplotaxis harra* and *Erucaria microcarpa* (rare Egyptian plants) on *E. coli* (HB101), *Azospirillum brasilense* (ATCC29710), *Bacillus subtilis* (CAIM1007), *Streptococcus lactis* (DSM20250), two yeasts-*Saccharomyces cerevisiae* (GT16034B) and *Kluyveromyces lactis*, and two fungi: *Fusarium oxysporium* and *Aspergillis niger*. Fatty acids of the plants were found to be very effective antimicrobials (between 1.1-1.5 cm zone of inhibition). The volatiles were still effective (1.0-1.2 with one value of 2.2 for *S. cervisiae*).

Wasabi (*Wasaba japonica* Matsum) stems were found to have a minimal bacte-ricidal concentration (MBC) of 0.1 g/ml of medium, while mustard seed was found to have a 1.0 MBC. It was found cidal for *E. coli* IFO 13275. The MBC of each methyl-sulfinylalkyl isothiocyanate is 0.1 for the alkyl ethyl, 0.1 for butyl, 0.1 for hexyl and 1.0 for octyl. The main component in wasabi was the hexyl homologue (Ono et al, 1997).

Brabban and Edwards (1995) found that the inhibitory concentration of sinigrin was species dependent. Sinigrin at 10 μg/ml caused growth-inhibition of *E. coli, B. sub-*

TABLE 4. Glucosinolates and their antifungal spectrum

Fungal species	Glucosinolate	MIC (ppm)	Reference
Aspergillis flavus (wheat isolate)	AITC (gas)	100 μg/L	Delaquis & Sholberg, 1997
Bortytis cinerea (apple isolate)	AITC (gas)	100 μg/L	Delaquis & Sholberg, 1997
Fusarium sambucinum	AITC	63 -360	Mayton et al., 1996
Hansenula mrakii Y1617	AITC	4	Shofran et al., 1998
Hansenula mrakii Y27	MMTSO	6	Kyung & Fleming, 1997
	AITC	4	
	$MMTSO_2$	500	
Penicillium expansum (apple isolate)	AITC (gas)	100 μg/L	Delaquis & Sholberg, 1997
Pichia membranefaciens Y1617	AITC	2	Shofran et al., 1998
Pichia membranefaciens Y20	MMTSO	8	Kyung & Fleming, 1997
	AITC	2	
	$MMTSO_2$	500	
Saccharomyces cerevisiae MD7	AITC	4	Shofran et al., 1998
Saccharomyces cerevisiae Y6	MMTSO	10	Kyung & Fleming, 1997
	AITC	4	
	$MMTSO_2$	50	
Torulaspora delbruekii Y2547	AITC	4	Shofran et al., 1998
Torulopsis etchellsii Y6651	AITC	1	Shofran et al., 1998
Torulopsis etchellsii Y24 (Y 6651)	MMTSO	10	Kyung & Fleming, 1997
	AITC	1	
	$MMTSO_2$	50	

tilis 168, *Saccharomyces cerevisiae*, *B. stearothermphilus* 170, and *Streptomyces thermoviolaceus* NCIMB 10076). Strains of *E. coli* 84 and *Saccharomyces cerevisiae* were more sensitive to sinigrin hydrolysis products at 10 μg/ml concentration.

Methyl thiocyanate seems to produce a negative chemotaxis response for *P. aeruginosa*. About 80% of the organisms were inhibited within the first 30 seconds of exposure (Ohga et al. 1993).

B. Antifungal effects

The inhibitory activity of glucosinolates against various fungal species is shown in *Table 4*. Bushwaldt and co-workers (1985) evaluated the effects of sinigrin obtained from horseradish (*Armoracia rusticana* G.) on common fungal diseases of oilseed rape (*Brassica napus* var. oleifera). The growth of *Alternaria brassicae* (causative agent of black spot) was not affected by sinigrin in Czapek Dox medium. However, as the sinigrin concentration increased, the colony morphology of the fungi was affected (changed to a darker color). When grown in the dark, sinigrin had no effect, however, in UV light the fungal growth was affected. *Pholma lingam* (causative agent of black leg) also demonstrated similar results with sinigrin. Furthermore, sinigrin did not influence the number of sclerotia formation in *Sclerotinia sclerotiorum* (causative agent of stem rot).

Kyung and Fleming (1997) reported that AITC is a potent antifungal agent effective against both oxidative and fermentative yeast with a MIC of ≤4 ppm. Vega et al. (1997) found that sinigrin had no effect against *Paecilomyces fumosoroseus*.

Mayton et al. (1996) determined the effects of green manure (*Brassica* plants) on *Fusarium sambucinum* (causative agent of tuber dry rot in potatoes). Results indicated that different species and cultivars or plant introduction accessions, had different susceptibilities, probably due to a varied amount of AITC in each plant. *B. juncea* cv. Cutlass was completely inhibitory at 0.18 ppm. In 5 cultivars of *B. napus*, no AITC was measured, as well as in *B. juncea* PI 470241. *B. nigra* PI 169067 and PI 169066 were also completely inhibitory at 0.36 and 0.288 ppm, respectively.

V. APPLICATIONS

A. Mustard

Common herbal preparations and remedies use mustard as a rubefacient stimulant, for stomach and pancreas problems, arthritis, inflammation and toothache. Mustard flour, wheat flour and water is sometimes made into a paste, spread between two pieces of cotton sheeting, wrapped in flannel and placed on the chest of a congested person. This poultice provides warmth, but does not irritate the chest. It is used for the relief of muscular, skeletal pains, bronchitis and rheumatism. Poultices are also made from the leaves of the mustard plant. Ground mustard is sometimes mixed into a bath. Ground mustard is made into a tea for the relief of sore throats, or to aid stomach and pancreatic problems. Pennycress (*Thlaspi arvense*) was used in a Mithridate Confection (antedote) for poisons such as snake poison.

B. Horseradish

Horseradish is used as a stimulant, rubefacient, diuretic and an antiseptic. As a stimulant, it is put on oily or high fat proteins such as fish and meats, served alone, or in vinegar. It is used to treat dropsy along with mustard. It promotes sweating, and is usually infused in wine to accomplish this effect. The juice is mixed with vinegar to bleach freckles, as well as a worm-expelling agent for children. The juice is boiled down and mixed with sugar to yield syrup for hoarseness or coughs.

C. Field applications

Thiocyanates have both antimicrobial and allelopathic effects. Methanethiol, and its oxidation product, CH_3SSCH_3, as well as methyl isothiocyanate were shown to suppress the pea root-rot disease (*Aphanomyces euteiches*) when used as a soil fumigant (Lewis & Papavizas, 1971). Since these compounds adversely affect the morphological properties, mycelial growth, and oospore development, it could possibly be used as a fungicide. Mayton et al. (1996) also reported that certain species of *Brassica* are very effective for complete inhibition of *Fusarium sambucinum* (potato tuber dry rot). Brassica containing high levels of AITC could be used a green manure or as a fungicide for *F. sambrucinum*.

Delaquis and Sholberg (1997) suggested possible applications of gaseous AITC as a preservative in packaged foods. AITC elicits a potent antimicrobial activity at very low concentrations (1 mg/L).

Leaf tissue of cabbage and other *Brassicas* when added to the soil, have a similar effect on pea, bean, sesame, wheat root-rot fungi (Kirkegaard et al., 1996; Angus et al., 1994). Brassica (disease-free) grown on unused farmland or during crop rotation could be recycled for pea, bean, sesame, and wheat crops and used as an organic fungicide.

Since volatile isothiocyanates are potent allelochemicals responsible for the inhibitory effects of *Brassica* on growth of other plants, it could be used as a weed-deterrent. It has already been shown that methyl isothiocyanate inhibits germination of a number of weeds such as pigweed, dandelion, crabgrass and common ragweed (Beekhuis, 1975; Teasdale & Taylorson, 1986).

Certain microorganisms may degrade sinigrin by myrosinase-like activity (Llanos Palop et al., 1995; Brabban & Edwards, 1994). This may be beneficial for producers of oil and meal. Degradation of glucosinolates in rapeseed meal could render it useful as an animal feed. (Llanos Palop et al., 1995).

D. Multifunctionality

Isothiocyanates are effective substrates for human glutathione transferases. These enzymes promote addition of a thiol group to the electrophillic central carbon of the isothiocyanate group to form dithiocarbamates (Kolm et al., 1995). Further studies on reverse catalysis and sequestration supports the view that transferases are catalysts and are involved in metabolic disposition of isocyanates (Meyer et al., 1995; Zhang et al., 1995).

Using flow cytometry, Hasegawa et al. (1993) found that isothiocyanates alter cell cycle progression and induce cell cycle arrest, particularly at G_2/M phase. BITC and PEITC have a stronger effect than AITC.

Yu et al. (1989) provided evidence that chemopreventive isothiocyanates induce apoptosis and caspase-3-like protease activity and stimulated proteolytic cleavage of ply (ADP) ribose polymerase.

1. Antitumor effects. Steinmetz and Potter (1996) summarized several epidemiological studies conducted on cruciferae. Studies have shown a protective association between intake of cruciferous vegetables and cancer of the colon and rectum. Broccoli, cabbage, and Brussels sprouts were protective against mammary carcinogenesis (Stoewsand et al., 1988).

The anticancer properties of organic isothiocyanates are attributed to their selective toxicity to tumor cells and their ability to block procarcinogen activation by inhibiting phase-I enzymes and cytochrome P450s and inducing phase-II enzymes, including glutathione-S-tranferases (GSTs) (Zhang & Talalay, 1994). Zhang et al. (1992) identified an isothiocyanate in broccoli, called sulforaphane, which is a primary inducer of phase II enzymes. The authors suggest that this ITC is responsible for the protective effect of yellow and green vegetables against carcinogenesis.

2. Enzyme regulation. Sulforaphane could cause a dose-dependent induction of mRNA for GST A1/A2 and P1 but not M1 in hepatocytes of rats. Induction did not persist and was no longer apparent after 48 h exposure. Sulforaphane induced subunits for GSTs A1, A2, A3, A4, GSTM1, and GSTM2. Enzyme activities for CYP1A1 and 2B1/2 were decreased in a dose-dependent manner (Maheo et al., 1997).

The effects of chemopreventive agents, and various isothiocyanates-sulfoaphane on the transcriptional regulation of rat glutathione-S-transferase Ya (rGSTya) were determined using constructs containing the 5'flanking region of rGSTya fused with the CAT reporter gene in HepG2 cells. Increasing concentrations of the chemopreventive agents resulted in corresponding increases in CAT gene expression. Sulforaphane induced the transcription factor at mRNA level (Fei et al., 1996).

Sulforaphane is a potent competitive inhibitor of CYP2E1 with a K_i of 37µM. CYP2E1 has a high conservation of catalytic and regulatory specificity between rodent and human forms. It inhibits the mutagenicity of NDMA in *Salmonella* Typhimurium after preincubation with cytosol extracts from mouse liver. Sulforaphane itself is neither mutagenic nor could induce UDS in mouse hepatocytes at concentrations up to 20 µM (40 µM was cytotoxic) (Barcelo et al., 1996).

Treatment of rat and human hepatocytes with sulforaphane induced glutathione transferases and was a transient inhibitor of cytochrome P450 (Morel et al., 1997).

VI. SAFETY AND EFFICACY

Hydrolysis products of glucosinolates are not only toxic to some animals and microbes, but also to the plant producing them, and therefore require detoxification. Plants attempt to recycle unusable metabolites to usable metabolites, or they are excreted. Some examples of excreted metabolites include methylisothiocyanate and methylcyanate (Brown & Morra, 1997).

A. Intake

Chung et al. (1998) described a rapid HPLC-based assay to determine total urinary isothiocyanates. This assay could be a useful biomarker to identify the uptake of dietary isothiocyanates in epidemiological studies. Sones et al. (1984) calculated mean daily intake of glucosinolates from cruciferous vegetables in the UK as 46.1 mg from fresh vegetables and 29.4 mg from cooked vegetables. Intake varied with geographic location, season and income level. Intake during the winter months was double that of summer months. The authors indicated that some individuals are likely to have a daily consumption of >300 mg total glucosinolates. Major contributors to the intake were cabbage, cauliflower, Brussels sprouts, and Swede-turnip. The average daily intake of AITC and its glucosinolate precursor sinigrin is in 2 to 7 mg range in the UK (Fenwick & Heaney, 1983; Musk & Johnson, 1993a).

B. Metabolism

Isothiocyanates and indoles induce detoxification by affecting the Phase 1 and Phase 2 biotransformation enzyme activities. Phase 1 involves oxidation-reduction and hydrolysis reactions, making the carcinogenic product more hydrophilic. The enzymes of Phase I are the cytochrome P450s. Phase 2 involves conjugation reactions that promote excretion of carcinogenic compounds. Some enzymes in the Phase 2 are glutathione-S-transferases and UDP-glucuronyl transferases. Some of the hydrolysis products can cause goiter. Goitrogenicity in this manner is caused by the thiocyanate ion (a pseudohalide) and 5-vinyloxazolidine. Thiocyanate acts similar to halides, and is taken up by the thyroid gland, instead of iodine. Thus it acts as a competitive inhibitor, and can not be transformed into thyroxine, a metabolic hormone (Hathcock & Rader, 1994).

Bruggeman (1985) suggested that glutathione and cysteine may be regarded as transporting agents for isothiocyanates in the body, and that initial detoxification may be followed by a release of the reactive compounds in other sites. Getahun and Chung (1999) found that 1,3-benodithiol-2-thione is a specific urinary marker for quantifying ITC after human ingestion of boiled watercress extract and mustard. It was shown that glucosinolates in boiled watercress are converted to free ITC's in humans.

C. Animal studies

The Galvage Study was a 2-year carcinogenesis bioassay of food-grade AITC (CAS No. 57-06-7) in rats (F344/N) and mice (B6C3F$_1$). Single dose studies (14 day with 50mg/kg and 13-week with 25 mg/kg) showed pathological findings that included a thickened mucosal surface of the stomach in both rats and mice, and a thickened urinary bladder wall in the male mice. The chronic study showed lower mean body weight in male rats, than in the controls, and higher mean body weight than controls in the females. Urinary bladder transitional-cell papillomas occurred in dosed male rats ($p < 0.05$), but not in the control F344/N rats. Other findings in the dosed rats showed epithelial hyperplasia occurring in increased incidence (0/49, 1/49, 6/49), but did not occur in rats with papillomas. Dosed male mice show an increased incidence of cytoplasmic vacuolization in the liver (controls 4%, low-dose 16%, and high dose 26%). The evidence of carcinogenicity results was: male rats positive, female rats equivocal, male mice negative, female rats negative. (USA Toxicology Program, 1990).

A five-week study done on Scottish Blackface castrated male lambs, fed *ad lib* with secondary compounds ACN (10 m.mols/d)(allyl cyanide), AITC (10 mM/d) and DMDS (25 mM/d) (dimethyl disulfide), showed some physiological effects. Animals were fed dried grass pellets and capsules containing each metabolite. Voluntary food intake was depressed ($p < 0.001$) in both the ACN and AITC groups, but not in the DMDS group. DMDS produced elevated Heinz body counts ($p < 0.001$). T4 concentrations rose in the presence of ACN ($p < 0.05$), and an increase in free thyroxine index values ($p < 0.05$). Reduced glutathione concentrations in the blood were depressed by ACN and AITC, but elevated by DMDS. Possible liver damage with ACN and AITC were determined by elevated plasma γ-glutamyl transpeptidase. However, plasma creatine and plasma alkaline phosphatase was unaffected by the treatments. Plasma concentrations of albumin and urea were depressed with AITC ($p<0.05$). Hemolytic anemia or cell lysis caused by DMDS was not significant in this study. Cytochrome oxidase activity, measured by muscle biopsy on day 25, was not significantly altered by any of the compounds (Duncan & Milne, 1993). Proposed explanations of decreased voluntary food intake include an effect on microorganisms in the rumen, decreased microbial activity, rate of food passage, inhibition of aerobic respiration and release of the free cyanide. AITC in addition to above may inhibit digestive enzymes. They also proposed that ACN's effect on the increased T4 concentration may be due to ACN's role as a depletor of glutathione. Decreased glutathione seems responsible for the increased concentration of T4.

Fischer rats fed human fecal flora and single strains of *E.coli* EMO and *B. vulgatus* showed that it was possible that human intestinal flora may be involved in glucosinolate metabolism when brassica species are consumed by humans. Rats fed rapeseed meal exhibited a lowered T3 and T4 plasma levels and developed goiters (Rabot et al., 1992).

D. Limitations

September 1998, Health Canada issued a warning to consumers not to use any mustard seed oils because they may be contaminated by argemone oil. Over one thousand people became ill from dropsy-induced argemone. Argemone can cause death when consumed or absorbed through the skin. It is found in the family Papavaraceae. Papavaraceae produces yellow flowers and black seeds, which resemble the flowers and seeds of the mustard plant. Argemone has also been found in adulterants, which are added to mustard oil to increase profits in developing countries.

Some of the hydrolysis products of glucosinolates can cause goiter. Goitrogenicity is caused by the thiocyanate ion (a pseudo-halide) and 5-vinyloxazolidine. Thiocyanate acts chemically similar to halides, and is taken up by the thyroid gland, instead of iodine. Thus it acts as a competitive inhibitor even if iodine is administered, and thiocyanate can not be transformed into thyroxine, a metabolic hormone (Hathcock & Rader, 1994).

Mutagenic intermediates of glucosinolate degradation may also occur. Neudecker and Henschler propose two metabolic pathways that might lead to mutagenic intermediates. Epoxidation of the carbon to carbon double bond may occur by microsomal oxygenase, which could result in a reactive exposide. This exposide may react with the cellular DNA. Another possible pathway could be the spontaneous or enzymatic cleavage by hydrolysis of the NCS-leaving group or an oxidized derivatative that possesses better leaving group capacities. This may give rise to an allyl alcohol, which can be oxidized to acrolein by alcohol dehydrogenase, and further to acrylic acid.

Listeria spp. have been found on *Brassica* spp. in fresh market produce. In 1981, an epidemic of *Listeria monocytogenes* was linked to raw cabbage. In one study, *Listeria* spp. were isolated from 6 of 10 types produce tested, with radishes and potatoes being the more affected crops (Heisick et al., 1989).

VII. SUMMARY

Glucosinolates are abundant in various cruciferous vegetables including cabbage, Brussel sprouts, broccoli, cauliflower, horseradish, mustard, watercress, turnips, radish, rutabaga and kohlrabi. These bioactive compounds have been used for centuries in common herbal preparations and remedies ranging from rheumatism to treating snake poison. As stimulants, mustard and horseradish were applied on oily or high fatty proteins such as fish and meats. Volatile glucosinolates such as the isothiocyanates elicit a wide range of antibacterial, and antifungal effects. Application of gaseous allyl isothiocyanates as a preservative in packaged foods has been suggested at low concentrations. Since volatile isothiocyanates are potent allelochemicals, a possible use as weed deterrents has also been suggested. A protective anti-tumor association between intake of cruciferous vegetables and cancer of the colon and rectum. Broccoli, cabbage and Brussel sprouts were also shown to be protective against mammary carcinogenesis. However, certain studies with purified glucosinolates indicated toxicity in to some animals.

VIII. REFERENCES

1. Angus, J.F., Gardner, P.A., Kirkegaard, J.A., and Desmarchelier, J.M. 1994. Isothiocyanates released from Brassica roots inhibit growth of the take-all fungus. *Plant and Soil.* 162.107-112.
2. Barcelo, S., Gardiner, J.M., Gescher, A., and Chipman, J.K. 1996. CYP2E1-mediated mechanism of anti-genotoxicity of the broccoli constituent sulforaphane. *Carcinogenesis* 17:277-282.
3. Beekhuis, H.A. 1975. Technology and industrial applications. In: *Chemistry and Biochemistry of Thiocyanic acid and its derivatives*, ed. A. S. Newman, pp. 222-225. New York:Academic Press.
4. Bones, A.M., and Rossiter, J.T. 1996. The myrosinase-glucosinolate system, its organization and biochemistry. *Physiol. Plant* 97:194-208.
5. Bones, A.M., and Slupphaug, G.J. 1989. Purification, characterization and partial amino acid sequencing of β-thioglucosidase from *B. napus* L. *Plant Physiology.* 134:722-729.
6. Brabban, A.D., and Edwards, C. 1994. Isolation of glucosinolate degrading microorganisms and their potential for reducing the glucosinolate content of rapemeal. *FEMS Microbiol. Lett.* 119:83-88.
7. Brabban, A.D., and Edwards, C. 1995. The effects of glucosinolates and their hydrolysis products on microbial growth. *J. Appl. Bacteriol.* 79:171-177.
8. Brown, P.D., and Morra, M.J. 1997. Control of soil-borne plant pests using glucosinolate-containing plants. *Adv. Agr.* 61:167-231.
9. Bruggeman, I.M., Temmink, J.H.M., and van Bladeren, P.J. 1986. Glutathione and cysteine mediated cytotoxicity of allyl and benzyl isothiocyanate. *Toxic. Appl. Pharm.* 83:349-359.
10. Bushwaldt, L., Nielsen, J.K., and Sorensen, H. 1985. Preliminary investigations of the effect of sinigrin on *in vitro* growth of three fungal pathogens of oilseed rape. *World Crops: Production, Utilization, Description* 11:260-267.
11. Choesin, D.N. and Boerner, R.E.J. 1991. Allyl isothiocyanate release and the alleopathic potential of *Brassica napus* (Brassicaceae). *Am. J. Bot.* 78:1083-1090.
12. Chung, F.L., Jiao, D., Getahun, S.M., and Yu, M.C. 1998. A urinary biomarker for uptake of dietary isothiocyanates in humans. *Cancer Epidemiol. Biomarkers Prev.* 2:103-108.
13. Cottax, S., Henrissat, B., and Driguez, H. 1996. Mechanism-based inhibition and stereochemistry of glucosinolate hydrolysis by myrosinase. *Biochemistry.* 25:15256-15259.
14. Cronquist, A. 1981. *An Integrative System of Classification of Flowering Plants.* New York: Columbia University Press.
15. Dahlgren, R.M.T. 1980. An updated angiosperm classification. *Bot. J. Linnean Soc.* 80:91-124.
16. David, W.A.L., and Gardner, B.O.C. 1962. Oviposition and the hatching of the eggs of *Pieris brassicae* (L) in a laboratory culture. *Bull. Ent. Res.* 53:91-109.
17. Delaquis, P.J., and Sholberg, P.L. 1997. Antimicrobial activity of gaseous allyl isothiocyanate. *J. Food Prot.* 60:943-947.

18. Duncan, A.J., and Milne, J.A. 1993. Effects of oral administration of brassica secondary metabolites, allyl cyanide, allyl isothiocyanate and dimethyl disulfide, on the voluntary food intake and metabolism of sheep. *Br. J. Nutr.* 70:631-645.
19. Ettlinger, M.G., and Kjaer, A . 1968. Sulfur compounds in plants. *Recent Adv. Phytochem.* 1:59-144.
20. Fahey, J.W., Zhang, Y., and Talalay, P. 1997. Broccoli sprouts: an exceptionally rich source of inducers of enzymes that protect against chemical carcinogens. *Proc. Natl. Acad. Sci. USA* 94:10367-10372.
21. Fei, P., Matwyshyn, G.A., Rushmore, T.H., and Kong, A.N. 1996. Transcription regulation of rat glutathione S-transferase Ya subunit gene expression by chemopreventive agents. *Pharm. Res.* 7:1043-1048.
22. Fenwick G.A., Heaney, R.K., and Mullin, W.J. 1983. Glucosinolates and their breakdown products in foods and food plants. *CRC Crit. Rev. Food Sci. Nutr.* 18:123-201.
23. Gethun, S.M., and Chung, F.L. 1999. Conversion of glucosiolates to isothiocyanates in humans after ingestion of cooked watercress. *Cancer Epidemiol. Biomark. Prevent.* 8:447-451.
24. Hanley, B.A., Parsley, K.R., Lewis, J.A., and Fenwick, R.G. 1990. Chemistry of indole glucosinolates: Intermediacy of indole-3-yl methyl isothiocyanates in the enzyme hydrolysis of indole glucosinolates. *J. Chem. Soc. Perkin Trans.* 1:2273-2276.
25. Hasegawa, T., Nishino, H., and Iwashima, A. 1993. Isothiocyanates inhibit cell cycle progression of HeLa cells at G2/M phase. *Anticancer Drugs* 2:273-279.
26. Hashem, F.A., and Saleh, M.M. 1999. Antimicrobial components of some Cruciferae Plants (*Diplotaxis harra* Forsk. and *Erucaria microcarpa* Boiss.). *Phytother. Res.* 13:329-332.
27. Hathcock, J.N., and Rader, J.I. 1994. Food Additives, Contaminants, and Natural Toxins. In: *Modern Nutrition in Health and Disease*. 8th ed., eds. Shils, M. E., Olson, J.A., and Shike, M., pp. 1605-1606. Philadelphia: Lea and Febiger.
28. Health Canada Online. Sept. 11, 1998. Warning not to use any products containing mustard seed/mustard oil. *Health Canada Press Release.*
29. Heisick, J.E., Wagner, D.E., Nierman, M.L., and Peeler, J.T. 1989. *Listeria* spp. found on fresh market produce. *Appl. Environ. Microbiol.* 55:1925-1927.
30. Hickman, J.C. (ed.). 1993. *The Jepson Manual Higher of Plants of California*, Berkeley: UC-Berkeley Press.
31. Hughes, P.R., and Renwick, J.A.A. 1997. New oviposition stimulants for the Diamondback moth in cabbage. *Ent. Exp. Et App.* 85:281-283.
32. Karcher, A., and El Rassi, Z. 1998. High performance liquid phase separation of glucosides. II. Calpillary electrophoresis of the fluorescently labeled acid hydrolysis products of glucosinolates-profiling of glucosinolates in white and red cabbages. *J. Liq. . Chrom. Rel. Technol.* 21:1411-1432.
33. Kawakishi, S., and Kaneko, T. 1987. Interaction of proteins with allyl isothiocyanate. *J. Agric. Food Chem.* 35:85-88.
34. Kelly P.J., Bones, A.M., and Rossiter, J.T. 1998. Subcellular immunolocalization of the glucosinolate sinigrin in seedlings of *Brassica juncea. Planta.* 206:370-377.
35. Kirkegaard, J.A., Wong, P.T.W., and Desmarchelier, J.M . 1996. *In vitro* suppression of fungal root pathogens of cereals by Brassica tissues. *Plant Pathol.* 45:593-603.
36. Kjaer, A. 1976. Glucosinolates in the Cruciferae. In: *The Biology and Chemistry of the Cruciferae*, eds. Vaughan, J.G., MacLeod, A.J., and Jones, B.M.G., pp. 207-219. London: Academic Press.
37. Kolm, R.H., Danielson, U.H., Zhang, Y., Talalay, P., and Mannervik, B. 1995. Isothiocyanates as substrates for human glutathione transferases: structure-activity studies. *Biochem. J.* 311:453-459.
38. Kyung, K.H., and Fleming, H.P. 1996. Antimicrobial activity of sulfur compounds derived from cabbage. *J. Food Prot.* 60:67-71.
39. Kyung, K.H., Han, D.C., and Fleming, H.P. 1997. Antibacterial activity of heated cabbage juice, S-methyl-L-cysteine sulfoxide and methyl methanethiosulfonate. *J. Food Sci.* 62:406-409.
40. Labague, L., Garrat, G., Coustille, J. L., Viaud, M. C., and Rollin, P. 1991. Identification of enzymatic degradation products from synthesized glucobrassicin by gas chromotography-mass spectrometry. *J. Chromatogr.* 586:166-170.
41. Lewis, J.A., and Papavizas, G.C. 1971. Effect of sulfur containing compounds and vapors from cabbage decomposition on *Aphanomyces eureiches. Phytopathology* 61:208-214.
42. Llanos Palop, M., Smiths, J.P., and Ten Brink, B. 1995. Degradation of sinigrin by *Lactobacillus agilis* strain R16. *Int. J. Food Microbiol.* 26:219-229.
43. Luthy, B., and Matile, P. 1984. The mustard oild bomb: Rectified analysis of the subcellular organization of the myrosinase system. *Biochem. Physiol. Pflanzen* 179:5-12.

44. Macleod, A.J., and Rossiter, J.T. 1985. The occurrence of activity of epithiospecifier protein in some cruciferae seeds. *Phytochemistry*. 24:1895-1898.
45. Maheo, K., Morel, F., Langouet, S., Kramer, H., Le Ferrec, E., Ketterer, B., and Guillouzo, A. 1997. Inhibition of cytochromes P-450 and induction of glutathione S-transferases by sulforaphane in primary human and rat hepatocytes. *Cancer Res.* 57:3649-3652.
46. Mayton, H.S. 1996. Correlation of fungicidal activity of Brassica species with allyl isothiocyanate production in macerated leaf tissue. *Phytopathology* 86:267-271.
47. Meyer, D.J., Crease, D.J., and Ketterer, B. 1995. Forward and reverse catalysis and product sequestration by human glutathion-S-transfererases in the reaction of GSH with dietary aralkyl isothiocyanates. *Biochem. J.* 306:565-569.
48. Mitchell, J., and Rook, A. 1979. *Botanical Dermatology*. Vancouver: Green Grass Ltd.
49. Morel, F., Langouet, S., Maheo, K., and Guillouzo, A. The use of primary hepatocyte cultures for the evaluation of chemoprotective agents. *Cell Biol. Toxicol.* 13:323-329.
50. Musk, S.R., and Johnson, I.T. 1993. The clastogenic effects of isothiocyanates. *Mutat. Res.* 300:111-117.
51. Neudecker, T., and Henschler, D. 1985. Allyl isothiocyanate is mutagenic in *Salmonella typhimurium*. *Mut. Res.* 156:33-37.
52. Ohga T., Masduki, A., Kato, J., and Ohtake, H. 1993. Chemotaxis away from thiocyanic and isothiocyanic esters in *Pseudomonas aeruginosa*. *FEMS Microbiol. Lett.* 113:63-66.
53. Olsson, K., and Jonasson, T. 1994. Leaf feeding by caterpillars on white cabbage cultivars with different 2-propenyl glucosinolate (sinigrin). *J. Appl. Entomol.* 118:197-202.
54. Ono, H., Tesaki, S., Tanabe, S. and Watanabe, M. 1997. 6-Methylsulfinylhexyl isothiocyanate and its homologues as food-originated compounds with antibacterial activity against *Escherichia coli* and *Staphylococcus aureus*. *Biosci. Biotechnol. Biochem.* 62:363-365.
55. Pulverer, G. 1969. Allyl isothiocyanate: A new broad-spectrum antibiotic from nasturtium. *German Medical Monthly*. 14:27-30.
56. Rabot, S., Nugon-Baudon, L., Raibaud, P., and Szylit, O. 1992. Rape-seed meal toxicity in gnotobiotic rats:influence of a whole human faecal flora or single human strains of *Escherichia coli and Bacteroides vulgatus*. *Br. J. Nutr.* 70:323-331.
57. Sapru, R.P. 1998. Epidemic dropsy. *Health Tribune India*. Sept. 02:2.
58. Shahidi, F., and Gabon, J.E. 1990. Fate of sinigrin in methanol/ammonia/water-hexane extraction of *B. juncea* mustard seed. *J. Food Sci.* 55:793-795.
59. Shikita, M., Fahey, J.W., Golden, T.R., Holtzclaw, W.D., and Talalay, P. 1999. An unusual case of 'uncompetitive activation' by ascorbic acid: purification and kinetic properties of a myrosinase from *Raphanus sativus* seedlings. *J. Biochem.* 341:725-732.
60. Shofran, B.G., Purrington, S.T., Breidt, F., and Fleming, H.P. 1998. Antimicrobial properties of sinigrin and its hydrolysis products. *J. Food Sci.* 63:621-624.
61. Sones, K., Heaney, R.K., and Fenwick, G.R. 1984. The glucosinolate content of UK vegetables -cabbage (*Brassica oleracea*), swede (*Brassica napus*) and turnip (*Brassica campestris*). *Food Additives and Contaminants* 1(3):289-296.
62. Steinmetz, K.A., and Potter, J.D. 1996. Vegetables, fruit, and cancer prevention: a review. *J. Am. Diet Assoc.* 96:1027-1039.
63. Stoewsand, G.S., Anderson, J.L., and Munson, L. 1988. Protective effect of dietary brussels sprouts against mammary carcinogenesis in Sprague-Dawley rats. *Cancer Lett.* 39:199-207.
64. Takhtajan, A.L. 1980. Outline of the classification of flowering plants (Magnoliophyta). *Bot. Rev.* 46: 225-359.
65. Tang, C.S. 1971. Benzyl isothiocyanate of papaya fruit. *Phytochem.* 10:117-121.
66. Teasdale, J.R., and Taylorson, R.B. 1986. Weed seed response to methyl isothiocyanate and metham. *Weed Sci.* 34:520-524.
67. Thorne, R.F. 1992. An updated phylogenetic classification of the flowering plants. *Aliso* 13:365-389.
68. Tookey, H.L. 1973. Crambe thioglucoside glucohydrolase (EC3.2.3.1): Separation of a protein required for epithiobutane formation. *Can. J. Biochem.* 51:1654-1660.
69. United States National Toxicology Program. 1990. Chemical Status Report. NTP Chemtrak System. (NIH TR-234). Research Triangle Park, NC.
70. VanEtten, C.H., Daxenbichler, M.E., Williams, P.H., and Kwolek, W.F. 1976. Glucosinolates and derived products in cruciferous vegetables. Analysis of the edible part from twenty-two varieties of cabbage. *J Agric. Food Chem.* 24:452-455.

71. Vega, F.E., Dowd, P.F., McGuire, M.R., Jackson, M.A., and Nelson, T.C. 1997. *In vitro* effects of secondary plant compunds on germination of blastospores of the entomopathogenic fungus *Paecilomyces fumosoroseus* (Deuteromycotina: Hyphomycetes). *J. Invertebrate Pathol.* 70:209-213.

72. Yu, R., Mandlekar, S., Harvey, K.J., Ucker, D.S., and Kong, A.N. 1998. Chemopreventive isothiocyanates induce apoptosis and caspase-3-like protease activity. *Cancer Res.* 58:402-408.

73. Zhang, Y., Kensler, T.W., Cho, C.G., Posner, G.H., and Talalay, P.1992. A major inducer of anticarcinogenic protective enzymes from broccoli: isolation and elucidation of structure. *Proc. Natl. Acad. Sci. USA.* 89:2399-2403.

74. Zhang, Y., and Talalay, P. 1998. Mechanism of differential potencies of isothiocyanates as inducers of anticarcinogenic Phase 2 enzymes. *Cancer Res.* 58:4632-4639.

75. Zhang, Y., Kolm, R.H., Mannervik, B., and Talalay, P. 1995. Reversible conjugation of isothiocyanates with glutathione catalyzed by human glutathione transferases. *Biochem. Biophys. Res. Commun.* 206:748-755.

A.S. Naidu

Agar

16

I. INTRODUCTION

Agar is a hydrophilic colloid extracted from certain marine algae of the class *Rhodophyceae*. It is insoluble in cold water but soluble in boiling water. A hot aqueous 1.5% solution is clear, and when cooled to 32-39°C forms a firm, resilient gel which melts at temperatures greater than 80°C. This difference in gelling and melting temperatures, known as hysteresis, makes agar of great importance in the microbiological/ medical, bio-pharmaceutical, and food industries. Agar is synonymous to agar-agar, *dai choy goh, kanten*, Japanese gelatin, Japanese isinglass, vegetable gelatin, and angel's hair.

Chemically, agar is a salt or a mixture of salts of anionic polysaccharides. Agar, *in situ*, is a cell wall constituent and probably serves as a structural, ion-exchange, and dialysis membrane. It is a neutral polymer of galactose with a low viscosity in water solution, stable gelling temperatures, and a strong gel structure. It has a melting point of 95°C and it solidifies at 45°C. Agar has a low sulfate content.

During investigation of the interaction of human lactoferrin with various bacteria, Gerlach et al. (1994) found that, in *Streptococcus pyogenes*, lactoferrin binding occurred to agar- rather than broth-grown cells irrespective of the nutrients used. Furthermore, binding of human lactoferrin to broth-grown, heat-killed bacteria was induced by overnight incubation on agar media or short-time exposure of the cells to water-soluble agar extract. The binding pattern was revealed in most of 92 strains of *S. pyogenes* representing various M- or T-types with no apparent type variation. The component thus bridging the attachment of human lactoferrin to the streptococcal cell surface was recovered in extracts of agar-grown cells and isolated by affinity chromatography on human lactoferrin-sepharose. By gel filtration in the presence of radiolabeled human lactoferrin, this component exhibited similar elution position as crude water-soluble agar extract. Chemical analysis identified the active human lactoferrin-binding agar component to be a galactose-rich polysaccharide (GRP). Further binding tests showed that the interaction between streptococci and GRP was stable in the presence of high molar NaCl, KSCN, or urea and was unaffected by various serum or matrix proteins or by streptococccal lipoteichoic acid; however, a moderate inhibition by heparin or bovine mucin was observed. Studies on isogenic mutants of *S. pyogenes* did not support the involvement of M-protein or the hyaluronate capsule in the binding of GRP. Sodium dodecyl sulfate-polyacrylamide gel electrophoresis and western blot analyses revealed a GRP-binding protein of approximately 70 kDa in the cell wall extracts of two strains of *S. pyogenes*, types M19 and M55. Finally, the adherence of (broth-grown) ^{3}H-thymidine-labeled *S. pyogenes*, type M19, to the pharyngeal epithelial cell line DT562 or to normal tonsillar epithelial cells was inhibited by GRP in a dose-related manner. Based on these data, Gerlach et al. (1994) proposed that the streptococcal GRP-binding component may represent a novel surface lectin acting as a mucosal adhesin for *S. pyogenes*, in accordance with previous data indicating that galactose-containing sugar moieties may serve as ligands for the adherence of streptococci to pharyngeal cells. These results also indicated that GRP-like components such as mucin or heparin might act to block epithelial adherence of *S. pyogenes* at the mucosal level.

The capacity of agar components to inhibit tissue adhesion-colonization of microbial pathogens such as *S. pyogenes* and the ability to interact with antimicrobial agents such as lactoferrin has unveiled a multifunctional role for agar ingredients as food additives.

II. DISTRIBUTION

Many marine algae accumulate significant quantities of sulfated polysaccharides in their cell walls or intracellular compartments. The nature of the polysaccharide sulfate from each alga is highly characteristic of their taxonomic division (Conway & Rees, 1962). In the *Rhodophyceae* (red seaweed) these polysaccharides usually contain galactose as the D- or as the D- and L-enantiomorphs, often together with such derivatives of each enantiomorph as the 3,6-anhydride, the 6-*O*-methyl ether and/or various sulfate esters.

The sulfated galactans of the red seaweed are a family of related molecules, each with properties appropriate for a particular species growing in a particular environment. Araki et al. (1958) have established that agarose is a repeating molecule (*FIGURE-1*). In this polysaccharide, the 1,4-linked galactose unit occurs as the 3,6-anhydride of the L-sugar, while the 1,3-linked galactose occurs as unmodified D-isomer. Later studies have indicated that agars from other species have similar structure (Anderson et al., 1965).

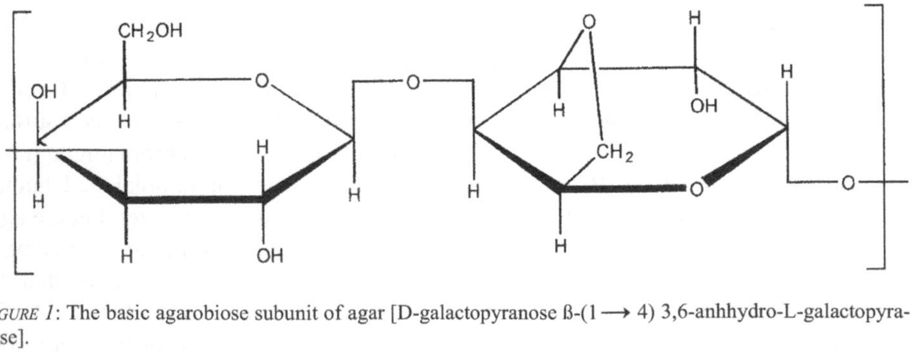

FIGURE 1: The basic agarobiose subunit of agar [D-galactopyranose ß-(1 ⟶ 4) 3,6-anhhydro-L-galactopyranose].

TABLE *1*: Structural variations in galactose units of *Rhodophyceae* polysaccharides

Species	Polysaccharide type	3-linked Gal	4-linked Gal
Gelidium amansii	Agarose	D	L
Gracilaria confervoides	Agar	D	L
Porphyra sp.	Porphyran	D	L
Gloiopeltis furcata	Funorin	D	L
Chondrus crispus	χ-Carrageenan	D	D
Chondrus crispus	λ-Carrageenan	D	D
Furcellaria fastigiata	Furcellaran	D	D
Hypnea specifera	Hypnean	D	D

Rees and Conway (1962) reported that porphyran (the polysaccharide of *Porphyra* species) is related structurally to agarose. The 1,4-linked unit is L-galactose, however, unlike agarose, only a fraction of this sugar occurs as the 3,6-anhydride; the remainder is present as the 6-sulfate.

The mucilage of *Gloipeltis furcata* consists of polysaccharide with similar alternating sequence of 1,4-linked L-galactose and 1,3-linked D-galactose units. It differs from porphyran in containing no 6-*O*-methyl-D-galactose units and is clearly distinguished from agarose by its high ester sulfate and L-galactose content (Turvey & Williams, 1963).

The basic structure of χ-carrageenan consists of alternative β-1-4- and α-1,3-linked D-galactose units, with 4-linked units occurring as 3,6-anhydro D-galactose and the 3-linked units as D-galactose-4-sulfate (Hirase et al., 1958). The λ–carrageenan also contains α-1,3- and β-1-4-linked D-galactose units. A small amount of the 4-linked units in this polysaccharide occur as the 3,6-anhydride; the majority occurs instead as the 2,6-disulfate. Structural variations on the theme $[\beta\text{-Gal}p1 \rightarrow 4\alpha\text{-Gal}p1 \rightarrow 3]_n$ in red seaweed polysaccharides are shown in TABLE *1*.

III. CHEMISTRY AND STRUCTURE

Agar is composed of two polysaccharides; agarose and agaropectin. Agarose consists of alternating 1,4 linked 3,6-anhydro-α-L-galactopyranose and 1,3-linked β-D-galactopyranose. Agaropectin is a more complicated structure that contains residues of sulfuric, pyruvic, and uronic acids, in addition to D-galactose and 3, 6-anhydro-L-galactose. In commercial agar, three molecular types are observed. These molecular types include neutral agarose, pyruvated agarose (4,6-*O*-1-carboxyethylidene-D-galactose, and a sulfated galactan of the agar type. Half-ester sulfates attached to D-galactopyranose units are present in varying amounts but are present on approximately every tenth D-galactopyranose subunit.

The agarose backbone is a diastereo-isomer with iota and kappa-carrageenans in that it contains 1- 4-linked 3-6-anhydro-α-L-galactose rather than the D-enantiomer. The agar backbone is substituted to various extents with neutral and charged groups.

The X-ray diffraction studies indicate that agarose is a double helix with parallel threefold chains and left-handed with a shorter helix pitch. Thus, with a pitch of 1.90 nm, the agarose helix is less extended than either iota-carrageenan (2.66 nm) or kappa-carrageenan (2.46 nm) with its double helical structure.

Agars could be fractionated by aqueous elution at various temperatures; the lower gel strength fractions are usually higher in organic sulfate and lower in gel strength.

TABLE 2: Chemical composition of water-soluble agar extract (GRP)

Analyte	Molar Proportions	Analyte	Molar Proportions
Carbohydrates		*Elements*	
Galactose	1.0	Carbon	36.8%
3,6 anhydrogalactose	0.2	Hydrogen	5.2%
2-OMe-galactose	trace[a]	Nitrogen	2.9%
4-Moe-galactose	trace	Oxygen	39.4%
		Sulfur	2.8%

[a]Trace = < 0.05; monosaccharides were assayed as alditol acetate derivatives according to Sawardeker et al. (1965). Based on Gerlach et al. (1994)

The molecular weight of agar has been estimated to be 5- to 30-kDa by osmotic pressure measurements. Estimates resulting from diffraction and polarization studies indicate average molecular weights as high as 160-kDa for laboratory-extracted agars. It appears desirable to use the word agar only in the generic sense to indicate a derivatized galactan possessing certain properties common to a group of substance, each of which may differ structurally from the others. Chain length may be reduced by ultrasonic vibrations and gamma radiation, as well as by intensive agitation, chemical means, and high temperature. Near neutral pH, agar is compatible with most other polysaccharide gums

Sodium alginate and starch decrease the strength of agar gels while dextrins and sucrose increase the gel strength of certain agars. Locust bean gum has a marked synergistic effect on the strength of agar gels. Incorporation of 0.15% locust bean gum could increase the rupture strength of an agar gel by 50-200%. Carboxyethylcellulose shows similar action to a lower degree. Most salts, glycerol, sorbitol, alkanolamines, and 1,2,6-hexanetriol have little effect on the strength of the gels.

IV. ISOLATION OF GRP

Ten grams of solid agar (bacteriological grade) is mixed with 100 ml of phosphate buffered saline containing 0.05% Tween-20. After constant stirring for 1 h, the soluble extract is passed through a scintered glass filter. The filtrate is dialyzed against distilled water, and the absorption maximum of the dialysate is determined. The agar extract solution is freeze-dried and stored at room temperature.

Carbohydrate analysis. The agar extract is hydrolyzed with 4 M trifluoroacetic acid at 100 °C for 4 h. Monosaccharides are quantitated as alditol acetates by gas-liquid chromatography. Chemical analysis revealed that the extract was composed mainly of galactose derivatives (*TABLE 2*); therefore, the term "galactose-rich polysaccharide (GRP)" has been used.

Molar contents of galactose and 3,6-anhydroglactose residues, 6-*O*-methyl and 6-sulfate groups and pyruvic acid residue in agar-type polysaccharides could be determined by the analysis of the 100 MHz NMR spectrum of their partial hydrolysates (Izumi, 1973). The NMR spectrum of D-galactopyranose 6-sulfate is shown in *FIGURE 2*. Partial assignments denoted in the figure were achieved by comparing spectral pattern with that of D-galactopyranose (Izumi, 1971), assuming that both patterns are similar except for the deshielding effects H-6 caused by the introduction of the O-sulfate group.

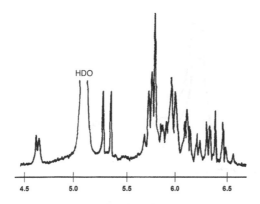

FIGURE 2: NMR spectrum of D-galactopyranose 6-sulfate in water at 100 MHz (redrawn from Izumi, 1973).

V. ANTIMICROBIAL EFFECTS OF GRP

Gerlach et al. (1994) demonstrated that galactose-rich polysaccharide (GRP), a water-soluble extract from agar could inhibit the adhesion of *Streptococcus pyogenes* to pharyngeal carcinoma cell line (DT-562). It is well known, however, that a cell line compared with its physiological origin, generally exhibits altered surface structures. In order to verify whether the effect also applied to normal tonsillar epithelia, the authors have also tested cells from healthy individuals and found a more potent inhibition by GRP. This finding seems to agree with earlier studies demonstrating an inhibitory effect of galactose on the adhesion of streptococci to pharyngeal cells, suggesting a lectin-type interaction (Tylewska et al., 1988). It appears possible, thus, that GRP-binding lectin may represent a streptococcal adhesin specific for pharyngeal cells.

Many bacteria, Gram negative in particular, are known to express surface lectins capable of recognizing simple sugars or complex saccharides, providing an important mechanism for adhesion/colonization to host tissue (Mirelman & Ofek, 1986). Thus, intestinal, respiratory, or urinary tract Gram-negative bacterial pathogens recognize various sugars such as D-galactose, D-GalNAc, D-mannose, and L-fucose in host mucosal colonization (Beachey, 1981). Among Gram-positive organisms, very few surface lectins have been reported to date; however, *Streptococcus mutans* and *Streptococcus cricetus* demonstrate lectins binding to dextran (Drake et al., 1988) and glucans (Gibbons & Fritzgerald, 1969), respectively, suggested in the adhesion of these oral streptococci to saliva-coated teeth and mucosa.

A. Streptococcal binding of GRP

The GRP-streptococci complex was stable in the presence of high molar NaCl, KSCN, or urea. Of the substances tested above, only heparin and bovine submaxillary mucin interfered with the binding of GRP to Todd-Hewitt broth-grown M19 cells; however, compared with the almost total inhibition of the uptake of [125]Iodine labeled-human lactoferrin to GRP-coated cells, the binding of GRP was moderately affected by heparin, and the possibility that heparin acted to inhibit the tracer substance, ^{125}I-human lactoferrin, rather than GRP could not be excluded. There was no effect of either human or bovine

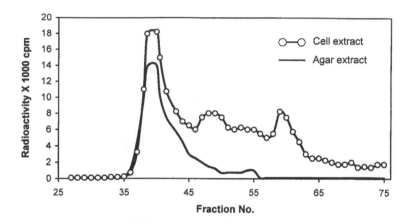

FIGURE 3: Gel filtration of GRP and solubilized, human lactoferrin-binding material from *Streptococcus pyogenes* after coupling to ^{125}I-labeled human lactoferrin. Either GRP or human lactoferrin affinity chromatography-purified alkaline extract of strain M19 (agar-grown) was incubated with ^{125}I-labeled-human lactoferrin at 20 °C for 1 h. Chromatography was performed with a Sephacryl S-200 HR column. Fractions were measured for radioactivity in a gamma counter (redrawn from Gerlach et al., 1994).

lactoferrin. Furthermore, lipoteichoic acid from *S. pyogenes* in a concentration of 0.1 mg/ml did not block the GRP binding to whole cells.

The heavily encapsulated *S. pyogenes* strain M50/873 and the acapsular transposon mutant M50/819, as well as M-positive strain M76/CS110, its M-negative mutant M76/CS112, and M-positive convertant M76/CS124, were examined for GRP binding. The presence or absence of hyaluronic acid capsule or M-protein did not appear to affect the whole-cell interaction with GRP.

The GRP interaction with human lactoferrin seemed to be specific, which provided a technical basis to identify a GRP-binding component in the cell extracts of streptococci. By SDS-PAGE and Western-blot analysis (with GRP as probe and horseradish peroxidase labeled hLF as tracer), a distinct GRP-binding protein with an estimated molecular mass of 70-kDa was identified in extracts of broth-grown *S. pyogenes* strains M19 (*FIGURE 3*) and M55. A slight amount of binding protein was also detected in the bacteria-free culture supernatant.

B. Inhibition of streptococcal adhesion to tissue by GRP

Gerlach et al. (1994) used human lactoferrin as a highly specific probe for the detection of GRP, in agreement with its known affinity for agar gels (Malmquist & Johansson, 1971). Based on this property the GRP interaction with *S. pyogenes* was studied. Furthermore, a 70-kDa GRP-binding, streptococcal lectin was identified and isolated. Studies on streptococcal mutants showed that GRP binding was independent of the presence of M protein or hyaluronate capsule. It was clearly distinct from streptococcal lipoteichoic acid. Since heparin was inhibitory for the binding of human lactoferrin to GRP rather than for the streptococcal binding of GRP, the GRP-binding moiety appeared distinct from a heparin-inhibitable surface protein binding to basement membranes (Bergey & Stinson, 1988); in addition, the reported molecular mass of this protein was much lower than that of the GRP-binding component.

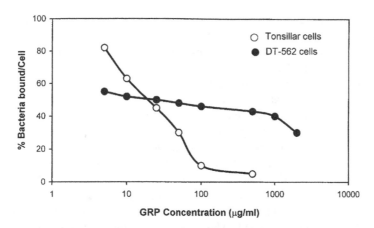

Figure 4: Effect of GRP on the interaction of strain M19 with DT-562 cells and normal tonsillar epithelial cells. ³H-thymidine-labeled bacteria were incubated with the indicated concentration of GRP for 10 min at 20 °C and then washed. The capacity of uncoated or GRP-coated bacteria to interact with eucaryotic cells was examined in an adhesion assay (redrawn from Gerlach et al., 1994).

The ratio between eucaryotic cells and ³H-thymidine-labeled bacteria (*S. pyogenes* strain M19) was maintained at 1:200 in the adhesion studies. Approximately 50 bacteria bound to each oropharyngeal (DT-562) cell. GRP moderately but in a dose-dependent manner decreased the streptococcal adherence and elicited 50% inhibition at ~250 μg/ml concentration. The adherence capacity of strain M19 to normal tonsillar epithelia was ~16 bacteria/cell, which was threefold lower than the adherence to DT-562 cells. This interaction was blocked by GRP also in a concentration-dependent fashion but required only ~30 μg/ml of GRP for 50% inhibition (*Figure 4*).

The 0.2:1 ratio of 3,6-anhydrogalactose: galactose in the GRP composition suggested the presence of galactan or a possible degradation of the 3,6-anhydro sugar to galactose during acid hydrolysis. In addition, the elementary analysis revealed the presence of nitrogen-containing nonsugar compounds in GRP, while the low sulfur content indicated that less than every second repeating unit was sulfated. Galactose-rich glycosides are abundant in mucins, while the mucosal layer is rich in sulfated polysaccharides such as heparan sulfate (Roussel et al., 1988). Since these substances may interact with human lactoferrin (Rejman et al., 1989), the inhibitory effect of bovine submaxillary mucin and of heparin on the binding of human lactoferrin to agar-grown *S. pyogenes* was expected.

VI. LARGE-SCALE PRODUCTION

Gelidium is common seaweed throughout the littoral zone. It is found under overhanging rocks and on the borders of rock pools. *Gelidium comeum* is a species used to produce agar in the United States. The frond has a rigid feel to it and a purplish color. The frond is forked into flat or cylindrical branches and grows generally in close colonies. The frond could get from one to three inches in height.

Another common species used for the production of agar is from the species, *Gracilaria verrucosa*, a branched cylindrical frond. Its frond ranges from six to twelve inches in height and is attached to a holdfast.

Agar and agaroids are obtained from various genera and species of red-purple seaweed found throughout the world's temperate zones. Seaweed of commercial interest is generally found in the intertidal zone to depths of more than forty meters in areas of surface turbulence and of marked top-to-bottom exchange. It grows from holdfasts and attains lengths of 0.1 to 2 meters. Propagation is by spores and stolons.

Newer methods for agar manufacture employ counter-current and cascade multiple extraction, centrifugation, plate-and-frames press filtration, artificial freezing, chemical bleaching, drying with hot air by drum and spray methods, and grinding. Although each manufacturer uses various locally developed modifications of the basic extraction-freezing-thawing-drying method, a common sequence of industrial extraction and purification is as follows - in the first step, the *Gelidium* is washed and soaked for twelve to fourteen hours. It is then transferred to pressure cookers and cooked for 6 h at a pressure of 15 lb/inch² in a dilute agar solution from the third and final cook. It then receives two more cookings before being discarded. The extract is clarified, filtered, poured into open tubes, and allowed to gel for 24 h. The gel is chopped and put into cans in a freezing chamber (14°C) for two days. After two days, it is thawed and placed in a dehydrator to remove water and soluble impurities. The flasks, containing about ninety percent moisture, are dried in hot air until the moisture is reduced by two-thirds (a final moisture content of approx. 35%). The flakes are bleached in 1% sodium hyphochlorite solution at room temperature. The excess bleaching agent is reduced by sodium sulfite, agar is removed, washed and dried to twenty percent moisture content.

The major commercial weeds contain 10-25% of recoverable agar and are frequently in short supply. The agar content of seaweed demonstrates intra-specific and inter-specific variation. Carbon dioxide concentration, oxygen tension, water temperature, and intensity of solar radiation are factors in the growth and agar content in a given species. The presence of epiphytes markedly affects the yield of commercially collected seaweed. Agar yield fluctuates through the year in a pattern correlating with the curve of solar ultraviolet intensity.

VII. APPLICATIONS

A. GRP applications

The capacity of GRP to block streptococcal adhesion to biosurfaces has potential implications. GRP based mouth washes could be beneficial in the control of streptococcal pharyngitis. GRP spray wash could also be useful in detaching streptococcal contaminants from food surfaces. The structural similarities of GRP with mucin components such as heparan sulfate and its ability to interact with mucosal defense factors such as lactoferrin gives rise various possibilities to develop specific antimicrobial delivery systems in medical and food sciences. GRP is also a good immobilizing agent for various bioactive compounds.

B. Laboratory applications

In 1882, Robert Koch published the use of agar as a solidifying agent for bacterial growth media. Since then, it has become the most common use for agar (Bridson, 1990). The melting point of agar is of great importance. Bacteriological agar remains liquid when cooled to 42 °C and most microorganisms can survive and be thoroughly homogenized in such a liquid phase. On the other hand, the agar-containing medium solidifies into a firm gel at 37 °C, the temperature most commonly used for microbial incubation. Characteristics such as resistance to microbial degradation, good gel firmness, elasticity, clarity, and stability makes agar an ideal media ingredient for microbiological applications. In addition, agar constitutes a natural non-toxic matrix for the formation of culture media. These unique properties have made agar a critical component in many microbiological applications. Agar is generally used at concentrations of one to two percent. It has also been used successfully at lower concentrations, for isolation and cultivation of anaerobes and testing motility of microorganisms.

Agarose applications as a supporting medium in the separation of bioactive molecules has revolutionized the fields of biochemistry, molecular biology and immunology. Agarose is used as packing bed for various chromatographic procedures, immunodiffusion and immunoelectrophoresis techniques.

C. Food applications

Agar has been used for many centuries as a high performance gelling agent. Its ability to produce clear, colorless, odorless, and natural gels without the support of other colloids has long been exploited by the food industry not only as a stabilizer and gelling agent but also in the manufacturing of confectionery, glazing, icing coating, piping jellies, salad dressings, etc. Agar jellifies at 40 °C and melts at 80 °C. This property offers a definite advantage particularly with regard to the shelf life of food preparations. Agar is used in dairy-based products when incorporation takes place at the pasteurization stage. It is considered as a cost-effective stabilizer for dairy products where water retention is of importance. It can also be mixed with other colloids to improve their final texture.

The carbohydrate moiety (mainly galactose) is digestible if hydrolyzed up to fifty percent. Therefore, agar is not used for nutritional value in food, but rather as an emulsifier. It forms a jelly to preserve food that is more suitable for consumption in the solid form. It is used as pectin in jellies, jams, and marmalade. Agar is also used as a clarifying agent in the production of beer, wine, and coffee. Agar is also used in the transport of cooked fish to prevent undesirable breakage of the product into pieces.

However, the gelling ability of agar is affected by the acidity or alkalinity of the ingredients in the food composition. Acidic foods, such as citrus fruits and strawberries, may require higher amounts of agar. Certain ingredients such as kiwi fruit (too acidic), pineapple, fresh figs, papaya, mango and peaches prevent agar gelling. Such foods contain enzymes which break down the gelling ability (although cooked fruit seems to lose this effect).

D. Medical applications

Agar is useful as a laxative. It is widely used as a treatment for constipation. Agar does not increase peristaltic action and its therapeutic value depends on its ability to absorb and retain moisture. The agar activity in the intestinal tract is mechanical and analogous to that of the cellulose of vegetable foods, aiding the regularity of the bowel movements. Upon hydration, it provides smooth, non-irritant bulk in digestive tract.

Agar is also used in preparation of emulsions, suspensions, capsules and suppositories in surgical lubricants and as a suspending agent for barium sulfate radiology. In prosthetic dentistry, agar is used in the preparation of dental casts. Agar is also used as a gelling ingredient in a wide range of cosmetic products.

VIII. SUMMARY

Agar has been a basic ingredient in the solid-state cultivation of bacteria over most of this century. Agar media are widely used in studies to elucidate the properties of bacterial surfaces. The possible adsorption of streptococci to GRP from agar warrants attention on interpreting data on bacterial cell surface charge of hydrophobic properties, bacterial adhesion to eucaryotic cells, and agglutination assays for rapid identification of bacteria. The ability to interact with specific mucosal defense factors such as lactoferrin as well as to bind various bioactive compounds makes GRP an excellent immobilizing agent and an antimicrobial delivery system.

IX. REFERENCES

1. Anderson, N.S., Dolan, T.C., and Rees, D.A. 1965. Evidence for a common structural pattern in the poly-saccharide sulphates of the Rhodophyceae. *Nature* 205:1060-1062.
2. Araki, C. 1959. *Proceeding of the 4th International Congress of Biochemistry*, Vol.I, pp. 15. London: Pergamon Press.
3. Beachey, E.H. 1981. Bacterial adherence: adhesin-receptor interactions mediating the attachment of bacteria to mucosal surfaces. *J. Infect Dis.* 143:325-345.
4. Bergey, E.J., and Stinson, M.E. 1988. Heparin-inhibitable basement membrane-binding protein of *Streptococcus pyogenes. Infect. Immun.* 56:1715-1721.
5. Bridson, E.Y. 1990. Media in microbiology. *Rev. Med. Microbiol.* 1:1-9.
6. Chapman, V.J., and Chapman, D.J. 1980. *Seaweeds and Their Uses.* New York: Chapman and Hall.
7. Christensen, O., and Trudsoe, J. 1980. Effect of other hydrocolloids on the texture of kappa carrageenan gels. *J. Texture Studies11*: 137-147.
8. Conway, E., and Rees, D.A. 1962. Water-soluble polysaccharides of *Porphyra* species: a note on the classification of *P. naidum. Nature* 195: 398-399.
9. Drake, D., Taylor, K.G., Bleiweis, A.S., and Doyle, R.J. 1988. Specificity of the glucan-binding lectin of *Streptococcus cricetus. Infect. Immun.* 56:1864-1872.
10. Gerlach, D., Schalen, C., Tigyi, Z., Nilsson, B., Forsgren, A., and Naidu, A.S. 1994. Identification of a novel lectin in *Streptococcus pyogenes* and its possible role in bacterial adherence to pharyngeal cells. *Curr. Microbiol.* 28:331-338.
11. Gibbons, R.J., and Fitzgerald, R.J. 1969. Dextran-induced agglutination of *Streptococcus mutans* and its potential role in the formation of microbial dental plaques. *J. Bacteriol.* 98:341-346.
12. Hirase, S., Araki, C., and Ito, T. 1958. *Bull. Chem. Soc. Japan.* 31:428-
13. Izumi, K. 1973. Structural analysis of agar-type polysaccharides by NMR spectroscopy. *Biochim. Biophys. Acta* 320:311-317.
14. Izumi, K. 1971. NMR spectra of some monosaccharides of galactopyranose series in deuterium oxide. *Agri. Biol. Chem.* 35:1816-1818.
15. Lee, R.E. 1980. *Phycology.* New York: Cambridge University Press.
16. Malmquist, J., and Johansson, B.G. 1971. Interaction of lactoferrin with agar gels and with trypan blue. *Biochim. Biophys. Acta* 236:38-46.
17. Mirelman, D., and Ofek, I. 1986. Introduction to microbial lectins and agglutinins. In *Microbial Lectins and Agglutinins*, ed. D. Mirelman, pp.1-20. New York: John Wiley & Sons.
18. Rees, D.A. 1961. The metabolism of isolated rat liver nuclei during chemical carcinogenesis. *Biochem. J.* 81: 347-352.
19. Rees, D.A. 1963. The carrageenan system of polysaccharides. Part I. The relation between the κ and λ compounds. *J. Chem. Soc.* 1821-1832.
20. Rees, D.A., and Conway, E. 1962. The structure and biosynthesis of porphyran: a comparison of some samples. *Biochem. J.* 84: 411-416.
21. Rejman, J.J., Hegarty, H.M., and Hurley, W.L. 1989. Purification and characterization of bovine lactoferrin from secretions of the involuting mammary gland: identification of multiple molecular weight forms. *Comp. Biochem. Physiol. [B]* 93:929-934.

22. Roussel, P., Lamblin, G., Lhermitte, M., Houdret, N., Lefitte, J.J., Perini, J.M., Klein, A., and Scharfman, A. 1988. The complexity of mucins. *Biochemie.* 70:1471-1482.
23. Sawardekar, J.S., Sloneker, H.H., and Jeanes, A. 1965. Quantitative determination of monosaccharides as their alditol acetates by gas-liquid chromatography. *Anal Biochem.* 37:1602-1604.
24. Selby, H.H., and Selby, T.A. 1959. *Industrial Gums: Polysaccharides and Their Derivatives,* Chapter-2. San Diego: American Agar and Chemical Company.
25. Stein, J., and Borden, C.A. 1984. Causative and beneficial algae in human disease conditions; a review. *Phycologia* 23:485-501.
26. Turvey, J.R., and Williams, T.P. 1963. *Proceeding of the 4th International Seaweed Symposium.* pp. 370. London: Pergamon Press.
27. Tylewska, S.K., Fischetti, V.A., and Gibbons, R.J. 1988. Binding selectivity of *Streptococcus pyogenes* and M protein to epithelial cells differ from that of lipoteichoic acid. *Curr. Microbiol.* 16:209-216.
28. Whyte, J.N.C., and Englar, J.R. 1980. The agar component of the red seaweed *Gelidium purpurascens. Phytochem.* 20:237-240.

Section-IV

BACTO-ANTIMICROBIALS

Probiotics
Nisin
Pediocin
Reuterin
Sakacin

A.S. Naidu
R.A. Clemens

Probiotics

17

I. INTRODUCTION

Lactic acid bacteria (LAB) have been widely used as probiotics in various fermented foods since antiquity. The preservative and health benefits of such traditional foods has been recognized for thousands of years, and accordingly, lactic acid fermentation played an important role in the early years of microbiology. After Louis Pasteur (1857) advocated germ theory for the fermentative changes, Joseph Lister attempted to prove the microbial nature of lactic acid fermentation. Using boiled milk as a nutrient medium, he isolated by chance the first bacterial pure culture described as *'Bacterium' lactis* (Lister, 1873). After decades of isolation and description of new bacteria, the genus *Lactobacillus* was proposed by Beijernick (1901); which remains a genus containing 64 valid species.

Metchnikoff (1907) proposed the health benefits related to the regular consumption of fermented milks based on his research finding with "Bulgarian bacillus" an organism closely related to *Lactobacillus delbreuckii* ssp. *bulgaricus*, a common LAB starter of yogurt. Lactobacilli continue to remain the most commonly used probiotic microorganisms. Currently available probiotic preparations contain *L. delbreuckii* ssp. *bulgaricus, L. acidophilus, L. casei, L. fermentum, L. plantarum, L. brevis, L. lactis* and *L. reuteri*.

A wide array of microorganisms colonize the human gastrointestinal tract, with some determined to be beneficial and others detrimental. Tissier (1905) isolated one such beneficial bacterium from the gut of breast-fed infants and designated the organism as *Bacillus bifidus*. The use of bifidobacteria as probiotics emerged from this finding, and currently used species are *B. adolescentis, B. animalis, B. bifidum, B. infantis, B. longum* and *B. thermophilum*. The first use of another genus of bacteria, *Streptococcus*, as probiotics was in the form of soured milk and yogurt (cited by Fuller, 1993). The yogurt starter *S. salivarius* ssp. *thermophilus* is still a common probiotic LAB.

LAB with probiotic activity are generally enteric flora, believed to play a beneficial role in the ecosystem of the human gastrointestinal tract. The probiotic spectrum of activity can be divided into nutritional, physiological, and antimicrobial effects. These observations have led to the development of a variety of foods and feeds containing LAB cells for probiotic use in man and animals. LAB are also potential adjuvants and their oral administration trigger(s) both mucosal and systemic immune responses (Gerritse et al., 1990). Various nutritional and therapeutic effects ascribed to LAB are: improvement of nutritional quality of food and feed; metabolic stimuli of vitamin synthesis and enzyme production; stabilization of gut microflora and competitive exclusion of enteric pathogens; enhancement of innate host defenses by production of antimicrobial substances; reduction of serum cholesterol by assimilation mechanisms; decreased risk of colon cancer by detoxification of carcinogens; and tumor suppression by modulation of cell mediated immunity.

Recent global marketing trends of probiotics are based on expectations of a prophylactic effect and in many cases as an alternative to more conventional pharmaceutical preparations. Although used in humans and animals for generations, probiotics have only recently been subject to clinical research. The most common use of probiotics is as food in the form of fermented milk products. However, the list of probiotic effects and health claims with the use of LAB is expanding.

II. ANTIMICROBIAL SYSTEMS

Several investigations have demonstrated that various species of LAB exert antagonistic action against intestinal and food-borne pathogens (Gibson et al., 1997). LAB are capable of preventing the adherence, establishment, replication, and/or pathogenic action of specific enteropathogens (Saavedra, 1995). These antagonist properties may be manifested by (a) decreasing the luminal pH through the production of volatile short-chain fatty acids (SCFA) such as acetic, lactic or proprionic acid; (b) rendering specific nutrients unavailable to pathogens; (c) decreasing the redox potential of the luminal environment; (d) producing hydrogen peroxide under anaerobic conditions; and/or (e) producing specific inhibitory compounds such as bacteriocins (Havenaar, et al. 1992).

A. Lactic acid and volatile acids

Fermentation involving LAB results in accumulation of organic acids, primarily lactic acid, as a major end product of carbohydrate metabolism, generated from pyruvate by lactic acid dehydrogenase. The accumulation of lactic acid and the concomitant reduction in pH of the milieu results in a broad-spectrum inhibitory activity against Gram-positive and Gram-negative bacteria. The acidic pH, dissociation constant (pK value), and mole concentration are factors that determine the inhibitory activity of lactic acid and acetic acid in the milieu (Ingram et al., 1956). Acetic acid has more antimicrobial activity than lactic acid (Rasic & Kurmann, 1983). Acetic acid has a pK of 4.756 while lactic acid has a pK of 3.860 (Lindgren & Dobrogosz, 1990). At an intestinal pH of 5.8, 8.4% of the acetic acid and 1.1% of the lactic acid are present in the undissociated form. Thus, production of acetic and lactic acids by bifidobacteria, in the ratio of 3:2, respectively, would yield approximately 11 times as much undissociated acetic acid as lactic acid. This is an important factor if one considers undissociated acid as antagonistic to the growth of many potential pathogens and putrefactive bacteria. Lipophilic acids such as lactic acid and acetic acid in undissociated form can penetrate the microbial cell membrane, and at higher intracellular pH, dissociate to produce hydrogen ions that interfere with essential metabolic functions such as substrate translocation and oxidative phosphorylation (Baird-Parker, 1980).

Lactic acid and acetic acid are known to inhibit *Staphylococcus aureus* (Kao & Frazier, 1996), however, this activity was observed only in early but not late stages of bacterial growth (Haines & Harmon, 1973). Goepfert and Hicks (1969) showed that *Salmonellae* are inhibited at pH values lower than 4.4 for lactic acid and 5.4 for acetic acid. Thus, small amounts of acetic acid produced by LAB could elicit a potent antimicrobial effect. Adams and Hall (1988) reported a synergism between lactic acid and acetic acid in the inhibition of *E.coli* and *Salmonella*. This phenomenon is due to the strong acidic effect of lactic acid that resulted increase in the mole ratio of inhibitory undissociated acetic acid. Wong and Chen (1988) reported that the growth of *Bacillus cereus* was blocked in the presence of LAB due to acetate production at pH 6.0 and spore germination was inhibited by, in decreasing order, formate, lactate and acetate at pH 4.4, 4.3, and 4.2, respectively.

Midolo et al. (1995) examined the *in vitro* effects of both pH and organic acids on the gastric pathogen *Helicobacter pylori*. Strains of *L. acidophilus, L. casei, L. bulgaricus, Pediococcus pentosaceus* and *Bifidobacterium bifidus* were assayed for their lactic acid production, pH and inhibition of *H. pylori* growth *in vitro*. Lactic, acetic and hydrochloric acids demonstrated inhibition of *H. pylori* growth in a concentration-dependent manner with the lactic acid resulting in highest inhibition. This inhibition was due both to the pH of the solution and concentration of the acids. Six strains of *L. acidophilus* and one strain of *L. casei* ssp. *rhamnosus* inhibited *H. pylori* and *L. bulgaricus* did not. Concentrations of lactic acid produced by these strains ranged from 50 to 156 mmol/L and correlated with *H. pylori* inhibition.

Bifidobacteria do not produce H_2O_2; however, they do produce both acetic acid and lactic acid (1986). The production of these acids reduces intestinal pH, which in turn limits the growth of many potential pathogens and putrefactive bacteria. By controlling intestinal pH, it is possible to restrict the production of phenols, ammonia, steroid metabolites, bacterial toxins as well as vasoconstriction amines such as histamine, tyramine, cadaverine and agmatine (Hidaka & Eida, 1988).

Control of intestinal pH also influences the degradation of aromatic amino acids (Mitsuoka, 1984). Phenol and *p*-cresol are putrefactive products of tyrosine, while indole and skatole are derived from tryptophan. These compounds are normally detoxified in the liver with acids, such as glucoronic or sulfuric and excreted in urine (Hidaka et al., 1986). Intestinal putrefactive products can cause diarrhea, liver disorders, and malfunction of the circulatory system (Mitsuoka, 1982). Products such as ammonia are normally detoxified by the liver; however, in instances where liver function is impaired, e.g. cirrhosis, ammonia levels may be elevated and interfere with peripheral and central nerve impulses. Bifidobacteria suppress ammonia production by inhibiting urea-splitting organisms and could conceivably be used to clinical advantage in the management of post-systemic encephalopathy (Modler et al., 1990b).

SCFA are widely used antimicrobials in food processing industry as preservatives and meat surface decontaminants. Lactic, acetic and other acid antimicrobials have been described in detail in chapters elsewhere in this volume.

B. Hydrogen peroxide

In the presence of oxygen, LAB produces hydrogen peroxide (H_2O_2) through electron transport via flavin enzymes. In the presence of H_2O_2, superoxide anions form destructive hydroxy radicals. This process may lead to peroxidation of membrane lipids (Morris, 1979), and increased membrane permeability (Kong & Davidson, 1980). The resulting bactericidal effect of these oxygen metabolites has been attributed to their strong oxidizing effect on the bacterial cell as well as destruction of nucleic acids and cell proteins (Dahl et al., 1989; Piard & Desmazeaud, 1992). In addition, H_2O_2 could react with other cellular and milieu components to form additional inhibitory substances.

H_2O_2 formation by LAB and its effect on various microorganisms is well documented for years (Klebanoff et al., 1966; Dahiya & Speck, 1968). LAB strains were reported to produce H_2O_2 under aerobic conditions in a complex glucose based media. The H_2O_2 accumulates during oxygen utilization by the cultures simultaneously with an increase in specific activity of NADH oxidase, pyruvate oxidase and NADH peroxidase (Murphy & Condon, 1984). H_2O_2 is not a product of NADH oxidase *in vitro*, but is formed in substantial quantities from oxygen during oxidation of pyruvate. The three enzymes were induced by oxygen and H_2O_2; the induction of NADH oxidase respond to lower levels of oxygen (but not of H_2O_2) than the pyruvate oxidase or the NADH peroxidase.

Nunez de Kairuz et al. (1988) reported fatty acid dependent H_2O_2 production in LAB. *Lactobacillus leichmanii* growing in complex medium supplemented with decanoic acid accumulate high concentrations of H_2O_2 in the culture. The H_2O_2-generating system was specifically induced by one of the saturated fatty acids from 4:0 to 16:0 or oleic acid. The induction of this system was associated with the presence of a fatty acyl-CoA-dependent H_2O_2-generating activity in the cell-free extracts. This activity is shown for the first time in a procaryote organism.

Fitzsimmons and Berry (1994) examined the inhibitory effect of H_2O_2 producing lactobacilli (LB+) on *Candida albicans*. A range of *Lactobacillus acidophilus* strains isolated from patients using oral, vaginal and endocervical swabs were investigated for their ability to (a) inhibit the growth of *Candida albicans*, and (b) generate peroxidase, H_2O_2 and hypothiocyanite. Inhibition of *Candida albicans* and H_2O_2 production was detected in nine out of twelve strains whereas peroxidase production was only detected in three out of twelve strains, all from oral swabs. Hypothiocyanite production was detected in two strains and it was only detected in these strains after growth in MRS medium in aerobic conditions.

C. Bacteriocins

The gastrointestinal tract contains many antimicrobial proteins such as colicins, defensins, cecropins, and magainins. These are low molecular weight, cationic, amphiphilic molecules, tend to aggregate and are benign to the producing organism. LAB also produce a wide range of similar antagonistic factors that include metabolic products, antibiotic-like substances and bactericidal proteins, collectively termed as bacteriocins. Bacteriocins vary in spectrum of activity, mode of action, molecular weight, genetic origin and biochemical properties. Bacteriocins can be produced spontaneously or induced.

TABLE 1: Types of bacteriocin production in lactic acid bacteria (adapted from Naidu et al., 1999)

LAB Strain	Bacteriocin	M$_r$	Reference
Small heat-stable bacteriocins			
Lc. lactis ssp. *lactis LMG2130*	Lactococcin A	5.8 kD	Holo et al. (1991)
Lc. lactis ssp. *cremoris 9B4*	Lactococcin B		van Belkum et al. (1992)
Lc. lactis ssp *lactis*	Lactococcin M		van Belkum et al. (1991)
Lb. acidophilus 11088	Lactacin F	6.3 kD	Klaenhammer (1993)
Leu. gelidium UAL187	Leucocin A	3.9 kD	Hastings & Stiles (1991)
Ped. acidilactici PAC1.0	Pediocin PA-1	4.6 kD	Gonzalez & Kunka (1987)
Large heat-labile bacteriocins			
Lb. helveticus 481	Helvetican J	37 kD	Joerger & Klaenhammer (1986)
Lb. delbrukii JCM1106	Lacticin A		Toba et al. (1991a)
Lb. delbrukii JCM 1248	Lacticin B		Toba et al. (1991a)
Lb. acidophilus LAPT1060	Acidophilucin A		Toba et al.(1991b)
Lb. casei B80	Caseicin 80	42 kD	Rammelsberg & Radler (1990)
Lantibiotics			
Lc. lactis ssp. *lactis ATCC114*	Nisin A	33.5 kD	Buchman et al. (1988)
Lc. lactis ssp. *lactis NIZ022186*	Nisin Z		Mulders et al. (1991)
Lb. sake L45	Lactosin S	3.8 kD	Mortvedt & Nes (1990)
Lc. lactis ssp. *lactis CNRZ481*	Lacticin 481	2.9 kD	Piard et al. (1990)
Carnobacterium ssp. *U149*	Carnocin U149	4.6 kD	Stoffels et al. (1992)
Lb. plantarum LPCO10	Plantaricin S		Jimenez-Diaz et al. (1995)
Ent. faecium T136	Enterocin A	4.8 kD	Casaus et al. (1997)
	Enterocin B		

The genetic determinants of most of the bacteriocins are located on the plasmids, with a few exceptions, which are chromosomal encoded. The release of bacteriocins requires the expression and activity of bacteriocin-release proteins and the presence of detergent-resistant phospholipase A in the bacterial outer membrane of the LAB. These antimicrobial agents are species specific and exert their lethal activity through adsorption to specific receptors located on the external surface of sensitive bacteria, followed by metabolic, biological and morphological changes resulting in the killing of such bacteria. There is increasing evidence that bacteriocins from many bacterial genera also share these characteristics.

Bacteriocins have been isolated from a variety of LAB (*TABLE 1*) A large number of new bacteriocins in LAB have been characterized and classified into four major classes: 1) lantibiotics, 2) small heat-stable peptides, 3) large heat-labile proteins, and 4) complex proteins whose activity requires the association of carbohydrate or lipid moieties (Nes et al., 1996). Most of the new bacteriocins belong to the class II bacteriocins that are small (30-100 amino acids) heat-stable and commonly not post-translationally modified. While most bacteriocin producers synthesize only one bacteriocin, it has been shown that several LAB produce multiple bacteriocins (2 or 3 bacteriocins). Based on common features, some of the class II bacteriocins can be divided into separate groups such as the pediocin-like (strong anti-listeria) bacteriocins, the two-peptide bacteriocins, and bacteriocins with a sec-dependent signal sequence.

In recent years, there has been a considerable increase in studies of bactericidal peptides produced by LAB, with particular emphasis upon their potential application as food preservatives. Several of these peptides contain lanthionine and other post-transla-

TABLE 2: Inhibitory spectrum of bacteriocins against non-lactic microorganisms[a]

Susceptible Organism	Bacteriocin
Aeromonas hydrophila	Sakacin A
Bacillus cereus	Lactocin-S, Lactostrepcin-5, Nisin, Pediocin-A, Pediocin-AcH, Sakacin-A
Bacillus coagulans	Nisin
Bacillus licheniformis	Nisin
Bacillus pumilis	Thermophillin
Bacillus stearothermophilus	Nisin
Bacillus subtilis	Lacticin-481, Nisin, Thermophillin
Bronchothrix thermospacta	Curvacin-A, Pediocin-AcH, Sakacin-A, Sakacin-P
Clostridium bifermentans	Nisin
Clostridium botulinum	Nisin, Pediocin-A, Reuterin, Sakacin-A
Clostridium butyricum	Nisin, Reuterin
Clostridium perfringens	Nisin, Pediocin-A, Pediocin-AcH, Pediocin-VTT, Reuterin, Thermophillin
Clostridium sporogenes	Nisin, Pediocin-A
Clostridium tyrobutricum	Lacticin-481, Lactocin-S, Pediocin-AcH
Escherichia coli	Reuterin, Thermophillin
Listeria innocua	Lacticin-481, Lactosin-S, Pediocin-A, Pediocin-AcH
Listeria ivanovii	Pediocin-A, Pediocin-AcH, Pediocin-PAC10
Listeria monocytogenes	Carnobacteriocin A&B, Curvacin-A, Enterocin-1146, Leucocin-A, Nisin, Pediocin-A, Pediocin-AcH, Pediocin-JD, Pediocin PA-1, Pediocin-PAC10, Pediocin-VTT, Piscicolin-61, Lactacin-B, Reuterin, Sakacin-A, Sakacin-P
Listeria seeligeri	Pediocin-A
Listeria welchii	Lacticin-481, Pediocin-A
Proteus mirabilis	Nisin
Pseudomonas aeruginosa	Thermophillin
Pseudomonas fluorescens	Thermophillin
Salmonella enteritidis	Reuterin, Thermophillin
Salmonella infantis	Pediocin-VTT, Reuterin
Salmonella typhimurium	Ruetuerin, Thermophillin
Shigella sp.	Reuterin, Thermophillin
Staphylococcus aureus	Nisin, Lacticin-481, Pediocin-A, Pediocin-AcH, Sakacin-A, Thermophillin, Plantaricin-SIK83,
Staphylococcus carnosus	Curvacin, Lacticin-481, Lactocin-S, Pediocin-AcH
Staphylococcus epidermis	Nisin
Staphylococcus simulans	Nisin
Mycobacterium tuberculosis	Nisin
Yersinia enterocolitica	Thermophillin

[a]Bacteriocin produced by LAB inhibits closely related bacteria belonging to the same genus, other LAB species, Gram positive bacteria, and /or less commonly Gram negative bacteria. (adapted from Naidu et al., 1999)

tionally modified amino acid residues. The lanthionine-containing molecules (lantibiotic) appear to have evolved in two quite different lineage, type A and type B. A wider diversity of compounds of type A lantibiotics has been described in the recent years. Novel features of some of the more recently described type A lantibiotics include a) new modifications such as D-Ala and 2-hydroxypropionyl residues both derived from serine; b) different types of pre-lantibiotic leader sequences; c) the apparent requirements for different numbers and types of genes for synthesis of some active type A lantibiotics; d) cytolysin functions as both a hemolysin and a bacteriocin; e) one of the newly-described lantibiotics (lactocin S) does not have any net charge at neutral pH; while carnocin UI49 is the largest of the lantibiotics discovered and the killing action of another (cytolysin) has been shown to depend on the interaction of two peptides.

In vitro studies have demonstrated that some bacteriocins have a broad inhibitory spectrum, including many Gram-positive and Gram-negative bacteria (*TABLE 2*). The best known bacteriocin from LAB is nisin, produced by *L. lactis* (Hurst, 1981). Gibson et al. (1997) reported that *L. reuteri* produces the bacteriocin reuterin which affects Gram-negative (*Salmonella* and *Shigella*) and Gram-positive (*Clostridia* and *Listeria*) bacteria.

Wolf et al. (1995) demonstrated that the oral administration of *L. reuteri*, the producer of the bacteriocin reuterin, was safe and well tolerated in adult male subjects. Ruiz-Palacios et al. (1996) reported similar results among children 12 to 36 months of age who were at high risk for rotaviral diarrhea. The children were fed a blend of *L. reuteri*, *L. acidophilus*, and *B. infantis* for 4 to 14 weeks without any adverse effects. Speck et al. (1993) indicated that *L. reuteri* is a common *Lactobacillus* used in the food industry and consumed without any known adverse effects. However, the potentially important role of reuterin in preventing diarrhea and controlling the gastrointestinal ecosystem remains to be established.

The well-characterized and most commonly used bacteriocins - nisin, pediocin, sakacin and reuterin are elaborately described in the following chapters.

D. Carbon dioxide

Carbon dioxide (CO_2) is a major end product of hexose fermentation by heterofermentative LAB. A number of LAB are capable of CO_2 production from malate and citrate (Fleming et al., 1986; London, 1990) and also by metabolizing arginine via the arginine deaminase pathway (Poolman, 1993). Finally, decarboxylation of amino acids (histidine, tyrosine) can also result in CO_2 formation.

The CO_2 also contributes to the antimicrobial activity of LAB. Its role in creating anaerobic environment by replacing existent molecular oxygen, its extra- and intracellular capability to decrease pH and its destructive effects on cell membranes (Eklund, 1984) make CO_2 a potent inhibitory system against a wide variety of microorganisms (Clark & Takacs, 1980). This protective role of CO_2 is critical, particularly in the fermentation of vegetables and silages to prevent growth of molds (Lindgren & Dobrogosz, 1990).

E. Diacetyl and acetaldehyde

Diacetyl (2,3-butanedione) is an end product of pyruvate metabolism (Condon, 1987) by citrate fermenting LAB (Hugenholtz, 1993). Diacetyl elicits a potent antimicrobial activity against various food-borne pathogens and spoilage microorganisms (Jay, 1982). Diacetyl is more effective against Gram-negative bacteria, yeasts and molds, than against Gram-positive organisms. Diacetyl interferes with arginine utilization by reacting with the arginine-binding protein of Gram-negative bacteria (Jay, 1996). High concentration of diacetyl (0.4 mg/ml) is required for antimicrobial effect against most organisms (Gupta et al., 1973). However, *E. coli* is susceptible to diacetyl at extremely low concentration (Spillman et al., 1978). Dose-dependent inhibition experiments established that concentrations of 0.2 and 0.3 mg/ml are required for eliciting antimicrobial activity against yeasts / Gram-negative bacteria, and non-lactic Gram-positive bacteria, respectively (Jay, 1982). Motlagh et al. (1991) reported bactericidal effect for diacetyl (344 ppm) against strains of *Yersinia enterocolitica*, *Aeromonas hydrophila*, *E. coli*, and *Salmonella anatum*, but not against *Listeria*.

Acetaldehyde formed during carbohydrate metabolism of heterofermentative LAB is reduced to ethanol by re-oxidation of pyridine nucleotides, catalyzed by an NAD-dependent alcohol dehydrogenase. Acetaldehyde imparts the typical aroma of yogurt. Kulshrestha and Marth (1974a; 1974b; 1974c) have reported antimicrobial activity of acetaldehyde (10-100 ppm) against food-borne pathogens, *E. coli, Salmonella typhimurium* and *S. aureus*.

III. ANTIMICROBIAL EFFECTS *IN VIVO*

A. Antibacterial effects

Savage (1969) studied *in vivo* microbial interference between LAB and indigenous yeast in rodents. Ducluzeau et al. (1971) reported the antagonist effect of a LAB strain on a *Ristella sp.* in the murine digestive tract.

Sato (1984) examined the *in vivo* antilisterial activity of 10 LAB species and observed that *L. casei* was the most potent inhibitor of *Listeria monocytogenes* infection in mice. The activity of *L. casei* differed with regard to the dose of administration. The study suggested that the enhanced host resistance to listeria infection was mediated possibly by macrophages migrating from the blood stream to the reticuloendothelial system in response to *L. casei* administration. The functions of liver macrophages and peritoneal macrophages obtained after injection of *L. casei* were examined by Sato et al. (1988a). Listericidal activity *in vivo* was enhanced in liver macrophages 13 days after *L. casei* injection but was suppressed in the macrophages 2 days after the injection. The listericidal activity *in vitro* was enhanced in peritoneal macrophages obtained 13 days after *L. casei* injection but was suppressed in cells obtained 2 days later. The PMA-triggered respiratory burst in the liver macrophages elicited by *L. casei* was higher than that of resident macrophages. Alkaline phosphodiesterase activity in the liver macrophages was decreased by *L. casei* injection, as was also the case with peritoneal macrophages. These observations indicate that *L. casei* enhanced cellular functions of both liver and peritoneal macrophages.

Itoh and Freter (1989) used gnotobiotic mice and a continuous-flow culture system to study the interactions between *E. coli* and (i) clostridia (in chloroform-treated cecal suspensions from conventional mice) and (ii) three strains of LAB isolated from conventional mice. In gnotobiotic mice, the LAB suppressed *E. coli* multiplication in the stomach and the small intestine, but had no demonstrable effect on *E. coli* multiplication in the large intestine. In contrast, clostridia were most effective in controlling *E.coli* multiplication in the large intestine. In the presence of both LAB and clostridia, *E. coli* populations in the various regions of the gastrointestinal tract resembled those found in conventionalized control animals. The control of *E. coli* populations was not related to changes in pH or intestinal motility. *In vitro* stimulation of the above-described *in vivo* interactions required a two-stage continuous-flow culture in which the effluent from the first stage represented the influx to the second. The first stage was inoculated with LAB, and the second stage was inoculated with either a pure culture of *E. coli* or *E. coli* and clostridia. In these instances, the *E. coli* populations in the second stage of the culture resembled in size those found in the large intestine of gnotobiotic mice harboring a similar flora.

Fichera et al. (1987) investigated the role of *B. bifidum* and *L. acidophilus*, in controlling a severe experimental infection by *Salmonella enteritidis* in a mouse model. The administration of *Bifidobacterium* and *Lactobacillus* markedly increased the animal's resistance against the lethal salmonella infection. The effect of a diet supplemented with yogurt containing live *L. bulgaricus* and *S. thermophilus* on the response of inbred mice to infection with *Salmonella typhimurium* was elaborated by De Simone et al. (1988). The relationship between modifications of the immune response following ingestion of yogurt with live LAB and increased defense mechanisms was confirmed by the bacterial counts in livers and spleens and by the reduced mortality to *Salmonella typhimurium* infection. The protective effect of feeding milk fermented with a mixture of *L. casei* and *L. acidophilus* against *Salmonella typhimurium* infection in mice (Perdigon, et al., 1990). The survival rate obtained after oral infection with *Salmonella typhimurium* was 100% in mice pretreated by feeding with fermented milk containing the LAB mixture. Elevated levels of anti-salmonellae antibodies in serum and in intestinal fluid were found in the group of mice fed with the mixture. When the mice were first infected with *Salmonella typhimurium* and then fed with the mixture of fermented milks, pathogen colonization was not prevented. The authors suggested that the augmentation of resistance to salmonellae was due to the anti-salmonellae protective immunity mainly mediated by the mucosal tissue induced by the treatment with fermented milk containing both the LAB strains. Perdigon et al. (1991) further demonstrated that the protective effect of *L. casei* against enteric pathogens *Salmonella typhimurium* and *E. coli* was due to elevated IgA antibodies in intestinal secretions. Beta-glucuronidase and beta-galactosidase activities in the intestinal fluid and histological samples showed a marked increase in intestinal inflammatory response. These results suggest that *L. casei* plays an important role in the prevention of enteric infections, a low dose being enough for protection against intestinal infections by increasing IgA secretion into the intestinal lumen, thus providing adequate defenses for the mucosal surface. Similar protective effect of feeding milk fermented with a mixture of *L. casei* and *L. acidophilus* against *Shigella sonnei* was also reported (Nader de Macias et al., 1992). There was a 100% survival rate in mice fed with fermented milk. High titer anti-shigella antibodies were found both in sera and in small intestinal fluid of mice treated with fermented milk, suggesting that the mucosal tissue could mediate the protective immunity.

The antagonistic activity of *L. casei GG* was examined in mice orally infected with *Salmonella typhimurium* C5 (Hudault et al., 1997). In germ-free mice, establishment of *L. casei GG* in the gut significantly delayed the occurrence of 100% mortality of the animals. Cecal colonization level and translocation rate of C5 to the mesenteric lymph nodes, spleen, and liver were significantly reduced in these animals.

A fermented mixture containing *L. acidophilus* (10^5 cells/ml) was shown to inhibit the growth of potent enteric pathogens including *Shigella dysenteriae, Salmonella typhimurium* and *E. coli* (Rani & Khetarpaul, 1998). A significant decline in pH with a corresponding increase in titratable acidity due to probiotic fermentation occurred in the developed food mixture. The fermented mixture has prevented *E. coli* induced diarrhea in mice.

Kabir et al. (1997) reported the ability of *L. salivarius* to inhibit both the attachment and IL-8 release of *Helicobacter pylori* (causative agent of gastric inflammatory diseases). *H. pylori* failed to colonize the stomach of *L. salivarius* infected gnotobiotic BALB/c mice, but colonized in large numbers and subsequently caused active gastritis in germ free mice. In addition, *L. salivarius* administered after *H. pylori* implantation could dissociate colonization by this gastric pathogen. These findings suggest the possibility of LAB application in the treatment of *H. pylori* gastritis.

B. Antifungal effects

Efficacy of *L. reuteri* as a probiotic for the control of *Cryptosporidium parvum* infection was evaluated in C57BL/6 female mice that were immunosuppressed by intraperitoneal inoculation with the LP-BM5 leukemia virus (Alak et al., 1997). Four months after inoculation, mice developed lymphadenopathy, splenomegaly, and suscepti- bility to *C. parvum* infection. After daily prefeeding with *L. reuteri* (10^8 cfu/day) for 10 days, mice were challenged with 6.5 x 10^6 *C. parvum* oocysts and fed *L. reuteri* during the entire study. Mice supplemented with *L. reuteri* and challenged with *C. parvum* cleared parasite loads from the gut epithelium. However, unsupplemented animals developed per- sistent cryptosporidiosis and shed high levels of oocysts in the feces. *L. reuteri* feeding increased its colonization of the intestinal tract, which was inversely related to the fecal shedding of oocysts. These findings suggest that *L. reuteri* may help prevent *C. parvum* infection in immunodeficient subjects.

Wagner et al. (1997) tested the antifungal ability of *L. acidophilus, L. reuteri, L. casei GG*, or *B. animalis* to protect athymic and euthymic mice from mucosal and sys- temic candidiasis. The presence of probiotic LAB in the gastrointestinal tracts prolonged the survival of adult and neonatal athymic mice compared to that of isogenic mice colo- nized with *C. albicans* alone. The incidence of systemic candidiasis in athymic mice was significantly reduced by each of the four probiotic bacterial species. The numbers of *C. albicans* present in the alimentary tracts of euthymic mice were significantly reduced by *L. casei GG* and *B. animalis*. None of the probiotic bacteria species completely prevent- ed mucosal candidiasis, but *B. animalis* reduced its incidence and severity. Probiotic bac- teria also modulated antibody- and cell-mediated immune responses to *C. albicans*. The prolonged survival of mice, decreased severity of mucosal and systemic candidiasis, mod- ulation of immune responses, decreased number of *C. albicans* in the alimentary tract, and reduced numbers of orogastric infections demonstrated not only that probiotic bacteria have biotherapeutic potential for prophylaxis against and therapy of this fungal disease but also that probiotic bacteria protect mice from candidiasis by a variety of immunolog- ic (thymic and extrathymic) and nonimmunologic mechanisms.

IV. ANTIMICROBIAL PROBIO-ACTIVE COMPOUNDS

Naidu et al. (1999) have defined probio-active substance as 'a cellular complex of LAB that has a capacity to interact with the host mucosa and may beneficially modu- late the immune system independent of LAB's viability'. Recent developments in the role of probiotics (live) and their probioactive (cellular) substances in the intestinal and extra- intestinal physiology of host are overwhelming.

A. Antibacterial effects

The Yakult Central Institute of Microbiological Research in Japan has initiated several studies on the protective effect of heat-killed *L. casei* LC9018 against microbial infections. These studies have led to the identification of various immunomodulating pro- bio-active compounds in the cell extracts of LAB.

Miake et al. (1985) observed that the survival of mice after intraperitoneal (i.p.) infection with *P. aeruginosa* was augmented in mice pretreated with heat-killed LC9018. Similar treatment of mice with *Corynebacterium parvum*, however, was not effective. The protective effect of LC9018 was observed in mice subjected to body irradiation but was abrogated when mice were treated with carrageenan. These results suggested that augmentation of the resistance of mice to *P. aeruginosa* was caused by the induction of activated macrophages. Lipoteichoic acid (LTA) from cell walls of *LC9018* or *L. fermentum YIT 0159* augmented the resistance of C57BL/6 mice to infection with *P. aeruginosa* (Setoyama et al., 1985).

Similar potent macrophage-activating effect by LC9018 was also demonstrated in experimental mouse infections with *Listeria monocytogenes* (Nomoto et al., 1985). The LC9018 induced peritoneal cells were composed mainly of macrophages and polymorphonuclear cells with enhanced phagocytosis and marked intracellular killing activity against *Listeria monocytogenes*. Küpffer cells of mice pretreated with LC9018 produced a high level of oxygen radicals. The antilisterial effect of lasted for three weeks after LC9018 administration. Sato et al. (1988b) examined the cell wall, cytoplasm, polysaccharide, and peptidoglycan fractions prepared from *L. casei*, *L. plantarum*, and *L. acidophilus* for their efficacies to enhance resistance of host mice against *Listeria monocytogenes* infection. Intraperitoneal injections of cellular fractions of *L. casei* led to elicitation of inflammatory cells in the peritoneal cavity and the effect was maximal with peptidoglycan. Macrophage ratio in the resultant peritoneal exudate cells was elevated in peptidoglycan-treated mice. Macrophages induced with cell wall fraction of *L. casei* showed the potent phorbol myristate acetic acid (PMA)-triggered respiratory burst. All the macrophages induced with cell wall of *L. casei* and *L. acidophilus* enhanced oxygen radical production in response to PMA but *L. plantarum* did not enhance similar oxygen radical-producing ability. The *L. casei*-cell wall also enhanced *in vitro* listericidal activity of mouse peritoneal macrophages, but such an activity was not observed with *L. acidophilus*-cell wall. When mice were intravenously given the cellular fractions 7 or 13 days before *L. monocytogenes* infection, cell wall fractions of *L. casei* elicited the most potent protective activity. A weak protective activity was also found in peptidoglycan of *L. casei*. The protective action of *L. casei* against *Listeria monocytogenes* infection in host mice may be attributed to cell wall compounds and partially to the peptidoglycan moiety.

Heat-killed LC9018 was also studied for its protective and therapeutic efficacies against *Mycobacterium fortuitum* and *M. chelonae* infections in mice (Saito et al., 1987). LC9018 reduced the incidence of spinning disease and gross renal lesions and enhanced the elimination of organisms at the site of infection in the host mice. The LC9018 administration caused a marked increase in the phagocytic function, oxygen radical-generating ability of the murine peritoneal macrophages. It also resulted in an enhancement of interleukin-1-producing function of the macrophages. These findings indicate the ability of LC9018 to activate macrophage functions, in particular those related to microbicidal activity and result in the protective and therapeutic efficacy of LC9018 against infection due to *M. fortuitum* complex.

After oral administration of *E. coli B41* (10^2 to 10^5 cfu) to 24-48 h old suckling mice, an 80 to 100% mortality rate is observed within three days. Fourniat et al. (1986) compared the effect of the oral treatment with a lyophilized preparation of heat-killed *L. acidophilus* and with sterile water on the mortality rate of newborn mice. In six out of seven assays, the heat-killed *L. acidophilus* administration extended survival of infected suckling mice. The influence of the oral administration of killed bifidobacteria, lactobacillus, bacteroides and fusobacteria on the anti- salmonella resistance of mice, infected orally with *S. dublin*, was studied (Shkarupeta et al., 1988). Bifidobacteria and lacto-

           Natural Food Antimicrobial Systems

TABLE 3: *In vivo* antimicrobial activity of probiotic and probio-active agents against microbial pathogens (adapted from Naidu et al., 1999)

Susceptible Pathogen	Antimicrobial Agent	Mode of Action	Reference
Antibacterial Activity			
Ristella sp.	LAB species	Antagonistic effect in digestive tract	Ducluzeau et al. (1971)
Mycobacterium intracellulare	*L.casei LC9018*	Immunopotentiation effect	Saito et al. (1983a)
Mycobacterium fortuitum	Heat-killed *LC9018*	Activated macrophage cidal activity	Saito et al. (1987)
Mycobacterium chelonae	Heat-killed *LC9018*	Activated macrophage cidal activity	Saito et al. (1987)
Listeria monocytogenes	*L. casei*	Macrophage-mediated host defense	Sato (1984)
Pseudomonas aeruginosa	Heat-killed *LC9018*	Induction of activated macrophages	Miake et al. (1985)
Streptococcus mutans	*S. faecium/S.equinus/L. fermentum/ L.salivarius*	Bacterial growth inhibition	Ishihara et al. (1985)
Escherichia coli	*L. casei*	Increased IgA secretion in lumen	Perdigon et al. (1991)
Enterotoxigenic *E. coli*	Heat-killed *L. acidophilus*	Protected the mouse challenge	Fourniat et al. (1986)
Uropathogenic *E.coli*	*L. fermentum CRL1058*	Inflammatory immune response	Silva de Ruiz et al. (1996)
Salmonella enteritidis	*L. acidophilus/ B. bifidum*	Increased resistance to lethality	Fichera et al. (1987)
Salmonella typhimurium	*L. bulgaricus/ S. thermophilus*	Increased defense mechanism	De Simone et al. (1988)
	L. casei/L. acidophilus	Prevention of intestinal infection	Perdigon et al. (1990)
	Killed LAB strains	Immunostimulating effect	Shkarupeta et al. (1988)
Salmonella dublin	*L. acidophilus*	Bacterial multiplication-inhibition	Rani & Khetarpaul (1998)
Salmonella typhi	*L. acidophilus*	Bacterial multiplication-inhibition	Rani & Khetarpaul (1998)
Shigella dysenteriae	*L.casei/ L. acidophilus*	Protective mucosal immunity	Nader de Macias, et al. (1992)
Shigella sonnei	*L. acidophilus* S/C supernate[a]	Microbial growth inhibition	Coconnier et al. (1997)
Shigella flexneri	*L. acidophilus* S/C supernate[a]	Microbial growth inhibition	Coconnier et al. (1997)
Klebsiella pneumoniae	*L. acidophilus* S/C supernate[a]	Microbial growth inhibition	Coconnier et al. (1997)
Enterobacter sp.	*L. acidophilus* S/C supernate[a]	Microbial growth inhibition	Coconnier et al. (1997)
Bacillus cereus	*L. salivarius*	Attachment inhibition/IL-8 release	Kabir et al. (1997)
Helicobacter pylori			
Antifungal Activity			
Candida albicans	*L.acidophilus/L.reuteri/L. casei GG/ B.animalis*	Cell-mediated immune modulation	Wagner et al. (1997)
Cryptosporidium parvum	*L. reuteri*	Reversal of immune suppression	Alak et al. (1997)
Antiviral Activity			
Herpes Simplex virus type-1	Heat-killed *LC9018*	Induction of peritoneal macrophages	Watanabe & Saito (1986)
	Heat-killed *LC9018*	Immunostimulant activity	Watanabe & Hotta (1996)
Murine Cytomegalo virus	Heat-killed *LC9018*	host-mediated antiviral activity	Ohashi et al. (1988)

[a]Spent culture (S/C) supernate; **IFN-γ**, interferon-gamma; **NK** cell, Natural Killer cell

bacillus were shown to produce a dose-dependent immunostimulating effect. The oral administration of killed bifidobacteria and lactobacillus led to the enhanced resistance of mice to salmonella infection. Bacteroids and fusobacteria were found to possess no such effect.

Ishihara et al. (1985) reported that the *in vitro* growth of *Streptococcus mutans* was completely inhibited by water-soluble extracts from cells of various intestinal LAB identified as *S. faecium, S. equinus, L. fermentum,* and *L. salivarius*. The growth inhibition was dependent on the concentrations of the extracts. In contrast, the extracts did not inhibit the growth of the major indigenous intestinal LAB isolated from humans. Coconnier et al. (1997) reported that the spent culture supernatant of the human *L. acidophilus* strain *LB* produces an antibacterial activity against a wide range of gram-negative and gram-positive pathogens. It decreased the *in vitro* viability of *Staphylococcus aureus, Listeria monocytogenes, Salmonella typhimurium, Shigella flexneri, Escherichia coli, Klebsiella pneumoniae, Bacillus cereus, Pseudomonas aeruginosa,* and *Enterobacter sp.* In contrast, it did not inhibit lactobacilli and bifidobacteria. The activity was heat stable and relatively sensitive to enzymatic treatments and developed under acidic conditions. Antimicrobial activity was independent of lactic acid production. Antimicrobial activity was observed against *S. typhimurium SL1344* infecting human cultured intestinal Caco-2 cells and also in a mouse model with *S. typhimurium C5* infection following oral treatment with the spent culture supernatant.

B. Antiviral effects

Heat-killed LC9018 cells were shown to enhance the resistance to herpes simplex virus type 1 (HSV-1) in adult mice (Watanabe et al., 1986). Protection of mice against HSV-1 infection and concurrent production of neutralizing antibodies were significantly enhanced by administration of LC9018 in combination with inactivated HSV-1 antigen. The optimal enhancement of protection was noticed in mice 14 days after the simultaneous administration of these substances. The immunity to HSV-1 infection in mice could be transferred with peritoneal exudate cells from syngeneic mice previously treated with LC9018 alone and LC9018 in combination with inactivated HSV-1 antigen or with thioglycollate broth. Conversely, transfer of peritoneal exudate cells induced by thioglycollate broth alone and of spleen cells induced by LC9018 in combination with thioglycollate broth or by thioglycollate broth alone was not effective. These results suggest that mouse peritoneal macrophages induced by the administration of LC9018 in combination with inactivated HSV-1 antigen may play an important role in host defense mechanisms against HSV-1 infection. Subsequently, Watanabe and Yamori (1989) suggested that the host defense mechanism of mice against intraperitoneal infection with HSV-1 may be mainly related to peritoneal macrophages activation by synergistic effect of the administration of LC9018 and inactivate HSV-1 antigen and interferon production.

The protective effect of heat-killed *L. casei* against murine cytomegalovirus (MCMV) infection was reported by Ohashi et al. (1988). The protective effect was evidenced by an increase in plaque-forming units (PFU) per 50% lethal dose (LD_{50}) and a decrease in titers of infectious viral replication in the target organs. This was further confirmed by severity of histopathological damage to the target organs, especially the liver. *L. casei* neither inactivated MCMV nor inhibited its replication in mouse embryonic fibroblasts (MEF). The spleen cells from *L. casei*-treated mice inhibited MCMV replication in MEF on co-cultivation. Augmentation of splenic NK cell activity by *L. casei* cor-

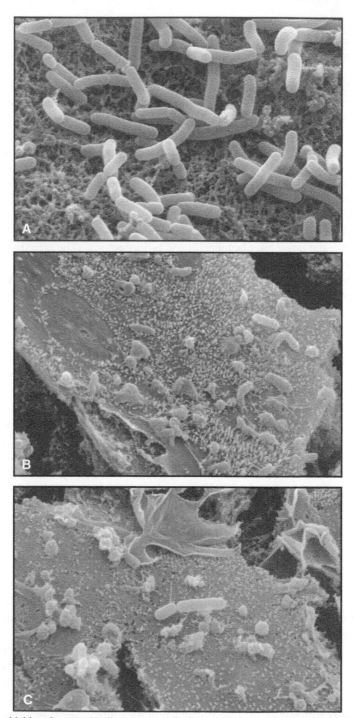

FIGURE 1. Microbial interference of Helicobacter pylori adhesion to eucaryotic cells by *L. johnsonii* (A) Adhesion of *L. johnsonii* on Caco-2 cells; (B) Adhesion of *H. pylori* on HT-29 cells; (C) Competition between *L. johnsonii* and *H. pylori* on HT-29 cells [reproduced with permission from the Nestlé Research Center, Lausanne, Switzerland].

related with mouse survival from otherwise lethal MCMV infection. Cytotoxic activity of peritoneal cells and level of serum interferon were elevated after MCMV infection, but they were not associated with survival of mice or with treatment of *L. casei*. The authors suggested that the *L. casei*-induced protection from MCMV infection was probably due to the augmentation of murine NK cell activity. A glycoprotein (GP) in the cell wall of *L. casei* was later identified as the potent inducer of the resistance (Ohashi et al., 1989). The protective effect of GP was evidenced by a decrease in titers of infectious viral replication in the target organs. Not only *L. casei* cells but also GP, to a lesser degree, enhanced NK cell activity in both uninfected mice and MCMV-infected mice. The activity of *L. casei* cells and GP to augment NK cell activity correlated with the protection index. GP treatment did not modify interferon production during MCMV infection. Thus, GP of LC cells seems to be the active principle to endow mice with resistance to MCMV.

Finally, the *in vivo* antimicrobial effects of live probiotic bacteria and probioactive compounds against various pathogens in experimental animal infection models are summarized in *TABLE 3*.

V. MICROBIAL INTERFERENCE

Probiotic LAB have been used to treat disturbed intestinal microflora and increased gut permeability which are characteristic of many intestinal disorders such as acute rotaviral diarrhea, food allergy, colon disorders, metabolic changes during pelvic radiotherapy and changes associated with colon cancer development. In all such conditions altered intestinal microflora, impaired gut barrier and different types of intestinal inflammation are present. Effective probiotic bacteria are able to survive gastric conditions and colonize the intestine, at least temporarily, by adhering to the intestinal epithelium. Such probiotic microorganisms appear to be promising candidates for the treatment of clinical conditions with abnormal gut microflora and altered gut mucosal barrier functions (Salminen et al., 1996).

Various laboratories have investigated LAB adhesion and its possible role in microbial ecology of the urogenital and intestinal tracts and the influence of host and microbial factors on bacterial interference and disease control. Although evidence is equivocal, *Lactobacillus acidophilus* appears to be beneficially involved in the interference with establishment of pathogens in the gastrointestinal tract. The mechanisms of action are believed to involve competitive exclusion and production of inhibitory substances, including bacteriocins (Reid et al., 1990). LAB strains that demonstrate wide spectrum of antimicrobial characteristics including acid and bile resistance, antimicrobial systems (i.e. bacteriocins, peroxide), adhesion to various types of mucosal cell are necessary for eliciting potent probiotic effect.

In veterinary studies, bacterial interference by administration of probiotics has also been beneficial in disease prevention in animals. Carefully selected bacterial mixtures integrated with the gastrointestinal flora of the animals can confer disease resistance and improve physiological function. Additional human and animal trials are needed to determine the practical, long-term usefulness of bacterial interference as a protective mechanism against infectious diseases.

Studies at the Nestlé Research Center, Switzerland revealed that *Lactobacillus johnsonii* could effectively interfere with the colonization of H. pylori to cultured eucaryotic cells such as HT-29 and Caco-2 (*FIGURE 1*; personal communication).

A. Human uropathogens

Chan et al. (1984) reported complete or partial inhibition of adherence of several Gram-negative uropathogens by preincubating the human uroepithelial cells *in vitro*, with several LAB species of the normal flora. Lactobacilli strains vary in their ability to interfere with colonization of uroepithelial cells by pathogenic organisms.

Microbial adherence to surfaces is a serious risk factor in the etiology of biomaterial-associated infections. The ability of organisms to adhere to and form biofilms on fibrous materials is believed to be an important initiating step in the induction of several diseases, such as the toxic shock syndrome. Hawthorn and Reid (1990) reported that precoating of polymers with certain LAB strains reduced the binding of uropathogenic coagulase-negative staphylococci and *E. coli* to biomaterial surface. Lactobacilli seem to detach from polymers of low surface tension and reattach to polymers with high surface tensions. Thus, effective exclusion of uropathogens was observed with precoating of lactobacilli on hydrophilic materials with high surface tension such as glass and sulfonated polystyrene.

Surfaces precoated with lactobacilli inhibit *Staphylococcus aureus* adhesion by 26- 97%, and *Candida albicans* by 0-67% (Reid et al., 1995). When the lactobacilli were used to challenge adherent pathogens, there was 99% displacement of the *S. aureus* and up to 91% displacement of *C. albicans*. Hydrophobic *L. acidophilus* strain 76 and strain T-13 were the most effective of five *Lactobacillus* isolates tested at interference by precoating. The moderately hydrophilic *L. casei* var. *rhamnosus* GR-1 was the most effective at displacing the candida. Experiments with uroepithelial cells also showed that the lactobacilli could significantly interfere with the adhesion of both pathogens to the cells. The results demonstrate the rapidity with which two pathogens adhered to fibers and epithelial cells, and raised the possibility that members of the normal female urogenital flora might interfere with infections caused by these organisms.

Velraeds et al. (1996) investigated the biosurfactants from *L. casei ssp. rhamnosus 36* and *ATCC 7469*, *L. fermentum B54*, and *L. acidophilus RC14* for their capacity to inhibit the initial adhesion of uropathogenic *Enterococcus faecalis 1131* to glass in a parallel-plate flow chamber. The initial deposition rate of *E. faecalis* to glass with an adsorbed biosurfactant layer from *L. acidophilus RC14* or *L. fermentum B54* was significantly decreased by approximately 70%, while the number of adhering enterococci after 4 h of adhesion was reduced by an average of 77%. The surface activity of the biosurfactants and their activity inhibiting the initial adhesion of *E. faecalis 1131* were retained after dialysis (molecular weight cutoff, 6 to 8 kD) and freeze-drying. Fourier transform infrared spectroscopy and X-ray photoelectron spectroscopy revealed that the freeze-dried biosurfactants from *L. acidophilus RC14* and *L. fermentum B54* were richest in protein, while those from *L. casei* ssp. *rhamnosus 36* and *ATCC 7469* had relatively high polysaccharide and phosphate contents. The ability of *L. acidophilus RC14* biosurfactant "surlactin" to reduce the initial adhesion of *Enterococcus faecalis 1131* was further characterized by Velraeds et al. (1997) on a hydrophilic and a hydrophobic substratum in a parallel-plate flow chamber, using phosphate-buffered saline and pooled human urine as a suspending fluid. The biosurfactant surlactin, as released by several *Lactobacillus* isolates, might open the way to the development of antiadhesive biologic coatings for catheter materials. It should be considered, however, that these results are preliminary and that the efficiency of the biosurfactant is probably affected not only by the hydrophobicity of the substratum and the suspending fluid, but also by the type of uropathogen involved.

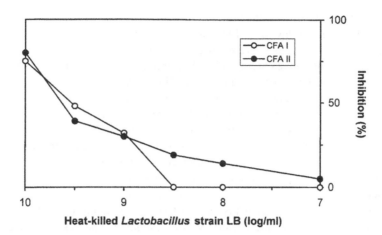

FIGURE 2: Competitive exclusion of diarrheagenic *Escherichia coli* (ETEC) from human enterocyte-like Caco-2 cells by heat-killed *Lactobacillus* (redrawn from Chauviere et al., 1992).

B. Human oral pathogens

The ability of oral lactobacilli to coaggregate with streptococci and actinomycetes was reported by Wilcox et al. (1993). Of the seven species of lactobacilli studied, only two were capable of coaggregation and the coaggregation was restricted to streptococci. *Lactobacillus salivarius* strains coaggregated with *Streptococcus salivarius*, *Streptococcus gordonii*, *Streptococcus crista* and tufted *Streptococcus sanguis II* strains. *Lactobacillus fermentum* coaggregated with *S. gordonii* and *S. sanguis*. The coaggregation between *L. salivarius* and *S. salivarius, S. gordonii* or tufted *S. sanguis II* strains was mediated by a protein on the surface of the lactobacilli and was not inhibited by lactose. The coaggregation between *L. fermentum* and the streptococci was mediated by protein on the surface of the streptococci and was inhibited by lactose.

C. Human enteropathogens

Escherichia coli B41 adheres to HeLa 229 cells in a diffuse pattern. Heat-killed *L. acidophilus* (Lacteol strain) was found to inhibit this adhesion in a dose-dependent manner (Fourniat et al., 1992). This inhibitory action was lost after lysis of the *L. acidophilus*, further suggested that the mechanism was steric hindrance of *E. coli* adhesion sites rather than competition for a common binding site. A thermostable factor excreted by *L. acidophilus* into the medium seems to mediate the adhesion of *L. acidophilus* to HeLa cells, and block adhesion of *E. coli* to these cells.

Enterotoxigenic *E. coli* (ETEC) bearing CFA/I or CFA/II adhesive factors specifically adhere to the brush border of the polarized epithelial human intestinal Caco-2 cells in culture. Heat-killed *L. acidophilus strain LB* also seem to inhibit ETEC adhesion to Caco-2 cells in a dose-dependent manner (*FIGURE 2*). The mechanism of this inhibition seems to be steric hindrance of the ETEC receptors on human enterocytes caused by heat-killed *L. acidophilus* strain LB (Chauviere et al., 1992).

Salmonella typhimurium and enteropathogenic *E. coli* (EPEC) adhere to the brush border of differentiated human intestinal epithelial Caco-2 cells in culture, whereas *Yersinia pseudotuberculosis* and *Listeria monocytogenes* bind to the periphery of undifferentiated Caco-2 cells. All these enterovirulent strains invade the Caco-2 cells. Live and heat-killed *L. acidophilus* strain LB, which strongly adheres both to undifferentiated and differentiated Caco-2 cells, were reported to inhibit both cell association and invasion of Caco-2 cells by the above enterovirulent bacteria in a dose-dependent manner (Coconnier et al., 1993a). The mechanism of inhibition of both adhesion and invasion seems to be due to steric hindrance of pathogen specific human enterocyte receptors by whole cells of lactobacilli rather than to a specific blockade of receptors. Bernet et al. (1994) tested four human *L. acidophilus* strains for their ability to adhere to human enterocyte like Caco-2 cells in culture. The LA1 strain exhibited calcium independent adhesive property. This adhesion to Caco-2 cells required a proteinaceous adhesion-promoting factor, which was present in the spent bacterial broth culture supernatant. Strain LA1 also strongly bound to the mucus secreted by the homogeneous cultured human goblet cell line HT29-MTX. The inhibitory effect of LA1 organisms against Caco-2 cell adhesion and cell invasion by a large variety of diarrheagenic bacteria was investigated. The dose-dependent inhibitions by strain LA1 were obtained for the following organisms: (a) cell association of enterotoxigenic, diffusely adhering and enteropathogenic *E. coli*, and *Salmonella typhimurium*; (b) cell invasion by enteropathogenic *E. coli, Yersinia pseudotuberculosis,* and *Salmonella typhimurium*. Incubations of *L. acidophilus* LA1 before and together with enterovirulent *E. coli* were more effective than incubation after infection by *E. coli*. Similar inhibitory effect by heat-killed *L. acidophilus*, strain LB, against the mucus-secreting HT29-MTX and human intestinal Caco-2 was also reported earlier by Coconnier *et al.* (1993b).

Oral inoculation of *H. pylori* usually results in a temporary colonization without a successful infection in the stomach of conventional mice in which lactobacilli are the predominant indigenous bacteria. Kabir et al. (1997) investigated the inhibitory effects of lactobacilli on *H. pylori* attachment to murine and human gastric epithelial cells and the *H. pylori* mediated release of interleukin-8 (IL-8) by these cells *in vitro*. The study included *Lactobacillus salivarius* infected gnotobiotic BALB/c mice and control germ free mice were inoculated orally with *H. pylori* pathogen. *L. salivarius* inhibited both the attachment and IL-8 release *in vitro*. *H. pylori* failed to colonize the stomach of infected with *L. salivarius*, but in large numbers colonized and subsequently caused active gastritis in germ free mice. In addition, *L salivarius* administration after *H. pylori* implantation could eliminate colonization by *H. pylori*. These findings suggest the possibility of LAB as an effective probiotic agent against *H. pylori*.

D. Porcine enteropathogens

Enteropathogenic *E. coli* (EPEC) colonizes the piglet ileum and adheres to the mucosa by K88 fimbrial appendages. Indigenous lactobacilli seem to suppress the colonization potential of EPEC to ileal mucus (Blomberg et al., 1992). An active component responsible for the blocking of K88 fimbrial interaction with ileal mucus was detected in the culture fluid. Inhibition occurred when mucus was pretreated with either spent culture dialysis retentate or the void volume (fraction of > 250 kDa molecular weight) after gel filtration. Three lactobacilli of porcine origin reduced adhesion of *E. coli K88* by approximately 50%.

E. Avian enteropathogens

Jin et al. (1996) examined the poultry isolates of *L. acidophilus* and *L. fermentum* for bacterial interference against chicken intestinal pathogens of *Salmonellae* in an ileal epithelial cell (IEC) radioactive assay. Exclusion, competition and displacement phenomena were investigated by incubating (a) LAB and IEC together, prior to addition of salmonellae, (b) LAB, IEC and salmonellae together, and (c) salmonellae and IEC, followed by the LAB. *L. acidophilus* significantly reduced the attachment of *Salmonella pullorum* to IEC in the tests for exclusion and competition, but not in the displacement tests. *L. fermentum* showed some ability to reduce the attachment of *Salmonella typhimurium* to IEC under the conditions of exclusion, competition, but not displacement. However, both *L. acidophilus* and *L. fermentum* were unable to reduce the adherence of *Salmonella enteritidis* to IEC under any of the conditions.

VI. APPLICATIONS

In recent years, the retail sales of bio-milks, bio-yogurts, and other probiotic products increased rapidly in the markets of Europe, North America and many other developed countries. Such products are made using a single genus, or in combination with other LAB as mixed starter cultures. In the absence of published surveys, it is impossible to predict the extent to which expected health benefits have contributed to this sales pitch, but it would seem unlikely that consumers would have acquired this intense probiotic-loyalty without a genuine belief in its therapeutic and/or prophylactic properties (Tamime et al., 1995)

A. Fermented foods

Fermented foods have a long history of safe usage and are found in diets throughout the world. Fermentation is a process in which foods are modified by the action of microorganisms or enzymes in order to achieve a desired biochemical change. Production and consumption of fermented foods dates back many thousands of years. Even today, fermentation continues to be used for household production and preservation of foods in many countries where mass-produced food is not widely available (Hull et al., 1992). With modern food processing technology, large-scale production of fermented products can be accomplished with precise controls over the microorganisms and enzymes used in the fermentation process.

Alcoholic fermentation of barley to beer and grapes to wine are early evidence of use of the process, dates back to 5,000 years (Borgstrom, 1968). In the Middle East and the Indian subcontinent, milk was fermented into yogurt and other products (Oberman, 1985). More than 1,000 different types of cheese have been produced from LAB in the Middle East and Europe (Campbell-Platt, 1994).

Fermented vegetables have been traditional foods in East Asia for thousands of years. *Kimchi* remains the traditional pickled vegetable food in Korea, served today at nearly every meal. In Indonesia, soybeans have been fermented to *tempe* and peanut press cake into *oncom*. Other fermented foods that have been used traditionally in Asian cuisine include fish sauces, fermented fish pastes, and soy products (soy sauce and *miso*). Production of these products began in China 3000 years ago and knowledge of these processes was later transferred to Japan (Campbell-Platt, 1994). In Africa, maize and sorghum are fermented to beers and products made from fermentation of beans from the tree legume *Parkia* (Campbell-Platt, 1994). Fermentation is also important in the detoxification of bitter varieties of cassava, which is one of the world's most important starch crops. Fermentation of meat is a traditional method for protecting it from spoilage. LAB, either alone or in combination with micrococci, yeasts or molds are essential for the fermentation process (Hammes & Tichaczek, 1994).

TABLE 4: LAB application in various fermented foods (adapted from Naidu et al., 1999)

Food Product	Origin	LAB Culture
Acidophilus milk	USA	*Lb. acidophilus*
Balao balao	Indonesia	*Lactobacillus sp.*
Breads	Egypt	*Lb. brevis, Lb. plantarum, Lb. fermentum, Lc. mesenteroides, S. thermophilus*
Bulgaria buttermilk	Bulgaria	*Lb. delbrueckii* spp. *bulgaricus*
Burong dalag	Southeast Asia	*Lc. mesenteroides, P. pentosaceus, Lb. plantarum*
Butter	Wordwide	*Lc. lactis* spp. *lactis, Lc. lactis* spp. *cremoris, Lc. lactis, Lc mesenteroides.*
Buttermilk	Worldwide	*Lc. lactis* ssp. *lactis*
Cacao	South America	Various LAB
Cheeses	Worldwide	*Lc. lactis* spp. *lactis, Lc. lactis* ssp. *cremoris, Lb. delbrueckii* spp. *bulgaricus, Lb. delbrueckii* ssp. *lactis, Lb. casei, Lb. helveticius, Lb. brevis, Lb. fermentum*
Coffee	Worldwide	Various LAB
Crackers	Europe /N. America	*Lb. plantarum, Lb. delbrueckii, Lb. casei, Lb. brevis, Lb. fermentum*
Cucumbers	Worldwide	*Lb. plantarum, Lb. brevis, P. pentosaceus*
Cured ham	Europe	*Lb. casei, Lb. plantarum*
Distillery mashes	Europe	*Lb. plantarum, Lb. delbrueckii,* ssp. *delbruckii, Lb. casei, Lb. fermentum*
Dahi	India	*S. thermophilus, Lb. bulgaricus, Lc. diacetylactuis, Leuconostoc* spp.
Dosa	India	*Lc. mesenteroides, E. faecalis, P. pentosaceus*
Feed additives	Europe / N.America	*Lb. acidophilus, Lb. delbrueckii, Lb. reuteri*
Fermented fish	China, Japan	Various LAB
Filmjölk	Sweden	*Lc. lactis, Lc. cremoris, Lc. diacetylactis*
Gari	Africa	*Leuconostoc sp.*
Green olives	Middle East	*Lb. plantarum, Lb. brevis, P. pentosaceus*
Idli	India	*Lc. mesenteroides, E. faecalis, P. pentosaceus*
Izushi	Japan	*Lactobacillus sp.*
Kefir	Caucasus	*Lc. lactis* ssp. *lactis, Lb. delbrueckii* ssp. *bulgaricus*
Kenkey	Africa	*Lactobacillus sp.*
Kimchi	Korea	Various LAB
Kishk	Saudi Arabia	*Lb. brevis, Lb. casei, Lb. plantarum*
Kisra	Middle East	*Lactobacillus sp.*
Koumiss	Mongolia	*Lb. delbrueckii* ssp. *bulgaricus, Lb. delbrueckii* ssp. *lactis*
Lactic acid	Europe	*Lb. delbrueckii* ssp. *delbruckii*
Lactic butter	Europe	*Lc. lactis* ssp. *cremoris*
Lassi	India	*Lactococcus* spp., *Lactobacillus* spp.
Lebanon bologna	Lebanon	*P. acidilactici*
Leben	Middle East	*S. thermophilus, Lb. bulgaricus, Lb. acidophilus, Lc. lactis*
Magon	Tunisia	*Lactobacillus sp.*
Miso	Japan	*Lactobacillus sp.*
Ogi	Africa	*Lb. plantarum, Lc. lactis* ssp. *lactis*
Oncom	Indonesia	*LAB ssp.*
Pulque	Mexico	*Lb. plantarum, Leuconostoc sp.*
Puto	Portugal	*Lc. mesenteroides, E. faecalis*
Sauerkraut	Europe	*Lb. plantarum, Lb. brevis, Lc. mesenteroides, P. pentosaceus*
Sausages, meats	Europe	*Lb. plantarum, Lb. casei, Lb. reuteri, Lb. sake, P. pentosaceus, P. acidilactii*
Shoyu	Indonesia	*Lactobacillus sp.*
Silage starters	North America	*Lb. plantarum, P.acidilactii*
Sour bread	France / W. Europe	*Lb. sanfrancisco, Lb.brevis*
Sour cream	Worldwide	*Lc. lactis* ssp. *cremoris*

TABLE 4 (Cont.):

Food Product	Origin	LAB Culture
Sour pumpernickel	North Africa	*Lc. mesenteroides*
Sour cream	Worldwide	*Lc. lactis* ssp. *cremoris*
Sour pumpernickel	North Africa	*Lc. mesenteroides*
Soy sauce	Southeast Asia	*Lb. delbrueckii* ssp. *delbruckii*
Sweet dough	France / W. Europe	*Lb. brevis, Lb. plantarum, Lb. fermentus, Lc. lactis, Leu. mesenteroides*
Taette	Netherlands	*Lc. lactis* ssp. *lactis* var. *taette*
Tempeh	Indonesia	*Lactobacillus sp.*
Uji	Japan	*Lb. plantarum, L. lactis* ssp. *lactis*
Villi	Finland	*Lc. lactis* ssp. *cremoris, Lc. diacetylactis*
Wine	N. America /Europe	*Lb. plantarum, Lb. hilgardii, P. damnosus, Lc. mesenteroides, Lc. oenos*
Yakult	Japan	*Lb. casei* Yogurt Asia, Balkans *Lb. delbrueckii* ssp. *bulgaricus, S. thermophilus*

Throughout the world, fermented foods continue to constitute a significant proportion of the diet. Production of fermented foods is highest in Europe, North America and sub-Saharan Africa, with significant production and consumption in South America, the Middle East, India, and Southeast Asia. In terms of total production and consumption worldwide, dairy foods, beverages and cereals constitute the majority of fermented foods (Campbell-Platt, 1994).

Traditionally, LAB have been used in the production of fermented foods (TABLE 4). These include species of *Lactobacillus* and *Bifidobacterium* and the species *S. thermophilus*. This may have begun as a process in nature in which nutrient availability, environmental conditions selected particular microorganisms, which modified, and preserved the food (Campbell-Platt, 1994). Approximately 80 bifid-containing products are now on the world market. Most are of dairy origin and include such products as yogurt, butter milk, sour cream, powdered milk, fortified milk, cookies and frozen desserts (Modler et al., 1990a).

LAB plays an important role in the production of raw fermented sausages (FIGURE 2). Starter cultures are therefore continuously subjected to re-evaluation and adapted to new environments. Protective cultures with the ability of bacteriocin production, especially to control *L. monocytogenes*, offer novel possibilities for the improvement of safety and shelf life extension of conventional meat products. Kröckel (1999) suggested a possible biopreservation with LAB cultures such as *L. sakei* or *L.curvatus* preferentially with the ability to produce bacteriocins or the direct use of purified anti-listerial bacteriocin as food additive for refrigerated ready-to eat meat products. During cold storage of vacuum-packed sliced Bologna-type sausage *L.sakei* Lb674 (sakacin P) produces sufficient amounts of bacteriocin when inoculated at a density of 10^5 to 10^6 bacteria/g sausage. The growth of *L.monocytogenes* was delayed and in certain cases completely inhibited. Similar results were obtained with *L.sakei* Lb706 (Sakacin A producer) (Kröckel, 1999).

B. Supplemented foods

The species of LAB used in the preparation of probiotic products include *L. bulgaricus, L. lactis, L. salivarius, L. plantarum, S. thermophilus, Enterococcus faecium, E. feacalis,* and *Bifidobacterium* sp.

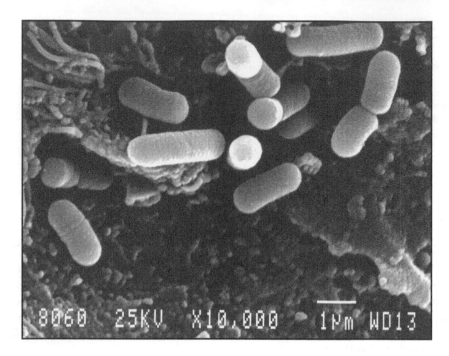

FIGURE 3: Rod-shaped lactobacilli in salami under scanning electron microscope (Reproduced with permission from Dr..Kröckel, Federal Center for Meat Research, Kulmbach, Germany).

Considerable attention has been given in recent years to the use of bifidobacteria in probiotic foods, particularly in Japan and Europe (Hughes & Hoover, 1991; Ishibashi & Shimamura, 1993). Hughes and Hoover (1991) reviewed and summarized a number of probiotic applications for bifidobacteria. In the 1940s, bifidus milk was used as a treatment for infants with nutritional deficiencies. In Japan, the first bifidus product (low-fat fresh milk containing *B. longum* and *L. acidophilus*) was developed by Morinaga Milk Industry Company in 1971; full-scale production began in 1977 when the company started a home-delivery service (Ishibashi & Shimamura, 1993). By 1984, there were 53 Bifidus products on the market in Japan. Today many products including yogurts have been reformulated to include bifidus cultures; total yogurt sales in Japan have nearly doubled from the 1980s to the 1990s (Hughes & Hoover, 1991). Today, many products containing LAB are available worldwide. Probiotic preparations are manufactured in various forms (tablets or powders) and used in a number of foods (milk, chewing gums, fiber preparations, sweets, cakes, beer, and soymilk).

C. Pharmaceutical products

Special dietary preparations containing viable cells of LAB are available in different markets as freeze-dried tablets. The primary objective of these products is to achieve persistant colonization of the bacteria in the gut during the treatment of a variety of conditions such as gastrointestinal disorders (eg. post antibiotic therapy, adjustment of microbial imbalances in the gut, liver diseases, chronic constipation, chronic duodenitis, peptic ulcers in children, and after irradiation therapy) (Tamime et al., 1995).

Pharmaceutical products made from cow milk and containing live bifidobacteria are Bifider® (in Japan), Bifidogène® (in France) and LioBif® (in Yugoslavia). Two probiotic products are in the German market: Eugalan Töpfer Forte® is a fat-free and glutanfree preparation that contains plant proteins, lactose, lactulose and 30 minerals; and Euga-Lein Töpfer Forte® a fat containing product with vitamin C. Ominflora® , which is also a German product contains *L. acidophilus* and *B.longum* and a saprophytic *E.coli* (Kurman et al., 1992).

VII. LARGE-SCALE PRODUCTION

Development of consistently effective probiotics is still at a very early stage. The probiotic concept will be accepted if the underlying mechanisms of claimed health benefits are elucidated, and appropriate selection criteria for probiotic microorganisms are defined. Probiotic strains must therefore fulfill a number of general and specific criteria (Havenaar et al., 1992). The general selection criteria are determined by biosafety aspects, methods of production and processing, and the method of administration of the probiotic and the location in the body where the microorganism must be active. Specific selection criteria must include phenotypic traits such as specific enzyme activities or stimulation of the immune system. A rational selection and validation of promising probiotic LAB is only possible from evidence obtained in validated in vitro models with a high predictive value and followed by *in vivo* studies, including human trials.

A. Strain selection

LAB strains for probiotic use must be representative of microorganisms that are Generally Recognized As Safe (GRAS microorganisms). The use of newer strains without a proven track record of biosafety, requires extensive toxicological and tolerance testing prior to acceptance.

The next important basic step in the selection procedure is the choice of origin of the strain. This choice will mainly be determined by the specific purpose of the probiotic. The generally desired properties of LAB for human probiotic use are:

- Human origin with a proven safety and tolerance record
- Ability to survive acidic conditions and bile in the gut
- Contribution to positive regulation of intestinal functions, correction of bowel disorders
- Ability to produce vitamins and release beneficial enzymes that would facilitate digestion of food, absorption of essential nutrients and inactivate carcinogens and toxic compounds
- Capacity to adhere to human intestinal epithelia, colonize the mucosal surface and proliferate in the human gut
- Ability to produce antimicrobial substances and demonstrate a wide spectrum inhibitory activity against human pathogens
- Possession of probio-active cellular components that could stimulate immune responses, augment host defense functions of reticuloendothelial cell cascade

B. Quality control

In recent years, consumers have become aware of probiotic properties of cultured milk and dosage specifications. A concentration of 1×10^5 cfu/g or ml of the final product has been suggested as the therapeutic minimum. Most probiotic LAB, including the common yogurt cultures *L. acidophilus* and *B.bifidum*, show a short stationary growth phase, followed by a rapid loss of cell viability, even in cold storage (Lee & Wong, 1993). This short shelf life represents a logistical problem for both manufacturers and retailers, and a technical challenge for researchers.

Thus, it is important to check the viability and resistance of the LAB during processing and storage. One of the first steps in probiotic production is large-scale culturing, washing and drying of the microorganisms. Most of the LAB can be cultured in large-scale fermentors and are rather resistant to centrifugation (Havennar et al., 1992). In general, LAB could withstand freezing and frozen storage at –20°C or lower (Klaenhammer & Kleeman, 1981), but are less resistant to freeze-drying and especially to spray-drying. In cases of storage over long periods and/or under unfavorable conditions, encapsulation of LAB cultures should be considered. The viability of LAB could dramatically decrease during pelleting, thus preservation or protective measures are warranted during such processing. Contamination of probiotic products with undesirable microorganisms is also possible, especially in uncontrolled fermentation procedures and a stringent microbiological quality control is necessary. The quality control should ensure the following stability and technical properties of probiotic LAB products:

- Ability to maintain viability through processing and storage
- Maintain good flavor, aroma profile and organoleptic qualities
- Maintain mild acidity throughout storage
- Retain intestinal colonizing properties throughout processing and storage
- Enhance the shelf life and storage stability of fermented products
- Demonstrate stability and functionality after freeze-drying and other drying methods
- Accurate strain identification and exclusion of undesirable contaminants
- Dose-response data for required effects

C. Biotechnology

It is possible that a potential probiotic LAB strain will fulfil a number of specific criteria for a product development, but lacks one important property. A preferable manipulation would be locating and isolating the genes that are responsible for a specific factor, such as adherence, and to transfer these genes to a probiotic LAB strain without altering other beneficial properties. Use of gene technology for the construction and improvement of microorganisms could provide great promise for future development of probiotic strains.

Natural gene transfer processes are of significance both in the evolution of LAB and as a tool for the construction of genetically manipulated strains. This approach may have advantages over recombinant DNA methods that require regulatory approval and consumer acceptance (Gasson & Fitzgerald, 1994). The advent of high-voltage electroporation greatly simplified the genetic manipulation of LAB. The tools for the genetic engineering of LAB are clearly available. The catalog of cloned genes and regulatory and expression signals available will allow thorough exploration of the expression and secretion signal sequence requirements and the potential of these bacteria (Mercenier et al., 1994). Knowledge thus gained could be put to rational application in the design and modification of strains.

VIII. SAFETY AND TOLERANCE

Food fermentation is one of the oldest known uses of biotechnology. Fermented foods and beverages continue to provide an important part of human diet and constitute 20-40% of food supply worldwide (Campbell-Platt, 1994). By tradition, LAB are involved in the production of fermented foods. These microorganisms constitute one quarter of our diet and are characterized by a safe history, certain beneficial health effects, and an extended shelf life when compared to raw materials. Since antiquity, people have regularly consumed large numbers of LAB, both in fermented and non-fermented foods, comprised as the material's natural flora. While the situation requires continued monitoring, it must be emphasized, however, had these organisms posed any serious health threat, it would have become apparent long time ago. The GRAS status of probiotic LAB should be considered based on the following four lines of scientific evidence:

A. Human trials

Oral administration of probiotic LAB was well tolerated and proven safe in 143 human clinical trials during 1961-1998 (Naidu et al., 1999). In these trials, 7,526 human subjects were orally administered with well-characterized or unknown LAB strains. These bacteria were used either as single or mixed doses (ranging from $10^7 - 10^{11}$ cfu/dose), live or heat-killed cultures, for prophylactic and/or therapeutic purposes in clinical management of various illnesses and physiological disorders. The number of clinical studies markedly increased during the past 40 years, with more than 50% trials in the recent 8 years alone. This trend in clinical trials indicate that clinicians, basic scientists, and corporate R&D personnel, all alike have developed a phenomenal interest in probiotic applications, a prospect in probiotic efficacy and a high degree of confidence in safe administration of LAB in formulations, diet or alone into humans. The new developments in molecular genetics and experimental biology has led to specific characterization of LAB strains such as *Lactobacillus GG, L. casei LC9018, L. acidophilus LC1, S. thermophilus Th4, B. lactis Bb12*, etc., and evidently clinical trials with well-defined species of single strains became dominant recently compared studies with mixed cultures or undefined LAB species in early years. All probiotic LAB strains administered in the 143 human trials were well-tolerated and not a single adverse effect or event was ever reported in any of the 7,526 subjects involved in these studies spanning a time-line of four decades.

B. Animal studies

LAB administration in experimental animal models *in vivo* i.e. mice, rats, guinea pigs, etc.; never elicited pyrogenicity, acute toxicity, or bacteremia. On the contrary, probiotic LAB appeared to prolong the survival of the animals.

C. *In vitro* studies

Probiotic LAB are shown to be non-invasive in eucaryotic cell interaction experiments with differentiated human intestinal epithelial cells (for microbial adhesion) or with human lymphoid carcinoma cell lines (for eliciting immunomodulatory effects). Various laboratories studied interactions of LAB with different eucaryotic cells and cell lines. LAB strains adhere to chicken crop epithelium (Brooker & Fuller, 1975), porcine gastric squamous epithelium (Henriksson et al., 1991), murine gastric squamous epithelium (Conway & Kjellberg, 1989), and chicken intestinal epithelial cells (Schneitz et al., 1993). These interactions are host-specific and establish LAB in the intestinal tract without any histopathological and toxicological outcomes. Chauviere et al. (1989) reported the binding of *Lactobacillus* strains to human colon carcinoma (Caco-2) or HT-29 cell lines. Elo et al. (1991) reported that *Lactobacillus casei* strain *GG* either freeze-dried or samples from dairy products bind to Caco-2 cell line with medium to strong affinity. It was observed that the common commercial dairy strains *L. bulgaricus* and *L. acidophilus NCFD1748* were not adhesive to the Caco-2 cell line, while *Bifidobacterium* strains demonstrated a moderate eucaryotic cell interaction. Unlike the non-binding strains, *Lactobacillus GG* shows rapid colonization and persistence in the human intestinal tract (Saxelin et al., 1991). Bernet et al. (1993) reported a high level of calcium-independent adherence of *B. breve 4, B. infantis,* and three fresh human isolates of bifidobacteria to Caco-2 cells and to mucus produced by the human mucus-secreting HT29-MTX cell line in culture. The bacteria interacted with the well-defined apical microvilli of Caco-2 cells without cell damage.

D. Probiotic Market

Metchnikoff's advocacy of fermented milk products as beneficial to human health has initiated the development of probiotic formulations as pharmaceuticals for therapy and prophylaxis. As early as 1913, a probiotic preparation of *L. bulgaricus* was successfully used in the treatment of 116 out of 117 cases of infantile diarrhea (Clock, 1913). In 1922, a probiotic-based tablet product, Paraghurt®, (Leo Laboratories, Denmark) was introduced into the market. Later this probiotic product underwent several improvisations and reformulation by adding chocolate and paraffin coating. Currently this product contains freeze-dried *Enterococcus faecium* only, without any coating. Today the global market is estimated to contain several thousands of different probiotic preparations in form of tablets, soft and hard gelatin capsules, powders, liquids, and pastes.

Probiotic bacteria are sold as food and dietary supplements in the U.S. market. The food products containing probiotic bacteria are almost exclusively dairy products such as fluid milk and yogurt, capitalizing on the historical association of LAB with fermented milk. *L. acidophilus* and *Bifidobacterium* species are the most commonly used cultures in probiotic foods in U.S. Approximately 0.6% of fluid milk products in the U.S. contain probiotic bacteria. These products use about 40% of the dairy-destined probiotic bacteria. In contrast, most major, and many minor, brands of yogurt contain probiotics, using the remaining 60% of the dairy-destined probiotic bacteria (Sanders, 1998). Sweet Acidophilus™ milk, a product containing *L.acidophilus* NCFM™ strain was introduced into U.S. market in 1970.

IX. SUMMARY

LAB with probiotic activity are generally enteric flora, believed to play a beneficial role in the ecosystem of the human gastrointestinal tract. Recent developments in the role of probiotics (live) and their probioactive (cellular) substances in the intestinal and extra-intestinal physiology of host are overwhelming. LAB are capable of preventing the adherence, establishment, replication, and/or pathogenic action of specific enteropathogens. Probiotic LAB have been used to treat disturbed intestinal microflora and increased gut permeability which are characteristic of many intestinal disorders such as acute rotaviral diarrhea, food allergy, colon disorders, metabolic changes during pelvic radiotherapy and changes associated with colon cancer development.

Food fermentation is one of the oldest known uses of biotechnology. Fermented foods and beverages continue to provide an important part of human diet and constitute 20-40% of food supply worldwide. The most common use of probiotics is as food in the form of fermented milk products. In recent years, consumers have become aware of probiotic properties of cultured milk and dosage specifications. A concentration of 1 x 10^5 cfu/g or ml of the final product has been suggested as the therapeutic minimum. However, the list of probiotic effects and health claims with the use of LAB is expanding. The global marketing trends of probiotics are based on expectations of a prophylactic effect and in many cases as an alternative to more conventional pharmaceutical preparations. The probiotic concept will come to acceptance if the underlying mechanisms of claimed health benefits are elucidated, and appropriate selection criteria for probiotic microorganisms are defined.

Acknowledgments: Thanks to Dr. M.F. Clerc, Nestlé Research Center, Lausanne, Switzerland; and to Dr. Luthar Kröckel, Federal Center for Meat Research, Kulmbach, Germany for providing the SEM figure.

X. REFERENCES

1. Adams, M. R., and Hall, C. J. 1988. Growth inhibition of food borne pathogens by lactic and acetic acids and their mixtures. *Int. J. Food Sci. Technol.* 23: 287-292.
2. Alak, J. I., Wolf, B. W., Mdurvwa, E. G., Pimentel-Smith, G. E., and Adeyemo, O. 1997. Effect of *Lactobacillus reuteri* on intestinal resistance of *Cryptosporidium parvum* infection in a murine model of acquired immunodeficiency syndrome. *J. Infect. Dis.* 175: 218-221.
3. Baird-Parker, A. C. 1980. Organic acids, In: *Microbial Ecology of Foods*, ed. J. H. Silliker, R. P. Elliott, A. C. Baird-Parker, F. L. Bryan, J. H. B. Christian, D. S. Clark, J. C. Olson, and T. R. Roberts, pp. 126-135. New York: Academic Press.
4. Beijernick, M. W. 1901. Anhaufungsversuche mit Ureumbakterien. Ureumspaltung durch Urease and durch Katabolismus. *Zentrubl. Bakteriol. Parasitenkd. Infektionskr. Hyg. II Abt* 7: 33-61.
5. Bernet, M. F., Brassart, D., Neeser, J. R., and Servin, A. L. 1994. *Lactobacillus acidophilus* LA 1 binds to cultured human intestinal cell lines and inhibits cell attachment and cell invasion by enterovirulent bacteria. *Gut* 35: 483-489.
6. Bernet, M. F., Brassart, D., Neeser, J. R., and Servin, A. L. 1993. Adhesion of human bifidobacterial strains to cultured human intestinal epithelial cells and inhibition of enteropathogen-cell interactions., *Appl. Environ. Microbiol.* 59: 4121-4128.
7. Blomberg, L., Henriksson, A., and Conway, P. L. 1992. Inhibition of adhesion of *Escherichia coli* K88 to piglet ileal mucus by *Lactobacillus* spp. *Appl. Environ. Microbiol.* 59: 34-39.
8. Borgstrom, G. 1968. Principles of Food Science. In: *Food Microbiology and Biochemistry, Vol. 2*, pp. 103-126. New York: Macmillan.
9. Brooker, B. F., and Fuller, R. 1975. Adhesion of lactobacilli to the chicken crop epithelium. *J. Ultrastruct. Res.* 52: 21-31.
10. Buchman, G. W., Banerjee, S., and Hansen, J. N. 1988. Structure, expression, and evolution of a gene encoding the precursor of nisin, a small protein antibiotic. *J. Biol. Chem.* 263: 16260-16266.
11. Campbell-Platt, G. 1994. Fermented foods - a world perspective. *Food Res. Int.* 27: 253-257.
12. Casaus, P., Nilsen, T., Cintas, L. M., Nes, I. F., Hernandez, P. E., and Holo, H. 1997. Enterocin B a new bacteriocin from *Enterococcus faecium* T136 which can act synergistically with enterocin A. *Microbiology* 143: 2287-2294.
13. Chan, R. C., Bruce, A. W., and Reid, G. 1984. Adherence of cervical, vaginal and distal urethral normal microbial flora to human uroepithelial cells and the inhibition of adherence of gram-negative uropathogens by competitive exclusion. *J. Urol.* 131: 596-601.

14. Chauviere, G., Barbat, A., Fourniat, J., and Servin, A. L. 1989. Adhesion of Lactobacillus on to the cultured human enterocyte-like cell lines Caco-2 and HT-29. Comparision between human and non-human strains. In: *Les laits fermentes. Actualite de la recherche*, pp. 287. John Libbey Eurotext Ltd.

15. Chauviere, G., Coconnier, M. H., Kerneis, S., Darfeuille-Micharud, A., Joly, B., and Servin, A. L. 1992, Competitive exclusion of diarrheagenic *Escherichia coli* (ETEC) from human enterocyte-like Caco-2 cells by heat-killed *Lactobacillus. FEMS Microbiol. Lett.* 70: 213-217.

16. Clark, D. S., and Takacs, J. 1980. Gases as preservatives. In: *Microbial Ecology of Foods*, ed. J. H. Sillikier, pp. 170-180. London, UK: Academic Press.

17. Clock, O.R. 1913. One hundred seventeen cases of infant diarrhea. *JAMA* 61: 164-168.

18. Coconnier, M. H., Bernet, M. F., Chauviere, G., and Servin, A. L. 1993b. Adhering heat-killed human *Lactobacillus acidophilus*, strain LB, inhibits the process of pathogenicity of diarrhoeagenic bacteria in cultured human intestinal cells. *J. Diarrhoeal Dis. Res.* 11: 235-242.

19. Coconnier, M. H., Bernet, M. F., Kerneis, S., Chauviere, G., Fourniat, J., and Servin, A. L. 1993a. Inhibition of adhesion of enteroinvasive pathogens to human intestinal Caco-2 cells by *Lactobacillus acidophilus* strain LB decreases bacterial invasion. *FEMS Microbiol. Lett.* 110: 299-305.

20. Coconnier, M. H., Lievin, V., Bernet-Camard, M. F., Hudault, S., and Servin, A. L. 1997. Antibacterial effect of the adhering human *Lactobacillus acidophilus* strain LB. *Antimicrob. Agents Chemother.* 41: 1046-1052.

21. Condon, S. 1987. Responses of lactic acid bacteria to oxygen. *FEMS Microbiol. Rev.* 46: 269-280.

22. Conway, P., and Kjellberg, S. 1989. Protein-mediated adhesion of *Lactobacillus fermentum* strain 737 to mouse stomach squamous epithelium. *J. Gen. Microbiol.* 135: 1175-1186.

23. Dahiya, R. S., and Speck, M. L. 1968. Hydrogen peroxide formation by lactobacilli and its effect on *Staphylococcus aureus. J. Dairy Sci.* 51(10): 1068-1072.

24. Dahl, T. A., Midden, W. R., and Hartman, P. E. 1989. Comparison of killing of Gram-negative and Gram-positive bacteria by pure singlet oxygen. *J. Bacteriol.* 171: 2188-2194.

25. De Simone, C., Tzantzoglou, S., Baldinelli, L., Di Fabio, S., Bianchi-Salvadori, B., Jirillo, E., and Vesely, R. 1988. Enhancement of host resistance against *Salmonella typhimurium* infection by a diet supplemented with yogurt. *Immunopharmacol. Immunotoxicol.* 10: 399-415.

26. Ducluzeau, R., Dubos, F., and Raibaud, P. 1971. Antagonist effect of a *Lactobacillus* strain on a *Ristella* sp. strain in the digestive tract of gnotoxenic mice ingesting lactose. *Ann. Inst. Pasteur (Paris)* 121: 777-794.

27. Eklund, T. 1984. The effect of carbon dioxide on bacterial growth and on uptake processes in the bacterial membrane vesicles. *Int. J. Food Microbiol.* 1: 179-185.

28. Elo, S., Saxelin, M., and Salminen, S. 1991. Attachment of *Lactobacillus casei* strain GG to human colon carcinoma cell line Caco-2: comparison with other dairy strains, *Lett. Appl. Microbiol.* 13: 154-156.

29. Fichera, G. A., Pappalardo, C., Furneri, P. M., Gismondo, M. R., and Nicoletti, G. 1987. Evaluation of antinfective activity of an association bifidobacterium / lactobacillus in mouse intestine. *G. Batteriol. Virol. Immunol.* 80: 189-205.

30. Fitzsimmons, N., and Berry, D. R. 1994. Inhibition of Candida albicans by *Lactobacillus acidophilus*: evidence for the involvement of a peroxidase system. *Microbios* 80(323): 125-133.

31. Fleming, H. P., McFeeters, R. F., and Daeschel, M. A. 1986. The lactobacilli, pediococci, and leuconostocs: vegetable products. In: *Bacterial Starter Cultures for Foods*, ed. S. E. Gilliland, pp. 97-118. Boca Raton, FL: CRC Press Inc.

32. Fourniat, J., Colomban, C., Linxe, C., and Karam, D. 1992. Heat-killed *Lactobacillus acidophilus* inhibits adhesion of *Escherichia coli* B41 to HeLa cells, *Ann. Rech. Vet.* 23: 361-370.

33. Fourniat, J., Djaballi, Z., Maccario, J., Bourlioux, P., and German, A. 1986. Effect of the administration of killed *Lactobacillus acidophilus* on the survival of suckling mice infected with a strain of enterotoxigenic *Escherichia coli. Ann. Rech. Vet.* 17: 401-407.

34. Fuller, R. 1993. Probiotic foods - current use and future developments. *IFI NR.3*: 23-26.

35. Gasson, M. J., and Fitzgerald, G. F. 1994. Gene transfer systems and transposition. In: *Genetics and Biotechnology of Lactic Acid Bacteria*, ed. M.J. Gasson, and W.M. de Vos, pp. 1-44. New York: Blackie Academic & Professional.

36. Gerritse, K., Posno, M., Schellekens, M. M., Boersma, W. J., and Claassen, E. 1990. Oral administration of TNP-Lactobacillus conjugates in mice: a model for evaluation of mucosal and systemic immune responses and memory formation elicited by transformed lactobacilli, *Res. Microbiol.* 141: 955-962.

37. Gibson, G. R., Saavedra, J. M., Macfarlane, S., and Macfarlane, G. T. 1997. Probiotics and intestinal infections. In: *Probiotics 2: Applications and Practical Aspects.* ed. R. Fuller, pp. 10-39. New York: Chapman & Hall.

38. Goepfert, J. M., and Hicks, R. 1969. Effect of volatile fatty acids on *Salmonella typhimurium. J. Bacteriol.* 97: 956-958.

39. Gonzalez, C. F., and Kunka, B. S. 1987. Plasmid-associated bacteriocin production and sucrose fermentation in *Pediococcus acidilactici. Appl. Environ. Microbiol.* 53: 2534-2538.

40. Gupta, K. G., Chandiok, L., and Bhatnagar, L. 1973. Antibacterial activity of diacetyl and its influence on the keeping quality of milk. *Zentralbl. Bakteriol. Parasitenkd. Infektionskr. Hyg. (Abt. 1 Orig. Reihe B)* 158: 202-205.

41. Haines, W. C., and Harmon, L. G. 1973. Effect of selected lactic acid bacteria on growth of *Staphylococcus aureus* and production of enterotoxin. *Appl. Microbiol.* 25: 436-444.

42. Hammes, W. P., and Tichaczek, P. S. 1994. The potential of lactic acid bacteria for the production of safe and wholesome food. *Z. Lebensm. Unters. Forsch.* 198: 193-201.

43. Hastings, J. W., and Stiles, M. E. 1991. Antibiosis of *Leuconostoc gelidum* isolated form meat. *J. Appl. Bacteriol.* 70: 127-134.

44. Havenaar, R., Brink, B., and Huis in't Veld, J. H. J. 1992. Selection of strains for probiotic use. In: *Probiotics, the Scientific Basis*, ed. R. Fuller, pp. 209-224. London, UK: Chapman and Hall.

45. Havenaar, R., Brink, B., and Huis in't Veld, J. H. J. 1992. Selection of strains for probiotic use. In: *Probiotics, the Scientific Basis*, ed. R. Fuller, pp. 209-224. London: Chapman and Hall.

46. Hawthorn, L. A., and Reid, G. 1990. Exclusion of uropathogen adhesion to polymer surfaces by *Lactobacillus acidophilus. J. Biomed. Mater. Res.* 24: 39-46.

47. Henriksson, A., Szewzyk, R., and Conway, P. L. 1991. Characteristics of the adhesive determinants of *Lactobacillus fermentum* 104. *Appl. Environ. Microbiol.*, 57: 499-502.

48. Hidaka, H., and Eida, T. 1988. Roles of fructooligosaccharides for the improvements of intestinal flora and suppression of production of putrid substances. *Shokuhim-Kogyo* 31:52.

49. Hidaka, H., Eida, T., Takizawa, T., Tokunaga, T., and Tashiro, Y. 1986. Effects of fructooligosaccharides on intestinal flora and human health. *Bifidobacteria Microflora.* 5: 37.

50. Holo, H., Nilssen, O., and Nes, I. F. 1991. Lactococcin A, a new bacteriocin from *Lactococcus lactis* subsp. *cremoris*: isolation and characterization of the protein and its gene. *J. Bacteriol.* 173: 3879-3887.

51. Hudault, S., Lievin, V., Bernet-Camard, M. F., and Servin, A. L. 1997. Antagonistic activity exerted in vitro and in vivo by *Lactobacillus casei* (strain GG) against *Salmonella typhimurium* C5 infection. *Appl. Environ. Microbiol.* 63: 513-518.

52. Hugenholtz, J. 1993. Citrate metabolism in lactic acid bacteria, *FEMS Microbiol. Rev.* 12: 165-178.

53. Hughes, D. B., and Hoover, D. G. 1991. Bifidobacteria: their potential for use in American diary products. *Food Tech.* (April) 74-83.

54. Hull, R. R., Conway, P. L., and Evans, A. J. 1992. Probiotic foods - a new opportunity, *Food Australia* 44(3): 112-113.

55. Hurst, A. 1981. Nisin. *Adv. Appl. Microbiol.* 27: 85-123.

56. Ingram, M., Ottoway, F. J. H., and Coppock, J. B. M. 1956. The preservative action of acid substances in food. *Chem. Ind.* 42: 1154-1165.

57. Ishibashi, N., and Shimamura, S. 1993. Bifidobacteria: research and development in Japan, *Food Tech.* (June) 126-135.

58. Ishihara, K., Miyakawa, H., Hasegawa, A., Takazoe, I., and Kawai, Y. 1985. Growth inhibition of *Streptococcus mutans* by cellular extracts of human intestinal lactic acid bacteria. *Infect. Immun.* 49: 692-694.

59. Itoh, K., and Freter, R. 1989. Control of *Escherichia coli* populations by a combination of indigenous clostridia and lactobacilli in gnotobiotic mice and continuous-flow cultures. *Infect. Immun.* 57: 559-565.

60. Jay, J. M. 1982, Antimicrobial properties of diacetyl. *Appl. Environ. Microbiol.* 44: 525-532.

61. Jay, J. M. 1996. *Modern Food Microbiology*. 5th Ed. New York: Van Nostrand Reinhold.

62. Jimenez-Diaz, R., Ruiz-Barba, J. L., Cathcart, D. P., Holo, H., Nes, I. F., Sletten, K. H., and Warner, P. J. 1995. Purification and partial amino acid sequence of plantaricin S, a bacteriocin produced by *Lactobacillus plantarum* LPCO10, the activity of which depends on the complementary action of two peptides. *Appl. Environ. Microbiol.* 61(12): 4459-4463.

63. Jin, L. Z., Ho, Y. W., Ali, M. A., Abdullah, N., and Jalaludin, S. 1996. Effect of adherent *Lactobacillus* spp. on in vitro adherence of salmonellae to the intestinal epithelial cells of chicken. *J. Appl. Bacteriol.* 81: 201-206.

64. Joerger, M. C., and Klaenhammer, T. R. 1986. Characterization and purification of helveticin J and evidence for a chromosomally determined bacteriocin produced by *Lactobacillus helveticus* 481. *J. Bacteriol.* 167: 439-446.

65. Kabir, A. M., Aiba, Y., Takagi, A., Kamiya, S., Miwa, T., and Koga, Y. 1997. Prevention of *Helicobacter pylori* infection by lactobacilli in a gnotobiotic murine model. *Gut* 41: 49-55.

66. Kao, C. T., and Frazier, W. C. 1996. Effect of lactic acid bacteria on growth of *Staphylococcus aureus*. *Appl. Microbiol.* 14: 251-255.

67. Klaenhammer, T. R. 1993. Genetics of bacteriocins produced by lactic acid bacteria. *FEMS Microbiol. Rev.* 12: 39-85.

68. Klaenhammer, T. R., and Kleeman, E. G. 1981. Growth characteristics, bile sensitivity, and freeze damage in colonial variants of *Lactobacillus acidophilus*. *Appl. Environ. Microbiol.* 41: 1461-1467.

69. Klebanoff, S. J., and Coombs, R. W. 1991. Virucidal effect of *Lactobacillus acidophilus* on human immunodeficiency virus type1: possible role in heterosexual transmission. *J. Exp. Med.* 174: 289-292.

70. Klebanoff, S. J., Clem, W. H., and Luebke, R. G. 1966. The peroxidase-thiocyanate-hydrogen peroxide antimicrobial system. *Biochim. Biophys. Acta* 117(1): 63-72.

71. Kong, S., and Davison, A. J. 1980. The role of interactions between O_2, H_2O_2, OH, e- and O_2^- in free radical damage to biological systems. *Arch. Biochem. Biophys.* 204: 13-29.

72. Kröckel, L. 1999. Natural barriers for use in biopreservation. Bacteriocinogenic lactic acid bacteria may inhibit pathogens. *Fleisch Wirtschaft* 2/99:36-38.

73. Kulshrestha, D. A., and Marth, E. H. 1974a. Inhibition of bacteria by some volatile and non-volatile compounds associated with milk. I. *Escherichia coli*. *J. Milk Food Technol.* 37: 510-516.

74. Kulshrestha, D. C., and Marth, E. H. 1974b. Inhibition of bacteria by some volatile and non-volatile compounds associated with milk. II. *Salmonella typhimurium*. *J. Milk Food Technol.* 37: 539-544.

75. Kulshrestha, D. C., and Marth, E. H. 1974c. Inhibition of bacteria by some volatile and non-volatile compounds associated with milk. III. *Stapylococcus aureus*. *J. Milk Food Technol.* 37: 545-550.

76. Kurmann, J. A., Rasic, J. L., and Kroger, M. 1992. *Encyclopedia of Fermented Fresh Milk Products*. New York: Van Nostrand Reinhold.

77. Lee, Y. K., and Wong, S. F. 1993. In: *Lactic Acid Bacteria*, ed. S. Salminen and A. von Wright, pp. 97-109. London: Marcel Dekker.

78. Lindgren, S. E., and Dobrogosz, W. J. 1990. Antagonistic activities of lactic acid bacteria in food and feed fermentations. *FEMS Microbiol. Rev.* 87: 149-164.

79. Lister, J. 1873. A further contribution to the natural history of bacteria and the germ theory of fermentative changes. *Quart. J. Microsc. Sci.* 13: 380-408.

80. London, J. 1990. Uncommon pathways of metabolism among lactic acid bacteria. *FEMS Microbiol. Rev.* 87: 103-112.

81. Mercenier, A., Pouwels, P. H., and Chassy, B. M. 1994. Genetic engineering of lactobacilli, leuconostocs and *Streptococcus thermophilus*. In: *Genetics and Biotechnology of Lactic Acid Bacteria*, ed. M.J. Gasson and W.M. DeVos, pp. 253-293. New York: Blackie Academic & Professional.

82. Metchnikoff, E. 1907. *The Prolongation of Life*. London: Heinemann.

83. Miake, S., Nomoto, K., Yokokura, T., Yoshikai, Y., Mutai, M., and Nomoto, K. 1985. Protective effect of *Lactobacillus casei* on *Pseudomonas aeruginosa* infection in mice. *Infect. Immun.* 48: 480-485.

84. Midolo, P. D., Lambert, J. R., Hull, R., Luo, F., and Grayson, M. L. 1995. In vitro inhibition of *Helicobacter pylori* NCTC 11637 by organic acids and lactic acid bacteria. *J. Appl. Bacteriol.* 79(4): 475-479.

85. Mitsuoka, T. 1982. Recent trends in research on intestinal flora. *Bifidobacteria Microflora* 1:3.

86. Mitsuoka, T. 1984. Taxonomy and ecology of bifidobacteria, *Bifidobacteria Microflora* 3: 11.

87. Modler, H. W., McKellar, R. C., and Yaguchi, M. 1990a. Bifidobacteria and bifidogenic factors. *Can. Inst. Food Sci. Tech.* 23: 29-41.

88. Modler, H. W., McKellar, R. C., Goff, H. D., and Mackie, D. A. 1990b. Using ice cream as a mechanism to incorporate bifidobacteria and fructooligosaccharides into the human diet. *Cultured Dairy Prod. J.* 23: 29-41.

89. Morris, J. G. 1979. Nature of oxygen toxicity in anaerobic microorganisms. In: *Strategies of Microbial Life in Extreme Environments*, ed. M. Shilo, pp. 149-162. Berlin: Verlag Chemi.

90. Mortvedt, C. I., and Nes, I. F. 1990. Plasmid-associated bacteriocin production by a *Lactobacillus sake* strain. *J. Gen. Microbiol.* 136: 1601-1607.

91. Motlagh, A. M., Johnson, M. C., and Ray, B. 1991. Viability loss of foodborne pathogens by starter culture metabolites. *J. Food Prot.* 54: 873-878, 884.

92. Mulders, J. W. M., Boerrigter, I. J., Rollema, H. S., Siezen, R. J., and de Vos, W. M. 1991. Identification and characterization of the lantibiotic nisin Z, a natural nisin variant. *Eur. J. Biochem.* 201: 581-584.

93. Murphy, M. G., and Condon, S. 1984. Correlation of oxygen utilization and hydrogen peroxide accumulation with oxygen induced enzymes in *Lactobacillus plantarum* cultures. *Arch. Microbiol.* 138(1): 44-48.

94. Nader de Macias, M. E., Apella, M. C., Romero, N. C., Gonzalez, S. N., and Oliver, G. 1992. Inhibition of *Shigella sonnei* by *Lactobacillus casei* and *Lactobacillus acidophilus*. *J. Appl. Bacteriol.* 73: 407-411.
95. Naidu, A. S., Bidlack, W. R., and Clemens, R. A. 1999. Probiotic spectra of lactic acid bacteria (LAB) *Crit. Rev. Food Sci. Nutr.*39: 3-126
96. Nes, I. F., Diep, D. B., Havarstein, L. S., Bruberg, M. B., Eijsink, V., and Holo, H. 1996. Biosynthesis of bacteriocins in lactic acid bacteria. *Antonie Van Leewenhoek* 70(2-4): 113-128.
97. Nomoto, K., Miake, S., Hashimoto, S., Yokokura, T., Mutai, M., Yoshikai, Y., and Nomoto, K. 1985. Augmentation of host resistance to *Listeria monocytogenes* infection by *Lactobacillus casei*. *J. Clin. Lab. Immunol.* 17: 91-97.
98. Nunez de Kairuz, M. S., Olazabal, M. E., Oliver, G., Pesce de Ruiz Holgado, A. A., Massa, E., and Farias, R. N. 1988. Fatty acid dependent hydrogen peroxide production in *Lactobacillus*. *Biochem. Biophys. Res. Commun.* 152(1): 113-121.
99. Oberman, H. 1985. Fermented milks. In: *Microbiology of Fermented Foods, Vol 1*, ed. B.J.B.Wood, pp.167-195. London: Elsevier Applied Sciences.
100. Ohashi, T., Minamishima, Y., Yokokura, T., and Mutai, M. 1989. Induction of resistance in mice against murine cytomegalovirus by cellular components of *Lactobacillus casei*. *Biotherapy* 1(2): 89-95.
101. Ohashi, T., Yoshida, A., and Minamishima, Y. 1988. Host-mediated antiviral activity of *Lactobacillus casei* against cytomegalovirus infection in mice. *Biotherapy* 1: 27-39.
102. Perdigon, G., Alvarez, S., and Pesce de Ruiz Holgado, A. 1991. Immunoadjuvant activity of oral *Lactobacillus casei*: influence of dose on the secretory immune response and protective capacity in intestinal infections. *J. Dairy Res.* 58: 485-496.
103. Perdigon, G., Nader de Macias, M. E., Alvarez, S., Oliver, G., and Pesce de Ruiz Holgado, A. A. 1990. Prevention of gastrointestinal infection using immunobiological methods with milk fermented with *Lactobacillus casei* and *Lactobacillus acidophilus*. *J. Dairy Res.* 57: 255-264.
104. Piard, J. C., and Desmazeaud, M. 1992. Inhibiting factors produced by lactic acid bacteria. 2. Bacteriocins and other antibacterial substances. *Lait* 71: 525-541.
105. Piard, J. C., Delorme, F., Giraffa, G., Commissaire, J., and Desmazeaud, M. 1990. Evidence for a bacteriocin produced by *Lactococcus lactis* CNRZ 481. *Neth. Milk Dairy J.* 44: 143-158.
106. Poolman, B. 1993. Energy transduction in lactic acid bacteria. *FEMS Microbiol. Rev.* 12: 125-147.
107. Rammelsberg, M., and Radler, F. 1990. Antibacterial polypeptides of *Lactobacillus* species. *J. Appl. Bacteriol.* 69: 177-184.
108. Rani, B., and Khetarpaul, N. 1998. Probiotic fermented food mixtures: possible applications in clinical anti-diarrhoea usage. *Nutr. Hlth.* 12: 97-105.
109. Rasic, J. L., and Kurmann, J. A. 1983. Bifidobacteria and their role. In: *Experientia Supplementum*, Vol. 39, Birkhauser: Verlag
110. Reid, G., Bruce, A. W., McGroarty, J. A., Cheng, K. J., and Costerton, J. W. 1990. Is there a role for lactobacilli in prevention or urogenital and intestinal infection? *Clin. Microbiol. Rev.* 3(4): 335-344.
111. Reid, G., Tieszer, C., and Lam, D. 1995. Influence of lactobacilli on the adhesion of *Staphylococcus aureus* and *Candida albicans* to fibers and epithelial cells. *J. Ind. Microbiol.* 15: 248-253.
112. Ruiz-Palacios, G., Tuz, F., Arteaga, F., Guerrero, M. L., Dohnalek, M., and Hilty, M. 1996. Tolerance and fecal colonization with *Lactobacillus reuteri* in children fed a beverage with a mixture of *Lactobacillus* spp. *Pediatr. Res.* 39(4): 184A.(Abstract #1090)
113. Saavedra, J. M. 1995. Microbes to fight microbes: a not so novel approach to controlling diarrheal disease. *J. Pediatr. Gastroenterol. Nutr.* 21(2): 125-129.
114. Saito, H., Nagashima, K., and Tomioka, H. 1983. Effects of bacterial immunopotentiators, LC 9018 and OK-432, on the resistance against *Mycobacterium* intracellular infection in mice. *Hiroshima J. Med. Sci.* 32: 145-148.
115. Saito, H., Tomioka, H., and Nagashima, K. 1987. Protective and therapeutic efficacy of *Lactobacillus casei* against experimental murine infections due to *Mycobacterium fortuitum* complex. *J. Gen. Microbiol.* 133: 2843-2851.
116. Salminen, S., Isolauri, E., and Salminen, E. 1996. Clinical uses of probiotics for stabilizing the gut mucosal barrier: successful strains and future challenges. *Antonie Van Leeuwenhoek*. 70: 347-358.
117. Sato, K. 1984. Enhancement of host resistance against *Listeria* infection by *Lactobacillus casei*: role of macrophages. *Infect. Immun.* 44: 445-451.
118. Sato, K., Saito, H., and Tomioka, H. 1988a. Enhancement of host resistance against *Listeria* infection by *Lactobacillus casei*: activation of liver macrophages and peritoneal macrophages by *Lactobacillus casei*. *Microbiol. Immunol.* 32: 689-698.
119. Sato, K., Saito, H., Tomioka, H., and Yokokura, T. 1988b. Enhancement of host resistance against *Listeria* infection by *Lactobacillus casei*: efficacy of cell wall preparation of *Lactobacillus casei*. *Microbiol. Immunol.* 32: 1189-1200.

120. Savage, D. C. 1969. Microbial interference between indigenous yeast and lactobacilli in the rodent stomach. *J Bacteriol.* 98: 1278-1283.
121. Saxelin, M., Elo, S., Salminen, S., and Vapaatalo, H. 1991. Dose response colonisation of faeces after oral administration of *Lactobacillus casei* strain GG., *Microbial Ecology in Health and Disease.* 6:119-122.
122. Schneitz, C., Nuotio, L., and Lounatma, K. 1993. Adhesion of *Lactobacillus acidophilus* to avian intestinal epithelial cells mediated by the crystalline bacterial cell surface layer (S-layer). *J. Appl. Bacteriol.* 74: 290-294.
123. Setoyama, T., Nomoto, K., Yokokura, T., and Mutai, M. 1985. Protective effect of lipoteichoic acid from *Lactobacillus casei* and *Lactobacillus fermentum* against *Pseudomonas aeruginosa* in mice. *J. Gen. Microbiol.* 131: 2501-2503.
124. Shkarupeta, M. M., Korshunov, V. M., Savenkova, V. T., and Pinegin, B. V. 1988. Influence of the oral administration of indigenous microorganisms on the resistance of mice to *Salmonella* infection. *Zh. Mikrobiol. Epidemiol. Immunobiol.* 7: 46-50.
125. Silva de Ruiz, C., Lopez de Bocanera, M. E., Nader de Macias, M. E., and Pesce de Ruiz Holgado, A. A. 1996. Effect of lactobacilli and antibiotics on *E. coli* urinary infections in mice. *Biol. Pharm. Bull.* 19: 88-93.
126. Speck, M. L., Dobrogosz, W. J., and Casas, I. A. 1993. *Lactobacillus reuteri* in food supplementation. *Food Technol* July: 90, 92-94.
127. Spillman, H., Puhan, Z., and Banhegyi, M. 1978. Antimikrobielle aktivitat *thermophiler Lactobazillen. Milchwissenschaft* 33: 148-153.
128. Stoffels, G., Nissen-Meyer, J., Gudmundsdottir, A., Sletten, K., Holo, H., and Nes, I. F. 1992. Purification and characterization of a new bacteriocin isolated from a *Carnobacterium* sp. *Appl. Environ. Microbiol.* 58: 1417-1422.
129. Tamime, A. Y., Marshall, V. M. E., and Robinson, R. K. 1995. Microbiological and technical aspects of milks fermented by bifidobacteria. *J Dairy Res.* 62: 151-187.
130. Tissier, H. 1905. Taxonomy and ecology of bifidobacteria. *Bifidobacteria Microflora* 3: 11-28.
131. Toba, T., Samant, S. K., Yoshioka, E., and Itoh, T. 1991a. Reuterin 6, a new bacteriocin produced by *Lactobacillus reuteri* LA 6. *Lett. Appl. Microbiol.* 13: 281-286.
132. Toba, T., Yoshioka, E., and Itoh, T. 1991b. Lacticin, a bacteriocin produced by *Lactobacillus delbrueckii* subsp. *Lactis. Lett. Appl. Microbiol.* 12: 43-45.
133. Van Belkum, M. J., Hayema, B. J., Jeeninga, R. E., Kok, J., and Venema, G. 1991. Organization and nucleotide sequences of two lactococcal bacteriocin operons. *Appl. Environ. Microbiol.* 57: 492-498.
134. Van Belkum, M. J., Kok, J., and Venema, G. 1992. Cloning, sequencing, and expression in *Escherichia coli* of Icn B, a third bacteriocin determinant from the lactococcal bacteriocin plasmid p9B4-6. *Appl. Environ. Microbiol.* 58: 572-577.
135. Velraeds, M. M., van der Mei, H. C., Reid, G., and Busscher, H. J. 1996. Inhibition of initial adhesion of uropathogenic *Enterococcus faecalis* by biosurfactants from *Lactobacillus* isolates. *Appl. Environ. Microbiol.* 62: 1958-1963.
136. Velraeds, M. M., van der Mei, H. C., Reid, G., and Busscher, H. J. 1997. Inhibition of initial adhesion of uropathogenic *Enterococcus faecalis* to solid substrata by an adsorbed biosurfactant layer from *Lactobacillus acidophilus. Urology* 49: 790-794.
137. Wagner, R. D., Pierson, C., Warner, T., Dohnalek, M., Farmer, J., Roberts, L., Hilty, M., and Balish, E. 1997. Biotherapeutic effects of probiotic bacteria on candidiasis in immunodeficient mice. *Infect. Immun.* 65: 4165-4172.
138. Watanabe, T., and Hotta, T. 1996. Suppressive effects of *Lactobacillus casei* cells, a bacterial immunostimulant, on the incidence of spontaneous thymic lymphoma in AKR mice. *Cancer Immunol. Immunother.* 42: 285-290.
139. Watanabe, T., and Saito, H. 1986. Protection of mice against herpes simplex virus infection by a *Lactobacillus casei* preparation (LC 9018) in combination with inactivated viral antigen. *Microbiol. Immunol.* 30: 111-122.
140. Watanabe, T., and Yamori, T. 1989. Primary resistance induced in mice by *Lactobacillus casei* following infection with herpes simplex virus. *Kansenshogaku Zasshi* 63: 182-188.
141. Wilcox, M. D., Patrikakis, M., Harty, D. W., Loo, C. Y., and Knox, K. W. 1993. Coaggregation of oral lactobacilli with streptococci from the oral cavity. *Oral Microbiol. Immunol.* 8: 319-321.
142. Wolf, B. W., Garleb, K. A., Ataya, D. G., and Casas, I. A. 1995. Safety and tolerance of *Lactobacillus reuteri* in healthy adult male subjects. *Micro. Ecol. Hlth. Dis.* 8: 41-50.
143. Wong, H. C., and Chen, Y. L. 1988. Effects of lactic acid bacteria and organic acids on growth and germination of *Bacillus cereus. Appl. Environ. Microbiol.* 54: 2179-2184.

L.V. Thomas
M.R. Clarkson
J. Delves-Broughton

18

Nisin

I. INTRODUCTION

Nisin was first discovered in England in 1928 as a result of difficulties experienced during cheese making. A 'lactic streptococci' was inhibiting growth of a milk starter culture (*Lactobacillus bulgaricus*) (Rogers, 1928; Rogers & Whittier, 1928). A few years later, similar problems were experienced in New Zealand. Occasionally, cheese-makers experienced failure of starter cultures to clot milk within an 18 h incubation period, and yet no bacteriophage could be detected. This only happened when, contrary to their usual practice, the cheese-makers had not used fresh milk but milk that had been stored overnight. It was realised that the starter culture was not being adversely affected by bacteriophage but by some unknown inhibitory proteinaceous substance that was present in the 'non-acid' milk (Whitehead, 1933; Whitehead & Riddet, 1933; Meanwell, 1943; Hunter & Whitehead, 1944). Storage of the milk had allowed the contaminant organisms to grow and produce the inhibitor. The protein was isolated and tested during World War II, in an effort not just to solve the cheese-making problems, but also to explore the potential of this compound in human or veterinary medicine (Meanwell, 1943; Mattick & Hirsch, 1944). Mattick and Hirsch (1947) characterized the compound and called it 'nisin', a name coined from its serogroup determined by the Lancefield serotyping scheme for streptococci. Nisin was a 'Group **N** **I**nhibitory **S**ubstance'. *Lactococcus lactis* subsp *lactis*, the producer organism, was formerly classified as a lactic streptococcus (Schleifer et al., 1985). The suffix '-**in**' was commonly used for antibiotics although nisin is now usually referred to as a bacteriocin or as an antimicrobial peptide and not as an antibiotic because it lacks the key attributes that make most antibiotics unsuitable for widespread food usage. Important differences include the facts that nisin does not persist in the body or the environment, nor is it associated with the development of bacterial resistance to nisin itself or to any medically important antibiotic.

Initial interest in nisin as a therapeutic agent focused on its use for the control of mastitis. It became apparent, however, that there were drawbacks associated with its use for the treatment of infections in animals or humans. Its antibacterial spectrum of activi-

0-8493-2047-X/00/$0.00+ $.50
© *2000 by CRC Press, LLC*

ty limited its usefulness. Moreover nisin had low solubility and was less stable at physiological pH. It could not be administered orally because of degradation by proteases in the digestive system (Hurst, 1983). Conversely, this latter characteristic, (together with some other attributes) made nisin suitable for use as a food preservative, and this was first demonstrated in 1951. A starter culture that produced nisin prevented clostridial gas formation and consequent spoilage of a Swiss cheese by *Clostridium butyricum* and *Cl. tyrobutyricum* (Hirsch et al., 1951). McClintock et al., (1952) successfully used nisin-producing cultures to inhibit the development of anaerobic spores in *Gruyere fondue* cheese. The difficulty of using nisin in foods made with starter cultures was then realized. Nisin will often inhibit starter cultures and thus prevent the ripening of the cheese, as was first shown by Winkler and Fröhlich (1957) with Emmenthal cheese.

Mattick and Hirsch (1947) first isolated nisin from culture broths of *Lac. lactis* subsp. *lactis*. The first commercial preparation of nisin (Nisaplin) was made in 1953 by Aplin and Barrett, Ltd (Cultor Food Science) and was originally intended for use solely within the company to prevent clostridial spoilage of processed cheese. This was a common problem in industry due to the poorer quality of raw cheese then available. The potential of the product was realized and before long it was being used by other cheese processors in other countries as well as the UK. Nisaplin, the nisin preparation developed by Aplin and Barrett, was developed between 1962 to 1965. In 1969 a joint FAO/WHO expert committee on food additives recognized nisin as a safe and legal biological food preservative (FAO/WHO, 1969).

Nisin is a small, heat stable antimicrobial peptide described as a class I bacteriocin, a group that comprises lantibiotics (Klaenhammer et al., 1993). Lantibiotics are a family of membrane active peptides that contain the unusual thioether amino acids lanthionine and β-methyl lanthionine as well as other modified amino acids, such as dehydrated serine and threonine (Jung, 1991a). Nisin was discovered over 50 years before most other bacteriocins and was the first compound of this type to be used in the food industry on a commercial scale. Its discovery prompted the search for other antimicrobial peptide products from lactic acid bacteria, but nisin is still the only purified bacteriocin commercially available and is unique in its acceptance worldwide as a food preservative.

Most bacteriocins have a limited spectrum of activity, generally effective only against bacteria related to the producer organism. Like other bacteriocins nisin is normally active against Gram positive and not against Gram negative bacteria, yeasts or molds. However, nisin has a much broader antimicrobial spectrum. Treatment of Gram negative bacteria with chelating agents renders these organisms nisin sensitive.

Nisin was the first lantibiotic whose structure was determined (Gross & Morell, 1971). Its unusual biosynthesis by post-translational modification of a gene-encoded protein distinguishes nisin (and other lantibiotics) from peptide antibiotics that are synthesized from amino acids by complexes of several enzymes.

Scientific literature on nisin is extensive. Major reviews have been written by Hurst (1981; 1983), Delves-Broughton (1990), Ray (1992), Hurst & Hoover (1993), de Vuyst & Vandamme (1994), Delves-Broughton & Gasson (1994), Delves-Broughton et al., (1996), Wessels et al., (1998), and Abee & Delves-Broughton (in press). In this CHAPTER, nisin concentrations are expressed as levels of pure nisin, e.g. µg/ml and µg/g (see *Section II D* for further explanation).

II. OCCURRENCE

A. Biosynthesis

Lantibiotics in general are synthesized on the ribosome as precursor or pre-peptides that are enzymatically modified following translation (de Vos et al., 1995). A general lantibiotic model described by Schnell et al., (1988) suggested that the synthesis and export of this group of bacteriocins occurs in 3 phases. The first phase involves intracellular post-translational modification that changes several of the amino acid residues. This includes dehydration of particular hydroxyl amino acids to form dehydroamino acids and the formation of thioethers by the addition of a sulfhydryl group. In the second phase, the modified peptide is exported across the cytoplasmic membrane. In the last phase, proteolytic cleavage of the leader peptide produces the final molecule with antimicrobial activity. The mechanism for nisin biosynthesis is illustrated in *Figure 1*.

The complete gene cluster for nisin biosynthesis in *Lac. lactis* is located on a 70 kb conjugative transposon located on the chromosome (Dodd et al., 1990; Rauch et al., 1990; Horn et al., 1991; Rauch & Vos, 1992). The transposon also codes for sucrose fermenting ability and reduced bacteriophage sensitivity (Rauch et al., 1994). The sequence and transcription of the nisin structural gene was identified by Buchman et al., (1988). A cluster of 11 contiguous genes, arranged in three multi-cistronic operons, are involved in nisin biosynthesis (de Ruyter et al., 1996; Siezen et al., 1996). This cluster contains the genes coding for nisin structure and immunity for the producer organism, as well as the proteins involved in nisin post-translational modifications, nisin translocation, cleavage of the leader protein, and regulation of the biosynthesis. The structural gene nisA is followed by *nisB*, *nisT*, *nisI*, *nisR* and *nisK*. There are also three additional open reading frames adjacent to *nisK* (*nisF*, *nisE*, and *nisG*) and in the opposite direction, three open reading frames downstream of *nisG* are also involved.

The nisin structural gene *nisA* gene codes for the intra-cellular synthesis of a 57 amino acid pre-peptide. This contains normal amino acids and comprises a C-terminal 23-residue leader peptide linked to an N-terminal 34-residue pro-peptide region from which the mature nisin is formed. The pre-peptide is modified by the gene products of *nisB*, (that dehydrates the serine and threonine residues) and *nisC* (that adds sulfur from neighboring cysteine residues), to form the lanthionine and β-methyllanthionine monosulfide bridges. Inactivation of both *nisB* and *nisC* genes results in complete loss of nisin production (Siegers et al., 1996). The maturation and secretion of nisin appears to take place at a multimeric lanthionine synthetase complex that is membrane associated. The complex is at least 350-kDa in size and contains the proteins NisB, NisC (at least 2 molecules) and a NisT dimer. Secretion of the modified peptide is carried out by an ATP binding cassette transporter enzyme; a 600 amino acid residue product of the *nisT* gene. The *nisP* gene encodes an extra-cellular protease that cleaves the leader peptide from the mature nisin molecule.

The self-protection or immune mechanism of nisin-producing organisms is controlled by the gene products of *nisI*, *nisE*, *nisF* and *nisG*. The immune mechanism is not fully understood but involves the immunity protein NisI that is attached to the outside of the cell membrane. Proteins NisE and NisF show strong homology with the ATP binding cassette transporters described for several colicins. The hydrophobic protein NisG appears to act similarly to immunity proteins described for some colicins. This has potential membrane spanning domains (Siegers & Entian, 1995).

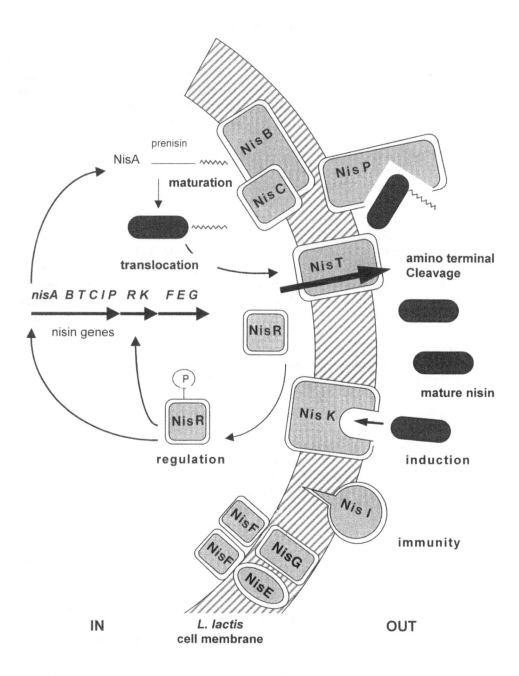

FIGURE 1. Nisin biosynthesis. NisA: pre nisin structural protein; NisB: maturation enzyme; NisT: transport protein; NisC: maturation enzyme; NisI: nisin immunity protein; NisP: proteinase cleaves leader; NisR: regulatory protein; NisK: regulatory protein; NisFEG: nisin immunity proteins. Reprinted with permission of Dr. Mike Gasson (Institute of Food Research Norwich, UK).

B. Screening and identification of nisin-producing strains

Two natural variants of nisin are known: nisin A and nisin Z (Gross & Morell, 1971; Mulders et al., 1991; de Vos et al., 1993) and particular lactococcus producer strains will only make one nisin variant. De Vos et al., (1993) examined 26 nisin-producing *Lac. lactis* strains, and found 12 strains produced nisin A and 14 strains produced nisin Z. Nisin-producing lactococci are frequently isolated from milk (Hoyle & Nichols, 1948). Nisin-producing strains can be selected by a series of growths in media containing increasing levels of nisin (Mattick & Hirsch, 1947; Hirsch 1952; Cheeseman & Berridge, 1959). These organisms share a common phenotype of ability to ferment glucose and sucrose, nisin production and immunity to nisin, but strains may vary in their ability to ferment lactose (LeBlanc et al., 1980; Gasson 1984; Gonzalez & Kunka, 1985; de Vuyst, 1994).

Commercial extracts such as Nisaplin and Chrisin are prepared using wild-type isolates of *Lac. lactis* subsp. *lactis*. Genetically modified organisms are not used, and strain selection is primarily based on good yield and phage resistance. It may also be preferable that producer strains are non-lysogenic and free from nisin-degrading enzymes (de Vuyst & Vandamme, 1994). There can be great variability in nisin yield among different strains, for instance levels of 0, 5, 12.5 and 40 µg nisin/ml culture broth have been reported (de Vuyst, 1990; de Vuyst & Vandamme, 1991a). More recently, Goulhen et al., (1999), using the nisin Z-producing strain *Lac. lactis* subsp. *lactis* biovar *diacetylactis* UL719s, claimed yields of approximately 250 µg/ml culture.

C. Biological advantage

Hirsch (1952) postulated that nisin-producing lactococci were of relatively recent origin in evolutionary terms, and their adaptation to the environment followed the domestication of mammals. The reasons proposed in favor of this theory were fourfold. Firstly these organisms are non-pathogenic and saprophytic. Secondly, the ability to ferment lactose is easily lost (suggesting adaptation to a new habitat). Thirdly their habitat is undefined, but they appear to occur ubiquitously in small numbers. Strains have been isolated from vegetables (Stark & Sherman, 1935; Mundt et al., 1967; Harris et al., 1992) silage, coats and udders of cows and in dung (Cox & Whitehead, 1936), and a number of other sources including human throats and feces (Hirsch & Wheater, 1951). In raw milk stored at ambient temperature, nisin-producing lactococci predominate (Rowlands et al., 1950). Hirsch (1952) proposed that milk has now become the true ecological source of lactococci. The possession of plasmid encoded metabolic activities such as fermentation of lactose and citrate, proteinase activity and bacteriophage resistance, support this theory (McKay, 1983). Lastly, nisin production would ensure a competitive advantage over similar lactic organisms occupying the same habitat (Hirsch, 1952; Hawley, 1955; Gross, 1977). Cross-inhibition of nisin and diplococcin produced by *Lac. lactis* subsp. *lactis* and *Lac. lactis* subsp. *cremoris*, may be evidence of competition in milk (Hirsch & Grinsted, 1951; Hirsch, 1952).

Nisin is produced only during the exponential phase of growth; production stops when the cells enter the stationary phase. If nisin is added to culture medium before inoculation, a short lag phase results but the growth rate and final cell yield is not affected (Hurst & Kruse, 1972). Kuipers et al., (1995) have shown that nisin biosynthesis is an autoregulating process. It is now thought that nisin is produced at low levels by uninduced cells and accumulates to a threshold concentration that indicates a specific cell density. This is an example of bacterial communication based on cell density known as 'quorum sensing'. Nisin participates in cell-cell communication and acts as a signal molecule indicating cell density (Sahl & Bierbaum, 1998). This is an input signal for the action of a nisin-binding sensor kinase (NisK) and a corresponding transcription activator (NisR), which upregulates nisin biosynthesis.

D. Quantitative assays

In the original bioassay to measure nisin potency, *Streptococcus agalactiae* was used as the test organism. For this test, the minimal nisin concentration required to inhibit the growth of one bacterial cell in 1 ml of broth was called a 'Reading Unit' (RU). The WHO Committee on Biological Standardization (1970) then established as an international nisin reference an International Unit (IU), which was defined as 0.001 mg of this preparation and was the same as the Reading Unit. Specific methods are now used to measure nisin potency in commercial concentrates and the concentration of nisin in treated food products (Joint FAO/WHO Expert Committee 1969). To avoid confusion in this review nisin is expressed as levels of pure nisin (μg/ml or μg/g). Multiplication by 40 will convert these values to IU (RU) (IU/ml, IU/g) or to equivalent levels of the commercial concentrate Nisaplin (mg/L, mg/kg). For example, 1 μg nisin/g is equivalent to 40 IU nisin/g or 40 mg Nisaplin/kg.

Bioassay methods are still routinely used to measure nisin potency, particularly the horizontal agar diffusion method using *Micrococcus luteus* as the target organism (Tramer & Fowler, 1964; Fowler et al., 1975). Nisin is extracted from the product by treatment with acid (boiling at pH 2.0). A set of controls and standards are included (such as nisin inactivation by alkali at 63 °C for 30 min) to ensure resulting zones of diffusion are due to nisin alone and not any other antimicrobial factors associated with the food. Dilutions of standard and test solutions are made in treated food extract in which any nisin has been destroyed. Nisin does not diffuse quickly through agar at pH 7.0 (Mattick & Hirsch, 1947), but this drawback has been overcome by using Tween 20 in the assay medium. The size of zones of inhibition is linearly related to the logarithm of nisin at concentrations between 0.03 - 0.5 mg/L. The method has been modified over the years (Rogers & Montville, 1991; Joosten & Nunez, 1995; Wolf & Gibbons, 1996). Problems have been reported due to the binding of nisin to food ingredients (Bell & de Lacy, 1986; Somers & Taylor, 1987). Bell and de Lacey (1986) reported that nisin extraction was adversely affected by the fat content of meat, but Davies et al., (1999) did not find this in a study of bologna sausage samples.

A method using the dye rezasurin (Friedmann & Epstein, 1951) is often used as a bioassay method due to its speed and accuracy. Standard nisin solutions and the unknown test solution are compared for the ability to inhibit the growth of *Lac. lactis* subsp. *cremoris* in reconstituted skim milk. This species is used because it has a fast growth rate. The dilution method is based on the reduction and subsequent color change of resazurin as a growth indicator.

A nisin-linked immunosorbent assay (ELISA) has been described that uses a sheep polyclonal antiserum conjugated to horseradish peroxidase (Falahee et al., 1990a; 1990b; Falahee & Adams, 1992). This assay has one disadvantage. Non-active nisin degradation products are measured as well as the entire biologically active molecule.

Bouksaim et al., (1998) described an immunodot detection technique using enhanced chemiluminescence that detected low levels of nisin Z in milk and whey.

A rapid and convenient method has been described involving spectro photometric analysis of 2,3,5-triphenyltetrazolium chloride, a colorless and water-soluble dye. Viable cells take up the dye and reduce it intracellularly to formazon, a red insoluble compound, which can be extracted and absorbance measured at 485 nm (Albayrak & Yousef, 1997).

ATP bioluminometry has been used to assay nisin in beer and wort (Waites & Ogden 1987). This technique measures the ATP release from *Lb. casei*, caused by nisin.

Mulders et al., (1991) described an assay using reverse phase HPLC. This analysis can be used to quantify nisin to a lower level of detection (10 mg/L). Nisin-containing liquid is assayed on a hydrophobic (C18) narrow base HPLC column by gradient elution. Nisin is detected by UV absorption at 225 nm. This method can be used to analyze broth samples and downstream process samples (Delves-Broughton & Friis, 1998).

III. MOLECULAR PROPERTIES

A. Physico-chemistry

Nisin is a cationic molecule due to the combination of three lysine residues and one or more histidine residues (depending on the nisin variant) together with a lack of glutamate and aspartate. The side chain groups in histidine have a pKa of 6.5, and those in lysine are 10.0, and this makes the net charge of nisin dependent on pH. Nisin is most soluble at acid pH, and becomes less soluble with increasing neutrality. Liu and Hansen (1990) found that at pH 2.2 solubility was approximately 56 mg/ml, at pH 5.0 it was 3 mg/ml and at pH 11.0 it was 1 mg/ml. Solubility is not a problem in food products since the nisin levels are usually much lower (< 0.25 mg/ml). Nisin is insoluble in non-polar solvents.

Nisin concentrates remain stable for long periods if they are kept dry, in the dark and at temperatures not exceeding 25 °C. It was originally thought that nisin stability improved as conditions became increasingly acidic and that nisin was stable to autoclaving at pH 2.0 (Tramer, 1964; 1966), but it has now been shown that optimum stability occurs at pH 3.0 (Davies et al., 1998). At pH 3.0, < 5% activity was reportedly lost after heat treatment of 115 °C for 20 min. At all pH values above and below this, a marked decrease in activity occurred: 28.5% loss occurring at pH 2.0 and 21.4% loss at pH 4.0. Greater losses were found in conditions nearer neutrality: 50% occurred at pH 5.0. At high pH it has been suggested that the Dha and Dhb residues become more susceptible to modification due to the presence of nucleophiles, thus decreasing the stability and solubility of the molecule (Liu & Hansen, 1990). In practical terms, nisin is usually protected by food ingredients and will not be damaged to such an extent by pasteurization or sterilization processing (Heinemann et al., 1965). This is illustrated by the successful use of nisin in high pH/heat-treated food products such as canned vegetables and pasteurized liquid egg (Delves-Broughton 1990; Delves-Broughton et al., 1992).

General characteristics of nisin including properties and applications are listed in *TABLE 1*.

TABLE 1 General characteristics of nisin

Characteristic	Description	
History	1928	Discovery
	1947	First isolation
	1951	First use in food
	1962-65	Development of Nisaplin
	1969	Recognised as food preservative by FAO/WHO
	1971	Primary structure determined
	1988	Granted GRAS status (USFDA)
Producer organism	*Lactococcus lactis* subsp. *lactis*	
Biosynthesis	Post-translational enzymatic modification	
Variants	Nisin A, Nisin Z	
EU number	E234	
Molecular weight	3353 Daltons (for the monomer)	
Structure	34 amino acid peptide with characteristic lanthionine rings. Contains unusual amino acids: lanthionine, ß methyl lanthionine, and dehydrated serine and threonine.	
Properties	Cationic with a positive charge of 3 (2 for nisin Z). Amphiphilic: hydrophobic at N-terminus and hydrophilic at C-terminus.	
Solubility	56 mg ml^{-1} at pH 2.2; 3 mg/ml at pH 5; 1 mg/ml at pH 11.	
Stability	Optimum at pH 3.0 (<5% loss after 115 °C for 20 min)	
Activity	Bactericidal and bacteriostatic against cells of a wide range of Gram positive bacteria. Sporostatic against spores of Gram-positive endospore formers (*Bacillus* and *Clostridium*).	
Methods of application	In solution or as a dry powder, mixed into food, often in combination with heat treatment. Also as surface treatment by spray, immersion or in packaging/casing material.	
Principal applications	Processed cheese and spreads, dairy desserts, cheeses (ricotta, cottage), milk, vegetable protein milk, liquid egg, canned foods, fresh soups, crumpets, beer and wine, dressings and sauces.	
Potential applications	Pasteurized fruit juice stored at ambient temperature, meat and meat products, fish products, rehydrated infant formula, vegetarian food.	

B. Structure

1. Primary structure: Lantibiotics have characteristic lanthionine rings with monosulfide bridges. Nisin is defined as a Group A lantibiotic mainly because it has a linear, rather than a circular (Group B) structure (Jung, 1991a). All Group A lantibiotics are cationic and amphiphilic polypeptides with elongated forms. Other lantibiotics of Group A include subtilin (very similar in structure to nisin), epidermin, gallidermin, Pep5 and lacticin 481 (de Vos et al., 1995). The structure of Group A lantibiotics has been closely examined in order to understand their ability to form pores in prokaryotic cytoplasmic membranes (Sahl, 1991). Group B lantibiotics such as duramycin are also amphiphilic but have a pH closer to neutral and have a more compact, globular structure.

The structure of nisin was investigated by several workers from 1949 onwards (Berridge, 1949; Berridge et al., 1952; Newton et al., 1953; Cheeseman & Berridge, 1959) but it was Gross and Morell (1971), using amino acid sequencing and cyanogen bromide cleavage, who first proposed the complete primary structure. Nisin is a 34 amino acid polypeptide that is described as a heterodetic pentacyclic subtype A lantibiotic (Gross & Morell, 1971, Shiba et al., 1991; de Vuyst & Vandamme, 1994.) It contains unusual thioether amino acids: lanthionine (ALA-S-ALA), at positions 3-7 and β-methyllanthion-

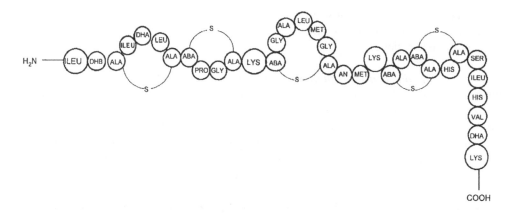

FIGURE 2. The structure of nisin A. ABA: Amino butyric acid; DHA: Dehydroalanine; DHB: Dehydrobutyrine (ß-Methyldehydroalanine); ALA-S-ALA: Lanthionine; ABA-S-ALA: ß-Methyllanthionine. (from Gross and Morell 1971. With permission.)

ine (ABA-S-ALA) at positions 8-11, 13-19, 23-36 and 25-28. It has a sulfur content of 5-6% due to these residues (Berridge et al., 1952). Five intramolecular sulfide bridges are present: ring A is formed by lanthionine and rings B-E by the four β-methyllanthionine residues. Nisin also contains three α,β unsaturated amino acids: dehydroalanine (DHA) at positions 5 and 33, and dehydrobutyrine (DHB) or β-methyl-dehydroalanine at position 2 (Gross & Morell, 1967, 1968; 1971). The integrity of ring A is critical to the antimicrobial activity. α,β dehydroalanine and α,β dehdyrobutyric acid are formed by dehydration of serine and threonine residues respectively. These compounds may then lead to the formation of lanthionine and β-methyllanthionine by making thioether bridges with nearby cysteine residues.

The α carbon atoms of the N-terminal units of the thioether amino acids are in the D-configuration (Gross & Morell, 1971) and the β carbon atoms are in the L-configuration (Knox & Keck, 1973; Morell & Gross, 1973; Knox & Keck, 1975; Fukase et al., 1988). The proposed structure of nisin A, based on a nisin preparation from *Lac. lactis* subsp. *lactis* NCFB 497 (Hirsch, 1951), is shown in FIGURE 2 (Gross & Morell 1971). The structural difference of the natural variant nisin Z is a substitution of histidine at position 27 for aspargine (Mulders et al., 1991). Nisin Z is produced by *Lac. lactis* subsp. *lactis* strain NIZO 22186, strain N8, strain SIK-83 and others (Graeffe et al., 1991; Kuipers et al., 1991a; Mulders et al., 1991; Nes, 1992). A molecule with the same substitution (His[27] to Asn[27]) with a molecular weight of 2700 was described (Araya et al., 1992). The primary structure proposed by Gross and Morell (*FIGURE 2*) has been supported by data derived from the chemical synthesis of nisin (Fukase et al., 1988; Bycroft et al., 1991; Shiba et al., 1991; Wakamiya et al., 1991) and the investigation of its spatial structure using mass and NMR spectroscopy (Barber et al., 1988; Nielsen & Roepstorff, 1988; Chan et al., 1989, Roepstorff et al., 1988; Craig, 1991).

The monomer of nisin has a molecular mass of 3353 Daltons (Jung, 1991 a; b). Nisin normally occurs as the more stable dimer form (7-kDa) and can also occur as a tetramer (14-kDa) (Cheeseman & Berridge, 1957; 1959; Ingram et al., 1967; Jarvis et al.,

1968). The multimers could arise from peptide bonds formed by intermolecular nucleophilic addition of an amino group from one residue and the double bond of one of the several unsaturated residues.

2. *Spatial configuration:* Several workers have studied the conformation of nisin in aqueous solution, mostly using two-dimensional NMR (Goodman et al., 1991; Kitazawa et al., 1991; van de Ven et al., 1991a, 1991b; Lian et al., 1991; 1992). In solution, nisin is flexible with no overall characteristic conformation. The thioether rings form the only defined structured regions, with nisin appearing as two fragments connected by a flexible region of variable structure (amino acids 20-22) (Sahl et al., 1995). In solvents that mimic membranes (e.g. trifluorothenol), nisin adopts a rod-shaped amphiphilic structure with clusters of charged and hydrophobic regions over the surface. The N terminal contains a high ratio of hydrophobic residues and the C terminal is more hydrophilic (van de Ven et al., 1991a). Similar conformations have been observed in zwitterionic dodecylphophocholine and anionic sodium dodecylsulphate micelles that mimic membranes (van den Hooven et al., 1996). The conformation of nisin complexed to both types of micelles was the same regardless of the different polar head-groups of the detergents.

IV. ANTIMICROBIAL ACTIVITY

A. Mechanism of action

1. Activity against vegetative cells: In vegetative cells the primary site of action for both nisin A and nisin Z is the cytoplasmic membrane where nisin forms pores, destroys the membrane integrity and acts as a voltage dependent polarizer (Ruhr & Sahl, 1985; Abee et al., 1994; Abee, 1995). This evidence has come from numerous studies on various physiological and artificial membrane systems. The flexible nature and amphiphilic properties of the nisin molecule are important for this mode of action.

The discovery that nisin inhibits N-acetylglucosamine synthesis in *B. subtilis*, and DNA, RNA and protein synthesis in *M. luteus* (Henning et al., 1986b) has prompted speculation that nisin interacts with the phospholipid components of the cytoplasmic membrane. Studies have indicated that isolated cytoplasmic membranes and phospholipid components of membranes show an antagonistic effect on nisin inhibition (Henning et al., 1986b). Nisin is a cationic molecule, which is believed to interact with the charged headgroups of membrane phospholipids via ionic forces. The C-terminal region is thought to be responsible for the initial interaction (Breukink et al., 1997). The initial association of nisin with the membrane may be charge-dependent (Garcia-Garcera et al., 1993; Driessen et al., 1995; Demel et al., 1996). Nisin was found to associate with non-energized liposomes, with the greatest interaction occurring with negatively charged phospholipids.

The pores formed by nisin across the cytoplasmic membrane are non-selective and short-lived. Pore size varies - a single molecule being too small to form a pore. This suggests that nisin molecules can join and leave the pore in a dynamic process (Jack et al., 1998). The energy that is needed for the formation and opening of the pores (Benz et al., 1991; Sahl, 1991), is mainly supplied by the trans-membrane potential of the microorganism. The lethal effect of nisin is optimum against actively growing cells whose membranes are fully energized (Sahl, 1991; Maisnier-Patin et al., 1992). Nisin activity could be elicited in the absence of a trans-membrane electric potential ($\Delta\Psi$) with high pH gradient across the membrane (ΔpH) (Moll et al., 1997). In such case, the rate of dissipation

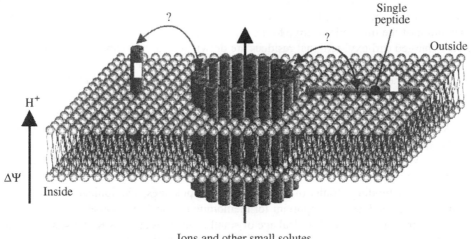

Single peptide

Outside

H⁺

ΔΨ

Inside

Ions and other small solutes
(eg. K⁺ATP, amino acids)

FIGURE 3. A general model of the 'barrel-stave' mechanism of pore formation by peptides. (reprinted with permission from *Lantibiotics and Related Peptides*, 1988, by Ralph W. Jack, Gabriele Bierbaum and Hans-Georg Sahl published by Springer-Verlag and Landes Bioscience).

of ΔpH is increased with the magnitude of ΔpH, and pores are formed only when the internal ΔpH is alkaline. The efficiency of $\Delta \Psi$-induced pore formation is strongly affected by the external pH, unlike that induced by ΔpH. At acidic pH, nisin dissipates the ΔpH more effectively than the $\Delta \Psi$.

The mechanism of pore formation is still unclear (Jack et al., 1998). The 'barrel-stave' mechanism of pore formation (Ojcius & Young, 1991; Nissen-Meyer et al., 1992) suggests that nisin monomers (the staves) bind to the membrane, insert into the membrane (dependent on the $\Delta \Psi$ and aggregate within the membrane to form a water-filled pore (the barrel). This is illustrated in *FIGURE 3*. The nisin molecule orients perpendicularly into the membrane with its hydrophobic sides facing the membrane and its hydrophilic regions facing into the lumen of the pore, forming the sides. If the membrane is sufficiently energized, nisin molecules travel through the membrane, adhering to the surface and carrying across phospholipids. Several molecules binding at the same site will result in the formation of a transient pore. Analysis of nisin orientation, using site-directed tryptophan fluorescence spectroscopy, revealed an overall parallel average orientation of nisin in the membrane with respect to the membrane surface, with the N-terminus inserted more deeply than the C-terminus (Breukink et al., 1998).

In the 'wedge' model (Driessen et al., 1995), nisin molecules are thought to bind by electrostatic interactions to the anionic membrane surface, leading to a high local concentration that disturbs the lipid dynamics and causes a localized strain which forces the nisin into the membrane. This model suggests wedge-like pores composed of multimeric nisin molecules that are formed once the $\Delta \Psi$ drives membrane insertion of the phospholipid surface bound molecules. The phospholipid composition of the membrane would affect this mode of action. Demel et al., (1996) showed that nisin penetrates more deeply into lipid monolayers of diphosphatidylglycerol compared to monolayers of other lipids including phosphatidylglycerol, phosphatidylcholine, phosphatidylethanolamine, monogalactosyldiacylglycerol and digalactosyldiacylglycerol. The spatial structure of nisin changes only in the flexible hinge region between the two structured domains, while the orientation of the molecule relative to the lipid headgroups remains unchanged. The positively charged carboxyl-terminal residues are drawn across the membrane in response to the $\Delta\Psi$, causing a structural defect in the biphospholipid bilayer.

Several unique features of nisin are vital for its ability to bind to the membrane and form pores: its amphipathic character, the flexibility of its central segment, the C-terminal and the overall negative surface charge of the membrane (Demel et al., 1996; Breukink et al., 1997; Giffard et al., 1997). Winkowski et al. (1996) obtained mechanistic information about nisin by measuring the efflux of a fluorescein from *Listeria monocytogenes*-derived liposomes. Initial leakage was dependent on the concentration of both lipid and nisin. Increased rates were observed at higher pH values, but this was not due to an increase in the binding affinity of the nisin for the liposomes. The ionization state of the nisin histidine residues was important for membrane permeabilization ability.

Pore formation leads to leakage of small molecules such as K^+ ions, solutes and metabolites such as ATP and amino acids. This rapidly depletes the proton motive force, starving the cell of energy and causing the cessation of certain biosynthetic processes (Ruhr & Sahl, 1985; Sahl & Bierbaum, 1998). Further evidence of this mechanism is based on studies such as that by Okereke and Montville (1992) where nisin-induced dissipation of the proton motive force of *Clostridium sporogenes*. Davies and Adams (1994) reported that *L. monocytogenes* cells exposed to nisin show an increased rate of loss of K^+ ions.

Nisin may have other mechanisms of activity against vegetative cells. Interference with cell wall biosynthesis has been observed, although this may simply be a consequence of energy loss and membrane depolarization resulting from pore formation. Linnett and Strominger (1973) reported *in vitro* evidence of nisin interference in cell wall biosynthesis, and nisin was also shown to inhibit murein synthesis in *Bacillus subtilis* (Henning et al., 1986b). Reisinger et al., (1980) used a cell-free system to show a formation of a nisin complex with the lipid-bound peptidoglycan precursor, undecaprenyl-pyrophosphate. However, Sahl and Bierbaum (1998) observed that the cidal effect of nisin occurs immediately after addition of the peptide, whereas inhibitors of cell wall biosynthesis, such as mersacidin, induce slow lysis and their effect needs at least one cell division cycle to become apparent. These data suggest that the inhibition of cell wall biosynthesis may not contribute to the cidal effect *in vivo*.

Another mechanism of action could be the induction of autolysis (Bierbaum & Sahl 1985; 1987; 1988). Nisin may adsorb to the teichoic acids of the cell walls and activate autolytic enzymes in the septum between two daughter cells. Extensive cell wall degradation has been observed by electron microscopy (Bierbaum & Sahl, 1991). This mechanism could be slower in effect than pore formation, but may contribute to an overall cidal effect. Studies on heat injured cells of *B. cereus* showed that nisin (as low as 10 µg/ml) prevented the repair of the cytoplasmic membrane of heat stressed cells (Beuchat et al., 1997).

2. Activity against endospores: (Endo) spore forming bacteria, such as *Bacillus* and *Clostridium* species are highly sensitive to nisin. In general, spores are more sensitive than cells, and the sensitivity is increased in acidic substrates, if the spores have suffered injury (e.g. heat treatment). Less is understood about the activity of nisin against spores, although it is known to be different from that against vegetative cells. The effect is normally sporostatic, but sporocidal effects have been reported depending on both the strain and severity of any heat treatment (de Vuyst & Vandamme, 1994). Germination is not affected by nisin. Instead, nisin prevents post-germination swelling and subsequent spore outgrowth (Hitchins et al., 1963; Gould, 1964). Spores that rupture their coats mechanically are more nisin sensitive than spores that lyse their coats (Gould & Hurst, 1962). Chan et al., (1996) produced a nisin mutant in which the dehydroalanine residue at position 5 was replaced by an alanine residue. This substitution caused a reduction in the nisin-induced inhibition of spore outgrowth, but did not affect activity against vegetative cells. This was also observed with subtilin (Liu & Hansen 1992; 1993). The double bond of the Dha5 residue is thought to provide a reactive group for an interaction between nisin and a spore-associated factor essential for outgrowth.

3. Sensitivity: There can be wide variation in nisin sensitivity among strains even of the same species (Gupta & Prasad, 1989a; Meghrous et al., 1999). Ramseier (1960) reported that approximately 10-100 times more nisin is needed to prevent the outgrowth of spores of *Cl. botulinum*, compared to *Cl. butyricum* spores. Both spores and cells of thermophilic species such as *B. stearothermophilus* and *Cl. thermosaccharolyticum* are more sensitive to nisin than mesophilic spore forming species (Jarvis, 1967). Meghrous et al. (1999) compared the activities of nisin A, nisin Z and pediocin, and found that lactic acid bacteria, *Clostridium* and *Bacillus* (spores and cells) are more effectively inhibited by the nisin variants compared to pediocin. Nisin A showed more anti-listerial activity than nisin Z. In other studies, *Listeria* species including *L. monocytogenes* have shown a wide variation in sensitivity (Ferreira & Lund, 1996; Ukuku & Shelef, 1997). Within a population of the same strain, certain cells may show greater resistance. Gravesen and Knochel (1998) examined 414 *L. monocytogenes* isolates from various sources and found natural resistance occurred rarely. Benkerroum and Sandine (1988) found that the nisin concentration required for complete inhibition of *L. monocytogenes* varied from 18.5 - 2500 μg/ml, depending on the culture medium and the target strain. This limitation could be overcome in foods by using nisin in combination with a starter culture producing a non-nisin anti-listerial bacteriocin (Schillinger et al., 1998). In practice, however, varied listerial sensitivities are overcome by standard hurdle technology, where nisin acts in combination with other control measures in the food matrix, such as reduced water activity.

The composition of the cell membrane may influence the sensitivity of vegetative cells (Verheul et al., 1997). Investigations of mutants that arise naturally from sensitive populations have shown that lower concentrations of cellular phospholipid as well as alterations in the fatty acid composition of the membrane are associated with reduced nisin sensitivity (Ming & Daeschel, 1995; Mazzotta & Montville, 1997; Crandall & Montville, 1998). Changes in the composition of the cell wall may also render cells less sensitive, as shown in a study comparing the activity of nisin against protoplasts and whole cells of strains of *L. monocytogenes* (Davies et al., 1996). Bennik et al., (1997) concluded that the association of nisin with the cell membrane and its subsequent insertion was similar for *Lb. sake* and *Pediococcus pentosaceus* strains with high or low nisin tolerance. For strains insensitive to nisin, overall membrane constitution as well as membrane-fluidity could preclude pore-formation. Verheul et al., (1997) isolated a nisin-resistant *L. monocytogenes* variant by stepwise exposure to increasing concentrations of nisin in BHI broth. The resistant isolate showed no significant change in the membrane fatty acyl composition, but produced more phosphatidylglyerol and less diphosphatidylglycerol.

Although mutation could render certain bacteria resistant to the low levels of nisin commonly found in foods, in practice, the physiological consequence of reduced nisin sensitivity often reduces cell fitness to the point where the bacterium is no longer a threat to food spoilage. A 20 % drop was observed in the growth rate of naturally occurring nisin-resistant *L. monocytogenes* strains (Anne Gravesen, personal communication). The metabolic cost of induced nisin resistance can be further inferred from the rapid loss of resistance when cells are cultured in the absence of nisin. Nisin resistance can therefore be seen as distinct from antibiotic resistance in terms of both mechanism and consequence.

Certain Gram-positive bacteria are nisin-resistant due to their ability to synthesize protease enzymes that could inactivate nisin (see *Section IV-B2*), although this has never emerged as a practical problem in food systems. Spontaneous resistance has been observed in strains in *Cl. butyricum, Lac. lactis* subsp. *cremoris*, *L. monocytogenes* and *Strep. agalactiae* (Hirsch, 1950; Ramseier, 1960; Hirsch & Grinsted, 1951; Harris et al., 1991). Relatively high frequencies of the occurrence of spontaneous *L. monocytogenes* mutants have been reported (Harris et al., 1991; Ming & Daeschel, 1993; Davies & Adams, 1994). In *Lb. casei* strains, induced nisin resistance was lost after one passage in nisin-free medium (Breuer & Radler 1996). Resistance was linked with the production of an anionic phosphate polysaccharide with subunits of rhamnose and galactose.

4. Antimicrobial kinetics of reaction mixture: Nisin works in a dose-dependent fashion, i.e. its effectiveness decreases in proportion to increasing numbers of target organisms (Rayman et al., 1981; Scott & Taylor, 1981). This is true for both vegetative cells and spores. Ramseier (1960) demonstrated that an increased nisin concentration of approximately 0.5 log cycle IU was required to control a ten-fold increase in spores of *Cl. butyricum*. Porretta et al., (1966) reported that twice as much nisin was required to achieve the same preservative effect in canned peas containing 810 *B. stearothermophilus* spores as in the same product with 141 spores. The overall environmental conditions of the test product dictate whether nisin activity against vegetative cells results in a predominantly bactericidal or bacteriostatic effect (Delves-Broughton et al., 1996). Optimum conditions for cell growth will give a morc bactericidal effect, as the cells will be more energized (see *Section IV-B*). Frequently other preservation hurdles might affect the overall conditions of a food product, e.g. acid pH, low water activity, chill temperature, etc. These conditions will result in the cells being in an non-energized state.

B. Factors affecting nisin action

Nisin activity in a food matrix may depend on several factors. These include: i) changes in nisin solubility and charge, ii) binding of nisin to food components (e.g. meat phospholipids), iii) inactivation of nisin (by proteases or food ingredients), and iv) changes in the cell envelope of the target organisms in response to environmental factors (Gänzle et al., 1999). The activity of nisin produced *in situ* may also be affected by food components. Hugas et al., (1996) showed that during sausage fermentation, the formulation of the sausage and the technology of the fermentation affected elimination of *L. monocytogenes* resulting from bacteriocins produced by starter cultures.

Thorough mixing is necessary for optimum results. The spatial heterogeneity of a food matrix may affect uniform distribution of nisin and thus its potential to protect a product (Wimpenny et al., 1995; Thomas et al., 1997). Due to its hydrophobic nature, nisin may preferentially associate with the fatty components of a food, making it unavailable for antimicrobial activity (Jung et al., 1992). Nisin works better in homogeneous liquid substrates rather than solid and heterogeneous systems, probably due to better diffusibility (Delves-Broughton et al., 1996) and better inhibition is often observed during *in vitro* tests conducted in liquid culture, compared to those in actual food products.

Blom et al., (1998) developed a model assay to investigate intrinsic factors that could affect diffusion of bacteriocins such as nisin. Factors with the potential to influence diffusion include pH, salt concentration, nitrite and nitrate, aqueous phase available for diffusion, fat content and fat surface available for solubilization, proteolytic activity, availability of growth substances and sugar concentration. The concentration of indicator strain, salt content and agar concentration were the most important factors affecting diffusion. Food ingredients may also benefit nisin by affording protecting to the molecule during processing treatments, particularly heating.

Other preservative hurdles, such as heat treatments, low water activity, modified atmosphere, low temperature, etc. can enhance the effect of nisin. For *L.monocytogenes*, decrease in temperature reduced the rate of K^+ efflux caused by nisin Z, an effect probably caused by reduced membrane fluidity (Abee et al., 1994). Szabo and Cahill (1998) examined different atmospheres, and found a combination of 4°C with a CO_2 enriched atmosphere controlled the growth of *L. monocytogenes* in a broth system (pH 6.0) for long periods. Ueckert et al., (1998) reported synergism of nisin activity against *Lactobacillus* as a result of exposure to a sequential treatment of mild heat (48 - 56°C) and then nisin (0.5 µg/ml) at 30°C. The combined variables of low pH and high NaCl concentration have been shown to boost nisin activity against *Lb. sake*, *Staph. aureus* and *L. monocytogenes* (Thomas & Wimpenny 1996; Blom et al., 1997; Datta & Benjamin, 1997).

Rogers and Montville (1994) investigated the factors influencing nisin inhibition of *Cl. botulinum* using a model food system containing yeast extract, proteose peptone, and glucose supplemented with various concentrations of protein, phospholipid or soluble starch. The experiments were conducted at pH 5.5, 6.0 or 6.5, and inoculated samples were incubated at 15, 25 or 35°C. Temperature was the most significant variable and nisin became less effective as the temperature increased. Nisin concentration had significant positive and phospholipid had significant negative linear effects. Beuchat et al., (1997) reported similar effects of temperature. Psychrophilic *B. cereus* strains in beef gravy were more sensitive at 8°C compared to 15°C. Spores were more sensitive than cells, and the sensitivity of the cells was increased by heat stress.

Currently of interest is the combined effect of nisin with novel preservation techniques such as ultra-high pressure or high hydrostatic pressure and pulsed electric field. High hydrostatic pressure kills or injures microorganisms mainly by damaging the membrane, and thus it was likely that its effect would be enhanced in the presence of a preservative such as nisin that targets the membrane. A four-dimensional process has been suggested: pressure x time x temperature x antimicrobial preservative (such as nisin). Inclusion of nisin would allow lower pressures to be employed, with less impairment to food quality whilst maintaining a high level of kill of foodborne pathogens (Kalchayanand et al., 1995; Ray, 1995). Kalchayanand et al., (1994) reported that cells of *L. monocytogenes* and the Gram negative pathogens *Escherichia coli* and *Salmonella typhimurium* that were injured by these treatments were sensitive to nisin. The antibacterial effect of the physical treatments was significantly increased with nisin. Ultra high pressure inactivation of *E. coli* and *Listeria innocua* inoculated in liquid whole egg was reportedly improved significantly by 1.25 and 5 mg/L nisin (Ponce et al., 1998). Hauben et al. (1996) suggested that pressure- treated *E. coli* became nisin sensitive because the pressure caused disruption of the outer membrane permeability. Evidence suggests that cells are least sensitive to pressurization between 20 - 30 ºC, but become extremely sensitive at temperatures greater than 35 ºC owing to phase transitions of membrane lipids (Carlez et al., 1992; Ludwig et al., 1992; Kalchayanand et al., 1996). Further investigations reported increased cell death by pressurization (345 Mpa at 25 ºC for 10 min) in the presence of a combination of nisin and pediocin AcH. Pressurization of *L. monocytogenes, Staph. aureus, E. coli* O157:H7 and *S. typhimurium* in the presence of the bacteriocins caused an increase of 1.3 - 5.1 log cycles in viability loss. Total loss in viability ranged from 5.6 - 8 log cycles (Kalchayanand et al., 1998).

Roberts and Hoover (1996) investigated the combined effects of ultra high pressure treatment, heat, acidity and nisin on *B. coagulans* spores, which were sensitive to a very low concentrations of nisin. A minimum of a 6 $\log_{10}$ reduction was achieved with 400 Mpa pressure in pH 4.0 buffer with a mild heat treatment (70 ºC for 30 min) and 0.8 IU/ml nisin.

1. Synergism and antagonism: Combinations of two or more preservative systems, or 'hurdles', are better at preventing the emergence of less sensitive organisms, and reduce the extent of a single extreme system or preservative (Leistner & Gorris, 1995). There has been extensive testing of nisin with other treatments to find synergistic combinations and to avoid those that are antagonistic. The main groups of compounds that enhance the antibacterial action of nisin are chelators (e.g. EDTA and citrate), non-ionic and amphoteric surfactants, and emulsifiers such as sucrose fatty acid esters and monoglycerides. Nisin is inactivated by oxidizers and certain proteases, and is inhibited by divalent cations such as Ca^{2+} and Mg^{2+} and strong anionic detergents.

Nisin is effective against Gram negative bacteria, if it is able to pass through the normally nisin-impermeable barrier of the cell wall, thus gaining access to its target: the cytoplasmic membrane. This could be achieved by disrupting the outer membrane (Kordel & Sahl, 1986), by using chelating agents (e.g. EDTA), organic acids, or treatments such as sublethal heat or freezing (Blackburn et al., 1989; Kalchayanand et al., 1992; Stevens et al., 1991; 1992; Delves-Broughton, 1993). Chelating agents remove divalent cations, such as Ca^{2+} and Mg^{2+}, from the Gram negative cell walls, releasing phospholipids and lipoproteins and increasing the cell wall permeability. Stevens et al. (1992a) found effective treatments that comprised 12.5 - 25 µg nisin/g with 20 mM EDTA or citric monohydrate at 30 - 42 ºC. However chelating agents are much less effective in food due to their preferential binding to free divalent ions present in the food matrix.

Another means of using nisin against both Gram negative and Gram positive bacteria was the subject of a patent application (Miles et al., 1997). This involved a treatment combining a solution of low concentration alkali metal orthophosphate (e.g. trisodium orthophosphate) with osmotic shock and/or lysozyme and/or nisin. Another synergistic combination, resulting in an antimicrobial effect against *Salmonella* spp., has been described between an antimicrobial polypeptide such as nisin, lactoperoxidase, thiocyanate and hydrogen peroxide applied at 30 - 40 °C at pH 3-5 (Bycroft et al., 1991). The lactoperoxidase-thiocyanate-hydrogen peroxide system (LPS) occurs naturally in milk. Its antimicrobial effect is due to hypothiocyanate ions: transient products of the oxidation of thiocyanate by hydrogen peroxide catalyzed by lactoperoxidase (Aune & Thomas, 1977; Bjork & Claesson, 1980; Reiter, 1985). Zapico et al., (1998) found synergistic activity of LPS and nisin against *L. monocytogenes* in UHT skim milk, causing a 5.6 log unit reduction in bacterial numbers compared to control milk. Higher *in situ* nisin production in raw milk by a *Lac. lactis* subsp *lactis* strains was achieved in the presence of LPS, compared to that without LPS activation (Rodriquez et al., 1997).

Mansour et al. (1999) described synergy between monolaurin and nisin against *B. licheniformis* in milk. Total inhibition of outgrown spores and vegetative cells within 7 days at 37 °C was achieved by a combination of nisin (0.75 μg/ml) and monolaurin (100 μg/ml) at pH 6.0. A synergist effect has also been reported between sucrose fatty acid esters (sucrose palmitate, sucrose stearate) and nisin (Thomas et al., 1998). Nisin activity could be stimulated by other neutral emulsifiers such as monolaurin, mono-oleate and Tween 80 (Jung et al., 1992) and this process has been patented (Blackburn et al., 1989).

Other commonly used preservatives have been reported to show synergism with nisin. For example, Scannell et al., (1997) found that 2% sodium lactate enhanced nisin inhibition of *S. kentucky* and *Staph. aureus* on fresh pork sausages. In another study, an agar diffusion method was used to show that a combination of lactic acid and whey permeate fermented by a nisin-producing *Lac. lactis* strain had a synergistic effect against fish isolates: *M. luteus*, *Pseudomonas aeruginosa*, *Ps. fluorescens*, and *Staph. hominis* (Nykänen et al., 1998b). Oscroft et al., (1990) found combinations of organic acids (citric acid, lactic acid, glucono-delta-lactone or acetic acid) with nisin improved inhibition of thermally-stressed *Bacillus* spores, particularly under less acidic conditions and ambient storage. The anti-listerial effect of a sorbate-nisin combination was assessed *in vitro* and on beef at refrigeration temperature (Avery & Buncic, 1997). The combined treatment greatly reduced initial *L. monocytogenes* counts on vacuum-packed and carbon dioxide packaged beef, and prevented growth during 4 weeks of refrigerated storage. Gänzle et al., (1999) reported an increase in nisin activity due to a combination of lower pH, and with EDTA, propyl-parabens or NaCl.

Synergism between lysozyme and nisin has been described (Monticello, 1989; 1990). Lysozyme is known to have antimicrobial capabilities, and is sometimes used to control growth of *Cl. tyrobutyricum* in cheese. Low levels of a combination of these compounds had a greater effect than either one alone at higher concentrations.

Hanlin et al., (1993) investigated combinations of bacteriocins, and found synergism between nisin and pediocin against *L. monocytogenes* and other bacteria. Mulet-Powell et al., (1998), also observed this and in addition reported antagonism between lacticin 481 and nisin. Walsh et al., (1996) reported that certain mixtures of antibiotics and nisin had an enhanced antibacterial effect in Cheddar cheese. Nisin (12.5 μg/g curd), penicillin G (12 U/g curd) and streptomycin (12 mg/g curd) reduced non-starter lactic acid bacteria by 4-5 log cycles compared with untreated control after 100-160 days. Apart from containing higher levels of free amino acids, no differences in the treated cheese were detected.

The food additives sodium metabisulphite and titanium dioxide are known to degrade nisin (Delves-Broughton et al., 1996). Sodium metabisulphite is used as an antioxidant, bleach and antimicrobial agent. Titanium dioxide is used as a white colorant in some salad dressings and processed cheese formulations. Adsorption of nisin (and other bacteriocins) to this compound resulted in the inactivation of the nisin (Wan et al., 1996), although there has been one report of titanium dioxide improving the antibacterial activity of nisin in low fat salad dressings (Yuan et al., 1998).

Divalent and trivalent cations (e.g. Mg^{2+}, Ca^{2+}, Gd^{3+}) have been shown to reduce the activity of nisin Z against *L. monocytogenes* (Abee et al., 1994). The ions may interact with the negatively charged phospholipid headgroups of phosphatidyl glycerol and cardiolipin of the cytoplasmic membrane (Harwood & Russell, 1984; O'Leary & Wilkinson 1988), reducing electrostatic interaction and/or causing the membrane to become more rigid (Abee et al., 1994).

Jung et al., (1992) found milk fat reduced nisin activity: approximately 33% decrease in nisin activity in skim milk was observed compared to >88% decrease in milk containing 12.9% fat. Conversely, Zapico et al., (1999) found no significant difference of nisin inhibition of *L. innocua* in whole milk and skim milk. In the study by Jung et al., (1992) the nonionic emulsifier Tween 80 partially counteracted the apparent decrease due to milk fat, whilst the anionic emulsifier lecithin had no effect. Henning (1984) found that ≥ 0.1% lecithin was antagonistic to nisin, an effect thought to be due to stable complexes forming between nisin and the zwitterionic phospholipid (Henning et al., 1986a). Gänzle et al., (1999) found that lecithin, as well as casein and divalent cations, were antagonistic to nisin activity. Casein could interact with nisin because of its amphiphilic nature, having negatively charged and hydrophobic domains. Richard (1993) observed that nisin added to milk heated to 72 ºC for a few seconds had less listericidal activity than the nisin addition to milk treated at 100 ºC/10 minutes. Thus, greater heat treatment seems to prevent binding of the nisin to milk proteins.

In studies with lactic acid bacteria and *Listeria*, calcium has been reported to be antagonistic towards nisin (Henning, 1984; Blackburn et al., 1989; Abee et al., 1994; Gänzle et al., 1999). A concentration of 5 n.mol/L Mg^{2+} caused a 31% reduction of the nisin Z induced efflux of K^+ ions from *L. monocytogenes cells*. A 60% reduction was caused by the same concentration of Ca^{2+} (Abee et al., 1994). Divalent cations could bind to anionic phospholipids, make the cytoplasmic membrane more rigid and reduce the affinity of the nisin towards the membrane (Abee et al., 1994; Demel et al., 1996; Crandall & Montville, 1998). The synergistic effect of EDTA on nisin activity against Gram negative bacteria (Delves-Broughton, 1993) is abolished if calcium and magnesium are added (Cutter & Siragusa, 1995a).

2. Enzyme inactivation: Nisin is inactivated by pancreatin, α-chymotrypsin and subtilopeptidase (Heinemann & Williams, 1966; Jarvis, 1967; Jarvis & Mahoney, 1969). The active component of pancreatin that inactivates nisin has been shown to be α-chymotrypsin. This enzyme is sometimes present in trypsin inactivates nisin (Hawley, 1958; Campbell & Sniff, 1959; Thorpe, 1960; Willimowska-Pelc et al., 1976)

Production of nisinases, the more specific enzymes which target nisin, have been reported from strains exhibiting reduced nisin sensitivity, including *B. cereus*, *B. polymyxa*, *B. megaterium*, *Enterococcus faecalis*, *Lb. plantarum*, *Lb. brevis*, *Lac. lactis*

TABLE 2. Nisin-sensitive bacterial species associated with food

Genus	Species	Comments
Alicyclobacillus	*acidoterrestris*	Gram positive spore former. Grows at pH 2.5 - 6, and 25 - 60°C. Spoilage organism of fresh/pasteurised fruit juice stored at ambient temperature.
Bacillus	*brevis, cereus, coagulans, licheniformis, macerans, megaterium, pumilus, subtilis, sporothermodurans, stearothermophilus*	Gram positive aerobes and facultative anaerobes. Spore-forming. Examples include spoilage organisms (cause of 'rope' in bread and flat sour spoilage of canned food), psychrophiles, thermophiles, acid-resistant spp., UHT-resistant sp., and food-poisoning pathogens.
Brochothrix thermosphacta		Gram-positive aerobes and facultative anaerobes. Heat sensitive. Spoilage organisms of raw and cooked meat. Growth between 0 - 30°C. Very sensitive to nisin.
Clostridium	*bifermentans, botulinum, butyricum, histolyticum, pasteurianum, perfringens, putrificum, sordelli, sporogenes, tertium, thermosaccharolyticum, tyrobutyricum*	Gram-positive obligate anaerobes. Spore forming. Examples include spoilage organisms (causing blowing, etc.), and food-poisoning pathogens.
Desulfotomaculum	*nigrificans*	Gram-variable obligate anaerobes. Spore-forming. Thermophile causing blackening of canned foods
Enterococcus	*faecalis, faecium*	Gram-positive aerobes or anaerobes. Varied response to nisin. Spoilage organisms
Lactobacillus	*spp., bulgaricus, brevis, buchneri, casei, helveticus, fermentum, lactis, plantarum.*	Gram-positive aerobes or anaerobes. Able to grow at low pH. Some species used in cheese production. Spoilage organisms of dressings, wine, beer, cider.
Leuconostoc	*oenos, mesenteroides*	Gram-positive aerobes and anaerobes. Spoilage organisms of wine and vegetables (slime-producing).
Listeria	*monocytogenes*	Gram-positive aerobes and anaerobes. Varied nisin sensitivity. Food poisoning organism able to grow at chill temperatures. Found in soft cheese, pate, frankfurters, sliced delicatessen meat, salad, smoked fish and seafood. Zero tolerance in the USA for food with no heat processing.
Micrococcus	*luteus, varians*	Gram-positive aerobes. *M. luteus* used in nisin bioassay
Pediococcus	*spp., damnosus, pentosaceus*	Gram-positive aerobes and anaerobes. Spoilage organisms of wine and beer.
Sporolactobacillus	*inulinus*	Gram-positive aerobe and anaerobe. Spore forming. Growth at low pH.
Staphylococcus	*aureus*	Gram positive aerobe and anaerobe. Varied nisin sensitivity. Food poisoning pathogen.

subsp. *lactis*, *Lac. lactis* subsp. *cremoris*, *Leuconostoc*, *Staph. aureus*, and *Strep. thermophilus* (Kooy, 1952; Carlson & Bauer, 1957; Lipinska & Strzalkowska, 1959; Alifax & Chevalier, 1962; Jarvis, 1966; Jarvis, 1967; Lipinska, 1977; Collins-Thompson et al., 1985). A nisinase from *B. cereus* was also capable of degrading subtilin (Jarvis, 1967). Enzymes isolated from several *Bacillus* species were active against nisin A. The site of action for nisinases seems to be the C-terminal dehydroalanine; the enzymes act either as reductases or dehydropeptide reductases (Gross et al., 1969; Jarvis 1970; Jarvis & Farr, 1971). Nisinase activity is a potential problem while using any nisin-producing starter culture (Galesloot, 1956; Lipinska, 1977).

C. Antimicrobial spectrum

Nisin is effective against Gram positive bacteria, including those that form heat-resistant endospores (see *TABLE 2*). Levels of nisin required to inhibit spore outgrowth are often less than those required to inhibit vegetative cells (Hurst, 1981). Gram negative bacteria are resistant to nisin because of their nisin-impermeable outer cell membrane, which prevents nisin reaching the cytoplasmic membrane. Schved et al., (1994) showed that if the cell wall were removed, spheroplasts of *E. coli* and *S. typhimurium* were sensitive to nisin. If Gram negative cells are exposed to sublethal treatments that damages the outer membrane, nisin can gain access to the cytoplasmic membrane and cause inhibition. Treatments that act in such a way, rendering Gram negative bacteria sensitive, include hydrostatic pressure, heat, freezing and thawing and chelating agents (described above) such as EDTA or ethyl maltol (Stevens et al., 1991; Kalchayanand et al., 1992; Delves-Broughton, 1993; Kalchayanand et al., 1994; Hauben et al., 1996; Schved et al., 1996). Yeasts are insensitive to nisin for exactly the same reason. Removal of the yeast cell wall allows the bacteriocin access to the yeast membrane and results in rupture of the cell (Dielbandhoesing et al., 1998). Variation in nisin sensitivity has been observed during the yeast cell cycle. However, the insensitivity of yeasts to nisin allows the use of the bacteriocin in yeast fermented products.

V. USE OF NISIN IN FOOD

A. Types of food suitable for nisin use

Nisin is suitable for use in a wide range of foods - liquid or solid, canned or packaged, chill or warm ambient storage. Based on target organisms, its usage falls into three broad categories: i) to prevent spoilage by Gram positive endospore formers (particularly in heat processed food) ii) to prevent spoilage by lactic acid bacteria and similar organisms such as *Brocothrix thermosphacta*, iii) to kill or inhibit Gram positive pathogens, e.g. *L. monocytogenes*, *B. cereus*, *Cl. botulinum*.

Nisin serves many purposes in food. It can extend shelf life of both chilled and ambient stored food. Typical examples of this are canned products stored at ambient temperatures in warm climates. Nisin also protects products, particularly chilled foods, from storage temperature abuse. Addition of nisin may contribute to the general preservation of a food so that thermal processing can be reduced. This has the double advantage of improving product quality (which may be impaired by harsh heat treatment) and reducing manufacturing costs, since both the time and temperature of thermal treatment can be cut. Distribution costs can also be reduced with nisin use, since this may allow products to be transported at ambient rather than chilled temperatures.

Bad manufacturing practice resulting in unacceptable levels of post processing contaminants is not masked by nisin use. Such contamination usually introduces Gram negative bacteria that are not controlled by nisin. As nisin works in a concentration-dependant manner, numbers of spores and/or vegetative cells contaminating a food product should not be at too high a level.

B. Methods of application

Nisin is best added as an aqueous solution, usually to the liquid portion of a product during its processing. For instance, in canned foods nisin can be combined with a small quantity of canning brine and then thoroughly mixed into the bulk product. In dairy desserts and milks, it can be added to a small quantity of milk, then mixed into the bulk milk, filled and processed (Fowler & McCann, 1971). It can also be added as a powder, but in all cases, it is essential to ensure uniform dispersal throughout the food matrix. The best time to add nisin is at the last practical stage before heat processing (if this is part of the manufacturing process). In the manufacture of processed cheese, for example, nisin is usually added to the heated cheese at the same time as the melting salts (Fowler & McCann, 1972). Nisin can also be used (at higher concentrations since this can be considered a processing aid) as a spray or dip for surface decontamination. As nisin is unstable at high pH, it is inadvisable to mix it with acid correctant emulsifying salts, which are typically > pH 9.0.

The addition level of nisin depends on the type of food, its heat process, pH, storage conditions and the required shelf life. Addition levels of a commercial extract such as Nisaplin vary from 10 – 750 mg/kg, which is equivalent to 0.25 - 18.7 µg nisin/g. Typical usage levels in foods are shown in *TABLE 3*. Nisin is often used in acidic foods, but is effective in products across a wide range of pH values (3.5 - 8.0). Nisin seems to be a very effective preservative in liquid egg, which generally has a pH of 7.3 to 7.8.

There have been several investigations into the application of nisin in packaging films and edible casings. Silicon surfaces coated with nisin have been shown to demonstrate antimicrobial activity (Daeschel et al., 1992; Bower et al., 1995a, b; Lakamraju et al., 1996; Wan et al., 1996). Ming et al., (1997) described the retention of nisin in food packaging films by dialysis. A 4.3 $\log_{10}$ reduction of *S. typhimurium* was achieved on inoculated broiler skin exposed to nisin-coated polyvinyl chloride film (Natrajan & Sheldon, 1994). Edible packaging films have also been considered because they have several advantages: they are biodegradable, they enhance the flavor and color of food, reduce moisture loss and can contribute to food safety and preservation if, for example, nisin is incorporated. Biodegradable films containing lysozyme, nisin and/or EDTA made from soy protein and corn zein prepared by heat-press and casting methods demonstrated good bactericidal activity against *Lb. plantarum* and *E. coli* (Padgett et al., 1998). Dawson et al., (1997) reported that decreasing the water vapor permeability of the protein films reduced the nisin activity as measured by a zone diffusion assay. Recent interest in the application of nisin in casings has mainly concentrated on meat products (Nettles-Cutter & Siragusa, 1996; Kassaify, 1998). Sheldon et al., (1996) examined the antimicrobial effectiveness of three commercial absorbent tray pads treated with a nisin-containing formulation (pH 3.5), comprising 50 µg/ml nisin, 5 mM EDTA, 3% citric acid and 5% Tween 20. The nisin solution was applied evenly across the pads, which were then air dried and inoculated with either *S. typhimurium* or the exudate from a pad taken from a purchased

TABLE 3. Typical addition levels of nisin (and Nisaplin) in examples of different foods

Type of food/application	Nisin (µg/g)	Nisaplin (mg/kg) or (mg/L)	Typical target organisms
Canned food	2.5 - 5	100 - 200	*Bacillus stearothermophilus* *Cl. thermosaccharolyticum* *Cl. botulinum*
Processed cheese	5 - 15	200 - 600	*Bacillus* spp. *Clostridium* spp.
Pasteurized chilled dairy desserts	1.25 - 3.75	50 - 150	*Bacillus* spp. *Clostridium* spp.
Ricotta cheese	2.5 - 5	100 - 200	*Listeria monocytogenes*
Dressings and sauces	1.25 - 5	50 - 200	LAB *Bacillus* spp. *Clostridium* spp.
Milk and milk products	0.25 - 1.25	10 - 50	*Bacillus* spp. e.g. *B. sporothermodurans*
Liquid egg	1.25 - 5	50 - 200	*Bacillus* spp., e.g. *B. cereus*
Beer:			LAB, e.g. *Lactobacillus, Pediococcus*
pitching yeast wash	25 - 37.5	1000 - 1500	
reduced pasteurization	0.25 - 1.25	10 - 50	
during fermentation	0.63 - 2.5	25 - 100	
post fermentation	0.25 - 1.25	10 - 50	
Crumpets	3.75 - 6.25	150 - 250	*Bacillus cereus*
Fruit juice (pasteurized, stored at ambient)	0.13 - 0.25	5 - 10	*Alicyclobacillus acidoterrestris*

tray of chicken portions. After incubation at 4°C for up to 168 h, reductions in bacterial numbers at pH 7.5 (compared to untreated controls) ranged from 1.4 to 3.7 log cycles and was greatly dependent on the type of pad used. At pH 3.5 the effect was even better. No mesophiles were detected in nisin treated pads after 168 h at 4°C, compared to a count of $\log_{10}$ 6.4 CFU ml[-1] in control pads. In two of the pads, no *Salmonella* were detected at the end, compared to an average of $\log_{10}$ 3.1 CFU ml[-1] in controls.

C. Limitations

1. Nisin retention after processing: A certain proportion of nisin is lost after addition as a direct result of processing treatment, and thus nisin addition levels must be adjusted to compensate for this. During the melt process used in manufacturing processed cheese, typically 15-20% nisin will be lost but this very much depends on the degree of heat treatment, the period of time that the cheese is kept molten and the pH of the product. Better retention is achieved at acidic pH values. Retention also increases if hot processed cheese is not kept for prolonged periods at raised temperatures before packing and if it is cooled quickly after packing. Nisin retention becomes worse if a UHT process is involved, with up to 40% loss occurring. However, after such a treatment surviving spores will be few in number, and thus approximately half the nisin level may be needed. Moreover UHT treatments damage endospores, rendering them more nisin sensitive (Delves-Broughton et al., 1996).

2. Nisin retention during shelf life: Decline of nisin levels during storage is slower at lower temperatures, as is shown in *FIGURE 4* (Delves-Broughton, 1990). In processed cheese, nisin is incorporated to prevent outgrowth of bacterial endospores. This is a sporostatic effect, and thus the level of nisin in the cheese must be maintained at an inhibitory level throughout the shelf life of the product. Higher levels of nisin (12.5 - 15.5 µg nisin/g) are necessary for products stored at high ambient temperatures for long periods. Manufacturers can avoid the necessity of storing products including processed cheese at chill temperatures by adding nisin. Zapico et al., (1999) found homogenization of milk at a pressure of 200 bar resulted in a high loss of nisin activity, with a consequent reduction in anti-listerial effect.

D. Safety and tolerance

Nisin-producing lactococci occur naturally in raw milk and cheese. A world-wide survey of raw milk samples found 43% contained nisin-producing lactococci, and these organisms were found in milk from all the countries examined (Delves-Broughton, 1990). Inadvertently and apparently harmlessly, humans and animals probably have consumed this bacteriocin for centuries. Toxicological studies by many groups have now confirmed that nisin is not toxic at levels much higher than those used in food (Frazer et al., 1962; Hara et al., 1962; Fowler, 1973; Shillinger et al., 1969; Bogorditskaya et al., 1970; Shtenberg & Igant'ev, 1970). Digestive enzymes rapidly inactivate nisin and consequently it does not alter the bacterial microflora in the intestinal tract (Barber et al., 1952; Heinemann & Williams, 1966; Jarvis & Mahoney, 1969; Tiemann & Belitz, 1969). Ten minutes after consumption of chocolate milk containing nisin, the bacteriocin could not be detected in human saliva (Claypool et al., 1966). The LD_{50} value is about 7 g/kg body weight; similar to that of common salt. As the preparation tested contained 75% salt, the toxicity can be attributed to this component alone (Hara et al., 1962). No ill effects were observed in pigs and poultry from feeding experiments (Barber et al., 1952; Coates et al., 1951). There is no evidence of any cross resistance with antibiotics used in medicine (Szybalski, 1953; Carlson & Bauer, 1957; Hossack et al., 1983).

In 1969 the FAO/WHO Expert Committee decided that the available evidence indicated a level of 3.3×10^6 units/kg body weight had no adverse effect, and permitted an unconditional acceptable daily intake (ADI) to be set at 3.3×10^4 units/kg body weight. (FAO/WHO 1969). In 1988, the US Food and Drug Administration affirmed nisin as GRAS (generally recognized as safe) for use as a direct ingredient in human food (FDA 1988).

E. International regulations

A guide to current food legislation on the use of nisin was recently reviewed (Turtell & Delves-Broughton, 1998), but changes occur all the time. *TABLE 4* shows an overview of products and countries where nisin is permitted. Nisin currently has self-affirmation status in the USA for use in 'fresh' soups that are heat-treated and stored at chill temperatures. It must be strongly emphasized that current legal situations should always be checked with the appropriate authorities.

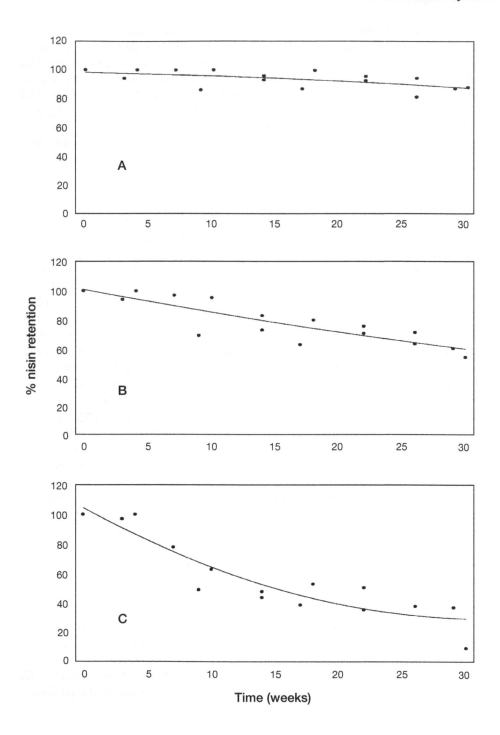

FIGURE 4. Retention of nisin in processed cheese spread at (a) 20˚C, (b) 25˚C and (c) 3˚C. Nisin addition level was 6.25 µg/g.

VI. EXISTING FOOD APPLICATIONS

A. Dairy products

1. Processed cheese and cheese spreads: Clostridium spp such as *Cl. butyricum*, *Cl. tyrobutyricum* and *Cl. sporogenes* are commonly associated with spoilage of these products (Meyer, 1973; Thomas, 1977), causing 'blowing' due to gas production, off-odors and liquefaction. Facultatively anaerobic *Bacillus* spp. may also cause problems. Nisin inhibits the outgrowth of the endospores of these species, which can survive the heat process of 85-105°C for 6-10 min that is used to melt and mix the cheese. Spores can be present in any of the raw ingredients of processed cheese (Meyer, 1973; Thomas, 1977; Zehren & Nusbaum 1992). Milk from herds fed on silage, for instance, is known to have a heavy spore load (Glaesser, 1989). For a desired shelf life of 6 months at 25°C, a nisin addition level of 2.5 µg nisin/g is required in a typical processed cheese product with a clostridial spore load of 10 spores/g. By comparison, 12.5 µg nisin/g would be required if the product contained 1000 spores/g. In general, twice as much nisin is required for each ten-fold increase in spore numbers.

The environmental conditions of processed cheese (particularly its low acidity, high moisture content and lack of oxygen) favor anaerobic growth. Moisture content varies according to the type of product. Approximate moisture levels are as follows: block cheese (44-46%), slices (48-50%), spreads (52-60%) and sauces or dips (56-65%) (Delves-Broughton, 1998). The main ingredients used in these products are raw cheese, butter, skim milk powder and whey powder. The preservative capability of nisin is affected by the type of emulsifying salt (phosphate or citrate) as well as flavor and color additives. Also of importance are the spore numbers, moisture and NaCl levels, the length and temperature of the heat process, and the length and realistic temperature of the storage period required (Delves-Broughton, 1998).

Suggested levels of nisin for effective preservation of processed cheese vary from 5 to 20 µg nisin/g (Delves-Broughton et al., 1996). Trials undertaken at Aplin & Barrett, found that 6.25 µg/g prevented spoilage of processed cheese products inoculated with a mixture of spores and then stored at 37°C. Samples not containing nisin showed rapid evidence of spoilage (Delves-Broughton et al., 1996). The samples had a range of moisture levels (40-60%) and onset of spoilage rates was significantly slower in samples with lower moisture content. Plocková et al., (1996) also found that nisin levels of 2.5 and 6.25 µg/g inhibited spore-forming species in processed cheese stored at 5 and 21°C for 3 months. Nisin (at 4 – 5.1 µg/g) has also been shown to reduce spoilage microflora (including aerobic and anaerobic sporeformers) in processed cheese analogues incorporating fat substitutes during a 4 months storage period at 5°C (Muir et al., 1999).

If processed cheese is not stored at the correct chill temperature there is a danger that *Cl. botulinum* can grow (Briozzo et al., 1983; Kosikowski, 1977). In processed cheese formulations, moisture, NaCl and phosphate levels as well as pH all combine to influence not only the growth rate of the pathogen, but also the protective ability of nisin (Tanaka et al., 1986). Higher levels of nisin are needed in conditions of high moisture and low levels of NaCl and phosphate. The level of nisin will also depend (as in every nisin application) on the numbers of contaminant endospores, the period and temperature of the shelf life (Hirsch & Grinsted, 1954; Scott & Taylor, 1981). An inoculation trial by Somers and

TABLE 4. Products and Countries in which nisin is approved for use

Product	Permitted level (μg/g food)	Country
Milk: pasteurized, flavoured, long life	no limit	Abu Dhabi, Bahrain, Dubai
Condensed milk products	12.5	Slovak Republic
Clotted cream	no limit	Cyprus, Gibraltar, Guyana, Hong Kong, Malta, Mauritius, Trinidad & Tobago
	10	EU
Cheese	no limit	Abu Dhabi, Australia, Costa Rica, Cyprus, Dubai, Gibraltar, Hong Kong, Malaysia, Malta, Mauritius, Papua New Guinea, Singapore, Trinidad & Tobago
	3	Czech Republic
	5	Bulgaria
	6.25	Taiwan
	12.5	Chile, EU, Slovak Republic, South Africa, Vietnam
	25	India
Cheese, excluding soft white cheese	no limit	Israel
Cheese, soft	25	India
Cheeses, very high, high medium and low moisture content	12.5	MERCOSUR[1]
Mascarpone	10	EU
Processed cheese	under review	Mexico
	no limit	Abu Dhabi, Bahrain
	2.5	Belgium
	12.5	Argentina, Brazil, Codex Alimentarius Commission, Colombia, Czech Republic, Egypt, EU, Jordan, Kuwait, Montenegro, Macedonia, New Zealand, Saudi Arabia, Serbia, South Africa, Venezuela.
	25	India
	100	Algeria, Thailand, Uruguay
	120	Poland [2]
	200	Russia
	4000	Philippines
Processed (pasteurized) cheese spread (and with vegetables, fruit, meats)	250	USA
Processed (pasteurized) cheese spread (and with vegetables, fruit, meats)	250	USA
Cheese preparations (spreadable)	12.5	Codex Alimentarius Commission
Cheese preparations	no limit	Zimbabwe
	12.5	Indonesia
Cheese products	no limit	Australia
Dairy products	no limit	Abu Dhabi, Bahrain, Dubai
	12.5	China
Puddings (semolina, tapioca, etc.)	3	EU, Czech Republic

Contd.

TABLE 4. (Contd.) Products and Countries in which nisin is approved for use

Product	Permitted level (μg/g food)	Country
Mayonnaise	12.5	Slovak Republic
Hermetically sealed food given botulinum process	no limit	Malaysia
Part prepared products	12.5	Slovak Republic
Canned food given botulinum process	no limit	Singapore, Trinidad & Tobago
Canned foods	no limit	Abu Dhabi, Bahrain, Cyprus, Dubai, Gibraltar, Guyana, Hong Kong, Malta, Mauritius
	5	China
	12.5	Slovak Republic
Canned tomatoes, paste, puree	no limit	Australia
Canned soups	no limit	Australia
Canned fruit, pH <4.5	no limit	Papua New Guinea
Tomato puree, canned tomato pulp, canned tomato paste, canned tomato juice, pH <4.5	no limit	Papua New Guinea
Coconut water	5000	India[3]
Plant protein foods	5	China
Breads	12.5	Slovak Republic
Crumpets, flapjacks	250	Australia
Beer	no limit	Australia, New Zealand
Ice for storing fresh fish	no limit	Bulgaria
Meat products	12.5	China
	200[4]	Brazil
Vegetables (raw, peeled, semi-preserved potatoes, cauliflower, green peas)	100	Russia
Vegetables: sterilized and pickled	12.5	Slovak Republic
Ready to eat meals	12.5	Slovak Republic
Branzcturi tapite	50	Romania
Requeijao	12.5	Brazil
Nisin as permitted preservative	not known	Bolivia, Ecuador, Qatar, Sri Lanka, Yugoslavia
	no limit	Peru

[1] Argentina, Brazil, Paraguay, Uruguay; [2] melted; [3] proposal for use, [4] as external treatment.
Adapted from Turtell & Delves-Broughton (1998).

Taylor (1987) demonstrated that 12.5 - 25 µg nisin/g prevented toxin production by *Cl. botulinum* strains types A and B, for 48 weeks in a cheese spread at pH with a moisture content of 52-54%, and 1.7% disodium phosphate. At least 12.5 µg nisin/g is needed to protect against *Cl. botulinum*, and other preservative hurdles should be effective and in place.

2. Dairy and other desserts: The quality of dairy desserts such as crème caramel and chocolate dairy desserts can be improved by reducing heat processing and using nisin to maintain the shelf life (Gregory et al., 1964; Heinemann et al., 1964; Gontzea et al., 1973; El-Sadek et al., 1976). Sterilization can be detrimental to the taste and appearance of these desserts. Nisin works well in these types of products that are usually homogeneous and pasteurized. Further examples include *kheer*, an Indian dessert made from milk, sugar and rice; nisin was reported to be a successful preservative in this product (Sukumar et al., 1976). Addition of 1.25 µg nisin/g to heat treated quarg desserts reportedly increased shelf-life by 5-7 days, depending on the storage temperature (Plocková et al., 1998).

Inclusion of rice, sago and semolina, common ingredients of canned milk puddings, increases the difficulty of preservation by heating alone. These puddings are made by adding dry cereal and sugar to empty cans, then filling with hot milk. This may be made up from powdered milk and may vary in fat content. The cans are sealed and heat-treated; thickening the contents and reducing heat transfer. As with other canned products, a heat treatment severe enough to kill all bacterial endospores will usually impair flavor. A nisin addition of 12.5 - 25 µg/g avoids the necessity for such high heat treatment whilst maintaining protection of the product from spoilage. Addition of 14 µg nisin/g to ice cream has been shown to have an anti-listerial effect during manufacture and subsequent storage at -18°C for 3 months (Dean & Zottola, 1996).

Although nisin inhibits many starter cultures, there is evidence that it could prolong the shelf life of yogurt. Over-acidity can be prevented by adding nisin after a suitable period of fermentation (Gupta & Prasad, 1988; 1989b; 1989c). Production of a good quality yogurt, containing 0.625 - 1.25 µg nisin/ml, was described by Bayoumi (1991). This had a shelf life at 10°C of 19 d, and 33 d at 6°C. The shelf life of *lassi*, an Indian yogurt drink stored at room temperature, was increased from < 7 days to 10 days by nisin addition, with little change in organoleptic quality (Kumar & Prasad, 1996). The shelf life of *nono*, a Nigerian fermented milk product, has also reportedly been extended by nisin (Olasupo et al., 1996).

3. Cottage cheese: Benkerroum and Sandine (1988) investigated nisin inhibition of *L. monocytogenes* ATCC 7644 in cottage cheese inoculated with 3.5 x 10^5 cells/g. Complete inhibition was achieved by 9.25 µg/ml nisin in cottage cheese at a pH level of ≥ 5.0 after 24 h, and within 24 h if the pH was between 3.5 to 4.5. No viable cells were found after 24 h at 37°C or 4°C. Ferreira and Lund (1996) also undertook *L. monocytogenes* challenge studies in cottage cheese, which was at pH 4.6-4.7. Numbers decreased by approximately 10-fold during storage at 20°C for 7 d; addition of 50 µg/g nisin increased this to a drop of 10^3 in 3 d.

4. *Soft white fresh cheese:* These cheeses are made without starter culture, normally by acidification or coagulation with rennet. They include ricotta, panir and Latin-American cheeses queso blanco and queso fresco. These products have minimal processing before packaging and are highly perishable. They have a short shelf life even at refrigerated temperatures, and at these temperatures *Listeria* spp. can grow. In 1985, an outbreak of more than 142 cases of listeriosis and 48 deaths was linked to consumption of contaminated cheese of this type (Linnan et al., 1988). Davies et al., (1997) showed that addition of 2.5 µg/L nisin effectively inhibited growth of a five strain cocktail of *L. monocytogenes* for ≥ 8 weeks. Cheese inoculated with the same strains contained unsafe levels of the organism after 1-2 weeks of incubation.

5. *Natural cheese:* The potential for the growth of the pathogens *L. monocytogenes* and *Staph. aureus* in soft cheeses has revived interest in the isolation or creation of nisin-resistant starter cultures. A nisin-producing starter culture suitable for producing Cheddar cheese was developed by combining a naturally occurring lactose-fermenting, nisin-producing, proteinase-positive *Lac. lactis* subsp *lactis* strain with a lactose-positive, nisin-positive, nisin-resistant, proteinase-positive transconjugant *Lac. lactis* subsp. *cremoris*. The culture was used to manufacture cheese containing 17.5 µg nisin/g (Roberts et al., 1992). Four of the nisin producing strains produced higher levels of nisin in pH controlled 10% reconstituted skim milk. Approximately 20% more nisin was detected in cheese prepared with a pH-controlled bulk starter compared to cheese made without pH control (Yezzi et al., 1993). Cheese made with such nisin-producing strains was later used to manufacture pasteurized processed cheese spreads at low and high moisture content (53 and 65%). The nisin content was 7.5 - 9.7 µg/g, and thermal processing did not inactivate. The cheese spread containing nisin had a longer shelf-life than controls not containing nisin, but there was no observed difference with regard to moisture content (Roberts & Zottola 1993).

Nisin-producing starter cultures were used in a challenge study by Ramsaran et al., (1998), in the manufacture of Feta and Camembert cheese. *L. monocytogenes* growth was inhibited in the Feta cheese, but not in the Camembert (possibly due to nisin breakdown by moulds).

6. *Milk and similar products:* Although nisin is not a permitted milk additive in countries with temperate climates such as the USA and UK, it is permitted in some countries in the Middle East (Anonymous 1989, *TABLE 4*). In warmer tropical and sub-tropical countries, facilities for cold storage, transport and distribution are often inadequate. Addition of low levels of nisin (at 0.25 - 1.25 µg/ml) to pasteurized milk stored at chilled, ambient and elevated temperatures can extend the shelf life up to more than double its normal period (Mahmoud et al., 1976; Fowler, 1979; 1981; Anonymous 1988; Delves-Broughton, 1990). Wirjantoro and Lewis (1996) monitored total microbial numbers, lactobacilli, calcium ion concentration, pH and total acidity in pasteurized milk containing nisin. 1 µg ml^{-1} nisin was added before a pasteurization of either 72°C for 15 s, 90°C for 15 s, or 115°C for 2 s. All of these combinations resulted in extension of the shelf life, and in the latter cases did not result in a 'cooked' flavor. Levels as low as 0.25 µg/ml added to milk prior to pasteurization extended the shelf life by at least 2 weeks. Use of nisin with pasteurization was suggested as particularly suitable for milk preservation in regions with poor refrigeration (Wirjantaro & Lewis, 1996).

UHT treatment (135-142°C for a few seconds) enables milk to be stored at ambient temperature for long periods, but the process often imparts a burnt flavor. Addition of nisin can prevent this by allowing heat treatments to be reduced without compromising the shelf life. 'Sterilized' milk may, contrary to its title, contain thermophilic heat-resistant spores (for discussion of *B. sporothermodurans* see below) whose growth can cause spoilage, especially in hot climates. Sterilized milk and flavored drinks such as chocolate milk are often kept at ambient temperature. Addition of flavorings contributes to the spoilage potential, since this usually increases the spore load. Nisin allows heat treatments of UHT milk to be reduced so that the flavor is not impaired and a good shelf life maintained. This has been shown for chocolate milk (Heinemann et al., 1964), whole buffalo milk (Shehata et al., 1976; 1977), and sterilized milk (Wajid & Kalra, 1976). Destruction of heat tolerant spores of *B. stearothermophilus* in a concentrated buffalo milk system was enhanced at 115-121°C by the addition of nisin; the maximum cidal effect was observed with 12.5 µg nisin/g (Rao & Mathur, 1996). Nisin has also been shown to increase the shelf life of canned evaporated milk (Gregory et al., 1964).

There have been many reports of the occurrence of *L. monocytogenes* in raw milk, its growth at refrigeration temperatures and subsequent survival and growth in soft cheese (examples: Rodriguez et al., 1985; Rosenow & Marth, 1987; Walker et al., 1990; Harvey & Gilmour, 1992; Sulzer & Busse, 1993). Doyle et al., (1987) showed that this organism survives pasteurization treatment of 71.7°C for 15 s. An investigation of *L. monocytogenes* by Maisnier-Patin et al., (1995) found that low levels of nisin in milk enhanced the lethal effect of a moderate heat treatment. It was suggested that this combined treatment would increase the safety margin of raw milk cheeses with no significant increase in processing costs and possible improvement in cheese quality. A challenge study using *L. innocua* undertaken by Zapico et al., (1999) found good recovery of nisin from milk and a corresponding effective anti-listerial activity.

Nisin has been shown to extend the shelf life of pasteurized cream. Low levels (0.625 - 2.5 µg/ml) were effective in the absence of post processing contamination (Philips et al., 1983).

B. Vegetable protein milks
Nisin works effectively in products such as soy milk and coconut milk. These products have usually been subjected to some kind of heat treatment, and may be sold in cans, packs or bottles. The use of titanium dioxide as a whitening agent should be avoided as this may degrade nisin and compromise its efficacy. Nisin prevents spoilage and can either extend the shelf life or allow a reduction in heat processing time.

C. Liquid egg
To ensure the complete elimination of *Salmonella*, in the UK for instance, liquid egg must be pasteurized for at least 2-5 min at a temperature of 64.4°C. Although this will kill most of the bacteria present, it will not destroy endospores. Vegetative cells of certain heat resistant species (Gram negative and Gram positive) may also survive. *B. cereus*, for example, is commonly found in liquid whole egg (Payne et al., 1979; Wood & Waites, 1988). Some surviving organisms may be capable of growth at 4 to 8°C, and the shelf life of this product is generally a maximum of 10-11 days at refrigeration temperatures. A nisin level of 5 mg/L was reported to significantly increase shelf life from 6-11 days to

17-20 days. Nisin also prevented the growth of *B. cereus* in the liquid egg (Delves-Broughton et al., 1992).

The temperature and duration of liquid egg pasteurization can be reduced with incorporation of nisin (Boziaris et al., 1998; Ponce et al., 1998). Nisin has also been reported to increase the effectiveness of heat treatment in killing *S. enteriditis* PT4 (Boziaris et al., 1998). Ponce et al., (1998) found that a combination of nisin and high hydrostatic pressure (450 Mpa) prevented the growth of *E. coli* and *L. innocua* in liquid egg stored at 4°C for 30 days.

D. Canned foods

A minimum heat treatment of $F_0=3$ is a mandatory requirement for the destruction of *Cl. botulinum* spores in canned foods of low acidity ($\leq$ pH 4.5). Heat resistant endospores of thermophilic bacterial species can survive this (and greater) heat treatment (Eyles & Richardson, 1988). Spoilage caused by species such as *B. stearothermophilus* and *Cl. thermosaccharolyticum* can be prevented by incorporation of nisin, which will help protect canned foods stored for prolonged periods at warm ambient temperatures. Nisin can also enable the heat treatment to be reduced to the minimum requirement without increasing the risk of spoilage. In high acid canned food (< pH 4.5) potential spoilage organisms are aciduric endospore-forming species (which are not pathogenic), such as *Cl. pasteurianum*, *B. macerans* and *B. coagulans*. The growth of these organisms will be controlled by nisin, and this has been shown in canned tomatoes (Vas 1963; Maslennikova et al., 1968) and tomato juice (Poretta et al., 1968).

Inoculation trials have shown excellent control of *B. stearothermophilus* in a variety of canned foods subjected to different processing. Examples include canned mushrooms processed at F=4.5 (Funan et al., 1990), okra at 120°C for 15 min (El-Samahy & Elias 1977), and peas at 118°C for 25 min and at F=12 (Gillespy, 1953; Nekhotenova, 1961; Vas et al., 1967). *Cl. sporogenes* control has been shown in canned asparagus at 115-117°C for 17 min (Hernandez et al., 1964) and potatoes at 100°C for 10 min (Maslennikova & Loshina 1968). In many of these studies 100% of the control cans not containing nisin became spoilt during the trial incubation period.

E. Fresh soups

Nisin is an effective preservative in fresh pasteurized 'home-made'-type soups, which are stored at refrigeration temperatures. This application has been achieved in the USA by a process of self-affirmation. Nisin successfully controls the spoilage microflora and extends the shelf life of these products.

F. Crumpets

Crumpets, a traditional British product, are also popular in Australia and New Zealand. They are made from flour batter mixed with yeast or a raising agent and are cooked on a hot plate. They are toasted or grilled before consumption. Crumpets have a moisture content of 48-54% and water activity of 0.95-0.97, and are not at acid pH (pH 6-8), characteristics that result in a short shelf life of up to 5 days at ambient temperature. Retailers would find it inconvenient to keep crumpets refrigerated.

Although not a problem in the cooler climes of the UK, there have been outbreaks of *B. cereus* food-poisoning associated with the growth of this organism in crumpets sold in Australia with its warmer climate (Murrell, 1978; Lee, 1988). *Bacillus* spores naturally present in the flour are not killed by the hot plate baking process. Although there is a high heat treatment at the bottom of the crumpet, the rest of the crumpet will receive a lower heat treatment that will not kill these bacterial endospores. During the shelf life, numbers of *B. cereus* may increase from undetectable levels to $>10^5$ CFU/g and toxin production will occur. Addition of at least 3.75 µg nisin/g (as 150 mg/kg Nisaplin) to the batter has been shown to reduce *B. cereus* numbers to approximately 10^3 cells/g after 5 days storage. In crumpets without nisin, counts of *B. cereus* reached numbers of $\geq 10^5$ cells/g in the same storage period (Jenson et al., 1994). In Australia, nisin is now permitted as a preservative in crumpets.

G. Alcoholic beverages – beer and wine

Alcoholic beverages such as beer and wine are susceptible to spoilage by lactic acid bacteria, in particular *Lactobacillus* and *Pediococcus* spp, which are major contaminants throughout the brewing process and prevalent in fermentations (Ogden 1986). The low pH and lack of fermentable sugar and oxygen in beer and wine prohibits the growth of most other bacteria, naturally selecting for lactic acid bacteria whose growth causes acidity, turbidity, slime and off-flavors. Nisin is an ideal preservative for these yeast-fermented beverages because it inhibits the unwanted lactic acid bacteria without affecting the growth or metabolic activity of the yeasts. Contamination problems in brewing are exacerbated by the practice of transferring the pitching yeast from one fermentation to the next. A survey of contaminant bacteria isolated from beer or wort found 92% of the Gram positive bacteria (mainly lactic acid bacteria) were inhibited by nisin (Ogden & Tubb 1985). Nisin can be used at several stages of the beer fermentation. A level of 25 µg nisin/ml proved a faster and more effective means of cleansing contaminated pitching yeast compared to acid washing with three different acids (Ogden 1987). At a lower concentration (0.625 - 2.5 µg nisin/ml) nisin could be added during the fermentation process; nisin activity survives kieslguhr filtration, fining and pasteurization (Ogden et al., 1988). An addition of 0.25 - 0.375 µg nisin/ml increases the shelf life of unpasteurized beer.

A similar application for nisin in the production of wine has been identified. In grape must, lactic acid bacteria, including the more resistant *Lb. casei*, are inhibited. In pilot scale experiments with fresh musts from several grape varieties, nisin added prior to fermentation did not cause any detectable change in the appearance or taste of the wine (Radler 1990a; 1990b). Nisin could be considered a suitable replacement for sulfur dioxide as a wine preservative; furthermore, nisin is effective above pH 4, unlike sulfur dioxide. Nisin-resistant *Ln. oenos* strains have been isolated that allow nisin use in wine for the suppression of unwanted malolactic acid fermentation by natural contaminant bacteria (Daeschel et al., 1991).

Distilled alcoholic beverages such as fruit brandy are produced by fermentation by yeasts and the naturally occurring lactic acid bacteria. If the latter are inhibited by nisin, more substrate is available for the yeast fermentation, which can increase the alcohol content in the distillate by over 10% (Hennings et al., 1986a). However, in a later study by Mawson and Costar (1993), this increase was not found.

H. Dressings and sauces

High concentrations of acetic acid are traditionally used in salad dressings and mayonnaise products to lower the pH sufficiently to inhibit spoilage by lactic acid bacteria. These emulsified semi-solid products are mainly composed of vegetable oil, acidulants, egg-yolk and starch (FDA, 1993). Water activity plays a part in preventing bacterial growth, but the low pH is the main preservative factor and current consumer demand is for less acidic products. The pH of commercial salad dressings is usually between pH 2.9 to 4.4 depending on the acetic acid concentration. Predominant spoilage bacteria are lactobacilli, particularly *Lb. fructivorans* (Kurtzman et al., 1971; Smittle & Flowers 1982). Muriana and Kanach (1995) undertook a microbiological analysis of nine commercial salad dressings and found bacterial numbers of 10^2 - 10^4 CFU/g but few yeasts or moulds. Nisin inhibited 27 of 30 bacterial isolates. 5 µg nisin/g inhibited the initial levels of an added lactobacillus inoculum in salad dressing (pH 4.2), and no significant growth occurred for a period of 90 days at 26˚C. In controls not containing nisin, both the added lactobacilli and the indigenous microflora increased in numbers from day 6, reaching a maximum at 30 days. Nisin (5 µg/g) was shown to inhibit the growth of spoilage lactobacilli in low fat salad dressings (pH 4.2 - 4.3), despite the inclusion of titanium dioxide (Yuan et al., 1998).

VII. POTENTIAL FOOD APPLICATIONS

A. New target organisms

In the 1980s a novel acid tolerant, endospore-forming bacterium was isolated from apple juice (Cerny et al., 1984). This organism, now called *Alicyclobacillus acidoterrestris*, (Wisototzkey et al., 1992), can grow in orange juice, grapefruit juice and apple juice to levels of 10^4 - 10^5 cfu ml⁻¹, producing detectable taint (Komitopoulou et al., 1999). Due to its ability to survive normal juice pasteurization conditions, this organism is a spoilage problem in pasteurized and heat-treated fruit juices stored at ambient temperature. Spores of this organism were very nisin sensitive: the MIC against spores was only 0.125 µg nisin/ml. Moreover the presence of nisin during heating decreased the D value by up to 40% (Komitopoulou et al., 1999).

Another novel *Bacillus* species, *B. sporothermodurans*, has recently been described (Pettersson et al., 1996). As its name suggests, this species produces highly heat-resistant endospores, which can survive sterilizing and ultra high temperature (UHT) treatments. UHT treated and sterilized milk are considered commercially sterile when the total count is ≤ 10 CFU per 0.1 ml (Anonymous, 1992). This species was first detected in UHT milk from southern Europe in 1985, where it caused the closure of a UHT milk operation. It was later detected in Germany in 1990, and has now been noted in several other European countries (Hammer et al., 1995) and is also believed to be prevalent in South America. The bacteria have been found in dried milk products (Pettersson et al., 1996). Endospores surviving the heat treatment can germinate and grow to a maximum of 10^5 per ml in milk stored at 30˚C for 5 days. (Duquet et al., 1987; Hammer et al., 1995; Pettersson et al., 1996). Spoilage, if detected, can be noticed by a slight pink color, off-flavors and structural differences (Klijn et al., 1997). Studies undertaken at our laboratories have found both vegetative cells and spores of *B. sporothermodurans* isolates from UHT milk were highly sensitive to nisin. An addition level of 1.25 - 2.5 µg nisin/ml would provide a wide margin of safety in the preservation of milk naturally contaminated with spores of this organism.

B. Meat products

Nisin has sometimes been described as demonstrating variable effectiveness in meat products, with reports of organisms that showed good nisin sensitivity *in vitro* apparently less sensitive in the presence of meat. For example, outgrowth of spores of a resistant *Cl. botulinum* strain was inhibited by 50 µg nisin/ml in trypticase peptone-yeast extract-glucose medium, but not by a higher nisin concentration (125 µg/ml) in cooked meat medium (Scott and Taylor 1981). In most studies investigating the activity of nisin in the presence of meat, there are many variables, all of which could influence the outcome of the antimicrobial effect. The nature of the meat system itself can affect activity, e.g. whether the meat is raw, cooked or cured. For example, Murray and Richard (1998) found nisin A to be more effective than pediocin AcH in inhibiting *L. innocua* on inoculated raw minced pork, evidently because nisin A was more stable to the activity of meat proteases than pediocin AcH. Other preservation hurdles are also influential, e.g. pH, NaCl content, vacuum packaging, modified atmosphere packaging, storage temperature, other ingredients (emulsifiers, flavorings, other preservatives). All these factors determine the nature of the spoilage microflora, and thus the target organisms for nisin activity (Bell & de Lacey, 1985; Calderon et al., 1985; Fowler & Gasson, 1991; Rayman et al., 1983; Taylor et al., 1984; Taylor & Somers, 1985). Poor activity has been attributed to nisin binding to meat proteins, uneven mixing, poor absorption into the meat matrix, interference of meat phospholipid with the nisin activity, or nisin heat lability in neutral pH conditions (de Vuyst & Vandamme, 1994; Henning et al., 1986b).

There have been reports of poor recovery of nisin from meats and associated products, thought to be due to its strong adsorption onto meat proteins, and this has given misleading and variable bioassay results. (Fowler, 1981; Tramer & Fowler, 1964). Bell and de Lacey (1986) showed that rate of recovery was not greatly affected by the curing salts, particle size or meat to extractant ratio, but they found that it was significantly affected by the fat content. An addition of 2.5 µg/g nisin to minced beef emulsions resulted in a nisin recovery ranging from 26% from samples with 3% fat content, to 76% nisin recovery from samples with 83% fat. However, a subsequent study by Davies et al., (1999) found that recovery of nisin from a simple pork meat emulsion and bologna-type sausage samples was not affected by the fat content. Good recovery of nisin was achieved at both the beginning and end of a storage period of 60 days at 8°C (and was reflected in control of lactic acid bacterial spoilage).

1. Raw meat: Nisin has been investigated as a surface treatment for beef and chicken carcass disinfection and on packaged raw meat. At refrigeration temperatures *L. monocytogenes* strains are able to grow on packaged raw and cooked meat in the presence of oxygen and under vacuum (Grau & Vanderline, 1990; Kaya & Schmidt, 1991; Hudson & Mott, 1993; Hudson et al., 1994; Barbosa et al., 1994). Treatment of packaged steak with a sorbate-nisin combination reduced counts of an inoculated cocktail of *L. monocytogenes* strains by 96.5% for vacuum-packed beef and 89.3% for carbon dioxide packaged beef. It also prevented growth of the organisms on vacuum-packaged beef stored for 4 weeks at 4°C (Avery & Buncic, 1997). Another suggested method for inhibiting bacterial growth on stored fresh meat is to incorporate nisin in plastic food-packaging films. Ming et al., (1997) successfully applied nisin as a powder in food packaging casings; dialysis showed good nisin retention.

Nisin protects against Gram positive organisms, but combinations of nisin with other preservatives can extend this protection to include Gram negative bacteria. In an investigation into the feasibility of replacing sulfur dioxide with organic acids and nisin in raw pork sausage, Scannell et al., (1997) found a combination of sodium lactate and nisin was effective in reducing the total bacterial count of the product. The combination of lactate and nisin protected against *Staph. aureus* and *Salmonella* species (common pathogenic contaminants). Jarvis and Burke (1976) found 10 µg nisin/g alone or in combination with 1000 ppm sorbic acid and 2.5% (wt/vol) polyphosphate delayed the spoilage of raw 'British fresh sausages'. In a trial of the antibacterial activity of biopreservatives in refrigerated vacuum-packaged raw beef, samples treated with 12.5µg nisin/g showed the greatest reduction of aerobic plate counts and LAB, and the bacterial population remained below 10^6 CFU g^{-1} for 8 weeks at 3°C. (Rozbeh et al., 1993).

Cutter and Siragusa (1994) tested a nisin spray as a means of decontaminating raw meat surfaces using a pilot scale model carcass washer. A significant reduction in the total bacterial count on both lean and adipose tissue was achieved. The incidence of *L. innocua* and the meat spoilage organism *Br. thermosphacta* was reduced using this treatment on vacuum-packed beef (Cutter & Siragusa, 1996b). Nisin may be less active in this situation due to its poor solubility and instability at the pH of raw meat (pH 5.8). Cut pieces of raw beef soaked in nisin solution (250 µg/ml) for 10 minutes at room temperature resulted in a rapid decrease of residual nisin activity after 4 days at room temperature, but the decrease was much slower at 5°C (Chung et al.,1989). The nisin loss was consistent with the observed antimicrobial activity. On meats artificially inoculated with *L. monocytogenes* or *Staph. aureus,* bacterial growth was delayed for at least 1 day at room temperature. If stored at 5°C, growth of *Listeria* was delayed for > 2 weeks and growth of *Staphylococcus* did not occur. A synergistic cidal effect was observed against *L. monocytogenes* due to a combined treatment of heat (52°C/ 3min) and 2.5 µg nisin/L, but a reduced effect was observed on turkey carcass skin (Mahadeo & Tahini, 1994). The stability of nisin on raw meat surfaces can be improved by immobilizing the preservative in a calcium alginate gel (Cutter & Siragusa, 1996a; Fang & Lin, 1995).

Nettles-Cutter and Siragusa (1996) inoculated post-rigor lean and fatty beef carcass tissues with *Br. thermosphacta* and tested the effect of 100 µg nisin/ml, calcium alginate edible gel and the same concentration of nisin immobilized in the same gel. The latter showed the greatest antimicrobial activity. Cutter and Siragusa (1998) have also experimented with incorporating nisin into a commercially available meat binding system. This system was applied to raw meat surfaces in order to study inhibition of *Br. thermosphacta* during extended refrigerated storage. All samples treated with nisin developed significantly lower numbers of the organism. A decrease in nisin activity was observed, particularly on vacuum packed samples but the reduction in bacterial numbers during the incubation period indicated nisin was still active. Kassaify (1998) investigated collagen and natural casings dipped in varying levels of nisin (as Nisaplin) and found good residual levels of nisin. The addition of nisin in a meat binding system has good potential in the preservation of meat products such as continental sausages and pastes, and meat used to prepare restructured meat products.

It has been suggested that carcass disinfection could be achieved by using nisin with a chelating agent (Cutter & Siragusa, 1995a), but promising results from this treatment achieved *in vitro* were not repeated when meat was treated by submerging in the same test solutions (Cutter & Siragusa, 1995b). An optimum treatment of nisin (100 µg/mg), a chelating agent (EDTA, citric acid) and a surfactant (Tween 20) was described by Shefet et al., (1995) for use on chicken drumsticks to inhibit *S. typhimurium* and extend the shelf life.

Combined treatments of high pressure and nisin at different pH values were applied to mechanically recovered raw poultry meat (Yuste et al., 1998). Due to its processing this type of meat has a high microbial load. A combination of 350 Mpa, 100 µg nisin/ml and 1% glucono-delta-lactone extended the shelf life during a 30 day chilled storage (2°C). In some samples a 3.4 - 5.4 log CFU/g reduction of mesophilic bacteria was achieved; psychotrophic bacteria were even more sensitive and were undetected after treatment.

2. Cooked meat products: Beuchat et al., (1997) examined the effects of nisin and temperature on survival, growth and enterotoxin production of psychrophilic *B. cereus* in beef gravy. These organisms are causing concern in ready-to-serve meats and meat products that are not subjected to sterilization treatments. Nisin was more effective on cells than spores, and at 8°C compared to 15°C. When cells were heat stressed they were more nisin sensitive. As little as 1 µg nisin/ml effectively controlled growth and enterotoxin production at 8°C.

Nisin has been suggested as a preservative in cooked meats, and as partial replacement for nitrite in cooked cured meat systems. A nisin/nitrite combination has been studied in bacon and chicken frankfurter emulsions (Calderon et al., 1985; Taylor et al., 1984; Taylor & Somers, 1985). Nitrite contributes to the flavor and color of cooked meat products but use of a natural betalainic pigment together with nisin was reported to reduce the formation of nitrosamines in ham whilst maintaining a satisfactory product appearance (Anaya et al., 1997). Outgrowth of *Cl. perfringens* spores was completely inhibited by a combination of 5 µg/g of nisin and 75 ppm of nitrite (Caserio et al., 1979b). Similarly, 1.875 - 2.5 µg/g nisin in combination with 40 ppm nitrite effectively inhibited outgrowth of *Cl. sporogenes* spores for 5 days in meat slurries incubated at 37°C (Rayman et al., 1981). This combination proved superior to 150 ppm nitrite alone. In a later study, 12.5 µg/g nisin in combination with 60 ppm nitrite failed to prevent outgrowth of *Cl. botulinum* spores in pork slurries adjusted to pH 5.8 and held at 25°C. If the pH of the slurry was reduced to pH 5.1, however, 7.5 µg/g nisin completely inhibited spore outgrowth (Rayman et al., 1983).

There is growing evidence that nisin is effective in preserving pasteurized, cured, vacuum-packaged, chilled-meat products where the major spoilage organisms are Gram positive lactic acid bacteria (LAB). The characteristics of these products (A_w, storage temperature, nitrite, curing salts, microaerophilic atmosphere) preferentially selects for these organisms. Predominant species are the facultative heterofermenters: *Lb. sake* and *Lb. curvatus*. *Lb. sake* constitute 68% and *Lb. curvatus* 16.9% of the lactic acid bacteria isolated from vacuum-packed processed meats (Holzapfel, 1998). Their growth results in surface slime production, off color, gas and odors. Other spoilage species include heterofermenting lactobacilli and leuconostocs, and *Br. thermosphacta* (Mukherji & Qvist 1981; Nielsen, 1983), (the latter when the packaging film is permeable). Investigations of lactic acid bacteria associated with spoilage of these products have found them sensitive to nisin (Collins-Thompson et al., 1985; Davies et al., 1999).

Continental sausages, patés and similar meat products are homogenized, and this allows nisin to be evenly mixed throughout the food. Incorporation of nisin into frankfurters prevented discoloration when the sausages were inoculated with LAB before pasteurization (Stankiewicz-Berger, 1969). 12.5 μg/g nisin was successfully used in combination with sorbic acid (0.125%) and monolaurin (0.25 - 0.5%) to control spoilage by *B. licheniformis* as well as the natural spoilage microflora in a pasteurized cured meat product (Bell & de Lacey, 1987). The fat: lean ratio in this product was 1:3. Control of *Cl. perfringens* was achieved in Italian wirstel sausage by incorporation of 5 μg/g nisin (Caserio et al., 1979a). Outgrowth and growth of inoculated spores of *B. licheniformis* was controlled by 12.5 μg/g nisin in cooked luncheon meat at 20°C for 10 days (Bell & de Lacy, 1986). *L. monocytogenes* was inhibited in liver paté by a combination of nisin, lysozyme and citrate (Ter Steeg, 1993) and also in cooked pork tenderloins by a surface treatment of 250 μg/ml nisin in combination with modified atmosphere (Fang & Lin 1994).

Cooked cured meat products differ considerably in their formulations. There can be wide variation in fat content, particle size, salt content, phosphate type, nitrite content, emulsifier, etc. The study by Davies et al., (1999) found nisin to be more effective in lean meat, and also that the type of phosphate used as emulsifier affected the nisin efficacy in inhibiting lactic acid bacteria. In this study, the optimum sausage formulation for nisin preservation of vacuum-packed bologna-type sausages incorporated 25% fat and diphosphate as an emulsifier. Nisin levels of 6.25 - 25 μg/g effectively inhibited the growth of a mixed LAB inoculum, and extended the shelf life of vacuum-packed bologna-type sausages stored at 8°C.

A different approach to nisin preservation of vacuum-packed frankfurters was reported by Wang et al., (1986). A nisin-producing strain was inoculated at >10⁹ CFU/g into raw beef which was stored for 5-7 days and then made into frankfurters. Nisin was produced *in situ* and this extended the shelf life of the product.

Exposure to high temperatures with microwave cooking may result in spore germination. Microwave heating of inoculated chicken samples was reported to stimulate germination and outgrowth *of Cl. perfringens* spores when the maximum internal meat temperature was 49- 84°C (Craven & Lillard, 1974). Hague et al., (1997) investigated the combination of microwave treatment and nisin on precooked beef inoculated with *Cl. sporogenes* spores, vacuum packaged and stored at 4 or 10°C for 21 or 70 d. Nisin alone or in combination with microwave treatment significantly reduced the viable vegetative cell count.

C. Fish products

In Bulgaria nisin is permitted (with no limit) in ice for storing fresh fish The use of nisin as a preservative of fish and fish products has not been studied extensively because spoilage is usually due to Gram negative bacteria (Gram & Huss, 1996). In a study by Zaki et al., (1976) bolti fish (*Tilapia nilotica*) were eviscerated, and soaked in brine solutions containing nisin and then either dehydrated or dried in the sun. After storage at room temperature for 3 months the total microbial load showed only a slight increase. In a similar study, El-Bedawey et al., (1985a; 1985b) found the shelf life of bolti fish could be extended by 9 days after dipping the gutted fish into a nisin solution. This allowed time for the fish to be distributed for retail.

An application has been identified for the use of nisin to eliminate or inhibit the psychotrophic pathogen *L. monocytogenes* in fresh and lightly preserved seafood products. Current preservation systems will not eliminate this problem. Cold smoked salmon, for example, is usually vacuum packed with a salt content less than 6% and a pH of ≥ 5.0. It is not usually heated before being consumed and *Listeria* can grow in this type of product. (Rørvik et al., 1991; Ben Embarek & Huss, 1992; Eklund et al., 1995). This is a potentially serious problem as these luxury products are often sent to customers by postal delivery, during which time they are at ambient temperature.

A study by Nilsson et al., (1997) reported that nisin in combination with a carbon dioxide atmosphere showed a pronounced anti-listerial effect *in vitro* as well as on cold-smoked salmon. Nisin produced *in situ* by inoculated *Lac. lactis* delayed the growth of *L. monocytogenes* in vacuum packed cold-smoked salmon stored at 10°C (Wessels & Huss, 1996). A combination of 2.5 g/kg lactic acid, 30 g/kg nisin whey permeate and 20 g/kg sodium chloride inhibited the total aerobic counts of minced rainbow trout (Nykänen et al., 1998a). The nisin whey did not affect the sensory quality of the fish but the lactic acid adversely affected the flavor, astringency and sourness.

Listeria has been isolated in many types of seafood (Ben Embarek, 1994; Jemmi, 1993; Jørgensen & Huss, 1998; Rørvik et al., 1995; Weagant et al., 1988), and moreover, seafoods have been implicated as the source of sporadic cases of listeriosis (Facinelli et al., 1989; Ericsson et al., 1997). In 1989 the U.S Food and Drug Administration set a zero tolerance level for this pathogen in ready-to-eat seafoods (Crawford, 1989; Klima & Montville, 1995). This has caused some problems in the industry and resulted in at least one product recall (cans of Canadian frozen shucked lobster). The use of nisin has been suggested to overcome this problem. For example, a combined treatment of nisin and moderate heat was shown to increase the killing of *L. monocytogenes* in cans of 'cold-pack' (frozen) lobster (Budu-Amoaka et al., 1999). In a further study, the combined effect of nisin and glycine caused a 2 and 3 log reduction in bacterial counts after incubation at 20 - 22°C for 0.5 and 1 h respectively. This combined treatment would enable a reduction of the heat processing, and thus improve the quality of the lobster meat in the cans (Budu-Amoaka et al., unpublished). Reduction of *L. monocytogenes* has also been achieved by washing crab meat in 2.5 - 5 μg nisin/ml (Degnan et al., 1994).

Concern has also been expressed about the potential of *Cl. botulinum* growth in chilled fish products that are either vacuum packed or in modified atmosphere. There is a fear that *Cl. botulinum* will be able to grow and produce toxin undetected in such products, because there would be no evidence that the product is spoiled. The normal indicator of spoilage, growth of aerobic microorganisms, would not occur due to the lack of oxygen. It has been reported that spraying with nisin may protect from this pathogen. A delay in toxin production on fish fillets (cod, herring, and smoked mackerel) inoculated with *Cl. botulinum type* E spores and stored at 10°C and 26°C, was achieved by spraying with nisin before packaging in 100% CO_2 (Taylor et al., 1990).

Addition of 12.5 μg/g nisin has been reported to extend the shelf life of caviar (Ushakova, 1972).

D. Miscellaneous potential applications

There is a potential use for nisin in the preservation of re-hydrated infant formulae to prevent the growth of the pathogen *B. cereus*. In inoculation trials using a 3-strain cocktail, growth of the pathogen occurred in three different re-hydrated infant formulas at 25°C but not at refrigeration temperatures. If the formula was kept at room temperature and then refrigerated, however, outgrowth of spores took place. Nisin was found to be the best bacteriocin active against *B. cereus in vitro* and at a concentration of 25 µg/ml inhibited 10^3 spores/ml in formula at 8, 12 and 25°C for up to 10 days (Wong & Luchansky 1999).

Vegetarian foods such as tofu show good potential for preservation with nisin because they are often pasteurized or heat-treated, and subsequent spoilage would largely be due to Gram-positive spore-formers. Fang et al., (1997) treated samples of vegetarian food made from soybeans by immersing in nisin and potassium sorbate solutions. The samples were inoculated with *Staph. aureus* or *B. cereus* spores before vacuum packing and storing at 4 or 30°C. A significant decrease in bacterial numbers was observed compared to untreated controls, achieving an extension of the shelf life.

Sikhae is a traditional rice beverage (also known as Danssol or Gamjoo) that has a strong malt flavor because of the amylolytic digestion of rice starch by enzymes in malt. It is now produced commercially in cans or retort pouches, which are processed at 121°C for 15 minutes. Spoilage is often due to *Bacillus* spp. A milder thermal process (110°C for 15 min) in combination with nisin resulted in a better quality product without loss of microbiological quality, estimated by storage at 30°C for 6 days or at ambient temperature for 3 months (Yoo et al., 1994).

The combination of EDTA and nisin has been suggested in vegetable or salad washing solutions. Bacterial soft rot is the leading biological cause of post-harvest loss of fruit and vegetable. It is caused by a diverse and mainly Gram negative range of bacteria. Combinations of EDTA and nisin were found synergistic in inhibiting the growth of 6 pectolytic soft-rotting bacteria in 80% trypticase soy broth. Unfortunately, *in vivo* experiments using freshly cut carrots did not support these observations (Wells et al., 1998).

Spoilage of fermented vegetable products can result from secondary fermentation by wild type contaminant bacteria. There have been several studies examining the use of nisin-resistant lactic acid bacteria starter cultures in the production of products such as sauerkraut, US fermented cabbage and kimchi (Harris et al., 1992; Breidt et al., 1993; Oh et al., 1994).

Liu (1997) investigated the efficacy of nisin against *L. monocytogenes* in roasted potatoes. Nisin decreased counts by $\log_{10}6.1$ after 21 days incubation at 5°C, and growth was delayed 48 h at 24°C.

VIII. BIOTECHNOLOGY

A. Fermentation

Fermentation is normally carried out as a batch (rather than continuous) process using *Lac. lactis* subsp. *lactis*. Optimum conditions for nisin production may be very strain and medium specific, but are generally maintained by holding the temperature at 28 - 37°C (Yoo et al., 1989; Matsusaki et al., 1996b) and adding alkali to keep the pH between pH 5.5 - 7 (Yoo et al., 1989; Meghrous et al., 1992; de Vuyst & Vandamme 1994; Liu, 1997). Ammonium or sodium hydroxide is generally used to control pH, which may be one of the most important factors influencing nisin production (Van't Hull, 1997; Matsusaki et al., 1996b). In the early days of interest in large-scale nisin production, Hirsch (1951) reported an optimum of pH 6.0 in a buffered system. De Vuyst et al., (1988) considered pH 6.8 the optimum level for production and used this in later studies (de Vuyst & Vandamme, 1992). With the producer strain *Lac. lactis* IFO 12007, Yoo et al.,

(1989) found that a sharp peak of increased nisin production occurred between pH 6.5 to pH 7.0, with a rapid decrease in production noted outside this optimum range. In the same study, the optimum temperature was 37°C. Matsusaki et al (1996b) reported a dramatic increase in nisin Z production between pH 5 - 5.5 in flasks, and the optimum temperature was 30°C. In continuous culture, maximum nisin activity levels were observed at pH 5.5 - 6.0 (Meghrous et al., 1992). Liu (1997), (using an immobilized cell packed-bed bioreactor) found optimum conditions for production by *Lac. lactis* ATCC 11454, were 31°C at pH 5.5.

The culture media is usually milk based, and growth promoters may be added to enhance production. Production of the commercial extract Nisaplin, for example, involves fermentation of a non-fat milk medium, followed by extraction and concentration. Complex media used include the following: stillage-based medium (a by-product of corn fermentation) (Van't Hull & Gibbons, 1997), natural rubber serum powder (Chinachoti et al., 1997c), whey enriched with potato starch syrup (Szadowska et al., 1974), potato juice (Kudryashov et al., 1995), whey permeate supplemented with casein hydrolyzate (Liu et al., 1997), and whey permeate supplemented with yeast extract and Tween 80 (Amiali et al., 1998; Goulhen et al., 1999).

Glucose, lactose and sucrose are frequently reported as carbon sources for nisin production. De Vuyst and Vandamme (1992) proposed that both cell growth and nisin biosynthesis was carbon-source regulated. Chandrapati and O'Sullivan (1998) assessed several nutritional parameters for their influence on nisin production using a modified rapid plate assay procedure. Of those tested, glucose was found to be the optimal carbon source, with glycerol having the greatest suppressive effect. Based on sugar composition, Chinachoti et al., (1998) found the yield of nisin Z from *Lac. lactis* subsp. *lactis* IO-1 was about 20% higher with xylose compared to glucose. Optimal conditions for nisin Z production were 4% xylose at pH 6.0 at 37°C. Addition of 0.1 M $CaCl_2$ increased production of nisin Z, without any increase in cell growth, acid production or xylose consumption (Matsusaki et al., 1996b; Chinachoti et al., 1998). The actual role of calcium ions in nisin production is still debated. In contrast, Yoo et al., (1989) had previously found xylose to be the poorest substrate for nisin production and glucose the best carbon source tested using *L. lactis* IFO 12007. Chen et al., (1995) found that soluble starch gave better nisin yields compared to a range of the more frequently used sugars.

The influence of phosphate and nitrogen source on nisin Z production by *Lac. lactis* subsp *lactis* NIZO 22186 was investigated by de Vuyst and Vandamme (1992; 1993). Potassium dihydrogen phosphate was found to be the best phosphate source, although this would also have created conditions of favorable pH. The best source of organic nitrogen was found to be cotton-seed mill. Baranova and Egorov (1967) found no noticeable difference of nisin production in media containing autolyzed yeast, casein hydrolyzate or peptone as principal nitrogen source. Using synthetic media (Kozak & Dobrzanski, 1977) found addition of either serine, proline, cysteine or cystine increased nisin production; addition of all four achieved yields similar to that using complex media such as skim milk. In later studies, serine, theonine, and cysteine were found to highly stimulate nisin production and, as expected, sulfur was essential for nisin production in a synthetic medium. Moreover cysteine was the best sulfur source tested, with a narrow optimum concentration of 0.03% (de Vuyst & Vandamme, 1991b; de Vuyst, 1995).

Chen et al., (1995) noted a stimulatory effect of Tween 80 (0.1%) that caused a 46% increase in nisin production by *Lac. lactis* subsp *lactis* CCRC 14016. Huot et al., (1996) also observed fourfold increase of another bacteriocin (J16) from *Lac. lactis* subsp. *cremoris* due to addition of 1% Tween 80. The non-ionic surfactant reduced the level of bound bacteriocin, but the increase in overall nisin yield was not fully explained.

Several groups have studied nisin production under conditions of continuous culture (Oberman & Jakubowska, 1969; Krasnikova et al., 1979; Yoo et al., 1989; Meghrous et al., 1992). The latter group reported that nisin biosynthesis was greatly influenced by the specific growth rate, and highest nisin activity was obtained at a specific growth rate of 0.25/h. This was in accordance with the observation of de Vuyst et al., (1990) that higher nisin yields could be achieved with lower specific growth rates. Taniguchi et al., (1994) described continuous production of nisin in a bioreactor coupled to a micro-filtration module. In this system, nisin productivity was approximately four times higher in continuous culture compared to that obtained in batch culture.

Kim et al., (1997), using a construct of *Lac. lactis* subsp. *lactis* C2SmPrt⁻ grown under non-nutrient limiting conditions, showed that high concentrations of nisin inhibited nisin production and observed that this phenomenon was host specific. This was further evidence that end-product inhibition could be occurring (de Vuyst, 1994). A system to partially overcome this limitation was proposed by Kim (1997), involving the use of a two-phase batch culture. Chinachoti et al., (1997a) described a system whereby the nisin Z fermentation process was integrated with a micro-filtration cartridge and a reverse-phase adsorbent system. Significant increases in nisin Z productivity were achieved compared to a system with no adsorbent. In a later study, Chinachoti et al., (1997b) investigated a number of materials to immobilize *Lac. lactis* cells by both entrapment and adsorption, as a means of improving productivity of nisin Z. Liu (1997) used cells of *Lac. lactis* immobilized by natural attachment to fiber surfaces in a packed-bed bioreactor. Various workers have immobilized *Lac. lactis* in calcium alginate beads and succeeded in demonstrating reasonable levels of nisin production (Zezza et al., 1993; Pasinin et al., 1995; Wan et al., 1995).

B. Recovery and purification

Mattick and Hirsch (1947) first purified nisin from acidified culture fluids, followed by concentration in a chloroform gel, dissolving in acid alcohol and precipitation at neutral pH. At pH 6.0, nisin is essentially associated with cell membranes, with significant release at values < pH 3 (Yang et al., 1992). Many methods of concentration and extraction (either from culture fluids or the cells of the producer organism) have since been used. They include fractional precipitation with salt and subsequent crystallization using ethanol, solvent-extraction, acid-extraction, frothing, gel filtration and ion-exchange chromatography (Berridge, 1947; 1949; Hirsch, 1951; Falconer, 1952, Cheeseman & Berridge, 1957; Hawley & Hall, 1960; Hurst, 1966; Bailey & Hurst, 1971; Lipinska et al., 1973; Wilimowska-Pelc et al., 1976; Lee & Kim, 1985). The commercial product Nisaplin is spray dried and then reduced to a fine particle size that is standardized with NaCl.

Suarez et al., (1997) described a rapid single step purification of nisin A using anti-nisin A specific monoclonal antibodies. Antibodies were bound to activated Sepharose. Nisin was purified in a single step by immunoaffinity chromatography using this immunoadsorption matrix.

TABLE 5. Specifications for the commercial nisin concentrate Nisaplin

General specifications	
Colour	Off white
Nisin content	Minimum of 900 IU nisin/mg (WHO)
Moisture	Not more than 3%
Shelf life	2 years from date of packing if stored below 25°C in dry conditions away from direct sunlight.
Microbiological specifications	
Total viable count	$<10^2$
Yeasts and moulds	$<10^2$
Coliforms	$<10^2$
Salmonellae	Absent in 10 g
Coagulase-positive Staphylococci	Absent in 10 g
Heavy metals	
Arsenic	Not more than 1 mg kg^{-1}
Lead	Not more than 2 mg kg^{-1}
Copper	Not more than 50 mg kg^{-1}
Zinc	Not more than 50 mg kg^{-1}
Mercury	Not more than 1 mg kg^{-1}
Heavy metals as lead	Not more than 10 mg kg^{-1}

Adapted from Delves-Broughton and Friis (1998).

Matsusaki et al., (1996a) described a purification strategy based on acid treatment, ammonium sulfate precipitation, cation-exchange chromatography and reverse phase HPLC. More recently Janes et al., (1998) proposed the use of rice hull ash or silicic acid as adsorbents for the partial purification of nisin.

C. Specifications for nisin preparation

The FAO/WHO Joint Expert Committee on Food Additives (1969) set specifications for the purity and identity of nisin. They ruled that pure nisin was too potent for use in food, but any commercial concentrates must contain at least 2.25% active nisin. The commercial preparation of nisin, Nisaplin, contains 2.5% nisin A. The preparation is made to a specific potency of 1 x 10^6 IU/g, and also contains added salts and milk solids derived from the fermentation of a modified milk medium (see *TABLE 5)*.

The specifications for commercial nisin concentrates such as Nisaplin and Chrisin are very similar. Those for Nisaplin are shown in *TABLE 5* (Delves-Broughton & Friis, 1998).

D. Protein engineering and gene technology

Due to public concern, manufacturers of commercial nisin consider it advisable not to use genetically modified organisms to produce nisin. The current level of understanding of the genetic control of nisin production, however, has made it possible to develop protein engineering strategies aimed at altering and possibly improving the activity of nisin. Protein engineering enables modified residues to be introduced or removed, or the hinge region of the molecule altered, in order to study their role in either biosynthesis or antimicrobial activity. Two different strategies have been employed: complementation of a deficiency in the structural gene of the host strain or substitution of the wild-type gene by a mutated copy using gene replacement technology. Nisin–producing strains carrying plasmid-encoded variant *nis-I* genes have been developed by Kuipers et al., (1992). Dodd et al., (1996) developed a lactococcal expression system that allowed the exclusive production of novel nisins encoded by mutated pre-nisin (*nisA* genes). This lead to the production and characterization of variant *nisA* genes containing various substitutions.

Investigations have also been aimed at altering or transferring the genes coding for nisin production in order to improve yield or reduce manufacturing costs. Alteration of the genes coding for nisin has been attempted by generating mutants using radiation treatment (Cziszar & Pulay, 1956; Kalra et al., 1973; Lipinska, 1977). Kuipers et al., (1991a; 1991b; 1992) made a range of nisin mutants using site-directed mutagenesis, purified by reverse phase HPLC. In most cases, reduced or similar activity was observed, although one strain produced nisin Z with increased activity specifically against *M. flavus*.

Conjugation and transformation techniques have been used to improve strains and nisin production (Gasson, 1984; Tsai & Sandine, 1987; Microlife-Technic, 1986). The transposon coding for nisin synthesis has been transferred into other bacteria (Rauch & de Vos, 1992), showing it could be possible to transfer nisin resistance to starter cultures in products preserved using nisin. Kleerebezem et al., (1997) introduced two plasmids that allowed nisin-inducible gene expression in *Lac. lactis*, *Ln. lactis* and *Lb. helveticus*. The beta-glucuronidase activity (used as a reporter) remained below the detection limits under non-inducible conditions and could be increased by adding sub-inhibitory amounts of nisin to the growth medium. The authors suggested this system could be used in other lactic acid bacteria, and could enable the overproduction of nisin during fermentation. Stoyanova and Egorov (1998) derived efficient nisin producers by fusing the protoplasts of two strains with low nisin-synthesizing abilities. The authors reported that this resulted in strains that were 10-14 times more efficient at producing nisin than the parent strains.

IX. SUMMARY

The polypeptide nisin, a bacteriocin produced by *Lac. lactis* subsp. *lactis*, exhibits antimicrobial activity against a wide range of Gram positive bacteria associated with food, and is particularly effective against heat-resistant bacterial spores. It is not active against Gram-negative bacteria, yeasts and molds because it is unable to penetrate these cells and gain access to the cytoplasmic membrane, its site of action. The unusual biosynthesis of nisin is by post translational modification of serine and threonine to form dehydro amino acids that react with cysteine to from thioether lanthionine rings, characteristic of this class of bacteriocin.

Toxicological studies have found nisin safe and it is now allowed as a food preservative in over 50 countries worldwide. The major use of nisin is in processed cheese, pasteurized dairy products and canned vegetables. In this type of product heat processing eliminates all spoilage microflora except for the nisin-sensitive Gram-positive spore forming bacteria such as *Bacillus* and *Clostridium*. Nisin is also used effectively in combination with heat in crumpets, liquid egg, and fresh soups. Spoilage by lactic acid bacteria can be controlled by nisin in products such as wine and beer, pasteurized vacuum packaged continental sausages, ricotta cheese and acidic foods such as salad dressings, sauces. As part of a hurdle system of preservation nisin can also be used to prevent the growth of pathogens such as *B. cereus*, *Cl. botulinum* and *L. monocytogenes*. Current interest is focussed on the combination of nisin with novel non-thermal preservation processes such as ultrahigh pressure and electroporation.

X. REFERENCES

1. Abee, T. 1995. Pore-forming bacteriocins of Gram-positive bacteria and self-protection mechanisms of producer organisms. *FEMS Microbiol. Letts.* 129: 1-10.
2. Abee, T., and Delves-Broughton, J. (in press). Bacteriocins - nisin. In *Food Preservatives.*
3. Abee, T., Rombouts, F. M., Hugenholtz, J., Guibard, G., and Letellier, L. 1994. Mode of action of nisin Z against *Listeria monocytogenes* Scott A grown at high and low temperatures. *Appl. Environ. Microbiol.* 60: 1962-1968.
4. Albayrak, N., and Yousef, A. E. 1997. A spectrophotmetric assay to determine bacteriocin activity. In *Abstracts of Papers Presented at IFT conference at Orlando, Florida, USA.*
5. Alifax, R., and Chevalier, R. 1962. Etude de la nisinase produite par *Streptococcus thermophilus. J. Dairy Res.* 29: 233-240.
6. Amiali, M. N., Lacroix, C., and Simard, R. E. 1998. High nisin Z production by *Lactococcus lactis* subsp. *lactis* biovar. *diacetlyactis* UL719 in whey permeate with aeration. *World J. Microbiol. Biotechnol.* 14: 887-894.
7. Anaya, M. A., Aviles, M., Castano, E., Alvarez, J., and Gonzalez de Mejia, E. 1997. Substitution of nitrites with natural additives in cooked ham. In *Abstracts of Papers Presented at IFT conference at Orlando, Florida, USA.*
8. Anonymous. 1988. Preservation of milk and milk products with Nisaplin. *Tech. Info. Leaflet No.* 11/88. Aplin & Barrett Ltd.
9. Anonymous. 1989. International acceptance of nisin as a food preservative. *Tech. Info. Leaflet No.* 4/89/11. Aplin & Barrett Ltd.
10. Anonymous. 1992. Directive 92/46. Council of the European Communities of 16th June. Health Rules for the production and trade of raw milk, heat treated milk, and products based on milk. *Official Journal of the European Community No.* L268, pp 1-32.
11. Araya, T., Ishibashi, N., and Shimamura, S. 1992. Genetic evidence that *Lactococcus lactis* JCM7638 produces a mutated form of nisin. *J. Gen. Appl. Microbiol.* 38: 271-278.
12. Aune, T. M., and Thomas, E. L. 1977. Accumulation of hypothiocyanate ion during peroxidase-catalysed oxidation of thiocyanate ion. *Eur. J. Biochem.* 80: 209-214.
13. Avery, S. M., and Buncic, S. 1997. Antilisterial effects of a sorbate-nisin combination *in vitro* and on packaged beef at refrigeration temperature. *J. Food Prot.* 60: 1075-1080.
14. Bailey, F. J., and Hurst, A. 1971. Preparation of a highly active form of nisin from *Streptococcus lactis. Can. J. Microbiol.* 17: 61-67.
15. Baranova, I. P., and Egorov, N. S. 1967. Influence of different nitrogen compounds on the growth of *Streptococcus lactis* and nisin formation. *Mikrobiologiya* 36: 804-808.
16. Barber, M., Elliot, G. J., Bordoli, R. S., Green, B. N., and Bycroft, B. W. 1988. Confirmation of the structure of nisin and its major degradation product by FAMS-MS and FAMS-MS/MS. *Experientia* 44: 266-270.
17. Barber, R. S., Braude, R., and Hirsch, A. 1952. Growth of pigs given skim milk soured with nisin-producing streptococci. *Nature* 169: 200.
18. Barbosa, W. B., Cabedo, L., Wederquist, H. J., Sofos, J. N., and Schmidt, G. R. 1994. Growth variation among species and strains of *Listeria* in culture broth. *J. Food Prot.* 57: 331-333, 336.
19. Bayoumi, S. 1991. Nisin and yoghurt manufacture. *Chem. Mikrobiol. Technol. Lebens.* 13:65-69.
20. Bell, R. G., and de Lacy, K. M. 1985. The effect of nisin-sodium chloride interactions on the outgrowth of *Bacillus licheniformis* spores. *J. Appl. Bacteriol.* 59: 127-132.
21. Bell, R. G., and de Lacy, K. M. 1986. Factors influencing the determination of nisin in meat products. *J. Food Technol.* 21: 1-7.
22. Bell, R. G., and de Lacy, K. M. 1987. The efficacy of nisin, sorbic acid and monlaurin as preservatives in pasteurized cured meat products. *Food Microbiol.* 4: 277-283.
23. Ben Embarek, P. K. 1994. Presence, detection and growth of *Listeria monocytogenes* in seafoods: a review. *Int. J. Food Microbiol.* 23: 17-34.
24. Ben Embarek, P. K., and Huss, H. H. 1992. Growth of *Listeria monocytogenes* in lightly preserved fish products. In *Quality Assurance in the Fish Industry. Proc. Int. Conf.,* 26-30th August 1991, Copenhagen, Denmark. eds. H. H. Huss, M. Jakobsnen, and J. Liston. pp. 293-303. Amsterdam: Elsevier.
25. Benkerroum, N., and Sandine, W. E. 1988. Inhibitory action of nisin against *Listeria monocytogenes. J. Dairy Sci.* 71: 3237-3245.

26. Bennik, M. H. J., Verheul, A., Abee, T., Naaktgeborne-Stoffels, Gorris, L. G. M., and Smid, E. 1997. Interactions of nisin and pediocin PA-1 with closely related lactic acid bacteria that manifest over 100-fold differences in bacteriocin sensitivity. *Appl. Environ. Microbiol.* 63: 3628-3636.

27. Benz, R., Jung, G., and Sahl, H. –G. 1991. Mechanism of channel formation by lantibiotics in black lipid membranes. In *Nisin and Novel Lantibiotics*. eds. G. Jung and H. –G. Sahl, pp 359-372. Leiden: Escom.

28. Berridge, N. J. 1947. Further purification of nisin. *Lancet* 2: 7-8.

29. Berridge, N. J. 1949. Preparation of the antibiotic nisin. *Biochem J.* 45: 486-493.

30. Berridge, N. J., Newton, G. G. G., and Abraham, E. P. 1952. Purification and nature of the antibiotic nisin. *Biochem. J.* 52: 529-535.

31. Beuchat, L. R., Clavero, M. R. S., and Jaquette, C. B. 1997. Effects of nisin and temperature on survival, growth, and enterotoxin production characteristics of psychotrophic *Bacillus cereus* in beef gravy. *Appl. Environ. Microbiol.* 63: 1953-1958.

32. Bierbaum, G., and Sahl, H. –G. 1985. Induction of autolysis of *Staphylococci* by the basic peptide antibiotics Pep5 and nisin and their influence on the activity of autolytic enzymes. *Arch. Microbiol. Lett.* 58: 223-228.

33. Bierbaum, G., and Sahl, H. –G. 1987. Autolytic system of *Staphylococcus simulans* 22: influence of cationic peptides on activity of N-acetylmuramoyl-L-alanine amidase. *J. Bact.* 169: 5452-5458.

34. Bierbaum, G., and Sahl, H. –G. 1988. Influence of cationic peptides on the activity of the autolytic endo--N-acetylglucosaminidase of *Staphylococcus simulans* 22. *FEMS Microbiol. Letts.* 58: 223-228.

35. Bierbaum, G., and Sahl, H. –G. 1991. Induction of autolysis of *Staphylococcus simulans* 22 by Pep5 and nisin and influence of the cationic peptides on the activity of the autolytic enzymes. In *Nisin and Novel Lantibiotics*. eds. G. Jung and H. –G. Sahl, pp 386-396. Escom: Leiden.

36. Björk, L., and Claesson, O. 1980. Correlation between concentration of hypothiocyanate and antibacterial effect of the lactoperoxidase system against *Escherichia coli*. *J. Dairy Sci.* 63: 919-922.

37. Blackburn, P., Polak, J., Gusik, S., Rubino, S. D. 1989. Nisin compositions for use as enhanced broad range bacteriocins. *International Patent No* PCT/US89/02625. *International Publication No.* WO89/12399. Applied Microbiology Inc., New York, USA.

38. Blom, H., Katla, T., Hagen, B. F., Axelsson, L. 1997. A model assay to demonstrate how intrinsic factors affect diffusion of bacteriocins. *Int. J. Food Microbiol.* 38: 103-109.

39. Bogorditsikaya, V. P., Shillinger, Y. I., and Osipova, I. N. 1970. Hygienic study of food products preserved with the antibiotic nisin. *Gig. Sanit.* 35: 37-40.

40. Bouksaim, M., Fliss, I., Meghrous, J., and Simard, R., and Lacroix, C. 1998. Immunodot detection of nisin Z in milk and whey using enhanced chemiluminescence. *J. Appl. Microbiol.* 84: 176-184.

41. Bower, C. K., McGuire, J., and Daeschel, M. A. 1995a. Influences on the antimicrobial activity of surface-adsorbed nisin. *J. Industrial Microbiol.* 15: 227-233.

42. Bower, C. K., McGuire, J., and Daeschel, M. A. 1995b. Suppression of *Listeria monocytogenes* colonization following adsorption of nisin onto silica surfaces. *Appl. Environ. Microbiol.* 61: 992-997.

43. Boziaris, I. S., Humpheson, L., and Adams, M. R. 1998. Effect of nisin on heat injury and inactivation of *Salmonella enteriditis* PT4. *Int. J. Food Microbiol.* 43: 7-13.

44. Breidt, F., Crowley, K. A., and Fleming, H. P. 1993. Isolation and characterization of nisin-resistant *Leuconostoc mesenteroides* for use in cabbage fermentations. *Appl. Environ. Microbiol.* 59: 3778-3783.

45. Breuer, B., and Radler, F. 1996. Inducible resistance against nisin in *Lactobacillus casei*. *Arch. Microbiol.* 165: 114-118.

46. Breukink, E., van Kraaij, C., Demel, R. A., Siezen, R. J., Kuipers, O. P., and de Kruijff, B. 1997. The C-terminal region of nisin is responsible for the initial interaction of nisin with the target membrane. *Biochem.* 36: 6968-6976.

47. Breukink, E., van Kraaij, C., van Dalen, A., Demel, R. A., Siezen, R. J., de Kruijff, B. and Kuipers, O. P. 1998. The orientation of nisin in membranes. *Biochem.* 37: 8153-8162.

48. Briozzo, J., de la Garde, E. A., Chirife, J., and Parada, J. L. 1983. *Clostridium botulinum* type A growth and toxin production in media and processed cheese spread. *Appl. Environ. Microbiol.* 45: 1150-1152.

49. Bycroft, B. W., Chan, W. C. , and Roberts, G. C. K. 1991. Synthesis and characterization of pro- and prepetides related to nisin and subtilin. In *Nisin and Novel Lantibiotics* eds. G. Jung and H. –G. Sahl, pp 204-217. Leiden: Escom.

50. Buchman, G. W., Banerjee, S., and Hansen, J. N. 1988. Structure, expression, and evolution of a gene encoding the precursor of nisin, a small protein antibiotic. *J. Biol. Chem.* 263: 9508-9514.

51. Budu-Amoaka, E., Ablett, R. F., Harris, J., and Delves-Broughton, J. 1999. Combined effect of nisin and moderate heat on the destruction of *Listeria monocytogenes* in cold-pack lobster meat. *J. Food Prot.* 62: 46-50.

52. Budu-Amoaka, E., Ablett, R. F., Thomas, L. V., and Delves-Broughton, J. Combined inhibitory effect of glycine and nisin on destruction of *Listeria monocytogenes* in "cold-pack" lobster meat. Submitted paper.

53. Bycroft, N. L., Byng, G. S., and Good, S. R. 1991. Synergistic antimicrobial compositions. *US Patent No.* 5, 043, 176. Haarman & Reimer Corp., USA.

54. Calderon, C., Collins-Thompson, D. L., and Usborne, W. R. 1985. Shelf-life studies of vacuum packaged bacon treated with nisin. *J. Food Prot.* 48: 330-333.

55. Campbell, L. L., and Sniff, E. E. 1959. Effect of subtilin and nisin on the spores of *Bacillus coagulans*. *J. Bacteriol.* 77: 766-770.

56. Carlez, A., Cheftel, J. –C., Rosec, J. P., Richard, N., Saldana, J. –L., and Balny, C. 1992. Effects of high pressure and bacteriostatic agents on the destruction of *Citrobacter freundii* in minced beef muscle. In *High Pressure and Biotechnology* Vol 224. eds. C. Balny, R. Hayashi, K. Heremand and P. Masson, pp. 365-368. London: Colloque Inserm/John Libbey and Co.

57. Carslon, S., and Bauer, H. M. 1957. Nisin, ein antibakterieller wirkstoff aus *Streptococcus lactis* unter Beruksichtigung des resistenzproblems. *Arch. Hyg. Bakteriol.* 141: 445-459.

58. Caserio, G., Ciampella, M., Gennari, M., and Barluzzi, A. M. 1979a. Utilisation of nisin in cooked sausages and other cured meat products. *Industrie Alimentaria* 18: 1-19.

59. Caserio, G., Stecchini, M., Pastore, M., and Gennari, M. 1979b. The individual and combined effect of nisin and nitrite on the spore germination of *Clostridium perfringens* in meat mixtures subjected to fermentation. *Industrie. Alimentaria* 18: 894-898.

60. Cerny, G., Hennlich, W., and Poralla, K. 1984. Spoilage of fruit juice by bacilli: isolation and characterization of the spoiling microorganism. *Zeitschrift für Lebensmittle Untersuchung und Forschung* 179: 224-227.

61. Chan, W. C., Leyland, M., Clark, J., Doff, H. M., Lian, L. –Y., Bycroft, B. W., Gasson, M. J., and Roberts, G. C. K. 1996. Structure-activity relationships in the peptide antibiotic nisin: antibacterial activity of fragments of nisin. *FEBS Lett.* 390: 129-132.

62. Chan, W. C., Lian, L. –Y., Bycroft, B.W., and Roberts, G. C. K. 1989. Confirmation of the structure of nisin by complete ¹H n.m.r. resonance assignment in aqueous and dimethyl sulphoxide solution. *J. Chem. Soc. Perkin. Trans.* I:2539-2567.

63. Chandrapati, S. R., and O'Sullivan, D. J. 1998. Procedure for quantifiable assessment of nutritional parameters influencing nisin production by *Lactococcus lactis* subsp. *lactis*. *J. Biotechnol.* 63: 229-233.

64. Cheeseman, G. C. and Berridge, N. J. 1957. An improved method of preparing nisin. *Biochem J.* 65: 603-608.

65. Cheeseman, G. C. and Berridge, N. J. 1959. Observations on the molecular weight and chemical composition of nisin A. *Biochem J.* 71: 185-194.

66. Chen, Y. –H., and Yu, R. –C. 1995. Production of nisin as a preservative by *Lactococcus lactis*. *Food Science, Taiwan.* 22: 77-85.

67. Chinachoti, N., Endo, N., Matsusaki, H., Sonomoto, K., and Ishizaki, A. 1997a. Bioreactor systems for efficient production and separation of nisin Z using *Lactococcus lactis* IO-1. *J. Fac. Agr., Kyushu Univ.* 43: 421-436.

68. Chinachoti, N., Matsusaki, H., Sonomoto, K., and Ishizaki, A. 1997b. Immobilization of *Lactococcus lactis* IO-1 for nisin Z production. *J. Fac. Agr., Kyushu Univ.* 42: 151-169.

69. Chinachoti, N., Matsusaki, H., Sonomoto, K., and Ishizaki, A. 1997c, Utilization of xylose as an alternative carbon source for nisin Z production. *J. Fac. Agr., Kyushu Univ.* 42: 171-181.

70. Chinachoti, N., Matsusaki, H., Sonomoto, K., and Ishizaki, A. 1998. Nisin Z production by *Lactococcus lactis* IO-1 using xylose as a carbon source. *Bioscience Biotechnol. Biochem.* 62: 1022-1024.

71. Chung, K. –T., Dickson, J. S., and Crouse, J. D. 1989. Effects of nisin on growth of bacteria attached to meat. *Appl. Environ. Microbiol.* 55: 1329-1333.

72. Claypool, L., Heinemann, B., Voris, L., and Stumbo, C. R. 1966. Residence time of nisin in the oral cavity following consumption of chocolate milk containing nisin. *J. Dairy Sci.* 49: 314-316.

73. Coates, M. E. Harrison, G. F., Kon, S. K., Mann, M. E., and Rose, C. D. 1951. Effect of antibiotics and vitamin B12 on the growth of normal and 'Animal Protein Factor' deficient chicks. *Proc. Biochem. Soc.,* xii-xiii.

74. Collins-Thompson, D. L., Calderon, C., and Usborne, W. R. 1985. Nisin sensitivity of lactic acid bacteria isolated from cured and fermented meat products. *J. Food Prot.* 48: 668-670.

75. Cox, G. A., and Whitehead, H. R. 1936. Streptococci which produce a substance inhibiting growth of lactic streptococci. *N. Z. J. Agric.* 52: 38.

76. Craig, A. G. 1991. Mass spectrometric fragmentation of the peptide chain and β-methyl lanthionine bridges of the polypeptide nisin. *Biol. Mass Spectrom.* 20: 195-202.

77. Crandall, A. D., and Montville, T. J. 1998. Nisin resistance in *Listeria monocytogenes* ATCC 700302 is a complex phenotype. *Appl. Environ. Microbiol.* 64: 231-237.

78. Craven, S. E., and Lillard, H. S. 1974. Effect of microwave heating of recooked chicken on *Clostridium perfringens. J. Food Sci.* 39: 211-212.

79. Crawford, L. M. 1989. Revised policy for controlling *Listeria monocytogenes. Fed. Regist.* 54: 22345-22346.

80. Cutter, C. N., and Siragusa, G. R. 1994. Decontamination of beef carcass tissue with nisin using a pilot scale carcass washer. *Food Microbiol.* 11: 481-489.

81. Cutter, C. N., and Siragusa, G. R. 1995a. Population reductions of Gram-negative pathogens following treatments with nisin and chelators under various conditions. *J. Food Prot.* 58: 977-983.

82. Cutter, C. N., and Siragusa, G. R. 1995b. Treatments with nisin and chelators to reduce *Salmonella* and *Escherichia coli* on beef. *J. Food Prot.* 58: 1028-1030.

83. Cutter, C. N., and Siragusa, G. R. 1996a. Reduction of *Brochothrix thermosphacta* on beef surfaces following immobilization of nisin in calcium alginate gels. *Letts. Appl. Microbiol.* 23: 9-12.

84. Cutter, C. N., and Siragusa, G. R. 1996b. Reductions of *Listeria innocua* and *Brochothrix thermosphacta* on beef following nisin spray treatments and vacuum packaging. *Food Microbiol.* 13: 23-33.

85. Cutter, C. N., and Siragusa, G. R. 1998. Incorporation of nisin into a meat binding system to inhibit bacteria on beef surfaces. *Letts. Appl. Microbiol.* 27: 19-23.

86. Cziszar, J., and Pulay, G. 1956. Studies on *Streptococcus lactis* which produces antibiotics effective against clostridia. I. The activity spectrum of antibiotic-producing strains, and how to increase their antibiotic production. *Proc. 14th Int. Dairy Congr.* 2: 423-430.

87. Daeschel, M. A., Jung, D. S., and Watson, B. T. 1991. Controlling wine malolactic fermentation with nisin and nisin-resistant strains of *Leuconostoc oenos. Appl. Environ. Microbiol.* 57: 601-603.

88. Daeschel, M. A., McGuire, J., and Al-Makhlafi, H. 1992. Antimicrobial activity of nisin adsorbed to hydrophilic and hydrophobic silicon surfaces. *J. Food Prot.* 55: 731-735.

89. Datta, A. T., and Benjamin, M. M. 1997. Factors controlling acid tolerance of *Listeria monocytogenes*: effects of nisin and other ionophores. *Appl. Environ. Microbiol.* 63: 4123-4126.

90. Davies, E. A., and Adams, M. R. 1994. Resistance of *Listeria monocytogenes* to the bacteriocin nisin. *Int. J. Food Microbiol.* 21: 341-347.

91. Davies, E. A., Bevis, H. E., and Delves-Broughton, J. 1997. The use of the bacteriocin, nisin, as a preservative in ricotta-type cheeses to control the food-borne pathogen *Listeria monocytogenes. Letts. Appl. Microbiol.* 24: 343-346.

92. Davies, E. H., Bevis, H. E., Potter, R., Harris, J., Williams, G. C., and Delves-Broughton, J. 1998. The effect of pH on the stability of nisin solution during autoclaving. *Letts. Appl. Microbiol.* 27: 186-187.

93. Davies, E. A., Falahee, M. B., and Adams, M. R. 1996. Involvement of the cell envelope of *Listeria monocytogenes* in the acquisition of nisin resistance. *J. Appl. Bacteriol.* 81: 139-146.

94. Davies, E. A., Milne, C. F., Bevis, H. E., Potter, R. W., Harris, J. M., Williams, G. C., Thomas, L. V., and Delves-Broughton, J. 1999. Effective use of nisin to control lactic acid bacterial spoilage in vacuum packed bologna-type sausage. *J. Food Prot.* 62:1004-1010.

95. Dawson, P. L., Han, I. Y., and Padgett, T. R. 1997. Effect of lauric acid on nisin activity on nisin activity in edible protein packaging films. *Poult. Sci.* 76 (Suppl): 74

96. Dean, J. P., and Zottola, E. A. 1996. Use of nisin in ice cream and effect on the survival of *Listeria monocytogenes. J. Food Prot.* 59: 476-480.

97. Degnan, A. J., Kaspar, C. W., Otwell, W. S., Tamplin, M. L., and Luchansky, J. B. 1994. Evaluation of lactic acid bacterium fermentation products and food grade chemicals to control *Listeria monocytogenes* in blue crab (*Callinectes sapidus*) meat. *Appl. Environ. Microbiol.* 60: 3198-3203.

98. Delves-Broughton, J. 1990. Nisin and its uses as a food preservative. *Food Technol.* 44: 100, 102, 104, 106, 108, 111-112 and 117.

99. Delves-Broughton, J. 1993. The use of EDTA to enhance the efficacy of nisin towards Gram negative bacteria. *Int. Biodet. Biodeg.* 32: 87-97.

100. Delves-Broughton, J. 1998. Use of nisin in processed cheese and natural cheese. *Bull. Int. Dairy Fed.* 329: 13-17.

101. Delves-Broughton, J., and Friis, M. 1998. Nisin preparations – production, specifications and assay procedures. *Bull. Int. Dairy Fed.* 329: 18-19.

102. Delves-Broughton, J. and Gasson, M. J. 1994. Nisin. In *Natural Antimicrobial Systems and Food Preservation*. eds. V. M. Dillon and R. G. Board, pp. 99-131. UK: CAB International.

103. Delves- Broughton, J., Blackburn, P., Evans, R. E., and Hugenholtz, J. 1996. Applications of the bacteriocin, nisin. *Anton. von Leeuwen.* 69: 193-202.

104. Delves-Broughton, J., Williams, G. C., and Wilkinson, S. 1992. The use of the bacteriocin, nisin, as a preservative in pasteurized liquid whole egg. *Letts. Appl. Microbiol.* 15: 133-136.

105. Demel, R. A., Peelen, T., Siezen, R. J., de Kruijff, B., and Kuipers, O. P. 1996. Nisin Z, mutant nisin Z and lacticin 481 interactions with anionic lipids correlate with antimicrobial acitvity. A monolayer study. *Eur. J. Biochem.* 235: 267-274.

106. de Ruyter, P. G. G. A., Kuipers, O. P., Beerthuyzeb, M. M., van Alen Boeerigter, I., and de Vos, W. M. 1996. Functional analysis of promoters in the nisin cluster of *Lactococcus lactis. J. Bacteriol.* 178:3434-3439.

107. de Vos, W. M., Kuipers, O. P., van der Meer, J. R., and Siezen, R. J. 1995. Maturation pathway of nisin and other lantibiotics: post-translationally modified antimicrobial peptides exported by Gram positive bacteria. *Mol. Microbiol.* 17: 427-437.

108. de Vos, W. M., Mulders, J. W. M., Hugenholtz, J., Siezen, R. J., and Kuipers, O. P. 1993. Properties of nisin Z and the distribution of its gene, *nisZ*, in *Lactococcus lactis. Appl Environ Microbiol* 59: 213-218.

109. de Vuyst, L. 1990. Biosynthese, Fermentatie en Genetica van het Lactococcus lactis subsp. lactis Lantibioticum Nisine. Ph. D. Thesis. Fac. Agr. Sci., University Ghent.

110. de Vuyst, L. 1994. Nisin production variability between natural *Lactococcus lactis* subsp. *lactis* strains. *Biotechnol. Lett.* 16: 287-292.

111. de Vuyst, L. 1995. Nutritional factors affecting nisin production by *Lactococcus lactis* subsp. *lactis* NIZO 221866 in a synthetic medium. *J. Appl. Bacteriol.* 78: 28-33.

112. de Vuyst, L., and Vandamme, E. J. 1991a. Localization and phenotypic expression of genes involved in the biosynthesis of the *Lactococcus lactis* subsp. *lactis* lantibiotic nisin. In *Bacteriocins, Microcins and Lantibiotics.* eds. R. James, C. Ladzunski and F. Pattus, pp. 449-462. Berlin: Springer-Verlag.

113. de Vuyst, L., and Vandamme, E. J. 1991b. Microbial manipulation of nisin biosynthesis and fermentation. In *Nisin and Novel Lantibiotics.* eds. G. Jung and H. –G. Sahl, pp. 397-409. Leiden: Escom.

114. de Vuyst, L., and Vandamme, E. J. 1992. Influence of the carbon source on nisin production in *Lactococcus lactis* subsp. *lactis* batch fermentations. *J. Gen. Microbiol.* 138: 571-578.

115. de Vuyst, L., and Vandamme, E. J. 1993. Influence of the phosphorous and nitrogen source on nisin production in *Lactococcus lactis* subsp. *lactis* batch fermentations using a complex medium. *Appl. Microbiol. Biotechnol.* 40: 12-22.

116. de Vuyst, L., and Vandamme, E. J. 1994. Nisin, a lantibiotic produced by Lactococcus lactis subsp. lactis: properties, biosynthesis, fermentation and applications. In *Bacteriocins of Lactic Acid Bacteria. Microbiology, Genetics and Applications*, eds. L. de Vuyst and E. J. Vandamme, pp. 151-221. London: Blackie Academic & Professional.

117. de Vuyst, L., de Poorter, G., and Vandamme, E. J. 1990. Metabolic control of nisin biosynthesis in *Lactococcus lactis* subsp. *lactis*. In *Fermentation Technologies Industrial Applications*. ed. P. –L. Yu, pp. 156-172. London: Elsevier Applied Science.

118. de Vuyst, L., Joris, K., Beel, C., and Vandamme, E. J. 1988. Physiological characterization of the nisin fermentation process. *Med. Fac. Landbouwwet.* 53: 2065-2069.

119. Dielbandhoesing, S. K., Zhang, H., Caro, L. H. P., van der Vaart, J. M., Klis, F. M., Verrips, C. T., and Brul, S. 1998. Specific cell wall proteins confer resistance to nisin upon yeast cells. *Appl. Environ. Microbiol.* 64: 4047-4052.

120. Dodd, H. M., Horn, N. and Gasson, M. J. 1990. Analysis of the genetic determinant for the production of the peptide antibiotic nisin. *J. Gen Microbiol.* 136: 555-566.

121. Dodd, H. M., Horn, N., Giffard, C. J., and Gasson, M. J. 1996. A gene replacement strategy for engineering nisin. *Microbiol.* 142: 47-55.

122. Doyle, M. P., Glass, K. A., Berry, J. T., Garcia, G. A., Pollard, D. J., and Schultz, R. D. 1987. Survival of *Listeria monocytogenes* in milk during high-temperature, short-time pasteurization. *Appl. Environ. Microbiol.* 53: 1433-1438.

123. Driessen, A. J., van den Hoven, H. W., Kaiper, W., van den Kamp, M., Sahl, H. –S., Konings, R. N. H., and Konings, W. N. 1995. Mechanistic studies of lantibiotic-induced permeabilization of phospholipid vesicles. *Biochem.* 34: 1606-1614.

124. Duquet, J. P., Trouvat, A., Mouniqua, G., Odet, G., and Cerf, O. 1987. Heat resistant spores in milk used in the manufacture of UHT milk. *Lait* 67: 393.

125. EEC. 1983. EEC Commission Directive 83/463/EEC.

126. Eklund, M. W., Poysky, F. T., Paranjpye, R. N., Lashbrook, L. C., Peterson, M. E., and Pelroy, G. A. 1995. Incidence and sources of *Listeria monocytogenes* in cold-smoked fishery products and processing plants. *J. Food Prot.* 58: 502-508.

127. El-Bedawey, A. E., El-Sherbiny, A. M., Zaki, M. S., and Khalil, A. H. 1985a. The effect of certain antibiotics on the keeping quality of bolti fish (*Tilapia nilotica*). *Nahr.* 29: 665-670.

128. El-Bedawey, A. E., Zaki, M. S., El-Sherbiny, A. M., and Khalil, A. H. 1985b. The effect of certain antibiotics on bolti fish (*Tilapia nilotica*) preservation. *Nahr.* 29: 303-308.
129. El-Sadek, G. M., Mahmoud, S. A., and Dawood, A. H. 1976. Effect of nisin on reducing the thermal processing process in commercially sterilized milk. *Zentralbl. Bakteriol. Parasitenkd. Infektionskr. Hyg.* 131: 285-290.
130. El-Samahy, S. K., and Elias, A. N. 1977. The use of nisin in canned okra. *Egyptian J. Food Sci.* 5: 78-81.
131. Ericsson, H., Eklow, A., Danielson-Tham, M. –L., Loncarevic, S., Mentzing, L. –O., Persson, I., Unnerstad, H., and Tham, W. 1997. An outbreak of Listeriosis suspected to have been caused by rainbow trout. *J. Clin. Microbiol.* 35: 2904-2907.
132. Eyles, M. J., and Richardson, K. C. 1988. Thermophilic bacteria and food spoilage. *CSIRO Food Res. Quart.* 48: 19-24.
133. Facinelli, B., Varaldo, P. E., Toni, M., Casolari, C., and Fabio, U. 1989. Ignorance about *Listeria. BMJ* 299: 738.
134. Falahee, M. B., and Adams, M. R. 1992. Cross-reactivity of bacteriocins from lactic acid bacteria and lantibiotics in a nisin bioassay and ELISA. *Letts. Appl. Microbiol.* 15: 214-216.
135. Falahee, M. B., Adams, M. R., Dale, J. W., and Morris, B. A. 1990a. An enzyme immunoassay for nisin. *Int. J. Food Sci. Technol.* 25: 590-595.
136. Falahee, M. B., Adams, M. R., and Dale, J. W. 1990b. Investigations of bacteriocins from lactic acid bacteria and lantibiotics with a nisin ELISA. *FEMS Microbiol. Rev.* 87: p88, E10.
137. Falconer, R. 1952. Improvements in and relating to antibiotic substances produced by *Streptococcus lactis. Brit. Patent.* 683423.
138. Fang, T. J., and Lin, C. –C. 1994. Growth of *Listeria monocytogenes* and *Pseudomonas fragi* on cooked pork in a modified atmosphere packaging/nisin combination system. *J. Food Prot.* 57: 479-485.
139. Fang, T. J., and Lin, C. –C. 1995. Inhibition of *Listeria monocytogenes* on pork tissue by immobilized nisin. *J. Food Drug Analys.* 3: 269-274.
140. Fang, T. J., Chen, C. –Y., and Chen, H. L. 1997. Inhibition of *Staphylococcus aureus* and *Bacillus cereus* on a vegetarian food treated with nisin combined with either potassium sorbate or sodium benzoate. *J. Food Safety* 17: 69-87.
141. FAO/WHO Expert Committee on Food Additives. 1969. Specifications for identity and purity of some antibiotics. 12th Report. WHO Technical Report Series, No. 430.
142. FDA (Food and Drug Administration). 1988. Nisin preparation: Affirmation of GRAS status as a direct human food ingredient. *Fed. Reg.* 53: 11247.
143. FDA (Food and Drug Administration) 1993. Salad dressings. Code of Federal Regulations. Title 21, part 169. Washington: US Government Printing Office.
144. Ferreira, M. A. S. S. and Lund, B. M. 1996. The effect of nisin on *Listeria monocytogenes* in culture medium and long-life cottage cheese. *Lett. Appl. Microbiol.* 22: 433-438.
145. Fowler, G. G. 1973. Toxicology of nisin. *Food Cosm. Toxicol.* 11: 351-352.
146. Fowler, G. G. 1979. The potential of nisin. *Food Manuf.* 54: 57-59.
147. Fowler, G. G. 1981. Nisin: Will it be used here? *Food Engineer.* 53: 82-83.
148. Fowler, G. G., and Gasson, M. J. 1991. Antibiotics – nisin. In *Food Preservatives.* eds. N. J. Russell, and G. W. Gould, pp. 134 – 152. Glasgow: Blackie and Son Ltd.
149. Fowler, G. G., and McCann, B. 1971. The growing use of nisin in the dairy industry. *Austr. J. Dairy Technol.* 26: 44-46.
150. Fowler, G. G., and McCann, B. 1972. The use of nisin in the food industry. *Food Ind. (South Africa)* 25: 49-55.
151. Fowler, G. G., Jarvis, B., and Tramer, J. 1975. The assay of nisin in foods. In *Some Methods for Microbiological Assay.* eds. R. G. Board and D. W. Lovelock, pp. 91-105. London: Academic Press.
151a. Frazer, A. C., Sharott, M., and Hickman, J. R. 1962. The biological effects of food additives. I. Nisin. *J. Sci. Food Agr.* 13: 32-42.
152. Friedmann, R., and Epstein, C. 1951. The assay of the antibiotic nisin by means of a reductase (resazurin) test. *J. Gen. Microbiol.* 5: 830-839.
153. Fukase, K. Kitazawa, M., Sano, A., Shimbo, K., Fujita, H., Horimoto, S., Wakamiya, T., and Shiba, T. 1988. Total synthesis of peptide antibiotic nisin. *Tetr. Lett.* 29: 795-798.
154. Funan, H., Meijun, S., Hebao, D., Linda, L., Jinza, C., Yichi, Y., and Xue, S. 1990. *Nisaplin Application Experiment in Sterilisation of Canned Mushrooms.* Report of Scientific Research Institute of Food and Fermentation Industry, Ministry of Light Industry, Shanghai, Peoples Republic of China 9 pp.
155. Galesloot, T. E. 1956. Melkzuurbacteriën die het antibioticum van *Streptococcus lactis* (nisine) onwerkzaam maken. *Ned. Melk Zuiveltijdschr.* 10: 143-155.

156. Gänzle, M. G., Weber, S., and Hammes, W. P. 1999. Effect of ecological factors on the inhibitory spectrum and activity of bacteriocins. *Int. J. Food Microbiol.* 46: 207-217.

157. Garcia-Garcera, M. J., Elferink, M. G. L., Driessen, A. J. M., and Konings, W. N. 1993. *In vitro* pore-forming activity of the lantibiotic nisin. Role of proton motive force and lipid composition. *Eur. J. Biochem.* 212: 417-422.

158. Gasson. M. J. 1984. Transfer of sucrose fermenting ability, nisin resistance and nisin production into *Streptococcus lactis* 712. *FEMS Microbiol. Lett.* 21: 7-10.

159. Giffard, C. J., Dodd, H. M., Horn, N., Ladha, S., Mackie, A. R. Parr, A., Gasson, M. J., and Sanders, D. 1997. Structure-function relations of variant and fragment nisins studied with model membrane systems. *Biochem.* 36: 3802-3810.

160. Giffard, C. J., Ladha, S., Mackie, A. R., Clark, D. C., and Sanders, D. 1996. Interaction of nisin with planar lipid bilayers monitored by fluorescence recovery after photobleaching. *J. Membrane Biol.* 151: 293-300.

161. Gillespy, T. G.1953. *Nisin Trials with Canned Beans in Tomato Sauce and Canned Garden Peas.* Research Leaflet No 3, Fruit and Vegetable Canning and Quick Freezing Research Association, Chipping Campden.

162. Glaeser, H. 1989. Use of nitrate in cheese production. *Dairy Ind. Int.* 54: 19-233.

163. Goodman, M., Palmer, D. E., Mierke, D., Ro, S., Nunami, K., Wakamiya, T., Fuakse, K., Hormioto, S., Gontzea, J., Toma, C., Barduta, Z., Moldovan, E., Calinescu, S., and Mavromati, E. 1973. L'effect de la nisine comparativement à celui du perhyrdrol sur la prévention du gonflement butyrique et sur la valeur nutrive du fromage Edam. *Le Lait* 53: 40-62.

164. Gonzalez, C. F., and Kunka, B. S. 1985. Transfer of sucrose-fermenting ability and nisin production phenotype among lactic streptococci. *Appl. Environ. Microbiol.* 49: 627-633.

165. Gould, G. W. 1964. Effect of food preservatives on the growth of bacteria from spores. In *Microbial Inhibitors in Foods.* ed. M. Molin, pp. 17-24. Stockholm: Almqvist and Wiksell.

166. Gould, G. W., and Hurst, A. 1962. Inhibition of *Bacillus* spore development. *Proc. 8th Internat. Conference Microbiol, Abstract* A2-A11.

167. Goulhen, F., Meghrous, J., and Lacroix, C. 1999. Production of a nisin Z/pediocin mixture by pH-controlled mixed-strains batch cultures in supplemented whey permeate. *J. Appl. Microbiol.* 86: 399-406.

168. Graeffe, T., Rintala, H., Paulin, L., and Saris, P. 1991. A natural nisin variant. In *Nisin and Novel Lantibiotics* ed. G. Jung and H. –G. Sahl, pp 260-268. Leiden: Escom.

169. Gram, L., and Huss, H. H. 1996. Microbiological spoilage of fish and fish products. *Int. J. Food Microbiol.* 33: 121-137.

170. Grau, F. H., and Vanderlinde, P. B. 1990. Growth of *Listeria monocytogenes* on vacuum-packaged beef. *J. Food Prot.* 53: 739-741.

171. Gravesen, A., and Knochel, S. 1998. Biopreservation of meat products: development of bacteriocin resistance in *Listeria monocytogenes. Third Int. Workshop on Lantibiotics, April 5-8,* p. 51.

172. Gregory, M. E., Henry, K. M., and Kon, S. K. 1964. Nutritive properties of freshly prepared and stored evaporated milks manufactured by a normal commercial procedure or by reduced thermal processes in the presence of nisin. *J. Dairy Res.* 31: 113-119.

173. Gross, E. 1977. α, β–Unsaturated and related amino acids in peptides and proteins. In *Protein Cross-Linking.* ed. M. Friedmann, pp 131-153. New York: Plenum Press.

174. Gross, E. and Morell, J. L. 1967. The presence of dehydroalanine in the antibiotic nisin and its relationship to activity. *J Am Chem Soc* 89: 2791 – 2792.

175. Gross, E. and Morell, J. L. 1968. The number and nature of α, β-unsaturated amino acids in nisin. *FEBS Lett.* 2: 61-64.

176. Gross, E. and Morell, J. L. 1971. The structure of nisin. *J Am Chem Soc* 93: 4634 – 4635.

177. Gross, E., Morell, J. L., and Craig, L. C. 1969. Dehydroalanyllysine: identical COOH-terminal structures in the peptide antibiotics nisin and subtilin. *Biochem.* 62: 952-956.

178. Gupta, R. K., and Prasad, D. N. 1988. Incorporation of nisin in stirred yoghurt I. Effect on lactic and non-lactic organisms during storage. *Cult. Prod. J. August:* 17-18.

179. Gupta, R. K., and Prasad, D. N. 1989a. Antibiotic activity of nisin in food preservation: a review. *Microbiol. Aliments. Nutrit.* 7: 199-208.

180. Gupta, R. K., and Prasad, D. N. 1989b. Incorporation of nisin in stirred yoghurt II. Effect on biochemical activities during storage. *Cult. Prod. J. February:* 9-10.

181. Gupta, R. K., and Prasad, D. N. 1989c. Incorporation of nisin in stirred yoghurt III. Quantitative estimation of residual nisin. *Cult. Prod. J. May:* 11.

182. Hague, M. A., Kastner, C. L., Fung, D. L., Kone, K., and Schwenke, J. R. 1997. Use of nisin and microwave treatment reduces *Clostridium sporogenes* outgrowth in precooked vacuum-packaged beef. *J. Food Prot.* 9: 1072-1074.

183. Hammer, P., Lembke, F., Suhren, G., and Heeschen, W. 1995. Characterization of a heat resistant mesophilic *Bacillus* species affecting quality of UHT-milk – a preliminary report. *Kiel. Milchwirtsch. Forschungsber.* 47: 303-311.

184. Hanlin, M. B., Kalchayanand, N., Ray, P., and Ray, B. 1993. Bacteriocins of lactic acid bacteria in combination have greater antibacterial activity. *J. Food Prot.* 56: 252-255.

185. Hara, S. Yakazu, K., Nakakawaji, K., Takenchi, T., Kobayashi, T., Sata, M., Imai, Z., and Shibuya, T. 1962. An investigation of toxicity of nisin with a particular reference to experimental studies of its oral administration and influences on digestive enzymes. *Tokyo Med. Univ. J.* 20: 175-207.

186. Harris, L. J., Fleming, H. P., and Klaenhammer, T. R. 1991. Sensitivity and resistance of *Listeria monocytogenes* ATCC 19115, Scott A, and UAL500 to nisin. *J. Food Prot.* 52: 384-387.

187. Harris, L. J., Fleming, H. P., and Klaenhammer, T. R. 1992. Characterization of two nisin producing *Lactococcus lactis* subsp. *lactis* strains isolated from a commercial sauerkraut fermentation. *Appl. Environ. Microbiol.* 58: 1477-1487.

188. Harvey, J., and Gilmour, A. 1992. Occurrence of *Listeria* species in raw milk and dairy products produced in Northern Ireland. *J Appl. Bacteriol.* 72: 119-125.

189. Harwood, J. L., and Russell, N. J. 1984. *Lipids in Plants and Microbes*. London: George Allen and Unwin Ltd.

190. Hauben, K. J.A., Wuytack, E. C., Soontjens, G. C. F., and Michiels, C. W. 1996. High pressure transient sensitization of *Escherichia coli* to lysozyme and nisin by disruption of outer membrane permeability. *J. Food Prot.* 59: 350-355.

191. Hawley, H. B. 1955. The development and use of nisin. *J. Appl. Bacteriol.* 18: 388-395.

192. Hawley, H. B. 1958. Uber die zulässigkeit und anerkennung von nisin als lebesmittelzusatz. *Milchwissenschaft* 6: 257-259.

193. Hawley, H. B., and Hall, R. H. 1960. The production of nisin. Patent No. GB844,782.

194. Heinemann, B., and Williams, R. 1966. The inactivation of nisin by pancreatin. *J. Dairy Sci.* 49. 312-313.

195. Heinemann, B., Stumbo, C. R., and Scurlock, A. 1964. Use of nisin in preparing beverage-quality sterile chocolate flavoured milk. *J. Dairy Sci* 47: 8-12.

196. Heinemann, B., Voris, L., and Stumbo, C. R. 1965. Use of nisin in processing food products. *Food Technol.* 19: 592-596.

197. Henning, S. 1984. Untersuchungen zur Wirkungsweiscs von Nisin und dessen Anwendung in der Lebensmitteltechnologies. *Dissertation, Universität Hohenheim, Fakultat fur Allgemeine und Angewande Naturwissenschaften*.

198. Henning, S., Metz, R., and Hammes, W. P. 1986a. New aspects for the application of nisin to foods based on its mode of action. *Int. J. Food Microbiol.* 3: 135-141.

199. Henning, S., Metz, R., and Hammes, W. P. 1986b. Studies on the mode of action of nisin. *Int. J. Food Microbiol.* 3: 121-134.

200. Hernandez, E., Duran, L., and Morell, J. 1964. Use of nisin in canned asparagus. *Agroquimica y Technologia de Alimentos* 4: 466-470.

201. Hirsch, A. 1950. The assay of the antibiotic nisin. *J. Gen. Microbiol.* 4: 70-83.

202. Hirsch, A. 1951. Growth and nisin production of a strain of *Streptococcus lactis*. *J. Gen. Microbiol.* 5: 208-221.

203. Hirsch, A. 1952. The evolution of the lactic streptococci. *J. Dairy Res.* 19: 290-293.

204. Hirsch, A., and Grinsted, E. 1951. The differentiation of the lactic streptococci and their antibiotics. *J. Dairy Res.* 18: 198-204.

205. Hirsch, A., and Grinsted, E. 1954. Methods for the growth and enumeration of anaerobic spore spore-formers from cheese, with observations on the effect of nisin. *J. Dairy Res.* 21: 101-110.

206. Hirsch, A., and Wheater, D. M. 1951. The production of antibiotics by streptococci. *J. Dairy Res.* 18: 193-197.

207. Hirsch, A., Grinsted, E., Chapman, H. R., and Mattick, A. 1951. A note on the inhibition of an anaerobic sporeformer in Swiss-type cheese by a nisin-producing *Streptococcus*. *J. Dairy Res.* 18: 205.

208. Hitchens, A. D., Gould, G. W., and Hurst, A. 1963. The swelling of bacterial spores during germination and outgrowth. *J. Gen. Microbiol.* 30: 445-453.

209. Holzapfel, W. H. 1998. The Gram-positive bacteria associated with meat and meat products. In: *The Microbiology of Meat and Poultry*. eds. A. Davies and R. Board, pp. 35-84. London: Blackie Academic & Professional.

210. Horn, N., Swindell, S., Dodd, H., and Gasson, M. 1991. Nisin biosynthesis genes are encoded by a novel conjugative transposon. *Mol Gen. Genet.* 228: 129-135.

211. Hossack, D. J. N., Bird, M. C., and Fowler, G. G. 1983. The effect of nisin on the sensitivity of microrgansism to antibiotics and other chemotherapeutic agents. In *Antimicrobials and Agriculture.* ed. M. Woodbine, pp. 425-433. London: Butterworths.

212. Hoyle, M., and Nichols, A. A. 1948. Inhibitory strains of lactic streptococci and their significance in the selection of cultures for starter. *J. Dairy Res.* 15: 398-408.

213. Hudson, J. A., and Mott, S. J. 1993. Growth of *Listeria monocytogenes, Aeromonas hydrophila* and *Yersinia enterocolitica* on cooked beef under refrigeration and mild temperature abuse. *Food Microbiol.* 10: 429-437.

214. Hudson, J. A., Mott, S. J., and Penney, N. 1994. Growth of *Listeria monocytogenes, Aeromonas hydrophila* and *Yersinia enterocolitica* on vacuum and saturated carbon dioxide controlled atmosphere-packaged sliced roast beef. *J. Food Prot.* 57: 204-208.

215. Hugas, M., Neumeyer, B., Pages, F., Garriga, M. and Hammes, W. P. 1996. Antimicrobial activity of bacteriocin-producing cultures in meat products. *Fleischwirtschaft* 76: 649-652.

216. Hunter, G. J. E., and Whitehead, H. R. 1944. *J. Dairy Res.* 13: 123.

217. Huot, E., Barrena-Gonzalez, C., and Petitdemange, H. 1996. Comparative effectiveness of nisin and bacteriocin J46 at different pH values. *Letts. Appl. Microbiol.* 22: 76-79.

218. Hurst, A. 1966. Biosynthesis of the antibiotic nisin by whole *Streptococcus lactis* organisms. *J. Gen. Microbiol.* 44: 209-220.

219. Hurst, A. 1981. Nisin. *Adv. Appl. Microbiol.* 27: 85-123.

220. Hurst, A. 1983. Nisin and other inhibitory substances from lactic acid bacteria. In *Antimicrobials in Foods,* eds. A. L. Branen and P. M. Davidson, pp. 327-351. New York: Marcel Dekker.

221. Hurst, A., and Hoover, D. G. 1993. Nisin. In *Antimicrobials in Foods.* eds. P. M. Davidson, and A. L. Branen, pp. 369-394. 2nd Edition. New York: Marcel Dekker.

222. Hurst, A., and Kruse, H. 1972. Effect of secondary metabolites on the organisms producing them: effect of nisin on *Streptococcus lactis* and enterotoxin B on *Staphylococcus aureus. Antimicrob. Agents Chemother.* 1: 277-279.

223. Ingram, L., Tombs, M. P., and Hurst, A. 1967. Mobility-molecular weight relationships of small proteins and peptides in acrylamide-gel electrophoresis. *Anal. Biochem.* 20: 24-29.

224. Jack, R. W., Bierbaum, G., and Sahl, H. –G. 1998. Biological activities. In *Lantibiotics and Related Peptides,* pp. 157-190. Georgetown, USA: Springer-Verlag and Landes Bioscience.

225. Janes, M. E., Nannapaneni, R., Proctor, A., and Johnson, M. G. 1998. Rice hull ash and silicic acid as adsorbents for concentration of bacteriocins. *Appl. Environ. Microbiol.* 64: 4403-4409.

226. Jarvis, B. 1966. Observations on the inactivation of nisin by enzymes from *Bacillus cereus* and *Bacillus polymyxa. Chem. Ind.* 882

227. Jarvis, B. 1967. Resistance to nisin and production of nisin-inactivating enzymes by several *Bacillus* species. *J. Gen Microbiol.* 47: 33-48.

228. Jarvis, B. 1970. Enzymic reduction of the C-terminal dehydroalanine sequence in nisin. *Proc. Biochem. Soc.* 119: 56P

229. Jarvis, B., and Burke, C. S. 1976. Practical and legislative aspects of the chemical preservation of food. In *Inhibition and Inactivation of Vegetative Microbes.* eds. F. A. Skinner and W. B. Hugo, pp.345-367. New York: Academic Press.

230. Jarvis, B., and Farr, J. 1971. Partial purification, specificity and mechanism of action of the nisin-inactivating enzyme from *Bacillus cereus. Biochim. Biophys. Acta.* 227: 232-240.

231. Jarvis, B., and Mahoney, R. R. 1969. Inactivation of nisin by alphacymotrypsin. *J. Dairy Sci.* 52: 1448-1450.

232. Jarvis, B., Jeffcoat, J., and Cheeseman, G. C. 1968. Molecular weight distribution of nisin. *Biochim. Biophys. Acta.* 168: 153-155.

233. Jemmi, T. 1993. *Listeria monocytogenes* in smoked fish: an overview. *Arch. Lebens. hyg.* 44: 10-13.

234. Jenson, I., Baird, L., and Delves-Broughton. 1994. The use of nisin as a preservative in crumpets. *J. Food Prot.* 57: 874-877.

235. Joint FAO/WHO Expert Committee on Food Additives. 1969. Specifications for identity and purity of some antibiotics: Nisin. *FAO Nutrition Meetings Report Series No 45A. 12th Report:* 53-67.

236. Joosten, H. M. L., and Nunez, M. 1995. Absorption of nisin and enterocin 4 to polypropylene and glass surfaces and its prevention by Tween 80. *Letts. Appl. Microbiol.* 21: 389-392.

237. Jørgensen, L. V., Huss, H. H. 1998. Prevalence and growth of *Listeria monocytogenes* in naturally contaminated seafood. *Int. J. Food Microbiol.* 42: 127-131.

238. Jung, G. 1991a. Lantibiotics - ribosomally synthesised biologically active polypeptides containing sulfide bridges and α, β–didehydroaminoacids. *Angew. Chem, Int. Ed. Engl.* 30: 1051-1068.

239. Jung, G. 1991b. Lantibiotics: a survey. In *Nisin and Novel Lantibiotics: Proceedings of the First International Workshop on Lantibiotics.* eds G. Jung and H. -G. Sahl, pp 1 -25. Leiden: Escom.

240. Jung, D. –S., Bodyfelt, F. W., and Daeschel, M. A. 1992. Influence of fat and emulsifiers on the efficacy of nisin in inhibiting *Listeria monocytogenes* in fluid milk. *J Dairy Sci.* 75: 387-393.

241. Kalchayanand, N., Hanlin, M. B., and Ray, B. 1992. Sublethal injury makes Gram negative and Gram positive bacteria sensitive to the bacteriocins, pediocin AcH, and nisin. *Letts. Appl. Microbiol.* 15: 239-243.

242. Kalchayanand, N., Ray, B., Sikes, T., and Dunne, C. P. 1995. Bacteriocin-based biopreservatives add an extra dimension in food preservation by hydrostatic pressure. In *Activities Report of the R & D Associates* 48: 280-286. San Antonio, TX, Research and Development Assocaites for Military Food and Packaging Systems, Inc.

243. Kalchayanand, N., Sikes, T., Dunne, C. P., and Ray, B. 1994. Hydrostatic pressure and electroporation have increased bactericidal efficiency in combination with bacteriocins. *Appl. Environ. Microbiol.* 60: 4174-4177.

244. Kalchayanand, N., Sikes, T., Dunne, C. P., and Ray, B., 1996.Viability loss kinetics of food spoilage and pathogenic bacteria at a moderate hydrostatic pressure. In *Activities Report of the R & D Associates* 49: 331-341. San Antonio, TX, Research and Development Associates for Military Food and Packaging Systems, Inc.

245. Kalchayanand, N., Sikes, A., Dunne, C. P., and Ray, B. 1998. Factors influencing death and injury of foodborne pathogens by hydrostatic pressure-pastuerization. *Food Microbiol.* 15: 207-214.

246. Kalra, M. S., Kuila, R. K., and Ranganathan, B. 1973. Activation of nisin production by UV-irradiation in a nisin-producing strain of *Streptococcus lactis. Experientia* 29: 624-625.

247. Kassaify, Z. G. 1998. The potential use of nisin on fresh sausages and onto sausage casings. *M. Sc. Thesis.* Oxford Brooks University, Oxford, UK.

248. Kaya, M., and Schmidt, U. 1991. Behaviour of *Listeria monocytogenes* on vacuum-packed beef. *Fleischwirtschaft* 71: 424-426.

249. Kim, W. S. 1997. Nisin production by *Lactocccus lactis* using two-phase batch culture. *Letts. Appl. Microbiol.* 25: 169-171.

250. Kim, W. S., Hall, R. J., and Dunn, N. W. 1997. The effect of concentration and nutrient depletion on nisin production of *Lactococcus lactis. Appl. Microbiol. Biotechnol.* 48: 449-453.

251. Kitazawa, M., Fujita, H., Kubo, A., and Shiba, T. 1991. Conformations of nisin and its fragments using synthesis, NMR and computer simulations. In *Nisin and Novel Lantibiotics.* eds. G. Jung and H. –G. Sahl, pp 59-75. Leiden: Escom.

252. Klaenhammer, T. R. 1993. Genetics of bacteriocins produced by lactic acid bacteria. *FEMS Microbiol. Revs.* 12: 39-86.

253. Kleerebezenm M., Beerthuyzen, M. M., Vaughan, E. E., de Vos, W. M., and Kuipers, O. P. 1997. Controlled gene expression systems for lactic acid bacteria: transferable nisin-inducible cassettes for *Lactococcus, Leuconostoc* and *Lactobacillus* spp. *Appl. Environ. Microbiol.* 63: 4581-4584.

254. Klijn, N., Herman, L., Langeveld, L., Varewijck, M., Wagendorp, A. A., Huemer, I., and Weerkamp, A. H. 1997. Genotypical and phenotypical characterization of *Bacillus sporothermodurans* strains, surviving heat sterilization. *Int. Dairy J.* 7: 421-428

255. Klima, R. A., and Montville, T. J. 1995. The regulatory and industrial response to listeriosis in the USA: a paradigm for dealing with foodborne pathogens. *Trends Food Sci. Technol.* 6: 87-93.

256. Knox, J. R., and Keck, P. C. 1973. Conformation and absolute configuration of β-methyllanthionine. *Biochem, Biophys. Res. Commun,* 53: 567-571.

257. Knox, J. R., and Keck, P. C. 1975. β-methyllanthionine.: a sulfur amino acid in subtilin and nisin antibiotics. *Acta Cryst.*B321: 698-700.

258. Komitopolou, E., Boziaris, I., Davies, E. A., Delves-Broughton, J., and Adams, M. R. 1999. *Alicyclobacillus acidoterrestris* in fruit juices and its control by nisin. *Int. J. Food Sci Technol.* 34:81-85.

259. Kooy, J. S. 1952. Strains of *Lactobacillus plantarum* which destroy the antibiotic made by *Streptococcus lactis. Neth. Milk Dairy J.* 6: 323-330.

260. Kordel, M., and Sahl, H. –G. 1986. Susceptibility of bacterial eukaryotic and artificial membranes to the disruptive action of the cationic peptides Pep 5 and nisin. *FEMS Microbiol. Letts.* 34: 139-144.

261. Kosikowski, F. V. 1977. *Cheese and Fermented Milk Foods*, 2nd Edition. Edward Brothers, Ann Arbor, MI, USA.

262. Kozak, W., and Dobrzanski, W. T. 1977. Growth requirements and the effect of organic components of the synthetic medium on the biosynthesis of the antibiotic nisin in *Streptococcus lactis* strain. *Acta Microbiol. Pol.* 26: 361-368.

263. Krasnikova, L. V., Vasil'ev, N. F., Sileva, M. I., Voronina, L. N., and Litvinova, M. N. 1979. Conditions of biosynthesis of nisin with lactic fermentation streptococci during continuous cultivation. *Izvestiya Vysshikh Uchebnykh Zavedenii, Pishchevaya Tekhnologiya* 5: 35-37. (In Russian).

264. Kudryashov, V. L., Sergeeva, I. D., Zadororzhnaya, T. I., and Duritskaya, L. I. 1995. Synthesis of the biopreservative nisin on wastes and secondary raw materials of a number of biotechnological plants. *Biotekhnologiya* 12: 25-28.

265. Kuipers, O. P., Beerthuyzen, M. M., de Ruyter, P. G. G. A., Luesink, E. J., and de Vos, W. M. 1995. Autoregulation of nisin biosynthesis in *Lactococcus lactis* by signal transduction. *J. Biol. Chem.* 207: 27299-27304.

266. Kuipers, O. P., Rollema, H. S., Yap, W. M. G. J., Boot, H. J., Siezen, R. J., and de Vos, W. M. 1992. Engineering dehydrated amino acid residues in the antimicrobial peptide nisin. *J. Biol. Chem.* 267: 24340-24346.

267. Kuipers, O. P., Yap, W. M. G., Rollema, H. S., Beerthyzen, M. M., Siezen, R. J., and de Vos, W. M. 1991a. Expression of wild-type and mutant nisin genes in *Lactococcus lactis*. In *Nisin and Novel Lantibiotics* eds. G. Jung and H-G. Sahl, pp 250-259. Leiden: Escom.

268. Kuipers, O. P., Boot, H. J., and de Vos, W. M. 1991b. Improved site-directed mutagenesis method using PCR. *Nucl. Acids Res.* 19: 4558.

269. Kumar, N., and Prasad, D. N. 1996. Preservative action of nisin in *lassi* under different storage temperatures. *Indian J. Animal. Sci.* 66: 525-528.

270. Kurtzman, C. P., Rogers, R., and Hesseltine, C. W. 1971. Microbiological spoilage of mayonnaise and salad dressings. *Appl. Microbiol.* 21: 870-874.

271. Lakamraju, M., McGuire, J., and Daeschel, M. 1996. Nisin adsorption and exchange with selected milk proteins at silanized silica surfaces. *J. Colloid and Interface Sci.* 178: 495-504.

272. LeBlanc, D. J., Crow, V. L., and Lee, L. N. 1980. Plasmid mediated carbohydrate catabolic enzymes among strains of *Streptococcus lactis*. In *Plasmids and Transposons: Environmental Effects and Maintenance Mechanisms*. eds. C. Stuttard and K. R. Rozee, pp. 31-41. New York: Academic Press.

273. Lee, H. W. 1988. *Bacillus cereus* food poisoning associated with crumpets. *Abstracts. Sixth Australian Food Microbiology Conference*, Sept 4-7th. Surfers Paradise.

274. Lee, S. H., and Kim, H. U. 1985. Studies on the antibiotic nisin produced by *Streptococcus lactis* IFO 12007. I. The production and purification of nisin. *Korean J. Anim. Sci.* 27: 480-483.

275. Leistner, L., and Gorris, L. G. M. 1995. Food preservation by hurdle technology. *Trends Food Sci Technol.* 6: 41-46.

276. Lian, L-Y., Chan, W. C., Morley, S. D., Roberts, G. C., Bycroft, B. W., and Jackson, D. 1991. NMR studies of the solution structure of nisin A and related peptides. In *Nisin and Novel Lantibiotics*. eds. G. Jung and H. –G. Sahl, pp 43-58. Leiden: Escom.

277. Lian, L-Y., Chan, W. C., Morley, S. D., Roberts, G. C., Bycroft, B. W., and Jackson, D. 1992. NMR studies of the solution structure of nisin A. *Biochem. J.* 283: 413-420.

278. Linnan, M. J., Mascola, L., Lou, X. D., Goulet, V., May, S., Saminen C., *et al.* 1988. Epidemic listeriosis associated with Mexican-style cheese. *N. Eng. J. Med.* 319: 823-828.

279. Linnett, P. E., and Strominger, J. L. 1973. Additional antibiotic inhibitors of peptidoglycan synthesis. *Antimicrob. Agents Chemother.* 4: 231-236.

280. Lipinska, E. 1977. Nisin and its applications. In *Antibiotics and Antibiosis in Agriculture*. ed. M. Woodbine, pp. 103-130. London: Butterworth.

281. Lipinska, E., and Strzalkowska, S. N. 1959. Experiments on training pure cheese cultures to resist antibiotics produced by *Streptococcus lactis*. *Dairy Congr. Proc.* 2: 572-579.

282. Lipinska, E., Szadowska, M., Jakubczyk, E., and Lipniewsko, D. 1973. Laboratory investigations on nisin concentration by means of microfiltration. *Ann. Inst. Dairy Ind.* 15: 25-35.

283. Liu, W., and Hansen, J. N. 1990. Some chemical and physical properties of nisin, a small antibiotic produced by *Lactococcus lactis*. *Appl. Environ.. Microbiol.* 56: 2551-2558.

284. Liu, W., and Hansen, J. N. 1992. Enhancement of the chemical and antimicrobial properties of subtilin by site-directed mutagenesis. *J. Biol. Chem.* 267: 25078-25085.

285. Liu, W., and Hansen, J. N. 1993. The antimicrobial effect of a structural variant of subtilin against outgrowing *Bacillus cereus* T spores and vegetative cells occurs by different mechanisms. *Appl. Environ. Microbiol.* 59: 648-651.

286. Liu, X. 1997. Continuous production of bacteriocins by lactic acid bacteria from whey permeate and applications in food systems. *Ph. D. Thesis. Ohio State University, USA.*

287. Liu, X., Yousef, A. E., and Yang, S. T. 1997. Continuous production of nisin by immobilized *L. lactis* subsp. *lactis* ATCC 11454 in a packed-bed bioreactor from whey permeate. *Abstracts of papers presented at IFT Annual Meeting, Orlando, Florida, USA.* 14-18th June.

288. Ludwig, H., Bieler, C., Hallbauer, K., and Scigalla, W. 1992. Inactivation of microoorganisms by hydrostatic pressure. In *High Pressure and Biotechnology* Vol 224. eds. C. Balny, R. Hayashi, K. Heremand and P. Masson, pp. 365-368. London: Colloque Inserm/John Libbey and Co.

289. Mahadeo, M., and Tatini, S. R. 1994. The potential use of nisin to control *Listeria monocytogenes* in poultry. *Letts. Appl. Microbiol.* 18: 323-326.

290. Mahmoud, S. A., El-Sadek, G. M., and Dawood, A. H. 1976. Effect of nisin on prolonging the keeping quality of pasteurized milk. *Zentralbl. Parasitenkd. Infektionskr. Hyg.*, 131: 277-284.

291. Maisnier-Patin, S. Deschamps, N., Tatini, S. R., and Richard, J. 1992. Inhibition of *Listeria monocytogenes* in Camembert cheese made with a nisin-producing starter. *Lait* 72: 249-263.

292. Maisnier-Patin, S., Tatini, S. R., and Richard, J. 1995. Combined effect of nisin and moderate heat on destruction of *Listeria monocytogenes* in milk. *Lait* 75: 81-91.

293. Mansour, M., Amri, D., Bouttefroy, A., Linder, M., and Milliere, J. B. 1999. Inhibition of *Bacillus licheniformis* spore growth in milk by nisin, monolaurin and pH combinations. *J. Appl. Microbiol.* 86: 311-324.

294. Maslennikova, N. M., and Loshina, P. B. 1968. The use of nisin in canned potatoes. *Konserv. i. Ovoshcheshusil* 23: 12-15. (Russian language).

295. Maslennikova, N. M., Shundova, Y. U., and Nekhotenova, T. 1968. The effect of nisin in the sterilisation procedure of canned whole tomatoes in brine. *Konserv. i Ovovshchesushil Prom.* 23: 7-9. (Russian language).

296. Matsusaki, H., Endo, N., Sonomoto, K., and Ishizaki, A. 1996a. Development of purification method and identification of a peptide antibiotic produced by *Lactococcus lactis* IO-1. *Food Sci. Technol.* 2: 157-162.

297. Matsusaki, H., Endo, N., Sonomoto, K., and Ishizaki, A. 1996b. Lantibiotic nisin Z fermentative production by *Lactococcus lactis* IO-1: relationship between production of the lantibiotic and lactate and cell growth. *Appl. Microbiol. Biotechnol.* 45: 36-40.

298. Mattick, A. T. R., and Hirsch, A. 1944. A powerful inhibitory substance produced by group N streptococci. *Nature* 154: 551-552.

299. Mattick, A. T. R., and Hirsch, A. 1947. Further observations on an inhibitory substance (nisin) from lactic streptococci. *Lancet* 2: 5-7.

300. Mawson, A. J., and Costar, K. 1993. Effects of nisin addition on the ethanol fermentation of casein whey permeate. *Letts. Appl. Microbiol.* 17: 256-258.

301. Mazzotta, A. S., and Montville, T. J. 1997. Nisin induces changes in membrane fatty acid composition of *Listeria monocytogenes* nisin-resistant strains at 10 ºC and 30 ºC. *J. Appl. Microbiol.* 82: 32-38.

302. McClintock, M., Serres, L., Marzolf, J. J., Hirsch, A., and Mocquot, G. 1952. Action inhibitrice des streptocoques producteurs de nisine sur le développement des sporulés anaérobies dans le fromage de Gruyère fondu. *J. Dairy Res.* 19: 187-193.

303. McKay, L. L. 1983. Functional properties of plasmids in lactic streptococci. *Antonie von Leeuwen.* 49: 259-274.

304. Meanwell, L. J. 1943. The influence of raw milk quality on 'slowness' in cheesemaking. *Proc. Soc. Agr. Bacteriol* (Abstracts) 19: 1-3.

305. Meghrous, J., Huot, E., and Quittelier, M. 1992. Regulation of nisin biosynthesis by continuous cultures and by resting cells of *Lactococcus lactis* subsp. *lactis*. *Res. Microbiol.* 143: 879-890.

306. Meghrous, J., Lacroix, C., and Simard, R. E. 1999. The effects on vegetative cells and spores of three bacteriocins from lactic acid bacteria. *Food Microbiol.* 16: 105-114.

307. Meyer, A. M. 1973. *Processed Cheese Manufacture.* Food Trade Press Ltd., London.

308. Microlife-Technics. 1986. Derived nisin producing microorganisms, method of production and use and products obtained thereby. *Europ. Patent* 0137869.

309. Miles, R. J., Cassar, C. A., da Silva Carneiro de Melo, A. M. 1997. Bacterial decontamination method. *International Patent No.* WO 97/23136. The Minister of Agriculture Fisheries and Food, London.

310. Ming, X., and Deaschel, M. A. 1993. Nisin resistance of foodborne bacteria and the specific resistance responses of *Listeria monocytogenes* Scott A. *J. Food Prot.* 11: 944-948.

311. Ming, X., and Deaschel, M. A. 1995. Correlation of cellular phospholipid content with nisin resistance of *Listeria monocytogenes* Scott A. *J. Food Prot.* 58: 416-420.

312. Ming, X, Weber, G. H., Ayres, J. W., and Sandine, W. E. 1997. Bacteriocins applied to food packaging materials to inhibit *Listeria monocytogenes* on meats. *J. Food Sci.* 62: 413-415.

313. Moll, G. N., Clark, J., Chan, W. C., Bycroft, B. W., Roberts, G. C. K., Konings, W. N., and Driessen, A. J. M. 1997. Role of transmembrane pH gradient and membrane binding in nisin pore formation. *J. Bacteriol.* 179: 135-140.

314. Monticello, D. J. 1989. Control of microbial growth with nisin/lysozyme formulations. *European Patent Application No.* 0374-823. Miles Inc., Eckhart, Indiana.

315. Monticello, D. J. 1990. Control of microbial growth with lantibiotic/lysozyme formulations. *European Patent Application No.* 90111939.6. Haarman & Reimer Corp.

316. Morell, J. L., and Gross, E. 1973. Configuration of the β-carbon atoms of the β-methyllanthionine residues in nisin. *J. Am. Chem. Soc.* 95: 6480-6481.

317. Muir, D. D., Tamime, A. Y., Shenan, M. E., and Dawood, A. H. 1999. Processed cheese analogues incorporating fat-substitutes. 1. Composition, microbiological quality and flavour changes during storage at 5°C. *Lebensm,. -Wiss. u -Technol.* 32: 41-49.

318. Mukherji, S., and Qvist, S. 1981. The impact of *Brochothrix thermosphacta* and lactic acid bacteria on the keeping quality of vacuum-packed vienna sausages. In *Psychotrophic Microorganisms in Spoilage and Pathogenicity.* eds. T. A. Roberts, E. Hobbs, J. H. B. Christian, and N. Skovgaard, p.231-239. London: Academic Press.

319. Mulders, J. W. M., Boerrigter, I. J., Rollema, H. S., Siezen, R. J., and de Vos, W. M. 1991. Identification and characterization of the lantibiotic nisin Z, a natural nisin variant. *Eur J Biochem* 201: 581-584.

320. Mulet-Powell, N., Lacoste-Armynot, A. M., Vinãs, M., and de Buochberg, M. S. 1998. Interactions between pairs of bacteriocins from lactic acid bacteria. *J. Food Prot.* 61: 1210-1212.

321. Mundt, J. O., Graham, W. F., and McCarty, I. E. 1967. Spherical lactic acid-producing bacteria of southern-grown raw and processed vegetables. *Appl. Microbiol.* 15: 1303-1308.

322. Muriana, P. M., and Kanach, L. 1995. Use of Nisaplin™ to inhibit spoilage bacteria in buttermilk ranch dressing. *J. Food Prot.* 58: 1109-1113.

323. Murray, M., and Richard, J. A. 1998. Comparative study of the antilisterial activity of nisin A and pediocin AcH in fresh ground pork stored aerobically at 5 °C. *J. Food Prot.* 60: 1534-1540.

324. Murrell, W. G. 1978. *Bacillus cereus* Food Poisoning from Pikelets. Communicable Disease Intelligence (Australia). Environmental Health Branch, Dept. of Health. Reporting period 20th April to 3rd May, 1978.

325. Natrajan, N., and Sheldon, B. W. 1995. Evaluations of bacteriocin-packaging and edible film delivery systems to reduce *Salmonella* in fresh poultry. *Poult. Sci.* 74 (Suppl): 31.

326. Nekhotenova, T. I. 1961. The possibility of modifying the sterilisation process of green peas by adding nisin. *Konserv. i. Ovoshchesushil Prom.* 16: 21-23 (Russian language).

327. Nes, I. F. 1992. Non-nisin-like bacteriocins in lactic acid bacteria. Paper presented at Bioteteen Päivät, Helsinki, Finland.

328. Nettles-Cutter, C., and Siragusa, G. R. 1996. Immobilization of nisin in an edible gel for reducing bacteria on the surface of beef and in ground beef. In Program and Abstracts Book for the 83rd Annual Meeting of IAMFES, Seattle, WA, USA.

329. Newton, G. G. F., Abraham, E. P., and Berridge, N. J. 1953. Sulphur-containing amino-acids of nisin. *Nature* 171: 606.

330. Nielsen, H. –J. S. 1983. Influence of temperature and gas permeability of packaging film on development and composition of microbial flora in vacuum-packed bologna sausage. *J. Food Prot.* 46: 693-698.

331. Nielsen, P., and Roepstorff, P. 1988. Sample preparation dependent fragmentation in 252-Cf plasma desorption mass spectrometry of the polycyclic antibiotic, nisin. *Biom. Environ. Mass Spectrom.* 17: 137-141.

332. Nilsson, L., Huss, H. H., and Gram, L. 1997. Inhibition of *Listeria monocytogenes* on cold-smoked salmon by nisin and carbon dioxide atmosphere. *Int. J. Food Microbiol.* 38: 217-227.

333. Nissen-Meyer, J., Holo, H., Havarsteuin, L. S., Sletten, K., and Nes, I. F. 1992. A novel bacteriocin whose activity depends on the complementary action of two peptides. *J. Bacteriol.* 174: 5686-5692.

334. Nykänen, A., Lapveteläinen, A., Hietnaen, R.-M., and Kallio, H. 1998a. The effect of lactic acid, nisin whey permeate, sodium chloride and related combinations on aerobic plate count and the sensory characteristics of rainbow trout. *Lebenum –Wiss u-Technol* 31: 286-290.

335. Nykänen, A., Vesanen, S., and Kallio, H. 1998b. Synergistic antimicrobial effect of nisin whey permeate and lactic acids on microbes isolated from fish. *Letts. Appl. Micro.* 27: 345-348.

336. Oberman, H., and Jakubowska, J. 1969. The nisin formation in continuous cultivation of *Streptococcus lactis* 91. *Chemia Spozywcza* 100: 511-516.

337. Ogden, K. 1986. Nisin: a bacteriocin with a potential use in brewing. *J. Inst. Brew.* 92: 379-383.

338. Ogden, K. 1987. Cleansing contaminated pitching yeast with nisin. *J. Inst. Brew.* 93: 302-307.

339. Ogden, K., and Tubb, R. S. 1985. Inhibition of beer-spoilage lactic acid bacteria by nisin. *J. Inst. Brew.* 91: 390-392.

340. Ogden, K., Waites, M. J., and Hammond, J. R. M. 1988. Nisin and brewing. *J. Inst. Brew.* 94: 233-238.

341. Oh, H. I., Leon, M. J., and Lee, W. J. 1994. Use of nisin to enhance storage stability of Kimchi. *36C-12 IFT Annual Meeting Technical Program* June 25-29. Abstracts.

342. Ojcius, D. M., and Young, J. D. –E. 1991. Cytolytic pore-forming proteins and peptides: is there a common structural motif? *Trends. Biochem. Sci.* 16: 225-229.

343. Okereke, A., and Montville, T. J. 1992. Nisin dissipates the proton motive force of the obligate anaerobe *Clostridium sporogenes* PA 3679. *Appl. Environ. Microbiol.* 58: 2463-2467.

344. Olasupo, N. A., Akinsanya, S. M., Oladele, O. F., and Azeez, M. K. 1996. Evaluation of nisin for the preservation of *nono* – a Nigerian fermented milk product. *J. Food Process. Preserv.* 20: 71-78.

345. O'Leary, W. M., and Wilkinson, S. G. 1988. Gram –positive bacteria. In *Microbial Lipids.* eds. R. Ratledge and S. G. Wilkinson, pp 127-201. London: Academic Press.

346. Oscroft, C. A., Banks, J. G., and McPhee, S. 1990. Inhibition of thermally-stressed Bacillus spores by combinations of nisin, pH and organic acids. *Lebensmittel-Wissenschaft und Technologie* 23: 538-544.

347. Padgett, T., Han, I., and Dawson, P. L. 1998. Incorporation of food-grade antimicrobial compounds into biodegradable packaging films. *J. Food Prot.* 61: 1330-1335.

348. Pasinin, G., Crapisi, A., Lante, A., Curioni, A., Zamorani, A., and Spettoli, P. 1995. Evaluation of bacteriocin activity produced in milk by *Lactococcus lactis* subsp. *lactis* immobilized in barium alginate beads. *Ann. N. Y. Acad. Sci.* 750: 465-468.

349. Payne, J., Gooch, J. E. T., and Barnes, E. M. 1979. Heat-resistant bacteria in pasteurised whole egg. *J. Appl. Bacteriol.* 46: 601-613.

350. Pettersson, B., Lembke, F., Hammer, P., Stackebrandt, E., and Priest, F. 1996. *Bacillus sporothermodurans*, a new species producing highly heat-resistant endospores. *Int. J. Syst. Bacteriol.* 46: 759-764.

351. Phillips, J. D., Griffiths, M. W., and Muir, D. D. 1983. Effect of nisin on the shelf-life of pasteurised double cream. *J. Soc. Dairy Technol.* 36: 17-21.

352. Plocková, M., Stepánek, M., Demnerová, K., Curda, L., and Sviráková, E. 1996. Effect of snisin for improvement in shelf-life and quality of processed cheese. *Adv. Food Sci.* (CMTL) 18: 78-83.s

353. Plockova, M., Siháková, Z., and Sviráková, E. 1998. The efficacy of nisin and nisin-producing strain *Lactococcus lactis* subsp. *lactis* in thermized quarg desserts. *Adv. Food Sci.* 21. 17-22, 23.

354. Ponce, E., Pla, R., Sendra, B., Guamis, B., and Mor-Mur, M. 1998. Combined effects of nisin and high hydrostatic pressure on destruction of *Listeria innocua* and *Escherichia coli* in liquid whole egg. *Int. J. Food Microbiol.* 43: 15-19.

355. Poretta, A., Casolari, A., and Cassara, A. 1968. Impiego della nisina nella preparazione dell succo di pomodora. *Industria Conserve* 1: 13-14.

356. Poretta, A., Giannone, L., and Casolari, A. 1966. Impiego della nisina nella lavorazione dei piselli. *Industria Conserve* 2: 89-91.

357. Proctor, V. A., and Cunningham, F. E. 1993. The antimicrobial properties of lysozyme alone and in combination with other additives *in vitro* and in selected meat products. *J. Rapid Methods and Automation in Microbiol* 1: 315-328.

358. Radler, F. 1990a. Possible use of nisin in wine making. I. Action of nisin against lactic acid bacteria and wine yeasts in solid and liquid media. *Am. J. Enol. Vit.* 41: 1-6.

359. Radler, F. 1990b. Possible use of nisin in wine making. II. Experiments to control lactic acid bacteria in the production of wine. *Am. J. Enol. Vit.* 41: 7-11.

360. Ramsaran, H., Chen, J., Brunke, B., Hill, A., and Griffiths, M. W. 1998. Survival of bioluminescent *Listeria monocytogenes* and *Escherichia coli* O157:H7 in soft cheeses. *J Dairy Sci.* 81: 1810-1817.

361. Ramseier, H. R. 1960. The action of nisin on *Clostridium butyricum. Arch. Mikrobiol.* 37: 57-94.

362. Rao, K. V. S. S., and Mathur, B. N. 1996. Thermal death kinetics of *Bacillus stearothermophilus* spores in a nisin supplemented acidified concentrated buffalo milk system. *Milchwissenschaft* 51: 186-191.

363. Rauch, P. J. G., and de Vos, W. M. 1992. Characterization of the novel nisin-sucrose conjugative transposon Tn5276 and its insertion in *Lactococcus lactis. J. Bacteriol.* 174: 1280-1287.

364. Rauch, P. J. G., Beerthuyzen, M. M., and de Vos, W. M. 1990. Nucleotide sequence of IS904 from *Lactococcus lactis* subsp. *lactis* strain NIZO R5. *Nucleic Acid Res.* 18: 4253-4254.

365. Rauch, P. J. G., Kuipers, O. P., Siezen, R. J., and de Vos, W. M. 1994. Genetics and protein engineering of nisin. In *Bacteriocins of Lactic Acid Bacteria: Microbiology, Genetics and Applications*, eds. L. de Vuyst and E. J. Vandamme, pp. 223-249. London: Blackie Academic & Professional.

366. Ray, B. 1992. Nisin of *Lactococcus lactis* subsp. *lactis* as a food preservative. In *Food Biopreservatives of Microbial Origin.* eds. B. Ray and M. A. Daeschel, pp. 207-264. USA: CRC Press.

367. Ray, B. 1995. Combined antibacterial efficiency of bacteriocins and ultra-high hydrostatic pressure. *Symposium on the Impact of High Pressure Effects on Food.* 9th World Congress of Food Science and Technology, 30 July - 4 August 1995. Budapest, Hungary. The Hungarian Science Society of Food Industry, Budapest (Abstract No. L078, Volume I).

368. Rayman, K., Aris, B., and Hurst, A. 1981. Nisin: a possible alternative or adjunct to nitrite in the preservation of meats. *Appl. Environ. Microbiol.* 41: 375-380.

369. Rayman, K., Naeem, M., and Hurst, A. 1983. Failure of nisin to inhibit outgrowth of *Clostridium botulinum* in a model cured meat system. *Appl. Environ. Microbiol.* 46: 1450-1452.

370. Reisinger, P., Seidel, H., Tschesche, H., and Hammes, W. P. 1980. The effect of nisin on murein synthesis. *Arch. Microbiol.* 127: 187-193.

371. Reiter, B. 1985. Lactoperoxidase system of bovine milk. In *The Lactoperoxidase System: Chemistry and Biological Significance. Immunology Series No.* 27. eds. K. M. Pruitt and J. O. Tenovuo, pp. 123-141. New York: Marcel Dekker.

372. Richard, J. 1993. Inhibition of *Listeria monocytogenes* during cheese manufacture by adding nisin to milk or using a nisin-producing starter. In *Food Ingredients Europe, Conference Proceedings*, pp. 59-64. Maarsens: Expoconsult Publishers.

373. Roberts, C. M., and Hoover, D. G. 1996. Sensitivity of *Bacillus coagulans* spores to combinations of high hydrostatic pressure, heat, acidity and nisin. *J. Appl. Bact.* 81: 363-368.

374. Roberts, R. F., and Zottola, E. A. 1993. Shelf-life of pasteurized process cheese spreads made from Cheddar cheese manufactured with a nisin-producing starter culture. *J. Dairy Sci.* 76: 1829-1836.

375. Roberts, R. F., Zottola, E. A., and McKay, L. L. 1992. Use of a nisin-producing starter culture suitable for Cheddar cheese manufacture. *J. Dairy Sci.* 75: 2353-2363.

376. Rodriguez, E., Tomillo, J., Nunez, M., and Medina, M. 1997. Combined effect of bacteriocin-producing lactic acid bacteria and lactoperoxidase system activation of *Listeria monocytogenes* in refrigerated raw milk. *J. Appl. Microbiol.* 83: 389-395.

377. Rodriquez, L. D., Garayzabal. J. F., Vasquez Boland, J. A., Ferri, E. R., and Fernandez, G. S. 1985. Isolation de microorganismses du genre *Listeria* à partir de lait cru destine a la consommation humaine. *Can J Microbiol.* 31: 938-941.

378. Roepstorff, P., Nielsen, P. F., Kamesnsky, L., Craig, A. G., and Self, R. 1988. Cf plasma desorption mass spectrometry of a polycyclic peptide antibiotic, nisin. *Biomed. Environ. Mass Spectrom.* 15: 305-310.

379. Rogers, L. A. 1928. The inhibitory effect of *Streptococcus lactis* on *Lactobacillus bulgaricus*. *J. Bacteriol.* 16: 321-325.

380. Rogers, A. M., and Montville, T. J. 1991. Improved agar diffusion assay for nisin quantification. *Food Biotechnol.* 5: 161-168.

381. Rogers, A. M., and Montville, T. J. 1994. Quantification of factors which influence nisin's inhibition of *Clostridium botulinum* 56A in a model food system. *J Food Sci* 59: 663-6668, 686.

382. Rogers, L. A., and Whittier, E. O. 1928. Limiting factors in lactic fermentation. *J. Bacteriol.* 16: 211-229.

383. Rørvik, L. M., Caugant, D. A., and Yndestad, M. 1995. Contamination pattern of *Listeria monocytogenes* and other *Listeria* spp. in a salmon slaughterhouse and smoked salmon processing plant. *Int J. Food Microbiol* 25: 19-27.

384. Rørvik, L. M., Yndestad, M., Skjerve, E. 1991. Growth of *Listeria monocytogenes* in vacuum-packed, smoked salmon during storage at 4°C. *Int J. Food Microbiol* 14: 111-118.

385. Rosenow, E. M., and Marth, E. H. 1987. Fate of *Listeria monocytogenes* in skim whole and chocolate milk and in whipping cream during incubation at 4, 8, 13, 21 and 35°C. *J Food Prot.* 50: 372-378.

386. Rowlands, A., Barkworth, H., Hosking, Z., and Kempthorne, O. 1950. *J. Dairy Res.* 17: 159.

387. Rozbeh, M., Kalchayanand, N., Field, R. A., Johnson, M. C., and Ray, B. 1993. The influence of biopreservatives on the bacterial level of refrigerated vacuum packaged beef. *J. Food Safety* 13: 99-111.

388. Ruhr, E., and Sahl, H. G. 1985. Mode of action of the peptide antibiotic nisin and influence on the membrane potential of whole cells and on cytoplasmic and artificial membrane vesicles. *Antimicrob. Agents Chemother.* 27: 841-845.

389. Sahl, H-G. 1991. Pore formation in bacterial membranes by cationic lantibiotics. In *Nisin and Novel Lantibiotics* eds. G. Jung and H. –G. Sahl, pp 347-358. Leiden: Escom.

390. Sahl, H-G., and Bierbaum, G. 1998. Lantibiotics: biosynthesis and biological activities of uniquely modified peptides from Gram-positive bacteria. *Annu. Rev. Microbiol.* 52: 41-79.

391. Sahl, H-G., Jack, R. W., and Bierbaum, G. 1995. Biosynthesis and biological activities of lantibiotics with unique post-translational modifications. *Eur. J. Biochem.* 230: 827-853.

392. Scannell, A. G. M., Hill, C., Buckley, D. J., and Arendt, E. K. 1997. Determination of the influence of organic acids and nisin on shelf-life and microbiological safety aspects of fresh pork sausage. *J. Appl. Microbiol.* 83: 407-412.

393. Schillinger, U., Chung, H. –S., and Holzapfel, W. H. 1998. Use of bacteriocinogenic lactic acid bacteria to inhibit spontaneous nisin-resistant mutants of *Listeria monocytogenes* Scott A. *J. Appl. Microbiol.* 85: 657-663.

394. Schleifer, K. H., Kraus, J., Dvorak, C., Kilpper-Bälz, R., Collins, M. D., and Fischer, W. 1985. Transfer of *Streptococcus lactis* and related streptococci to the genus *Lactococcus* gen. nov. *System. Appl. Microbiol.* 6: 183-195.

395. Schnell, N., Entian, K. D., Schneider, U., Gotz, F., Zahner, H., Kellner, R., and Jung, G. 1988. Prepeptide sequence of epidermin, a ribosomally synthesized antibiotic with four sulphide rings. *Nature* 333: 276-278.

396. Schved, F., Henis, Y., Juven, B. J. 1994. Response of spheroplasts and chelator-permeabilized cells of Gram-negative bacteria to the action of bacteriocins pediocin SJ-1 and nisin. *Int. J. Food Microbiol.* 21: 305-314.

397. Schved, F., Pierson, M. D., and Juven, B. J. 1996. Sensitization of *Escherichia coli* to nisin by maltol and ethyl maltol. *Letts. Appl. Microbiol.* 22: 189-191.

398. Scott, V. N., and Taylor, S. L. 1981. Temperature, pH, and spore load effects on the ability of nisin to prevent the outgrowth of *Clostridium botulinum* spores. *J Food Sci* 46: 121-126.

399. Shefet, S. M., Sheldon, B. W., and Klaenhammer, T. R. 1995. Efficacy of optimized nisin-based treatments to inhibit *Salmonella typhimurium* and extend shelf life of broiler carcasses. *J. Food Prot.* 58: 1077-1082.

400. Shehata, A. E., Khalafalla, S. M., Magdoub, M. N., and Hofi, A. A. 1976. The use of nisin in the production of sterilized milk drinks. *Egyptian J. Dairy Sci.* 4: 37-42.

401. Shehata, A. E., Khalafalla, S. M., El-Magdoub, M. N., and Hofi, A. A. 1977. The use of nisin in the production of sterilized milk drinks. *Milchwissenschaft* 32: 412-416.

402. Sheldon, B. W., Hale, S. A., and Beard, B. M. 1996. Efficacy of incorporating nisin-based formulations into absorbent meat tray pads to control pathogenic and spoilage microorganisms. *Poultry Sci.* 75 (Suppl.): 97.

403. Shiba, T., Wakamiya, T., Fukase, K., Ueki, Y., Teshima, T., and Nishikawa, M. 1991. Structure of the lanthionine peptides nisin, ancovenin and lanthiopeptin. In In *Nisin and Novel Lantibiotics.* eds. G. Jung and H. –G. Sahl, pp 113-122. Leiden: Escom.

404. Shillinger, Y. I., Bogoroditska, V. P., and Osipova, I. N. 1969. Hygienic characteristics of a Soviet-made preparation of nisin, an antibiotic employed for preservation of food products. *Vop. Pitan.* 28: 44-48.

405. Shtenberg , A. J., and Ignat'ev, A. D. 1970. Toxicological evaluation of some combinations of food preservatives. *Food Cosmet. Toxicol.* 8: 369-380.

406. Siegers, K., and Entian, K. D. 1995. Genes involved in immunity to the lantibiotic nisin produced by *Lactococcus lactis* 6F3. *Appl. Environ. Microbiol* 61: 1082-1089.

407. Siegers, K., Heinzmann, S., and Entian, K.-D. 1996. Biosynthesis of lantibiotic nisin. Post-translational modification of its prepeptide occurs at a multimeric membrane-associated lanthionine synthetase complex. *J. Biol. Chem.* 271: 12294-12301.

408. Siezen, R. J., Kuipers, O. P., and de Vos, W. M. 1996. Comparison of lantibiotic gene clusters and encoded proteins. *Antonie von Leeuwenhoek* 69: 171-184.

408a. Smittle, R. B., and Flowers, R. S. 1982. Acid tolerant microorganisms involved in the spoilage of salad dressings. *J. Food Prot.* 45: 977-983.

409. Somers, E. B., and Taylor, S. L. 1987. Antibotulinal effectiveness of nisin in pasteurized process cheese spreads. *J. Food Prot.* 50: 842-848.

410. Stankiewicz-Berger, H. 1969. Effect of nisin on the *lactobacilli* that cause greening of cured meat products. *Acta Microbiol. Polon. Ser. B* 13: 117-120.

411. Stark, P. and Sherman, J. M. 1935. Concerning the habitat of *Streptococcus lactis. J. Bacteriol.* 30: 639-646.

412. Stevens, K. A., Sheldon, B. W., Klapes, N. A., and Klaenhammer, T. R. 1991. Nisin treatment for the inactivation of *Salmonella* species and other Gram negative bacteria. *Appl. Environ. Microbiol.* 57: 3613-3615.

413. Stevens K. A., Sheldon, B. W., Klapes, N. A., and Klaenhammer, T. R. 1992. Effects of treatment conditions on nisin inactivation of Gram negative bacteria. *J. Food Prot.* 55: 763-766.

414. Stoyanova, L. G., and Egorov, N. S. 1998. Derivation of efficient nisin producers by fusing protoplasts of tow allied strains of *Lactococcus lactis* ssp. *lactis* with low nisin-synthesizing abilities. *Mikrobiologiya* 67:47-54.

415. Suarez, A. M., Azcona, J. I., Rodriguez, J. M., Sanz, B., and Hernandez, P. E. 1997. One-step purification of nisin A by immunoaffinity chromatography. *Appl. Environ. Microbiol.* 63: 4990-4992.

416. Sukumar, G., and Thompkinson, D. K., Gahlot, D. P., and Mathur, O. N. 1976. Studies on a method of preparation and preservation of kheer. *Indian J. Dairy Sci.* 29: 316.

417. Sulzer, G., and Busse, M. 1993. Behaviour of *Listeria* spp. during the production of camembert cheese under various conditions of inoculation and ripening. *Milchwissenschaft* 48: 196-200.

418. Szabo, E. A. and Cahill, M. E. 1998. The combined effects of modified atmosphere, temperature, nisin and ALTA 2341 on the growth of *Listeria monocytogenes. Int. J. Food Microbiol.* 43: 21-31.

419. Szadowska, M., Lipinska, E., Jakubczyk, E., and Lipniewska, D. 1974. Whey media enriched with substances stimulating the synthesis of nisin. *Roczniki Instytutu Przemyslu Mleczarskiego*, 16: 5-13.
420. Szybalski, W. 1953. Cross resistance of *Micrococcus pyogenes* var. *aureus* to thirty-four antimicrobial drugs. *Antibiot. Chemother.* 3: 1095-1103.
421. Tanaka, N., Traisman, E., Plantinga, P., Finn, L., Flom, W., Meske, L., and Guggisberg, J. 1986. Evaluation of factors involved in antibotulinal properties of pasteurized cheese spreads. *J. Food Prot.* 49:526-531.
423. Taniguchi M., Hoshino, K., Urasaki, H., and Fujii, M. 1994. Continuous production of an antibiotic polypeptide (nisin) by *Lactococcus lactis* using a bioreactor coupled to a microfiltration module. *J. Ferment. Biotechnol.* 77: 704-708.
424. Taylor, L. Y., Cann, D. D., and Welch, B. J. 1990. Antibotulinal properties of nisin in fresh fish packaged in an atmosphere of carbon dioxide. *J. Food Prot.* 53: 953-957.
425. Taylor, S. L., and Somers, E. B. 1985. Evaluation of the antibotulinal effectiveness of nisin in bacon. *J. Food Prot.* 48: 949-952.
426. Taylor, S. L., Somers, E. B., and Krueger, L. A. 1984. Antibotulinal effectiveness of nisin-nitrite combinations in culture medium and chicken frankfurter emulsions. *J. Food Prot.* 234-239.
427. Ter Steeg, P. F. 1993. Interacties tussen Nisine, Lysozym en Citrat in Bioconserving. *De Ware (n) Chemicus* 23: 183-190.
428. Thomas, L. V., and Wimpenny, J. W. T. 1996. Investigation of the effect of combined variations in temperature, pH and NaCl concentration on nisin inhibition of *Listeria monocytogenes* and *Staphylococcus aureus. Appl. Environ. Microbiol.* 62: 2006-2012.
429. Thomas, L. V., Davies, E. A., Delves-Broughton, J., and Wimpenny, J. W. T. 1998. Synergist effect of sucrose fatty acid esters on nisin inhibition of Gram-positive bacteria. *J. Appl. Microbiol.* 85: 1013-1022.
430. Thomas, L. V., Wimpenny, J. W. T., and Barker, G. C. 1997. Spatial interactions between subsurface bacterial colonies in a model system: a territory model describing the inhibition of *Listeria monocytogenes* by a nisin-producing lactic acid bacterium. *Microbiol.* 143: 2575-2582.
431. Thomas, M. A. 1977. *The Processed Cheese Industry.* Dept. of Agriculture, New South Wales, Australia.
432. Thorpe, R. H. 1960. The action of nisin on spoilage bacteria. I. The effect of nisin on the heat resistance of *Bacillus stearothermophilus* spores. *J. Appl. Bacteriol.* 23: 136-143.
433. Tiemann, U., and Belitz, H. D. 1969. About nisin: I. Effect of proteases on nisin activity. *Z. Lebesm. Unters. Forsch.* 141: 260-262.
434. Tramer, J. 1964. The inhibitory action of nisin on *Bacillus stearothermophilus.* In *The Action, Use and Natural Occurrence of Microbial Inhibitors in Food.* ed. N. Molin, pp. 25-33. Stockholm: Almqvist & Wiksell.
435. Tramer, J. 1966. Nisin in food preservation. *Chem. Ind.* 11: 446-450.
436. Tramer, J., and Fowler, G. G. 1964. Estimation of nisin in foods. *J. Sci. Food Agric.* 15: 522-528.
437. Tsai, H. J., and Sandine, W. E. 1987. Conjugal transfer of nisin plasmid genes from *Streptococcus lactis* 7962 to *Leuconostoc dextranicum* 181. *Appl. Environ. Microbiol.* 53: 352-357.
438. Turtell, A., and Delves-Broughton, J. 1998. International acceptance of nisin as a food preservative. *Bull. Int. Dairy Fed.* 329: 20-23.
439. Ueckert, J. E., ter Steeg, and Cole, P. J. 1998. Synergistic antibacterial action of heat in combination with nisin and magainin II amide. *J. Appl. Microbiol.* 85: 487-494.
440. Ukuku, D. O. and Shelef, L. A. 1997. Sensitivity of six strains of *Listeria monocytogenes* to nisin. *J. Food Prot.* 60: 867-869.
441. Ushakova, R. F. 1972. Effect of nisin (English) on the microbial count of grainy caviar. *Trudy Vsesoyuznogo Nauchno-Issledovatel'skogo Instituta Morskogo Rybnogo Khozyaistva i Okeanografi (VNIRO)* 88: 24-32 .
442. van de Ven, Ven, F. J. M., van den Hooven, H. W., Konings, R. N. H., and Hilbers, C. W. 1991a. NMR studies of lantibiotics. The structure of nisin in aqueous solution. *Eur. J. Biochem.* 202: 1181-1188.
443. van de Ven, Ven, F. J. M., van den Hooven, H. W., Konings, R. N. H., and Hilbers, C. W. 1991b. The spatial structure of nisin in aqueous solution. In *Nisin and Novel Lantibiotics.* eds. G. Jung and H. –G. Sahl, pp 35-42. Leiden: Escom.
444. van den Hooven, H. W., Doeland, C. C. M., van de Kamp, M., Konings, R. N. H., Hilbers, C. W., van de Ven, F. J. M. 1996. Three-dimensional structure of the lantibiotic micelles of dodecylphocholine and of sodium dodecylsulphate. *Eur. J. Biochem.* 235: 382-393.
445. Van't Hul, J. S., and Gibbons, W. R. 1997. Neutralization/recovery of lactic acid from *Lactococcus lactis*: effects on biomass, lactic acid, and nisin production. *World J. Microbiol. Biotechnol.* 13: 527-532.
446. Vas, K. 1963. Use of nisin in the thermal preservation of tomato products. *Fruchtasft-Industrie ver Confructa* 8: 73-77.

447. Vas, K., Kiss, I., and Kiss, N. 1967. Use of nisin for shortening the heat treatment in the sterilisation of green peas. *Zeitschrift fur Lebensmittle-Untersuchung und Forschung* 133: 141-144.

448. Verheul, A., Russell, N. J., van 't Hof, R., Rombouts, F. M., and Abee, T. 1997. Modifications of membrane phospholipid composition in nisin-resistant *Listeria monocytogenes* Scott A. *Appl. Environ. Microbiol.* 63: 3451-3457.

449. Waites, M. J., and Ogden, K. 1987. The estimation of nisin using ATP-bioluminometry. *J. Inst. Brewing* 93: 30-32.

450. Wajid, H. R. A., and Kalra, M. S. 1976. Nisin as an aid for extending the shelf-life of sterilised milk. *J. Food Sci. Technol. (Mysore)* 13: 6-8.

451. Wakamiya, T., Fukase, K., Sano, A., Shimbo, K., Kitazawa, M., Hormoto, S., Fujita, H., Kubo, A., Maeshiro, Y., and Shiba, T. 1991. Studies on the chemical synthesis of the lanthionine peptide nisin. In *Nisin and Novel Lantibiotics.* eds. G. Jung and H. –G. Sahl, pp 189-203. Leiden: Escom.

452. Walker, S. J., Archer, P., and Banks, J. G. 1990. Growth of *Listeria monocytogenes* at refrigeration temperatures. *J. Appl. Bacteriol.* 68: 157-162.

453. Walsh, E. M., McSweeney, P. L. H., and Fox, P. F. 1996. Use of antibiotics to inhibit non-starter lactic acid bacteria in Cheddar cheese. *Int. Dairy J.* 6: 425-431.

454. Wan, J., Gordon, J., Hickey, M. W., Mawson, R. F., and Coventry, M. J. 1996. Adsorption of bacteriocins by ingestible silica compounds. *J. Appl. Bacteriol.* 81: 167-173.

455. Wan, J., Hickey, M. W., and Coventry, M. J. 1995. Continuous production of bacteriocins, brevicin, nisin, and pediocin using calcium alginate immobilized bacteria. *J. Appl. Bacteriol.* 79: 671-676.

456. Wang, S. Y., Dockerty, T. R., Ledford, R. A., and Stouffer, J. R. 1986. Shelf-life extension of vacuum packaged frankfurters made from beef inoculated with *Streptococcus lactis*. *J. Food Prot* 49: 130-134.

457. Weagant, S. D., Sado, P. N., Colburn, K. G., Torkelson, J. D., Stanley, F. A., Krane, M. H., Shields, S. C., Thayer, C. F. 1988. The incidence of *Listeria* species in frozen seafood products. *J. Food Prot.* 51: 655-657.

458. Wells, J. M., Liao, C. –H, and Hotchkiss, A. T. 1998. *In vitro* inhibition of soft-rotting bacteria by EDTA and nisin and *in vivo* response on inoculated fresh carrots. *Plant Dis* 82: 491-495.

459. Wessels, S., and Huss, H. H. 1996. Suitability of *Lactococcus lactis* subsp. *lactis* ATCC 11454 as a protective culture for lightly preserved fish products. *Food Microbiol.* 13: 323-332.

460. Wessels, S., Jelle, B., and Nes, I. F. 1998. *Bacteriocins of the Lactic Acid Bacteria: An Overlooked Benefit for Food.* Danish Toxicology Centre, Denmark.

461. Whitehead, H. R. 1933. A substance inhibiting bacterial growth, produced by certain strains of lactic streptococci. *Biochem. J.* 27: 1793-1800.

462. Whitehead, H. R., and Riddet, W. 1933. Slow development of acidity in cheese manufacture. *N. Z. J. Agric.* 46: 225-229.

463. WHO Expert Committee on Biological Standardization. 1970. Twenty-second report. *Wld. Hlth. Org. Techn. Rep. Ser.* No 444.

464. Wilimowska-Pelc, A., Olichwier, Z., Malicka-Blaszkiewicz, M., and Mejbaum-Katzenellenbogen, W. 1976. The use of gel-filtration for the isolation of pure nisin from commercial products. *Acta Microbiol. Pol.* 25: 71-77.

465. Wimpenny, J. W. T., Leistner, L., Thomas, L. V., Mitchell, A. J., Katsaras, K., and Peetz, P. 1995. Submerged bacterial colonies within food and model systems: their growth, distribution and interactions. *Int. J. Food Microbiol.* 28: 299-315.

466. Winkler, S., and Fröhlich, M. 1957. To prove the influence of nisin on the most important organisms which promote ripening in Emmenthal cheese. *Milchwissenschaft* 8: 87-96.

467. Winkowski, K., Ludescher, R. D., and Montville, T. J. 1996. Physicochemical characterization of the nisin-membrane interaction with liposomes derived from *Listeria monocytogenes*. *Appl. Environ. Microbiol.* 62: 323-327.

468. Wirjantaro, T., and Lewis, M. J. 1996. Effect of nisin and high temperature pasteurization on the shelf-life of whole milk. *J. Soc. Dairy Technol.* 49: 99-102.

469. Wisotzkey, J. D., Jurshuk, P., Fox, G. E., Deinhard, G. and Poralla, K. 1992. Comparative sequence analysis on the 16S rRNA DNA of *Bacillus acidocaldarius, Bacillus acidoterrestris* and *Bacillus cycloheptanicus* and proposal for creation of new genus *Alicyclobacillus*. *Int. J. Syst. Bacteriol.* 44: 263-269.

470. Wolf, C. E., and Gibbons, W. R. 1996. Improved method for quantification of the bacteriocin nisin. *J. Appl. Bacteriol.* 80: 453-457.

471. Wong, A., and Luchansky, J. 1999. Growth and biocontrol of enterotoxigenic *Bacillus cereus* in infant formula. *FRI Newsletter* (University of Wisconsin-Madison, USA) 11: 2-3.

472. Wood, S. L., and Waites, W. M. 1988. Factors affecting the occurrence of *Bacillus cereus* in liquid whole egg. *Food Microbiol.* 5: 103-107.
473. Yang, R., Johnson, M. C., and Ray, B. 1992. Novel method to extract large amounts of bacteriocins from lactic acid bacteria. *Appl. Environ. Microbiol.* 58: 3353-3359.
474. Yezzi, T. L., Ajao, A. B., and Zottola, E. A. 1993. Increased nisin in Cheddar-type cheese prepared with pH control of the bulk starter culture system. *J. Dairy Sci* 76: 2827-2831.
475. Yoo, J.-Y., Choi, S.-Y., Jin, Y.-O., Koo, Y. –J., and Chung, K. –S. 1989. Process kinetics of nisin production in batch and continuous culture. *Kor. J. Appl. Microbiol. Bioeng.* 17: 504-509.
476. Yoo, J. -Y., Kwon, D. J., Park, J. H., and Koo, Y. J. 1994. Use of nisin as an aid in reduction of thermal process of bottled *sikhae* (rice beverage). *J. Microbiol. Biotechnol* 4: 141-145.
477. Yuan, A., Cousin, M. A., and Muriana, P. M. 1998. Microbial activity after the use of nisin in salad dressings. In *Book of Abstracts. 1998 IFT Annual Meeting*. 20-24 June, 1998. Atlanta, Georgia, USA. Institute of Food Technologists.
478. Yuste, J., Mor-Mur, M., Capellas, M., Guamis, B., and Pla, R. 1998. Microbiological quality of mechanically recovered poultry meat treated with high hydrostatic pressure and nisin. *Food Microbiol.* 15: 407-414.
479. Zaki, M. S., El Mansy, A. H., Hassan, Y. M., and Rahma, E. H. A. 1976. Effect of nisin in saturated brine and storage on the quality of dried bolti fish (*Tilapia nilotica*). *Die Nahrung* 20: 691-697.
480. Zapico, P., de Paz, M., Medina, M., and Nunez, M. 1999. The effect of homogenization of whole milk, skim milk and milk fat on nisin activity against *Listeria innocua. Int. J. Food Microbiol.* 46: 151-157.
481. Zapico, P., Medina, M., Gaya, P., and Nunez, M. 1998. Synergistic effect of nisin and the lactoperoxidase system on *Listeria monocytogenes* in skim milk. *Int. J. Food Microbiol.* 40: 35-42.
482. Zehren, V. L., and Nusbaum, D. D. 1992. *Processed Cheese*. Cheese Reporter Publishing Co. Inc., Madison, Wisconsin, USA. 364 pp.
483. Zezza, N., Pasinin, G., Lombardi, A., Mercenier, A., Spettoti, P., Zamorani, A., and Nuti, M. P. 1993. Production of a bacteriocin active on lactate fermenting clostridia by *Lactococcus lactis* subsp. *lactis* immobilized in coated alginate beads. *J. Dairy Res.* 60:581-591.

B. Ray
K. W. Miller

19

Pediocin

I. INTRODUCTION

The term "pediocins" is used here for the ribosomally synthesized antibacterial peptides or bacteriocins produced by certain strains among the species that are currently included in the genus *Pediococcus* (Garvie, 1986). At present, *Pediococcus cerevisiae* is no longer considered a separate species and *Pediococcus halophilus* has been assigned a new genus, *Tetragenococcus* (Collins et al., 1990). These two species are not included among the pediocin-producers in this chapter. A *Lactobacillus plantarum*, strain WHE92, that was isolated from cheese and found to be a pediocin producer (Ennahar et al., 1996) as well as several strains of *Lactococcus lactis* (Chikindas et al., 1995; Horn et al., 1998) and *Escherichia coli* (Marugg et al., 1992; Bukhtiyarova et al., 1994; Miller et al., 1998a,b) genetically engineered to produce pediocins are included here.

The genus *Pediococcus* is included with the group collectively designated as lactic acid bacteria (LAB). Strains of two species in this genus, *Pediococcus pentosaceus* and *Pediococcus acidilactici*, naturally found in many plant sources, have been used for a long time, both in natural and pure culture (as starter cultures) fermentations to produce fermented foods of plant, meat, fish, cereal and dairy origin (Ray, 1995). Their cells are spherical, appear as tetrads and pairs, but never single cells or chains. They are Gram-positive, nonmotile, facultative anaerobes and homolactic fermentative. Several strains of *Ped. pentosaceus* and *Ped. acidilactici* have many identical and overlapping biochemical and physiological characteristics that differ from the "type/strains" of both species and thus, it is often difficult to accurately identify an isolate at a species level without performing DNA/DNA hybridization. The two species differ by only 20 to 30% in DNA homology. However, based on the characteristics of the respective "type strains", *Ped. acidilactici* can arbitrarily be differentiated from *Ped. pentosaceus* species by its ability to grow at 50°C and survive 70°C for 10 min, but inability to ferment maltose (Garvie, 1986; Anonymous, 1996; Tanasupawat et al., 1993).

Several strains of *Ped. acidilactici* and *Ped. pentosaceus* have been reported to produce pediocins. In recent years, several chapters and reviews have been published that mostly narrated the early studies on pediocins (Ray, 1992; 1994; 1995; 1996; Ray & Hoover, 1993; De Vuyst, 1994; Jack et al., 1995). In the current chapter, more emphasis will be given to the recent studies of pediocins.

Early studies in the 1960s identified that two strains of *Ped. pentosaceus*, FBB61 and L7230, isolated from cucumber fermentation brine produced antagonist effects against associated bacteria; subsequently the antibacterial components were suggested to be bacteriocins (Fleming et al., 1975). Later, Daeschel and Klaenhammer (1985) reported that both strains produced the same bacteriocin, which they designated as pediocin A. In 1987, Bhunia et al. (1987) reported that *Ped. acidilactici* strain H produced a bacteriocin, which later they designated as pediocin AcH (Bhunia et al., 1991). Gonzalez and Kunka (1987) also reported in the same year that *Ped. acidilactici* PAC1.0 [originally designated as #B5627 in the NRRL culture collection, Peoria, IL; Gonzalez & Kunka, 1983] produced a bacteriocin that was designated as pediocin PA-1. Later, independent studies revealed that both pediocin PA-1 and pediocin AcH have the same nucleotide and amino acid sequences and thus are essentially the same (Henderson et al., 1992; Motlagh et al., 1992a). It has been suggested before to use a common designation, pediocin PA-1/AcH for the two to avoid confusion (Ray, 1996). Several researchers have used this designation (pediocin PA-1/AcH), while there are others who are using pediocin PA-1 or pediocin AcH. Prior to the 1990's, several researchers also have reported pediocin production by different strains of *Ped. pentosaceus* (Graham & McKay, 1985) and *Ped. acidilactici* (Hoover et al., 1988; 1989; some of these strains were found not to be pediocin producers in our laboratory; Ray et al., 1989a). Amino acid sequences of these pediocins were not determined. During the 1990's pediocin production by other *Ped. acidilactici* strains were reported. These included: strain JD 1-23 (Berry et al., 1990; 1991), strain M (Kim et al., 1992: later designated as KFRI168 by Kim et al., 1993; Elegado et al., 1997), strain PC (Jager & Harlander, 1992), strains AB6, RS2, etc. (a total of 18 strains from foods; Bhunia et al., 1994), strain SJ-1 (Schved et al., 1993), strain UL5 (Daba et al., 1994: originally reported as *Leuconostoc mesenteroides* UL5; Daba et al., 1993),and strain PO$_2$ (Coventry et al., 1995; *TABLE 1*). Among these strains, nucleotide and/or amino acid sequences of four other pediocins from strains, UL5 (Daba et al., 1994), F (Ray et al., 1989; Miller, unpublished data), RS2 (Yang & Ray, unpublished data) and a commercial strain (Lozano et al., 1992), besides strains PAC 1.0 and H (and LB42-923), have been studied and all have the same sequence as pediocin PA-1/AcH. In addition, two of our collaborators, one in India (Central Food Research Institute, Mysore: strain isolated from natural vegetable fermentation) and another in Indonesia (Udayana University, Denpasar, Bali: strain isolated from naturally fermented meat) reported that their isolates also have the same nucleotide base sequence and amino acid sequence as pediocin PA-1/AcH. A careful analysis of the reported pediocin-producing strains in the USA revealed the strains were obtained either from reputed culture collections or from commercial culture houses or from fermented meat products. Many of them, if not all, probably originated from the same strain (Ray, 1994, 1995; Ray & Hoover, 1993). Those strains have been carefully studied and found to have the same plasmid harboring the pediocin-production phenotype. Besides *Ped. acidilactici* strains, two strains of *Pediococcus parvulus*, isolated from vegetables, were found to produce pediocins and on analysis were found to have an identical amino acid sequence as pediocin PA-1/AcH (Bennik et al., 1997). It was mentioned earlier that *Lab. plantarum* WHE92, isolated from cheese, was also found to produce pediocin PA-1/AcH (Ennahar et al., 1996).

TABLE 1: Pediocin-Producing strains used in most studies.

Strains	Other designations	Original source/comments
Ped. acidilactici		
H[a]	JBL 1095	Fermented beef stick
E, F[a]	None	Fermented beef stick
LB42-923[a]	None	Transconjugant of H
PAC1.0[a]	NRRL-B-5627, NRRL-B-18050	NRRL Collection, Peoria, IL
	B 4627, JLB 1096	
PC	M, KFRI 161	Marschall Div (WI)
		Diversitech (FL)
		Marschall Div (WI)
PO$_2$	JBL 1097	Diversitech (FL)
MC-03	None	Pepperoni
Ab6, RS-2[a]	None	Fermented sausage
JD 1-23	None	Commercial starter
SJ1	None	Fermented sausage
UL5[a,b]	*Leu. mesenteroides* UL5	Cheddar cheese. Produces mesenterocin
Not designated[a,c]		CHR Hansen
L 50[b]	*Enterococcus faecium* L50	Spanish dry sausage. Produces enterocin
Ped. parvulus		
ATO34[a]		
ATO77[a]		
Lab. plantarum		
WHE92[a]	None	Munster cheese
Ped. pentosaceus		
FBB61	*Ped. cerevisiae,* ATCC43200	Cucumber brine
FBB63	*Ped. cerevisiae*	Not known
Pediococcus ISK-1[b]	*Staphylococcus warnari*	Fermented rice bran

[a]Amino acid sequence same as pediocin PA-1/AcH. For more information see Ray, B., 1995, 1996.
[b]Initially the species were incorrectly identified.
[c]Lozano et al., (1992).

During the 1990s, pediocin production by several strains of *Ped. pentosaceus* was also reported (Skytta et al., 1993; Strasser de Sadd & Monca de Narda, 1993), but their amino acid sequence were not studied. Although the amino acid sequence of pediocin A of *Ped. pentosaceus* FBB61 is not known, Piva and Headon (1994) reported that it has an Mr of about 80 kDa; thus, it cannot have the same amino acid sequence as pediocin PA-1/AcH which has a Mr of 4.6 kDa.

Several strains have also been wrongly identified and later recognized to be different (*TABLE 1*). Daba et al., (1993) initially designated a bacteriocin producing strain to be *Leuconostoc mesenteroides*, but later reported it to be *Ped. acidilactici* UL5 (Daba et al., 1994). Cintas et al. (1995) also reported that they have isolated a broad-spectrum pediocin-producing strain *Ped. acidilactici* L50. However, later studies revealed the isolate to be a strain of *Enterococcus faecium* (Cintas et al., 1998). Kim et al. (1998) reported to have isolated a *Lactococcus lactis* strain that produced pediocin PA-1/AcH. They recently found the isolate to be a *Pediococcus* species. Kimura et al. (1998) observed that a salt-tolerant *Pediococcus* species, ISK-1, produced bacteriocin ISK-1, which could be a lantibiotic. However, morphological and biochemical characteristics of the strain were not reported. Recently, it has been identified as a strain of *Staphylococcus warneri* (personal communication A. Ishizaki). These reports stress the importance of proper identification of an isolate (microscopic, Gram-staining, physiological and biochemical profiles, and if necessary, 16S rRNA and DNA homology studies).

Among the pediocins, pediocin PA-1/AcH has been extensively studied. The materials in this chapter will reflect that. Where appropriate, information on pediocins of other strains will also be included. However, in the absence of amino acid sequence, reporting that a specific strain is producing a different pediocin (such as pediocin JD, pediocin 5) probably is not correct. A specific name/designation should be given only when the amino acid sequence of a bacteriocin, including a pediocin, is correctly known (Jack et al., 1995).

II. OCCURRENCE

A. Biosynthesis

1. Plasmid linkage of pediocin production phenotype. Bacteriocins of Gram-positive bacteria, including *Pediococcus* spp., are ribosomally synthesized peptides. Among LAB strains, the structural gene encoding a specific bacteriocin can be linked to a plasmid, a transposon, or the chromosome (Jack et al., 1995). Among the pediocin-producing strains so far studied, the genetic determinants for pediocin-production phenotype have been identified to be plasmid-linked. In general, most of the plasmids encoding the pediocin-phenotype are smaller than 20 Kb. Although among strains, especially in *Ped. acidilactici*, there are some differences in molecular weight of the plasmids linked to pediocin production. This may be due to differences in the techniques of gel electrophoresis and the selection of MW-standards. Limited studies have also revealed that another gene, encoding for an immunity protein that confers immunity to the producer cells against their own pediocin, is also located in the same plasmid that encodes the pediocin gene, i.e., the two genes are linked.

Daeschel and Klaenhammer (1985) reported that in the two *Ped. pentosaceus* strains, FBB61 and L7230, the pediocin and the immunity genes are encoded in a 13.6 MDa (about 20.4 Kb) plasmid, pMD136. Graham and McKay (1985) also showed that in *Ped. pentosaceus* FBB63 (originally designated as *Pediococcus cerevisiae* FBB63) the pediocin gene is located in a 10.5 MDa (about 15.8 Kb) plasmid. They did not study the linkage of the immunity gene.

Plasmid-linkage of pediocins in *Ped. acidilactici* strains has been reported by several researchers. Gonzalez and Kunka (1987) showed that in *Ped. acidilactici* PAC 1.0 the gene encoding for pediocin PA-1 is located in a 6.2 MDa (about 9.3 kb) plasmid, pSRQ11. At that time, they reported that the related immunity gene was not located in pSRQ11. However, later studies proved that both genes were encoded by pSRQ11 (Marugg et al., 1992). Hoover et al. (1988) reported the plasmid linkage of pediocin production in *Ped. acidilactici* PO$_2$, PC, B5627 (same as strain PAC 1.0) and *Ped. pentosaceus* MC-03 (later found to be a *Ped. acidilactici* strain; Ray, B. unpublished data). In all four strains, the pediocin gene was located in a 5.5 kDa (about 8.3 kb) plasmid. Jager and Harlander (1992) also found that the pediocin-gene in strains PO$_2$, PC and PAC 1.0 was linked to 5.5 MDa (8.3 kb) plasmids. The differences in plasmid size (9.3 kb vs. 8.3 kb) encoding for pediocin PA-1 in strains PAC 1.0 could be due to differences in analytical methods used by the researchers. Ray et al. (1989a) and Kim et al. (1992) reported that in *Ped. acidilactici* H, E, F and M, the pediocin gene and immunity gene are linked to plasmids of MW between 11 to 12 kb. Later studies, however, revealed that in all four strains, both genes are located in the same plasmid, pSMB74, containing 8877 bp (8.9 kb) and encoding for the same pediocin, designated as pediocin AcH (Ray et al., 1992; Motlagh et al., 1992a, 1994; Bukhtiyarova et al., 1994). Subsequent studies showed that both pediocin PA-1 and pediocin AcH have the same amino acid sequence (Henderson et al., 1992; Motlagh et al., 1992) and most probably the two plasmids, pSMB74 and pSRQ11, are the same. Schved et al. (1993) observed that in pediocin-producing *Ped. acidilactici* SJ-1, isolated from naturally fermented sausage in Israel, the pediocin-gene is located in a 4.6 MDa (about 6.2 kb) plasmid. Daba et al. (1994) reported that in *Ped. acidilactici* UL5 (isolated from local Cheddar cheese), the pediocin-gene was located in a 12.5 kb plasmid, and have the amino acid sequence of the pediocin (designated as pediocin 5) was the same as pediocin PA-1/AcH. As mentioned earlier, one researcher in India and one researcher at Bali in Indonesia found that two strains of *Ped. acidilactici*, isolated from food sources, had the same amino acid sequence the same as pediocin PA-1/AcH.

An unrelated study by Luchansky et al. (1992) showed that three pediocin-producing *Ped. acidilactici* strains, H, PAC1.0 and PO$_2$ (designated respectively as JLB1095, JLB1096 and JLB1097) harbored the same DNA fragment of 1,861 kb, indicating the likelihood of all three strains originating from a single strain. Bhunia and Johnson (1992) observed that a specific monoclonal antibody, Ped-2B2, obtained following injecting broken heat-killed cells of *Ped. acidilactici* H into BALB/c mice, cross-reacted only with 16 pediocin-producing *Ped. acidilactici* strains isolated from different brands of fermented sausages. In a later study, Bhunia (1994) found that another monoclonal antibody, RS2-AR, produced against pediocin RS2 from *Ped. acidilactici* RS2, reacted only with pediocin RS2 and pediocin AcH, but not with bacteriocins from several other LAB. These observations raise the possibility that all pediocin-producing *Ped. acidilactici* strains, so far studied, might have the same origin and probably produce the same pediocin, pediocin PA-1/AcH.

Among pediocin PA-1/AcH producing non-*Ped. acidilactici* strains, *Lab. plantarum* WHE92 seems to harbor an 11 kb plasmid that encodes a pediocin gene (Ennahar et al., 1996). The two *Ped. parvulus* strains ATO34 and ATO77 appeared to have the plasmid, pRSQ11 (pSMB74) encoding for pediocin PA-1/AcH (Bennick et al., 1997).

2. Genetics of pediocins. The location of the genes involved in pediocin PA-1/AcH production has been studied separately using plasmids pRSQ11 (in *Ped. acidilactici* PAC1.0) and pSMB74 (in *Ped. acidilactici* H and LB42-923). Like other low molecular weight bacteriocins of Gram-positive bacteria, pediocin PA-1/AcH initially is translated in the cytoplasm as 62 amino acid containing prepediocin. Although prepediocin has bactericidal activity, in the reduced cytoplasmic environment, it may remain inactive due to the inability of the four cysteines in the molecule to form two disulfide bonds ñ which are essential for bactericidal activity (Ray et al., 1999). As prepediocin molecules are sensitive to the hydrolytic action of cytoplasmic proteases, they may need to be rapidly transported through the cytoplasmic membrane by a specific ABC-transport system. An 18 amino acid segment in the N-terminus of prepediocin, designated the leader peptide, helps in the recognition of the specific transport system. During membrane translocation, the leader peptide is enzymatically removed at a double glycine (Gly^{-2}-Gly^{-1}) cleavage site by the transporter proteins and the pediocin PA-1/AcH containing 44 amino acids is released from the cell to the environment. Thus, the transporter protein(s) have two functions, namely, enzymatic cleavage of the leader peptide and the transport and export of prepediocin-pediocin PA-1/AcH molecules. Previous studies have predicted that the structural gene for pediocin (actually prepediocin) and host cell immunity gene are linked and located in the same plasmid (pSMB74 in *Ped. acidilactici* H). Attempts were made to locate the sites of these two genes in the related plasmids. Details of the methods used in locating the genes associated with pediocin PA-1/AcH have been reviewed earlier (Ray, 1994, 1996; Ray & Hoover, 1993; Jack et al., 1995). A brief description is given here summarizing the findings.

Analysis of the complete nucleotide sequence of the plasmid pSMB74 revealed that it has a total of 8,877 bp with 16 unique restriction enzyme (RE) cleavage sites (*TABLE 2*; Motlagh et al., 1994; Ray et al., 1992). Analysis of the cloned RE-cleaved fragments generated from pSRQ11 and pSMB74 revealed that in a segment of 3,500 bp, there are four open reading frames (ORF 1 to 4), each separated by a ribosome binding site. There is a promoter upstream of ORF1 and a terminator downstream of ORF4. Thus, it appears to be an operon-system and is designated as the *ped* (or *pap*) operon. The ORF's are designated as *ped A, B, C, D* in pSRQ11 and as *pap A, B, C, D* in pSMB74 (Marugg et al., 1992; Motlagh et al., 1992a, 1994; Bukhtiyarova et al., 1994; *FIGURE 1*). The first gene containing 186 bp is the structural gene for the 62 amino acid containing prepediocin. The other three ORF's, respectively, encode for proteins containing 112, 174 and 724 amino acids. Mutation and deletion analyses of the four genes revealed that active pediocin PA-1/AcH could be produced by the transformed host cells carrying the intact *ped A* and *D* (*pap A* and *D*) genes. Since *ped A* (*pap A*) encodes for prepediocin, it was concluded that *ped D* (or *pap D*) protein has the dual functions of removing the N-terminal leader peptide as well as transporting prepediocin-pediocin through the membrane resulting in the release of pediocin PA-1/AcH into the environment (Bukhtiyarova et al., 1994; Marugg et

TABLE 2: Unique restriction enzyme cutting sites in pSMB74.

Restriction enzymes	Cutting sites at bp[a]
Aoc1, Bsu36I, Cvn1, Eco81I, Mst2, Sau1	380
Bam H1	782
Sal 1	2134
Afl 2	2284
Bst X1	2329
Xma 3, Eag 1, Eco 52I	2874
Bal 1	3246
Acy 1, Aha 2	3838
Bgl 1	5783
Xba 1	5854
EcoR 1	7724
Rsr 2	7976
Hgi A1, Sac 1, Sst 2	8104
EcoT 221, Nsi 1, Ava 3	8345

[a]Total bp 8877. The positions in relation to one of the three cutting sites of Hind III. The three Hind III sites are at 1, 3470 and 6210 bp. The *pap* operon of 3500 bp is between bp 3136 and 6636 including the promoter, four open reading frames and the terminator (Motlagh et al., 1994; and unpublished data, Ray, B.).

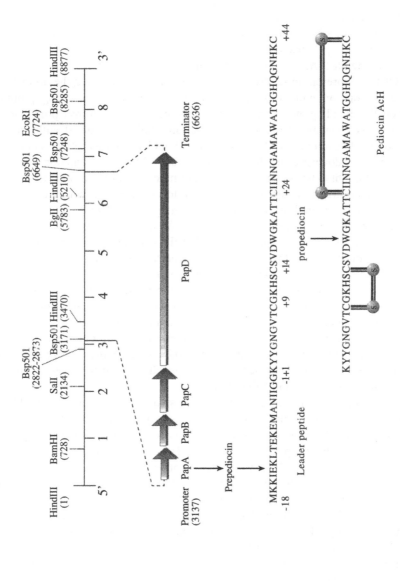

FIGURE 1. Physical map of the *pap* (*ped*) operon carried by the 8.6 kb *Ped. acidilactici* AcH plasmid, pSMB74. The 3.6 kb operon is encoded on a Bsp501 restriction fragment. The functions of the four genes in the *pap* (*ped*) cluster are described in the text. The amino acid sequences of the 62 amino acid prepediocin PA-1/AcH primary translation product and the 44 amino acid propediocin PA-1/AcH species produced by leader peptide processing also are shown. (From Motlagh et al., 1994. With permission.)

al., 1992). Bukhtiyarova et al. (1994) also suggested that Pap C protein, although not essential, probably acts as a helper for the efficient transportation function of Ped D protein. However, in a subsequent study, Venema et al. (1995), reported that both Ped C and Ped D proteins are required for the transportation of pediocin and Ped D, in addition, is involved in the removal of the leader peptide from prepediocin. Further investigation showed Ped D (Pap D) is essentially similar to some other known ATP-binding cassette-(ABC)-transporter proteins. Ped D (Pap D) has a proteolytic domain in its N-terminal part, an ATP-binding cassette at the C-terminal part and as many as six membrane-spanning segments located in between. For efficient transport of the prepediocin-pediocin molecules, it may form a homodimer (Havarstein et al., 1995). RNA-transcript investigations revealed that the *ped*-operon is transcribed in two separate mRNAs, one of about 1.2 kb, related to the ORF A, B and C, is abundant and the other of about 3.5 kb encoding all four ORF's, is present in relatively small amounts. By replacing the normal promoter of the *ped*-operon with a strong promoter or introducing a strong promoter upstream of ORF 4, Ped D protein can be over-expressed, resulting in over-production of pediocin PA-1/AcH (Venema et al., 1995). Deletion studies conducted with the four ORF's showed that deletion of one ORF upstream did not have a polar effect on the transcription of the genes downstream (Bukhtiyarova et al., 1994). The function of protein Ped B (Pap B) is thought to be associated with immunity of the host cells against pediocin PA-1/AcH, since there was no other ORF in the plasmid and also pediocin-production and immunity against pediocin are linked phenotypes (Motlagh et al., 1994). A later study by Venema et al. (1995) also suggested that Ped B is the immunity protein necessary for the protection of producer cells against pediocin PA-1/AcH. However, the mechanism(s) by which it provides protection to the cytoplasmic membrane of the producer cells is not clear. In a study with a separate bacteriocin, lactococcin A, of a *Lactococcus lactis* strain, the immunity protein did not react directly with lactococcin A. It was hypothesized that the immunity protein probably blocks the lactococcin A-specific receptor protein on the membrane of the producer cells and protects the membrane from bactericidal action of lactococcin A (Nissen-Myer et al., 1993). The existence of a pediocin-specific receptor protein in the cytoplasmic membrane of pediocin-producing *Ped. acidilactici* strains has not been conclusively proved. At present, the exact mechanism of producer cell protection by the immunity protein is not definitely known.

In their study, Venema et al. (1995) suggested that a region located 135 bp upstream of Ped A may carry some element(s) that controls production of pediocin PA-1/AcH and the product of one of the genes in the *ped*-operon regulates expression of the operon by interacting with this upstream element(s). However, Bukhtiyarova et al. (1994) cleaved the *pap*-operon from pSMB74 using restriction enzyme *Bsp* 501 which cuts very close to the *pap A* gene, down stream of its promoter, and cloned it in pHSP9. The resultant vector pMBR1.0, when transformed into both *Escherichia coli* JM109 and *Ped. acidilactici* LB42, the transformed cells produced active pediocin PA-1/AcH in relatively large amounts. In a later study, Miller et al. (1998a,b) cloned only the nucleotide segment coding for the 44 amino acids of pediocin PA-1/AcH into a vector and the transformed *Esc. coli* JM109 produced pediocin PA-1/AcH in large amounts. Thus, the role of control elements on pediocin production is not clear (Venema et al., 1995).

Multiple functions for the 18 amino acid N-terminal leader peptide of prepediocin have been suggested by several researchers. Its main function is probably to recognize the specific Ped C-Ped D (Pap C-Pap D) transporter protein complex in the cyto-

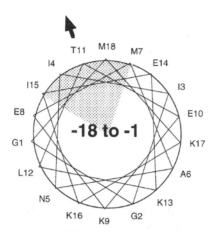

FIGURE 2. Schiffer-Edmundson helical wheel plot of the 18 amino acid leader peptide region of pediocin PA-1/AcH. The diagram indicates that this putative α-helical sequence is amphipatic. The calculated average sequence hydrophobic moment (<μ>) of the sequence is 4.0 (Eisenberg et al., 1984). The hydrophobic face of the helix is shaded (unpublished, Miller, K. W.).

plasmic membrane and direct a prepediocin molecule through this specific ABC-transporter. During its passage, the proteolytic domain in the N-terminus of Ped D (Pap D) recognizes the -Gly^{-2}-Gly^{-1}- region in the molecule and cleaves it, changing the molecule to pediocin PA-1/AcH. As indicated earlier, prepediocin molecules are potentially bactericidal, but in the reduced environment of the cytoplasm they remain inactive. Following release into the environment, the pediocin molecules are exposed to an oxidizing environment and the four cysteine form two disulfide bridges, between cys 9 and cys 14 and cys 24 and cys 44 and the molecules become biologically active (Ray et al., 1999). It was also found that prepediocin molecules, as compared to pediocin PA-1/AcH, are more prone to proteolysis. The other function of the leader peptide is to facilitate passage of the prepediocin through the transporter. In addition, the leader peptide is more hydrophilic than the rest of the molecules. This may prevent it from interacting with to the hydrophobic membrane (*FIGURE 2*).

3. Optimum conditions of production of pediocins. Pediocin producing strains of *Pediococcus* spp., while growing in a broth, an agar medium or a food system, synthesize pediocin and release it in the growth environment. The amount of pediocin produced by a strain in a growth medium can be quantitatively determined by a suitable bioassay system. Optimum growth conditions of pediocin production by the *Pediococcus* strains have been studied by different researchers in several bacteriological broths, and have been reviewed before (Ray, 1992, 1994, 1995, 1996; Ray & Hoover 1993). Some of these methods are briefly discussed here.

Pediocin production studies have been mainly conducted with *Ped. acidilactici* strains and many researchers used the complex lactobacilli-MRS-broth (Gonzalez & Kunka, 1987; Christensen & Hutkins, 1992; Lozano et al., 1992; Chikindas et al., 1993; Cho et al., 1996; Huang et al., 1996). Bhunia et al. (1987, 1988, 1991) used casein-glucose broth containing casein (1%), glucose (1%), yeast extract (0.5%), Tween 80 (0.1%), sodium acetate (0.5%), MgSO$_4$ (0.01%), MnSO$_4$ (0.005%) and sodium phosphate (0.2%). Biswas et al. (1991) compared, for the first time, the influences of different growth parameters, including several growth media on pediocin PA-1/AcH production by *Ped. acidilactici* H. The results showed that this strain produced relatively much higher levels of pediocin in non-buffered TGE broth, containing trypticase (or tryptone), glucose, yeast extract (each at 1%), Tween 80 (0.2%), MgSO$_4$ and MnSO$_4$ (each 0.005%), pH 6.8, than in MRS broth and several modifications of MRS broth, as well as buffered TGE broth (TGE broth + 0.2% each of sodium acetate, sodium citrate and sodium phosphate, pH 6.8). Influences of several carbohydrate sources, incubation temperatures, initial and terminal pH's, several B vitamins and trace elements were also evaluated. Their studies showed that the highest level of pediocin was produced in TGE broth, especially when the concentrations of tryptone, glucose and yeast extract were increased to the 2% level each. In addition, at a terminal pH for the culture broth of 4.0 or below, more pediocin was present than at pH above 4.0. The highest level of pediocin was produced by growing the cells in TGE broth at 35°C to 37°C within 16 to 18 h starting with 1% initial inoculum. In general, the level of pediocin production appeared to be directly related to the amount of cell mass. Later studies by this group revealed that *Ped. acidilactici* strains differ greatly in the level of pediocin production (*TABLE 3*). However, all strains tested produced more pediocin in TGE broth than in MRS broth. One such variant, *Ped. acidilactici* LB42-923 (Ray et al., 1989b) produced much higher levels of pediocin than the parental strain H (Yang et al., 1992; Yang & Ray, 1994a). Extending the incubation time beyond 24 h reduced the pediocin level, probably due to the action of intracellular proteases from dead and lysed cells. Two other studies on optimum conditions of pediocin production by *Ped. acidilactici* were reported by Cho et al. (1996) and Huang et al. (1996). The highest levels of pediocin production in their studies were much lower than that obtained by Biswas et al. (1992) and Yang et al. (1992, 1994a). The differences in the amount of pediocin production by the different research groups could be due to the use of buffered vs. non-buffered growth media as well as due to the use of different producer strains, different indicator strains and differences in assay methods used to determine the activity or arbitrary unit (AU) of pediocin/ml of culture broth.

In contrast to *Ped. acidilactici* strains, *Lactobacillus plantarum* WHE92 produced pediocin PA-1/AcH at a higher level in MRS broth than in TGE broth and at a terminal pH of 5.0 and above. These differences could be due to greater suitability of *Lab. plantarum* to grow in MRS broth than in TGE broth and differences in some control systems for the expression of gene(s) associated in pediocin-production between *Lab. plantarum* WHE92 and pediocin-producing *Ped. acidilactici* strains. At present, no genetic information on pediocin production in *Lab. plantarum* WHE92 is known.

B. Screening/identification of pediocin-producing strains

Several direct and indirect bioassay methods have been used to screen bacteriocin production potentiality of *Pediococcus* strains as well as different LAB. Some of these have been designated as: "Flip-streak method", "spot-on-the-lawn method" and "well-diffusion method" (Lewus & Montville, 1991; Ray, 1992, 1994). The ability of a strain to produce bacteriocin is identified by the presence of a distinct zone of growth inhibition of the indicator bacteria used. Unfortunately, many researchers have used acid sensitive Gram-positive bacteria (like some *Listeria* species) and even Gram-negative bacte-

TABLE 3: Pediocin PA-1/AcH production level among *Ped. acidilactici* strains[a]

Producer Strains	Activity unit (AU)/ml of culture broth against indicators				
	Lab. plantarum NCDO 955	*Ped. acidilactici* LB42	*Leu. mesenteroides* Ly	*Ent. faecalis* MB1	*Lis. innocua* LIN 11
H	30×10^3	30×10^3	20×10^3	80×10^3	60×10^3
E	30×10^3	20×10^3	20×10^3	50×10^3	50×10^3
F	30×10^3	40×10^3	30×10^3	80×10^3	80×10^3
M	50×10^3	40×10^3	20×10^3	80×10^3	80×10^3
TR	80×10^3	60×10^3	50×10^3	80×10^3	$80 \times 10^3+$
HL	30×10^3	20×10^3	20×10^3	80×10^3	60×10^3
AB6	50×10^3	40×10^3	40×10^3	80×10^3	80×10^3
RS2	30×10^3	30×10^3	20×10^3	70×10^3	60×10^3
MC03	30×10^3	30×10^3	20×10^3	70×10^3	80×10^3

[a]Heated culture broths were diluted and assayed against soft agar overlays seeded with indicator cells from early stationary growth phase by the methods of Biswas et al. (1991) and Yang et al. (1992). The results are the average of three separate tests. Strains H,E,F,M are isolated from fermented sausage (Ray et al, 1989; Kim et al., 1991. M is the same as PC). TR and HL are variants isolated from H. AB6 and RS2 are from Bhunia and Johnson (1992), MC03 from Hoover et al. (1988; it is a *Ped. acidilactici*, not *Ped. pentosaceus*) (unpublished data, Ray, B.).

ria as indicator strains. In many studies, highly concentrated (10x) culture broth supernatants in large volumes (100 ml) have been used. Even after neutralization of the culture broths, inhibitory metabolites produced by some LAB, other than bacteriocins, can produce zone of growth inhibition and can be misinterpreted as bacteriocin.

We have been using a direct method with a high degree of success to isolate bacteriocin-producing LAB, including *Pediococcus* strains, and other Gram-positive bacteria from natural sources (Yang & Ray, 1994b). We have selected six species/strains of LAB from five genera that are capable of growing at pH 4 and above and are highly sensitive to bacteriocins produced by different LAB. We routinely use four of these six strains in isolating bacteriocin producing LAB: *Ped. acidilactici* LB42, *Lab. plantarum* NCDO 955, *Leuconostoc mesenteroides* Ly, and *Enterococcus faecalis* MB1. All four strains are sensitive to pediocin PA-1/AcH and a pediocin produced by a *Ped. pentosaceus* strain 34, which is different from pediocin PA-1/AcH.

To isolate pediocin producing *Pediococcus* strains from natural sources, a product sample is serially diluted and aliquots are pour plated with 3 to 4 ml of TGE agar (TGE broth + 1.5% agar) and MRS agar (MRS broth + 1.5% agar), using four plates for each medium in each dilution. Following solidification of the medium, another 3 ml/plate of the same agar medium is poured to cover the surface of the first layer of agar medium. The plates are incubated at 30°C for 18 to 24 h, or longer if necessary, for the formation of small colony forming units (cfu). A plate with about 100 colonies is then overlaid with 5 ml of the same melted soft agar medium (containing 0.7 to 1.0 agar) containing 5 ml (about 10^5 to 10^6 cells) of a freshly grown indicator bacteria. Four indicators are used in four plates of both types of agar media. The plates are incubated overnight at 30°C and examined for a clear and circular zone of inhibition, which is not irregular and fuzzy, around a cfu. Cells from such a colony are purified, grown in TGE broth or MRS broth and the heat treated culture broth in 5 ml volume is examined to determine if it produces a clear zone of inhibition of growth of an indicator strain inoculated on the surface of an agar media plate. A known pediocin-producing strain is used as a control in each step to reduce confusion.

An immunological method for the identification of pediocin-producing *Ped. acidilactici* strains has been developed by Bhunia and Johnson (1992). A monoclonal antibody, ped-2B2, was produced by injecting mice with broken and heat-killed cells of *Ped. acidilactici* H. When cells of bacterial isolates from food samples were examined by enzyme-linked immunosorbsant assay methods, only those producing pediocin (probably pediocin PA-1/AcH) gave a positive reaction. Bhunia et al. (1994) also developed a plasmid probe containing the *pap-A* gene from plasmid pSMB74 in *Ped. acidilactici* H and used Southern hybridization techniques to screen for plasmids encoding the *pap-A* gene in the pediococci isolates from natural sources. This technique facilitates differentiation between pediocin PA-1/AcH producing and non-producing isolates. Recently, Martinez et al. (1998) also developed a colony hybridization method to differentiate pediocin-producing *Ped. acidilactici* colonies from non-producing colonies. They generated a specific probe by PCR and used it to identify *Ped. acidilactici* strains capable of producing pediocin PA-1/AcH.

C. Quantitative assay

Three methods have been developed to determine the concentrations of pediocin PA-1/AcH in a preparation. These are "dilution-to-extinction", "minimum-inhibitory concentrations" and an "immunological assay". Although some researchers have used the diameter of zones of growth inhibition of an indicator strain to compare relative strengths of pediocin preparations, it has not been used for quantitative assay of pediocin concentrations, as it has for the determination of IU of nisin.

In the most used dilution-to-extinction procedure, a pediocin preparation is serially diluted in a buffer or water and an aliquot from each dilution is spotted over a lawn of indicator bacteria. We use dilutions over narrow range, such as 1:10, 1:20, etc. then two-fold dilution, since the latter at higher range (such as 1:128, 1:256, etc.) does not give an accurate concentration. We add 5 ml of a freshly grown indicator bacteria (about 10^5 to 10^6 cells total) to 5 µl of melted TGE-soft and pour it over a prepoured TGE agar plate. The plates are usually kept at 4°C for 1 h so the indicator cells go into lag phase and the surface dries. Then 5 ml portions from each dilution, in duplicate, are spotted over the lawn. The plates are usually kept at 4°C to 10°C for 1 to 2 h for the pediocin molecules to come in contact with the indicator cells and kill them, before the cells start to multiply. The plates are then incubated at 30°C overnight and examined for a definite circular clear zone around the area of the pediocin spot. The highest dilution that gave a definite zone of 1 to 2 mm is considered as one activity unit (1 AU). Pediocin concentration in the preparation is calculated by multiplying this dilution factor by 200 (Biswas et al., 1992; Ray, 1992; Yang et al., 1992). Some researchers, especially those who used two-fold dilutions, expressed the concentration as arbitrary units (AU; Pucci et al., 1988; Chikindas et al., 1995; Chen et al., 1997a,b). We have seen that the former gives reproducible results for the same indicator strain. To assay pediocin concentration in a culture broth we heat-treat the broth at 90°C for 5 min to kill the producer cells and go through dilution procedures, usually in sterile water. Since we use LAB as indicators and very high ranges of dilutions (generally 1:250 to 1:300 or more), we do not neutralize the broth. For a dried preparation, we usually dissolve 1 mg powder in 1 ml of a 0.02% SDS solution. This is done to prevent clumping of pediocin molecules. After about 30 min, with occasional mixing in a vortex mixer, the preparation is diluted serially in sterile water and used for spotting against an indicator on a plate. Due to the high range of dilutions used, indicator strains, other than LAB, such as *Listeria* species, can also be used in this method. By using different indicators against a pediocin preparation, the relative sensitivity of the strains can be calculated. Also, using the same procedure, relative bactericidal potency of different bacteriocins can be compared and expressed in terms of AU (such as 1 AU of pediocin is equal to "X" AU of nisin, "Y" AU of sakacin and "Z" AU of leuconocin; Yang & Ray, 1994b).

In the minimum-inhibitory concentration (MIC) assay procedure, a pediocin-preparation is diluted through two-fold dilution procedures in a suitable broth and a 200 ml portion from each dilution is transferred into the wells of a microtitre plate. A culture broth of an indicator is diluted with the same broth to OD 0.1 at 600nm and a specific volume of this is added to each well. Following incubation for 3 to 5 h at 30°C, the OD at 600nm of culture broth in the wells is measured spectrophotometrically with a microtitre plate reader. One pediocin unit is defined as the amount of a pediocin preparation (AU or mg) that produced 50% growth inhibition, as compared to controls, of the indicator (Finland et al., 1996; Ennahar et al., 1996; Lozano et al., 1992). This method has not been standardized with respect to indicator strain(s), indicator cell concentration, growth conditions (time, temperature), and dilution preparation of pediocin. It has not been compared with other methods (determination of AU) with respect to its reproducibility and efficiency.

Martinez et al. (1998) described several variations of enzyme-linked assays (ELISA) methods that seemed to have a high degree of sensitivity against pediocin PA-1/AcH. According to them, pediocin concentrations ranging between 0.025 mg/ml to 2.5 mg/ml in a preparation can be detected. The method has not been studied extensively and the relationship between mg and AU of pediocin has not been determined.

III. MOLECULAR PROPERTIES

A. Isolation and purification

Pediocins of *Ped. acidilactici* strains have been purified from the culture broths, for two reasons. One is for quantitative studies and usually requires relatively large amounts of partially purified products. Such preparations are used to determine bactericidal host range against different bacterial species and strains, to study bactericidal effectiveness in food systems, determine storage stability, and to use as a starting material for further purification. The other method is for qualitative studies, such as to determine molecular weight by SDS-PAGE, amino acid composition and amino acid sequence of pediocins, and to study structure/function of the molecules in relation to their bactericidal action. The highly purified materials necessary for these studies are obtained from the partially purified materials obtained for quantitative studies.

In most methods used to prepare partially purified pediocin PA-1/AcH, a producer strain is grown in a suitable broth, such as MRS broth, casein-glucose-yeast extract broth or TGE broth and the cells are removed by centrifugation (Ray, 1992, 1994; Hoover & Ray, 1993). The supernatant liquid is generally filter sterilized and the proteins, including pediocin PA-1/AcH, are precipitated by adding NH_4-sulfate to supersaturation. The precipitated proteins are collected by centrifugation, solubilized in water, and dialyzed to remove NH$_4$-sulfate. If a dried product is desired, it is freeze-dried (Gonzalez & Kunka, 1987; Pucci et al., 1988; Bhunia et al., 1987, 1988; Coventry et al., 1995). In a separate method, the proteins in a cell-free supernatant are precipitated by adding ethanol (1:1 by volume), usually after first concentrating the supernatant, and the precipitated proteins are collected by centrifugation. The precipitate is then resuspended in sterile deionized water and used for further studies (Bhunia et al., 1991). In another method, the cell-free supernatant is lyophilized, the powder is dissolved in a small volume of sterile deionized water (1:250) and dialyzed at 25°C for 48 h. The pediocins (with other proteins) precipitate in the small volume of dialysate, which are collected by centrifugation and the pellet is resuspended in 0.5 ml deionized water for further studies (Bhunia & Johnson, 1992).

A quite different method was developed in our laboratory from the observation that pediocin PA-1/AcH molecules adsorb to and desorp from the cell surface of producer strains (as well as other Gram-positive bacterial cells) depending on the pH of the environment (Bhunia et al., 1991). Based on this observation, Yang et al. (1992) developed the partial purification of large amounts of pediocin (and other bacteriocins from LAB) starting from culture broths. In this method, a producer strain of *Ped. acidilactici* is grown at 35°C for 16 to 18 h in TGE broth with the terminal pH of culture broth reaching 3.6 to 3.8. The culture broth is then heat-treated at ≥75°C for 30 min (or a temperature-time combination to kill all cells and destroy intracellular proteases and peptidases). The pH of the culture broth is then adjusted to pH 6.5 by adding 10 M NaOH and stirred slowly with a magnetic stirrer for 4 h at room temperature (22° to 25°C) to facilitate absorption of the pediocin molecules on the cell surface. The cells are harvested by centrifugation and resuspended in a small volume (usually 1:20) of a solution containing 100 mM NaCl and 100 mM lactic acid or HCl, pH 2.0. The cell suspension is then stirred for at least 2 h at room temperature (or overnight at 10°C) for the desorption of pediocin molecules from the cell surface into the liquid. Following another centrifugation, the supernatant liquid is collected and dialyzed against deionized water for at least 4 h in a 1000 MW cut-off dialysis bag with at least two changes of water, to remove the salt and acid. If a higher purification (but low yield) is desired, the material can be dialyzed for 48 h at 10°C with changes of water made every 12 h. The dialysate is then freeze-dried (producing a white fluffy powder) and stored at -20°C for future use. In this procedure, the loss of pediocin at each stage, such as in the culture broth supernatant and in the cells following desorption, can be determined to calculate the final yield in the freeze-dried powder. For pediocin PA-1/AcH (and nisin), with producer cells grown in TGE broth the yield is 90% and over, of that present in the initial culture broth (Yang et al., 1992). However, for two other bacteriocins, leuconocin Lcm1 and sakacin, the yield is 50% or lower. We suspect this is due to the use of buffered TGE broth containing phosphate, acetate and citrate. It should be recognized that cationic bacteriocin molecules absorb on the cell surface through ionic interaction with lipoteichoic and teichoic acid molecules to the cell wall of Gram-positive bacteria (Bhunia et al., 1991). If a culture broth contains large amounts of anions, they will interfere with the adsorption of the bacteriocin molecules on the cell surface. Following centrifugation, the molecules will remain in the supernatant, instead of with the cells and will produce a low final yield. This could be the reason why low yields were obtained when researchers grew producer *Ped. acidilactici* strains in MRS broth that contains many anions (Daba et al., 1994). This procedure has been used with some modifications by Elegado et al. (1997), Daba et al. (1994) and Janes et al. (1998). Janes et al.

(1998) first removed the producer cells from the culture broth, then added rice hull ash or silicic acid to the supernatant liquid and adjusted its pH to 8 to 9 for the adsorption of pediocin PA-1/AcH. The rice hull ash or silicic acids with adsorbed pediocin were removed by centrifugation and then resuspended in 5% phosphoric acid at pH 2.5 to facilitate the release of pediocin. The liquid is collected by centrifugation and the rice hull ash and silicic acid removed in the pellets. The final yield of pediocin was about 50% of the original amount present in the culture broth. This method also was used to purify bacteriocins of other LAB.

The methods developed to produce highly purified pediocin PA-1/AcH involve multi-steps, require a relatively long time and give very low yields. The initial step, as described above, consists of preparation of partially purified pediocin either by ammonium sulfate or ethanol precipitation or by pH-dependent cell adsorption-desorption from the culture broth. The subsequent steps are based on the cationic nature and hydrophobicity of pediocin molecules. These involve separation and concentration of the molecules by different chromatographic methods, including hydrophobic, ion exchange, and size exclusion chromatography (Jack et al., 1995; Henderson et al., 1992). Venema et al. (1995) used a method in which following ethanol precipitation of pediocin PA-1/AcH from culture broth supernatant, the preparation was subjected to isoelectric focusing and ultra-filtration. Miller (1999, unpublished data) used a procedure in which freeze-dried pediocin PA-1/AcH obtained by pH-dependent cell adsorption-desorption, is further purified by HPLC using a hydrophobic column. The method produced a product which is over 95% pure with relatively high yield. Ennahar et al. (1997) purified pediocin PA-1/AcH produced by *Lab. plantarum* WHE92 by a four-step procedure consisting of ammonium sulfate precipitation, cation exchange chromatography, then reverse-phase chromatography and finally stationary-phase chromatography. The final yield was less than 20%.

Only one study by Piva et al. (1994) reports on the purification of pediocin A of *Ped. pentosaceus* BFF61. The culture broth supernatant was first dialyzed against polyethanol glycol-2000 and then extracted with butanol. The final step involves electro-endosmotic preparative electrophoresis. The procedure gave less than 20% final yield.

B. Physico-chemical properties

The physical and chemical characteristics of pediocin are very similar to the characteristics of other ribosomally synthesized bacteriocins of Gram-positive bacteria and in particular to those which are grouped as "pediocin-like" or of the "pediocin-family" (Jack et al., 1995). Some minor differences have been reported by some researchers, but may be mostly due to the specific techniques and the source of pediocin used in a study, instead of actual differences in the molecules. Depending upon a study, pediocin has been used in a heated culture broth or in a cell-free supernatant, or as a partially purified product or a highly purified product. Some of the characteristics of pediocin PA-1/AcH are listed in *TABLE 4*. From amino acid sequence and/or corresponding nucleotide sequence analysis of pediocins from nine *Ped. acidilactici* strains from different sources and different countries, it is evident the molecule has 44 amino acids with a molecular mass of 4,628 Da. It is cationic due to the presence of 4 lysines and 3 histidines and has an isoelectric point of 9.6. The net charge varies with the pH of the environment, especially at the acidic range, due to histidine. Pediocin PA-1/AcH has four cysteine residues which retain the thiol (-SH) group in a reducing environment but in the oxidized state form two disulfide bonds (-S-S-) between -Cys 9-Cys 14- and -Cys24-Cys44-. This oxidized state is essential for the bactericidal property of pediocin (Henderson et al., 1992; Marugg et al., 1992; Motlagh et al., 1992, 1994; Bukhtiyarova et al., 1994; Lozano et al., 1992). In the N-terminus, there are five amino acids, -Tyr 3, Gly 4, Asn 5, Gly 6 and Val 7 (-YGNGV-) which are found in all bacteriocins of the pediocin-family and are important for its bactericidal property (Jack et al., 1995). Oxidation of the molecules one methionine to methionine sulfoxide ($-CH_2-CH_2-S-CH_3 \rightarrow -CH_2-CH_2-S-CH_3$) makes the molecule biologically inactive. The molecule is hydrophobic due to the presence of relative high numbers of non-polar amino acids. The N-terminal half is proportionately more hydrophilic than the C-terminal half of the molecule. Due to their hydrophobic nature, molecules tend to form aggregates, especially in a concentrated state and at pH above 6.0. The aggregates can be easily dispersed in the presence of 0.02% SDS solution.

The influence of several factors on the bactericidal properties of pediocin PA-1/AcH are described below.

1. Enzymes. Being a peptide, pediocins, like other bacteriocins of Gram-positive bacteria, are hydrolyzed by proteolytic enzymes and lose bactericidal activity. Pediocin PA-1/AcH from *Ped. acidilactici* strains so far studied, are in general, sensitive to chymotrypsin, trypsin, ficin, papain, and proteases: IV, IX, X, XIV, XV and XIX (Gonzalez & Kunka, 1987; Bhunia et al., 1987, 1988; Coventry et al., 1995; Schved et al., 1993). In general, they are not sensitive to lipases, amylases, RNases or DNases. However, several inconsistencies have been reported by some researchers and in those situations, neither the amino acid sequence of the pediocin has been determined nor a purified pediocin preparation used. Jorger and Harlander (1992) reported that pediocins from several *Ped. acidilactici* and *Ped. pentosaceus* strains are inactivated by chymotrypsin but not by pepsin. One of the strains they used is *Ped. acidilactici* PAC 1.0 that produces pediocin PA-1/AcH. Gonzalez and Kunka (1987) observed pediocin PA-1/AcH to be sensitive to pepsin. Similarly, pediocin from *Ped. acidilactici* PO_2 was found by Coventry et al. (1995) to be pepsin resistant although it is sensitive to other proteolytic enzymes studied. Pediocin produced by *Ped. acidilactici* SJ1 was reported to be sensitive to proteolytic

TABLE 4; Some physio-chemical characteristics of Pediocin PA-1/AcH[a].

Molecular mass: 4.6 kDa		**Amino acid composition:**	
Isoelectric point: 9.6		Nonpolar:	25.0%
Charges: (pH >6): 3+			Absent: Leu, Pro, Phe
(pH <6): 6+		Polar:	56.8% (Gly 18.2%)
Enzyme cuts:		Acidic:	2.3%
Chymotrypsin: 2, 3, 18, 33			Absent: Glu
Trypsin/Trypsin K: 1, 11, 20, 43		Basic:	15.9%
Thermolysin: 6, 15, 24, 25			Absent: Arg
Staph-V8: 17			
		Cysteine: 4	
Non-cutting enzymes:		Methionine: 1	
Trypsin R			
Thrombin			

[a]Jack et al. (1995) and unpublished data (Ray, B).

enzymes as well as to amylase. Although the amino acid sequence of this pediocin is not known, from the available information on the structures of pediocin from six different strains of *Ped. acidilactici* and bacteriocins from many Gram-positive bacteria, it is not likely that this pediocin has a carbohydrate moiety, particularly one that is crucial for its bactericidal activity. Pediocin A from *Ped. pentosaceus* FBB61 is also sensitive to proteolytic enzymes (Piva & Headon, 1994).

2. *Heat*. The bactericidal effect of pediocins from the *Ped. acidilactici* strains so far studied remains stable after heating up to 60 min at 80°C. Partial inactivation occurs following heating at 121°C for 15 min; the level of inactivation is higher in culture broth than in a purified preparation. When heating is conducted at pH between 5 to 6, the level of inactivation is less (Gonzalez & Kunka, 1987; Bhunia et al., 1988; Schved et al., 1993; Coventry et al., 1995). Partial inactivation at very high heat could be due to the interaction of pediocin with some component of the culture broth and due to oxidation of methionine, especially in purified pediocin PA-1/AcH. In contrast, pediocin A from *Ped. pentosaceus* FBB61 is sensitive to high heat losing over 95% of activity at 100°C in 10 min (Piva & Headon, 1994). This could be due to the high molecular weight (about 80 kDa) of pediocin A.

3. *pH*. Pediocins from *Ped. acidilactici* strains retain their bactericidal activity at a pH range between 3 to 9. At higher pH values, pediocin PA-1/AcH lost activity following storage at 25°C at pH 10.0 for 24 h, but with 2 h at pH 12.0 (Bhunia et al., 1988). Schved et al. (1993) showed that 50% of the activity of pediocin from *Ped. acidilactici* SJ1 is lost within 24 h at pH 2.5 and 12.0. However, in our adsorption-desorption purification method, we reduced the pH to 2.0 for overnight at 10°C without observing any loss of activity. In general, bacteriocins are stable at lower pH. It is possible that during storage at pH 2.5 the pediocin molecules form aggregates and produce a situation of apparent loss of activity as observed by Schved et al. (1993).

4. *Organic solvents and chemicals*. Pediocin PA-1/AcH retained full activity following exposure to several organic chemicals, namely acetone (100%), chloroform (40%), isopropanol (80%), ethylalcohol (80%), hexane (80%), acetonitrile (70%), EDTA (10 mM), SDS (1%) and Tris (0.6 M) (Bhunia et al., 1988, 1991). Treatment with 10% formaldehyde reduced the activity considerably (Coventry et al., 1995).

5. *Ions*. Cations, such as Ca^{2+}, Mg^{2+}, Mn^{2+}, K^+ and Na^+ did not reduce the bactericidal activity of pediocin PA-1/AcH. In contrast, anions, especially Cl^- and $(PO_4)^{3-}$ reduced bactericidal efficiency to a great extent. However, this effect is dependent upon the relative concentrations of pediocin vs. anions; at a higher concentration of pediocin, the bactericidal effect is reversed (Bhunia et al., 1991). The anions, probably through ionic interaction with the cationic pediocin molecules, prevent them from binding to anionic teichoic and lipoteichoic acids in the cell wall of sensitive indicator bacteria. This binding is important for concentrating of pediocin on the cell surface and making it available for destabilization of the cytoplasmic membrane resulting in cell death. This aspect is discussed in detail in a later section related to "mode of action" of pediocin.

6. Hydrostatic pressure. Pediocin PA-1/AcH, both purified and in culture broth, has been subjected to high hydrostatic pressure of 345 MPa (50 kPsi) for 30 min at 25°C. Although hydrostatic pressure is known to alter secondary and tertiary structures of proteins, pediocin retained full activity following pressurization. During subsequent storage for 5 months at 2°C, the preparations retained full activity (Kalchayanand et al., 1995).

7. Storage time and temperature. Pediocin PA-1/AcH in culture broth, and partial and highly purified preparations, in liquid and dried states, retains full activity at -20°C and below for over 6 months. In contrast, the vacuum-packaged dried product is relatively more stable at room temperature (Ray, 1994, 1996). The loss of activity of pediocin PA-1/AcH during storage in the presence of air could be due to oxidation of methionine. In a liquid preparation, aggregation of the molecules can also produce an apparent loss which could be reversed by adding SDS (0.02%) during assaying as indicated previously. In general, a dried pediocin preparation stored in the absence of air at -20°C or below will have a longer storage life.

C. Structure

The primary structure, i.e. the amino acid sequence, of pediocins from at least nine *Ped. acidilactici* strains have been elucidated (*TABLE 1*). All of them have the same amino acid sequence as pediocin PA-1/AcH. These strains are: PAC1.0 (Henderson et al., 1992; Marugg et al., 1992), H and LB42-923 (Motlagh et al., 1992, 1994; Bukhtiyarova et al., 1994), a commercial strain (Lozana et al., 1992), UL5 (Daba et al., 1994; Huang et al., 1996), F (Miller, 1997, unpublished data) RS$_2$ (Yang & Ray, 1994, unpublished data) and one strain from cucumber brine (Central Food Research Institute, India) and one strain from naturally fermented sausage (Udayana University, Bali, Indonesia). Data for the last two strains have been communicated to us. In two pediocin producing *Ped. parvulus* strains, isolated from vegetables, the primary structure was found to be the same as pediocin PA-1/AcH. The amino acid sequence in these strains is determined from the nucleotide sequence of the structural gene (Bennik et al., 1997). The bacteriocin producing *Lab. plantarum* WHE92, isolated from cheese, was also found to produce a pediocin with the same amino acid sequence as pediocin PA-1/AcH (Ennahar et al., 1996). The amino acid sequence of pediocin A and *Ped. pentosaceus* FBB61 is not known, but it has a Mr of 80 kDa which is much larger than pediocin PA-1/AcH with a Mr of 4.6 kDa.

Several indirect studies tend to indicate that most other pediocin-producing *Ped. acidilactici* strains so far studied, probably originate from a single strain. The data are based on plasmid profiles and MWs of plasmids coding for pediocin (Ray et al., 1989a; Kim et al., 1991), immunological similarities (Bhunia & Johnson, 1992; Bhunia et al., 1994; Bhunia, 1994), genomic analysis (Luchansky et al., 1992) and DNA finger printing and ribotyping (Jager & Harlander, 1992). The probability of their having the same origin became more apparent due to the fact they are all isolated from fermented meat products or initially received from one source and then given a different designation (*TABLE 1*).

As has been indicated before, the cytoplasmic form of the molecule has 62 amino acids and is designated as prepediocin. During passage through the membrane, the 18 amino acid leader peptide at the N-terminus is removed and the 44 amino acid pediocin PA-1/AcH is excreted into the environment. Once outside, the molecule forms secondary and tertiary structures that are essential for its bactericidal property. The four cysteine

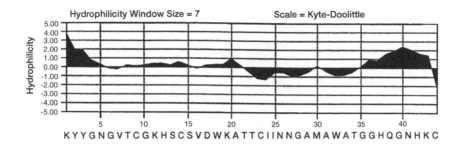

FIGURE 3. Hydropathy plot of deduced amino acid sequence of pediocin PA-1/AcH encoded by *pap A* in plasmid pSMB74. The X axis shows amino acid residues and the Y axis shows hydropathy index. Hydrophobicity is below and hydrophilicity is above the baseline on the X axis (unpublished, Ray, B.).

residues form two disulfide bonds and may cause the molecule to fold. The hydropathy plot shows the pediocin PA-1/AcH molecule to be moderately amphipathic (*FIGURE 3*). The N-terminal half (up to amino acid 21 to 22) is relatively polar while the C-terminal half is relatively non-polar. The N-terminal half is predicted to form a 3-strand antiparallel β-sheet in which the second and third strands are linked by a disulfide bond between Cys 9 and Cys 14 and the first and second strands are connected by a β-turn (-Gly 4-Ans 5-Gly 4-Val 7-; Chen et al., 1998; Montville & Chen, 1998). A helical wheel plot of the C-terminal region of the molecule indicates that it could form amphipathic α-helix (*FIGURE 4*). Possibly the C-terminus may loop back and become crosslinked to the α-helix via the Cys 24 and Cys 44 disulfide bond (Miller et al., 1998a, b; *FIGURE 5*).

Miller et al. (1998a,b), through random amino acid substitution, produced mutants of pediocin PA-1/AcH that differed in single amino acids and studied the relative importance of the amino acids for bactericidal activity. Analysis of these mutants and other variants indicated that bactericidal property of pediocin PA-1/AcH is not confined to any specific segment on the primary structure. The five amino acid sequence -Tyr 3-Gly 4-Asn 5-Gly 6-Val 7- is essential for the bactericidal action and so are the two -S-S-bridges between -Cys 9-Cys 14- and -Cys 24-Cys 44. In addition, the bactericidal property of the molecule is completely lost in 8 of the following substitution mutants: Asn 5 Lys, Cys 9 Arg, Cys 14 Ser, Cys 14 Tyr, Cys 24 Ser, Gly 37 Glu, Gly 37 Arg and Cys 44 Trp. In nine other mutants (Lys 1 Asn, Trp 18 Arg, Iso 26 Thr, Met 31 Thr, Ala 34 Asp, Asn 41 Lys, His 42 Leu, Lys 43 Asn and Lys 43 Glu) the activity was reduced to between <1% and 60% of the wild type. One mutant (Lys 11 Glu), however, had 2.8 fold higher activity than the wild type. It is evident from these data that the entire 44 amino acid sequence is necessary to confer the specific secondary and tertiary structures to the molecule that are essential for the high level of bactericidal activity (Miller et al., 1998b). It is also evident that a free N-terminus of pediocin PA-1/AcH is not necessary for its bactericidal action. The molecule retains bactericidal efficiency with its N-terminus linked to a large foreign protein (such as maltose binding protein) or to its leader peptide (Miller et al., 1998a,b; Ray et al., 1999).

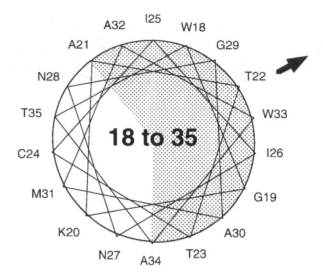

FIGURE 4; Schiffer-edmundson helical wheel plot of the putative 18 amino acid amphipathic α-helical mem brane-binding sequence of pediocin PA-1/AcH. The location of the membrane binding sequence is proposed based on comparison to the three-dimensional structure of the corresponding region of leucocin A (Gallagher et. al., 1997). The hydrophobic face of the helix is shaded, and $<\mu> = 0.3$ (Eisenberg et al., 1984; unpublished, Miller, K. W.).

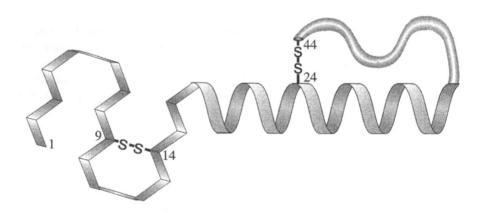

FIGURE 5; Hypothetical secondary structure model of the membrane-bound form of pediocin PA-1/AcH. The structure is based on computer modeling of the N-terminal half of pediocin PA-1 (Chen et al., 1997) and on the three-dimensional structure of leucocin A determined by NMR spectroscopy in membrane-mimicking environ ments (Gallagher et al., 1997). The N-terminal region is proposed to fold as a 3-strand ß-sheet stabilized by the Cys 9 -Cys 14 disulfide bond. The C-terminal half is proposed to fold as an amphipathic a-helix (refer to *FIGURE 3*), and may be stabilized by fold-back of the C-terminus and crosslinking via the Cys 24 -Cys 44 disulfide bond (unpublished, Miller, K. W.).

IV. ANTIMICROBIAL ACTIVITY

A. Mechanism(s) of antibacterial action

1. Antibacterial effect and target site(s). Early studies, Bhunia et al. (1988), indicated that the antibacterial effect of pediocin PA-1/AcH towards sensitive Gram-positive bacterial cells is caused by its bactericidal and not by bacteriostatic action. The effect is very fast initially and then continues to increase slowly over a period of time, and is dependent directly on the level of pediocin concentration for a given indicator bacterium. They also showed that in a population of a sensitive strain there are always some resistant variant cells that are capable of growth in a suitable broth and retain the characteristics of resistance only while growing in the presence of pediocin. In a subsequent study it was shown that pediocin-induced cell death occurred from impairment of the cytoplasmic membrane of sensitive bacteria resulting in the leakage of cellular ions and small molecules and UV absorbing material (Bhunia et al., 1991). They also reported that the cationic pediocin molecule binds in a pH dependent manner to anionic lipoteichoic acids (and probably teichoic acids) in the wall of producer cells and other Gram-positive bacterial cells, and that the adsorption is relatively higher for sensitive than resistant cells. Some anions, especially Cl^-, I^- and $(PO_4)^{3-}$, but not acetate or citrate, reduce both adsorption and bactericidal effect of pediocin but the effects are dependent on the relative concentrations of pediocin and anions. Cells of some sensitive strains also are lysed, probably due to a secondary effect. From these observations they resolved that in sensitive bacteria adsorption of pediocin PA-1/AcH on the cell wall in higher concentrations facilitates molecules to bringing about changes in membrane integrity resulting in loss of important intracellular small molecules and subsequent cell death. In resistant Gram-positive cells, adsorption of pediocin at a lower level may not allow the molecules to act on the cell membrane. In Gram-negative cells, due to the lack of lipoteichoic and teichoic acid, pediocin molecules are not absorbed (or adsorbed at very low concentrations, probably to lipopolysaccharides that have a barrier function) and the inner membrane remains unaffected. Limited studies have shown that following a sublethal stress, the resistant Gram-positive and Gram-negative bacteria become sensitive to pediocin PA-1/AcH. In the injured cells, due to loss of the barrier function of the wall or outer membrane of the cells, pediocin can come in contact with the membrane and bring about changes to cause cell death (Ray, 1995). Bacterial spores of the *Bacillus* and *Clostridium* strains are resistant to pediocin PA-1/AcH. However, after the spores are induced to germinate and outgrow they are killed by pediocin, provided the cells are sensitive to pediocin (Kalchayanand, 1995).

2. Mechanisms of bactericidal action. Christensen and Hutkins (1992, 1994) reported that treatment of *Listeria* strains with pediocin, from *Ped. acidilactici* JD 1-23, caused rapid collapse of the membrane potential and proton motive force in the cell membrane. It also increased loss of protons and inhibited transport of glucose. These changes produced pediocin-induced cell death. Chikindas et al. (1993) also confirmed that pediocin PA-1/AcH killed cells of sensitive strains of *Pediococcus* spp. by dissipating transmembrane electrical potential. They exposed cytoplasmic membrane vesicles from sensitive cells as well as liposomes fused with membrane vesicles from sensitive cells to pediocin and observed a rapid efflux of cytoplasmic molecules in a voltage independent manner. They suggested that the pediocin molecules form small pores in the cytoplasmic membrane allowing loss of small molecules (not exceeding 9400 Da) and causing cell death. They also postulated that the pediocin molecules initially bind to a specific receptor protein on the membrane then aggregate and form pores in the sensitive, but not resistant cells. The size and life span of the pores could not be determined, but they form immediately following exposure of cells to pediocin. Based on several models, Chen et al. (1997a,b) postulated that the N-terminal half of pediocin PA-1/AcH form a 3-strand antiparallel β-sheet, consisting of a patch of positive residues (Lys 11 and His 12) that binds electrostatically to anionic phospholipids in the cytoplasmic membrane. Using membrane lipid vesicles rich in phosphotidylglycerol from a *Lis. monocytogenes* strain as well as synthetic lipid vesicles, prepared from phosphotidylglycerol, they showed the existence of strong electrostatic interactions between pediocin and both types of lipid vesicles. The affinity of pediocin for synthetic lipid vesicles indicates that a membrane protein receptor is not essential for the pediocin-induced pore formation in the membrane. The binding affinity of Lys 11 and His 12 to phosphotidylglycerol vesicles is greater than that of the -Tyr3-Gly 4-Asn 5-Gly 6-Val7- motif at the N-terminus of pediocin PA-1/AcH. Shifting the pH of the environment above 6 with concomitant deprotonation of His 12 reduced the binding affinity of pediocin fragments to lipid vesicles. This suggests electrostatic interactions are important for the binding of pediocin to the target membrane. They also observed that the presence of a higher concentration of negatively charged phospholipids increased the affinity of pediocin for the membrane. This also supports the necessity of electrostatic interaction for the initial binding of the N-terminal region of pediocin to the membrane of sensitive cells (Chen et al., 1998). The role of the C-terminal hydrophobic half of pediocin PA-1/AcH has not been elucidated yet. It is suspected that the ability of this half to form amphiphilic α-helical conformation may be important in pore formation. The -S-S- bond between -Cys 24-Cys 44- may help stabilize the conformation of this part of the molecule (*FIGURE 5*). This also can facilitate pore formation in the cytoplasmic membrane.

The mechanism of bactericidal action of pediocin PA-1/AcH is not yet completely understood. From the available information, a hypothesis for the mechanism of action can be formed as follows (Montville & Chen 1998; Miller et al., 1998b):

- The N-terminal half of the molecule adopts a b-sheet conformation, probably after membrane binding, which is stabilized by a -S-S- bond between -Cys 9 and Cys 14.
- The cationic amino acids (Lys 11 and His 12) bind to the anionic phospholipids in the membrane through electrostatic interactions. This interaction can be reduced by the presence of certain anions (Cl-, I-, PO$_4^{3-}$) in the environment, at pH above 6, and increased levels of neutral phospholipids.
- A protein receptor in the membrane is not essential for the binding of pediocin, but if such a receptor is present, it might enhance the action of pediocin.
- The C-terminal half also may have a random conformation in the aqueous environment, but the -S-S- bond between Cys 24-Cys 44, and b-sheet may help transform it to a defined a-helical amphiphilic structure following coming in contact with the lipid bilayer.
- Pediocin molecules subsequently may assemble in the membrane causing pore formation. This causes the dissipation of proton motive force due to the collapse of the pH gradient and membrane potential in a voltage independent manner.

- The cells lose intracellular ions, UV absorbing materials, ATP and can no longer transport amino acids and glucose.
- These changes finally bring about cell death.

The role of binding of pediocin PA-1/AcH to cell wall anions in the subsequent binding of the N-terminal region to anions in the membrane and pore formation by the C-terminal region is not known. It is slowly becoming evident that the binding of pediocin (and other bacteriocins) to the cell wall of sensitive Gram-positive bacteria is important for the bactericidal effectiveness of the molecules (Bhunia et al., 1981; Fimland et al., 1998; Peschel et al., 1999). The role of the cell wall in controlling bactericidal properties of pediocin PA-1/AcH (and nisin) is particularly evidenced from the higher sensitivity of sublethally-injured sensitive Gram-positive bacteria and developed sensitivity of sub-lethally-stressed resistant Gram-positive and Gram-negative bacterial cells (Kalchayanand et al., 1992, 1998a, b; Ray, 1995).

A recent report has shown that in the presence of a 15-mer peptide fragment, derived from the C-terminal half of pediocin PA-1/AcH (Lys 20 to Ala 34), the bactericidal effect of the molecules on target cells is inhibited. The authors have suggested that the fragment, by interacting with an entity either on the cell surface binding site or in the membrane of the target cells prevents binding and inhibits bactericidal action of pediocin PA-1/AcH. When sensitive cells are exposed to pediocin PA-1/AcH, this part of the molecule probably is associated with increased membrane permeation and cytotoxicity in the target cells (Fimland et al., 1999). To perform the studies, they mixed the sensitive cells, pediocin PA-1/AcH and the 15-mer peptide fragment in a broth, incubated the materials for 10 to 16 h at 30°C and measured the OD (at 610nm) to compare growth with appropriate controls. However, they neither measured the relative adsorption of pediocin to sensitive cells or the level of death due to pediocin in the presence and absence of this 15-mer peptide. Bhunia et al. (1991) reported that when sensitive cells, pediocin and NaCl or sodium phosphate are mixed together both pediocin adsorption on the surface and death of sensitive cells are reduced in a concentration dependent way. They suggested that anions of the salt bind to the cations in the pediocin molecules and make them unavailable for subsequent binding to the cell wall and probably to the cell membrane. It will be important to directly determine if the 15-mer peptide is involved in preventing adsorption of pediocin to target cells and thereby neutralizing its bactericidal function.

3. Lysis of sensitive cells. Initially, Pucci et al. (1988) reported that when pediocin PA-1/AcH was added to growing cultures of *Lis. monocytogenes* and *Ped. pentosaceus*, the strains not only stopped growing, but also lysed, as evident from the reduction in OD at 660 nm. At 200 AU/ml of pediocin the cells lysed at a rate of 0.19 OD unit/h for *Lis monocyt*ogenes and 0.29 OD unit/h for *Ped. pentosaceus*. Bhunia et al. (1991) observed that addition of pediocin PA-1/AcH to growing cultures inhibited growth of both *Lab. plantarum* NCDO955 and *Leu. mesenteroides* Ly but caused lysis only in *Leu. mesenteroides* Ly. Transmission electron micrographs showed the presence of lysed ghost cells only in *Leu. mesenteroides* Ly. Motlagh et al. (1991) reported that exposing growing cells of five strains of *Lis. monocytogenes* to pediocin PA-1/AcH reduced viability of all strains but caused lysis in only two strains. It was suspected that while cell death results from the direct action of pediocin, lysis occurs by an indirect mechanism. Cell

death probably triggers activation of the autolytic enzyme(s) in some strains, which causes the cells to lyse.

Recently, we have studied the relationship between pediocin related cell death
and lysis in *Leu. mesenteroides* Ly (*TABLE 5*). Washed cells taken during the exponential
growth phase were suspended in 0.1% peptone, mixed with 2,000 AU/ml of purified
pediocin PA-1/AcH, incubated at 25°C and at selected intervals samples were analyzed for
cfu/ml and OD at 600nm. The rate of viability loss was much higher than the rate of cell
lysis, especially during the initial 30 min of incubation. In a separate study, when exponentially growing cells of strain Ly were incubated at ≤4°C with pediocin, there was a
high level of cell death but no lysis. Thus, cell lysis appears to require the action of the
lytic enzyme(s) of the target cells, and pediocin may have an indirect role in triggering
lysis of some strains.

4. Immunity and resistance against pediocin. Compared to most bacteriocins
produced by the pediocin-family of LAB, both pediocin PA-1/AcH and pediocin A have
a relatively wide host range against Gram-positive bacteria. Yet, there are many species
and strains of Gram-positive bacteria that are not sensitive to them. A producer strain is
also not sensitive to its own pediocin, although it can be sensitive to other bacteriocins.
These two types of insensitivity to pediocins are manifested by two separate mechanisms
and are discussed here.

The insensitivity of a producer strain to its own pediocin is attributed to an
immunity protein, the genetic information of which is encoded in a specific immunity
gene. In *Ped. acidilactici* strains PAC 1.0, LB42-923 (a derivative of strain H) and F, this
is the *ped B* (or *pap B*) gene in the *ped* (or *pap*) operon (*FIGURE 1*). The Ped B (or Pap B)
protein has 112 amino acids with an estimated MW of 12,993 Da and a pI of 7.4 and contains 37.5% non-polar, 34.6% polar, 11.6% acidic and 16.1% basic amino acids. It does
not contain Arg and Cys. It is principally hydrophilic and contains no large hydrophobic
segments that could span the cytoplasmic membrane (Motlagh et al., 1994; Jack et al.,
1995; Marugg et al., 1992). The mechanism(s) by which this protein protects the cytoplasmic membrane of the producer cells against pediocin PA-1/AcH has not been eluci-

TABLE 5: Viability loss and reduction in OD of *Leuconostoc Mesenteroides* Ly in the presence of
Pediocin PA-1/AcH[a]

Min at 25°C	Control		Pediocin (2,000 AU/ml)	
	OD 600nm	cfu/ml	OD 600nm	cfu/ml
0	0.48	16×10^8	0.48	16×10^8
10	0.48	14×10^8	0.47 (2%)	9×10^7 (94%)
20	0.47	13×10^8	0.43	8×10^7
30	0.45	11×10^8	0.27	6×10^7
60	0.44	10×10^8	0.19	8×10^7
120	0.41	10×10^8	0.12	2×10^7
180	0.38	10×10^8	0.08	1×10^3
240	0.35 (27%)	10×10^8 (38%)	0.06 (87%)	2×10^2 (>99%)

[a]Cells from exponential phase were suspended in 0.1% peptone water treated with a purified preparation of
pediocin PA-1/AcH at 2,000 AU/ml. The results in parentheses indicate reduction in cfu and OD at incubation
time indicated (unpublished data, Ray, B.).

dated. Pediocin molecules are adsorbed on the cell surface of the producer strains. Also, as indicated before, the producer strains can be sensitive to other bacteriocin(s) of the pediocin family, whose N-terminal halves are fairly conserved and as in the case of pediocin PA-1/AcH, bind to the cytoplasmic membrane by the C-terminal regions in the initial stage of pore formation (Montville & Chen, 1998). It may be that the immunity protein, which is specific against pediocin PA-1/AcH, interact with the C-terminal half of pediocin and prevents formation of pores in the cytoplasmic membrane. In that case, Ped B (Pap B) molecules could have to be transported from the inside to the outside of the cytoplasmic membrane by some mechanism(s).

A producer strain becomes sensitive to its pediocin following the loss of the plasmid encoding the immunity protein (Ray et al., 1989a). However, there are several reports that in some strains the plasmid-cured cells still retained their resistance to pediocin as the two genes are not linked (Gonzalez & Kunka, 1987; Ray et al., 1989a; Schved et al., 1993). Noerlis and Ray (1994) showed that this resistance resulted from growing the cured cells in the presence of pediocin. After several transfers and growth in a pediocin-free medium, the cells became sensitive to pediocin.

In the population of a strain of Gram-positive bacteria, sensitive to pediocin PA-1/AcH, some variant cells are resistant to pediocin and are capable of multiplying in pediocin-containing broth. The percentage of such variant cells differs greatly with strains. Also, their proportion increases in old cultures and cultures maintained at low pH and low temperature (Bhunia et al., 1991; Motlagh et al., 1991, 1992b), as long as they are maintained in the presence of pediocin. These resistant cells also adsorb pediocin PA-1/AcH on the cell surface. Perhaps due to selective pressure they have low levels of anionic phospholipids in the membrane, reducing the affinity of the cationic N-terminus for the membrane (Montville & Chen, 1998). Alternatively, these cells (and stressed cells) could be induced to synthesize a stress protein, that acts as an immunity protein and protect the cells against bacteriocins.

Some Gram-positive bacterial species and strains are inherently resistant to pediocin PA-1/AcH (Bhunia et al., 1988, 1991; Hanlin et al., 1993). Limited studies have shown that they also adsorb pediocin on the cell surface, but usually at a lower amount than a sensitive strain. This could be due to a low level of anionic molecules in the cell wall (Bhunia et al., 1991; Fimland et al., 1998; Peschel et al., 1999). They could also have lower quantities of anionic phospholipids in the membrane (Chen et al., 1998; Montville & Chen, 1998). Following sublethal stresses these cells become sensitive to pediocin (Ray, 1993, 1995; Kalchayanand et al., 1992). This could be due to the loss of barrier functions of the cell wall in the injured cells allowing pediocin molecules to encounter the cytoplasmic cell membrane.

Gram-negative bacteria are, in general, resistant to pediocin PA-1/AcH and other bacteriocins of Gram-positive bacteria. Lipopolysaccharide (LPS) molecules on the outer membrane either may not adsorb pediocin or the outer membrane itself may simply prevent pediocin from coming in contact with the cytoplasmic membrane. However, sublethally stressed cells with impaired barrier functions of LPS-molecules become sensitive to pediocin PA-1/AcH (Kalchayanand et al., 1992, 1994, 1998a,b). Limited studies have shown that *Esc. coli* strains, transformed with plasmids carrying either the pediocin operon (Bukhtiyarova et al., 1994) or the matured pediocin gene (Miller et al., 1998b) but lacking the specific immunity gene, failed to grow well. It is suspected that the pediocin molecules ultimately gain access to the periplasmic area, and destabilize the membrane causing cell death.

B. Influencing Factors for Antibacterial Action

1. Synergism. The bactericidal efficiency of pediocin PA-1/AcH can be enhanced when used in combination with nisin A and several other suitable bacteriocins of LAB. Hanlin et al. (1993) showed that a combination of pediocin PA-1/AcH and nisin killed more cells than either molecule alone, several Gram-positive food spoilage and pathogenic bacterial strains. These included strains of *Lactobacillus* spp., *Leuconostoc* spp., *Clostridium laramie*, that all are associated with spoilage of refrigerated vacuum-packaged processed and fresh meat, and pathogens like *Lis. monocytogenes*, *Clo. perfringens* and *Clo. botulinum*. Other studies by this group revealed that a combination of pediocin and nisin also had synergistic bactericidal effect against sublethally stressed but normally resistant Gram-positive and Gram-negative foodborne pathogenic and spoilage bacteria (Kalchayanand et al., 1992, 1994, 1998b; Ray, 1993). Pediocin also shows synergistic/additive bactericidal effects against *Lis. innocua* Lin 11, when used in combination with leuconocin Lcm 1 (from *Leu. carnosum* Lcm) and sakacin Z (from *Lab. sake* Z) (*TABLE 6*). Recently, Mulet-Powell et al. (1998) observed synergistic/additive bactericidal effects for the combination of pediocin PA-1/AcH and nisin against *Listeria* spp., including *Lis. monocytogenes*. They also reported that pediocin showed synergistic/additive bactericidal properties with several other bacteriocins of LAB. Limited studies in our laboratory have indicated that the bactericidal spectrum of pediocin is enhanced by combining pediocin with other suitable bacteriocins and other antibacterial compounds like lactic acid, EDTA and lysozyme.

2. Antagonism. Pediocins, being protein, are hydrolyzed by proteolytic enzymes causing them to lose their bactericidal activity (Ray, 1992, 1994). Also, as mentioned above, several anions, like Cl^-, I^-, and $(PO_4)^{3-}$, reduce the bactericidal property of pediocin PA-1/AcH in a direct concentration-dependent way (Bhunia et al., 1991). Zwitterionic lipids may also antagonize the action against cells (Chen et al., 1998). Recently, Fimland et al. (1998) reported that in the presence of a 15-mer fragment in the C-terminal half of pediocin PA-1/AcH (Lys 20 to Ala 34) inhibited bactericidal effect of pediocin. It probably interacts with a specific pediocin-binding site in the membrane of the target cells and prevents pediocin-membrane interaction. In general, the bacteriocins of Gram-positive bacteria are hydrophobic, will accumulate in the lipid phase in an environment, such as a fatty food, and will not be available for bactericidal action. This should be considered while planning to use pediocins in a food as a preservative.

3. Miscellaneous factors. The antibacterial efficiency of pediocin PA-1/AcH is influenced by several other factors. These include pH and temperature of the environment, species, strains, age of target bacteria, relative concentrations of pediocin vs. target cells and storage conditions (Ray, 1992). In general, the bactericidal efficiency of pediocin is higher in an acidic environment (pH 4 to 6), and at ≥20°C. At pH 6.0 and above the molecules become less cationic due to loss of positive charges in the three histidine side-chains. At lower temperatures, especially at 4°C and below, adsorption effi-

TABLE 6: Enhanced bactericidal efficiency of Pediocin PA-1/AcH with other bacteriocins against *Listeria monocytogenes* LIN 11.

Bacteriocins[a]	Log$_{10}$ cfu/ml after storage at 4°C in 0.1% peptone for					
	2 h	1 d	3 d	7 d	14 d	21 d
Control	8.8	8.8	8.5	8.4	8.0	7.0
Nisin A (Nis)	7.1	5.8	4.0	2.9	NT[b]	NT
Pediocin PA-1/AcH (Ped)	5.5	5.4	5.1	4.9	NT	NT
Leuconocin Lcm1 (Leu)[d]	8.6	8.5	8.4	8.3	NT	NT
Sakacin Z (Sak)[d]	7.0	6.4	6.0	4.7	NT	NT
Nis + Ped	3.1	1.4	ND[c]	<1	ND	ND
Nis + Ped + Leu	3.2	2.9	1.9	ND	ND	ND
Nis + Ped + Leu + Sak	2.3	1.4	1.0	ND	ND	ND

[a]Purified bacteriocins were used at 2,000 AU/ml. Results are the average of duplicate samples.
[b]NT, not tested.
[c]ND, no cfu detected in 0.4 ml of the cell suspension.
[d]Leuconocin Lcm1 and Sakacin Z are respectively from *Leu. mesenteroides* Lcm1 and *Lab. sake* Z from our collection (Unpublished data, B. Ray).

ciency of pediocin molecules to the cell wall (and possibly to the membrane) may be reduced. The bactericidal efficiency against a given number of cells of a sensitive strain is higher with higher pediocin concentration, up to a certain level (Bhunia et al., 1991). Gram-positive bacterial strains of the same species also differ greatly in sensitivity to pediocin (Motlagh et al., 1991; Bhunia et al., 1991; Eijsink et al., 1998).

During storage in a concentrated solution, pediocin molecules, due to their hydrophobic nature, have a tendency to form aggregates that can produce an apparent loss of potency. In addition, the single methionine residue can undergo oxidation in the presence of oxygen causing the molecules to lose activity. The molecules have a tendency to form dimers, probably due to breakage and reformation of -S-S- bonds with other molecules under certain conditions such as high heating and cooling. The bactericidal efficiency of such molecules is not known.

C. Bactericidal spectrum

Compared to other bacteriocins in the pediocin-family and to some lantibiotics, pediocin A of *Ped. pentosaceus* FBB61 and pediocins from *Ped. acidilactici* strains have a relatively wider bactericidal spectrum against Gram-positive bacteria. Daeschel and Klaenhammer (1985) reported that pediocin A inhibits growth of several strains of *Ped. pentosaceus, Ped. acidilactici, Staphylococcus aureus, Lactococcus lactis, Lactobacillus* spp., *Clo. botulinum, Clo. perfringens* and *Clo. sporogenes*. Hoover et al. (1989) also found pediocin A inhibited growth of six out of eight strains of *Lis. monocytogenes*. Piva and Headon (1994) used a butanol extracted, partially purified pediocin A preparation to compare the relative sensitivity of Gram-positive bacteria from several genera. A standard *Lac. lactis* strain was the most sensitive of the test strains, and one strain each of *Bacillus cereus, Lis. innocua, Clo. sporogenes* and two strains of *Clo. tyrobutyricum* and several strains of *Lactobacillus* spp. were sensitive to pediocin A.

The bactericidal spectra of pediocins (probably all are pediocin PA-1/AcH) from several *Ped. acidilactici* strains have been studied using neutralized culture broths and partially purified materials. The bactericidal spectrum of pediocin PA-1/AcH has been reviewed before (Ray 1992, 1994, 1996; Ray & Hoover, 1993). There are some conflict-ing reports, particularly when culture broths were used as a pediocin source. Probably even after neutralization, the culture broths of certain strains can contain metabolites other than organic acids, that produce a fuzzy zone of inhibition against an indicator strain, especially when used in concentrated form or in large volumes. This zone characteristic is quite different from the clear circular zone produced by relatively pure bacteriocins, including pediocin. In the following discussion, studies conducted with partially or fully purified pediocin will be reviewed. Bhunia et al. (1981) observed that a ammonium sul-fate precipitated fraction of pediocin PA-1/AcH, formed zones of growth inhibition against *Bac. cereus, Clo. perfringens, Sta. aureus* and several species and strains of *Lactobacillus* and *Leuconostoc*. Later, using a higher purity preparation, this group observed that pediocin PA-1/AcH is able to form zones of growth inhibition against strains of *Lis. monocytogenes, Bac. cereus, Clo. perfringens, Clo. botulinum. Clo. laramie*, and many *Lactobacillus, Leuconostoc, Enterococcus, Pediococcus* and *Lactococcus* species. However, for each species studied strains were found that are resis-tant to pediocin PA-1/AcH. Similar observations were also made by others (Eijsink et al., 1998; Schved et al., 1993; Jager and Harlander, 1992; Coventry et al., 1995). Although bacterial spores are not sensitive to pediocin PA-1/AcH and other bacteriocins, no growth was observed following inoculation of spores of *Clo. laramie, Clo. perfringens* and *Clo. botulinum* into suitable broths containing pediocin. A later study showed that pediocin produced a bactericidal effect on the germinated and outgrowing spores. Limited studies have also indicated that resistant but stressed Gram-positive bacterial strains of *Sta. aureus, Lac. lactis* and Gram-negative bacterial strains of *Salmonella, Esc. coli* O157:H7, *Serratia liquefaciens* and *Pseudomonas* sp. became susceptible to pediocin PA-1/AcH (Ray, 1992; Kalchayanand et al., 1992, 1994, 1998a, b).

V. BIOLOGICAL ADVANTAGE

Gram-positive bacteria that produce pediocins and other bacteriocins probably have growth advantages by completely killing bacteria that are likely to be growing in the same ecological niche (Jack et al., 1995; Yang & Ray, 1994b). Because of this, many bac-teriocins have relatively narrow bactericidal spectrums; some are effective only against a few strains of the same species of producer strain and maybe a few closely related species. In contrast, pediocins of both *Ped. pentosaceus* FBB61 and *Ped. acidilactici* strains have a relatively wide spectrum that includes strains and species from many genera of Gram-positive bacteria. This includes many foodborne spoilage and pathogenic bacteria. Pediocin PA-1/AcH is also effective against many strains of sublethally stressed Gram-positive and Gram-negative spoilage and pathogenic bacteria. Such injured bacteria can be present in foods that have a pH below 6, water activity (A_w) below 0.9, or have been given low heat treatment, subjected to hydrostatic pressure, or stored at low temperature, including long storage at refrigerated temperature (for mesophilic and thermophilic bac-teria). Incorporation of pediocins as preservatives in such foods can help in killing the normally sensitive and resistant but injured cells of spoilage and pathogenic bacteria and ensure longer product shelf-life and greater consumer safety. In the production of certain fermented foods, especially in controlled fermentation where specific strains of starter cultures are used, pediocin PA-1/AcH has a specific application to control *Lis. monocyto-*

genes. Milk used in cheese processing, especially where raw milk is used for unique flavor characteristics (such as many French cheeses), may contain different species of *Listeria*, including *Lis. monocytogenes*. During the initial processing stage, when starter cultures are allowed to grow in milk, *Lis. monocytogenes* can also grow, making the final product unsafe for consumption. Pediocin PA-1/AcH is extremely effective in killing *Listeria*, but does not kill the starter cultures. Thus, pediocin can be added to milk to control *Listeria* in the final product. As *Ped. acidilactici* strains are not capable of metabolizing lactose, the carbohydrate in milk, pediocin-producing *Ped. acidilactici* strains cannot be used for their growth and pediocin production in the starter culture. Efforts are being made to develop *Lactococcus* and *Lactobacillus* starter cultures capable of producing pediocin PA-1/AcH that can be used both for fermentation and for *Listeria* control in the production of cheese. In meat fermentation, where *Pediococcus* spp. are normally used, a pediocin-producing strain can be used to control *Lis. monocytogenes* in the products. Many refrigerated vacuum-packaged processed food products from meat, dairy, fish and vegetable groups contain normally psychrotrophic Gram-positive bacterial strains from genera: *Leuconostoc, Lactobacillus, Carnobacterium, Brochothrix* and *Clostridium*. They are capable of multiplying at refrigerated temperature and causing spoilage of the product. By incorporating pediocin PA-1/AcH during the formulation of the raw ingredients, spoilage problems in the final products can be reduced (Yang & Ray, 1994b; Bennik et al, 1997; Ennahar et al., 1998). This aspect is further discussed below.

VI. APPLICATION

A. Suitability and adaptability

1. As food biopreservatives. It has already been mentioned that pediocin A from *Ped. pentosaceus* FBB61 and pediocin PA-1/AcH from *Ped. acidilactici* strains as well as from *Ped. parvulus* strains and *Lab. plantarum* WHE92 are bactericidal against many normal as well as injured foodborne spoilage and pathogenic bacteria. Thus, these bacteriocins can be used to control these bacteria in foods. Several challenge studies have been conducted, either by using pediocin PA-1/AcH producing strains or by using pediocin itself. Some of them have already been reviewed (Ray, 1992, 1994, 1995, 1996; Ray & Hoover, 1993). A brief discussion is presented here, first for the spoilage bacteria and then pathogenic bacteria.

Motlagh et al. (1992b) showed that pediocin PA-1/AcH at a level of 1,350 AU/g or ml killed over 99% of meat spoilage psychrotrophic *Leu. mesenteroides* Ly in sterile food systems, namely ground beef, sausage mix, milk and ice cream mix. However, the resistant survivors multiplied during storage at 4°C. Application of pediocin PA-1/AcH at a level of 3,000 AU/g also reduced the populations of *Clo. laramie, Lactobacillus* spp. and *Leuconostoc* spp. during storage at 4°C for 12 wk (Ray, 1992; Rozbeh et al., 1993). Coventry et al. (1995) observed that incorporating pediocin (from *Ped. acidilactici* PO$_2$) reduced the population and controlled growth of *Lab. cervatus* in a meat paste during 15 d storage at refrigerated temperature. The effectiveness of pediocin PA-1/AcH in controlling spoilage of salad dressing and salad caused by *Lab. bifermentum* and also in extending the shelf-life of coleslaw, macaroni salad and potato salad was also reported by Gonzalez (1999). The possible use of pediocin-producing *Ped. parvatus* strains to control spoilage and pathogenic bacteria in fresh ready-to-eat vegetables has been suggested by Bennik et al. (1997).

The effectiveness of pediocin from *Ped. acidilactici* strains in controlling *Lis. monocytogenes* in processed meat products has been studied by several researchers. In ground beef and sausage mix, pediocin PA-1/AcH reduced populations of several strains of *Lis. monocytogenes* in a concentration-dependent manner. At a level of 1,350 AU/g the populations were reduced by 3.5 log cycles for strain $Ohio_2$ 3 log cycles for strain Scott A and 1.5 log cycles for strain CA. The resistant survivors multiplied during refrigerated storage (Motlagh et al., 1992b). Pediocin PA-1/AcH application also reduced the populations of *Lis. monocytogenes* in fresh beef (Nielsen et al., 1990) and beef by-products, such as heart, kidney, liver and cheek muscle (Ray, 1992). In liver, the bactericidal effect of pediocin is lost relatively quickly during storage at 4°C, probably due to degradation of pediocin by proteolytic enzymes. Pediocin PA-1/AcH, at a level of 5,000 AU/g in roast beef, Cotto salami and summer sausage inoculated with *Sta. aureus, Lis. monocytogenes, Sal. typhimurium* and *Esc. coli* O157:H7, in combination with hydrostatic pressure at 345 MPa (50 kPsi) at 50°C for 5 min reduced the population by >7 log cycles, 2 log cycles more as compared to pressure-treated products without pediocin (Kalchayanand et al., 1998a,b,c).

Berry et al. (1990) and Foegeding et al. (1992) used pediocin-producing strains of *Ped. acidilactici* strain JD 1-23 and PAC1.0, respectively, as starter cultures to produce dry fermented sausage with meat inoculated with *Lis. monocytogenes*. The *Pediococcus* strains produced pediocin during fermentation and reduced the level of *Listeria* in the final products. Pediocin PA-1/AcH producing *Ped. acidilactici* cells when introduced in frankfurter packages along with *Lis. monocytogenes* cells and stored at 4°C for 60 d or more, greatly reduced the population of *Listeria* reduced greatly (Berry et al., 1991; Degnan et al., 1992).

Pucci et al. (1988) showed that a dried preparation of pediocin PA-1/AcH from *Ped. acidilactici* PAC1.0 effectively controlled the growth of *Lis. monocytogenes* in cottage cheese, half-and-half and cheese sauce during storage at 4°C for 14 d. By spraying a pediocin PA-1/AcH producing cell suspension of *Lab. plantarum* WHE92 on the surface of Munster cheese, Ennahar et al. (1998) were able to prevent the appearance of *Lis. monocytogenes* during ripening for 21 d. The results of these limited studies suggest that both pediocin-producing strains and pediocins have potential application as preservatives to control foodborne spoilage and pathogenic bacteria.

2. Other uses. Very few studies have been conducted on the applications of pediocins, other than a few challenge studies against spoilage and pathogenic bacteria in several food systems. Piva et al. (1995) observed that pediocin A producing *Ped. pentosaceus* reduced proteolysis and ammonia formation in caecal fermentation without interfering with the activity of normal caecal microflora. They suggested that this strain could be used as a probiotic bacterium to maintain intestinal health. Pediocin-producing *Pediococcus* strains can also be used to control undesirable fermentation during silage fermentation (Ray, 1995). Hanlin et al. (1995) developed a procedure in which pediocin PA-1/AcH, along with a low concentration of SDS, was used to control growth of Gram-positive and Gram-negative bacteria that are associated with spoilage of cattle hides. They showed that the hide preserved this way did not undergo sloughening and after tanning produced very good quality leather. Pediocin PA-1/AcH is inhibitory to *Streptococcus mutans* strains, and thus could be used to control dental cavities by incorporating in mouthwash, tooth paste and chewing gum. Other suggested applications of pediocin PA-1/AcH, especially in combination with other bacteriocins, are in items like cosmetics, personal hygiene products, disinfectants and cleaners and animal feed (Ray & Hoover, 1993; Ray, 1995, 1996).

B. Limitations

1. Inherent to pediocins: Several factors have to be recognized when applying pediocin as a bactericidal agent. Pediocin is inactivated by proteolytic enzymes; thus, it might have limited application in raw food systems and in the formulation of raw ingredients used to produce heat processed products, such as heat processed meat products. This problem can be reduced by adding a higher concentration of pediocin or by using encapsulated pediocin, which will be released during heating. Also, certain anions in foods, especially Cl^- and $(PO_4)^{3-}$ either present naturally or added, can reduce the bactericidal property in a concentration-dependent way. This can be overcome by increasing the concentration of pediocin. Being hydrophobic, pediocin can accumulate preferentially at the lipid-water interface in a fatty food, instead of being distributed uniformly in the food system. We found pediocin PA-1/AcH effectively reduced populations of *Listeria* strains in ice cream mix, sausage mix, ground beef and whole milk with lipid percentages ranging between 3.5% and 30%, especially when used in a relatively higher concentration (Motlagh et al., 1992b). The pediocin PA-1/AcH molecules (and other bacteriocins) have a tendency to aggregate during storage, especially in a concentrated liquid. This can be reduced by either maintaining the preparation at a low pH or by adding a food grade surface active compound in small concentrations during storage. Finally, pediocinPA-1/AcH has a methionine which, in the presence of oxygen, can oxidize to methionine sulfoxide, leading to loss of bactericidal activity. This can be avoided by storing the liquid and dried preparations of pediocin under vacuum or in the presence of an inert gas, like nitrogen.

2. Inherent to bacteria: It has already been mentioned that pediocins have relatively wider bactericidal spectra compared to other currently known bacteriocins from the pediocin-family and several lantibiotics. They are effective against many Gram-positive bacteria. The effectiveness of pediocin PA-1/AcH can also be increased to cover many other Gram-positive as well as Gram-negative bacteria, that are normally resistant, by inflicting sublethal injury to the cells. Many of the currently used methods of processing and preservation of foods can induce sublethal injury, and in the presence of pediocin, the injured cells will be killed. Similarly, spores of pediocin-sensitive bacteria, although not sensitive to pediocins, following germination and outgrowth will be killed by pediocin PA-1/AcH.

There are foodborne spores of resistant bacteria, yeasts, molds and pathogenic viruses that are not killed by pediocins, indicating that they have some limitations as food preservatives. In addition, sensitive bacterial strains contain natural variants that are resistant to pediocin PA-1/AcH and thus, are capable of multiplying and reaching spoilage or disease-producing levels in foods with a long shelf life. Both of these problems can be reduced by using one or more suitable bacteriocins with pediocin PA-1/AcH. Pediocin can also be hydrolyzed and inactivated by the extracellular proteolytic enzymes produced by some bacteria. If such bacteria are growing in a food, the potency of pediocin can be reduced depending upon the conditions. Some of these enzymes are heat stable and will retain catalytic properties in heat-treated foods, and pediocin added to such foods can be hydrolyzed during storage. Finally, pediocin-sensitive bacterial cells, after being killed, are capable of releasing intracellular proteases, which in turn can hydrolyze and reduce the potency of pediocin used in a food. Many of these problems are hypothetical and need to be studied properly. An understanding of the actual situation will help design procedures to overcome the problems.

VII. SAFETY AND LEGAL STATUS

A. Safety in food

Both *Ped. pentosaceus* and *Ped. acidilactici* strains have been used as starter cultures in controlled and natural fermentation of meat, fish, vegetables, cereals and dairy products (Ray, 1995). Cells of beneficial bacteria and their metabolites have been consumed by humans for thousands of years without any known adverse effects, and these bacteria are regarded as safe and food grade. Pediocin A producing *Ped. pentosaceus* FBB61 was isolated from a brine of cucumber fermentation. Most pediocin-producing *Ped. acidilactici* strains were obtained from companies that provide commercial starter cultures or from fermented sausages. Pediocin-producing *Ped. parvulus* strains and *Lab. plantarum* WHE92 were isolated from vegetables and cheese, respectively (*TABLE 1*). Thus, pediocin(s), at least from *Ped. acidilactici* have been consumed by humans without any ill effects.

In several studies, pediocin PA-1/AcH was injected in mice and rabbits subcutaneously, intraperitoneally and intravenously. The animals did not show any adverse reactions. Pediocin PA-1/AcH also did not kill hybridoma cells (Bhunia et al., 1990; Bhunia & Johnson, 1992). These studies suggest that pediocin PA-1/AcH will be safe when consumed with food. Also, when consumed, it will be hydrolyzed by the proteolytic enzymes in the digestive tract and unusual products will not be produced because the molecule does not have any unusual amino acids. There is almost no chance for pediocin PA-1/AcH, when consumed, to cause any adverse effects in the gastrointestinal tract or changes to the normal profile of intestinal microorganisms.

Pediococcus species are included with the LAB and are not regarded as pathogens, especially *Ped. pentosaceus* and *Ped. acidilactici*. There have been several reports indicating the isolation of *Ped. acidilactici* strains from human infections. However, they have been considered as secondary contaminants rather than associated with the primary infection. Considering that other bacteria can be mistakenly identified as *Pediococcus*, as indicated in the "Introduction", it will be important to examine such isolates to see if infections are really caused by *Pediococcus*.

B. Legal status

Pediocin PA-1/AcH, besides nisin, is the most studied bacteriocin of LAB (Ray, 1992, 1994, 1995, 1996; Ray & Hoover, 1993). Researchers in several countries have recognized its potential as a food preservative, especially for use in certain specific foods. However, it has not yet been legally approved by the regulatory agencies in the USA, Europe and countries where there is interest for its use as a food biopreservative. It has been stated by individuals associated with US regulatory agencies that use of bacteriocins in food, even from food grade starter culture bacteria, is regulated by the US Federal

Food, Drug and Cosmetic Act and will require integrated reviews by the Food and Drug Administration and Food Safety Inspection Service of the United Stated Department of Agriculture, depending on the nature of a food (Field, 1996; Post, 1996). This includes the use of purified bacteriocins and bacteriocin-producing strains in prepared foods as they will be classified as food additives; if used in fruits and vegetables they will be considered as pesticides. At present, at least one product is being sold commercially that contains pediocin PA-1/AcH. However, the product does not indicate that it has pediocin.

Interestingly, both FDA and USDA/FSIS research laboratories have been studying the bactericidal effectiveness of bacteriocins, especially nisin from the starter culture bacterium, *Lactococcus lactis*. It is expected that if the results of the studies currently being conducted in many laboratories are promising, pediocin PA-1/AcH and other bacteriocins of LAB could be approved for use in food the regulatory agencies. This is possible, since nisin has been approved for use in some specific foods in many countries, including the USA.

VIII. BIOTECHNOLOGY

A. Large-scale production

Pediocin A is produced by *Ped. pentosaceus* FBB61 in very low concentration and probably no work has been conducted to achieve its large scale production. Pediocin production by *Ped. acidilactici* strains are relatively high and several methods have been tested for large scale production. Results presented in *TABLE 3* indicate that strains of *Ped. acidilactici* differ greatly under identical conditions with respect to the level of pediocin PA-1/AcH production. We have isolated several of these strains from several parental strains through selective procedures. Our experience is that pediocin PA-1/AcH production is much higher in a simple non-buffered broth, such as TGE broth, than in a complex buffered broth, like MRS broth (Biswas et al., 1991). Even in buffered TGE broth, the production of pediocin is much lower than TGE broth. While it is not clearly understood, we have observed the highest production of pediocin at pH 3.6 to 3.8. Originally, it was thought that the processing of prepediocin to pediocin is probably dependent on low pH (Ray et al., 1993), this may not be the case. By increasing tryptone (or trypticase), glucose and yeast extract to the 2% level in TGE broth, yield can be increased by another 30% or higher. We have used up to 14L TGE broth in a fermenter and obtained as high as 80,000 AU of pediocin/ml of culture broth by assaying against *Lab. plantarum* NCDO955. Pediocin PA-1/AcH is currently being produced from *Ped. acidilactici* PAC1.0 by a commercial firm and the product is being sold as a flavor enhancer; their production process has not been made available.

Cho et al. (1996) studied the feasibility of continuous pediocin production by *Ped. acidilactici* PO$_2$ using immobilized cells in MRS broth in a packed bed reactor. At optimum conditions, maximum production was 1,000 AU/ml/d. The bioreactor was operated for three months without any difficulties or contamination. Huang et al. (1996) also described a method they used to study continuous production of pediocin by *Ped. acidilactici* UL5. They used both free cells and immobilized cells and obtained a maximum of 4,915 AU/ml with the free cells and 1,024 AU/ml with the immobilized cells. The level of pediocin production by continuous methods, as reported in these two studies, appeared to be much less than that obtained in a static culture in TGE broth (Biswas et al., 1991; Yang et al., 1992; Yang & Ray, 1994a).

B. Gene technology/protein engineering

1. Expression of pediocin PA-1/AcH in heterologous hosts. Since pediocin-producing *Ped. acidilactici* strains do not ferment lactose but pediocin PA-1/AcH is bactericidal to *Lis. monocytogenes*, but not to many *Lactococcus lactis* and *Lactobacillus* spp. used in dairy fermentations, efforts are being made to use gene-technology to develop starter culture strains that will produce pediocin. These or suitable strains can then be used to both control *Listeria* in the product and to ferment milk. Chikindas et al. (1995) cloned the whole *ped* (or *pap*) operon (*FIGURE 1*) from the plasmid pSRQ11 of *Ped. acidilactici* PAC1.0 in a vector containing a strong lactococcal promoter and a erythromycin-resistant (Em^r) selective marker. This cloned plasmid (pMC117), when introduced by electroporation into a *Lactococcus lactis* strain, produced pediocin PA-1/AcH at about 50 AU/ml. Raising the copy number of pMC117 by introducing it into another *Lactococcus lactis* host, increased pediocin production to 3,000 to 6,000 AU/ml, which is the same as produced by *Ped. acidilactici* PAC1.0, but much lower than the 40,000 to 80,000 AU/ml we obtained strains we have been using. Plasmid pMC117, when introduced into non-bacteriocin-producing strains of *Ped. pentosaceus* and *Ped. acidilactici*, also produced pediocin. Buyong et al. (1998) introduced pMC117 into a starter culture strain of *Lactococcus lactis* ssp. *lactis*, that is used in Cheddar cheese production. They inoculated this strain along with *Lis. monocytogenes* into milk and produced Cheddar cheese. During various steps of processing and ripening of the cheese, they measured the pediocin level and *Lis. monocytogenes* population. The pediocin PA-1/AcH level was as high as 64,000 AU/g after 14 d of ripening and the *Listeria* population was reduced from 3.2 log cycles/g on day 1 to 0.9 log cycles/g on day 184. In contrast, *Listeria* populations were 4.8 log cycles/g on day 1 and 4.1 log cycles/g on day 184 during ripening in cheese made using the same starter strain lacking pMC117.

Horn et al. (1998) modified a plasmid from a *Lactococcus lactis* strain that had the promoter and leader segment encoding for lactococcin A fused to the nucleotide sequence of mature pediocin PA-1/AcH. When this plasmid was introduced into a *Lac. lactis* strain that is normally resistant to pediocin PA-1/AcH, but encode the lactococcin A-translocation apparatus, it produced pediocin, but only at a level of about 25% of that produced by the parental *Ped. acidilactici* strain. It is suggested that this strain can also be used in dairy fermentation to control *Lis. monocytogenes* or in applications where *Ped. acidilactici* cannot be used for fermentation.

2. Synthetic and hybrid pediocin. There are many conserved areas in the primary structure of pediocin-like bacteriocins. The similarities are more apparent in the N-terminal half than the rather diverse in the C-terminal half that is considered to determine the bactericidal specificity of a bacteriocin. Attempts are being made to exchange these halves between bacteriocins to study if such an exchange reduces or increases bactericidal efficiency compared to normal bacteriocins. Fimland et al. (1996) synthetically prepared normal pediocin PA-1/AcH and sakacin P as well as two hybrid-molecules Ped-Sak and Sak-Ped. In the Ped-Sak hybrid, the N-terminal segment of pediocin from Lys 1 to Ala 21 was fused with the C-terminal part of sakacin from Ile 22 to Lys 43. Similarly, the Sak-Ped hybrid was constructed by fusing Lys 1 to Ala 21 of sakacin with the Thr 22 to Cys 44 region of pediocin (*FIGURE 6*). The yields obtained for the synthetic normal bacteriocins, were about 1% for pediocin PA-1/AcH and 10% for sakacin P with purity of

```
            1      5       10        15        20      25        30        35        40    44
Pediocin PA-1/AcH  K Y Y G N G V T C G K H S C S V D W G K A T T C I I N N G A M A W A T G G H Q G N H K C

Sakacin P          K Y Y G N G V H C G K H S C T V D W G T A I G N I G N N A A A N W A T G G N A G W N K -

Ped-Sak            K Y Y G N G V T C G K H S C S V D W G K A I G N I G N N A A A N W A T G G N A G W N K -

Sak-Ped            K Y Y G N G V H C G K H S C T V D W G T A T T C I I N N G A M A W A T G G H Q G N H K C
```

FIGURE 6; Amino acid sequences of pediocin PA-1/AcH, sakacin P and their two hybrids. The dissimilar amino acids in the sequences of parental types are underlined in sakacin P. Only 14 were different, 11 of which are at C-terminal regions.

about 85 to 95% for pediocin and 70% for sakacin. The two hybrids had several characteristics that are different from the two normal bacteriocins, especially for the number of cysteines and net charges (*TABLE 7*). The two synthetically prepared normal bacteriocins had specific activities comparable to those of their natural types. The two hybrid bacteriocins, however, displayed similar activity spectrum and bactericidal effectiveness to what normally is specified by the C-terminal portion of the natural bacteriocins. Thus, the Ped-Sak hybrid showed the same specificity pattern as that of sakacin P and the Sak-Ped hybrid had the same specificity pattern as pediocin PA-1/AcH. These results tend to agree with the idea that bactericidal efficiency of a bacteriocin, at least in the pediocin-family, is determined by the primary structure of the C-terminal half.

3. Chimeric and mutant pediocins. Among the four genes in the *ped* (or *pap*) operon, the minimum requirement for the production of pediocin PA-1/AcH is the presence of the *ped* (or *pap*) A, and ped (or pap) D genes (Bukhtiyarova et al., 1994). Venema et al. (1995), however, suggested that *ped* (or *pap*) C is also essential for pediocin production (*FIGURE 1*). This has been shown by cloning the operon in suitable plasmids and expressing the genes in *Esc. coli* hosts. Miller et al. (1998a) constructed a plasmid in which only the nucleotide segment of the matured pediocin (without the leader segment) was fused to the COOH-terminus of the secretory protein, maltose binding protein (MBP). The *sec* machinery of the host *Esc. coli* secreted the chimeric MBP-pediocin PA-1/AcH molecule which proved to be bactericidal to bacterial strains which are sensitive to the normal pediocin PA-1/AcH.

TABLE 7: Some characteristics of Pediocin PA-1/AcH, Sakacin P and their hybrids.

Characteristics	Bacteriocins[a]			
	Pediocin PA-1/AcH	Sakacin P	Ped-Sak	Sak-Ped
Total amino acids	44	43	43	44
Cysteines	4	2	2	4
Lysine	4	3	4	3
Histidine	3	2	1	4
Aspartic acid	1	1	1	1
Methionine	1	0	0	1
Net charge: pH <6	6	4	4	6
pH >6	3	2	3	2

[a]The hybrids have more similarities with the C-terminal halves of the parental types.

Using the same system, Miller et al. (1998b) subsequently generated, by PCR random mutagenesis of DNA encoding for MBP-pediocin PA-1/AcH, 17 substitution mutants at 14 of the 44 amino acids of the molecule. Seven of these mutants (Asn 5 Lys, Cys 9 Arg, Cys 14 Ser, Cys 14 Tyr, Gly 37 Asp, Gly 37 Arg, Cys 44 Trp) were biologically inactive. The other nine mutants (Lys 1 Asn, Trp 18 Arg, Ile 26 Thr, Met 31 Thr, Ala 34 Asp, Asn 41 Lys, His 42 Leu, Lys 43 Asn, Lys 43 Glu) retained <1% to about 60% of the activity of the normal MBP-pediocin PA-1/AcH. The results suggest that for full activity, the amino acid sequences in both the N- and C-terminal parts are important. Surprisingly, one mutant (Lys 11 Glu) had about 2.8-fold higher activity than wild type, revealing the possibility that substitution of suitable amino acid(s) in the primary structure may enable production of mutant pediocin PA-1/AcH variants with higher potency and maybe a wider bactericidal spectrum.

4. Bactericidal efficiency of prepediocin. The removal of the leader peptide from prepediocin during translocations, resulting in the production of biologically active pediocin PA-1/AcH, lead to the suspicion that prepediocin (and other pediocin-like bacteriocins) are biologically inactive. A recent study, however, showed that the 62 amino acid prepediocin is bactericidal and has a similar activity pattern, but perhaps only about 80% of the potency of the normal pediocin (Ray et al., 1999). The prepediocin is relatively more sensitive to proteolytic enzymes than pediocin; by removing the leader peptide during translocation, the matured pediocin is made more resistant to these enzymes.

IX. SUMMARY

Pediocins from several strains of *Ped. acidilactici* and *Ped. pentosaceus*, due to their bactericidal property against many Gram-positive bacteria are being studied for their potential use as food preservatives and antibacterial agents. Pediocin A, of *Ped. pentosaceus*, is 80 kDa, pediocins of *Ped. acidilactici* strains are about 4.6 kDa. From the available information, it is anticipated that the *Ped. acidilactici* strains currently studied may have the same origin and they all have 44 amino acids as pediocin PA-1/AcH. Pediocin PA-1/AcH production phenotype is encoded in the first of the four open reading frames in an operon; the other three open reading frames encode for immunity protein against pediocin PA-1/AcH and for the translocation and proteolytic cleavage of the propediocin. At the translation level, prepediocin has 62 amino acids, during translocation an 18 amino acid leader segment is removed from the N-terminal part, leaving the 44 amino acid pediocin PA-1/AcH. Two *Ped. parvulus* strains and one *Lactobacillus plantarum* strain also produce pediocin PA-1/AcH. In general, the bactericidal property of pediocins is relatively resistant to heat, storage conditions, pH, organic solvents and hydrostatic pressure, but are destroyed by proteolytic enzymes and neutralized by some anions. Pediocin PA-1/AcH by *Ped. acidilactici* strains is produced in large amounts in a non-buffered nutritional medium, especially at a low terminal pH. The bactericidal action of pediocin PA-1/AcH against target cells is produced through the ability of the molecules to form pores in the membrane and disrupt the proton motive force. Most probably, the N-terminal half of the molecule interacts with the membrane through electrostatic interaction while the C-terminal half causes destabilization of the membrane. Studies with amino acid mutants have indicated that the bactericidal activity of pediocin PA-1/AcH is not controlled by a specific segment of the molecule. Limited studies have shown that pediocin PA-1/AcH, or the producer strain(s), can be effectively used in food systems to control *Lis. monocytogenes*. Under appropriate processing conditions it can also be used to control Gram-positive and Gram-negative spoilage bacteria in foods.

X. REFERENCES

1. Anonymous. 1996. Designation of strain B213c (DSM20284) in place of strain NCDO1859 as the type strain of *Pediococcus acidilactici* Lindner 1887. *Int. J. Systemic Bacteriol.* 46:835.
2. Bennik, M. H. J., Smid, E. J., and Gorris, L. G. M. 1997. Vegetable-associated *Pediococcus parvulus* produces pediocin PA-1. *Appl. Environ. Microbiol.* 63:2074-2076.
3. Berry, E. D., Liewen, M. B., Mandiog, R. W., and Hutkins, R. W. 1990. Inhibition of *Listeria monocyto genes* by bacteriocin-producing *Pediococcus* during the manufacture of fermented semidry sausage. *J. Food Prot.* 53:194-197.
4. Berry, E. D., Hutkins, R. W., and Mandigo, R. W. 1991. The use of bacteriocin-producing *Pediococcus acidilactici* to control postprocessing *Listeria monocytogenes* conamination of frankfurters. *J. Food Prot.* 54:681-686.
5. Bhunia, A. K., Kim, W.-J., Johnson, M. C., and Ray, B. 1987a. Partial purification and characterization of antimicrobial substances from *Pediococcus acidilactici* and *Lactobacillus plantarum*. Abstract #143. *Inst. Food Technol. Ann. Meeting*, June 16-19, Las Vegas.
6. Bhunia, A. K., Johnson, M. C., and Ray, B. 1987b. Direct detection of an antimicrobial peptide of *Pediococcus acidilactici* in sodium dodecyl sulfate-polyacrylamide gel electrophoresis. *J. Indust. Microbiol.* 2:319-322.
7. Bhunia, A. K., Johnson, M. C., and Ray, B. 1988. Purification, characterization, and antimicrobial spectrum of a bacteriocin produced by *Pediococcus acidilactici*. *J. Appl. Bacteriol.* 65:261-268.
8. Bhunia, A. K., Johnson, M. C., Ray, B., and Belden, E. L. 1990. Antigenic property of pediocin AcH produced by *Pediococcus acidilactici* H. *J. Appl. Bacteriol.* 69:211-215.
9. Bhunia, A. K., Johnson, M. C., Ray, B., and Kalchayanand, N. 1991. Mode of action of pediocin AcH from *Pediococcus acidilactici* H on sensitive bacterial strain. *J. Appl. Bacteriol.* 70:25-33.
10. Bhunia, A. K., and Johnson, M. G. 1992. Monoclonal antibody-colony immunoblot method specific for isolation of *Pediococcus acidilactici* from foods and correlation with pediocin (bacteriocin) production. *Appl. Environ. Microbiol.* 58:2315-2320.
11. Bhunia, A. K. 1994. Monoclonal antibody-based enzyme immunoassay for pediocins of *Pediococcus acidilactici*. *Appl. Environ. Microbiol.* 60:2692-2696.
12. Bhunia, A. K., Bhowmik, T. K., and Johnson, M. G. 1994. Determination of bacteriocin-encoding plasmids of *Pediococcus acidilactici* strains by Southern hybridization. *Lett. Appl. Microbiol.* 18:168-170.
13. Biswas, S. R., Ray, P., Johnson, M. C., and Ray, B. 1991. Influence of growth conditions on the production of bacteriocin, pediocin AcH, by *Pediococcus acidilactici* H. *Appl. Environ. Microbiol.* 57:1265-1267.
14. Bukhtiyarova, M., Yang, R., and Ray, B. 1994. Analysis of the pediocin AcII gene cluster from plasmid pSMB74 and its expression in a pediocin negative *Pediococcus acidilactici* strain. *Appl. Environ. Microbiol.* 60:3405-3408.
15. Buyong, N., Kok, J., and Luchansky, J. B. 1998. Use of a genetically enhanced pediocin-producing starter culture, *Lactococcus lactis* subsp. *lactis* MM217, to control *Listeria monocytogenes* in Cheddar cheese. *Appl. Environ. Microbiol.* 64:4842-4845.
16. Chen, Y., Shapira, R., Eisenstein, M., and Montville, T. J. 1997a. Functional characterization of pediocin PA-1 binding to liposomes in the absence of a protein receptor and its relationship to a predicted tertiary structure. *Appl. Environ. Microbiol.* 63:524-431.
17. Chen, Y., Ludescher, R. D., and Montville, T. J. 1997b. Electrostatic interactions, but not the YGNGV consensus motif, govern the binding of pediocin PA-1 and its fragments to phospholipid vesicles. *Appl. Environ. Microbiol.* 63:4770-4777.
18. Chen, Y., Ludescher, R. D., and Montville, T. J. 1998. Influence of lipid composition on pediocin PA-1 binding to phospholipid vesicles. *Appl. Environ. Microbiol.* 64:3530-3532.
19. Chikindas, M. L., Garcia-Garcera, M. J., Driessen, A. J. M., Ledeboer, A. M., Nissen-Meyer, J., Nes, I. F., Abee, T., Konings, W. N., and Venema, G. 1993. Pediocin PA-1, a bacteriocin of *Pediococcus acidilactici* PAC1.0, forms hydrophilic pores in the cytoplasmic membrane of target cells. *Appl. Environ. Microbiol.* 59:3577-3584.

20. Chikindas, M. L., Venema, K., Ledeboer, A. M., Venema G., and Kok, J. 1995. Expression of lactococcin A and pediocin PA-1 in heterologous hosts. *Lett. Appl. Microbiol.* 21:183-189.

21. Cho, H.-Y., Yousef, A. E., and Yang, S.-T. 1996. Continuous production of pediocin by immobilized *Pediococcus acidilactici* PO₂ in a packed-bed reactor. *Appl. Microbiol. Biotechnol.* 45:589-594.

22. Christensen, D. P., and Hutkins, R. W. 1992. Collapse of the proton motive force in *Listeria monocytogenes* caused by bacteriocin produced by *Pediococcus acidilactici*. *Appl. Environ. Microbiol.* 58:3312-3315.

23. Christensen, D. P., and Hutkins, R. W. 1994. Glucose uptake by *Listeria monocytogenes* Scott A and inhibition by pediocin JD. *Appl. Environ. Microbiol.* 60:3870-3873.

24. Cintas, L. M., Rodriguez, J. M., Fernandez, M. F., Sletten, K., Nes, I. F., Hernandez, P. E., and Holo, H. 1995. Isolation and characterization of pediocin L50, a new bacteriocin from *Pediococcus acidilactici* with a broad inhibitory spectrum. *Appl. Environ. Microbiol.* 61:2643-2648.

25. Cintas, L. M., Casaus, P., Hollo, H., Hernandez, P. E., Nes, I. F., and Havarstein, L. S. 1998. Enterocins L50A and L50B, two novel bacteriocins from *Enterococcus faecium* L50, are related to staphyloccal hemolysins. *J. Bacteriol.* 180:1988-1994.

26. Collins, M. D., Williams, A. M., and Wallbanks, S. 1990. The phylogeny of *Aerococcus* and *Pediococcus* as determined by 16S rRNA sequence analysis: description of *Tetragenococcus* gen. nov. *FEMS Microbiol. Lett.* 70:255-262.

27. Coventry, M. J., Muirhead, K., and Hickey, M. W. 1995. Partial characteristics of pediocin PO₂ and comparison with nisin for biopreservation of meat products. *Int. J. Food Microbiol.* 26:133-145.

28. Daba, H., Lacroix, C., Huang, J., and Simard, R. E. 1993. Influence of growth conditions on production and activity of mesenterocin 5 by a strain of *Leuconostoc mesenteroides*. *Appl. Microbiol. Biotechnol.* 39:166-173.

29. Daba, H., Lacroix, C., Huang, J., Simard, R. E., and Lemieux, L. 1994. Simple method of purification and sequencing of a bacteriocin produced by *Pediococcus acidilactici* UL5. *J. Appl. Bacteriol.* 77:682-688.

30. Daeschel, M. A., and Klaenhammer, T. R. 1985. Association of a 13.6-megadalton plasmid in *Pediococcus pentosaceus* with bacteriocin activity. *Appl. Environ. Microbiol.* 50:1538-1541.

31. De Vuyst, L. 1994. Bacteriocins of *Pediococcus*. In *Bacteriocins of Lactic Acid Bacteria*, eds. L. De Vuyst and E. J. Vandamme, pp. 461-464. Blackie Academic and Professional, Inc. New York.

32. Degnan, A. J., Yousef, A. E., and Luchansky, J. B. 1992. Use of *Pediococcus acidilactici* to control *Listeria monocytogenes* in temperature-abused vacuum-packaged wieners. *J. Food Prot.* 55:98-103.

33. Eijsink, V. G., Skeie, M., Middelhoven, P. H., Brurberg, M. B., and Nes, I. F. 1998. Comparative studies of class IIa bacteriocins of lactic acid bacteria. *J. Appl. Microbiol.* 64:3275-3281.

34. Eisenberg, D., Schwarz, E., Komaromy, M., and Wall, R. 1984. Analysis of membrane and surface protein sequences with the hydrophobic moment plot. *J. Mol. Biol.* 179:125-142.

35. Elegado, F. B., Kim. W.-J., and Kwon, D. Y. 1997. Rapid purification, partial characterization and antimicrobial spectrum of the bacteriocin, pediocin AcH, from *Pediococcus acidilactici* M. *Int. J. Food Microbiol.* 37:1-11.

36. Ennahar, S., Aoude-Werner, D., Sorokine, O., Dorsselaer, A. V., Bringel, F., Hubert, J.-C., and Hasselmann, C. 1996. Production of pediocin AcH by *Lactobacillus plantarum* WHE92 isolated from cheese. *Appl. Environ. Microbiol.* 62:4381-4387.

37. Ennahar, S., Assobhel, D., and Hasselmann, C. 1998. Inhibition of *Listeria monocytogenes* in a smear-surface soft cheese by *Lactobacillus plantarum* WHE92, a pediocin AcH producer. *J. Food Prot.* 61:186-191.

38. Field, F. A. 1996. Use of bacteriocins in food: regulatory considerations. *J. Food Prot.* Supplement: 72-77.

39. Fimland, G., Blingsmo, O. R., Sletten, K., Jung, G., Nes, I. F., and Nissen-Meyer, J. 1996. New biologically active hybrid bacteriocins constructed by combining regions from various pediocin-like bacteriocins: the C-terminal region is important for determining specificity. *Appl. Environ. Microbiol.* 62:3313-3318.

40. Fimland, G., Jack, R., Jung, G., Nes, I. F., and Nissen-Meyer, J. 1998. The bacterial activity of pediocin PA-1 is specifically inhibited by a 15-mer fragment than spans the bacteriocin from the center toward the C-terminus. *Appl. Environ. Microbiol.* 64:5057-5060.

41. Fleming, H. P., Etchells, J. L., and Costilow, R. N. 1975. Microbial inhibition by an isolate of *Pediococcus* from cucumber brines. *Appl. Microbiol.* 30:1040-1042.

42. Foegeding, P. M., Thomas, A. B., Pilkington, D. H., and Klaenhammer, T. R. 1992. Enhanced control of *Listeria monocytogenes* by in situ-produced pediocin during dry fermented sausage production. *Appl. Environ. Microbiol.* 58:884-890.

43. Gallagher, N. L. F., Sailer, M., Niemczura, W. P., Nakashima, T. T., Stiles, M. E., and Vederas. 1997. Three-dimensional structure of leucocon A in trifuoroethanol and dodecylphosphocholine micelles: spatial location of residues critical for biological activity in type IIa bacteriocins from lactic acid bacteria. *Biochemistry* 36:15062-15072.

44. Garvie, E. I. 1986. Genus *Pediococcus* Claussen 1903. In *Bergey's Manual of Systematic Bacteriology*, ed. P. H. A. Sneath, Vol. 2, pp. 1075-1079. Willams and Wilkins, Baltimore.

45. Gonzalez, C. F., and Kunka, B. S. 1983. Plasmid transfer in *Pediococcus* spp: intergeneric and intrageneric transfer of pIP501. *Appl. Environ. Microbiol.* 46:81-89.

46. Gonzalez, C. F., and Kunka, B. S. 1987. Plasmid-associated bacteriocin production and sucrose fermentation in *Pediococcus acidilactici*. *Appl. Environ. Microbiol.* 53:2534-2538.

47. Gonzalez, C. F. 1989. Methods for inhibiting bacterial spoilage and resulting composition. *U. S. Patent 4883673*.

48. Graham, D. C., and McKay, L. L. 1985. Plasmid DNA in strains of *Pediococcus cerevisiae* and *Pediococcus pentosaceus*. *Appl. Environ. Microbiol.* 50:532-534.

49. Hanlin, M. B., Kalchayanand, N., Ray, P., and Ray, B. 1993. Bacteriocins of lactic acid bacteria in combination have greater antibacterial activity. *J. Food Prot.* 56:252-255.

50. Hanlin, M. B., Field, R. A., Ray, B., and Bailey, D. G. 1995. Characterization of predominant bacteria in cattle hides and their control by a bacteriocin based preservative. *J. Am. Leather Chemist Assoc.* 90:308-321.

51. Havarstein, L. S., Diep, D. B., and Nes, I. F. 1995. A family of ABC transporters carry out proteolytic processing of their substrates concomitant with export. *Mol. Microbiol.* 16:229-240.

52. Henderson, J. T., Chopko, A. L., and van Wassenaar, P. D. D. 1992. Purification and primary structure of pediocin PA-1 produced by *Pediococcus acidilactici* PAC1.0. *Arch. Biochem. Biophys.* 295:5-12.

53. Hoover, D. G., Walsh, P. M., Kolaetis, K. M., and Daly, M. M. 1988. A bacteriocin produced by *Pediococcus* species associated with 5.5-metadalton plasmid. *J. Food Prot.* 51.29-31.

54. Hoover, D. G., Dishart, K. J., and Hermes, M. A. 1989. Antagonistic effect of *Pediococcus* spp. against *Listeria monocytogenes*. *Food Biotechnol.* 3:183-196.

55. Horn, N., Martinez, M. I., Martinez, J. M., Hernandez, P. E., Gasson, M. J., Rodriguez, J. M., and Dodd, H. M. 1998. Production of pediocin PA-1 by *Lactococcus lactis* using the lactococcin secretory apparatus. *Appl. Environ. Microbiol.* 64:818-823.

56. Huang, J., Lacroix, C., Daba, H., and Simard, R. E. 1996. Pediocin 5 production and plasmid stability during continuous free and immobilized cell cultures of *Pediococcus acidilactici* UL5. *J. Appl. Bacteriol.* 80:635-644.

57. Jack, W. R., Tagg, J. R., and Ray. B. 1995. Bacteriocins of Gram-positive bacteria. *Microbiol. Rev.* 59:171-200.

58. Jager, K., and Harlander, S. 1992. Characterization of a bacteriocin from *Pediococcus acidilactici* PC and comparison of bacteriocin-producing strains using molecular typing procedures. *Appl. Microbiol. Biotechnol.* 37:631-637.

59. Janes, M. E., Nannapaneni, R., Proctor, A., and Johnson, M. J. 1998. Rice hull ash and silicic acid as adsorbents for concentration of bacteriocins. *Appl. Environ. Microbiol.* 64:4403-4409.

60. Kalchayanand, N., Hanlin, M. B., and Ray, B. 1992. Sublethal injury makes Gram-negative and resistant Gram-positive bacteria sensitive to the bacteriocins, pediocin AcH and nisin. *Lett. Appl. Microbiol.* 15:239-243.

61. Kalchayanand, N., Sikes, T., Dunne, C. P., and Ray, B. 1994. Hydrostatic pressure and electroporation have increased bactericidal efficiency in combination with bacteriocins. *Appl. Environ. Microbiol.* 60:4174-4177.

62. Kalchayanand, N., Ray, B., Sikes, T., and Dunne, C. P. 1995. Bacteriocin-based biopreservatives add an extra dimension in food preservation by hydrostatic pressure. *Proceedings of the Spring Meeting*, April 24-26, 1995, pp. 282-286. Research and Development Associates for Military Food and Packaging System, Inc. San Antonio, TX.

63. Kalchayanand, N., Sikes, A., Dunne, C. P., and Ray, B. 1998a. Interaction of hydrostatic pressure, time and temperature of pressurization and pediocin AcH on inactivation of foodborne bacteria. *J. Food Prot.* 61:425-431.

64. Kalchayanand, N., Sikes, A., Dunne, C. P., and Ray, B. 1998b. Factors influencing death and injury of foodborne pathogens by hydrostatic pressure pasteurization. *Food Microbiol.* 15:207-214.

65. Kalchayanand, N., Ray, B., Sikes, A., and Dunne, C. P. 1998c. Enhancement of safety of processed meat products by hydrostatic pressure in combination with temperature and bacteriocins. *Proceedings, 44th International Congress of Meat Science and Technology*. pp. 522-523, August 30 to September 4, 1998, Barcelona, Spain. Inst. Food Ag. Res. and Technol. (IRTA), Spain.

66. Kim, W.-J., Ray, B., and Johnson, M. C. 1992. Plasmid transfers by conjugation and electroporation in *Pediococcus acidilactici*. *J. Appl. Bacteriol.* 72:201-207.

67. Kim, W.-J., Hong, S.-S., Cha, S.-K., and Koo, Y.-J. 1993. Use of bacteriocinogenic *Pediococcus acidilactici* in sausage fermentation. *J. Microbiol. Biotechnol.* 3:199-203.

68. Kim, W.-J., Ryoo, C. R., Lee, Y. J. and Kwon, D. Y. 1998. *Lactococcus lactis* sp. isolated from kimchi produces pediocin AcH. Abstract #29-2, p. 55. *Book of Abstracts*, 1998 IFT Annual Meeting, Atlanta, June 29-24. Inst. Food Technol. Chicago, IL.

69. Kimura, H., Matsusaki, H., Sashihara, T., Sonomoto, K., and Ishizaki, A. 1998. Purification and partial identification of bacteriocin ISK-1, a new lantibiotic produced by *Pediococcus* sp. ISK-1. *Biosci. Biotechnol. Biochem.* 62:2341-2345.

70. Lewus, C. B., and Montville, T. J. 1991. Detection of bacteriocins produced by lactic acid bacteria. *J. Microbiol. Methods* 13:145-150.

71. Lozano, J. C. N., Nissen-Meyer, J., Sletten, K., Pelaz, C., and Nes, I. F. 1992. Purification and amino acid sequence of a bacteriocin produced by *Pediococcus acidilactici*. *J. Gen. Microbiol.* 138:1985-1990.

72. Luchansky, J. B., Glass, K. A., Harsono, K. D., Degnan, A. J., Faith, N. G., Cauvin, B., Baccus-Taylor, G., Arihara, K., Bater, B., Maurer, A. J., and Cassens, R. G. 1992. Genomic analysis of *Pediococcus* starter cultures used to control *Listeria monocytogenes* in turkey summer sausage. *Appl. Environ. Microbiol.* 58:3053-3059.

73. Martinez, M. I., Rodriguez, E., Medina, M., Hernandez, P. E., and Rodriguez, J. M. 1998. Detection of specific bacteriocin-producing lactic acid bacteria by colony hybridization. *J. Appl. Microbiol.* 84:1099-1103.

74. Marugg, J. D., Gonzalez, C. F., Kunka, B. S., Ledeboer, A. M., Pucci, M. J., Toonen, M. Y., Walker, S. A., Zoetmulder, L. C. M., and Vandenbergh, P. A. 1992. Cloning, expression and nucleotide sequence of genes involved in production of pediocin PA-1, a bacteriocin from *Pediococcus acidilactici* PAC1.0. *Appl. Environ. Microbiol.* 58:2360-2367.

75. Miller, K. W., Schamber, R., Chen, Y., and Ray, B. 1998a. Production of active chimeric pediocin AcH in *Escherichia coli* in the absence of processing and secretion genes from the *Pediococcus pap* operon. *Appl. Environ. Microbiol.* 64:14-20.

76. Miller, K. W., Schamber, R., Osmanagaoglu, O., and Ray, B. 1998b. Isolation and characterization of pediocin AcH chimeric protein mutants with altered bactericidal activity. *Appl. Environ. Microbiol.* 64:1197-2005.

77. Montville, T. J., and Chen, Y. 1998. Mechanistic action of pediocin and nisin: recent progress and unresolved questions. *Appl. Microbiol. Biotechnol.* 50:511-519.

78. Motlagh, A. M., Johnson, M. C., and Ray, B. 1991. Viability loss of foodborne pathogens by starter culture metabolites. *J. Food Prot.* 54:873-878, 884.

79. Motlagh, A. M., Bhunia, A. K., Szostek, F., Hansen, T. R., Johnson, M. G., and Ray, B. 1992a. Nucleotide and amino acid sequence of *pap*-gene (pediocin AcH production) in *Pediococcus acidilactici* H. *Lett. Appl. Microbiol.* 15:45-48.

80. Motlagh, A. M., Holla, S., Johnson, M. C., Ray, B., and Field, R. A. 1992b. Inhibition of *Listeria* spp. in sterile food systems by pediocin AcH, a bacteriocin produced by *Pediococcus acidilactici* H. *J. Food Prot.* 55:337-343.

81. Motlagh, A. M., Bukhtiyarova, M., and Ray, B. 1994. Complete nucleotide sequences of pSMB74, a plasmid encoding production of pediocin AcH in *Pediococcus acidilactici*. *Lett. Appl. Microbiol.* 18:305-312.

82. Mulet-Powell, N., Lacoste-Armynot, A. M., Vinas, M., and Simeon De Buochberg, M. 1998. Interactions between pairs of bacteriocins from lactic bacteria. *J. Food Prot.* 61:1210-1212.

83. Nielsen, J. W., Dickson, J. S., and Crouse, J. D. 1990. Use of bacteriocin produced by *Pediococcus acidilactici* to inhibit *Listeria monocytogenes* associated with fresh meat. *Appl. Environ. Microbiol.* 56:2142-2145.

84. Nissen-Meyer, J., Havarstein, L. S., Holo, H., Sletten, K., and Nes, I. F. 1993a. Association of the lactococcin A immunity factor with the cell membrane: purification and characterization of the immunity factor. *J. Gen. Microbiol.* 139:1503-1509.

85. Nissen-Meyer, J., Larsen, A. G., Sletten, K., Daeschel, M., and Nes, I. F. 1993b. Purification and characterization of plantaricin A, a *Lactobacillus plantarum* bacteriocin whose activity depends on the action of two peptides. *J. Gen. Microbiol.* 139:1973-1978.

86. Noerlis, Y., and Ray, B. 1994. Factors influencing immunity and resistance of *Pediococcus acidilactici* to the bacteriocin, pediocin AcH. *Lett. Appl. Microbiol.* 18:138-143.

87. Peschel, A., Otto., M., Jack, R. W., Kalbacher, H., Jung, G., and Gotz, F. 1999. Inactivation of the *dlt* operon in *Staphylococcus aureus* confer sensitivity to defensins, protegrins and other antimicrobial peptides. *J. Biol. Chem.* 274:8405-8410.

88. Piva, A., and Headon, D. R. 1994. Pediocin A, a bacteriocin produced by *Pediococcus pentosaceus* FBB61. *Microbiol.* 140:697-702.

89. Piva, A., Meola, E., and Panciroll, A. 1995. Effect of *Pediococcus pentosaceus* FBB61, pediocin A producer strain, in caecal fermentations. *J. Appl. Bacteriol.* 78:616-610.

90. Post, R. C. 1996. Regulatory perspective of the USDA on the use of antimicrobials and inhibitors in foods. *J. Food Prot.* Supplement 1996:78-81.

91. Pucci, M. J., Vedamuthu, E. R., Kunka, B. S., and Vandenbergh, P. A. 1988. Inhibition of *Listeria monocytogenes* by using bacteriocin PA-1 produced by *Pediococcus acidilactici* PAC1.0. *Appl. Environ. Microbiol.* 54:2349-2353.

92. Ray, S. K., Johnson, M. C., and Ray. B. 1989a. Bacteriocin plasmids of *Pediococcus acidilactici*. *J. Ind. Microbiol.* 4:163-171.

93. Ray, S. K., Kim, W.-J., Johnson, M. C., and Ray, B. 1989b. Conjugal transfer of a plasmid encoding bacteriocin production and immunity in *Pediococcus acidilactici* H. *J. Appl. Bacteriol.* 66:393-399.

94. Ray, B. 1992. Pediocin(s) of *Pediococcus acidilactici* as a food biopreservative. In *Food Biopreservatives of Microbial Origin*, eds. B. Ray and M. A. Daeschel, pp. 265-322. CRC Press, Inc. Boca Raton, FL.

95. Ray, B., Motlagh, A. M., Johnson, M. C., and Bozoglu, F. 1992. Mapping of pSMB74, a plasmid encoding bacteriocin, pediocin AcH, production (Pap +) by *Pediococcus acidilactici* H. *Lett. Appl. Microbiol.* 15:35-37.

96. Ray, B. 1993. Sublethal injury, bacteriocins, and food microbiology. *ASM-News* 59:285-291.

97. Ray, B., and Hoover, D. G. 1993. Pediocins. In *Bacteriocins of Lactic Acid Bacteria*, eds. D. G. Hoover and L. R. Steenson. pp. 108-210. Academic Press, Inc. San Diego, CA.

98. Ray, B., Motlagh, A. M., and Johnson, M. C. 1993. Processing of pediocin of *Pediococcus acidilactici*. Abstract G5. *FEMS Microbial Rev.* 12:P119.

99. Ray, B. 1994. Pediocins of *Pediococcus* species. In *Bacteriocins of Lactic Acid Bacteria*, eds. L. De Vuyst and E. J. Vandamme, pp. 465-496. Blackie Academic and Professional, Inc. New York.

100. Ray, B. 1995. *Pediococcus* in fermented foods. In *Food Biotechnology Microorganisms*, eds. Y. H. Hui and G. G. Khachatourians, pp. 745-796. VCH Publishers, Inc. New York.

101. Ray, B. 1996. Characteristics and applications of pediocin(s) of *Pediococcus acidilactici*: pediocin PA-1/AcH. In *Lactic Acid Bacteria*, eds. T. F. Bozoglu and B. Ray, pp. 155-204. NATO ASI Series, Vol. 98. Springer-Verlag, Inc. New York.

102. Ray, B., Schamber, R., and Miller, K. W. 1999. The pediocin AcH precursor is biologically active. *Appl. Environ. Microbiol.* 65:2281-2286.

103. Rozbeh, M., Kalchayanand, N., Field, R. A., Johnson, M. C., and Ray, B. 1993. The influence of biopreservatives on the bacterial level of refrigerated vacuum-packaged beef. *J. Food Safety* 13:99-111.

104. Schved, F., Lalazar, A., Henis, Y., and Juven, B. J. 1993. Purification, partial characterization and plasmid-linkage of pediocin SJ-1, a bacteriocin produced by *Pediococcus acidilactici*. *J. Appl. Bacteriol.* 74:67-77.

105. Skytta, E., Haikara, A., and Mattila-Sandholm, T. 1993. Production and characterization of antibacterial compounds produced by *Pediococcus damnosus* and *Pediococcus pentosaceus*. *J. Appl. Bacteriol.* 74:134-142.

106. Strasser de Saad, A. M., and Manca de Nadra, M. C. 1993. Characterization of bacteriocin produced by *Pediococcus pentosaceus*. *J. Appl. Bacteriol.* 74:406-410.

107. Tanasupawat, S., Okada, S., Kozaki, M., and Komagata, K. 1993. Characterization of *Pediococcus pentosaceus* and *Pediococcus acidilactici* strains and replacement of the type strain of *P. acidilactici* with a proposed neotype DSM 20284. *Int. J. System. Bacteriol.* 43:860-863.

108. Venema, K., Kok, J., Marugg, J. D., Toonen, M. Y., Ledeboer, A. M., Venema, G., and Chikindas, M. L. 1995. Functional analysis of the pediocin operon of *Pediococcus acidilactici* PAC1.0: Ped B is the immunity protein and Ped D is the precursor processing enzyme. *Mol. Microbiol.* 17:516-522.

109. Waite, B. L., Siragusa, G. R., and Hutkins, R. W. 1998. Bacteriocin inhibition of two glucose transport systems in *Listeria monocytogenes*. *J. Appl. Microbiol.* 84:715-721.

110. Yang, R., and Ray, B. 1994a. Factors influencing production of bacteriocins by lactic acid bacteria. *Food Microbiol.* 11:281-191.

111. Yang, R., and Ray, B. 1994b. Prevalence and biological control of bacteriocin-producing psychrotrophic leuconostocs associated with spoilage of vacuum-packaged processed meats. *J. Food Prot.* 57:209-217.

112. Yang, R., Johnson, M. C., and Ray, B. 1992. Novel method to extract large amounts of bacteriocins from lactic acid bacteria. *Appl. Environ. Microbiol.* 58:3355-3359.

M.G. El-Ziney

J.M. Debevere

M. Jakobsen

20

Reuterin

I. INTRODUCTION

Lactobacilli produce a number of antimicrobial compounds like short chain organic acids, mainly lactate and acetate, hydrogen peroxide, diacetyl, carbon dioxide, bacteriocins and reuterin. Reuterin, β-hydroxypropionaldehyde, has a broad-spectrum of antimicrobial activity towards prokaryotes and eukaryotes (Chung et al., 1989). This non-peptide compound is produced as an intermediate during anaerobic metabolism of glycerol by *Lactobacillus reuteri* (Axelsson et al., 1989, El-Ziney 1997, El-Ziney et al., 1998). *L. reuteri* was first isolated by Lerche and Reuter (1962) but classified at the time as *Lactobacillus fermentum* Type II. Since 1980 (Kandler et al., 1980) it has been recognized as a species of heterofermentative lactobacilli inhabiting the gastrointestinal tract of humans and animals. *L. reuteri* is also found in a range of foods particularly dairy products, meats and sourdough (Dellaglio et al., 1981; Kandler & Weiss, 1986; Vogel et al., 1994). The fact that *L. reuteri* strains are natural inhabitants of the gastriointestinal tract of human and animals combined with the broad antimicrobial effect of reuterin, its resistance to heat, proteolytic and lipolytic enzymes, its stability over wide range of pH values and high solubility in water and lipid, makes *L. reuteri* as well as reuterin ideal as bio-preservatives for foods and other non-food applications.

II. OCCURRENCE

A. Biosynthesis

Growing and non-growing cells of *L. reuteri* can produce reuterin, from glycerol under anaerobic conditions (Axelsson et al., 1989; El-Ziney et al., 1998). The substance itself is active both under aerobic and anaerobic conditions, indicating that only the synthesis is affected by the presence of oxygen (Axelsson et al., 1989; Chung et al, 1989). The synthesis of reuterin from glycerol is also reported to occur in *Aerobacter aerogenes* (Paweekiewicz & Zagalak, 1965) *Klebsiella pneumoniae* (Forage & Foster, 1982; Forage & Foster, 1982; Ahrens et al., 1998), *Clostridium* spp. (Forsberg, 1987) and also in other

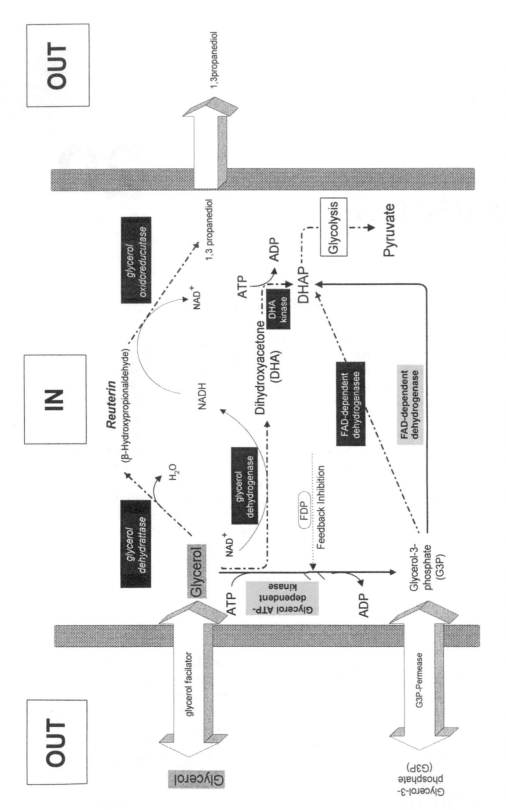

FIGURE 1. Schematic representation of the dissimilation pathways of glycerol in *Klebsiella. pneumoniae*. Fermentative dissimilation of the glycerol occurs through pathways controlled by the *dha* regulon (dashed arrows). Respiratory dissimilation of glycerol occurs through pathways specified by the *glp* regulon (solid arrows) (adapted after Johnson & Lin, 1987; Ahrens et al., 1998)

lactobacilli (Sobolov & Smiley, 1960; Schutz & Radler, 1984). Studies on the glycerol metabolism in *K. pneumoniae* have shown that under anaerobic conditions glycerol can be used as sole carbon and energy source. This is accomplished by a set of enzymes which are genetically controlled by the so called '*dha* regulon' (Forage & Lin, 1982). The energy-yielding part of this fermentation is an NAD+-linked oxidation of glycerol to dihydroxyacetone (DHA) and phosphorylation of DHA to DHAP, which enters the glycolytic pathway (*FIGURE 1*). The NADH generated during the oxidation of glycerol to DHA is reoxidized to NAD+ by dehydration of glycerol to β-hydroxypropionaldehyde (β-HPA) and subsequent reduction to 1,3-propanediol (PPD) (Forage & Lin, 1982; Johnson & Lin, 1987). Thus, glycerol is used as a carbon and energy source as well as an electron acceptor (*FIGURE 1*). Although the anaerobic utilization of glycerol in *L. reuteri* is very similar to the glycerol dissimilation in *K. pneumoniae*, *L. brevis* and *L. buchneri*, neither growing nor resting cells of these bacteria accumulate or excrete reuterin (Schutz & Radler, 1984). An immediate dismutation and/or isomerization of reuterin could explain this phenomenon (Sobolov & Smiley 1960; Schutz & Radler, 1984).

It has been demonstrated that *L. brevis* and *L. buchneri* can not grow on glycerol alone, but require glycerol as an external electron acceptor during growth on glucose (Schutz & Radler, 1984). In fact, many of these lactobacilli grow very poorly on glucose only and require glycerol or another electron acceptor for growth. This co-fermentation of glycerol with sugars leads to a higher growth rate and alters the end product profile by suppressing ethanol formation and increasing the acetate production (Schutz & Radler, 1984). Studies of *L. reuteri* ATCC 1063, a reuterin producing strain, (Talarico et al., 1990) and an isolate of *L. reuteri* from sourdough (Ragout et al., 1996), supported this finding. This can be explained by the inability to reduce acetylphosphate to ethanol or a lack of acetaldehyde dehydrogenase (Eltz & Vandemark, 1960; Stamer & Stoyla, 1967). As mentioned above it has been established that in the presence of glucose and glycerol, other lactobacilli than *L. reuteri* ferment glycerol to reuterin (Sobolov & Smiley, 1960; Schutz & Radler, 1984) but it is immediately converted to equimolar amounts of β-hydroxypropionic acid and 1,3 propanediol, or used as a hydrogen acceptor, resulting only in accumulation of 1,3 propanediol (Sobolov & Smiley, 1960; Schutz & Radler, 1984). In fermentation experiments, with medium containing only radiolabeled (^{14}C) glycerol, it was reported that *L. reuteri* 1063, converted glycerol into reuterin, 1,3 propanediol and to a lesser extent b-hydroxypropionic acid (Talarico et al., 1988). It was concluded that 1 mol glycerol is required to produce 1 mol of reuterin. The role of glycerol as a hydrogen acceptor during glycolysis in lactobacilli is well established (*FIGURE 2*). *L. reuteri* as a heterofermentative lactic acid bacterium is able to ferment glucose to lactic acid, acetic acid, ethanol and CO_2. The shift from ethanol to acetic acid production in the presence of glycerol compared to the fermentation of solely glucose, clearly points at another pathway of regenerating NAD+, different from the reduction of acetylphosphate to ethanol. This alternative pathway is the dehydration of glycerol to form reuterin and subsequent reduction to 1,3-propanediol (Schutz & Radler, 1984; Talarico et al., 1988). As long as NADH is generated during fermentation of glucose, the reuterin is immediately used as the main electron acceptor. When the supply of NADH is depleted (i.e. glucose), which is the situation for resting cells, reuterin accumulates and is excreted (Talarico et al., 1990). Hence, it is concluded that reuterin is a key factor for recycling NADH generated during hexose catabolism via the 6-phosphogluconate pathway, resulting in accumulation of 1,3

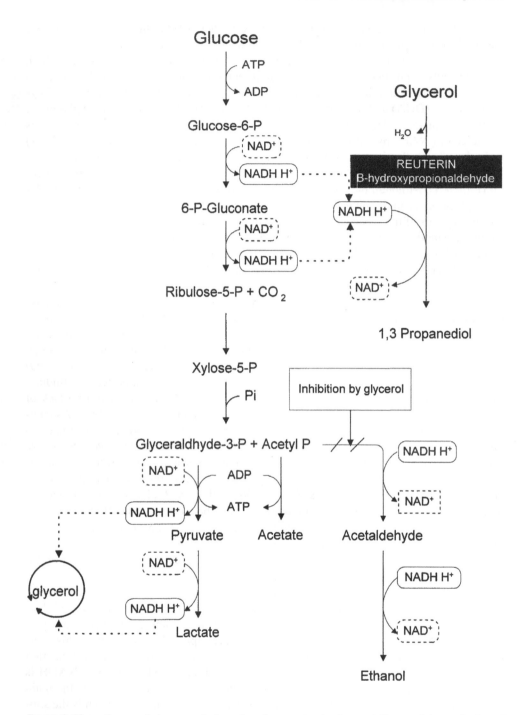

FIGURE 2. The pathways of glucose and glycerol co-fermentation by *L reuteri* illustrated the production of reuterin and its role to recycle of NADH.

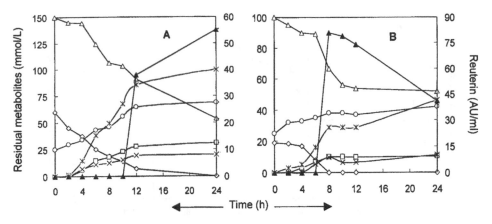

FIGURE 3. Glycerol-glucose co-fermentation batch culture of *L. reuteri* 12002 in m-MRS broth at pH 5.5 at 60 mmol\L glucose and 150 mmol/L glycerol (A) and at 20 mmol/L glucose and 100 mmol/L glycerol (B). Δ, glycerol residual; ◇, glucose residual; ▲, reuterin; ○, acetate; □, lactate; ×, ethanol; ∗, 1,3 PDD (El-Ziney et al., 1998).

propanediol (Schutz & Radler, 1984; Sobolov & Smily, 1960; Talarico et al., 1990). In batch fermentation studies (El-Ziney et al., 1998), the co-fermentation of glycerol and glucose by *L. reuteri* strain 12002 (DSM 12246), a reuterin producing strain, was shown to provide the growing cells with a highly effective alternative hydrogen acceptor system. This resulted in increased growth rate and cell yield. The fermentation shifted towards a high accumulation of acetate, which is a high-energy metabolite for ATP synthesis via the acetate kinase reaction. At the same time it resulted in reduced ethanol production which is produced through the utilization of acetyl phosphate as a hydrogen acceptor. Increasing the initial glycerol/glucose molar ratio from 2.5 to 5 did not completely eliminate the ethanol synthesis (*FIGURE 3A & 3B*). This indicates that both hydrogen acceptor systems, reuterin and acetyl phosphate, are acting competitively to reduce NADH in *L. reuteri*.

It is established that reuterin is accumulated by resting or non-growing cells of *L. reuteri* in the presence of glycerol in a medium free from sugars (Chung et al., 1989; Talarico & Dobrogosz, 1989). In batch fermentation studies (El-Ziney et al., 1998), using different initial molar ratio between glycerol and glucose, it was shown that reuterin is produced solely by non-growing cells (stationary phase) of *L. reuteri* 12002 only when the glucose is very low (8 mmol/L) (*FIGURE 3B*). This indicates a glucose repressing effect on the reuterin production. In the same study, chemostat experiments, run under non-glucose repressing conditions, i.e. 20 mmol/L glucose, it was clearly demonstrated for the first time that growing cells of *L. reuteri* are able to produce reuterin during the co-fermentation of glycerol and glucose (*FIGURE 4*). These results indicate that the reuterin production observed during glycerol and glucose co-fermentation in the batch cultures is controlled by the glucose concentration in the growth medium (El-Ziney et al., 1998). To explain this mechanism it should be mentioned that in both Gram-positive and Gram-negative bacteria, the uptake of sugar via phosphotransferase system (PTS) leads to repression of the enzymes responsible for glycerol uptake and metabolism (Deutscher et al., 1993; Oh et al., 1973; Postma & Lengeler, 1985). It has been reported that, in *Enterobacter faecalis*, glucose and other sugars metabolized via the Embden-Meyerhof pathway strongly repress the synthesis of glycerol kinase which is responsible for respiratory dissimilation of glyc-

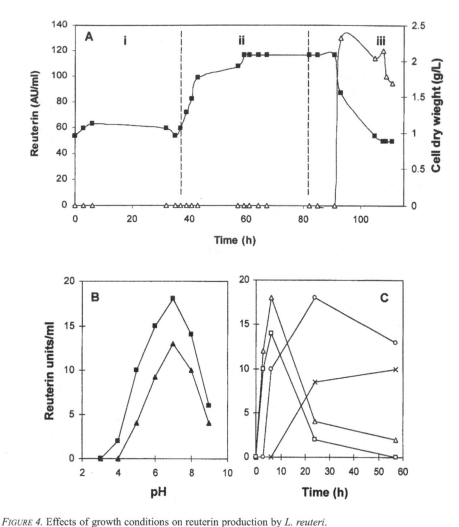

FIGURE 4. Effects of growth conditions on reuterin production by *L. reuteri*.
(A) Growth and reuterin production by *L. reuteri* 12002 in chemostat cultures in m-MRS broth maintained at pH 5.5. At 20 mmol/L glucose (i), at 20 mmol/L glucose and 50 mmol/L glycerol (ii) and at 20 mmol/L glucose and 150 mmol/L glycerol (iii). ■, Cell dry weight (g/L); Δ, reuterin (El-Ziney et al., 1998).
(B) Effect of pH on reuterin production by *L. reuteri* in a co-cultured system with *E. coli* LMG 8223 in BM medium containing 40 mmol/L glycerol after 3h; (▲), and 6h; (■), of incubation at 37°C (El-Ziney, 1997).
(C) Effect of temperature (✕, 9°C; ○, 25°C; Δ, 37°C; ▫, 45°C) on reuterin production by *L. reuteri* in a co-cultured system with *E. coli* LMG 8223 in BM medium containing 40 mmol/L glycerol after 3 h of incubation at 37°C (El-Ziney, 1997).

erol to glycerol-3-phosphate (Deutscher et al., 1993). Under non-glucose repressing conditions, it appears that reuterin production is not started until a certain level of glycerol is exceeded i.e. 150 mmol/L for *L. reuteri* 12002 (El-Ziney et al., 1998). It has been observed in chemostat trials that a high level of reuterin production is followed by a decline in the growth of *L. reuteri* coinciding with a reduction in the glycerol uptake and oscillatory behavior (El-Ziney et al., 1998) (*FIGURE 4A*). This leads to a reduction of the amount of reuterin in the growth medium and *L. reuteri* initiates growth again (unpublished data). This observation has shown that the accumulation of reuterin to a toxic level in the growth medium suppresses the growth of the producer strain (*L. reuteri* 12002). Veiga da Cunha and Foster (1992) suggested that the presence of a high level of reuterin might have a feedback inhibition towards the glycerol metabolism.

Curing studies on *L. reuteri* 12002 by acriflavine resulted in a mutant that was unable to produce reuterin but still resistant as parent cells to the compound. This *Reu⁻* variant of *L. reuteri* 12002 showed the same sugar fermentation pattern and the same growth rate compared to the parental strain but it became erythromycin sensitive (Em^s). Plasmid profile analysis of the *Reu⁻* revealed a loss of all four plasmids. These plasmids may therefore encode for both reuterin production and regulation and carry copies of the gene of the glycerol dehydratase and regulator gene that are responsible for the dissimilation of glycerol and the accumulation of the reuterin (El-Ziney, 1997).

Two enzymes are reported to control the glycerol metabolism in *L. reuteri* (Talarico & Dobrogosz, 1990; Talarico et al., 1990). The first is glycerol dehydratase or coenzyme B^{12} -dependent glycerol dehydratase that seems to be the 'reuterin-enzyme' (Talarico & Dobrogosz, 1990). This enzyme has a molecular weight of 200-kD, determined by gel filtration, while SDS electrophoresis showed a single major band with a molecular weight of 52-kD and it was concluded that the dehydratase is composed of four identical subunits (Talarico & Dobrogosz, 1990). However, genetic studies should be carried out to complement fully the biochemical analysis. The optimum activity of the enzyme was observed at pH 7.2 and it declined sharply outside a pH range of 6.5 to 8.0 Potassium and 1,2 propanediol are required to maintain the activity (Talarico & Dobrogosz, 1990). The second enzyme is 1,3 propanediol:NAD$^+$ oxidoreductase which is able to carry out a number of oxidations and reductions (Talarico et al., 1990). The enzyme oxidizes glycerol to dihydroxyacetone (DHA) and reduces DHA to glycerol. It is also capable of catalyzing an irreversible reaction, which reduces reuterin to 1,3 propanediol (Talarico et al., 1990). But as *L. reuteri* can not grow on glycerol as sole carbon source, probably due to a lack of DHA kinase, no physiological role other than the reduction of can presently be ascribed to this enzyme (Talarico et al., 1990). 1,3 propanediol:NAD$^+$ oxidoreductase has a molecular weight of 180-kD; but it is assumed to exist as a tetramer of 42-kD subunits (Talarico & Dobrogosz, 1990). Compared to the glycerol enzymatic dissimilation system in *K. pneumoniae*, it is likely that the expression of these two enzymes is constitutive in *L. reuteri*, since the purification of both enzymes was done from cells grown in a medium only containing glucose (Talarico & Dobrogosz, 1990).

The pH optimum growth of *L. reuteri* 12002 occurs at pH 5.0-4.6 at 37 °C (El-Ziney, 1997) and the maximum reuterin production took place within the same pH range (El-Ziney 1997). The effect of pH on reuterin production was carried out in co-cultured system of *L. reuteri* 12002 and *E. coli* K12 inoculated in BM medium containing glycerol (40 mmol/L) at pH values ranging between 3 and 9 (El-Ziney 1997). Under these non-growing conditions, it was observed that there were no significant change in the viable cells of *L. reuteri* 12002 at all the tested pH values after 6 h of incubation (El-Ziney, 1997). The overall yield of reuterin exhibited an optimum at pH 7 but it was still appreciable at values as low as pH 5 and as high as pH 9 (*FIGURE 4B*). In the same study, reuterin production controlled at temperatures ranging from 9 to 45 °C was compared (*FIGURE 4C*). As reuterin production is an enzyme-catalyzed process, the production increases as the incubation temperature is increased from 9 to 45 °C but the maximum occurred at 37 °C. Increasing incubation time reduced the reuterin level in the media, which was concomitant with increasing the 1,3 propanediol (El-Ziney, 1997).

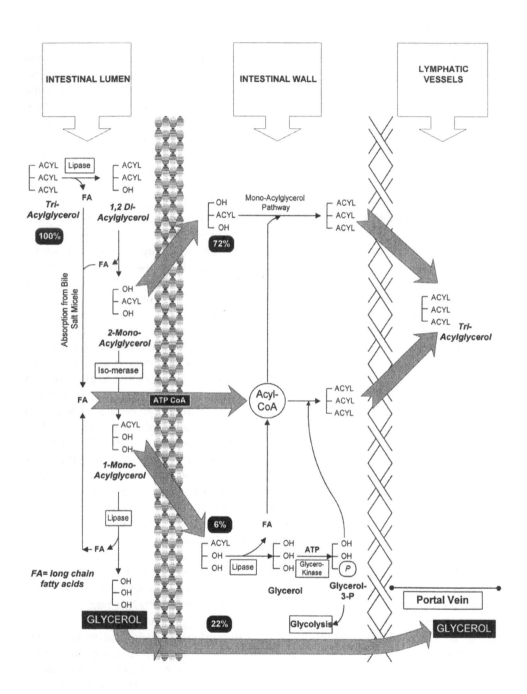

FIGURE 5. Schematic of chemical mechanisms of digestion and absorption of triacylglycerols and generation of glycerol in digestive tract (adapted after Mayes, 1977).

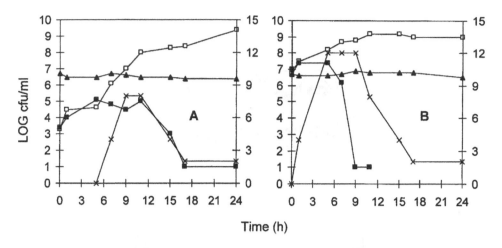

FIGURE 6. Heterologous effect on reuterin production ; ✕, by *L. reuteri* 12002 in co-culture with *E. coli* LMG 8223 at two ratios of 1000:1 (A) and 1:3 (B) in BM medium supplemented with 40 mmol/L glycerol. The viable counts of *E. coli*; ☐, alone; ■, in the presence of *L. reuteri*; ▲, *L. reuteri* in co-culture (El-Ziney, 1997).

B. Formation and possible role in ecosystem balance

Reuterin is produced *in vitro* under conditions similar to those of the intestines e.g. at low oxygen tension, pH values ranging from acid to slightly alkaline and a temperature of 37°C (Dobrogosz et al., 1989). The glycerol required for reuterin production is presumed to be available in the gasterointestinal tract (GI) as a product of host and microbial lipolytic activities (*FIGURE 5*). The free glycerol released in the intestinal lumen represents approximately 22% of the total amount of triacylglycerol originally present in the diet (Mayes, 1977). Dobrogosz et al. (1989) showed that reuterin-producing *L. reuteri* strains were present in relatively large numbers in the stomach and small intestine of animals, especially in newborn animals fed with colostrum. Actually, it is not fully explained why *L. reuteri* is able to accumulate reuterin in surrounding medium. The possible explanation could be that reuterin plays a significant role to enable the species to compete and to grow to significant levels in their environment as confirmed by results of associative growth experiments (*FIGURE 6*). Chung et al. (1989) and El-Ziney (1997) demonstrated that the production of reuterin is stimulated in associative cultures of *E. coli*. Two explanations for this heterologous effect have been suggested (Chung et al. 1989). One is that increasing numbers of heterologous cells results in a high oxygen consumption that creates absolute anaerobic conditions which simulate and enhance reuterin production. However, they found that *Clostridium sporogenes*, an obligate anaerobic microorganism, also stimulated reuterin production. As second explanation Chung et al. (1989) proposed that *L. reuteri* has a mechanism for sensing increased numbers of heterologous cells, and responds to this stimulus by producing reuterin which decreases the heterologous cell population without seriously affecting the producer cells. Recent studies have indicated that diverse bacteria exploit a cell-cell communication device, known as 'quorum sensing', to regulate the transcription of multiple target genes (Hardman et al., 1998). This quorum sensing acts depending on the production of diffusible signal molecules termed 'pheromones' which make the bacterium able to monitor its own cell population. Many

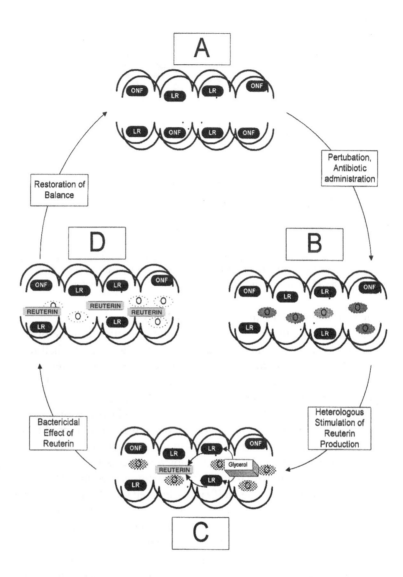

FIGURE 7. Schematic representation of proposed regulatory scheme of the *L. reuteri*-reuterin system to restore the balance in the gasterointestinal tract. LR, *L. reuteri*; ONF, other natural flora, O, opportunistic organism (adapted after Axelsson, 1990).

examples of quorum sensing have been studied in prokaryotes including antibiotic biosynthesis, and the production of virulence characters in animals and fish (Hardman et al., 1998).

A working hypothesis was formulated (Axelsson, 1990) on the role of *L. reuteri* in the regulation of microbiota in the gastrointestinal tract. It is suggested that in the healthy animal, *L. reuteri* constitutes, together with other bacteria, the indigenous microbiota (*FIGURE 7A*). Perturbation or administration of antibiotics results in increased numbers of opportunistic organisms, that could lead to disturbing the equilibrium in the intestine (*Figure 7B*). Hence, *L. reuteri* cells respond to the increased number of such opportunistic organisms (through heterologous stimulation) by increasing in reuterin production (*Figure 7C*). At certain levels of reuterin, the number of the opportunistic organisms will decrease, while *L. reuteri* cells (and other LAB) are not affected. Finally, this will result in a gradual reestablishment of microbial balance (*Figure 7D*).

C. Screening for reuterin producing LAB strains

Lactobacillus reuteri is the only member of the LAB reported to have the ability to produce and excrete reuterin. This phenomenon has not been statistically investigated or even considered when a new isolateof LAB is characterized. The ability to produce reuterin is only reported for strains of *L. reuteri* isolated from human and animal intestine.

The protocol normally used to examine the ability of any LAB strain to produce reuterin includes two main steps (Chung et al., 1989, El-Ziney, 1997). First, the cultures are propagated in MRS broth (De Man et al., 1960) overnight at 37°C, centrifuged and washed twice with phosphate buffer (50 mmol/L, pH 7.0). In the second step, the cells are incubated anaerobically for 3-h at 37°C in a water-glycerol solution (250 mmol/L) for production and accumulation of reuterin. The optimum cell density required for reuterin production was found to be ~10 mg cell dry cell/ml (Axelsson et al., 1989; El-Ziney, 1997).

After fermentation for 3 h, cells are removed by centrifugation, the cell free extract is filter-sterilized and stored at 4°C. The presence of reuterin in the supernatant can be demonstrated using the bioassay method with *E. coli* K-12 as an indicator strain (Chung et al., 1989; El-Ziney, 1997). The presence of reuterin can be confirmed by the HPLC analysis (El-Ziney & Debevere, 1998). El-Ziney (1997) reported that out of fifty-nine strains of LAB, isolated from swine intestine, one *L. reuteri* 12002 (DSM 12246) showed the ability to produce reuterin by anaerobic fermentation of glycerol.

D. Quantitative assays

1. Bioassay method: Reuterin activity is quantified by the modified minimum inhibitory concentration (MIC) method using microtiter plates (Chung et al., 1989; El-Ziney & Debevere 1998). In general, the culture of *E. coli* K-12 (indicator strain) is grown overnight and harvested by centrifugation (7000 x g), washed twice with phosphate buffer (pH 7.2, 50 mmol/L), suspended in the same buffer and diluted to give a final concentration of $\log_{10}$ 6 CFU/ml. For the producer strain, *L. reuteri*, the resulting supernatant of the water-glycerol from the two-step fermentation process are serially diluted in Muller-Hinton (El-Ziney et al., 1999) in microtiter wells (200 ml). Then it is inoculated with the bacterial suspension of *E. coli* K-12 (50 ml) prepared as described above. The plates are incubated in a humidity chamber at 37°C, and the growth is determined visually or measured as absorbance at 550 nm in an automated microplate reader after 24 and 48 h. The MIC is recorded as the lowest concentration that does not permit visible growth or an absorbance exceeding 0.06 for the indicator strain. Reuterin concentration, expressed as AU/ml, is defined as the reciprocal of the highest dilution that does not permit growth of the indicator strain.

2. High-performance liquid chromatography (HPLC): The reuterin production can be confirmed and determined by HPLC (El-Ziney & Debevere, 1998). In brief, an Aminex HPX-87H column (300 by 7.8 mm, Bio-rad, Richmond, CA) protected by a cation H+ Micro-Guard column (30 by 4.6 mm, Bio-rad) is used at a flow-rate of 0.6

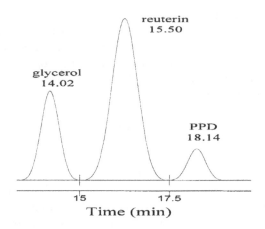

FIGURE 8. HPLC chromatogram showing the separation of residual glycerol, reuterin and 1,3 propanediol (1,3 PPD) produced during glycerol fermentation by *L. reuteri*.

ml/min (H_2SO_4 10 mmol/L); at 30°C. Reuterin is detected by refractive index measurement with a model RI 1047 (Hewlett-Packard). The retention time of reuterin under these conditions is 15 min and is eluted between glycerol (14 min) and 1,3 propanediol (18 min) as shown in *FIGURE 8*. The metabolites are identified by comparison with well-characterized standards by HPLC and by mass spectrometry (Talarico & Dobrogosz, 1989). The known producer strains such as DSM 20016 (type strain), and DSM 12246 can be used as a reference for reuterin.

III. MOLECULAR PROPERTIES

A. Isolation and purification

Reuterin is preferably produced using a two-step fermentation process, which reduces the purification steps and excludes a lot of extraneous materials. Thus reuterin, 1,3 propanediol, and unconsumed glycerol are the only residual metabolites found in the fermentation pool. Nevertheless, the following purification method has also been used efficiently for the isolation and purification of reuterin produced by *L. reuteri* in continuous culture. The purification scheme includes semipreparative HPLC using ion-exchange column, followed by analytical liquid chromatography by applying an Aminex HPX-87H analytical column. Talarico and Dobrogosz (1989) purified reuterin using a polar solvent system composed of 60% acetonitrile and 40% deionized water containing 1.1-g of trifluoroacetic acid. El-Ziney and Debevere (1998) purified reuterin with solvent composed of 10 mmol/L H_2SO_4. Activity in the fractions is monitored by the MIC assay as described above.

B. Physico-chemistry

As described earlier, reuterin is an equilibrium mixture of monomeric, hydrated monomeric and cyclic dimeric forms of β-hydroxypropionaldehyde (*FIGURE 9*). On the basis of Liquid Chromatography-Mass Spectrometry (LC-MS) data, the cyclic form of reuterin was determined to have a molecular weight of 148, while, it is predicted to be 74 for monomeric reuterin (Talarico & Dobrogosz, 1989). Reuterin is soluble in water as well

dimeric form   dehydrated form   hydrated form

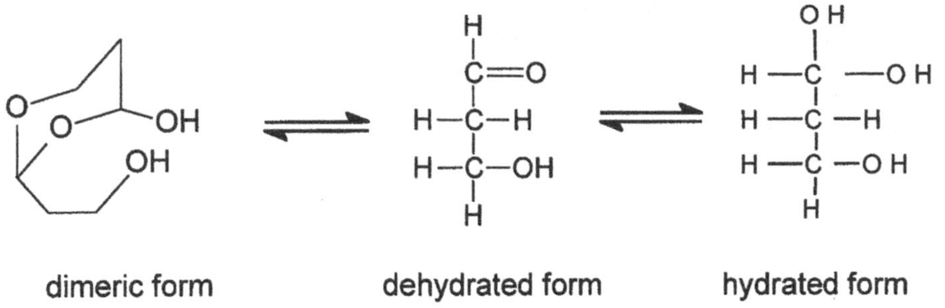

FIGURE 9. Chemical structure of reuterin in aqueous solution (Talarico & Dobrogosz, 1989).

as lipid and stable in a pH range from 2 to 9 (El-Ziney, 1997) with an immediate irreversible breakdown at pH 11 (Talarico et al., 1988). Purified reuterin incubated at 37°C with the pH adjusted to 6.5 had a half-life of approximately 2 weeks, but when incubated under acidic conditions it had a half-life of approximately twice that. Heat treatments of crude reuterin at 72°C for 15 sec and at 100°C for 5 min, did not reduce the reuterin activity compared with untreated samples (El-Ziney, 1997). Reuterin stored below 5°C at neutral and acidic pH showed no appreciable breakdown over a 6-month period (Talarico et al., 1988). Freeze-dried reuterin stored at 4°C was shown to be stable over one year without any changes in the activity (El-Ziney, unpublished data).

C. Structure

The major functional groups of reuterin molecule were determined using Fourier Transform Infrared Spectrophotometry (FTIR) (Talarico & Dobrogosz, 1989). Spectra contained a peak due to (C-O) stretching and a broad (O-H) stretch band, indicating hydroxyl functionality. This is in addition to a (C=O) stretch which illustrated the presence of aldehydes, along with (C-H) stretches (Talarico & Dobrogosz, 1989). Proton and Carbon 13 nuclear magnetic resonance (^{1}H NMR, ^{13}C NMR) analysis and homonuclear coupling experiments indicated that two molecules of a three-carbon compound were present in aqueous solutions; as predominantly hydrated and dehydrated form of β-hydroxy propionaldehyde (FIGURE 9). A proton spectrum showed several weak signals, which revealed the presence of third molecule in small amount, a cycle dimer of the aldehyde (Talarico & Dobrogosz, 1989). Mass spectroscopy data predicted that the reuterin has a molecular weight of 148, which confirmed the presence of cyclic dimeric structure of reuterin. Talarico and Dobrogosz (1989) suggested that the solvent background effects mask detection of the lower-molecular-weight monomer form of reuterin. Hence, it was predicted that the monomeric reuterin has a molecular weight of 74.

IV. ANTIMICROBIAL ACTIVITY

A. Mechanisms of action

Reuterin is a broad-spectrum antimicrobial agent active against Gram-positive and Gram-negative bacteria, yeasts, fungi and protozoa (Chung et al., 1989; El-Ziney, 1997; El-Ziney et al., 1998). For viruses, it is reported that reuterin inhibits replication of B2 and lambda phages in *L. plantarum* and *E. coli* respectively (Chung et al., 1989). The

TABLE 1. Minimum inhibitory concentration of reuterin against different bacterial, yeast, fungal and protozoan species (Chung et al., 1989; El-Ziney 1997).

Species	Tested strain	MIC (AU/ml)
Gram-positive bacteria		
Bacillus cereus	3	2
Bacillus megaterium	1	5
Staphylococcus aureus	2	2
Staphylococcus epidermidis	1	5
Listeria monocytogenes	6	4-8
Lactobacillus bulgaricus	1	9
Lactobacillus plantarum	1	12
Lactobacillus acidophilus	6	12-40
Leuconostoc mesenteroides	1	16
Pediococcus cerevisiae	1	12
Gram-negative bacteria		
Escherichia coli	ETEC (1), VETC (1), K12 (2), EHEC (2)	4
Pseudomonas aeruginosa	1	2
Pseudomonas alcaligenes	1	4
pseudomonas fluorescens	2	2-4
Shewanella putrefacines	1	2
Yersinia enterocolitica	0:3 (1), 0:8 (1), 0:9 (1)	2-4
Salmonella typhimurium	1	4
Yeast		
Candida albicans	2	2
Saccharomyces cerevisiae	1	12
Saccharomycopsis fibuligera	1	16
Torulopsis glabrata	1	4
Fungi		
Aspergillus flavus	1	8
Fusarium samfucienum	1	8
Protozoa		
Trypanosoma cruzi	1	5

results of a survey carried out to determine the sensitivity of different microorganisms towards reuterin from *L. reuteri* 1063 *and L. reuteri* 12002 are presented in TABLE 1 (Chung et al., 1989; El-Ziney, 1997). It was observed that 4 to 5 AU/ml of reuterin are sufficient to inhibit the growth of the various species tested. LAB appeared to be slightly more resistant compared to other bacterial group. The yeast *Candida albicans* was as sensitive as bacterial species. The yeast *Saccharomycopsis fibuligera*, and plant pathogenic *Fusarium* spp. were slightly less sensitive. Reuterin from *L. reuteri* 12002 had a bactericidal effect towards different pathogens and spoilage microorganisms whether Gram-positive or Gram-negative bacteria (*TABLE 1*). The MIC of reuterin was 2 to 4 AU/ml for most species including spore-forming species such as *Bacillus cereus* (2 AU/ml). *L. monocytogenes* appeared to be relatively more resistant and had a MIC of 8 AU/ml (El-Ziney, 1997). This could be explained by the fact that *L. monocytogenes* is phylogenetically close to the genus *Lactobacillus* (Wilkinson & Jones, 1977; Ruhland & Fieldler, 1987). Therefore, it is possible that this related group of organisms has a common mechanism responsible for the resistance to reuterin. The effect of reuterin against *E. coli*, including

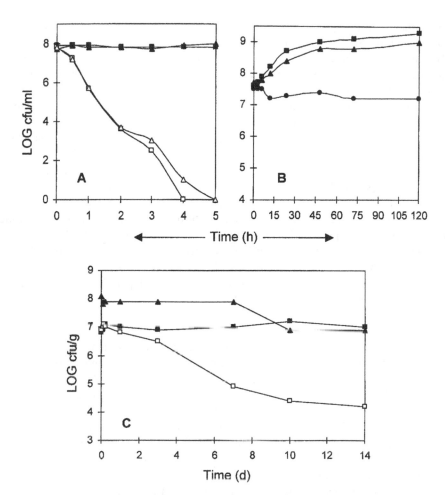

FIGURE 10. Antimicrobial effects of reuterin (adapted from El-Ziney, 1997).
(A) Bactericidal effect of the reuterin; 100 AU/ml, against exponential cells of *E. coli* LMG 8223. Viable count of *E. coli* in phosphate buffer with; □, and without; ■, reuterin. In phosphate buffer containing 20 mmol/L glucose with; △, and without; ▲, reuterin.
(B) Inhibitory effect of *L. reuteri* 12002-glycerol system on *L. monocytogenes* (Ohio) in skimmed milk stored at 7°C under vacuum. Viable count of listeria; ■, alone; ●, in the presence of *L. reuteri* 12002, a reuterin producing strain, (Reu⁺) and; ▲, its negative clone, *L. reuteri* 12003 (Reu⁻).
(C) Bactericidal effect of the *L. reuteri* 12002-glycerol system on *E. coli* O157:H7 (strain 932) in ground beef incubated under vacuum at 7°C. Viable count of *E. coli*; ■, in control and; □, in the treatment. ▲, *L. reuteri* were determined during 14 days of storage.

the enterohemorrhagic serotype O157:H7, has been demonstrated. Reuterin had a marked bactericidal effect on growing and stationary phase cells of enteropathogenic *E. coli* LMG 8223 (indicator strain) suspended in phosphate buffer (50 mmol/L, pH 7) with and without glucose and the cells were undetectable after 6 h (*FIGURE 10A*) (El-Ziney, 1997). After 1h of incubation the initial population of indicator strain was reduced by 99.4%. The bactericidal effect of reuterin was not affected by the physiological state of the indicator strain i.e. exponential or stationary phase cells (El-Ziney, 1997). The absorbancy ($A_{600 nm}$) of the reuterin treated cultures did not change, indicating that the killing effect of reuterin proceeded without cell lysis (El-Ziney 1997). Reuterin is considered an analog of D-ribose capable of inhibiting DNA synthesis by inhibiting ribonucleotide reductase (NRD), involved in the first step in *de novo* synthesis of deoxyribonucleotide for DNA synthesis (Talarico & Dobrogosz, 1989). This was observed by testing the inhibitory effect of reuterin against purified B1 and B2 subunits of the ribonucleotide reductase isolated from *E. coli*. Reuterin caused a 50% inhibition of the activity of B1 subunits. This inhibition of the ribonucleolide reductase activity could explain the broad antimicrobial effect. In these experiments it was also shown that thioredoxin, another sulfhydryl enzyme, was inhibited by reuterin. This suggests that the mechanism of action of reuterin might be directed towards sulfhydryl enzymes (Talarico & Dobrogosz, 1989). Furthermore Yunmbam and Roberts (1992) reported that reuterin inhibited the DNA and protein synthesis, the motility and the viability of *Trypanosoma brucei* ssp. *brucei*.

Until now it is not established which one of the three forms of reuterin or which combination is the biological active entity (Talarico & Dobrogosz, 1989). It is also difficult to ascertain changes that may occur in the relative concentration or biological efficacy of these entities under various environmental conditions. Further, the higher resistance of the LAB and to less extent *Listeria monocytogenes* as compared to other bacteria, to reuterin still needs to be explained.

B. Influencing factors

There is limited information about the interaction of reuterin and intrinsic factors (e.g. food composition, pH, a_w and oxidation-reduction potential) and extrinsic factors (e.g. temperature, humidity, atmosphere, and food preservatives). El-Ziney and Debevere (1998) reported that milk fat in the range of 0.5 to 3% did not significantly affect the reuterin activity against *Listeria monocytogenes*. After 3 days of storage at 7°C, the decimal reduction values (*D*) were 17.14 h in skimmed milk (0.5% fat), 20.20-h in half-fat milk (1.5% fat) and 21.6 h in whole-fat milk (3% fat). However, after 5 days, *L. monocytogenes* was not detected in the three types of milk. In the same study, a synergistic effect between reuterin and sodium chloride was observed. After one day of storage at 7°C of *L. monocytogenes* in skimmed milk with reuterin (150 AU/ml), the viability decreased by 0.7 $\log_{10}$ and 1.6 $\log_{10}$ in milk without and with 3% (w/v) of salt. Adding lactic acid (5% v/v) to reuterin (500 AU/ml) enhanced significantly the efficacy of reuterin in decontamination of meat infected with *E. coli* O 157:H7 (El-Ziney et al., 1999). It was found that the addition of ethanol (40% v/v) to the reuterin-lactic acid mixture did not improve the bactericidal effect of reuterin.

V. APPLICATIONS

In Sweden, *Lactobacillus reuteri* is used as an additional culture to supplement *L. acidophilus* and *Bifidobacterium* spp. in fermented milk named 'BRA' (Speck et al., 1993). *In vivo* studies in animal models have indicated that *L. reuteri* is able to improve the feed flavor and adds nutritional benefits (Speck et al., 1993). In USA, many applications are reported for *L. reuteri* as a feed additive (Casas-Perez, 1996). The use of *L. reuteri* as a protective culture in food is also proposed for the production of dry sausage but its hetero-fermentative characteristics should be taken into account (Weber 1994). It has also been used in the production of Emmental-like cheese in France (Delespaul et al.,

1994). Recently, in Finland, it is reported that *L. reuteri* has been added to orange juice, drinkable yogurt, and cottage cheese as a probiotic culture (Dairy Food Magazine OnLine, 1999). Evidence is being accumulated to support the role of *L. reuteri* as a probiotic culture, able to maintain health and prevent diseases. Administration of *L. reuteri* in both chicks and poults has increased the inhibition of colonization of *Salmonella* spp. and *E. coli* (Edens et al., 1997). This has led to reduced mortality due to in-hatcher exposure to pathogens (Edens et al., 1997). The probiotic properties of *L. reuteri* have also been indicated for humans. The organism showed a high efficacy in inhibiting rotavirus and significantly shortened the duration of watery diarrhea in young children (Shornikova el al., 1997a). Furthermore, it has been suggested that *L. reuteri* be used as a therapeutic agent in acute rotavirus diarrhea in children (Shornikova et al 1997b). In addition to these data, evidence for hypocholesterolemic effect of *L. reuteri* is recently reported (Taranto et al., 1998).

The potential application of reuterin and *L. reuteri*-glycerol-system in food preservation was first demonstrated in 1990 by Lindgren and Dobrogosz, who utilized a reuterin producing strain of *L. reuteri* 16003 along with glycerol to extend the shelf-life of herring. The filets were dipped into a solution containing 1 x 10⁹ cfu/ml of *L. reuteri* and 250 mmol/L glycerol and routinely examined for Gram-negative spoilage bacteria. The growth of Gram-negative bacteria was retarded which resulted in a reduced the accumulation of total volatile basic nitrogen (TVB-N) during storage in a controlled atmosphere (100% nitrogen) at 2°C. Recently, the combined addition of *L. reuteri* and glycerol was applied to inhibit *E. coli* O157: H7 in ground beef and *L. monocytogenes* in skimmed milk (El-Ziney, 1997). Addition of *L. reuteri* (12002) in combination with 150 mmol/L glycerol to inhibit *E. coli* O157: H7 (932) in vacuum packed ground beef at 7°C was shown to be a successful approach to enhance the safety of this food product (*FIGURE 10C*). During the first three days of storage this system exhibited a slight inhibitory effect, raised to a significant level from the third day. Hence, after 7 days the initial population of *E. coli* O157:H7 decreased by 2 log_{10} while, it remained unchanged in the control. It is interesting to note that at the end of incubation period, the pH of ground meat treated with reuterin did not change significantly (5.5± 0.1). In skim milk, *L. reuteri*-glycerol system inhibited the growth of *L. monocytogenes* while the reuterin negative variant (*Reu-*), strain 12003 did not inhibit the growth of the pathogen as compared to the untreated samples (*FIGURE 10B*). The system elicited a bacteriostatic effect against *Listeria monocytogenes* while in the control the pathogen counts were increased by 1 log_{10} after one day. The pH of the milk, in the presence of parent and reuterin-negative variant cells of *L. reuteri* did not drop below 5.8 indicating that reuterin and not the acidification, was the cause of the inhibitory effect. As mentioned earlier, the production of reuterin was reduced when sugars e.g. glucose, lactose, sucrose were introduced into the growth media, which was the case when the reuterin producing organism was added to skimmed milk (El-Ziney, 1997).

Daeschel (1989) cited that Dobrogosz (1988) applied reuterin at concentrations of 50 and 100 AU/g in ground beef kept at 4°C for 6 days. It was observed that the viability of coliforms was reduced by 0.5 and 2 log_{10} respectively, while it was increased by 4 log_{10} in the control. The bactericidal action of reuterin against *E. coli* O157:H7 and *L. monocytogenes* in minced meat was further demonstrated (El-Ziney et al., 1999). In vacuum packed ground ham, addition of reuterin caused a pronounced reduction in the viability of *E. coli* O157:H7. After one day at 7°C, reuterin at a level of 100 AU/g reduced the population by 5 $\log_{10}$. For *L. monocytogenes* the effect of reuterin was significant but less pronounced. At a reuterin concentration 150 AU/g the number of viable cells decreased by log 1.3 $\log_{10}$ after 7 days and at a concentration of 250 AU/g the reduction was 3.2 $\log_{10}$ for the same period at 7°C. During the 14 days of storage at 7°C, no growth was observed for surviving cells of *L. monocytogenes*. In creamed-style cottage cheese, pH 5.4 and 5.5 % fat, addition of reuterin in concentration of 50-250 AU/g, was bactericidal effect to *E. coli* O157:H7 and *L. monocytogenes* (El-Ziney & Debevere, 1998). The inactivation rate was more pronounced with *E. coli* O157:H7 than *L. monocytogenes* and it was dependent on reuterin concentration. At reuterin concentration of 50 AU/g *Listeria* decreased by 1.5 $\log_{10}$ after three weeks, whereas *E. coli* O157:H7 was reduced by 3.5 $\log_{10}$ during the same period. Increasing reuterin concentration to 150 AU/g killed *E. coli* O157:H7 within 7 days whereas, 250 AU/g of reuterin was required for *L. monocytogenes* to obtain a comparable effect (El-Ziney et al., 1999).

In UHT skim milk with 150 AU of reuterin/ml, stored at 7°C, the decline in the cell numbers of *L. monocytogenes* was higher than in cottage cheese (El-Ziney & Debevere, 1998). After 3 days of incubation, in skim milk, the initial *Listeria* population, 4.8 $\log_{10}$, was eliminated. While in creamed-cottage a concentration of 250 AU of reuterin/g was required to obtain a similar effect within 7 days of storage at 7°C. In the same study, it was observed that the *E. coli* O157:H7 was more susceptible to reuterin. In creamed-cottage cheese, a concentration of 150 AU of reuterin/g was sufficient to eliminate the *E. coli* O157:H7 after 7 days at refrigeration (El-Ziney & Debevere, 1998).

Reuterin has been successfully used in decontaminating different surfaces. Reuterin (500 AU/ml) was applied, at 7°C for 15-s, to decontaminate cooked pork surfaces which were artificially inoculated by *E. coli* O157:H7 or *L. monocytogenes* (El-Ziney et al., 1999). The results showed that both organisms significantly decreased in viability immediately after treatment. The immediate inactivation of *E. coli* O157:H7 (0.45 $\log_{10}$ cfu/cm^2) was more pronounced than for *L. monocytogenes* (0.3 $\log_{10}$ cfu/cm^2). After 24 h of storage at 7°C, the bactericidal effect of reuterin led to 2.78 and 0.63 $\log_{10}$ reduction for *E. coli* O157:H7 and *L. monocytogenes*, respectively (El-Ziney et al., 1999).

The germicidal activity of reuterin against *L. monocytogenes* adhered to cows teat skin was studied (El-Ziney, 1997). In model experiments where teats were disinfected by dipping in reuterin (200 AU/ml in 10 mmol/L buffer lactate, pH 6.0) for varying periods up to 10 min and kept for 8 h at room temperature, the viable cells on skin decreased by 0.3 $\log_{10}$ after 5 min of dipping.

VI. SUMMARY

Reuterin has been proven to be an effective broad-spectrum antimicrobial substance able to play a powerful role to improve food safety and prolong shelf life. Further, *L. reuteri* combined with glycerol represent an example of a protective culture which could be added in combination with starter cultures in the manufacturing of fermented fish, sausage, cheese and pickles. This system could control pathogen and spoilage microorganisms during fermentation process and storage. Hence, it is believed that the efficacy of reuterin as biopreservative agent and *L. reuteri* as a protective culture will be explored over a wide range of applications. For example, it would be interesting to apply the *L. reuteri*-glycerol system in highly enriched glycerol foods such as shrimps and other seafoods to extend shelf life. The production of reuterin in such foods needs to be followed and quantified at different storage times and temperatures to provide the food industry with the applicable models for different food systems. In this regard, the interaction between reuterin and intrinsic (a_w, pH, natural antimicrobial compounds, etc.) or extrinsic (storage temperature, humidity, packaging) factors will represent another important research area. These studies will allow us to evaluate and quantify the synergistic effect of reuterin in combination with those factors in order to optimize the bactericidal effect of reuterin.

The successful use of reuterin as a decontaminating agent for meat and cows' teat surfaces also suggests more surface disinfectant applications. Reuterin-based formulations could be applied as a disinfectant for different surfaces including stainless steel, plastics, etc., and even for medical equipment.

It is reported that *L. reuteri* has probiotic effects in animals and humans. But still further studies are needed to confirm such findings and to explore and elucidate the full probiotic and therapeutic potential of *L. reuteri*. Hence, it is believed that future research will focus on the probiotic role of this species and the possible therapeutic applications. This development will coincide with more surveys to locate, isolate and select the organism in different ecosystems and to conduct advanced genetic and phenotypic analysis.

Basic research on the genes encoding for reuterin production and their regulations is still lacking. Elucidation of these gaps in our knowledge will improve significantly the possibilities of applying reuterin-producing strains of *L. reuteri* and reuterin in food industry as well as in other industries.

VII. REFERENCES

1. Ahrens, K., Menzel, K., Zeng, A. P. and Deckwer, W. D. 1998. Kinetic, dynamic, and pathway studies of glycerol metabolism by *Klebsiella pneumoniae* in anaerobic continuous culture: III. Enzymes and fluxes of glycerol dissimilation and 1,3 propanediol formation. *Biotechnol. Bioeng.* 59:544-552.
2. Axelsson, L. 1990. *Lactobacillus reuteri*, u member of the gut bacterial flora: studies on antagonism, metabolism, and genetics. Ph.D. thesis. Swedish University of Agricultural Sciences.
3. Axelsson, L., Chung, T.C., Debrogosz, W.J. and Lindgren S. 1989. Production of a broad spectrum antimicrobial substance by *Lactobacillus reuteri*. *Microb. Ecol. Hlth. Dis.* 2:131-136.
4. Casas-Perez, I.A. 1996. Feed additive which consists of whey and *Lactobacillus reuteri* and a method of delivering *Lactobacillus reuteri* to the gastrointestinal tract. *US Patent* : 5,480,641.
5. Chung, T.C., Axelsson, L., Lindgren, S. E. and Dobrogosz, W.J. 1989. *In vitro* studies on reuterin synthesis by *Lactobacillus reuteri*. *Microb. Ecol. Illth. Dis.* 2:137-144.
6. Daeschel, M. A. 1989. Antimicrobial substances from lactic acid bacteria. *Food Technol.* 43:164-167.
7. Dairy Foods Magazine onLine. 1999. http://www.dairyfoods.com/cultured.html.
8. Deutscher, J., Bauer, B., and Sauerwald, H. 1993. Regulation of glycerol metabolism in *Enterococcus faecalis* by phosphoenolpyruvate-dependent phosphorylation of glycerol kinase catalyzed by enzyme I and HPr of the phosphotransferase system. *J. Bacteriol.* 175:3730-3733.
9. De Man, J.C., Rogosa, M. and Sharpe, M.E. (1960) A medium for the cultivation of lactobacilli. *J. Appl. Bacteriol.* 23:130-135.
10. Delespaul, G., Leclerc, G., Lepeltier, M., and Chassagne, M.H. 1994. Cheese-making process and the cheese obtained. *French Patent FR* 2,696,620.
11. Dellagio, F., Arriza, F.S. and Leda, A. 1981. Classification of citrate-fermenting lactobacilli isolated from lamb stomach, sheep milk and pecorino romano cheese. *Zentralbl. Bakteriol. Hyg., Abt. Orig.* C2:349-356.
12. Dobrogosz, W. J., Casas, I. A., Pagano, G. A., Talarico, T. L. Sjoberg and Karlsson, K. 1989. *Lactobacillus reuteri* and the enteric microbiota. In *The Regulatory and Protective Role of the Normal Flora*, ed. R. Grub, T. Midtvedt, and E. Norin. A Wenner-Gren International Symposium. The Macmillan Press Ltd., London.
13. Edens, F. W., Parkhust, C. R., Casas, I. A. and Dobrogosz, W. J. 1997. Principles of *ex-ovo* competitive exclusion and *in-ovo* administration of *Lactobacillus reuteri*. *Poult. Sci.* 76:179-196.

14. Eltz, R. W. and Vandemark, P. J. 1960. Fructose dissimulation by *Lactobacillus brevis*. *J. Bacteriol.* 79: 763-776.

15. El-Ziney, M. G. 1997. Antimicrobial activity of lactic acid bacteria metabolites: the role of lactic acid, enterocin 5701 and reuterin. Ph.D. thesis. University of Ghent.

16. El-Ziney, M. G. and Debevere, J. M. 1998. The effect reuterin on *Listeria monocytogenes* and *Escherichia coli* O157:H7 in milk and cottage cheese. *J. Food Prot.* 61:1275-1280.

17. El-Ziney, M. G., Arneborg, N., Uyttendaele, M., Debevere, J. and Jakobsen, M. 1998. Characterization of growth and metabolites production of *Lactobacillus reuteri* during glucose/glycerol co-fermentation in batch and continuous cultures. *Biotechnol. Lett.* 20:913-916.

18. El-Ziney, M., van den Tempel, T., Debevere, J. and Jakobsen, M. 1999. Application of reuterin produced by *Lactobacillus reuteri* 12002 for meat decontamination and preservation. *J. Food Prot.* 62:257-261.

19. Forage, R. G. and Foster, M. A. 1982. Glycerol fermentation in *Klebsiella pneumoniae*: functions of the coenzyme B_{12}-dependent glycerol and diol dehydratases. *J. Bacteriol.* 149:413-419.

20. Forage, R. G. and Lin, E. C. C. 1982. *dha* system mediating aerobic and anaerobic dissimilation of glycerol in *Klebsiella pneumoniae* NCIB 418. *J. Bacteriol.* 151:591-599.

21. Forsberg, C. W. 1987. Production of 1,3 propanediol from glycerol by *Clostridium acetobutylicum* and other *Clostridium* species. *Appl. Environ. Microbiol.* 53: 639-643.

22. Hardman, A. M., Stewart, G. S. and Williams P. 1998. Quorum sensing and the cell-cell communication dependent regulation of gene expression in pathogenic and non-pathogenic bacteria. *Antonie van Leeuwenhock.* 74:199-210.

23. Johnson, E. A. and Lin, E. C. C. 1987. *Klebsiella pneumoniae* 1,3-propanediol: NAD⁺ oxidoreductase. *J. Bacteriol.* 169:2050-2054.

24. Kandler, O., Steter, K. O., and Köhl, R. 1980. *Lactobacillus reuteri* sp. nov., a new species of heterofermentative lactobacilli. *Zentralbl. Bakteriol. Hyg. Abt., Orig.* C1:264-269.

25. Kandler, O. and Weiss, N. 1986. Genus *Lactobacillus*. In: Sneath, P.H., Mair, N. S., Sharpe, M.E. and Holt, J.G. (eds.). *Bergey's Manual of Systematic Bacteriology*. Vol. 2. Williams and Wilkins Co., Baltimore, pp.1209-1234.

26. Lerche, M. and Reuter, G. 1962. Das Verkommen aerob wachsander grampositiver Stabchendes Genus *Lactobacillus beijerinck* im Darminhalt erwachsener Menschen. *Zbl. Bakt. Parasit. Infekt. Hyg.* 1 Orig. 185.s:446-481.

27. Lindgren, S. E. and Dobrogosz, W. J. 1990. Antagonistic activities of lactic acid bacteria in food and feed fermentation. *FEMS Microbiol. Rev.* 87:149-164.

28. Mayes, P. A. 1977. Digestion and absorption from the gastrointestinal tract. In *Review of Physiological Chemistry*, ed. H. A. Harper, V. W. Rodwell, and P. A. Mayes (Eds.), pp. 202-217, 16th ed, Lange.

29. Oh, Y. K., Freese, E. B. and Freese, E. 1973. Abnormal septation and inhibition of sporulation by accumulation of L-μ-glycerophosphate in *Bacillus subtilis* mutants. *J. Bacteriol.* 1131034-1045.

30. Paweekiewicz, J. and Zagalak, B. 1965. Enzymic conversion of glycerol into β-hydroxypropionaldehyde in a cell-free extract from *Aerobacter aerogenes*. *Acta Biochim. Polonica* 12:207-218.

31. Postma, P. W. and Lengeler, J. W. 1985. Phosphoenolpyruvate: carbohydrate phosphotransferase system of bacteria. *Microbiol.* 49:232-269.

32. Ragout, A., Sineriz, F., Diekmann, H. and de Valdez, G.F. 1996. Shifts in the fermentation balance of *Lactobacillus reuteri* in the presence of glycerol. *Biotechnol. Lett.* 18:1105-1108.

33. Ruhland, G. J. and Fiedler, F. 1987. Occurrence and biochemistry of lipoteichoic acids in the genus *Listeria*. *Syst. Appl. Microbiol.* 9:40-46.

34. Shornikova, A. V., Casas, I. A., Isolauri, E., Mykkanen H. and Vesikari, T. 1997a. *Lactobacillus reuteri* as a therapeutic agent in acute diarrhea in young children. *Pediatr Infect. Dis. J.* 24:399-404.

35. Shornikova, A. V., Casas, I. A., Mykkanen H., Salo, E. and Vesikari, T. 1997b. Bacteriotherapy with *Lactobacillus reuteri* in rotavirus gastroenteritis. *Pediatr Infect. Dis. J.* 16:1103-1107.

36. Schutz, H., and Radler, F. 1984. Anaerobic reduction of glycerol to propanediol-1,3 by *Lactobacillus. brevis* and *Lactobacillus. bucheneri*. *Syst. Appl. Microbiol.* 5:169-178.

37. Stamer, H. R. and Stoyla, B. O. 1967. Growth response of *Lactobacillus brevis* to aeration and organic catalysts. *Appl. Microbiol.* 15:1025-1030.

38. Sobolov, M. and Smiley K., L. 1960. Metabolism of glycerol by an acrolein-forming *Lactobacillus*. *J. Bacteriol.* 79:261-266.

39. Speck, M. L., Dobrogosz, W.J. and Casas, I. A. 1993. *Lactobacillus reuteri* in food supplementation. *Food Technol.* 47:90-94.

40. Talarico, T. L., Casas, I.A., Chung, T.C. and Dobrogosz, W.J. 1988. Production and isolation of reuterin, a growth inhibitor produced by *Lactobacillus reuteri*. *Antimicrob. Agents Chemother.* 32:1854-1858.

41. Talarico, T. L. and Dobrogosz, W. J. 1989. Chemical characterization of an antimicrobial substance produced by *Lactobacillus reuteri*. *Antimicrob. Agents Chemother.* 33:674-679.
42. Talarico, T. L. and Dobrogosz, W. J. 1990. Purification and characterization of glycerol dehydratase from *Lactobacillus reuteri*. *Appl. Environ. Microbiol.* 56:1195-1197.
43. Talarico, T. L., Axelsson, L., Novotny, J., Fiuzat, M. and Dobrogosz, W. J. 1990. Utilization of glycerol as a hydrogen acceptor by *Lactobacillus reuteri*: purification of 1,3-propanediol: NAD$^+$ oxidoreductase. *Appl. Environ. Microbiol.* 56:943-948.
44. Taranto, M. P., Medici, M., Perdigon, A., Ruiz Holgado, P. and Valdez, G. F. 1998. Evidence for hypocholesterolemic effect of *Lactobacillus reuteri* in hypercholesterolemic mice. *J. Dairy Sci.* 81: 2336-2340.
45. Veiga da Cunha, M. and Foster, M. A. 1992. Sugar-glycerol co-fermentations in Lactobacilli: the fate of lactate. *J. Bacteriol.* 174:1013-1019.
46. Vogel, R. F., Boecker, G., Stolz, P., Ehrmann, M., Fanta, D., Ludwig, W., Pot, B., Kersters, K., Schleifer, K. H. and Hammes, W, P. 1994. Identification of lactobacilli from sourdough and description of *Lactobacillus pontis* sp. nov. *Int. J. Systm. Bacteriol.* 44:223-229.
47. Weber, H. 1994. Dry sausage manufacture. The importance of protective cultures and their metabolic products. *Fleisch.* 74:278-282.
48. Wilkinson, B. J. and Jones, D. 1977. A numerical taxonomic survey of *Listeria* and related bacteria. *J. General Microbiol.* 98:399-421.
49. Yunmbam, M. K. and Roberts, J. F. 1992. The *IN VITRO* efficacy of reuterin on the culture and bloodstream forms of *Trypanosoma brucei brucei*. *Compen. Biochem. Physiol.* 101:235-238.

F. Leroy
L. De Vuyst

21

Sakacins

I. INTRODUCTION

Currently, natural antimicrobials have become increasingly attractive for application in food products. Bacteriocins produced by lactic acid bacteria (LAB) may be considered as such. Lactic acid bacteria are an important group of industrial microorganisms which are involved in various food fermentation processes, and contribute to the enhancement of organoleptic attributes of the food, as well as to the preservation and microbial safety of the end-product (Stiles & Holzapfel, 1997). They are able to inhibit the growth of many undesirable microbes because of the production of lactic acid and other metabolites such as acetic acid, carbon dioxide, hydrogen peroxide, diacetyl, and bacteriocins (De Vuyst & Vandamme, 1994a). Bacteriocins produced by LAB are low-molecular-mass peptides or proteins with an antibacterial mode of action towards sensitive species that are closely related to the producer cell (De Vuyst & Vandamme, 1994b). They are secreted into the microbial environment where they may cause the inhibition of spoilage bacteria and food-borne pathogens (Nettles & Barefoot, 1993), and hence contribute to the competitiveness of the producing cells (Vogel et al., 1993; Ruiz-Barba et al., 1994). Bacteriocins of LAB may be classified as follows (Klaenhammer, 1993; Nes et al., 1996):

Class I. Lantibiotics, bacteriocins that contain the unusual amino acids lanthionine and 3-methyl-lanthionine (e.g. nisin).

Class II. Small, heat stable, non-lantibiotic peptides, subdivided as:

- Pediocin-like, one-peptide bacteriocins with a double-glycine leader peptide and a dedicated secretion and processing machinery. Consist of a relatively well-conserved, hydrophilic N-terminal sequence, containing the distinctive YGNGVXC motif (X means any amino acid residue), and a more diverse, relatively hydrophobic C-terminal half. Display inhibition against *Listeria* (e.g. pediocin PA-1).
- Bacteriocins as in a) but require combination of two polypeptide chains for full activity (e.g. lactococcin G).
- Bacteriocins with a *sec*-dependent signal sequence (e.g. divergicin A).

Class III. Large, heat labile protein bacteriocins (e.g. helveticin J).

Class IV. Complex bacteriocins with carbohydrate or lipid moiety (not fully accepted).

0-8493-2047-X/00/$0.00+$.50

Sakacins are class IIa bacteriocins produced by certain strains of *Lactobacillus sakei*. As all bacteriocins of this subclass, they are small, cationic and hydrophobic peptides that contain an N-terminal leader sequence of the double-glycine type, and possess strong antilisterial activity. **Sakacin A** was the first sakacin described and is produced by *Lb. sakei* Lb706, a bacterium that was isolated from raw meat (Schillinger & Lücke, 1989). Subsequently, other sakacins were discovered, including **sakacin M** produced by *Lb. sakei* 148 (Sobrino et al., 1991, 1992), **sakacin P** by *Lb. sakei* LTH673 (Tichaczek et al., 1992), **sakacin 674** by *Lb. sakei* Lb674 and *Lb. sakei* Lb16 (Holck et al., 1994), **sakacin B** by *Lb. sakei* 251 (Samelis et al., 1994) and **sakacin K** by *Lb. sakei* CTC 494 (Hugas et al., 1995). It was further shown that *Lb. sakei* LTH673 produces at least one other bacteriocin in addition to sakacin P (Eijsink et al., 1998). All known sakacin producers were isolated from non-fermented meat (*Lb. sakei* Lb706 from raw meat, *Lb. sakei* Lb674 from vacuum-packed lamb, and *Lb. sakei* Lb16 from vacuum-packed beef), or from dry fermented sausage (*Lb. sakei* 148, *Lb. sakei* LTH673, *Lb. sakei* 251, and *Lb. sakei* CTC 494). Still other bacteriocinogenic *Lb. sakei* strains were isolated from meat sources, such as *Lb. sakei* CTC 372 (fermented sausage; Garriga et al., 1993), and *Lb. sakei* 265 (chilled meat; Buncic et al., 1997), but their bacteriocins were not or only partially characterized. In addition, *Lb. sakei* L45, producer of the lantibiotic lactocin S (Mørtvedt & Nes, 1990), and *Lb. curvatus* LTH1174, producer of curvacin A (Tichaczek et al., 1992) which is identical to sakacin A (Axelsson & Holck, 1995), were both isolated from fermented sausage. Strains of *Lb. sakei* and *Lb. curvatus* are highly adapted to the meat habitat and predominate in European naturally fermented sausages because of their competence at lower ripening temperatures (Hammes et al., 1990; Hugas et al., 1993). In American-style fermented meat products, *Pediococcus* usually is the main starter culture.

II. OCCURRENCE

A. Biosynthesis

1. Production by fermentation. Sakacin production, as bacteriocin production by LAB in general, is a growth-related process and hence follows primary metabolite kinetics (*FIGURE 1*) (Leroy & De Vuyst, 1999a). The bacteriocin molecules are released into the cell environment while cells are growing exponentially, but production ceases from the time the cell growth starts to level off. From that point bacteriocin activity decreases, due to the adsorption of bacteriocin molecules to the cells, in particular at high pH values (De Vuyst et al., 1996b; Leroy & De Vuyst, 1999a).

The microbial environment, besides interfering with bio-availability and efficiency (see *Section IVB - Influencing factors*), will strongly determine sakacin production. Acidity and temperature strongly influence cell growth and consequently affect the growth-related sakacin production, as shown in the production of sakacin A (Yang & Ray, 1994), sakacin M (Sobrino et al., 1992) and sakacin K (Hugas et al., 1998b; Leroy & De Vuyst, 1999a). At temperatures between 20 and 25°C, high amounts of both biomass and sakacin K were produced by *Lb. sakei* CTC 494 in MRS broth. However, at 35°C cells grew well but no sakacin K could be detected (Leroy & De Vuyst, 1999a). Abundant production of sakacin K was also detected at refrigerated temperatures in MRS broth, although the growth rate was very low (Hugas et al., 1998b). Curvacin A, identical to sakacin A and K, was preferably produced at temperatures below 25°C (Vogel et al.,

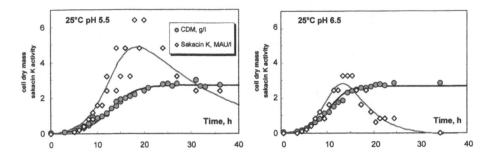

FIGURE 1. Growth-associated production of sakacin K by *Lactobacillus sakei* CTC 494 in de Man-Rogosa-Sharpe (MRS) broth at constant conditions of temperature and pH. CDM = Cell Dry Mass; MAU = Mega Arbitrary Units.

1993). Although sakacin K production is clearly linked to the growth of its producer, the optimal pH for cell growth (pH 6.0) did not correspond to the pH for maximal sakacin K production (pH 5.0) (Leroy & De Vuyst, 1999*a*). Likewise, production of bacteriocins from *Lb. sakei* 265 and *Lb. sakei* Lb706 was more efficient at lower pH (in buffered MRS broth of pH 5.4 and in non-buffered MRS broth) than at a neutral pH (in buffered broth of pH 6.5) (Buncic et al., 1997). These data suggest that the physical factors temperature and pH have, besides their growth-related effect on sakacin production, also a pronounced influence on the specific sakacin production by the cells, meaning the production per gram of cells (as may be deduced from *FIGURE 1*). In this matter, it is interesting to mention that, at optimal conditions for sakacin K production (23°C and pH 5.0), the specific bacteriocin production, being lower than expected from predictive modeling, appeared to have reached its upper limit (*FIGURE 2*) (Leroy & De Vuyst, 1999*a*). This could be due to limited immunity of the producer to its own bacteriocin. A bacteriocin production saturation level has also been observed for *Enterococcus faecium* CTC 492 (Nilsen et al., 1998).

Some mineral compounds such as nitrite, salt and cations, present in the cell environment, may also interfere with sakacin production (Leroy & De Vuyst, 1999*b*). This is important while considering *in situ* applications of sakacin-producing strains. Nitrite (up to 0.04%) when used in meat fermentations for color stability, flavor and antimicrobial action, did not affect the sakacin K production per cell of *Lb. sakei* CTC 494 but lowered the overall bacteriocin production slightly by its reductive effect on the cell growth of the producer cells. Sodium chloride, which is added for flavor and microbial selection, reduced the sakacin K production extensively by lowering both the growth and the specific bacteriocin production of the producer cells. Sakacin K production was possible up to a concentration of 6% sodium chloride, while at 8% concentration, sakacin K could no longer be detected. Curvacin A, identical to sakacin A and K, was produced in the presence of up to 4% sodium chloride (Vogel et al., 1993). It has been suggested that sodium chloride molecules influence the binding of the induction factor, necessary to start transcription of the bacteriocin operon, to its receptor (Nilsen et al., 1998). However, fermentation experiments with glycerol indicate that a decrease in water activity might be the main reason for the reduced bacteriocin production at high salt concentrations (Leroy & De Vuyst, 1999*b*).

TABLE 1. Amino acids (AA) and molecular mass of mature sakacins produced by *Lb. sakei* strains.

Bacteriocin	Producer	AA residues	Mass (Da)	Reference
Sakacin A	*Lb. sakei* Lb706	41	4308.7	Holck et al., 1992
= sakacin K	*Lb. sakei* CTC 494	41	4308.7	Aymerich et al., 1996
= curvacin A	*Lb. curvatus* LTH1174	41	4308.7	Tichaczek et al., 1993

```
              -10        -1 +1        +10         +20        +30        +40
       MNNVKELSMTELQTITGG ARSYGNGVYCNNKKCWVNRGEATQSIIGGMISGWASGLAGM
```

Bacteriocin	Producer	AA residues	Mass (Da)	Reference
Sakacin B	*Lb. sakei* 251	not available	6300*	Samelis et al., 1994
Sakacin M	*Lb. sakei* 148	not available	4640*	Sobrino et al., 1992
Sakacin P	*Lb. sakei* LTH673	43	4436.6	Tichaczek et al., 1994
= sakacin 674	*Lb. sakei* Lb674	43	4436.6	Holck et al., 1994

```
              -10        -1 +1        +10         +20        +30        +40
       MEKFIELSLKEVTAITGG KYYGNGVHCGKHSCTVDWGTAIGNIGNNAAANWATGGNAGWNK
```

* estimated by gel electrophoresis
Sequencing indicated for sakacin A (data from Holck et al., 1992) and sakacin P (data from Tichaczek et al., 1994), consisting of the leader sequence, the proteolytic cleavage site (↓) and the mature bacteriocin.

Other factors that might interfere with bacteriocin production of LAB strains in a food environment include the presence of sugars (De Vuyst & Vandamme, 1992), nutrients and several growth factors (Parente & Hill, 1992; Yang & Ray, 1994; Vignolo et al., 1995), fat and emulsifiers (Parente & Hill, 1992; Huot et al., 1996), ethanol (De Vuyst et al., 1996*b*; Nilsen et al., 1998; Callewaert et al., 1999), spices (Hugas et al., 1998*a*), and atmospheric conditions (De Vuyst et al., 1996*b*; Hugas et al., 1998*b*).

2. Molecular organization and regulation. Sakacins, and class IIa bacteriocins in general, are synthesized as prebacteriocins (*TABLE 1*). Sakacin prepeptides possess a leader sequence of 18 amino acid residues, which is cleaved off during externalization of the bacteriocin (Holck et al., 1992, 1994; Tichaczek et al., 1994). Characteristic of all class IIa bacteriocins is the presence of glycine residues at positions -2 and -1 relative to the proteolytic cleavage site. The functions of the leader sequence probably consist of maintaining the bacteriocin biologically inactive while detained inside the producer cell, and providing a recognition signal for the transporter system (see below) (Nes et al., 1996). All proteins involved in the biosynthesis of sakacins are encoded by a genetic structure which is organized in several operons (*FIGURE 2*) (Nes et al., 1996):
- a structural gene that encodes the prebacteriocin;
- an immunity gene that is located next to and downstream of the structural gene, on the same operon, encoding for the immunity protein;
- a gene encoding for the ABC transporter needed for externalization of the bacteriocin and the concomitant proteolytic cleavage of the leader sequence;
- a gene encoding for an accessory protein needed for the externalization;
- a set of regulatory genes, encoding for an inducing factor, a histidine protein kinase, and a response regulator.

3. Sakacin A. The biosynthesis of sakacin A has been described in detail. All information necessary for sakacin A production and immunity is present on a 8,668-bp fragment, part of a 60-kb plasmid (Schillinger & Lücke, 1989), consisting of two diver-

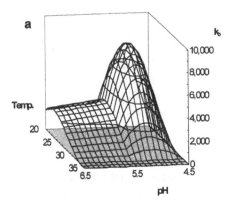 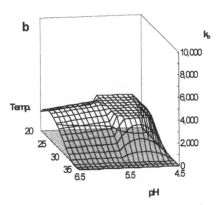

FIGURE 2. Specific sakacin K production, represented by the biokinetic parameter k_B, by *Lactobacillus sakei* CTC 494 in MRS broth, (a) as expected from predictive modeling and (b) after introduction of an upper limit (with permission from the American Society for Microbiology; from Leroy & De Vuyst, 1999a).

gently transcribed operons, *sapA-saiA* and *sapKRTE*, separated by an insertion element IS*1163* (Axelsson & Holck, 1995) (*FIGURE 3A*). The insertion element did not appear to be involved in expression of the *sap* genes. The first operon contains the structural gene *sapA* (originally designated *sakA*; Holck et al., 1992) followed by *saiA* (originally designated *sakB*; Axelsson et al., 1993), the gene which encodes a putative immunity peptide. Sequencing of the structural gene *sapA* revealed a primary translation product of 59 amino acid residues which is cleaved between amino acids 18 and 19 to yield the active sakacin A (*TABLE 1*) (Holck et al., 1992). The immunity peptide SaiA is synthesized together with the bacteriocin and provides protection to the producer strain against its own bacteriocin. SaiA consists of 90 amino acid residues, is largely hydrophilic, rich in lysine and possesses a calculated isoelectric point of 10.3 (Axelsson & Holck, 1995). It is surprising that the immunity proteins of the two identical bacteriocins sakacin A (from *Lb. sakei* Lb706) and curvacin A (from *Lb. curvatus* LTH1174) share little homology; this suggests that immunity proteins do not directly interact with their bacteriocin (Nes et al., 1996). The second operon involved in sakacin A production contains genes that encode a histidine kinase (*sapK*) and a response regulator (*sapR*), as well as proteins that are involved in the signal sequence-independent transport and processing mechanism (*sapT* and *sapE*). Frameshift mutations and deletion analyses have indicated that the *sapK/sapR* system is essential for production and immunity, constituting a classical bacterial two-component signal-transducing system. In response to environmental signals this system acts as a transcriptional activator on both operons (*sapKRTE* and *sapA-saiA*). The histidine kinase SapK monitors the signal and the response regulator SapR mediates an adaptive response (Fabret et al., 1999), *i.e.* a change in the expression of the genes for sakacin production. The genes *sapT* and *sapE* are necessary for production only. SapT, a dedicated ABC transporter, is responsible for the simultaneous secretion and processing of the prebacteriocin and is assisted by an accessory protein SapE. The latter is essential for the externalization of the bacteriocin but its specific role remains unclear. Recent evidence indicates that sakacin production, including the secretion/processing machinery, is regulated by an induction mechanism which is affected by the growth conditions (Eijsink et al., 1996; Nes et al., 1996; Brurberg et al., 1997). The inducing factor is a basic, non-bactericidal, bacteriocin-like peptide with a double-glycine leader, processed and externalized most probably by the dedicated ABC transporter associated with the bacteriocin. The open reading frame preceding the regulatory genes in the sakacin A gene cluster, named *sapIA*, analogus to the *sppIP* gene for sakacin P (see below), is the most likely factor regulating sakacin A production (*FIGURE 3A*) (Eijsink et al., 1996). It encodes a protein of 23 residues containing a double-glycine leader. The inducing factor, capable of adopting a highly amphifilic helical structure in membrane environments, may function as a cell density signal (*quorum sensing*).

4. Sakacin P. The six consecutive genes (*sppK, sppR, sppA, spiA, sppT* and *sppE*) that are present on a 7.6 kb chromosomal DNA fragment responsible for sakacin P production and immunity constitute a comparable genetic system (*FIGURE 3B*) (Tichaczek et al., 1994; Hühne et al., 1996). The structural gene encoding sakacin 674 (*sakR*) was also located on the chromosome and its sequencing indicated that sakacin 674 is equal to sakacin P (Holck et al., 1994). Sakacin P, and its equivalent sakacin 674, are synthesized as prepeptides of 61 amino acid residues (*TABLE 1*). Interestingly, a copy of the immunity gene *spiA* has also been discovered on the chromosome of *Lb. sakei* Lb706, the sakacin A producer (Hühne et al., 1996). It has been shown for *Lb. sakei* LTH673 that the secretion of an inducing factor (a 19-residue peptide) is essential for production and immunity (Eijsink et al., 1996). The gene encoding the inducing factor (*sppIP*) is located within a few kilobases of the structural gene and precedes the histidine kinase gene, as it was the case for sakacin A, and hence is the first gene in the regulatory operon (*FIGURE 3B*). The gene is co-transcribed with *sppK* and *sppR*. The inducing factor tightly regulates the promoter in front of the structural gene *sppA* and also, but less strictly, the promoter in front of *sppIP* (Eijsink et al., 1996; Brurberg et al, 1997). This molecular organization provides an attractive genetic explanation for growth-phase dependent autoinduction of sakacin production. However, some physiological data indicate that sakacin P production by *Lb. sakei* Lb674 takes place constitutively (Hühne et al., 1996). An explanation might be that the induction peptide sticks to the cell surface of the sakacin-producing cells and continuously turns on production (Callewaert et al., 1999).

B. Screening / Identification

The production of bacteriocins is a widespread bacterial feature (Tagg et al., 1976; Jack et al., 1995). However, their detection poses many difficulties, as production depends on the growth medium and the overall environmental conditions, while activity determination is dependent on the indicator strains used. Moreover, there is a great variation in sensitivity between species and even between strains, which makes it difficult to find the correct indicator strains.

Methods to test a strain for sakacin production consist of causing inhibition zones on a soft agar overlay containing the indicator organism. This may be obtained by overlaying a plated bacteriocinogenic culture, that grows on a medium containing little sugar in order to avoid strong acidification (Schillinger & Lücke, 1989; Sobrino et al., 1991). Diversely, cell-free supernatant of the producing culture may be added to the overlay via spotting (agar spot test; Hugas et al., 1995) or by placing it into wells (well-diffusion assay; Schillinger & Lücke, 1989; Papathanasopoulos et al., 1991; Tichaczek et al.,

1992; Samelis et al., 1994; Hugas et al., 1995). Another possibility is to dip paper discs into cell-free culture supernatant and place them on the overlay (Sobrino et al., 1991, 1992; Rodríguez et al., 1994). Neutralization of the culture supernatant excludes inhibition of the indicator due to lactic acid secretion. Treatment of the supernatant with catalase or incubating under anaerobic conditions makes it possible to exclude inhibition due to peroxide formation (Schillinger & Lücke 1989; Papathanasopoulos et al., 1991; Sobrino et al., 1991; Samelis et al., 1994). Treatment with proteases, amylases, glucosidases, *etc.* further enables the confirmation of the proteinaceous nature of the antimicrobial agents (Schillinger & Lücke, 1989; Sobrino et al., 1991, 1992; Tichaczek et al., 1992; Samelis et al., 1994; Hugas et al., 1995).

C. Quantitative assays

Sakacin activity may be determined with a critical dilution method (Barefoot & Klaenhammer, 1983; De Vuyst et al., 1996*a,b*; Leroy & De Vuyst, 1999*a*). Briefly, serial twofold dilutions of cell-free culture supernatant in sodium phosphate buffer (pH 6.5) are placed into wells (100 μl) or spotted (10 μl) onto an indicator lawn. The lawn is prepared by adding fresh, exponentially growing cultures of a sensitive strain to an overlay agar. After incubation of the plates, activity is expressed in arbitrary units (AU), corresponding with 10 μl of the highest dilution causing a definite zone of inhibition on the lawn of the indicator organism.

The soluble sakacin activity may also be quantified using a microtiter plate assay system (Mørtvedt & Nes, 1990; Holck et al., 1992; Eijsink et al., 1998). A sensitive indicator organism is added to the wells and after incubation culture turbidity is measured spectrophotometrically. One bacteriocin unit is then defined as the amount of sakacin that inhibits the growth of the indicator by 50% as compared to a control culture.

All methods mentioned above are indirect bioassays. Disadvantages are therefore the dependence on a sensitive indicator organism, which varies among laboratories, and the use of arbitrary units, which vary with experimental conditions. Future possibilities for the detection of sakacins could include immunoassays (Martínez et al., 1998) and mass spectrometry (Rose et al., 1999). However, such methods only detect 'mass' and not biological activity.

III. MOLECULAR PROPERTIES

A. Isolation and purification

Usually, purification to homogencity of class IIa bacteriocins is done by subsequent steps of ammonium sulfate precipitation, ion exchange, hydrophobic interaction and reversed-phase HPLC (Tichaczek et al., 1992; Holck et al., 1992, 1994). Ion exchange and hydrophobic interaction chromatography are included because of a surplus of positively charged amino acid residues and the hydrophobic nature of sakacins. Producer cells are grown to the early stationary phase and are removed by centrifugation. The bacteriocin present in the cell-free culture supernatant is precipitated by ammonium sulfate addition (40%), and harvested through a centrifugation step. The precipitated pellet is then dissolved in a buffer solution (e.g. pH 4.2) and applied to an ion exchange chromatography column. After this step, loss of bacteriocin activity might take place. Next, the bacteriocin is eluted from the column in another buffer solution (e.g. pH 5.7) and applied to a column

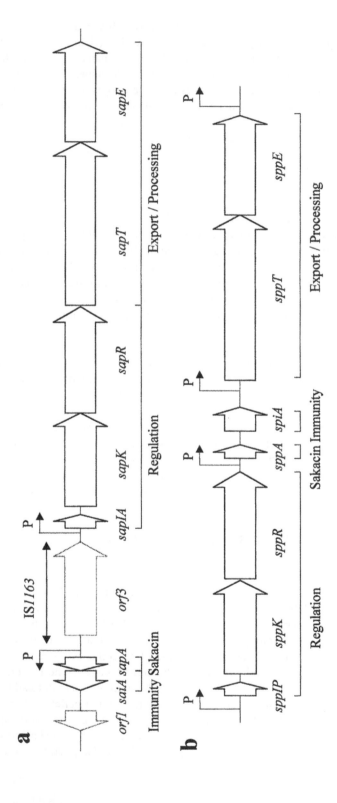

Figure 3. Genetic organization of the gene cluster of (a) the sakacin A (*sap*) bacteriocin system (data obtained from Axelsson & Holck, 1995) and (b) the sakacin P (*spp*) bacteriocin system (data obtained from Hühne et al., 1996; and Brurberg et al., 1997). The functions or potential functions of the different gene products are indicated. P indicates the promoter site.

for hydrophobic interaction chromatography. For reasons that remain unknown, an increase of sakacin activity was observed after this hydrophobic interaction step. The eluate is then diluted in 0.1% trifluoroacetic acid (TFA) and applied to a reversed-phase column, equilibrated with 10% (v/v) 2-propanol containing 0.1% TFA. A linear gradient of 15 to 50% 2-propanol, containing 0.1% TFA, is then imposed to elute the bound sakacin. The diluted sakacin is rechromatographed on the same column. This method was successfully applied for the purification of sakacin A (> 9000-fold; Holck et al., 1992), sakacin P (>5000-fold, Tichaczek et al., 1992) and sakacin 674 (Holck et al., 1994). Sakacin B, however, is more difficult to purify since it is not precipitated by the addition of up to 70% ammonium sulfate. Although sakacin B may be purified by ion exchange, 97.5% of the activity is lost during this step. The purification becomes even more difficult with hydrophobic interaction chromatography since sakacin B becomes totally inactivated due to irreversible reaction with the gel matrix (Samelis et al., 1994). Sakacin M was partially purified by gel filtration, following concentration of the cell-free culture supernatant by lyophilization (Sobrino et al., 1992).

Isolation procedures which rely on salt precipitation of the sakacin molecules from the cell-free culture supernatant have the disadvantage that only a portion of the total sakacin present in the culture broth is recovered since the portion adsorbed to the cells will be lost. To overcome the problem Yang et al. (1992) established a simple method based on the reduced bacteriocin adsorption under acidic conditions. After fermentation, the culture broth of the producer strain is adjusted to pH 6.5 for maximum adsorption of sakacin to the producer cells. The producer cells are heat-killed (70°C for 25 min) to avoid proteolytic degradation by microbial enzymes. The cells are then removed from the culture broth by centrifugation and, after washing and suspending in a sodium chloride solution, the bacteriocin is released from the cells by lowering the pH to 2.0. Collection of the supernatant after a final centrifugation step to remove the cells permits a concentrated preparation of sakacin to be obtained. Further purification to homogeneity may be done by reversed-phase HPLC.

B. Physico-chemistry

Sakacins are quite stable bacteriocins which are resistant to heat and acidity. Heat stability has been demonstrated for sakacin A (100°C for 20 min; Schillinger & Lücke, 1989), sakacin P (100°C for 7 min; Tichaczek et al., 1992), sakacin M (80°C for 1 h and 150°C for 9 min; Sobrino et al., 1992), sakacin B (100°C for 20 min; Samelis et al., 1994) and sakacin K (100°C for 20 min, Hugas et al., 1995). Stability in a wide pH range has been reported for sakacin B (pH 2.0-9.0; Samelis et al., 1994).

However, despite their stable nature, their proteinaceous structure makes them very sensitive to the action of certain proteases. Trypsin and pepsin severely inhibit the activity of sakacin A (Schillinger & Lücke, 1989; Lewus et al., 1991) and sakacin M (Sobrino et al., 1992). Sakacin M was, to a lesser degree, also inhibited by papain and the Sigma proteases II and XIV (Sobrino et al., 1992). Sakacin P and K were found sensitive to proteinase K and trypsin but insensitive to pepsin (Tichaczek et al., 1992; Hugas et al., 1995).

Because of their hydrophobic nature sakacins tend to form aggregates or adsorb to cells, and consequently become unavailable for inhibitory activity (De Vuyst et al., 1996b). The adsorption of bacteriocin molecules to the bacterial cell surface is pH-dependent (Yang et al., 1992). In general, between 93 and 100% of bacteriocin molecules are adsorbed at pH near 6.0, and the lowest adsorption (≤ 5%) occurs at pH 1.5 to 2.0. Sakacin A adsorption to the cells is maximal at pH values above 5.5 while at pH 2.0 no sakacin A is adsorbed to the *Lb. sakei* Lb706 producer cells but 75% remains adsorbed onto *E. faecalis* MB1 indicator cells.

The deamidation and oxidation of, respectively, the asparagine residues (positions 24, 32, 38, and 42) in sakacin P and the two methionine residues (positions 30 and 41) in sakacin A (see *TABLE 1*), may lead to less active compounds and hence contribute to the instability of bacteriocins (Fimland et al., 1996).

C. Structure

Sakacins are, as all other class IIa bacteriocins, small peptides consisting of one polypeptide chain (*TABLE 1*). They are amphiphilic (rather hydrophobic), cationic, and posses a high content of small amino acids (glycine, in particular). The charged amino acid residues are all located in the N-terminal hydrophilic part of the polypeptide chain, whilst the more hydrophobic C-terminal part consists of pairs of hydrophobic amino acid residues which are regularly spaced by pairs of hydrophilic ones (Holck et al., 1992). The complete amino acid sequence (*TABLE 1*) of sakacin A (Holck et al., 1992) and sakacin 674 (Holck et al., 1994) was determined by Edman degradation. Sakacin A consists of 41 amino acid residues and has a calculated molecular mass of 4308.7 Da, which is in good agreement with the value determined by plasma desorption mass spectrometry (4309.6 Da). Sakacin 674, and its equivalent sakacin P, consists of 43 amino acid residues, and has a calculated molecular mass of 4436.6 Da, corresponding with the molecular mass of 4437.2 Da as determined by plasma desorption mass spectrometry (Holck et al., 1994; Tichaczek et al., 1994). Sakacin A shares 33% amino acid identity with sakacin P, and is equal to sakacin K and curvacin A (Aymerich et al., 1996). Sakacin 674 is identical to sakacin P. These data are of considerable importance if one assumes that the C-terminal part of the class IIa bacteriocins is an important determinant of the target cell specificity (Fimland et al., 1996). In addition, the presence of a disulfide bond would be important but not crucial for sakacin activity (Chikindas et al., 1993; Eijsink et al., 1998).

Up to now, the 3D NMR structure of sakacins has not been studied. In general, bacteriocins are unstructured in aqueous solutions but adopt a partly helical structure in membrane-mimicking environments (Eijsink et al., 1998).

IV. ANTIMICROBIAL ACTIVITY

A. Mechanism of action

In cases where the mode of action of class IIa bacteriocins has been investigated, the site of action appears to be the cytoplasmatic membrane. The lethal action of the bacteriocin results from a permeabilization of the membrane by hydrophilic pore formation with a consequent leakage of vital components as well as a depletion of the proton motive force (Montville & Bruno, 1994; Abee, 1995; Jack et al., 1995). The mode of action of sakacins resulted to be bactericidal for sakacin A (Schillinger & Lücke, 1989) and sakacin B (Samelis et al., 1994), even bacteriolytic for sakacin P (Tichaczek et al., 1992) and sakacin K (Hugas et al., 1995), but only bacteriostatic for sakacin M (Sobrino et al., 1991, Rodríguez et al., 1994). Moreover, *Lb. sakei* CTC 382 displayed a bacteriostatic mode of action against *L. monocytogenes* but was bactericidal against *Lb. plantarum* (Garriga et al., 1993). However, the antibacterial mode of action seems to be dependent on several factors such as the concentration and purity of the bacteriocin preparation, the type of buffer or broth, the sensitivity of the indicator strain and the density of the cell suspension used.

Sensitivity to class IIa bacteriocins is very much strain-dependent (see *Section IVC - Spectrum*). Moreover, it is remarkable that *Lb. sakei* Lb706 and *Lb. curvatus* LTH1174, which produce the identical bacteriocins sakacin A and curvacin A, respectively, show different sensitivity behavior to several class IIa bacteriocins. Eijsink et al. (1998) have postulated that differences in bacteriocin sensitivity of closely related lactic acid bacteria may be due to the fact that these bacteria possess one or more immunity genes for class IIa bacteriocins which may be expressed to various extents. However, it has been suggested that specific surface receptors are responsible for the initial binding of the bacteriocin and its lethal action, but the nature of these receptors remains unclear and highly hypothetical (Nes et al., 1996). It has even been shown that antimicrobial activity of class IIa bacteriocins does not necessarily require specific receptors (Kaiser & Montville, 1996; Chen et al., 1997). In addition, the C-terminal part of the bacteriocin molecule, which shows diversity amongst class IIa bacteriocins, was shown to be important in determining target cell specificity (Fimland et al., 1996). The C-terminal part of sakacin P becomes amphiphilic if it adapts an α-helical secondary structure, whereas sakacin A has a more hydrophobic character. A reasonable explanation for sensitivity behavior might be that the hydrophobicity of the rather diverse C-terminal half determines the degree of interaction with membrane entities of the target cell, and hence the degree of permeabilization (Bhugaloo-Vial et al., 1999).

Gram-negative bacteria are generally insensitive to bacteriocins from LAB since their outer membrane provides them with a permeability barrier. Sublethal injury, in order to disrupt the outer membrane, may however lead to sensitiveness (Stevens et al., 1991; Kalchanayand et al., 1992). Yeasts and fungi are not inhibited at all.

B. Influencing factors

The bacteriocin environment strongly determines bioavailability and efficiency of antibacterial action and hence stability of bacteriocin activity. This is of utmost importance when considering applications of sakacins in food products. As an example, it has been found that class IIa bacteriocins have an increased activity at low pH values, as it was the case for sakacin P (Gänzle et al., 1996). Possible explanations are an enhanced capacity to pass through the cell wall of the target cells or inhibition of the binding with putative membrane receptors at higher pHs (Jack et al., 1995). Low acidity will also reduce bacteriocin losses due to adsorption or aggregation phenomena (Yang et al., 1992), and hence favor stability over time.

In addition, temperature might play an important role in the mode of action and stability of sakacins. Low temperatures result in the ordering of the lipid hydrocarbon chains and hence decrease membrane fluidity, which may interfere with bacteriocin efficiency (Abee et al., 1994). Furthermore, low temperatures decreased the inactivation rate of sakacin K and hence extended bioavailability (Leroy & De Vuyst, 1999*a*).

Chemicals, such as nitrite and salt, and cations might as well influence the antibacterial action of sakacins. Although nitrite does not affect specific sakacin K production (Leroy & De Vuyst, 1999*b*), it seems to reinforce the *in situ* antimicrobial efficiency of this bacteriocin (Hugas et al., 1998*a*). It was also shown that the presence of nitrite has no remarkable effect on the activity of the bioavailable sakacin P, while an increase of activity is observed upon addition of salt to sakacin P solutions at acidic conditions (pH 5.5) (Gänzle et al, 1996). It has further been suggested that di- and trivalent cations, such as magnesium and calcium, might interfere with the electrostatic interactions between the negatively charged phospholipid headgroups of the cytoplasmic membrane of the target cell and the positively charged bacteriocin (Abee et al., 1994; Abee, 1995).

Other conditions, such as the presence of spices and particular atmospheric conditions (Hugas et al., 1998*b*), may also influence sakacin activity and stability. Moreover, bacteriocin bioavailability may be reduced due to diffusion limitations (Blom et al., 1997) and inactivation by proteases or adsorption to food particles (e.g. fat; Jung et al., 1992). The presence of ethanol and emulsifiers, on the other hand, may partially prevent adsorption losses (Jung et al., 1992; Callewaert et al., 1999).

C. Spectrum

The inhibitory spectrum of bacteriocins is restricted to species that are closely related to the producing cell. Sensitivity is strain- rather than species-related. Sakacins are characterized by a narrow inhibitory spectrum, being active against various strains of LAB, especially *Lactobacillus sp.*. As is common for class IIa bacteriocins, they are also active against *Listeria* (Schillinger & Lücke, 1989; Lewus et al., 1991; Motlagh et al., 1991; Sobrino et al., 1991; Tichaczek et al., 1992; Hugas et al., 1995); only sakacin B does not seem to inhibit *Listeriae* (Samelis et al., 1994).

Sakacin A is active against various LAB species (mainly *Lactobacilli* but also *Pediococci, Carnobacterium piscicola, Leuconostoc paramesenteroides, Enterococcus faecium* and *E. faecalis*) and *Listeria monocytogenes* (Schillinger & Lücke, 1989; Lewus et al., 1991). Surprisingly, activity against the Gram-negative psychrotroph *Aeromonas hydrophila* was observed as well (Lewus et al., 1991). There are considerable differences in the sensitivity of different strains of a species. Whereas Schillinger and Lücke (1989) found *Staphylococcus aureus* 14458 to remain unaffected by the presence of sakacin A, Lewus et al. (1991) detected activity against *S. aureus* 25923 and, to a lesser extent, also against *S. aureus* 196E. The sakacin A analog, sakacin K, was shown to be active against selected strains of *L. monocytogenes* and *L. innocua* (Hugas et al., 1995). Sakacin M is active against other LAB (*Lactobacillus, Leuconostoc, Carnobacterium*), and against the foodborne pathogens *S. aureus* and *L. monocytogenes* (Sobrino et al., 1991). Sakacin B inhibits several strains of *Lactobacillus* and *Leuconostoc* but is not effective against *L. monocytogenes* or *S. aureus* (Samelis et al., 1994).

Sakacin P is inhibitory towards many strains of *Lactobacillus* and against strains of *Leuconostoc, Carnobacterium, E. faecalis, Brochothrix thermosphacta* and *L. monocytogenes* (Tichaczek et al., 1992). It was further shown that pure sakacin P was five to ten times more active towards *Listeria* than curvacin A (= sakacin A), although ten times less active than the strongly antilisterial enterocin A. On the other hand, the activity of pure sakacin P towards lactic acid bacteria is rather modest (Eijsink et al., 1998). Because of this combination of high antilisterial activity and a restricted inhibitory spectrum towards lactic acid bacteria, sakacin P is a promising bacteriocin to combat *Listeria* contamination of foods.

Since sakacins do not seem to affect the growth of selected strains of *Micrococci* (Sobrino et al., 1991; Tichaczek et al., 1992) and *Staphylococcus carnosus* (Schillinger & Lücke, 1989; Tichaczek et al., 1992), which are important components of meat starter cultures, they are suitable for practical use in the production of fermented meat products, for instance to combat *Listeriae, E. faecalis* and *S. aureus*.

V. BIOLOGICAL ADVANTAGE

A. Additive advantage to foods

The use of sakacin or sakacin-producing LAB in a food matrix, especially in meat products, is currently a point of interest (Stiles & Hastings, 1991; Eckner, 1992; McMullen & Stiles, 1996; Muriana, 1996; Hugas, 1997). Besides the inhibitory effects towards foodborne pathogens, the presence of sakacin in the food will contribute to the inhibition of the outgrowth of spoilage bacteria that are for instance rope-inducing, gas-producing or flavor-disturbing (heterofermentative LAB). For this reason, the overall sensorial and organoleptic qualities of the food may be more efficiently preserved.

B. Physiological advantage to host

Since sakacins are often inhibitory towards foodborne pathogens such as *Listeria monocytogenes* and *Staphylococcus aureus*, their use either as additive or using sakacin-producing strains as starter or protective cultures, may contribute to safer food products. Additionally, they may lead to a reduction of some traditional, chemical preservatives that are being questioned because of their possible harmful effects towards the consumer (e.g. nitrates).

VI. APPLICATIONS

A. Suitability / Adaptability

The *in vitro* inhibitory activity of sakacins towards *Listeria monocytogenes*, using liquid laboratory media, has been demonstrated. In the presence of *Lb. sakei* Lb706 the *L. monocytogenes* count in MRS broth dropped sharply from 6×10^5 to 4×10^3 bacterial counts (CFU) per ml after 24 hours at 15°C, while a bacteriocin-negative mutant permitted further *Listeria* outgrowth up to approximately 10^8 CFU per ml (Schillinger & Lücke, 1990; Schillinger et al., 1991). Similar results were obtained for the indicator strain *Lb. sakei* Lb790. Whereas the cell density of *Lb. sakei* Lb790 continued to increase in the presence of the bacteriocin-negative mutant, more than 99% of the indicator cells were killed by the sakacin A-producing *Lb. sakei* Lb706 (Schillinger & Lücke, 1989). In APT broth, *Lb. sakei* 148 prevented an initial load of *L. monocytogenes* (10^3 CFU ml^{-1}) from attaining the numbers they reached in pure cultures (10^9 CFU ml^{-1}). Maximal loads were around 10^5 CFU ml^{-1} but the inhibition was mainly due to the production of lactic acid, since the non-bacteriocinogenic *Lb. sakei* 23 yielded similar results (Rodríguez et al., 1994).

The *in vitro* production of sakacin K by *Lb. sakei* CTC 494 is optimal at the temperature and pH conditions that prevail during the fermentation stage of European sausage manufacturing (22-25°C and pH 5.8-4.9, respectively) (Leroy & De Vuyst, 1999a). Although the addition of salt and, to a lesser extent, of nitrite lowers the *in vitro* production of sakacin K by *Lb. sakei* CTC 494, results suggest that under sausage conditions the bacteriocin activity will nevertheless be sufficient (Leroy & De Vuyst, 1999*b*). Hence, the sakacin K producer *Lb. sakei* CTC 494 seems a promising candidate for meat fermentation processes. Although sakacin activity *in situ* is less pronounced, experiments have indeed confirmed its application potential (Hugas et al., 1995).

B. Existing applications

Nisin, a bacteriocin which is produced by certain strains of *Lactococcus lactis* subsp. *lactis*, and Alta™ 2341, a pediocin preparation from *Pediococcus acidilactici*, are commercially applied as food additives (Anonymous, 1986, 1992). The allowed application fields of both bacteriocins are however limited. In addition, the pH of meat is not suited to nisin solubility and stability. On the other hand, *P. acidilactici* is not a psychrotroph, so it is not suited for *in situ* production of pediocins in meats. This leaves potential for sakacin-producing *Lb. sakei* strains that grow and produce bacteriocin at low temperatures, and hence can be used for the biopreservation of meats. Although sakacins display no antibacterial spectra equivalent to nisin or pediocin, they are active against *Listeria* species that are of concern in meat technology.

In meat technology, sakacin could be used as added (partially) purified bacteriocin preparation or through *in situ* production by the use of sakacin-producing strains. The latter may serve two purposes. When applied to non-fermented meat or meat products they may suppress growth of spoilage organisms or food pathogens (use as protective culture, for instance in fresh meat, high pH meat products, minimally processed meat products like vacuum- or modified-atmosphere-packaged meat). Secondly, they may be used as starter culture or coculture in fermented meat products to control the fermentation process by inhibiting the growth of the competing flora, including foodborne pathogens. The bacteriocin indeed provides an additional and acceptable barrier to prevent the outgrowth of undesirable bacteria in a given food, but, of course, can not substitute good manufacturing practice. Applications of purified sakacin as an additive are currently absent since the compound has not been officially approved yet. Therefore, *in situ* applications by means of 'food grade' producers are more suitable, as special labeling is not necessary (Holzapfel et al., 1995). Furthermore, it is likely that the starter flora of several fermented meat products already contains sakacin producers. Patents have been deposited for the specific use of bacteriocin-producing LAB as commercial starter cultures for meat fermentations (Hugas et al., 1996*a*; Tichaczek et al., 1995).

C. Possible applications

The effect of adding supernatant containing sakacin A to control the growth of psychrotrophic foodborne pathogens in refrigerated foods is rather small (a decrease of 0.6 log for *L. monocytogenes*), probably due to the low concentration of sakacin A in the supernatant (Motlagh et al., 1991). The addition of purified sakacin is however not legally accepted. As long as sakacin has not obtained the 'Generally Regarded As Safe'-status (GRAS), the *in situ* application of sakacin producers in food processing may be an acceptable alternative to direct addition of purified sakacin. Since the sakacin-producing LAB are GRAS microorganisms, and therefore have a legal status, their application as starter, co- or protective cultures escapes the additives regulation. Moreover, it has been shown that the sakacin K-producing *Lb. sakei* CTC 494 provided good organoleptic and sensorial qualities to the end-product (Hugas et al., 1995). The application of sakacins in food products other than meat has not been cited in the literature up to now.

1. Starter or co-cultures. The *in situ* application of sakacins in meat may be exploited using sakacin-producing starter cultures for meat fermentation *(FIGURE 4)*. In German 'Mettwurst', made of pork meat with a normal pH (5.6-5.8), *L. monocytogenes* did not grow but was able to survive. While the addition of its bacteriocin-negative variant had no antilisterial effect, *Lb. sakei* Lb706 was able to reduce *Listeria* counts with about one log cycle, although both the bacteriocin producer and non-producer acidified the meat to the same extent. Higher risk of *L. monocytogenes* outgrowth may be expected from sausages made of meat with a high initial pH (6.3), since *Listeria* was able to grow rapidly from 5×10^3 CFU g^{-1} to 4×10^6 CFU g^{-1} within 8 days. Growth was only slightly affected by the bacteriocin-negative variant of *Lb. sakei* Lb706 while the bacteriocin producer prevented multiplication during the first 2 days. After this initial delay, however, *Listeriae* started to grow until 5×10^4 CFU g^{-1} after 8 days (Schillinger et al., 1991).

Likewise, in Spanish dry fermented sausages the use of the sakacin K-producing *Lb. sakei* CTC 494 as a starter culture permitted to diminish the number of artificially added *Listeriae* (10^3 CFU g^{-1}) by 2.64 log, which was 1.25 log more compared to a non-bacteriocinogenic control strain (Hugas et al., 1995). In another experiment, following the same recipe but with *Staphylococcus carnosus* as additional starter culture, *Lb. sakei* CTC 494 reduced listerial counts from 10^4 CFU g^{-1} to 2-4 MPN (most probable number) g^{-1} after 8 days, while the sakacin A-producing *Lb. sakei* Lb706 hardly had any effect. However, when the recipe was modified (lower amounts of added sugar and a nitrate cure instead of a nitrate/nitrite cure), *Lb. sakei* CTC 494 failed to show any result, but, surprisingly, *Lb. sakei* Lb706 was now capable of a slight antilisterial effect. The sakacin P producer, *Lb. sakei* LTH673, was ineffective for both recipes (Hugas et al., 1996*b*).

Another possibility is the use of *Lb. sakei* cocultures that may be added to a classical starter culture in order to reinforce microbial selection. The release of bacteriocin by the co-culture will then assist the starter in its antibacterial activity.

2. Protective cultures. Sakacin producers may also be applied in non-fermented foods where they can act as a protective culture. Protective cultures represent an additional safety factor, reducing the risk of growth and survival of foodborne pathogenic bacteria (Holzapfel et al., 1995). A potential application is the inoculation of fresh meat with sakacin-producing strains to prevent the outgrowth of the psychrotrophic pathogen *L. monocytogenes*. In chilled beef mince (8°C), the pathogen is able to increase its viable numbers from 10^5 CFU g^{-1} to 10^7-10^8 CFU g^{-1} within six days (Schillinger et al., 1991) to two weeks (Schillinger & Lücke, 1990). In samples inoculated with a non-bacteriocinogenic *Lb. sakei* Lb706 (10^5 CFU g^{-1}), the *Listeria* counts were slightly reduced to 10^4 CFU g^{-1} (Schillinger & Lücke, 1990) or increased to 10^6 CFU g^{-1} in pasteurized meat (Schillinger et al., 1991). Addition of sakacin-producing *Lb. sakei* Lb706 (10^5 CFU g^{-1}) permitted to decrease the listerial count to 10^3 CFU g^{-1} (Schillinger & Lücke, 1990) or to stabilize it throughout 7 days in pasteurized meat, before reaching final counts of 10^6 CFU g^{-1} after two weeks (Schillinger et al., 1991). The inhibitory effect on *L. monocytogenes* was also assayed in sliced vacuum-packed Bologna-type sausages for *Lb. sakei* Lb706 (Schmidt & Kaya, 1990). Compared to the bacteriocin non-producing variant, the presence of a protective culture of *Lb. sakei* Lb706 (10^3 CFU g^{-1}) delayed the time for *Listeria* to develop from 10^3 to 10^4 CFU g^{-1} from one week to more than three weeks at a storage temperature of 7°C. Likewise, inoculation of *Lb. sakei* CTC 494 (10^6 CFU g^{-1}) (or addi-

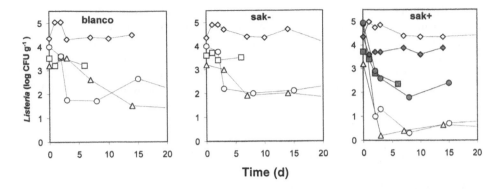

Time (d)

FIGURE 4. Growth of *Listeria* in the absence of a starter culture (blanco), in the presence of a non-bacteriocino-genic control strain (sak-) and in the presence of *Lactobacillus sakei* Lb706 (closed symbols) or *Lb. sakei* CTC 494 (open symbols) as a sakacin-producing starter culture (sak+), during fermentation of Spanish-style dry sausages (triangles, after Hugas et al., 1995; circles, after Hugas et al., 1996*b*), German-style dry sausages (diamonds, after Hugas et al., 1996*b*), and Mettwurst (rectangles, after Schillinger et al., 1991).

tion of 400 AU g^{-1} of sakacin K) inhibited the growth of *Listeria* in raw minced pork, poultry breasts and cooked pork under vacuum or a modified atmosphere (Hugas et al., 1998*b*). The most efficient inhibition was obtained in vacuum-packaged samples of poultry breasts and cooked pork and modified-atmosphere-packaged samples of raw minced pork.

Protective cultures may also be applied to prevent the outgrowth of spoilage bacteria. For instance, in cooked pork, *Lb. sakei* CTC 494 was able to prevent slime formation due to *Lb. sakei* CTC746 up to 21 days of storage, while control samples were already slimy after one week. The strain was however ineffective against *Leuconostoc* CTC 747, another slime producer (Garriga et al., 1998).

High initial numbers of protective strains may however lead to an unacceptable end-product. The presence of *Lb. sakei* Lb674 in sliced vacuum-packed Bologna-type sausages was effective to control listerial growth at inoculation numbers of 10^5-10^6 CFU g^{-1} but resulted in a too acid product. At lower inoculation levels (10-100 CFU g^{-1}), however, *Lb. sakei* Lb674 offered no advantage (Kröckel, 1997). Likewise, amounts of bacteriocin produced by low initial numbers (10^3 g^{-1}) of *Lb. sakei* 265 and *Lb. sakei* Lb706 were not sufficient to inhibit listerial growth on chilled, vacuum-packaged, raw beef (pH 5.3-5.4) and sausages (pH 6.4) (Buncic et al., 1997). In this matter it is interesting to mention that some strains of *Lb. sakei* produce strong sulfur odors when inoculated onto anaerobically stored meat (Shay et al., 1988).

D. Limitations

The sakacin producers to be used as bacteriocin-producing starter or protective cultures in meat technology must meet some essential criteria (*TABLE 2*).

One of the main bottlenecks is the limited effectiveness of bacteriocins in a food environment because of low bioavailability and stability of the molecules. For instance, the antilisterial effectiveness of *Lb. sakei* Lb706 in a food system is less pronounced than in broth (Schillinger et al., 1991). Besides, most of the inhibition of *Listeria* in dry fermented sausage by the sakacin K-producing *Lb. sakei* CTC 494 occurred in the beginning of the fermentation process, with little additional inhibition after that (Hugas et al., 1995). The reduced efficacy of sakacin in a meat environment may possibly be attributed to a limited diffusion throughout the product resulting in an uneven distribution, an inactivation by meat proteases or binding to meat components such as protein and fat particles (Schillinger et al., 1991). Moreover, the efficacy of sakacin producers in a food environment seems to be dependent on the ingredients used (sugars, curing agents, spices, etc.) and the food manufacturing process applied (Hugas et al., 1996b; 1998a). It has been shown that the presence of elevated salt concentrations in the sausage batter is harmful for sakacin K production by *Lb. sakei* CTC 494 (Leroy & De Vuyst, 1999b).

Another important limitation is the fact that sakacins are not active against all spoilage or pathogenic organisms of concern in foods, particularly in meat products. Gram-negative bacteria are normally not affected since their outer membrane forms a protective barrier. However, rapid acidification by the starter culture may overcome this problem; the presence of lactic acid and the low pH will avoid their outgrowth. On the other hand, the aciduric character of sakacin-producing *Lb. sakei* strains, producing a low pH in the presence of a fermentable carbohydrate, may limit their desirability as a protective culture in non-fermented cured meat.

Finally, spontaneous resistance development of the target organisms may reduce the application potential of sakacin. However, it was shown that a sakacin A-resistant phenotype of *L. monocytogenes* was unlikely to become stable in a natural population since it had a lower growth rate than the sensitive phenotype and was unable to invade populations of the sensitive strain (Dykes & Hastings, 1998).

VII. SAFETY / TOLERANCE AND EFFICACY

LAB are microorganisms which are categorized as 'Generally Regarded as Safe' (GRAS). They have a long history of safe use in foods either as indigenous natural contaminants or as starter cultures. Moreover, bacteriocin-producing lactic acid bacteria are omnipresent in foods and hence consumed by men for thousands of years. In this matter, 10% of the *Lactobacilli* strains that were isolated from meat and meat products by Schillinger and Lücke (1989) were identified as bacteriocin producers. This percentage may be as high as 20 to 30% (Garriga et al., 1993; McMullen & Stiles, 1996). Finally, since sakacins are sensitive to trypsin and pepsin, they will be degraded in the gastrointestinal tract and hence be harmless to the consumer.

VIII. BIOTECHNOLOGY

Starter strains may be improved by gene technology and at the same time maintain their desired features for which they were originally selected. Recombinant DNA technology may lead to the construction of improved bacteriocinogenic starter or protective cultures to combine technically desirable features (Geisen & Holzapfel, 1996). The transfer of the genes responsible for sakacin production into other bacteria may contribute to the enhancement of the industrial interest of these bacteria. In this matter, after introduction by electroporation of all genes responsible for sakacin P production and immunity, the bacteriocin-negative strains *Lb. sakei* Lb790 and *Lb. sakei* Lb706X produced sakacin P in comparable amounts to the parental strain, *Lb. sakei* Lb674 (Hühne et al.,

TABLE 2. Criteria for the use of sakacin-producing starter or protective cultures in meat products.

Sakacin producing starter or protective cultures in meat products must:	
Be adaptive to meat environment	• Growth in meat
	• Competitive features
	• Psychrotrophic nature (chilled meat)
	• Salt tolerance (sausage production)
Be functional in meat environment	• Production of sakacin
	• Availability of sakacin
	• Stability of sakacin
	• Diffusion of sakacin
	• Activity spectrum
	• Lactic acid prodution and sensory qualities (starters)
Not provoke negative effects	• No gas or slime production
	• No discoloration effects
	• No production of toxins or biogenic amines
	• Homofermentative starters
	• Limited acidification and proteolytic activity (protective cultures)

1996). Also, since the activity of sakacin does not include all of the Gram-positive spoilage or pathogenic bacteria of concern in meats, strains may be engineered to produce a variety of bacteriocins to enlarge their activity spectrum and to overcome potential full resistance development of target strains. This might also be accomplished by further screening for wide spectrum bacteriocins next to genetic and/or protein engineering of existing compounds. For instance, by targeting specific spoilers or pathogens associated with meats, it must be possible to obtain bacteriocin molecules suitable for use in bacteriocin gene cassettes for the production of multiple bacteriocins by one appropriate producer strain. In addition, with the use of a heterologous expression system, bacteriocin may be produced at levels equal to or higher than those obtained with the corresponding wild-type producer strains (Axelsson et al., 1998). This was obtained by introducing two plasmids involved in bacteriocin production in a bacteriocin-negative *Lactobacillus sakei* strain. The first plasmid contains the genes for transcriptional activation of the sakacin A promoter and for the export and processing of bacteriocin precursors. The second plasmid contains the structural and immunity genes for the class II bacteriocin of interest, fused to the sakacin A promoter. Optimizing the combination of the leader peptide and the appropriate secretion machinery, may further lead to a facilitated multiple bacteriocin expression and increased bacteriocin yields (Allison et al., 1995; Biet et al., 1998). Moreover, a better knowledge of the structure of bacteriocins in relation to their mode of action may lead to the development of hybrid molecules with a very broad spectrum or, on the contrary, a specific lethality (Fimland et al., 1996).

IX. SUMMARY

Sakacins are bacteriocins produced by several strains of *Lactobacillus sakei*. They are small, cationic, hydrophobic peptides with an antibacterial mode of action against species that are closely related to the producer, especially *Lactobacilli*, but also including other lactic acid bacteria and often foodborne pathogens such as *Listeria monocytogenes*. Sakacins are produced as primary metabolites, following cell growth of the producer, and become inactivated over time, mainly because of adsorption or aggregation phenomena. Environmental conditions strongly determine sakacin activity levels, because of their effect on cell growth as well as on the specific production by the cells. Moreover, the environmental conditions, particularly the pH, will affect sakacin stability and efficacy. Application possibilities are predominantly situated in the field of sakacin-producing starter or co-cultures for meat fermentations since addition of purified sakacin is not (yet) accepted by the food additives legislation. Continued study of the physical and chemical properties and structure-function relationships of sakacins and similar compounds, as well as their *in situ* functionality either as food additive or being produced by starter, co- or protective cultures is necessary if their potential for future use in meat preservation is to be exploited.

Acknowledgments. The authors acknowledge their financial support from the Research Council of the Vrije Universiteit Brussel, the Fund for Scientific Research-Flanders (FWO), the Institute for the Encouragement of Scientific and Technological Research in the Industry (IWT), in particular the STWW-project 'Functionality of novel starter cultures in traditional fermentation processes', and the European Commission (grant FAIR-CT97-3227). FL was supported by a grant of the IWT.

X. REFERENCES

1. Abee, T. 1995. Pore-forming bacteriocins of Gram-positive bacteria and self-protection mechanisms of producer organisms. *FEMS Microbiol. Lett.* 129:1-10.
2. Abee, T., Rombouts, F., Hugenholtz, J., Guihard, G., and Letellier, L. 1994. Mode of action of nisin Z against *Listeria monocytogenes* Scott A grown at high and low temperatures. *Appl. Environ. Microbiol.* 60:1962-1968.
3. Allison, G.E., Ahn, C., Stiles, M.E., and Klaenhammer, T.R. 1995. Utilisation of leucocin A export system in *Leuconostoc gelidum* for production of a *Lactobacillus* bacteriocin. *FEMS Microbiol. Lett.* 131:87-93.
4. Anonymous. 1986. International acceptance of nisin as a food additive. Issue No. 1/86. Aplin & Barrett Ltd, Trowbridge, Wiltshire, England.
5. Anonymous. 1992. Product information sheet for ALTA™ 2341. Quest International, Sarasota, FL.
6. Axelsson, L., and Holck, A. 1995. The genes involved in production of and immunity to sakacin A, a bacteriocin from *Lactobacillus sake* Lb706. *J. Bacteriol.* 177:2125-37.
7. Axelsson, L., Holck, A., Birkeland, S.E., Aukrust, T., and Blom, H. 1993. Cloning and nucleotide sequence of a gene from *Lactobacillus sake* Lb706 necessary for sakacin A production and immunity. *Appl. Environ. Microbiol.* 59:2868-2875.
8. Axelsson, L., Katla, T., Björnslett, M., Eijsink, V.G., and Holck, A. 1998. A system for heterologous expression of bacteriocins in *Lactobacillus sake*. *FEMS Microbiol. Lett.* 168:137-43.
9. Aymerich, M.T., Nes, I.F., and Vogel, R.F. 1996. Unpublished data. Cited in: M.E. Stiles. Biopreservation by lactic acid bacteria. In *Lactic Acid Bacteria: Genetics, Metabolism and Applications*, eds. G. Venema, J.H.J Huis in 't Veld, and J. Hugenholtz, pp. 243. Dordrecht, The Netherlands: Kluwer Academic Publishers.
10. Barefoot, S.F., and Klaenhammer, T.R. 1983. Detection and activity of lactacin B, a bacteriocin produced by *Lactobacillus acidophilus*. *Appl. Environ. Microbiol.* 45:1808-1815.
11. Bhugaloo-Vial, P., Dousliez, J.-P., Mollé, D., Dousset, X., Boyaval, P., and Marion, D. 1999. Delineation of key amino acid side chains and peptide domains for antimicrobial properties of divercin V41, a pediocin-like bacteriocin secreted by *Carnobacterium divergens* V41. *Appl. Environ. Microbiol.* 65:2895-2900.
12. Biet, F., Berjeaud, J.M., Worobo, R.W., Cenatiempo, Y., and Fremaux, C. 1998. Heterologous expression of the bacteriocin mesenterocin Y105 using the dedicated transport system and the general secretion pathway. *Microbiol.* 144:2845-2854.
13. Blom, H., Katla, T., Hagen, B.F., and Axelsson, L. 1997. A model assay to demonstrate how intrinsic factors affect diffusion of bacteriocins. *Int. J. Food Microbiol.* 38:103-109.
14. Brurberg, M.B., Nes, I.F., and Eijsink, V.G. 1997. Pheromone-induced production of antimicrobial peptides in *Lactobacillus*. *Mol. Microbiol.* 26:347-60.
15. Buncic, S., Avery, S.M., and Moorhead, S.M. 1997. Insufficient antilisterial capacity of low inoculum *Lactobacillus* cultures on long-term stored meats at 4 °C. *Int. J. Food Microbiol.* 34:157-170.

16. Callewaert, R., Holo, H., Devreese, B., Van Beeumen, J., Nes, I.F., and De Vuyst, L. 1999. A three-step method to purify amylovorin L471, a bacteriocin produced by *Lactobacillus amylovorus* DCE 471, and bacteriocin characterization and production. *Microbiol.* 145:2559-2568.

17. Chen, Y., Shapira, R., Eisenstein, M., and Montville, T.J. 1997. Functional characterization of pediocin PA-1 binding to liposomes in the absence of a protein receptor and its relationship to a predicted tertiary structure. *Appl. Environ. Microbiol.* 63:524-531.

18. Chikindas, M.L., García-Garcerá, M.J., Driessen, A.J.M., Ledeboer, A.M., Nissen-Meyer, J., Nes, I.F., Abee, T., Konings, W.N., and Venema, G. 1993. Pediocin PA-1, a bacteriocin from *Pediococcus acidilactici* PAC1.0, forms hydrophilic pores in the cytoplasmic membrane of target cells. *Appl. Environ. Microbiol.* 59:3577-3584.

19. De Vuyst, L., Callewaert, R., and Pot, B. 1996a. Characterisation and antagonistic activity of *Lactobacillus amylovorus* DCE 471 and large scale isolation of its bacteriocin amylovorin L471. *Syst. Appl. Microbiol.* 19:9-20.

20. De Vuyst, L., Callewaert, R., and Crabbé, K. 1996b. Primary metabolite kinetics of bacteriocin biosynthesis by *Lactobacillus amylovorus* and evidence for stimulation of bacteriocin production under unfavourable growth conditions. *Microbiol.* 142:817-827.

21. De Vuyst, L., and Vandamme, E.J. 1992. Influence of the carbon source on nisin production in *Lactococcus lactis* subsp. *lactis* batch fermentations. *J. Gen. Microbiol.* 138:571-578.

22. De Vuyst, L., and Vandamme, E.J. 1994a. Antimicrobial potential of lactic acid bacteria. In *Bacteriocins of Lactic Acid Bacteria: Microbiology, Genetics and Applications*, eds. L. De Vuyst and E.J. Vandamme, pp. 91-142. London: Blackie Academic & Professional.

23. De Vuyst, L., and Vandamme, E.J. 1994b. *Bacteriocins of Lactic Acid Bacteria: Microbiology, Genetics and Applications*. London: Blackie Academic & Professional, 536 pp.

24. Dykes, G.A., and Hastings, J.W. 1998. Fitness costs associated with class IIa bacteriocin resistance in *Listeria monocytogenes* B73. *Lett. Appl. Microbiol.* 26:5-8.

25. Eckner, K.F. 1992. Bacteriocins and food applications. *Dairy Food Environ. Sanit.* 12:204-209.

26. Eijsink, V.G., Brurberg, M.B., Middelhoven, P.H., and Nes, I.F. 1996. Induction of bacteriocin production in *Lactobacillus sake* by a secreted peptide. *J. Bacteriol.* 178:2232-7.

27. Eijsink, V.G., Skeie, M., Middelhoven, P.H., Brurberg, M.B., and Nes, I.F. 1998. Comparative studies of class IIa bacteriocins of lactic acid bacteria. *Appl. Environ. Microbiol.* 64:3275-81.

28. Fabret, C., Feher, V.A., and Hoch, J.A. 1999. Two-component signal transduction in *Bacillus subtilis*: how one organism sees its world. *J. Bacteriol.* 181:1975-1983.

29. Fimland, G., Blingsmo, O.R., Sletten, K., Jung, G., Nes, I.F., and Nissen-Meyer, J. 1996. New biologically active hybrid bacteriocins constructed by combining regions from various pediocin-like bacteriocins: the C-terminal region is important for determining specificity. *Appl. Environ. Microbiol.* 62:3313-3318.

30. Gänzle, M.G., Hertel, C., and Hammes, W.P. 1996. Die antimikrobielle Wirkung von Bakteriozin-bildenden Kulturen in Fleischwaren. Modellhafte Beschreibung des Effektes von Sakacin P auf *Listeria ivanovii* DSM20750 in Abhängigkeit von pH-Wert, NaCl- und Nitritkonzentration. *Fleischwirtsch.* 76:409-412.

31. Garriga, M., Aymerich, M.T., Costa, S., Gou, P., Monfort, J.M., and Hugas, M. 1998. Bioprotective cultures in order to prevent slime in cooked meat products. Congress Proceedings of the 44th International Congress of Meat Science and Technology, Barcelona, Spain, pp. 328-329.

32. Garriga, M., Hugas, M., Aymerich, T., and Monfort, J.M. 1993. Bacteriocinogenic activity of lactobacilli from fermented sausages. *J. Appl. Bacteriol.* 75:142-148.

33. Geisen, R., and Holzapfel, W.H. 1996. Genetically modified starter and protective cultures. *Int. J. Food Microbiol.* 30:315-324.

34. Hammes, W.P., Bantleon, A., and Min, S. 1990. Lactic acid bacteria in meat fermentation. *FEMS Microbiol. Rev.*, 87: 165-174.

35. Holck, A., Axelsson, L., Birkeland, S.E., Aukrust, T., and Blom, H. 1992. Purification and amino acid sequence of sakacin A, a bacteriocin from *Lactobacillus sake* Lb706. *J. Gen. Microbiol.* 138:2715-20.

36. Holck, A., Axelsson, L., Hühne, K., and Kröckel, L. 1994. Purification and cloning of sakacin 674, a bacteriocin from *Lactobacillus sake* Lb674. *FEMS Microbiol. Lett.* 115:143-9.

37. Holzapfel, W.H., Geisen, R., Schillinger, U. 1995. Biological preservation of foods with reference to protective cultures, bacteriocins and food-grade enzymes. *Int. J. Food Microbiol.* 24:343-362.

38. Hugas, M. 1997. Biopreservation of meat and meat products. *Actes du Colloque Lactic* 97, Caen, France, pp. 213-227.

39. Hugas, M., Garriga, M., Aymerich, M.T., and Monfort, J.M. 1993. Biochemical characterization of lactobacilli from dry fermented sausages. *Int. J. Food Microbiol.* 18:107-113.

40. Hugas, M., Garriga, M., Aymerich, M.T., and Monfort, J.M. 1995. Inhibition of *Listeria* in dry fermented sausages by the bacteriocinogenic *Lactobacillus sake* CTC 494. *J. Appl. Bacteriol.* 79:322-330.

41. Hugas, M., Garriga, M., Pascual, M., Aymerich, M.T., and Monfort, J.M. 1998a. Influence of technological ingredients on the inhibition of *Listeria monocytogenes* by a bacteriocinogenic starter culture in dry fermented sausages. Congress Proceedings of the 44th International Congress of Meat Science and Technology, Barcelona, Spain, pp. 338-339.

42. Hugas, M., Garriga, M., Monfort, J.M., and Ylla, J. 1996a. Bacteriocin from *Enterococcus faecium* active against *Listeria monocytogenes*. European Patent Office, EP0705843.

43. Hugas, M., Neumeyer, B., Pagés, F., Garriga, M., and Hammes, W.P. 1996b. Comparison of bacteriocin-producing lactobacilli on *Listeria* growth in fermented sausages. *Fleischwirtsch.* 76:649-652.

44. Hugas, M., Pagés, F., Garriga, M., and Monfort, J.M. 1998b. Application of the bacteriocinogenic *Lactobacillus sakei* CTC 494 to prevent growth of *Listeria* in fresh and cooked meat products packed with different atmospheres. *Food Microbiol.* 15:639-650.

45. Hühne, K., Axelsson, L., Holck, A., and Kröckel, L. 1996. Analysis of the sakacin P gene cluster from *Lactobacillus sake* Lb674 and its expression in sakacin-negative *Lb. sake* strains. *Microbiol.* 142:1437-48.

46. Huot, E., Barrena-Gonzales, C., and Petitdemange, H. 1996. Tween 80 effect on bacteriocin synthesis by *Lactococcus lactis* subsp. *cremoris* J46. *Lett. Appl. Microbiol.* 22:307-310.

47. Jack, R.W., Tagg, J.R., and Ray, B. 1995. Bacteriocins of Gram-positive bacteria. *Microbiol. Rev.* 59:171-200.

48. Jung, D.-S., Bodyfelt, F.W., and Daeshel, M.A. 1992. Influence of fat and emulsifiers on the efficacy of nisin in inhibiting *Listeria monocytogenes* in fluid milk. *J. Dairy Sci.* 75:387-393.

49. Kaiser, A.L., and Montville, T.J. 1996. Purification of the bacteriocin bavaricin MN and characterization of its mode of action against *Listeria monocytogenes* Scott A cells and lipid vesicles. *Appl. Environ. Microbiol.* 62:4529-4535.

50. Kalchayanand, N., Hanlin, M.B., and Ray, B. 1992. Sublethal injury makes Gram-negative and resistant Gram-positive bacteria sensitive to the bacteriocins pediocin AcH and nisin. *Lett. Appl. Microbiol.* 15:239-243.

51. Klaenhammer, T.R. 1993. Genetics of bacteriocins produced by lactic acid bacteria. FEMS *Microbiol. Rev.* 12:39-86.

52. Kröckel, L. 1997. Use of sakacin P and *Lactobacillus sake* Lb674 to suppress growth of *Listeria monocytogenes* in vacuum-packed sliced bologna-type sausages. *Actes du Colloque Lactic* 97, Caen, France, pp. 213-227.

53. Leroy, F., and De Vuyst, L. 1999a. Temperature and pH conditions that prevail during the fermentation of sausages are optimal for the production of the antilisterial bacteriocin sakacin K. *Appl. Environ. Microbiol.* 65:974-981.

54. Leroy, F., and De Vuyst, L. 1999b. The presence of salt and curing agents reduces the bacteriocin production by *Lactobacillus sakei* CTC 494, a potential starter culture for sausage fermentation. *Submitted for publication.*

55. Lewus, C.B., Kaiser, A., and Montville, T.J. 1991. Inhibition of food-borne bacterial pathogens by bacteriocins from lactic acid bacteria isolated from meat. *Appl. Environ. Microbiol.* 57:1683-1688.

56. Martínez, J.M., Martínez, M.I., Surez, A.M., Herranz, C., Casaus, P., Cintas, L.M., Rodríguez, J.M., and Hernandez, P.E. 1998. Generation of polyclonal antibodies of predetermined specificity against pediocin PA-1. *Appl. Environ. Microbiol.* 64:4536-4545.

57. McMullen, L.M., and Stiles, M.E. 1996. Potential use of bacteriocin-producing lactic acid bacteria in the preservation of meats. *J. Food Prot. 1996 (Supplement):* 64-71.

58. Montville, T.J., and Bruno, M.E.C. 1994. Evidence that dissipation of proton motive force is a common mechanism of action for bacteriocins and other antimicrobial proteins. *Int. J. Food Microbiol.* 24:53-74.

59. Mørtvedt, C.I., and Nes, I.F. 1990. Plasmid-associated bacteriocin production by a *Lactobacillus sake* strain. *J. Gen. Microbiol.* 136:1601-1607.

60. Motlagh, A.M., Johnson, M.C., and Ray, B. 1991. Viability loss of foodborne pathogens by starter culture metabolites. *J. Food Prot.* 54:873-884.

61. Muriana, P.M. 1996. Bacteriocins for control of *Listeria* spp. in food. *J. Food Prot. 1996 (Supplement):* 54-63.

62. Nes, I.F., Diep, D.B., Håvarstein, L.S., Brurberg, M.B., Eijsink, V., and Holo, H. 1996. Biosynthesis of bacteriocins in lactic acid bacteria. *Anton. Leeuw. Int. J. Gen. Molec. Microbiol.* 70:113-128.

63. Nettles, C.G., and Barefoot, S.F. 1993. Biochemical and genetic characteristics of bacteriocins of food-associated lactic acid bacteria. *J. Food. Protect.* 56:338-356.

64. Nilsen, T., Nes, I.F., and Holo, H. 1998. An exported inducer peptide regulates bacteriocin production in *Enterococcus faecium* CTC 492. *J. Bacteriol.* 180:1848-1854.

65. Papathanasopoulos, M.A., Franz, C.M.A.P., Dykes, G.A., and von Holy, A. 1991. Antimicrobial activity of meat spoilage lactic acid bacteria. *S. Afr. J. Sc.* 87:243-246.

66. Parente, E., and Hill, C. 1992. A comparison of factors affecting the production of two bacteriocins from lactic acid bacteria. *J. Appl. Bacteriol.* 73:290-298.

67. Rodríguez, J.M., Sobrino, O.J., Moreira, W.L., Fernández, M.F., Cintas, L.M., Casaus, P., Sanz, B., and Hernández, P.E. 1994. Inhibition of *Listeria monocytogenes* by *Lactobacillus sake* strains of meat origin. *Meat Sci.* 38:17-26.

68. Rose, N.L., Sporns, P., and McMullen, L.M. 1999. Detection of bacteriocins by matrix-assisted laser desorption/ionization time-of-flight mass spectrometry. *Appl. Environ. Microbiol.* 65:2238-2242.

69. Ruiz-Barba, J.L., Cathcart, D.P., Warner, P.J., and Jiménez-Díaz, R. 1994. Use of *Lactobacillus plantarum* LPCO10, a bacteriocin producer, as a starter culture in Spanish-style green olive fermentations. *Appl. Environ. Microbiol.* 60:2059-2064.

70. Samelis, J., Roller, S., and Metaxopoulos, J. 1994. Sakacin B, a bacteriocin produced by *Lactobacillus sake* isolated from Greek dry fermented sausages. *J. Appl. Bacteriol.* 76:475-486.

71. Schillinger, U., Kaya, M., and Lücke, F.-K. 1991. Behaviour of *Listeria monocytogenes* in meat and its control by a bacteriocin-producing strain of *Lactobacillus sake. J. Appl. Bacteriol.* 70:473-478.

72. Schillinger, U., and Lücke, F.-K. 1989. Antibacterial activity of *Lactobacillus sake* isolated from meat. *Appl. Environ. Microbiol.* 55:1901-1906.

73. Schillinger, U., and Lücke, F.-K. 1990. Lactic acid bacteria as protective cultures in meat products. *Fleischwirtsch.* 70:1296-1299.

74. Schmidt, U., and Kaya, M. 1990. Verhalten von *Listeria monocytogenes* in vakuum verpactem Brühwurstaufschnitt. *Fleischwirtsch.* 70:236-240.

75. Shay, B.J., Egan, A.F., Wright, M., and Rogers, P.J. 1998. Cysteine metabolsim in an isolate of *Lactobacillus sake*: plasmid composition and cysteine transport. *FEMS Microbiol. Lett.* 56:183-188.

76. Sobrino, O. J., Rodríguez, J.M., Moreira, W.L., Fernández, M.F., Sanz, B., and Hernández, P.E. 1991. Antibacterial activity of *Lactobacillus sake* isolated from dry fermented sausages. *Int. J. Food Microbiol.* 13:1-10.

77. Sobrino, O. J., Rodríguez, J.M., Moreira, W.L., Fernández, M.F., Sanz, B., and Hernández, P.E. 1992. Sakacin M, a bacteriocin-like substance from *Lactobacillus sake* 148. *Int. J. Food Microbiol.* 16:215-225.

78. Stevens, K.A., Sheldon, B.W., Klapes, N.A., and Klaenhammer, T.R. 1991. Nisin treatment for inactivation of *Salmonella* species and other Gram-negative bacteria. *Appl. Environ. Microbiol.* 57:3613-3615.

79. Stiles, M.E., and Hastings, J.W. 1991. Bacteriocin production by lactic acid bacteria: potential for use in meat preservation. *Trends Food Sci. Technol.* 2:247-251.

80. Stiles, M.E. and Holzapfel, W.H. 1997. Lactic acid bacteria of foods and their current taxonomy. *Int. J. Food Microbiol.* 36:1-29.

81. Tagg, J.R., Dajani, A.S., Wannamaker, L.W. 1976. Bacteriocins of Gram-positive bacteria. *Bacteriol. Rev.* 40:722-756.

82. Tichaczek, P. S., Meyer, J.N., Nes, I.F., Vogel, R.F., and Hammes, W.P. 1992. Characterization of the bacteriocins curvacin A from *Lactobacillus curvatus* LTH1174 and sakacin P from *L. sake* LTH673. *Syst. Appl. Microbiol.* 15:460-468.

83. Tichaczek, P.S., Pohle, S.B., Vogel, R.F., and Hammes, W.P. 1995. Use of bacteriocin-producing organisms to cure raw sausage. European Patent Office, EP0640291.

84. Tichaczek P.S., Vogel R.F., and Hammes, W.P. 1993. Cloning and sequencing of *curA* encoding curvacin A, the bacteriocin produced by *Lactobacillus curvatus* LTH1174. *Arch. Microbiol.* 160:279-283.

85. Tichaczek P.S., Vogel R.F., and Hammes, W.P. 1994. Cloning and sequencing of *sakP* encoding sakacin P, the bacteriocin produced by *Lactobacillus sake* LTH 673. *Microbiol.* 140:361-7.

86. Vignolo, G.M. de Kairuz, M.N., de Ruiz Holgado, A.A.P., and Oliver, G. 1995. Influence of growth conditions on the production of lactocin 705, a bacteriocin produced by *Lactobacillus casei* CRL 705. *J. Appl. Bacteriol.* 78:5-10.

87. Vogel, R.F., Pohle, B.S., Tichaczek, P.S., and Hammes, W.P. 1993. The competitive advantage of *Lactobacillus curvatus* LTH 1174 in sausage fermentations is caused by formation of curvacin A. *System. Appl. Microbiol.* 16:457-462.

88. Yang, R., Johnson, M.C., and Ray, B. 1992. Novel method to extract large amounts of bacteriocins from lactic acid bacteria. *Appl. Environ. Microbiol.* 58:3355-3359.

89. Yang, R., and Ray, B. 1994. Factors influencing production of bacteriocins by lactic acid bacteria. *Food Microbiol.* 11:281-291.

Section-V

ACID-ANTIMICROBIALS

Lactic acid
Sorbic acid
Acetic acid
Citric acid

J-C. Bogaert
A.S. Naidu

Lactic acid

22

I. INTRODUCTION

Lactic acid, also known as 2-hydroxypropionic acid, is present in almost all life forms and was probably also a constitutent of primitive life forms in the beginning of evolution (Brin, 1965). Lactic acid, although it is relatively abundant, is a relatively recent discovery.

Lactic acid, as the term implies, is a milk-associated acid, or rather is found in curdled milk. In 1780, Carl Wilhelm Scheele, a Swedish chemist, isolated an acid by precipitation with a calcium salt and named it *mjölksyra* - or 'milk acid' - assuming that it was a normal component of milk. He did not realize that it was a fermentation product of rancid milk (Dobbin, 1931). Later in 1813, Henri Braconnot at the French Nancy University, working with the acidic components of fermented food, found a product which he named 'nanceic acid', after the town of Nancy. He concluded that during the rapid fermentation alcohol is first produced and vinegar accumulated, whereas nanceic acid was produced during the slow fermentation process (Braconnot, 1813). In 1817, Vogel, a German chemist clearly showed that nanceic and lactic acid are identical (Holten, 1971).

The fermentative mechanism of lactic acid synthesis was first described by Cogniard-Latour, that small 'globules' were able to multiply in a medium containing sugar and produced gas and alcohol. In 1857, Louis Pasteur, during distillery experiments with beet juice fermentation, first described 'lactic yeast', a greyish substance, made of small globules aligned in short segmented ligaments, as producers of lactic acid ('lactic fermentation') as an essential ingredient for the growth of yeast cells and in alcohol fermentation.

During the early 19th century, the use of lactic acid was limited merely as a pharmaceutical speciality. In 1881, Charles Avery established the 'Lactate Company' near Boston, in the United States, with an objective to replace the use of tartric acid in bread making. Around 1900, the Boeringher Company in Germany successfully marketed lactic acid. Currently, the market for lactic acid and its derivatives amounts to 100,000 metric tons annually with a steady growth rate of 12 to 15% per year. Due to the perceived potential, many large industrial groups are now entering the field of acid technology.

II. OCCURRENCE

Lactic acid is frequently found in various life forms including animals, plants, and microorganisms. It is present as an intermediary of the carbohydrate and amino acid metabolism. Quantitatively, the L(+) form of lactic acid is a more predominant optical isomert in the nature, however, the D(-) form is also prevalent. In certain milieu, recemic mixtures of both optical isomers may exist in variable proportions. Lactic acid is found, in virtually all tissues, physiological fluids and biological secretions. In humans, lactic acid is found in blood (1.4 μmol/ml), in sperm (4.1 μmol/ml), in sweat (4.0 - 40.0 μmol/ml), in the cerebrospinal fluid (1.6 μmol/ml) and in urine (0.3 μmol/ml) (Geigy, 1968). In the human body the L(+) form is predominant, however, in urine, the D(-) form could be detected. The D(-) form of lactic acid is metabolized by the liver, and it is generally recommended not to add this form to food for children under 3 years due to its possible role in hepatic dysfunction. However, adverse physiological consequences of D(-) lactic acid ingestion have not been proved (Vrese, 1990).

In biological systems, lactic acid L(+) is formed by reduction of pyruvic acid. This hydrogen transfer is done by a NADH-dependent enzyme, lactate dehydrogenase. Lactic acid levels may change depending on variations in the pyruvate concentration (for instance, intravenous administration of glucose or fructose during hyperventilation; in certain cases of diabetes or after administration of adrenalin) or when the NADH/NAD$^+$ ratio fluctuates (improper oxygenation during intense muscular exercise or in altitude) (Huckabee, 1958; Lissac, 1966).

With an exception of skeletal muscles, the lactate/pyruvate ratio is remarkably constant in all tissues, and plasma although the concentration of both of these components varies according to specific tissues. In normal oxygenation conditions, this ratio is stable at 12 moles of lactate for 1 mole of pyruvate and is maintained at this level by incitation of hepatic cells (Schimassek, 1965). The hydrophobic nature of lactate and pyruvate molecules allows free diffusion through the cell membranes which establishes a direct relation between these two substances' concentration in the blood stream and the redox equilibrium of the cells' NADH/NAD$^+$ system (Hohorst, 1965).

In aerobic conditions, the energy feed to muscle cells is provided by oxidative degradation of glucose that generates energy compounds such as the adenosine triphosphate (ATP) and creatin phosphate. During anaerobic conditions or in case of increased energy demands (i.e. intense physical exertion), the required energy is supplied by glycolysis, which induces the accumulation of lactic acid in the muscles. This excess acid diffuses into the blood stream and is absorbed in the liver and kidneys to re-synthesize glucose ('gluconeogenesis') and glycogen (glyconeo genesis). Evidently, this metabolic path uses energy produced by oxidative degradation of lactic acid. In 1930, Meyerhof (1930) established that the ratio between used lactic acid and oxidized lactic acid was 5 to 1.

When re-synthesized in the liver or in kidneys, the glucose molecule is circulated into the blood stream and reaches the muscular cells where it is transformed into energy or lactic acid thus completing the 'Cori cycle' (Alpert, 1965; Brecht, 1967). Reichard *et al.* (1963) determined that at least 40% of the lactic pool exists as glucose and about 6% of the total glucose volume is recycled via the Cori cycle. The average turnover of lactic acid in an adult male is about 120 to 150 g per day (US-FDA, 1978).

Lactic acid is also found in a wide array of microorganisms. Several bacterial species produce large quantities of lactic acid, among which the best known are

Lactobacillus, Sporolactobacillus, Enterococcus, Lactococcus, Bacillus, Streptococcus, Pediococcus, Leuconostoc and bifidobacteria. Bifidobacteria are strict anaerobes while the other species are generally regarded as microaerophiles. The two lactic acid isomers can be produced by the members of *Lactobacillus* genus. *Lb. maltaromicus, Lb. agilis, Lb. sharpae, Lb. amylophilus, Lb. murinus* are known to produce only the L(+) form of acid whereas *Lb. delbrueckii, Lb. vitulinus, Lb. coryniformis* produce the D(-) isomer only and *Lb. plantarum or Lb. helveticus* are producers of both the molecular forms. The pH conditions, and age of cultures may influence the L(+)/D(-) ratio (Garvie, 1967). The stereospecificity of the acid is influenced by the presence of one and/or the other specific lactate dehydrogenase that converts pyruvate into lactate. Lactate hydrogenase that could simultaneously produce both lactic isomers has not been detected. Therefore, the existence of the L(+) and D(-) forms in a pure culture indicates the presence of independent genes for the two lactate dehydrogenases. However, the enzyme racemase that converts an isomer into the other form was reported in certain microorganisms such as *Clostridium butylicum* or *Lactobacillus sake* (Hiyama, 1968).

Fermented substrates used in the production of lactic acid are generally hexoses. Bifidobacterial fermentation, however, is a special process, where lactic synthesis may occur in two different pathways, a homolactic or a heterolactic fermentation; both conditions produce lactic acid as the only product or with various co-products such as ethanol, carbon dioxide and acetate.

Homolactic fermentation follows the metabolic path of glycolysis according to the classic EMP pathway. The following equation describes the general equilibria of this pathway:

ADP (2) + PO₄- (2)   ATP (2)

Glucose (1)   **Lactic Acid (2)**

This equation shows an anaerobic and exothermic reaction with a lactic acid yield often higher than 90%.

The heterolactic fermentation does not work along this path due to the lack of specific enzymes the aldolase and the triose phosphate isomerase for glucose catabolism according to the EMP scheme. The oxidative degradation of pentoses phosphates allows glucose metabolism via the hexose monophosphate shunt. Microorganisms adapted to this path ferment the D-xylose, the L-arabinose and the D-ribose to form lactic acid and acetic acid (Bernstein, 1953). The following equation describes the general equilibria of glucose heterofermenting catabolism.

ADP (2) + PO₄⁻ (2)   ATP (2)

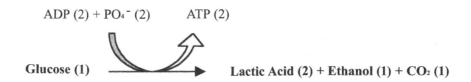

Glucose (1) ────────▶ **Lactic Acid (2) + Ethanol (1) + CO₂ (1)**

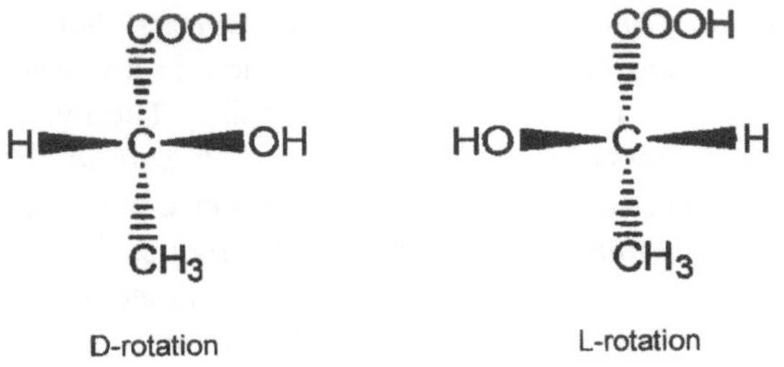

FIGURE 1: Structure of L(+) and D(-) optical isomers of lactic acid.

Industrial uses of lactic acid bacteria (LAB) are numerous and diversified (ensilage, dairy industry, bread making, brewery, vine-growing, probiotic, etc., (see *Chapter 17: Probiotics*). This chapter is limited to discussion of the use of LAB for the industrial production of lactic acid. Nearly 90% of the annual 100,000 mT output of lactic acid is produced by a fermentation process, the only one which produces the necessary isomeric purity. This acid may also be produced by chemical synthesis of petroleum derivatives; however, these products could be a racemic mix of the two lactic acid optical isomers.

The only chemical process currently in production is based on a catalytic transformation of acetaldehyde into lactonitrile using hydrocyanic acid, a by-product of the plastic industry. This step is followed by an acid hydrolysis, at about 100°C, of the nitrile to recover lactic acid, which is then purified by distillation of its methylic ester. Other chemical processes have been developed based on either propionic acid chlorination plus hydrolysis, or the reaction of propene with liquid dinitrogen tetroxide giving the 1-nitropropane-2-ol, which is then hydrolyzed into lactic acid (Chahal, 1990).

III. MOLECULAR PROPERTIES

Lactic acid is a weak organic acid and one of the smallest molecules with an asymetrical carbon in location α of the carboxylic function. This feature allows its existence with two different stereoisomers as the L(+) and D(-) forms (*FIGURE 1*). The plus and minus signs show the polarized light rotation, respectively to the right and to the left, known as dextro- and levo-rotatory effects. The L and D symbols indicate the spatial configuration of the molecule with reference to the two glyceraldehyde isomers. When containing no water, both these optically active acids form white solid crystals which melt at higher temperature (52.7 - 52.8°C) than their racemic mix containing 50% of each isomer (16.4°C) (Walter, 1988).

In addition to optical activity, a lactic acid molecule has simultaneous carboxylic and hydroxyl groups allowing it to react either as an acid or as an alcohol. This dual functionality leads to the spontaneous formation of intermolecular esters (oligomers) with discharge of water leading to a thermodynamic equilibrium, of concentration and age of the solution. The kinetics of this condensation reaction is dependent on temperature, possible presence of catalyst species (Lewis acid) and the differences in equilibrium of the solutions. As for every esterification reaction, it is reversible by adding water (hydrolysis of formed oligomers). In addition, the formation of lactic acid oligomers during heating and concentration of the lactic acid solution make it impossible to precisely measure the boiling point of the pure product at atmospheric pressure.

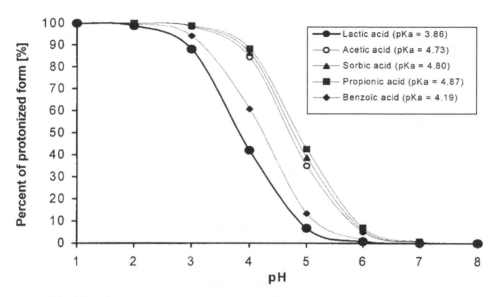

Figure 2: Variation of the percentage of protonized molecules as a function of pH for some monoacids.

Lactic acid is a weak acid with a pKa of 3.86 at 25°C. However, the tendency of every monoacid to dissociate is a function of pH of the medium as compared to the pK_a of the contemplated acid. Indeed, higher the pH increase, the more the acid molecule will be dissociated. On the other hand, with the same pH, different acids will be more dissociated if their pKa is low, as indicated in *Figure 2*. Such characteristics are of prime importance for the microbiostasis or microbicidal potential. The antimicrobial effects are larger when the acid is protonized, i.e. not dissociated. The main thermodynamic characteristics of lactic acid are shown in *Table 1*.

The salts of lactic acid are numerous. Holten (1971) has mentioned more than fifty metallic salts and close to twenty double salts containing the lactate ion. This chapter will discuss the properties of only sodium and potassium lactates, due to their wide use as preservatives by the food industry (*Table 2*). These two salts are generally produced by the reaction of lactic acid with the hydroxide or the carbonate of the desired metallic counter-ion; they are marketed either as a solid powder or as a liquid solution of 50%wt (by weight) or 60%wt.

Generally, lactic acid and lactates are used by the food industry for one of the following properties: (1) the acidification potential of lactic acid, (2) pH regulation property of sodium and potassium lactates, (3) reduction of water activity by sodium lactate, (4) synergism with common antioxidants such as ascorbic acid and (5) antimicrobial activity, which will be the prime focus of this chapter.

Assays. Methods for determining lactic acid content in foods may be found in the Official Methods of Analysis of the AOAC (Helrich, 1990). Volatile acids can be separated from foods using steam distillation. For a mixture of acids, the distillate is separated on a silicic acid column and the component acids can be identified with paper chromatography (Young et al., 1965) or thin layer chromatography (Tijan & Jansen, 1971).

TABLE 1 : Main thermodynamic characteristics of lactic acid (adapted from Holten, 1971).

Dissociation constant (Ka) :	0.000137 (at 25 °C)
Heat of dissociation (ΔH) :	- 63 cal/mol (at 25 °C)
Free energy of dissociation (ΔF) :	~5000 cal/mol
Heat of solution (ΔH) :	+ 1868 cal/mol (for crystalline L(+) lactic acid at 25 °C)
Heat of dilution (ΔH) :	-1000 cal/mol (for dilution with a large volume of water)
Heat of fusion (ΔH) :	+ 2710 cal/mol (for racemic lactic acid)
	+ 4030 cal/mol (for L(+) lactic acid)
Entropy of solution (ΔS) :	+ 6.2 cal/mol/ °C
Entropy of dilution (ΔS) :	- 3.6 cal/mol/ °C
Entropy of fusion (ΔS) :	+ 9.4 cal/mol/ °C (for racemic lactic acid)
	+ 12.2 cal/mol/ °C (for L(+) lactic acid)
Heat of combustion (ΔH_c^0) :	- 321220 cal/mol (for crystalline L(+) lactic acid at 25 °C)
	- 325600 cal/mol (for liquid racemic lactic acid at 25 °C)
Heat of formation (ΔH_f^0) :	- 165890 cal/mol (for crystalline L(+) lactic acid at 25 °C)
	- 163000 cal/mol (for liquid lactic acid)
	- 164020 cal/mol (for lactic acid in dilute solution)
	- 164080 cal/mol (for dissociated and diluted lactic acid)
Heat capacity (C_p) :	0.338 cal/g/°C (for crystalline lactic acid at 25 °C)
	0.559 cal/g/°C (for liquid lactic acid at 25 °C)
Absolute entropy (S^0) :	34.0 cal/mol/°C (for crystalline L(+) lactic acid at 25 °C)
	45.9 cal/mol/°C (for liquid racemic lactic acid at 25 °C)
Entropy of formation (ΔS_f^0) :	- 137.2 cal/mol/°C (for crystalline lactic acid at 25 °C)
	- 125.3 cal/mol/°C (for liquid lactic acid at 25 °C)
Free energy of formation (ΔF_f^0) :	- 124980 cal/mol (for crystalline L(+) lactic acid at 25 °C)
	- 126500 cal/mol (for liquid racemic lactic acid at 25 °C)

Derivatives of the acids can be identified using gas chromatography (Martin et al., 1971; Tonogai et al., 1983; Tsuji et al., 1986), and it is possible to identify components by retention times alone (James & Martin, 1952). Several modifications of HPLC methods have been developed for particular acids in foods (Bevilacqua & Califano, 1989; Bouzas et al., 1991; Bushway et al., 1984; Marsili et al., 1981; Nassos et al., 1984). Ion-exclusion chromatography coupled to HPLC has also been used (Togami et al., 1990). Lactic acid can also be identified using enzymatic assays.

IV. ANTIMICROBIAL ACTIVITY

The purpose of most antimicrobial agents is to reduce or eliminate microbial activity by stasis or cidal mechanisms. In the food industry, the antimicrobial application of lactic acid and lactates is generally considered at two stages: 1) decontamination of food such as the beef carcasses at slaughterhouses, and 2) shelf life enhancement of fresh or semi-processed foods. Consumer acceptance of such processed foods is largely based on objective considerations such as visual (appearance, color) and organoleptic (odor, flavor) qualities.

Although different inhibition mechanisms may be considered (Shelef, 1994), the two main ones are: (A) the capacity of lactic acid to reduce pH of the medium and the lipophilic characteristic of its protonized (undissociated) form, which facilitates its pene-

TABLE 2: Density, viscosity, freezing and boiling points of aqueous of sodium and potassium lactates (adapted from Dietz et al., 1944).

Concentration [%wt]	Density [g/ml at 25°C]	Freezing point [°C]	Boiling point [°C at 742 mmHg]	Viscosity [cP at 25°C]
Sodium Lactate				
5	1.0225	-2.2	99.8	1.060
10	1.0511	-4.1	100.1	1.240
20	1.1002	-9.7	101.4	2.038
30	1.1542	-18.2	103.3	3.638
40	1.2078	-32.5	105.5	7.291
50	1.2629		108.7	17.979
60	1.3131		112.9	55.063
Potassium Lactate				
5	1.0217	-1.6	99.7	0.999
10	1.0465	-3.7	100.1	1.128
20	1.0985	-8.6	101.4	1.490
30	1.1528	-16.1	103.1	2.105
40	1.2102	-28.1	105.7	3.289
50	1.2699	-51.0	109.4	5.492
60	1.3307		114.6	11.320

tration through the cell wall (Freese, 1973; Hunter, 1973; Salmond, 1984) and (B) the effect of lactates (mainly sodium lactate) on water activity in foods with intermediate humidity (Debevere, 1989; Loncin, 1975).

A. Effects on intracellular pH

It has been observed for a long time that most microorganisms are susceptible to antimicrobial effects in the presence of organic acids compared to inorganic acids. It is also well known that such inhibition increases by lowereing pH of the media or milieu (Sorrell, 1989; Farber, 1989; Conner, 1990; Ita, 1991). This phenomenon is due to the hydrophobic feature of most organic acids, often refered to as 'volatile fatty acids', which allows free diffusion of the protonized form through the cell membrane. This diffusion process takes place spontaneously due to the pH and osmolarity gradients that exist between the inner and outer sides of the cell. Generally, the intracellular pH is higher than the extracellular, and the acid undergoes dissociation as soon as it enters the cellular cytoplasm and then decreases the intracellular pH by releasing the proton. It seems however, that there is no common optimum intracellular pH for microorganisms. Acidophiles have generally an intracellular pH range of about 6.5 - 7.0; the comparable values for neutrophiles is about pH 7.5 - 8.0 range and for alkalophiles about pH 8.4 - 9.0 range (Eklund, 1983). This dissociation effect is described, for a monoacid such as lactic acid, by the relation shown hereunder which shows the ratio between the dissociated form [A-] and the non-dissociated one (or protonized form, [HA]) as an exponential function of the neighboring pH.

$$[A\text{-}] / [HA] = 10^{\,pH - pKa}$$

In order to counter the decrease of cellular cytoplasmic pH, resulting from the ionization of the entered acid, the cell allocates the main part of its energy content to eliminate these newly formed protons which results in slower growth kinetics (Cassio, 1987; Ten Brink, 1980; Ten Brink, 1985).

The ability of microorganisms to keep a constant intracellular pH varies (Booth, 1985). Fermentative bacteria, compared to respiratory microorganisms, often show a larger intracellular interval where growth rate remains stable. This logic is based on the chemiosmotic assumption initially suggested by Mitchell in 1961 (Mitchell, 1961; Mitchell & Moyle, 1969) which describes the transmembrane transport phenomenon and the energy generating phenomenon as a function of the electrochemical gradient existing in the membrane and resulting from the proton extrusion bringing about the intracellular pH decrease. Indeed, according to this chemiosmotic assumption, the proton translocation from the intracellular to the extracellular is of prime importance based on the assumption that the membrane is entirely inert and protons could only pass through by following a spatially oriented electron transportation chain. This transfer generates an electro-chemical potential, called proton-motive force (Dp) which may be represented by following mathematical formula: Dp = Dy - Z . DpH (where Dy and DpH are electropotential differential and pH differential across the membrane, respectively). Z is a constant which is a function of the temperature (Z = 2.3 RT/F, where R is the gas constant; T the absolute temperature and F the Faraday constant). The ionization of an organic acid in the cytoplasm will therefore be immediately and systematically balanced by the export of a proton under the impulse of the promotive force which is established at the cell membrane level. When this system is continuously activated or at least continued, one presumes that the ATP regeneration system will be affected (Sheu et al., 1975) as is the active nutrient transportation in the cell. Other cell functions, in addition to the energy metabolism, will also be affected. Freese (1973) and Eklund (1980) indicated that the occurrence of perturbations in the transmembrane transportation of amino acids is due to the occuring proton gradient (Hunter, 1973). Protein structural changes, nucleic acids, phospholipids and perturbations in the enzymatic system may also be affected.

The inhibition of active transportation of nutrients by neutralization of the promotive force only partially explains the microbial growth inhibition effect of organic acids (Eklund, 1985; Ramos, 1977). Measuring the influence of such acids on Dy and DpH, the occurence of an effect on the pH gradient while establishing an electric charge differential across the membrane which is high enough to allow the active transportation of amino acids. Moon (1983) reported that lactic acid does not react as one may expect from its dissociation constant. Studies on *Listeria monocytogenes* indicated that the inhibition of this bacterium is not associated with the decrease of its intracellular pH (Young, 1993; Ita, 1991).

However, various studies showed that bacterial growth is affected by the anion as well as by the proton (Eklund, 1983) and that lactate anion acts itself differently from other anions produced by fermentative processes such as propionate, sorbate or acetate. In non-fermented foodstuffs that pH is close to neutral; thus a ratio between dissociated and undissociated forms clearly favors the latter form. In addition, for lactic acid salts such as sodium and potassium lactates, an inhibition phenomenon can not be assigned to the acid's dissociation and to the subsequent production of protons, since these compounds totally dissociate in aqueous phase. Studies on calcium lactate indicated a bacteriostatic activity, however, the cation may not be implicated because it does not affect the balance of the Na/K transmembrane equilibrium (Weaver, 1993).

B. Effects on water activity

Loncin (1975) found that sodium lactate has a deflationary effect on water activity (a_w) higher than other organic acids or sodium chloride at equimolar concentrations, Chirife and Ferro Fontan (1980) reported a much lesser effect. These studies found that sodium lactate decreased water activity less than sodium chloride, however, this effect was higher than tartarate or sodium citrate at equimolar concentrations. Houstma *et al.* (1983) reported that sodium lactate and chloride have similar effect on water activity at concentrations ranging from 0.089 to 0.89 mol/L.

Houstma *et al.* (1993) found that the inhibition effect of sodium lactate is more intense than for the corresponding chloride and suggested a specific inhibition effect for the lactate ion on cell growth. From studies on gram-positive, gram-negative bacteria and yeasts, it appears that the sodium lactate minimal inhibition concentration is far lower than chloride with the exception of *Campylobacter jejuni*, *Trichosporon beigelii* and *Torulaspa delbrueckii* strains. The difference in efficiency between the two salts significantly varied between the test strains. Certain microorganisms such as *Yersinia enterocolitica*, *Listeria monocytogens*, *Staphylococcus aureus* and *Brochotrix thermosphacta* which are frequent contaminants of meat products are more sensitive to lactate than chloride. Similar obervations have been reported by Chen and Shelef (1992) for *Listeria monocytogens*. The microbial growth on cooked beef meat was unaffected by sodium chloride at 4% concentration while sodium lactate caused stasis at the same concentration.

The specific inhibition of sodium lactate seems stronger for gram-positive bacteria compared to Gram-negatives (Houstma, 1993). While interpreting these data it should be considered that the two salts tested are at equal molar concentrations, but the difference in molecular weights (respectively 58.5 and 112.1 for sodium chloride and lactate) and mass concentrations nearly doubles for sodium lactate compared to sodium chloride (Dacosta, 1997). The conditions in the growth media could also significantly influence the antimicrobial outcome (Shelef, 1994). Thus, a lesser sodium lactate concentration is needed to inhibit *Listeria monocytogens* in cooked food compared to the amout needed for the same strain grown in culture medium. Indeed, if one compares minimal values of A_W (between 0.90 and 0.94) which allow *Listeria* to grow in several media with an average of about 0.92, it may be noted that these values are significantly lower than those reported for meat products processed with sodium lactate, whose average water activity is about 0.95 (Miller, 1992; Nolan, 1992).

Therefore, assuming that the A_W limit is the same in a culture medium and in food, such results can only be explained by a specific effect of the lactate ion which is separate from its action on A_W. The variations in the inhibition concentration values in culture media as compared to meat could be explained by the fact that the majority of lactate remains in the aqeous part of the meat where concentration is higher than 7% for a 4% sodium lactate treatment in meat with 55% moisture (Shelef, 1994).

De Wit and Rombouts (1990) studied the influence of oxygenation conditions in media and found similar results under aerobic and anaerobic conditions. In contrast, Grau (1981) observed that 150 mol/L of sodium lactate inhibits the anerobic growth of enterobacteria *Yersinia enterocolitica*, *Enterobacter clocae*, *Aeromanas hydrophila* and *Serrata liquefaciens* when only the growth of *Aeromonas hydrophila* is affected in aerobic conditions at 25°C, pH 5.5 and pH 6.1. However, at 5°C, similar obervations could be made at pH 6.1 when the strains are grown under aerobic conditions at pH 5.5. The influence of temperature was also studied by De Wit and Rombouts (1990). The sensitivity of *Streptococcus faecalis* for sodium lactate was increased at low temperature (10°C) compared to the observed effect at higher temperature (30°C).

It should also be mentioned that the organoleptic impact of the two salts is different, as sodium lactate has a salted flavor which is five times lower than for sodium chloride for similar mass concentrations (Dacosta, 1997).

C. Other effects

Various effects of lactic acid other than the two described above have been reported. Hydroxycarboxylic acids are known for their chelating properties and lactic acid, in particular, is commonly used by the food industry for this purpose as are polyphosphates, citrates and ethylenediaminetetracetic (EDTA) acids (Kabara, 1991). Iron chelation property may possibly contribute to the antilisterial effect of lactate (Shelef, 1994) and to the stabilization of food lipids by the antioxidizing (Nnanna, 1994).

On the other hand, certain fungi, particularly *Aspergillus parasiticus* and *Aspergillus flavus* produce aflatoxins, extremely potent toxic substances under certain conditions. The most common aflatoxins types are B1, B2, G1 and G2. Moulds producing aflatoxins may proliferate in many foodstuffs (Hesseltine, 1966, Bullerman, 1986) and in various environmental conditions (Holmquist, 1983). Different studies have reported the effects of lactic acid on growth and production of aflatoxins in foods such as cheese (Lie, 1967; Bullerman, 1981) and certain meat (Bullerman, 1969a; Strzelecki, 1969; Sutic, 1972). El-Gazzar (1987) showed that lactic acid does not limit the growth of mycelium but could reduce the production of aflatoxins B1 and G1 when the initial pH of the media is maintained at 4.5, whereas the toxin production increase when the pH is set at 3.5. Luchese and Harrigan (1990) demonstrated that the total production of aflatoxins is increased when *Aspergillus parasiticus* is grown in a mixed culture in the presence of *Lactococcus lactis* as compared to a direct addition of lactic acid.

D. Comparison of lactate with other organic acids

Lactic acid is generally considered less efficient than sorbic, acetic, propionic and benzoic acids (Baird-Parker, 1980; Minor, 1970; Chung, 1970). Based only on its pKa, one might expect that lactic acid displays a higher pH deflationary effect but a more limited growth inhibition effect. However, at a given pH, lactic acid would be ionized more than the other acids. Brackett (1987) suggested the following classification as a function of the type of effect observed on the growth of *Yersinia enterocolitica*: (1) antimicrobial activity based on pH : propionic > lactic > acetic > citric; (2) antimicrobial activity based on the concentration of the non-dissociated form (for monoprotonic acids only): lactic > propionic > acetic; (3) antimicrobial activity based on equal molar concentrations: citric > lactic > propionic > acetic. It was concluded that there are no significant differences between these acids. The only possible difference seems to be related to the concentrations used. Comparable conclusions were obtained in studies with *Salmonella typhimurium* in milk where a: few differences in effects on growth were observed between hydrochloric, lactic and citric acids at 37°C, but a very clear hierarchy at 22°C: citric > lactic > hydrochloric (Subramanian, 1968). Conne et al. (1990) have studied the effects of different acids on the growth of *Listeria monocytogenes* and concluded that, at 10°C (in tryptic soy broth yeast extract media), the survival was 11 to 12 weeks in presence of

TABLE 3: Impact of lactic acid on organoleptic profile of foodstuffs.

Processed product	Effect	Reference
Color/Appearance		
Veal	Discoloration at concentration > 1.25%	Woolthuis, 1985
Veal	Discoloration at concentration > 1.25%	Snijders, 1985
Veal (head)	No effect on color at concentration of 1%	Cudjoe, 1988
Veal (tongue)	Improved appearance at concentration of 2%	Visser, 1988
Beef	No effect on color at concentration of 2%	Hamby, 1987
Beef	No effect on color at concentration of 2%	Prasai, 1991
Beef	Discoloration at concentration > 1%	Snijders, 1985
Pork	Discoloration at concentration > 1.5%	
Pork	Color darker with 1% lactic + 1% acetic acids	Mendonca, 1989
Chicken	Color lighter at concentration ranging from 1-5%	Van Der Marel, 1989
Chicken	Appearance deteriorates at concentration of 3%	Kolsarici, 1995
Crayfish	No effect on color at concentration of 1%	Dorsa, 1995
Catfish	Appearance deteriorates at concentration > 2%	Marshall, 1996
Flavor		
Veal	Little effect on flavor at concentrations < 2%	Woolthuis, 1985
Veal	No effect on flavor at concentrations < 2%, Alters flavor at concentration of 4%	Snijders, 1985
Veal (tongue)	No effect on flavor at concentration of 2%	Visser, 1988
Beef	No effect on flavor at concentration of 1%	Acuff, 1987
Beef	No effect on flavor at concentration of 1%	Dixon, 1987
Beef	Improvement of flavor at concentration of 1%	Hamby, 1987
Chicken	No effect on flavor at concentration 1 to 5%	Van Der Marel, 1989
Pork	No effect on flavor at concentration < 2%	Epling, 1993
Crayfish	No effect on flavor at concentration of 1%	Dorsa, 1995
Odor, aroma		
Beef	Improvement of aroma for concentration of 1%	Prasai, 1991
Chicken	Aroma deteriorates for concentration of 3%	Kolsarici, 1995
Catfish	Odor deterirates for concentrations > 2%	Marshall, 1996

acetic, citric or propionic acid but only 6 weeks in presence of lactic or hydrochloric acid. At 30°C, the listerial population was reduced to less than 10 cfu/ml within 1 to 3 weeks for all tested acids. Woolford (1975) showed that lactic acid is an excellent growth inhibitor for sporulating bacteria. Rice (1954) found an effect about four times more potent for lactic acid on the growth of *Bacillus coagulans* when compared to malic, citric, propionic and acetic acids.

In addition, numerous publications reported synergestic antimicrobial effects between organic acids, acetic and lactic acids, in particular, thus suggested to use con-cominantly (Rubin, 1978; Adams, 1988; Dickson, 1992). Various mixtures of lactic acid, citric acid and potassium sorbate showed synergy either by slowing the growth of *Salmonella, Yersinia enterocolitica, Pseudomonas fluorescens,* two strains of LAB (Restaino, 1981) and four strains of osmophilic yeasts (Restaino, 1982) or by total inhibition of their growth.

E. Effects on sensory characteristics of foods

In addition to food safety and shelflife attributes it is of prime importance that microbial interventions do not alter sensory characters such as flavor, color or odor when used as preserving agents in foods. Indeed, it is obvious that a food product with a perfect

TABLE 4: Impact of sodium and potassium lactates on the organoleptic profile of foodstuffs.

Processed product	Effect	Reference
Color, Appearance		
Beef	Sodium lactate emphasizes red color (specially at 2% concentration)	Papadopoulos, 1991
Beef (minced)	Sodium lactate does not alter color at concentration of 2 to 3%	Egbert, 1992
Pork (sausages)	Sodium lactate improves color for concentration of 1.8%	Lamkey, 1991
Pork (sausages)	Sodium lactate keeps red color unaltered at concentration of 2 to 3%	Brewer, 1991
Pork (sausages)	Sodium lactate does not alter color	Brewer, 1993
Pork (sausages)	Potassium lactate does not alter color at concentration of 1.2%	Bradford, 1993
Pork (minced)	Sodium lactate keeps red color at concentrations of 2 to 3%	Brewer, 1995
Flavor		
Beef	Sodium lactate improves flavor	Papadopoulos, 1991
Beef (minced)	Potassium lactate does not alter flavor at conc of 2 to 3%	Egbert, 1992
Pork (sausages)	Sodium lactate gives a less acidic, more salted taste at 2 to 3%	Brewer, 1991
Pork (sausages)	Sodium lactate prevents weakening of flavor at 1-2% concentrations. At 3% increases flavor substantially. High conc increases salty taste. Bitter/acidic tastes unaltered	Brewer, 1993
Pork (sausages)	Potassium lactate does not alter flavor for concentration of 1.2%	Bradford, 1993
Pork (minced)	Sodium lactate emphasizes flavor up to 3% concentration. Salty taste of sodium lactate is not as much as with sod chloride	O'Connor
Odor, aroma		
Pork (sausages)	Sodium lactate slows the appearance of unwanted odors	Lamkey, 1991
Pork (sausages)	Sodium lactate prevents off odors for concentrations of 2 to 3%	Brewer, 1991

microbiological keeping quality will not be consumed if it does not appeal to the consumers. TABLES 3 and TABLE 4 show a few of the numerous publications dealing with the organoleptic impact of the use of lactic acid, sodium lactate or potassium lactate to process various foodstuffs.

V. ANTIMICROBIAL SPECTRUM

Lactic acid elicits antimicrobial effects on a variety of microorgansims (TABLE 5). This antimicrobial activity is based on the ability of lactic acid to reduce pH of the milieu. Thus, in fermented foods, lactic acid in combination with other antimicrobial factors such as bacteriocins produced by LAB inhibit the competing microorganisms. Nykanen et al. (1998) have recently shown that the antimicrobial activity of a combination of lactic acid and whey permeat fermented by a nisin-producing *Lactococcus lactis* strain against fish-borne bacteria. Lactic acid inhibited all bacterial strains studied, but nisin whey permeate inhibited Gram-positive bacteria only. The combination was more effective than lactic acid alone against *Pseudomonas fluorescens* and *Staphylococcus hominis* isolated from fish, and *Pseudomonas aeruginosa* ATCC9721 and *Micrococcus luteus* ATCC9341.

Lactic acid is an excellent inhibitor of spore-forming bacteria at pH 5.0, but is ineffective against yeast and molds (Woolford, 1975a). Lactic acid is about four times more effective than malic, citric, propionic, or acetic in preventing the proliferation of *Bacillus coagulans* in tomato juice (Rice & Penderson, 1954). Lactic acid is more inhibitory to *Mycobacterium tuberculosis* at lower pH (Dubos, 1950).

TABLE 5: Antimicrobial spectrum of lactic acid and lactates on various microorganisms

Susceptible microorganism	Antimicrobial effect	Reference
Mycobacterium tuberculosis	Inhibits proliferation at low pH	Dubos, 1950
Bacillus coagulans	Inhibits growth in tomato juice	Rice & Penderson, 1954
Yersinia enterocolitica	Inhibits anaerobic growth (100 mM)	Grau, 1981
Aeromonas hydrophila	Inhibits aerobic growth at pH 5.5	Grau, 1981
Cl. botulinum A, B & E	Synergistic inhibition with monolaurin	Notermans & Dufrenne, 1981
Clostridium botulinum	Prevents growth in raw turkey	Maas et al., 1989
Enterobacteriaceae	2% acid spray of beef caracasses	Smulders & Woolthus, 1983
Lactobacillaceae	0.2% acid dip of porcine livers	Woolthus & Smulders, 1984
Asperigillus sp.	Inhibits aflatoxins B1 & G1 production	El-Gazzar, 1987
Pseudomonas fragi	Synergistic inhibition with Na-alginate	Harmanyi et al., 1991
Clostridium sporogenes	Inhibits growth at 2% concentration	Unda et al., 1991
Listeria monocytogenes	5% Na-lactate inhibits growth in TSB	Shelef & Yand, 1991
	With modified atmospheric packaging	Zeitoun & Debevere, 1991
	Synergistic inhibition with monolaurin	Oh & Marshall, 1993
Vibrio vulnificus	Cidal activity at 300 – 400 ppm	Sun & Oliver,1994
Helicobacter pylori	PH-dependent growth inhibition	Midolo et al., 1995
Escherichia coli O157:H7	Acid wash with hot water & TSP spray	Dorsa et al., 1998
Pseudomonas sp.	Synergistic inhibition with nisin	Nykanen et al., 1998
Salmonella typhimurium	2% acid with 0.5% cetylpyridinium	Yang et al., 1998

Maas et al. (1989) demonstrated that sodium lactate was effective in preventing growth of *Clostridium botulinum* in raw turkey using 0, 2, 2.5, 3,or 3.5% sodium lactate. The meat was cooked to an internal temperature of 71.1°C and vacuum packaged. Toxicity occurred at 3, 4-5, 4-6, 7, or 7-8 days for each concentration, respectively, at 27°C. Organoleptically, the 2.5 % concentration was acceptable; off flavors developed in meat containing the 3.5 % concentration. Unda et al. (1991) also reported that 2 % lactate inhibited *C. sporogenes*.

Lactate influence the growth of LAB in meat, depending on the pH and atmospheric conditions (Grau, 1980). A 100 mM concentration of lactate buffered to pH 5.5 prevented the anaerobic growth of *Serratia liquefaciens*, *Yersinia enterocolitica*, *Enterobacter cloacae*, and *Aeromonas hydrophila* and aerobic growth of the aeromonads (Grau, 1981). At pH 6.1 the amount of lactate ordinarily found in muscle was not inhibitory to any of the strains.

Papadopoulos et al. (1991a; 1991b; 1991c) injected beef top rounds with 0-4 % sodium lactate, cooked and vacuum packaged the roasts. The aerobic plate counts for control meat increased from 2.7 to 8.0 log CFU/cm^2 after 84 days of storage, but the plate counts of treated roasts increased to 6.5 log CFU/cm^2 in the same period. The microflora changed from predominantly staphylococci and micrococci at 0 days to coryneform, micrococci, lactobacilli, and yeasts at 42 days and to homo- and heterofermentative lactobacilli and streptococci at 84 days. Lactate increased cooking yields and enhanced flavor at the 1 % level by adding a salty taste, but produced a mild throat irritation at a 4 % concentration. Shelef and Yang (1991) suggested a possible a_w effect for sodium lactate, but Papadopoulos et al. (1991b) did not find this effect in their studies.

Lactates have been combined with other compounds in multiple-barrier studies. Lactic acid at 25-30 mEq in combination with sodium nitrite, potassium sorbate, or glycerol monolaurate enhanced inhibition of *S. aureus* grown anaerobically (Smith & Palumbo, 1980). Sodium lactate at the concentration had no effect. Notermans and

Dufrenne (1981), however, found that glycerol monolaurate at a concentration of 5g/kg of a meat slurry inhibited *C. botulinum* Types A, B, and E only when lactic acid was used to reduce the pH to below 5.2.

Harmayani et al. (1991) studied the interactive effects of sodium alginate and calcium or sodium lactate used in restructured meats on *Pseudomonas fragi* and *S. typhimurium*. Inoculated ground beef was stored at 5°C for 5 days. Sodium alginate at concentrations upto 0.8 %, calcium lactate upto 0.3 % or a combination of the two did not affect growth of either organism. A 3 % concentration of sodium lactate moderately affected the growth of *S. typhimurium* but did inhibit *P. fragi* at the 3 and 4 % levels.

Fresh pork sausage formulated with 0-3 % sodium lactate was vacuum packaged and stored for up to 28 days at 4°C (Brewer et al., 1991). Using 10^8 CFU/g as a microbial end point, 1% sodium lactate delayed growth for 10 days, but a 2 % level extended the time to 24 days. Using 4% potassium lactate and 3% sodium chloride reduced growth at 5°C. No synergistic effect was observed between a 4 % concentration of lactate combined with 140 ppm potassium nitrite.

Enteropathogens. *Vibrio vulnificus* is a bacterium indigenous to estuarine waters and is a potent human pathogen. Infections are generally associated with the consumption of raw oyster. Sun and Oliver (1994) found that lactic acid and butylated hydroxyanisole are cidal at 300 ppm and 400 ppm, respectively, within 3 h. The mode of action of lactic acid against *V. vulnificus* appears to be a pH effect.

Helicobacter pylori is a gastric pathogen and etiological agent of peptic ulcers in humans. Lactic, acetic and hydrochloric acids were shown to inhibit the growth of *H. pylori* in a concentration-dependent manner with the lactic acid demonstrating the greatest inhibition (Midolo et al., 1995). This inhibition was due both to the pH of the solution and its concentration. Six strains of *Lb. acidophilus* and one strain of *Lb. casei ssp. rhamnosus* also inhibited *H. pylori* growth where as *Bifidobacterium bifidus*, *Ped. pentosaceus* and *Lb. bulgaricus* did not. Concentrations of lactic acid produced by these strains ranged from 50 to 156 mmol/L and correlated with *H. pylori* inhibition.

Listeria monocytogenes. Shelef and Yand (1991) found that a 5.0 % concentration of sodium or potassium lactate delayed the growth of *L. monocytogenes* is tryptic soy broth. Growth of *L. monocytogenes* decreased by using 4 % lactate in comminuted chicken or beef when held at 35, 20, or 5°C. The authors concluded that the two salts gave comparable results and could be used interchangeably, which suggested that the lactate moiety is the effective component. Lactate contributed to water-holding capacity and increased cooking yields.

Zeitoun and Debevere (1991) studied the effect of treatments with various concentrations (2%, 5% and 10% w/v) of lactic acid/sodium lactate buffer (pH 3.0), modified atmosphere packaging (MAP) (90% CO_2 and 10% O_2) and 10% (w/v) lactic acid/sodium lactate buffer (pH 3.0) combined with MAP on *L. monocytogenes* Z7 serotype 1 and on the shelf life of chicken legs stored at 6°C. The antimicrobial effect of lactic acid buffer systems (pH 3.0) increased with higher concentrations of lactic acid in the buffered system. The best results were obtained by the combined use of 10% acid/sodium lactate buffer (pH 3.0) and modified atmosphere packing.

Oh and Marshall (1993) investigated the antimicrobial effects of glycerol mono-laurate (monolaurin), ethanol and lactic acid, either alone or in combination, against *L. monocytogenes* in tryptic soy broth. Ethanol at concentrations up to 1.25% did not inhibit growth, but a strong inhibition was observed in the presence of 5% ethanol. The MIC values of monolaurin and ethanol alone were 10 μg/ml (0.001%) and 50 mg/ml (5%), respectively. However, MIC values were not changed when monolaurin was combined with ethanol. When 5 μg/ml monolaurin was combined with 5% ethanol, the inhibitory effect of the combination was similar to the most active compound alone after 24 h incubation. These data indicated an interaction between monolaurin and ethanol against *L. monocytogenes*. MIC value of lactic acid alone was 50 mg/ml (0.5%), but was lower when 1.25% ethanol was combined with 0.25% lactic acid. When 2.5% ethanol was combined with 0.25% lactic acid, the combination did not increase the inhibitory effect of the most active single compound alone. This result also indicated an interaction between ethanol and lactic acid.

VI. APPLICATIONS

Lactic acid is a hygroscopic, syrupy liquid with a moderately strong acid taste. Lactic acid is used in the manufacture of jams, jellies, sherbets, confectionery products, and beverages. It is used to adjust acidity in brines for pickles and olives. Calcium lactate is used as a firming agent for apple slices, to prevent discoloration in fruit, and in baking powders.

A. Beef processing

Lactic acid sprays have been effective in limiting microbial growth on meat carcasses under a variety of storage conditions. A 1-1.25% concentration of acid sprayed on veal carcasses, followed by vacuum packing, lowered microbial counts after storage for 14 days at 2°C, but not at 7 days (Smulders & Woolthus, 1983). A 2% concentration discolored the surface of the carcasses (Woolthus & Smulders, 1985). A 2% lactic acid treatment at pH 2.3 of veal tongues, combined with vacuum packing and storage at 3°C, reduced aerobic mesophilic plate counts from 5.6 to 2.7 log CFU/cm^2 (Visser et al., 1988).

Reductions in microbial counts were reported with a 1.25% acid spray of beef carcasses followed by a treatment of hot-boned cuts with 2% acid, vacuum packaging, and storage for 7 and 10 days. Acid treatment when coupled with vacuum packaging was more effective in prolonging shelf life than vacuum packaging alone (Smulders & Woolthus, 1985). *Enterobacteriaceae* contaminated 50% of the untreated samples, which was reduced to 10% after treatment. A study using lactic acid sprays on skinned cow heads found that 1% acid was effective in significantly reducing total counts when product was stored at 4, 15, and 20°C (Cudjoe, 1988). Lactic acid extended shelf life by 3 days at 4°C and 1 day at 15 and 20°C.

Osthold et al. (1984) found that spraying beef and sheep carcasses with a combination of 1% lactic acid, 2% acetic, 0.25% citric, and 0.1% ascorbic acids selectively inhibited *Enterobacteriaeae* at 10°C. Acuff et al. (1987) compared this combination adjusted to pH 2.6 with either 1% lactic acid at pH 2.9 or 1% acetic acid at pH 3.3. There was little difference in bactericidal activity between the combination and either acid alone on subprimal cuts of beef.

Dixon et al. (1987) confirmed their findings using beef strip loins stored in polyvinyl chloride (PVC) film for 6 days or high-oxygen barrier film for 28 days. When temperature changed from 20 to 70°C or lactic and acetic acid concentration increased individually to 3%, aerobic bacterial counts and *S. typhimurium* levels decreased approx. by 1-log for *Enterobacteriaceae*, and 0.5 log for *E. coli* (Anderson & Marshall,1990). The combination of acids were effective as well as 1% lactic or acetic acid alone. Lactic acid was most effective against *S. typhimurium* at 70°C, reducing numbers by approx. 2 log and *Enterobacteriaceae* by 1.5 log (Anderson et al., 1992).

Woolthus et al. (1984) effectively reduced total plate counts and *Enterobacteriaceae* and *Lactobacillaceae* counts by 2-3 log by immersing porcine livers in 0.2% lactic acid for 5 min, vacuum packing, and storing for 5 days at 3°C.

Dorsa et al. (1998) evaluated the long-term efficacy of several beef-carcass surface-tissue wash interventions on the microbiological quality of ground beef. Beef tissue was inoculated with bovine feces containing 4 or 6 log CFU/ml of *E. coli* O157:H7, *Listeria innocua, S. typhimurium*, and *Clostridium sporogenes*. The tissue was subjected to wash treatments with: 2% lactic acid, 2% acetic acid, 12% trisodium phosphate, hot water (74 +/- 2°C at the tissue surface), or water (32 +/- 2°C at the tissue surface). A control group was left untreated. After treatments, the samples were held at 4°C for 24 h and then ground. The ground beef was packaged and incubated at 4°C for 21 days or 12°C for 3 days. Samples treated with acid wash, held at 12°C for 3 days yielded significantly lower aerobic plate counts than the control and also yielded the lowest levels of pseudomonads when compared to other sample groups. After being held at 4°C for 21 days or 12°C for 3 days, samples treated with antimicrobial compounds had lower or no detectable (< 1 CFU/g) levels of *E. coli* O157:H7, *L. innocua, S. typhimurium*, and *C. sporogenes* than beef treated with a water or the control. Ground beef produced from tissue treated with hot water yielded lower populations of these bacteria when compared to water or untreated control beef, but the populations were generally higher than those observed in any of the antimicrobial chemical-treated samples. Results from this study indicated that the use of carcass interventions, especially acid antimicrobial compounds, presently available to the slaughter industry would lower bacterial counts in ground beef.

B. Poultry processing

Lactic acid dips and spray washes of prechilled birds have also been used successfully in the poultry industry. Total microbial numbers from skin of birds immersed for 15 s at 19°C in 1 or 2% lactic acid at pH 2.2 decreased from 5.2 to 3.7 log CFU/g (van der Marel et al., 1988). Levels of psychotrophs decreased from 3.9 to 2.7 log CFU/g and *Enterobacteriaceae* from 3.3 to 2.6 log CFU/g. Higher concentrations of the acid did not ensure greater decontamination. A 10% lactic acid and sodium lactate-buffered acid spray (pH 3.0) for chicken legs increased the shelf life from 6 to 12 days at 60°C, and a 2% lactic acid dip at pH 2.3 prolonged shelf life to 8 days (Zeitoun & Debevere, 1990). These treatments inhibited hydrogen sulfide-producing spoilage bacteria, such as the *Pseudomonas spp.*

Lactic acid, chlorine, lactic acid, trisodium phosphate (TSP) and a commercial phosphate blend (Avgard) were evaluated for their potential bactericidal effects on fecally contaminated turkey carcasses (Bautista et al., 1997). Carcasses were sprayed for 10 seconds with each antimicrobial, at various concentrations and pressure combinations.

Lactic acid at various concentrations showed a significant effect in reducing total bacteria and coliform counts. The results indicate that lactic acid at 4.25% (w/w) has the potential for reducing the total microbial load and coliforms by more than 95%. Chlorine, TSP and Avgard concentration did not affect the microbial load when compared with a water spray (control). Preliminary presumptive testing indicated that lactic acid and Avgard had some effect against *Salmonella* spp. These findings suggest that lactic acid is an effective bactericide for reducing microbial contamination and improving the safety of poultry meat.

The antimicrobial efficacy of lactic acid (2%), trisodium phosphate (TSP, 10%), cetylpyridinium chloride (CPC, 0.5%), and sodium bisulfate (SBS, 5%) to control *Salmonella typhimurium* in prechilled poultry spray was evaluated (Yang et al., 1998). Each chicken carcass was inoculated by spraying the outside and inside of each carcass with *S. typhimurium* at 10^5 CFU/carcass. The inoculated carcasses then were passed through the birdwasher and sprayed with selected chemicals at 35 °C for 17 seconds. After a 60-second setting time on a shackle line, the carcasses were sprayed with tap water to rinse off chemical residue. All the chemical treatments reduced *Salmonella* on the chicken carcasses by approximately 2 $\log_{10}$ CFU per carcass. Total aerobes on the chicken carcasses, however, were reduced by 2.16, 1.66, 1.03, and 0.74 $\log_{10}$ CFU per carcass after spraying with 0.5% CPC, 5% SBS, 2% lactic acid, or 10% TSP, respectively. Spray treatments of both SBS and lactic acid caused slight discoloration in part of the chicken skin.

VII. BIOTECHNOLOGY

As indicated earlier, numerous microorganisms produce lactic acid as the main metabolite (homofermentation) or mixed with other by-products (heterofermentation). The utilization of these strains has led to the commercial production of large quantities of lactic acid. This method is considered superior to chemical synthesis process since the latter method inevitably produces a racemic mix of the two optical isomers in equal volumes. Fungi (*Rhyzopus Oryzae*) and lactic acid bacteria are now being used worldwide to manufacture bulk qunatities of lactic acid. The basic industrial operation and certain important technological choices for bulk synthesis of lactic acid is briefly described.

Lactic acid is secreted into the milieu during the log phase as well as the stationary phase of microbial growth as shown by the Luedeking and Piret (1959) kinetic model according to the equation below:

$$\delta P / \delta t = \alpha . \delta X / \delta t + \beta . X$$

This equation indicates that the lactic acid production velocity, or instant productivity (dP/dt) is proportional to the instant growth speed (dX / dt) but also to the cell population (X) whether multiplying or stationary. Factors **a** and **b** are specific constant coefficients for the strain under consideration for temperature and pH conditions. This model has been modified in several instances (Friedman, 1970; Hanson, 1972; Rogers, 1978; Jorgensen, 1987). In addition, a very complete model was suggested by Aborhey and Williamson (1977) which takes into account, among others, the lactic acid inhibition.

The above kinetic model is suitable for batch fermentation systems. However, various laboratories have investigated continuous fermentation techniques to attempt to shorten non-productive operations which are inevitably related to batch systems (filling, lag phase, emptying, etc). Continuous systems make use of different processes such as the immobilization or inclusion of entire microorganism in or at the surface of a substructure

(Stenroos, 1982; Boyaval, 1985; Roy, 1986), cell recycling using centrifugation or making use of membrane techniques (ultra-filtration, micro-filtration)(Prigent, 1984; Boyaval, 1988; de Raucourt, 1989).

Lactic acid is purified either by precipitation as metal lactates followed by a neutralization with sulfuric acid (Maesato, 1985) or by esterification with alcohol, distillation and hydrolysis of the formed ester (Boroda, 1966), or by electro-dialysis (Jacquement, 1968). A more recent purification process to extract lactic acid by liquid-liquid extraction using an organic solvent not miscible in water with a Lewis base such as a tertinary amine has been developed. With this process, lactic acid could be recovered in a second step with a liquid-liquid back-extraction. This step allows to re-transfer the lactic acid to water (Baniel, 1972; Baniel, 1980). Finally, lactic acid in acid- and/or ammonium lactate or metal lactate-form may be purified by processing on cationic and/or anionic ion-exchange columns (Napierala, 1972; Maesato, 1985; Shkurino, 1986; Obara, 1987a; Obara, 1987b; Zelenava, 1982).

It should be emphasized that these purification steps generally start with solutions of diluted lactic acid in water. As explained earlier, this step is introduced because the bearing molecule structure is made of an hydroxyl function and a carboxylic acid group. Indeed this double functionality is at the origin of condensation reactions producing lactic acid oligomers. This explains why lactic acid was, for a long time, thought of as a weak volatile substance and one which can not be distilled under atmospheric pressure. In fact, lactic acid condenses to form oligomers with a high boiling point at this pressure. Certain studies on lactic acid distillation by vapor at 160-200 °C show that it is possible to distillate with yields of about 75 to 85%. However such drastic conditions are detrimental to the product quality, degradation and possible racemization. Certain variants of this process have been suggested (Sepitka, 1962; Shishkini, 1977). Current industrial-scale processes are however based on esterification followed by distillation and hydrolysis as well as on liquid-liquid extraction using a tertinary amine and calcium lactate crystallization with sulfuric acid acidification plus gypsum (calcium sulfate) filtration.

VIII. SAFETY AND TOLERANCE

Most mammals, including humans, tolerate daily oral intakes of lactic acid higher than 1500 mg/kg of body weight. Values for LD_{50} vary from 1810 mg/kg for the guinea pig and up to more than 4800 mg/kg for the mouse (WHO, 1967). Various studies show that no lactic acid accumulation is recorded in the rats fed with 1 to 2 g/kg of body weight during 14 to 16 days. Similarly, dogs have tolerated oral doses of 600 to 1600 mg of lactic acid per kg of body weight when ingested 42 times over a 10-weeks time-span (WHO, 1967).

Due to the absence of toxicological contra-indication, lactic acid and lactates have been long accepted as GRAS (*Generally Recognized As Safe*) food additives. In the United States, lactic acid is approved as a GRAS substance for miscellaneous or general purpose usage (21 CFR 184.1061). Calcium (21 CFR 184.1207), potassium (21 CFR 184.1639), and sodium lactates (21 CFR 184.1768) are also approved. The United States Food and Drug Administration is in the process of developing specifications for sodium lactate. The acceptable daily human intake of lactic acid and calcium, potassium, ammonium and sodium salts of lactate is unlimited, however, for recemic mixtures of D/L-lactic acid [upper limit: D(-) lactic acid] is limited to 0-100 mg/kg body weight (FAO, 1965; FAO, 1973).

In Europe, the 95/2/CE Directive, published in the March 18th, 1995 issue of the Official Gazette, deals with the use of additives by the food industry. In the case of lactic acid, sodium, potassium and calcium lactates, the *quantum satis* principle is prescribed i.e. the necessary and adequate quantities may be added to reach the desired effect but not more.

IX. SUMMARY

Lactic acid and its derivatives elicit a broad-spectrum of antimicrobial activity against Gram-positive bacteria including spore-forming *Clostridia* and *Bacillus* sp., as well as Gram-negative pathogens such as *E.coli O157:H7* and *Salmonella* sp. Lactate compounds also demonstrate antifungal activity against aflatoxin-producing *Aspergillus* sp. Considering these attributes, lactic acid and lactates are widely used in the food industry for decontamination of meat foods such as beef, poultry and pork during processing and packaging. They are also used in shelflife enhancement of fresh and semi-processed foods. Consumer acceptance of such processed foods, however, is largely based on visual and organoleptic qualities. Currently, the worldwide utility of lactic acid and its derivatives amounts to 100,000 metric tons per year.

X. REFERENCES

1. Aborhey, S. and Williamson, D. 1977. Modelling of lactic acid production by *Streptococcus cremoris J. Gen. Appl. Microbiol.* 23:7-21.
2. Acuff, G.R., Vanderzant, C., Savel, J.W., Jones, D.K., Griffin, D.B. and Ehlers, J.G. 1987. Effect of acid decontamination of beef subprimal cuts on the microbiological and sensory characteristics of steaks. *Meat Sci.* 19:217-226.
3. Adams, M.R. and Hall, C.J. 1988. Growth inhibition of food-borne pathogens by lactic and acetic acids and their mixtures. *Int. J. Food Sci. Technol.* 23:287-292.
4. Alpert, N.R. 1965. Lactate production and removal and the regulation of metabolism. *Ann. N.Y. Acad. Sci.* 119:995-1012.
5. Anderson, M.E. and Marshall, R.T. 1990. Reducing microbial populations on beef tissues: Concentration and temperature of an acid mixture. *J. Food Sci.* 55:903.
6. Anderson, M.E., Marshall, R.T. and Dickson, J. S. 1992. Efficacies of acetic, lactic and two mixed acids in reducing numbers of bacteria on surfaces of lean meat. *J. Food Safety* 12:139.
7. Anonymous. 1987. Lactic acid §184.1061, Sodium lactate §184.1768; Potassium lactate §184.1639; Calcium lactate §184.1061. *Code of Federal Regulations* No. 21. U.S. Gov. Printing Office, Washington, DC.
8. Baird-Parker, A.C. 1980. Organic acids. In *Microbial Ecology of Food, Vol. 1, Factors Affecting Life and Death of Microorganisms,* ed. Silliker, J.H., Elliot, R.P., and A.C. Baird-Parker, pp.126-135. London: Academic Press.
9. Ballabriga, A., Conde, C. and Gallart-Catala, A. 1970. Metabolic response of prematures to milk formulas with different lactic acid isomers or citric acid. *Helv. Paediatr. Acta.* 25:25.
10. Baniel, A.M. 1980. MILES Lab. EP 49:429.
11. Baniel, A.M., Blumberg, R. and Hadju, K. 1972. IMI. DE 2,329,480.
12. Bautista, D.A., Sylvester, N., Barbut, S., and Griffiths, M.W. 1997. Determination of efficacy of antimicrobial rinses on turkey carcasses using response surface designs. *Int. J. Food Microbiol.* 34:279-292.
13. Bergmeyer, H.U. 1974. *Methods of Enzymatic Analysis,* Vols. 1 and 2, Academic Press, New York.
14. Berntein, I.A. 1953. Fermentation of Ribose-C[14] by *Lactobacillus pentosus. J. Biol. Chem.* 205:309-316.
15. Bevilacqua, A.E. and Califano, A.N. 1989. Determination of organic acids in dairy products by high performance liquid chromatography. *J. Food Sci.* 54:1076

16. Booth, I.R. 1985. Regulation of cytoplasmic pH in bacteria. *Microbiol. Rev.* 49:359-378.

17. Bouzas, J., Kantt, C.A., Bodyfelt, F. and Torres, J.A. 1991. Simultaneous determination of sugars and organic acids in chedder cheese by high-performance liquid chromatography. *J. Food Sci.* 56:276.

18. Boyaval, P. and Goulet, J. 1988. Optimal conditions for production of lactic acid from cheese whey permeate by calcium alginate-entrapped *Lactobacillus helveticus*. *Enzyme Microb. Technol.* 10:725-728.

19. Boyaval, P., Lebrun, A., and Goulet, J. 1985. Immobilization of *Lactobacillus helveticus* in calcium alginate beads. *Lait.* 65:185-199.

20. Brackett, E. R. 1987. Effects of various acids on growth and survival of *Yersinia enterocolitica*. *J. Food Protection*. 50:598-601.

21. Bradford, D.D., Huffman, D.L., Egbert, W.R. and Jones,W.R. 1993. Low-fat fresh pork sausage patty stability in refrigerated storage with potassium lactate, *J. Food Sci.* 58:488-491.

22. Brecht, K. 1967. Muskelphysiologie. In KEIDEL: *Physiologie* pp. 262-294. Stuttgart:Thieme Verlag.

23. Brewer, M.S., McKeith, F., Martin, S.E., Dall-Mier, A.W. and Meyer, J. 1991. Sodium lactate effects on shelf-life, sensory and physical characteristics of fresh pork sausages. *J.Food Sci.* 56:1176-1178.

24. Brewer, M.S., McKeith, F.K. and Sprouls, G. 1993. Sodium lactate effects on microbial, sensory, and physical characteristics of vacuum-packaged pork sausages. *J. Muscle Foods*. 4:179-192.

25. Brewer, M.S., Rostogi, B.K., Argoudelis, L. and Sprouls, G.K. 1995. Sodium lactates/sodium chloride effects on aerobic plate counts and color of aerobically packaged ground pork. *Food Sci.* 60:58-62.

26. Brin, M. 1965. Lactic acid - some definitions. *Ann. N.Y. Acad. Sci.* 107:1084-1090.

27. Bullerman, L. B. 1981. Public health significance of molds and mycotoxins in fermented dairy products. *J. Dairy Sci.* 64:2439-2452.

28. Bullerman, L. B. 1986. Mycotoxins and food safety. *Food Technol.* 40: 59-66.

29. Bullerman, L.B., Hartman, P.A., and Ayres, J.C. 1969a. Aflatoxin production in meats. I. Stored meats. *Appl. Microbiol.* 18:714-717.

30. Bullerman, L.B., P. A. Hartman, P.A., and Ayres, J.C. 1969b. Aflatoxin production in meats. II. Aged dry salami and aged country cured hams. *Appl. Microbiol.* 18:718-722.

31. Bushway, R.J., Bureau, J.L.and McGann D.F. 1984. Determinations of organic acids in potatoes by high performance liquid chromatography. *J. Food Sci.* 49:75

32. Cassio, F., Leao, C. and Van Uden, N. 1987. Transport of lactate and other short-chain monocarboxylates in the yeast *Saccharomyces cerevisiae*. *Appl. Environ. Microbiol.* 53:509-513.

33. Chahl, S.P.1990. Lactic Acid. *Ullmann's Encyclopedia of Industrial Chemistry*.5th ed., Vol. 15A, pp.97. Stuttgart: VCH.

34. Chen, N. and Shelef, L.A. 1992. Relationship between water activity, lactate and growth of *Listeria monocytogenes* in a meat model system. *J. Food Prot.* 55:574-578.

35. Chirife, J. and Ferro Fontan, C. 1980. Prediction of water activity of aqueous solutions in connection with intermedite moisture foods: Experimental investigation of the a_w lowering behavior of sodium lactate and some related compounds. *J. Food Sci.* 45:802-804.

36. Chung, K.C. and Goepfert, J.M. 1970. Growth of Salmonella at low pH. *J. Food Sci.* 35:326-328.

37. Conner, D.E., Scott V.N., and Bernard, D.T. 1990. Growth inhibition and survival of *Listeria monocytogenes* as affected by acidic conditions. *J. Food Prot.* 53:652-655.

38. Cudjoe, K.S. 1988. The effect of lactic acid sprays on the keeping qualities of meat during storage. *Int. J. Food Microbiol.* 7:1

39. Cudjoe, K.S. 1988. The effect of lactic acid sprays on the keeping qualities of meat during storage. *Int. J. Food Microbol.* 7:1-7.

40. De Raucourt, A., Girard, D., Prigent, Y., and Boyaval,P. 1989b. Lactose continuous fermentation with cells recycled by ultrafiltration and lactate separation by electrodialysis model identification. *Appl. Microbiol. Biotechnol.* 30:528-534.

41. De Raucourt, A.., Girard, D., Prigent,Y., and Boyaval, P. 1989a. Lactose continuous fermentation with cells recycled by ultrafiltration and lactate separation by electrodialysis modelling and simulation. *Appl. Microbiol. Biotechnol.* 30:521-527.

42. De Vrese, M., Koppenhoefer, B., and Barth, C.A. 1990. D-lactic acid metabolism after an oral load of DL-lactate. *Clin. Nutr.* 9:23-28.

43. De Wit, J.C. and Rombouts, F.M. 1990. Antimicrobial activity of sodium lactate. *Food Microbiol.* 7:113-120.

44. Debevere, J.M. 1989. The effect of sodium lactate on the shelf-life of vacuum-packed coarse liver pâté. *Fleischwirtsch.* 69:223-224.

45. Dickson, J.S. and Maynard, E.A. 1992. Microbiological decontamination of food animal carcasses by washing and sanitizing systems : A review, *J. Food Protection*. 55:133-140.

46. Dietz, A.A., Derening. E.F. and Schopmeyer, H.H. 1941. Physical properties of sodium, potassium, and ammonium lactate solutions. *Ind. Eng. Chem.* 33:1444-1447.

47. Dixon, Z.R., Vanderzant, C., Acuff, G.R., Savell, J.W. and Jones, D.K.1987. Effect of acid treatment of beef trip loin steaks on microbiological and sensory characteristics, Int. *J. Food Microbiol.* 5:181-186.

48. Dobbin, L. 1931. The collected papers of Carl Wilhelm Scheele. London:G. Bell and Sons.

49. Dorsa, W.J. and Marshall, D.L. 1995. Influence of lactic acid and modified atmosphere on thermal destruction of *Listeria monocytogenes* in crawfish tall meat homogenate. *J. Food Saf.* 15:1-9.

50. Dorsa, W.J., Cutter, C.N., and Siragusa, G.R. 1998. Bacterial profile of ground beef made from carcass tissue experimentally contaminated with pathogenic and spoilage bacteria before being washed with hot water, alkaline solution, or organic acid and then stored at 4 or 12 degrees C. *J. Food Prot.* 61:1109-1118.

51. Dubus, R.J. 1950. The effect of organic acids on mammalian tubercle bacilli. *J. Exp. Med.* 92:319.

52. Egnert, W.R., Huffman, D.L., Bradford, D.D. and Jones, W.R. 1992. Properties of low-fat ground beef containing potassium lactate during aerobic refrigerated storage. *J. Food Sci.* 57:1033-1037.

53. Eklund, T. 1980. Inhibition of growth and uptake processes in bacteria by some chemical food preservatives. *J. Bacteriol.* 48:423-432.

54. Eklund, T. 1983. The antimicrobial effect of dissociated and undissociated sorbic acid at different pH levels. *J. Appl. Bacteriol.* 54:383-389.

55. El–Gazzar, E.F., Rusul, G. and Marth, E.H. 1987. Growth and aflatoxin production by *Aspergillus parasiticus* NRRL 2999 in the presence of lactic acid and at different initial pH values. *J. Food Protection.* 50:940-944.

56. Epling, L.K., Carpenter, J.A. and Blankenship, L.C. 1993. Prevalence of *Campylobacter spp.* and *Salmonella spp.* on pork carcasses and the reduction effected by spraying with lactic acid. *J. Food Prot.* 53:536-540.

57. Farber, J.M., Sanders, G.W., Dunfield, S. and Prescott, R. 1989. The effect of various acidulants on the growth of *Listeria monocytogenes*. *Lett. Appl. Microbiol.* 9:181-183.

58. Freese, E., Sheu C.W. and Galliers, E. 1973. Function of lipophilic acids as antimicrobial food additives. *Nature.* 241:321-325.

59. Friedman, M.R. and Gaden, E.J. 1970. Growth and acid production by *Lactobacillus delbrueckii* in a dialysis culture system. *Biotechnol. Bioeng.* 12:961.

60. Garvie, E.I. 1967. The production of L - (+) and D - (-) Lactic acid in cultures of some lactic acid bacteria,with a special study of *lactobacillus acidophilus*. *J. Dairy Res.* 34:31-37.

61. Geigy, J.R. 1968. A-G/ documenta Geigy. *Wissenschaftliche tsabellen* 7.

62. Granados, H., Glavind, J.and Dam, H. 1949. Observations on experimental dental caries. III. The effect of dietary lactic acid. *J. Dent. Res.* 28:282.

63. Grau, F. H. 1981. Role of pH, Lactate, and anaerobiosis in controlling the growth of some fermentative gram-negative bacteria on beef. *Appl. Environ. Microbiol.* 42:1043.

64. Grau, F.H. 1980. Inhibition of the anaerobic growth of *Brochothrix thermosphacta* by lactic acid. *Appl. Environ. Microbiol.* 40:433.

65. Hamby, P.L., Savell, J.W., Acuff, G.R., Vanderzant, C. and Cross, H.R . 1987. Spray-chilling and carcass decontamination systems using lactic and acetic acid. *Meat Sci.* 21:1-14.

66. Hanson, T.P. and Tsao, G.T. 1972. Kinectic studies of the lactic acid fermentation in batch and continuous cultures. *Biotechnol. Bioeng.* 14:233-252.

67. Harmayani, E. Sofos, J. N. and Schmidt. G.R. 1991. Effect of sodium lactate, calcium lactate and sodium alginate on bacterial growth and animopeptidase activity. *J. Food Safety.* 11:269.

68. Helrich, K., ed. 1990. *Official Methods of Analysis of the Association of Official Analytical Chemists*, 15th ed., Association of Official Analytical Chemists. Washington, D.C.

69. Hesseltine, C.W., Shotwell, O.L., Ellis, J.J. and Stubblefield, R.D.1966. Aflatoxin formation by *Aspergillus flavus. Bacteriol. Rev.* 30:795-805.

70. Hiyama, T., Fukui, S. and Kitihara, K. 1968. Purification and properties of lactate racemase from *Lactobacillus sake*. *J. Biochem. Tokyo.* 64:99-107.

71. Hohorst, H.J., Arese, P., Bartels, H., Stratmann, D. and Talke, H. 1965. Lactic acid and the steady state of cellular red/ox-systems. *Ann. N.Y. Acad. Sci.* 119:974-994.

72. Holmquist, G.U., Walker, H.W., and Stahr, H.M. 1983. Influence of temperature, pH; water activity and antifungal agents on growth of *Aspergillus flavus* and *A. parasiticus. J. Food Sci.* 43:778-782.

73. Holten, C.H. 1971. Lactic Acid. Weinheim:Verlag Chemie.

74. Houtsma, P.C., De Witt, J.C. and Rombouts, F.M. 1993. Minimum inhibitory concentration (MIC) of sodium lactate for pathogens and spoilage organims occuring in meat products. *Int. J. Food Microbiol.* 20:247-257.

75. Huckabee, W.E. 1958. Relationships of Pyruvate and lactate during anaerobic metabolism, I. Effects of infusion of pyruvate or glucose and of hyperventilation. *J. Clin. Invest.* 37:244-254.

76. Hunter, D.R. and Segal, I.H. 1973. Effect of weak acids on amino acid transport by *Penicillium chrysogenum*: Evidence for proton or charge gradient as the driving force. *J. Bacteriol.* 113:1184-1192.

77. Ita, P. and Hutkins, R.W. 1991. Intracellular pH and survival of *Listeria monocytogenes* Scott A. in tryptic soy broth containing acetic, lactic, citric and hydrochloric acids. *J. Food Prot.* 54:15-19

78. James, A.T. and Martin, A.J.P. 1952. Gas-liquid partition chromatography: The separation and micro-estimation of volatile fatty acids from formic acid to dodecanoic acid. *Biochem. J.* 50:679.

79. Kabara, J.J. 1991. Phenols and chelators. In *Food Preservatives,* ed. Russell, N. J. and Gould, G.W., pp. 200-214. Glasgow:Blackie Publishers.

80. Kolsarici, N. and Candogan, K. 1995. The effects of potassium sorbate and lactic acid on the shelf-life of vacuum-packed chicken meat. *Poult. Sci.* 74:1884-1893.

81. Lamkey, J.W., Leak, F.W., Tuley, W.B., Johnson, D.D. and West, R.L. 1991. Assessment of sodium lactate addition to fresh pork sausage. *J. Food Sci.* 56:220-223.

82. Lie, J.L. and Marth, E.H. 1967. Formation of aflatoxin in cheddar cheese by *Aspergillus flavus* and *Aspergillus parasiticus*. *J. Dairy Sci.* 50:1708,1710.

83. Lissac, J. and Amstutz, P.H. 1966. L'acidose lactique. Acidoses metab. *Meningitis reanim.* 25-48.

84. Loncin, M. 1975. Basic principles of moisture equilibria In *Freeze Drying and Advanced Food Technology,* ed. Goldblith, S.A., Rey, L. and Rothmayr, W.W., pp. 599-617. New York:Academic Press.

85. Luchese, R.H. and Harrigan, W.F. 1990. Growth of, and aflatoxin production by *Aspergillus parasiticus* when in the presence of either *Lactococcus lactis* or lactic acid and at different initial pH values. *J. Appl. Bacteriol.* 69:512-519.

86. Lück, E.1980. *Antimicrobial Food Preservatives.* Berlin:Springer-Verlag.

87. Maas, M.R., Glass, K.A. and Doyle, M.P. 1989. Sodium lactate delays toxin production by *Clostridium botulinum* in cook-in-bag turkey products. *Appl. Environ. Microbiol.* 55:2226.

88. Marshall, D.L. and Kim, C.R. 1996. Microbiological and sensory analyses of refrigerated catfish fillets treated with acetic and lactic acids. *J. Food Qual.* 19:317-329.

89. Marsili, R.T., Ostapenko, H., Simmons, R.E. and Green, D.E. 1981. High performance liquid chromatography determination of organic acids in dairy products. *J. Food Sci.* 46:52.

90. Martin, G.E., Sullo J. G. and Schoeneman, R.L. 1971. Determination of fixed acids in commercial wines by gas-liquid chromatography. *J. Agric. Food chem.* 19:995.

91. Mendonca, A.F., Molins, R.A., Kraft, A.A. and Walker, H.W. 1989. Microbiological, chemical, and physical changes in fresh, vacuum-packaged pork treated with organic acids and salts. *J. Food Sci.* 54:18-21.

92. Meyerhof, O. 1930. *Die chemischen Vorgänge im Muskel.* Berlin: Springer Verlag.

93. Midolo, P.D., Lambert, J.R., Hull, R., Luo, F., and Grayson, M.L. 1995. In vitro inhibition of *Helicobacter pylori* NCTC 11637 by organic acids and lactic acid bacteria. *J. Appl. Bacteriol.* 79:475-479.

94. Miller, A.J. 1992. Combined water activity and solute effects on growth and survival of *Listeria monocytogenes* Scott A. *J. Food Protect.* 55:414-418.

95. Minor, T.E. and Marth, E.H. 1970. Growth of *Staphylococcus aureus* in acidified pasteurized milk. *J. Mild Food Technol.* 33:516-520.

96. Mitchell, P. 1961. Coupling of phosphorylation to electron and hydrogen transfer by a chemiosmotic type of mechanism. *Nature.* 191:144-148.

97. Mitchell, P. and Moyle, J. 1969. Estimation of membrane potential and pH difference across the cristae membrane of rat liver mitochondria. *Eur. J. Biochem.* 7:471-484.

98. Napierala, W. and Siminski, M.1972. Production of alimentary lactic acid of high purity. *Przem. Ferment. Rolny.* 16:4-10.

99. Nassos, P.S., Schade, J.E., King, A.D., Jr. and Stafford, A.E. 1984. Comparison of HPLC and GC methods for measuring lactic acid in ground beef. *J. Food. Sci.* 49:671.

100. Nnanna, I.A., Ukuku, D.O., Mc Vann, K.B. and Shelef, L.A. 1994. Antioxidant activity of sodium lactate in meat and a model system. *Lebensm-Wissensch und Technol.* 27:78-85.

101. Nolan, D.A., Chamblin, D.C. and Troller, J.A. 1992. Minimal water activity levels for growth and survival of *Listeria monocytogenes* and *Listeria innocua*. *Int. J. Food Microbiol.* 16:323-325.

102. Notermans, S. and Dufrenne, J. 1981. Effect of glyceryl monolaurate by *Clostridium botulinum* in meat slurry. *J.Food Safety.* 3:83.

103. Nykanen, A., Vesanen, S., and Kallio, H. 1998. Synergistic antimicrobial effect of nisin whey permeate and lactic acid on microbes isolated from fish. *Lett. Appl. Microbiol.* 27:345-348.

104. O'Connor, P.L., Brewer, M.S., McKeith, F.K., Novakofski, J.E. and Carr, T.R. 1993. Sodium lactate-sodium chloride effects on sensory characteristics and shelf-life of fresh ground pork. *J. Food Sci.* 58:978-986.

105. Obara, H. 1987 Shimadzu corp., JP 0,191,788.

106. Obara, H. 1987. Shimadzu corp., JP 63,188,632.

107. Oh, D.H., and Marshall, D.L. 1993. Antimicrobial activity of ethanol, glycerol monolaurate or lactic acid against *Listeria monocytogenes. Int. J. Food Microbiol.* 20:239-246.

108. Osthold, W., Shin, H.K., Dresel, J. and Leistner, L. 1984. Improving the storage life of carcasses by treating their surfaces with an acid spray. *Fleischwirtschaft.* 64:828.

109. Papadopoulos, L.S., Miller, R.K., Acuff, G.R,. Vanderzant, C. and Cross, H.R. 1991b. Effect of sodium lactate on microbial and chemical composition of cooked beef during storage. *J. Food Sci.* 56:341.

110. Papadopoulos, L.S., Miller, R.K., Ringer, L.J. and Cross, H.R. 1991a. Sodium lactate effect on sensory characteristics, cooked meat color and chemical composition. *J. Food Sci.* 56:621.

111. Papdopoulos, L.S., Miller, R.K., Acuff, G.R., Lucia, L.M., Vanderzant, C. and Cross, H.R. 1991. Consumer and trained sensory comparisons of cooked beef top, rounds treated with sodium lactate. *J. Food Sci.* 56:1141-1153.

112. Park, H.S., Marth, E.H. and Olson, N.F. 1970. Survival of *Salmonella typhimurium* in cold-pack cheese food during refrigerated storage. *J. Milk Food Technol.* 33:383.

113. Pitkin, C.E. 1935. Lactic acid stricture of esophagus. *Ann. Otol. Rhinol. Laryngol.* 44:842.

114. Prasai, R.K., Acuff, G.R., Lucia, L.M., Hale, D.S., Savell, J.W. and Morgan, J.B. 1991. Microbiological effects of acid decontamination of beef carcasses at various locations in processing. *J. Food Prot.* 54:868-872.

115. Prigent, Y. and Franco, A. 1984. Study of lactic membrane fermenter - extraction of sodium lactate included in supplemented lactose fermentation liquor using ultrafiltration and electrodialysis. *Lait.* 64:217-238.

116. Reichard, G.A., Moury, N.P., Hochella, N.J., Patterson, A.L. and Weinhouse, E. 1963. Quantative estimation of the cori cycle in the human. *J. Biol. Chem.* 238: 495-501.

117. Restaino, L., Komatsu, K.K., and Syracuse, J. 1981. Effects of acids on potassium sorbate inhibition of food-related microorganisms in culture media. *J. Food Sci.* 47:134-143.

118. Restaino, L., Lenvich, L.M., and Bills, S. 1982. Effect of acids and sorbate combinations on the growth of four osmophilic yeasts. *J. Food Prot.* 45: 1138-1142.

119. Rice, A.C., and Pederson, C.S. 1954. Factors affecting growth of *Bacillus coagulans* in canned tomato juice. II. Acidic constituents of tomato juice and specific organic acids. *Food Res.* 19:124-131.

120. Rogers, P.L., Bramall L. and Mc Donald, I.J. 1978. Kinetic analysis of batch and continuous culture of *Streptococcus cremoris. Can. J. Microbiol.* 24:372-380.

121. Roy D. 1986. Ph. D. Thesis. University of Laval, Quebec, Canada.

122. Rubin H.E. 1978. Toxicological model for a two-acid system. *Appl. Environ. Microbiol.* 36:623-624.

123. Salmond, C.V., Kroll R.G. and Booth, I.R.1984. The effects of food preservatives on pH homeostasis in *Escherichia coli. J. Gen. Microbiol.* 130: 2845-2850.

124. Schimassek, H. 1965. Lactate metabolism in the isolated perfused rat liver. *Ann. Sci.* 119:1013-1028.

125. Shelef, A.L. 1994. Antimicrobial effects of lactates : A Review. *J. Food Protection.* 57:445-450.

126. Shelef, L.A. and Yang, Q. 1991. Growth suppression of *Listeria monocytogenes* by lactates in broth, chicken and beef. *J. Food Prot.* 54:283.

127. Sheu, C.W., Salomon, D., Simmons, J.L., Sreevalsan, T. and Freese, E. 1975. Inhibitory effect of lipophilic acids and related compounds on bacteria and mammalian cells. *Antimicrob. Agents Chemother.* 7:349-63.

128. Shkurina, O.V. and Dauksha, V.E. 1986. Investigation of calcium lactate mother liquor conversion with ionites. *Khim-Farm.Zh.* 20:1375-1377.

129. Smith, J.L. and Palumbo, S.A. 1980. Inhibition of aerobic and anaerobic growth of *Staphylococcus aureus* in a model sausage system. *J. Food Safety.* 2:221.

130. Smulders, F.J.M. and Woolthus, C.H.J. 1983. Influence of two levels of hygiene in the microbiological condition of veal as a product of two slaughtering/processing sequences. *J. Food Prot.* 46:1032.

131. Smulders, F.J.M. and Woolthus, C.H.J. 1985. Immediate and delayed microbiological effects of lactic acid decontamination of calf carcasses-influence on conventionally boned versus hot-boned and vacuum-packaged cuts. *J. Food Prot.* 48:838.

132. Snijders, J.M.A., Van Logtestijn, J.G., Mossel, D.A.A. and Smulders, F.J.M. 1985. Lactic acid as a decontaminant in slaughter and processing procedures. *The Quarterly Veterinary* 277-282.

133. Stenroos, S.L., Linko, Y.Y. and Linko, P. 1982. Lactic acid fermentation with immobilized *Lactobacillus* sp. *Biotechnol.* Lett. 4:159.

134. Strzelecki, E., Lillard, H. S. and Ayres, J.C. 1969. Country cured hams as a possible source of aflatoxin. *Appl. Microbiol.* 18:938-939.

135. Subramanian, C.S. and Marth, E.H. 1968. Growth of *Salmonella typhimurium* in acidified milks. *J. Dairy Sci.* 51:934.

136. Sun, Y., and Oliver, J.D. 1994. Antimicrobial action of some GRAS compounds against *Vibrio vulnificus*. *Food Addit. Contam*. 11:549-558.
137. Sutic, M.J., Ayres, J.C., and Koehler, P.E. 1972. Isolation and aflatoxin production of molds isolated from country cured hams. *Appl. Microbiol*. 23: 656-658.
138. Ten Brink, B. and Konings, W.N. 1980. Generation of an electrochemical proton gradient by lactate efflux in membrane vesicles of *Escherichia coli*. *Eur. J. Biochem*. 11:59-66.
139. Ten Brink, B., Otto, R., Hansen, U.P and Konings, W.N. 1985. Energy recycling by lactate efflux in growing and nongrowing cells of *Streptococcus cremoris*. *J. Bacteriol*. 162:383-390.
140. Tijan, G.H. and Jansen, J.T.A. 1970. Identification of acetic, propionic, and sorbic acids in bakery products by thin layer chromatography. *J. Assoc. Off. Anal. Chem*. 54:1150
141. Togami, D.W., Treat-Clemons, L.G. and Hometchko, D.J. 1990. Separation of organic acids in food products. *Am. Lab*. June:15.
142. Tonogai, Y., Kingkate, A., Thanissorn, W. and Punthanaprated, U. 1983. Gas chromatographic determination of propionic acid and sodium and calcium propionate in bread and cake. *J. Food Prot*. 46:284.
143. Trainer, J.B., Krippaehne, W.W., Hunter, W.C. and Lagozzino, D.A. 1945. Esophageal stenosis due to lactic acid. *Am. J. Dis. Child*. 69:173.
144. Tsuji, S., Tonogai, Y. and Ito, Y. 1986. Rapid determination of mono, di- and tri- isopropyl citrate in foods by gas chromatography. *J. Food Prot*. 49:914.
145. Unda, J.R., Molins, R.A. and Walker, H.W. 1991. *Clostridium sporogenes* and *Listeria monocytogenes*: Survival and inhibition in microwave-ready beef roasts containing selected antimicrobials. *J. Food Sci*. 56:198.
146. Van Der Marel, G.M., De Vries, A.W., Van Logtestijn, J.G. and Mossel, D.A.A. 1989. Effect of lactic acid treatment during processing on the sensory quality and lactic acid content of fresh broiler chickens. *Int. J. Food Sci. Technol*. 24:11-16.
147. Van der Marel, G.M., van Logtestijn, J.G. and Mossel, D.A.A. 1988. Bacteriological quality of broiler carcasses as affected by in-plant lactic acid decontamination. *Int. J. Food Microbiol*. 6:31.
148. Vanderzant, C. and Nickelson, R. 1969. A microbiological examination of muscle tissue of beef, pork and lamb carcasses. *J. Milk Food Technol*. 32:357.
149. Visser, I.J.R., Koolmees, P.A. and Bijker, P.G.H. 1988. Microbiological conditions and keeping quality of veal tongues as affected by lactic acid decontamination and vacuum packaging. *J. Food Prot*. 51:208-213.
150. Walter, A. 1988. Milchsaüre, Lehrbuchd. *Org. Chemie*, Stuttgart:Verlag..
151. Weaver, R.A. and Shelef, L.A. 1993. Antilisterial activity of sodium, potassium or calcium lactate in pork liver sausage. *J. Food Safety*. 13:133-146.
152. WHO 1967. Food and Agricult. Organ. of the United Nations, FAO Nutrition Meetings, Report Series No.40 A, B, C, Toxicological evaluation of some antimicrobials, antioxidants, emulsifieres, stabilisers, fluor-treatment agents, acids and bases. *WHO Food Add*. 67:29-144.
153. Woolford, M.K. 1975. Microbiological screening of the straight chain fatty acids (C_1-C_{12}) as potential silage additives. *J. Sci. Food Agric*. 28:219.
154. Woolford, M.K. 1975a. Microbiological screening of food preservatives, cold sterilants and specific antimicrobial agents as potential silage additives. *J. Sci Food Agric*. 26:229.
155. Woolthuis, C.H.J., and Smulders, F.J.M. 1985. Microbial decontamination of calf carcasses by lactic acid sprays. *J. Food Prot*. 48:832-837.
156. Woolthus, C.H.J., Mossel, D.A.A., van Logtestijn, J.D., DeKruiff, J.M. and Smulders, F.J.M. 1984. Microbial decontamination of porcine liver with lactic acid and hot water. *J. Food Prot*. 47:220.
157. Yang, Z., Li, Y., and Slavik, M. 1998. Use of antimicrobial spray applied with an inside-outside birdwasher to reduce bacterial contamination on prechilled chicken carcasses. *J. Food Prot*. 61:829-832.
158. Young, J.A., Schwartzman, G. and Melton, A.L. 1965. Separation and identification of sodium salts of acetic, propionic, butyric, and valeric acids by paper chromatography. *J. Assoc. Off. Anal. Chem*. 48:622.
159. Zeitoun, A.A., and Debevere, J.M. 1991. Inhibition, survival and growth of *Listeria monocytogenes* on poultry as influenced by buffered lactic acid treatment and modified atmosphere packaging. *Int. J. Food Microbiol*. 14:161-169.
160. Zeitoun, A.A.M. and Debevere, J.M. 1990. The effect of treatment with buffered lactic acid in microbial decontamination and on shelf life of poultry. *Int. J. Food Microbiol*. 11:305.

J.N. Sofos

Sorbic acid

23

I. INTRODUCTION

Sorbic acid was first isolated from the oil of rowanberries (sorbapple or mountain ash tree) by A.W. von Hofmann in 1859, while its antimicrobial properties were discovered in the late 1930s and 1940s both in Germany and in the United States by E. Miller and C. M. Gooding, respectively (Lück, 1976, 1980; Sofos, 1989; Sofos & Busta, 1981). The first United States patent on sorbic acid was awarded in 1945 to C. M. Gooding and Best Foods, Inc. (Gooding, 1945) for its fungistatic properties in foods and food wrappers (Sofos, 1992; Sofos & Busta, 1993). The evaluation of sorbic acid as a preservative increased in the late 1940s and in the 1950s when it became commercially available. The biological properties and health effects of the compound were also evaluated during that time. These efforts have led to the extensive use of sorbic acid and its salts as preservatives in foods and other materials throughout the world (Lück, 1976; 1980; Sofos, 1989; 199; 1994; Sofos & Busta, 1981; 1993; Sofos et al., 1979a). Further research, performed during the late 1950s and in the 1960s, examined the mechanism of sorbic acid activity against microbial growth and the potential for use of the compound in additional food products (Sofos & Busta, 1981; 1993). In the 1970s and early 1980s, sorbic acid and its potassium salt were evaluated for use as inhibitors of *Clostridium botulinum* in meat products, especially in combination with reduced nitrite levels to minimize the potential for formation of carcinogenic nitrosamines during product cooking (Sofos, 1981; 1989; 1992; Robach & Sofos, 1982; Sofos & Busta, 1993).

Currently, sorbic acid and its more water soluble salts, especially potassium sorbate, are well established safe antimicrobial agents, and are used throughout the world as preservatives for various foods, animal feeds, pharmaceuticals, cosmetics, and in various industrial applications (Sofos, 1989). This chapter discusses their commercial production, analytical methodologies, molecular properties, antimicrobial activity, degradation by microorganisms, applications in food preservation, and their safety and tolerance in the human diet.

II. COMMERCIAL PRODUCTION

As indicated, sorbic acid is a natural component of the berries of the mountain ash tree (*Sorbus aucuparia,* L. *Rocaceae)*, in which it was first detected, as well as of aphids where it is biosynthesized through an acetogenic pathway (Walters & Mullin, 1994). Commercial production of the compound has involved either oxidation of 2.4-hexadienal or more commonly, the condensation reaction of ketene and crotonaldehyde. The crude sorbic acid produced by decomposition of the ketene-crotonaldehyde reaction polymer is purified and stabilized for storage, distribution, and use as an antimicrobial agent. Purification may involve treatment with sodium hydroxide, hydrochloric acid, and activated carbon; distillation and crystallization; or washing, distillation, and vacuum drying. Granulation is desirable for uniform mixing in foods and for improved solubility, and it is accomplished by extrusion and pelletization of slurry (Sofos, 1989; 1992).

In addition to alkaline salts formed by neutralization of sorbic acid, other derivatives of the compound manufactured and tested for antimicrobial activity include esters, alcohols, aldehydes, amine salts, and amide derivatives. The derivative with the widest commercial application as an antimicrobial agent is potassium sorbate. Its high solubility allows preparation of concentrated stock solutions to be used in dipping or spraying applications (Sofos, 1989; 1992).

III. DETECTION AND ANALYSIS

Detection methods require quantitative extraction and separation of sorbic acid from the food material without food ingredient interference (DeLuca et al., 1995; Mandrou et al., 1999; Montano et al., 1995). Extraction may involve acid-steam distillation, selective gas diffusion, and solvent extraction using ethyl or petroleum ether, dichloromethane, and isooctane, while in some foods, filtration, dialysis, or direct analysis has been used (Sofos, 1989; Sofos & Busta, 1993). Steam distillation has been used extensively, but it is considered time consuming and compounds present in the food or generated by decomposition of lipid materials may interfere with colorimetric or spectrophotometric detection of sorbic acid. Several modifications have been proposed as useful in avoiding interference and in improving the accuracy of the steam distillation procedure, while combinations of various treatments have also been used to improve extraction and reduce interference. Such combinations have involved extraction under acid conditions from the steam distillate and reextraction with sodium hydroxide; dialysis and solvent extraction; successive extractions with either, sodium hydroxide, and methylene chloride; and double distillation and ether extraction (Sofos, 1989).

Methods for qualitative and quantitative detection of sorbic acid in foods include acidimetry, bromometry, colorimetry, spectrophotometry, polarography, and chromatography (Sofos, 1989). Colorimetric and spectrophotometric methods are widely used, while chromatographic methods have also been applied (Sofos & Busta, 1993).

The colorimetric detection of sorbic acid at an absorbance of 532 nm is an official method for quantitative determination in foods and beverages (AOAC Int. method 975.31), cheese (AOAC Int. methods 971.15 & 975.22) and wine (AOAC Int. method 975.10) by the color reaction with α-thiobarbituric acid (AOAC Int., 1995). The method is simple and usually very specific (Lück, 1980). The spectrophotometric (ultraviolet absorption) procedure has also been used in many foods, including fruit products, bakery items, wine, cheese, and sausage products and it involves measurement of absorbance at 250 nm. Special extraction and purification steps have been proposed to reduce interference problems (Sofos, 1989). The method is an official final action (AOAC Int. method 974.10) procedure for dairy products (AOAC, Int., 1995; Sofos & Busta, 1993).

Numerous reports of chromatographic methods for determination of sorbic acid in food products are found in the literature (Sofos, 1989). They include gas chromatography, high-performance liquid chromatography (HPLC), thin-layer chromatography, paper chromatography, and micellar electrokinetic capillary chromatography (MECC). A gas chromatographic method is a first action procedure (AOAC Int. method 983.16) for sorbic and benzoic acids in foods (AOAC, Int., 1995). Solid extraction has been used as a clean-up procedure for its determination by liquid chromotography in fruit products (Mandrou et al., 1999). Others, however, have reported that clean-up procedures did not improve liquid chromatography determination (Benassi & Cecchi, 1998). In addition, microdialysis was used to extract sorbic acid and benzoic acid from food to be separated and detected by HPLC (Mannino & Cosio, 1996). A rapid method for the identification and quantitation of sorbic and benzoic acids in beverages and foods by MECC has been reported (Pant & Trenerry, 1995). Capillary electrophoresis was also applied for detection in citrus juices and other substrates (Cancalon, 1999; Mercier et al., 1998). Inhibition of microorganisms has also been evaluated as a procedure for the qualitative detection of sorbates. It is obvious, however, that in addition to sorbates, such inhibition may be caused by a number of other inhibitors (Sofos & Busta, 1993).

IV. MOLECULAR PROPERTIES

Sorbic acid is a naturally occurring *trans-trans* unsaturated fatty acid (2.4-hexadienoic; $CH_3-CH=CH-CH=CH-COOH$), with its highly reactive carboxylic group forming salts and esters, while its conjugated double bonds are also reactive. The reactivitiy of sorbate in food systems may influence its activity as an antimicrobial agent as well as the safety of the products. The volatility of the compound in steam is useful in its isolation for quantitative detection in foods or other materials. Commercially important salts of sorbic acid include calcium, sodium, and potassium sorbates (Sofos & Busta, 1993).

The solubility of sorbic acid is higher in lipids than in water (0.15% w/v at 20°C), whereas its salts, especially potassium sorbate, are more soluble in water. Solubility is higher with increasing temperature and with increasing concentration of ethanol, propylene glycol, and acetic acid, whereas it decreases with increasing concentrations of sucrose and sodium chloride. Partitioning of sorbic acid between the lipid and aqueous phases of foods depends on the pH of the food, the amount and type of lipids, and other ingredients present such as sugar and ethanol. The high solubility of potassium sorbate in water (58.2% w/v at 20°C) makes it ideal for use as a solution. In water, potassium sorbate hydrolyzes to yield 74% of its weight as sorbic acid. Sodium sorbate is soluble in water at the rate of approximately 30%, whereas calcium sorbate is soluble at 1.2% in water and insoluble in fats, which makes it valuable as a delayed-release form of sorbic acid (Sofos, 1989; 1992).

In the pure and dry form, sorbic acid and its derivatives are remarkably resistant to oxidation, but in aqueous solutions they are relatively unstable and degrade by first-order reaction kinetics (Sofos, 1989). The olefinic bonds of sorbate are attacked by oxidizing agents, resulting in formation of peroxides, followed by their degradation or polymerization. Products of sorbic acid oxidation include carbonyl compounds such as crotonaldehyde, malonaldehyde, acrolein, formic acid, and malonic acid (Sofos & Busta, 1993). Factors influencing the rate of oxidation include pH, sugars, salt, amino acids, antioxidants, metal ions, light, temperature, time, types of food materials present, and properties of the packaging materials. Low or no oxygen permeability is desirable in packaging materials to avoid oxidation of sorbic acid (Sofos, 1992).

Oxidation is more rapid at decreasing pH values and in the presence of acetic, sulfuric and hydrochloric, whereas citric and malic acids appear to have no major effect on degradation of sorbic acid. Oxidation may also be accelerated by certain concentrations of copper, iron, cobalt, and manganese. Chloride salts, antioxidants and alcohols appear to inhibit oxidation, whereas sulfates and phosphates exert pro-oxidant activity. Oxidation is also inhibited by antioxidants and alcohols. In general, oxidative degradation of sorbic acid in foods, is less extensive than in aqueous solutions, and it is influenced by the nature of the food and its components and by processing, handling, and storage conditions (Sofos, 1989; 1992; 1994; Sofos & Busta, 1993).

Diffusion of sorbic acid into foods is desirable when preservation of the core of the product is necessary instead of only the surface layer. When there is no need for microbial inhibition in the interior of the product, then it is desirable that sorbate remains on the surface where its activity is needed. Uptake and diffusion of sorbic acid into model systems or foods is affected by moisture content and water activity of the substrate, its structure, and its hydration properties. Diffusivity appears to decrease with increasing concentrations of gelling agents, decreasing concentrations of sorbic acid, increasing concentration and molecular weight of solutes, and decreasing temperature (Sofos, 1989; 1992; Sofos & Busta, 1993).

Sorbates may be applied by direct mixing into the food formulation, by treating food wrappers and packaging materials that come in contact with the food, by immersing, dipping, or spraying of foods with solutions, or by dusting foods with the dry form of the compound. Direct mixing of sorbic acid or its salts into a food product during processing is the ideal application procedure for uniform distribution and control of the amount. The other methods of application may result in variable residual levels of sorbic acid present in different parts of the food. The residual levels will vary not only with method of application, but also with concentration, time/length of exposure, type of substrate and handling of the substrate after exposure to sorbate. Variations can be avoided by experimentation before an application to determine the appropriate solution concentration, exposure time, and handling for the desirable levels of residual sorbate in the product (Sofos, 1989; 1992; Sofos & Busta, 1993).

V. ANTIMICROBIAL ACTIVITY

A. Spectrum

Sorbic acid is active against yeasts and molds, as well as against many bacteria. Extensive research during the 1950s demonstrated the antimicrobial effectiveness of sorbates against yeasts and molds and led to the use of the compounds as fungistatic agents in many foods. Effective antimicrobial concentrations of sorbic acid in most foods are in the range of 0.02-0.30% (Sofos & Busta, 1993).

Inhibition of yeasts, first demonstrated in the 1950s in fermented vegetable products, includes species of the generea *Brettanomyces, Candida, Cryptococcus, Debaryomyces, Endomycopsis, Hansenula, Kloeckera, Pichia, Rhodotorula, Saccharomyces, Sporobolomyces, Torulaspora, Torulopsis* and *Zygosaccharomyces* (Sofos, 1989). In addition to its effectiveness in fermented vegetables, sorbic acid inhibits yeasts in fruit juices, wines, cottage cheese, dried fruits, meat and fish products. Use of sorbates for inhibition of yeasts is especially important in low-pH and/or intermediate water activity products, such as carbonated beverages, salad dressings, syrups, tomato products, jams, candy, jellies and chocolate syrup (Liewen & Marth, 1985a; Restaino et al., 1982; Sofos & Busta, 1993).

A major application of sorbates in food products is their use for inhibition of molds in cheeses (Chichester & Tanner, 1972). Sorbates also inhibit molds in butter, sausages, fruits and juices, cakes, grains, breads and smoked fish (Liewen & Marth, 1985a; Sofos, 1989; Sofos & Busta, 1993). The antimicrobial activity of sorbic acid against molds includes species of the genera *Alternaria, Ascochyta, Aspergillus, Botrytis, Cephalosporium, Chaetomium, Cladosporium, Colletotrichum, Cunninghamella, Curvularia, Fusarium, Geotrichum, Gliocladium, Helminthospolaria, Pestalotiopsis, Pullularia, Rhizoctonia, Rhizopus, Rosellinia, Sporotrichum, Trichoderma, Truncatella* and others (Sofos, 1989).

Although sorbates inhibit the formation of mycotoxins by various molds, under certain conditions, subinhibitory levels of sorbate may stimulate production of mycotoxins (Liewen & Marth, 1985a-c; Monnet al., 1988). Stimulation of mycotoxin formation by low levels of sorbate depends on species and strains of molds, storage temperature and other factors (Sofos, 1989; Sofos & Busta, 1993).

Sorbates have been reported to inhibit bacterial species belonging to the genera *Acetobacter, Achromobacter, Acinetobacter, Aeromonas, Alcaligenes, Alteromonas, Arthrobacter, Bacillus, Campylobacter, Clostridium, Enterobacter, Escherichia,* including *E. coli* O157:H7, *Klebsiella, Lactobacillus, Listeria, Micrococcus, Moraxella, Mycobacterium, Pediococcus, Proteus, Pseudomonas, Salmonella, Serratia, Staphylococcus, Vibrio, Yersinia,* and others. Depending on pH and concentration, sorbate inhibited or inactivated *Listeria monocytogenes* in a broth (El-Shenawy & Marth, 1988) and in a cold-pack cheese food (Ryser & Marth, 1988). In other studies (Kouassi & Shelef, 1995a,b), although sorbate did not affect growth of *L. monocytogenes* in broth, it suppressed cysteine activation of listeriolysin O secretion. It was concluded that combinations of sorbate with propionate or lactate, which inhibited growth, could extend shelf-life and increase safety.

Overall, sorbates can inhibit Gram-positive and Gram-negative, catalase-positive and catalase-negative, aerobic and anaerobic, and mesophilic and psychrotorphic microorganisms, as well as spoilage and pathogenic bacteria. Inhibition of bacteria by sorbate appears to effect an extension of the lag phase, with a lesser influence on rate and extent of growth. The effect of sorbate on spore-forming bacteria may be exerted on spore germination, outgrowth and/or vegetative cell division (Blocher & Busta, 1985; Sofos, 1992; Sofos et al., 1979d; 1986; Sofos & Busta, 1993). However, various species and strains of microorganisms exhibit different sensitivities to inhibition by sorbate. Studies in the 1950s indicated that sorbic acid could be used as a selective agent for catalase-negative lactic acid-producing bacteria and clostridia because it was highly inhibitory against catalase-positive organisms. Subsequent studies, however, reported either no effect or inhibition of lactics and clostridia by sorbate (Sofos, 1989). The overall inhibitory action of sorbate against lactics is less than that against yeasts, which explains the usefulness of the compound as a preservative in vegetable fermentations. Another bacterium that appears to be more resistant to inhibition by sorbate than other sporeformers is *Sporolactobacillus*, while growth of *Gluconobacter oxydans* in the presence of sublethal concentrations of sorbic acid before studies to determine the minimal inhibitory concentration (MIC) resulted in a substantial increase in the MIC within 1 h. Varying sensitivities of bacterial species and strains to sorbate may lead to shifts in the microbial population during storage of foods (Blocher & Busta, 1983; 1985; Sofos & Busta, 1993).

Under certain conditions some yeasts and molds may also be resistant to inhibition by sorbate. Yeast strains resistant to sorbate belong to the genera *Zygosaccharomyces, Saccharomyces, Pichia, Debaryomyces, Torulopsis, Brettanomyces, Candida* and *Triganopsis* (Cole et al., 1987; Lenovich et al., 1988; Mihyar et al., 1997; Restaino et al., 1981, 1982; Sousa et al., 1996; Warth 1977, 1985). Of 100 yeast strains isolated from spoiled foods and beverages, most tolerated 150 ppm sorbic acid, 40% tolerated 500 ppm, and two strains of *Zygosaccharomyces bailii* tolerated 800 ppm of sorbic acid (Neves et al., 1994). Resistance of yeasts to inhibition by sorbate depends on species and strains, sorbate concentration, pH, inoculum level, storage temperature and previous exposure of the organism to low levels of sorbate (Sofos, 1989; Sofos & Busta, 1993). When the yeast cells have been previously adapted to sorbate in media containing glucose or sucrose, subsequent exposure of the cells shows little effect of solute types on sorbate resistance (Lenovich et al., 1988). However, potassium sorbate suppressed growth of *Z. bailii* in salsa mayonnaise more than sodium benzoate (Wind & Restaino, 1995). Potassium sorbate or sodium benzoate resulted in complete inhibition of *Zygosaccharomyces rouxii* in high moisture prunes (El Halouat et al., 1998).

Resistance of osmotolerant yeasts to inhibition by sorbate was acquired by preconditioning the yeast to sorbate (Bills et al., 1982). One proposed mechanism of resistance of osmotolerant yeasts has involved an inducible, energy-requiring system that transports the preservative out of the cell (Warth, 1977). Other proposed mechanisms of yeast resistance to sorbate at reduced water activities have been related to yeast cell shrinkage and decreases in membrane pore size retarding the flow of sorbate into the cell, or protection of enzyme systems from inhibition by sorbate through production of compatible solutes, such as polyols (Sofos & Busta, 1993). Exposure of *S. cerevisiae* to sorbic acid caused strong induction of two plasma membrane proteins, one of which was identified as ATP-binding cassette transporter (Pdr12), which is essential for the adaptation of yeast to growth under weak acid stress and confers weak acid resistance by mediating energy-dependent extrusion of water soluble carboxylate ions (Holyoak et al., 1996; Piper et al., 1998).

B. Influencing factors

The antimicrobial activity of sorbic acid is influenced by compositional, processing and environmental factors, such as ingredients, pH, concentration, water activity, temperature, gas atmosphere, packaging, microbial flora, and additives. These factors can act synergistically or be antagonistic and either enhance or negate the antimicrobial activity (Sofos, 1989; 1992; Sofos & Busta, 1993).

The antimicrobial activity increases as the pH of the substrate decreases and approaches the dissociation constant (pK_a = 4.76), but sorbic acid shows effectiveness at pH values as high as 6.5 while certain studies have indicated small antimicrobial activity by sorbate at pH values as high as 7.0 (Draughon et al., 1982; Sofos & Busta, 1980; 1981; Statham & McMeekin, 1988). In contrast, the maximum pH for antimicrobial activity by other common lipophilic food preservatives is lower, (e.g., 5.0-5.5 and 4.0-4.5 for propionate and benzoate, respectively). Antimicrobial activity at pH values above 5.5 is useful in foods in which preservatives, such as parabens, might not be effective because of their increased solubility in fat. In certain instances, however, sorbates can partially or totally replace benzoate in foods even at lower pH in order to avoid possible off-flavors due to higher benzoate levels and to extend the range of microorganisms inhibited compared to benzoate or propionate (Sofos & Busta, 1981, 1993). Potassium sorbate was found more effective than chitosan in inhibiting growth of *Aspergillus niger* on low-sugar candied 'kumquat' (Fang et al., 1994).

The higher antimicrobial activity at lower pH values is attributed to the increased levels of undissociated acid present, which is the effective antimicrobial form (Lück, 1980; Sofos, 1989; Sofos & Busta, 1981). However, the dissociated sorbic acid was also found to have antimicrobial activity, but it was 10-600 times less inhibitory than the undissocated acid (Eklund, 1983; Statham & McMeekin, 1988). In environments of pH above 6.0, however, more than 50% of the inhibition was due to dissociated compound (Eklund, 1983).

In products with high fat content, the amount of dissociated sorbic acid in the water phase, where it is needed for microbial control, is decreased, while food ingredients such as salt and sugars also reduce the concentration of sorbate in the aqueous phase, but they act synergistically to enhance the antimicrobial activity (Beuchat, 1981c). In general, solutes should increase the inhibitory activity of sorbate by reducing the water activity of the substrate (Cerruti et al., 1990; Sofos, 1989). Sucrose, glucose, and sodium chloride, however, have reduced the synergistic effect of sorbate and heat on thermal inactivation of microorganisms (Beuchat, 1981a-c; Cerruti et al., 1988), and sodium chloride (1.25 and 2.5%) reduced inhibition of *C. botulinum* by sorbate in a nutrient broth (Wagner & Busta, 1984). Exposure of *Sacharomyces rouxii* cells in a 60% sucrose + 0.1% sorbate environment enhanced their inhibition by sorbate compared to preconditioning in 0% sucrose + 0.1% sorbate (Bills et al., 1982). The resistance of the same organism to increasing concentrations of sorbate was also increased at lower water activities (Restaino et al., 1981). Antimicrobial activity is also reduced in presence of increased levels of microbial contamination confirming that sorbates, similar to any microbial inhibitor, should be used to preserve foods processed using good manufacturing practices and not as substitutes for appropriate sanitation and hygienic practices (Sofos & Busta, 1993).

Presence of sorbic acid in food formulations may affect the rate and extent of microbial destruction during heating, as well as dormancy and recovery of heated microoganisms (Sofos, 1989). Sorbate may enhance heat activation and destruction of spores and may inhibit repair and growth of thermally injured cells (Banks et al., 1998; Beuchat, 1981a-d; Lopez et al., 1996; Oloyede et al., 1994), but its influence on thermal inactivation and recovery of injured microorganisms is variable among species and strains. Low concentrations of sorbic and fumaric acids in the heating medium had little effect on the heat resistance of *Eurotium herbariorum,* a mold involved in the spoilage of grape preserves (Splittstoesser et al., 1989). Concentrations of sorbate as high as 0.1% had little effect on the thermal resistance of ascospores of *Neosartorya fischeri*, but the growth of surviving spores that had been exposed to high temperatures was greatly inhibited by sorbate concentrations as low as 0.007% (Splittstoesser & Churey, 1989). Inclusion of sorbic acid in the formulation may also eliminate the protective effect of sucrose against thermal inactivation of yeasts and molds (Sofos, 1989; Sofos & Busta, 1993). Sorbic and benzoic acids affected the thermotolerance and heat shock response of *S. severisiae* depending on pH (Cheng & Piper, 1994). At low pH, sorbate inhibited induction of thermotolerance by sublethal heat shock, but at pH above 5.5 it acted as a powerful inducer of thermotolerance in the absence of sublethal heat treatment. Sorbic acid was found to induce thermotolerance without inducing the heat shock response through accumulation of trehalose in *S. cerevisiae* (Cheng et al., 1999). Potassium sorbate sensitized cells of *L. monocytogenes* and *Z. bailii* to inactivation by high hydrostatic pressure (Mackey et al., 1995; Palou et al., 1997). Sorbic acid (0.1%) enhanced destruction of *E. coli* O157:H7 cells at pH 3.4 but not at pH 6.4 (Liu et al., 1997). Sorbic acid (0.1%) reduced the $D_{50^\circ C}$ of *E. coli* O157:H7 in apple juice from 18 to 5.2 min, while benzoic acid reduced it to 0.64 min (Splittstosser et al., 1996). Potassium sorbate or sodium benzoate did not affect survival of *E. coli* O157:H7 during storage of apple cider (Miller & Kaspar, 1994). Reductions of 5-10 log-units of *E. coli* O157:H7 or *Salmonella typhimurium* DT104 in apple cider were achieved through freeze-thaw treatments in the presence of sorbic acid (Uljas & Ingham, 1999).

Reduced storage temperature enhances inhibition of microbial growth by sorbic acid, which indicates that the compound should be more useful as a preservative in refrigerated foods (Roland & Beuchat, 1984). However, when combined with acidification, treatment with sorbic acid may enhance the storage stability of fruit juices even at temperatures higher than refrigeration (Alli & Kermasha, 1989).

Vacuum or modified gas atmosphere conditions usually enhance the antimicrobial activity of sorbic acid, as indicated with meat and fish products (Statham et al., 1985). Combinations of sorbic acid with carbon dioxide have also been reported as effective inhibitors of microbial growth (Elliot, et al., 1985; Gray et al., 1984). Sorbic acid (0.2 - 0.4%), however, did not enhance the bactericidal activity of cellulose-based edible films against *Salmonella montevideo* (Zhuang et al., 1996).

Food acids may reduce the water solubility of sorbic acid, but they can enhance antimicrobial activity by increasing the concentration of the undissociated form of the compound. In addition, the specific anion itself may contribute to the antimicrobial activity but effects vary with substrate, microorganisms, and types of acids (Huhtanen et al., 1983; Restaino et al., 1981; 1982; Sofos, 1989).

Sorbic acid combinations with antioxidants, such as butylated hydroxyanisole (BHA), butylated hydroxytoluene (BHT), tertiary butyl hydroxyguinone (TBHQ), and propyl gallate (PG), have shown increased antimicrobial activity compared to individual components (Davidson et al., 1981). Although results varied with types of microorganisms, antioxidants and substrates, these combinations offer the advantage of simultaneous inhibition of microbial growth and rancidity (Sofos & Busta, 1993). Potassium sorbate was found more inhibitory against yeasts than hydroxycinnamic acids (Stead, 1995).

Increased antimicrobial effects have also been reported with combinations of sorbate and polyphosphates (Nelson et al., 1983; Sofos, 1986a-c; Thomas & Wagner, 1987; Wagner & Busta, 1983). Furthermore, use of sorbate in combination with benzoate or propionate may expand the range of microorganisms inhibited with reduced concentrations of each preservative. Increased antimicrobial effects were also observed when sorbic acid was combined with propionate, ascorbate, certain amino acids, fatty acids, sucrose fatty acid esters, sulfur dioxide, propylene glycol, glucose oxidase and many other factors or compounds (Marshall & Bullerman, 1986; Sofos, 1989; Sofos & Busta, 1993; Tellez-Giron et al., 1988). The microbial stability of shelf-stable banana puree was enhanced through inhibition of inoculated osmophilic and nonosmophilic yeasts, molds, *Bacillus coagulans, Clostridium pasteurianum* and *Clostridium butyricum* by adjustment of water activity to 0.97, pH to 3.4 and addition of 250 ppm ascorbic acid, 100 ppm potassium sorbate, 400 ppm sodium bisulfite and mild heat (Guerrero et al., 1994). However, the extent of the antimicrobial activity of sorbic acid under commercial conditions is difficult to predict because it is affected by and interacts with factors such as product pH and composition (fat and moisture), presence of other ingredients and preservatives, product processing, sanitation and extent of contamination, packaging, atmosphere and storage temperature (Sofos, 1989; Sofos & Busta, 1981; 1993).

C. Mechanisms of action

Sorbic acid has been described as a competitive and reversible inhibitor of amino acid inducted bacterial spore germination by acting on the connecting reactions that follow initiation of the germination process. Because the nature of the connecting reactions of spore germination is largely unknown, however, it is difficult to present an exact mechanism of their inhibition by compounds such as sorbate. Indirect evidence suggests that inhibition could involve alteration of permeability of spore membranes or inhibition of spore lytic enzymes that may be involved in germination by causing cortex hydrolysis and loss of refractility (Blocher & Busta, 1985; Sofos, 1989, 1992; Sofos & Busta, 1993; Sofos et al., 1979d, 1986).

Generally, the amounts of sorbic acid employed in food preservation (≤0.3%) inhibit or delay growth of vegetative microorganisms without causing permanent or lethal damage to their cells, while higher concentrations may cause inactivation. Inhibition of microbial cell metabolic function may result from alterations in cell membranes and cell transport functions, inhibition of enzymes involved in transport function or metabolic activity, alteration of the morphological structure of the cell, or changes in the genetic material. Each one of these modes of inhibition may be operational in specific microbial species and in different substrates, and environmental as well as processing parameters (Sofos, 1989, 1992; Sofos & Busta, 1993; Sofos et al., 1986).

Even though information on the effect of sorbic acid on the genetic material of microbial cells is limited, the compound is believed not to possess mutagenic activity, but it may form mutagenic products when reacting under certain conditions with compounds such as sulfur dioxide and sodium nitrite. Sorbates have been reported to cause morphological changes in microbial cells including development of yeasts cells with dense lipoprotein granules, irregular nuclei, numerous mitochrondria of various sizes, and vacuoles. Spores of *C. botulinum* treated with sorbate have become long and bulbous in shape and have exhibited defective division and elongation of emerging cells. The mechanism causing such morphological changes may involve incorporation of sorbate into specific cell structures and/or inhibition or enhancement of specific biosynthetic processes. Sorbic acid has also been reported to be associated with microbial cell alterations in the morphology, integrity, permeability, and function of cell membranes. Elongation and aberrant cell formation are also associated with changes in cell walls and membranes (Sofos, 1989; 1992).

Inhibition of microbial cell metabolism in the presence of sorbic acid has been detected as a decrease in assimilation of carbon from substrates such as glucose, acetate, lactate, ethanol, succinate, pyruvate, and oxaloacetate. Inhibition of cell metabolism, however, may be the result of adverse effects on enzymes, or transport mechanisms. Sorbate is known to inhibit the *in vitro* activity of many enzymes, especially sulfhydryl-containing enzymes (Kouassi & Shelef, 1995a,b). Although exact mechanisms of inhibition and inactivation of specific enzymes are unknown, indirect evidence supports the postulation that binding of sorbic acid with sulfhydryl groups inhibits their activity. Inhibition of the enzyme catalase, however, was attributed to autoxidation of sorbic acid and formation of sorbyl peroxide. It has also been suggested that sorbate combines with coenzyme A forming sorbyl coenzyme A, which causes microbial inhibition through interference with oxygen uptake (Sofos, 1992).

Sorbic acid may also inhibit microorganisms by interfering with mechanisms of electron and substrate transport in the cell, leading to growth-inhibition by starvation. Inhibition of nutrient uptake may be due to neutralization of the proton-motive force needed for substrate transport, inhibition of transport enzyme systems, or inhibition of synthesis and/or depletion of ATP. Inhibition of amino acid uptake has been attributed to inhibition of amino acid permease enzymes or the metabolic energy utilization of amino acid transport mechanisms. This effect may be occurring through incorporation of sorbic acid, or unsaturated fatty acid, into the cell membrane, where it may cause steric disorganization of active membrane transport proteins (Sofos, 1989, 1992; Sofos & Busta, 1993). Cells of *Z. rouxii* cultured in media supplemented with sorbate contain higher percentages of C18:1 fatty acids than cells cultured in media without sorbate (Golden et al., 1994).

Sorbic acid has been reported as an inhibitor of mold growth through the depletion of ATP levels in conidia. This could potentially take place because ATP levels may be depleted as the cell attempts to maintain ion balance when dissociation of sorbic acid in the cytoplasm increases the intercellular cation concentration and because the primary sodium/hydrogen pump is directly linked to hydrolysis of ATP. As the hydrogen influx exceeds the pumped efflux, a shift in charge may potentially take place and lead to a decrease in the net negative intercellular change. This could then discharge the pH gradient required for ATP formation according to the chemostatic theory of oxidative phosphorylation. Studies with bacteria, however, found no decreases in ATP in the presence of sorbic acid (Sofos, 1989; 1992).

Inhibition of microbial growth by sorbic acid through cell starvation may be due to neutralization of the transport driving proton-motive force that exists across cell membranes. In lower pH environments, the difference across the cell membrane is large, and the amount of acid entering and dissociating in the cytoplasm is higher. This accumulation of hydrogen ions inside the cell acts as an inhibitor by interfering with metabolic processes and causing a dissipation of the transmembrane proton gradient, which is one of the components of the proton-motive force. Elimination of the transport driving pH difference across the cell membrane has been presented as a major mechanism of microbial inhibition by compounds such as sorbic acid. According to this hypothesis, undissociated sorbic acid acts as a protonophore, which decreases the intracellular pH and dissipates the proton-motive force of the membrane that energizes transport of compounds such as amino and keto acids. Inhibition of uptake of such components is believed to induce a stringent-type regulatory response in the cells, resulting in inhibition of growth but in maintenance of cell viability. A stringent response involves readjustments occurring in bacteria when amino acids become limiting or their specific ratios are disturbed (Eklund, 1985, 1989; Salmond et al., 1984; Sofos, 1989, 1992; Sofos & Busta, 1993).

It is believed that neutralization of the pH difference across the cell membrane is not the only mechanism of microbial inhibition by sorbate. Reasons for this conclusion include (1) although dissociation of the acid inside the cell may eliminate the pH difference across the membrane, its effect on the other component of the proton-motive force, which is the difference in electrical potential, is smaller. The remaining difference in electrical potential is smaller. The remaining difference in electrical potential, however, is believed to be adequate to energize the uptake of substances needed for cell maintenance and growth. (2) Sorbic acid has been shown to inhibit microbial growth, although less efficiently, even at pH values near neutrality. In such situations, uptake of the compound based on pH difference across the cell membrane would not be adequate to explain the observed antimicrobial activity. (3) Neutralization of the proton-motive force is probably not involved in growth inhibition in the presence of carbohydrates, because they do not depend on this force for their transport. Therefore, mechanisms involved in inhibition of microbial metabolism and proliferation by sorbic and other similar lipophilic acid preservatives appear to be different, depending on microbial types, substrates, and environmental conditions (Sofos, 1989; 1992).

VI. DEGRADATION

As a fatty acid, sorbate may be metabolized through β-oxidation or Ω-oxidation by animals and certain microorganisms (Lück, 1980). Like caproic and butyric acids, under normal conditions of alimentation, sorbate is completely oxidized to carbon dioxide and water yielding 6.6 kcal/g, of which 50% is biologically utilizable (Sofos & Busta, 1993). In addition, strains of molds have been found to metabolize sorbate under certain conditions, in cheeses and fruit products (Sofos, 1989), as shown with *Penicillium* strains isolated from cheeses treated with sorbate, which were able to degrade high (0.18-1.20%) sorbate levels (Finol et al., 1982; Marth et al., 1966). It should be noted that 0.1% sorbate is usually sufficient to inhibit sensitive molds but it appears that selection may occur in sorbate-treated cheeses for certain molds tolerant to the compound (Liewen and Marth, 1985a,b). Degradation of sorbic acid by molds yields 1,3-pentadiene, which is volatile, has an odor like kerosene, plastic paint, or hydrocarbon and is formed through a decarboxylation reaction (Liewen & Marth, 1985a,b; Marth et al., 1966). Other strains of molds that may metabolize sorbate belong to the genera *Aspergillus, Fusarium, Mucor,* and *Geotrichum* (Sofos, 1989), but it appears that there is no apparent relationship between sorbate resistance and the toxigenic properties of molds (Tsai et al., 1988). In general, many molds are sensitive to inhibition by sorbate, but some strains may be resistant and able to metabolize the compound, using it as a carbon source. Degradation of sorbate by molds depends on species and strains, prior exposure to subinhibitory levels of sorbate, level of inoculum, amount of sorbate present, and type of substrate (Sofos, 1989; Sofos & Busta, 1993).

Sorbic acid may also be degraded by certain bacterial strains, under appropriate conditions. This metabolism is mostly associated with lactic acid-producing bacterial strains present as high inocula in sublethal concentrations of sorbate (Horwood et al., 1981; Liewen & Marth, 1985a). Degradation of sorbate by lactic acid bacteria has been associated with geranium-type off-odors in wines and fermented vegetables, caused by ethyl sorbate, 4-hexenoic acid, 1-ethoxyhexa-2,4-diene, and 2-enthoxyhexa-3,5-diene (Edinger & Splittstoesser, 1986a,b; Sofos, 1989). In general, a geranium-like odor is usually associated with wines treated with sorbate and contaminated with high microbial loads (Sofos & Busta, 1993).

VII. APPLICATIONS

Sorbic acid and its potassium salt are widely used as antimicrobials throughout the world in the preservation of human foods, animal feeds, pharmaceuticals, cosmetic products, packaging materials and technical preparations that come in contact with foods or the human body. Legally, sorbates may be used in any food product in the United States that allows generally recognized as safe (GRAS) food additives and in about 80 or more food products that have federal standards and identity (Sofos, 1989; 1992; Sofos & Busta, 1993).

Groups of foods that may be preserved with sorbates include dairy products, bakery items, fruit and vegetable products, edible fat emulsion products, certain meat and fish products, sugar and confectionery items (Sofos, 1989; Sofos & Busta, 1993). As should be the case with any food preservative, sorbic acid is not a substitute for good sanitation or an agent to be used to improve the quality of partially spoiled or degraded foods, but it should be considered as adjunct to good sanitation and hygiene.

As indicated, application of sorbates to foods includes direct addition in the formulation, spraying or immersing the food material in a solution, dusting with a powder, or addition in a coating or packaging material. Selection of the most appropriate method depends on processing procedures, type of food, objectives to be accomplished, equipment available, and convenience (Sofos, 1989). Concentrations of sorbic acid used in food preservation are in the range of 0.02-0.3% while higher levels may cause undesirable changes in the taste of foods. In general, amounts of 0.1-0.3% are tolerated, but levels as low as 0.1% may be detectable in some foods (Sofos & Busta, 1993).

Preservation of cheeses is a common application of sorbates, and United States federal standards of identity permit their use in more than 40 types of cheese, cheese spreads, and cheese food. Reasons for the wide use of sorbates in preserving cheese include the susceptibility of all types of cheeses to microbial deterioration, especially surface mold growth during storage, aging, and distribution; the effectiveness of sorbate against molds; and the increased activity of sorbates at high pH values compared to other common preservatives. In addition to mold spoilage, sorbates inhibit formation of mycotoxins, but they should not be used in mold-ripened cheeses, such as Roquefort or blue and Camembert, in which they can interfere with the desirable molds involved in their manufacture (Sofos & Busta, 1993).

The application of sorbates is generally limited to the surface of cheese, where mold development usually occurs, while in products such as cottage cheese, sorbic acid may be incorporated into the curd. The method of application varies with the type of cheese, preservation objectives, available facilities and equipment, and regulations, which may vary among countries (Lück, 1976, 1980). Procedures for application on the surface of cheese include one of the following: (1) potassium sorbate in solution or in a brine, which can be used to immerse, spray or wash the cheese; (2) a sorbic acid powder for dusting the cheese; and (3) sorbic acid, potassium or calcium sorbate or other derivatives impregnated into the packaging material or other coatings (e.g., wax) used in cheese storage and marketing (Han et al., 1998; Sofos, 1989; Weng & Chen, 1997). Amounts of sorbic acid permitted and applied in cheese preservation range from 0.05 to 0.30% (Lück, 1980; Sofos & Busta, 1981). Levels of 0.05-0.07% sorbic acid are generally used in direct addition to cheese. Concentrations applied to cheese surfaces range from 0.1 to 0.3 g/dm^2; when applied to packaging films, amounts of 2-4 g/m^2 are used (Lück, 1980; Sofos & Busta, 1993). Inclusion of sorbic acid in cottage cheese was effective as a preservative but not against *L. monocytogenes* (Piccinin & Shelef, 1995). However potassium sorbate (0.3%) or sodium benzoate (0.3%) had a significant effect in delaying growth of *E. coli* O157:H7 in Queso Fresco, soft cheese (Kasrazadeh & Genigeorgis, 1995)

Absorption of sorbic acid into the cheese is dependent on the porosity of the surface and the amount and distribution of fat in the cheese. The solubility properties of calcium sorbate may make it the preferred type for cheese preservation, since it is insoluble in fat and only slightly soluble in water. Thus, most of the preservative remains on the surface of hard, ripened cheese without migrating inside the product or dissolving in the water on the shelves during the long maturation period (Lück, 1976). In addition to reducing cheese spoilage and the possibility of aflatoxin formation, use of sorbates during cheese aging also reduces labor and other costs associated with cheese trimming and washing (Sofos & Busta, 1993).

Water-soluble salts of sorbic acid are used widely in the preservation of fresh, fermented, and pickled vegetables (Sofos, 1989). Sorbate levels of 0.05 – 0.20% inhibit growth of organisms not desired in vegetable fermentations, such as yeasts, molds, and putrefactive bacteria while their selective action allows growth of the desirable lactic acid-producing organisms. Selection of precise levels of sorbate, however, depends upon products and specific fermentation conditions (e.g., salt concentration). Inclusion of sorbate at the initial stage of the process results in a 'clean' fermentation and prevents turbidity development in the final product (Lück, 1980). Mold and yeast spoilage is also prevented by sorbates in pickled vegetable products (e.g., cucumbers and olives). Addition of 0.02-0.05% sorbate to a 15-20%, salt brine retards yeast scum formation. Levels of 0.02-0.10% sorbate reduce the number of bloaters and poor seed cavities. In sweet cucumber packs, 0.1% sorbate and 0.5% acetic acid completely inhibit microbial spoilage at sucrose concentrations ranging from 2 to 40%. Sorbate levels in the range of 0.05-0.10% may be used to preserve refrigerated fresh salads, while other vegetable products preserved with sorbates include tomato products and fermented Asian sauces (Chichester & Tanner, 1972; Lück, 1980; Sofos, 1989; Sofos & Busta, 1993).

Sorbates are used to preserve fruit products such as dried fruits, fruit juices and syrups, fruit cocktails, jams, jellies, preserves, beverages and wines. Depending on the particular product, sorbate acts as an inhibitor of yeasts, molds, or bacteria. Sorbate concentrations as low as 0.02-0.05% are adequate for preserving high-moisture dried fruits (e.g., prunes, raisins and figs). The higher moisture content in these products is preferred by the consumers, but it makes them susceptible to mold and yeast spoilage. In low moisture content (e.g., dried fruits) and in high-sugar products (e.g. jams and jellies), small quantities of sorbic acid (0.05-0.10%) are adequate for preservation because of the synergistic antimicrobial effects (Lück, 1980). In these products, sorbate is either added directly into the product or applied to the surface of the product or packaging material. Other fruit products that may be preserved with sorbates include maraschino cherries and strawberry puree (Sofos & Busta, 1993).

Sorbates are used in the preprocessing steps of fruit juice and drink processing, together with sulfur dioxide and pasteurization, to inhibit chemical, enzymatic, or microbiological deterioration (e.g. fermentation). Concentrations as low as 0.02-0.10% are adequate in preservation of soft drinks, while combinations of sorbic acid and sulfur-dioxide are used in the preservation of high-pulp fruit juices. In these products sorbic acid acts as the microbial inhibitor and sulfur-dioxide prevents oxidation and enzymatic spoilage (Lück, 1980). Potassium sorbate and sodium benzoate were highly effective in reducing yeast and mold populations in tomato juice (Bizri & Wahem, 1994).

In wine processing, sorbates are used to prevent refermentation. Potassium sorbate is preferred because of its solubility, but sodium sorbate may be used to avoid potassium bitartate precipitation in some wines caused by potassium sorbate. Acceptable wine preservation can be achieved with concentrations of 0.02% sorbic acid in combination with 0.002-0.004% sulfur-dioxide. Sorbate acts as the inhibitor of yeast refermentation during storage, and sulfur-dioxide protects against enzymatic and oxidative changes and bacterial fermentation. Sorbate levels above 0.03% may impart an off-flavor to the delicate taste of wine. Sorbate can also be added before bottling the wine and is valuable in the preservation of sweet wines and champagne (Sofos, 1989; Sofos & Busta, 1993).

Sorbates, in general, may be more suitable for fruit products than other common preservatives because of their neutral taste, which is especially important in preservation of juices, soft drinks and other beverages. Often, however, sorbates are used in combination with benzoate in preserving fruit juices and drinks. In fruit preserves (e.g., jams, jellies and marmalades), mold inhibition may be achieved by sprinkling sorbate on the surface of the product. In dried products with an irregular conformation (e.g., figs), the immersion method of application is more effective (Sofos, 1989).

In addition to propionic acid-based preservatives, which are the most widely used compounds in preserving bakery goods, sorbates are also valuable because they are effective at higher pH and against molds (Chichester & Tanner, 1972; Lück, 1976, 1980). Concentrations of 0.03-0.30% sorbic acid are adequate for preserving baking products, and in addition to undesirable molds, they inhibit rope-forming bacteria and pathogens, such as *Staphylococcus aureus*, in cream pies (Sofos, 1989). Potassium sorbate also extended the shelf-life of acidified tortillas, especially when applied in combination with calcium propionate (Tellez-Giron et al., 1988).

Use of sorbates must be avoided in products raised by yeasts, unless the amount of yeast and the leavening time are increased, or if yeast-leavened products are treated with sorbate by spraying after baking. In some applications sorbate may be used in the coating oil or the pan grease (e.g. brown-and-serve products). In addition, to avoid the effect of sorbates on desirable yeasts in yeast leavened products, the derivative sorboyl palmitate may be used, which is a mixed anhydride of sorbic and palmitic acids. Sorboyl palmitate has no direct antimicorbial activity, but when heated during baking, it is chemically degraded and releases sorbic acid after the yeast-leavened dough is raised. Use of this compound has the advantage of preserving yeast-leavened bread without interfering with the fermentation process (Lück, 1976, 1980; Sofos & Busta, 1993).

Sorbates should be valuable in baking powder-raised products, including cakes and confectionery. Concentrations of 0.05-0.10% sorbate inhibit the spoilage of cake icings, toppings and fillings. Pie crusts and fillings, refrigerated dough products, pizza shells and toppings, muffins and doughnut mixes may be preserved with sorbates (Sofos, 1989; Sofos & Busta, 1993, 1999).

Potassium sorbate solutions (10%) are approved for use in meat products in the United States for the suppression of mold growth on the surface of dry sausages during the drying period are more common meat preservatives in other countries such as Japan and Korea. Limited sporadic studies in the 1950s and 1960s indicated the potential of sorbic acid as a preservative for meat products (Sofos, 1989). Tompkin et al. (1974) showed that sorbate retarded growth of *Salmonella* spp. and *S. aureus* and delayed growth and toxin production by *C. botulinum* in dry sausage. Extensive studies that followed examined the action of sorbates in a variety of fresh, uncured and processed meat products against clostridia and other pathogens (National Academy of Sciences 1982; Robach & Sofos, 1982; Sofos, 1981; Sofos, 1989, 1992; Sofos & Busta, 1993). The reason for extensive testing of the antimicrobial properties of sorbate in meat products was the perceived need for an alternative to sodium nitrite as an inhibitor of *C. botulinum* in cured meat products. Nitrite may be a potential carcinogen and, more importantly, it can react with amines to form carcinogenic nitrosamines during meat product processing (Sofos et al., 1979a; National Academy of Sciences, 1981). The antimicrobial activity of sorbate was demonstrated in bacon, comminuted pork products, beef and pork frankfurters, pork slurries, uncured cooked sausage, sliced bologna, raw and cooked pork chops, beef steaks, and poultry products (Chang et al., 1983; Draughon et al., 1982; Nelson et al., 1983; Robach et al., 1980c; Sofos et al., 1979b,c, 1980a,b; Tompkin et al., 1974; Unda et al., 1990; U.S. Department of Agriculture, 1979a,b; Wagner & Busta, 1983). In addition to *C. botulinum*, other pathogenic and spoilage bacteria inhibited by sorbate in various meat products include *S. aureus, L. monocytogenes, Salmonella, Clostridium sporogenes, Bacillus cereus, Bacillus licheniformis, Pseudomonas*, mesophiles, psychotroph and lipolytic organisms (Chang et al., 1983; Elliot et al., 1985; Gray et al., 1984; Morad et al., 1982; Petaja et al., 1979; Rice & Person, 1982; Robach, 1979; Sofos et al., 1980b; Tompkin et al., 1974; Wederquist et al., 1994; Zamora & Zaritzky, 1987). Products in which these microorganisms are inhibited include bacon and other cured meats, cured poultry products, raw beef and pork and uncured poultry meat and carcasses. The antimicrobial activity of sorbate in several products was enhanced when combined with nitrite, sodium chloride, phosphates, antioxidants, acids, low pH, low storage temperature, low oxygen, and increased carbon dioxide atmospheres (Sofos, 1989; Sofos & Busta, 1993).

Sorbate (<0.3%) had no major adverse effects on sensory qualities, such as color and flavor of meat products (Amundson et al., 1981; Berry et al., 1981; Chambers et al., 1981; Huhtanen et al., 1983; Kemp et al., 1983; Paquette et al., 1980; Vareltzis et al., 1984), while the reduction of nitrite levels in bacon-pumping brine solutions resulted in reduced levels of nitrosamines in the fried product (Robach et al., 1980a,b; U.S. Department of Agriculture, 1979b). However, sorbate was not approved for use in cured meat products in the United States following a report that consumption of experimental bacon resulted in allergic-type reactions in certain individuals (Berry & Blumer, 1981). It should be noted, however, that no other studies reported undesirable flavors or allergic reactions (Sofos, 1989) and that a National Academy of Sciences committee concluded that although the flavor of bacon treated with combinations of sorbate and nitrite was not the same as the flavor of bacon treated only with nitrite, the flavors of both types of bacon were equally desirable (National Academy of Sciences, 1982).

Sorbates, alone or in combination with benzoates are effective preservatives of oil-in-water or water-in-oil emulsion-based products, such as margarine, mayonnaise, salad dressings, delicatessen, and other similar products (Sofos & Busta, 1993). Other uses of sorbates in foods include sugar-based and confectionery products, at concentrations of 0.05-0.20% to prevent growth of molds and osmophilic yeasts in items such as fillings for chocolate and pralines (Chichester & Tanner, 1972). Sorbates are also used to inhibit molds on dried and smoked fish (e.g., dried cod), while in Asian countries they are commonly used in combination with other preservatives to preserve fish and meat sausages, as well as fermented plant foods (Chichester & Tanner, 1972; Lück, 1976; Robach & Sofos, 1982; Sofos, 1989). Although several studies have found improved preservation of various types of fish with sorbate, one indicated that toxin production by *C. botulinum* in vacuum packaged shucked scallops was slightly more rapid (27 °C) when sorbate (0.1%) was used (Fletcher et al., 1988). A sorbate (0.1%) - benzoate (0.1%) solution preserved brind shrimp for 59 days, which is longer than the preservation (31 days) achieved by the most effective bacteriocin (nisin) tested (Einarsson & Lauzon, 1995). Application of potassium sorbate with bifidobacteria extended the shelf-life of whole and peeled shrimp by three days (Al-Dagal & Bazaraa, 1999). Several other food products may be preserved with sorbates and include high-moisture marshmallow candy, sour cream, tangerine sherbet base, orange peels, strawberry puree, pineapples, tomato juice, Mexican-type hot sauces, prepeeled carrots, potato chips, cut squash, shredded cabbage, cucumbers, garlic oil and pancake batter (Sofos, 1989; Sofos & Busta, 1993).

Additional uses of sorbic acid and its salts as preservatives include pharmaceuticals, cosmetics, and animal feeds (Sofos, 1989). In addition, the esters of sorbic acid (e.g., ethyl and propyl) may also be used in cosmetics and pharmaceuticals, especially those of the emulsion type, in which the emulsifiers may be inactivated by nonionic compounds (Lück, 1976). Sorbates may be particularly important in intermediate-moisture (25% moisture content) pet foods, in which levels of up to 0.30% can be effective mold inhibitors. Sorbic acid levels in the range of 0.03-0.10%, depending on moisture content, are also effective in preventing mold growth on animal feeds during storage. This also assures mycotoxin-free animal feeds and thus eliminates any hazards associated with mycotoxin-contaminated feed. Other reported uses of sorbates include preservation of tobacco products, adhesives and glues (Lück, 1976; Sofos, 1989; Sofos & Busta, 1993).

VIII. SAFETY AND TOLERANCE

Sorbic acid and its salts are popular preservatives for many foods throughout the world because they are considered as generally safe food additives. Extensive testing and feeding to test animals have indicated that they are among the safest of all food preservatives with oral lethal doses (LD_{50}) of 7.4-10.5 g of sorbic acid/kg body weight in rats and 4.2-6.2 g of potassium sorbate/kg body weight compared to the sodium chloride LD_{50} of 5 g/kg body weight. Thus, based on acute toxicity, sorbic acid is considered one of the least harmful preservatives in use (Sofos, 1989; 1992).

Increases in liver or other organ weights associated with long-term feeding studies involving diets high in sorbic acid have been attributed to functional hypertrophy through caloric utilization of sorbic acid, while the histopathological appearance of the liver has been normal. As indicated, being an aliphatic carboxylic acid, sorbate can be used in the animal body as a source of calories (Sofos, 1989; 1992; Sofos & Busta, 1993). No carcinogenic or mutagenic effects have been observed with sorbate (Dickens et al., 1968; Gaunt et al., 1975; Hendy et al., 1976; Lück, 1980; Shtenberg & Ignat'ev, 1970; Sofos, 1994; Sofos & Busta, 1981, 1993).

An allergic-type response has been reported for sorbic acid in one study in which bacon samples with sorbate-nitrite combinations were tested against *C. botulinum*. Some taste panelists reported certain 'allergic-type' symptoms after tasting uninoculated experimental bacon (U.S. Department of Agriculture, 1979b; Berry & Blumer, 1981). Even though it was implied that sorbate might have been involved in producing those symptoms, no direct relation could be proven between symptoms and specific ingredients used in formulating the bacon (U.S. Department of Agriculture, 1979b). In addition, no such symptoms were observed by other individuals who tasted sorbate-nitrite experimental bacon from other studies (Robach & Adam, 1980). This, of course, does not rule out the possibility that high concentrations of sorbate may act as irritants to certain susceptible individuals (Sofos, 1989).

As preservatives for cosmetics and pharmaceutical products, sorbates have been extensively examined for skin tolerance (Lück, 1976; Patrizi et al., 1999). The literature is contradictory, but it indicates that most people are not affected by sorbic acid applied to the skin. Some sensitive individuals, however, show skin irritations when exposed to sorbic acid (National Academy of Sciences, 1982). Average sorbic acid concentrations for skin irrigation are in the range of 1%, while some very sensitive individuals may show irritations at lower levels (Lück, 1976). Considering the average use levels of 0.10-0.30% in food processing, the potential for such irritations in commercial products is minor (Sofos & Busta, 1993).

There has been some concern about possible sorbate-nitrite reactions when both are added to the same substrate (Sofos, 1981; National Academy of Sciences, 1982). Since there is evidence of reactions between nitrite and fatty acids (Benedict, 1980) and because sorbic acid is an unsaturated fatty acid, it may react with nitrite in products such as cured meats. The products of such reactions have the potential of being mutagenic (Sofos, 1989). In general, however, various studies have indicated that nitrite reacts with sorbic acid to form mutagens optimally at pH 3.5, not at 6.0. Mutagen formation of the c-nitroso mutagens requires the presence of excess nitrite. Mutagen formation is inhibited by ingredients such as ascorbic acid, cysteine, and vegetable juices and is inactivated by heat (Kada, 1974; Kito et al., 1978; Robach et al., 1980a). Thus, the requirements for the reactions to proceed and the instability of the products make it unlikely for mutagens to be formed at detectable or harmful concentrations in foods treated with nitrite and sorbate (Sofos, 1981, 1989). It should be noted, however, that certain studies have reported inhibition by sorbate of carcinogenic nitrosamine formation in model systems, but sorbic acid had no effect on nitrosamine formation from aminopyrine and sodium nitrite in animal stomachs (Sofos, 1989; Sofos & Busta, 1993). Potential injury in cell hepatocytes induced by potassium sorbate was prevented by antioxidants (Sugihara et al., 1997).

No synergism in acute toxicity has been detected for combinations of sorbic acid with various other additives, including parabens, benzoate, and propionate (Sofos, 1989). Sorbate, however, has not protected against the toxic effects of other substances, while the derivates of sorbic acid are also relatively nontoxic (Sofos & Busta, 1993).

IX. SUMMARY

Overall, sorbic acid appears to be one of the safest food preservatives available. In addition to being considered less toxic, sorbic acid is also more effective than benzoate or propionate in preserving foods such as cheese, fish and bakery products. The usefulness of sorbates as preservatives relies on their ability to inhibit the growth of numerous yeasts, molds, and bacteria. Inhibition of microorganisms by sorbates, however, is variable with microbial types, species, strains, food, and environmental conditions. Some microbial strains are resistant to inhibition by sorbate or even metabolize the compound under certain conditions. In general, however, sorbates are considered effective food preservatives when used under sanitary conditions and in products processed following good manufacturing practices. While the mechanisms of antimicrobial activity by sorbic acid are not fully defined, inhibition of bacterial spore germination is believed to occur at the connecting reactions, probably through action on spore membranes or protease enzymes involved in germination. Inhibition of microbial metabolic function is probably due to morphological alterations in the structure of the cells, changes in the genetic material, alterations in cell membranes, and inhibition of enzymes or transport functions.

X. REFERENCES

1. Al-Dagal, M.M., and Bazaraa, W.A. 1999. Extension of shelf life of whole and peeled shrimp with organic acid salts and bifidobacteria. *J. Food Prot.* 62:51-56.
2. Alli, Z.L.I. and Kermasha, S. 1989. Use of acidification, low temperature and sorbates for storage of orange juice. *J. Food Sci.* 54:674-678.
3. Amundson, C.M., Sebranek, J.G., Kraft, A.A., Rust, R.E., Wagner, M.K. and Robach, M.C. 1981. Effect of belly handling (chillled, frozen, prerigor) and smoking delay on regular and sorbate-cured bacon. *J. Food Sci.* 47:222-225.
4. AOAC Int.. 1995. *Official Methods of Analysis,* 16th ed., AOAC International, Gaithersburg, Maryland.
5. Banks, J.G., Morgan, S., and Stringer, M.F. 1988. Inhibition of heated *Bacillus* spores by combinations of potassium sorbate, sodium, benzoate, pH and organic acids. *Legensm. Wiss. Technol.* 21:250-255.
6. Benassi, M.T., Cecchi, H.M. 1998. Method development for the simultaneous determination of carboxylic acids, phenolic compounds, and sorbid acid in white wines. *J. Liq. Chromatogr. Rel. Technol.* 21:491-501.
7. Benedict, R.C. 1980. Biochemical basis for nitrite-inhibition of *Clostridium botulinum* in cured meats. *J. Food Prot.* 43:877-891.
8. Berry, B.W. and Blumer, T.N. 1981. Sensory, physical and cooking characteristics of bacon processed with varying levels of sodium nitrite and potassium sorbate. *J. Food Prot.* 46:321-327.
9. Berry, B.W., Quick, J.A., Douglass, K.W., and Tennent, I.N. 1981. Shelf-life characteristics of bacon processed with various levels of sodium nitrite and potassium sorbate. *J. Food Prot.* 46:596-600.
10. Beuchat, L.R. 1981a. Effects of potassium sorbate and sodium benzoate on inactivating yeasts heated in broths containing sodium chloride and sucrose. *J. Food Prot.* 44:765-769.
11. Beuchat, L.R. 1981b. Synergistic effects of potassium sorbate and sodium benzoate on thermal inactivation of yeasts. *J. Food Sci.* 46:771-777.
12. Beuchat, L.R. 1981c. Combined effects of solutes and food preservatives on rates of inactivation of an colony formation by heated spores and vegetative cells of molds. *Appl. Environ. Microbiol.* 41:472-477.
13. Beuchat, L.R. 1981d. Effects of potassium sorbate and sodium benzoate on inactivating yeasts in broths containing NaCl and sucrose. *J. Food Prot.* 44:765-769.

14. Bills, S., Restaino, L., and Lenovich, L.M. 1982. Growth response of an osmotolerant sorbate-resistant yeast, *Saccharomyces rouxii*, at different sucrose and sorbate levels. *J. Food Prot.* 45:1120-1124, 1128.

15. Bizri, J.N., and Wahem, I.A. 1994. Citric acid and antimicrobials affect microbiological stability and quality of tomato juice. *J. Food Sci.* 59:130-134.

16. Blocher, J.C. and Busta, F.F. 1983. Influence of potassium sorbate and reduced pH on the growth of vegetative cells of four strains of type A and B *Clostridium botulinum. J. Food Sci.* 48:574-575, 580.

17. Blocher, J.C. and Busta, F.F. 1985. Multiple modes of inhibition of spore germination and outgrowth by reduced pH and sorbate. *J. Appl. Bacteriol.* 59:469-478.

18. Cancalon, P.F. 1999. Analytical monitoring of citrus juices by using capillary electrophoresis. *J. AOAC Int.* 82:95-106.

19. Cerruti, P., Alzamora, S.M., and Chirife, J. 1988. Effect of potassium sorbate and sodium bisulfite on thermal inactivation of *Saccharomyces cerevisiae* in media or lowered water activity. *J. Food Sci.* 53:1911-1912.

20. Cerruti, P., Alzamora, S.M., and Chirife, J. 1990. A multi-parameter approach to control the growth of *Saccharomyces cerevisiae* in laboratory media. *J. Food Sci.* 55:837-840.

21. Chambers, E., IV, Bowers, J.A., Prusa, K. and Craig, J. 1981. Sensory attributes and instron measurements of reduced-nitrite poultry frankfurters with sorbic acid or potassium sorbate. *J. Food Sci.* 47:207-209.

22. Chang, C.-W., Sebranek, J.G., Walker, H.W., and Galloway, D.E. 1983. Packaging film permeability in conjunction with sodium nitrite, potassium sorbate or lactic acid starter culture for control of *Clostridium sporogenes* (PA 3679) growth in sliced bologna. *J. Food Sci.* 48:861-864.

23. Cheng, L., and Piper, P.W. 1994. Weak acid preservatives block the heat shock response and heat-shock-element-directed lacZ expression of low pH *Saccharomyces cerevisiae* cultures, an inhibitory action partially relieved by respiratory deficiency. *Microbiology* 140:1085-1096.

24. Cheng, L., Moghraby, J., and Piper, P.W. 1999. Weak organic acid treatment causes a trehalose accumulation in low-pH cultures of *Saccharomyces cerevisiae*, not displayed by the more preservative-resistant *Zygosaccharomyces bailii. FEMS Microbiol. Lett.* 170:89-95.

25. Chichester, D.F. and Tanner, F.W. 1972. Antimicrobial food additives. In *Handbook of Food Additives*, ed. T.E. Furia, *pp.* 115-184. Boca Raton: CRC Press.

26. Cole, M.B., Franklin, J.G. and Keenan, M.H.J. 1987. Probability of growth of the spoilage yeast *Zygosaccharomyces bailii* in a model fruit drink system. *Food Microbiol.* 4:115-119.

27. Davidson, P.M., Brekke, C.J. and Branen, A.L. 1981. Antimicrobial activity of butylated hydroxyanisole, tertiary butylhydroquinone, and potassium sorbate in combination. *J. Food Sci.* 46:314-316.

28. De Luca, C., Passi, S., and Quattrucci, E. 1995. Simultaneous determination of sorbic acid, benzoic acid and parabens in foods: a new gas chromatography-mass spectrometry technique adopted in a survey on Italian foods and beverages. *Food Addit. Contam.* 112:1-7.

29. Dickens, F., Jones, H.E.H., and Waynforth, H.B. 1968. Further tests on the carcinogenicity of sorbic acid in the rat. *Br. J. Cancer* 22:762-768.

30. Draughon, F.A., Sung, S.C., Mount, J.R., and Davidson, P.M. 1982. Effect of the parabens with and without nitrite on *Clostridium botulinum* growth and toxin production in canned pork slurry. *J. Food Sci.* 47:1635-1637, 1642.

31. Edinger, W.D. and Splittstoesser, D.F. 1986a. Production by lactic acid bacteria of sorbic alcohol, the precursor of the geranium odor compound. *Am. J. Enol. Vitic.* 37:34-38.

32. Edinger, W.D. and Splittstoesser, D.F. 1986b. Sorbate tolerance by lactic acid bacteria associated with grapes and wine. *J. Food Sci.* 51:1077-1078.

33. Einarsson, H., and Lauzon, H.L. 1995. Biopreservation of brined shrimp (*Pandalus borealis)* by bacteriocins from lactic acid bacteria. *Appl. Environ. Microbiol.* 61:669-676.

34. Eklund, T. 1985. The effect of sorbic acid and esters of P-hydroxybenzoic acid on the protonmotive force in *Escherichia coli* membrane vesicles. *J. Gen. Microbiol.* 13:73-76.

35. Eklund, T. 1989. Organic acids and esters. In *Mechanisms of Action of Food Preservation Proceedings*, ed. G.W. Gould, *pp.* 161-200. London: Elsevier Applied Science.

36. Eklund, T. 1983. The antimicrobial effect of dissociated and undissociated sorbic acid at different pH levels. *J. Appl. Bacteriol.* 54:383-389.

37. El Halouat, A., Gourama, H., Uyttendaele, M., and Debevere, J.M. 1998. Effects of modified atmosphere packaging and preservatives on the shelf-life of high moisture prunes and raisins. *Int. J. Food Microbiol.* 41:177-184.

38. Elliott, P.H., Tomlins, R.I., and Gray, R.J.H. 1985. Control of microbial spoilage on fresh poultry using a combination potassium sorbate/carbon dioxide packaging system. *J. Food Sci.* 50:1360-1363.

39. El-Shenawy, M.A. and Marth, E.H. 1988. Inhibition and inactivation of *Listeria monocytogenes* by sorbic acid. *J. Food Prot.* 51:842-847.

40. Fang, Shao W., Li, Chin F., and Shih, D.Y.C. 1994. Antifungal activity of chitosan and its preservative effect on low-sugar candied kumquat. *J. Food Prot.* 57:136-140.

41. Finol, M., Marth, E.H., and Lindsay, R.C. 1982. Depletion of sorbate from different media during growth of *Penicillium* species. *J. Food Prot.* 45:398-404, 409.

42. Fletcher, G.C., Murrell, W.G., Statham, J.A., Stewart, B.J., and Bremner, H.A. 1988. Packaging of scallops with sorbate: An assessment of the hazard from *Clostridium botulinum*. *J. Food Sci.* 53:349-352, 358.

43. Gaunt, I.F., Butterworth, K.R. Hardy, J., and Gangolli, S.D. 1975. Long-term toxicity of sorbic acid in the rat. *Food Cosmet. Toxicol.* 13:31-45.

44. Golden, D.A., Beuchat, L.R., and Hitchcock, H.L. 1994. Changes in fatty acid composition of various lipid components of *Zygosaccharomyces rouxii* as influenced by solutes, potassium sorbate and incubation temperature. *Int. J. Food Microbiol.* 21:293-303.

45. Gooding, C.M. 1945. Process of inhibiting growth of molds. U.S. Patent 2,379,294.

46. Gray, R.J.H., Elliott, P.H., and Tomlins, R.I. 1984. Control of two major pathogens on fresh poultry using a combination potassium sorbate/carbon dioxide packaging treatment. *J. Food Sci.* 49:142-145, 179.

47. Guerrero, S., Alzamora, S.M., and Gerschenson, L.N. 1994. Development of a shelf-stable banana puree by combined factors: Microbial stability. *J. Food Prot.* 57:902-907.

48. Han, Jung Hoon, and Floros, J.D. 1998. Simulating diffusion model and determining diffusivity of potassium sorbate through plastics to develop antimicrobial packaging films. *J. Food Process. Preserv.* 22:107-122.

49. Hendy, R.J., Hardy, J., Gaunt, I.F., Kiss, I.S., and Butterworth, K.R. 1976. Long-term toxicity studies of sorbic acid in mice. *Food Cosmet. Toxicol.* 14:381-386.

50. Holyoak, D.C., Stratford, M., McMullin, Z., Cole, M.B., Crimmins, K., Brown, A.J.P., and Coote, P.J. 1996. Activity of the plasma membrane H super (+) -TPase and optimal glycolytic flux are required for rapid adaptation and growth of *Saccharomyces cerevisiae* in the presence of the weak-acid preservative sorbic acid. *Appl. Environ. Microbiol.* 62:3158-3164.

51. Horwood, J.F., Lloyd, G.T., Ramshaw, E.H., and Stark, W. 1981. An off-flavour associated with the use of sorbic acid during feta cheese maturation. *Aust. J. Dairy Technol.* 36:38-40.

52. Huhtanen, C.N., Feinberg, J.I., Trenchard, H., and Phillips, J.G. 1983. Acid enhancement of *Clostridium botulinum* inhibition in ham and bacon prepared with potassium sorbate and sorbic acid. *J. Food Prot.* 46:807-810.

53. Kada, T. 1974. DNA-damaging products from reaction between sodium nitrite and sorbic acid. *Rep. Natl. Inst. Genet. Jpn.* 24:43-44.

54. Kasrazadeh, J. and Genigeorgis, C. 1995. Potential growth and control of *Escherichia coli* O157:H7 in soft Hispanic type cheese. *Int. J. Food Microbiol.* 25:289-300.

55. Kemp, J.D., Langlois, B.E., and Fox, J.D. 1983. Effect of potassium sorbate and vacuum packaging on the quality and microflora of dry-cured intact and boneless hams. *J. Food Sci.* 48:1709-1714.

56. Kito, Y., Namiki, M., and Tsuji, K. 1978. A new N-nitropyrrole, 1,4-dinitro-2-methylpyrrole, formed by the reaction of sorbic acid with sodium nitrite. *Tetrahedron* 34:505-508.

57. Kouassi, Y., and Shelef, L.A. 1995a. Listeriolysin O secretion by *Listeria monocytogenes* in the presence of cysteine and sorbate. *Lett. Appl. Microbiol.* 20:295-299.

58. Kouassi, Y., and Shelef, L.A. 1995b. Listeriolysin O secretion by *Listeria monocytogenes* in broth containing salts of organic acids. *J. Food Prot.* 58:1314-1319.

59. Lenovich, L.M., Buchanan, R.L., Worley, N.J., and Restaino, L. 1988. Effect of solute type on sorbate resistance in *Zygosaccharomyces rouxii*. *J. Food Sci.* 43:914-915.

60. Liewen, M.B. and Marth, E.H. 1985b. Production of mycotoxins by sorbate-resistant molds. *J. Food Prot.* 48:156-157.

61. Liewen, M.B. and Marth, E.H. 1985a. Growth and inhibition of microorganisms in the presence of sorbic acid: A review. *J. Food Prot.* 48:364-375.

62. Liewen, M.B. and Marth, E.H. 1985c. Growth of sorbate-resistant and –sensitive strains of *Penicillum roqueforti* in the presence of sorbate. *J. Food Prot.* 48:525-529.

63. Liu, X., Yousef, A.E., and Chism, G.W. 1997. Inactivation of *Escherichia coli* O157:H7 by the combination of organic acids and pulsed electric field. *J. Food Saf.* 16:287-299.

64. Lopez, M., Mazas, M., Gonzalez, I., Gonzalez, J., and Bernardo, A. 1996. Thermal resistance of *Bacillus stearothermophilus* spores in different heating systems containing some approved food additives. *Lett. Appl. Microbiol.* 23:187-191.

65. Lück, E. 1976. Sorbic acid as a food preservative. *Int. Flavors Food Addit.* 7:122-124, 127.

66. Lück, E. 1980. *Antimicrobial Food Additives, Characteristics, Uses, Effects.* Berlin: Springer-Verlag.

67. Mackey, B.M., Forestiere, K., and Isaacs, N. 1995. Factors affecting the resistance of *Listeria monocytogenes* to high hydrostatic pressure. *Food Biotech.* 9:1-11.

68. Mandrou, B., Nolleau, V., Gastaldi, E., and Fabre, H. 1999. Solid phase extraction as a clean up procedure for the liquid chromatographic determination of benzoic and sorbic acids in fruit derived products. *J. Liq. Chromatogr. Rel. Technol.* 21:829-842.

69. Mannino, S., Cosio, M.S. 1996. Determination of benzoic and sorbic acids in food by microdialysis sampling coupled with HPLC and UV detection. *Italian J. Food Sci.* 8:311-316.

70. Marshall, D.L. and Bullerman, L.R. 1986. Effect of sucrose esters in combination with selected mold inhibitors in growth and aflatoxin production by *Aspergillus parasiticus. J. Food Prot.* 49:378-382, 388.

71. Marth, E.H., Capp, C.M. Hasenzahl, L., Jackson, H.W., and Hussang, R.V. 1966. Degradation of potassium sorbate by *Penicillium* species. *J. Dairy Sci.* 49:1197-1205.

72. Mercier, J.P., Chaimbault, P., Morin, P., Dreux, M. and Tambute, A. 1998. Identification of phosphonic acids by capillary electrophoresis-ionspray mass spectrometry. *J. Chromatogr. A.* 825:71-80.

73. Mihyar, G.F., Yamani, M.I., and Al-Sa'Ed, A.K. 1997. Resistance of yeast flora of labaneh to potassium sorbate and sodium benzoate. *J. Dairy Sci.* 80:2304-2309.

74. Miller, L.G., and Kaspar, C.W. 1994. *Escherichia coli* O157:H7 acid tolerance and survival in apple cider. *J. Food Prot.* 57:460-464.

75. Monnet, D., Vidal, D., and Creach, O. 1988. Influence of metabolic and physical factors on production of diacetoxyscirpendol *Fusarium sambucinum. Appl. Environ. Microbiol.* 54:2167-2173.

76. Montano, A., Sanchez, A.H., and Rejano, J. 1995. Determination of benzoic and sorbic acids in packaged vegetable products. Comparative evaluation of methods. *Analyst* 120:2483-2487.

77. Morad, M.M., Branen, A.L., and Brekke, C.J. 1982. Antimicrobial activity of butylated hydroxyanisole and potassium sorbate against natural microflora in raw turkey meat and *Salmonella typhimurium* in cooked turkey meat. *J. Food Prot.* 45:1038-1043.

78. National Academy of Sciences. 1981. *The Heath Effects of Nitrate, Nitrite and N-nitroso Compounds,* Washington DC: National Academy Press.

79. National Academy of Sciences. 1982. *Alternatives to the Current Use of Nitrite in Foods,* Washington DC: National Academy Press.

80. Nelson, K.A., Busta, F.F., Sofos, J.N., and Wagner, M.K. 1983. Effect of polyphosphates in combination with nitrite-sorbate on *Clostridium botulinum* growth and toxin production in chicken frankfurther emulsions. *J. Food Prot.* 46:846-850.

81. Neves, L., Pampulha, M.E., and Loureiro-Dias, M.C. 1994. Resistance of food spoilage yeasts to sorbic acid. *Lett. Appl. Microbiol.* 19:8-11.

82. Oloyede, O.B., and Scholefield, J. 1994. Inhibition of *Bacillus* spores by combinations of heat, potassium sorbate, NaCl and pH. *World J. Microbiol. Biotechnol.* 10:579-582.

83. Palou, E., Lopez-Malo, A., Barbosa-Canovas, G.V., Welti-Chanes, J., and Swanson, B.G. 1997. High hydrostatic pressure as a hurdle for *Zygosaccharomyces bailii* inactivation. *J. Food Sci.* 62:855-857.

84. Pant, I., and Trenerry, V.C. 1995. The determination of sorbic acid and benzoic acid in a variety of beverages and foods by micellar electrokinetic capillary chromatography. *Food Chem.* 53:219-236.

85. Paquette, M.W., Robach, M.C., Sofos, J.N., and Busta, F.F. 1980. Effects of various concentrations of sodium nitrite and potassium sorbate on color and organoleptic qualities of commercially prepared bacon. *J. Food Sci.* 45:1293-1296.

86. Patrizi, A., Orlandi, C., Vincenzi, C., and Bardazzi, F. 1999. Allergic contact dermatitis caused by sorbic acid: rare occurrence. *Am. J. Contact Dermat.* 10:52.

87. Piccinin, D.M., and Shelef, L.A. 1995. Survival of *Listeria monocytogenes* in cottage cheese. *J. Food Prot.* 58:128-131.

88. Piper, P., Mahe, Y., Thompson, S., Pandjaitan, R., Holyoak, C., Egner, R., Muehlbauer, J., Coote, P., and Kuchler, K. 1998. The Pdr12 ABC transporter is required for the development of weak organic acid resistance in yeast. *EMBO J.* 17:4257-4265.

89. Restaino, L., Komatsu, K.K., and Syracuse, M.J. 1981. Effects of acids on potassium sorbate inhibition of food-related microorganisms in culture media. *J. Food Sci.* 47:134-138.

90. Restaino, L., Lenovich, L.M., and Bills, S. 1982. Effect of acids and sorbate combinations on the growth of four osmophillic yeasts. *J. Food Prot.* 45:1138-1142.

91. Rice, K.M., and Pierson, M.D. 1982. Inhibition of *Salmonella* by sodium nitrite and potssium sorbate in frankfurters. *J. Food Sci.* 47:1615-1617.

92. Robach, M.C. 1979. Extension of shelf-life of fresh, whole broilers using a potassium sorbate dip. *J. Food Prot.* 42:855-857.

93. Robach, M.C. and Adam, K. 1980. Preliminary observations on human reactions related to sensory evaluation of cooked, commercial bacon. *J. Food Prot.* 43:933-935, 938.

94. Robach, M.C. and Sofos, J.N. 1982. Use of sorbates in meat products, fresh poultry and poultry products: A review. *J. Food Prot.* 44:614-622.

95. Robach, M.C., DiFate, V.G., Adams, K., and Kier, L.D. 1980a. Evaluation of the mutagenicity of sorbic acid-sodium nitrite reaction products produced in bacon-curing brines. *Food Cosmet. Toxicol.* 18:237-240.

96. Robach, M.C., Owens, J.L., Paquette, M.W., Sofos, J.N., and Busta, F.F. 1980b. Effects of various concentrations of sodium nitrite and potassium sorbate on nitrosamine formation in commercially prepared bacon. *J. Food Sci.* 45:1280-1284.

97. Robach, M.C., To, E.C., Meydav, S., and Cook, C.F. 1980c. Effect of sorbates on microbiological growth in cooked turkey products. *J. Food Sci.* 45:638-640.

98. Roland, J.O. and Beuchat, L.R. 1984. Biomass and patulin production by *Byssochlamys nivea* in apple juice as affected by sorbate, benzoate, SO_2 and temperature. *J. Food Sci.* 49:402-406.

99. Ryser, E.T. and Marth, E.H. 1988. Survival of *Listeria monocytogenes* in cold-pack cheese food during refrigerated storage *J. Food Prot.* 51:615-621.

100. Salmond, C.V., Knoll, R.G., and Booth, I.R. 1984. The effect of food preservatives on pH homeostasis in *Escherichia coli. J. Gen. Microbiol.* 130-2845-2850.

101. Shtenberg, A.J. and Ignat'ev, A.D. 1970. Toxicological evaluation of some combinations of food preservatives. *Food Cosmet. Toxicol.* 8:369-380.

102. Sofos, J.N. 1994. Antimicrobial agents. In *Food Additive Toxicology*, pp. 501-529. New york: Marcel Dekker, Inc.

103. Sofos, J.N. 1981. Nitrite, sorbate and pH interaction in cured meat products. *Proc. Rec. Meat Conf.* American Meat Science Association, Chicago, IL. 34:104-120.

104. Sofos, J.N. 1986a. Use of phosphates in low-sodium meat products. *Food Technol.* 40(9):52-69.

105. Sofos, J.N. 1986b. Antimicrobial activity and functionality of reduced sodium chloride and potassium sorbate in uncured poultry products. *J. Food Sci.* 51:16-19, 23.

106. Sofos, J.N. 1986c. Influence of sodium chloride, sorbic acid and potassium sorbate on weight losses and gas production in emulsified meat products. *Proc. Eur. Meet. Meat Res. Work* 32:303.

107. Sofos, J.N. 1989. *Sorbate Food Preservatives*. Boca Raton, FL: CRC Press.

108. Sofos, J.N. 1992. Sorbic acid, mode of action. In *Encyclopedia of Microbiology*, Volume 4, ed. J. Lederberg, *pp.* 43-52. San Diego, CA: Academic Press, Inc.

110. Sofos, J.N. and Busta, F.F. 1980. Alternatives to the use of nitrite as an antibotulinal agent. *Food Technol.* 34(5):244-251.

111. Sofos, J.N. and Busta, F.F. 1981. Antimicrobial activity of sorbate. *J. Food Prot.* 44:614-622.

112. Sofos, J.N. and Busta, F.F. 1993. Sorbic acid and sorbates. In *Antimicrobials in Food*, ed. P.M. Davidson and A.L. Branen, pp. 49-94. New York: Marcel Dekker, Inc.

113. Sofos, J.N. and Busta, F.F. 1999. Chemical food preservatives. In *Principles and Practice of Disinfection, Preservation and Sterilization*, 3rd edition, eds. A.D. Russell, W., B. Hugo and G.A.J. Ayiffe, pp.485-541. Oxford: Blackwell Scientific.

114. Sofos, J.N., Busta, F.F., and Allen, C.E. 1979a. Botulism control by nitrite and sorbate in cured meats: A review. *J. Food Prot.* 42:739-770.

115. Sofos, J.N., Busta, F.F., and Allen, C.E. 1979c. *Clostridium botulinum* control by sodium nitrite and sorbic acid in various meat and soy protein formulations. *J. Food Sci.* 44:1662-1671.

116. Sofos, J.N., Busta, F.F., and Allen, C.E. 1979d. Sodium nitrite and sorbic acid effects on *Clostridium botulinum* spore germination and total microbial growth in chicken frankfurther emulsions during temperature abuse. *Appl. Environ. Microbiol.* 37:1103-1109.

117. Sofos, J.N., Busta, F.F., and Allen, C.E. 1980a. Influence of pH on *Clostridium botulinum* control by sodium nitrite and sorbic acid in chicken emulsions. *J. Food Sci.* 45:7-12.

118. Sofos, J.N., Busta, F.F., Bhothipaksa, K., Allen, C.E. Robach, M.C., and Paquette, M.W. 1980b. Effects of various concentrations of sodium nitrite and potassium sorbate on *Clostridium botulinum* toxin production in commercially prepared bacon. *J. Food Sci.* 45:1285-1292.

119. Sofos, J.N., Busta, F.F., Bhothipaksa, K., and Allen, C.E. 1979b. Sodium nitrite and sorbic acid effects on *Clostridium botulinum* toxin formation in chicken frankfurter-type emulsions. *J. Food Sci.* 44:668-672, 675.

120. Sofos, J.N., Pierson, M.D. Blocher, J.C., and Busta, F.F. 1986. Mode of action of sorbic acid on bacterial cells and spores. *J. Food Microbiol.* 3:1-17.

121. Sousa, M.J., Miranda, L., Corta-Real, M., and Leao, C. 1996. Transport of acetic acid in *Zygosaccharomyces bailii*: Effects of ethanol and their implications on the resistance of the yeast to acidic environments. *Appl. Environ. Microbiol.* 62:3152-3157.

122. Splittstoesser, D.F., and Churey, J.J. 1989. Effect of low concentrations of sorbic acid on the heat resistance and viable recovery of *Neosartorya fisheri* ascospores. *J. Food Prot.* 52:821-822.

123. Splittstoesser, D.F., Lammers, J.M., Downing, D.L., and Churey, J.J. 1989. Heat resistance of *Eurotium herbariorum*, a xerophilic mold. *J. Food Sci.* 54:683-685.
124. Splittstoesser, D.F., McLellan, M.R., and Churey, J.J. 1996. Heat resistance of *Escherichia coli* O157:H7 in apple juice. *J. Food Prot.* 59:226-229.
125. Statham, J.A., and McMeekin, T.A. 1988. The effect of potassium sorbate on the structural integrity of *Alteromonas putrefaciens*. *J. Appl. Bacteriol.* 65:469-476.
126. Statham, J.A., Bremner, H.A. and Quarmby A.R. 1985. Storage of Morwong (*Nemadactylus macropterus* Block and Schneider) in combination of polyphosphate, potassium sorbate and carbon dioxide at 4°C. *J. Food Sci.* 50:1580-1584, 1587.
127. Stead, D. 1995. The effect of hydroxycinnamic acids and potassium sorbate on the growth of 11 strains of spoilage yeasts. *J. Appl. Bacteriol.* 78:82-87.
128. Sugihara, N., Shimomichi, K., and Furuno, K. 1997. Cytotoxicity of food preservatives in cultured rat hepatocytes loaded with linolenic acid. *Toxicology.* 120:29-36.
129. Tellez-Giron, A., Acuff, G.R., Vanderzant, C., Rooney, W., and Waniska R.D. 1988. Microbiological characteristics and shelf life of corn tortillas with and without antimicrobial agents. *J. Food Prot.* 51:945-948.
130. Thomas, D.J. and Wagner, M.K. 1987. Effect of sodium acid pyrophosphate and/or potassium sorbate on *Staphyloccoccus aureus* FRI-100 growth and toxin production. *J. Food Sci.* 52:793-784, 800.
131. Tompkin, R.B., Christiansen, L.N., Shaparis, A.B., and Bolin, H. 1974. Effect of potassium sorbate on *salmonellae, Staphylococcus aureus, Clostridium perfringens* and *Clostridium botulinum* in cooked, uncured sausage. *Appl. Microbiol.* 28:262-264.
132. Tsai, W.J., Liewen, M.B., and Bullerman, L. B. 1988. Toxicity and sorbate sensitivity of molds isolated from surplus commodity cheese. *J. Food Prot.* 51:457-462.
133. U.S. Department of Agriculture. 1979a. Food Safety and Quality Service four-plant study to investigate the use of 40 ppm sodium nitrite and 0.26% potassium sorbate in bacon, Final Report, July.
134. U.S. Department of Agriculture. 1979b. Final report: Shelf-life, sensory, cooking and physical characteristics of bacon cured with varying levels of sodium nitrite and potassium sorbate, July.
135. Uljas, H.E., and Ingham, S.C. 1999. Combinations of intervention treatments resulting in 5-$\log_{10}$-unit reductions in numbers of *Escherichia coli* O157:H7 and *Salmonella typhimurium* DT104 organisms in apple cider. *Appl. Environ Microbiol.* 65:1924-1929.
136. Unda, J.R., Molins, R.A., and Walker, H.W. 1990. Microbiological and some physical and chemical changes in vacuum-packaged beef steaks treated with combinations of potassium sorbate, phosphate, sodium chloride and sodium acetate. *J. Food Sci.* 55:323-326.
137. Vareltzis, K., Buck, E.M., and Labbe, R.G. 1984. Effectiveness of betalain/potassium sorbate system versus sodium nitrite for color development and control of total aerobes. *Clostridium perfringens* and *Clostridium sporogenes* in chicken frankfurters. *J. Food Prot.* 47:532-536.
138. Wagner, M.K. and Busta, F.F. 1983. Effect of sodium acid pyrophosphate in combination with sodium nitrite or sodium nitrite/potassium sorbate on *Clostridium botulinum* growth and toxin production in beef/pork frankfurter emulsions. *J. Food Sci.* 48:990-991, 993.
139. Wagner, M.K., and Busta, F. F. 1984. Inhibition of *Clostrium botulinum* growth from spore inocula in media containing sodium acid pyrophosphate and potassium sorbate with or without added sodium chloride. *J. Food Sci.* 49:1588-1594.
140. Walters, F.S., and Mullin, C.A. 1994. Biosynthetic alternatives for acetogenic production of sorbic acid in the potato aphid (*Homoptera: Aphididae*). *Arch. Insect Biochem. Physiol.* 27:249-264.
141. Warth, A.D. 1977. Mechanism of resistance of *Saccharomyces bailii* to benzoic, sorbic and other weak acids used as food preservatives. *J. Appl. Baceteriol.* 43:215-230.
142. Warth, A.D. 1985. Resistance of yeast species to benzoic and sorbic acids and to sulfur dioxide. *J. Food Prot.* 48:564-569.
143. Wederquist, H.J., Sofos, J.N., and Schmidt, G.R. 1994. *Listeria monocytogenes* inhibition in refrigerated vacuum packaged turkey bologna by chemical additives. *J. Food Sci.* 59:498-500.
144. Weng, Y.M., Chen, M.J. 1997. Sorbic anhydride as antimycotic additive in polyethylene food packaging films. *Leben. Wiss. Tech. J. Food Sci.* 30:485-487.
145. Wind, C.E., and Restaino, L. 1995. Antimicrobial effectiveness of potassium sorbate and sodium benzoate against *Zygosaccharomyces bailii* in a salsa mayonnaise. *J. Food Prot.* 58:1257-1259.
146. Zamora, M.C., and Zaritzky, N.E. 1987. Antimicrobial activity of undissociated sorbic acid in vacuum packaged beef. *J. Food Sci.* 52:1449-1454.
147. Zhuang, R., Beuchat, L.R., Chinnan, M.S., Shewfelt, R.L., and Huang, Y-W. 1996. Inactivation of *Salmonella montevideo* on tomatoes by applying cellulose-based edible films. *J. Food Prot.* 59:808-812.

D.L. Marshall
L.N. Cotton
F.A. Bal'a

Acetic acid

24

I. INTRODUCTION

Naturally-occurring and relatively inexpensive, organic acids traditionally have been used as food preservatives. Whether naturally produced during fermentation or intentionally added, these acids retard microbial growth and contribute desirable sensory properties to a number of foods. Acetic acid, historically diluted in the form of vinegar, has been utilized perhaps longer than any other preservative for its antimicrobial effect, that influences food keeping-quality, wholesomeness, and safety. Reviews of vinegar contain historical information that couples the discovery of acetic acid to wine spoilage, and since wine has been used for at least 1,000 years, vinegar probably has also been used that long (Nickol, 1979; Pederson, 1979). Early uses of vinegar were for medicinal purposes, such as wet compresses or consumption as a drink elixir (Nickol, 1979).

Because acetic acid is the predominant flavoring and antimicrobial component in vinegars, the following review will focus on the importance of this acid as a direct food additive or more recently as a food processing aid, to decontaminate foods prior to distribution and consumption.

II. SYNTHESIS

A. Vinegar production

Earlier processes used for making vinegar (Nickol, 1979) were: i) the Let-alone process, ii) the Field process, iii) the Orleans process (which is also known as the slow process), iv) the Quick process, and v) the Submerged culture process. Of these, the quick process and submerged culture process were developed and used for commercial vinegar production (Nickol, 1979). Industrial developments and advances in vinegar manufacturing and use have been reviewed (Allgeier & Hildebrandt, 1960; Conner & Allgeier, 1976; Nickol, 1979; Pederson, 1979; Ebner, 1982; Adams, 1985).

Acetic acid is formed in a four-step reaction involving conversion of starch to sugar by amylases, anaerobic conversion of sugars to ethanol by yeast fermentation, con-

version of ethanol to hydrated acetaldehyde, and dehydrogenation to acetic acid by aldehyde dehydrogenase (Nickol, 1979; Canning, 1985). The last two steps are performed aerobically with the aid of acetic acid forming bacteria. Acetic acid yield from fermented
sugar is approximately 40%, with the remaining sugar metabolites either lost by
volatilization or converted into other compounds. Yield improvements can be achieved
with high aeration rates during continuous production (Ghommidh et al., 1986).

In the development of vinegar, microbial action by acetic acid forming bacteria
has been difficult to achieve with use of pure cultures (Nickol, 1979), and reviews have
described ecological aspects of aerobic fermentation of vinegar by *Acetobacter aceti* or
Gluconobacter suboxydans (Conner & Allgeier, 1976; Ebner, 1982; Ito et al., 1991; Park
et al., 1991). Additionally, Cheryan et al. (1997) reviewed the anaerobic process of acetic
acid production by *Clostridium thermoaceticum,* which is more economical and typically
is used for manufacture of industrial acetic acid. Furthermore, an unusual commensal relationship has been described between ethanol forming yeasts and acetic acid forming bacteria during the fermentation of acetic acid containing Haipao tea beverage (Liu et al.,
1996).

Main types of vinegar include white distilled vinegar, cider vinegar, wine vine-
gar, rice vinegar, and malt vinegar. Further processing of vinegar, following actual substrate conversion to acetic acid involves filtration, clarification, pasteurization (74°C), and
bottling. Concentration from 10-13% to 20-30% acetic acid can be achieved by freezing
vinegar to a slush and isolating the liquid portion by centrifugation (Ebner, 1982; Nickol,
1979) or by electrodialysis (Chukwu & Cheryan, 1996). Acetic acid concentration in vinegar may be expressed in 'grains'. For example, a 100 grain distilled vinegar is a 10%
acetic acid solution (Nickol, 1979). Regulations in the United States require vinegar to
contain at least 4% acetic acid resulting from acetic acid fermentation of ethanol-contain-
ing substrates. Labels stating the diluent used to meet the listed concentration of acid also
are needed.

B. Vinegar defects

Vinegar defects that occur during processing include off-odors associated with
presence of vinegar eels (actually a nematode *Anguilula aceti*), vinegar mites (*Iphidulus*
spp.) and/or fruit flies (*Drosophila melanogaster*), slime formation in generators caused
by *Acetobacter xylinum* (Nickol, 1979; Pederson, 1979), blackening from contaminating
metals (usually iron) (Pederson, 1979), and natural enzymatic browning from use of fruit
juices (Pederson, 1979). Occasionally yeasts can destroy acetic acid during the fermentation process. Further interest in vinegar should be directed to cited reviews, particularly
those by Conner and Allgeier (1976), Ebner (1982), and Adams (1985).

III. MOLECULAR PROPERTIES

A. Physicochemical properties

Physicochemical properties of acetic acid (CAS Reg. No. 64-19-7) and some
derivatives are listed in *TABLE 1*. The basic structure of acetic acid is shown in *FIGURE 1*.
Compounds related to acetic acid that are used as functional ingredients in foods include
sodium and calcium salt derivatives, collectively known as acetates. Acetic acid (ethanoic
acid; CH_3COOH) is a colorless liquid having a pungent vinegar-like odor and sour taste.

TABLE 1. Physical and chemical properties of acetic acid and related GRAS compounds

Property	Acetic Acid[a]	Calcium Acetate[b]	Sodium Acetate[c]
Molecular weight	60.05	158.17	82.04
Molecular formula	$C_2H_4O_2$	$C_4H_6CaO_4$	$C_2H_3NaO_2$
Dissociation constant (pKa)	4.74	NA[d]	NA
Surface tension (in contact with vapor) dynes/cm	10 °C=28.8 20 °C=27.8 50 °C=24.8	solid	solid
Melting point °C	16.6	>160 decomposes	58 for trihydrate
Boiling point °C	117.9	NA	>120 decomposes
Density	1.05	1.50	1.45 trihydrate
Solubility	Miscible with water, alcohol, glycerol, ether, and carbon tetrachloride	Water-soluble/anhydrate insoluble in ethanol, acetone, and benzene monohydrate slightly soluble in alcohol	Soluble in water; moderately soluble in alcohol
Color / Form	Liquid, pungent odor	Rod, needle-shaped crystals; granules; powder	Colorless crystals, white crystalline powder or flakes pH of solutions
pH of solutions	0.1M=2.9	0.2M=7.6	0.1M=8.9
LD_{50} Orally in rats	3.53 g/kg	4.28 g/kg	3.53 g/kg anhydrous
Safety	Causes burns / irritant to eyes skin and bronchi / erodes dental enamel	Acetone vapor at >160 °C, acrid fumes, and possibly carbon monoxide	May cause irritation to respiratory tract

[a] The Merck Index, 11th Edition. 1989. Merck and Co. Inc. Rahway, NJ.
[b] J.T. Baker. MSDS Number C0266. Phillipsburg, NJ.
[c] J.T. Baker. MSDS Number S2666. Phillipsburg, NJ.
[d] NA= Not available

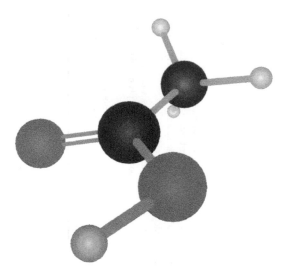

FIGURE 1. Three-dimensional illustration of acetic acid.

It is weakly ionized in aqueous solution (K_a 1.8 x 10^{-5}; pK_a 4.74; pH of aqueous solution: 1.0 M = 2.4; 0.1 M = 2.9; 0.01 M = 3.4). It is an excellent solvent for organic compounds and is miscible with water, alcohol, glycerol, and lipids. Acetic acid is incompatible with carbonate and phosphate. It is synthesized by: i) conversion of ethanol to vinegar by acetic acid producing bacteria, ii) destructive distillation of wood, or iii) acetylene and water oxidation (Ray & Sandine, 1992). Lin and Sen (1994) described the direct catalytic conversion of methane to acetic acid in an aqueous medium using rhodium trichloride as a catalyst. Product from the latter three conversions is glacial acetic acid, which is not used in foods.

Acetic acid (21 CFR 184.1005) and its salts, calcium acetate (CAS Reg. No. 62-54-4; 21 CFR 184.1185), sodium acetate (CAS Reg. No. 6131-90-4; 21 CFR 184.1721), and sodium diacetate (CAS Reg. No. 126-96-5; 21 CFR 184.1754) are affirmed as generally recognized as safe (GRAS), while calcium diacetate (21 CFR 182.6197) is GRAS. Toxicological information for acetates also is summarized in *TABLE 1.*

Advantages of using acetic acid and acetates include relatively low cost, GRAS status, and relatively low toxicity. The primary disadvantage of acetic acid use in food is its acidic taste and strong sourness intensity. Based on sensory evaluation of organic acid solutions, acetic acid is more sour than lactic, malic, or citric acids, with perceived sourness related to the pK_a of the acid (Hartwig & McDaniel, 1995). In fact, acetic acid solutions at pH 4.5 were rated more sour than citric acid solutions at pH 3.5 (Hartwig & McDaniel, 1995); however, earlier sensory research reported tartaric acid as more sour than acetic acid (Pangborn, 1963). Another potential problem with using acetic acid as an ingredient is its potential to cause delamination of packaging materials. For example, it has been shown that delamination of a polyethylene-aluminum foil occurs with exposure to 3% acetic acid (Olafsson et al., 1993).

B. Assays

Published reports on methods for the detection and quantification of acetic acid in foods have shown that various techniques can be used besides traditional pH and titratable acidity tests. Liquid chromatography has been used to determine acetic acid concentration in cucumbers (Lazaro et al., 1989). Differentiation of the organic acid content of cheese can be accomplished by high performance liquid chromatography (Bevilaqua & Califano, 1992). Miwa & Yamamoto (1993) used 2-nitrophenylhydrazide derivatization and reversed-phase ion-pair chromatography to profile acetic acid and other carboxylic acids in beverages. Corn-derived acetic acid used as an adulterant of apple cider vinegar can be detected using a stable carbon isotope ratio method (Krueger, 1992).

IV. ANTIMICROBIAL PROPERTIES

A. Antimicrobial activity

Acetic acid has been evaluated for *in vitro* antimicrobial efficacy against specific pathogens using microbial media and/or foods (Doores, 1993). Antimicrobial activity of acetates applied to foods for shelf-life extension or improving keeping quality has been evaluated as well. Important variables, such as medium composition (Zayaitz & Ledford, 1985) and incubation or storage temperature, are limiting or controlling factors influencing antimicrobial activity of acetic acid. Published reports often are difficult to compare because acetic acid concentrations have been variably expressed as percentage or weight, molarity of amount added, or as final pH of test medium acidified.

Existing literature contains extensive discussions pertaining to the causal nature and possible mechanisms underlying antimicrobial activity of acetic acid. It is commonly agreed that antimicrobial activity depends upon exposure time, temperature, type of acid, concentration of acid, dissociation level, and/or pH (Levine & Fellers, 1940; Hentges, 1967; Parish & Higgins, 1989; Sorrells et al., 1989; Conner et al., 1990; Brown & Booth, 1991; Cherrington et al., 1991a; Ouattara et al., 1997). General findings show that effectiveness of acetic acid increases as concentration increases, pH decreases, temperature increases, and microbial load decreases (Ray & Sandine, 1992). *FIGURE 2A* shows the relationship between increasing sodium acetate concentration and decreasing growth of aerobic spoilage bacteria on catfish fillets (Kim et al., 1995b). *FIGURE 2B* demonstrates that increasing duration of exposure to acetic acid also affects antimicrobial activity by increasing microbial generation time (decreasing growth rates) (Marshall & Kim, 1996). Because fungi (yeasts and molds) can grow at low pH they demonstrate more resistance to acetic acid than bacteria. Many fungi are also inhibited by acetic acid (Moon, 1983). Among bacteria, Gram-positives are typically more resistant than Gram-negatives and fermentative bacteria are more resistant than respiratory bacteria (Ray & Sandine, 1992). Bacterial spores and viruses are more resistant than vegetative bacterial cells. Anaerobic conditions increase the antimicrobial effect of acetic acid (Ray & Sandine, 1992). Also, food composition can influence antimicrobial activity. For example, the buffering capacity of a food influences pH, which in turn influences the degree of acetic acid dissociation. It has been shown that prior exposure to sublethal concentrations of acetic acid can increase the tolerance of bacteria to additional acid exposure (Abee & Wouter, 1999).

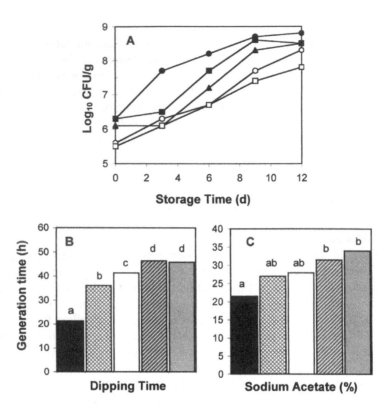

FIGURE 2A. Total aerobic bacterial counts on catfish fillets treated with sodium acetate during storage at 4°C. Control (●), and sodium acetate at concentrations of 0.25% (■), 0.50% (▲), 0.75% (○) and 1.0% (□) [Redrawn from Kim et al., 1995b)].

FIGURE 2B. Generation times of aerobic bacteria at 4°C on catfish fillets dipped in 3% acetic acid for different time. Bars from *Left* to *Right* represent time of 0, 5, 15, 30 and 60 seconds, respectively. Means having the same letter are not significantly different (P>0.05) [Redrawn from Marshal & Kim, 1996].

FIGURE 2C. Generation times of aerobic bacteria at 4°C on catfish fillets treated with different levels of sodium acetate. Bars from *Left* to *Right* represent levels of 0%, 0.25%, 0.5%, 0.75% and 1.0%, respectively. Means having the same letter are not significantly different (P>0.05) [Redrawn from Kim et al., 1995b].

B. Mechanism of action

At equimolar levels, organic acids (weak acids) have greater antimicrobial activity than strong inorganic acids. In addition, the antimicrobial effect of weak acids increases as pH decreases. This implies that the activity of organic acids is directly related to the amount of undissociated molecules (protonated acids), which increase as pH decreases due to increasing amounts of protons (Winslow & Lochridge, 1906; Levine & Fellers, 1940; Ingram et al., 1956; Eklund, 1989; Brown & Booth, 1991; Ray & Sandine, 1992).

When acetic acid is dissolved in solution, it dissociates to release free protons, which decrease solution pH. The increased number of protons on the outer surfaces of microorganisms can disrupt membrane function by denaturing enzymes and by altering permeability leading to membrane destabilization (Booth, 1985; Cassio et al., 1987; Ray & Sandine, 1992; Casal et al., 1998). Undissociated acetic acid can also traverse the lipid bilayer of bacteria and yeasts and release protons into the cytoplasm (Young & Foegeding, 1993; Casal et al., 1998; Guldfeldt & Arneborg, 1998). Excess intracellular protons can acidify the cytoplasm and cause protein denaturation and energy loss due to activation of ATP-dependant proton pumps located in the cell membrane (Booth, 1985; Pampulha & Loureiro-Dias, 1989; Ray & Sandine, 1992). Because ATP is needed for active transport of nutrients across membranes, cell starvation also can occur due to acetic acid exposure (Ray & Sandine, 1992). Furthermore, the data suggests that bacterial ribosomes are target sites for antimicrobial action of acids (Zayaitz & Ledford, 1985). Thus, it is generally thought that the activity of acetic acid is due to the action of both undissociated and dissociated molecules. When pH is high, the amount of undissociated molecules will be relatively small, which suggests that acetic acid can possibly cause extracellular damage around neutral pH. In contrast, at low pH the amount of protons is higher and the amount of undissociated acetic acid is higher, which results in significantly more internal and external cell damage. This damage can be caused by protons, undissociated molecules, and by acetic acid anions (Ray & Sandine, 1992). The damage caused may be permanent leading to cell death (bactericidal) or may be transient leading to cell injury that either prevents cell multiplication until repair or dramatically slows the rate of cell multiplication (bacteriostatic) (Foster & Hall, 1990). *FIGURE 2C* shows the increase in generation time (reduction in growth rates) of aerobic spoilage bacteria on catfish exposed to sublethal concentrations of sodium acetate (Kim et al., 1995b).

Physiological responses of certain microorganisms to varying concentrations of acetic acid have been investigated (Huang et al., 1986; Diez-Gonzalez & Russell, 1997). At bacteriostatic concentrations of acetic acid or sodium acetate, sublethal injury has been observed in *Escherichia coli* (Roth & Keenan, 1971; Przybylski & Witter, 1979; Buchanan & Edelson, 1999), *Salmonella bareilly* (Blankenship, 1981), and *Listeria monocytogenes* (Ahamed & Marth, 1990). For *Clostridium acetobutylicum* at pH 4.5, Huang et al. (1986) observed an intracellular concentration of acetic acid that was 13-fold higher than extracellular concentration. This high acid concentration intracellularly resulted in lower cytoplasm pH. Differences in regulation of metabolic processes occurring in response to increases in acetic acid concentration have been reported (Diez-Gonzalez & Russell, 1997; Ita & Hutkins, 1991). Ita & Hutkins (1991) compared intracellular pH effects of acetic, lactic, citric, and hydrochloric acids on *L. monocytogenes*. In their study, lactic and citric acids reduced intracellular pH more than acetic acid. In contrast, Young & Foegeding (1993) found that acetic acid lowered internal pH of *L. monocytogenes* more than lactic or citric acids. In both studies, acetic acid reduced numbers of *L. monocytogenes* more effectively than the other acids tested. In a study comparing responses of non-pathogenic *E. coli* K-12 to pathogenic *E. coli* O157:H7, intracellular pH was lower for *E. coli* O157:H7 as was the observed extracellular to intracellular pH difference, reportedly due to increased lactate production (Diez-Gonzalez & Russell, 1997). The presence of alternative modes of pH regulation in *E. coli* O157:H7 compared to *E. coli* K-12 represents an adaptive strategy supporting concerns about the pathogenic strain's ability to persist in certain foods and cause illness following their consumption. Fermentation acid-resistance in bacteria has been characterized by the ability to generate ATP, grow, and/or maintain a low pH membrane gradient under conditions generally bacteriostatic (Russell & Diez-Gonzalez, 1998).

C. Survival of microorganisms exposed to acetic acid

Many studies have demonstrated the ability of acetic acid to inhibit a variety of microorganisms (Osterling & Lindgren, 1993). For example, growth of the peptic ulcer causing bacterium *Helicobacter pylori* was inhibited by acetic acid (above 0.2 mol/L) and by sodium acetate (above 2.5 mol/L) (Midolo et al., 1995). Chung and Goepfert (1970) demonstrated that acetic acid was more lethal than lactic acid to *Salmonella* spp. Similarly, acetic acid had greater activity against *Yersinia enterocolitica* during low temperature incubation than lactic, citric, or sulfuric acids (Brackett, 1987; Little et al., 1992). Ryu et al. (1999) showed that acetic acid inhibited *E. coli* O157:H7 better than lactic, malic, or citric acids. Vegetative cells of *Bacillus cereus* were inhibited more by acetic acid than by formic acid (Wong & Chen, 1988). In contrast, spore germination of *B. cereus* was inhibited more by formic acid than acetic acid (Wong & Chen, 1988). Other bacilli (*B. subtilis* and *B. licheniformis*) known to cause rope spoilage of bread were inhibited by 0.1% acetic acid (Rosenquist & Hansen, 1998). Bacteriostatic activity of acetic acid against *L. monocytogenes* was reported at 0.2% and bactericidal activity was found at concentrations exceeding >0.3% (Ahamad & Marth, 1987; Farber et al., 1989). Activity against this pathogen increased as incubation temperature increased (Ahamad & Marth, 1987, Sorrells et al., 1989). In addition, the effects of pH and salt on sodium acetate inhibition of *L. monocytogenes* has been reported (Nerbrink et al., 1999). Greater than 5 log reductions in the numbers of *Campylobacter jejuni* could be achieved with a 2.5 minute exposure to 1% acetic acid at 50°C (Stern et al., 1985). One and two log reductions were observed in the population of *Staphylococcus aureus* inoculated into milk acidified to pH 5.2 and 5.0 with acetic acid, respectively (Minor & Marth, 1970). Growth and aflatoxin production by *Aspergillus parasiticus* were reduced as concentration of acetic acid increased from 0 to 0.75%, with activity of acetic acid increasing as pH decreased (Rusul et al., 1987).

Implication of acidic foods like apple cider (Besser et al., 1993) and apple juice (Steele et al., 1982) as transmission vehicles in outbreaks of *E. coli* O157:H7 infections, has shifted research focus on acetic acid. Emphasis on the acid-tolerance of *E. coli* O157:H7 has generated numerous reports investigating growth and survival of this bacterium in foods typically acidified with acetic acid, such as apple cider (Zhao et al., 1993; Miller & Kaspar, 1994; Semanchek & Golden, 1996), mayonnaise (Weagant et al., 1994; Zhao & Doyle, 1994; Erickson et al., 1995; Hathcox et al., 1995; Raghubeer et al, 1995), and even vinegar (Tsujihata et al., 1998). Acid tolerance has been determined to be a characteristic trait of *E. coli* O157:H7 rather than a response (Leyer et al., 1995). This trait is enhanced at refrigeration temperatures (Miller & Kaspar, 1994). High survival rates have been reported for *E. coli* O157:H7 during refrigerated storage of apple ciders containing preservatives (Miller & Kaspar, 1994). Also, *Salmonella typhimurium*, *S. aureus*, and *E. coli* O157:H7 survival in mayonnaise and mayonnaise-based products has been investigated (Wethington & Fabian, 1950; Doyle et al., 1982; Gomez-Lucia et al., 1987; Weagant et al., 1994; Zhao & Doyle, 1994; Raghubeer et al., 1995). Mayonnaise acidified to pH <4.1 with 0.7% acetic acid was able to reduce 10^4 *L. monocytogenes* and 10^7 *Salmonella* per gram below detectable levels at 24°C after 72 hours (Glass & Doyle, 1991). Erickson and Jenkins (1991) also compared survival rates of these two bacteria in four commercial mayonnaise products.

Conner and Kotrola (1995) reported enhanced survival of *E. coli* O157:H7 in tryptic soy broth acidified to pH 5.2 with various organic acids, including acetic acid, with storage at 4 or 10°C. Growth was observed at 25°C except for one acid treatment, mandelic acid. Abdul-Raouf et al. (1993) reported survival of *E. coli* O157:H7 in beef salads containing up to 40% mayonnaise (pH 5.4) stored at 5°C. Beef salads containing less mayonnaise and incubated at 21 or 30°C supported growth. Additionally, growth responses of *E. coli* O157:H7 in beef slurries acidified to various pH levels with acetic, citric, or lactic acids was tested (Abdul-Raouf et al., 1993). Importantly, population decreases were observed in acetic acid-acidified slurries stored at 21°C with pH levels of 5.0 and 5.4, and in slurries acidified to pH 4.7 with acetic, citric, or lactic acid. With 30°C storage of acetic acid-acidified beef slurries, numbers decreased for pH levels of 4.7 and 5.0. Conversely, acidification to pH 4.7 using lactic or citric acids did not result in decreases in numbers of *E. coli* O157:H7 at 30°C (Abdul-Raouf et al., 1993). The general observation of this study was that acetic acid was better than lactic or citric acid in controlling the growth of the bacterium. Likewise, at a given pH, acetic acid has greater ability than HCl to reduce the thermal tolerance of *Salmonella enteritidis* phage type 4 and *E. coli* O157:H7 (Perales & Garcia, 1990; Blackburn et al, 1997).

Acetic acid-resistant microorganisms have been associated with spoilage of acid-containing food products such as mayonnaise (Kurtzman et al., 1971; Smittle, 1977), salad dressings (Kurtzman et al., 1971; Smittle, 1977; Smittle & Flowers, 1982), tomato ketchup and juice (Rice & Pederson, 1954; Sorrells & Leonard, 1988; Bjorkroth & Korkeala, 1997), and mustard (Buchta et al., 1996). These studies have characterized predominant spoilage microorganisms for listed food products; however, characterization was largely limited to descriptive observations. Identification of primary spoilers was inaccurate due to difficulty in subculturing lactobacilli (Smittle & Flowers, 1982). Regardless, lactobacilli, such as *Lactobacillus fructivorans* (Kurtzman et al., 1971; Bjorkroth & Korkeala, 1997), and yeasts, particularly *Zygosaccharomyces bailii* (previously named *Saccharomyces bailii*) (Kurtzman et al., 1971; Buchta et al., 1996) or *Saccharomyces acidifaciens* (Smittle, 1977), are considered primary spoilers of food products containing appreciable amounts of vinegar. *Acetobacter* spp. and some lactic acid bacteria are resistant to acetic acid found in acetic acid preserved foods (Dankin & Radwell, 1971).

Resistance or tolerance of *Z. bailii* to acetic acid is due to an inductive, ATP-requiring physiological response (Warth, 1977). Further work by Sorrels and Leonard (1988) confirmed that acid tolerance by a *Saccharomyces* spp. involved an energy requiring system. They also demonstrated the involvement of cell walls and transport as factors influencing acetic acid resistance. *E. coli* O157:H7 previously adapted to acetic acid at pH 5.0 showed nearly a 1,000-fold higher tolerance to pH 4.0 conditions than did nonadapted cells (Brudzinski & Harrison, 1998). Acetic acid injured *L. monocytogenes* was able to survive nine times longer during acid exposure at 13°C compared to 35°C (Ahamad & Marth, 1990).

The ability of microorganisms to survive acetic acid exposure is related to a number of factors besides the previously mentioned extrinsic factors. Intrinsic inducible acid tolerance mechanisms have been reviewed (Abee & Wouters, 1999; Gahan & Hill, 1999).

A number of gene regulatory features are induced upon exposure to organic acids. Many of these code for the production of acid shock proteins (Foster, 1991; Lee et al., 1995). Using acid-repair mutants, Sinha (1986) found that the polymerase repair system was responsible for acid resistance, while excision repair was not. The SOS response is apparently not induced by exposure to acetic acid (Cherrington et al., 1991b). An important consequence of the production of acid-resistance products is the potential for development of resistance to other food processing hurdles such as low pH, heat, and low water activity (Leyer & Johnson, 1993; Garren et al., 1998; Ryu & Beuchat, 1998; Abee and Wouters, 1999). Acid-adapted *E. coli* O157:H7 was more resistant to low pH caused by acetic acid than was its nonadapted counterpart (Deng et al., 1999; Ryu et al., 1999). However, evidence suggests that thermal tolerance of *E. coli* O157:H7 is not enhanced by prior exposure to 1.5% acetic acid (Williams & Ingham, 1998). An additional concern is the potential for acid stressed or shocked foodborne pathogens previously exposed to acetic acid to better survive acid barriers encountered in the stomach, epithelial cells, and macrophages (Gahan & Hill, 1999). Thus, prior exposure to acetic acid in foods, if not bactericidal, may lead to increased virulence of pathogens. When *L. monocytogenes* was exposed to acetic acid, slightly higher levels of virulence factors superoxide dismutase and listeriolysin O were produced (Dimmig et al., 1994). The same study revealed no change in the amount of catalase produced, which is another *L. monocytogenes* virulence factor.

D. Antimicrobial combinations

The concept of hurdle technology, which uses multiple antimicrobial barriers, emerged from the practice of employing combinations of biological, chemical, and/or physical food preservation techniques (Marshall & Dickson, 1998). Organic acids are utilized frequently as acidifying components in such food preservation systems. Hence, acetic acid, used in combination with other antimicrobial compounds and/or physical preservation treatments (usually involving refrigeration or a thermal treatment) can improve food quality characteristics, such as shelf-life extension and maintenance of desirable sensory properties. Pathogen control or elimination has been an additional applied research approach of studies using acetic acid in combination with other antimicrobial treatments (Minor & Marth, 1972a; 1972b; Ito et al., 1976).

Antimicrobial synergism against *L. monocytogenes* was reported with a combination of acetic acid and monolaurin (Oh & Marshall, 1994). The combined effect of this treatment was synergistic (*Figure 3*). Tamblyn & Conner (1997) tested the effects of 125 ppm sodium lauryl sulfate, 2% ethanol, 100 ppm sorbitan monolaurate, or 15% dimethyl sulfoxide combined with 1% acetic acid to decontaminate *S. typhimurium* from poultry skin. Only sorbitan monolaurate had the ability to increase the antimicrobial effect of acetic acid. All other compounds either failed to elicit a response or acted as antagonists. A spray wash treatment consisting of 1% acetic acid combined with 3% hydrogen peroxide and 1% sodium bicarbonate was able to reduce *E. coli*, *Listeria innocua*, and *Salmonella wentworth* counts by greater than 3 logs compared to unsprayed samples (Bell et al., 1997). Dickson (1990) found that the effectiveness of a 2% acetic acid treatment on reducing the numbers of *S. typhimurium* and *L. monocytogenes* on beef carcass tissues was increased when combined with osmotic stress. The osmotic stress included either surface drying or prewashing with 20% NaCl prior to acetic acid exposure. An additional 1.5 to 2.0 log reduction in counts was seen with the combined acetic acid-osmotic stress treat-

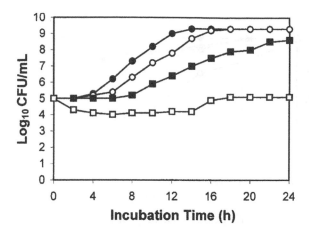

FIGURE 3. Effect of acetic acid combined with monolaurin on inhibition of *Listeria monocytogenes* at 37°C. Control (●), 18 μM monolaurin (○), 21.5 mM acetic acid (■), and combination of both (□). [Redrawn from Oh & Marshall, 1994].

ment. Pickled herring fillets require 9% salt in combination with 2.6% acetic acid to achieve destruction of *Anisakis* sp. larvae within 5 to 6 weeks of storage (Karl et al., 1994). Of course, the combination of acetic acid and salt provides pickled products with an important safety barrier against outgrowth of *Clostridium botulinum* and *S. aureus* (Nunheimer & Fabian, 1940; Minor & Marth, 1972b; Ito et al., 1976). When 1% acetic acid was combined with 0.145 to 1.1 kGy electron beam irradiation a synergistic lethal effect was observed against *E. coli* and *Lactobacillus curvatus* (Fielding et al., 1997). Thermal resistance of *Bacillus stearothermophilus* and *Bacillus coagulans* spores in a frankfurter emulsion slurry was reduced by over 50% with addition of acetic acid (Lynch & Potter, 1988). Acetic acid combined with the antimicrobial peptide nisin produces synergistic antimicrobial activity against several heat stressed *Bacillus* species (Oscroft et al., 1990). When combined with nisin, acetic acid performed better than treatments consisting of nisin combined with lactic and citric acids and performed similarly to nisin combined with glucono-delta-lactone.

A synergistic antimicrobial effect in systems containing acid mixtures has been reported for acetic and lactic acid combinations (Minor & Marth, 1972a; Adams & Hall, 1988). A combination of 2% lactic acid, 1% acetic acid, and 0.25% citric acid was more effective in reducing bacterial numbers on beef than was either 3% acetic acid alone or 3% lactic acid alone at treatment temperatures of 45 and 70°C (Anderson et al, 1992). Growth of *L. monocytogenes* could be inhibited when broth containing 0.3% sodium propionate was adjusted with acetic acid to pH 5.6 or lower (El-Shenawy & Marth, 1992). Likewise, growth of *L. monocytogenes* was prevented when 0.3% potassium sorbate was combined with acetic acid at pH below 5.0 (El-Shenawy & Marth, 1991). Refrigerated shelf-life of pork chops treated with a 60:40 mixture of acetic:propionic acid (1.36 M) was extended 2 days longer than chops treated with 250 ppm sodium hypochlorite (Carpenter et al., 1986). Buffalo meat treated with a combination of acetic acid with lactic or propionic acids had a longer refrigerated shelf-life than untreated controls (Surve et al., 1991). Sodium benzoate has been tested in combination with organic acids, with acetic acid

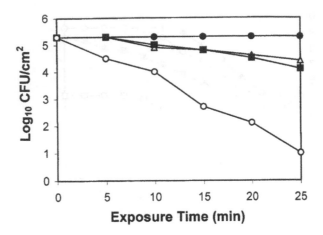

FIGURE 4. Inactivation of 1-day adherent *Listeria monocytogenes* cells on stainless steel exposed to acetic acid and monolaurin at 25°C. Control (●), 0.1 mg/ml monolaurin (△), 1% acetic acid (■), and combination of both (○). [Redrawn from Oh & Marshall, 1996].

enhancing benzoate activity against *L. monocytogenes* more so than lactic or citric acids (El-Shenawy & Marth, 1989). In this study, however, the acids were added to achieve desired pH levels (5.0 or 5.6) and were not compared by weight, molarity, or dissociation. Growth of *S. aureus* during tempeh manufacture could be prevented by acidification of soy beans during water soaking (Tuneel & Goktan, 1990). Parameters needed were addition of 0.6% lactic acid, 0.06% acetic acid, and a pH of 4.0.

V. APPLICATIONS

A. General, non-commodity-based usage

Use of acetic acid as a sanitizing agent of food contact surfaces has shown promise. One percent acetic acid combined with 50 to 100 μg/ml monolaurin effectively eliminated a 10^5 cfu/cm^2 biofilm of *Listeria monocytogenes* on stainless steel (*FIGURE 4*) (Oh & Marshall, 1996). Entani et al. (1997) reported effective reductions (>3 logs) in numbers of *E. coli* O157:H7 on cutting boards after surface treatment with vinegar solutions (1 to 2% acetic acid) containing added salt (3 to 7%).

B. Decontamination of meats

Effectiveness of acetic acid for decontamination of beef and pork carcasses has been reviewed (Dickson & Anderson, 1992; Frederick, 1994; Siragusa, 1995; Bolder, 1997; Smulders & Greer, 1998). The effectiveness of acetic acid as a carcass decontaminating treatment depends on a number of factors. In general, the efficacy of acetic acid increases with increasing treatment temperature, increasing acid concentration, and decreasing level of carcass contamination (Greer & Dilts, 1992). Meat-associated pathogens, such as *Y. enterocolitica, L. monocytogenes, S. typhimurium, E. coli, Campylobacter jejuni,* and *S. aureus*, are more resistant to acetic acid than are spoilers, such as *Pseudomonas fragi* and *Brochothrix thermosphacta* (Greer & Dilts, 1992).

Of the many carcass decontamination procedures, treatment with acetic acid has been shown to be the best chemical treatment over steam, hot water, hypochlorite, hydrogen peroxide, or trisodium phosphate (Anderson et al., 1979; Cabedo et al., 1996). Acetic acid treated beef plates had at least a 16 day shelf-life extension compared to untreated plates as determined by total plate counts (Anderson et al., 1979) presumably due to the 1.5 log decrease in total microbial counts achieved with 3% acetic acid treatment (Anderson et al., 1980). Beef carcasses spray washed with 1% acetic acid and then stored under vacuum had lower aerobic plate counts than untreated samples, with lactobacilli predominating (Hamby et al., 1987). Pork products treated with 2-3% acetic acid and vacuum packaged had lower aerobic, anaerobic, psychrotrophic, lactic acid, and Enterobacteriaceae bacterial counts, which yielded products having much longer refrigerated shelf-life (Mendonca et al., 1989; Cacciarelli et al., 1983). Hot (52°C) acetic acid (1.5%) carcass wash solution was more effective in reducing microbial counts than was cooler solution (14.4°C) (Anderson et al., 1987). Similar findings were found using 55°C acetic acid dips compared with 25°C dips to decontaminate lamb carcasses (Anderson et al., 1988). Increasing temperature of dip solution to 70°C allowed for substantial reduction in total aerobic plate counts, Enterobacteriaceae counts, *E. coli* counts, and *Salmonella* counts even with low (1%) acetic acid concentrations (Anderson & Marshall, 1989).

Beef carcasses contaminated with feces were effectively decontaminated by washing followed by organic acid treatment than by washing alone or trimming (Hardin et al., 1995). Aerobic plate counts, coliform counts, and the incidence of *Salmonella* contamination all decreased on pork cheek meat treated with 2% acetic acid (Frederick et al., 1994). Dickson (1992) demonstrated that 65% of a *S. typhimurium* population was sublethally injured on lean and fat tissues of beef when treated with 2% acetic acid. When acetic acid was combined with calcium alginate gel and applied to beef tissue, counts of *S. typhimurium*, *L. monocytogenes*, and *E. coli* O157:H7 were generally reduced to a similar level as seen with acetic acid alone (Siragusa & Dickson, 1993). A 2% acetic acid spray wash was superior to spray washing with 0.003% chlorine, 5% hydrogen peroxide, or 12% trisodium phosphate in reducing the number of aerobic plate counts on lamb carcasses (Kochevar et al., 1997). Effectiveness of 2% acetic acid can be increased when carcass surfaces are dried or treated with 20% NaCl to cause osmotic stress of bacteria (Dickson, 1990). A combination of 15 ml/L acetic acid dip (55°C) for 10 sec with vacuum packaging significantly increased refrigerated shelf-life of pork (Shay & Egan, 1986). A further refinement of acetic acid carcass washing has shown dramatic reductions in total counts and *E. coli* counts when carcasses were washed twice with 2% acetic acid during processing (Graves Delmore et al., 1998). In this study, carcasses were washed pre-evisceration and after final carcass fabrication. Count reductions achieved were 2.9 and 4.3 logs for high inoculum levels of aerobes and *E. coli*, respectively. It has been shown that bacterial reduction of carcasses by acetic acid is better on adipose tissue than on lean tissue (Cutter & Siragusa, 1994). Saponins (plant-derived triterpenoid glycosides) added as foaming agents to acetic acid spray washes did not increase effectiveness of acetic acid in reducing the levels of pathogens and spoilers on beef (Cutter, 1999).

Dorsa et al. (1997; 1998a; 1998b) assessed effectiveness of several beef carcass wash treatments (2% lactic acid, 2% acetic acid, 12% trisodium phosphate, water at 78°C, water at 32°C, and untreated as experimental control) for beef subprimals and ground beef control or elimination of *S. typhimurium*, *L. monocytogenes*, *E. coli* O157:H7, or *Clostridium sporogenes*. Their work demonstrated that acetic acid offered residual antimicrobial activity on subprimals during vacuum packaged chill storage (Dorsa et al., 1998a). For the ground beef studies (Dorsa et al., 1998b), treated beef carcass tissue was inoculated with low or high numbers of pathogens, and microbiological sampling was performed i) less than 30 minutes after treatment and ii) after 24 h storage at 4°C for treated carcasses, iii) immediately following processing to ground beef, and iv) after storage at 12°C for 3 days or 4°C for 21 days. Surface pH of acetic acid-treated carcasses increased from 4.3 (following acetic acid exposure) to 4.9 after 24 h at 4°C. Following processing to ground beef, an additional increase to pH 5.7 was reported for acetic acid treatment (Dorsa et al., 1998b). Ground beef processed from carcasses treated with acetic acid, lactic acid, TSP, or hot water (78°C) had lower numbers of recovered *E. coli* O157:H7 (high inoculum) than did ground beef made from carcasses treated with water (32°C) or left untreated (as controls). For samples inoculated with low levels of *E. coli* O157:H7, only acetic acid treatment resulted in significantly lower numbers of *E. coli* O157:H7 following storage at 12°C for 3 days. In general, carcass wash treatments were effective in reducing numbers of pathogens (low level inoculum) subsequently recovered in ground beef after storage for 3 days at 12°C (Dorsa et al., 1998b). Additionally, acetic acid carcass treatment reduced total aerobic and pseudomonad counts on ground beef stored for 3 days at 12°C (Dorsa et al., 1998b) compared to controls. No difference was reported for these counts following refrigeration storage (4°C) for 21 days (Dorsa et al., 1998b); however, the authors reported a trend of gradually decreasing numbers of pseudomonads and total aerobic bacterial counts for ground beef processed from acetic acid-washed carcasses. The primary approach of Dorsa et al. (1998a; 1998b) was microbial profiling to test whether or not reduction of overall bacterial numbers resulted with concurrent increases in numbers of pathogens due to lack of competition. Therefore, the main conclusion was that antimicrobial treatments, while not dramatically reducing total microbial counts, were effective for pathogen control, and initial reductions of total microbial numbers did not result in increases in pathogen levels (Dorsa et al., 1998a; 1998b).

Others have found that acetic acid treatment of low count carcasses has little benefit. For example, Ockerman et al. (1974) found small reductions in microbial counts on lamb carcasses sprayed with 6% acetic acid. Beef carcasses treated with 5% acetic acid showed less than a 1 log reduction in inoculated *E. coli* O157:H7 (Delazari et al., 1998). Likewise, Brackett et al. (1994) showed that 1.5% acetic acid was not effective in reducing the levels of *E. coli* O157:H7 on beef. In addition, Conner et al., (1997) found that 2-4% acetic sprays of beef trim destined for ground beef manufacture had little impact on the survival of inoculated *E. coli* O157:H7 and *L. monocytogenes*. These studies differed from those described previously because the beef chosen for use was post-chill retail cuts rather than pre-chill carcasses. Fu et al. (1994) showed that pork carcasses sprayed with 1.5% acetic acid had lower coliform counts initially but when carcasses were processed into loins this effect was minimal during storage. Likewise, in a commercial beef slaughter plant, dilute acetic acid spray was ineffective in reducing microbial counts of low count carcasses (Avens et al., 1996). The latter experiment failed to adequately describe the concentration of acetic acid used. Thus, given the abundance of evidence showing efficacy of acetic acid as a pre-chill carcass decontaminating agent, use of acetic acid would be best in this circumstance. Use on post-chill beef cuts seems to render acetic acid less potent.

Unfortunately, direct acetic acid treatment of meat pieces can lead to discoloration (Kotula & Thelappurate, 1994) and off flavors (Bell et al., 1986), thus making acetic acid decontamination treatments only valid for carcass decontamination. Although treatment of animal carcasses and consumer cuts with acetic acid can reduce microbial loads, there remains the question of whether such treatments yield marketable products. Kotula & Thelappurate (1994) showed that sensory panelists were generally unable to differentiate 1.2% acetic acid treated retail beef cuts from those that were untreated.

C. Decontamination of poultry

As with treatment conditions described with meats, acetic acid can also increase the shelf-life of poultry by inhibiting growth of spoilage bacteria (Mountney & O'Malley, 1965). Acetic acid can aid in controlling *Salmonella* cross-contamination from scald tank water to bird carcasses; however, treatment with acetic acid may not result in lower aerobic microbial counts on picked carcasses (Lillard et al., 1987; Dickens & Whittemore, 1994). This suggests that carcass cross-contamination can occur during picking. Others, however, have shown that 1% acetic acid treatment for 30 seconds produced a 0.6 log reduction in total aerobic plate counts (Dickens & Whittemore, 1997) and 0.6% acetic acid treatment for 10 min gave a 0.7 log reduction in *Enterobacteriaceae* counts (Dickens et al., 1994). Five percent acetic acid was effective in reducing the number of *S. typhimurium* attached to chicken skin by greater than 2 logs in both chiller and scalder applications (Tamblyn et al., 1997). A 0.8 log reduction in the numbers of *C. jejuni* was achieved with a 0.5% acetic acid immersion at 50°C for 90 seconds (Stern et al., 1985). Addition of acetic acid to chill tank water or scald tank water may be useful to control pathogens during immersion processing and might reduce the rate of carcass cross-contamination. Indeed, Okrend et al. (1986) found dramatic reductions in D_{52} of *S. typhimurium*, *Salmonella newport*, and *C. jejuni* when scald water contained as little as 0.1% acetic acid. Increasing acetic acid concentration to 1% caused immediate bacterial death at 52 °C. For control of *L. monocytogenes* growth on vacuum packaged turkey bologna, 0.5% sodium acetate proved slightly superior to 1.0% sodium bicarbonate, 2.0% sodium lactate, or 0.26% potassium sorbate as formulation additives (Wederquist et al., 1994).

Even though microbial count reductions on chickens following immersion in acetic acid solutions largely have not been significant, reported decreases in numbers of Enterobacteriaceae and/or *Salmonella* spp. recovered from immersion solution samples have potentially significant ramifications in cross-contamination control. Evaluation of poultry processing steps in order to identify contamination points for *Salmonella* or factors influencing pathogen incidence from fresh poultry products have resulted in development and evaluation of various treatments for pathogen control (Lillard, 1990).

D. Decontamination of seafoods

Catfish fillet aerobic plate counts were reduced by treatment with sodium acetate (*FIGURE 2A*) (Kim et al., 1995b). This resultant inhibition of microbial spoilers gave a 6 day shelf-life increase of fillets during refrigerated storage presumably due to slowing the growth rate of survivors (*FIGURE 2C*). Combining 0.7% sodium acetate with 0.4% monopotassium phosphate gave microbial growth rate retardation (*FIGURE 5B*) that was similar to that seen with use of 1% sodium acetate alone (*FIGURE 2C*) (Kim et al., 1995b). Sensory panelists found that fillets treated with the combined chemicals were preferred over untreated controls during storage (Kim et al., 1995b). It was concluded that most of

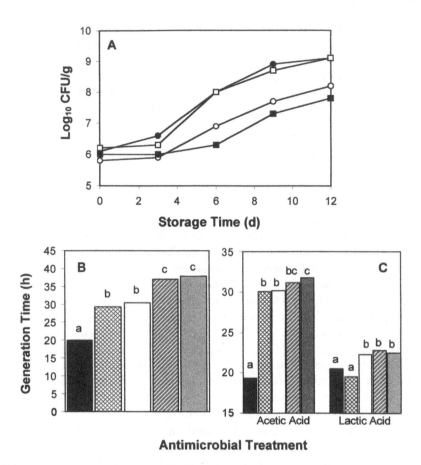

FIGURE 5A. Aerobic plate counts of catfish fillets treated with *Bifidobacterium infantis* culture and sodium acetate, either alone or combined, during storage at 4°C. Control (●), 2.5% *B. infantis* (□), 0.5% sodium acetate (○), and a combination of both (■). [Redrawn from Kim et al., 1995a].

FIGURE 5B. Generation times of aerobic bacteria at 4°C on catfish fillets treated with combinations of sodium acetate (0.5%) and different concentrations of monopotassium phosphate. Bars from *Left* to *Right* represent control, 0.1%, 0.2%, 0.3% and 0.4% monopotassium phosphate, respectively, in combination with 0.5% sodium acetate. [Redrawn from Kim et al., 1995b].

FIGURE 5C. Generation times of aerobic bacteria at 4°C on catfish fillets treated with different concentrations of acetic acid or lactic acid. Bars from *Left* to *Right* represent acid concentrations, 0%, 1%, 2%, 3% and 4%, respectively. [Redrawn from Marshal & Kim, 1996].

the antimicrobial effect was associated with sodium acetate and the combined effect with monopotassium phosphate was additive. Growth of Gram negative bacteria was prevented for 6 days on catfish fillets treated with 1% sodium acetate (Kim & Hearnsberger, 1994). In this study sodium acetate was also combined with 0.25% potassium sorbate and 2.5% *Lactococcus lactis* culture to test for interactions between the treatments. Results demonstrated that antimicrobial activity resided primarily with acetate. The addition of sorbate or lactic culture had little impact on inhibiting Gram negative bacteria. In a similar study, Kim et al. (1995a) showed that a combination of 0.5% sodium acetate and a 2.5% culture of *Bifidobacterium infantis* gave greater inhibition of aerobic microorganisms than did either acetate or culture alone (*FIGURE 5A*). Thus, the interaction between

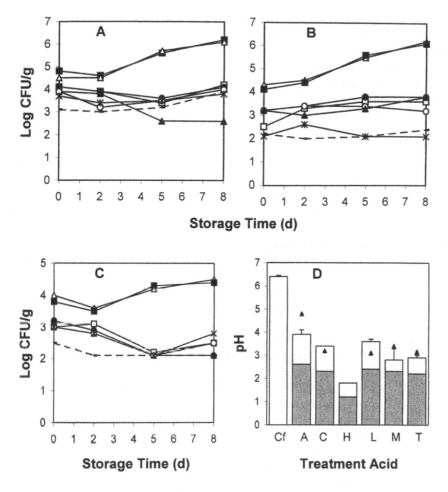

FIGURE 6. 2% acid-treated catfish fillets (Cf) during storage at 4 C. Unrinsed (■), rinsed (△) or treated with acetic (A: ▲), citric (C: ◻), hydrochloric (H: --), lactic (L: ●), malic (M: ○), and tartaric (T: ✶) acids. *FIGURE 6A.* Change in aerobic plate counts; *FIGURE 6B.* Change in coliform counts; *FIGURE 6C.* Change in *Listeria monocytogenes* counts; *FIGURE 6D.* Composite representation of dip solution pH (solid bar), catfish fillet surface pH (open bar), and pKa (▲) of treatment acids. [Redrawn from Bal'a & Marshall, 1998].

acetate and *B. infantis* culture was deemed an additive response. Acetate treatment, whether alone or combined with culture, increased refrigerated shelf-life of fillets by 3 days (Kim et al., 1995a). Like many previous studies comparing the activity of acetic acid with other acids, Marshall & Kim (1996) demonstrated that acetic acid slowed the rate of growth of aerobic spoilage bacteria more than did lactic acid (*FIGURE 5C*). More recently, Bal'a and Marshall (1998) demonstrated that 2% acetic acid had greatest inhibitory ability towards catfish fillet aerobic (*FIGURE 6A*), coliform (*FIGURE 6B*), and *L. monocytogenes* (*FIGURE 6C*) bacteria compared to 2% concentrations of citric, lactic, malic, tartaric, or hydrochloric acids. They also demonstrated that among the acids tested, acetic acid-treated fillets had the highest surface pH (*FIGURE 6D*). Thus, the greater activity of acetic acid was not due to its ability to reduce surface pH lower than the other acids. These results were subsequently confirmed by Fernandes et al. (1998). Marshall & Kim (1996) demonstrated that catfish fillets treated with greater than 2% acetic acid were objectionable to sensory panelists in terms of off-color and off-odor problems.

Crabmeat contaminated with *L. monocytogenes* showed 0.8 log or 2.0 log reductions in counts after washing with 4 M sodium acetate or 2 M sodium diacetate solutions for 1 minute (Degnan et al., 1994). Acetic acid was superior to formaline and orthrophosphoric, formic, tartaric, citric, hydrochloric, and sulfuric acids in preserving mackerel fish destined for fish meal production (Borquez et al., 1994). In addition, processing performance of acetic acid treated mackerel during fish-meal production was superior to untreated controls (Borquez & Gonzalez, 1994). Treatment of whole and peeled shrimp with sodium acetate improved initial microbiological quality and extended shelf-life (Zhuang et al., 1996; Al-Dagal & Bazaraa, 1999). Sodium acetate was superior to treatment with potassium sorbate, sodium lactate, trisodium citrate, or bifidobacteria culture.

E. Decontamination of produce and grains

Fruit decay could be prevented by fumigation with >2 mg acetic acid vapor per liter of air (Sholberg & Gaunce, 1995). Apples and pears inoculated with the molds *Botrytis cinerea* and *Penicillium expansum* failed to decay. Similarly, tomatoes, grapes, and kiwifruit contaminated with *B. cinerea* and oranges contaminated with *Penicillium italicum* failed to decay when fumigated with acetic acid at 5°C. Greater activity of acetic acid was obtained at high relative humidity (98%) (Sholberg & Gaunce, 1995). Table grapes fumigated with 0.27% acetic acid yielded results similar to fumigation with sulfur dioxide, with effective control of both *Botrytis* and *Penicillium* decay during refrigerated storage (Sholberg et al., 1996). Decay of peaches by *Monilinia fructicola* and *Rhizopus stolonifer* could be prevented by fumigation with 1.4 to 2.7 mg/L acetic acid (Sholberg & Gaunce, 1996a). Additionally, *Alternaria* decay of cherries and brown rot (*M. fructicola*) of apricots could be reduced with acetic acid fumigation (Sholberg & Gaunce, 1996a). An additional study has shown that nectarines, oranges, lemons, and grapefruits fumigated with 1.9 µl/L acetic acid could reduce decay from 98% to 16% for nectarines and from 86% to 11% for citrus during storage at 20°C (Sholberg, 1998).

Acetic acid fumigation combined with low carbon dioxide modified atmosphere packaging increased strawberry shelf-life 2-3 fold (Moyles et al., 1996). Fumigated strawberries were free of decay for 14 days at 5°C while untreated strawberries had 89% rotted fruits by this time. Proliferation of aflatoxigenic *Aspergillus flavus* in high moisture rape, rice, and wheat grains could be controlled by fumigation with acetic acid (Sholberg & Gaunce, 1996b). A fumigation treatment with 0.50 ml acetic acid/100 g cabbage resulted in undetectable levels of microorganisms on prepared coleslaw after 22 days of storage at 5°C (Delaquis et al., 1997). This fumigation treatment of shredded cabbage prior to processing to coleslaw proved to reduce the numbers of spoilage bacteria and extend refrigerated shelf-life. In contrast, numbers of *L. monocytogenes* on cabbage and lettuce could not be effectively reduced by a 10 min exposure to 1% acetic acid (Zhang & Farber, 1996). Entani et al. (1997) reported effective reductions (>3 logs) in numbers of *E. coli* O157:H7 on cabbage and cucumbers after surface treatment with vinegar solutions (1 to 2% acetic acid) containing added salt (3 to 7%). Mung bean seeds used to produce bean sprouts were fumigated with 242 µl acetic acid per liter of air for 12 hours at 45°C

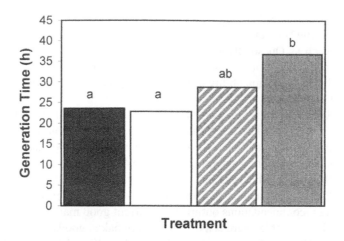

FIGURE 7. Generation times of aerobic bacteria at 4°C on catfish fillets treated with sodium acetate and *Bifidobacterium infantis*. Bars from *Left* to *Right* represent control, 2.5% *B. infantis*, 0.5% sodium acetate and a combination of both, respectively. Means having the same letter are not significantly different (P>0.05) [Redrawn from Kim et al., 1995a].

(Delaquis et al., 1999). This treatment yielded 5, 6, and 4 log reductions in contaminating *S. typhimurium*, *E. coli* O157:H7, and *L. monocytogenes* without adversely affecting sprouting performance of the seeds.

F. Natural preservation systems

Natural preservation systems for food have incorporated cultures of lactic acid bacteria, bifidobacteria, and propionic bacteria (Kim & Hearnsberger, 1994; Kim et al., 1995a). For example, bifidobacteria culture combined with sodium acetate slowed the rate of growth of aerobic spoilage bacteria on catfish fillets greater than either treatment alone (*FIGURE 7*). Use of desirable microorganisms in natural systems-approaches works because they form and/or accumulate starter culture-derived compounds, such as lactic, acetic, or propionic acids (Ray & Sandine, 1992), nisin (Ray, 1992a), and/or acetaldehyde (Ray, 1992b). These compounds individually or combined have antimicrobial activities that can extend shelf-life of certain food products or inhibit foodborne pathogens (Daeschel, 1989; Ray, 1992a). Shelf-life extension of various products has been approached by incorporating bacteriocins of lactic acid bacteria (Davies et al., 1997). The effects of combining acetic acid with bacteriocins or other naturally occurring antimicrobial compounds, such as chitosan, protamine, lysozyme, and lactoferrin have not been reported.

VI. REGULATORY STATUS

Acetic acid, used and label-listed generally as vinegar, has been employed as a flavoring ingredient but also has functioned primarily as an acidulant or preservative in foods (Gardner, 1968; Stroup et al., 1985). Excluding vinegar, the foremost food types important when considering acetic acid are mayonnaise, salad dressings, ketchups/catsups, sauces or condiments, and pickles and relishes (Fabian & Wadsworth, 1939; Gardner, 1968; Kabara & Eklund, 1991; Davidson, 1997). Other foods containing acetic acid include cheeses and non-dairy equivalents, baked products, oils, gravies, and meat products (Davidson, 1997). Acetic acid is the preferred acidulant used to make direct set soft-style cheeses such as Queso Blanco (Farkye et al, 1995). Acetic acid is frequently used to prevent ropy spoilage of breads (Rosenquist & Hansen, 1998); however, levels used must not inhibit fermentation ability of the leavening yeast *Saccharomyces cerevisiae* (Phowchinda et al., 1995).

Principle U.S. regulations for use of acetic acid and its salts are found in the Code of Federal Regulations (CFR), Title 21, Volume 3, Parts 170-199. In addition, these food ingredients meet specifications outlined in Food Chemicals Codex. Defined uses of acetic acid are as a pickling and curing agent (21 CFR 184.1005 Sec. 170.3(o)(5)), flavor enhancer (Sec. 170.3(o)(11)), flavoring agent and adjuvant (Sec. 170.3(o)(12)), pH control agent (Sec. 170.3(o)(23)), and carrier solvent (Sec. 170.3(o)(27)). It may be used at levels not exceeding recommendations outlined for current good manufacturing practices (Sec. 184.1(b)(1)). For example, there is a 0.25% limit in baked goods (Sec. 170.3(n)(1)), 0.8% in cheese (Sec. 170.3(n)(5)) and dairy product analogs (Sec. 170.3(n)(10)), 0.5% in chewing gum (Sec. 170.3(n)(6)), 9.0% in condiments and relishes (Sec. 170.3(n)(8)), 0.5% in fats and oils (Sec. 170.3(n)(12)), 3.0% in gravies and sauces (Sec. 170.3(n)(24)), 0.6% in meat products (Sec. 170.3(n)(29)), and 0.15% or less in other food product categories.

Calcium acetate is used as a firming agent (21 CFR 184.1185 Sec. 170.3(o)(10)), pH control agent (Sec. 170.3(o)(23)), processing aid (Sec. 170.3(o)(24)), sequestrant (Sec. 170.3(o)(26)), stabilizer and thickener (Sec. 170.3(o)(28)), and texturizer (Sec. 170.3(o)(32)). Usage limits for this compound are 0.2% in baked goods (Sec. 170.3(n)(1)), 0.02% in cheese (Sec. 170.3(n)(5)), 0.2% in gelatins, puddings, and fillings (Sec. 170.3(n)(22)), 0.15% in sweet sauces, toppings, and syrups (Sec. 170.3(n)(43)), and 0.0001% for all other food categories.

Sodium acetate is used as a flavoring agent and adjuvant (21 CFR 184.1721 Sec. 170.3(o)(12)) and as a pH control agent (Sec. 170.3(o)(23)). It may be used at levels not exceeding 0.007% in breakfast cereals (Sec. 170.3(n)(4)), 0.5% in fats and oils (Sec. 170.3(n)(12)), 0.6% in grain products and pastas (Sec. 170.3(n)(23)) and snack foods (Sec. 170.3(n)(37)), 0.15% in hard candy (Sec. 170.3(n)(25)), 0.12% in jams and jellies (Sec. 170.3(n)(28)) and meat products (Sec. 170.3(n)(29)), 0.2% in soft candy (Sec. 170.3(n)(38)), 0.05% in soups and soup mixes (Sec. 170.3(n)(40)) and sweet sauces (Sec. 170.3(n)(43)).

Sodium diacetate is used as an antimicrobial agent (21 CFR 184.1754 Sec. 170.3(o)(2)), flavoring agent and adjuvant (Sec. 170.3(o)(12)), and pH control agent (Sec. 170.3(o)(23)). Maximum usage levels are 0.4% in baked goods (Sec. 170.3(n)(1)), 0.1% in fats and oils (Sec. 170.3(n)(12)), meat products (Sec. 170.3(n)(29)), and soft candy (Sec. 170.3(n)(38)), 0.25% in gravies and sauces (Sec. 170.3(n)(24)), and 0.05% in snack foods (Sec. 170.3(n)(37)) and soups and soup mixes (Sec. 170.3(n)(40)).

VII. SUMMARY

Acetic acid and its GRAS salts have a long history of functional use in foods. The ability to lower pH with acetic acid or vinegar provides pickled foods with unsurpassed keeping quality, flavor, and safety. Aside from this traditional use, acetic acid use as a decontaminating agent of fresh and minimally processed foods remains under exploited. Few other naturally occurring antimicrobial agents have such a wide spectrum of activity. Newer uses of acetic acid will likely be discovered. Food processors should find acetic acid a desirable functional ingredient or processing aid because of its wide availability, low cost, consumer acceptance, and clean label requirements.

VIII. REFERENCES

1. Abdul-Raouf, U.M., Beuchat, L.R., and Ammar, M.S. 1993. Survival and growth of *Escherichia coli* O157:H7 in ground, roasted beef as affected by pH, acidulants, and temperature. *Appl. Environ. Microbiol.* 59: 2364-2368.
2. Abee, T. and Wouters, J.A. 1999. Microbial stress response in minimal processing. *Int. J. Food Microbiol.* 50: 65-91.
3. Adams, M.R. 1985. Vinegar. In *Microbiology of Fermented Foods, Vol 1.*, ed. B.J.B. Wood, pp. 1-47. New York: Elsevier Applied Science Publishers.
4. Adams, M.R. and Hall, C.J. 1988. Growth inhibition of foodborne pathogens by lactic and acetic acids and their mixtures. *Int. J. Food Sci. Technol.* 23: 287-292.
5. Ahamad, N. and Marth, E.H. 1987. Behavior of *Listeria monocytogenes* at 7, 13, 21 and 35 °C in tryptic soy broth acidified with acetic, citric or lactic acid. *J. Food Prot.* 52: 688-695.
6. Ahamad, N. and Marth, E.H. 1990. Acid-injury of *Listeria monocytogenes*. *J. Food Prot.* 53: 26-29.
7. Al-Dagal, M.M. and Bazaraa, W.A. 1999. Extension of shelf life of whole and peeled shrimp with organic acid salts and bifidobacteria. *J. Food Prot.* 62: 51-56.
8. Allgeier, R.J. and Hildebrandt, F.M. 1960. Newer developments in vinegar manufacture. *Adv. Appl. Microbiol.* 11: 163-181.
9. Anderson, M.E. and Marshall, R.T. 1989. Interaction of concentration and temperature of acetic acid solution on reduction of various species of microorganisms on beef surfaces. *J. Food Prot.* 52: 312-315.
10. Anderson, M.E., Marshall, R.T., Stringer, W.C., and Naumann, H.D. 1979. Microbial growth on plate beef during extended storage after washing and sanitizing. *J. Food Prot.* 42: 389-392.
11. Anderson, M.E., Marshall, R.T., Stringer, W.C., and Naumann, H.D. 1980. In-plant evaluation of a prototype carcass cleaning and sanitizing unit. *J. Food Prot.* 43: 568-570.
12. Anderson, M.E., Huff, H.E., Naumann, H.D., Marshall, R.T., Damare, J.M., Pratt, M., and Johnston, R. 1987. Evaluation of an automated beef carcass washing and sanitizing system under production conditions. *J. Food Prot.* 50: 562-566.
13. Anderson, M.E., Huff, H.E., Naumann, H.D., and Marshall, R.T. 1988. Counts of six types of bacteria on lamb carcasses dipped or sprayed with acetic acid at 25° or 55°C and stored vacuum packaged at 0 °C. *J. Food Prot.* 51: 874-877.
14. Anderson, M.E., Marshall, R.T., and Dickson, J.S. 1992. Efficacies of acetic, lactic and two mixed acids in reducing numbers of bacteria on surfaces of lean meat. *J. Food Safety* 12: 139-147.
15. Avens, J.S., Clayton, P., Jones, D.K., Bolin, R., Lloyd, W., and Jankow, D. 1996. Acetic acid spray ineffective on beef carcasses with low bacteria counts. *Lebensm.-Wiss. u.-Technol.* 29: 28-32.
16. Bal'a, M.F.A. and Marshall, D.L. 1998. Organic acid dipping of catfish fillets: effect on color, microbial load, and *Listeria monocytogenes*. *J. Food Prot.* 61: 1470-1474.
17. Bell, M.F., Marshall, R.T., and Anderson, M.E. 1986. Microbiological and sensory tests of beef treated with acetic and formic acids. *J. Food Prot.* 49: 207-210.
18. Bell, K.Y., Cutter, C.N., and Sumner, S.S. 1997. Reduction of foodborne microorganisms on beef carcass tissue using acetic acid, sodium bicarbonate, and hydrogen peroxide spray washes. *Food Microbiol.* 14: 439-448.
19. Besser, R.E., Lett, S.M., Weber, J.T., Doyle, M.P., Barret, T.J., Wells, J.G., and Griffin, P.M. 1993. An outbreak of diarrhea and hemolytic uremic syndrome from *Escherichia coli* O157:H7 in fresh pressed apple cider. *J. Am. Med. Assoc.* 269: 2217-2220.
20. Bevilacqua, A.E. and Califano, A.N. 1992. Changes in organic acids during ripening of Port Salut Argentino cheese. *Food Chem.* 43: 354-349.
21. Bjorkroth, K.J. and Korkeala, H.J. 1997. *Lactobacillus fructivorans* spoilage of tomato ketchup. *J. Food Prot.* 60: 505-509.

22. Blackburn, C.W., Curtis, L.M., Humpheson, L., Billon, C., and McClure, P.J. 1997. Development of thermal inactivation models for *Salmonella enteritidis* and *Escherichia coli* O157:H7 with temperature, pH and NaCl as controlling factors. *Int. J. Food Microbiol.* 38: 31-44.

23. Blankenship, L.C. 1981. Some characteristics of acid injury and recovery of *Salmonella bareilly* in a model system. *J. Food Prot.* 44: 73-77.

24. Bolder, N.M. 1997. Decontamination of meat and poultry carcasses. *Trends Food Sci. Technol.* 8: 221-227.

25. Booth, I.R. 1985. Regulation of cytoplasmic pH in bacteria. *Microbiol. Rev.* 49: 359-378.

26. Borquez, R. and Gonzalez, C. 1994. Influence of acetic acid preservation of Chilean mackerel (*Trachurus murphy*) on fish meal production. *J. Sci. Food Agric.* 66: 187-192.

27. Borquez, R., Espinoza, M., and Ormeno, R. 1994. Effects of storage time and chemical preservatives on the total volatile basic nitrogen content in Chilean mackerel (*Trachurus murphy*) prior to fish meal production. *J. Sci. Food Agric.* 66: 181-186.

28. Brackett, R.E. 1987. Effects of various acids on growth and survival of *Yersinia enterocolitica*. *J. Food Prot.* 50: 598-601,607.

29. Brackett, R.E., Hao, Y.Y., and Doyle, M.P. 1994. Ineffectiveness of hot acid sprays to decontaminate *Escherichia coli* O157:H7 on beef. *J. Food Prot.* 57: 198-203.

30. Brown, M.H. and Booth, I.R. 1991. Acidulants and low pH. In *Food Preservatives*, ed. N.J. Russell and G.W. Gould, pp. 22-43. New York: AVI Publishing Co.

31. Brudzinski, L. and Harrison, M.A. 1998. Influence of incubation conditions on survival and acid tolerance response of *Escherichia coli* O157:H7 and non-O157:H7 isolates exposed to acetic acid. *J. Food Prot.* 61: 542-546.

32. Buchanan, R.L. and Edelson, S.G. 1999. pH-dependent stationary-phase acid resistance response of enterohemorrhagic *Escherichia coli* in the presence of various acidulants. *J. Food Prot.* 62: 211-218.

33. Buchta, V., Slavikova, E., Vadkartiova, R., Alt, S., and Jilek, P. 1996. *Zygosaccharomyces bailii* as a potential spoiler of mustard. *Food Microbiol.* 13: 133-135.

34. Cabedo, L., Sofos, J.N., and Smith, G.C. 1996. Removal of bacteria from beef tissue by spray washing after different times of exposure to fecal material. *J. Food Prot.* 59: 1284-1287.

35. Cacciarelli, M.A., Springer, W.C., Anderson, M.E., and Naumann, H.D. 1983. Effects of washing and sanitizing on the bacterial flora on vacuum-packaged pork loins. *J. Food Prot.* 46: 231-234.

36. Canning, A. 1985. Vinegar brewing. *Nutr. Food Sci.* Sept/Oct, pp. 20-21.

37. Carpenter, J.A., Reagan, J.O., Bauer, F.T., and Harrison, M.A. 1986. The effect of carcass decontamination on retail case life of pork chops. *J. Food Qual.* 9: 311-317.

38. Casal, M., Cardoso, H., and Leão, C. 1998. Effects of ethanol and other alkanols on transport of acetic acid in *Saccharomyces cerevisiae*. *Appl. Environ. Microbiol.* 64: 665-668.

39. Cassio, F., Leao, C., and van Uden, N. 1987. Transport of lactate and other short-chain monocarboxylates in the yeast *Saccharomyces cerevisiae*. *Appl. Environ. Microbiol.* 53: 509-513.

40. Cherrington, C.A., Hinton, M., Mead, G.C., and Chopra, I. 1991a. Organic acids: chemistry, antibacterial activity and practical applications. *Adv. Micro. Phys.* 32: 87-108.

41. Cherrington, C.A., Hinton, M., Pearson, G.R. and Chopra, I. 1991b. Inhibition of *Escherichia coli* K12 by short-chain organic acids: lack of evidence for induction of the SOS response. *J. Appl. Bacteriol.* 70: 156-160.

42. Cheryan, M., Parekh, S., Shah, M., and Witjitra, K. 1997. Production of acetic acid by *Clostridium thermoaceticum*. *Adv. Appl. Microbiol.* 43: 1-33.

43. Chukwu, U. and Cheryan, M. 1996. Concentration of vinegar by electrodialysis. *J. Food Sci.* 61: 1223-1226.

44. Chung, K.C. and Goepfert, J.M. 1970. Growth of *Salmonella* at low pH. *J. Food Sci.* 35: 326-328.

45. Conner, H.A. and Allgeier, R.J. 1976. Vinegar: its history and development. *Adv. Appl. Microbiol.* 20: 81-133.

46. Conner, D.E. and Kotrola, J.S. 1995. Growth and survival of *Escherichia coli* O157:H7 under acidic conditions. *Appl. Environ. Microbiol.* 61: 382-385.

47. Conner, D.E., Scott, V.N., and Bernard, D.T. 1990. Growth, inhibition, and survival of *Listeria monocytogenes* as affected by acidic conditions. *J. Food Prot.* 53: 652-655.

48. Conner, D.E., Kotrola, J.S., Mikel, W.B., and Tamblyn, K.C. 1997. Effects of acetic-lactic acid treatments applied to beef trim on populations of *Escherichia coli* O157:H7 and *Listeria monocytogenes* in ground beef. *J. Food Prot.* 60: 1560-1563.

49. Cutter, C.N. 1999. Combination spray washes of saponin with water or acetic acid to reduce aerobic and pathogenic bacteria on lean beef surfaces. *J. Food Prot.* 62: 280-283.

50. Cutter, C.N. and Siragusa, G.R. 1994. Efficacy of organic acids against *Escherichia coli* O157:H7 attached to beef carcass tissue using a pilot scale model carcass washer. *J. Food Prot.* 57: 97-103.

51. Daeschel, M. 1989. Antimicrobial substances from lactic acid bacteria for use as food preservatives. *Food Technol.* 43(1): 164-167.

52. Dankin, J.C. and Radwell, J.Y. 1971. Lactobacilli causing spoilage of acetic acid preserves. *J. Appl. Bacteriol.* 34: 541-545.

53. Davidson, P.M. 1997. Chemical preservatives and natural antimicrobial compounds. In *Food Microbiology Fundamentals and Frontiers*, ed. M.P. Doyle, L.R. Beuchat, and T.J. Montville, pp. 520-556. Washington, D.C.: ASM Press.

54. Davies, E.A., Bevis, H.E., and Delves-Broughton, J. 1997. The use of the bacteriocin, nisin, as a preservative in ricotta-type cheeses to control the food-borne pathogen *Listeria monocytogenes*. *Lett. Appl. Microbiol.* 24: 343-346.

55. Degnan, A.J., Kaspar, C.W., Otwell, W.S., Tamplin, M.L., and Luchansky, J.B. 1994. Evaluation of lactic acid bacterium fermentation products and food-grade chemicals to control *Listeria monocytogenes* in blue crab (*Callinectes sapidus*) meat. *Appl. Environ. Microbiol.* 60: 3198-3203.

56. Delaquis, P.J., Graham, H.S., and Hocking, R. 1997. Shelf-life of coleslaw made from cabbage treated with gaseous acetic acid. *J. Food Process. Preserv.* 21: 129-140.

57. Delaquis, P.J., Sholberg, P.L., and Stanich, K. 1999. Disinfection of mung bean seed with gaseous acetic acid. *J. Food Prot.* 62:953-957.

58. Delazari, I., Iaria, S.T., Riemann, H.P., Cliver, D.O., and Mori, T. 1998. Decontaminating beef for *Escherichia coli* O157:H7. *J. Food Prot.* 61: 547-550.

59. Deng, Y., Ryu, J.-H., and Beuchat, L.R. 1999. Tolerance of acid-adapted and non-adapted *Escherichia coli* O157:H7 cells to reduced pH as affected by type of acidulant. *J. Appl. Microbiol.* 86: 203-210.

60. Dickens, J.A. and Whittemore, A.D. 1994. The effect of acetic acid and air injection on appearance, moisture, pick-up, microbiological quality, and *Salmonella* incidence on processed poultry carcasses. *Poultry Sci.* 73: 582-586.

61. Dickens, J.A. and Whittemore, A.D. 1997. Effects of acetic acid and hydrogen peroxide application during defeathering on the microbiological quality of broiler carcasses prior to evisceration. *Poultry Sci.* 76:657-660.

62. Dickens, J.A., Lyon, A.D., Whittemore, A.D., and Lyon, C.E. 1994. The effect of and acetic acid dip on carcass appearance, microbiological quality, and cooked breast meat texture and flavor. *Poultry Sci.* 73:576-581.

63. Dickson, J.S. 1990. Surface moisture and osmotic stress as factors that affect the sanitizing of beef tissue surfaces. *J. Food Prot.* 53: 674-679.

64. Dickson, J.S. 1992. Acetic acid action on beef tissue surfaces contaminated with *Salmonella typhimurium*. *J. Food Sci.* 57: 297-301.

65. Dickson, J.S. and Anderson, M.E. 1992. Microbiological decontamination of food animal carcasses by washing and sanitizing systems: a review. *J. Food Prot.* 55: 133-140.

66. Diez-Gonzalez, F. and Russell, J.B. 1997. The ability of *Escherichia coli* O157:H7 to decrease its intracellular pH and resist the toxicity of acetic acid. *Microbiol.* 143: 1175-1180.

67. Dimmig, L.K., Myers, E.R., and Martin, S.E. 1994. Catalase, superoxide dismutase and listeriolysin O production by *Listeria monocytogenes* in broth containing acetic and hydrochloric acids. *J. Food Prot.* 57: 626-628.

68. Doores, S. 1993. Organic acids. In *Antimicrobials in Foods, 2nd ed.*, ed. P.M. Davidson and A.L. Branen, pp. 95-136. New York: Marcel Dekker, Inc.

69. Dorsa, W.J., Cutter, C.N., and Siragusa, G.R. 1997. Effects of acetic acid, lactic acid, and TSP on the microflora of refrigerated beef carcass surface tissue with *Escherichia coli* O157:H7, *Listeria monocytogenes*, and *Clostridium sporogenes*. *J. Food Prot.* 60: 619-624.

70. Dorsa, W.J., Cutter, C.N., and Siragusa, G.R. 1998a. Long-term effect of alkaline, organic acid, or hot water washes on the microbial profile of refrigerated beef contaminated with bacterial pathogens after washing. *J. Food Prot.* 61: 300-306.

71. Dorsa, W.J., Cutter, C.N., and Siragusa, G.R. 1998b. Bacterial profile of ground beef made from carcass tissue experimentally contaminated with pathogenic and spoilage bacteria before being washed with hot water, alkaline solution, or organic acid and then stored at 4 or 12 °C. *J. Food Prot.* 61: 1109-1118.

72. Doyle, M.P., Bains, N.J., Schoeni, J.L., and Foster, E.M. 1982. Fate of *Salmonella typhimurium* and *Staphylococcus aureus* in meat salads prepared with mayonnaise. *J. Food Prot.* 45: 152-156.

73. Ebner, H. 1982. Vinegar. In *Prescott & Dunn's Industrial Microbiology 4th ed.*, ed. G. Reed, pp. 802-834.

74.   Eklund, T. 1989. Organic acids and esters. In *Mechanisms of Action of Food Preservation Proceedings*, ed. A.W. Gould, pp. 161-200. London: Elsevier Applied Science.

75.   El-Shenawy, M.A. and Marth, E.H. 1989. Inhibition or inactivation of *Listeria monocytogenes* by sodium benzoate together with some organic acids. *J. Food Prot.* 52: 771-776.

76.   El-Shenawy, M.A. and Marth, E.H. 1991. Organic acids enhance the antilisterial activity of potassium sorbate. *J. Food Prot.* 54: 593-597.

77.   El-Shenawy, M.A. and Marth, E.H. 1992. Behavior of *Listeria monocytogenes* in the presence of sodium propionate together with food acids. *J. Food Prot.* 55: 241-245.

78.   Entani, E., Asai, M., Tsujihata, S., Tsukamoto, T., and Ohta, M. 1997. Antibacterial action of vinegar against food-borne pathogenic bacteria including *Escherichia coli* O157:H7 (part 2). Effect of sodium chloride and temperature on bactericidal activity. *Kansenshogaku Zasshi* 71: 451-458.

79.   Erickson, J.P. and Jenkins, P. 1991. Comparative *Salmonella* spp. and *Listeria monocytogenes* inactivation rates in four commercial mayonnaise products. *J. Food Prot.* 54: 913-916.

80.   Erickson, J.P., Stamer, J.W., Hayes, M., McKenna, D.N., and van Alstine, L.A. 1995. An assessment of *Escherichia coli* O157:H7 contamination risks in commercial mayonnaise from pasteurized eggs and environmental sources, and behavior in low-pH dressings. *J. Food Prot.* 58: 1059-1064.

81.   Farber, J.M., Sanders, G.W., Dunfield, S., and Prescott, R. 1989. The effects of various acidulants on the growth of *Listeria monocytogenes*. *Lett. Appl. Microbiol.* 9: 181-183.

82.   Fabian, F.W. and Wadsworth, C.K. 1939. Experimental work on lactic acid in preserving pickles and pickle products. II. Preserving value of acetic and lactic acids in the presence of sucrose. *Food Res.* 4: 511-519.

83.   Farkye, N.Y., Prasad, B.B., Rossi, R., and Noyes, O.R. 1995. Sensory and textural properties of Queso Blanco-type cheese influenced by acid type. *J. Dairy Sci.* 78: 1649-1656.

84.   Fernandes, C.F., Flick, G.J., Cohen, J., and Thomas, T.B. 1998. Role of organic acids during processing to improve quality of channel catfish fillets. *J. Food Prot.* 61: 495-498.

85.   Fielding, L.M., Cook, P.E., and Grandison, A.S. 1997. The effect of electron beam irradiation, combined with acetic acid, on the survival and recovery of *Escherichia coli* and *Lactobacillus curvatus*. *Int. J. Food Microbiol.* 35: 259-265.

86.   Foster, J.W. 1991. *Salmonella* acid shock proteins are required for adaptive acid tolerance response. *J. Bacteriol.* 173: 6896-6902.

87.   Foster, J.W. and Hall, H.K. 1990. Adaptive acidification tolerance response of *Salmonella typhimurium*. *J. Bacteriol.* 172: 771-778.

88.   Frederick, T.L. 1994. The effects of organic acid and temperature on the microbiological properties of pork. *J. Food Sci.* 59: 300-302.

89.   Frederick, T.L., Miller, M.F., Thompson, L.D., and Ramsey, C.B. 1994. Microbiological properties of pork cheek meat as affected by acetic acid and temperature. *J. Food Sci.* 59: 300-302.

90.   Fu, A.-H., Sebranek, J.G., and Murano, E.A. 1994. Microbial quality characteristics of pork cuts from carcasses treated with sanitizing sprays. *J. Food Sci.* 59:306-309.

91.   Gahan, C.G.M. and Hill, C. 1999. The relationship between acid stress responses and virulence in *Salmonella typhimurium* and *Listeria monocytogenes*. *Int. J. Food Microbiol.* 50: 93-100.

92.   Gardner, W.H. 1968. Acidulants in food processing. In *Handbook of Food Additives*, ed. T.E. Furia, pp. 247-287. Cleveland, OH: The Chemical Rubber Co.

93.   Garren, D.M., Harrison, M.A., and Russell, S.M. 1998. Acid tolerance and acid shock response in *Escherichia coli* O157:H7 and non-O157:H7 isolates provide cross protection to sodium lactate and sodium chloride. *J. Food Prot.* 61: 158-161.

94.   Ghommidh, C., Cutayar, J.M., and Navvarro, J.M. 1986. Continuous production of vinegar I. research strategy. *Biotechnol. Lett.* 8: 13-18.

95.   Glass, K.A. and Doyle, M.P. 1991. Fate of *Salmonella* and *Listeria monocytogenes* in commercial, reduced-calorie mayonnaise. *J. Food Prot.* 54: 691-695.

96.   Gomez-Lucia, E., Goyache, J., Blanco, J.L., Garayzabal, J.F.F., Orden, J.A., and Suarez, G. 1987. Growth of *Staphylococcus aureus* and enterotoxin production in homemade mayonnaise prepared with different pH values. *J. Food Prot.* 50: 872-875.

97.   Graves Delmore, L.R., Sofos, J.N., Schmidt, G.R., and Smith, G.C. 1998. Decontamination of inoculated beef with sequential spraying treatments. *J. Food Sci.* 63: 890-893.

98.   Greer, G.G. and Dilts, B.D. 1992. Factors affecting the susceptibility of meatborne pathogens and spoilage bacteria to organic acids. *Food Res. Int.* 25: 355-364.

99.   Guldfeldt, L.U. and Arneborg, N. 1998. Measurement of the effects of acetic acid and extracellular pH on intracellular pH of nonfermenting, individual *Saccharomyces cerevisiae* cells by fluorescence microscopy. *Appl. Environ. Microbiol.* 64: 530-534.

100. Hamby, P.L., Savell, J.W., Acuff, G.R., Vanderzant, C., and Cross, H.R. 1987. Spray-chilling and carcass decontamination systems using lactic acid and acetic acid. *Meat Sci.* 21: 1-14.

101. Hardin, M.D., Acuff, G.R., Lucia, L.M., Oman, J.S., and Savell, J.W. 1995. Comparison of methods for decontamination from beef carcass surfaces. *J. Food Prot.* 58: 368-374.

102. Hartwig, P. and McDaniel, M. R. 1995. Flavor characteristics of lactic, malic, citric, and acetic acids at various pH levels. *J. Food Sci.* 60: 384-388.

103. Hathcox, A.L., Beuchat, L.R., and Doyle, M.P. 1995. Death of enterohemorrhagic *Escherichia coli* O157:H7 in real mayonnaise and reduced-calorie mayonnaise dressings as influenced by initial population and storage temperature. *Appl. Environ. Microbiol.* 61: 4172-4177.

104. Hentges, D.J. 1967. Influence of pH on the inhibitory activity of formic and acetic acids for *Shigella*. *J. Bacteriol.* 93: 956-958.

105. Huang, L., Forsberg, C.W., and Gibbins, L.N. 1986. Influence of external pH and fermentation products of *Clostridium acetobutylicum* intracellular pH and cellular distribution of fermentation products. *Appl. Environ. Microbiol.* 51: 1230-1234.

106. Ingram, M., Ottoway, F.J.H., and Coppock, J.B.M. 1956. The preservative action of acid substances in food. *Chem. Ind. (London)* 42: 1154-163.

107. Ita, P.S. and Hutkins, R.W. 1991. Intracellular pH and survival of *Listeria monocytogenes* Scott A in tryptic soy broth containing acetic, lactic, citric, and hydrochloric acids. *J. Food Prot.* 54: 15-19.

108. Ito, K.A., Chen, J.K., Lerke, P.A., Seeger, M.L., and Unverferth, J.A. 1976. Effect of acid and salt concentration in fresh-pack pickles on the growth of *Clostridium botulinum* spores. *Appl. Environ. Microbiol.* 32: 121-124.

109. Ito, T., Sota, H., Honda, H., Shimizu, K., and Kobayashi, T. 1991. Efficient acetic acid production by repeated fed-batch fermentation using two fermentors. *Appl. Microbiol. Biotech.* 36: 295-299.

110. Kabara, J.J. and Eklund, T. 1991. Organic acids and esters. In *Food Preservatives*, ed. N.J. Russell and G.W. Gould, pp. 44-71. Glasgow: Blackie and Son Ltd.

111. Karl, H., Roepstorff, A., Huss, H.H., and Bloemsma, B. 1994. Survival of *Anisakis* larvae in marinated herring fillets. *Int. J. Food Sci. Technol.* 29: 661-670.

112. Kim, C.R. and Hearnsberger, J.O. 1994. Gram negative bacteria inhibition by lactic acid culture and food preservatives on catfish fillets during refrigerated storage. *J. Food Sci.* 59: 513-516.

113. Kim, C.R., Hearnsberger, J.O., Vickery, A.P., White, C.H., and Marshall, D.L. 1995a. Sodium acetate and bifidobacteria increase shelf-life of refrigerated catfish fillets. *J. Food Sci.* 60: 25-27.

114. Kim, C.R., Hearnsberger, J.O., Vickery, A.P., White, C.H., and Marshall, D.L. 1995b. Extending shelf life of refrigerated catfish fillets using sodium acetate and monopotassium phosphate. *J. Food Prot.* 58: 644-647.

115. Kochevar, S.L., Sofos, J.N., LeValley, S.B., and Smith, G.C. 1997. Effect of water temperature, pressure and chemical solution on removal of fecal material and bacteria from lamb adipose tissue by spray-washing. *Meat Sci.* 45: 377-388.

116. Kotula, K.L. and Thelappurate, R. 1994. Microbiological and sensory attributes of retail cuts of beef treated with acetic and lactic acid solutions. *J. Food Prot.* 57: 665-670.

117. Krueger, D.A. 1992. Stable carbon isotope ratio method for detection of corn-derived acetic acid in apple cider vinegar: collaborative study. *J. AOAC Int.* 74: 725-728.

118. Kurtzman, C.P., Hesseltine, C.W., and Rogers, R. 1971. Microbiological spoilage of mayonnaise and salad dressing. *Appl. Microbiol.* 21: 870-874.

119. Lazaro, M.J., Carbonell, E., Aristoy, M.C., Safon, J., and Rodrigo, M. 1989. Liquid chromatographic determination of acids and sugars in homolactic cucumber fermentations. *J. Assoc. Off. Anal. Chem.* 72: 52-55.

120. Lee, I.S., Lin, J., Hall, H.K., Bearson, B., and Foster, J.W. 1995. The stationary-phase sigma factor (RpoS) is required for a sustained acid tolerance response in virulent *Salmonella typhimurium*. *Mol. Microbiol.* 17: 155-167.

121. Levine, A.S. and Fellers, C.R. 1940. Action of acetic acid on food spoilage microorganisms. *J. Bacteriol.* 39: 499-515.

122. Leyer, G.J. and Johnson, E.A. 1993. Acid adaptation induces cross-protection against environmental stresses in *Salmonella typhimurium*. *Appl. Environ. Microbiol.* 59: 1842-1847.

123. Leyer, G.J., Wang, L.-L., and Johnson, E.A. 1995. Acid adaptation of *Escherichia coli* O157:H7 increases survival in acidic foods. *Appl. Environ. Microbiol.* 61: 3752-3755.

124. Lillard, H.S. 1990. The impact of commercial processing procedures on the bacterial contamination and cross-contamination of broiler carcasses. *J. Food Prot.* 53: 202-204, 207.

125. Lillard, H.S., Blankenship, L.C., Dickens, J.A., Craven, S.E., and Shackelford, A.D. 1987. Effect of acetic acid on the microbiological quality of scalded picked and unpicked broiler carcasses. *J. Food Prot.* 50: 112-114.

126. Lin, M. and Sen, A. 1994. Direct catalytic conversion of methane to acetic acid in an aqueous medium. *Nature* 368: 613-615.

127. Little, C.L., Adams, M.R., and Easter, M.C. 1992. The effect of pH, acidulant and temperature on the survival of *Yersinia enterocolitica*. *Lett. Appl. Microbiol.* 14: 148-152.

128. Liu, C.H., Hsu, W.H., Lee, F.L., and Liao, C.C. 1996. The isolation and identification of microbes from a fermented tea beverage, Haipao, and their interactions during Haipao fermentation. *Food Microbiol.* 131: 407-415.

129. Lynch, D.J. and Potter, N.N. 1988. Effects of organic acids on thermal inactivation of *Bacillus stearothermophilus* and *Bacillus coagulans* spores in frankfurter emulsion slurry. *J. Food Prot.* 51: 475-480.

130. Marshall, D.L. and Dickson, J.S. 1998. Ensuring food safety, In *Maxcy-Rosenau-Last Public Health and Preventive Medicine, 14 ed.,* ed. R.B. Wallace, pp. 723-736. Stamford, CT: Appleton and Lange.

131. Marshall, D.L. and Kim, C.R. 1996. Microbiological and sensory analyses of refrigerated catfish fillets treated with acetic and lactic acids. *J. Food Qual.* 19: 317-329.

132. Mendonca, A.F., Molins, R.A., Kraft, A.A., and Walker, H.W. 1989. Microbiological, chemical and physical changes in fresh, vacuum-packaged pork treated with organic acids and salts. *J. Food Sci.* 54: 18-21.

133. Midolo, P.D., Lambert, J.R., Hull, R., Luo, F., and Grayson, M.L. 1995. *In vitro* inhibition of *Helicobacter pylori* NCTC 11637 by organic acids and lactic acid bacteria. *J. Appl. Bacteriol.* 79: 475-479.

134. Miller, L.G. and Kaspar, C.W. 1994. *Escherichia coli* O157:H7 acid tolerance and survival in apple cider. *J. Food Prot.* 57: 460-464.

135. Minor, T.E. and Marth, E.H. 1970. Growth of *Staphylococcus aureus* in acidified pasteurized milk. *J. Milk Food Technol.* 33: 516-520

136. Minor, T.E. and Marth, E.H. 1972a. Loss of viability by *Staphylococcus aureus* in acidified media. I. Inactivation by several acids, mixtures of acids, and salts of acids. *J. Milk Food Technol.* 35:191-196.

137. Minor, T.E. and Marth, E.H. 1972b. Loss of viability by *Staphylococcus aureus* in acidified media. II. Inactivation by acids in combination with sodium chloride, freezing, and heat. *J. Milk Food Technol.* 35:548-555.

138. Miwa, H., and Yamamoto, M. 1993. Determination of mono-, poly-, and hydroxy-carboxylic acid profiles of beverages as their 2-nitrophenylhydrazides by reversed-phase ion-pair chromatography. *J. Chromatogr.* 721: 261-268.

139. Moon, N.J. 1983. Inhibition of the growth of acid tolerant yeasts by acetate, lactate and propionate and their synergistic mixtures. *J. Appl. Bacteriol.* 55: 453-460.

140. Mountney, G.J. and O'Malley, J. 1965. Acids as poultry meat preservatives. *Poultry Sci.* 44: 582-586.

141. Moyles, A.L., Sholberg, P.L., and Gaunce, A.P. 1996. Modified-atmosphere packaging of grapes and strawberries fumigated with acetic acid. *HortSci.* 31: 414-416.

142. Nerbrink, E., Borch, E., Blom, H., and Nesbakken, T. 1999. A model based on absorbance data on the growth rate of *Listeria monocytogenes* and including the effects of pH, NaCl, Na-lactate and Na-acetate. *Int. J. Food Microbiol.* 47: 99-109.

143. Nickol, G.B. 1979. Vinegar. In *Microbial Technology, 2nd ed.,* ed. H.J. Peppler and D. Perlman, pp. 155-172. New York: Academic Press.

144. Nunheimer, T.D. and Fabian, F.W. 1940. Influence of organic acids, sugars, and sodium chloride upon strains of food poisoning staphylococci. *Am. J. Public Health* 30: 1040-1049.

145. Ockerman, H.W., Borton, R.J., Cahill, V.R., Parrett, N.A., and Hoffman, H.D. 1974. Use of acetic acid and lactic acid to control the quantity of microorganisms on lamb carcasses. *J. Milk Food Technol.* 37: 203-204.

146. Oh, D.-H. and Marshall, D.L. 1994. Enhanced inhibition of *Listeria monocytogenes* by glycerol monolaurate with organic acids. *J. Food Sci.* 59: 1258-1261.

147. Oh, D.-H. and Marshall, D.L. 1996. Monolaurin and acetic acid inactivation of *Listeria monocytogenes* attached to stainless steel. *J. Food Prot.* 59: 249-252.

148. Okrend, A.J., Johnston, R.W., and Moran, A.B. 1986. Effect of acetic acid on the death rates at 52°C of *Salmonella newport, Salmonella typhimurium* and *Campylobacter jejuni* in poultry scald water. *J. Food Prot.* 49: 500-503.

149. Olafsson, G., Jagerstad, M., Oste, R., and Wesslen, B. 1993. Delamination of polyethylene and aluminum foil layers of laminated packaging material by acetic acid. *J. Food Sci.* 58: 215-219.

150. Ouattara, B., Simard, R.E., Holley, R.A., Piette, G.J.-P., and Bégin, A.. 1997. Inhibitory effect of organic acids upon meat spoilage bacteria. *J. Food Prot.* 60: 246-253.

151. Oscroft, C.A., Banks, J.G., and McPhee, S. 1990. Inhibition of thermally-stressed *Bacillus* spores by combinations of nisin, pH and organic acids. *Lebensm.-Wiss. u.-Technol.* 23:538-544.

152. Osterling, C.E. and Lindgren, S.E. 1993. Inhibition of enterobacteria and *Listeria* growth by lactic, acetic and formic acids. *J. Appl. Bacteriol.* 75: 18-24.

153. Pampulha, M.E. and Loureiro-Dias, M.C. 1989. Combined effect of acetic acid, pH and ethanol on intracellular pH of fermenting yeast. *Appl. Microbiol. Biotechnol.* 31: 547-550.
154. Pangborn, R.M. 1963. Relative taste intensities of selected sugars and organic acids. *J. Food Sci.* 28: 726-733.
155. Parish, M.E. and Higgins, D.P. 1989. Survival of *Listeria monocytogenes* in low pH model broth systems. *J. Food Prot.* 52: 144-147.
156. Park, Y.S., Toda, K., Fukaya, M., Okumura, H., and Kawamura, Y. 1991. Production of a high concentration of acetic acid by *Acetobacter aceti* using a repeated fed-batch culture with cell recycling. *Appl. Microbiol. Biotech.* 35: 149-153.
157. Pederson, C.S. 1979. *Microbiology of Food Fermentations, 2nd Ed.* Westport, CT: AVI Publishing Co.
158. Perales, I. and Garcia, M.I. 1990. The influence of pH and temperature on the behavior of *Salmonella enteritidis* phage type 4 in homemade mayonnaise. *Lett. Appl. Microbiol.* 10: 19-22.
159. Phowchinda, O., Delia-Dupuy, M.L., and Strehaiano, P. 1995. Effects of acetic acid on growth and fermentative activity of *Saccharomyces cerevisiae*. *Biotechnol. Lett.* 17:237-242.
160. Przybylski, K.S. and Witter, L.D. 1979. Injury and recovery of *Escherichia coli* after sublethal acidification. *Appl. Environ. Microbiol.* 37: 261-265.
161. Raghubeer, E.V., Ke, J.S., Campbell, M.L., and Meyer, R.S. 1995. Fate of *Escherichia coli* O157:H7 and other coliforms in commercial mayonnaise and refrigerated salad dressing. *J. Food Prot.* 58: 13-18.
162. Ray, B. 1992a. Cells of lactic acid bacteria as food biopreservatives. In *Food Biopreservatives of Microbial Origin*, ed. B. Ray, pp. 81-101. Boca Raton, FL: CRC Press, Inc.
163. Ray, B. 1992b. Diacetyl of lactic acid bacteria as a food biopreservative. In *Food Biopreservatives of Microbial Origin*, ed. B. Ray, pp. 137-153. Boca Raton, FL: CRC Press, Inc.
164. Ray, B. and Sandine, W.E. 1992. Acetic, propionic, and lactic acids of starter culture bacteria as biopreservatives. In *Food Biopreservatives of Microbial Origin*, ed B. Ray, pp. 103-136. Boca Raton, FL: CRC Press, Inc.
165. Rice, A.C. and Pederson, C.S. 1954. Factors influencing growth of *Bacillus coagulans* in canned tomato juice. II. Acidic constituents of tomato juice and specific organic acids. *Food Res.* 19: 124-133.
166. Rosenquist, H. and Hansen, A. 1998. The antimicrobial effect of organic acids, sour dough and nisin against *Bacillus subtilis* and *B. licheniformis* isolated from wheat bread. *J. Appl. Microbiol.* 85: 621-631.
167. Roth, L.A. and Keenan, D. 1971. Acid injury of *Escherichia coli*. *Can. J. Microbiol.* 17: 1005-1008.
168. Russell, J.B. and Diez-Gonzalez, F. 1998. The effects of fermentation acids on bacterial growth. *Adv. Microbiol. Physiol.* 39: 205-234.
169. Rusul, G., El-Gazzar, F.E., and Marth, E.H. 1987. Growth and aflatoxin production by *Aspergillus parasiticus* NRRL 2999 in the presence of acetic or propionic acid and at different initial pH values. *J. Food Prot.* 50, 909-914.
170. Ryu, J.-H. and Beuchat, L.R. 1998. Influence of acid tolerance responses on survival, growth, and thermal cross-protection of *Escherichia coli* O157:H7 in acidified media and fruit juices. *Int. J. Food Microbiol.* 45: 185-193.
171. Ryu, J.-H., Deng, Y., and Beuchat, L.R. 1999. Behavior of acid-adapted and unadapted *Escherichia coli* O157:H7 when exposed to reduced pH achieved with various organic acids. *J. Food Prot.* 62: 451-455.
172. Semanchek, J.J. and Golden, D.A. 1996. Survival of *Escherichia coli* O157:H7 during fermentation of apple cider. *J. Food Prot.* 59: 1256-1259.
173. Shay, B.J. and Egan, A.F. 1986. Studies of possible techniques for extending the storage life of chilled pork. *Food Technol. Aust.* 38: 144-146.
174. Sholberg, P.L. 1998. Fumigation of fruit with short-chain organic acids to reduce the potential of postharvest decay. *Plant Dis.* 82: 689-693.
175. Sholberg, P.L. and Gaunce, A.P. 1995. Fumigation of fruit with acetic acid to prevent postharvest decay. *Hortsci.* 30: 1271-1275.
176. Sholberg, P.L. and Gaunce, A.P. 1996a. Fumigation of stone fruit with acetic acid to prevent postharvest decay. *Crop Prot.* 15: 681-686.
177. Sholberg, P.L. and Gaunce, A.P. 1996b. Fumigation of high moisture seed with acetic acid to control storage mold. *Can. J. Plant Sci.* 76: 551-555.
178. Sholberg, P.L., Reynolds, A.G., and Gaunce, A.P. 1996. Fumigation of table grapes with acetic acid to prevent postharvest decay. *Plant Dis.* 80: 1425-1428.
179. Sinha, R.P. 1986. Toxicity of organic acids for repair-deficient strains of *Escherichia coli*. *Appl. Environ. Microbiol.* 51: 1364-1366.
180. Siragusa, G.R. 1995. The effectiveness of carcass decontamination systems for controlling the presence of pathogens on the surfaces of meat animal carcasses. *J. Food Safety* 15: 229-238.

181. Siragusa, G.R. and Dickson, J.S. 1993. Inhibition of *Listeria monocytogenes*, *Salmonella typhimurium* and *Escherichia coli* O157:H7 on beef muscle tissue by lactic or acetic acid contained in calcium alginate gels. *J. Food Safety* 13: 147-158.
182. Smittle, R.B. 1977. Microbiology of mayonnaise and salad dressing: a review. *J. Food Prot.* 40: 415-422.
183. Smittle, R.B. and Flowers, R.S. 1982. Acid tolerant microorganisms involved in the spoilage of salad dressings *Lactobacillus fructivorans*, *Saccharomyces bailii*. *J. Food Prot.* 45:977-983.
184. Smulders, F.J.M. and Greer, G.G. 1998. Integrating microbial decontamination with organic acids in HACCP programs for muscle foods: prospects and controversies. *Int. J. Food Microbiol.* 44: 149-169.
185. Sorrells, K.M. and Leonard, B. 1988. Mechanisms of acid tolerance by a yeast isolated from spoiled ketchup. *J. Food Prot.* 51: 489-490.
186. Sorrels, K.M., Enigl, D.C., and Hatfield, J.R. 1989. Effect of pH, acidulant, time and temperature on the growth and survival of *L. monocytogenes*. *J. Food Prot.* 52: 571-573.
187. Steele, B.T., Murphy, N., Arbus, G.S., and Rance, C.P. 1982. An outbreak of hemolytic uremic syndrome associated with ingestion of fresh apple juice. *J. Pediatr.* 101: 963-965.
188. Stern, N.J., Rothenberg, P.J., and Stone, J.M. 1985. Enumeration and reduction of *Campylobacter jejuni* in poultry and red meats. *J. Food Prot.* 48: 606-610.
189. Stroup, W.H., Dickerson, R.W., Jr., and Johnston, M.R. 1985. Acid equilibrium development in mushrooms, pearl onions and cherry peppers. *J. Food Prot.* 48: 590-594.
190. Surve, A.N., Sherikar, A.T., Bhilegaonkar, K.N., and Karkare, U.D. 1991. Preservative effect of combinations of acetic acid with lactic or propionic acid on buffalo meat stored at refrigeration temperature. *Meat Sci.* 29: 309-322.
191. Tamblyn, K.C. and Conner, D.E. 1997. Bactericidal activity of organic acids in combination with transdermal compounds against *Salmonella typhimurium* attached to broiler skin. *Food Microbiol.* 14: 477-484.
192. Tamblyn, K.C., Conner, D.E., and Bilgili, S.F. 1997. Utilization of the skin attachment model to determine the antibacterial efficacy of potential carcass treatments. *Poultry Sci.* 76: 1318-1323.
193. Tsujihata, S., Entani, E., Asai, M., Tsukamoto, Y., Ohta, M. 1998. Mathematical modeling to predict the bactericidal effect of processed vinegar on *Escherichia coli* O157:H7. *Int. J. Food Microbiol.* 43: 135-138.
194. Tuneel, G. and Goktan, D. 1990. Effect of different methods of soaking soya beans on the growth of *Bacillus cereus, Klebsiella pneumoniae,* and *Staphylococcus aureus* in tempeh. *J. Sci. Food Agric.* 53: 287-296.
195. Warth, A.D. 1977. Mechanism of resistance of *Saccharomyces bailii* to benzoic, sorbic and other weak acids used as food preservatives. *J. Appl. Bacteriol.* 43: 215-230.
196. Weagant, S.T., Bryant, J.L., and Bark, D.H. 1994. Survival of *Escherichia coli* O157:H7 in mayonnaise-based sauces at room and refrigerated temperatures. *J. Food Prot.* 57: 629-631.
197. Wederquist, H.J., Sofos, J.N., and Schmidt, G.R. 1994. *Listeria monocytogenes* inhibition in refrigerated vacuum packaged turkey bologna by chemical additives. *J. Food Sci.* 59: 498-500,516.
198. Wethington, M.C. and Fabian, F.W. 1950. Viability of food poisoning *Staphylococci* and *Salmonellae* in salad dressing and mayonnaise. *Food Res.* 15: 125-134.
199. Williams, N.C. and Ingham, S.C. 1998. Thermotolerance of *Escherichia coli* O157:H7 ATCC 43894, *Escherichia coli* B, and an rpoS-deficient mutant of *Escherichia coli* O157:H7 ATCC 43895 following exposure to 1.5% acetic acid. *J. Food Prot.* 61:1184-1186.
200. Winslow, C.E. and Lochridge, E.E. 1906. The toxic effect of certain acids upon typhoid and colon bacilli in relation to degree of dissociation. *J. Infect. Dis.* 3: 547-571.
201. Wong, H.-C. and Chen, Y.-L. 1988. Effects of lactic acid bacteria and organic acids on growth and germination of *Bacillus cereus*. *Appl. Environ. Microbiol.* 54: 2179-2184.
202. Young, K.M. and Foegeding, P.M. 1993. Acetic, lactic and citric acids and pH inhibition of *Listeria monocytogenes* Scott A and the effect on intracellular pH. *J. Appl. Bacteriol.* 74: 515-520.
203. Zayaitz, A.E.K. and Ledford, R.A. 1985. Characteristics of acid-injury and recovery of *Staphylococcus aureus* in a model system. *J. Food Prot.* 48: 616-620.
204. Zhang, S. and Farber, J.M. 1996. The effects of various disinfectants against *Listeria monocytogenes* on fresh-cut vegetables. *Food Microbiol.* 13: 311-321.
205. Zhao, T. and Doyle, M.P. 1994. Fate of enterohemorrhagic *Escherichia coli* O157:H7 in commercial mayonnaise. *J. Food Prot.* 57: 780-783.
206. Zhao, T., Doyle, M.P., and Besser, R.E. 1993. Fate of enterohemorrhagic *Escherichia coli* O157:H7 in apple cider with and without preservatives. *Appl. Environ. Microbiol.* 59: 2526-2530.
207. Zhuang, R.Y., Huang, Y.W., and Beuchat, L.R. 1996. Quality changes during refrigerated storage of packaged shrimp and catfish fillets treated with sodium acetate, sodium lactate or propyl gallate. *J. Food Sci.* 61: 241-244,261.

R.K. Sharma

25

Citric acid

I. INTRODUCTION

Citric acid is used in industry for a variety of purposes. It plays an important role in the food and beverage industry, as well as in the manufacture of pharmaceuticals and cosmetics. The global production of citrate during 1990s was estimated at around 500 metric tons with an increasing demand.

Citric acid was first isolated from lemon juice by Swedish chemist Scheele in 1784. The John and Edmund Sturge Company successfully introduced the early commercial production of citric acid. Although citric acid was produced initially from citrus fruits, due to increasing demand, alternative methods of production were explored.

Grimoux and Adam (1880) synthesized citric acid from glycerol. Wehmer (1919) produced citric acid by a fungal fermentation process using *Penicillium* species. In 1917, Currie used *Aspergillus niger* to produce citric acid in a high yielding fermentation process. Currie transferred this technology to Pfizer Company to establish the first commercial fermentation process for citric acid. Many changes have taken place in the technology since its inception. Among various substrates, beet molasses was found economical for large-scale production. Doelger and Prescott (1939) used static medium containing sucrose and inorganic salts to grow *Aspergillus*. In 1949, Perlman reported development of submerged fermentation processes using *Aspergillus* and molasses. Since1965, different methods including yeast-based fermentation processes were developed. It is interesting to note that only a small amount of citric acid is produced from citrus fruits.

Citrate is widely used as an acidulant in beverages, confectionery, and effervescent salts; in pharmaceutical syrups, elixirs, in effervescent powders and tablets, to adjust the pH of foods and as synergistic antioxidant, in processing cheese. Also used as an additive in beverages, jellies, jams, preserves and candy to provide tartness. For non-dietary applications, citrate is used in the manufacture of alkylated resins; in esterified form as plasticizer, and foam inhibitor. Citrate is also used for sequestering color; in electroplating; in special inks; in analytical chemistry for determining citrate-soluble P_2O_5; as a reagent for albumin, mucin, lactoferrin, glucose, and bile pigments. In hematology, citrate

0-8493-2047-X/00/$0.00+$.50

is widely used as a component of anticoagulant solutions (citrate dextrose solution; citrate phosphate dextrose solution; and citric acid syrup).

II. OCCURRENCE AND PROPERTIES

A. Biosynthesis

Citric acid is produced as a bio-active molecule by plants and animals during metabolism. Acetyl-CoA derived from the oxidation of carbohydrates, fatty acids and amino acids during catabolism enters the tricarboxylic acid cycle. In this cycle, at each turn one molecule of acetic acid enters as acetyl CoA and condenses with a molecule of 4-Carbon oxaloacetic acid to form the 6-Carbon tricarboxylic compound 'citric acid'.

Acetyl-CoA + Oxaloacetic acid → Citroyl-CoA → Citric acid + CoASH

The rate of this reaction is largely determined by the availability of acetyl-CoA, oxaloacetate and the concentration of succinyl-CoA, which competes with acetyl-CoA and inhibits citrate synthase.

Aspergillus, yeasts, and certain bacteria accumulate citric acid in large quantities. Although such accumulation does not occur in normal physiological processes, it is found among mutants or strains with altered metabolic pathways. The addition of specific nutrients as well as selection of substrates paves the way for such production (Clark, 1962; Clark et al., 1965).

B. Fermentation

It is generally accepted that a rapid flux of carbon through glycolysis to pyruvate and a restriction in tricarboxylic acid cycle prevents the breakdown of citric acid once it is formed and results in the rapid accumulation of citric acid.

Netik et al. (1997) reported that the cellular uptake and export of citric acid by *Aspergillus niger* occurs by active, H(+)-symport dependent systems. These systems are inhibited by metabolizable tricarboxylic acid analogs and pthalic acid, and by several other mono- di-, and tri-basic organic acids. Citrate export is demonstrated in mycelia cultivated under manganese-deficient conditions, whereas the uptake of citrate from media was only detectable during pre-cultivation of *A. niger* in media supplemented with manganese ions.

C. Properties

Chemical nomenclature: 2-Hydroxy-1,2,3-propanetricarboxylic acid; β-hydroxytricarballylic acid. Chemical formula: $C_6H_8O_7$; molecular weight: 192.13 (Carbon, 37.51%; Hydrogen, 4.20%; Oxygen, 58.29%). Citric acid is widely distributed in plants and in animal tissues and fluids. As a monohydrate, citrate forms ortho-rhombic crystals in cold aqueous solution. Citrate has a pleasant, sour taste. The monohydrate crystals lose water in dry air or when heated at about 40 to 50°C, and are slightly deliquescent in moist air. It softens at 75°C with melting point ≈100°C. A 0.1N citrate solution has a pH of 2.2. Citric acid is a pure white crystalline solid. Citrate salts of ammonium, sodium and potassium are all highly soluble in water and are weak alkalines. The physico-chemical properties of citric acid are listed in *TABLE 1*.

TABLE 1. Citric acid - Physico-chemical properties

Molecular weight		
	Amorphous form	192.1
	Hydrated form	210.1
Melting point		153°C
Solubility		
	146 g dissolves in 1 L water at 20°C	
	525 g at dissolves in 1 liter water at 100°C	
	62 g dissolves in 1 liter ethanol at 25°C	
	2.25 g at 15°C in esters	
Dissociation constant		
	pK_1	3.14
	pK_2	4.77
	pK_3	6.40

```
          COOH
          |
HO — C — COOH
          |
          COOH
```

D. Assays

The methods of analysis according to the Association of Official Analytical Chemists (Williams, 1984) describes chemical assays for detection and quantitation of various acids in foods. Assays for organic acids in foods are generally carried out for three reasons: i) organic acid is a natural component of the product and therefore the assay is a quantitative or qualitative measure of wholesomeness or lack of adulteration or is a confirmation of the standard of identity for that product; ii) acid is not normally present in the food product or is present at levels lower than that normally detected in a standard assay and serves as a measure of adulteration either through addition of the acid or fermentation of the substrate to form acidic by-products; and iii) organic acid is added to achieve a desired effect as regulated by Good Manufacturing Practice (GMP).

In general, methods in practice are dependent on ease of separation of citrate from the food, the food product itself, and the desire for quantitative or qualitative results. It is desirable to estimate the levels of acetic and propionic acids in breads and cakes, eggs, and seafood; citric acid in wines, cheese, and dried milk. In the case of organic acids, enzymatic analysis for acetic, citric, isocitric, L-and D-lactic, L-malic, and succinic acids can be performed to identify citric acid (Bergmeyer, 1974). Acids are also quantitated using paper chromatography (Tijan & Jansen, 1971), gas chromatography (Martin et al., 1971), and high performance liquid chromatography (HPLC) (Marsili et al., 1981).

III. ANTIMICROBIAL ACTIVITY

Citric acid, like many other organic acids, elicits antimicrobial activity by various mechanisms. The antimicrobial outcome is influenced by various factors including pH, concentration, temperature, chain length etc. In addition to the antimicrobial property, the choice of the acidulant may depend upon secondary activity. Citric acid contributes to the taste and tartness of a product. It may create a synergistic relationship with antioxidants by chelating metal ions. It can control pectin gel formation, aid in the inversion of sucrose, prevent browning, and protect color. Certain citrate formulations find use as chemical leavening agents. Salts of acids may be important in regulating the acidity of foods.

Although citric acid is not used solely for its antimicrobial activity, it is shown to exhibit certain antimicrobial properties against molds and bacteria. In 1950, Murdock reported that some of the organisms isolated from tomato juice were inhibited by citric acid. Fabian and Graham (1953) found that citric acid was the more inhibitory to thermophilic bacteria than acetic and lactic acids. Further studies showed that pH was not a reliable indicator to measure preservation efficacy (Fabian & Wadsworth, 1939). Canned tomatoes acidified with citric, fumaric and malic acids did not show significant differences in their physical or chemical quality (Schoenemann & Lopez, 1973; Schoenemann et al., 1974).

Concentrations of 12-12.5% sodium citrate were inhibitory to *Salmonella* Anatum and *Salmonella* Oranienburg (Davis & Barnes, 1952). Thomson et al. (1967) found that 0.3% citric acid could reduce viable *Salmonella* on poultry carcasses. Skim milk acidified with citric acid was the most inhibitory to *Salmonella* Typhimurium followed by lactic and hydrochloric acids (Subramanian & Marth, 1968). *Staphylococcus aureus* is inhibited by increasing concentrations of citric acid and decreasing pH (Minor & Marth, 1970).

Sodium citrate could elicit bacteriostatic activity. Sodium citrate in concentrations of 0.1-4.0% was not as inhibitory to *Streptococcus agalactiae* when added to skim milk or fresh milk as citric acid at concentrations of 1, 2, and 4. This phenomenon seems due to a reduced pH effect rather than chelation of magnesium ions in the milieu (Sinha et al., 1968).

The antimicrobial activity of citrates has been attributed to its ability to chelate metal ions (Barnen & Keenan, 1970). *Lactobacillus casei* is inhibited by sodium citrate induced metal-chelation at 12-18 μM concentrations. Chelation effect also appears to be an influencing factor in the growth-inhibition of *Staphylococcus aureus* (Rammell, 1962) and *Arthrobacter pascens* (Imai et al., 1970).

Ahamad and Marth (1990) examined the ability of acetic, citric and lactic acid to prevent growth of *Listeria monocytogenes* in TB during extended incubation at 7-35°C. Citric aid was less inhibitory than acetic acid, with growth of listeriae occurring in all samples with 0.1% citric acid regardless of temperature. Based on average minimum pH values permitting growth of four *L. monocytogenes* strains at 10, 20, and 35°C, Sorrells and others found that acetic acid was again most inhibitory (pH 5.04) followed by lactic (pH 4.73), citric (pH 4.53), malic (pH 4.46) and hydrochloric acids (pH 4.46).

Citric and lactic acids have different effects on survival of listeriae. Although high levels of both acids inactivated listeriae in a similar pattern, low levels (0.1-0.5 M) of citric acid, especially at pH 5-6, protected listeriae from death (El-Gazzar & Marth, 1991; Buchanan et al., 1993; 1994). From the studies of Young and Foegeding (1993), it can be inferred that at the same pH levels, growth of listeria was affected most by acetic acid, followed by lactic and citric acids. The inhibitory effects of different salts of weak acids also were similar, with acetates being more inhibitory than lactates and citrates exhibiting the weakest action (Kouassi & Shelef, 1996).

Post et al. (1985) observed no change in numbers of *Clostridium botulinum* in tomato puree, shrimp puree acidified with citric or acetic acids to pH 4.2 or 4.6. No toxins were detected in any of these foods for up to 8 weeks. Similarly Tsang et al. (1985) found neither growth nor toxin production in TPGY medium acidified to pH <4.6. Notermans et al. (1985) examined the effects of dipping potatoes in a solution of 1% citric acid and 2% ascorbic acid for 2 minutes before vacuum packing and cooking. This process inhibited *C. botulinum* type B for 70 days at 15°C or 14 days at 20°C.

Cole et al. (1990) determined that the minimum pH values that allowed survival of *L. monocytogenes* after 4 weeks were pH 4.66 at 30°C, 4.36 at 10°C, and 4.19 at 5°C. The minimum pH that allowed growth within 60 days (>100-fold increase in cell numbers) was the same at 30°C but elevated to 4.83 at 10°C and 7.0 at 5°C. *L. monocytogenes* survived low pH values better at lower temperatures. These experiments used citrate phosphate buffer in which the citric acid content was calculated as 1.1% at pH 4.4 and 0.7% at pH 6.0. The authors suggested that the inhibitory activity might be due to the chelating ability of citric acid. Additional interactive effects were also observed in the presence of salt.

Using the multiple-barrier concept, hard-cooked eggs allowed to equilibrate in 0.5, 0.75, or 1.0% concentrations of citric acid and 0.2% sodium benzoate were held for 30 days at 4°C, or in 0.75% citric acid alone for 21 days at 4°C. A concentration of 0.75% citric acid was sufficient to reduce inoculated populations of *Salmonella* Typhimurium, *Yersinia enterocolitica*, *Escherichia coli*, and *S. aureus* (Fischer et al., 1985). Daly et al. (1973) used a combination of 0.1% citric acid and 0.75% gluconolactone incorporated into sausage before fermentation. This combination inhibited the growth of *S. aureus* early in the fermentation, and the increased acidity of the meat mixture allowed the fermentation to proceed more quickly.

Restaino et al. (1982) used a combination of 0.05% potassium sorbate and citric acid at pH 5.0 to reduce the growth rate of *Saccharomyces rouxii* and *Saccharomyces bisporus*. Potassium sorbate at 0.2% concentration in combination with citric and lactic acid at pH 5.5 suppressed the growth of *Lactobacillus plantarum* but did not effect *Pseudomonas aeruginosa* (Resatino et al., 1981).

Citric acid is also known to limit growth production of *Aspergillus parasiticus* and *A. versicolor*. At a level of 0.75% both citric acid and lactic acids had a slight growth retarding effect on *A. parasiticus*. At 0.25% toxin production was greatly reduced. 0.5% citric acid inhibited the growth and at 0.25% toxin production was suppressed. Even a concentration of 0.75% of citric or lactic acids did not affect the growth of *Penicillium expansum* (Reiss, 1976)

Beelman et al. (1989) developed an acid-blanch-chelate (ABC) process designed to reduce mesophilic and thermophilic bacterial spoilage in canned mushrooms. They demonstrated that acid blanching in a citric acid (0.05 M) solution buffered to pH 3.5 enhanced inhibitory activity during processing. This process also reduced thermophilic spoilage from 68 to 23.9% by adding citric acid to can brine, which was further reduced to 16.8% by the addition of EDTA.

Some studies have demonstrated a synergism between citric acid, polyphosphoric and EDTA with antimicrobials such as monolaurin, monocarpin, sucrose dicaprylate and sucrose monolaurate, thus exhibiting a high antibacterial activity against *E. coli* and other Gram-negative bacteria. It was concluded that the transport of monolaurin into the cell membrane of *E. coli* was enhanced by the action of citric, polyphosphoric acids and EDTA (Shibsaki & Kato, 1978; Blaszyk & Holley, 1998).

Nickerson and Darak (1978) described a composition and method for safely extending the storage life of refrigerated foods, such as shelled, hard-cooked eggs, cooked peeled shrimp, cooked and uncooked shrimp and cooked mushroom. The technique is based on applying mixtures of preservatives with a buffering agent such as citric and/or phosphoric acids. A buffering agent is needed in amounts to maintain the pH of the solution in the range of approximately 4.5 to 5.5 and selected from the group consisting of both citric acid and sodium or potassium hydroxide, or a mixture of sodium or potassium monohydrogen phosphate, dihydrogen phosphate or phosphoric acid.

The fate of *Salmonella* Enteridis PT4 in home-made mayonnaise prepared with citric acid solution [citric acid concentration of $\geq 4.98\%$ (w/v)] was investigated (Xiong et al., 1999). It was found that the pH of mayonnaise is closely related to the ratio of egg yolk to citric acid, and the inactivation rate of the microorganisms increased as the ratio decreased and/or incubation temperature increased. To achieve a *Salmonella* Enteritidis PT4-free home-made mayonnaise, it was recommended to prepare with pure lemon juice [citric acid concentration $\geq 5\%$(w/v)], and the pH should be 3.30 or below, or, in practice, at least 20 ml pure lemon juice per fresh egg yolk should be used.

Silla-Santos et al. (1995; 1996) found that citric acid has a stronger inhibiting effect than glucano-delta-lactone on the recovery of *Clostridium sporogenes* PA 3679 spores. In an earlier experiment these authors compared glucono-delta-lactone and citric acid as acidulants for lowering the heat resistance of *Clostridium sporogenes* PA 3679 in HTST working conditions. Citric acid was more effective in reducing the heat resistance of spores than GDL at all of the temperatures. The reduction in pH diminished the value of the 'z' parameter, although it was necessary to lower the pH to 4.5 to obtain a significant reduction.

El-Shenawy and Marth (1989; 1991; 1992) investigated the anti-listerial activity at 13 and 35°C of sodium propionate in combination with common organic acids. TB was prepared to contain 0, 500, 1500, or 3000 ppm sodium propionate and the pH of the medium was adjusted to 5.0 or 5.6 with HCl, or one of four common organic acids (acetic, tartaric, lactic, and citric). Decreasing the pH from 5.6 to 5.0 enhanced the anti-listerial activity of propionate. Organic acids, when compared with HCl, greatly enhanced the anti-listerial activity of propionate, with acetic acid being the most effective followed by tartaric, lactic and citric acids. In a subsequent study, the same authors investigated the antibacterial activity of sorbate in the presence of other organic acids. TB was prepared to contain 500, 1500, and 3000 ppm potassium sorbate and pH of the medium was adjusted to 5.0 or 5.6 using HCl or organic acids (acetic, tartaric, lactic or citric), inoculated with *L. monocytogenes*, and incubated at 13 or 35°C. When compared with HCl as an acidulant, the anti-listerial activity of sorbate was enhanced more by organic acids, with acetic and tartaric acids being more effective than lactic and citric acids.

Sheldon and Schuman (1996) reported that when stored at 4°C, *L. monocytogenes* populations in an inoculated commercially available reduced-cholesterol liquid whole egg product could be reduced as much as 3.9 orders of magnitude by adjusting the pH of the product to 6.6 with citric acid and adding nisin at a level of 1000 IU/mL.

Yamaguchi et al. (1996) investigated various properties of citric acid and EDTA solution as decalcifying and cleansing agents in root canal irrigation, as well their antibacterial effects. Citric acid solution was found to exhibit antibacterial effects on the bacteria used.

IV. APPLICATIONS

A. Clinical applications

A unique application of citric acid was studied by Marazova et al (1997). They studied the effect of newly synthesized agent MX1, a salt of active metabolite of the H2-blocker roxatidine with a complex of bismuth and citric acid against restraint ulcers. The experiments, conducted on rats, showed that MX1 has a gastroprotective effect, but also showed that bismuth subcitrate significantly reduced the ulcer size when higher doses are employed (Marzova et al, 1997). Topical citric acid produces notable changes very similar to those produced in response to glycolic acid, ammonium lactate, and retinoic acid including increases in epidermal and dermal glycosaminoglycans and viable epidermal thickness (Bernstein et al, 1997).

Renvert et al. (1997) evaluated subgingival irrigation with citric acid using an enzyme immunoassay. Subgingival irrigation is sought to be a supplement to scaling and root planing. Citric acid irrigation had no assurable microbiological effects. In another study Domiguez (1997) demonstrated that citric acid solution, used as a test drink, is superior to the previously proposed semi-liquid test meals in terms of the production of $^{13}CO_2$-urea breath for the diagnosis of *Helicobacter pylori* infection.

Nagoba et al. (1998) recently studied a possible treatment of superficial pseudomonal infections with citric acid. The authors feel that the treatment was both effective and cheaper than the conventional methods.

Most current dentin bonding procedures use acid etchants to partially demineralize the dentin structure and provide pathways for resin infiltration. On rehydration, the dehydrated surfaces underwent an expansion up to level seen after etching and tubule diameters returned to the etched values. These results indicate that the collapse of demineralized matrix is almost totally recoverable on rehydration (Marshall et al, 1998).

B. Dietary applications

The effect of dietary citric acid supplementation on calcium and phosphorus bioavailability in rats showed that citric acid increased intestinal absorption of calcium and phosphorus (Lacour et al., 1997). The study also showed that prolonged administration of calcium citrate supplements may help increase bone mineral concentration.

C. Pharmaceutical applications

In pharmaceutical industry citric acid is included in the formulation for many types of effervescent tablets, producing effervescent effects when it reacts with carbonates or bicarbonates. Citric acid is also used for its weak acidic nature, when the active agent is a basic substance. Trisodium citrate is used routinely as a blood anticoagulant. It complexes with calcium, one of the requirement to complete clotting. In cosmetics industry, citric acid is added in several creams, ointments, and shampoos. Zinc citrate is used in toothpastes for its antibacterial and antiplaque properties.

V. INDUSTRIAL PRODUCTION

A. Production processes

Currently, there are two organisms that are widely used in the production of citric acid, i.e. yeast and *Aspergillus niger*. Many other fungi are also able to produce citric acid such as *A. clavatus, A fumigatus, A. citrinum* and *Candida* species. Yeast has certain advantages over *Aspergillus* due to its independence with trace metals and higher fermentation rates. It also has the potential for higher sugar concentrations using osmophilic variants (Pfizer Inc. 1974).

Citric acid is produced by a number of different processes mainly by submerged liquid fermentation, liquid-surface and koji methods (Kubicek & Rohr, 1986; Kristiansen & Sinclair, 1979).

1. Submerged liquid fermentation process. Submerged liquid fermentation is one of the important and probably the most popular method used for the production of citric acid (Hustede & Rudy, 1976a; 1976b). A suitable substrate such as beet molasses, glucose syrup is sterilized and inoculated with a strain of *Aspergillus*. The inoculum can be introduced directly as spores or in vegetative form. The vital part lies in the actual morphology of the hyphae and the subsequent aggregation to form small spherical pellets. The hyphae should be abnormally short arising out of the deficiency of the manganese (Kisser et al., 1980). Factors resulting in the production of such pellets are the correct ferrocyanide level (Clark, 1964) and the correct concentration of manganese (Clark, 1966).

The medium is aerated or aerated and agitated depending on the type of fermenter tank used. A period of growth phase, during which almost no citric acid is produced, is followed by a production phase. In this phase the cells do not grow any longer and the entire substrate is now made available for citric acid production. The fermentation time is usually 2 - 5 days. Temperature and pH are accurately controlled. Initial pH is usually between 5 and 7, although final pH is around 2.0. Oxygen content is maintained at 10%, air saturation at 12%.

2. Liquid surface process. This process was first developed by Pfizer (1974). In this method, the fermentation is carried out in shallow trays with suitable medium. Beet molasses are used as a source for carbohydrates. pH is regulated to around 5.0, and additional nutrients are added, before the medium is sterilized. It is then inoculated with *Aspergillus* and left at a temperature of 30°C. The room is kept aerated so that oxygen supply is unhindered. Since the absorption of oxygen in to the cells is by surface aeration only, the whole process is completed in 7-12 days. At the end of the process, the mycelial mat formed is separated from the liquid. This liquid contains citric acid which is then purified.

3. Koji process. This process is almost exclusively employed in Japan. The main features are the use of sterilized wheat bran or sweet potato waste, giving a final water concentration of 70 - 80%, which is then inoculated with spores of *A. niger* and spread out in trays. The place is aerated and temperature maintained at around 28°C. The entire process is finished in a week, and after the mash is removed citric acid is extracted.

The effect of pH value, methanol, and salt concentration on the production of citric acid from cheese whey by two strains of *Aspergillus niger* (CAIM 111 and CAIM 167) was investigated by Samragy et al. (1996). The cumulative effect of fermentation medium pH (3.5), methanol concentration (4% v/v) and salt concentration (10%, w/v) during the fermentation process of whey did not enhance the production of citric acid by *A. niger* CAIM 111, while it did increase the production of citric acid by *A. niger* CAIM 167 by 4 times. Influence of calcium on fungal growth, hyphal morphology and citric acid production in *Aspergillus niger* was studied by Pera and Callieri (1997). Addition of 0.5g/L $CaCl_2$ to the fermentation medium lowered the final biomass dry mass by 35% and increased the uptake of phosphate and sucrose, and the production of citric acid by 15, 35 and 50% respectively.

Changes in the process which alter the *Aspergillus niger* metabolism during citric acid fermentation was studied by Legisa et al. (1995), by altering the substrate and by application of different substances at different concentrations. Initially, a higher sucrose concentration was used for the germination of spores, which caused a higher intracellular level of the osmo-regulator, glycerol, to be present. When citric acid started to be excreted into the medium, the substrate was suddenly diluted. This, the authors reasoned, induced the increased citric acid production.

Kirimura et al. (1999) described a method for the production of citric acid from xylan hydrolysate cultivated in a semi-solid culture of *Aspergillus niger*. They were able to directly produce 39.6 g/L of citric acid maximally in 3 d from 140 g/L of xylan.

B. Product recovery

After the organism synthesizes and accumulates citric acid and it is removed from the final broth, the product can be recovered by one of the two methods that are generally applied—The classical recovery method or the solvent extraction method (Marison, 1988; Sirman et al., 1990).

1. Classical recovery method. When molasses-based fermentation is used for manufacture of citric acid, impurities originate from the substrate. So, it is essential to purify and recover citric acid. The fermented broth is heated and lime is added. This precipitates insoluble calcium citrate trihydrate, which is then washed and treated with sulfuric acid. It releases citric acid and the calcium sulfate is removed.

2. Solvent extraction method. This method is employed when cleaner feedstocks are used in manufacturing process. This process involves removal of citric acid from the broth by treating it with the solvent at a low temperature followed by subsequent removal of the acids at a high temperature. Extractants used include butan-2-ol and tributyl phosphate diluted with small quantity of kerosene.

VI. SAFETY AND TOLERANCE

Toxicological data in animal models show different tolerance levels depending on the route of administration. Yokotani et al. (1971) tested the toxicity of commercial citric acid. The LD_{50} for mice administered by oral route is 5.0-5.8 g, subcutaneous is 2.7 g and for intravenous is around 0.95 g per kg body weight. For rat toxicity through oral administration the LD_{50} is 11.7 g and subcutaneous is 5.5 g per kg body weight. Death resulted due to respiratory or cardiac failure and hemorrhage of gastric mucosa. Gruber and Halbeisen (1948) inferred that the symptoms were similar to that of calcium deficiency. Studies conducted by Horn et al. (1957) showed that death of mice resulted from acute acidosis and hemorrhages of respiratory organs when citrate was administered by intravenous route.

Short-term studies on rats fed 0.2, 2.4, and 4.8% citric acid in a commercial diet showed lowered weight gains as a result of decreased intake of food, and slight blood chemistry abnormalities at the 4.8% level; slight atrophy occurred in regions of the thymus and spleen (Yokotani et al., 1971).

Long-term feeding studies with rats fed quantities up to 5% citric acid showed no significant changes from control groups (Bonting & Jansen, 1956; Horn et al., 1957). Cramer and co-workers (1956) investigated the relationship of the intake of sodium citrate and citric acid in vitamin D free diets containing low levels of phosphorus but adequate levels of calcium. When vitamin D was not given, the citrate completely prevented the absorption of calcium. Although there was no effect on weight gain, the urine contained calcium citrate. The authors concluded that citrate had a rachitogenic effect on the test animals.

Gomari and Gulyas (1944) administered 0.32-1.20 g/kg body weight of sodium citrate to dogs, which left the blood calcium levels unchanged but at the same time increased the urine calcium. Gruber and Halbeisen (1948) concluded that when citrate salts were administrated the acid part of the molecule was responsible for toxicity rather than the citrate moiety.

VII. REGULATORY STATUS

Citric acid is universally accepted as a safe food ingredient. With a pleasant sour taste, it has found its way into several foods and beverages. It is highly soluble in water and enhances the flavor of citrus-based foods. The Food and Drug Administration lists citric acid and its sodium, potassium and calcium salts, as multipurpose Generally Recognized As Safe (GRAS) food additives.

The joint FAO/WHO (JEFCA) has allocated to citric acid an Acceptable Daily Intake (ADI) of 'Non-specified' category. This means that, on the basis of the available data (chemical, biochemical, toxicological and others), the total daily intake of the substance, arising from its use at the levels required to achieve the desired effect, does not represent a hazard to health.

Citric is also used in countering oxidative discoloration of meats and fruits and other foods. When used in accordance with good manufacturing practice, citric acid may be used as a sequestrant. (21 CFR§182.6033). The calcium, sodium and potassium salts of citric acid, and isopropyl citrate at less than 0.02%, monoisopropyl citrate and stearyl citrate are commonly used as sequestrants. Monoglyceride citrate is used in antioxidant formulations for addition to oils and fats at less than ppm of the combined weight of the additive and fat or oil. Triethyl citrate is used in dried egg whites at levels less than 0.25%. An iron-choline-citrate complex may be used as a source of iron in foods for special dietary use. Iron ammonium citrate is used as an anti-caking agent in salt for human consumption when the level of the additive does not exceed 25 ppm in the finished product. Citric acid is also used as an acidulant in carbonated beverages for citrus flavor. Sodium citrate and tartrate are used as chelating agents to prepare noncaking, meat-salt mixtures (Gardner, 1966; 1972).

VIII. SUMMARY

Most plants and animals produce citric acid during metabolism. The tartness of the citrus fruits is due to the citric acid. Lemon juice contains 4 - 8% and orange has about 1% citric acid. Citric acid is widely used in foods, pharmaceuticals, and cosmetics. Citric acid has gained universal acceptance as a GRAS ingredient and the Joint FAO/WHO committee (JEFCA) has not set any ADI value. Although not a direct antimicrobial agent, citrate is recognized for enhancing the antimicrobial property of many other substances, either by reducing pH or by chelation of metal ions. Its use in controlling thermophiles, in particular, is well established. Citric acid is also used as an antioxidant to prevent oxidative discoloration in foods such as meats, fruits, and potatoes. Citrates of sodium, potassium and calcium also are recognized as GRAS ingredients with a wide range of food applications.

IX. REFERENCES

1. Ahamad, N., and Marth, E.H. 1987. Behavior of *Listeria monocytogenes* at 7, 13, 21, and 35°C in tryptic soy broth acidified with acetic, citric or lactic acid. *J. Food Prot.* 52:688-695.
2. Beelman, R.B., Witowski, M. E., Doores, S. Kilara, A., and Kuhn, G. D. 1989. Acidification process technology to control thermophilic spoilage in canned mushrooms. *J. Food Prot.* 52:178.
3. Bergmeyer, H.U.1974. *Methods of Enzymatic Analysis,* vols.1 and 2. New York: Academic Press.
4. Bernstein, E.F., Underhill, C.B., Lakkakorpi, J., Ditre, C.M., Uitto, J., Yu, R.J., and Scott, E.V. 1997. Citric acid increases viable epidermal thickness and glycosaminoglycan content of sun-damaged skin. *Dermatol. Surg.* 23:689-694.
5. Blaszyk, M., and Holley, R.A. 1998. Interaction of monolaurin, eugenol and sodium citrate on growth of common meat spoilage and pathogenic organisms. *Int. J. Food Microbiol.* 39:175-183.
6. Bonting, S.L., and Jansen, B. C. P. 1956. The effect of a prolonged intake of phosphoric acid and citric acid in rats. *Voeding* 17:137.
7. Branen, A.L., and Keenan, T.W. 1970. Growth stimulation of *Lactobacillus casei* by sodium citrate. *J. Dairy Sci.* 53:593-597.
8. Buchanan, R.L., and Golden, M.H. 1994. Interaction of citric acid concentration and pH on the kinetics of *Listeria monocytogenes* inactivation. *J. Food Prot.* 57:567-570
9. Buchanan, R.L., Golden, M.H., and Whiting, R.C. 1993. Differentiation of the effects of pH and lactic or acetic acid concentration on the kinetics of *Listeria monocytogenes* inactivation. *J. Food Prot.* 56:474-478.
10. Clark, D.S. 1962. Submerged citric acid fermentation of ferrocyanide treated beet molasses: Morphology of pellets of *A. niger. Can J. Microbiol.* 8:133-136.
11. Clark, D.S. 1964. Citric acid production, United States Patent No.3,118,821. *Chem. Abstr.* 61 4926d.
12. Clark, D.S., Ito, K., and Horitsu, H. 1966. Effect of manganese and other heavy metals on submerged citric acid fermentation of molasses. *Biotechnol. Bioeng.* 8:465-471.
13. Clark, D.S., Ito, K., and Tymchuck, P. 1965. Effect of potassium ferrocyanide addition on the chemical composition of molasses mash used in the citric acid fermentation. *Biotechnol. Bioeng.* 7:269-278.
14. Cole, M.B., Jones, M.V., and Holyoak C. 1990. The effect of pH, salt concentration and temperature on the survival and growth of *Listeria monocytogenes. J. Appl. Bacteriol.* 69:63.
15. Cramer, J.W., Porrata-Doria, E. I., and Steenbock, H. 1956. A rachitogenic and growth-promoting effect of citrate. *Arch. Biochem. Biophys.* 60:58-63.
16. Currie, J.N. 1917. The citric acid fermentation of *A.niger. J. Biol. Chem.* 31:15-17.
17. Daly, C., Lachance, M., Sandine, W. E., and Elliker, P. R. 1973. Control of *Staphylococcus aureus* in sausage by starter cultures and chemical acidulation. *J. Food Sci.* 38:426.
18. Davis, F., and Barnes, L.A. 1952. Suppression of growth of *Salmonella anatum* and *Salmonella oranienburg* by concentration variation of energy sources in a synthetic basal medium. *J. Bacteriol.* 63:33-38.
19. Dominguez, M. J. E., Leodolter, A., Sauerbruch, T., and Malfertheiner, P. 1997. A citric acid solution is an optimal drink in the ^{13}C-urea breath test for the diagnosis of *Helicobacter pylori. Gut* 40:459-62.
20. Doulger, W.P., and Prescot, S.C. 1939. Citric acid fermentation. *Ind. Eng. Chem.* 26:1142-1149.
21. El-Gazzar, F.E., and Marth, E.H. 1991. Survival of *Listeria monocytogenes* in food colorant derived from red beets. *J. Dairy Sci.* 74:81-85.
22. El-Shenawy, M.A., and Marth, E.H. 1989. Inhibition and inactivation of *Listeria monocytogenes* by sodium benzoate together with some organic acids. *J. Food Prot.* 52:771-776.
23. El-Shenawy, M.A., and Marth, E.H. 1991. Organic acids enhance the antilisterial activity of potassium sorbate. *J. Food Prot.* 54:593-597.
24. El-Shenawy, M.A., and Marth, E.H. 1992. Behavior of *Listeria monoytogenes* in the presence of sodium propionate together with food acids. *J. Food Prot.* 55:241-245.
25. Fabian, F. W., and Graham, H. T. 1953. Viability of thermophilic bacteria in the presence of varying concentrations of acids, sodium chloride and sugars. *Food Technol.* 7:212-217.
26. Fabian, F. W., and Wadsworth, C. K. 1939b. Experimental work on lactic acid in preserving pickles and pickle products. II. Preserving value of acetic acid and lactic acid in the presence of sucrose. *Food Res.* 4: 511-519.

27. FAO/WHO. 1963. Specifications for the Identity and Purity of Food Additives and Their Toxicological Evaluation: Emulsifiers, Stabilizers, Bleaching and Maturing agents. 7th Report of the Joint Food and Agriculture Organization of the United Nations/World Health Organization Expert Committee on Food Additives. WHO Tech. Report Ser. No. 281. FAO Nutrition Meetings Report Ser No.35.

28. FAO/WHO. 1973. Toxicological Evaluation of Certain Food Additives with a Review of General Principles and of Specifications. 17th Report of the Joint Food and Agriculture Organization of the United Nations/World Health Organization Expert Committee on Food Additives. WHO Tech. Report Ser. No.539. FAO Nutrition Meetings Report Ser. No. 53.

29. Fischer, J. R., Fletcher, D.L., Cox, N. A., and Bailey, J.S. 1985. Microbiological properties of hard-cooked eggs in a citric acid-based preservation solution. *J. Food Prot.* 48:252.

30. Gardner, W. H. 1972. Acidulants in food processing. In *Handbook of Food Additives*, 2nd ed., ed. T.E. Furia, pp.225. Cleveland, OH: CRC Press.

31. Gardner, W.H. 1966. Food Acidulants. New York: Allied Chemical Corporation.

32. Gomori, G., and Gulyas, E. 1944. Effect of parenterally administered citrate on the renal excretion of calcium. *Proc. Soc. Exp. Biol. Med.* 56:226-228.

33. Grimoux, E., and Adam, P. 1880. Synthase de l'acide citrique. *C.R. Hebd Séance Acad. Sci.* 90:1252-1255.

34. Gruber, C.M., and Halbeisen W. A. 1948. A Study on the comparative toxic effects of citric acid and its sodium salts. *J. Pharmacol. Exp. Ther.* 94:65-67.

35. Horn, H.J., Holland, E.G., and Hazelton, L. W. 1957. Safety of adipic acid as compared with citric and tartaric acid. *J. Agr. Food Chem.* 5:759-761.

36. Hustede, H., and Rudy, H. 1976a. Manufacture of citric acid by submerged fermentation. *United States Patent No.* 3,941,656.

37. Hustede, H., and Rudy, H. 1976b. Manufacture of citric acid by submerged fermentation. *United States Patent No.* 3,940,315.

38. Imai, K., Banno, I., and Ujima, T. 1970. Inhibition of bacterial growth by citrate. *J. Gen. Appl. Microbiol.* 16:479-487.

39. Kirimura, K., Watanabe, K., Sunagawa, T., and Usami, S. 1999. Citric acid production from xylan and xylan hydrolysate by semi-solid culture of *Aspergillus niger*. *Biosci. Biotechnol. Biochem.* 63:226-8.

40. Kisser, M., Kubicek, C.P and Rohr, M. 1980. Influence of manganese on morphology and cell wall composition of *A.niger* during citric acid fermentation. *Arch. Microbiol.* 128:26-33.

41. Kouassi, Y., and Shelef, L.A. 1996. Metabolic activities of *Listeria monocytogenes* in the presence of sodium propionate, acetate, lactate and citrate. 81:147-153.

42. Kristiansen, B., and Sinclair, C.G. 1979. Production of citric acid in continuous culture. *Biotechnol. Bioeng.* 21:297-315.

43. Kubicek, C.P., and Rohr, M. 1986. Citric acid fermentation. *Crit. Rev. Biotechnol.* 3:331.

44. Lacour, B., Tardivel, S., and Druke, T. 1997. Stimulation by citric acid of calcium and phosphorous bioavailability in rats fed a calcium-rich diet. *Miner. Electrolyte Metab.* 23:79-87.

45. Legisa, M., Gradisnik, G., and Grapulin, M. 1995. Sudden substrate dilution induces a higher rate of citric acid production by *Aspergillus niger*. *Appl. Environ. Microbiol.* 61:2732-2737.

46. Marazova, K., Kloucheck, E., Popov, A., Ivanov, C.H., and Krushkov, I. 1997. Gastroprotective effect of MX1 (a novel salt of the active metabolite of roxatidine with a complex of bismuth and citric acid) against stress ulcers in rats. *J. Pharm. Pharmacol.* 49: 791-795.

47. Marison, I.W. 1988. Citric acid production. In: *Biotechnology for Engnieers: Biological Systems in Technological Processes*, ed. A.H. Scragg, pp. 323-326. Chichester, England: Ellis Horwood.

48. Marshall, G.W. Jr., Wu, Maigidi, I. C., Watanabe, L. G., Inai, N., Balooch, M., Kinney, J.H., and Marshall, S.J. 1998. Effect of citric acid concentration on dentin demineralization, dehydration, and rehydration: atomic force microscopy study. *J. Biomed. Mater. Res.* 42:500-507.

49. Marsili, R.T., Ostapenko, H., Simons, R.E., and Green, D. E. 1981. High performance liquid chromatography determination of organic acids in dairy products. *J. Food Sci.* 46:52-57.

50. Martin, S.M., and Waters, W.R. 1952. Production of citric acid by submerged fermentation. *Ind. Eng. Chem.* 44:2229-2240.

51. Minor, T.E., and Marth, E. H. 1970. Growth of *Staphylococcus aureus* in acidified pasteurized milk. *J. Milk Food Technol.* 33:516.

52. Murdock, D. I. 1950. Inhibitory action of citric acid on tomato juice flat sour organisms. *Food Res.* 15:107.

53. Nagoba, B. S., Deshmukh, S. R., Wadher, B. J., Mahabaleshwar, L., Gandhi, R.C., Kulkarni, P.B., Mane, V.A., and Deshmukh, J. S. 1998. Treatment of superficial pseudomonal infections with citric acid. *J. Hosp. Infect.* 40:155-157.

54. Netik, A., Torres, N.V., Riol, J.M., and Kubieck, C.P. 1997. Uptake and export of citric acid by *Aspergillus niger* is reciprocally regulated by manganese ions. *Biochim. Biophys. Acta* 1326:287-294.

55. Nickerson, J.T.R., and Darack, J.R. 1978. *US Patent* 4076850.

56. Notermans, S., Dufrenne, J., and Keybets, M.J.H. 1985. Use of preservatives to delay toxin formation by *Clostridium botulinum* (type B strain okra) in vacuum - packed, cooked potatoes. *J. Food Prot.* 48:851.

57. Pera, L.M., and Callieri, D.A.1997. Influence of calcium on fungal growth, hyphal morphology and citric acid production in *Aspergillus niger. Folia-Microbiol.* 42:551-556.

58. Perlman, D. 1949. Mycological production of citric acid - the submerged culture method. *Econ. Bot.* 3:360-374.

59. Pfizer Inc. 1974 Fermentation process for the production of citric acid. Br. Pat. 1,364,094. *Chem. Abstr.* 78: 70192y.

60. Post, L. S., Amoroso, T.L., and Solberg, M. 1985. Inhibition of *Clostridium botulinum* type E in acidified food systems. *J. Food Sci.* 50:966.

61. Rammell, C. G. 1962. Inhibition by citrate of the growth of coagulase positive staphylococci. *J. Bacteriol.* 84:1123.

62. Reiss, J. 1976. Prevention of the formation of mycotoxins in whole wheat bread by citric acid and lactic acid. *Experimentia* 32:168.

63. Renvert, S., Dahlen, G., and Snyder, B. 1997. Clinical and microbiological effects of subgingival antimicrobial irrigation with citric acid as evaluated by an enzyme immunoassay and culture analysis. *J. Periodontol.* 68:346-352.

64. Restaino, L., Lenovich, L.M., and Bills, S. 1982. Effects of acids on potassium sorbate inhibition of food-related microorganisms in culture media. *J. Food Sci.* 47:134.

65. Restaino., L., Komatsu, K.K., and Syracuse, M.J. 1981. Effects of acids and sorbate combinations on the growth of four osmophilic yeasts. *J. Food Prot.* 45:1138.

66. Rohr, M., Zenhentgruber, O., and Kubicek, C.P., 1981, Kinetics of citric acid production by *A.niger* pilot plant scale. *Biotechnol. Bioeng.* 23:2433-2455.

67. Samragy, Y. A., Khorshid, M. A., Foda, M.I., and Shehata, A.E., 1996. Effect of fermentation conditions on the production of citric acid from cheese whey by *Aspergillus niger. Int. J. Food Microbiol.* 29: 411-416.

68. Schoenemann, D.R., Lopez, A., and Cooler, F. W. 1974. pH and acidic stability during storage of acidified and nonacidified canned tomatoes. *J. Food Sci.* 39:257.

69. Schoenemannn, D.R., and Lopez, A. 1973. Heat processing effects on physical and chemical characteristics of acidified canned tomatoes. *J. Food Sci.* 38:195.

70. Sheldon, B.W., and Schuman, J.D. 1996. Thermal and biological treatments to control psychrotrophic pathogens. *Poultry Sci.* 75:1126-1132.

71. Shibasaki, I., and Kato, N. 1978. In: *The Pharmacological Effect of Lipids*, ed. J.J. Kabara, pp.15. Champaign, IL: The American Oil Chemists Society.

72. Silla, Santos, M.H., and Torres, Z. J. 1995. Glucono-delta-lactone and citric acid as acidulants for lowering the heat resistance of *Clostridium sporogenes* PA3679 in HTST working conditions. *Int. J. Food Microbiol.* 25:191-7.

73. Silla, Santos, M.H., and Torres, Z. J. 1996. Evaluation of citric acid and GDL in the recovery at different pH levels of *Clostridium sporogenes* PA3679 spores subjected to HTST treatment conditions. *Int. J. Food Microbiol.* 29:241-54.

74. Sinha, D.P., Drury, A.R., and Conner, G. H. 1968. The *in vitro* effect of citric acid and sodium citrate acid and sodium citrate on *Streptococcus agalactiae* in milk. *Indian Vet. J.* 45:805-810.

75. Sirman, T., Pyle, D.L., and Grandison, A.S. 1990. Extraction of citric acid using a supported liquid membrane. In *Separations for Biotechnology*, ed. D. L. Pyle, pp. 245-254. London: Elsevier.

76. Subramanian, C.S., and Marth, E. H. 1968. Multiplication of *Salmonella typhimurium* in skim milk with and without added hydrochloric, lactic and citric acids. *J. Milk Food Technol.* 31:323-326.

77. Thomson, J.E., Banwart, G.J., Sanders, D.H., and Mercuri, A. J. 1967. Effect of chlorine, antibodies, propiolactone, acids and washing on Salmonella typhimurium on eviscerated fryer chickens. *Poul. Sci.* 46:146-151.

78. Tijan, G.H., and Jamen, J.T.A. 1971. Identification of acetic, propionic, and sorbic acids in bakery products by thin layer chromatography. *J. Assoc. Off. Anal. Chemists.* 54:1150-1151.

79. Tsang, N., Post, L. S. and Solberg, M. 1985. Growth and toxin production by *Clostridium botulinum* in model acidified systems. *J. Food Sci.* 50:961.

80. Wehmer, C. 1919. Verlust des oxalsaure-bildungs-vermogens bei einem degenerierten *Aspergillus niger. Zentrabl. Bacteriol.* 49:145-148.

81. Williams, S. 1984. *Official Methods Analysis of the Association of Official Analytical Chemists,* 14th ed. A.O.A.C., Arlington, Va.
82. Xiong, R., Xie, G., and Edmondson, A. S. 1999. The fate of *Salmonella enteritidis* PT4 in homemade mayonnaise prepared with citric acid. *Lett. Appl. Microbiol.* 28:36-40.
83. Yamaguchi, M., Yoshida, K., Suzuki, R., and Nakamura, H. 1996. Root canal irrigation with citric acid solution. *J. Endodontics.* 22:27-9.
84. Yokotani, H., Usui, T., Nakaguchi, T., Kanabayashi, T., Tanda, M., and Aramaki, Y. 1971. Acute and subacute toxicological studies of TAKEDA-citric acid in mice and rats. *J. Takeda Res. Lab.* 30:25-35.
85. Young, K.M., and P.M. Foegeding. 1993. Acetic, lactic and citric acids and pH inhibition of *Listeria monocytogenes* Scott A and the effect on intracellular pH. *J. Appl. Bacteriol.* 74: 515-520.

Section-VI

MILIEU-ANTIMICROBIALS

Sodium chloride
Polyphosphates
Chloro-cides
Ozone

S. Ravishankar

V.K. Juneja

Sodium chloride

26

I. INTRODUCTION

Sodium chloride, commonly called salt, table salt, or rock salt, is a vital part of human life. The food that we eat is tasteless without salt. An average American consumes approximately 8-10 g of salt per day (Dickinson, 1980). Salt enhances the flavor of foods and plays a functional role in food processing. For instance, salt controls microbial growth and shifts the fermentation in a desirable direction in products like pickles and sauerkraut; it controls yeast activity, strengthens the gluten and enhances crust color in baking goods; it controls the lactic acid fermentation rate as well as enhances flavor, texture, ripening and shelf life extension in cheese; it lowers water activity, strengthens gel structure and enhances color in processed meats (Lynch, 1987). Apart from its use in the food industry, salt aids in road transportation during severe winter weather, where it is still the best method of ice removal. Salt is also widely used in the chemical industry.

The history of salt dates back to the beginning of civilization, about 2700 BC, when one of the earliest known treatises on pharmacology, published in Chinese, contained information on several types of salt (Salt Institute, 1999). It has been referred to as the essence of life in Biblical teachings (Dickinson, 1980). Salt has been of economical importance in ancient history. In ancient Greece, goods were exchanged for salt. Salt tax has been an important issue of human history in many parts of the world. In Chinese history the salt tax was a major source of revenue (Salt Institute, 1999). Salt was a cause of the French Revolution. The salt tax supported British monarchs and people were imprisoned for smuggling salt. During the British rule in India, Mahatma Gandhi opposed the salt tax during the struggle for freedom. The desire for salt is believed to be an evolutionary process (Forsyth & Miller, 1980). It has been suggested that the neural-endocrine interactions in higher mammals play a role in behavior towards salt emergence and salinity control, and this relates to their parallel evolution (Denton, 1982). Studies on kangaroos living on the snowy mountains of Australia show a specific sodium appetite organized in the animals' brains (Denton, 1982).

0-8493-2047-X/00/$0.00+ $.50

Other related compounds of sodium that are used in food processing and preservation include sodium ascorbate, sodium benzoate, sodium bicarbonate, sodium citrate, sodium nitrate, sodium nitrite, sodium phosphate and sodium propionate. Some other salts that have been tried as a substitute for sodium chloride include potassium chloride, calcium chloride, ammonium chloride, magnesium chloride and lithium chloride. Each of these has limitations. Potassium chloride is the most closely related compound to sodium chloride, especially with regard to its physical and functional properties, but the limiting factor is its extreme bitter taste. Calcium and magnesium chlorides are very bitter too. Ammonium chloride is a highly unstable compound and lithium chloride is toxic to humans.

Replacing sodium chloride partially with a mixture of potassium, magnesium and calcium chlorides in dry fermented sausage caused a decrease in saltiness (decrease in sensory acceptability) as well as an increase in acidity and water activity (Gimeno et al, 1998). The lactic acid bacteria were not affected, but the counts of Micrococcaceae decreased. However, partial replacement of sodium chloride with potassium lactate in dry fermented sausages did not cause significant sensory defects (Gou et al., 1996; Askar et al., 1993). Partial replacement of sodium chloride with magnesium chloride at low levels did not produce any sensorial defects in bologna (Seman et al., 1980) while in frankfurters magnesium chloride was not a suitable substitute (Hand et al., 1982). Substituting sodium chloride with glycine (above 40%) produced an unacceptable sweet taste and non-uniform texture (Gou et al., 1996). Potassium chloride has been used as a substitute in some formulations (Ibanez et al., 1995; 1996; 1997) while substitution of the same has caused significant bitterness in some others (Leak et al., 1987; Keeton, 1984). A 1:1 mixture of sodium and potassium chloride when fed to human subjects showed a reduction of sodium intake by 44-55% (Mickelson et al., 1972).

Several seasonings and flavoring blends are also being tried as substitutes for sodium chloride. One such substitute included a mixture of potassium chloride, hydrolyzed vegetable protein, dipotassium orthophosphate, glucose, 5'-inosinic acid and 5'-guanosinic acid (Mohlenkamp & Miller, 1981). This blend was free of the bitter taste associated with potassium chloride. Autolyzed yeast blended with potassium chloride reduced the bitter taste due to potassium chloride (Shackelford, 1981). Another low sodium seasoning blend consisted of potassium chloride, monopotassium glutamate, potassium citrate, potassium phosphate, L-glutamic acid and silicon dioxide (Allen & Day, 1980). Glycinamide hydrochloride formed a good blend with monosodium glutamate (Sternberg et al., 1980). Addition of certain amino acids and their esters with sodium chloride has been found to enhance saltiness. L- ornithyl-B-alanine and Glycine ethyl ester hydrochloride enhanced the salty effect of sodium chloride (Okai et al., 1989). Some of the salty peptides included glycine methyl, ethyl ester hydrochlorides (Kawasaki et al., 1988) and L-ornithyltaurine (Tada et al., 1984).

II. OCCURRENCE AND PROPERTIES

Salt is abundantly available in nature and has been designated as the fifth element, next to air, water, fire and earth (Dickinson, 1980). In the United States, salt deposits are found in the states of Michigan, Ohio, New York, Pennsylvania, Louisiana and Texas. The country has an annual production of about 22 millions of dry salt (Dickinson, 1980). Historically, salt was made by boiling brine from salt springs. In 1800, large-scale production of salt began and the first underground salt mine was found in 1869 (Salt Institute, 1999).

Salt is produced by three methods (Dickinson, 1980). Recovering the rock salt by digging 600 to 2000 feet below the surface of the earth is called mining. Explosives are used to loosen the salt embedded in the rocks in the mines. Another method is removing the brine from brine wells using pipes, wherein, a pipe flushes water into a salt deposit and the brine formed is removed and brought to the surface using another pipe. The water is then evaporated or removed by heat, which leaves salt granules or flakes. The third method is natural evaporation of water from concentrated brines by sunshine, producing solar salt. The brine is concentrated in ponds to a point where it crystallizes and evaporates to form a salt bed. The crystallized salt is then harvested, washed and screened. This method can be only used in areas of heavy sunshine, very little rain and good air motion. In the United States, the sources of solar salt include the great Salt Lake in Utah, as well as the San Francisco Bay and San Diego areas in California.

Salt as a chemical compound is made of two elemental substances, the cationic sodium ion (Na^+) and the anionic chloride ion (Cl^-), which react together forming the halide salt, sodium chloride. The role of sodium ion in the human body is to maintain the blood volume and cellular osmotic pressure as well as aiding in the transmission of nerve impulses (Institute of Food Technologists, 1980). With regard to the physical properties of sodium chloride, it is present either as colorless transparent crystals or as white crystalline powder, with a molecular weight of 58.44, melting point of 80°C, boiling point of 1413°C, pH of 6.7 to 7.3 and a density of 2.165 (Lewis, 1980). It is soluble in water and glycerine. One gram of sodium chloride will dissolve in 2.8 ml of water at 25°C and a saturated solution will contain 26.5 g sodium chloride in 100 g of water (Shelef & Seiter, 1993). The solution is clear, colorless and odorless. It is stable under ordinary conditions of use and storage. Sodium constitutes 40% of sodium chloride by weight (Bodyfelt, 1982). To determine the salt content of a product, the food or its ash is extracted in warm water and the sodium and chloride contents determined by appropriate techniques. The concentration of sodium present in various bodily secretions can be measured through either flame photometry or by ion-specific electrodes (Michell, 1995). The chloride can be measured by titration using Morh's or Volhards's method (Luick, 1980). For assays of sodium chloride in various foods refer to 'Official Methods of Analysis' of the American Association of Analytical Chemists (1990).

III. ANTIMICROBIAL ACTIVITY

The antimicrobial activity of sodium chloride may be called either direct or indirect depending upon the purpose it serves and the amount added in a food product (Sofos, 1983). In case of dried and smoked meats, a large amount of sodium chloride is added, which makes these products shelf stable. These products were popular in earlier days and in certain parts of the world. They depended solely on sodium chloride for their preservation, and hence the effect could be called direct. In recent years sodium chloride is added in minimal amounts and is combined with other preservatives or hurdles to prevent microbial growth. The amount of sodium chloride that needs to be added in foods required to prevent microbial growth is large (16.54% salt solution to bring the water activity to 0.90 (Robinson & Stokes, 1959)) and will cause an unacceptable taste and hence salt is usually combined with other preservation techniques. In certain instances, sodium chloride is added mainly as a flavoring and functional ingredient and hence in these cases the effect could be called indirect. In some fermented food products, sodium chloride in appropriate concentrations extracts nutrients from vegetables, which in turn allows the growth of lactic acid bacteria, thereby preventing growth of certain spoilage organisms and even pathogens such as *Staphylococcus* species. In these cases, the antimicrobial effect could be called indirect. Another reason that the antimicrobial effect of sodium chloride may be called indirect is that it reduces the water activity in many foods (Sperber, 1983) and thereby indirectly prevents microbial growth. The reduction in water activity of the food causes an osmotic shock to the bacterial cell. Plasmolysis occurs and the cell loses fluids and either dies or enters dormancy. Some other possible mechanisms of sodium chloride inhibition are limiting oxygen solubility to the microbial cell, alteration of pH, toxicity of sodium and chloride ions, loss of magnesium ions (Banwart, 1979) and interference with the cellular enzymes (Shelef & Seiter, 1993).

Studies on the effect of sodium chloride on various organisms have indicated that sodium chloride could have a role on interfering with substrate utilization in these organisms. In *Staphylococcus aureus* sodium chloride was found to inhibit respiration, glucose utilization, phospho-β-galactosidase induction, and staphylococcal enterotoxin-A synthesis as well as hydrolysis of O-nitrophenyl-β-galactoside (ONPG), thereby inhibiting substrate transport into the bacterial cell (Smith et al., 1987). Erecinska and Deutsch (1985) studied the effect of sodium chloride on washed cells of *Paracoccus denitrificans* and found that increasing the concentration of sodium chloride (from 0.2 to 1.8%) gradually decreased the respiration rate and uptake of α-isoaminobutyrate with the highest concentration, completely inhibiting the uptake of the amino acid. In *Clostridium sporogenes*, glucose utilization was gradually inhibited and the level of intracellular ATP progressively decreased with increasing concentration of sodium chloride (Woods & Wood, 1982). However, the degree of glucose conversion to ethanol remained unchanged, indicating that sodium chloride interfered only with uptake and not with metabolism. In *Pseudomonas fluorescens* an increasing concentration of sodium chloride inhibited oxygen uptake (Prior, 1978). Protein synthesis was not always inhibited in different *S. aureus* cultures (Troller and Stinson, 1978).

Bacteria can be classified into different groups based on their salt tolerance (Ayers et al., 1980). Those that grow well at concentrations ranging from 0 to 0.5% as well as tolerating higher levels are called nonhalophiles. Some bacteria belonging to the genera *Staphylococcus*, *Micrococcus* and *Clostridium* belong to this group. Those that can tolerate and grow in the range of 1.5 to 5.0% sodium chloride are called slight halophiles and some members of *Achromobacter*, *Flavobacterium*, *Pseudomonas* and *Vibrio* fall into this category. The third group, which includes those that can tolerate up to 5 to 20% sodium chloride, are called moderate halophiles. Some of the lactic acid bacteria and some spore formers fall under this category. Some of the moderate halophiles were studied for their susceptibility to a wide range of antibiotics at different salt concentrations and a group of them were found to have increased susceptibility at low salt concentrations (Coronado et al., 1995). Organisms that require and can grow in high concentrations of salt are called halophiles and those that can tolerate but not grow in high concentrations of salt are termed halodurics (Jay, 1992). The extreme halophiles require at least 9% sodium chloride for growth, with the optimum concentration for growth ranging between 12 to 23% and the maximum being 32% (Madigan et al., 1997). The cell wall of halophilic strain *Halobacterium salinarum* is stabilized by sodium ions, to such an extent that in low sodium environments or where the concentration of sodium is insufficient the cell wall breaks down causing cell lysis (Madigan et al., 1997). The need for sodium cannot be replaced by any other ion, including potassium. Obligate halophiles require and tolerate 15% sodium chloride and can be found in saturated brines. Bacteria that can grow in high salt concentrations however, are of less importance as food spoilage or pathogenic agents (Lueck, 1980).

Sufficiently high amounts of sodium chloride are needed to cause bactericidal effects in foods and such amounts would not be permissible in foods, and such foods may not even be palatable. Hence, sodium chloride is generally used with other preservation methods. The antimicrobial effect of sodium chloride is thus dependent upon on several other factors present in foods as well as during food processing. Some of these factors include pH of the food, temperature and time during processing and storage, type of substrate, water activity of the substrate, type and populations of microorganisms, other preservatives/antimicrobial present in the substrate etc. Hence, most of the research done on sodium chloride is on its combinations with other hurdles and these studies will be discussed.

As mentioned earlier, the mechanism of salt inhibition is by lowering the water activity of the substrate. Most of the pathogenic organisms will not grow below water activity of 0.90 to 0.92. One exception is *Staphylococcus aureus*, which has a generally recognized minimum water activity of 0.86, but has been reported to grow in conditions of water activity as low as 0.83 (Jay, 1992). This organism can grow in the presence of 7 to 10% sodium chloride with some strains being able to grow at 20% sodium chloride. Concentrations of sodium chloride up to 10% had very slight effects on the growth of *S. aureus*, but concentrations more than 3% caused a decline in the production of enterotoxin-B (McLean et al., 1968). In another study, both the cell growth and enterotoxin-B production were inhibited with increasing sodium concentration (0 to 5.3%), but there was no effect on enterotoxin-A production (Troller & Stinson, 1978). However, other researchers have shown an inhibitory effect of sodium chloride on enterotoxin-A production by *S. aureus*. Increasing the concentration from 0 to 4% (Pereira et al., 1982) and from 0 to about 10% (Smith et al., 1987) caused a decline in the enterotoxin-A production by the organism. An increase in the concentration of sodium chloride in tryptic soy broth caused an increase in the effectiveness of butylated hydroxyanisole (BHA) in inhibiting *S. aureus*. Accordingly, the strongest effect was observed at a combination of 100 ppm BHA and 5% or 10% sodium chloride (Stern et al., 1979). Sodium chloride (7%) exhibited a synergistic effect with potassium sorbate (0.2%) in inhibiting *S. aureus* strains in TSB at 37°C (Robach & Stateler, 1980). However, there was no synergistic effect of sodium chloride (2, 3 and 7%) and potassium sorbate (0.2 and 0.3%) observed after prolonged exposure (15 days at 22°C and 48 h at 35°C) (LaRocco & Martin, 1987). An alternative sigma factor s[B], produced by *S. aureus* as a response to environmental stresses was repressed in the presence of 1 M sodium chloride (Chan et al., 1998).

There are reports in the literature showing that the heat resistance of *S. aureus* increases with increasing concentrations of sodium chloride due to a decrease in the water activity of the medium. Sodium chloride at different concentrations (3, 5 and 9 % w/v) protected *S. aureus* from heat injury, with the highest concentration affording the maximum protection (Smith et al., 1982). The authors suggested that sodium chloride might be involved in stabilizing membrane protein structures such as nucleic acids and nucleotides as well as increasing the melting temperatures of membrane phospholipids.

Thus, the damage to the cell membrane and leakage of cell components from the cytoplasm is prevented. In another study, 4 and 8% sodium chloride protected *S. aureus* cells from heat injury at pH 7.0 while at pH 6.5 a concentration of 8% gave protection (Bean & Roberts, 1975). Different substrates with reduced water activity were inoculated with *S. aureus* and the organism showed increased heat resistance (Troller, 1973). The effects of salt on thermal resistance have been examined by determining the relationships between thermal resistance and either solute concentration or water activity of the heating menstruum (Juneja et al., 1999). The heat resistance of *S. aureus* in sodium chloride increased as the degree of salt-water association increased (Tuncan & Martin, 1990). The authors explained that the effects of salt on thermal inactivation of microorganisms are mainly related to reduced water activity and increased osmotic pressure of the heating menstruum. For a given solute a certain optimum concentration of the solute gives a maximum heat protection, whereas levels outside this optimum solute concentration result in an increase in the heat sensitivity of the organism (Leistner & Russell, 1991).

The influence of sodium chloride on spore formers has been the subject of interest for many researchers. Sodium chloride has been found to inhibit toxin production in *Clostridium botulinum*. Greenberg et al. (1959) studied the inhibitory effect of sodium chloride on growth and toxin production in *C. botulinum* types A and B in cured meat. With less than 6.25% sodium chloride there was no inhibition of toxin production as well as putrefactive changes. Between 6.25 and 9% there was no inhibition on toxin production but the putrefactive changes did not occur. Above 9% growth was inhibited. In laboratory medium, 3 to 4 % sodium chloride inhibited toxin production by *C. botulinum* type E (Segner et al., 1966). The influence of sodium chloride on toxin production by *C. botulinum* types A, B and C at different temperatures in cooked meats vacuum packed in air-impermeable plastic pouches was studied (Pivnick & Barnett, 1965). Toxin production in both types was inhibited by 3% salt during 4 weeks at 30°C. However, 2.2% salt did not inhibit toxin production at 30°C in 1 week, while it did in 1-month at 25°C and in 2 months duration at 20°C. Sodium chloride at concentrations 2.1 and 3.4% inhibited toxin production in cooked ham within a month at 15°C. In the case of *C. botulinum* type E, 1.9% sodium chloride failed to inhibit toxin production in jellied pork tongue, while 2.7% sodium chloride was very effective in inhibiting toxin production. The authors suggested adequate refrigeration and an increase in salt concentration to acceptable limits for types A and B and heat and low salt concentration for type E. Sodium chloride at 1.4 and 2.3% concentrations did not inhibit toxin production by *C. botulinum* E in jellied ox tongue (Pivnick & Bird, 1965). A concentration of 5.0% sodium chloride at pH values lower than 5.03 was needed to inhibit growth of *C. botulinum* type E (Segner et al., 1966).

The effect of sodium chloride on growth of *C. botulinum* spores at different temperatures and pH levels in the presence of sodium nitrite was investigated (Emodi & Lechowich, 1969). The substrate was laboratory media and the pH range varied from 5.2 to 6.6. About 4.87% sodium chloride was sufficient to inhibit outgrowth at all temperatures (30, 15.6, 10, 7.2, 5.0 and 3.4) with lower temperatures requiring slightly lower concentrations. At lower temperatures, inhibition occurred at high pH levels. In bottled lumpfish caviar toxin production of *C. botulinum* was inhibited at combinations of sodium chloride concentrations $\geq 5.56\%$ and pH <5.0 at an abusive temperature of 30°C (Hauschild & Hilsheimer, 1979). Products with 4.5% sodium chloride were capable of supporting growth of proteolytic *C. botulinum* strains even at 15°C, even though they may have an extended lag time (Gibson et al., 1987).

The effect of sodium chloride on the heat resistance and recovery of non-proteolytic *C. botulinum* type B strain in turkey meat slurry was studied (Juneja & Eblen, 1995). Increasing the salt concentration from 1 to 2 and 3% reduced the heat resistance of the spores (the spore D-values declined). The measured heat resistance in the heating medium as well as the recovery medium (reduced recovery of heat-injured spores) was lower with increasing salt concentrations. A similar phenomenon of reduction in recovery with increasing salt concentrations has been observed in other spore formers such as proteolytic *C. botulinum* (Pivnick & Thacker, 1970), *C. sporogenes* (Roberts et al., 1966) and *Bacillus stearothermophilus* (Cook & Gilbert, 1969). Interactive effects of temperature, pH, sodium chloride and phosphate on the thermal inactivation of non-proteolytic *C. botulinum* type B strains have been studied (Juneja et al., 1995). Combined with heat treatment sodium chloride was able to prevent growth from spores of non-proteolytic *C. botulinum*. (Stringer & Peck, 1997). Heating at 90˚C for 30 min alone or incubating at 10˚C with 4.0% salt by itself did not prevent growth, while heat treatment at the same conditions following by incubation with the same amount of salt did prevent growth. Sodium chloride at 3% concentration along with low pH (5.5) reduced the nisin resistance of *C. botulinum* strains (Mazzotta et al., 1997).

The effect of sodium chloride on spore outgrowth of *C. perfringens* in cook-in-bag ground beef at different temperatures, pH and sodium pyrophosphate levels was investigated (Juneja & Majka, 1995). At 28˚C, a combination of 3% sodium chloride and pH 5.5 delayed growth for 24 h. At 15°C and pH 7.0 there was growth within 6 days, but with the same conditions when 3% salt was added growth was delayed until 8 days. At pH 5.5 and 15˚C in the presence of 3% salt, there was no growth of vegetative cells even after 21 days. The temperature abuse of this product for more than 15 h without sodium chloride led to growth of *C. perfringens*. In case of vacuum packaged cook-in-bag ground turkey, at 28˚C 3% salt delayed growth of *C. perfringens* from spore inoculum for 12 h (Juneja & Marmer, 1996). At 15˚C with 1-2% salt growth was very slow. With 3% salt no vegetative cells were observed even after 28 days. The D-value was 23.2 min with no salt, which decreased to 17.7 min with 3% salt. Gibson and Roberts (1986a) studied the influence of increasing concentration of salt in combination with sodium nitrite, pH and temperature variables, and found that the growth rate of *C. perfringens* decreased with increasing concentrations of salt. Salt showed synergistic effects with other variables. Juneja et al. (1996) studied the interactive effects of sodium chloride, temperature, initial pH and sodium pyrophosphate on growth kinetics of *C. perfringens*.

The effect of sodium chloride on heat and radiation resistance and on recovery of injured spores of *Bacillus* species was investigated (Briggs & Yazdany, 1970). An increase in the concentration of sodium chloride caused a decrease in the heat resistance of *B. sterothermophilus*, but not the radiation resistance of the same organism as well as the heat and radiation resistance of other species of *Bacillus*. Sodium chloride presence during heating increased the sodium chloride sensitivity of *B. stearothermophilus*, but not that of other species. An increase in the irradiation dose increased sodium chloride sensitivity of *B. stearothermophilus* in the recovery medium while the effect was less severe for other species. The growth of *B. stearothermophilus* in salty carrot medium as a function of pH, temperature and sodium chloride has been characterized and mathematical models derived (Ng & Schaffner, 1997). *B. subtilis* when exposed to mild concentrations of salt was able to survive subsequent toxic concentrations (Volker et al., 1992). A mild heat shock produced cross-protection against lethal salt stress, but not vice-versa. A set of general stress proteins were believed to be induced in response to salt or heat stress.

The effect of sodium chloride in combination with other factors on *Escherichia coli* has been investigated. A concentration of 8% or more of sodium chloride completely inhibited growth of enteropathogenic *E. coli* at different temperature and pH levels, while a concentration of 4% in combination with pH 5.6 and 200 ppm of nitrite did not (Gibson & Roberts, 1986b). *E. coli* O157:H7 was inhibited by 8.5% or more sodium chloride in TSB (Glass et al., 1992). A concentration of 2.5% did not have any inhibiting effect, while at 4.5% the generation time was longer and at 6.5% the lag time was very long (36 h) which the authors attributed to the presence of a salt tolerant population of the organism. In fermented sausage with 3.5% sodium chloride, 69 ppm. sodium nitrite and pH 4.8 the organism population was reduced but was not inhibited completely (Glass et al., 1992). The survival of bioluminescent *E. coli* O157:H7 in brain heart infusion medium (BHI) containing sodium chloride with other factors as well as in model systems representing fermented sausage was studied (Tomicka et al., 1997). Sodium chloride at concentrations up to 3.5% did not inhibit growth in BHI. In American-style fermentation (high temperature short time) with 2% sodium chloride, starter culture, dextrose and sodium nitrite, the organism survived for more than 51 days and the authors explained that sodium chloride and sodium nitrite enhanced the survival of the organism in this system. In the European-style fermentation, at low inoculum levels of the organism there was inhibition after 9 days, while with high initial inoculum levels, there was survival for more than 30 days. The reason for this survival was attributed to possible inhibition of starter cultures by the salts added and hence lesser competition for *E. coli*. In skim milk (10% rehydrated non-fat dry milk), *E. coli* O157:H7 had up to a 3-log reduction with 4% salt at pH 4.7 at different times and temperatures, while 6% sodium chloride caused complete inhibition at all treatment conditions (Guraya & Frank, 1998). It was concluded that sodium chloride at increasing concentration enhanced inactivation at pH levels between 4.1 and 4.7.

The interactive effects of temperature (55 to 62.5 °C), pH (4 to 8), sodium chloride (0 to 6% w/v) and sodium pyrophosphate (0 to 0.3% w/v) on the heat resisnce of *E. coli* O157:H7 was examined (Juneja et al., 1999). All four factors interactively affected the inactivation of the pathogen, and a mathematical model for the interactive effects of the four variables on the thermal inactivation was developed. The protective effect of salt against lethality was only observed at pH 4.0. The influence of sodium chloride (8%) combined with sodium lactate (4.0%) and polyphosphate (0.5%) on the heat resistance of turkey meat at 50, 55, 57 and 60°C was studied (Kotrola & Conner, 1997). The D- values for turkey meat with these additives were higher than turkey meat without additives, indicating that the additives enhanced survival of the organism. A predictive model fitted using the Gompertz equation for the growth of *E. coli* O157:H7 as a function of temperature, pH and sodium chloride was developed by Sutherland et al. (1995) and has been extrapolated to a range of foods including meat, poultry, milk, cheese and tempeh. Acid adaptation of some strains of *E. coli* O157:H7 exhibited cross-protection against increased osmolarity (Garren et al., 1998; Cheville et al., 1996). Salt, heat and acid tolerances in *E. coli* O157:H7 were found to be regulated by the *rpo*S sigma factor (Cheville et al., 1996).

Another organism studied with regard to its sensitivity to sodium chloride is *Listeria monocytogenes*. The influence of sodium chloride, temperature and pH on the growth of *L. monocytogenes* in cabbage juice was investigated (Conner et al., 1986). A 2% and higher concentration of sodium chloride in the juice inhibited the organism. With 5% sodium chloride the survival of one strain (Scott A) declined 90% over 70 days at 5°C. Another strain (coleslaw outbreak strain) survived in cabbage juice containing 3.5% or less sodium chloride over a 70 day period. At 1.5 and 2.0% sodium chloride, both the strains had extended lag times. Initial exposure to heat and ethanol significantly increased the tolerance of *L. monocytogenes* to sodium chloride (Lou & Yousef, 1997). Sodium chloride inhibited the repair of thermal injury in *L. monocytogenes* cells (Pagan et al., 1997; Linton et al., 1992; Smith & Hunter, 1988; Golden et al., 1988).

Models to fit the interactive effects of sodium chloride with other variables on *L monocytogenes* have been the subject of interest for many investigators. The heat resistance of *L. monocytogenes* at different temperatures, pH levels and sodium chloride concentrations was studied in phosphate buffer (Linton et al., 1995) and in infant formula (Linton et al., 1996) and models were fit using the Gompertz equation. Interactive effects of temperature, lactic acid, sodium chloride and sodium nitrite on the survival of *L. monocytogens* under anaerobic conditions (Buchanan & Golden, 1995) and the interactive effects of these same factors on the time to achieve a 4-log reduction in aerobic acidic conditions (Buchanan et al., 1997) were studied and models fit for the same. Another model included the effect of CO_2, pH, temperature and sodium chloride on the growth of *L. monocytogenes* (Fernandez et al., 1997). Juneja and Eblen (1999) studied the interactive effects of temperature, pH, sodium chloride and sodium pyrophosphate on the heat inactivation of *L. monocytogenes* in beef gravy. In their findings, sodium chloride protected the organism against the lethal effects of heat. A predictive model describing the combined effects of the factors was developed. Another model describing the effects and interactions of temperature, pH, growth atmosphere (aerobic and anaerobic), sodium chloride and sodium nitrite on the growth of *L. monocytogenes* was developed by Buchanan and co-workers (1989).

Effects of sodium chloride on other food-borne pathogens have also been documented. Thermally processed as well as fresh food products are chilled using brine that could contain up to 20% sodium chloride. The growth, survival and injury potential of *Yersinia enterocolitica*, *S. aureus* and *L. monocytogenes* in chiller brine with different sodium chloride concentrations was investigated (Miller et al., 1997). Growth of *Y. enterocolitica* was observed in 0.5 and 5% brine at -2 and 5°C, respectively. *S. aureus* grew in 5% sodium chloride at 12°C and *L. monocytogenes* was able to grow in 5 and 9% sodium chloride at 5 and 12°C, respectively. Sodium chloride (9%) at -2°C was either bacteriostatic or -cidal depending on the organism. *L. monocytogenes* was able to survive 20% sodium chloride at -12°C for 30 days. Caution should be taken in recycling brine, if such survival could occur.

Li et al. (1997) tested the efficacy of sodium chloride spray to reduce *Salmonella typhimurim* in prechill chicken carcasses. Prechill chicken carcasses inoculated with *S. typhimurium* were sprayed with tap water, 0.85% sodium chloride solution and other antimicrobial sprays. Compared to tap water, sodium chloride did not significantly reduce *S. typhimurium* population and hence did not prove as effective an antimicrobial spray. The efficacy of sodium chloride and sodium carbonate solutions on removing *E. coli* O157:H7 from surfaces of chopped lettuce was studied (Janes et al., 1999). A 2-log reduction was observed with a concentration lower than 0.8% and about 2.5-log reduction with 0.8 to 1% solutions of each of these salts at pH 8.0, whereas a 3-log reduction was observed at pH 2.0 and 10. *Shigella* is a foodborne pathogen of concern that can survive in seawater (Nakamura et al., 1964). *Shigella flexneri* strains were able to grow in 3.78% sodium chloride (highest concentration allowing growth in nutrient agar) at 37°C (Fehlhaber, 1981). The interactive effects of sodium chloride, pH and temperature on the growth of *S. flexneri* was examined (Zaika et al., 1989). Combinations of low temperature, low pH and high sodium chloride concentration proved inhibitory to the organism. The inhibitory effect of a combination of antimicrobial compounds (lysozyme, monolaurin, triglycerol 1,2 laurate and BHA) against a number of spoilage and pathogenic microroganisms (*Bacilus* sp., *Pseudomonos* sp., *Enterobacter aerogenes, Lactococcus lactis, L. monocytogenes, E. coli* O157:H7, *Staphylococcus* sp., *S. typhimurium*) in the presence of sodium chloride and EDTA was studied (Razavi-Rohani & Griffiths, 1996). Sodium chloride exhibited a synergistic effect with these combinations in the presence of EDTA and low pH in inhibiting the organisms. The effect of sodium chloride and nitrite on the antimicrobial activity of lysozyme, nisin, and EDTA combination treatments against a variety of microroganisms of concern in cured meats was tested (Gill & Holley, 1999). Sodium chloride alone was inhibitory against *E. coli* O157:H7, *Salmonella typhimurium, Serratia grimesii* and *Shewanella putrifaciens*. A combination of sodium chloride and EDTA was found to be inhibitory to *E. coli, Salmonella* and *Serratia*, while in combination with sodium nitrite, sodium chloride inhibited *Salmonella* and *Shewanella*. Water containing electrolytic products of sodium chloride (electrolytic water) was tested for its bactericidal and hand washing efficiency and compared with hypochlorite solution (Hitomi et al., 1998). The electrolytic water was equally effective in its bactericidal and handwashing efficacy and is recommended for handwashing in place of running water.

IV. BIOLOGICAL ADVANTAGE

Sodium is a nutrient essential for all animals including humans. It is a dominant ion of human extracellular body fluid both quantitatively and functionally and is important for maintaining an appropriate blood volume and pressure, cell osmotic pressure and for transmitting nervous impulses (Institute of Food Technologists, 1980). Both sodium and chloride ions predominate in the extracellular body fluid whereas in the intracellular body fluid, sodium is present in very small quantities with almost no chloride present. Chloride is essential for maintaining tissue osmolarity, acid-base balance in the blood and the electrolyte balance of the body, for activating some of the essential enzymes in the stomach and for the formation of hydrochloric acid in the stomach, for transmitting nervous impulses and for the passage of water across the cell walls (Schreiber & Harner, 1983; Institute of Food Technologists, 1980). According to Schreiber and Harner (1983) the physiological roles of sodium chloride in the human body include: i) regulation of extracellular fluid volume. A decrease or increase in the bodily sodium chloride will cause a corresponding change (increase or decrease) in the extracellular body fluid. ii) regulation of intracellular fluid. Based on the amount of salt the intracellular water maintains equilibrium with the extracellular body fluid. iii) regulation of neutrality of body fluids. The body fluids have a neutral pH and their neutrality is maintained by sodium buffer systems and the excretory action of kidneys and lungs. iv) other physiological processes such as cardiovascular, intestinal muscles and the stomach require exact proportions of sodium and chloride. Any of these ions if excessively consumed, the amount other than needed will be excreted by the body in the urine and a balance between the two ions is always maintained by the body unless there are other disorders. Sodium chloride has its benefits in the clinical field. Intra-dermal bacteriostatic sodium chloride (0.9%) containing the

TABLE 1. Sodium content of some natural and processed foods

Food product	Sodium content (mg/100g)
Celery	140
Carrot	50
Apple	2
Milk	50
Lightly salted butter	870
Stilton cheese	1150
White bread	540
Soy sauce	6082
Tomato ketchup	1120
Boiled eggs	140
Corned beef	950
Steak	54

Adapted from Paul & Southgate (1978)

preservative benzyl alcohol was found to be as effective as 1% lidocaine hydrochloride (an amide local anasthetic) for the attenuation of intravenous cannulation pain (McNelis, 1998).

The minimum daily requirement of sodium chloride for normal individuals is less than 2 g/day and the recommendation of health agencies is not to exceed 6 g/day. The daily dietary intake of sodium is estimated to be 1.1 and 3.3 g for adults and between 115 and 750 mg for infants (NAS, NRC, 1980). For chloride, the minimum requirement daily set by the food and nutrition board is 750 mg for adults and 180 to 350 mg for infants. There are several sources through which sodium chloride is consumed by humans. The potable water contains sodium chloride that accounts for less than 1% of the daily intake. Sodium chloride is also present in some natural food products such as fruits and vegetables. Another source of salt is many processed foods. The sodium content of some natural and processed foods is summarized in TABLE 1. Another major source of sodium chloride is the salt added during cooking or at the dining table. Drugs (medicines) taken for ailments may also contain sodium and hence people who are taking medications regularly or for ailments should consider such intake. Excessive sodium chloride intake has been identified to be one of the causes for developing hypertension and the related cardiovascular problems and stroke (Pearson & Wolzak, 1982), however, a clear link between the two has not been established. Because of such possible health risk, there are many consumer concerns and the food industry is trying to minimize the salt content of food products.

V. APPLICATIONS

Salt is used not only by the food industry but also by the agricultural, chemical and transportation industries. TABLE 2 summarizes the average amount and cost of salt by the various industries (adapted from Dickinson, 1980). In many food formulations salt is added for flavoring, functional as well as preservative effects. Some of these food products include butter, cheese, fermented vegetables and fish, cured meats, bread etc. In butter and margarine, salt is used as a flavoring agent but also serves as a preservative. In these products, emulsified fats in the water phase are susceptible to microorganisms. Up to 2% and 3% by weight of salt is added in butter and margarine respectively after wash-

TABLE 2. Average amount of salt used and the cost incurred by various industries in the US

Industry	Amount of salt used (tons)	Cost incurred ($)
Food	1,002,800	98,302,900
Chemical	4,114,200	37,450,000
Highway	10,927,000	97,779,600
Water conditioning	2,304,800	78,947,000
Agriculture	1,980,000	73,944,300
TOTAL	20,328,800	386,423,800

Adapted from Dickinson (1980)

ing the grains before kneading (Lueck, 1980). The addition of 1-4% of salt to butter caus-es a significant increase in shelf life (Schreiber & Harner, 1983). There is about 16% moisture in these products, which allows for up to 12.5% sodium chloride, the concentra-tion that can be inhibitory to most microorganisms at refrigeration temperatures (Shelef & Seiter, 1993).

In the case of cheeses, salt acts as a flavoring and functional ingredient as well as helping preservation when combined with other antimicrobials such as sorbates. Salt is added to cheese curd either in the dry form or as a solution, based on the type of cheese. Up to 5% salt, relative to the water content of cheese is considered optimum (Lueck, 1980). Salt inhibits the development of bitter flavor in cheddar cheese by inhibiting the proteolysis of β-casein (Fox & Walley, 1971). Sodium chloride introduced in the serum phase of mozzarella cheese, promoted microstructure swelling, caused an increase in water holding capacity, and promoted solubilization of intact caseins from the paracasein matrix (Guo et al., 1997). Sodium chloride seems to play a role aiding/affecting the func-tional properties of some proteins and carbohydrates. Addition of sodium chloride to process whey protein solution increased the viscosity and aided in gelation (Kinekawa et al., 1998). Sodium chloride inhibited the *in vitro* proteolysis/digestibility of phaseolin proteins from dry beans such as the Great Northern bean and tepary bean (Sathe & Sze-tao, 1997; Sathe et al., 1994; Sathe et al., 1984). Sodium chloride aided in the clarity of starch and amylopectin (amaranth, waxy corn, normal corn) pastes, with the % transmit-tance values increasing with increasing salt concentration (Bello-Perez & Paredes-Lopez, 1996). Sodium chloride induces aggregation and gelation of some polysaccharides. 200 mM sodium chloride aided carageenan to form a strong gel, while in the presence of 100 mM salt a weak gel was formed with a mixture of carageenan and locust bean gum (Goncalves et al, 1997).

In fish, salting has been one of the oldest methods of preservation. Soon after harvest, the fish are cleaned and salt is added. Fish is one of the commodities to which a large amount of salt is sometimes added and if lesser salt is added, it is usually combined with other methods of preservation. Dipping fish in sodium chloride solution preserves the texture and color combined with modified atmospheric packaging (MAP) and storage (Mitsuda et al., 1980). Hake slices were dipped in sodium chloride solution (5 min in 5% brine) and MAP stored and these were compared with MAP stored slices of hake without sodium chloride dipping (Pastoriza et al., 1998). In sodium chloride dipped slices, bio-chemical, microbiological and sensory deterioration changes were inhibited, shelf life extended and the total volatile bases and total viable microbial counts were significantly lower than those of non dipped slices. The postmortem changes (rigor mortis) of Atlantic salmon influenced the salt uptake of the fish muscle (Wang et al., 1998). The equilibrium salt concentration of pre-rigor fillet was much lower (0.53 g/g salt-free solids) than that of in-rigor (0.66 g/g salt-free solids) and post-rigor mortis (0.75 g/g salt-free solids) salmon fillets in 20% (w/v) sodium chloride solution at 10°C.

In the case of liquid whole egg and egg yolk, 5 to 8 % of sodium chloride is used for preservation. Sodium chloride had a cryoprotective effect on egg yolk stored at -24°C by inhibiting its gelation at concentrations of 4 to 8% but not at 10% (Telis & Kieckbusch, 1998). Sodium chloride at concentrations of 0.1 and 1.0 M and at pH 7.2 and 9.0 protected egg white lysozyme, a natural antimicrobial enzyme, against heat inactivation at temperatures between 73-100°C (Makki & Durance, 1996). Sodium chloride was found to have an influence on egg white lysozyme solubility, in that the enzyme dissolved at increasing temperatures and decreasing salt concentrations in sodium chloride solutions (Forsythe et al., 1999). The effect of sodium chloride and sucrose on the bactericidal activity of egg white lysozyme (HL80/6) as well as its synergy with glycine was investigated (Ibrahim et al., 1996). The bactericidal effect of the enzyme against *Escherichia coli* K12 was found to decrease with increasing sodium chloride and sucrose concentrations. Sodium chloride at 1% concentration was needed to suppress inhibition of *Staphylococcus aureus*. However, even at inhibitory doses of sodium chloride and sucrose, the enzyme exhibited good synergy with glycine with regard to its action against Gram-positive bacteria, and this according to the authors suggest a possible food preservative application in the industry.

Salt in fermented vegetables such as pickles, sauerkraut, soy sauce etc., plays a preservative role as well as a functional role controlling the rate of lactic acid bacterial growth and thereby the fermentation. Salt helps in creating an anaerobic condition in the fermentation vessels, thereby promoting the growth of lactic acid bacteria (Buckenhuskes, 1997). Salt is also added in canned vegetables mainly to provide flavor. About 0.5 to 2% salt is added to the hot blanching water during canning to make certain vegetables like peas and lima beans tender, and about 10 % is needed to soften the cucumber pickles for ease of packing (Kaufmann, 1960). Sodium chloride concentration had an influence on the flavor compounds produced by yeast in soy sauce (Sasaki, 1996). The glycoproteins produced by various yeast species under sodium chloride stress acted as cryoprotective agents (Breierova, 1997). Some of the flavor compounds were produced in the largest amounts at a sodium chloride concentration of 17-18%, which is the common amount used for soy sauce. Sodium chloride is rarely used as a preservative in fruit products. One product in which 6 to 8% brine is added at the intermediate stage before preserving with sugar is the raw material used for making succades (Lueck, 1980). However, research has shown that sodium chloride can have some indirect beneficial roles in the fruit industry. Polyphenol oxidase is the browning enzyme, which causes an undesirable color on many fruits. Sodium chloride has been effective in inhibiting this enzyme isolated from grapes at pH values less than 5.0 (Valero & Gracia-carmona, 1998). Polygalacturonase is an enzyme present in some fruits, which is used by the food industry in the extraction and clarification of fruit juices. This enzyme is thermo-labile and attempts have been made to increase its thermostability using various additives. Sodium chloride enhances the thermostability of this enzyme even at low pH values below the optimum for this enzyme (Devi & Rao, 1998).

In the case of meats, salt is mainly used in fermented sausages where it acts as a preservative as well as aids in the development of flavor and texture. With regard to other types of meats, ground meats have a concentration of 2-4%, bacon around 2.25%, hams between 3 to 6% and corned beef about 6.25% of salt (Kaufmann, 1960). Salt lowers the water activity and in conjunction with nitrite can inhibit the growth of *Clostridium botulinum* (Terrell, 1983). Sodium chloride was found to show inhibitory effects against *C. botulinum* at a concentration of 4.5 to 4.8% (w/v) in wiener sausage even in the absence of nitrite (Hauschild, 1982). Concentrations lower than 4.5 % have been found ineffective against *C. botulinum*. Salt is also used in the curing and tanning of hides and skin. About 1 lb. of salt is added per lb. of hide and in curing, hides are immersed up to 24 h in saturated brine (Kaufmann, 1960). Salt is also used in the manufacture of animal by-products such as oleo stock, oleo oil and stearine. About 60 to 75 lbs of salt per 5000 lbs of the raw material is added at the melting stage and the final product contains about 2.5% salt added as a flavor enhancer and preservative (Kaufmann, 1960). Salt is also used in animal feeds, in fertilizers and weedicides, in soil stabilizers, in textile manufacture and in drilling oil wells.

VI. SAFETY AND TOLERANCE

The LD_{50} of sodium chloride when fed orally to fasted rats was determined to be 3.75 g/kg body weight (Boyd & Boyd, 1973). When administered to fasted rats over 100 days the LD_{50} was estimated to be 2.7 g/kg body weight and the respective value for non-fasted rats was 6.14 g/kg body weight (Lueck, 1980). A concentration of 2.8 to 5.6% sodium chloride given in the feed resulted in growth retardation and shortening of life-span of rats (Meneely et al., 1953). The LD_{50} for rats via oral route was estimated to be 3000 mg/kg, while via inhalation was >42 gm/m^3/1H1 (J. T. Baker, Material safety data sheet, 1996). The same reference cites LD_{50} for rabbits via skin to be >10 gm/kg and estimated it to be a mutagen and terratogen. For a detailed toxicity data and safety profile, refer to a food safety additives handbook.

About 1 to 3.3 g/day is the level of sodium that is considered safe as well as adequate for the body (Darby, 1980). Owing to its solubility in water, salt is excreted relatively easily and quickly from the body through the kidneys as well as the skin (in the form of sweat). Renal losses are about 6.0 to 12.5 g/day (Darby, 1980).

The effect of sodium chloride on cell division of rats was studied (Lugli & Lutz, 1999). Sodium chloride was given as a supplement at 2 and 4% concentrations in the feed for 4 weeks. This feed resulted in a significant stimulation of cell division in the stomach and liver with both concentrations of sodium chloride and in the bladder with 4% sodium chloride. Feed with ascorbic acid (2g/kg feed) or β-carotene (12.5 mg/kg feed) for 1 week before sodium chloride supplementation inhibited the stimulation of cell division.

VII. SUMMARY

Sodium chloride, commonly called table salt or salt, is a vital part of human life. Salt enhances the flavor of foods and plays a preservative as well as functional role in food processing. Apart from its use in the food industry, salt has uses in the agricultural and chemical industries as well as in water conditioning and transportation. Sodium chloride has been designated the fifth element and is abundantly available in nature. The production and recovery methods of sodium chloride are described. Two elemental substances, the cationic sodium and the anionic chloride, react to form the halide salt 'sodium chloride'. Sodium chloride occurs in the form of colorless transparent crystals or white crystalline powder, with a molecular weight of 58.44. To determine the salt content of a food product, the food or its ash is extracted in warm water and the sodium and chloride contents determined by appropriate techniques. The antimicrobial activity of salt can be both direct or indirect depending on the amount added and the purpose it serves. Since the amount of sodium chloride needed to be added to foods to prevent microbial growth is large and will cause an unacceptable taste, it is usually added in combination with other hurdles. The mechanism of inhibition of microorganisms by sodium chloride is mainly by lowering the water activity of the substrate. Studies have also indicated that sodium chloride could have a role in interfering with substrate utilization in microorganisms. The influence of sodium chloride alone as well as its interactive effects with other factors on various microorganisms such as *Staphylococcus aureus*, spore formers, *Escherichia coli* O157:H7, *Listeria monocytogenes* etc. in laboratory media as well as foods has been discussed in detail. The effect of sodium chloride on the heat resistance of microorganisms is discussed as well. The functional applications of sodium chloride in various products are discussed. Excessive sodium intake in humans has been linked to hypertension and the related cardiovascular problems and stroke. Hence there are many consumer concerns and the food industry is trying to minimize the salt content of food products.

VIII. REFERENCES

1. Allen, A. E. Jr., and Day, J. W. 1980. Low sodium salt seasoning. U. S. Patent. 4,216,244. August, 5.
2. Askar, A., El-Samahy, S. K., Shehata, H. A., and Tawfik, M. 1993. Pasterma and beef bouillon. The effect of substituting KCl and K-lactate for sodium chloride. *Fleischwirtschaft.* 73:289-292.
3. Ayres, J. C., Mundt, J. O., and Sandine, W. E. 1980. *Microbiology of Foods.* San Francisco: W. H. Freeman and Company.
4. Baker, J. T. 1996. Material Safety Data Sheet on sodium chloride.
5. Banwart, G. J. 1979. *Basic Food Microbiology.* Connecticut: AVI Publishers.
6. Bean, P. G., and Roberts, T. A. 1975. Effect of sodium chloride and sodium nitrite on the heat resistance of *Staphylococcus aureus* NCTC 10652 in buffer and meat macerate. *J. Food Technol.* 10:327-332.
7. Bello-Perez, L. A., and Paredes-Lopez, O. 1996. Starch and amylopectin. Effects of solutes on clarity of pastes. *Starch.* 48:205-207.
8. Bodyfelt, F. W. 1982. Processors should put some pinch on salt. *Dairy Rec.* 82:83-84.
9. Boyd, E. M., and Boyd, C. E. 1973. *Toxicity of Pure Foods.* Cleveland: CRC Press.
10. Breierova, E. 1997. Yeast exoglycoproteins produced under NaCl-stress conditions as efficient cryoprotective agents. *Lett. Appl. Microbiol.* 25:254-256.
11. Briggs, A., and Yazdany, S. 1970. Effect of sodium chloride on the heat and radiation resistance and on the recovery of heated or irradiated spores of the genus *Bacillus. J. Appl. Bacteriol.* 33:621-632.
12. Buchanan, R. L., and Golden. M. H. 1995. Model for the non-thermal inactivation of *Listeria monocytogenes* in a reduced oxygen environment. *Food Microbiol.* 12:203-212.
13. Buchanan, R. L., Golden. M. H., and Phillips, J. G. 1997. Expanded models for the thermal inactivation of *Listeria monocytogenes. J. Appl. Microbiol.* 82:567-577.
14. Buchanan, R. L., Stahl, H. G., and Whiting, R. C. 1989. Effects and interactions of temperature, pH, atmosphere, sodium chloride and sodium nitrite on the growth of *Listeria monocytogenes. J. Food Prot.* 52:844-851.
15. Buckenhuskes, H. J. 1997. Fermented vegetables. In *Food Microbiology- Fundamentals and Frontiers,* ed. M. P. Doyle, L. R. Beuchat and T. J. Montville, pp 600-601. Washington, D. C.: ASM Press.
16. Chan, P. F., Foster, S. J., Ingham, E., and Clements, M. O. 1998. The *Staphylococcus aureus* alternative sigma factor s[B] controls the environmental stress response but not starvation survival or pathogenicity in a mouse abscess model. *J. Bacteriol.* 180:6082-6089.
17. Cheville, A. M., Arnold, K. W., Buchrieser, Cheng, C.-M., and Kaspar, C. W. 1996. *rpoS* regulation of acid, heat and salt tolerance in *Escherichia coli* O157:H7. *Appl. Environ. Microbiol.* 62:1822-1824.

18. Conner, D. E., Brackett, R. E., and Beuchat, L. R. 1986. Effect of temperature, sodium chloride, and pH on growth of *Listeria monocytogenes* in cabbage juice. *Appl. Environ. Microbiol.* 52:59-63.

19. Cook, A. M., and Gilbert, R. J. 1969. The effect of sodium chloride on heat resistance and recovery of heated spores of *Bacillus stearothermophilus. J. Appl. Bacteriol.* 49:96-102.

20. Coronado, M. J. Vargas, C., Kunte, H. J., Galinski, E. A., Ventosa, A., and Nieto, J. J. 1995. Influence of salt concentrations on the susceptibility of moderately halophilic bacteria to antimicrobials and its potential use for genetic transfer studies. *Curr. Microbiol.* 31:365-371.

21. Darby, W. J. 1980. What salt?. How much?. *Contemp. Nutr.* 5: 1.

22. Denton, D. 1982. *The Hunger for Salt. An Anthropological, Physiological and Medical Analysis.* New York: Springer-Verlag.

23. Devi, N. A., and Rao, A. G. A. 1998. Effect of additives on kinetic thermal stability of polygalacturonase II from *Aspergillus carbonarius*: Mechanism of stabilization by sucrose. *J. Agric. Food Chem.* 46:3540-3545.

24. Dickinson, W. E. 1980. Salt sources and markets. In *Biological and Behavioral Aspects of Salt Intake,* ed M. Kare, R., Fregly, M.J., and R. A. Bernard, pp. 49-52. New York: Academic Press.

25. Emodi, A. S., and Lechowich, R. V. 1969. Low temperature growth of type E *Clostridium botulinum* spores. 1. Effects of sodium chloride, sodium nitrite and pH. *J. Food Sci.* 34:78-81.

26. Erecinska, M., and Deutsch, C. J. 1985. Osmotic effects on bacterial transport and energetics. *FEBS letters.* 188:145-149.

27. Fehlhaber, K. 1981. Untersuchungen uber lebensmittel hygienisch bedeutsame Eigenschsften von Shigellan. *Arch. Exper. Vet. Med. Leipzig.* 35:955-964.

28. Fernandez, P. S., George, S. M., Sills, C. C., and Peck, M. W. 1997. Predictive model of the effect of CO_2, pH, temperature and NaCl on the growth of *Listeria monocytogenes. Int. J. Food Microbiol.* 37:37-45.

29. Forsythe, E. L., Judge, R. A., and Pusey, M. L. 1999. Tetragonal chicken egg white lysozyme solubility in sodium chloride solutions. *J. Chem. Eng. Data.* 44:637-640.

30. Forsythe, R. H., and Miller, R. A. 1980. Salt in processed foods. In *Biological and Behavioral Aspects of Salt Intake,* ed, Kare, M.R., Fregly, M.J., and Bernard, R.A., pp. 221-228. New York: Academic Press.

31. Fox, P. F., and Walley, B. F. 1971. Influence of sodium chloride on the proteolysis of casein by rennet and by pepsin. *J. Dairy Res.* 28:165-170.

32. Garren, D. M., Harrison, M.A., and Russell, S. M. 1998. Acid tolerance and acid shock response of *Escherichia coli* O157:H7 and non-O157:H7 isolates provide cross-protection to sodium lactate and sodium chloride. *J. Food Prot.* 61:158-161.

33. Gibson, A. M. Bratchell, N., and Roberts, T. A. 1987. The effect of sodium chloride and temperature on the rate and extent of growth of *Clostridium botulinum* type A in pasteurized pork slurry. *J. Appl. Bacteriol.* 62:479-490.

34. Gibson, A. M., and Roberts, T. A. 1986a. The effect of pH, sodium chloride, sodium nitrite and storage temperature on the growth of *Clostridium perfringens* and fecal streptococci in laboratory media. *Int. J. Food Microbiol.* 3:195-210.

35. Gibson, A. M., and Roberts, T. A. 1986b. The effect of pH, water activity, sodium nitrite and storage temperature on the growth of enteropathogenic *Escherichia coli* and *Salmonella* in a laboratory medium. *Int. J. Food Microbiol.* 3:183-194.

36. Gill, A. O., and Holley, R. A. 1999. In Vitro evaluation on the effects of nitrite and NaCl on the antimicrobial activity of lysozyme, nisin and EDTA combination treatments. Presented at the 86th annual meeting of the International Association of Milk, Food and Environmental Sanitarians, Aug 1-4, Dearborn, MI.

37. Gimeno, O., Astiasaran, I., and Bello, J. 1998. A mixture of potassium, magnesium and calcium chlorides as a partial replacement of sodium chloride in dry fermented sausages. *J. Agric.Food Chem.* 46:4372-4375.

38. Glass, K. A., Loeffelholz, J. M., Ford, J. P., and Doyle, M. P. 1992. Fate of *Escherichia coli* O157:H7 as affected by pH or sodium chloride and in fermented, dry sausage. *Appl. Environ. Microbiol.* 58:2513-2516.

39. Golden, D. A., Beuchat, L. R., and Brackett, R. E. 1988. Inactivation and injury of *Listeria monocytogenes* as affected by heating and freezing. *Food Microbiol.* 5:17-23.

40. Goncalves, M. P., Gomes, C., Langdon, M. J., Viebke, C., and Williams, P. A. 1997. Studies on k-Carageenan/Locust bean gum mixtures in the presence of sodium chloride and sodium iodide. *Biopoly.* 41:657-671.

41. Greenberg, R. A., Silliker, J. H., and Fata, L. D. 1959. The influence of sodium chloride on toxin production and organoleptic breakdown in perishable cured meat inoculated with *Clostridium botulinum. Food Technol.* 13:509.

42. Guo, M. R., Gilmore, J. A., and Kindstedt, P. S. 1997. Effect of sodium chloride on the serum phase of Mozzarella cheese. *J. Dairy Sci.* 80:3092-3098.

43. Gou, P., Guerrero, L., Gelabert, J., and Arnau, J. 1996. Potasium chloride, potasium lactate and glycine as sodium chloride substitutes in fermented sausages and in dry-cured pork loin. *Meat Sci.* 42:37-48.

44. Guraya, R., Frank, J. F., and Hassan, A. N. 1998. Effectiveness of salt, pH and diacetyl as inhibitors for *Escherichia coli* O157:H7 in dairy foods stored at refrigeration temperatures. *J. Food Prot.* 61:1098-1102.

45. Hand, L.W., Terrell, R. N., and Smith, G. C. 1982. Effects of chloride salts on physical, chemical and sensory properties of frankfurters. *J. Food Sci.* 47:1800-1802.

46. Hauschild, A. H. W. 1982. Assessment of botulism hazards from cured meat products. *Food Technol.* 36:95-104.

47. Hauschild, A. H. W., and Hilsheimer, R. 1979. Effect of salt content and pH on toxinogenesis by *Clostridium botulinum* in caviar. *J. Food Prot.* 42:245-248.

48. Hitomi. S. Baba, S. Yano, H. Morisawa, Y., and Kimura, S. 1998. Antimicrobial effects of electrolytic products of sodium chloride- comparative evaluation with sodium hypochlorite solution and efficacy in handwashing. *Kansenshogaku Zasshi.* 1998. 72:1176-1181.

49. Ibanez, C., Quintanilla, L., Astiasaran, I., and Bello, J. 1997. Dry fermented sausages elaborated with *Lactobacillus plantarum- Staphylococcus carnosus.* Part II. Effect of partial replacement of NaCl with KCl on the proteolytic and insolubilization processes. *Meat Sci.* 46:277-283.

50. Ibanez, C., Quintanilla, L., Cid, C., Astiasaran, I., and Bello, J. 1996. Part 1. Effect of partial replacement of NaCl with KCl on the stability and the nitrosation process. *Meat Sci.* 44:227-234.

51. Ibanez, C., Quintanilla, L., Irigoyen, A., Garcia-Jalon, L., Cid, C., Astiasaran, I., and Bello, J. 1995. Partial replacement of sodium chloride with potassium chloride in dry fermented sausages: Influence on carbohydrate fermentation and the nitrosation process. *Meat Sci.* 40:45-53.

52. Ibrahim, H. R., Higashiguchi, S., Sugimoto, Y., and Aoki, T. 1996. Antimicrobial synergism of partially-denatured lysozyme with glycine: effect of sucrose and sodium chloride. *Food Res. Int.* 29:771-777.

53. Institute of Food Technologists. 1980. Dietery salt- A scientific status summary by the Institute of Food 65.Technologists's expert panel on food safety and nutrition and the committee on public information. *Food Technol.* 34:85-91.

54. Janes, M. E., Nannapaneni, R., Howard, L., and Johnson, M. G. 1999. Sodium chloride and sodium bicarbonate washing solution for removal of enterohaemorrhagic *Escherichia coli* O157:H7 from the surfaces of chopped lettuce. Presented at the 86th annual meeting of the International Association of Milk, Food and Environmental Sanitarians, Aug 1-4, Dearborn, MI.

55. Jay, J. M. 1992. *Modern Food Microbiology.* New York: Van Nostrand Reinhold.

56. Juneja, V. K., and Eblen, B. S. 1995. Influence of sodium chloride on thermal inactivation and recovery of nonproteolytic *Clostridium botulinum* type B strain KAP B5 spores. *J. Food Prot.* 58:813-816.

57. Juneja, V. K., and Eblen, B. S. 1999. Predictive thermal inactivation model for *Listeria monocytogenes* with temperature, pH, NaCl and sodium pyrophosphate as controlling factors. *J. Food Prot.* 62:986-993.

58. Juneja, V. K., and Majka, W. M. 1995. Outgrowth of *Clostridium perfringens* spores in cook-in-bag beef products. *J. Food Safety.* 15:21-34.

59. Juneja, V. K., and Marmer, B. S. 1996. Growth of *Clostridium perfringens* from spore inocula in *sous-vide* turkey products. *Int. J. Food Microbiol.* 32:115-123.

60. Juneja, V. K., Marmer, B. S.,and Eblen, B. S. 1999. Predictive model for the combined effect of temperature, pH, sodium chloride and sodium pyrophosphate on the heat resistance of *Escherichia coli* O157:H7. *J. Food Safety.* 19:147-160.

61. Juneja, V. K., Marmer, B. S., Phillips, J. G., and Miller, A. J. 1995. Influence of the intrinsic properties of food on thermal inactivation of spores of nonproteolytic *Clostridium botulinum*: Development of a predictive model. *J. Food Safety.* 15:349-364.

62. Juneja, V. K., Marmer, B. S., Phillips, J. G., and Palumbo, S. A. 1996. Interactive effects of temperature, initial pH, sodium chloride, and sodium pyrophosphate on the growth kinetics of *Clostridium perfringens.* *J. Food Prot.* 59:963-968.

63. Kaufman, D. W. 1960. *Production and Properties of Salt and Brine.* New York: Reinhold Public Corporation.

64. Kawasaki, Y., Seki, T., Tamura, M., Kikuchi, E., Tada, M., and Okai, H. 1988. Glycine methyl or ethyl ester hydrochloride as the simplest examples of salty peptides and their derivatives. *Ag. Biol. Chem.* 52:2679-2681.

65. Keeton, J. T. Effects of potassium chloride on properties of country-style hams. 1984. *J. Food Sci.* 49:146-148.

66. Kinekawa, Y-I., Fuyuki, T., and Kitabatake, N. 1998. Effects of salts on the properties of sols and gels prepared from whey protein isolate and process whey protein. *J. Dairy Sci.* 81:1532-1544.

67. Kotrola, J. S., and Conner, D. E. 1997. Heat inactivation of *Escherichia coli* O157:H7 in turkey meat as affected by sodium chloride, sodium lactate, polyphosphate and fat content. *J. Food Prot.* 60:898-902.

68. LaRocco, K. A. and Martin, S. E. 1987. Effects of potassium sorbate alone and in combination with sodium chloride on growth of *Staphylococcus aureus* MF-31. *J. Food Prot.* 50:750-

69. Leak, F. W., Kemp, J. D., Fox, J. D. and Langlois, B. E. 1987. Effects of boning time, mechanical tenderization and partial replacement of sodium chloride on the quality and microflora of boneless dry-cured ham. *J. Food Sci.* 52:263-266.

70. Leistner, L., and Russell, N. J. 1991. Solutes and low water activity. In *Food Preservatives*, ed. Rusell, N.J., and Gould,G.W., pp.11-134. London: Blackie and Son Ltd.

71. Lewis, R. J. 1989. *Food Additives Handbook*. New York: Van Nostrand Reinhold.

72. Li, Y., Slavik, M. F., Walker, J. T., and Xiong, H. 1997. Pre-chill spray of chicken carcasses to reduce *Salmonella typhimurium*. *J. Food Sci.* 62:605-607.

73. Linton, R. H., Carter, W. H., Pierson, M. D., and Hackney, C. R. 1995. Use of a modified Gompertz equation to model nonlinear survival curves for *Listeria monocytogenes* Scott A. *J. Food Prot.* 58:946-954.

74. Linton, R. H., Carter, W. H., Pierson, M. D., Hackney, C. R., and Eifert, J. D. 1996. Use of a modified Gompertz equation to predict the effects of temperature, pH and NaCl on the inactivation of *Listeria monocytogenes* Scott A heated in infant formula. *J. Food Prot.* 59:16-23.

75. Linton, R. H., Webster, J. B., Pierson, M. D., Bishop, J. R., and Hackney, C. R. 1992. The effect of sublethal heat shock and growth atmosphere on the heat resistance of *Listeria monocytogenes* Scott A. *J. Food Prot.* 55:84-87.

76. Lou, Y., and Yousef, A. E. 1997. Adaptation to sublethal environmental stresses protects *Listeria monocytogenes* against lethal preservation factors. *Appl. Environ. Microbiol.* 63:1252-1255.

77. Lueck, E. 1980. *Antimicrobial Food Additives*. Berlin: Springer-Verlag.

78. Lugli, S. M., and Lutz, W. K. 1999. Stimulation of cell division in the rat by NaCl, KCl, MgCl$_2$, and CaCl$_2$, and inhibition of the sodium chloride effect on the glandular stomach by ascorbic acid and β-carotene. *J. Cancer Res. Clin. Oncol.* 125:209-213.

79. Lynch, N. M. 1987. In search of the salty taste. *Food Technol.* 41:82-86.

80. Madigan, M. T., Martinko, J. M., and Parker, J. 1997. *Brock's Biology of Microorganisms*. New Jersey: Prentice Hall.

81. Makki, F., and Durance, T. D. 1996. Thermal inactivation of lysozyme as influenced by pH, sucrose and sodium chloride and inactivation and preservative effect in beer. *Food Res. Int.* 29:635-645.

82. Mazzotta, A. S., Crandall, A. D., and Montville, T. J. 1997. Nisin resistance in *Clostridium botulinum* spores and vegetative cells. *Appl. Environ. Microbiol.* 63:2654-2659.

83. McLean, R. A., Lilly, H. D., and Alford, J. A. 1968. Effects of meat-curing salts and temperature on production of Staphylococcal enterotoxin B. *J. Bacteriol.* 95:1207-1211.

84. McNelis, K. A. 1998. Intradermal bacteriostatic 0.9% sodium chloride containing the preservative benzyl alcohol compared with intradermal lidocaine hydrochloride 1% for attenuation of intravenous cannulation pain. *J. Amer. Assoc. Nurse Anes.* 66:583-585.

85. Meneely, G. R., Tucker, R. G., Darby, W. J., and Auerbach, S. H. 1953. Chronic sodium chloride toxicity: hypertension, renal and vascular lesions. *Ann. Int. Med.* 39: 991-998

86. Michell, A. R. 1995. *The Clinical Biology of Sodium*. New York: Pergamon.

87. Mickelson. O., Nakadani, D., Gill, J.L., and Frank, R. L. 1972. Sodium and potassium intakes and excretions of normal men consuming sodium chloride or a 1:1 mixture of sodium and potassium chloride. *Am. J. Clin. Nutr.* 30:2033-2040.

88. Miller, A. J., Call, J. E., and Eblen, B. S. 1997. Growth, injury and survival potential of *Yersinia enterocolitica, Listeria monocytogenes* and *Staphylococcus aureus* in brine chiller conditions. *J. Food Prot.* 60:1334-1340.

89. Mitsuda, H., Nakajima, K., Mizuno, H., and Kawai, F. 1980. Use of sodium chloride solution and carbon dioxide for extending shelf-life of fish fillets. *J. Food Sci.* 45: 661-666.

90. Mohlenkamp, M. J. Jr., and Miller, G. D. 1981. Sodium free salt substitute. U. S. Patent 4,243,691. Jan, 6.

91. Nakamura, M., Stone, R. L., Krubsack, J. E., and Pauls, F. P. 1964. Survival of Shigella in seawater. *Nature.* 203:213-214.

92. National Academy of Sciences. 1980. Recommended Dietery Allowances. National Research Council, Food and Nutrition Board. 9th ed. Washington, D. C.

93. Ng, T. M., and Schaffner, D. W. 1997. Mathematical models for the effects of pH, temperature and sodium chloride on the growth of *Bacillus stearothermophilus* in salty carrots. *Appl. Environ. Microbiol.* 63:1237-1243.

94. Official Methods of Analysis, 1980. Assn. Of Analytical Chemists, Washington, D. C.

95. Okai, H., Tamura, M., Seki, T., Tada, M., and Kawasaki, Y. 1989. An enhanced effect of sodium chloride using amino acid esters and amino acids. *Chem. Senses.* 14:320.
96. Pagan, R., Condon, S., and Sala, F. J. 1997. Effects of several factors on the heat shock-induced thermo-tolerance of *Listeria monocytogenes*. *Appl. Environ. Microbiol.* 63:3225-3232.
97. Paul, A. A., and Southgate, D. 1978. In: *McCance and Widowson's -Composition of Foods*. H. M. S. O.
98. Pastoriza, L., Sampedro, G., Herrera, J. J., and Cabo, M. L. 1998. Influence of sodium chloride and mod-ified atmosphere packaging on microbiological, chemical and sensorial properties in ice storage of slices of hake (*Merluccius merluccius*). *Food Chem.* 61:23-28.
99. Pearson, A. M., and Wolzak, A. M. 1982. Salt- its use in animal products, a human health dilemma. *J. Animal Sci.* 54:1263-1278.
100. Pereira, J. L., Salzberg, S. P., and Bergdoll, M. S. 1982. Effect of temperature, pH and sodium chloride concentrations on production of staphylococcal enterotoxin A and B. *J. Food Prot.* 45:1306-1309.
101. Pivnick, H., and Barnett, H. 1965. Effect of salt and temperature on toxinogenesis by *Clostridium botu-linum* in perishable cooked meats vacuum-packed in air-impermeable plastic pouches. *Food Technol.* 142:1164-1167.
102. Pivnick, H., and Bird, H. 1965. Toxinogenesis by *Clostridium botulinum* types A and E in perishable cooked meats vacuum-packed in plastic pouches. *Food Technol.* 19:132-140.
103. Pivnick, H., and Thacker, C. 1970. Effect of sodium chloride and pH in initiation of growth by heat-dam-aged spores of *Clostridium botulinum*. *Can. Inst. Food Technol. J.* 3:70-75.
104. Prior, B. A. 1978. The effect of water activity on growth and respiration of *Pseudomonos fluorescens*. *J. Appl. Microbiol.* 4:97-106.
105. Razavi-Rohani, S. M., and Griffiths, M. W. 1996. Inhibition of spoilage and pathogenic bacteria associat-ed with foods by combinations of antimicrobial agents. *J. Food Safety* 16:87-104.
106. Reddy K. A., and Marth, E. H. 1991. Reducing the sodium content of foods: A review. *J. Food Prot.* 54: 138-150.
107. Robach, M.C., and Stateler, C. L. 1980. Inhibition of *Staphylococcus aureus* by potassium sorbate in com-bination with sodium chloride, tertiary butylhydroquinone, butylated hydroxyanisole or ethylenediamine tetraacetic acid. *J. Food Prot.* 43:208-211.
108. Roberts, T. A., Gilbert, R. J., and Ingram, M. 1966. The effect of sodium chloride on heat resistance and recovery of heated spores of *Clostridium sporogenes* (PA 3679/S₂). *J. Appl. Bacteriol.* 29:549-555.
109. Robinson, R. A., and Stokes, R. H. 1959. *Electrolyte Solutions. The Measurement and Interpretation of Conductance, Chemical Potential and Diffusion in Solutions of Simple Electrolytes*. London: Butterworths.
110. Salt Institute, 1999. History of salt. Internet website publication.
111. Sasaki, M. 1996. Influence of sodium chloride on the levels of flavor compounds produced by Shoyu yeast. *J. Agric. Food Chem.* 44:3273-3275.
112. Sathe, S. K., Deshpande, S. S., and Salunkhe, D. K. 1984. Dry beans of *Phaseolus*. A review. Part 1. Chemical composition:proteins. *CRC Crit. Rev. Food Sci. Nutr.* 20:1-46.
113. Sathe, S. K., Idouraine, A., and Weber, C. W. 1994. Purification and biochemical characterization of tepary bean (*Phaseolus acutifolius*) major globulin. *Food Chem.* 50:261.
114. Sathe, S. K., and Sze-tao, K. W. C. 1997. Effects of sodium chloride, phytate and tannin on *in vitro* prote-olysis of phaseolin. *Food Chem.* 59:253-259.
115. Schreiber, B. C., and Harner, H. L. 19. Sixth international symposium on salt. Vol 2. Virginia: The Salt Institute.
116. Segner, W. P., Schmidt, C. F., and Boltz, J. K. 1966. Effect of sodium chloride and pH on outgrowth of spores of type E *Clostridum botulinum* at optimal and suboptimal temperatures. *Appl. Microbiol.* 14:49-54.
117. Seman, D. L., Olson, D. G., and Mandigo, R. W. 1980. Effect of reduction and partial replacement of sodi-um on bologna characteristics and acceptability. *J. Food Sci.* 45:1116-1121.
118. Shackelford, J. R. 1981. Salt substitutes having reduced bitterness. U. S. Patent. 4,297,375. Oct, 27.
119. Shelef, L. A., and Seiter, J. A. 1993. Indirect antimicrobial. In *Antimicrobial in Foods*, ed. P. M. Davidson and A. L. Branen. pp. 539-569. New York: Marcel Dekker.
120. Smith, J. L. Benedict, R. C.,and Palumbo, S. A. 1982. Protection against heat-injury in *Staphylococcus aureus* by solutes. *J. Food Prot.* 45:54-58.
121. Smith, J. L., and Hunter, S. E. 1988. Heat injury in *Listeria monocytogenes*. Prevention by solutes. *Lebensm. Wiss. Technol.* 21:307-311.
122. Smith, J. L., Maurer, M. J., Bencivengo, M. M., and Kunsch, C. A. 1987. Effect of sodium chloride on uptake of substrate by *Staphylococcus aureus* 196E. *J. Food Prot.* 50:968-974.

123. Sofos, J. N. 1983. Antimicrobial effects of sodium and other ions in foods: A review. *J. Food Safety*. 6: 45-78.
124. Sperber, W. H. 1983. Influences of water activity on foodborne bacteria: a review. *J. Food Prot*. 46:142-150.
125. Stern, N. J., Smoot, L. A., and Pierson, M. D. 1979. Inhibition of *Staphylococcus aureus* growth by combinations of butylated hydroxyanisole, sodium chloride and pH. *J. Food Sci*. 44:710-712.
126. Sternberg, M., Cornelius, D. A., Eberts, N. J., Schwende, F. J., and Chiang, J. P. C. 1980. Glycinamide hydrochloride, a compound with common salt flavor. In *Biological and Behavioral Aspects of Salt Intake*, ed M. R. Kare, M. J. Fregly and R. A. Bernard, pp 319-329. New York: Academic Press.
127. Stringer, S. C., and Peck, M. W. 1997. Combinations of heat treatment and sodium chloride that prevent growth from spores of nonproteolytic *Clostridium botulinum*. *J. Food Prot*. 60:1553-1559.
128. Sutherland, J. P., Bayliss, A. J., and Braxton, D. S. 1995. Predictive modeling of growth of *Escherichia coli* O157:H7. The effects of temperature, pH and sodium chloride. *Int. J. Food Microbiol*. 25:29-49.
129. Tada, M., Shinoda, I., and Okai, H. 1984. L-Ornithyltaurine, a new salty peptide. *J. Ag. Food Chem*. 32:992-996.
130. Telis, V. R. N., and Kieckbusch, T. G. 1998. Viscoelasticity of frozen/thawed egg yolk as affected by salts, sucrose and glycerol. *J. Food Sci*. 63:20-24.
131. Terrell, R. N. 1983. Reducing the sodium content of processed meats. *Food Technol*. 37:66-71.
132. Tomicka, A., Chen, J., Barbut, S., and Griffiths, M. W. 1997. Survival of bioluminescent *Escherichia coli* O157:H7 in a model system representing fermented sausage production. *J. Food Prot*. 60:1487-1492.
133. Troller, J. A. 1973. The water relations of foodborne bacterial pathogens: a review. *J. Milk Food Technol*. 36:276-288.
134. Troller, J. A., and Stinson, J. V. 1978. Influence of water activity on the production of extracellular enzymes by *Staphylococcus aureus*. *Appl. Environ. Microbiol*. 35:521-526.
135. Tuncan, E. U., and Martin, S. E. 1990. Combined effects of salts and temperature on the thermal destruction of *Staphylococcus aureus* MF-31. *J. Food Sci*. 55:833-836.
136. Valero, E., and Garcia-Carmona, F. 1998. pH-dependant effect of sodium chloride on latent grape polyphenol oxidase. *J. Agric. Food Chem*. 46:2447-2451.
137. Volker, U., Mach, H., Schmid, R., and Hecker, M. 1992. Stress proteins and cross-protection by heat shock and salt stress in *Bacillus subtilis*. *J. Gen. Microbiol*. 138:2125-2135.
138. Wang, D., Tang, J., Correia, L. R. and Gill, T. A. 1998. Postmortem changes of cultivated Atlantic salmon and their effects on salt uptake. *J. Food Sci*. 63:634-637.
139. Woods, L. F. J., and Wood, J. M. 1982. The mechanism of the inhibition of *Clostridium sporogenes* by sodium chloride. British Food Manufacturing Industries Research Assn. Research Report. No. 382.
140. Zaika, L. L., Engel, L. S., Kim, A. H., and Palumbo, S. A. 1989. Effect of sodium chloride, pH and temperature on growth of *Shigella flexneri*. *J. Food Prot*. 52:356-359.

A. Prakash

Polyphosphates 27

I. INTRODUCTION

Phosphorus, as the phosphate anion, is essential to sustain life. It is a component of many enzyme systems involved in the metabolism of various nutrients, and the formation and structure of various tissues (Ellinger, 1972). Thus, phosphates are natural constituents of almost every food we eat. To produce food grade phosphates, mined phosphates are converted to phosphoric acid, followed by partial or complete neutralization with alkalis such as sodium, potassium, or calcium. Their buffering capacity, ability to sequester metal ions, water-holding capacity, and ability to interact with long chain polyelectrolytes such as protein, make phosphates widely used as food additives. The classification, and various applications of polyphosphates in food products are summarized in TABLE 1. Corresponding chemical structures are shown in FIGURE 1. Inclusion of polyphosphates as functional additives in many food formulations and observations of preservative activity have prompted researchers to elucidate the mechanism of antimicrobial action. Regardless, polyphosphates continue to be added to foods as functional additives, with their antimicrobial effect being regarded as an added benefit rather than their primary function. The only approved antimicrobial use of an orthophosphate is in a pre-chill or post-chill dip to reduce microbial levels of raw poultry carcasses.

II. STRUCTURE/CHEMISTRY

Phosphates are derived from phosphoric acid neutralized with alkali metal ions, such as sodium, potassium, or calcium. There are two classes of phosphates: orthophosphates and condensed phosphates. Orthophosphates, prepared from orthophosphoric acid, H_3PO_4, consist of a phosphorus atom surrounded by four oxygens. This molecule has three replaceable hydrogen molecules giving rise to a number of combinations of hydrogen and metal cations such as monobasic orthophosphates (one alkali metal ion and two hydrogen molecules), dibasic orthophosphates (two alkali metal ions and one hydrogen), and tribasic orthophosphates (three metal ions). Orthophosphates when heated (a process

TABLE 1: Classes, Formulas, pH, Solubility, and Functions of Several Phosphates

Class of phosphates	Phosphate name	Formula	pH[1]	Solubility[2]	Functions
ORTHOPHOSPHATES	Monosodium phosphate	NaH_2PO_4	4.6	87	Emulsifier, buffer
	Disodium phosphate	Na_2HPO_4	9.2	12	Emulsifier, buffer
	Disodium phosphate dihydrate	$Na_2HPO_4\ 2H_2O$	0.9	15	Emulsifier, buffer
	Trisodium phosphate	Na_3PO_4	11.8	14	Emulsifier, buffer
	Monopotassium phosphate	KH_2PO_4	4.6	25	Water binding in meats
	Dipotassium phosphate	K_2HPO_4	9.3	168	Emulsifier, buffer
	Tripotassium phosphate	K_3PO_4	11.9	107	Emulsifier, buffer
	Monocalcium phosphate	$Ca(H_2PO_4)_2\ H_2O$	3.8		Acidulant, leavening acid, dough conditioner, yeast food, nutrient
CONDENSED PHOSPHATES					
Pyrophosphates	Sodium acid pyrophosphate	$Na_2H_2P_2O_7$	4.3	15	Emulsifier, buffer, sequestrant, water binding agent in meats
	Tetrasodium pyrophosphate	$Na_4P_2O_7$	10	3.8	Dispersant, coagulant, crystallization inhibitor in canned tuna
	Tetrapotassium pyrophosphate	$K_4P_2O_7$	10.5	187	Emulsifier, water binding agent in meats, suspending agent
Tripolyphosphates	Sodium tripolyphosphate	$Na_5P_3O_{10}$	9.9	15	Emulsifier, water binding agent in meats
	Potassium tripolyphosphate	$K_5P_3O_{10}$	9.6	193	Emulsifier, water binding agent in meats
Long-chain polyphosphates	Sodium polyphosphates, glassy, or Graham's Salt; three chain lengths; sodium hexametaphosphate has an average chain length of 13	$(NaPO_3)_{13}Na_2O$ $(NaPO_3)_{21}Na_2O$ $(NaPO_3)_5\ Na_2O$	7.7 6.9 6.3	40	Sequestrant, emulsifier, water binding agent in meats, suspending agent
Metaphosphates	Sodium trimetaphosphate	$(NaPO_3)_3$	6.7	23	
	Sodium tetrametaphosphate	$(NaPO_3)_4\ 4H_2O$	6.2	18	

[1] pH measurement represents a 1% solution; [2] Solubility measurement at 25°C (g/100g water)

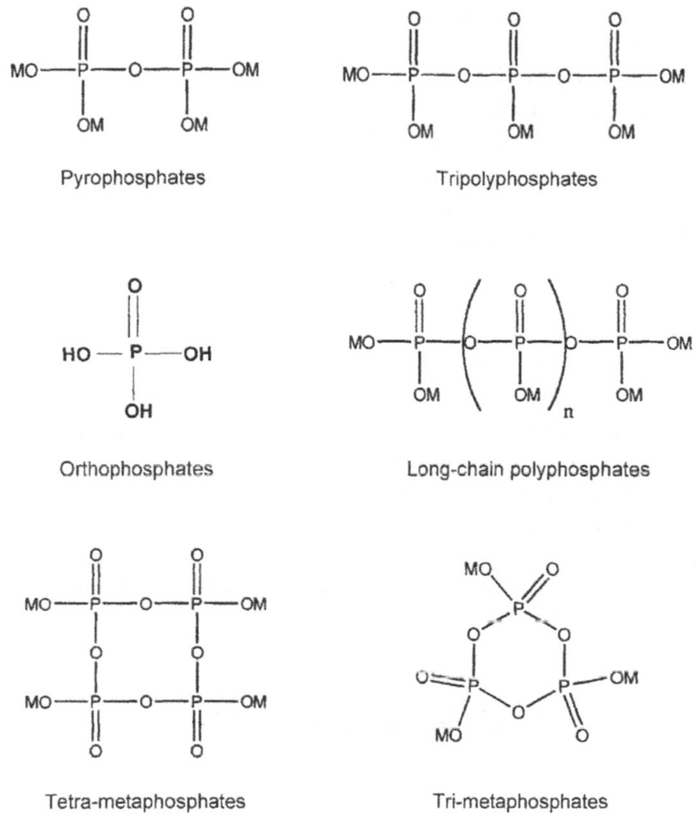

FIGURE 1. Stereochemical structures of various polyphosphates

called calcining), condense to form straight chain compounds (polyphosphates) and cyclic compounds (metaphosphates). Straight chain compounds consist of pyrophosphates with two monomers, tripolyphosphates with three orthophosphates, and polyphosphates with four or more orthophosphates. While the pyrophosphates and tripolyphosphates are crystalline materials, the straight chain polyphosphates form amorphous or glassy particles. The cyclic metaphosphates are not commonly used in the food industry, although sodium trimetaphosphate is approved by the FDA for modification of food starches. Another seldom used form is referred to as ultraphosphates, which incorporate branching PO_4 groups in the structure resulting in branched chains and rings as well as combinations of chains and rings. Branched phosphates rapidly hydrolyze to structures with no branching points (Van Wazer, 1971).

A. Properties

Orthophosphates have a pH range of 4.0-12.0, thus considered to be compounds with superior buffering capacity (Prasad & Prasad, 1989). They are also used in foods to adjust pH to optimum values. While pyrophosphates are good buffers in the pH range (5.5-7.5) of most foods longer chain polyphosphates are less effective (Sofos, 1989). Long chain polyphosphates are polyelectrolytes and react with other polyelectrolytes such as proteins, pectins, and starches to function as dispersive and emulsifying agents. They can form strong complexes with metal ions such as calcium and magnesium; higher pH increases such complexing efficiency. Short-chain polyphosphates can efficiently sequester heavy metal ions such as iron and copper, in this case, increasing pH decreases complexing efficiency. Ring phosphates form weaker complexes with cations and orthophosphates form complexes only at very low concentrations and precipitates at higher concentrations. (Wagner, 1986). The ability to sequester metal ions (form soluble complexes) is the basis of the mechanism of antimicrobial action.

B. Assay

The level of orthophosphate obtained from hydrolysis of polyphosphates can be determined by a colorimetric method. This method is based on the rapid formation of molybdenum blue color by the reaction of orthophosphate with molybdenum ions in the presence of ascorbic acid-trichloroacetic acid and citrate-arsenite reagents (Murphy & Riley, 1962; Dick & Tabatabai, 1977). NMR and gel filtration have been used to calculate the average length of polyphosphate chains (Wood & Clark, 1988). Tsuji et al. (1994) and Yasuno et al. (1996) determined the orthophosphate content in various raw and processed foods by ion chromatography. A modified thin-layer chromatographic (TLC) method for condensed polyphosphate determination in frozen seafood is described by Krzynowek and Panunzio (1995).

III. ANTIMICROBIAL ACTIVITY

A. Mechanism(s)

A number of mechanisms for the antimicrobial activity of polyphosphates have been suggested. The foremost mechanism suggested is the chelation of metal ions in cell membranes which leads to cation deficiency with resulting loss of membrane integrity and inhibition of normal cell division (Wagner, 1986). Other mechanisms include effect of pH (Shelef et al., 1990), increase in ionic strength, interactions with cell walls and membranes (Kohl, 1971), and interference with various transport functions (Knabel et al., 1991).

Polyphosphates can form stable complexes with metallic ions thus preventing precipitation or redissolving precipitates of alkaline earth metals (Van Wazer & Callis, 1958). Cobalt, copper, and iron are bound strongly by polyphosphates while calcium and magnesium are less strongly bound; these bound ions then become unavailable for metabolic functions. Because polyphosphates are polyvalent anions, they have a high degree of surface activity causing them to aggregate at cell surfaces when dissolved in water. Microbial cell membranes contain metal chelators, which are responsible for ferrying metal ions across the membrane. Metal chelating agents such as polyphosphates may compete with cell membrane-chelators for available ions and selectively remove calcium and magnesium or other ions from the cell wall, membranes, or cytoplasm. Post et al. (1968), suggested that this mechanism accounts for the observed lysis of gram-negative bacteria and the growth inhibition of gram-positive bacteria as well as fungi. Seward et al. (1982), examined the germination of *C. botulinum* type E spores in media supplemented with several antibotulinum agents, including sodium hypophosphite and sodium tripolyphosphate. At high pH levels, 0.5 or 1.0% hypophosphite and the combination of 1.5% potassium sorbate and 0.5% sodium tripolyphosphate resulted in germination and outgrowth of abnormally shaped cells culminating in cell lysis and defective cell division. Support for the metal chelation theory is provided by the fact that in foods that contain higher concentrations of divalent cations, polyphosphates are not as effective as in culture media (Shelef et al., 1990; Rajkowski, 1994; Zaika et al., 1997). Knabel et al. (1991), also showed that orthophosphates that lack the metal chelating potential did not inhibit *Aspergillus flavus* or any of the several gram-positive as well as gram-negative bacteria tested. Molins and co-workers (1987) reported that bacterial inhibition did not correlate with soluble orthophosphate content in ground pork. Addition of free cations to polyphosphate-containing cultures has been shown to reverse the inhibitory effects of polyphosphates on bacterial growth (Elliott et al., 1964; Jen & Shelef, 1986; Knabel et al., 1991).

Knabel et al. (1991) sought to explain the difference in inhibition of polyphosphates on growth of gram-negative versus gram-positive bacteria. Walls of most gram-positive bacteria consist of a thick layer of peptidoglycan and large amounts of teichoic and/or teichuronic acids. These anionic polymers can bind metal ions and are responsible for maintaining a high concentration of bivalent cations such as Ca^{2+}, Mg^{2+}, Mn^{2+}, and Fe^{3+} for cation-dependent membrane systems. Polyphosphates are able to remove metal ions from the cell membranes because of their higher affinity for the metal cations, thus inhibiting growth. Because gram-negative bacteria have more efficient systems for binding and transporting metal ions across cell membranes, polyphosphates are not as successful in chelating the metal ions, and thus not as effective in inhibiting gram-negative bacteria. The authors also suggested that fungi are inhibited by polyphosphates due to polymers such as chitin, chitosan, and glycoproteins present in their cell walls and which are involved in the passive uptake of metals.

Working with pure cultures of *Staphylococcus aureus*, Lee et al. (1994a) attributed the antibacterial effects of polyphosphates to damage to the cell wall or cell membrane which resulted in the leakage of intracellular nucleotides and proteins. The addition of Ca^{2+} and Mg^{2+} reversed the bactericidal and bacteriolytic effects (Lee et al., 1994b). It was also determined that polyphosphates bound to cell walls did not release metals from the cell walls into the medium leading to the conclusion that polyphosphates remain bound to the cell walls even after chelating metal ions from the cell walls (Lee et al., 1994c). The authors offered the hypothesis that structurally essential metals such as magnesium form bridges across phosphate groups in adjacent ribitol teichoic acid chains in *S. aureus* cell walls and that polyphosphates bound to the cell walls chelate these metals resulting in bactericidal and bacteriolytic effects. Matsuoka et al. (1995), suggested that the antibacterial action of sodium hexametaphosphate on *S. aureus* was caused by a weak effect on cell membranes resulting in loss of osmoregulation and selective permeability due to damage to the membrane, as well as loss of metabolic function caused by substrate leakage.

Other researchers have proposed that enzyme inhibition may be responsible for the antimicrobial mechanism. Wagner and Busta (1985) found that sodium acid pyrophosphate (SAPP) in culture media inhibited the production or activity of the protease responsible for activating *C. botulinum* toxin resulting in delayed toxigenicity in meat formulations inoculated with *C. botulinum*. Using fish muscle, Tarr (1951) showed that SAPP inhibited 5'-nucleotidase which hydrolyses adenosine monophosphate (AMP) to adenosine and orthophosphate thus demonstrating that inorganic pyrophosphates can regulate enzyme activity. In the previous year, Vishniac (1950) investigated the effect of sodium tripolyphosphate on yeast hexokinase, and reported that the compound inhibited the anaerobic conversion of glucose by intact yeast cells, yeast zymase preparations, as well as crystalline hexokinase preparations and that the inhibition was reversed by addition of ATP. It was concluded that polyphosphate inhibition was due to chelation of Mg^{2+}, which is essential for hexokinase activity and that ATP competes with polyphosphate for the metal ions.

B. Spectrum

The following is a review of studies published in the past ten years. Prior studies have been reviewed by Tompkin, 1984; Wagner, 1986; Sofos, 1989; Molins, 1991 and Shelef & Seiter, 1993. In the eighties there was particular emphasis on the effect of polyphosphates on *C. botulinum* and *S. aureus* growth and toxin production in culture media as well as cured and restructured meat products. The emergence of pathogens such as *Salmonella*, *E. coli* O157:H7, and *Listeria monocytogenes*, has prompted recent investigations into the suppression of growth of these bacteria by polyphosphates.

1. Clostridia: Pasteurized process cheese spread containing one of three types of phosphate emulsifiers was inoculated with *Clostridium botulinum* and analyzed over 30 weeks for growth of *C. botulinum* and toxigenesis (Eckner et al., 1994). Toxin formation was observed at 8 weeks in the 60% moisture cheese containing disodium orthophosphate, while the polyphosphates containing cheeses tested positive for toxin at 20 weeks. Toxin formation did not occur in cheese spreads with lower moisture containing phosphates. The results indicate that polyphosphates in high moisture cheese spreads can impart a margin of safety by delaying toxin formation, although toxigenesis is not prevented.

Loessner et al. (1997) investigated the effects of polyphosphates on the growth and development of *C. tyrobutyricum* in processed cheese spreads. Butyric blowing is a common defect observed in such products caused by outgrowth of *C. tyrobutyricum* spores. They found that 0.5% sodium polyphosphate glassy was sufficient to control *C. tyrobutyricum* growth under normal conditions, while 1% totally inhibited the organism.

2. Staphyloccocus aureus: At pH 7.0, 1.2% sodium acid pyrophosphate (SAPP), 1% sodium tripolyphosphate (STPP), or 0.4% sodium hexametaphosphate (also known as sodium polyphosphate, glassy or SPG) exhibited no antibacterial action against *S. aureus* strain 196E in N-Z amine broth plus 1% yeast extract (Shelef & Wang, 1989), although a bacteriostatic effect was observed at slightly higher concentrations. Without pH adjustment, 1% SAPP decreased the pH of the media to 5.5 at which level significant inhibition of the bacterium was observed. In contrast, 1% SAPP in BHI broth (pH 6.0) did not significantly inhibit *S. aureus*, whereas alkaline phosphates (pH 7.8) exhibited higher antimicrobial effectiveness (Knabel et al., 1991). Lee et al. (1994a) compared the effects of different phosphates on the growth of *S. aureus* ISP40 8325 in a synthetic medium. The minimum inhibitory concentrations were found to be 0.1% for sodium ultraphosphate and sodium polyphosphate glassy and 0.5% for SAPP and tetrasodium pyrophosphate (TSPP), indicating that long chain polyphosphates exerted greater inhibition (*FIGURE 2*).

Addition of 1% SAPP to custard reduced counts of *S. aureus* strain 196E by about 0.5 log, and in beef, 0.5 and 1.0% SAPP reduced counts by 1.5-2 logs at 22°C (Shelef et al., 1990). This effect was eliminated when pH was adjusted to that of the control or when the storage temperature was increased to 30°C. When the pH of beef was adjusted to the same levels as produced by 0.5 and 1.0% SAPP, no significant differences

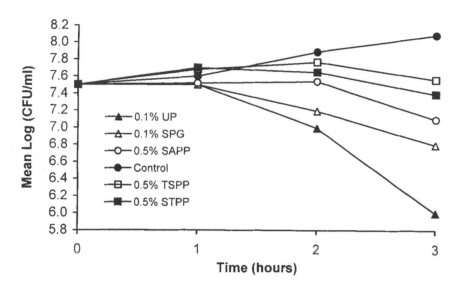

FIGURE 2. Effect of phosphates at MIC on growth (mean log CFU/ml) of early-exponential-phase cells of *Staphylococcus aureus* (redrawn from Lee et al., 1994a).

between the treatments was apparent. The authors surmised that the reduction of pH due to SAPP addition was a major factor in the observed growth inhibition. STP and SHMP had no effect on bacterial growth. Likewise, addition of 0.5 or 1.0% SHMP to UHT-sterilized milk had no activity against *S. aureus* 196E at 19, 28, and 37°C (Rajkowski et al., 1994), although SHMP enhanced the inhibitory effect of NaCl. The high levels of Ca^{2+} and Mg^{2+} in custard (Shelef et al., 1990) as well as UHT milk (Rajkowski et al., 1994) were most likely responsible for elimination of the inhibitory effects of polyphosphate.

It appears that SAPP compared to other phosphates has a greater inhibitory effect on *S. aureus* in food, although the inhibition is small. The inhibition seems to be related to the decrease in pH by SAPP and is enhanced in the presence of other antimicrobials.

3. Listeria monocytogenes: While SAPP appears to be most effective against *S. aureus*, the higher polymers are more effective in inhibiting growth of *L. monocytogenes*. The effect of three sodium polyphosphate polymers (average chain lengths = 6, 13, and 21) on the growth of *L. monocytogenes* in BHI containing 0.3% glucose, pH 6.0 at 28, 19, 10 and 5°C were evaluated (Zaika & Kim, 1993). The longer polymers were more inhibitory. Higher concentrations, lower temperatures, decreasing pH, as well as addition of NaCl enhanced the inhibitory effect (Zaika et al., 1997). Growth inhibition was reversed by addition of certain polyvalent metal cations, especially Ca^{2+}, Mg^{2+}, and Mn^{2+}. Thus, in foods with high levels of metal cations such as UHT-sterilized milk and single-ingredient baby foods, sodium polyphosphate exerted little or no inhibition on the growth of *L. monocytogenes* (Rajkowski et al., 1994; Zaika et al., 1997). Knabel et al. (1991) found that suppression of *L. monocytogenes* by 1% tetrasodium pyrophosphate (TSPP) could be reversed by iron supplementation. Addition of a 0.5% commercial phosphate blend to smoked sausage and cured smoked ham inoculated with *L. monocytogenes* had minimal or no effect (Flores et al., 1996).

Surface disinfection of chicken skin inoculated with *L. monocytogenes* with 1% trisodium phosphate (TSP) was more effective than solutions of 10% monosodium phosphate (MSP), SAPP, SPG, and STPP (Hwang & Buechat, 1995). Addition of Tween 80 to TSP did not enhance removal of *Listeria*.

4. Salmonella: In general, gram-negative bacteria are less inhibited by phosphates than are gram-positive bacteria. Nevertheless, the effectiveness of polyphosphates in inhibiting the growth of *Salmonella* in egg whites, chicken breast muscle, and culture media, has been demonstrated (Garibaldi et al., 1969; Foster & Mead, 1976; Molins et al., 1984). Tetrasodium pyrophosphate (TSPP) in BHI broth inhibited growth of *S. typhimurium*, and addition of Fe^{3+} enhanced the inhibition (Knabel et al., 1991). The authors theorized that formation of the polyphosphate-iron chelate complex blocked siderophore receptors on the surface of the cells preventing siderophore-mediated Fe^{3+} transport.

In 1992, the U.S. Department of Agriculture approved the use of trisodium phosphate (TSP) as a postchill processing aid to remove bacteria from the surfaces of raw poultry carcasses (Geise, 1993). It is hypothesized that TSP enables the removal of a thin layer of fat on the surface of poultry skin allowing the removal of bacteria embedded in the tissue (Geise, 1994). Two patents describe the reduction in counts in broiler carcasses inoculated with nalidixic acid-resistant (NAL) *Salmonella typhimurium* with a 10% TSP postchill dip (Bender & Brotsky, 1991; Bender & Brotsky, 1992). Kim & Slavik, 1994 achieved a 2-log reduction on salmonella inoculated chickens by treating with TSP. A 1% TSP was significantly more effective than solutions of 10% monosodium phosphate (MSP), SAPP, SPG, and STPP in reducing the levels of *Salmonella* spp. inoculated on the surface of chicken skin (Hwang & Buechat, 1995). Even though the levels of TSP used in this experiment were lower than the 10% used by Bender and Brotsky (1992), the authors found that the low levels applied for longer times were just as effective. Li et al. (1994) investigated the combined effect of low voltage, low current pulsed electrical signals and several food salts on *S. typhimurium* attached to chicken skins. Treatment with 1% TSP reduced *Salmonella* counts by 76% while the combination of electrical and TSP treatment reduced counts by 96% (1.4 logs).

The bactericidal effect of a commercial blend of SAPP and orthophosphoric acid on raw poultry carcasses inoculated with NAL *Salmonella typhimurium* was evaluated by Rathgeber & Waldroup (1995). Addition of a 1.0 or 1.5% SAPP blend significantly reduced salmonella counts, although the researchers cautioned that the extent of reduction was related to the low pH of the rinse solution. When the pH of the rinse solution was adjusted closer to neutral, no reduction in salmonella counts was observed, although *E. coli* and coliform counts were significantly reduced.

In addition to poultry, *S. typhimurium* is also a beef contaminant. Kim and Slavik (1994) inoculated fat and fascia surfaces of beef with *S. typhimurium* followed by rinse with 10% TSP for 15 seconds. A reduction of 0.91 and 0.51 logs was obtained on the TSP-treated fat and fascia surfaces, respectively. The greater reduction on the fat surfaces can be attributed to the removal of the fat layer by TSP and stronger entrapment of bacteria by the collagen fiber networks of the fascia surface.

5. Escherichia coli: The successful reduction in *Salmonella* counts by TSP on poultry surfaces led to the investigation of the effect on *E.coli* O157:H7. Kim and Slavik (1994) found that 1% TSP decreased *E. coli* counts by 1.35 and 0.92 logs on TSP-treated beef fat and fascia surfaces, respectively. 1% trisodium phosphate significantly reduced the levels of inoculated *E. coli* O157:H7 on alfalfa seeds, and concentrations greater than 4% precluded recovery of *E. coli* by direct plating, however, the bacterium was recovered by selective enrichment (Taormina & Buechat, 1999). The added heat stress (55°C) did not reduce counts further. Although TSP was effective in reducing counts without effect on seed germination, the reduction was not sufficient to justify its use since surviving cells would continue to grow during the sprouting process.

Although TSP was found to be effective against surface bound *E. coli* in the above-mentioned studies, polyphosphates are more effective in ground meat products. Conflicting results have been obtained with different phosphate forms, with meat type and presence of spices and other additives influencing effectiveness. A commercial blend of sodium poly-, meta-, and ortho-phosphate, was ineffective on the growth and survival of *E. coli* O157:H7 in ground beef, and mildly effective in fresh pork sausage (Flores et al., 1996). The authors attributed these results to the lack of heat treatment to inactivate phosphatases present in meat and a possible synergistic action of the phosphate blend with the spices added to the fresh pork sausage.

6. *Aeromonas hydrophila:* Although 2% polyphosphate exhibited no growth inhibition of *Aeromonas hydrophila*, a combination of 2% polyphosphate and 3.5% NaCl completely inhibited growth in BHI media (Palumbo et al., 1995). However, this combination treatment had only a limited effect on *A. hydrophila* in scallops and prevented outgrowth of the organism in ground pork during refrigerated storage.

7. *Fungi:* Inhibition of mold growth and mycotoxin production in high-moisture corn varieties treated with various phosphates was investigated (Lebron et al., 1989). The most effective treatments were 2.0% TSPP (tetrasodium pyrophosphate), and 1 and 2% acidic sodium polyphosphate glassy (same as SHMP) added to corn in powder form. These compounds protected undamaged corn kernels against infection but also prevented infection in damaged kernels. 2% TSPP and 2% SHMP were also effective when sprayed on the corn. The treatments were effective against the inoculated mold species and also naturally occurring species. Mycotoxin production was significantly inhibited by phosphate treatment. Inhibition of *Aspergillus flavus* by several phosphates was observed by Knabel et al., 1991, in culture media. Addition of Mg^{2+} was found to reverse inhibition of *A. flavus* growth by TSPP, whereas addition of Mn^{2+} reversed the growth inhibition by SHMP. Phosphate salts have been found to be effective against powdery mildew in cucumbers (Reuveni et al., 1992a and 1993) and clusters of winegrapes (Reuveni & Reuveni, 1995), as well as northern leaf blight and common rust in maize (Reuveni et al., 1992b; 1994).

IV. SAFETY AND TOLERANCE

A. Regulatory status

The FDA lists nearly all food phosphates as "GRAS" (Generally Recognized As Safe). Thus, phosphates are extensively used in food products around the world, including baby formulas and health food. Food grade phosphates and phosphoric acid are approved for use by the FDA in Title 21 of the Code of Federal Regulations (CFR). In

Parts 182 and 184, they are identified as GRAS. Under part 182, they are listed with functional groupings as follows:

- Subpart B: Multiple Purpose GRAS Food Substances
- Subpart F: Dietary Supplements
- Subpart G: Sequestrants
- Subpart I: Nutrients

Applications in the Meat and Poultry industry are regulated by the USDA and are listed in Title 9 of the CFR. Specific approvals are as follows:

- Part 318.7: Use in Meat Products
- Part 381.147: Use in Poultry Products

The USDA limits the use of phosphates in these products to 0.5% by weight of the final product. The USDA specifically prohibits the use of phosphates in fresh meat and poultry products. Meat and poultry products processed with phosphates should be labeled appropriately, and the labels must be approved by the USDA. Only clear solutions may be injected into meat and poultry.

Current regulations by the FDA limit the level of phosphates in seafood to Good Manufacturing Practice (GMP) and must be labeled accordingly. Many food phosphates are also approved for use as indirect ingredients and for other applications. Specific references follow:

- 21 CFR 172.892: Use in preparation of modified food starches.
- 21 CFR 173.310: Use in boiler water.
- 21 CFR 173.315: Use in washing or to assist in lye peeling of fruits and vegetables.
- 21 CFR 175: Subpart B - For use only as components of adhesive.
- 21 CFR 175: Subpart C - For use as components of cleaning.
- 21 CFR 176: Indirect Food Additives – Paper and paperboard components.

Since regulation of alcoholic beverages is the responsibility of the Bureau of Alcohol, Tobacco and Firearms, the approval of ammonium phosphates for treatment of wine and alcoholic juices is listed in 27 CFR 24.246.

Phosphates are included in the Standards of Identity of many standardized foods, including processed cheese, processed cheese food, processed cheese spread, evaporated milk, baking powder, phosphated flour, self-rising flour, enriched self-rising flour, self-rising white corn meal, self-rising yellow corn meal, and bread, rolls and buns.

B. Physiological role, RDA, and dietary intake

Phosphorus is the second most abundant mineral in the body, after calcium (Sizer & Whitney, 1997). About 85% of the phosphorus occurs bound to calcium in the bones and teeth in the form of calcium phosphate ($Ca_3[PO_4]_2$) and the crystal, hydroxyapatite ($Ca_{10}[PO_4]_6[OH]_2$). The remaining phosphorus is found in extracellular fluid or inside cells. As an important component of the nucleic acids DNA and RNA, phosphorus is part of the genetic material of each cell. It occurs in cell membranes as phospholipids and plays a vital role in energy metabolism through its inclusion in high energy phosphate bonds in molecules such as ATP, and through the phosphorylation of substrates (Groff et al., 1995). Phosphorus also plays an important role as a buffering agent for cellular fluids. The 1989 U.S. RDA for phosphorus ranges from 300 to 1200 mg/day depending on age (U.S. Food and Nutrition Board, 1989). Since excess phosphorus can cause calcium excretion, a 1:1 or higher ratio of calcium to phosphorus in the diet is considered optimal (Feldheim, 1983). Although phosphates are used extensively as additives in food products, their estimated contribution to the total phosphorus content of the diet is varied. Based on data from the 1977 Survey of Industry on the Use of Food Additives (Committee on the GRAS List Survey-Phase III, 1979), Greger & Krystofiak (1982) estimated that food additives contributed 20-30% of the phosphorus in the diet of an average adult (assuming a daily phosphorus consumption of 1200-1700 mg). Zemel & Bidari (1983) have estimated that phosphate additives in the American diet can increase phosphorus intake by 250 to 1000 mg/day (25-100%). Although toxicity from phosphorus is rare, Yates et al. (1998) suggest a tolerable upper intake level of 4000 mg for ages 9-70 and 3000 mg for ages 70 and up.

C. Nutritional effects

Bell et al. (1977) determined that a high phosphate diet increased parathyroid hormone synthesis, increased urinary secretion of cyclic AMP, and contributed to bone calcium loss. Phosphate supplementation with calcium has been shown to decrease iron (Monsen et al., 1976; Steinhardt Bour, 1984) and zinc (Cabell &Earle, 1965) absorption and utilization. A detailed review of the nutritional and toxicological effects of phosphates is provided by Molins (1991).

V. SUMMARY

The effects of polyphosphates on microorganisms in food products depend on various factors which influence their activity such as the category of phosphate and concentration, pH, nature of substrate, heat treatment, type and level of microbial contamination, formulation, and storage temperature (Sofos, 1989). For these reasons, food-based experiments often do not mimic the results obtained in culture media. Moreover, results from one kind of food cannot be extrapolated to other food products.

Long chain alkaline polyphosphates are more effective as antimicrobial agents compared to short chain polyphosphates, although sodium acid pyrophosphate, a short chain acidic polyphosphate and TSP, an orthophosphate are very effective in certain applications. The inhibition of SAPP is related to the decrease in pH, although pH reduction by itself does not account for suppression of microbial growth. TSP, an orthophosphate, is highly effective as a surface disinfectant. For long chain polyphosphates, inhibition increases with an increase in pH. In general, higher concentrations of polyphosphates have enhanced effects. Toxin production by bacteria such as *C. botulinum* and *S. aureus* generally occurs under condition that support the growth of the bacteria, although delayed toxin synthesis has been reported under conditions which allow the bacteria to grow.

The nature of the substrate, ground meat versus whole muscle, raw versus processed, affects polyphosphate activity. Hydrolysis of polyphosphates due to heating can lead to a loss in activity, thus in certain applications involving long-chain polyphosphates these compounds should be added after heat treatment. However, many foods, especially meat products, contain phosphatases, which must be inactivated by heat to maximize polyphosphate effectiveness. The presence of polyphosphates also appears to increase the sensitivity of several microorganisms to heat. Since polyphosphates function by chelating metal ions essential for microbial growth, in foods with high levels of cations, polyphosphates are not very effective. Interactions of polyphosphates with other polyelectrolytes such as protein, starch, and pectin must also be considered.

Bacterial species differ in their sensitivity to polyphosphates. Gram-positive bacteria and mold are sensitive to polyphosphates and gram-negative bacteria are resistant to the action of polyphosphates. Level of contamination and age of the culture can also affect polyphosphate action.

Synergistic effects have been reported with combinations of polyphosphates and other preservatives such as nitrites and sorbate. Sodium chloride, in some cases, enhances antimicrobial action, and in other cases, decreases effectiveness. Presence of spices may enhance activity. Certain combinations of polyphosphates and other additives such as salt have an additive effect, while other combinations elicit synergism. In the presence of other preservative agents, polyphosphates can provide added protection during storage at abuse temperatures.

Polyphosphates lack the broad spectrum of activity exhibited by primary antimicrobials such as sorbate and benzoate. However, as indirect antimicrobials, polyphosphates can impart considerable protection against microbial growth and spoilage while at the same time providing physical and chemical functionality.

VI. REFERENCES

1.   Bell, R.R., Draper, H.H., Tzeng, D.Y.M., Shin, H.K., and Schmidt, G.R. 1977. Physiological responses of human adults to foods containing phosphate additives. *J. Nutr*. 107:42-50.
2.   Bender, F.G., and Brotsky, E. 1991. Process for treating poultry carcasses to control salmonellae growth. U.S. Patent 5,069,922.
3.   Bender, F.G., and Brotsky, E. 1992. Process for treating poultry carcasses to control salmonellae growth. U.S. Patent 5,143,739.
4.   Cabell, C.A., and Earle, I.P. 1965. Additive effect of calcium and phosphorus on utilization of dietary zinc. *J. An. Sci*. 24:800.
5.   Committee on the GRAS List Survey-Phase III. 1979. The 1977 Survey on the Use of Food Additives. National Academy of Sciences. Washington, D.C.
6.   Dick, W. A., and Tabatabai, M.A. 1977. Determination of orthophosphates in aqueous solutions containing organic and inorganic phosphorous compounds. *J. Environ. Qual*. 6:82-85.
7.   Eckner, K.F., Dustman, W.A., Rys-Rodriguez, A.A. 1994. Contribution of composition, physicochemical characteristics and polyphosphates to the microbial safety of pasteurized cheese spreads. *J. Food Prot*. 57:259-300.
8.   Ellinger, R.H. 1972. *Phosphates as Food Ingredients*. Boca Raton, FL: CRC Press.
9.   Elliott, R.P., Straka, R.P., and Garibaldi, J.A. 1964. Polyphosphate inhibition of growth of pseudomonads from poultry meat. *Appl. Microbiol*. 12:517- 522.
10. Feldheim, W. 1983. On the ratio of calcium and phosphorus in human nutrition. *Milchwiss*. 38: 284-286.
11. Flores, L.M., Summer, S.S., Peters, D.L., Mandigo, R. 1996. Evaluation of a phosphate to control pathogen growth in fresh and processed meat products. *J. Food Prot*. 59:356-359.
12. Foster, R.D., and Mead, G.C. 1976. Effect of temperature and added polyphosphate on the survival of poultry meat during cold storage. *J. Appl. Bacteriol*. 41:505-510.
13. Garibaldi, J.A., Ijichi, K., and Bayne, H.G. 1969. Effect of pH and chelating agents on the heat resistance and viability of *Salmonella typhimurium* TM-1 and *Salmonella senftenberg* 775 W in egg white. *Appl. Microbiology*. 18:318-322.
14. Geise, J. 1993. *Salmonella* reduction process receives approval. *Food Tech*. 47:110.
15. Giese, J. 1994. Antimicrobials: Assuring food safety. *Food Tech*. 48 : 101-110.
16. Greger, J.L., and Krystofiak, M. 1982. Phosphorus intake of Americans. *Food Tech*. 36:78-84.
16a. Groff, J.L., and Gropper, S.S. 1999. *Advanced Nutrition and Human Metabolism*, 3rd ed. Belmont, California: Wadsworth/Thomson Learning.
17. Hwang, C., and Beuchat, L.R. 1995. Efficacy of selected chemicals for killing pathogenic and spoilage microorganisms on chicken skin. *J. Food Prot*. 58:19-23.

18. Jen, C. M. C., and Shelef, L.A. 1986. Factors affecting sensitivity of *Staphylococcus aureus* 196E to polyphosphates. *Appl. Microbiol.* 52: 842-846.

19. Kim, J.W., and Slavik, M.F. 1994. Trisodium (TSP) treatment of beef surfaces to reduce *Escherichia coli* O157:H7 and *Salmonella typhimurium. J. Food Sci.* 59:20-22.

20. Knabel, S.J., Walker, H.W., Hartman, P.A. 1991. Inhibition of Aspergillus flavus and selected gram-positive bacteria by chelation of essential metal cations by polyphosphates. *J. Food Prot.* 54:360-365.

21. Kohl, W.F. 1971. A new process for pasteurizing egg whites. *Food Technol.* 25:1176-1184.

22. Krzynowek, J., and Panunzio, L.J. 1995. Practical application of thin-layer chromatography for detection of polyphosphates in seafood. *J. AOAC Int.* 78:1328-1332.

23. Lebron, C.I., Molins, R.A., Walker, H.W., Kraft, A.A., and Stahr, H.M. 1989. Inhibition of mold growth and mycotoxin production in high- moisture corn treated with phosphates. *J. Food Prot.* 52:329-336.

24. Lee, R.M., Hartman, P.A., Olson, D.G., Williams, F.D. 1994a. Bacterial and bacteriolytic effects of selected food-grade phosphates using *Staphylococcus aureus* as a model system. *J. Food Prot.* 57:276-283.

25. Lee, R.M., Hartman, P.A., Olson, D.G., Williams, F.D. 1994b. Metal ions reverse the inhibitory effects of selected food-grade phosphates in *Staphylococcus aureus. J. Food Prot.* 57:284-288, 300.

26. Lee, R.M., Hartman, P.A., Stahr, H.M., Olson, D.G., Williams, F.D. 1994c. Antibacterial mechanism of long-chain polyphosphates in *Staphylococcus aureus. J. Food Prot.* 57:289-294.

27. Li, Y., Kim, J-.W., Slavik, M.F., Griffis, C.L., Walker, J.T., and Wang, H. 1994. *Salmonella typhimurium* attached to chicken skin reduced using electrical stimulation and inorganic salts. *J. Food Sci.* 59:23-29.

28. Loessner, M.J, Maier, S.K., Schiwek, P., and Scherer, S. 1997. Long-chain polyphosphates inhibit growth of *Clostridium tyrobutyricum* in processed cheese spreads. *J. Food Prot.* 60:493-498.

29. Matsuoka, A., Tsutsumi, M., Watanabe, T. 1995. Inhibitory effect of hexametaphosphate on growth of *Staphylococcus aureus. J. Food Hyg. Soc. Japan.* 36:588-594.

30. Molins, R.A. 1991. *Phosphates in Foods.* Boca Raton, FL: CRC Press.

31. Molins, R.A., Kraft, A.A., and Marcy, J.A. 1987. Extension of the shelf-life of fresh ground pork with phosphates. *J. Food Sci.* 52:513-514.

32. Molins, R.A., Kraft, A.A, Olson, D.G., and Hotchkiss, D.K. 1984. Recovery of selected bacteria in media containing 0.5% food grade poly- and pyrophosphates. *J. Food Sci.* 49:948-949.

33. Monsen, E.R. and Cook, J.D. 1976. Food iron absorption in human subjects. IV. The effects of calcium and phosphate salts on the absorption of nonheme iron. *J. Clin. Nutr.* 29:1142-1148.

34. Murphy, J., and Riley, J.P. 1962. A modified single solution method for the determination of phosphate in natural waters. *Anal. Chim. Acta.* 27:13-36.

35. Palumbo, S.A., Call, J.E., Cooke, P.H., Williams, A.C. 1995. Effect of polyphosphates and NaCl on *Aeromonas hydrophila* K144. *J. Food Safety.* 15:77-87.

36. Post, F.J., Colbentz, W.S., Chou, T.W., and Salunkhe, D.K. 1968. Influence of phosphate compounds on certain fungi and their preservative effect on fresh cherry fruit. *Appl. Microbiol.* 16:138-142.

37. Prasad, S., and Prasad, C. 1989. Phosphates in dairy and food industries. *Indian Dairyman* 41:121- 126.

38. Rajkowski, K.T, Calderone, S.M., and Jones, E. 1994. Effect on polyphosphate and sodium chloride on the growth of *Listeria monocytogenes* and *Staphylococcus aureus* in ultra- high temperature milk. *J. Dairy Sci.* 77:1503-1508.

39. Rathgeber, B.M., and Waldroup, A.L. 1995. Antibacterial activity of sodium acid pyrophosphate product in chiller water against selected bacteria on broiler carcasses. *J. Food Prot.* 58:530-534.

40. Reuveni, M., Agapov, V., and Reuveni, R. 1992a. Local and systemic resistance against powdery mildew and growth increase in cucumber plants induced by phosphate salts. (Abstr.) *Phytopath.* 82:1179.

41. Reuveni, M., Agapov, V., and Reuveni, R. 1992b. Systemic resistance against northern leaf blight and common rust in maize induced by foliar spray of phosphates. (Abstr.) *Phytopath.* 82:1179.

42. Reuveni, M., Agapov, V., and Reuveni, R. 1993. Induction of systemic resistance to powdery mildew and growth increase in cucumber by phosphates. *Biol. Agric. Hort.* 9:305-315.

43. Reuveni, M., Agapov, V., and Reuveni, R. 1994. Foliar spray of phosphates induces growth increase and systemic resistance to Puccinia sorghi in maize. *Plant Path.* 43:245-250.

44. Reuveni, M., and Reuveni, R. 1995. Efficacy of foliar application of phosphates in controlling powdery mildew fungus on field-grown wine grapes: effects on cluster yield and peroxidase activity in berries. *J. Phytopath.* 143:21-25.

45. Seward, R.A., Deibel, R.H., and Lindsay, R.C. 1982. Effects of potassium sorbate and other antibotulinal agents on germination and outgrowth of *Clostridum botulinum* type E spores in microcultures. *Appl. Environ. Microbiol.* 44:1212-1221.

46. Shelef, L.A., and Seiter, J.A. 1993. Indirect antimicrobials. In *Antimicrobials in Foods*, eds. P.M. Davidson and A.L. Branen, pp. 539-570. New York: Marcel Dekker, Inc.

47. Shelef, L.A., and Wang, Z.L. 1989. Effects of polyphosphates on cell numbers, enterotoxins A, and extra-celluar protein in *Staphylococcus aureus* 196E. J. Food Sci. 54:1550-1552.

48. Shelef, L.A, Wang, Z.L., and Udeogu, A.C. 1990. Growth of *Staphylococcus aureus* and enterotoxin A production in foods containing polyphosphates. *J. Food Safety*. 10:201-208.

49. Sizer, F.S., and Whitney, E.N. 1997. *Nutrition: Concepts and Controversies*, 7th ed. California: Wadsworth Publishing Company.

50. Sofos, J.N. 1989. Phosphates in meat products. In *Developments in Foods Preservation,* ed. S. Thorne, pp.207-252 . Barking, U.K: Elsevier.

51. Steinhardt Bour, N.J., Soullier, B.A., and Zemel, M.B. 1984. Effect of level and form of phosphorus and level of calcium intake on zinc, iron and copper bioavailability in man. *Nutr. Res*. 4:371.

52. Taormina, P.J., Beuchat, L.R. 1999. Comparison of chemical treatments to eliminate enterohemorrhagic *Escherichia coli* O157: H7 on alfalfa seeds. *J. Food Prot*. 62:318-324.

53. Tarr, H.L.A., Gardner, L.J., and Ingram, P. 1969. Pacific cod muscle 5' nucleotidase.   *J. Food Sci*. 34: 637-640.

54. Tompkin, R.B. 1984. Indirect antimicrobial effects on foods: Phosphates. *J. Food Safety*. 6:13-27.

55. Tsuji, S., Shibata, T., Uchibori, N., Kobayashi, T., Suzuki, H., Uchibori H.S., Muroi, J., Kaneda, N., and Ito, Y. 1994. *J. Food Hyg. Soc. Jap*. 35(1): 56-65.

56. U.S. Food and Nutrition Board. 1989. Recommended Dietary Allowances, 10th ed., National Academy of Sciences, National Research Council, Washington, D.C.

57. Van Wazer, J.R. 1971. Chemistry of phosphates and condensed phosphates. In *Symposium: Phosphates in Food Processing,* eds. J.M. Deman, and P. Melnychyn, chap. 1, Westport, CN: AVI Publishing.

58. Van Wazer, J.R., and Callis, C.F. 1958. Metal complexing by phosphates. *Chem. Rev*. 58:1011-1046.

59. Vishniac, W. 1950. The antagonism of sodium tripolyphosphate and adenosine tryphosphate in yeast. *Arch. Biochem*. 26:167-172.

60. Wagner, M.K. 1986. Phosphates as antibotulinal agents in cured meats: A review. *J. Food Prot*. 49:482-487.

61. Wagner, M.K., and Busta, F.F. 1985. Inhibition of *Clostridium botulinum* 52A toxicity and protease activity by sodium acid pyrophosphate in media systems. *Appl. Environ. Micribiol*. 50:16-20.

62. Wood, H.G., and Clark, J.E. 1988. Biological aspects of inorganic polyphosphates. *Ann. Rev. Biochem*. 57:235-260.

63. Yasuno, T., Funayama, K., Hagiwara, T., and Suzuki, S. 1996. Determination of orthophosphates in condensed phosphates used as food additives by ion chromatography. *Ann. Rep. Tokyo Metro. Res. Lab. Pub. Health*. 47:189-193.

64. Yates, A., Schlicker, S., and Suitor, C. 1998. Dietary reference intakes: the new basis for recommendations for calcium and related nutrients, B vitamins and choline. *J. Am. Diet. Assoc*. 98:699-706.

65. Zaika, L.L., and Kim, A.H. 1993. Effect of sodium polyphosphates on growth of *Listeria monocytogenes*. *J. Food Prot*. 56:577-580.

66. Zaika, L.L, Scullen, O.J, and Fanelli, J.S. 1997. Growth inhibition of *Listeria monocytogenes* by sodium polyphosphate as affected by polyvalent metal ions. *J. Food Sci*. 62:867-896, 872.

67. Zemel, M.E. and Bidari, M.T. 1983. Zinc, iron, and copper availability as affected by orthophosphates, polyphosphates and calcium. *J. Food Sci*. 48:567-573.

N. Khanna
A.S. Naidu

Chloro-cides

28

I. INTRODUCTION

Chlorine, a member of the halogen family, serves as a building block for several synthetic molecules that exhibit potent antimicrobial properties. These molecules range from simple elemental forms of diatomic chlorine, to complex organic chloramines. In this chapter, these molecules are referred to as 'chloro-cides'.

Among the various chloro-cides available today, elemental chlorine was the first to be discovered (Baldwin, 1927). At ambient conditions, elemental chlorine exists as a greenish-yellow gas with a penetrating odor. In 1774, Swedish chemist Karl W. Scheele obtained the gas by oxidizing nascent HCl with MnO_2. Scheele observed that the gas was soluble in water, and demonstrated a permanent bleaching effect on paper, vegetables, and flowers. These properties brought chlorine to its first commercial use, as bleaching agent for the textile industry. Later in 1789, in a French chemical plant in Javel, Berthollet improved the bleaching action of chlorine solution by mixing it with potassium hydroxide. The resulting compound was a hypochlorite (OCl^-) solution termed 'Javelle water' after the name of the town where it was discovered. Later, Labarraque modified the manufacturing process of hypochlorite by replacing potassium hydroxide with inexpensive sodium hydroxide, commonly known as caustic soda. Chlorination of sodium hydroxide remains the most popular method for the production of hypochlorite. Shortly after the discovery of sodium hypochlorite, the discovery of lithium- and calcium-hypochlorites followed.

Later, various antimicrobial properties of hypochlorite were discovered. In 1881, a German bacteriologist, Robert Koch, demonstrated the activity of hypochlorites to destroy pure cultures of bacteria under controlled laboratory conditions. By end of the nineteenth century, the American Public Health Association (APHA) recognized hypochlorites as potent broad-spectrum antimicrobial agents (Hadfield, 1957). The first commercial use of chlorine compound for disinfection was in the form of hypochlorite. It was used in London during 1850 to treat a contaminated water supply that had led to an outbreak of cholera.

Since the discovery of chlorine and hypochlorites, many other chlorine compounds were discovered that had distinct advantages as antimicrobial agents. A variety of organic chloro-cides were discovered that were not as susceptible to photo-decomposition as elemental chlorine or hypochlorite. They provided better disinfection properties for treating exposed water environments such as swimming pools, ponds, and open reservoirs. The organic chlorinating agents were also helpful in reducing metal corrosion on equipment used in the processing of water or food. Moreover, these compounds released hypochlorous acid in a sustained manner, preventing the liberation of chlorine gas. Many organic chloro-cides were also used in medical applications. Dakin et al. (1916) used Chloromine-T during World War I for treating infected wounds. Thereafter, Leech (1923) reported the germicidal properties of various organic chloro-cides.

The discovery of chlorine dioxide (ClO_2) was reported by Humphry Davy, soon after he declared chlorine to be an element. The antimicrobial effectiveness of ClO_2 was soon elucidated. However, due to its energetic nature, the use of ClO_2 did not become widespread. Greater than 15% concentration of ClO_2 in confined spaces is a potential explosive. Therefore, compressed ClO_2 could not be packaged into cylinders for shipping. It must be produced at the site of application. The first on-site generation of ClO_2 was reported in 1944, when it was used for the disinfection of water at the Niagara Falls (Aston, 1947). Since then, the generation technology for ClO_2 has undergone major breakthroughs and currently it is widely used for disinfection and sanitation in various applications.

In recent times, the use of certain chlorine compounds has faced opposition from environmentalists. A number of studies indicated that chlorine-based compounds form carcinogenic by-products (Richardson et al. 1998). A chloro-cide that is exempted from these controversial issues is ClO_2. The oxidation mechanism of ClO_2 is fundamentally different than other chlorine based compounds and therefore, it does not produce any detectable levels of chlorinated organic by-products. Due to the continued pressure to control disinfection by-products (DBP) by the Environmental Protection Agency (EPA), ClO_2 has received a great deal of attention. The focus on ClO_2 is increasing due to its potent sanitizing efficacy and for not forming any carcinogenic by-products. Therefore, this chapter will discuss the molecular mechanisms of ClO_2 more in detail than the other chloro-cides.

II. STRUCTURE AND CHEMISTRY

Natural forms of chlorine display only a single oxidation state where it exists as a monovalent chloride anion in the form of carnallite ($KMgCl_3$, $6H_2O$), sylvite (KCl) and common salt ($NaCl$). None of the natural forms have any unique antimicrobial properties. The synthetic compounds of chlorine, however, could demonstrate oxidation states ranging from -1 to +7 (Taube et al., 1949). Such a wide range of possible oxidation states of chlorine results in a plethora of chloro-cides that satisfy the diverse needs of the sanitation and disinfection industry. A list of these compounds and their structures are listed in *TABLE 1* and *FIGURE 1*. Further details regarding the commercially important compounds are discussed under four subsections: elemental chlorine, inorganic chloro-cides, organic chloro-cides and chlorine dioxide. ClO_2 being an inorganic chloro-cide is elaborated under a separate header, as it is considered different from most chlorinating agent in terms of activity and residual byproducts.

TABLE 1. Antimicrobial chloro-cides commonly used in the industry

Chemical Name [CAS No.]	Formula
Inorganic compounds	
Chloramines	NH_xCl_y
Sodium hypochlorite [7681-52-9]	NaOCl
Calcium hypochlorite [7778-54-3]	$Ca(OCl)_2$
Lithium hypochlorite [13840-33-0]	LiOCl
Chlorinated trisodium phosphate [56802-99-4]	$4(Na_2PO_4.11H_2O)NaOCl$
Chlorine dioxide [10049-04-4]	ClO_2
Organic compounds	
Sodium benzenesulfonchloramide [127-52-6] *(Chloramine-B)*	$C_6H_4ClNNaO_2S$
Sodium *p*-toluene sulfonchloramide [127-65-1] *(Chloramine-T)*	$p\text{-}CH_3C_6H_4SO_2NClNa$
p-Toluene sulfonchloramide [473-34-7] *(Dichloramine-T)*	$p\text{-}CH_3C_6H_4Cl_2NO_2S$
p-Sulfondichloramidobenzoic acid [80-13-7] *(Halazone)*	$p\text{-}C_7H_5Cl_2NO_4S$
Trichloroisocyanuric acid [87-90-1] *(Symclosene)*	$C_3Cl_2N_3O_3$
Sodium dichloroisocyanurate [51580-86-0]	$Na[C_3Cl_2N_3O_3]$
Potassium dichloroisocyanurate [2244-21-5]	$K[C_3Cl_2N_3O_3]$
Trichloromelamine [7673-09-8]	$C_3H_3Cl_2N_6$
1,3-Dichloro-5,5-dimethyl hydantoin [118-52-5] (Halane)	$C_5H_6Cl_2N_2O_2$
N-Chlorosuccinimide [128-09-6]	$C_4H_4ClNO_2$
N,N''-dichlorodiazenedicaboximidamide [502-98-7] *(Chloroazodin)*	$C_2H_4Cl_2N_2$

The availability of several different chloro-cides is often overwhelming to an end user. In general, a user considers replacement of one sanitizer with another for reasons related to efficacy, convenience, or cost. In such a case, it is useful to compare the biocidal efficiencies of chloro-cides by calculating their 'available chlorine' content. Available chlorine is a measure of the oxidizing capacity of a compound as compared to elemental chlorine. It is calculated by determining the electrochemical equivalent amount of Cl_2 to that compound. Available chorine content provides a normalized scale for comparing the efficiencies of various sanitizers. For example, compare NaOCl with $Ca(OCl)_2$. It could be deduced from *Equation 1* and *Equation 2* that the electrochemical reactions of elemental chlorine and the hypochlorite ion, each proceed with a consumption of two electrons. Therefore, it may be concluded that electrochemically, one mole of elemental chlorine is equivalent to one mole of hypochlorite. Since the molecular weight of Cl_2 is 70.91 g, one mole of hypochlorite shall contain 70.91 g of available chlorine. Accordingly, one mole of NaOCl contains 70.91 g of available chlorine, whereas, one mole of $Ca(OCl)_2$ contains 141.8 g of available chlorine.

$$Cl_2 + 2e^- \rightarrow 2Cl^- \qquad \text{(Equation 1)}$$

$$OCl^- + 2e^- + 2H^+ \rightarrow Cl^- + H_2O \qquad \text{(Equation 2)}$$

Although the 'available chlorine' scale provides a good general tool for comparing the biocidal efficacy of chloro-cides, one must be cautious in using this concept when ClO_2 is compared with another chloro-cide. According to *Equation 3*, the electrochemical reaction of ClO_2 consumes 5 electrons which is 2.5 times more than that of elemental chlorine. Therefore, the available chlorine content in one mole of ClO_2 is 177.3 g (obtained from 70.91 x 2.5). Since the molecular weight of one mole of ClO_2 is 67.45, it is expected to be 2.63 times more potent than elemental chlorine (2.63 is obtained from

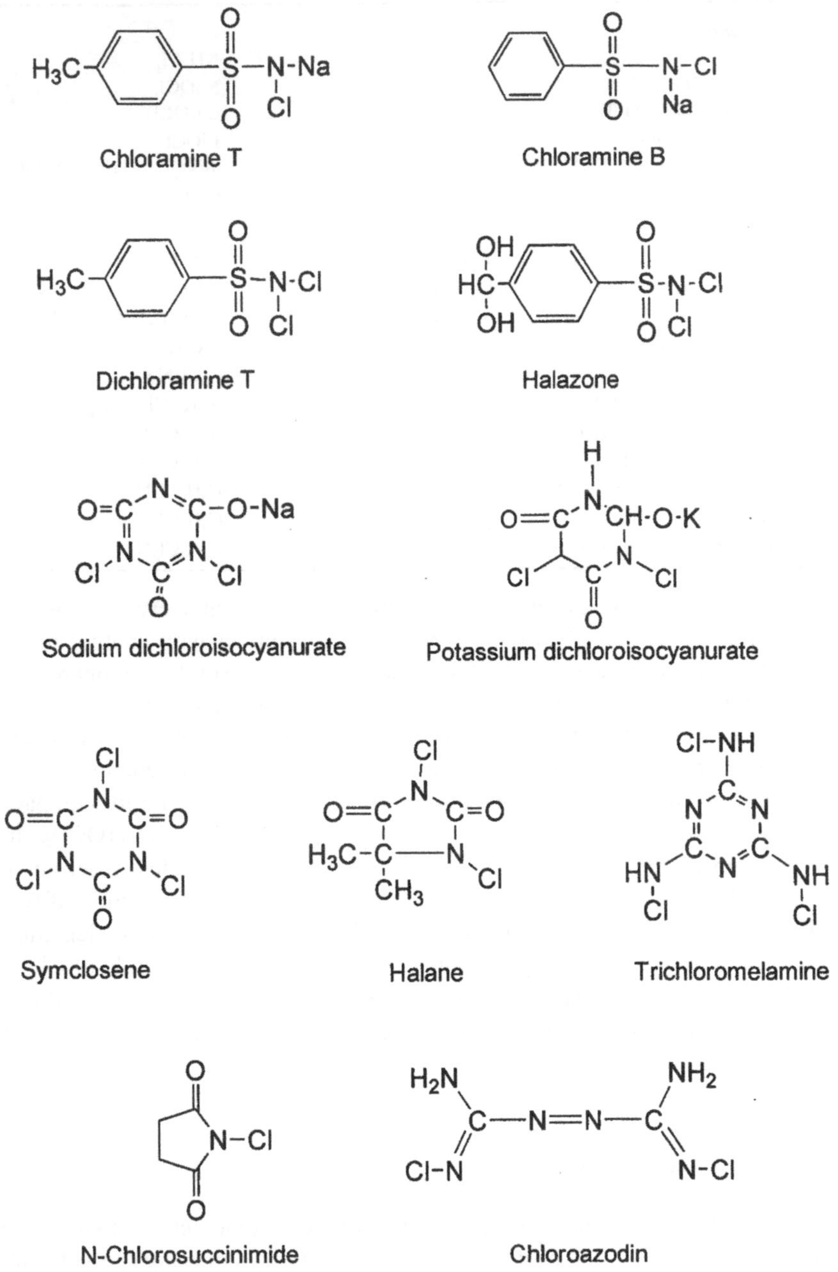

FIGURE 1. Structures of commercially produced organic chloro-cide compounds.

177.3/67.45). However, it has been observed that the biocidal efficacy of ClO_2 is greater than two orders of magnitude when compared to that of chlorine (Tanner, 1989). This may be due to the difference in the mechanism of the two biocides.

$$ClO_2 + 5e^- + 4H^+ \rightarrow Cl^- + 2H_2O \qquad \text{(Equation 3)}$$

A. Elemental chlorine

Elemental chlorine is a widely used disinfectant (White, 1992). Although at ambient conditions it exists as a gas, it can be safely compressed into a liquid form. Liquid chlorine is easier to handle and cheaper to transport. The first liquification process of chlorine was demonstrated in 1805 by Thomas Northemore by compressing the gas into a clear amber-colored liquid. It was noted that upon release of pressure, the liquid rapidly volatilized into gaseous form. The gaseous form of chlorine is yellowish-green in color. Green translates as '*chloros*' in Greek, and is the root of the name 'chlorine' as nomenclatured by Humphry Davy. He also declared chlorine to be an element in 1811 after unsuccessful attempts to effect its decomposition.

Elemental chlorine proves economical in high volume uses such as in the treatment of water and wastewater. A substantial amount of chlorine is used in the power industry to prevent biofouling of heat exchangers. For majority of antimicrobial applications, chlorine is converted into other chlorine compounds such as hypochlorites, which provide ease of use and safe handling.

On a commercial scale, chlorine is manufactured by electrolysis of brine with an overall reaction according to *Equation 4*.

$$NaCl + H_2O + \text{electric current} \rightarrow NaOH + 1/2\ Cl_2 + 1/2\ H_2 \qquad \text{(Equation 4)}$$

Differences in the manufacturing processes of chlorine are primarily based in the types of electrolytic cells used, i.e. diaphragm, mercury and membrane (White, 1992). In 1998, about 12.8 million tons of chlorine was produced in the United States (C&E News, 1999). An estimated one million tons have been used in the treatment of drinking water and wastewater.

In aqueous solutions, chlorine disproportionates according to *Equation 5*.

$$Cl_2 + H_2O \rightarrow HOCl + H^+ + Cl^- \qquad \text{(Equation 5)}$$

The major pathway for the above reaction is via hydroxyl ion where the rate approaches to that of the diffusion control reactions (Morris, 1692). This is of great practical significance since the rapid kinetics of this reaction enable high passage of aqueous chlorine solutions from chlorine generators without degassing problems. The properties of gas and liquid phase chlorine are listed in *TABLE 2*.

B. Inorganic chloro-cides

The two primary chlorine compounds that fall under this category are hypochlorites and chloramines.

1. Hypochlorites. Hypochlorites are the oldest and the most widely used antimicrobial agents for sanitation and disinfection. They are commonly known as bleach and their biggest application is in the treatment of potable water and wastewater. Although the

TABLE 2. Properties of chlorine

Symbol	Cl_2
Molecular weight	70.914 g/mol
Solubility in water (68°F)	7.29 g/L
Molar absorptivity in water (325 nm)	75 cm^{-1} M^{-1}
Gas:	
Density (34°F)	0.00321 g/ml
Viscosity (68°F)	0.01325 centipoise
Liquefying point	-30.1°F (-34.5°C)
Liquid:	
Boiling point	-30.1°F (-34.5°C)
Freezing point	-149.8°F (-101.0°C)
Density (32°F)	1.47 g/ml
Viscosity (68°F)	0.345 centipose
1 volume liquid at 32°F,1 atm	457.6 vol gas
1 kg liquid at 32°F, 1 atm	310.2 L gas

cost of hypochlorites is approximately two to four times higher than that of chlorine gas, use of the former is growing due to the risk of chlorine accidents. This is particularly true in large metropolitan areas where high volume of disinfectant is required for water treatment.

The hypochlorite ion is comprised of chlorine and an oxygen molecule with an overall negative charge, ClO⁻. The common counter cations of OCl⁻ that are used in the industry are Na⁺, Li⁺ and Ca²⁺. Sodium hypochlorite (NaOCl), also known as soda bleach liquor or liquid bleach, is the most widely used of all the chlorinated bleaches. In the United States, more than 150 tons of sodium chlorite is used every day (White, 1992). The use of sodium hypochlorite is preferred over other hypochlorites due to its low cost and ease of handling. Sodium hypochlorite has the least maintenance problems with pumping and metering equipment. Moreover, there is no fire hazard associated with its storage. Sodium hypochlorite is often supplied as a mixture with trisodium phosphate (TSP). This product is known as chlorinated TSP. Chlorinated TSP provides the benefits of disinfection along with the cleaning action of TSP.

Sodium hypochlorite is manufactured by reacting gaseous chlorine with the solution of caustic soda as shown in *Equation 6*.

$Cl_2 + NaOH \rightarrow NaOCl + NaCl + H_2O + Heat$ *(Equation 6)*

To improve the stability of the product, more caustic soda is added than suggested by the reaction stoichiometry. The commercially available solutions of sodium hypochlorite are as high as 15% available chlorine in concentration. More concentrated solutions cause purification problems due to solidification.

Calcium hypochlorite solutions are made by passing chlorine into milk of lime suspension according to *Equation 7*.

$2Cl_2 + 2Ca(OH)_2 \rightarrow Ca(OCl)_2 + CaCl_2 + 2 H_2O$ *(Equation 7)*

The maximum strength of these solutions is limited to about 35% available chlorine. These solutions are prone to sludge formation caused by the presence of impurities such as magnesium or aluminum oxides. Therefore, calcium hypochlorite is not desired in operations where sophisticated metering equipment is involved. The potential insolubles result in maintenance problems. Due to the much higher available chlorine content of calcium hypochlorite, it is used in remote areas where the transportation cost becomes significant. Calcium hypochlorite is also commercially available in granular form where the available chlorine content is 65 to 70%. Granular calcium hypochlorite is the most common form of dry bleach used in the industry. Granular calcium hypochlorite in concentrations greater than 35% is a fire hazard and must be stored away from heat and oxidizable organic compounds. Although manufacturers use different approaches for the production of granulated calcium hypochlorite, the basic process involves chlorination of lime slurry.

Among the granulated forms of hypochlorites, lithium hypochlorite is another prevalent form. It is prepared by mixing strong solutions of lithium chlorite and sodium hypochlorite, followed by evaporation of the reaction mixture. Lithium hypochlorite is a clean and free-flowing form of granulated hypochlorite, which is readily soluble in water. Therefore, it has distinct advantages over calcium hypochlorite such as that the solution 1) does not contain any sludge, 2) does not significantly affect the alkalinity or the pH of water and 3) is relatively more stable in the solution form.

Degradation of active components during storage is a major problem with hypochlorite solutions (Gordon, 1997). The loss is proportional to the exposure to light and heat. The presence of heavy metal cations such as iron, copper, nickel, and cobalt, adversely affect the stability of hypochlorite solutions. For maximum shelf life of hypochlorite solutions, dilution to a low concentration (~10%) is recommended. The presence of heavy metal ions should be minimized (preferably <0.5 ppm) and the solutions should be stored away from light and heat.

2. Inorganic chloramines. Inorganic chloramines are produced when ammonia is introduced in an aqueous solution of chlorine or hypochlorite. Specifically, the compounds produced are monochloramine (NH_2Cl), dichloramine ($NHCl_2$) and trichloramine (NCl_3). Trichloramine is commonly known as nitrogen trichloride. Production of chloromines is a complex phenomenon that involves a series of coupled reactions. During water-processing, this is known as the 'breakpoint phenomenon'. The final concentrations of chloramine are dependent on pH, temperature and initial reactant concentrations.

Inorganic chloramines were used in the 1920s for treatment of water supplies. They were generated in the water stream by introduction of ammonia. The primary objective was to improve the taste of water. Prolonged stability of chorine was an added benefit. However, by early 1940s the germicidal activity of chloramine was found approximately 25 times less than that of free available chlorine (Kereluk et al. 1983; Wattie, 1944). It was also reported that the contact times required for chloramines are about 100 times longer than the same residual concentration of free available chlorine. Additionally, it was discovered that the presence of monochloramine suppresses the release of hypochlorous acid rendering chlorine ineffective against microbes. Thus, the use of ammonia to generate chloramines was discontinued in the United States. Chloramines are not produced on a commercial scale, however, some interest in these compounds has emerged due to their potential use in prevention of trihalomethane formation during chlorination of water (Norman et al., 1966).

C. Organic chloro-cides

Most organic chloro-cides are N-chloroderivatives of one of the following four groups: sulfonamides, heterocyclic nitrogen compounds, condensed amines from guanidines derivatives, and anilides (Dychdala, 1991) (*FIGURE 1*). They all have a faint chlorine odor, exist as solids, and fall under the broad category of chloramines. As compared with hypochlorites, organic chloramines are less corrosive on skin and metal. They do not react as rapidly with organic load and are relatively more stable when exposed to heat and light. All organic chloro-cides, in general, have low solubility in water.

Chloramine-B, chloramine-T and dichloramine-T are crystalline powders. Due to their less irritating nature compared to hypochlorites, these compounds have been used for wound treatment and other antiseptic medicinal preparations (Dakin et al., 1916; Leech, 1923; Guiteras et al., 1934). They have limited use as sanitizers and are used only in applications that require low pH and long contact times. Halazone, a white crystalline powder, has been used for the treatment of potable water and for sanitizing equipment in the food and dairy industry. Geiger et al. (1952) demonstrated that the presence of organic load inhibits the biocidal efficacy of halazone. Concentration of halazone equivalent to 1 ppm available chlorine requires 3 minutes to kill a population of *E.coli* in the absence of organic load, whereas, the contact time required to achieve the same activity in the presence of organic load was one hour.

Chloroisocyanurates, a subclass of organic chloro-cides are widely used as sanitizers in dairies, restaurants and swimming pools. They consist of three commercial products: 1) trichloroisocyanuric acid, 2) sodium dichloroisocyanurate, and 3) potassium dichloroisocyanurate. After dissolving in water, these compounds hydrolyze to form HOCl and cyanuric acid. Ionic surfactants such as sodium dodecylbenzene sulfonate could be added to these chloro-cides for improved cleaning power. Trichloromelamine (TCM), a white powder, is a relatively stable organic chloro-cide. Very little chlorine loss is experienced when the powder is heated at 100°C for a day or boiled as aqueous solutions. Thus, this product can be easily blended with cleaning agents.

Several commercial sanitizing products are based on 1,3-dichloro-5,5-dimethyl hydantoin as the main active ingredient. Ortenzio et al. (1957) showed that the biocidal efficacy of 65 ppm of hydantoin solution at pH 6.0 is equivalent to 200 ppm of hypochlorite. N-Chlorosuccinimide is a heterocyclic chlorine compound that is not effected by the organic load in the system. This product has been primarily used in the purification of water supplies. Due to low water solubility of this compound, no HOCl was detected in the aqueous solution (Chang, 1944). *N, N″*-dichlorodiazenedicarboximid amide, also known as chlorozodin or azochloramid, appears as bright yellow flakes. This material has been used extensively for treating wounds. Guiteras et al. (1934) showed that azochloramid had a very low reactivity with tissue and body fluids and thus serves as a mild topical applicant.

D. Chlorine dioxide (ClO$_2$)

Chlorine dioxide was discovered in 1811 by Sir Humphry Davy, soon after he declared the chlorine element. This molecule turned out to be a very powerful antimicrobial agent. The oxidative capacity (not to be confused with oxidation potential) of ClO$_2$ is 2.6 times greater than that of a chlorine molecule. Although the antimicrobial proper-

TABLE 3: Properties of chlorine-dioxide

Symbol	ClO_2
Molecular weight	67.452 g/mol
Solubility in water (77°F, 34.5 mm Hg)	~3 g/L
Solubility in water (32 - 41°F, 70 - 100 mm Hg)	~20 g/L
Solubility in hexane (71.6°F)	~60 g/L
Molar absorptivity in water (360nm)	1,250 cm^{-1} M^{-1}
Boiling point	51.8°F (11°C)
Freezing point	-74.2°F (-59°C)
Vapor pressure (32°F)	490 mm Hg

ties of ClO_2 started to become known soon after its discovery, it did not become a popular choice for a long time because of its energetic nature.

In recent years, ClO_2 has found widespread use in the disinfection market (Gates, 1996). Approximately 361 metric tons of ClO_2 was produced in the United States in 1991; since then the production has significantly increased. Among the chlorine-based antimicrobial compounds, this compound has a great future because unlike chlorine or hypochlorites, ClO_2 does not produce carcinogenic byproducts such as trihalomethanes (Lykins et al., 1986; Myers, 1990; Glaze et al., 1990; Li et al., 1996; Richardson et al., 1998). Moreover, compared to chlorinating compounds, ClO_2 has a much higher antimicrobial efficacy. Narkis et al. (1992) have demonstrated that for treatment of wastewater, ClO_2 is superior to chlorine in eliminating indicator microorganisms and viruses.

Like chlorine, ClO_2 also exists as a greenish-yellow gas at ambient temperatures. However, it cannot be liquified and supplied ready for use like chlorine. It is estimated that more than 15% concentration of ClO_2 in air is explosive (Masschelein, 1979). Currently, the Department of Transportation (DOT) regulations do not allow shipment of ClO_2 in any form except for small quantities of frozen aqueous solutions. Therefore, ClO_2 is usually generated at the site of application. The properties of ClO_2 are summarized in *TABLE 3*.

1. Generation using acids. The most common precursors used for the generation of ClO_2 are sodium chlorate ($NaClO_3$) and sodium chlorite ($NaClO_2$). Mixing either of these precursors with acid generate ClO_2 (*Equation 8* and *Equation 9*). Any mineral or organic acid may be used for this purpose. In food processing facilities, the acids used are those with the GRAS status by the FDA. The most commonly used acids are hydrochloric or phosphoric. Often, citric acid granules are used where acid is required in solid state. The reaction with $NaClO_2$ is much more efficient than that with $NaClO_3$. The process of generation with $NaClO_3$ is primarily used in applications where purity of ClO_2 is not a critical factor. The reaction of $NaClO_2$ is sluggish and requires excess acid to drive the kinetics (Equation 9). If the acid concentration is high, a competing reaction is triggered (Equation 10) which produces unwanted chlorine. Excessive acid concentration in the effluent is undesired due to its high corrosion potential.

$$5NaClO_2 + 4HCl \rightarrow 4ClO_2 + H_2O + NaCl \qquad \text{(Equation 8)}$$

$$NaClO_3 + 2HCl \rightarrow ClO_2 + 1/2\ Cl_2 + NaCl + H_2O \qquad \text{(Equation 9)}$$

$NaClO_3 + 6\ HCl \rightarrow 3Cl_2 + NaCl + 3\ H_2O$ *(Equation 10)*

Many commercial operations use the 'Solvay' process for the generation of ClO_2 (Julien et al., 1959). This process is carried out by reacting $NaClO_3$ with H_2SO_4 with methanol as the reducing agent (*Equation 11*). Some manufacturers have modified this process by replacing methanol with hydrogen peroxide (*Equation 12*).

$2NaClO_3 + 2H_2SO_4 + CH_3OH \rightarrow 2ClO_2 + 2NaHSO_4 + HCHO + 2H_2O$
                                                                                       (Equation 11)

$2NaClO_3 + H_2SO_4 + H_2O_2 \rightarrow 2ClO_2 + O_2 + Na_2SO_4 + 2H_2O$ *(Equation 12)*

The Solvay process is also used in the manufacturing of speciality formulations of sodium chlorite sometimes referred as 'stabilized chlorine dioxide' products. For their synthesis, ClO_2 is first produced as above. It is then purified by soublizing in chilled water. The solubility of ClO_2 in water is approximately ten times greater than that of chlorine. Extracted ClO_2 is reacted with hydrogen peroxide and caustic to yield high-grade sodium chlorite. The optimum concentrations for production are between 9 to 10%. The product synthesized by this method is generally shipped as a liquid.

The solid form of sodium chlorite is obtained by evaporation of electrochemically generated commodity-grade sodium chlorite. Because of its high reactivity, the solid product is blended with 20% of inert salts that commonly consist of 13% NaCl, 5% Na_2CO_3 and 2% $NaClO_3$ (Masschelein, 1979). The commodity-grade sodium chlorite solutions are commercially available at concentrations up to 25%.

Although sodium chlorite is considerably more expensive than sodium chlorate, the ease of handling and purity of the ClO_2 obtained makes it by far the precursor of choice for small-scale systems. Besides the acid generation of ClO_2 described above, there are some alternative approaches to produce ClO_2 using chlorite solutions that are discussed next..

2. Generation using chlorine or hypochlorite. High purity ClO_2 can be produced by reacting chlorine gas with chlorite ion as shown in *Equation 13* (Pitochelli, 1995). This method takes advantage of the rapid reaction rate of molecular chlorine with the chlorite ion (Aieta et al., 1986). When dry chlorine gas is directly contacted with aqueous concentrate of chlorite ion, the yield of this process could be achieved up to 95 to 98%. In order to avoid any carry-over of chlorine impurities in the product, chlorine gas may be diluted with nitrogen or air that ensures a complete reaction.

$Cl_2 + 2NaClO_2 \rightarrow 2\ ClO_2 + 2NaCl$ *(Equation 13)*

ClO_2 can also be produced using aqueous solutions of chlorine or hypochlorites at low pH. Under low pH conditions, both chlorine and hypochlorites produce HOCl that reacts with chlorite salt as shown in *Equation 14*.

$HOCl + 2NaClO_2 \rightarrow 2ClO_2 + NaCl + NaOH$ *(Equation 14)*

3. Generation using UV-Vis light. When solutions of chlorite ion are subject to sunlight, they turn yellow within a few minutes. The color change occurs due to the conversion of the chlorite ion into ClO_2 upon exposure to light. Although the ultraviolet light is much more efficient for this conversion, the visible light is also capable of driving this reaction at a sluggish rate. This principle is used to design an apparatus that provides a source of pure ClO_2 solutions (Callerame, 1989). ClO_2 can also be produced by subjecting a mixture of chlorine gas and oxygen to ultraviolet radiation (Callerame, 1989).

4. Electrolytic generation. Several approaches are possible to generate ClO_2 electrochemically that use chlorite or chlorate ions as precursors (Griese et al., 1992; Kaczur et al., 1993; Gates, 1998). One method is the electrolysis of an aqueous solution of chlorite ion in the presence of chloride ion. The process could be carried out in an electrolytic cell consisting a platinum anode, a diaphragm and a lead cathode. The heat evolved in the anode compartment is preferably dissipated by a cooling device.

5. Catalytic generation. Many transition metals in ionic or elemental form are capable of catalyzing the conversion of chlorite ion into ClO_2 (Ringo, 1991). Particularly notable among these are: vanadium, cobalt, copper, ruthenium, palladium, and platinum. The generation of ClO_2 is a function of chlorite ion concentration, pH, ionic strength, temperature, and the presence of buffers in the system. The buffer systems that are known to increase the rate of generation are phosphate, citrate and borate.

6. Generation in dry media. Chlorine dioxide may be generated in a dry media that consists of chlorite ion and a source of proton. The latter may be achieved by using diatomaceous earth or zeolite crystals (Klatte, 1996). The mixture of these two components in an inert media such as 'attapulgite' clay forms a system that is capable of providing a sustain release of low concentrations of ClO_2 gas (~0.1 to 0.5 ppm/m³/hr). The shelf life of these formulations can be controlled by changing the quantity of different ingredients.

III. ANTIMICROBIAL EFFECTS

Several theories have been advanced, including cell membrane effects (Baker, 1926; Chang, 1944) and inhibition of sulfhydryl enzymes and enzymes involved in glucose metabolism (Green & Stumpf, 1946; Knox et al., 1948). Haas and Engelbrecht (1980) suggested that chlorine elicits cidal activity at or near the cell membrane, as well as affects the cellular DNA. It has also been indicated that NaOCl reacts with the DNA of living cells and cause mutations by oxidation of purine and pyrimidine bases (Rosenkranz, 1973; Wlodkowski & Rosenkranz, 1975). Bernarde et al. (1967b) attributed inactivation by ClO_2 to disruption of protein synthesis. The sporicidal effects of chlorine have been attributed to spore coat disruption by either chlorine combining with and removing protein (Wyatt & Waites, 1975). Kulikovsdy et al. (1975) implicated spore inactivation to permeability damage resulting in a loss of calcium ion, dipicolinic acid, RNA, and DNA.

A. Disinfectant mechanism(s)

The need to elucidate the biocidal action of chlorine-based compounds is interesting from academic standpoint and lucrative from industrial standpoint. However, despite enormous amount of research, the antimicrobial mechanism has not yet been fully unraveled. It is generally believed that oxidative killing is the cidal mechanism. Based on the antimicrobial mechanism, chloro-cides could be classified into two categories.

The first category comprises of compounds with antimicrobial mechanism similar to that of elemental chlorine. Hypochlorites and chloramines fall under this category. It was earlier believed that the germicidal effect of chlorine was due to the liberation of nascent oxygen. In 1944, Chang established that liberation of oxygen is not involved in the reaction of chlorine and that the disinfecting agent in fact was hypochlorous acid (HOCl) (*Equations 15*). Both, hypochlorites and chloramines, produce HOCl as a result of hydrolysis (*Equations 16 - 18*) which is believed to be the lethal component. HOCl is also known as the free available chlorine residual (White, 1992).

$$Cl_2 + H_2O \rightarrow HOCl + H^+ + C^- \qquad\qquad \textit{(Equation 15)}$$

$$NaOCl + H_2O \rightarrow HOCl + Na^+ (OH)^- \qquad\qquad \textit{(Equation 16)}$$

$$Ca(OCl)_2 + H_2O \rightarrow 2HOCl + Ca^{2+} (OH)_2^- \qquad\qquad \textit{(Equation 17)}$$

$$NH_2Cl + H_2O \rightarrow HOCl + NH_3 \qquad\qquad \textit{(Equation 18)}$$

HOCl is an electrochemically neutral molecule with a structure similar to that of the water molecule (HOH). Its modest size, low molecular weight and electro-neutrality allow penetration into the cell wall and to approach an enzyme residue with ease. It is assumed that once the penetration of the cell wall is accomplished, the disinfecting compound has the ability to attack the enzyme group and such destruction results in death of the organism.

The second category comprises of compounds with biocidal mechanism other than that of chlorine (Benarde, 1965). A widely used member of this category is ClO_2. It is hypothesized that the biocidal component of ClO_2 is chlorous acid ($HClO_2$) (Gordon, 1972). Although there are differences in the mechanisms of the two categories of compounds, the underlying basic mechanism for both is oxidation.

Most studies in this area were performed on bacterial or fungal systems and a limited information is available on the mode of virus inactivation. A virus is composed of nucleic acid contained in a protein shell. It is speculated that the biocide penetrates the outer shell and reacts with the nucleic acid rendering it inactive for multiplication.

Rate of destruction could be described as a first order process as in *Equation 19*.

$$N_f = N_0\, e^{-kt} \qquad\qquad \textit{(Equation 19)}$$

A plot of 't' versus log 'N_f' is linear where the slope 'k' represents the rate-constant and the intercept represents the logarithm of bacterial population. The bacterial population at time zero is given by the inverse natural log of the intercept of this plot. The time taken to reduce the microbial population to half ($t_{1/2}$) is equal to 0.693k.

It is generally believed that the relative efficiency of various disinfecting compounds is a function of the rate of diffusion of the active agent through the cell wall. The mode of action of chloro-cides have been studied by different groups (Green et al., 1946; Benarde et al., 1967; Noss et al. 1986, Berg et al. 1986). The destruction of microorganisms may occur via protein denaturation or implosion of the cell.

B. Influencing factors

Various factors affect the efficiency of the sanitizing agent. These factors are summarized below.

1. Concentration and contact time: It is essential to apply the appropriate concentration of sanitizer for adequate duration of time known as the "contact time". Generally, the appropriate concentration of a sanitizer for a specific application is recommended by the regulatory agencies such as FDA, EPA or USDA. The economics of a sanitizing program is driven by the concentration of the sanitizer. The requisite concentrations to kill some indicator organisms for chlorine based sanitizers are summarized in TABLE 2

To a certain extent, the concentration and contact time could compensate for each other. These parameters are expected to follow the relationship expressed by *Equation 20*, where, 'C' is the concentration of the biocide, 'n' is the coefficient of dilution 't' is the time, and 'K' is an empirical parameter that varies with sanitizer, microorganism, and milieu (Wickramanayake et al., 1991).

$$C^n \times t = K \qquad \textit{(Equation 20)}$$

2. pH: Milieu pH is a critical factor for efficacy of a sanitizer. In general, chlorinating sanitizers lose effect rapidly at pH levels >9.0. Hypochlorites are sold at alkaline pH and are to be mixed with acids for effective sanitation. Lowering of pH facilitates hydrolysis that produces the primary biocidal component, HOCl. The addition of acids in the hypochlorite systems must be performed with caution since at low pH they start liberating hazardous chlorine gas. Depending on the type application, the chlorinating systems are typically operated between the pH of 6.5 and 7.5. The role of pH in antimicrobial efficacy of chlorine is well documented (Dychdala, 1983). The pH also plays a critical role in the sporicidal efficacy of hypochlorite. Most commercial hypochlorite solutions produce a slightly alkaline pH at their use dilution. The antimicrobial properties of chlorine are not favorable under alkaline conditions.

As compared to various chlorinating systems, the pH range of ClO_2 is more flexible. ClO_2 does not ionize in water, and therefore its bactericidal efficiency remains essentially constant over the normal range of pH values in natural water (Ridenour & Ingols, 1947; Bernarde et al., 1965). Owing to its high solubility, the release of gas from the solution at low pH is much less as compare to chlorine. Once generated and introduced into the use solution, ClO_2 is effective up to pH 10.0. At pH values >10.0, ClO_2 slowly starts converting into ClO_2^-. Bernarde et al. (1965) found that equivalent concentrations of ClO_2 and hypochlorite at pH 8.5 produced an equivalent degree of cidal activity against *E. coli* in 15 and 300 s, respectively. At pH 6.5, the two compounds were almost equally effective at equivalent concentrations.

Berman et al. (1984) studied the effect of chlorine, ClO_2, and monochloramine on viruses at 5°C at pH 6 and 10. The results indicated that 0.5 ppm (as Chlorox), inactivated virions in less than 15-s at pH 6.0, resulting in 4-log reduction, but not at pH 10. With chlorine dioxide, however, 0.5 ppm was more efficient at pH 10 than at pH 6.0, with a 15-s inactivation. Monochloramine at 10 ppm concentration and pH 8.0 required more than 6 h to produce an equivalent inactivation of virions.

3. Temperature: The destruction of microorganisms follows first order kinetics. According to the Arrhenius relationship the reaction rate (biocidal activity) doubles with every 10°C (18°F) change in temperature (Wilkins, 1991). However, higher temperatures adversely affect the solubility, and gassing-off becomes a concern. The effects of temperature on the antimicrobial activity of chlorine is well documented (Trueman, 1971; Dychdala, 1983). At high concentrations of chlorine, the temperature effect is not readily apparent; however, at low concentrations the effect is multifold. Johns (1954) found no difference in the rate of kill for hypochlorite against *Micrococcus pyogenes* var. *aureus, E. coli*, or *Pseudomonas aeruginosa* at approximately 10 ppm and temperatures of 5, 20, and 40°C. These results also indicate that although hypochlorite is sensitive to temperature effects, it is significantly less sensitive than the other commonly used sanitizers. Rudolph and Levine (1941) found a pronounced temperature effect on the spores of *B. metiens*.

Temperature sensitivity data for the organic chlorine compounds are not readily available. Johns (1954) found that a dichlorodimethylhydantoin product was more temperature sensitive than hypochlorites, especially for *P. aeruginosa*. Decrease in temperature from 45 to 5°C necessitated a 4-fold increase in organic chlorine content to achieve the same rate of kill. Bernarde et al. (1967a) evaluated the effect of temperature on ClO_2 disinfection. A substantial temperature effect was observed over the temperature range of 5-32°C; however, no direct comparison to temperature effects on hypochlorite was established. Ridenour and Ingols (1947) showed a lower antimicrobial activity for ClO_2 at lower temperature and demonstrated that pH in range of 6-10 had no special effect on its activity.

4. Organic load: Organic matter in the milieu significantly affect the bactericidal efficacy of chlorine compounds (Trueman, 1971; Dychdala, 1983). Johns (1954) studied the activity of various sanitizing solutions in the presence of 0.5% skim milk at varying temperatures. The organo-chlorine compound dichlorodimethylhydantoin was less sensitive to the addition of organic material than hypochlorite solution. Bloomfield and Miles (1979b) compared the antimicrobial effect of sodium hypochlorite and sodium dichloroisocyanurate in the presence of organic material. Both chlorine compounds demonstrated equivalent cidal activity in the absence of organic material, however, an abrupt change in the antimicrobial efficacy was observed as the milk concentration increased from 0.5 to 2.0%. ClO_2 is a highly reactive compound, but as chlorine, it does not react with ammonia and other nitrogenous compounds and in general, is not as susceptible to the effects of organic matter (White, 1972). Lillard (1979) found that ClO_2 was significantly more effective than chlorine in reducing the aerobic and fecal coliform counts in poultry processing water.

5. Water quality: Alkalinity, hardness and organic load content of water affect the efficacy of a sanitizer. The effect of alkalinity could be seen particularly in cases where sanitizers are used for stasis of microorganisms. The mechanism of stasis involves reaction of acidic metabolites of the organism with the biocide precursor. High alkalinity (CO_3^{2-} and HCO_3^- ions) neutralize the acid produced by microorganism preventing the production of the active ingredient. With extremely hard water (>500 ppm), the hypochlorites form precipitates. Water hardness *per se* does not exert a marked effect on the antimicrobial activity of chlorine. However, in many cases, hard water could cause an upward drift in pH and diminish chlorine efficacy (Hays et al., 1967; Mosley et al., 1979)

6. Mode of application: The sanitizer could be applied by various delivery mechanisms including circulating, submerging, brushing, fogging, or spraying systems. Automation of the beverage industry has led to the increased use of clean-in-place (CIP) systems managed by computer-controlled valves and pumps. Sanitation of these systems entails circulating sanitizer through pipelines. For effective sanitation of tanks and vats, scrubbing is required to detach food particles and biofilm. Fogging is an effective way to reduce microorganisms from air that are potential risk for product contamination. Jeng and Woodworth (1990) studied ClO_2 gas sterilization under square-wave conditions. By using controlled humidity, gas concentration, and temperature at atmospheric pressure, standard biological indicators and spore disks of environmental isolates were exposed to ClO_2 gas. Prehumidification enhanced the ClO_2 activity. The D values (time required for 90% inactivation) of *Bacillus subtilis* ssp.*niger* were estimated to be 1.5, 2.5 and 4.2 min when exposed to ClO_2 concentration of 30, 15, and 7 mg/L, respectively, at 23°C and ambient (20-40%) relative humidity (*Figure 2*). The study indicated that ClO_2 gas on a molar basis was 1,075 times more potent than ethylene dioxide as a sterilant at 30°C.

C. Microbial resistance

Vegetative bacteria are generally more susceptible to chlorine inactivation than microorganisms that form spores. Based on free available chlorine concentration and contact time, bacterial spores are 10-10,000 times more resistant to destruction by chlorine than vegetative cells (Ito & Seeger, 1980; Odlaug, 1981).

Several authors (Johns, 1934, 1948; Trueman, 1971; Mosely et al., 1976) have observed a greater resistance in certain strains of *Staphylococcus aureus* compared to pseudomonads and other Gram-negative bacteria. With other commonly used sanitizers and disinfectants, notably quaternary ammonium compounds, the reverse is generally true. Haas and Engelbrecht (1980) suggested that yeasts and acid-fast bacteria would be more appropriate indicators of water disinfection than *E. coli* because of their greater resistance to free available chlorine. A comprehensive report published by the EPA (1979) ranked *Mycobacterium fortuitum* and *Candida parapsilosis* well ahead of *E. coli* in resistance to chlorine.

Bolton (1988) showed that strains of *S. aureus* isolated from biofilms and collected from the defeathering equipment were almost 8-fold more resistant than strains from the natural skin flora of poultry. It was concluded that the resistance of these strains was due to their ability to form macroclumps by production of an extracellular slime.

Caldwell (1990) reported that low chlorine levels (0.5-5 ppm) are only inhibitory to biofilms and their cells. Higher chlorine level of 50 ppm and up was needed for the reduction of biofilms by biocidal action. El-Kest and Marth (1988) tested *L. monocto-*

genes against sodium hypochlorite and concluded that a good program of cleaning and sanitizing should be able to control non-spore-forming organisms, including *L. monocytogenes*.

Acquired resistance to chlorine has not been demonstrated; however, Scheusner et al. (1971) provided evidence for chlorine-induced sublethal injury in *E. coli* by exposure to 1 µg/ml of hypochlorite.

Stress resulting from a variety of chemical and physical environments has been recognized in indicator bacteria. Indicator microorganisms are sublethally impaired due to a variety of causes associated with foods. Workers in the area of water microbiology are also gaining an appreciation of the importance of these stressed cells in the assessment of water quality using bacterial indicators. Chemical agents, including chlorine, which are employed in water disinfection processes, are important causes of bacterial stress injury.

Ojajarvi and Makela (1976) studied the disinfecting properties of chloramine and compounds containing chlorinated trisodium phosphate and potassium bromide or sodium dichloroisocyanurate and detergents. They found that the effectiveness of chlorine-bromine disinfectant substantially decreased in the presence of organic material. Irritation of the skin and the mucus membranes when using chlorine disinfectants was also observed.

Bacteria located within microbial aggregates (formed by flocculation during water treatment) are protected against chlorine. *Enterobacter cloacae* attached to drinking water distribution particles are also protected from chlorine disinfection. This effect was dependent on the level of chlorine in the system and attachment time (Herson et al., 1987). Indigenous coliforms associated with the particle fractions were tested for resistance to chlorine and monochloramine (Berman et al., 1988). Coliforms associated with < 7-µ fraction were inactivated more rapidly by 0.5 mg of chlorine per liter at 5°C and pH 7 than coliforms associated with > 7-µ fraction. The time required for 99% inactivation of the particle fractions with monochloramine at pH 7 or 8 was 20- to 50-fold greater than the time required for the same amount of inactivation with chlorine at pH 7. These results indicate that coliforms associated with sewage effluent particles are inactivated more rapidly with 0.5 mg of chlorine per liter than with 1.0 mg of monochloramine per liter. However, > 7-µ particles have a protective effect against the disinfecting action of chlorine.

Lisle and co-workers (1998) reported that chlorine resistance in *E. coli* O157:H7 progressively increased through the starvation period. After 29 days of starvation, there was no significant difference in chlorine resistance between control cultures that had not been exposed to the disinfectant and cultures that had been exposed. This study suggested that *E. coli* O157:H7 adapts to starvation conditions by developing a chlorine resistance phenotype.

D. Storage stability

The stability of sodium hypochlorite (diluted household bleach) when stored for 30 days in various types of containers and the efficacy of low concentrations of free available chlorine to inactivate bacteria was tested (Rutala et al., 1998). Solutions of standard household bleach were prepared using tap water or sterile distilled water at dilutions of 1:100, 1:50, and 1:5. Chlorine concentrations were measured, and the solutions were left in polyethylene containers at room temperature (20°C) under various conditions (translucent containers with light exposure and with or without air; brown opaque container without light or air exposure). Samples for chlorine and pH determinations were taken at time 0 and on days 7, 14, 21, 30, and 40. Bactericidal activity of chlorine solutions was assessed against strains of *Pseudomonas aeruginosa, Staphylococcus aureus*, and *Salmonella choleraesuis*. Chlorine concentrations at 30 days varied from the 40% to 50% range for 1:50 or 1:100 dilutions stored in containers other than closed brown containers to 83% to 85% for the 1:5 dilution stored in closed but non-opaque containers to 97% to 100% for 1:50 or 1:5 solutions stored in closed brown containers. The lowest concentration of sodium hypochlorite solution that reliably inactivated all the test organisms was 100 ppm. These data suggest that chlorine solutions do not need to be prepared fresh daily and 100 ppm concentration of chlorine effectively inactivates *S. aureus, S. choleraesuis*, and *P. aeruginosa*.

IV. ANTIMICROBIAL SPECTRUM

Chloro-cides can be considered a broad-spectrum germicide. Hypochlorites have documented antimicrobial activity against viruses, non-acid-fast bacteria, acid-fast bacilli, bacterial spores, fungi, algae, and protozoa (Dychdala, 1983a). The antimicrobial spectrum of chlorine is shown in *TABLE 4*.

A. Antibacterial activity

Many *Legionella* infections are acquired through inhalation or aspiration of drinking water and about 25% of municipalities in the USA use monochloramine for disinfection of drinking water. In a case study, Kool and coworkers (1999) concluded that 90% of outbreaks of Legionnaires' disease associated with drinking water might not have occurred if monochloramine had been used instead of free chlorine for residual disinfection. The authors suggested that chloramination of drinking water may be a cost-effective method for control of Legionnaires' disease at the municipal level or in individual hospitals, and widespread implementation could prevent thousands of cases.

However, *Legionellae* are more resistant to chlorine than coliforms (Kuchta et al., 1983). At 21°C, pH 7.6, and 0.1 mg of free chlorine residual per liter, a 99% kill of *Legionella pneumophila* was achieved within 40 min, compared with less than 1 min for *E. coli*. The observed resistance is enhanced as conditions for disinfection become less optimal. The required contact time for the removal of *L. pneumophilia* was twice as long at 4°C than it was at 21°C. These data suggest that legionellae can survive low levels of chlorine for relatively long periods of time. Lopes (1986) reported that 100 ppm available chlorine as sodium hypochlorite and dichloroisocyanurate was effective for the 5-log reduction of *L. monocytogenes* and *Salmonella typhimurium* within 30-s. The susceptibility of a strain of *Legionella pneumophila* to disinfection by an organic halamine, free chlorine, and a mixture of organic halamine and free chlorine was reported (Swango et al., 1987). The organic halamine had superior stability in solution and exhibited adequate disinfectant potential over a period of 1 month of repeated reinoculations of fresh bacteria. The combined halamine exhibited great potential for use in maintaining closed-cycle cooling water systems free of *L. pneumophila*. Water disinfection systems utilizing electrolytically generated copper and silver ions (200 and 20, 400 and 40, or 800 and 80 μg/L) and low levels of free chlorine (0.1 to 0.4 mg/L) were evaluated at room (21 to 23°C) and

TABLE 4. Antimicrobial spectrum of chlorine.

Bacterial species	Free chlorine	Contact time	Reference
Antibacterial activity			
Aerobacter aerogenes	0.01 ppm	5 min	Ridenour & Ingols, 1947
Aeromonas hydrophila / A. caviae	n/a	n/a	Sisti et al., 1998
Alcaligenes faecalis	n/a	n/a	Greene et al., 1993
Bacillus cereus/ B. cereus	100 ppm	5 min / 60 min	Cousins & Allan, 1967
Bacillus macerans	7.5 ppm	8 min	Seeger, 1978
Bacillus stearothermophilus	200 ppm	9 min	Seeger, 1978
Bacillus coagulans	5 ppm	27 min	Labree et al., 1960
Campylobacter jejuni	0.1 mg	5 min	Blaser et al., 1986
Clostridium botulinum types A & E	4.5 ppm	10.5 & 6.0 min	Ito & Seeger, 1980
C. perfringens / C. histolyticum	5 ppm	60 min / 10 min	Dye & Mead, 1972
C. tertium / C. bifermentans	5 ppm	20 min	Dye & Mead, 1972
C. sporogenes	5 ppm	35 min	Dye & Mead, 1972
Corynebacterium bovis	n/a	n/a	Oliver et al., 1989
Escherichia coli	12.5 ppm	15 sec	Mosley et al., 1976
E. coli O157:H7	2000 ppm	1 min	Beuchat et al., 1998
Helicobacter pylori	n/a	n/a	Johnson et al., 1997
Lactobacillus plantarum	5.0 ppm	15 sec	Hays et al., 1967
Legionella pneumophila	0.1 mg	40 min	Kuchta et al., 1983
Listeria monocytogenes	2000 ppm	1 min	Beuchat et al., 1998
	25 ppm	n/a	Lin et al., 1996
Pseudomonas aeruginosa	25 mg/ml	2 hours	Samrakandi et al., 1997
Pseudomonas fluorescens	n/a	n/a	Greene et al., 1993
Salmonella typhimurium	50 mg/L	1 min	Williams et al., 1990
Salmonella paratyphi B	0.02 ppm	5 min	Ridenour & Ingols, 1947
Salmonella derby	12.5 ppm	15 sec	Mosley et al., 1976
Salmonella stanley	0.1 mg/ ml	5-10 min	Jaquette et al., 1996
Staphylococcus aures	1 g/L	n/a	King et al., 1977
Streptococcus lactis	6 ppm	15 sec	Hays et al., 1967
Strep. dysgalactiae / Strep. uberis	n/a	n/a	Oliver et al., 1989
Yersinia enterocolitica	10 mg/L	120 sec	Paz et al., 1993
	500 ppm	n/a	Escudero et al., 1999
Antiviral activity			
Polio virus type I	0.03 mg/L	n/a	Thraenhart & Kuwert, 1975
Human adeno virus	0.3 - 0.6 mg/L	15-30 min	Carlson et al., 1976
	1 mg/L	n/a	Abad et al., 1994
Simian virus 40 / Kilham rat virus	n/a	n/a	Engelbrecht et al., 1980
Coxasackie virus CB3 & CB5	n/a	n/a	Jensen et al., 1980
Human immunodeficiency virus	50 ppm	2 min	Bloomfield et al, 1990
Human rotavirus	0.5 mg/L	n/a	Abad et al., 1994
Hepatitis A virus	n/a	n/a	Grabow et al., 1983
	0.5 - 1.0 mg/L	n/a	Abad et al., 1994
Antiparasitic activity			
Cryptosporidium parvum	0.4 mg/L	15 min	Peeters et al., 1989
Giardia sp.	n/a	n/a	Haas & Aturaliye, 1999

(n/a, data not available)

elevated (39 to 40°C) temperatures in filtered well water (pH 7.3) for their efficacy in inactivating *Legionella pneumophila* ATCC 33155 (Landeen et al., 1989). At room temperature, a contact time of at least 24 h was necessary for copper and silver (400 and 40 µg/L) to achieve a 3-$\log_{10}$ reduction in bacterial numbers. As the copper and silver concentration increased to 800 and 80 µg/L, the inactivation rate significantly increased from $K = 2.87 \times 10^{-3}$ to $K = 7.50 \times 10^{-3}$ ($\log_{10}$ reduction per minute). In water systems with and without copper and silver (400 and 40 µg/L), the inactivation rates significantly increased as the free chlorine concentration increased from 0.1 mg/L ($K = 0.397$ $\log_{10}$ reduction per min) to 0.4 mg/L ($K = 1.047$ $\log_{10}$ reduction per min). Compared to room temperature, no significant differences were observed when 0.2 mg of free chlorine per liter with and without 400 and 40 µg of copper and silver per liter was tested at 39 to 40°C. All disinfection systems, regardless of temperature or free chlorine concentration, showed increase inactivation rates when 400 and 40 µg of copper and silver per liter was added; however, this trend was significant only at 0.4 mg of free chlorine per liter. Orth and Mrozek (1989) evaluated the effectiveness of sodium hypochlorite against a number of pathogenic organisms, including *L. monocytogenes, Campylobacter jejuni*, and *Yersinia enterocolitica*. The results indicated that different organisms were destroyed by chlorine under practical use conditions of time, temperature, and concentration. However, certain strains displayed varying degrees of resistance to chlorine.

Campylobacter jejuni and closely related organisms are important bacterial causes of acute diarrheal illness. Both endemic and epidemic infections have been associated with consuming untreated or improperly treated surface water. Blaser et al. (1986) compared susceptibility of three *C. jejuni* strains and *E. coli* with standard procedures used to disinfect water. Inactivation of bacterial preparations with 0.1 mg of chlorine and 1.0 mg of monochloramine per liter was determined at pH 6 and 8 and at 4 and 25°C. Under virtually every condition tested, each of the three *C. jejuni* strains was more susceptible than the *E. coli* control strain, with greater than 99% inactivation after 15 min of contact with 1.0 mg of monochloramine per liter or 5 min of contact with 0.1 mg of free chlorine per liter. Results of experiments in which an antibiotic-containing medium was used suggest that a high proportion of the remaining cells were injured. An animal-passaged *C. jejuni* strain was as susceptible to chlorine disinfection as were laboratory-passaged strains. These results suggest that disinfection procedures commonly used for treatment of drinking water to remove coliform bacteria are adequate to eliminate *C. jejuni* and further correlate with the absence of outbreaks associated with properly treated water.

The effects of chlorine at varying pH, culture media and incubation temperatures on one type and two wild type strains of *Yersinia enterocolitica* were studied (Paz et al., 1993). Exposure to 1 and 5 mg/L did not diminish viability, even after prolonged exposure. A level of 10 mg/L was required to achieve a 5-log reduction in 120 s for the type strain and 80 s for the wild strains. There was an increase of more than 30% in the rate of disinfection with a 10°C rise, a remarkable increase in antimicrobial activity at pH 5-log reduction in 20 s, as well as marked neutralization of the effect in the presence of 0.1% peptone. Cells at exponential growth-phase were more susceptible than stationary-phase ones. Cells from liquid medium were more resistant than those from solid medium.

Johnson et al. (1997) demonstrated that *Helicobacter pylori* was readily inactivated by free chlorine and suggested the control of this gastric pathogen by disinfection practices normally employed in the treatment of drinking water.

The susceptibility of toxigenic *Aeromonas* spp. to free chlorine in drinking water supplies, and the influence of environmental temperature on the bactericidal activity of the oxidant, were evaluated (Sisti et al., 1998). The inactivation curves were characterized by an initial phase of rapid reduction of viable cells followed by a slow inactivation of bac-

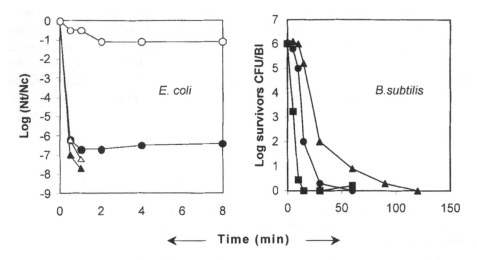

FIGURE 2. Antimicrobial activity of chlorine dioxide against *E.coli* and *B. subtilis*. Inactivation of *E.coli* cells in aqueous suspension at ClO_2 concentrations (mg/L): 0.7 (O); 1.4 (●); 2.8 (△); and 3.5 (▲) (redrawn from Foschino et al., 1998). Sporicidal kinetics of ClO_2 gas sterilization in ambient humidity - *B. subtilis* ssp. *niger* cells were exposed to 7 (▲); 15 (●) or 30 (■) mg/L of ClO_2 gas under square-wave conditions at 23°C (redrawn from Jeng & Woodworth, 1990).

teria. The effect of a chlorine compound is markedly influenced by water temperature. At summer water temperature (20°C), the efficacy of the chlorine concentrations tested was found two to three times lower compared to that found at a winter temperature (5°C). Resistance was higher in *A. hydrophila* than *A. caviae* and *A. sobria*, but all *Aeromonas* spp. were more susceptible than *E. coli*. Selective pressure with free chlorine did not produce *Aeromonas* cells with higher levels of chlorine resistance.

The efficacy of ClO_2 against cells of *E. coli* in aqueous suspension and adhering to the surfaces of stainless steel AISI 304 and PVC was evaluated (Foschino et al. 1998). The concentrations tested ranged from 0.7 to 14 mg/L; at 30s and 1, 2, 4, and 8 min exposure times. When the bacteria were suspended in water with 1.4 mg/liter of ClO_2, a 5-log reduction occurred within 30s (*FIGURE 2*); when cells were attached to the steel surface, similar inactivation took place only after 6 min with 7 mg/L or 4 min with 14 mg/L of ClO_2. A 5-log reduction was not obtained when organisms adhered to polyvinyl chloride (PVC).

Effect of chlorine and ozonation on *E. coli* cells resuspended in wastewater was compared (Arana et al., 1999). Selected chlorination and ozonation conditions produced a similar decrease in viability (2-2.5 log). Under such conditions, differences in membrane permeability and cell surface hydrophobicity were detected, depending on the disinfectant tested. No changes in cell surface hydrophobicity were observed after ozonation, however, approximately 95.5% of cells showed altered membrane permeability. The effect of chlorine was not linked to changes in membrane permeability. After chlorination, *E. coli* cells showed a tendency to aggregate. The degree of toxicity was unrelated to the effect on cellular activity.

B. Antiviral activity

Different types of enteric viruses show marked variations in the degree of resistance to free chlorine. Poliovirus, coxsackievirus, and some echoviruses are more resistant than coliform or enteric pathogenic bacteria (Dunham, 1977; Kabler et al., 1961).

Poliovirus type I is a highly stable pathogen in drinking-water and sewage lines. Thraenhart and Kuwert (1975) evaluated a comparative inactivation of poliovirus type I strains (wild type and attenuated) with chlorine and ozone treatments. Defined quantities of disinfectants were examined for viral inactivation in water without redox-potential (double-distilled water), water with low defined redox-potential (double-distilled water + KOH), previously chlorinated water with a residual chlorine content of 0.03 mg chlorine per liter (tap water) and water with a high redox-potential (well water from the drinking-water plant). Time-course studies were performed, with both chlorine and ozone, in order to evaluate the characteristics of the inactivation procedure. Chlorine and ozone demonstrated similar virus-disinfection profile. The initial rate and kinetics of virus disinfection were identical. The antiviral activities of both disinfectants were dependent on redox-potential and pH of the water. In contrast to the amount of free chlorine, the value of the oxidation-reduction potential (ORP) was found to be a criterion of virus inactivation (Carlson et al., 1976). For virus inactivation, higher ORP values and longer periods of contact was necessary than for the killing of bacteria. To ensure the inactivation of poliovirus in water contaminated with organic substances, an ORP of +780 mV (0.3-0.6 mg/L free chlorine) should be maintained for 15-30 min. Adenovirus also demonstrated similar resistance to inactivation. The rate of inactivation of poliovirus in water by chlorine is strongly influenced by the pH, which in turn influences the relative amounts of HOCl and OCl⁻ that are present and acting on the virus in the region of pH 6 to 10. The distribution of HOCl and OCl⁻ is influenced to a lesser extent by the addition of NaCl. The major part of the sharp increase in disinfection rate seen with this salt is thought to be due to its effect on the virus itself resulting in an increased chlorine sensitivity, especially at high pH (Sharp et al., 1980). ClO_2, bromine chloride and iodine were compared with chlorine as virucidal agents (Taylor & Butler, 1982). Under optimal conditions, all disinfectants were effective at low concentrations, but each disinfectant responded differently to acidity and alkalinity. Disinfection by chlorine was impaired by the presence of ammonia, but the other disinfectants retained much of their potency. Disinfection of poliovirus by iodine resulted in structural changes in the virions as seen by electron microscopy, but the other disinfectants were able to inactivate poliovirus without causing any apparent structural changes. Berg et al. (1989) reported a rapid inactivation of poliovirus by free chlorine at 5 °C and pH 9 in drinking water. Different ions from many salts present in the water seem to potentiate the virucidal activity of chlorine.

The kinetics of inactivation of six enteric viruses, simian virus 40 and Kilham rat virus by free available chlorine under controlled laboratory conditions was reported (Engelbrecht et al., 1980). Different virus types demonstrated a wide range of susceptibility to chlorine disinfection. The rate of inactivation was greater at pH 6 than at pH 10; however, the relative susceptibilities of the viruses were affected differently by a change in pH. The presence of potassium chloride also affected the susceptibility of viruses to chlorine.

The inactivation rates of coxsackievirus B3 (CB3) and B5 (CB5) by chlorine in dilute buffer at pH 6 were comparable and about half that of poliovirus (Mahoney) under similar conditions (Jensen et al., 1980). Purified CB3, like the poliovirus, aggregated in the acid range but not at pH 7 and above. Purified CB5 aggregated rapidly at all the pH values tested. Addition of 0.1 M NaCl to the buffer at pH 6 did not influence the aggregation of CB5 or the rate of chlorine action on either of the coxsackie-viruses, but at pH 10, the disinfection activity of OCl⁻ for both viruses was increased by 20-fold. Cesium chloride had a similar but smaller effect. KCl was the most active causing inactivation of OCl⁻ at pH 10 about equal to that of HOCl at pH 6.

In a study by Grabow et al. (1983). hepatitis A virus (HAV) and selected indicator organisms were mixed together in chlorine-demand-free buffers at pH 6, 8, or 10 and exposed to free chlorine residuals and the survival kinetics of individual organisms were compared. HAV was enumerated by a most-probable-number dilution assay, using PLC/PRF/5 liver cells for propagation of the virus and radioimmunoassay for its detection. At all pH levels, HAV was more sensitive than *Mycobacterium fortuitum*, coliphage V1 (representing a type of phage common in some sewage-polluted waters), and poliovirus type 2. Under certain conditions, HAV was more resistant than *E. coli*, *Streptococcus faecalis*, coliphage MS2, and reovirus type 3. It remained more resistant than SA-11 rotavirus. However, conditions generally specified for the chlorine disinfection of drinking-water supplies effectively inactivated HAV.

Using a quantitative suspension test method, the antiviral activity of sodium hypochlorite (NaOCl) and sodium dichloroisocyanurate (NaDCC) against human immunodeficiency virus (HIV) was investigated (Bloomfield et al., 1990). Viral suspensions were prepared containing 10^4-10^5 syncitial forming units/ml in 0.9% saline or 0.9% saline containing 10% v/v plasma to simulate clean and dirty conditions. A syncitial inhibition assay on C8166 lymphoblastoid line was performed to determine viral titer. Results indicate that satisfactory disinfection (3-4 log reduction in 2 min) could be achieved with NaDCC and NaOCl at concentrations of 50 ppm and 2500 ppm available chlorine for clean and soiled conditions, respectively. For treatment of blood spillage, the addition of NaDCC and NaOCl solutions (10,000 ppm) to equal volumes of contaminated blood (giving a final available chlorine concentration of 5000 ppm of blood) was sufficient to produce total kill within 2 min. For treatment of spillage material, chlorine-releasing powder formulations that produce higher available chlorine concentrations and achieve containment of spillage material offer an effective alternative.

The efficacy of copper and silver ions, in combination with low levels of free chlorine (FC), was evaluated for the disinfection of hepatitis A virus (HAV), human rotavirus (HRV), human adenovirus, and poliovirus (PV) in water (Abad et al., 1994). HAV and HRV showed little inactivation in conditions tested. PV showed more than a 4 $\log_{10}$ titer reduction in the presence of copper and silver combined with 0.5 mg of FC per liter or in the presence of 1 mg of FC per liter alone. Human adenovirus persisted longer than PV with the same treatments, although it persisted significantly less than HRV or HAV. The addition of 700 µg of copper and 70 µg of silver per liter did not enhance the inactivation rates after the exposure to 0.5 or 0.2 mg of FC per liter, although on some occasions it produced a level of inactivation similar to that induced by a higher dose of FC alone. Virus aggregates were observed in the presence of copper and silver ions, although not in the presence of FC alone.

Chlorine compounds have demonstrated cidal activity against bacteriophage. Sing et al. (1964a, b) compared the destructive activity of several germicidal aerosols against *Streptococcus cremoris* phage 144F. The levels of 500-2000 ppm available chlorine were highly effective against bacteriophage on a variety of surfaces. McCoy and Irwin (1974) examined the effect of various disinfectants on *E.coli* phage →X174 and found similar concentrations of chlorine to be effective.

Finichiu et al. (1986) studied the resistance of different microorganisms to chlorine to assure good drinking water quality and concluded that the resistance of bacteria to chlorine was much lower than that of viruses.

C. Antifungal activity

The fungicidal activity of chlorine has not been as extensively investigated as its bactericidal or sporicidal activity. Cheng and Levin (1970) studied the inactivation of *Aspergillus niger* conidiospores in the presence of 1-20 ppm chlorine. Fungal spores are slightly more resistant than vegetative bacteria since the germicide must mobilize farther to penetrate the coat of a fungal spore compared to a vegetative cell. Ver Kuilen and Marth (1980) reported the sporicidal effect of hypochlorite on *Aspergillus parasiticus*.

Brown and Wardowski (1986) reported the use of chlorine in citrus packaging houses for the reduction of decay on citrus fruits caused by *Penicillium digitatum* and *Geotrichum candidum*. The use of chlorine or ClO_2 under different plant conditions demonstrated an effective control of these organisms.

Although 0.4% chlorine for 2 min has been recommended for surface disinfection of food samples before direct plating for fungal enumeration, this procedure may not be adequate for highly contaminated products. The effectiveness of a range of chlorine solutions was investigated using barley samples artificially contaminated with four different concentrations of *Aspergillus flavus. A. niger, A. ochraceus, Eurotium repens, Penicillium brevicompactum, P. chrysogenum* and *Cladosporium cladosporioides* (Andrews et al., 1997). At initial contamination levels >10^4/g, 0.4% chlorine did not inactivate sufficient spores to produce less than 20% contamination. Of the test fungi, ascospores of *E. repens* were the most resistant to chlorine inactivation, whereas the conidia of *C. cladosporioides* were the most susceptible.

D. Sporicidal activity

The sporicidal activity of some of the organic chlorine compounds was reviewed by Trueman (1971). For trichloroisocyanuric acid, the sporicidal activity was significantly lower, although comparable, if not superior, results were obtained against sensitive vegetative bacteria compared to hypochlorite. Chloramine is also less sporicidal than hypochlorite (Odlaug, 1981).

Dichloroisocyanuric acid was found to be comparable, to sodium hypochlorite in the destruction of bacteriophage (Sing et., 1964a, 1964b). Trichloroisocyanuric acid has also been shown to inactivate bacteriophage at low levels (10 ppm) in aqueous solution (Fortney, 1958).

Sodium and potassium hypochlorites may be used on some polymeric and metallic membranes. Because of material restrictions, only a few types of membranes are compatible with chlorine, but even these require restrictions of temperature, concentration, pH, and exposure time. Bragulla and Liutner (1987) suggested the use of chlorine on different membranes and some of its limitations.

Russel (1990) showed that active chlorine compounds are not only bactericidal but also sporicidal. Hypochlorites in combination with sodium hydroxide were found more efficient sporicidal mixtures than by themselves.

E. Antiparasitic activity

Effect of disinfection of drinking water with ozone or chlorine dioxide on survival of *Cryptosporidium parvum* oocysts was reported (Peeters et al., 1989). Preliminary trials indicated that a minimum infection level of 1,000 oocysts (0.1-ml inoculum) per mouse was necessary to induce 100% infection. Treatment of water containing 10^4 oocysts per ml with 1.11 mg of ozone per liter (concentration at time zero [C0]) for 6 min totally eliminated the infectivity of the oocysts for neonatal mice. A level of 2.27 mg of ozone per liter (C0) was necessary to inactivate water containing 5×10^5 oocysts per ml within 8 min. Also, 0.4 mg of ClO_2 per liter (C0) significantly reduced infectivity within 15 min of contact, although some oocysts remained viable.

The effect of electroporation (pulses of high voltage electricity for very short duration) on the viability of *Giardia* cysts and *Cryptosporidium* oocysts, and on the viability of these organisms in the presence of free chlorine, combined chlorine, hydrogen peroxide and potassium permanganate, was examined (Haas & Aturaliye, 1999). While electroporation itself had only a minor effect on survival, the combination of electrical and chemical treatment produced superior inactivation, particularly with combined chlorine, hydrogen peroxide and potassium permanganate. This enhancement may provide a practical way of achieving enhanced inactivation of resistant protozoa by water disinfection processes.

V. APPLICATIONS

Among the various sanitizers used in the industry today, such as quaternary ammonium salts, ozone, iodophores, gluteraldehyde, and ethylene oxide, chloro-cides have the largest market-share. The key features of these compounds that have made them a popular choice are: 1) high antimicrobial efficacy, 2) low toxicity to humans, 3) versatile application, 4) ease of use, 5) low-cost, and 6) safe handling. Some of the drawbacks of chloro-cide use that are faced in the industry today are related to their irritant nature and their corrosive potential. Nevertheless, these problems could be circumvented by following appropriate directions for use and by adding anti-corrosion agents. The commonly used salt additives that diminish metal corrosion are nitrates, phosphates, disodium phosphate or sodium pyrophosphates. Corrosion can be prevented with 0.5 to 4 g/L of acid or neutral form of tartrates (Na, K or NH_4), or organic amino compounds such as amines, amides or heterocyclic bases (Masschelein, 1979).

Chlorine in its various forms is the most widely used chemical sanitizer in the food industry. In general, the organo-chlorines are slower acting bactericides than the inorganic forms, but they offer the advantage of stability and are relatively less irritating to personnel and less corrosive to equipment. Chlorine compounds are utilized: As adjuncts to water used for conveying raw food products, as well as water used for cooling of heat-sterilized cans. As sanitizing solution for food contact surfaces. In the treatment of raw meat, poultry, and fish to reduce microbial load and to extend shelf life.

The following *Section* describes the specific applications of chloro-cides in different industry sectors.

A. Dairy sanitation

The efficacy of chlorine disinfectant on teats artificially contaminated with a milk suspension of *Staphylococcus aureus* was reported (King et al., 1977). A solution of sodium hypochlorite with 40 g/L available chlorine was significantly more bactericidal than one containing 1 g/L available chlorine and more bactericidal than most other disinfectants tested. However, there were no distinguishable differences in efficacy between solutions containing 40 g/L and 10 g/L available chlorine and some of the iodophors containing 5 g/L available iodine. The addition of 190-416 g/L (15-33% v/v) glycerol significantly reduced the bactericidal properties of 3 iodophors (5 g/L available iodine), but soluble lanolin at approximately 20 g/L did not appear to lower the efficiency of sodium hypochlorite (45 g/L available chlorine) or of an iodophor (5 g/L available iodine).

Ozonated water and chlorinated sanitizer were compared for effectiveness against biofilms of milk spoilage bacteria (Greene et al., 1993). Stainless steel plates were incubated in UHT-pasteurized milk inoculated with pure cultures of either *Pseudomonas fluorescens* or *Alcaligenes faecalis*. Both ozonation and chlorination reduced bacteria populations by > 99% at initial cell densities in the range of approximately 1.24×10^5 to 8.56×10^5 cfu/cm^2 for *P. fluorescens* and 1.53×10^4 to 8.56×10^5 cfu/cm^2 for *A. faecalis* in milk films on stainless steel surfaces. Chlorine and monochloramine show an equal biocidal activity on lactose medium-grown *E. coli* and glycerol-ammonium nitrate medium-grown nonmucoid *Pseudomonas aeruginosa* biofilms (Samrakandi et al., 1997). In contrast, the effect of monochloramine is greater compared with that of chlorine on *E. coli* and mucoid *P. aeruginosa* biofilms grown in sucrose and glycerol-ammonium nitrate media, respectively. In these culture conditions, treatment with 25 mg monochloramine/L for 2 h reduced viable cells by 4.5 logs for *E. coli* and about 3 logs for mucoid *P. aeruginosa* while the similar treatment with chlorine reduced viable cells by 4.5 logs for *E. coli* and about 3 logs for mucoid *P. aeruginosa* while the similar treatment with chlorine reduced viable cells in these biofilms by 2.2 logs and 1 log, respectively. The decrease of chlorine disinfection efficacy on sucrose and glycerol-ammonium nitrate medium-grown biofilms is postulated to be linked to the higher polysaccharide production observed in these media. It seems likely that monochloramine produces a high leakage of material absorbing at 260 nm from sucrose medium-grown *E. coli* biofilm, which could indicate its better penetration into biofilms.

Prevention of bovine mastitis by a postmilking teat disinfectant containing chlorous acid and chlorine dioxide in a soluble polymer gel was reported by Oliver et al. (1989). A natural exposure study was conducted in a herd of 150 lactating dairy cows for 18 months to determine the effectiveness of chlorous acid and chlorine dioxide in a soluble polymer gel as a postmilking teat disinfectant for the prevention of bovine mastitis. Right quarters of cows were dipped in the experimental teat dip after milking machine removal. Left quarters were not dipped and served as within-cow negative controls. The experimental teat dip reduced *Staphylococcus aureus* infections 67.4%, *Streptococcus dysgalactiae* infections 63.8%, and *Streptococcus uberis* infections 27.8%. Overall efficacy of the chlorous acid and chlorine dioxide teat dip against major mastitis pathogens was 52.2%. The experimental teat dip reduced *Corynebacterium bovis* infections and coagulase-negative staphylococcal infections also by 45.8 and 38.7%, respectively. Overall efficacy against minor mastitis pathogens was 43.4%. Under conditions of this trial, the experimental teat dip containing chlorous acid and chlorine dioxide was effective in preventing new intramammary infections against a variety of mastitis pathogens.

B. Beef and poultry sanitation

The Food and Drug Administration has recently approved the use of acidified sodium chlorite solutions for the decontamination of red meat carcasses and poultry (FDA, 1998). Castillo et al., (1999) studied the effect of acidified sodium chlorite on beef carcasses that were inoculated with *Escherichia coli* O157:H7 and *Salmonella typhimurium*. The sodium chlorite solutions were activated with phosphoric or citric acid, and were applied on the carcasses as spray wash subsequent to a water wash. The initial numbers for both pathogens were reduced by 3.8 to 3.9 log cycles when a phosphoric acid activated sodium chlorite was used. The reduction in pathogen count was 4.5 to 4.6 log cycles when a citric acid activated sodium chlorite solution was used. The study concluded that acidified sodium chlorite sprays are very effective for decontaminating beef carcass surfaces. The study also demonstrated that the choice of acid used for the activation of sodium chlorite can have a significant influence on the biocidal activity of this sanitizer.

The efficacy of ClO_2 as an alternative sanitizing agent for hatching eggs was investigated to overcome health risks with formaldehyde fumigation (Patterson et al., 1990). Hatchability of chicken eggs was reduced when the eggs were dipped in the ClO_2 solutions for more than 5 minutes or in concentrations greater than 100 ppm chlorine. However, treatment of hatching eggs with a ClO_2 foam or fumigating with formaldehyde had no adverse effect on hatchability compared with untreated control eggs. Sanitizing soiled duck eggs with ClO_2 foam improved hatchability by more than 10% and hatch by more than 6% compared with untreated eggs. A novel method for assessing bactericidal potential of egg-sanitizing agents was developed. Using this technique, both chlorine dioxide foam and formaldehyde fumigation reduced the number of egg-contaminant bacteria inoculated on sterile chicken eggs compared with the number of bacteria on untreated eggs. These findings suggested that sanitizing hatching eggs with ClO_2 foam might be a viable alternative to fumigating with formaldehyde.

Three organic N-halamine compounds (combined halogen disinfectants) were compared with free chlorine (as calcium hypochlorite) as bactericides against *Salmonella typhimurium* and unidentified normal poultry bacterial flora under controlled conditions of pH, temperature, and halogen demand similar to those encountered in poultry processing (Williams et al., 1990). Two of the compounds (3-chloro-4,4-dimethyl-2-oxazolidinone and 1,3-dichloro-4,4,5,5-tetramethyl-2-imidazolidinone) at a concentration of 50 mg/L were found to cause a significant decline in viable organisms in less than 1 min at 48 °C, whereas a third compound (1-bromo-3-chloro-4,4,5,5-tetramethyl-2-imidazolidinone) was found to be less suitable.

C. Produce sanitation

According to the US National Advisory Committee on Microbiological Criteria for Foods, in the past two decades there has been a considerable increase in the consumption of fresh fruits and vegetables in the world (Roever, 1999). In the US, there has been a 432% increase in the average number of commodities offered for sale in supermarket produce departments. Subsequently, the occurrence of foodborne illnesses from these foods has also gone up. Therefore, effective compounds for the sanitation of produce are needed.

Chlorine has been used as a disinfectant in wash, spray and flume waters in the raw fruit and vegetable industry. To disinfect produce, chlorine is commonly used at concentrations of 50-200 ppm with a contact time of 1-2 minutes. Maximum solubility of chlorine is achieved in water at about 4°C. However, the temperature of the chlorinated water should ideally be at least 10°C higher than that of fruits or vegetables to achieve a positive temperature differential, thereby minimizing the uptake of wash-water through stem tissues (Bartz & Showalter, 1981; Zhuang et al., 1995) and open areas in the skin or leaves, whether due to mechanical assault or natural presence (e.g. lenticels and stomata). Elimination of uptake of wash-water that may contain microorganisms, including those that may cause human illnesses, should be considered as a critical control point in handling, processing and disinfection of raw fruits and vegetables.

The possible use of chlorinated water in packing-houses and during washing, cooling and transport for the purpose of controlling post-harvest diseases of fruits and vegetables has also been suggested (Eckert & Ogawa, 1988). The effects of chlorine concentration on aerobic microorganisms and fecal coliforms present on leafy salad greens was studied by Mazollier (1988). Populations of pathogens were markedly reduced with increased concentrations of chlorine to 50 ppm, but further increases in concentration to 200 ppm did not have a substantial additional effect. A standard procedure for washing lettuce leaves in tap water reduced population (ca. 10^7/g) of microflora by 92% (Adams et al., 1989). Inclusion of 10 ppm chlorine (pH 9.0) reduced the count by 97.8%, indicating that this concentration of chlorine in treatment water was only slightly more effective than using water with no chlorine. The pH adjustment from 9.0 to 4.5-5.0 with inorganic and organic acids resulted in a 1.5-4.0-fold increase in microbicidal effect. Increasing the washing time in hypochlorite solution from 5 to 30 minutes did not decrease numbers of microbes further, whereas extended washing in tap-water resulted in a reduction comparable to hypochlorite. Addition of 100 ppm of a surfactant (Tween 800) to hypochlorite washing solution enhanced lethality by enhancing surface contact but adversely affected sensory qualities of lettuce. Somers (1963) reported that wash-water with about 5-ppm chlorine reduced microbial populations on several fruits and vegetables by <90% from initial populations of 10^4-10^6 CFU/g.

Various studies have examined the efficacy of chlorine in killing bacterial pathogens inoculated onto the surface of raw vegetables. Dipping Brussels sprouts into a 200-ppm chlorine solution for 10 seconds decreased the viable *L. monocytogenes* cell counts (10^6 CFU/g) by about 100-fold (Brackett, 1987). However, dipping inoculated sprouts in sterile water containing no chlorine reduced the number of viable cells by about 10-fold. The maximum log_{10} reduction of *L. monoctogenes* on shredded lettuce and cabbage treated with 200 ppm chlorine for 10 minutes was reported to be 1.3-1.7 log_{10} CFU/g and 0.9-1.2 log_{10} CFU/g respectively (Zhang & Farber, 1996). Initial populations ranged from log_{10} 5.4 to 5.7 CFU/g. Reductions were greater when treatment was at 22°C than at 4°C. The above studies indicated that the antimicrobial efficacy of chlorine is generally greater when the temperature of the solution is higher than the temperature of the fruit or vegetable. Treatment was more effective at both temperatures for *L. monocytogenes* on lettuce than on cabbage, indicating that the antimicrobial activity is influenced by the nature of the vegetable being treated and/or the microflora it harbors. Numbers decreased only marginally with increased exposure time from 1 to 10 minutes, which supports the observations by Brackett (1987) that the action of chlorine against *L. monocytogenes* occurs primarily during the first 30 seconds of exposure. Nguyen-the and Carlin (1994) concluded that the reduction of *L. monocytogenes* from the surface of vegetables by chlorine is both unpredictable and limited.

The efficacy of chlorine treatment on inactivation of *Salmonella montevideo* on mature green tomatoes has been studied. Populations on the surface and in the stem core tissue were significantly reduced by dipping tomatoes for 2 minutes in a solution containing 60 ppm or 110 ppm chlorine respectively. However, treatment in a solution containing 320 ppm chlorine did not result in complete inactivation (Zhuang et al., 1995); the ineffectiveness of 100 ppm chlorine against *S. montevideo* inoculated into cracks in the skin of mature green tomatoes was demonstrated by Wei et al. (1995). Immersion of warm (26-40°C) tomatoes for 10 minutes or longer in cool (20-22°C) suspensions of bacteria resulted in infiltration of cells into the stem tissue (Bartz & Showalter, 1981). Uptake of bacterial cells was associated with a negative temperature difference between the water and the tomato, i.e. the water temperature was less than the tomato temperature. When the differential was shifted to a positive relationship, i.e. when the water temperature was higher than the tomato temperature, the extent of infiltration was reduced. A significantly higher number of *S. montevideo* cells were taken up by the core tissue when tomatoes at 25°C are dipped in suspensions at 10°C compared with the number of cells taken up by tomatoes dipped in suspensions at 25 or 37°C (Zhuang et al., 1995). Thus, the concentration of free chlorine reaching viable cells of *S. montevideo* that have infiltrated the core tissues is reduced to the point that lethality is substantially diminished. The uptake of pathogens and spoilage microorganisms by other fruits and vegetables during washing has not been investigated. However, infiltration of microbial cells due to a negative temperature differential between the water and the fruit or vegetable would appear to be possible.

The efficacy of chlorine and hot water treatments in killing *Salmonella stanley* inoculated onto alfalfa seeds was determined (Jaquette et al., 1996). Treatment of seeds containing 10^2 to 10^3 CFU/g in 100 µg/ml active chlorine solution for 5 or 10 min caused a significant reduction in population, and treatment in 290 µg/ml chlorine solution resulted in a significant reduction compared with treatment in 100 µg/ml chlorine. However, concentrations of chlorine of up to 1 mg/ml failed to result in further significant reductions. Treatment of seeds containing 10^1 to 10^2 CFU of *S. stanley* per g for 5 min in a solution containing 2 mg/ml chlorine reduced the population to undetectable levels. Treatment of seeds in water for 5 or 10 min at 54°C caused a significant reduction in the *S. stanley* population, and treatment at $\geq 57°C$ reduced populations to ≤ 1 CFU/g. However, treatment at $\geq 54°C$ for 10 min caused a substantial reduction in viability of the seeds. Treatment at 57 or 60°C for 5 min appears to be effective in killing *S. stanley* without substantially decreasing germination ability of seeds. Storage of seeds for 8 to 9 weeks at 8 and 21°C resulted in reductions in populations of *S. stanley* of about 1 $\log_{10}$ and 2 $\log_{10}$ CFU/g, respectively. The behavior of *S. stanely* on seeds during soaking, germination, sprouting, and refrigerated storage of sprouts was determined. An initial population of 3.29 $\log_{10}$ CFU/g increased slightly during 6 h of soaking, by about 10^3 CFU/g during a 24-h germination period, and by an additional 10 CFU/g during a 72-h sprouting stage. A population of 10^7 CFU/g of mature alfalfa sprouts was detected throughout a subsequent 10-day storage period at 5°C. These studies indicate that while populations of *S. stanley* can be greatly reduced, elimination of this organism from alfalfa seeds may not be reliably achieved with traditional disinfection procedures. If *S. stanley* is present on seeds at the initiation of the sprout production process, populations exceeding 10^7 CFU/g can develop and survive on mature sprouts exposed to handling practices used in commercial production and marketing.

In another study, alfalfa sprouts inoculated with a five-serovar mixture of *Salmonella* (*S. agona, S. enteritidis, S. hartford, S. poona,* and *S. montevideo*) were dipped in 200,500 or 2000 ppm chlorine solutions for 2 minutes (Beuchat, 1997). The pathogen was reduced by about 3.4 $\log_{10}$ CFU/g after treatment with 500 ppm chlorine and to an undetectable level (<1 CFU/g) after treatment with 2000 ppm chlorine. Chlorine treatment (2000 ppm) of cantaloupe cubes inoculated with the same *Salmonella* serovars resulted in <90% reduction in viable cells. The very high level of organic matter in the juice released from cut cantaloupe tissue apparently neutralized the chlorine prior to manifestation of its lethal effects.

Failure to maintain adequate chlorine in wash-water may lead to increases in microbial populations on fruits and vegetables. In a study designed to determine microbiological changes in fresh market tomatoes during packing operations, Senter and co-workers (1985) have observed that total plate counts and populations of *Enterobacteriaceae* were higher, compared to controls, on tomatoes washed in water containing an average of 114 ppm (range 90-140) chlorine; decreased were noted when tomatoes were treated in water containing 226 ppm chlorine (range 120-280). Recontamination of tomatoes occurred in the waxing operation as evidenced by increased total plate counts and mould populations.

Fruit and vegetable tissue components neutralize chlorine, rendering it inactive against microorganisms. The inaccessibility of HOCl to microbial cells in cracks, creases, crevices, pockets and natural openings in the skin undoubtedly also contributes to the overall lack of effectiveness of chlorine. The hydrophobic nature of the waxy cuticle on the surface of fruits and vegetables protects microbial cells from exposure to chlorine and, undoubtedly, other chemicals used as disinfectants that do not penetrate or dissolve these waxes. Surface-active agents such as detergents and ethanol reduce the hydrophobicity of fruit and vegetable skins as well as the surfaces of edible leaves, stems and flowers, but also tend to cause deterioration of sensory qualities (Adams et al., 1989; Zhang & Farber, 1996). Disinfectants that contain a solvent that would remove the waxy cuticle layer, and with it surface contaminants, without adversely affecting sensory characteristics would hold greater potential in reducing microbial populations on the surface of raw fruits and vegetables. Such disinfectants may be limited to use on fruits and vegetables that are to be further processed into juice or cut products, or on whole fruits, vegetables or plant parts that are destined for immediate consumption, since removal of cuticle material will also hasten deterioration of sensory quality. Application of such disinfectants at point of use diminishes the importance of this drawback.

Washing whole and cut produce by dipping or submerging in chlorinated water has a sanitizing effect, although reduction in microbial populations is minimal and is usually less than 100-fold. A study was undertaken by Beuchat and co-workers (1998) to evaluate the efficacy of a spray application of chlorine in killing Salmonella, *E. coli* 0157:H7, *L. monocytogenes*, yeasts and molds, and total aerobic mesophilic microorganisms on whole apples, tomatoes, and lettuce leaves. Inoculated produce was treated (sprayed and then soaked) with water (control) or solutions containing 200 or 2,000 ppm of chlorine for 0,1,3,5, or 10 min, rinsed with sterile water, and analyzed for populations (CFU/cm^2) of target microorganism. Compared to the control treatment, further reductions in numbers of pathogens of 0.35 to 2.30 log CFU/cm^2 were achieved by treatment with chlorine. Chlorine was generally more effective at 2,000 ppm than at 200 ppm.

Inactivation of microorganisms occurred essentially within 1 min after application of chlorine. These reductions are significant relative to populations of pathogenic microorganisms that may be present on produce. Spray application of chlorine to raw produce at food service or household levels may be a suitable, and more convenient, alternative to treatment by dipping or submersion.

Attachment of *E.coli* O157:H7 to lettuce leaf surface and bacterial viability in response to chlorine treatment was demonstrated by using confocal scanning laser microscopy (Seo & Frank, 1999). Sections of leaves (ca. 0.5 by 0.5 cm) were inoculated by submersion in a suspension of *E.coli* O157:H7 (ca. 10^7 to 10^8 CFU/ml) at 7°C overnight. Fluorescein isothiocyanate-labeled antibody was used to visualize the attached bacteria. *E.coli* O157:H7 was found attached to the surface, trichomes, stomata, and cut edges. Three-dimensional volume reconstruction of interior portions of leaves showed that *E. coli* O157:H7 was entrapped 20 to 100 µm below the surface in stomata and cut edges. Agar plate culturing and microscopic observation indicated that *E. coli* O157:H7 preferentially attached to cut edges, as opposed to the intact leaf surface. Dual staining with fluorescein isothiocyanate-labeled antibody and propidium iodide was used to determine viability of cells on artificially contaminated lettuce leaves after treatment with 20 mg/L chlorine solution for 5 min. Many live cells were found in stomata and on cut edges following chlorine treatment. *E. coli* O157:H7 did not preferentially adhere to biofilm produced by *Pseudomonas fluorescens* on the leaf surface. In contrast *to E.coli* O157:H7, *Pseudomonas* adhered to and grew mainly on the intact leaf surface rather than on the cut edges.

Escudero et al. (1999) studied the effect of temperature, storage and the efficacy of different free chlorine concentrations in washing solutions on the survival of *Yersinia enterocolitica* on surface of inoculated fresh tomatoes. At 6°C temperature, *Y. enterocolitica* did not proliferate during the first 4 days, but the counts increased until day 15. *Yersiniae* were able to grow on tomatoes stored at 22°C and 35°C. No detectable levels of viable cells were observed by using 500 ppm free chlorine washing solution.

Chlorine dioxide (ClO_2) has received attention as a disinfectant for fruits and vegetables, largely because its efficacy is less affected by pH and organic matter and it does not reach with ammonia to form chloramines, as do liquid chlorine and hypochlorites. In the United States, a maximum of 200 ppm ClO_2, is permitted for sanitizing equipment for fruit and vegetable processing. Chlorine dioxide is authorized for use in washing whole fresh fruits, vegetables, shelled beans, and peas with intact cuticles at a concentration not exceeding 5 ppm. The use of ClO_2 to disinfect fresh-cut fruits and vegetables was recently approved in the United States at up to 3 ppm (FDA, 1998).

Compared to the information that is available on the effectiveness of chlorine, as a disinfectant for fruits and vegetables, much less is known about the efficacy of ClO_2. Control of post-harvest fungal pathogens on pears (Spotts & Peters,1980) and protozoa in water (Chen *et al.*, 1985) has been studied. *In vitro* tests with conidia and sporangiospores of several fungal pathogens of apples and other fruits demonstrated >99% mortality from a 1-minute treatment in water containing 3 or 5 ppm ClO_2 (Roberts & Reymond, 1994). Longer exposure times were necessary to achieve similar mortalities by treatment with 1 ppm. Of the moulds tested, *Botrytis cinerea* and *Penicillium expansum* were least sensitive to ClO_2. Treatment of belts and pads in a commercial apple and pear packing-house with 14-18 ppm ClO_2, in a foam formulation resulted in significantly lower numbers of fungi. It was concluded that ClO_2 has desirable properties as a sanitizing agent for post-harvest decay management when residues of post-harvest fungicides are not desired or not allowed.

The efficacy of ClO_2 in preventing build-up of microorganisms in water for handling cucumbers and on the microorganisms present in fresh cucumbers has been studied (Costilow et al., 1984). At 2.5 ppm, ClO_2 was effective in killing microorganisms in wash-water but, at concentrations up to 105 ppm, failed to reduce the population of microorganisms present in or on fresh cucumbers. It was concluded that many microorganisms were so intimately associated with the cucumber fruit that they were unaffected by chlorine and ClO_2. Reina et al. (1995) evaluated the efficacy of ClO_2 in controlling microorganisms in recycled water in a spray-type hydro-cooler used to treat pickling cucumbers. Residual ClO_2 at 1.3 ppm was found to optimally control (2-6 $\log_{10}$ CFU/ml reduction) the number of microorganisms in the water. At 0.95 ppm ClO_2, the population was static, while at 2.8 and 5.1 ppm the odor became excessive. Populations of microorganisms on and in cucumbers were not greatly influenced by ClO_2, even at 5.1 ppm. It was concluded that the use of ClO_2 in water used to cool cucumbers seems to be an effective means of controlling microbial build-up, but that it has little effect on the viability of microorganisms on cucumbers.

The effectiveness of ClO_2, in killing *L. monocytogenes* inoculated onto the surface of shredded lettuce and cabbage leaves has been studied (Zhang & Farber, 1996). A 10-minute exposure of lettuce to 5 ppm ClO_2 caused a maximum reduction of 1.1 and 0.8 logs in numbers of *L. monocytogenes* at 4°C and 22°C, respectively, compared with a tap-water control. Similar results were obtained with cabbage. Thus, the maximum reduction in populations of *L. monocytogenes* on shredded lettuce and cabbage treated with ClO_2 at target concentrations up to 5 ppm was only slightly more than 90% which conforms with observations on the lack of effectiveness of ClO_2 in killing microorganisms on and in cucumbers (Costilow et al., 1984; Reina et al., 1995). As observed with chlorine and other disinfectant, microorganisms differ greatly in their sensitivity to ClO_2, and environmental conditions under which ClO_2 is applied can greatly influence efficacy. The effectiveness of ClO_2 in killing particular pathogens on specific types of fruits and vegetables deserves further research attention.

D. Seafood sanitation

The FDA amended on August 13, 1999 the food additive regulations (21 CFR Part 173.325) to provide the use of 40 to 50 ppm of acidified sodium chlorite for controlling the microbial population on seafood (FDA, 1999). The additive can be used as an antimicrobial agent to rinse, wash, thaw, transport, or store seafood. As stated above, the acidified sodium chlorite solutions are the source of chlorine dioxide that is generated when the acid is mixed with the sodium chlorite solution. The final solution is a mixture of ClO_2 and ClO_2^-. Lin et al. (1996) demonstrated that ClO_2/ClO_2^- mixtures are more effective in controlling pathogenic growth on seafood as compared to pure solutions of chlorine dioxide that are made by dissolving gaseous ClO_2 in water. Acidified sodium chlorite killed significant population of *E. coli*, *L. monocytogenes* Scott A and its streptomycin strain at 15, 10 and 7.5 ppm, respectively. Chlorine and ClO_2 solutions achieved the same results at 25 and 20 ppm, respectively.

E. Treatment of potatoes

Recently, EPA has granted exemption for the use of acidified sodium chlorite formulations on potatoes for the control of a fungal disease known as 'late blight'. This disease is caused by the organism *Phytophthora infestans*. Late blight causes black spots on potatoes and eventually leads to a secondary infection known as soft rot. In 1997, late blight resulted in a colossal loss of stored potatoes.

Significant reduction in spoilage is observed when 400 ppm of acidified sodium chlorite solutions is sprayed on the potatoes going into storage. Potatoes are typically stored under 90 to 95% relative humidity. For potatoes that are under higher risk of infection, it is recommended that 200 ppm of acidified sodium chlorite is misted through the humidification system. The amount and frequency of application via humidified system depends on the potential risk. The EPA approval is brand specific and is granted to only a few states in the US for a limited period. Studies are underway for a permanent approval that requires detailed information regarding the residual by-products.

F. Pharmaceutical preservative

A burgeoning use of chlorite salt is in the preservation of pharmaceutical formulations. Sodium chlorite is an excellent replacement for the old generation preservatives such as bezalkonium chloride or thimersol, which cause severe irritation of the skin in some patients. Several formulations are currently in the market that contain sodium chlorite preservative at concentrations between 50 to 100 ppm. It is speculated that the mechanism of preservation involves the conversion of chlorite ion into chlorine dioxide on demand. The demand is created by the presence of spoilage bacteria which generate acidic metabolites. The availability of acid triggers the production of ClO_2 form ClO_2^- as shown by *Equation 21*.

$$5ClO_2^- + 4\ H^+ \rightarrow 4\ ClO_2 + Cl^- + 2H_2O \qquad\qquad (Equation\ 21)$$

Similar mechanism is operative for the action of chlorine dioxide based mouthwash formulations that are commercially available. Upon contact with the metabolites of the odor-causing bacteria in the mouth, the chlorite ion converts into chlorine dioxide molecule that kills the bacteria. Moreover, because of its properties to readily react with sulfur compounds, ClO_2 neutralizes the mercaptans that are the primary source of odor. The commercially available mouthwash formulations contain approximately 1000 ppm of sodium chlorite solution.

G. Biofilm control

Microorganisms in a aquatic environment have a marked tendency to colonize and proliferate on submerged surfaces, such as vessels or pipes carrying a flow of water. Furthermore, bacteria exposed to surfaces could attach readily and irreversibly and form a thick matting of microcolonies. These adherent microorganisms secrete an extra cellular matrix (ECM) which is usually composed of a polysaccharide material called 'glycocalyx'. The matrix structure formed by the adherent bacteria and the glycocalyx are together referred to as 'biofilm'. Formation of biofilms have been observed on various surfaces including glass, rubber, stainless steel, different types metals and plastics (Mafu et al., 1990; Frank & Koffi, 1990; Herson et al., 1987). The type of surface markedly influences the behavior of the biofilm cells. The biofilm structure allows the inhabitant microorganisms to remain viable during periods of environmental stress such as nutrient limitation, adverse temperature, drying, etc. Moreover, microorganisms within the biofilm are reported to be less susceptible to antimicrobial agents (Anwar et al., 1990; Czechowski, 1990).

Ronner and Wong (1992) examined the efficacy of four types of detergent and non-detergent sanitizers to inactivate biofilm-associated *L. moncytogenes* and *S. typhimurium* on stainless steel and buna-n rubber (a gasket material commonly used in food processing). Bacterial biofilm populations on stainless steel were reduced 3-5 log by all the sanitizers, but those on rubber were resistant and were reduced less than 1-2 log. In contrast, planktonic (free-suspended) bacteria were reduced 7-8 log by these sanitizers. Chlorine and anionic sanitizers generally removed ECM from biofilms better than iodine and quaternary ammonium detergent sanitizers (*FIGURE 3*).

Attachment of unencapsulated *Klebsiella pneumoniae* grown in medium to glass with excess nutrients caused a 150-fold increase in chlorine disinfection resistance (LeChevallier et al., 1988). Other mechanisms which increased disinfection resistance included the age of the biofilm, bacterial encapsulation, and previous growth conditions (e.g., medium and temperature). These factors increased resistance to chlorine from 2- to 10-fold. Disinfection by free chlorine was affected by surfaces, age of the biofilm, encapsulation, and nutrient effects. Disinfection by monochloramine, however, was only affected by surfaces. LeChevallier et al. (1988b) further examined the inactivation of biofilm bacteria and characterized the interaction of biocides with pipe surfaces. Biofilm bacteria grown on surfaces of granular activated carbon particles, metal coupons, or glass microscope slides more resistant to hypochlorous acid (free chlorine, pH 7.0) than were unattached cells. In contrast, resistance of biofilm bacteria to monochloramine disinfection ranged from 2- to 100-fold more than that of unattached cells. The results suggested that, relative to inactivation of unattached bacteria, monochloramine was better able to penetrate and kill biofilm bacteria than free chlorine. For free chlorine, the data indicated that transport of the disinfectant into the biofilm was a major rate-limiting factor. Because of this phenomenon, increasing the level of free chlorine did not increase disinfection efficiency. Experiments where equal weights of disinfectants were used suggested that the greater penetrating power of monochloramine compensated for its limited disinfection activity. These studies showed that monochloramine was as effective as free chlorine for the inactivation of biofilm bacteria.

Transient chlorine concentration profiles were measured in biofilms during disinfection by use of a microelectrode (De Beer et al., 1994). The electrode had a tip diameter of ca. 10 μm and was sensitive to chlorine in the micromolar range. The biofilms contained *Pseudomonas aeruginosa* and *Klebsiella pneumoniae*. Chlorine concentrations measured in biofilms were typically only 20% or less of the concentration in the bulk liquid. Complete equilibration with the bulk liquid did not occur during the incubation time of 1 to 2 h. The penetration depth of chlorine into the biofilm and rate of penetration varied depending on the measurement location, reflecting heterogeneity in the distribution of biomass and in local hydrodynamics. The shape of the chlorine profiles, the long equilibration times, and the dependence on the bulk chlorine concentration showed that the penetration was a function of simultaneous reaction and diffusion of chlorine in the biofilm matrix. Frozen cross sections of biofilms, stained with a redox dye and a DNA stain, showed that the area of chlorine penetration overlapped with nonrespiring zones near the biofilm-bulk fluid interface. These data indicate that the limited penetration of chlorine

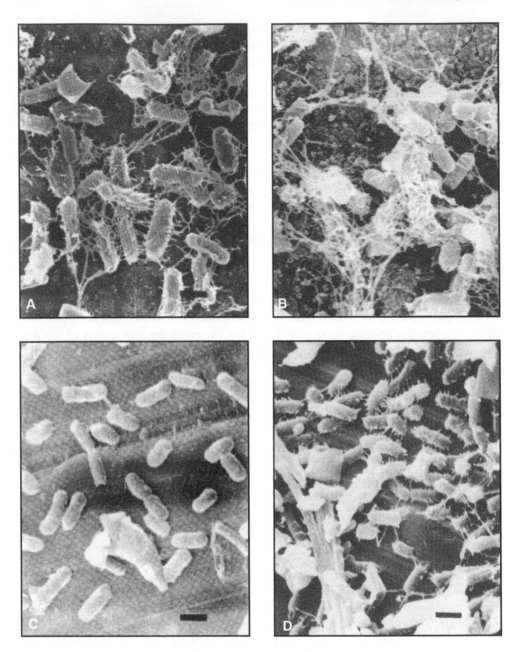

FIGURE 3. Sanitizer treatment of *Salmonella typhimurium* biofilm. Bar = 1 μm. (A) Bacteria with extra cellular matrix (ECM) on Buna-n rubber; (B) ECM remained on Buna-n rubber after chloro-cide treatment; (C) ECM removed from stainless steel after chloro-cide treatment; (D) ECM remained after iodine treatment. (Reproduced with permission from Ronner & Wong, 1993; and the Journal of Food Protection, IAMFES).

into the biofilm matrix is likely to be an important factor influencing the reduced efficacy of this biocide against biofilms as compared with its action against planktonic cells.

ClO_2 has several efficacy characteristics suited for microbiological control of biofilms in dispersing, dissolving and removing inhabitant microorganisms. Clark and Langley (1990) described a method for the removal of biofilms from submerged surfaces in an aqueous medium with an aqueous stabilized ClO_2.

VI. SAFETY AND TOLERANCE

Chlorine has been successfully used for the control of waterborne infectious diseases for nearly a century. In the 1970s, it was found that chlorine reacted with natural organic matter present in surface waters to produce disinfection by-products (DBP).

Over the past few years, concerns regarding the use of chlorine as a drinking water disinfectant have received a significant amount of attention (Rook, 1974; Bellar et al., 1974; Maugh, 1981). The concern is based upon the finding that chlorine reacts with the organic materials (principally humic acids) present in water, with the resultant formation of trihalomethane (THM) compounds. Regulatory agencies have expressed concern over the adequacy of safety data to support the use of chlorine, especially when employed for direct food contact (meat, poultry, and shellfish).

THM are organic contaminants found in drinking water that includes chloroform, bromodichloromethane, dibromochloromethane, and bromoform. Since drinking water could possibly be contaminated with THM and that THM produce carcinogenic effects in test animals (mice and rats), the Environmental Protection Agency (EPA) in 1979 promulgated regulations limiting the permissible levels of THM in drinking water to a maximum of 0.10 mg/L. The act requires water companies to monitor for THM levels in drinking water, and if present, these chemicals must be removed by a specific treatment because exposure to THM in drinking water poses a risk to human health.

Concern focused initially on the THM, but a variety of DBPs are now known to result from chlorination. Chlorination of drinking water has been one of the most effective public health measures ever undertaken. There are a number of alternatives to chlorination that are in active use in many parts of the world, but the risks associated with their by-products are even less well established than for chlorination. Moreover, the uses of these alternatives vary in their effectiveness and some require greater sophistication in their application. This can mean less protection to public health because of inappropriate application and control. Therefore, hazards associated with the use of such a clearly beneficial process, as chlorination, must be carefully considered not only in an absolute sense, but also in the context of alternative approaches for producing safe drinking water. The important question is whether the hazards associated with by-products have been sufficiently well established to warrant regulations that will undoubtedly have both positive and negative impacts on the public health.

Following the November 1979 amendment to the National Interim Primary Drinking water Regulations, many utilities in the US needed to change their disinfection practices to comply with the 0.1 mg/L maximum contaminant level for THMs. In 1983, an amendment to the THM regulation listed ClO_2 as an alternative or supplemental oxidant-disinfectant as one of the most suitable treatment technologies for control of THMs

(Rav Acha et al., 1983). The total organic halides formed with ClO_2 was only 1.25% of that formed with chlorine under the same reaction conditions (Fleischacker & Randtke, 1983). Therefore, ClO_2 is widely used in the US, Canada and Europe to control tastes, odors and microbiological pollution. Fukayama et al. (1986) reviewed the potential health risks associated with the use of chlorine with some food products. Solutions containing chlorine should not be combined with acid solutions or adjusted to low pH because of the formation of highly toxic chlorine gas.

Human exposure: An epidemiological study was conducted in a rural village on 198 persons who were exposed for 3 months to drinking water disinfected with ClO_2 (Michael et al., 1981). A control population of 118 non-exposed persons was also studied. Pre-exposure hematological and serum chemical parameters were compared with test results after 115 days of exposure. Statistical analysis of the data failed to identify any significant exposure-related effects. The occupational exposure limit (OSHA ceiling) for ClO_2 in the United States is 0.1 ppm expressed as Permissible Exposure Limit (PEL, 8 hours) and 0.3 ppm expressed as Short Term Exposure Limit (STEL, 15 minutes).

Animal trials: Exon et al. (1987) performed immunotoxicological evaluation of chlorine-based drinking water disinfectants. Male Sprague-Dawley rats were exposed to chlorine-based disinfectants in the drinking water from weaning to 12 weeks of age, at which time they were terminated and assessed for immune competence. Chlorine-based drinking water disinfectants sodium hypochlorite (5, 15 and 30 ppm) and monochloramine (9, 19 and 38 ppm) were used. Parameters of immunity measured were spleen and thymus weights, antibody production, delayed-type hypersensitivity (DTH) reactions, natural killer cell (NKC) cytotoxicity, oxidative metabolism response (i.e chemiluminescence-CL) and phagocytosis by macrophages, and production of 2 immunoregulatory cytokines, interleukin 2 (IL2) and prostaglandin E2 (PGE2). Significant (P less than or equal to 0.05) reductions of spleen weight, DTH reactions, and oxidative metabolism by macrophages were observed only in groups of rats exposed to high levels (30 ppm) of sodium hypochlorite, while PGE2 production was elevated. Rats exposed to the higher doses of monochloramine had reduced spleen weights (38 ppm), decreased antibody synthesis (9 and 19 ppm) and augmented PGE2 production (19 and 38 ppm). These results extend the earlier observations of others that macrophage function of laboratory rodents may be impaired by exposure to high concentrations of chlorinated drinking water. Furthermore, the function of other major populations of immunocytes and types of immune responses may also be altered following subchronic exposure to high concentrations of chlorinated drinking water. These types of effects on the immune system are a previously unrecognized potential side effect of the ubiquitous practice of disinfection of water with chlorine compounds. Alteration of immune function of chlorine-based disinfectant-exposed rats in this study was only evident at relatively high doses, and only selected immune responses were altered. Therefore, these chlorine-based disinfectants are not particularly strong immunodepressants.

Dichloroacetonitrile (DCAN), a by-product of drinking water disinfection formed by reaction of chlorine with background organic materials, was evaluated for its developmental effects in pregnant Long-Evans rats (Smith et al., 1989). Animals were dosed by oral intubation on Gestation Days 6-18 (plug = 0) with 0, 5, 15, 25, or 45 mg/kg/day. Tricaprylin was used as a vehicle. The highest dose tested (45 mg/kg) was lethal in 9% of the dams and caused resorption of the entire litter in 60% of the survivors. Embryolethality averaged 6% per litter at the low dose and 80% at the high dose and was statistically significant at 25 and 45 mg/kg/day. The incidence of soft tissue malformations was dose related and was statistically significant at doses toxic to the dam (45 mg/kg). These anomalies were principally in the cardiovascular (interventricular septal defect, levocardia, and abnormalities of the major vessels) and urogenital (hydronephrosis, rudimentary bladder and kidney, fused ureters, pelvic hernia, cryptorchidism) systems. The frequency of skeletal malformations (fused and cervical ribs) was also dose related and significantly increased at 45 mg/kg. The no-observed-adverse-effect dose for toxicity in pregnant Long-Evans rats was established by statistical analysis to be 15 mg/kg/day.

VII. SUMMARY

Chloro-cides, in a variety of forms, are the most common sanitizing agents used in the industry today. Numerous studies have reported their potent germicidal spectrum. A long history of usage has established their acceptance in the field as safe and relatively non-toxic products. Chloro-cides are broad-spectrum antimicrobial compounds and oxidative damage is their fundamental mechanism for biocidal action. The biocidal component of most chloro-cides is hypochlorous acid. Currently, the compounds that produce hypochlorous acid are subject to scrutiny because of risks to form undesired by-products such as THM.

With major advancements in chemical technology, ClO_2 has emerged as safe and effective class of chloro-cides. The use of ClO_2 as a disinfectant and sanitizer is growing due to its high antimicrobial efficacy and absence of any hazardous by-products. It is extensively used for the treatment of drinking water without adding chlorine's taste to the water. ClO_2 is also a popular choice of antimicrobial used for the sanitation of produce, poultry and beef. It is highly effective for the prevention of late blight on potatoes. The recent FDA approval for the use of ClO_2 on seafood had provided a useful tool to the food-processing industry for controlling persisting microbial pathogens such as *E. coli, Salmonella sp.* and *L. monocytogens*.

Oxidative damage has been nature's way of killing microorganisms in phagosomes and an integrated mechanism for antigen processing in various life forms. Various enzyme systems such as oxidases and halide-based catalysts contribute both additive and synergistic effects to numerous natural antimicrobials in the innate host defense. Thus, chloro-cides could be considered as one of the natural hurdle mechanism in the physiological milieu. As potent milieu-antimicrobials, chloro-cides could open a powerful hurdle-approch to food safety if integrated with other natural food antimicrobial systems.

Acknowledgments: The authors wish to thank Dr. Amy Wong, Food Research Institute, University of Wisconsin, Madison, WI for the SEM figure, and Bio-Cide International, Inc., Norman, Oklahoma for supporting the preparation of this manuscript.

VIII. REFERENCES

1.   Abad, F.X., Pinto, R.M., Diez, J.M., and Bosch, A. 1994. Disinfection of human enteric viruses in water by copper and silver in combination with low levels of chlorine. *Appl. Environ. Microbiol.* 60:2377-2383.

2.   Adams, M.R., Hartley A.D.,and Cox, L.J. 1989. Factors affecting the efficiency of washing procedures used in the production of prepared salads. *Food Microbiol.* 6:69-77.

3.   Andrew, S., Pardoel, D., Harun, A., and Treloar, T. 1997. Chlorine inactivation of fungal spores on cereal grains. *Int. J. Food Microbiol.* 35:153-162.

4.   Anwar, H., Dasgupta, M.K., and Costerton, J.W. 1990. Testing the susceptibility of bacteria in biofilms to antibacterial agents. *Antimicrob. Agents Chemother.* 34:2043-2046.

5.   Arana, I., Santorum, P., Muela, A., and Barcina, I. 1999. Chlorination and ozonation of waste-water:comparative analysis of efficacy through the effect on *Escherichia coli* membrances. *J. Appl. Microbiol.* 86:883-888.

6.   Aston, R.N. 1947. Chlorine Dioxide use in plants on Niagara border. *J. Am. Water Works Assoc.,* 40: 687-690.

7.   Baker, J.C. 1926. Chlorine in sewage and waste dispoal. *Can. Eng. Water Sew.* 50:127.

8.   Baldwin, R.T., 1927. History of the Chlorine Industry, *J. Chem. Ed.* 4:313-319.

9.   Bartz, J.A., and Showalter, R.K. 1981. Infiltration of tomatoes by bacteria in aqueous suspension. *Phytopathology.* 71:515-518.

10.   Bellar, T.A., Lichtenberg, J.J., and Kroner, R.C. 1974. The occurrence of organochlorides in chlorinated drinking water. *J. Am. Water Works Assoc.* 66:703-706.

11.   Benarde, M.A., Israel, B.M., Olivieri, V.P., and Granstrom, M.L. 1965. Efficiency of chlorine dioxide as a bacteriocide. *Applied Microbiology,* 13: 776-780.

12.   Berg, G., Sanjaghsaz, H., and Wangwongwatana, S. 1989. Potentiation of the virucidal effectiveness of free chlorine by substances in drinking water. *Appl. Environ. Microbiol.* 55:390-393.

13.   Berg, J. D., Roberts, P. V., and Matin, A. 1986. *J. Appl. Bact.* 60:213-220.

14.   Bergman, D., and Hoff, J.C. 1984. Inactivation of simian rotavirus SA 11 by chlorine, chlorine dioxide, and monochloramine. *Appl. Environ. Microbiol.* 48 (2)317-323.

15.   Berman, D., Rice, E.W., and Hoff, J.C. 1988. Inactivation of particle-associated coliforms by chlorine and monochloramine. *Appl. Environ. Microbiol.* 54:507-512.

16.   Bernarde, M.A., Israel, B.M., Olivieri, V.P., and Grandstram, M.L. 1965. Efficiency of chlorine dioxide as a bactericide. *App. Microbiol.* 13:776-780.

17.   Bernarde, M.A., Snow, W.B., and Olivieri, V.P. 1967a. Chlorine dioxide disinfection temperature effects. *J. Appl. Bacteriol.* 30:159-167.

18.   Bernarde, M.A., Snow, W.B., Olivieri, V.P., and Davidson, B. 1967b. Kinetics and mechanism of bacterial disinfection by chlorine dioxide. *Appl. Microbiol.* 15:257-265.

19.   Beuchat, L.R., Nail, B.V., Adler, B.B., and Clavero, M.R. 1998. Efficacy of spray application of chlorinated water in killing pathogenic bacteria on raw apples, tomatoes, and lettuce. *J. Food. Prot.* 61:1305-1311.

20.   Blaser, M.J., Smith, P.F., Wang, W.L., and Hoff, J.C. 1986. Inactivation of *Campylobacter jejuni* by chlorine and monochloramine. *Appl. Environ. Microbiol.* 51:307-311.

21.   Bloomfield, S.F., and Miles, G.A. 1979a. The antimicrobial properties of sodium dichloroisocyanurate and sodium hypochlorite formulations. *J. Appl. Bacteriol.* 46:65-73.

22.   Bloomfield, S.F., and Miles, G.A. 1979b. The relationship between residual chlorine and disinfection capacity of sodium hypochlorite and sodium dichloroisocyanurate solutions in the presence of *Escherichia coli* and of milk. *Microb. Lett.* 10:33.

23.   Bloomfield, S.F., Smith-Burchnell, C.A., and Dalgleish, A.G. 1990. Evaluation of hypochlorite-releasing disinfectants against the human immunodeficiency virus (HIV). *J. Hosp. Infect.* 15:273-278.

24.   Bolton, K.J., Doss, C.E.R., Mead, G.C., and Waits, W.M. 1988. Chlorine resistance of strains of *Staphylococcus aureus* isolated form poultry processing plants. *Appl. Microbiol.* 6:31-34.

25.   Brackett, R.E. 1987a. Antimicrobial effect of chlorine on *Listeria monocytogenes. J. Food Prot.* 50:999-1003.

26.   Bragulla, S., and Lintner, K. 1987. Basics in cleaning and disinfection for ultrafiltration, reverse osmosis and electro-dialysis. *Henkel Technical Information.* March, 1-4.

27.   Brown, G.E., and Wardowski, W.F. 1986. Use of chlorine and chlorine dioxide in Florida citrus packing houses to reduce inoculum of decay pathogens. *Citrus Ind.* 67(6):48-56.

28.   Bull, R.J. 1980. Health effects of alternative disinfectants and their reaction products. *J. Am. Water Works Assoc.* 72:299-303.

29. Caldwell, D.R. 1990. Analysis of biofilm formation: confocal laser microscopy and computer image analysis. *Abstract of Papers presented at the 77th Annual Meeting of the International Assoc. of Milk, Food and Environ. Sanitarians, Inc.* p.11.

30. Callerame, J. 1989a. Process for production of chlorine dioxide, *US Patent* 4,874,489.

31. Callerame, J. 1989b. Process for production of chlorine dioxide, *US Patent* 4,877,500.

32. Carlson, S., Hasselbarth, U., and Sohn, F.W. 1976. Studies on virus inactivation by chlorine during water disinfection. *Zentralbl Bakteriol* [Orig B] 162:320-329.

32a. Castillo, A. Lucia, M. L., Kemp, G.K., Acuff, G.R. 1999. Reduction of *Escherichia coli O157: H7* an *Salmonella typhimurium* on beef carcass surfaces using acidified sodium chlorite. *J. Food Prot.* 62: 580-584.

33. Chang, S.L. 1944. Destruction of microorganisms. *J. Am. Water Works Assoc.* 36:1192-1207.

34. Chemical & Engineering News, June 1999.

35. Chen, Y.S.R., Sproul, O.J., and Rubin, A.J. 1985. Inactivation of *Naegleria gruberi* cysts by chlorine dioxide. *Water Res.* 19:783-790.

36. Cheng, M.K.C., and Levin, R.E. 1970. Chemical destruction of *Aspergillus niger* conidiospores. *J. Food Sci.* 35:62-66.

37. Clark, J.B., and Langley, D.E. 1990. *United State Patent:* dated May 29, Number: 4,929,365.

38. Costilow, R.N., Uebersax, M.A., and Ward, P.J. 1984. Use of chlorine dioxide for controlling microorganisms during handling and storage of fresh cucumbers. *J. Food Sci.* 49:396-401.

39. Czechowski, M.H. 1990. Bacterial attachment to Buna-n gaskets in milk processing equipment. *Aust. J. Dairy Technol.* 45:113-114.

40. Dakin, H.D., and Cohen, J.B. 1916. On Chloromine antiseptics, *Br. Med. J.*, 1, 160-162.

41. De Beer, D., Srinivasan, R., and Stewart, P.S. 1994. Direct measurement of chlorine penetration into biofilms during disinfection. *Appl. Environ. Microbiol.* 60:4339-4344.

42. Dunham, W.B. 1977. Virucial agents. In *Disinfection, Sterilization and Preservation*, 2nd ed., edited by S.S Block, pp. 426-441. Philadelphia: Lea and Febiger.

43. Dychdala, G.R. 1983. Chlorine and chlorine compounds. In *Disinfection, Sterilization and Preservation*, 3rd ed., edited by S.S Block, pp. 157-182. Philadelphia: Lea and Febiger.

44. Dychdala, G.R. 1991. Chapter 7: Chlorine and Chlorine compounds In *Disinfection, Sterilization, and Preservation*, 4th edition, ed. S.S. Block, pp. 131-151, Philadelphia: Lea & Febiger.

45. Eckert, C.T., and Ogawa, J.M. 1988. The chemical control of post- harvest diseases: deciduous fruits, berries, vegetables and root/tuber crops. *Annu. Rev. Phytopathol.* 26:433-469.

46. El-Kest, S.E., and Marth, E.H. 1988. Inactivation of *Listeria monocytogenes* by chlorine. *J. Food Prot.* 51:520-524.

47. Engelbrecht, R.S., Weber, M.J., Salter, B.L., and Schmidt, C.A. 1980. Comparative inactivation of viruses by chlorine. *Appl. Environ. Microbiol.* 40:249-256.

48. Environmental Protection Agency. 1979. Acid-fast bacteria and yeasts as indicators of disinfection efficiency. EPA-600/2-79-091. U.S. Environmental Protection Agency, Cincinnati.

49. Escudero, M.E., Velazquez, L., DiGenaro, M.S., de Cortinez, Y.M., and de Guzman, A.M. 1999. Elimination of *Yersinia enterocolitica* by chlorine on fresh tomatoes. *Cent. Eur. J. Public Health* 7:24-26.

50. Exon, J.H., Koller, L.D., O'Reilly, C.A., and Bercz, J.P. 1987. Immunotoxicologic evaluation of chlorine-based drinking water disinfectants, sodium hypochlorite and monochloramine. *Toxicology* 44:257-269.

51. Finichiu, M., Strainer, M., Cotor, F., and Diaconu, R. 1986. Study of resistance to chlorine of certain microorganisms from water, with a view to improving sanitary and bacteriological criteria for evaluating qualities of drinking water. *Igiena* 35:93-100.

52. Fleischaker, S.J., and Randtke, S.J. 1983. Formation of organic chlorine in public water supplies. *J. Am. Wat. Works. Assoc.* 75:132-138.

53. Food and Drug Administration, Department of Health and Human Services. 1999. Secondary direct food additives permitted in food for human consumption. *Fed Reg.* 64: 44122-44123.

54. Food and Drug Administration. 1967. Permitted uses of chlorine in food processing. Invitation to submit data information and reviews. *Fed. Reg.* 41:27856-27857.

54a. Food and Drug Administration, Department of Health and Human Services. 1998. Secondary direct food additives permitted in food for human consumption. *Fed Reg.* 63: 11118-11119.

54b. Food and Drug Administration, Department of Health and Human Services. 1998. Secondary direct food additives permitted in food for human consumption. *Fed Reg.* 63: 38746-38747.

55. Fortney, K.F. 1958. The germicidal properties of trichlorocyanuric acid. Master's Thesis, Oregon State University, Corvallis.

56. Foschino, R., Nervegna, I., Motta, A., and Galli, A. 1998. Bactericidal activity of chlorine dioxide against *Escherichia coli* in water and on hard surfaces. *J. Food. Prot.* 61:668-672.
57. Frank, J.F., and Koffi, R.A. 1990. Surface-adherent growth of *Listeria monocytogenes* is associated with increased resistance to surfactant sanitizers and heat. *J. Food Prot.* 53:550-554.
58. Fukayama, M.Y., Tan, H. Wheeler, W.B., and Wei, C.I. 1986. Reactions of aqueous chlorine and chlorine dioxide with model food compounds. *Environ. Health Perspect.* 69:267-274.
59. Garg, N., Churey, J.J., and Splittstoesser, D.F. 1990. Effect of processing conditions on the microflora of fresh-cut vegetables. *J. Food Prot.* 53:701-703.
60. Gates, D. 1998. The Chlorine Dioxide Handbook, Water Disinfection Series, American Water Works Association: Denver.
61. Geiger, K.H., and Moloney, P.J. 1952. Enhanced effectiveness of chlorination. *Can. J. Pub. Health,* 43:359-367.
62. Gilbert, G., Kieffer, R. G., and Rosenblatt, D. H. 1972. The chemistry of chlorine dioxide. *Prog. Inorg. Chem.* 15: 201-286.
63. Glaze, W.H., Schep, R., Chauncey, W., Ruth, E.C., Zarnoch, J.J., Aieta, E.M., Tate, C.H., and McGuire, M.J. 1990. Evaluating oxidants for the removal of model taste and odor compounds from a municipal water supply. *J. AWWA.* 82 79-84.
64. Gordon, G., Adam, L.C., Bubnis, B.P., Kuo, C., Cushing, R.S., and Sakaji, R.H. 1997. Predicting liquid bleach decomposition. *J. AWWA.* 89: 142-149.
65. Grabow, W.O., Gauss-Muller, V., Prozesky, O.W., and Deinhardt, F. 1983. Inactivation of hepatitis A virus and indicator organisms in water by free chlorine residuals. *Appl. Environ. Microbiol.* 46:619-624.
66. Greene, A.K., Few, B.K., and Serafini, J.C. 1993. A comparison of ozonation and chlorination for the disinfection of stainless steel surfaces. *J Dairy Sci* 76:3617-3620.
67. Greene, D.E., and Stumpf, P.K. 1946. The mode of action of chlorine. *J. Am. Water Works Assoc.* 38:1301.
68. Griese, M.H., Kaczur, J.K., and Gordon, G. 1992. Combining methods for the reduction of oxychlorine residuals in drinking water. *J. AWWA* 84: 69-77.
69. Guiteras, A.F., and Schmelkes, F.C. 1934. The comparative action of sodium hypochlorite, chloramine-T, and azochloramid on organic substrates. *J. Biol. Chem.* 107:235-239.
70. Haas, C.N., and Aturaliye, D. 1999. Semi-quantitative characterization of electroporation-assisted disinfection processes for inactivation of Giardia and Cryptosporidium. *J. Appl. Microbiol.* 53:1178-1180.
71. Haas, C.N., and Engelbrecht, R.S. 1980. Physiological alterations of vegetative microorganisms resulting from chlorination. *J. Water Pollut. Control Fed.* 52:1976.
72. Hadfield, W.A. 1957. Chlorine and Chlorine compounds. In *Antiseptics, Disinfectants, Fungicides, Chemical and Physical Sterilization*. 2nd edition, pp 558-580. Philadelphia: Lea & Febiger.
73. Harakeh, M.S., Berg, J.D., Hoff, J.C., and Matiac A. 1958. Susceptibility of chemostat-grown *Yersinia enteroclitica* and *Klebiella pneumoniae* to chlorine dioxide. *Appl. Environ. Microbiol.* 49: 67-72.
74. Hays, H.A., Elliker, P.R., and Sandine, W.E. 1967. Microbial destruction by low concentrations of hypochlorite and iodophor germicides in alkaline and acidified water. *Appl. Microbiol.* 15:575-581.
75. Hekmati, M., and Bradley, R.L. 1979. Effect of milk constituents on the persistence of sodium hypochlorite sanitizer. *J. Dairy Sci.* 62:47.
76. Herson, D.S., McGonigle, B., Payer, M.A., and Baker, K.H. 1987. Attachment as a factor in the protection of *Enterobacter cloacae* from chlorination. *Appl Environ Microbiol* 53:1178-1180.
77. Ito, K.A., and Seeger, M.L. 1980. Effects of germicides on microorganisms in can cooling water. *J. Food Prot.* 43:484-487.
78. Jaquette, C.B., Beuchat, L.R., and Mahon, B.E. 1996. Efficacy of chlorine and heat treatment in killing *Salmonella stanley* inoculated onto alfalfa seeds and growth and survival of the pathogen during sprouting and storage. *Appl. Environ. Microbiol.* 62:2212-2215.
79. Jensen, H., Thomas, K., and Sharp, D.G. 1980. Inactivation of coxsackieviruses B3 and B5 in water by chlorine. *Appl. Environ. Microbiol.* 40:633-640.
80. Johns, C.K. 1934. Germicidal power of sodium hypochlorite. *Ind. Eng. Chem.* 26:787.
81. Johns, C.K. 1948. Influence of organic matter on the germicidal efficacy of quaternary and hypochlorite compounds. *Can. J. Res.* 26:91.
82. Johns, C.K. 1954. Iodophors as sanitizing agents. *Can. J. Technol.* 32:71-77.
83. Johnson, C.H., Rice, E.W., and Reasoner, D.J. 1997. Inactivation *of Helicobacter pylori* by chlorination. *Appl. Environ. Microbiol.* 63:4969-4970.
84. Julien, A.P., and Roger, S.M., 1959. Production of chlorine dioxide. *US Patent 2,881,052.*
85. Junli, H., Li W., Nenqi, R., Li, L.X., Fun, S.R., and Guanle, Y. 1997. Disinfection effect of chlorine dioxide on viruses, algae and animal planktons in water *Wat. Res.* 31:455-460.

86. Kabler, P.W., Clarke, N.A., Berg, G., and Chang, S.L. 1961. Virucidal efficacy of disinfectants in water. *Public Health Rep.* 76:565.

87. Kaczur, J.J., and Cawlfield, D.W. 1993. Chlorous acid, chlorites and chlorine dioxide (ClO₂, HClO₂). *Kirk-Othmer Encyclopedia of Chemical Technology,* 4ᵗʰ ed., pp. 968-997. New York:John Wiley.

88. Kereluk, K., and Borisenok, W.S. 1983. The antimicrobial activity of monochloramine, *Dev. Ind. Microbiol.* 24:401-408.

89. King, J.S., Neave, F.K., and Westgarth, D.R. 1977. Disinfection properties of some bovine teat dips. *J. Dairy Res.* 44:47-55

90. Knox, W.E., Stumpf, P.D., Green, E.D., and Auerbach, V.H. 1948. The inhibition of sulfhydryl enzymes as the basis of the bactericidal action of chlorine. *J. Bacteriol.* 55:451-458.

91. Kool, J.L., Carpenter, J.C., and Fields, B.S. 1999. Effect of monochloramine disinfection of municipal drinking water on risk of nosocomial Legionnaires' disease. *Lancet* 353:272-277.

92. Kuchta, J.M., States, S.J., McNamara, A.M., Wadowsky, R.M., and Yee, R.B. 1983. Susceptibility of *Legionella pneumophila* to chlorine in tap water. *Appl. Environ. Microbiol.* 46:1134-1139.

93. Kulikoysky, A., Pankratz, H.S., and Sadoff, H.L. 1975. Ultra-structural and chemical changes in spores of *Bacillus cereus* after action of chlorine. *J. Appl. Bacteriol.* 38:39-46.

94. Landeen, L.K., Yahya, M.T., and Gerba, C.P. 1989. Efficacy of copper and silver ions and reduced levels of free chlorine in inactivation of *Legionella pneumophila. Appl. Environ. Microbiol.* 55:3045-3050.

95. LeChevallier, M.W., Cawthon, C.D., and Lee, R.G. 1988a. Factors promoting survival of bacteria in chlorinated water supplies. *Appl. Environ. Microbiol.* 54:649-654.

96. LeChevallier, M.W., Cawthon, C.D., and Lee, R.G. 1988b. Inactivation of biofilm bacteria. *Appl Environ Microbiol* 54:2492-2499.

97. Leech, P.N. 1923. Examination of American-made Chloroamine-T, Dichloroamine-T, Halazone and preparation. *J. Am. Pharm. Asso.* 12, 592-602.

98. Li, J.W., Yu, Z., Cai, X., Gao, M., and Chao, F. Trihalomethane formation in water treated with chlorine dioxide. *Wat. Res.* 30: 2371-2376.

99. Lillard, H.S. 1979. Levels of chlorine and chlorine dioxide of equivalent bactericidal effect in poultry processing water. *J. Food Sci.* 44:1597.

100. Lin, W., Huang, T., Cornell, J.A., Lin, C., and Wei, C. 1996. Bactericidal activity of aqueous chlorine and chlorine dioxide solutions in a fish model system. *J. Food Sci.* 61: 1030-1034.

101. Lisle, J.T., Broadaway, S.C., Prescott, A.M., Pyle, B.H., Fricker, C., and McFeters, G.A. 1998. Effects of starvation on physiological activity and chlorine disinfection resistance in *Escherichia coli* 0157:H7. *Appl. Environ. Microbiol.* 64:4658-4662.

102. Lopes, J.A. 1986. Evaluation of dairy and food plant sanitizers against *Salmonella typhimurium* and *Listeria monocytogenes. J. Dairy Sci.* 69:2791-2796.

103. Lykins Jr. B.W., and Griese, M.H. 1986. Using chlorine dioxide for trihalomethane control. *J. AWWA* June: 88-93.

104. Mafu, A.A., Roy, D., Goulet, J., and Magny. P. 1990. Attachment of *Listeria monocytogenes* to stainless steel, glass, polypropylene, and rubber surfaces after short contact times. *J. Food Prot.* 53:742-746.

105. Masschelein, W.J. 1979. *Chlorine Dioxide, Chemistry and Environmental Impact of Oxychlorine Compounds,* Ann Arbor Science: Ann Arbor.

106. Maugh, T.H. 1981. New study links chlorination and cancer. *Science* 211:694.

107. Mazollier, J.R.. 1988. Ive gamme. Lavage-desinfection des salades. *Infros-Ctift.*41:19.

108. McCoy, Z., and Irwin, J. 1974. The effect of disinfectants on bacteriophage →X174. *Lab. Anim. Sci.* 24:630.

109. Michael, G.E., Miday, R.K., Bercz, J.P., Miller, R.G., Greathouse, D.G., Kraemer, D.F., and Lucas, J.B. 1981. Chlorine dioxide water disinfection: a prospective epidemiology study. *Arch. Environ. Health* 36:20-27.

110. Morris, J. C. 1946. The mechanism of the hydrolysis of chlorine. *J. Am. Chem. Soc.* 68:1692.

111. Mosley, E.B., Elliker, P.R., and Hays, H. 1976. Destruction of food spoilage indicator and pathogenic organisms by various germicides in solution and on a stainless steel surface. *J. Food Prot.* 39:830-836.

112. Myers, A.G. 1990. Evaluating Alternative disinfectants for THM control in small systems. *J. AWWA* 82: 77-84.

113. Narkis, N., and Kott, Y. 1992. Comparison between chlorine dioxide and chlorine for use as a disinfectant of wastewater effluents. *Wat. Sci. Tech.* 26: 1483-1492.

114. Nguyen-the, C., and Carlin, F. 1994. The microbiology of minimally processed fresh fruits and vegetables. *Crit. Rev. Food Sci. Nutr.* 34:371-401.

115. Norman, T.S., Harms, L.L., and Looyenga, R.W. 1966. The use of chloramines to prevent trihalomethane formation. *J. Am. Water Works Assoc.* 58:1475-1482.
116. Noss, C.I.; Hauchman, F.S.; and Olivieri, V.P. 1986. Chlorine dioxide reactivity with proteins. *Wat. Res.* 20: 351-356.
117. Odlaug, T.E. 1981. Antimicrobial activity of halogens. *J. Food Prot.* 44:608-613.
118. Ojajarvi, J., and Makela, P. 1976. Evaluation of chlorine compounds for surface disinfection by laboratory and in-use testing. *Scand. J. Infect. Dis.* 8:267-270.
119. Oliver, S.P., King, S.H., Torre, P.M., Shull, E.P., Dowlen, H.H., Lewis, M.J., and Sordillo, L.M. 1989. Prevention of bovine mastitis by a postmilking teat disinfectant containing chlorous acid and chlorine dioxide in a soluble polymer gel. *J Dairy Sci* 72:3091-3097.
120. Ortenzio, L.F. 1957. Available chlorine gemicidal equivalent concentration test. *J. Assoc. Off. Agric. Chem.,* 40:755-759.
121. Orth, R., and Mrozek, H. 1989. Is the control of *Listeria, Camphylobacter* and *Yersinia* a disinfection problem? *Fleischwirtshaft* 69(10):1575-1576.
122. Patterson, P.H., Ricke, S.C., Sunde, M.L., and Schaefer, D.M. 1990. Hatching eggs sanitized with chlorine dioxide foam: egg hatchability and bactericidal properties. *Avian Dis* 34:1-6.
123. Paz, M.L., Duaigues, M.V., Hanashiro, A., D'Aquino, M., and Santini, P. 1993. Antimicrobial effect of chlorine on *Yersinia enterocolitica*. *J Appl Bacteriol* 75:220-225.
124. Peeters, J.E., Mazas, E.A., Masschelein, W.J., Villacorta Martiez de Maturana, I., and Debacker, E. 1989. Effect of disinfection of drinking water with ozone or chlorine dioxide on survival of *Cryptosporidium parvum* oocysts. *Appl. Environ. Microbiol.* 55:1519-1522.
125. Pieterse, M.J. 1988. The potential risk of trihalomethanes in drinking water: A perspective. *S. Afr. J. Sci.* 84:166-170.
126. Rav Acha, C.H., Choshen, E., and Limoni, B. 1983. The action of chlorine dioxide on aquatic organic materials during disinfection of drinking water. *J. Environ. Sci. Health.* A18:651.
127. Reina, L.D., Fleming, H.P., and Humphries, E.G. 1995. Microbiological control of cucumber hydrocooling water with chlorine dioxide. *J. Food Prot.* 58:541-546.
128. Richardson, D.S., Thruston Jr., A.D., Caughran, T.V., Collette, T.W., Patterson, K.S., and Lykins Jr. B.W. 1998 Chemical by-products of chlorine and alternative disinfectants. *Food Technol.* 52: 5861.
129. Ridenour, G.M., and Ingols, R.S. 1947. Bactericidal properties of chlorine dioxide. *J. Am. Water Works Assoc.* 39:561-567.
129a.Ringo, J., 1991. Catalyst enhanced generation of chlorine dioxide. *US Patent* 5,008,096
130. Roberts, R.G., and Reymond, S.T. 1994. Chlorine dioxide for reduction of post-harvest pathogen inoculum during handling of tree fruits. *Appl. Environ. Microbiol.* 60:2864-2868.
131. Ronner, A.B., and Wong, A.C.L. 1993. Biofilm development and sanitizer inactivation of *Listeria monocytogenes* and *Salmonella typhimurium* on stainless steel and buna-n rubber. *J. Food Prot.* 56:750-758.
132. Rook, J.J. 1974. Formation of haloforms during chlorination of natural water. *Water Treat. Exam.* 23:234.
133. Rosenkranz, H.S.1973. Sodium hypochlorite and sodium perborate: Preferential inhibitors of DNA polymerase deficient bacteria. *Mutat. Res.* 21:171-174.
134. Rudolph, A.S., and Levine, M. 1941. Factors affecting the germicidal efficiency of hypochlorite solutions. Bull. 150, Eng. Exp. Sta., Iowa State College, Ames.
135. Russell, A.D. 1971. The destruction of bacterial spores. In *Inhibition and Destruction of the Microbial Cell,* edited by W.B. Hugo, pp.451-612. New York: Academic Press.
136. Russell, A.D.1990. Bacterial spores and chemical sporicidal agents. *Clin. Microbiol. Rev.* 3:99-119.
137. Rutala, W.A., Cole, E.C., Thomann, C.A., and Weber, D.J. 1998. Stability and bactericidal activity of chlorine solutions. *Infect. Control Hosp. Epidemiol.* 19:323-327.
138. Samrakandi, M.M., Roques, C., and Michel, G. 1997. Influence of trophic conditions on exopolysaccharide production:bacterial biofilm susceptibility to chlorine and monochloramine. *Can. J. Microbiol.* 43:751-758.
139. Scheusner, D.L., Busta, F.F., and Speck, M.L. 1971. Injury of bacteria by sanitizers. *Appl. Microbiol.* 21:41-45.
140. Senter, S.D., Cox, N.A., Bailey, J.S., and Forbus, W.R. 1985. Microbilogical changes in fresh market tomatoes during packing operations. *J. Food Sci.* 50:254-255.
141. Seo, K.H., and Frank, J.F. 1999. Attachment of *Escherichia coli O157:H7* to lettuce leaf surface and bacterial viability in response to chlorine treatment as demonstrated by using confocal scanning laser microscopy. *J. Food. Prot.* 62:3-9.
142. Sharp, D.G., Young, D.C., Floyd, R., and Johnson, J.D. 1980. Effect of ionic environment on the inactivation of poliovirus in water by chlorine. *Appl. Environ. Microbiol.* 39:530-534.

143. Sing, E.L., Elliker, P.R., and Sandine, W.E. 1964a. Comparative destruction of airborne lactic bacteriophages by various germicides applies as aerosols. *J. Milk Food Technol.* 27:129-134.

144. Sing, E.L., Elliker, P.R., and Sandine, W.E. 1964b. Comparative destruction of lactic *Streptococcus* bacteriophages on various surfaces by germicidal aerosols. *J. Milk Food Technol.*27:161-165.

145. Sisti, M., Albano, A., and Brandi, G. 1998. Bactericidal effect of chlorine on motile *Aeromonas* spp. in drinking water supplies and influence of temperature on disinfection efficacy. *Lett. Appl. Microbiol.* 26:347-351.

146. Smith, K.E., and Bradley, R.L. 1987. Efficacy of sanitizers using unsoiled sprialwound polysulfone ultrafiltration membranes. *J. Food Prot.* 50:567-572.

147. Smith, M.K., George, E.L., Zenick, H., Manson, J.M., and Stober, J.A. 1987. Developmental toxicity of halogenated acetonitriles: drinking water by-products of chlorine disinfection. *Toxicology* 46:83-93

148. Smith, M.K., Randall, J.L., Stober, J.A., and Read, E.J. 1989. Developmental toxicity of dichloroacetonitrile: a by-product of drinking water disinfection. *Fundam. Appl. Toxicol.* 12:765-772.

149. Somers, II.1963. Studies on in-plant chlorination. *Food Technol.* 5:46-51.

150. Spotts, R.A., and Peters, B.B. 1980. Chlorine and chlorine dioxide for control of d 'Anjou pear decay. *Plant Dis.* 64:1095-1097.

151. Swango, L.J., Wilt, G.R., Killen, A.D., Williams, D.E., and Worley, S.D. 1987. Inactivation of *Legionella pneumophila* by hypochlorite and an organic chloramine. *Appl. Environ. Microbiol.* 53:2983-2986.

152. Tanner, R.S., 1989. Comparative testing and evaluation of hard-surface disinfectants. *J. Ind. Microbiol.* 4:145-154.

153. Taube, H., and Dodgen, H. 1949. Applications of radioactive chlorine to the study mechanisms of reactions involving changes in the oxidation state of chlorine. *J. Am. Chem. Soc.* 71: 3330-3336.

154. Taylor, G.R., and Butler, M. 1982. A comparison of the virucidal properties of chlorine, chlorine dioxide, bromine chloride and iodine. *J. Hyg. (Lond).* 89:321-328.

155. Thraenhart, O., and Kuwert, E. 1975. Comparative studies on the action of chlorine and ozone on polioviruses in the reprocessing of drinking water in Essen. *Zentralbl. Bakteriol.* [Orig B] 160:305-341.

156. Trueman, J.R. 1971. The halogens. In *Inhibition and Destruction of the Microbial Cell*, edited by W.B. Hugo, pp. 137-183. New York: Academic Press.

157. Ver Kuilen, S.D., and Marth, E.H. 1980. Sporicidal action of hypochlorite on conidia of *Aspergillus parasiticus*. *J. Food Prot.* 43:784-788.

158. Wattie, E., and Butterfield, C.T. 1944. Relative resistance of *Escherichia coli* and *Eberthella typhosa* to chlorine and chloramines. *Public Health Reports* 59:1661-1671.

159. Wei, C.I., Huang, T.S., Kim, J.M., Lin, W.F., Tamplin, M.L., and Bartz, J.A. 1995. Growth and survival of *Salmonella montevideo* on tomatoes and disinfection with chlorinated water. *J. Food Prot.* 58:829-836.

160. White, G. C. 1992. *The Handbook of Chlorination and Alternative Disinfectants*, 3rd edition, New York:Van Nostrand.

161. White, G.C. 1972. *Handbook of Clorination*, New York: Van Nostrand-Reinhold.

162. Wickramanayake, G.B., and Sproul, O.J. 1991 Kinetics of the inactivation of microorganisms In *Disinfection, Sterilization, and Preservation*, 4th edition, eds. S.S. Block, S. S.; pp 131-151. Philadelphia: Lea & Febiger.

163. Wilkins, R.G. 1991. The deduction of mechanism In *Kinetics and Mechanism of Reactions of Transition Metal Complexes*. 2nd edition, pp 87-89.New York:VCH.

164. Williams, D.E., Smith, M.S., and Worley, S.D. 1990. Combined halogen disinfectants in poultry processing. *Poult. Sci.* 69:2248-2251.

165. Wlodkowski, T.J., and Rosenkranz, H.S. 1975. Mutagenicity of sodium hypochlorite for *Salmonella typhimurium*. *Mutat. Res.* 31:39-42.

166. Wyatt, L.R., and Waites, W.M. 1975. The effects of chlorine on spores of *Clostridium bifermentans*, *Bacillus subtilis* and *Bacillus cereus*. *J. Gen. Microbiol.* 89:337-344.

167. Zhang, S., and Farber, J.M. 1996. The effects of various disinfectants against *Listeria monocytogenes* on fresh-cut vegetables. *Food Microbiol.* 13:311-321.

168. Zhuang, R.T., Beuchat, L.R., and Angulo, F.J. 1995. Fate of *Salmonella montevideo* on and in raw tomatoes as affected by temperature and treatment with chlorine. *Appl. Environ. Microbiol.* 61:2127-2131.

K. Muthukumarappan
F. Halaweish
A.S. Naidu

Ozone

29

I. INTRODUCTION

Ozone is formed in the earth's atmosphere as result of lightning or high-energy ultraviolet radiation. Oxygen molecules are ruptured, producing oxygen fragments that combine with other oxygen molecules to produce ozone (O_3). The characteristic fresh, clean smell in air following a thunderstorm represents freshly generated ozone in nature. Ozone is also a by-product of various photochemical oxidation processes involving hydrocarbons, oxygen, and nitrogen.

The history of ozone and its applications has been extensively reviewed (Easton, 1951; Hill & Rice, 1982; Graham, 1997). In the United States, in 1888, Fewson invented a generator to produce ozone for deodorizing sewer gases. In Germany, in 1902, Siemens and Halske established the first full-scale ozone-generating plant for water treatment (cited by Graham, 1997). Later, in 1904, De la Coux reported extensive use of ozone in gelatin, casein and albumin production plants. During the same year, the city of Nice (France) was the first to implement commercial-scale disinfection of potable water using ozone (Lebout, 1959). Various European countries followed this lead and adapted ozonation as a standard practice for water treatment and disinfection. In the United States, ozonation of potable water was first installed in Whiting, Indiana, in 1940. By 1987, over 200 potable water treatment plants in the U.S. were using ozone (Rice, 1986). During 1953-56, ozone-containing air under pressure was found very effective for sterilizing empty food containers and was adopted for use on glass bottles in Switzerland.

Demand for effective antimicrobial sanitizers to control infection and disease as well as the need to reduce accumulation of chemical residues to maintain safe air, water, and food is ever-growing. Antimicrobial sanitizers such as heavy metal salts, halogen compounds, reducing gases, oxidizers, and alcohol have been used in many specific applications. The accumulation of toxic chemicals in our environment has increased the national focus on the safe use of sanitizers, bleaching agents, pesticides, and other chemicals in industrial processing.

There are several processing methods available for inactivation of microorganisms in foods including thermal, high pressure, pulsed electric field, oscillating magnetic field, irradiation, and ozonation. Ozonation is a relatively new method for food process-ing. Ozone has been used safely and effectively in water treatment for nine decades, at scales from a few gallons per minute to millions of gallons per day. It is approved in the US as generally recognized as safe (GRAS) for treatment of bottled water and as a sani-tizer for process trains in bottled water plants (FDA, 1995). In June 1997, ozone was deemed the GRAS status as a disinfectant for foods by an independent panel of experts, sponsored by the Electric Power Research Institute (EPRI, 1997). This action cleared the way for the use of ozone in the \$430 billion food processing industry. However, the FDA's final ruling has not been published yet.

Foods of muscle origin provide favorable conditions for microbial growth. The muscle foods are exposed to microbial contamination during slaughter, and handling, and this can lead to microbial spoilage and food-borne illness. Newly emerging pathogens have prompted a need to improve the microbiological status of ready to eat meat products, and the Food Safety and Inspection Service (FSIS) of the USDA has responded with establishment of 'zero tolerance' and 'clean meat' programs. Numerous studies have been conducted in the application of different processing methods to inactivate microorganisms in various food materials. However, the application of ozone in inactivating the microor-ganisms in various food materials is still in the developmental stage.

II. CHEMISTRY

Ozone (O_3) is unstable gas at ambient temperature and pressure. It has a charac-teristic pungent odor, which is not offensive. Pure gas is blue-colored; it is hard to notice the color at the low concentration in ozone applications. Ozone melts at -192.5°C and boils at 111.9 °C (Manley & Niegowski, 1967) and condenses at -112 °C to form a dark blue and highly explosive liquid. However, ozone applications mainly use about 10 % concentration; so, it is safe to handle under the current industrial applications.

Ozone is a naturally occurring substance found in our atmosphere. It can also be produced synthetically and used as an important industrial chemical. Ozone is formed in nature through photochemical reactions in the Earth's stratosphere and produced in low concentration at the Earth's surface by photochemical decomposition of air pollutants. Ozone is relatively stable at normal temperature and its half-life is about 12 hours, but in aqueous solution its half-life depends on the purity of water and/or the other constituents of water, or wastewater. For example, the half-life of ozone in distilled water is about 22 minutes, in tap water is about 20 minutes and filtered water from lake Zurich is about 20 minutes (Rosenthal, 1974).

A. Solubility

It is partially soluble in water and like most gases, the solubility increases as the water temperature decreases. It has the unique property of autodecomposition, producing numerous free radical species, the most prominent being the hydroxyl free radical (OH-). Although ozone is more soluble in water than oxygen [13 times more], its applications in aqueous systems require techniques to overcome the gas/liquid interface. For this reason, most of the methods of contacting ozone with water involve passing the ozone-containing gas into liquid. This forms bubbles of gas in the liquid. If the gas is passed through a dif-fuser, relatively large bubbles are formed.

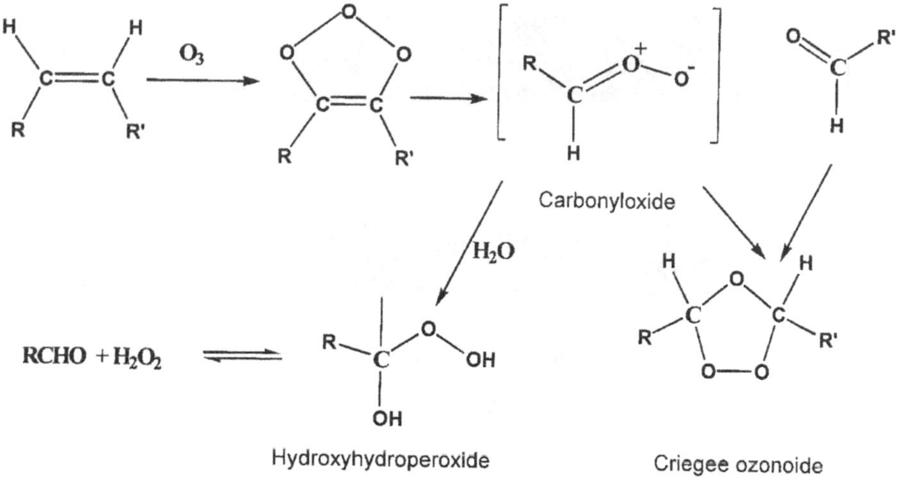

FIGURE 1. Ozone chemistry - generation of intermediate oxidative compounds.

B. Reactivity

Ozone reacts with olefins to produce the primary ozonoids, which decompose into carbonyl oxide and carbonyl compounds (Uppu et al., 1995). In nonparticipating organic solvents, these two species recombine to become Criegee ozonoids. If water is present, the carbonyl can react to give hydroxyhydroperoxide that can decompose to give hydrogen peroxide and a second mole of aldehyde (*FIGURE 1*). Ozone reacts easily with amines, sulfides, and ethers (electrophilic reaction), which contain heteroatoms and form oxidation products of the alkyl side chain (Bailey et al., 1972). This reaction contributes to the oxidation of most organic contaminants in ozone applications.

III. SYNTHESIS

Ozone gas can be generated using various methods. The most common methods are corona discharges (CD), photochemical, electrolytic, and radiochemical. The first industrial ozone generator was developed by Siemens in 1857, based on corona discharge (Siemens, 1857). Corona discharge in a dry process gas containing oxygen is presently the most widely used method for ozone generation. The formation of ozone through electrical discharge in a process gas is based on the nonhomogeneous corona discharge in air or oxygen. There are numerous distributed microdischarges by which the ozone is effectively generated. According to Kogelschatz (1987), each individual microdischarge lasts only several nanoseconds, lasting about 2.5 to 3 times longer in air than in oxygen. Passing air through a high voltage electrical discharge or corona produces ozone (*FIGURE 2*). A minimum of 5,000 volts of electricity is necessary to create the corona. Air or concentrated oxygen dried to a minimum of -60°C dewpoint passes through the corona which causes the O_2 bond to split, freeing two O_1 atoms which then collide with other O_2 molecules to create ozone (*refer to the equation*). The ozone/gas mixture discharged from the CD ozone generator normally contains from 1% to 3% ozone when using dry air, and 3% to 6% ozone when using high purity oxygen as the feed gas.

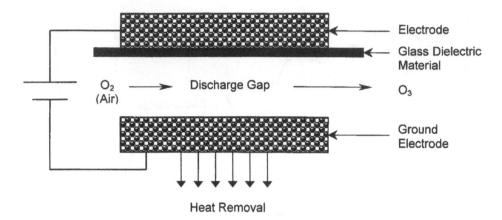

FIGURE 2. Schematic design of Corona Discharge (CD) ozone generation system.

$$\text{Equation:} \qquad O_2 \xrightarrow{\quad e \quad} 2O$$

$$O + O_2 \longrightarrow O_3$$

In the photochemical ozone generation method, the ozone is formed when oxygen is exposed to UV light of 140-190 nm wavelength. The method has been reviewed thoroughly by Langlais et al (1990). So far, due to poor yields, this method has been used only in small-scale applications. In the Electrolytic method, ozone is generated by the electrolysis of sulfuric acid. There are many advantages associated with this method, including: the use of low-voltage DC current, no feed gas preparation, reduced equipment size, possible generation of ozone at high concentration, and generation in the water. Some of the disadvantages of the method are corrosion and erosion of electrodes and the need for special electrolytes or water with low conductivity. In the Radiochemical method, the ozone is formed from the high-energy irradiation of oxygen by radioactive rays. Due to its complicated process requirements, this method has not been employed in water or wastewater treatment.

IV. ANTIMICROBIAL EFFECTS

A. Mechanism of action

Ozone is effective in killing bacteria (cidal effect) by lysis mechanism(s). Bacterial cell walls and outer membrane are the primary target sites for antimicrobial action. Ozone is a highly unstable molecule with a chemical composition of three oxygen atoms. The third atom is loosely bound and could detach easily from the molecule. This unstable third atom could quickly break, and the free single oxygen atom readily reacts with other molecules. This process generates a highly reactive oxidant system which affects the structural integrity of the reactant. Such alterations and disruption of cellular components could lead to oxidative damage and lysis of microorganisms. Similar changes at the molecular level also occur in various inorganic materials upon contact with ozone (Horváth et al., 1985). The contact times and dosages for ozone disinfection are much lower and very effective compared to many other sanitizers.

The half-life of ozone is about 12 hours in air, thus, one-half of the ozone will disappear after 12 hours, particularly the molecules that do not encounter an oxidizable substance. In such cases, ozone is depleted almost instantaneously. The natural degradation of ozone is a function of a very weak bond on the third oxygen atom. Elemental oxygen (comprised of two oxygen atoms) is held together by a double bond, each atom contributing a single bond. The addition of a third atom could only contribute one bond which is extremely weak, and is not sustained over time. However, when this bond is broken, the ozone molecule is degraded into elemental oxygen, and a single free oxygen atom.

B. Influencing factors

Ozone is a potent disinfecting agent which has been used successfully in water treatment, as well as to heal wounds. It can also be used in dental clinics, in laboratories, and in hospitals to clean and disinfect work surfaces and instruments.

The stability of ozone in water depends on water temperature, initial ozone concentration and length of holding time. The germicidal effects of ozone are affected by contact time, temperature, pH, and presence of inorganic and organic materials in the solution. Longer contact time, lower pH and temperature results in enhanced bactericidal effect (Yang & Chen, 1979).

Antimicrobial activity in ozonated water on bacterial suspensions and contaminated materials was effective and the efficacy depends on concentration and time of exposure. On buccal flora, one rinse alone had no effect, but various successive rinses led to substantial reductions in the number of colonies of bacteria. Ozone had a similar effect, although more pronounced, on the flora of the hands. Ozonated water placed in an open dish kept up antimicrobial activity for the first 20 minutes, but after 30 minutes this activity substantially decreased (Minguez et al., 1990).

C. Oxidative effects on food

Activated oxygen species such as ozone and others, including singlet oxygen, hydroxyl radical, superoxide anion (perhydroxyl radical at low pH), and hydrogen peroxide do not only elicit potent cidal activity against a broad-spectrum of microorganisms but also cause a myriad of oxidative changes in foods (Korycka-Dahl & Richardson, 1978). Chemical and enzymatic reactions known to occur in biological materials could generate singlet oxygen, hydroxyl radical, superoxide anion, and hydrogen peroxide. Reactions involving ozone, singlet oxygen, and hydroxyl radicals with food constituents could ultimately yield peroxides which decompose to initiate oxidative chain reactions. Superoxide anion and hydrogen peroxide are relatively inert toward organic molecules but could decompose to produce the more reactive singlet oxygen and hydroxyl radical. Inhibition of reactions initiated by reactive oxygen species in foods is very important in preserving the oxidative stability of foods.

Initiation of lipid peroxidation in foods may be accomplished by a variety of mechanisms. Two principal initiation reactions involve homolytic scission of preformed peroxides as catalyzed by metal ions and heme proteins and the reaction of activated oxygen species with the lipid substrate to yield peroxides and free radicals. Copper and

TABLE 1. Ozone - antimicrobial spectrum and sanitary applications in food industry

Sanitation	Dosage	Susceptible microorganisms	Reference
Water	0.35 mg/L	*A. hydrophila/ B. subtilis/ E. coli/ V. cholerae/ P.aeruginosa/ L. monocytogenes/ Salm. typhi/ Staph. aureus / Y. enterocolitica*	Korol et al., 1995
Dairy	5 ppm	*Alcaligens faecalis/ P. fluorescens*	Greene et al., 1993
Fish	0.27 mg/L	*P. putida/ B. thermospacta/ L. plantarum / Shewanella putrefaciens/ Enterobacter sp.*	Silva et al., 1998
	0.111 mg/L	*Enterococcus seriolicida*	Sugita et al., 1992
	0.064 mg/L	*Pasteurella piscicida/ Vibrio anguillarum*	Sugita et al., 1992
Shrimp	1.4 ml/L	*E.coli/ Salmonella typhimurium*	Chen et al., 1992
Poultry	0.2-0.4 ppm	*Salmonella sp./ Enterobacteriaceae*	Bailey et al., 1996
Animal	>100 ppm	*HVJ virus/ TME virus/ Reo type 3 virus/ murine hepatitis virus*	Sato et al., 1990
Potatoes	20-25 mg/m^3	Shelf life extension	Enshina & Voitik, 1989
Cabbage	7-13 mg/m^3	Shelf life extension	Enshina & Voitik, 1989
Carrot	5-15 mg/m^3	Shelf life extension	Enshina & Voitik, 1989
	60 µl/L	*Botrytis cinerea/ Scerotinia sclerotiorum*	Liew & Prange, 1994
Peppercorn	6.7 mg/L	3-6 log reduction of microbial load	Zhao & Cranston, 1995
Black berries	0.3 ppm	*Botrytis cinerea*	Barth et al., 1995
Media	3-18 ppm	*E.coli* O157:H7	Byun et al., 1998

cytochromes in the milk fat globule membrane may serve as focal points for initiation of lipid peroxidation by catalyzing homolytic scission of peroxides (Korycka-Dahl & Richardson, 1980).

The effect of ozone stress on polyamine metabolism and membrane lipid peroxidation in lentil seedlings through the amine oxidase and lipoxygenase activity and expression has been investigated (Maccarrone et al., 1997). Ozone could control the expression of these enzymes at the transcriptional level, down-regulating the amine oxidase gene and up-regulating the lipoxygenase gene. The decrease of amine oxidase activity was correlated with the increase of putrescine concentration in the ozone-treated plantlets, whereas the increase of lipoxygenase activity was paralleled by enhanced membrane lipid peroxidation.

V. ANTIMICROBIAL SPECTRUM

Ozone is a potent antimicrobial system that could elicit a broad-spectrum of cidal activity on a variety of microorganisms including bacteria, viruses and fungi (*TABLE 1*). This property has found applications of ozone system in a wide array of public and commercial sanitation processes.

A. Antibacterial activity

Ozone and chlorine are agents that disinfect by destroying, neutralizing or inhibiting the growth of pathogenic microorganisms. Korol et al. (1995) compared the efficacy of ozone and chlorine treatment of water and its antimicrobial spectrum on a variety of bacterial pathogens. The treatment of drinking water with ozone has shown to be more efficient against spores of *Bacillus subtilis*. It was observed that the ozone already in dose of 0.35 mg/L produced the reduction of at least 5 log in populations of approximately 10^6 cells/ml of *E. coli, Vibrio cholerae, Salmonella typhi, Yersinia enterocolitica,*

Pseudomonas aeruginosa, Aeromonas hydrophila, Listeria monocytogenes and *Staphylococcus aureus*. With a dose of 0.50 mg/L of chlorine, the reduction was much smaller for the tested microorganisms (except *Vibrio cholerae*), while the effect of 2 mg/L of chlorine was similar to the ozone treatment. For spores of *Bacillus subtilis*, the reduction observed with ozone concentrations of 0.35 and 0.70 mg/L was of almost 3 log, while no considerable effect was obtained with chlorine in the tested conditions. These results have indicated that both disinfectants were consumed during the treatment period, probably because of the water demand and the added bacterial mass. The ability of ozone to disinfect spores of *Bacillus spp.* and *Clostridium spp.* has been reported (Foegeding, 1985).

The antimicrobial effects of ozonated water in a recirculating concurrent reactor were evaluated against four gram-positive and four gram-negative bacteria, two yeasts, and spores of *Aspergillus niger* (Restaino et al., 1995). More than 5-log units each of *Salmonella typhimurium* and *Escherichia coli* cells were killed instantaneously in ozonated water with or without addition of 20 ppm of soluble starch. In ozonated water, death rates among the gram-negative bacteria (*S. typhimurium, E. coli, Pseudomonas aeruginosa, and Yersinia enterocolitica*) were similar. Among gram-positive bacteria, *Listeria monocytogenes* was more sensitive than *Staphylococcus aureus* or *Enterococcus faecalis*. In the presence of organic material, death rates of *S. aureus* compared with *L. monocytogenes* and *E. coli* compared with *S. typhimurium* in ozonated water were not affected by addition of soluble starch but were markedly reduced by addition of 20 ppm of bovine serum albumin. More than 4.5-log units each of *Candida albicans* and *Zygosaccharomyces bailii* cells were killed instantaneously in ozonated water, whereas <1-log unit of *Aspergillus niger* spores was killed after a 5-min exposure. The average ozone output levels in the deionized water (0.188 mg/ml) or water with soluble starch (0.198 mg/ml) did not differ but were significantly lower in water containing bovine serum albumin (0.149 mg/ml).

Methicillin-resistant *Staphylococcus aureus* (MRSA) has become an important problem in chemotherapy in the past decade because of its resistance to commonly used disinfectants. The usefulness of ozone solutions to disinfect MRSA has been evaluated (Yamayoshi & Tatsumi, 1993). Ozone strongly oxidized the cell walls and cytoplasmic membranes of bacteria. The microbicidal effect of ozone was elicited within the first five seconds of treatment. An ozone concentration 1.5 times that needed to kill the one MRSA strain tested was sufficient to kill the resistant clinical isolates. Such an ozone solution could be used to disinfect medical instruments and similar equipment. Ozone solutions should be useful in reducing the number of bacterial infections caused by inadequate disinfection against new resistant strains.

Ozonated water and chlorinated sanitizer were compared for effectiveness against biofilms of milk spoilage bacteria (Greene et al., 1993). Stainless steel plates were incubated in UHT-pasteurized milk inoculated with pure cultures of either *Pseudomonas fluorescens* or *Alcaligenes faecalis*. After incubation, the plates were removed and rinsed in sterile PBS. A control rinsed stainless steel plate was swabbed and plated on standard plate count agar. A second rinsed stainless steel plate was covered and treated for 2 min with a commercial chlorinated sanitizer (dichloro-s-triazinetrione). After treatment, the plate was rinsed twice in sterile PBS, swabbed, and plated on standard plate count agar. A third rinsed stainless steel plate from the culture was placed in ozonated deionized H_2O (5 ppm of ozone) for 10 min, rinsed twice as described, swabbed, and plated. Both ozonation and chlorination reduced bacteria populations by >99% at initial cell densities in the range of approximately 1.24 to 8.56 x 10^5 CFU/cm^2 for *P. fluorescens* and 1.53 x 10^4 to 8.56 x 10^5 CFU/cm^2 for *A. faecalis* in milk films on stainless steel surfaces.

The bactericidal activity of gaseous ozone was examined on five species of fish bacteria, *Pseudomonas putida, Shewanella putrefaciens, Brochothrix thermospacta, Enterobacter sp.,* and *Lactobacillus plantarum* (Silva et al., 1998). Test strains were inoculated on agar surfaces and exposed to different ozonation times in a gas chamber. Relatively low concentrations (<0.27 mg/L) of ozone has elicited potent cidal activity against vegetative cells of all five fish bacteria. The age of the cell culture was shown to influence the cell response following exposure. Survival rate was not linearly related to ozonation time, but exhibited biphasic death over an extended period. Similar bactericidal effects were observed on fish skin treated with ozone, with decreases of 1.0 log CFU/cm^2. Treatment of the whole fish using a commercial ozone generator showed improved scores for sensory analyses compared with the controls.

Reduction and elimination of verotoxigenic *E. coli* O157:H7 by gamma irradiation and ozone treatment was investigated (Byun et al., 1998). Log-phase cells were found to be more sensitive to gamma irradiation than stationary-phase cells. *E. coli* O157:H7 was more resistant to irradiation at -18°C than at 20°C. The D-values for this organism for treatment with ozone in trypticase soy agar were higher than those for treatment with ozone in phosphate buffer. Gamma irradiation at a dose of 1.5 kGy or ozone treatment at a concentration of 3 to 18 ppm for 20 to 50 min was required to assure the elimination of *E. coli* O157:H7.

B. Antiviral activity

Katzenelson et al (1974) reported a two-step ozonation process to inactivate viruses that are known to be more resistant to disinfectants than bacteria. Step-1 lasts < 10 seconds with a 99% viral inactivation and step-2 for several minutes to complete particle destruction. Ozone was applied at seven intermediate levels between 0.07 and 2.5 mg dm -3 but the phenomenon was independent of the changes in the concentration. The higher resistance of viruses is caused by the formation of clumps. Thus, an ultrasound of 100w for 2 minutes at 20mhz of poliovirus caused the breakdown of such virus clumps and made them highly susceptible to ozone.

Ozonation has been successfully used to inactivate the viral pathogens of laboratory animals (Sato et al., 1990). Ozone (> 100 ppm) at high humidity was highly virucidal against four RNA viruses including HVJ, Theiler's murine encephalomyelitis virus (TMEV), Reo type 3 virus (RV) and murine hepatitis virus (MHV). Antiviral tests with ozone were performed with 0.1 ml of virus suspension in deionized water or saline, placed in 35-mm dishes. The titer of 10^6 plaque-forming units of TMEV in a liquid-phase, which was highly stable against physical treatments, was reduced within 1 h to a level of 0 by 300 ppm of ozone at 80% humidity and 22-25°C. HVJ and MHV were more susceptible than TMEV to the ozone treatment. RV was the most resistant of the 4 viruses tested. The authors suggested that ozonation is a promising method to disinfect not only the laboratory animal RNA-viruses (both of enveloped and non-enveloped viruses) but also animal rooms, clean rooms and safety cabinets.

VI. APPLICATIONS

In recent years, ozone technology has found uses in agriculture and the food industry; however, it has been used in other industrial applications for a long time. The industrial applications of ozone have been focused on ozone's ability to oxidize and sterilize, without leaving residues. Applications of ozone in the water, wastewater treatment, and food processing appear highly promising. These include preserving raw agricultural commodities during storage and transit, controlling odor, retarding metabolic processes associated with ripening, and sanitizing water utilized for washing food equipment, foods, and packaging materials.

A. Treatment of drinking water

During the early 1900s, ozone was suggested as a disinfectant for drinking water. Ozone is still used in the disinfection of water supplies in many European countries. Ozone acts with efficacy even at lower concentration under controlled conditions with few known side effects such as taste, odor and toxic by-products. Ozone applications in water treatment can be classified into three types:

1. Disinfection. Ozonation is generally performed at the end of the water treatment process, prior to chemical addition, flocculation, sedimentation and filtration. Such treated water is clean and free of high concentrations of oxidizable organic materials, other than the lysed bacterial cells or inactivated viruses (Rice, 1986). A residual ozone standard for water treatment plants has been adopted first by the French Public Health Authorities. A residual ozone concentration of 0.4 mg/L is recommended for at least 4 minutes and maintained for an additional 4 minutes of contact time. A contact time of 8 minutes has a 99.9% efficacy to inactivate viruses and kill bacteria. In operational drinking water treatment plants, ozone disinfection/viral inactivation normally is accomplished using two or more side-by-side contact chamber. The 4 mg/L level is attained in the first chamber, and is maintained in the second or subsequent chambers. In normal practice, the first four minutes contact time is at least doubled, to eight to 12 minutes, to ensure that the require disinfection conditions are achieved (Rice, 1986).

2. Oxidation. Impurities such as iron, manganese, organic contaminants, phenols, inorganic impurities such as cyanide sulfide and nitrite, instantly react with ozone. It is hard at this point to provide a standard contact time and ozone concentration due to the variability and level of contaminants.

Ozonation at this stage normally produces water which is more turbid than the prior treatments. Products from oxidation reaction combines with polyvalent cations, hydrogen bonding and produces higher molecular weight adducts which are insoluble and could be precipitated and removed from the water.

3. Bio-processing. The purpose of water treatment in this case is to increase dissolved oxygen and to oxidize complex organic contaminants for easier degradation. The optimum ozone concentration for this treatment is 1 mg/L per mg/L of organic contaminants (Nagel et al., 1982).

B. Wastewater treatment

Ozone has been successfully used in sewage treatment plants for several decades and in many countries. Fox example, ozone could be used instead of chlorine in the final disinfection process of effluent prior to discharge into a receiving stream. Other applications include pretreatment and odor control of influent (Rice, 1986).

C. Food processing

The possible formation of carcinogenic compounds in water (Brungs, 1973) has raised questions about the use of certain chlorine compounds in food processing. The food industry is searching for alternative compounds for cooling water and wastewater disinfection. Numerous investigations on disinfection of foods (including the usage of gaseous ozone for shelf life enhancement, ozone dissolved in water for sanitizing surfaces of vegetables, fruits, and other agricultural products) strongly support the idea that ozone is a powerful disinfectant. Ozone has been used in the European food industry for decades, in certain instances for almost a century. However, the food industry in the U.S. has little or no experience with application of ozone to foods. This may be due to lack of studies on the efficacy of ozone in food processing, safety issues, toxicology, and impact on nutrients. Applications for ozone in the food processing/packaging industries are extremely varied (Graham, 1997).

Based on potent disinfection and oxidation properties, various possibilities for use of ozone in food industry and agriculture have been developed. There are few published applications of ozone, e.g., cleaning shellfish (Anonymous, 1972), preserving fish (Haraguchi et al., 1969), aflatoxin reduction in peanut and cottonseed meals (Dwankanth, 1968), sterilization of bacon, bananas, butter, eggs, mushrooms, cheese, fruits (Gammon & Kerelak, 1973), and beef (Kaess & Weidemann, 1968a; 1968b), disinfection of poultry carcasses and chill water (Sheldon & Brown, 1986; Bolder, 1997).

Sheldon and Brown (1986) investigated the efficacy of ozone as a disinfectant for poultry carcasses. The microbial counts of ozone treated carcasses stored at 4°C were markedly lower than carcasses chilled under non-ozonated conditions. Ozone also has been evaluated against various sanitizing agents (5% hydrogen peroxide, 0.5% ozone, 12% trisodium phosphate, 2% acetic acid, and 0.3% commercial sanitizer), and water (16 to 74°C) spray-washing interventions for their ability to reduce bacterial contamination on beef samples in a model spray-washing cabinet (Gorman et al., 1995) Under the conditions of their study, hydrogen peroxide and ozonated water were more effective than other sanitizing agents. However, the study conducted by Reagan et al (1996) on the effects of different treatments (74°C hot-water washing, 5% hydrogen peroxide, and 0.5% ozone) in reducing bacterial populations on beef carcasses showed that the ozone and hydrogen peroxide treatments had minor effects and were equivalent to conventional washing in reducing bacterial populations on beef.

1. Preservation and storage. Ozone is also being shown effective for increasing the storage life, ranging from minimally processed to fully processed commodities, by reducing the level of food spoilage microorganisms contained on the food surface. The use of ozone has become popular in several major cold storage plants in Europe (Horváth et al, 1985).

Practical operations for such preservation is based on the sterilization of air entering the storage room contain a sufficient amount of ozone to destroy microorganisms. Billion (1975) conducted a detailed investigation on the storage life of beef, veal, lamb, pork, chicken, and rabbit in ozonated atmospheres. Ozonated atmosphere increased the storage life of all the foods studied by 7 days compared to normal atmosphere. In general the growth of the surface microflora (*Pseudomonas* sp., spore-formers, salmonellae, and staphylococci) was retarded in refrigerated atmospheres and in the presence of ozone.

2. Effect on odors. Ozone has a characteristic odor, yet the result of application does not mask odors. Atomic oxygen formed by decomposition of ozone immediately oxidize various odor-causing materials. The characteristic putrid odor, however, remains and is difficult to eliminate with ozonation. In general, lower the temperature and larger the molecules in the reaction, weaker is the oxidizing effect. The moisture content in the air has no effect on the process. At very low concentration (0.01 to0.04 ppm ozone), air of the storage facility is felt fresh and pleasant without any stuffy odor. Studies showed that the odor of aromatic fruits such as strawberries is enhanced in the presence of ozone (Gammon & Kerelak, 1973). The formation of fragrances and odors that impart characteristic flavor to fruit is seemed to be enhanced by ozone. The sterilization of air in fruit storage by ozone prevents the odors of packaging materials from transfer to the commodities, a phenomena which frequently takes place otherwise, particularly when wooden crates are used in refrigerated stores at relative humidity of 85% to 90% (Horváth et al., 1985).

Storage places, warehouses and refrigerated stores could be disinfected in most cases by administering ozonated air. Besides disinfection, such a process could remove unpleasant odors of packaging materials and could retain the original flavor. The oxidation of compounds creating odors in such premises has the advantage of creating an atmosphere resembling pleasant fresh air. For such purposes, a low concentration of ozone (0.01 to 0.04 ppm) would be sufficient.

3. Effect on metabolism. Effects of ozone on the metabolism is also a consequence of its strong oxidizing ability. No deterioration of fruit is observed, since ozone only affects the surface. During storage, the respiration and ripening process of the fruit is enhanced. Ethylene produced during the storage affects the other fruit and initiates even more intensive ripening. The external signs of this process are the turning brown of the skin, softening of the flesh and, finally, the decay of the fruit. This process is controlled by ozone by the oxidation of metabolic products, thereby reducing the process of back action on other fruits. Moreover, ozone can also promote the healing of wounds and enhance resistance of fruit to infection (Horváth et al., 1985).

D. Sanitation of produce

The food value of potatoes, cabbage and carrot was studied after ozone treatment to prolong shelf life (Enshina & Voitik, 1989). The regimen of ozone-treatment of the vegetables under laboratory conditions was: 20-40 mg/m^3/4-h/day/25 times. Ozone concentration for potatoes treatment was 20-25 mg/m^3, cabbage 7-13 mg/m^3, carrot 5-15 mg/m^3. Ozone treatment did not significantly alter the protein and starch in potatoes; sugar and carotene in cabbage or carrot. Organoleptic evaluation indicated no changes in the properties of potatoes and vegetables as compared to non-treated products. However, ozonation has decreased the ascorbic acid content by 16 to 25%.

Effects of ozone and storage temperature on carrots and two post-harvest pathogens, *Botrytis cinerea* Pers. and *Sclerotinia sclerotiorum* de Bary were investigated (Liew & Prange, 1994). Pathogen-inoculated and uninoculated whole carrots were exposed to an ozone concentration of 0 (control), 7.5, 15, 30, or 60 µl/L. Treatment chambers were flushed with a total flow rate of 0.5 liters/min (air and ozone) for 8 h daily for 28 days. The experiment was repeated twice at storage temperatures of 2, 8, and 16°C. The residual ozone concentration (ozone supplied-exhausted and reacted ozone) increased with ozone supply concentration but was less at higher storage temperatures. A 50% reduction of daily growth rates of both fungi at the highest ozone concentration indicated that ozone is a potent fungistatic system. Carrot respiration rate, electrolyte leakage, and total color differences increased with ozone concentration. Ozone-treated carrots were lighter and less intense in color than control carrots.

The use of ozone as a substitute for ethylene oxide to decontaminate whole black peppercorn and ground black pepper and the effects of ozone on the volatile oil constituents of the spice were studied (Zhao & Cranston, 1995). Black peppercorns were immersed in water and sparged with ozonized air (ozone conc., 6.7 mg/L) for 10 min at an air flow rate of 6 L/min. This treatment reduced the microbial population of peppercorn by 3-4 log numbers. Ground black pepper with various moisture levels was sparged with ozonized air for upto 6 h. This treatment reduced the microbial population by 3-6 log numbers, depending on the moisture content of the spice. Higher moisture content led to greater reduction in microbial load. The volatile oil constituents of the spice with and without ozone treatment were extracted with isopropyl ether and analyzed by gas chromatography (GC) and GC-mass spectrometry. Ozone treatment of ground black pepper resulted in the oxidation of certain volatile oil constituents while the treatment had no significant effect on the volatile oil constituents of whole peppercorn.

Ozone exposure was assessed for storage of thornless blackberries which are prone to fungal decay (Barth et al., 1995). Blackberries were harvested and stored for 12 days at 2°C in 0, 0.1. and 0.3 ppm ozone. Berries were evaluated for fungal decay, anthocyanins, color and peroxidase activity. Ozone storage suppressed fungal development for 12 days, while 20% of control fruits showed decay. The main mold was *Botrytis cinerea*. Ozone storage did not cause observable injury or defects. By 12 days, anthocyanin content of juice was similar to initial levels of all treatments. Surface color was better retained in 0.1 ppm-stored berries by 5 days and in 0.3 ppm-stored berries by 12 days, by hue angle values. Peroxidase was greater in controls and 0.1 ppm samples and was lowest in 0.3 ppm fruits by 12 days. Ozone storage resulted in market quality extension. Ozone has also been used in the control of post-harvest decay of table grapes caused by *Rhizopus stolonifer* (Sarig et al., 1996)

Earlier studies have demonstrated that a novel source of ozone gas could be used to chemically degrade numerous mycotoxins, including aflatoxin (AF) B1. Subsequent *in vitro* analyses demonstrated detoxification of AFB1, suggesting a potential method of remediate AF-contaminated grain. McKenzie and co-workers (1998) investigated the ability of electrochemically produced ozone to degrade AFB1 in naturally contaminated whole kernel corn. Corn was procured from the southern coastal areas of Texas and HPLC analysis revealed 1,220 +/- 73.3 ppb AFB1. Control and contaminated corn were treated for 92 h with ozone at 200 mg/min in 30 kg batches; greater than 95% reduction of AFB1 in contaminated corn was achieved.

The effects of gamma irradiation and ozone treatment on microbiological and physico-chemical properties of bee pollen were investigated (Yook et al., 1998). Gamma irradiation at 7.5 kGy reduced the total microbial loads below detection levels (>10^2 CFU/g), but after ozone treatment of upto 18 ppm for 8 h the total aerobic bacteria were found in concentrations of >10^3 CFU/g. Physico-chemical properties such as amino acid and fatty acid composition, thiobarbituric acid value, mineral content, and pigment were not significantly changed by gamma irradiation, whereas ozone treatment caused significant changes in fatty acid composition, thiobarbituric acid value and pigment by lipid oxidation and discoloration.

E. Sanitation of poultry

A pilot plant study of the effect of using an ozone contact system on the quality of poultry broiler carcasses and recirculating poultry chill water revealed lower carcass microbial levels and a destruction of all microbes (>99%) washed from the carcass in the chill water (Sheldon & Brown, 1986). Ozonation also reduced the chemical oxygen demand by a third and increased the light transmission of the treated chill water. The authors indicated that ozonation had no adverse effect on carcass color or lipid peroxidation, and did not produce a carcass off-flavor.

Whistler and Sheldon (1989) evaluated ozone and formaldehyde as disinfectants in a prototype laboratory setter against microorganisms that are naturally present on fertile, freshly laid, broiler hatching eggs. Microbial counts were reduced by 2.5 $\log_{10}$ for water-misted and ozonated eggs or formaldehyde-fumigated eggs than for control and water-misted eggs. Eggshell conductance studies as measured by egg moisture losses in a desiccator showed no differences among the treatments. Hatchability was significantly reduced (26.5 to 37.5%) following ozonation (3.03% /wt for 2 h) in comparison with no treatment or water misting. Misting with ozonation was equally as effective as formaldehyde fumigation in reducing microbial counts. However, ozone treatment at the concentrations tested significantly reduced hatchability when compared with results of either no treatment, water misting, or an average hatchability figure for formaldehyde fumigation. These findings indicated that ozone is a good disinfectant yet may adversely affect embryo development when given in the gaseous form.

Bailey and co-workers (1996) have conducted four trials to evaluate the efficacy of hatcher air sanitation utilizing ultraviolet light (UV), ozone, or hydrogen peroxide on bacterial populations, the spread of *Salmonella*, and hatchability of broiler eggs. The UV light (254 nm, 146 μ) and ozone (0.2 or 0.4 ppm) treatments were continuously applied through the final three days of hatch, the hydrogen peroxide treatment (2.5%) was administered 1 or 2 min of each 10 min at rates of 500 or 100 ml/h. Hatchability was not significantly reduced by sanitizing treatments when compared with the untreated control (94 versus 95.6%). As compared to controls, all sanitizing treatments reduced 75 to 99% of the total bacteria, *Enterobacteriaceae*, and *Salmonella* in the hatching cabinet air samples. The use of hydrogen peroxide resulted in greater reduction of bacteria than ozone or UV light. Only hydrogen peroxide significantly reduced *Salmonella* levels on eggshell fragments. Significant reductions in the number of *Salmonella*-positive chicks occurred using the ozone and hydrogen peroxide treatments. Hydrogen peroxide significantly reduced the magnitude of *Salmonella* colonization in chicken ceca. These trials demonstrated that the spread of bacteria could be effectively reduced in the hatching cabinet by air sanitation

TABLE 2. Approved levels of ozone application

Exposure	Ozone level (ppm)
Detectable odor	0.01 - 0.05
OSHA 8 hr limit	0.1
OSHA 1.5 min limit	0.3
Lethal in few minutes	>1700

using UV light, ozone, and hydrogen peroxide. The potential to reduce bacterial cross contamination in the hatcher seems to be achievable without compromising the hatchability.

F. Sanitation of sea-food

Chen and co-workers (1992) have examined solubility and stability of ozone in shrimp-meat extract (SME), bactericidal effects on shrimp-meat organisms, mutagenicity of ozonated shrimp meat, and ozonolysis of DNA. The saturated concentration of ozone (1.4 ml/L) in SME was lower than in 2% saline or distilled water at 5 and 25˚C. Upon standing for 25 min after ozonation, ozone exhibited the same decomposition rate (2.7% /min) in SME at 5 and 25˚C. Among nine bacterial strains tested, *Salmonella typhimurium* was more resistant to ozone in shrimp meat. Mutagens were not detected in shrimp meat after ozonation in saline. Ozone in saline (<5 mg/L) has lysed M13 RF DNA in *E. coli* and single-stranded DNA in phage M13 outside bacterial cell within 30 min.

Survival of bacterial fish pathogens, including *Enterococcus seriolicida*, *Vibrio anguillarum*, and *Pasteurella piscicida*, in ozonated seawater was determined in a batch system (Sugita et al., 1992). Bacterial counts of all fish pathogens decreased at more than 0.040 to 0.060 mg of total residual oxidants (TROs) per liter, whereas no decrease in viable counts was observed at less than 0.018 to 0.028 mg of TROs per liter. The 99% inactivation point was achieved at concentrations of 0.111 mg/liter for *E. seriolicida*, 0.063 mg/liter for *P. piscicida*, and 0.064 mg/L for *V. anguillarum* within 1 min. Moreover, the mean 99 and 99.9% killing concentration-contact time (C_t) products were 0.123 and 0.186 mg/min/L for *E. seriolicida*, 0.056 and 0.084 mg/min/L for *P. piscicida*, and 0.081 and 0.123 mg/min/L for *V. anguillarum*, respectively. However, the mean 99 and 99.9% C_t products for the mixed population in coastal seawater were 0.200 and 0.621 mg.min/L. These results strongly suggest that ozone treatment at more than 1.0 mg of TROs per liter for several minutes could efficiently disinfect seawater for aqua-culture.

VII. SAFETY AND TOXICITY

Ozone is a toxic gas, and like chlorine can cause severe illness and death if inhaled in high quantity. However, ozone systems have various safety features. Ozone is generated *in situ*, thus eliminating transportation hazards. Also, the generation system could be shut down if an ozone leak developed. Another safety advantage is the physical characteristic of ozone that allows its detection (odor) at concentrations much lower than harmful levels (*TABLE 2*). These ozone levels have been recommended by the Occupational Safety and Health Administration (OSHA), the American National Standards Institute/American Society for Testing and Materials (ANSI/ASTM), the American Conference of Governmental Industrial Hygienists (ACGIH), and the American Industrial Hygiene Association (AIHA).

Ozone is one of the highly active oxidants with strong toxicity to animals, plants and living organisms. A large quantity of the photochemical oxidants in air pollution are created from ozone. The generation of ozone in work places such as from copying machines, in office environments, has been reported. Ozone toxicity depends on the level of ozone in the environment and/or working condition as well as the duration of exposure. Toxicity symptoms such as sharp irritation to the nose and throat could result instantly at 0.1 ppm dose. Loss of vision could arise from 0.1-0.5 ppm after exposure for 3-6 hours. Ozone toxicity of 1-2 ppm could cause distinct irritation on the upper part of the throat, headache, pain in the chest, cough and drying of the throat. Higher levels of ozone (5-10 ppm) could cause increase in pulse, edema of lung. Ozone level of 50 ppm or more is potentially fatal (Masaki, 1986).

The main toxic effects of ozone occur at the lung-air interface when it crosses the air/pulmonary tissue barrier and by the products formed during such reactions. The high reactivity of ozone prevents its penetration deep into the air-tissue boundary before it can cause toxic reactions. For this reason, the principal target for the reaction of ozone probably is the fluid layers covering the internal surface of the lung (Pryor, 1992 & 1993). The lung lining fluid layer is a patchy and highly dynamic material of variable thickness (0.1-0.2 μm) consisting of lipids (ca. 90%), proteins (ca. 10%), and antioxidants such as ascorbate, glutathione and uric acid (Hawgood, 1991). Ozone displays a high degree of toxicity and/or reactivity towards unsaturated lipids, certain amino acids residue in proteins and antioxidants. Studies on reactivity of ozone with lipids have identified lipid ozonation products that could cause toxic effects to deeper tissue where ozone itself could not reach (Pryor, 1991; Uppu et al., 1995). Similar studies with proteins also indicated oxidative damage during ozonation.

Studies indicated that Criegee ozonoides (*FIGURE 1*) are formed in synthetic bilayer membranes and that ozonoides have a long halflife (3-6 weeks) that could initiate lipid peroxidation (Squadrito et al., 1992). Recent studies showed that ozone exposure increases aldehydes in epithelial linning fluid in human lung and the ozonation products activate phospholipases (Frampton et al., 1999; Kafoury et al., 1998). The latter results suggest that the activation of these phospholipases may play a role in the development of lung inflammation during ozone exposure.

VIII. SUMMARY

The use of ozone in water purification for decades was mainly due to its potential disinfectant effect on a wide-variety of water-borne pathogens. Numerous studies have indicated that ozone can destroy with extreme efficiency the spores of molds, amoebae, and viruses, and bacteria as well as various other pathogenic and saprophytic organims. Such broad-spectrum and successful use of ozone has created a renewed interest in its utilization as a general germicidal agent.

Ozone shows much promise in the food and agriculture industry. However, more research and development is required before its multipurpose usage. Recent GRAS status has encouraged food processors to use ozone for sanitation. Ozone eliminates a significant problem associated with chlorine-based disinfection, in particular, the potential accumulation of toxic residues.

Limited data are available on the effect of different concentrations of ozone in reducing bacterial contamination of various food materials and the subsequent changes in physical and biochemical characteristics of these foods. Future studies should focus on the quantitative estimation of the changes in sensory, physical and nutritional characteristics of various food materials resulting from ozone application.

IX. REFERENCES

1. Anonymous, 1972. Use of ozone in sea water for cleansing shellfish. *Effluent Water Treatment Journal* 12: 260-264.
2. Bailey, J.S., Buhr, R.J., Cox, N.A., and Berrang, M.E. 1996. Effect of hatching cabinet sanitation treatments on *Salmonella* cross-contamination and hatchability of broiler eggs. *Poult. Sci.* 75:191-196.
3. Bailey, W.S., Ward, J.W., Hornish, R.E., and Potte II, F.E., 1972, Complexes and radicals produced during ozonation of olefins, In *Ozone Reactions with Organic Compounds*, ed. P.S. Baily, pp. 1-8. Washington D.C.: American Chemical Society.
4. Barth, M.M., Zhou, C., Mercier, J., and Payne, F.A. 1995. Ozone storage effects on anthocyanin content and fungal growth in blackberries. *J. Food Sci.* 60:1286-1288.
5. Billion, M. 1975. Storage life of foods in ozonized atmospheres. *J. Food Prot. Preserv.* 45: 625-635.
6. Bolder, N. M. 1997. Decontamination of meat and poultry carcasses. *Trends Food Sci. Technol.* 4:625-528.
7. Brungs, W. A. 1973. Effects of residual chlorine on aquatic life. *J. Water Poll. Control Fed.* 45: 2180.
8. Byun, M.W., Kwon, O.J., Yook, H.S., and Kim, K.S. 1998. Gamma irradiation and ozone treatment for inactivation of *Escherichia coli* O157:H7 in culture media. *J. Food Prot.* 61:728-730.
9. Chen, H.C., Huang, S.H., Moody, M.W., and Jiang, S.T. 1992. Bacteriocidal and mutagenic effects of ozone on shrimp (*Penaeus mondon*) meat. *J. Food Sci.* 57:923-927.
10. De la Coux., H. 1904. L'Ozone et ses applications industrielles. *Vve. Ch. Dunod, Paris.* Pp.49.
11. Dwankanth, C. T. 1968. Reduction of aflatoxin levels in cottonseed and peanut meals by ozonation. *J. Am. Oil Chem.* 45:93-96.
12. Easton, T. 1951. The uses of ozone. *Austral. J. Dairy Technol.* 4:142-143.
13. Enshina, A.N., and Voitik, N.P. 1989. The effect of regular treatment with ozone on the chemical composition of potatoes and vegetables. *Vopr. Pitan* 6:61-64.
14. EPRI. 1997. Expert Panel Report: Evaluation of the history and safety of ozone in processing foods for human consumption: Volumes 1-3. EPRI, CA.
15. FDA. 1995. Beverages: Bottled Water; Final Rule, Food and Drug Admin., *Fed. Reg.* 60:57075-57130.
16. Fewson, H. 1888. Apparatus for treating noxious gases with electricity. *U.S. Patent* 387,286.
17. Foegeding, P.M. 1985. Ozone inactivation of *Bacillus* and *Clostridium* spore populations and the importance of the spore coat to resistance. *Food Microbiol.* 2:123-134.
18. Frampton, M. W., Pryor, W. A., Cueto, R., Cox, C., Morrow, P. E., and Utell, M. J. 1999. Ozone exposure increases aldehydes in epithelial lining fluid in human lung. *Am. J. Respir. Crit. Care Med.* 159:1134-1137.
19. Gammon, R. and Kerelak, K. 1973. Gaseous sterilization of foods. *Am. Inst. Chem. Engineering* 69:91-94.
20. Gorman, B. M., Sofos, J. N., Morgan, J. B., Schmidt, G. R., and Smith, G. C. 1995. Evaluation of hand-trimming, various sanitizing agents, and hot water spray-washing as decontamination interventions for beef brisket adipose tissue. *J. Food Prot.* 58:899-907.
21. Graham, D.M. 1997. Use of ozone in food processing. *Food Tech.* 51:72-75.
22. Greene, A.K., Few, B.K., and Serafini, J.C. 1993. A comparison of ozonation and chlorination for the disinfection of stainless steel surfaces. *J. Dairy Sci.* 76:3617-3620.
23. Haraguchi, T., Simidu, U., and Aiso, K. 1969. Preserving effect of ozone to fish. *Bulletin of Japan Society of Fish* 35:915-919.
24. Hawgood, S. 1991. In *The Lung: Scientific Foundations*, eds. R.G. Crystal, J.B. West, P.J. Barnes, N.S. Cherniack, and E.R. Weibel, vol. 1, pp.247. New York: Raven Press.
25. Hill, A.G., and Rice, R.G. 1982. Historical background properties and applications of ozone. In *Handbook of Ozone Technology and Applications,* ed. R.G. Rice, and A. Netzer, Vol.1, pp.1-37. Michigan: Ann Arbor Science Publishers.
26. Horváth, M., Bilitzky, L., and Huttner, J. 1985. Ozone. In *Topics in Inorganic and General Chemistry-* monograph 20, Amsterdam: Elsevier.
27. Kaess, G., and Weidemann, J. F. 1968a. Ozone treatment of chilled beef. I. Effect of low concentrations of ozone on microbial spoilage and surface color of beef. *J. Food Technol.* 3:325-334.

28. Kaess, G., and Weidemann, J. F. 1968b. Ozone treatment of chilled beef. I. Interaction between ozone and muscle. *J. Food Technol.* 3:335-343.

29. Kafoury, R. M., Pryor, W. A., Squadrito, G. L., Salgo, M. G., Zuo, X., and Friedman, M. 1998. Lipid ozonation products activate phospholipases A2, C and D, Toxicol. *Appl. Pharmacol.* 150:338-349.

30. Katzenelson, E., Kletter, B., and Shuval, H. I. 1974. Inactivation kinetics of viruses and bacteria in water by use of ozone. *J. AWWA* 66:725-729.

31. Kogelschatz, U. 1987. Ozone generation from oxygen and air: Discharge physics and reaction mechanisms. Proc. 8th Ozone World Congress, IOA, Zurich, Switzerland.

32. Korol, S., Fortunato, M.S., Paz, M., Sanahuja, M.C., Lazaro, E., Santini, P., and D'Aquino, M. 1995. Water disinfection: comparative activities of ozone and chlorine on a wide spectrum of bacteria. *Rev. Argent. Microbiol.* 27:175-183.

33. Korycka-Dahl, M., and Richardson, T. 1980. Initiation of oxidative changes in foods. *J. Dairy Sci.* 63:1181-1198.

34. Korycka-Dahl, M.B., and Richardson, T. 1978. Activated oxygen species and oxidation of food constituents. *CRC Crit. Rev. Food Sci. Nutr.* 10:209-241.

35. Langlais, B., Reckhow, D. A., and Brink, D. R. 1990. Fundamental Aspects. In *Ozone in Water Treatment, Application and Engineering*. Cooperative Research Report. Michigan: Lewis Publishers.

36. Lebout, H. 1959. Fifty years of ozonation at Nice. In: *Ozone Chemistry and Technology, Adv. Chemistry Series, Vol.21,* ed. R.G. Rice, and A. Netzer, pp. 450-452. Washington DC: Americal Chemical Society.

37. Liew, C.L., and Prange, R.K. 1994. Effects of ozone and storage temperature on post harvest diseases and physiology of carrots (*Daucus carota L*). *J. Am. Soc. Hort. Sci.* 119:563-567.

38. Maccarrone, M., Veldink, G.A., Vliegenthart, F.G., and Finazzi Agro, A. 1997.Ozone stress modulates amine oxidase and lipoxygenase expression in lentil (*Lens culinaris*) seedlings. *FEBS Lett.* 408:241-244.

39. Manley, T. C., and Niegowski, S. J. 1967. Ozone. In *Encyclopedia of Chemical Technology*, Vol. 14, 2nd Edition. New york: J. Wiley & Sons, Inc.

40. Masaki, M. 1986. Ozonation. In *Environmental Chemistry of Dyes and Pigments*, eds. A. Reife and H.S. Freeman, pp 43-60, New York: John Wiley.

41. McKenzie, K.S., Kubena, L.F., Denvir, A.J., Rogers, T.D., Hitchens, G.D., Bailey, R.H., Harvey, R.B., Buckley, S.A., and Phillips, T.D. 1998. Aflatoxicosis in turkey poults is prevented by treatment of naturally contaminated corn with ozone generated by electrolysis. *Poult. Sci.* 77:1094-1102.

42. Minguez, F., Gomez-Lus, M.L., Andre, J., Cabronero, M.J., and Prieto, J. 1990. Antimicrobial activity of ozonized water in determined experimental conditions. *Rev. Sanid. Hig. Publica (Madr)* 64:415-423.

43. Nagel, G., Kuhn, W., Werner, P. and Sontheimer, H. 1982. Purification of groundwater by infiltration of ozone-treated water. *Wasser-Abwasser*, 123:399-407.

44. Pryor W.A. 1992. How far does ozone penetrate into the pulmonary air/tissue boundary before it reacts? *Free-Radic. Biol. Med.* 12:83-88.

45. Pryor, W. A. 1991. The ozonation of unsaturated fatty acids: aldehydes and hydrogen peroxide as products and possible mediators of ozone toxicity. *Chem. Res. Toxicol.* 4:341-348

46. Pryor, W.A. 1993. Ozone in all its reactive splendor. *J. Lab. Clin. Med.* 122:483-486.

47. Reagan, J.O., Acuff, G. R., Bueye, D. R., Buyck, M. J., Dickson, J. S., Kastner, C. L., Marsden, J. L., Morgan, J. B., Nickelson, R., Smith, G.C., and Sofos, J. N. 1996. Trimming and washing of beef carcasses as a method of improving the microbiological quality of meat. *J. Food Prot.* 59:751-756.

48. Restaino, L., Frampton, E.W., Hemphill, J.B., and Palnikar, P. 1995. Efficacy of ozonated water against various food-related microorganisms. *Appl. Environ. Microbiol.* 61:3471-3475.

49. Rice, R. G. 1986. Applications of ozone in water and wastewater treatment. In *Analytical Aspects of Ozone Treatment of Water and Wastewater*. Eds. R.G. Rice, L.J. Bollyky and W.J. Lacy, pp 7-26, Lewis Publishers.

50. Rosenthal, H. 1974. Selected bibliography of ozone, its biological effects and technical applications. Fisheries research board of Canada technical report No. 456, Fisheries and Marine Service, Pacific Biological Station, Nanaimo, BC.

51. Sarig, P., Zahavi, T., Zutkhi, Y., Yannai, S., Lisher, N., and Ben-Arie, R. 1996. Ozone for control and post harvest decay of table grapes caused by *Rhizopus stolonifer. Physiol. Molec. Plant Pathol.* 48:403-415.

52. Sato, H., Wananabe, Y., and Miyata, H. 1990. Virucidal effect of ozone treatment of laboratory animal viruses. *Jikken Dobutsu* 39:223-229.

53. Sheldon, B.W. and Brown, A. L. 1986. Efficacy of ozone as a disinfectant for poultry carcasses and chill water. *J. Food Sci.* 51:305-309.

54. Siemens, V. 1857. Uber die electrostatische Induktion und die verzogerung des storms in Flaschendrahten. *Poggendorff's Ann.*, 102: 120-124.

55. Silva, M.V., Gibbs, P.A., and Kirby, R.M. 1998. Sensorial and microbial effects of gaseous ozone on fresh scad (*Trachurus trachurus*). *J. Appl. Microbiol.* 85:802-810.

56. Squadrito, G. L., Uppo, R. M., Cueto, R. and Pryor, W. A. 1992. Production of Criegee ozonoide during the ozonation of 1-palmitoyl-2-oleolyl-sn-glycro-3-phosphocholine liposomes, *Lipids* 27:955-958.

57. Sugita, H., Asai, T., Hayashi, K., Mitsuya, T., Amanuma, K., Maruyama, C., and Deguchi, Y. 1992. Application of ozone disinfection to remove *Enterococcus seriolicida, Pasteurella piscicida*, and *Vibrio anguillarum* from seawater. *Appl. Environ. Microbiol.* 58:4072-4075.

58. Uppu, R. M., Cueto, R., Squadrito, G. L., and Pryor, W. A. 1995. What does ozone react with at the air/lung interface? Model studies using human red blood cell membranes. *Arch. Biochem.* 319:257-266.

59. Whistler, P.E., and Sheldon, B.W. 1989. Bactericidal activity, eggshell conductance, and hatchability effects of ozone versus formaldehyde disinfection. *Poult. Sci.* 68:1074-1077.

60. Yamayoshi, T., and Tatsumi, N. 1993. Microbicidal effects of ozone solution on methicillin-resistant *Staphylococcus aureus. Drugs Exp. Clin. Res.* 19:59-64.

61. Yang, P.P.W., and Chen, T.C. 1979. Stability of ozone and its germicidal properties on poultry meat microorganisms in liquid phase. *J. Food Sci.* 44:501-504.

62. Yook, H-S, Lim, S-I, and Byun, M-W. 1998. Changes in microbiological and physicochemical properties of bee pollen by application of gamma irradiation and ozone treatment. *J. Food Prot.* 61:217-220.

63. Zhao, J., and Cranston, P.M. 1995. Microbial decontamination of black pepper by ozone and the effect of the treatment on volatile oil constituents of the spice. *J. Sci. Food Agri.* 68:11-18.

Index

A

Acetaldehyde, 437
ACETIC ACID, 661-688
  antimicrobial activity, 665-666
  antimicrobial combinations, 670-672
  meat decontamination, 672-675
  mechanism of action, 666-667
  poultry decontamination, 675
  produce decontamination, 678
  seafood decontamination, 675-678
  vinegar defects, 662
  vinegar production, 661-662
Acetobacter aceti, 662
Acetyl-CoA, 690
Achromobacter spp. 333
Acid resistance, 669-670
Acid shock proteins, 669
Acid-blanch-chelate (ABC), 693
Acid-fast bacilli, 755
Acidic polysaccharides, 83
Acinetobacter calcoacetica, 286
Acquired immunity, 227
Actinobacillus actinomycetemcomitans,
  36, 38
Acute phase host response, 20
Adenosine triphosphate, 17
Aerobacter aerogenes, 31, 567
Aerobactin, 57
Aeromonas hydrophila, 46, 49, 65, 196,
  278, 283, 600, 621, 733, 789
Aeromonas spp., 757
Affinity chromatography, 23, 141
Aflatoxin, 278, 622, 649
AGAR, 417-427
  adhesion-inhibition, 422-423
  antimicrobial effects, 421-423
  isolation of GRP, 420-421
  streptococcal binding, 421-422

  structure, 419-420
Agaropectin, 419
Agarose, 419
Agglutination, 143
Aglycones, 298, 312, 325-326
Ajoene, 351, 355-358, 363-365, 370, 372
Alamar-Blue™ cytotoxicity assay, 163
Alcaligenes faecalis, 763
Algal lipids, 160
Allelopathic effects, 404
Allicin, 350-353, 356-357, 361-364, 370
Alliin lyase (alliinase), 350-353, 355-358
Alliin, 350-353, 356
Allium species, 296
Allyl disulfide, 351
Allyl isothiocyanate (AITC), 400
Allyl thiocyanate (ATC), 403
Amastigote killing, 43
Amastigotes, 52
Amyloid fibril protein, 205
Anise oil, 284
Anthocyanidins, 326
Anthocyanins, 325, 331-332
Antibacterial peptide, OTAP-92, 219
Antibody-antigen complexes, 138
Anti-bovine rotavirus IgY, 237
Anticancer drugs, 176
Anti-candida activity, 42-43, 353,
  356-357
Anti-candidal component, 351
Anti-chlamydial activity, 168
Anti-ETEC IgY, 238
Anti-fimbrial antibody, 235
Anti-human rotavirus IgY, 236
Anti-inflammatory, 195, 273, 275, 280
Anti-listerial activity, 622, 694
Antimicrobial peptides, 39, 160
Anti-OMP antibodies, 235

Printed and bound by CPI Group (UK) Ltd, Croydon, CR0 4YY

23/10/2024

01778259-0012